PRINCIPLES OF
Radiation Interaction
IN **Matter and Detection**
2nd Edition

P R I N C I P L E S   O F
Radiation Interaction
IN Matter and Detection
2nd Edition

Claude Leroy
Université de Montréal, Canada

Pier-Giorgio Rancoita
Istituto Nazionale di Fisica Nucleare, Milan, Italy

World Scientific

NEW JERSEY · LONDON · SINGAPORE · BEIJING · SHANGHAI · HONG KONG · TAIPEI · CHENNAI

Published by

World Scientific Publishing Co. Pte. Ltd.

5 Toh Tuck Link, Singapore 596224

USA office: 27 Warren Street, Suite 401-402, Hackensack, NJ 07601

UK office: 57 Shelton Street, Covent Garden, London WC2H 9HE

British Library Cataloguing-in-Publication Data
A catalogue record for this book is available from the British Library.

Cover images: The space image is provided by NASA.
The curves are from Figure 2.71 of this book.

PRINCIPLES OF RADIATION INTERACTION IN MATTER AND DETECTION
(2nd Edition)

ISBN-13 978-981-281-827-0
ISBN-10 981-281-827-8
ISBN-13 978-981-281-828-7 (pbk)
ISBN-10 981-281-828-6 (pbk)

Printed by FuIsland Offset Printing (S) Pte Ltd, Singapore

To Our Families

For their patience and encouragement

Preface to the Second Edition

The second edition of this book takes its origin from the positive comments of readers received by the authors following the publication of the first edition. It includes additional material covering the fundamental mechanisms of energy deposition and particle interactions resulting in a) permanent damage (like the displacement damage) in silicon semiconductor devices and b) single event effects due to individual events caused by the interactions of particles inside the active volume of silicon devices. This treatment also includes a description of radiation environments, in which these mechanisms are expected to operate. The extensive coverage of the displacement damage is discussed in the framework of the electromagnetic and nuclear interactions treated in the book. Furthermore, the electromagnetic interaction resulting in energy-loss processes is extended to cover low energy Coulomb scatterings with atomic electrons and nuclei of the medium, thus introducing processes depending on the sign of the incoming particle charge and the nuclear energy-losses. The applications of silicon devices in particle physics experiments, reactor physics, nuclear medicine and space possibly occur in adverse (or, even, very adverse) radiation environments that may affect the operation of the devices. These environments, which are described in this edition of the book, are generated by i) the operation of the high-luminosity machines for particles physics experiments , ii) the cosmic rays and trapped particles of various origins in the interplanetary space and/or the Earth magnetosphere and iii) the operation of nuclear reactors.

In addition to people and Institutions acknowledged in the first edition of the book, we wish to thank the library staff of the Department of Physics of the University of Milano-Bicocca for the help and assistance received. We are also grateful to Profs. Nathan Croitoru (Tel-Aviv University) for his suggestions and Stanislav Pospisil (Czech Technical University in Prague) for discussions about interpretation of results of spectrometry methods applied to the study of semiconductor detectors and neutron detection. Marie-Hélène Genest and Céline Lebel of the University of Montreal have provided help for many figures presented in several chapters of the book. The help of Andrea Gutiérrez of the University of Montreal for the sections on neutron detection with silicon detectors is gratefully acknowledged. We are indebted to Drs. Cristina Consolandi and Davide Grandi for the their help and Dr.

vii

Monica Rattaggi for her careful reading of the text regarding the radiation effects in silicon devices. We wish to thank those Editors who permitted us to reproduce or adapt figures from their articles or books for the text added in this edition. We acknowledge and wish to thank again the Institute of Physics (IoP) for the permission in reproducing and adapting text material, figures and tables from the Author's review article published in *Rep. Prog. in Phys.* **70** (2007) 403 for this revised version, in addition to that already permitted, published in *Rep. Prog. in Phys.* **63** (2000) 505, for the previous Edition. American Geophysical Union (AGU), Annual Reviews Inc., Elsevier and IEEE organization are acknowledged for their permissions to reproduce and adapt figures from their articles or books. The permissions are indicated in the figure captions according to indications from Editors.

Claude Leroy   *Pier-Giorgio Rancoita*

Université de Montréal (Québec)   *Istituto Nazionale di Fisica Nucleare*

Canada H3C3J7   *I-20126 Milan Italy*

15 August 2008

Preface to the First Edition

This book originates from lectures given to undergraduate and graduate students over several academic years. Students questions and interests have driven the need to make systematic and comprehensive (we hope) the presentation of the basic principles of a field which is under continuous development. The physics principles of radiation interaction with matter are introduced as a general knowledge background needed to understand how radiation can be detected. Technical developments are making available detectors and detecting media of increasing complexity. Historically, the first nuclear particle detectors (like those based on X-rays films) were very simple. In the course of time, the detectors have become more and more sophisticated. In addition, complex systems of detectors generally targeting a wide range of physics goals led to large experimental apparata often constituted by several sub-detectors. These large detector assemblies require dedicated methods of reconstruction and analysis of data to decrease the experimental errors. Therefore, both detectors and detection methods are fields of developments and investigations. To be detected, radiation and particles have to interact during their passage through a medium. Therefore, the first chapters are dealing with collision and radiation energy losses by charged particles, photon absorption and nuclear collision in matter. A particular attention has been given to the discussion of both the energy loss and the energy straggling, and the absorption of photons and hadrons in media. The second part of the book covers the particle energy determination, solid state, wire chambers and droplet detectors, and applications in the field of nuclear medicine. Detailed examples are presented which illustrate the operation of the various types of detectors, and help the understanding of the optimization factors.

We are grateful for the help received from individuals and groups of students in writing this book. The chapters on electromagnetic and hadron interactions in matter have taken advantage of discussions with undergraduate and graduate students of the University of Milan and Montreal. Their questions have helped the shaping of the content of these chapters. Help for the drawing of some of the figures and assistance have been provided by Pasquale D'Angelo from the National Institute of Nuclear Physics (Milan) and Dr. Simonetta Pensotti from the University of Milano-Bicocca. The chapters on solid state and nuclear medicine benefitted from the input of Céline Lebel PhD student at the University of Montreal and Dr. Patrick Roy former PhD student at the Montreal University. We have to acknowledge our collaborators of the SICAPO collaboration for the scientific achievements in the field of high energy electromagnetic and hadronic shower propagation in matter presented in the chapter on particle energy determination. Sections of the chapter on droplet detectors present results obtained in the framework of the PICASSO experiment in Montreal. They are the result of collaboration with Profs. Louis Lessard and Viktor Zacek of Montreal University. Input on this chapter has also been provided by Marie-Hélène Genest. The part of the chapter on wire chambers dealing with ionization chambers and their application in the measurement of liquid argon purity borrows material developed with our Dubna colleagues, in particular Drs. Alexander Tcheplakov and Victor Kukhtin.

We wish to thank many Authors and Editors who permitted us to reproduce adapt figures from their articles or books. For their permission in reproducing materials and figures, we acknowledge the Annual Review of Nuclear Science, the American Physical Society (APS), Cambridge University Press, European Organization for Nuclear Research (CERN), Elsevier, the International Atomic Energy Agency (IAEA), the Institute of Physics (IoP), the National Academic Press (NAS), Zeitschrift für Naturforschung, the Oxford University Press, Physica Scripta, the Italian Physical Society (IPS), and Springer-Verlag. We wish to thank for their collaboration Profs. A. Bohr, A. Fassò, R. Fernow, B. Mottelson, B. Povh, J.O. Rasmussen, K. Rith, G.B. Yodh, F. Zetsche, the Particle Data Group at Lawrence Berkeley National Laboratory, and the American Institute of Physics responsible for the succession of E. Segrè. The permissions are indicated in figure captions according to the indications from Editors.

C. Leroy　　　　　　　　　　　　*P.G. Rancoita*
Université de Montréal (Québec)　*Istituto Nazionale di Fisica Nucleare*
Canada H3C3J7　　　　　　　　　*I-20126 Milan Italy*

19 March 2004

Contents

Chapter 1

Introduction

The phenomena associated with the radiation interaction in matter are commonly understood to include a wide variety of physical effects. Moreover, the nature of interactions in matter depends on the incoming type of radiation and energy.

Furthermore, the incoming radiation also generates permanent or temporary damage. In most cases, detectors are not prevented from operating normally, but, in presence of large irradiations and/or as a result of large accumulated fluences, radiation effects may degrade devices performance. The second edition of this book is augmented by new chapters and additional sections have been added to some of the other chapters. Among these are treated radiation environments and basic mechanisms resulting in temporary and/or permanent radiation damages in devices, in particular those based on silicon semiconductor.

Historically, the first nuclear particle detectors (like those based on X-rays films) were very simple. In the course of time, the detectors have evolved toward higher specificity and at the same time toward more complexity. In addition, complex systems of detectors leading to large experimental apparata often consist of several sub-detectors and require sophisticated methods of reconstruction and analysis of data to decrease the experimental errors. Therefore, both detectors and detection methods are fields of development and investigation.

In each application field, the detector configuration has to be designed to respond to the collision geometry, i.e., the experimental apparatus has to be capable to process the particles emerging from the interaction or production volume. For instance, for medical applications, the detection system needs to be adapted to the part of the patient's body under examination. In most cases, the large amount of data to analyze, their variety and often the large energy range to be covered can only be handled by the construction of detectors composed of many sub-detectors assigned to dedicated tasks. For example, in high energy physics experiments, collisions between two particles beams (*collider type of collision*) or by a beam hitting a fixed target (*fixed target type of collision*) produce a variety of particles, most of the time with very complicated configurations of events. Space-based experiments are another example of application where detector development is needed. These experiments aim at the study of astrophysics, gamma rays astronomy and cosmic rays, interactions of cosmic rays with the space matter or with the Earth atmosphere or the detection of Galactic and extra-Galactic photons. The achievement of such vast research programs requires very reliable apparata.

In this chapter (Chapter 1), an overview of the types of interactions in matter and the physical meaning of interaction cross section are presented. These interactions and their properties will be extensively discussed in the two following chapters (Chapters 2 and 3) from the phenomenological point of view: these features constitute fundamental principles for designing detection systems. In addition, the basic equations and relations of relativistic kinematics will be introduced. Finally, an overview of detection methods and detecting media will be given.

Radiation environments and processes resulting in displacement damage are addressed in Chapter 4.

In the following chapters, i) the scintillation processes and scintillating media (Chapter 5), ii) the *electron–hole* carriers formation, charge transport and damage effects in semiconductors employed as radiation detectors (Chapter 6), iii) radiation effects in silicon devices (Chapter 7), and iv) the signal generation in ionization chambers (Chapter 8) will be dealt with. The principles of particle energy determination for both electromagnetic and hadronic particles are discussed in Chapter 9. In Chapter 10, droplet detectors, their response to neutrons and α-particles, and, in addition, their usage for a search of cold dark matter are treated. In Chapter 11, detector applications for Medical Physics are considered.

A collection of *General Properties and Constants* is included in Appendices A.1–A.8. Elements of *Mathematics and Statistics* for detection systems are summarized in Appendices B.1–B.2.

Finally, excellent books, complementary to the current *Second Edition* (for the first edition see [Leroy and Rancoita (2004)]), are available on the topics of radiation interactions in matter and techniques in particle detection and measurements: for instance, without being comprehensive, the more recent are [Fernow (1986); Gilmore (1992); Leo (1994); Kleinknecht (1998); Knoll (1999); Green (2000); Wigmans (2000); Grupen and Shwartz (2008); PDB (2008)].

1.1   Radiation and Particle Interactions

Radiation is detected by its interaction in matter. Every detection system has the same structure: it starts with the interaction of the radiation with the detection medium; the result of the interaction is transformed into signals, which are readout and usually recorded. The interaction processes depend on both the type and energy of the incoming particles* (or photons). The energy ranges encountered also vary by

---

*An incoming particle is sometimes referred to as primary, while particles (or photons) produced or emitted in the interaction are sometimes referred to as secondary.

orders of magnitude. For example, the electromagnetic spectrum[†] (photons) covers many decades of frequencies. The situation is similar for charged particles. Their energy ranges from fractions of eV to 10^{20} eV, in the case of ultra-high energy cosmic rays. The detecting media to be used in a particular application have to be carefully selected as function of particle type and energy.

Instruments for radiation detection evolve as new technologies become available and, as a consequence, more and more sophisticated devices are made available to users, nowadays. For instance, complex, large and advanced instrumentation can be found in usage for applications ranging from nuclear medicine and health physics to experimental high energy particle and space physics.

In practice, books on instrumentation have to be oriented to application fields. Nevertheless, in order to understand the principle of operation of radiation detectors, a deep knowledge of the interaction of radiation with matter is required. Physical phenomena allowing detection often involve soft electrons or photons, or atomic and molecular excitations. This is the case even for energetic and very energetic incoming charged particles. Furthermore, except for the case of total absorption of the incoming particle like in high-energy physics calorimetry, the particle is assumed to lose only a small fraction of its energy inside the detecting medium.

The loss of energy by a charged particle is caused by the interaction of the electric field associated with the moving charge and the one generated by the electronic (and, in few cases, nuclear) structure of detecting media. This process is referred to as *energy-loss process* and allows the dissipation of energy inside the detecting medium itself.

For sufficiently extended absorbers, high-energy hadronic particles (see Sect. 1.2) deposit their energy through a series of strong interactions with nuclei or nuclear constituents of the detector absorbing medium. The emerging particles will lose their energy by subsequent strong interactions but mostly by energy-loss processes. Thus, a cascade of particles is generated and their energies are fully absorbed in the detecting medium. Similarly, electromagnetic cascades are generated by primary electrons, positrons and photons whose energy is degraded by electromagnetic interactions and energy-loss processes.

The fundamental mechanism, on which radiation detectors are based, is the dissipation of a fraction[‡‡] of the incoming radiation-energy inside the detecting material. The transferred energy is distributed among excited states, which are capable of generating *carriers*, for instance, electrons-holes in semiconductors, ion pairs in gaseous devices, photons in scintillating media, etc.. These *carriers* are processed by appropriate readout elements, for instance, front-end electronics for semiconductor detectors and for gaseous devices, or photomultipliers for scintillating materials,

---

[†] The electromagnetic spectrum is subdivided into frequency regions, i.e., radiowaves, microwaves, infrared radiation, visible radiation, violet and ultraviolet radiation, X-rays and γ-rays.

[‡‡] As mentioned above, in some type of large detectors the whole amount of energy is deposited.

etc.. Hence, the required radiation information (such as momentum, energy, velocity) can be obtained. For example, in gas based detectors, the energy dissipation process results in creating *ion pairs* (i.e., electrons and positive ions) which are separated and move under the influence of an applied external electric field. Typically about 30 eV are required to create an ion pair. However, due to the limited number of ion pairs generated in a low density medium like a gas, a multiplication is usually needed in order to have enough carriers to induce a charge signal in the readout electronics. In semiconductor detectors, the medium is denser and an *electron–hole pair* requires about 3.6 eV to be generated; usually, no multiplication is needed inside these devices. In scintillating materials, whose densities are typically about half of semiconductor densities, the energy dissipation process results in emitting photons (about an order of 100 eV are needed to emit a photon), a fraction of which can be conveyed onto the photodiode of a photomultiplier where, in turn, *photoelectrons* are emitted and subsequently multiplied.

It is worthwhile to mention that these topics are based on both progresses in understanding of physical phenomena and discoveries of physical effects. These advancements have been recognized by a number of Nobel Prizes, e.g., like those awarded to W.C. Röntgen (1901), H. Lorentz and P. Zeeman (1902), J.J. Thomson (1906), A.A. Michelson (1907), J.D. van der Waals (1910), M. von Laue (1914), W.H. Bragg and W.L. Bragg (1915), C.G. Barkla (1917), M. Planck (1918), A. Einstein (1921), N. Bohr (1922), R.A. Millikan (1923), M. Siegbahn (1924), J. Franck and G. Hertz (1925), A.H. Compton and C.T.R. Wilson (1927), L. de Broglie (1929), E. Fermi (1939), P.M.S. Blackett (1948), C.F. Powell (1950), W.B. Shockley, J. Bardeen and W.H. Brattain (1956), P.A. Čerenkov, I. Frank and I.Y. Tamm (1958), D.A. Glaser (1960), L.D. Landau (1962), L.W. Alvarez (1968) and G. Charpak (1992).

1.2   Particles and Types of Interaction

Nowadays, it is usual to omit the term *elementary* while referring to the so-called *elementary particles* in fields like particle and nuclear physics. In the first half of the past century, only a few particles[††] were already discovered. At present, we know that these particles are final products from the interactions and decays of a very large number of particle states. This multitude of particles is proven to derive from i) a few fundamental constituent *fermions* of *spin* $\frac{1}{2}$, i.e., the *quarks* with fractional electric charges ($+\frac{2}{3}e$ and $-\frac{1}{3}e$, where e is the electron charge) and ii) the *leptons* (like the electron and its corresponding neutrino) with integral electric charge or neutral. For instance, neutrons and protons are built from a set of three quarks. These constituents interact by exchanging *spin* 1 *bosons*, which mediate three types of fundamental interactions: strong, electromagnetic and weak

---

[††]Protons, neutrons, electrons, neutrinos and photons were among the particles already discovered.

interactions. A fourth interaction, gravitation, is mediated by a *spin 2 boson* (*graviton*). The *photon* mediates the electromagnetic interaction, $W^\pm$ and Z the weak interaction, and the *gluon* the strong interaction.

Particles interacting via the *strong interaction* are known as *hadrons*. There are two main classes of hadrons: the *baryons*, with half-integral spin values, and the *mesons*, with integral spin values. For example, protons and neutrons are baryons, while pions are mesons. Protons and neutrons are constituents of nuclei and often referred to as *nucleons*. The strong interaction also provides the necessary binding forces to hold together nucleons inside the nucleus. Most of hadrons are unstable and are called *hadronic resonances*.

The *electromagnetic interaction* is usually responsible for most of non-nuclear interactions in physics beyond the *gravitational attraction*, and generates bound states in atoms and molecules. The *Quantum Electrodynamics Theory* (QED), one of the most successful theory in physics, allows extremely precise calculations of electromagnetic interactions of particles. *Weak interactions* are, for instance, responsible for processes like radioactive β-decays in nuclear physics. The gravitational force is the interaction involving massive bodies at very large distances. However, the gravitational force has negligible effect in particle–particle interaction at short distances.

The relative strengths of interactions at distances of $\simeq 10^{-18}$ cm are (see [Fernow (1986)] and references therein):

- strong interaction: 1
- electromagnetic interaction: $\approx 10^{-2}$
- weak interaction: $\approx 10^{-5}$
- gravitational interaction: $\approx 10^{-39}$

There is solid evidence that part of, if not all, the interactions are *unified*, i.e., are different aspects or manifestations of one single interaction. For instance, the electromagnetic and weak interactions have been unified in the *electroweak theory*, whose prediction of the existence of massive *gauge bosons*, $W^\pm$ and Z, has been experimentally verified.

Another remarkable feature of nature is the existence of *antimatter*. For every particle (fermion or boson), there exists an antiparticle, which has the same mass and spin, but opposite values of electric charge and magnetic momentum. An example of antiparticle is the positron, which is the antiparticle of the electron.

It is customary (see for instance [Fernow (1986); Perkins (1986)] and Section 38 of [PDB (2008)]) to measure energies in *Mega* or *Giga* electron Volt (MeV or GeV) and to use conversion factors in such a way the *speed of light c* and the *reduced Planck constant*$^\parallel$ $\hbar$ are set to 1 to simplify relativistic calculations.

$^\parallel$The reduced Planck constant (also known as *Dirac constant*) $\hbar$ is given by

$$\hbar \equiv \frac{h}{2\pi},$$

Nevertheless, in the present book, we will use the momenta in units of MeV/c or in GeV/c and the masses in units of MeV/c² or GeV/c², except when otherwise explicitly indicated.

1.2.1   *Quarks and Leptons*

At present, experimental data support the picture in which the matter is built from two basic types of fermions, i.e., quarks and leptons. As mentioned before, quarks carry fractional electric charges $(+\frac{2}{3}e$ and $-\frac{1}{3}e)$. Antiquarks carry opposite electric charges. There are different types of quarks distinguished by their *flavor*, i.e., u, d, s, c, b and t quarks. Their masses range from $\simeq 5\,\mathrm{MeV/c^2}$ for the lightest quark (u) to $\approx 171.2\,\mathrm{GeV/c^2}$ for the heaviest quark* (t). Baryons are built from three quarks, while mesons from quark-antiquark pairs. Ordinary matter is usually constituted of baryonic particles, like protons (stable) and neutrons (unstable). Mesons are unstable.

Leptons carry integral electric charges. Three types of negatively charged leptons are known: the electron (e), the muon (μ) and the tau (τ), whose masses are 0.511, 105.7 and 1777 MeV/c², respectively. Their antiparticles are positively charged. Associated with negative leptons are neutrinos, which are neutral leptons: ν_e (with mass $< 2.2\,\mathrm{eV}$), ν_μ (with mass $< 170\,\mathrm{keV}$) and ν_τ (with mass $< 18.2\,\mathrm{MeV}$). Neutrinos are longitudinally polarized with helicity $-\frac{1}{2}$ with regard to the direction of motion (they are *left handed*), while their correspondent antiparticles are *right handed* with helicity $+\frac{1}{2}$. The recent observation of neutrino oscillations implies that the neutrino has non-zero mass. This phenomenon, predicted by Pontecorvo whereby a neutrino created with a specific lepton flavor $(\nu_e, \nu_\mu, \nu_\tau)$ can later be observed to have a different flavor, is beyond the Standard Model of particle physics. At present, the experiments are only sensitive to the difference in the squares of the masses of neutrinos. These differences are known to be very small, i.e., $\lesssim 0.05\,\mathrm{eV/c^2}$.

Charged leptons undergo both electromagnetic and weak interactions, while neutrinos only undergo weak interactions. Quarks can interact via electromagnetic, weak and strong interactions.

1.3   Relativistic Kinematics

In this section, we will recall a few basic formulae of relativistic kinematics. In addition, we will discuss the relativistic kinematics of the two-body collision process and, also, the invariant mass of a many-particle system.

In relativistic mechanics, the *momentum* $\vec{p}$ of a material point of mass m_r and

---

where $h = 6.6260693 \times 10^{-34}\,\mathrm{J\,s}$ is the Planck constant (Appendix A.2).

*This is the value reported by the particle data group (e.g., see [PDB (2008)] and the web site: http://pdg.lbl.gov).

velocity $\vec{v}$ is

$$\vec{p} = m_r \vec{v}$$
$$= m_r \vec{\beta} c, \tag{1.1}$$

where

$$\vec{\beta} = \frac{\vec{v}}{c}, \tag{1.2}$$

is the ratio between the material-point velocity and the speed of light c. The mass m_r is also referred to as the *relativistic mass* and is related to the *rest mass, m_0,* by the so-called *Lorentz factor γ*:

$$m_r = \gamma m_0, \tag{1.3}$$

where

$$\gamma = \frac{1}{\sqrt{1 - \beta^2}} \tag{1.4}$$

and, conversely,

$$\beta = \frac{\sqrt{\gamma^2 - 1}}{\gamma}. \tag{1.5}$$

Thus, using Eq. (1.3), we can rewrite Eq. (1.1) as

$$\vec{p} = \vec{\beta} \gamma m_0 c \tag{1.6}$$

and consequently

$$\vec{\beta} \gamma = \frac{\vec{p}}{m_0 c}. \tag{1.7}$$

For the *total energy, E,* both the rest mass and the momentum have to be taken into account:

$$E = \sqrt{m_0^2 c^4 + p^2 c^2}, \tag{1.8}$$

or, equivalently from Eqs. (1.2–1.4)

$$E = \sqrt{m_0^2 c^4 + m_r^2 \beta^2 c^4} \tag{1.9}$$

$$= \sqrt{m_0^2 c^4 + (\gamma m_0)^2 \beta^2 c^4}$$

$$= m_0 c^2 \sqrt{1 + \beta^2 \gamma^2}$$

$$= \gamma m_0 c^2 \tag{1.10}$$

$$= m_r c^2. \tag{1.11}$$

The *kinetic energy* is given by

$$E_k = E - m_0 c^2 \tag{1.12}$$

and using Eq. (1.10) one finds:

$$E_k = m_0 (\gamma - 1) c^2, \tag{1.13}$$

or, equivalently, using Eq. (1.4)

$$E_k = m_0 c^2 \left(\frac{1}{\sqrt{1 - \beta^2}} - 1 \right). \tag{1.14}$$

Furthermore, from Eqs. (1.8, 1.12), one can also express the momentum (p) as:

$$m_0^2 c^4 + p^2 c^2 = \left(E_k + m_0 c^2 \right)^2$$
$$m_0^2 c^4 + p^2 c^2 = E_k^2 + 2 m_0 c^2 E_k + m_0^2 c^4,$$

from which one obtains

$$p = \frac{\sqrt{E_k \left(E_k + 2 m_0 c^2 \right)}}{c} \tag{1.15}$$

Let us indicate with $\mathcal{E}$ the quantity determined by the ratio

$$\mathcal{E} = \frac{E_k}{m_0 c^2}.$$

From Eqs. (1.10, 1.12), we have

$$\gamma = \mathcal{E} + 1; \tag{1.16}$$

thus, Eq. (1.5) can be rewritten as

$$\beta = \frac{\sqrt{\mathcal{E}(\mathcal{E} + 2)}}{\mathcal{E} + 1}, \tag{1.17}$$

and, finally, we get

$$\beta\gamma = \sqrt{\mathcal{E}(\mathcal{E} + 2)}. \tag{1.18}$$

The energy and the momentum of a particle in a second reference system, whose constant velocity is $-\vec{\beta}_0 c$ with respect to the original system, are obtained by the so-called *Lorentz transformations* (see for instance [Hagedorn (1964)]):

$$\vec{p} = \vec{p}' + \vec{\beta}_0 \gamma \left(\frac{\gamma}{\gamma + 1} \vec{\beta}_0 \cdot \vec{p}' + \frac{E'}{c} \right), \tag{1.19}$$

$$\frac{E}{c} = \gamma \left(\frac{E'}{c} + \vec{\beta}_0 \cdot \vec{p}' \right), \tag{1.20}$$

where E' and $\vec{p}'$ are the energy and the momentum in the original reference system. Furthermore, under Lorentz transformations, a time interval τ elapsed in a coordinate system, where the particle is at rest, is dilated by the Lorentz factor in a coordinate system moving with a velocity $-\beta c$ with respect to the particle (namely in the system in which the particle moves with a speed βc):

$$t = \gamma\tau. \tag{1.21}$$

For example, in its rest frame, a charged pion with a rest mass ≈ 139.57 MeV/c^2 has a mean-life of $\approx 2.6 \times 10^{-8}$ s. From Eq. (1.21), at 10 GeV in the laboratory frame (i.e., $\gamma \approx 72$ and $\beta \approx 1$), it becomes $t \approx 72 \times 2.6 \times 10^{-8} \approx 1.87 \times 10^{-6}$ s. Before decaying, the pion path is $l = c\,t \approx 562$ m.

1.3.1 *The Two-Body Scattering*

Radiation processes, like the ones resulting in energy losses by collision, take place in matter and can be considered (see following chapters) as two-body scatterings in which the target particle is almost at rest. In this section, we will study the kinematics of these processes and, in particular, derive equations regarding the maximum energy transfer.

Let us consider an incident particle [e.g., a proton (p), pion (π), kaon (K), etc.] of mass m and momentum $\vec{p}$, and a target particle of mass m_e [Rossi (1964)] at rest. For collision energy-loss processes, the target particle in matter is usually an atomic electron (see Fig. 1.1). After the interaction, two scattered particles emerge: the former with mass m and momentum $\vec{p}''$ and the latter with mass m_e and momentum $\vec{p}'$. The latter one has the direction of motion (i.e., the direction of the three-vector $\vec{p}'$) forming an angle θ with the incoming particle direction. θ is the angle at which the target particle is *scattered*. The kinetic energy [see Eq. (1.12)] of the scattered particle is related to its momentum by

$$E_k + m_e c^2 = \sqrt{p'^2 c^2 + m_e^2 c^4}, \tag{1.22}$$

from which we get

$$p'^2 = \frac{\left(E_k + m_e c^2\right)^2 - m_e^2 c^4}{c^2}. \tag{1.23}$$

The total energy before and after scattering is conserved. Thus, we have

$$\sqrt{p^2 c^2 + m^2 c^4} + m_e c^2 = \sqrt{p''^2 c^2 + m^2 c^4} + E_k + m_e c^2$$

and, consequently,

$$\sqrt{p''^2 c^2 + m^2 c^4} = \sqrt{p^2 c^2 + m^2 c^4} - E_k, \tag{1.24}$$

while from momentum conservation:

$$\vec{p}'' = \vec{p} - \vec{p}' \quad \Longrightarrow \quad p''^2 = p^2 + p'^2 - 2p\,p' \cos\theta. \tag{1.25}$$

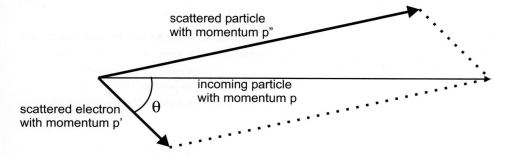

Fig. 1.1 Incident particle of mass m and momentum $\vec{p}$ emerges with momentum p'' = $|\vec{p}''|$, while the scattered electron emerges with momentum p' = $|\vec{p}'|$.

Equation (1.25) can be rewritten taking into account Eq. (1.23):

$$p''^2 = p^2 + \frac{\left(E_k + m_e c^2\right)^2 - m_e^2 c^4}{c^2} - 2p\cos\theta\sqrt{\frac{\left(E_k + m_e c^2\right)^2 - m_e^2 c^4}{c^2}},$$

which becomes, after substituting p'' obtained from Eq. (1.24) and squaring both sides of that equation,

$$E_k\sqrt{p^2 c^2 + m^2 c^4} = -E_k m_e c^2 + pc\cos\theta\sqrt{\left(E_k + m_e c^2\right)^2 - m_e^2 c^4},$$

from which we get

$$pc\cos\theta\sqrt{\frac{E_k^2 + 2E_k m_e c^2}{E_k^2}} = m_e c^2 + \sqrt{p^2 c^2 + m^2 c^4},$$

and, finally, by squaring both sides of the equation we can derive the expression for the kinetic energy E_k of the scattered target particle, i.e.,

$$E_k = \frac{2m_e c^4 p^2 \cos^2\theta}{\left(m_e c^2 + \sqrt{p^2 c^2 + m^2 c^4}\right)^2 - p^2 c^2 \cos^2\theta}. \qquad (1.26)$$

The kinetic energy E_k of the recoiling target particle is the amount of transferred energy in the interaction. From Eq. (1.26), we note that the maximum energy transfer W_m is for $\theta = 0$, i.e., when a head-on collision occurs. For $\theta = 0$, Eq. (1.26) becomes:

$$W_m = \frac{p^2 c^2}{\frac{1}{2}m_e c^2 + \frac{1}{2}\left(m^2/m_e\right)c^2 + \sqrt{p^2 c^2 + m^2 c^4}}. \qquad (1.27)$$

From Eq. (1.10), the incoming particle energy E_i is given by

$$E_i = m\gamma c^2 = \sqrt{p^2 c^2 + m^2 c^4}.$$

We can rewrite Eq. (1.27) as:

$$W_m = 2m_e c^2 \beta^2 \gamma^2 \left[1 + \left(\frac{m_e}{m}\right)^2 + 2\gamma\frac{m_e}{m}\right]^{-1}. \qquad (1.28)$$

Massive particles (e.g., proton[§], K, π etc.) are particles whose masses are much larger than the electron (or positron) mass m_e, i.e.,

$$m \gg m_e \ (\approx 0.511\,\text{MeV}/c^2).$$

For massive particles, at sufficiently high energies[‡], i.e., when the incoming momentum p is

$$p \gg \frac{m^2}{m_e}c,$$

[§]The rest mass of the proton is $\approx 938.27\,\text{MeV}/c^2$.

[‡]For instance, this condition is satisfied by an incoming π with momentum $\gg 36\,\text{GeV/c}$ or an incoming proton with momentum $\gg 1.7\,\text{TeV/c}$.

Eq. (1.27) becomes

$$W_m \approx pc \approx E_i.$$

In the extreme relativistic case, a massive particle can transfer all its energy to the target electron in a head-on collision, i.e., a proton can be stopped by interacting with an electron. At lower energies[†], i.e., when

$$p \ll \frac{m^2}{m_e}c,$$

the maximum energy transfer [see Eq. (1.27)] by particles with $m \gg m_e$ is approximated by

$$W_m \approx 2m_e c^2 \left(\frac{p}{mc}\right)^2$$

and, because $p = m\beta\gamma c$, we have:

$$W_m \approx 2m_e c^2 \frac{\beta^2}{1 - \beta^2} = 2m_e c^2 \beta^2 \gamma^2. \qquad (1.29)$$

For instance, a proton of $10\,\text{GeV}$ has a Lorentz factor $\gamma \approx 10$ and $\beta \approx 1$. Thus, its maximum energy transfer is $W_m \approx 100\,\text{MeV}$.

1.3.2 *The Invariant Mass*

The four-momentum of a particle with rest mass[‡] m_0 is defined as

$$\tilde{q} = \left(\frac{E}{c}, \vec{p}\right).$$

The scalar product between two four-momenta $\tilde{q}$ and $\tilde{q}'$ is an invariant[**] quantity and is given by (e.g., Section 38 of [PDB (2008)])

$$\tilde{q} \cdot \tilde{q}' = \frac{E\,E'}{c^2} - \vec{p} \cdot \vec{p}'. \qquad (1.30)$$

The *invariant mass* of a particle is related to the scalar product of its four-momentum by:

$$\begin{aligned}
\tilde{q} \cdot \tilde{q} &= q^2 \\
&= \frac{E^2}{c^2} - \vec{p} \cdot \vec{p} \\
&= \frac{E^2}{c^2} - p^2 \\
&= m_0^2 c^2
\end{aligned}$$

[†]For instance, this condition is satisfied by an incoming π with momentum $\ll 36\,\text{GeV/c}$ or an incoming proton with momentum $\ll 1.7\,\text{TeV/c}$.

[‡]The *rest mass* is the mass of a body that is isolated (free) and at rest relative to the observer.

[**]The *invariant mass* or *intrinsic mass* or *proper mass* or just *mass* is the mass of an object that is the same for all frames of reference.

and, finally,

$$m_0 = \sqrt{\frac{\tilde{q} \cdot \tilde{q}}{c^2}}. \tag{1.31}$$

The invariant mass of a single particle is its rest mass.

The invariant mass, M, of a set of particles is the energy available in their center-of-mass system. It is given by

$$M = \sqrt{\frac{\tilde{q}_s^2}{c^2}} = \sqrt{\frac{\left[\sum_i (E_i/c)\right]^2 - \left(\sum_i \vec{p}_i\right) \cdot \left(\sum_i \vec{p}_i\right)}{c^2}}, \tag{1.32}$$

where

$$\tilde{q}_s = \sum_i \tilde{q}_i$$

is the total four-momentum. Let us consider two particles with masses m_1 and m_2 and momenta $\vec{p}_1$ and $\vec{p}_2$. From Eq. (1.32), we have that their invariant mass $M_{1,2}$ is:

$$M_{1,2} = \frac{1}{c}\sqrt{\left(\frac{E_1 + E_2}{c}\right)^2 - p_1^2 - p_2^2 - 2p_1 p_2 \cos\theta}$$

$$= \frac{1}{c}\sqrt{2\frac{E_1 E_2}{c^2} + m_1^2 c^2 + m_2^2 c^2 - 2p_1 p_2 \cos\theta}, \tag{1.33}$$

where θ is the angle between the three-vectors $\vec{p}_1$ and $\vec{p}_2$. For example, let us take a proton of 100 GeV incident on a target proton at rest in a high-energy physics fixed target experiment. From Eq. (1.33), because $p_2 = 0$, $E_2 = m_2 c^2$, $m_1 = m_2 \approx 1$ GeV/c^2, the available center-of-mass energy (i.e., the invariant mass) becomes

$$M_{1,2} \approx \sqrt{200 + 1 + 1} \approx 14.2 \text{ GeV}/c^2.$$

Furthermore, in the scattering between an incoming particle 1 and a target particle 2, we define the invariant quantity s as:

$$s = \left(\tilde{q}_1 + \tilde{q}_2\right)^2$$

$$= m_1^2 c^2 + m_2^2 c^2 + 2\frac{E_1 E_2}{c^2} - 2\vec{p}_1 \cdot \vec{p}_2. \tag{1.34}$$

If the particle 1 (2) emerges as particle 3 (4), the invariant quantity s is also given by

$$s = \left(\tilde{q}_3 + \tilde{q}_4\right)^2.$$

From Eq. (1.33), s is the invariant mass square of the system 1,2 (3,4) times c^2, i.e., the square of the total energy in the center-of-mass system divided by c^2. For the same reaction, we define the invariant quantity t as the *square of four-momentum transfer*:

$$t = \left(\tilde{q}_1 - \tilde{q}_3\right)^2 = \left(\tilde{q}_2 - \tilde{q}_4\right)^2 \tag{1.35}$$

$$= m_1^2 c^2 + m_3^2 c^2 - 2\frac{E_1 E_3}{c^2} + 2\vec{p}_1 \cdot \vec{p}_3. \tag{1.36}$$

Both s and t are called *Mandelstam variables.* There is a third Mandelstam variable

$$u = \left(\tilde{q}_1 - \tilde{q}_4\right)^2 = \left(\tilde{q}_3 - \tilde{q}_2\right)^2.$$

However, it is not independent of s and t, as

$$s + t + u = m_1^2 c^2 + m_2^2 c^2 + m_3^2 c^2 + m_4^2 c^2.$$

1.4 Cross Section and Differential Cross Section

The cross section, σ, for a physical process is derived from the reaction probability for the occurrence of such an interaction. More precisely, when a collimated particle beam impinges on a target (see Fig. 1.2), some particles are removed by the physical reactions, resulting in an attenuated beam. The physical reactions occurring between the beam and the target particles include for example elastic scattering and particle production. A net difference between the incoming and outgoing particles can be measured and the removal probability of beam particles can be determined.

The simplest way of representing such a reaction is to imagine the incoming beam made of a uniform distribution of particles and the target as made of a disk onto which a certain amount of beam particles interact. This way, particles impinging onto the disk surface interact, while particle outside this surface continue their trajectory unaffected. However, this naive point of view has to account for the finite dimensions of both projectiles and target. This disk does not coincide with the *geometrical cross section* presented by the target and depicted by

$$\sigma_g = \pi R_g^2$$

in Fig. 1.2, where R_g is the physical (i.e., geometrical) radius of the target. It means the *effective area* struck by incoming point-like particles. This effective area is the so-called *total reaction cross section* (σ_{total} in Fig. 1.2) and takes into account the different types of reactions (often referred to as *partial cross sections*) between the projectile and the target. In interactions among particles, or particles and nuclei, or particles and atoms, the cross section size is usually expressed in units of *barn* indicated by b (see Appendix A.1):

$$1\,\text{b} = 10^{-24}\,\text{cm}^2 = 10^{-28}\,\text{m}^2.$$

Let us have a monochromatic beam of F_0 particles, for which σ_{total} is the total atomic cross section for all interaction processes between incoming particles

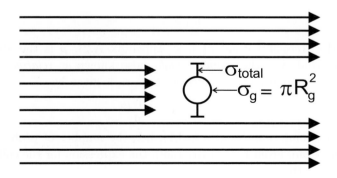

Fig. 1.2 Reaction and geometrical cross sections (see for instance [Marmier and Sheldon (1969)]).

and target atoms inside the absorber. In addition, we suppose that the overall absorber thickness is such that the probability of double particle interaction can be neglected. In the passage through a thickness dx' of the medium, the number of removed particles $-dF$ (the minus sign indicates that particles are removed from the beam) is proportional to the number of the beam particles F' at depth x' and to the number of target atoms per unit of volume, n_A, of the traversed material:

$$-dF = F' P_{\text{int}},$$

where $P_{\text{int}} = (\sigma_{\text{total}} dx') n_A$ is the probability of removing a particle in the thickness dx'. It has to be noted that n_A is the reciprocal of the atomic volume. We have:

$$\begin{aligned} -dF &= F'(\sigma_{\text{total}} dx') n_A \\ &= F'(\sigma_{\text{total}} n_A) dx' \\ &= \frac{F'}{\lambda_{\text{col}}} dx', \end{aligned}$$

where

$$\lambda_{\text{col}} = \frac{1}{n_A \sigma_{\text{total}}}. \tag{1.37}$$

The coefficient λ_{col} has the dimension of a length and is the so-called *collision length*, i.e., it is the *mean free path* between successive collisions. As a consequence, by traversing a thickness x of the absorber, we have:

$$\frac{dF}{F'} = -\frac{1}{\lambda_{\text{col}}} dx' \Rightarrow \int_{F_0}^{F} \frac{dF}{F'} = \int_0^x -\frac{dx'}{\lambda_{\text{col}}} \Rightarrow \ln \frac{F}{F_0} = -\frac{x}{\lambda_{\text{col}}},$$

and, finally,

$$F = F_0 \exp\left[-\left(\frac{x}{\lambda_{\text{col}}}\right)\right]. \tag{1.38}$$

Thus, there is an exponential decrease of the number of particles upon the passage in the absorbing medium.

The value of n_A, in units of cm^{-3}, is given by

$$n_A = \frac{N\rho}{A}, \tag{1.39}$$

where N is the *Avogadro constant* (see Appendix A.2), ρ is the absorber density, in g/cm^3, and A is the *atomic weight* [also known as *relative atomic mass* (Sect. 1.4.1)]. The *number of electrons* per cm^3, n, is given by

$$n = Z n_A = \frac{Z N \rho}{A}, \tag{1.40}$$

where Z is the *atomic number* (Sect. 3.1), i.e., the number of protons inside the nucleus of that atom. From Eqs. (1.37, 1.39), the expression for the collision length in cm can be rewritten as:

$$\lambda_{\text{col}} = \frac{A}{N\rho\, \sigma_{\text{total}}}, \tag{1.41}$$

where σ_{total} is in cm^2.

An interaction, which results in the emission of a reaction product, can depend on parameters like the incoming energy or the emission angle. Therefore, we can introduce the so-called *differential cross section* to express the dependence of the emission probability on these parameters. For instance, the differential cross section per unit of solid angle $\frac{d\sigma}{d\Omega}$ gives, once multiplied by the solid angle element $d\Omega$, the incoming particle cross section to yield the reaction product into the element of solid angle $d\Omega$ lying at a mean angle θ with respect to the incident beam direction (the so-called *scattering angle*) and at a mean *azimuthal angle* ϕ. We have:

$$\sigma = \int_0^\Omega \frac{d\sigma}{d\Omega}\, d\Omega = \int_{\phi=0}^{2\pi} \int_{\theta=0}^{\pi} \frac{d\sigma}{d\Omega} \sin\theta\, d\theta\, d\phi,$$

where σ is the cross section for the reaction and $d\Omega = \sin\theta\, d\theta\, d\phi$.

1.4.1 *Atomic Mass, Weight, Standard Weight and Atomic Mass Unit*

The *atomic mass*** is the rest mass of an atom in its ground state. The commonly used unit is the unified atomic mass unit (indicated by the symbol u, see Appendix A.2). The *unified*[††] *atomic mass unit*, as adopted by the International Union of Pure and Applied Chemistry (IUPAC) in 1966, is used to express masses of atomic particles and is defined to be exactly one twelfth of the mass of a ^{12}C atom in its ground state. The unified atomic mass unit replaced the *atomic mass unit* (chemical scale) and the *atomic mass unit* (physical scale), both having the symbol amu. The amu (physical scale) was one sixteenth of the mass of an atom of ^{16}O. The amu (chemical scale) was one sixteenth of the average mass of oxygen atoms as found in nature.

The *atomic weight* (also known as *relative atomic mass*[‡]) of an element can be determined from the knowledge of the isotopic abundances and corresponding atomic masses of the nuclides (e.g., see [Tuli (2000); IUPAC (2006); NNDC (2008a)]) of that element as found in a particular environment: it is expressed by the ratio of the average, weighted by isotopic abundance, of atomic masses of all its isotopes to the unified atomic mass unit.

The *standard atomic weights* are the recommended values of relative atomic masses of the elements determined by their isotopic abundances in the surface and atmosphere of the Earth and are revised biennially by IUPAC. For instance, hydrogen has a standard atomic weight of 1.00794 [IUPAC (2006)] (see Appendix A.3 and references therein).

[**]As it is defined in the IUPAC Compendium of Chemical Terminology. The reader can see the web site: *http://www.iupac.org/goldbook/A00496.pdf*.

[††]One unified atomic mass unit (u) is equal to $(1/N)$ gram, where N is the Avogadro constant.

[‡]As it is defined in the IUPAC Compendium of Chemical Terminology. The reader can see the web site: *http://www.iupac.org/goldbook/R05258.pdf*.

It has to be noted that, in nuclear-physics, the symbol amu is the standard notation for particle masses expressed in relative atomic masses when, for example, ion kinetic-energies are given in MeV/amu or in GeV/amu. It is also of common usage, for instance in space physics, that these kinetic-energies (E_k) using Eq. (1.13) are expressed in MeV/nucleon or in GeV/nucleon (also termed *specific energy*, e.g., see Section 2.5.3 in [ICRUM (2005)]) as:

$$\frac{E_k}{M_A} = u\,c^2\,(\gamma - 1),$$

where M_A is the mass number of the atom (Sect. 3.1). As noted in Section 2.5.3 of [ICRUM (2005)], in general no distinction is made between energy per nucleon and energy per atomic mass unit, because the small numerical difference[§] (e.g., see discussion in Sect. 3.1). Furthermore using Eq. (1.14), the specific energy of an ion at the Bohr velocity, $v_0 = c\,\alpha$ (see page 74 and Appendix A.2), is given by:

$$\left(\frac{E_k}{M_A}\right)_{v=v_0} = u\,c^2\left(\frac{1}{\sqrt{1-\alpha^2}} - 1\right) = 0.02489\,\text{MeV}.$$

1.5   Classical Elastic Coulomb Scattering Cross Section

Rutherford (1911) derived the classical differential cross section for the elastic Coulomb scattering of charged particles in connection with his proposal of atomic model. In this model, the atom consisted of a very small (i.e., almost point-like) nucleus surrounded by a more diffuse electron distribution. The mass and charge were supposed to be concentrated in the nucleus. This theory explained successfully experimental results[‡‡] from the scattering of α-particles upon gold target.

A complete derivation of the *Rutherford differential cross section* (also termed *Coulomb cross section* or *classical cross section*) can be found, for instance, in Section 2.1 of [Melissinos (1966)], in Section 3.5.7 of [Marmier and Sheldon (1969)] and in Section 2.2 of [Segre (1977)]. In the treatment, the incoming particle with charge ze^*, rest mass m and non-relativistic velocity[†] $v = \beta c$ scatters at an angle θ upon a target particle initially at rest with charge Ze and rest mass M under the action of a repulsive Coulomb force in the laboratory system. In the center-of-mass system (CoMS) for the reaction and assuming that the azimuthal distribution is

---

[§]The difference amounts to at most 0.25% for stable isotopes of all elements from lithium upward (Section 2.5.3 in [ICRUM (2005)]).
[‡‡]These results were obtained by Geiger and Marsden (1913).
[*]e is the electron charge.
[†]Under this assumption, we have $\gamma \simeq 1$, thus the kinetic energy and momentum of the incoming particle are $E_k \simeq mv^2/2$ and $p \simeq mv = m\beta c$, respectively.

isotropic, the differential cross section for scattering into a solid angle[**]

$$d\Omega' = \int_0^{2\pi} d\varphi' d\cos\theta' \; (\varphi' \text{ is the azimuthal angle})$$

$$= 2\pi \sin\theta' d\theta'$$

is given in cgs esu units by:

$$\frac{d\sigma^{\text{Rut}}}{d\Omega'} = \left\{\frac{zZe^2}{(mMv^2)/[2(m+M)]}\right\}^2 \left[\frac{1}{4\sin^2(\theta'/2)}\right]^2 \tag{1.42}$$

$$= \left\{\frac{zZe^2}{(4ME_k)/(m+M)}\right\}^2 \frac{1}{\sin^4(\theta'/2)}, \tag{1.43}$$

where θ' is the scattering angle[‡] {see Equation (3.124) of [Marmier and Sheldon (1969)]}.

Furthermore, for $M \gg m$ the distinction between laboratory (i.e., the system in which the target particle is initially at rest) and center-of-mass system disappears and $\theta \approx \theta'$; thus, Eqs. (1.42, 1.43) reduce to *Rutherford's formula*

$$\frac{d\sigma^{\text{Rut}}}{d\Omega} \simeq \left(\frac{zZe^2}{2p\beta c}\right)^2 \frac{1}{\sin^4(\theta/2)} \tag{1.44}$$

$$= \left(\frac{zZe^2}{4E_k}\right)^2 \frac{1}{\sin^4(\theta/2)}$$

$$= \left(\frac{zZe^2}{2}\right)^2 \frac{1}{4\,E_k^2 \sin^4(\theta/2)}. \tag{1.45}$$

For E_k in MeV, Eq. (1.45) is re-expressed as

$$d\sigma^{\text{Rut}} \simeq 0.12951 \times 10^{-26} \times \left(\frac{zZ}{E_k}\right)^2 \frac{1}{\sin^4(\theta/2)} \, d\Omega \; [\text{cm}^2/\text{nucleus}]. \tag{1.46}$$

Rutherford's formula is non-relativistic and is valid for a fixed scattering center. In addition, it does not account for nuclear forces. Nevertheless, in practice it can be considered an appropriate approximation for describing particle scattering, when this occurs at a distance to the target particle larger than its nuclear radius [see Eq. (3.12)). It has to be noted that, for a small angle scattering, Eq. (1.44) becomes

$$\frac{d\sigma^{\text{Rut}}}{d\Omega} \simeq \left(\frac{2\,zZe^2}{p\beta c}\right)^2 \frac{1}{\theta^4}. \tag{1.47}$$

This equation is also valid at relativistic velocities (e.g., see discussion in Section 2.2 of [Segre (1977)]). Rutherford's formula can also be obtained using, for instance, the *Born approximation*[*] in a quantum mechanical approach (e.g., see Section 2.2 and Appendix A of [Segre (1977)]).

[**]This corresponds to an angular aperture between θ' and $\theta' + d\theta'$.

[‡]For the scattering angle we have $0° < \theta' < 180°$.

[*]In quantum mechanical potential scattering, the scattered wave may be obtained by the so-called Born expansion. The Born approximation is the first term of the Born expansion (e.g., see Sections 3-1–3-2a of [Roman (1965)] and Section 7.2 of [Sakurai (1994)]).

Let us now derive the expression for the differential cross section with respect to the transferred energy in the scattering. We can compute the quantity t, i.e., the square of the four momentum transfer[‡‡], in i) the laboratory system where the target particle, initially at rest, recoils with a kinetic energy T equal to the amount of energy transferred in the reaction and ii) the CoMS where the incoming particle with momentum p_{cm} is scattered[§] at an angle θ'. Because the square of the four momentum transfer is an invariant quantity, from Eqs. (1.35, 1.36) we have

$$2M^2c^2 - \frac{2Mc^2(Mc^2+T)}{c^2} = 2mc^2 - \frac{2(p_{\mathrm{cm}}^2c^2+m^2c^4)}{c^2} + 2p_{\mathrm{cm}}^2\cos\theta'$$

$$2M^2c^4 - 2Mc^2(Mc^2+T) = 2(p_{\mathrm{cm}}^2c^2\cos\theta' + mc^4) - 2(p_{\mathrm{cm}}^2c^2+m^2c^4)$$

$$-2MT = -2p_{\mathrm{cm}}^2(1-\cos\theta')$$

and, consequently,

$$T = \frac{p_{\mathrm{cm}}^2}{M}\,(1-\cos\theta')\,. \tag{1.48}$$

The incoming particle momentum in the CoMS can be expressed in terms of incoming-particle laboratory momentum and rest mass of the particles (e.g., see Equation 38.6 of [PDB (2008)]):

$$p_{\mathrm{cm}} = p\frac{M}{M_{\mathrm{cm}}},$$

where M_{cm} is the reaction invariant-mass, which can be computed using Eq. (1.33); thus, we obtain

$$p_{\mathrm{cm}} = p\frac{M}{\sqrt{m^2 + M^2 + 2\,M\sqrt{(p/c)^2+m^2}}}\,. \tag{1.49}$$

The quantity $\cos\theta'$ can be rewritten as:

$$\cos\theta' = \cos^2(\theta'/2) - \sin^2(\theta'/2)$$

$$= 1 - 2\sin^2(\theta'/2). \tag{1.50}$$

Using Eqs. (1.49, 1.50), Eq. (1.48) becomes

$$T = \left[\frac{p^2M}{m^2+M^2+2\,M\sqrt{(p/c)^2+m^2}}\right]\left[2\sin^2(\theta'/2)\right]$$

$$= \frac{2\,p^2M}{m^2+M^2+2\,M\sqrt{(p/c)^2+m^2}}\sin^2(\theta'/2)\,. \tag{1.51}$$

In a non-relativistic scattering for which $pc \ll mc^2$, Eq. (1.51) reduces to

$$T \approx \frac{2\,p^2M}{m^2+M^2+2\,Mm}\sin^2(\theta'/2)$$

$$= \frac{2\,p^2M}{(m+M)^2}\sin^2(\theta'/2)$$

$$= \frac{4\,mME_k}{(m+M)^2}\sin^2(\theta'/2) \tag{1.52}$$

[‡‡]The reader can see page 12 and, also, discussion in Sect. 1.3.2.

[§]In the CoMS, the scattering angle is the rotation angle of the particle momentum.

{e.g, see Equation (2-62) of [Ziegler, Biersack and Littmark (1985a)] or, equivalently, Equation (2-75) of [Ziegler, J.F. and M.D. and Biersack (2008a)]}. It has to be noted that the maximum energy transfer$^{\parallel}$ $T_{\max}$ occurs for $\theta' = 180°$, i.e., in case of a *head-on repulsive collision*, and is given by

$$T_{\max} = \frac{4\,mME_k}{(m+M)^2};$$ (1.53)

thus, the recoil energy can be written as

$$T = T_{\max}\sin^2(\theta'/2).$$ (1.54)

Since [e.g., see Eq. (1.50)]

$$d\Omega' = 2\pi\sin\theta'\,d\theta'$$
$$= 2\pi\,d\cos\theta'$$
$$= 4\pi\,d[-\sin^2(\theta'/2)],$$

using Eq. (1.52) we can rewrite the Rutherford cross section [Eq. (1.43)] in terms of the recoil kinetic energy of the target particle (i.e., in terms of the energy transferred in the Coulomb interaction) as:

$$d\sigma^{\text{Rut}} = \left\{\frac{zZe^2}{(4ME_k)/(m+M)}\right\}^2\frac{1}{\sin^4(\theta'/2)}\,d\Omega'$$
$$= \left\{\frac{zZe^2}{(4ME_k)/(m+M)}\right\}^2\left[\frac{4\,mME_k}{T(m+M)^2}\right]^2 4\pi\,d[-\sin^2(\theta'/2)].$$

In the latter equation we can introduce Eq. (1.43), thus, we get:

$$d\sigma^{\text{Rut}} = 4\pi\left[\frac{mzZe^2}{T(m+M)}\right]^2\frac{(m+M)^2}{4\,mME_k}\,|d(-T)|$$
$$= \pi\,\frac{m(zZe^2)^2}{ME_k}\frac{1}{T^2}\,dT,$$

where the negative sign in the term $d(-T)$ indicates that the incoming particle looses energy as a result of the interaction§; this energy is absorbed via the target recoil. Finally, the differential cross section corresponding to a transferred energy between T and $T+dT$ (e.g, see Equation 7 of [Bakale, Sowada and Schmidt (1976)]) is given by

$$\frac{d\sigma^{\text{Rut}}}{dT} = \pi\,\frac{m(zZe^2)^2}{ME_k}\frac{1}{T^2}$$
$$= 2\pi\,\frac{(zZe^2)^2}{Mv^2}\frac{1}{T^2}.$$ (1.55)

In Coulomb interactions, for instance those resulting in displacement damage for silicon devices (e.g., see discussions in Sects. 4.2.1.3 and 7.1.3), the average energy

---

$^{\parallel}$$T_{\max}$ can also be obtained using Eq. (1.27) for $\gamma = 1$.

§The Coulomb scattering of charged particles on atomic electrons is the dominant mechanism of *collision* (also termed *electronic*) *energy-loss process* discussed, for instance, in Sect. 2.1.1.

transferred in the reaction determines the extension of the subsequent cascade development of atomic displacements. Assuming that the interaction is purely electrostatic and neglecting the screening effect[*] of nuclear Coulomb potentials, to a first approximation the average energy transferred $\langle T \rangle$ above a minimum energy threshold $E_{\min}$ up the maximum energy-transfer can be obtained by means of Eq. (1.55) and it is expressed as:

$$
\begin{aligned}
\langle T \rangle &= \int_{E_{\min}}^{T^{\max}} \frac{d\sigma^{\mathrm{Rut}}}{dT} T \, dT \bigg/ \left(\int_{E_{\min}}^{T^{\max}} \frac{d\sigma^{\mathrm{Rut}}}{dT} \, dT \right) \\
&= \int_{E_{\min}}^{T^{\max}} \frac{1}{T} \, dT \bigg/ \left(\int_{E_{\min}}^{T^{\max}} \frac{1}{T^2} \, dT \right) \\
&= \frac{E_{\min} T^{\max}}{T^{\max} - E_{\min}} \ln \left(\frac{T^{\max}}{E_{\min}} \right).
\end{aligned}
\tag{1.56}
$$

1.6   Detectors and Large Experimental Apparata

Radiation detectors use *detecting* (sometime referred to as *active*) *media* and *readout* systems. The operation of any detecting device is based on specific effects of radiation interaction in matter. These effects are exploited in order to produce quantitative measurable signals in the readout system associated with the detection system itself.

The basic features of radiation detectors can be understood once fundamental processes of radiation interaction in matter are considered (see chapters on *Electromagnetic Interaction of Radiation in Matter* and on *Nuclear Interactions in Matter*). For instance, the collision energy-loss is the mechanism generating primary electrons (which in turn can generate secondary electrons) in gaseous detectors, or excited molecules (which decay emitting photons) in scintillating devices, or electron–hole pairs in semiconductors, etc.. The primary mechanism of transferring energy from an incoming charged particle to a medium has to enable the generation of secondary detectable particles, whose number or flux of energy has to be as much as possible linearly related to the incoming number of particles or incoming particle energy. Radiation detectors, or combinations of them, can provide a wide range of information as spatial locations, particle momentum, velocity, energy, etc..

For instance, in nuclear medicine, the image formation is based on the spatial reconstruction of the radiation emitted from a patient. The detecting system needs to be complex and consists of many sub-detectors in order to cover large emission area and provide a 3-dimensional reconstruction of the internal volume of the patient under investigation. In high-energy physics experiments, detectors are located downstream a fixed target or surrounding the collision point in colliding beams ma-

---

[*]The screening of the Coulomb potential cannot be neglected at low energy (e.g., see discussion in Sect. 4.2.1.3).

chines. In space experiments, sets of highly reliable detectors are located on board of satellites and recently on board of the International Space Station for photons and energetic cosmic rays detection. Particles created in high energy collisions or impinging on a space detector, pass through various types of detectors with dedicated tasks. Examples are:

- tracking, with capability of momentum analysis in *magnetic spectrometers*,
- electron and hadron separation,
- particle identification,
- energy determination,
- triggering,
- data acquisition,
- monitoring.

Each sub-detector has very specific characteristics and its functionality has to be optimized taking into account the features of other sub-detectors.

Particles passing through *tracking detectors* (or *trackers*) have their trajectories reconstructed by dedicated computer software codes. The accuracy of reconstructed trajectories depends on the spatial resolution of the tracker, which can be of a few μm's for *semiconductor detectors* (see the chapter on *Solid State Detectors*), ≈ 100 μm for *drift chambers* and a few hundreds μm's for *MultiWire Proportional Chambers* (MWPC). The two latter ones are gas based devices. The accuracy also depends on the multiple Coulomb scattering inside the tracker. In *magnetic spectrometers*, the particle momentum can be determined once the particle trajectory has been reconstructed.

As will be discussed in details in this book, electrons, photons and hadrons can create electromagnetic and hadronic showers in matter (see Sects. 2.4 and 3.3). The characteristics of the shower can be determined with calorimeters, as discussed in the chapter on *Principles of Particle Energy Determination*. These detectors can measure both the energy and the impact position of the incoming particle. In addition, the distribution of energy deposition along and perpendicularly to the incoming particle direction is used to discriminate among incoming electrons/photons and hadrons.

Other devices, such as the *Transition Radiation Detectors* (TRD), can provide signatures for discriminating electrons from hadrons at high energy. These devices are classified among the so-called *particle identification detectors*, like the *Čerenkov detectors*.

All the detectors or sub-detectors in an experiment or in a large, complex detection system must be carefully integrated. The design system does not need to exploit all the best sub-detector features. It only needs to make them suited to the purpose of the overall resulting apparatus. Fast outputs from sub-detectors or dedicated detectors provide *trigger signals*, which indicate that a particular type of *event* has occurred inside the apparatus. The trigger looks for spatial and/or temporal correlation of detectors or sub-detector signals, which have been set in designing the apparatus. The trigger can also be based on threshold, such as energy threshold.

Calibration and on-line monitoring are tasks which have to be particularly investigated and implemented in large and complex apparata. Data are usually collected via a *data acquisition system* (DAQ), which sends them or part of them to monitoring computer codes and the recording system.

Computers play a central role in processing data and afterwards in displaying processed data. Much of the software code used in the reconstruction of physical events is detector or apparatus dependent. Graphical routines for data display are by themselves an important field of continuous development and require more and more powerful processors.

1.6.1   *Trigger, Monitoring, Data Acquisition, Slow Control*

A trigger signal is an electronic signal which indicates the occurrence of an *event* which has to be processed and, possibly, collected by the data acquisition part of the apparatus. A well designed trigger avoids the data collection system to be swamped by similar but background events. This way, expected events together with a minimal amount of background events will be collected and processed. For instance, the trigger can be constructed to identify particles, to separate electrons from hadrons, to count the event multiplicity, etc.. For large apparata, the trigger is organized with a few levels following a hierarchical sequence. Programmable devices are commonly employed such as look-up tables, hard-wired processors, microprocessors, emulators.

In recent years, the pipelined processing has been often adopted in order to avoid loss of information. It is typically employed for event recording and storage in high performance computing systems. There are different types of pipelines for storing signals coming from various parts of the apparatus. The pipeline can be both analog and digital.

Particularly interesting events, as well as events sampled on a statistical basis, can be passed to the on-line event display task. This latter task can be integrated inside the on-line monitoring and calibration tasks, which allow us to verify that the whole detector is properly functioning during the data collection. Within this framework, automatic processes search for anomalous behaviors of sub-detectors and send warning messages or try to readjust remote controlled sub-detector elements to restore the working condition. Sets of calibration constants are also collected in systematically updated databases and used in reconstruction procedures.

The term "data acquisition" includes the data collection from the whole apparatus, data storage on accessible media and subsequently to external software codes for reconstruction and display.

Complex detector systems usually require that sub-detector operations and readjustments be carried out by dedicated remote software codes. Independently, if the event has to be considered or not as a background event, sub-detector operations are expected to be performed properly when some detector parameters are within predefined value ranges. These quantities are for instance high voltages and currents of sub-detectors, flow of gases (for gaseous detectors), temperature distribution, etc.. The slow control includes the recording, monitoring and control of all parameters which are expected to be within predefined ranges during data collection.

1.6.2   *General Features of Particle Detectors and Detection Media*

The spatial resolution depends on the detector type and, among detectors using similar media, on the type of readout system. The *resolution time* and the *dead time*¶ also depend both on the detector type. Fast *drift chambers* have resolution times of $\simeq 2$ ns and dead times of $\simeq 100$ ns. Scintillator devices can reach resolution and dead times of about (100–150) ps and 10 ns, respectively. *Photographic emulsions* have spatial resolution of $\simeq 1\,\mu$m. *Silicon strip** and *silicon pixel detectors* can achieve spatial resolutions of $\simeq$ a few mm and $2\,\mu$m (and less), respectively.

The *photomultiplier* is a suitable vacuum tube designed as a source of primary electrons (*photoelectrons*) emitted by the photocathode and as a tool of electron amplification by secondary emission process. The amplification factor can be as large as 10^7 (for 12 stages). Photomultipliers are widely used in association with detection media in which photons are emitted, like scintillators and Čerenkov detectors. Photomultiplier operations can be limited, or even completely impaired, inside strong magnetic fields.

Organic scintillators are classified as organic crystals (i.e., anthracene, naphthalene, etc.), liquids‖ or plastics** depending on the type of the scintillating medium in usage and their densities are between $\simeq (1.03$–$1.25)\,$g/cm^3. The scintillation mechanism is particularly noticeable in organic substances containing aromatic rings, for instance polystyrene, polyvinyltoluene and naphtalene. In liquid scintillator, molecules of toluene and xylene are typically included. About 3% of the deposited energy is re-emitted as optical photons. By far, the plastic scintillators are the most widely employed. Photon emission yields are approximately 100 eV of energy deposited inside the scintillating medium (see chapter on *Scintillating Media and*

¶The dead time is the time during which a detector is not capable of detecting a next coming particle.

*One of the first examples of the usage of silicon strip and (in a second time also) microstrip detectors was for the active Si-target of the NA14 experiment at CERN-SPS [Rancoita and Seidman (1982); Barate et al. (1985)]; another was the microstrip detectors assembled as a silicon counter telescope for the target of the NA11 experiment [Rancoita and Seidman (1982); Belau, Klanner, Lutz, Neugebauer and Wylie (1983)].

‖In a liquid scintillator an organic crystal (solute) is dissolved in a solvent.

**Plastic scintillators are similar in composition to liquids. Polystyrene and polyvinyl toluene are commonly used as base plastics to replace the solvent.

Scintillator Detectors). A *minimum ionizing particle** can generate up to 2×10^4 photons traversing $\sim 1\,$cm of plastic scintillator. However, only a small fraction, typically less than 10%, of the generated photons arrive on the photomultiplier photocathode, whose *quantum efficiency*[†] does not exceed $\sim 30\%$ for the most favorable photon frequency. In addition, local ionizations much larger than those generated by minimum ionizing particles emit less light. Plastic scintillators are reliable and robust. With their high hydrogen content, they are particularly suited for neutron detection. However, their light yield degrades due to aging effects. For instance, aging can be enhanced by exposure to solvent vapors, irradiation and mechanical flexing. The radiation damage depends not only on the integrated dose, but on the dose rate, and environmental factors (like temperature and atmosphere) before and after the irradiation, as well as on material properties. Commonly used inorganic scintillators have larger densities, between $\simeq (3.67\text{–}8.28)\,\mathrm{g/cm^3}$. They are very efficient for the detection of electrons and photons; they are employed when large densities and good energy resolution are required.

Detectors using a gas, or more likely a mixture of gases, have undergone a great deal of development since the first planar detector geometry: the MWPC (Multi-Wire Proportional Chamber), realized a few decades ago. This planar geometry has allowed one to exploit the cylindrical gas detector properties (see the chapter on *Ionization Chambers*) regarding both the small electron multiplication volume around the anode wire and the large drift volume available to positive ions at almost constant electric field towards the cathode. In the detector, ion pairs are generated in a gas layer, whose thickness is typically between a few cm's and a fraction of a cm. Anodes are regularly separated by less than 2 mm, but not much less than 1 mm, because there are practical difficulties of precisely stringing wires below 1 mm and in addition the mechanical tension, balancing the electrostatic force between wires, cannot exceed its critical value. This allows the achievement of spatial resolutions up to a few hundreds μm. These detectors need an individual readout channel per anode. Similar or even better spatial resolutions, with a small readout channel density, can be achieved with the so-called *drift chambers*. In these devices, the position of the passing particle is determined by the time difference between the passage of the particle and the arrival of electrons at the wire. Detectors (the so-called *time projection chamber*) with long drift distances perpendicularly to a multi-anode proportional plane provide three-dimensional information. Large volume chambers up to a few tens of m^3 and several thousands of wires have been successfully operated in high energy physics experiments. The anode spacing limitation can be overcome by using lithographic technique, which have been able to produce a miniaturized version of a MWPC with thin aluminum strips engraved in an insulating support and an anode spacing reduced to $(0.1\text{–}0.2)\,$mm.

Silicon detectors are the most widely used semiconductor detectors. They are p–n

---

*The reader can see the definition at page 47.

[†]The quantum efficiency is the probability of emitting a photoelectron per impinging photon.

junction diodes operated in reverse bias (see chapter on *Solid State Detectors*). Their substrate is typically n-type high resistivity silicon with thickness between (300–500) μm. Full depletion voltages are usually between (50–150) V for 300 μm thick detectors, depending on their resistivity. The energy needed to be deposited inside the detector active volume to create an *electron–hole (e-h) pair* is about 3.6 eV. Electrons and holes are referred to as *carriers*. A minimum ionizing massive particle[‡] loses on average approximately 30 keV per 100 μm in a silicon detector,[§] i.e., it generates approximately 83 e-h pairs/μm. The *transit time*, i.e., the time needed by carriers to drift towards the electrodes and, consequently, to induce electric signals on them, decreases as the reverse bias voltage necessary for full depletion increases. Typical transit times are $\leq$ 10 ns for electrons and $\leq$ 25 ns for holes in the case of a fully depleted 300 μm thick device. The transit time is mainly limited by the carrier mobility saturation for electric field larger than $\simeq 10^4$ V/cm. In the beginning of the 1960s, first demonstrations were made that silicon detectors operated at room temperature could be used for nuclear reaction studies and spectroscopy. Since then, a great deal of work and combined efforts have been carried out, making this type of detectors more and more reliable and easy to operate. Up to the beginning of the 1980s, these devices were expensive and with active areas not exceeding a fraction of 1 cm^2. At that time (see for instance [Rancoita and Seidman (1982)]), the need to use them in high-energy physics experiments has resulted in developing large area *strip silicon detectors*, at relatively low cost. A further important step was the development of monolithic front-end electronics. Thus, a high density of readout channels (for instance one readout channel every 50 μm) could be achieved. Their usage as tracking devices has allowed the construction of complex detectors with some m^2 of overall active area. A subsequent development has achieved a three-dimensional readout by employing both the so-called *pixel silicon detectors* (Sect. 6.5) and the *double sided silicon detectors*.

As discussed in the chapter on *Principles of Particle Energy Determination* (see for instance [Leroy and Rancoita (2000)] and references therein) around mid-eighties, high energy sampling calorimeters using silicon detectors as active medium were developed for particle physics experiments. Very large active areas of silicon detectors have been realized for both high energy and space physics applications.

Silicon diodes can be used as photodiodes. Their quantum efficiencies are > 70% for photon wavelengths between $\sim$ 600 nm and 1 μm. In practical applications, for instance in detecting photons from a scintillator, attention has to be paid to make sure that the detected signal due to photon conversion is larger than that due to the energy loss of particles traversing simultaneously the photodiode.

Both silicon detectors and photodiodes can work properly even in strong magnetic fields.

[‡]A massive particle is at the minimum of energy-loss for $\beta\gamma \approx 3$ (see page 47).

[§]Since fast δ-rays are not fully absorbed in thin absorbers, the average energy-loss per 100 μm can depend on the detector thickness, see Sect. 2.1.1.4.

Radiation damage can impair the silicon device performance. Nevertheless, if properly designed and associated with well suited front-end electronics, it can maintain its performance in large radiation fluence environments. Radiation damage causes the creation of *Frenkel pairs*, i.e., pairs consisting of a displaced atom from a lattice site and the corresponding vacancy in the previously occupied lattice site. This leads to i) a dose dependent increased leakage-current, ii) the creation of deep and shallow defects which can act as carrier trapping centers, iii) the build up of space charge able to change (i.e., to highly increase) the required reverse bias for achieving full depletion, etc.. In addition, there are surface damages resulting in an increase of the surface leakage current. In strip detectors, the inter-strip isolation is usually affected.

Particle identification detectors are designed to determine the particle rest mass m_0 once the measurement of particle momentum $p = m_0\beta\gamma c$ (see page 6) has been performed by other detectors. At low energy, e.g., for nuclear and particle physics applications, the particle velocity ($= \beta c$) can be measured by using the *time-of-flight* technique, i.e., by determining how much time it takes for the particle to pass through two subsequent detectors. For instance, time-of-flight systems using two plastic scintillators 1 m apart are able to provide a good particle mass identification for electrons, pions, kaons and protons up to particle momenta of $\approx (1.5\text{–}2)\,\mathrm{GeV/c}$ for time resolutions of $\approx (120\text{–}150)\,\mathrm{ps}$.

Čerenkov detectors exploit the properties of the Čerenkov radiation (described in Sect. 2.2.2), which depends on the particle velocity. Threshold Čerenkov counters provide an information whether the particle is above or below the threshold velocity for emitting Čerenkov radiation in the radiator medium. Differential Čerenkov counters exploit the dependence on both the particle velocity and the emission angle of the emitted radiation. Finally *Ring Imaging Čerenkov Detectors* (RICH) exploit the properties of the Čerenkov radiation in geometries up to the full solid angle. These devices typically include more than one radiator medium.

Transition radiation (described in Sect. 2.2.3) is emitted when a charged particle crosses the boundaries of two media with different dielectric constants. Emitted photon energies depend both on the medium plasma frequency and on the *Lorentz factor* γ of the particle. In practice, the transition radiation becomes useful for particle detection when Lorentz factors are larger than 1000. The transition radiation devices are usually employed to provide electron/hadron separation in the energy range $0.5 \leq p \leq 100\,\mathrm{GeV/c}$, when soft X-rays radiated by electrons have energies of several keV's and can be detected inside wire chambers operated with gas mixtures containing xenon. A *Transition Radiation Detector* (TRD) is typically composed of several modules, each made of an X-ray detector and a radiator. The radiator is subdivided in order to have several hundred boundaries, because the photon probability emission is of $\simeq 1\%$ per boundary crossing. The TRD performance depends on its overall length.

The development of electromagnetic and hadronic showers is described in

Sects. 2.4 and 3.3, respectively. These showers differ largely both for their longitudinal (i.e., along the incoming particle direction) and transversal (i.e., on the plane perpendicular to the incoming particle direction) shapes. By exploiting these characteristics, the electron/hadron identification is achieved in calorimetry. *Calorimeters*, as discussed in the chapter on *Principles of Particle Energy Determination*, are devices in which the total incoming particle energy is deposited by a multiplicative process called the *cascading shower development*. In homogeneous calorimeters, the incoming particle releases its energy in a medium which is at the same time the *passive absorber* shower generator and the *active detection medium*. These calorimeters can achieve the best energy resolution and are typically employed for particles depositing their energy by electromagnetic cascades. Sampling calorimeters, mostly used for high energy electromagnetic and hadronic showers, consist in passive absorber layers interspaced with active detection media layers. This way, only a small fraction of the incoming particle energy, usually less than a few percents or even a fraction of percent, is deposited in the active part of the detection system. The *sampling fluctuations* are dominating the electromagnetic calorimeter energy resolution and are largely contributing to the overall hadronic calorimeter resolution. Because physical mechanisms by which energy is deposited in matter by electromagnetic and hadronic showers are different, care has to be given in hadronic calorimetry in equalizing the hadronic and the electromagnetic responses of the calorimeter, i.e., by achieving the so-called *compensation condition* (e.g., to achieve the ratio $e/\pi = 1$). In fact, contrary to electromagnetic cascades initiated by electrons and photons, cascades initiated by hadrons will proceed by generating both hadronic particles and particles showering via electromagnetic cascades, i.e., a hadronic shower will always contain some electromagnetic sub-cascades, due to the production in cascading of neutral particles (like π^0, η, ...) decaying into photons. Electromagnetic sampling calorimeters typically have a response which is proportional to the incoming particle energy, E, and the energy resolution varies as $1/\sqrt{E}$. These features are present in compensating hadronic calorimeters. In this latter case, the energy resolution is worsened by the so-called *intrinsic fluctuations*, which take into account that a non negligible fraction of the incoming hadron energy is spent in breaking nuclear bounds, as for instance in nuclear spallation processes or in emitting largely undetected neutrons. In the calorimeter resolution, the extent of sampling fluctuations depends on both the type of passive absorbers and active media, as well as on the type of calorimetric structure realized, for instance thicknesses of passive absorbers. Active media commonly employed are scintillators, liquid argon, silicon detectors, gas detectors, etc.. Calorimeters with very a large volume (a few tens of m^3) have been constructed for high energy physics experiments. Specialized and compact electromagnetic calorimeters with imaging capabilities have been flown in balloon experiments. For all these systems, calibration procedures have been designed and set into operation in order to keep constantly calibrated and controlled many thousands of readout channels.

The very low interaction cross section between *Weakly Interacting Massive Particles*, WIMP, and the nuclei of the detector's active medium requires the use of very massive detectors to achieve a sensitivity level allowing the detection of WIMPs particles in the galactic environment. The effective cost of these detectors has to be minimized. Superheated droplet detectors, referred to as "bubble detectors" (see chapter on *Superheated Droplet (Bubble) Detectors*), of low cost, offer an attractive solution to the problem of WIMP detection. These detectors use superheated freon liquid droplets (active material) dispersed and trapped in a polymerized gel. This detection technique is based on the phase transition of superheated droplets at room or moderate temperatures. The phase transitions are induced by nuclear recoils when undergoing interactions with particles. These detectors are threshold detectors, a minimal energy deposition has to be achieved for inducing a phase transition. Their sensitivity to various types of radiation strongly depends on the operating temperature and pressure. Over the years, bubble detectors have been developed using detector formulations that are appropriate for a range of applications such as the direct measurement of neutralinos predicted by minimal supersymmetric models of cold dark matter, or portable neutron dosimeters for personal dosimetry or the measurement of the radiation fields in irradiation zones near particle accelerators or reactors.

Most of the instrumentations and experimental techniques developed for particle, nuclear and space physics have been used in medical applications. Furthermore, in this book, particular attention is paid to the physics mechanisms of radiation interaction with matter. These mechanisms, treated in the chapters on *Electromagnetic Interaction of Radiation in Matter* and on *Nuclear Interactions in Matter*, are of very important for the understanding of detectors and detector systems commonly used in nuclear medicine and in general for *medical applications*. In the chapter on *Medical Physics Applications*, imaging techniques based on *Magnetic Resonance Imaging* (MRI), *Single Photon Emission Computed Tomography* (SPECT) and *Positron Electron Tomography* (PET) are treated. The main advantage of MRI is that no radioactive material is needed, i.e., it exploits the non-zero spin property of some nuclei. MRI uses magnetic fields varying from 0.2 to 2 T and radio frequency waves to observe the magnetization change of the non-zero spin nuclei. The isotope of hydrogen, ^{1}H, which has a nuclear spin of $\frac{1}{2}$, is a major component of the human body and is used as the main source of information. Both SPECT and PET make use of properties of photon interactions in matter. For SPECT, the gamma-ray imaging technique proceeds through the injection into the patient of a radioactive substance which emits photons of well-defined energy. The distribution of radionuclides, position and concentration, inside patient's body is monitored externally through the emitted radiation deposited in a photon detector array rotating around the body and which allows the acquisition of data from multiple angles. PET is a nuclear medical imaging technique which relies on the measurement of the distribution of a radioactive tracer or radiopharmaceutical labeled with a positron emitting isotope injected into a patient. The positron emitted by the radioactive tracer or radiopharmaceutical annihilates very close to the emission point (≤ 1 mm) with an electron of the body to produce a pair of 511 keV photons emitted back-to-back, which in turn are detected by the PET camera.

It is important to note that many detectors rely partially or crucially on low-noise electronics. Most of the more visible and useful detector applications have been achieved through the usage of cheap monolithic electronics, designed and developed for fundamental physics researches and afterwards adjusted to other applications.

1.6.3 *Radiation Environments and Silicon Devices*

In the revised edition of this book, basic mechanisms resulting in temporary and/or permanent radiation-damages to silicon detectors and devices are presented and discussed for i) low- and high-resistivity bulk material and ii) in most cases up to cryogenic temperature. A review on permanent damage inflicted to silicon devices operated in radiation environments was provided by Leroy and Rancoita (2007).

Silicon is the active material of radiation detectors and in electronic devices used in the fabrication and development of electronic circuits of large to very large scale integration (VLSI) applications. As discussed in the previous section, these devices are commonly used for a large variety of applications in many fields including particle physics experiments, reactor physics, nuclear medicine, and space. These fields generally present adverse (or, even, very adverse) radiation environments that may affect the operation of the devices and are described in the chapter on *Radiation Environments and Radiation Damage in Silicon Semiconductor*. These environments are generated by i) the operation of the high-luminosity machines [e.g., the Large Hadron Collider¶ (LHC)] for particles physics experiments , ii) the cosmic rays and trapped particles of various origins in interplanetary space and/or Earth magnetosphere and iii) the operation of nuclear reactors. Their potential threats to the devices operation, in terms of radiation-induced damage by displacement are also described in that chapter. Furthermore to understand the way it may affect the devices operation, the particles energy deposition mechanisms are discussed with i) the radiation-induced damage and ii) resulting effect on bulk and device parameters evolution. These latter modifications depend on the type of irradiation particles, their energy and irradiation fluence. A treatment is done of damage inflicted by total ionizing dose and non-ionization energy-loss (NIEL) effects for which the accumulated fluence causes permanent degradation because of the induced displacement damage.

In the case of irradiated detectors, the study of the peak evolution with bias voltage in spectroscopy measurement and irradiation fluence allows the measurement of the charge collection efficiency degradation (in various regions of the diodes) and structure alteration. The chapter on *Solid State Detectors* also includes a discussion

---

¶This machine is located at CERN (Geneva, Switzerland).

of the violation of non-ionizing energy-loss (NIEL) scaling. This *scaling property*, which expresses the proportionality between NIEL-value and resulting damage effects, is *violated* for low energy ($\lesssim 10\,\mathrm{MeV}$) protons.

Integrated radiation dose delivered by a high particle flux may permanently damage silicon detectors and electronics. This is a total dose effect which leads to degradation due to displacement damage and particle interaction and is reviewed in the chapter on *Displacement Damage and Particle Interactions in Silicon Devices*, where features (under large irradiations) of VLSI bipolar transistors are treated. These devices are mostly affected by the displacement damage generated by non-ionizing energy-loss processes. For instance, at large cumulative irradiation this mechanism was found to be responsible i) for the decrease of the gain of bipolar transistors mostly as a result of the decrease of the minority-carrier lifetime in the transistor base and ii) for the degradation of the series-noise performance of charge-sensitive-preamplifiers with bipolar junction transistors in the input stage, mainly because of the increase of the base spreading-resistance. The gain degradation depends almost linearly on the amount of the displacement damage generated (e.g., the amount of energy deposited by NIEL processes) independently of the type of incoming particle, thus following an approximate *NIEL-scaling*.

However, temporary or permanent damage may be inflicted by a single particle (*single event effect* - SEE) to electronic devices or integrated circuits. Chapter 7 contains also a large section on temporary and permanent damage inflicted by a single particle (single event effect) to electronic devices or integrated circuits. The generation of SEE in the various radiation field environments, and calculation of their rate of occurrence using data and simulation techniques are outlined. It is also emphasized that the understanding of radiation effects (not only SEE) on the silicon devices combined with simulation techniques has an impact on device design and allows the prediction of the behavior of specific devices, when exposed to a radiation field of interest. On several occasions, emphasis is put on the fact that the understanding of the principles of radiation effects on silicon devices combined with simulation techniques has an impact on device design and allows one to predict the behavior of specific devices in radiation environments.

Chapter 2

Electromagnetic Interaction of Radiation in Matter

The study of radiation detection requires a deep understanding of the interaction of particles and photons with matter, where they deposit energy via electromagnetic or nuclear processes. As a result of such processes, a fraction[‡‡] of the incoming particle energy is released inside the medium. However, there are detectors, like the so-called *calorimeter detectors*, in which almost the whole particle energy is degraded and deposited via a series of successive electromagnetic and/or nuclear processes.

Inside matter, any type of moving charged particle (α-particle, p, K, π, μ, e, etc) will lose energy. Particles heavier than electrons will undergo energy losses mainly due to the excitation and ionization of atoms of the medium, close to their trajectory. With increasing energy, some of scattered electrons can escape from the (thin) absorber or detecting medium. Thus, the deposited energy may be smaller than the energy lost by a charged particle. For heavy ions, with velocity of the order (or lower) of the Bohr orbital velocity[††] of electrons in the hydrogen atom, the energy loss due to collisions with target nuclei is no longer negligible. Electrons (and positrons) can also lose energy by radiating photons. This latter process is dominant above the so-called *critical energy*.

Photons can be fully absorbed by matter in a single scattering or by a few subsequent interactions. Emerging electrons will mainly experience collision losses at sufficiently low energy (i.e., below the critical energy) or will radiate other photons. The most relevant processes for photon absorption in matter are the photoelectric, Compton and pair production of electrons and positrons.

In this chapter, we will discuss the various electromagnetic processes which take place when charged particles - electrons, light and heavy ions, etc. - or photons pass through matter, while in the next chapter nuclear interactions will be considered; for convenience, the units are expressed in cgs esu and e is the electron charge if not otherwise explicitly indicated.

[‡‡]In (fast) particle detectors, the deposited energy is, usually, a small fraction of the incoming particle energy.

[††]The reader can see page 74 for a discussion on the Bohr orbital velocity.

2.1   Passage of Ionizing Particles through Matter

Charged particles passing through matter lose their kinetic energy by dominant electromagnetic interactions which consist of excitation* and ionization[†] atoms along their passage. This process is usually called *collision loss process* or *collision process*[‡]. Subsequent collision processes mediated by the electromagnetic field associated with charged incoming particles and medium targets (i.e., bound electrons and nuclei) lead to the formation of the *primary ionization*. Fast electrons resulting from ionization processes are called *δ-rays* (Sect. 2.1.2.1).

If sufficiently energetic, these electrons will also excite and ionize atoms. Thus, a *secondary ionization* process will take place. However, the deposited energy per unit path inside the medium is usually lower than the energy lost by collisions, because the fastest δ-rays can be completely absorbed far away from where they were generated or escape from the medium.

Furthermore, deflections from the incoming direction are usually small, but become relevant in the final stage of fully absorbing processes. It has to be noted that a detailed energy-loss computation for a charged particle traversing a medium is beyond the scope of this book and can be found, for instance, in [ICRUM (1993a, 2005)] (e.g., see also databases available on web in [Berger, Coursey, Zucker and Chang (2005)]).

2.1.1   *The Collision Energy-Loss of Massive Charged Particles*

The collision process induced by massive charged particles (i.e., particles with rest mass much larger than the electron mass) was studied by Bohr (1913), Bethe (1930), Bloch (1933) and others [Møller (1932); Williams (1932); Heitler (1954); Sternheimer (1961); Andersen and Ziegler (1977); Ahlen (1980); Lindhard and Sørensen (1996); Sigmund (1997)] (see also references therein). For an incoming particle of mass m_p, velocity $v = \beta c$, *charge number*[††] z and, thus, charge ze, the theoretical expression for the energy loss by collision, dE/dx, is given by the *energy-loss formula*:

$$-\frac{dE}{dx} = \frac{2\pi n z^2 e^4}{mv^2} \left\{ \ln\left[\frac{2mv^2 W_m}{I^2(1-\beta^2)}\right] - 2\beta^2 - \delta - U \right\}, \tag{2.1}$$

where m is the electron mass (as will be indicated throughout this chapter instead of the usual notation m_e); n is the number of electrons per cm^3 of the traversed material, I is the mean excitation energy of the atoms of the material, W_m is the

---

*These are electron transitions from their initial states to higher discrete bound states.

[†]These are electron transitions from their initial states to states in the continuum, where electrons are no longer bounded.

[‡]The process is also termed *electronic energy-loss*, e.g., see Sects. 2.1.4, 4.2.1.1 and Chapter 3 of [Ziegler, Biersack and Littmark (1985a); Ziegler, J.F. and M.D. and Biersack (2008a)].

[††]The *charge number* is the coefficient that, when multiplied by the elementary charge, gives the charge of an incoming particle [Wikipedia (2008a)], i.e., it is the atomic number of the fully-ionized incoming particle.

incoming particle direction

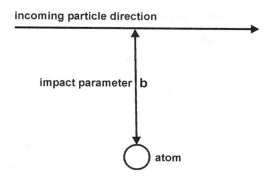

impact parameter b

atom

Fig. 2.1   The impact parameter b is the minimum distance between the incoming particle and the target by which it is scattered.

maximum transferable energy from the incident particle to atomic electrons, δ is the correction for the *density-effect*, and finally U is the term related to the non-participation of electrons of inner shells (K, L, ...) for very low incoming kinetic energies (i.e., the *shell correction term*). The number of electrons per cm^3 (n) of the traversed material is given by $(Z\rho N)/A$ [Eq. (1.40)], where ρ is the material density in g/cm^3, N is the Avogadro number (see Appendix A.2), Z and A are the *atomic number* (Sect. 3.1) and *atomic weight* (see page 14 and Sect. 1.4.1) of the material, respectively. The atomic number Z is the number of protons inside the nucleus of that atom.

The minus sign for dE/dx, in Eq. (2.1), indicates that the energy is lost by the particle. For a heavy particle, the collision energy-loss, dE/dx, is also referred to as the *stopping power*.

In Sect. 1.3.1, we have derived the expression of the maximum energy transfer W_m for the relativistic scattering of a massive particle onto an electron at rest. Usually, because the maximum energy transfer is much larger than the electron binding energy, this latter can be neglected. Thus, the value of W_m is given by formula (1.28), or, in most practical cases, by its approximate expression given in Eq. (1.29): we will make use of this latter equation in the present chapter. Once the approximate expression of W_m is used, Eq. (2.1) (i.e., the *energy-loss formula*) can be rewritten in an equivalent way as

$$-\frac{dE}{dx} = \frac{4\pi n z^2 e^4}{mv^2} \left[\ln\left(\frac{2mv^2\gamma^2}{I}\right) - \beta^2 - \frac{\delta}{2} - \frac{U}{2} \right], \qquad (2.2)$$

or

$$-\frac{dE}{dx} = \frac{4\pi n z^2 e^4}{mv^2} L, \qquad (2.3)$$

where L is a dimensionless parameter called the *stopping number*, which contains the essential physics of the process. In Eq. (2.3), L is given by the term in brackets in Eq. (2.2). As discussed later, the stopping number can be modified by adding

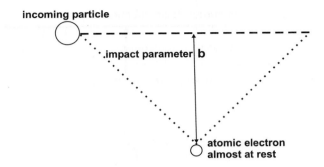

Fig. 2.2   Incoming fast particle of charge ze scattered by an atomic electron almost at rest: for small energy transfer, the particle trajectory is not deflected.

correction terms which also depend on the particle velocity v, charge number z, atomic target number Z and excitation energy I. Equation (2.1) and its equivalent expression given by Eq. (2.2) are also termed *Bethe–Bloch formula*. In literature, the Bethe–Bloch formula may also be found without the shell correction term. In this latter form, it describes the energy-loss of a charged massive particle like a proton with kinetic energy larger than a few MeV (e.g., see Sect. 2.1.1.2 and Section 27.2.1 of [PDB (2008)]).

Let us discuss an approximate derivation, i.e., without entering into complex calculations, of the energy-loss formula following closely previous approaches [Fermi (1950); Sternheimer (1961); Fernow (1986)]. In this way, the physical meaning of the terms appearing in the formula and their behavior as a function of incoming velocity become more evident. We restrict ourselves to cases where only a small fraction of the incoming kinetic energy is transferred to atomic electrons, so that the incoming particle trajectory is not deviated.

Now, we introduce the *impact parameter b* describing how close the collision is (see Fig. 2.1): b is the minimal distance of the incoming particle to the target electron. In general, large values of b correspond to the so-called *distant collisions*, conversely small values to *close collisions*. Both kinds of collisions are important for determining the *average energy-loss* [Eq. (2.1)], the energy straggling (i.e., the energy-loss distribution) and the *most probable energy-loss*.

When a particle of charge ze interacts with an electron *almost at rest*[‡‡], we assume that, to a first approximation, the electron will emerge only after the particle passage so that we can consider the electron essentially at rest throughout the interaction. This way, for symmetry reason (see Fig. 2.2), the transferred momentum $I_\perp$ will be almost along the direction perpendicular to the particle trajectory. In addition, the order of magnitude of the maximum strength of the Coulomb force acting along the perpendicular direction $F_\perp$ is $\approx ze^2/b^2$. Thus, the maximum electric

---

[‡‡]An electron is almost at rest, when its velocity is much smaller than the incoming particle velocity v.

field strength, at which the Coulomb field affects the target electron, decreases as the impact parameter increases. Furthermore, the interaction time between the two particles is inversely proportional to the incoming particle velocity, v, and directly proportional to b, namely the interaction time is $\approx b/v$. Therefore, the transferred momentum $I_\perp$ is given by:

$$I_\perp = \int F_\perp \, dt$$
$$\sim \frac{ze^2}{b^2} \left(\frac{b}{v} \right) = \frac{ze^2}{bv}.$$

This estimate differs by a factor 2 from a more refined calculation (see for instance [Fernow (1986)]), in which relativistic corrections are also considered while estimating the perpendicular electric field. Thus, we can finally write:

$$I_\perp = \frac{2ze^2}{bv}. \tag{2.4}$$

Since the electron before the interaction was supposed to be at rest, its recoil momentum is $I_\perp$. Its kinetic energy W, which is usually so small that we do not need to deal with a relativistic formula, is given by:

$$W = \frac{I_\perp^2}{2m} \tag{2.5}$$

$$= \frac{2z^2e^4}{mb^2v^2}. \tag{2.6}$$

Equation (2.6) shows the relationship between the impact parameter b and the transferred energy W: distant collisions are typically soft ones, while close collisions allow large transfers of kinetic energy, which goes as $1/b^2$. As a consequence, for sufficiently large b values the *shell binding energies* of electrons have to be taken into account. Close collisions may happen with very large energy transfers, i.e., with the emission of fast outgoing δ-rays. From Eq. (1.26), we see that, as the kinetic energy of the δ-ray increases, the emission angle (the one formed with the incoming particle direction) decreases: very fast δ-rays are emitted close to the particle trajectory. Finally, we have to consider that the energy loss will fluctuate from one collision to the next, depending on collision distances.

Furthermore, the energy transferred to recoiling nuclei can be usually neglected* with respect to the energy of recoiling electrons: their ratio is, typically, of the order of a few 10^{-4}. For any given value of b the kinetic energy (W) acquired by a recoiling electron is expressed by Eq. (2.6); while, if the interaction occurs on a nucleus with charge Ze and mass m_A, we have to rewrite Eq. (2.4) as:

$$I_{A\perp} = \frac{2Zze^2}{bv}. \tag{2.7}$$

---

*For heavy ions with velocity of the order of (or lower than) the Bohr orbital velocity of electrons in the hydrogen atom (defined at page 74), the *nuclear stopping power* cannot be neglected (see discussion in Sects. 2.1.4 and 2.1.4.1).

Using Eq. (2.7), we can re-express Eq. (2.5) for estimating the recoil nuclear energy as:

$$W_A = \frac{2Z^2 z^2 e^4}{m_A b^2 v^2}.$$

The ratio between the two recoil energies is

$$\frac{W_A}{W} = \left(\frac{2Z^2 z^2 e^4}{m_A b^2 v^2}\right) \Big/ \left(\frac{2z^2 e^4}{m b^2 v^2}\right)$$

$$= \frac{Z^2 m}{m_A}.$$

When a particle interacts with an atom, the impact parameter is almost the same for the Z atomic electrons and the nucleus. The overall transferred energy (W_e) to the electrons of the medium is about ZW and we have

$$\frac{W_A}{W_e} = \frac{Zm}{m_A}.$$

On hydrogen $(Z = 1)$, this ratio is $\simeq 5.44 \times 10^{-4}$, for heavier nuclei like Pb and U it becomes $\simeq 2.17 \times 10^{-4}$ and 2.12×10^{-4}, respectively. It has to be noted that the accumulated energy deposited in processes resulting from interactions on nuclei is the dominant mechanism for (permanent) radiation damage by displacement in Coulomb scatterings (e.g., see Chapter 4 and, for silicon devices, Chapter 6 and 7); although, usually, it is negligible with respect to the overall energy lost by collisions in traversing a medium (e.g., see Sects. 4.2.3 and 4.2.3.1).

Let us consider (see Fig. 2.3) a particle of charge ze traversing a material, in which the number of electrons per cm^3 is n [Eq. (1.40)]. The number of electrons encountered by the particle along a path dx at impact parameter between b and

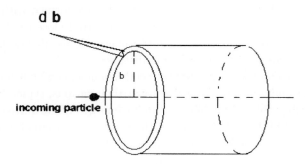

Fig. 2.3   An incoming fast particle of charge ze interacts with electrons at impact parameter between b and $b + db$.

$b+db$ is $n(2\pi b)\,db\,dx$. From Eq. (2.6), the overall kinetic energy transferred to atomic electrons is

$$W_b = \frac{2z^2e^4}{mb^2v^2}n(2\pi b)\,db\,dx$$
$$= \left(\frac{4\pi nz^2e^4}{mv^2}\right)\frac{db}{b}\,dx.$$

Thus, for impact parameters between b and $b + db$, the energy lost in a length dx becomes:

$$-\frac{dE_b}{dx} = \left(\frac{4\pi nz^2e^4}{mv^2}\right)\frac{db}{b}\;,$$

and the overall energy loss by collisions calculated by integrating from $b_{\min}$ up to $b_{\max}$, i.e., over the range of the impact parameter for which this approach is valid, becomes

$$-\frac{dE}{dx} = \int_{b_{\min}}^{b_{\max}}\left(\frac{4\pi nz^2e^4}{mv^2}\right)\frac{db}{b}$$
$$= \frac{4\pi nz^2e^4}{mv^2}\ln\left(\frac{b_{\max}}{b_{\min}}\right). \tag{2.8}$$

The upper limit $b_{\max}$ can be estimated by considering that the collision time τ cannot exceed the typical time period associated with bound electrons, namely $\tau \simeq (1/\bar{\nu})$ where $\bar{\nu}$ is the characteristic mean frequency of excitation of electrons. In fact, if the collision time were much larger than the typical revolution period, the passage of the particle could be considered as similar to an adiabatic process which does not affect the electron energy. In addition, at relativistic energies the region of space at the maximum electric field strength is contracted by the Lorentz factor γ and, consequently, the collision time becomes $\simeq b_{\max}/(\gamma v)$. Thus, for $b_{\max}$ we have:

$$\tau \simeq \left(\frac{1}{\bar{\nu}}\right) \simeq \left(\frac{b_{\max}}{\gamma}\right)\frac{1}{v} \qquad \text{and} \qquad b_{\max} \simeq \frac{v\gamma}{\bar{\nu}}.$$

Introducing the mean excitation energy $I = h\bar{\nu}$, we obtain:

$$b_{\max} \simeq \frac{v\gamma h}{I}. \tag{2.9}$$

The lower limit $b_{\min}$ is evaluated considering the extent to which the classical treatment can be employed. In the framework of the classical approach, the wave characteristics of particles are neglected. This assumption is valid as long as the impact parameter is larger than the de Broglie wavelength** of the electron in the center-of-mass system (CoMS) of the interaction. For instance, we can assume

$$b_{\min} \simeq \frac{h}{2P_{ecm}}, \tag{2.10}$$

_____

**The de Broglie wavelength of a particle with momentum p is

$$\lambda = \frac{h}{p},$$

where h is the Planck constant.

where P_{ecm} is the electron momentum in the CoMS. Because the electron mass is much smaller than the mass of the incoming heavy-particle, the CoMS is approximately associated with the incoming particle and conversely the electron velocity in the CoMS is opposite and almost equal in absolute value to that of the incoming particle, v. Thus, we have that

$$|P_{ecm}| \simeq m\gamma v = m\gamma \beta c$$

and Eq. (2.10) becomes

$$b_{\min} \simeq h/(2m\gamma\beta c). \tag{2.11}$$

Substituting the values of $b_{\min}$ and $b_{\max}$ in Eq. (2.8), we obtain:

$$-\frac{dE}{dx} = \frac{4\pi n z^2 e^4}{mv^2} \ln\left[\left(\frac{v\gamma h}{I}\right)\left(\frac{2m\gamma\beta c}{h}\right)\right]$$

$$= \frac{2\pi n z^2 e^4}{mv^2} \ln\left(\frac{2m\gamma^2 v^2}{I}\right)^2.$$

Finally, using the value of the maximum energy transfer W_m from Eq. (1.29), we get:

$$-\frac{dE}{dx} = \frac{2\pi n z^2 e^4}{mv^2} \ln\left[\frac{2mv^2 W_m}{I^2(1-\beta^2)}\right]. \tag{2.12}$$

Table 2.1 Values of Z, Z/A, I, ρ, $h\nu_p$ and density-effect parameters S_0, S_1, a, md, and δ_0 for some elemental substances.

El.	Z	Z/A	I eV	ρ g/cm^3	$h\nu_p$ eV	S_0	S_1	a	md	δ_0
He	2	0.500	41.8	1.66×10^{-4}	0.26	2.202	3.612	0.134	5.835	0.00
Li	3	0.432	40.0	0.53	13.84	0.130	1.640	0.951	2.500	0.14
O	8	0.500	95.0	1.33×10^{-3}	0.74	1.754	4.321	0.118	3.291	0.00
Ne	10	0.496	137.0	8.36×10^{-4}	0.59	2.074	4.642	0.081	3.577	0.00
Al	13	0.482	166.0	2.70	32.86	0.171	3.013	0.080	3.635	0.12
Si	14	0.498	173.0	2.33	31.06	0.201	2.872	0.149	3.255	0.14
Ar	18	0.451	188.0	1.66×10^{-3}	0.79	1.764	4.486	0.197	2.962	0.00
Fe	26	0.466	286.0	7.87	55.17	-0.001	3.153	0.147	2.963	0.12
Cu	29	0.456	322.0	8.96	58.27	-0.025	3.279	0.143	2.904	0.08
Ge	32	0.441	350.0	5.32	44.14	0.338	3.610	0.072	3.331	0.14
Kr	36	0.430	352.0	3.48×10^{-3}	1.11	1.716	5.075	0.074	3.405	0.00
Ag	47	0.436	470.0	10.50	61.64	0.066	3.107	0.246	2.690	0.14
Xe	54	0.411	482.0	5.49×10^{-3}	1.37	1.563	4.737	0.233	2.741	0.0
Ta	73	0.403	718.0	16.65	74.69	0.212	3.481	0.178	2.762	0.14
W	74	0.403	727.0	19.30	80.32	0.217	3.496	0.155	2.845	0.14
Au	79	0.401	790.0	19.32	80.22	0.202	3.698	0.098	3.110	0.14
Pb	82	0.396	823.0	11.35	61.07	0.378	3.807	0.094	3.161	0.14
U	92	0.387	890.0	18.95	77.99	0.226	3.372	0.197	2.817	0.14

Data are from [Sternheimer, Berger and Seltzer (1984)]

This equation is equivalent to the *energy-loss formula* [see Eq. (2.1)] except for the terms $-2\beta^2$, $-\delta$ due to the density-effect and the shell correction term $(-U)$. When the first of these latter terms is added to Eq. (2.12), using Eq. (1.29) we obtain the expression (termed *Bethe relativistic formula*)

$$-\frac{dE}{dx} = \frac{4\pi n z^2 e^4}{mv^2} \left\{ \ln\left[\frac{2mv^2}{I(1-\beta^2)}\right] - \beta^2 \right\}$$

(2.13)

derived in the quantum treatment of energy loss by collisions of a heavy, spin 0 incident particle [e.g., see the discussion in [Ahlen (1980)] and references therein; see also [Fano (1963)]]. Furthermore, it has to be noted that spin plays an important role when the transferred energy is almost equal to the incoming energy (this occurs with very limited statistical probability). The other terms $(-U$ and $-\delta)$ to be added to Eq. (2.13) are discussed in Sects. 2.1.1.2 and 2.1.1.3, respectively. At low particle velocity[††], additional corrections are added: for instance, corrections accounting for the *Barkas effect*[‡‡] discussed in Sect. 2.1.1.1 and the *Bloch correction*[**] discussed later.

The Bloch correction (e.g., see [Northcliffe (1963)], Section D of [Ahlen (1980)], [de Ferraiis and Arista (1984); Lindhard and Sørensen (1996)], Section 3.3.3 of [ICRUM (2005)] and Chapter 6 of [Sigmund (2006)]) derives from the Bloch quantum-mechanical approach in which he did not assume, unlike Bethe, that it is valid to consider the electrons[*] to be represented by plane waves in the center-of-momentum reference frame. It results in adding the quantity (e.g., see Equation 6.4 at page 184 of [Sigmund (2006)], [Northcliffe (1963)], Section D of [Ahlen (1980)], Section 2.2 of [ICRUM (1993a)], Section 3.3.3 of [ICRUM (2005)] and references therein)

$$\Delta L_{\text{Bloch}} = \Psi(1) - \text{Re}\left[\Psi\left(1 + j\frac{z\,v_0}{v}\right)\right]$$

(2.14)

to the stopping number [see Eq. (2.3)] of the energy-loss expression; in Eq. (2.14) v_0 is the Bohr velocity (defined at page 74, see also Appendix A.2), the function

$$\Psi(\iota) = \frac{d\ln\Gamma(\iota)}{d\iota}$$

(called *digamma function*) is the logarithmic derivative of the gamma function and Re denotes the real part. An accurate approximation for Eq. (2.14) can be obtained from [de Ferraiis and Arista (1984)] (an approximated expression[†] can also be found

---

[††]At very low velocity, the energy-loss process due to Coulomb interactions on nuclei (termed *nuclear energy-loss* and resulting in the *nuclear stopping power*) cannot be neglected (e.g., see Sects. 2.1.4, 2.1.4.1 and 2.1.4.1).

[‡‡]For the Barkas effect, the reader can see [Barkas, Dyer and Heckman (1963); Ashley, Ritchie and Brandt (1972); Lindhard (1976); Bichsel (1990); Arista and Lifschitz (1999)]. This effect is also referred to as *Barkas–Andersen effect* after the systematic investigations carried out by Andersen, Simonsen and Sørensen (1969) (e.g., see Section 2.3.4 of [ICRUM (2005)]).

[**]It is determined by the difference between the Bloch non-relativistic expression and that one of Bethe, see Section III of [Ahlen (1980)].

[*]The Bloch correction originates in close collisions (e.g., see Section 3.3.3 of [ICRUM (2005)]).

[†]In Section 2.2 of [ICRUM (1993a)], the Block correction is approximated by the expression $-y^2\,[1.20206 - y^2\,(1.042 - 0.8549\,y^2 + 0.343\,y^4)]$, where $y = z\alpha/\beta$.

in Section 37 of [Heitler (1954)], in [Bloch (1933); Bichsel (1990)] and in Section 2.2 of [ICRUM (1993a)]). The Bloch approach is meaningful for charged particles with velocities

$$\frac{1}{z} \lesssim \frac{v}{z\,v_0} \lesssim \frac{1}{z\alpha}$$

(e.g., see [Sigmund (1997, 1998)] and references therein), where the quantity $2z\,v_0/v$ is referred to as the *Bohr k parameter* of a particle with charge ze and velocity v. However, the correction can be neglected at large velocities (e.g., see [Northcliffe (1963)] and Section 3.3.3 of [ICRUM (2005)]).

The energy-loss formula, as re-expressed in Eq. (2.2), shows that the logarith-

Table 2.2   Values of Z/A, I, ρ, $h\nu_p$ and density-effect parameters S_0, S_1, a, and md for some compounds and mixtures.

Material	Z/A	I eV	ρ g/cm^3	$h\nu_p$ eV	S_0	S_1	a	md
(dry) Air at sea level	0.499	85.7	1.21 $\times 10^{-3}$	0.71	1.742	4.276	0.109	3.399
Anthracene	0.527	69.5	1.28	23.70	0.115	2.521	0.147	3.283
Ethane	0.599	45.4	1.25 $\times 10^{-3}$	0.79	1.511	3.874	0.096	3.610
Ethyl Alcohol	0.564	62.9	0.79	19.23	0.222	2.705	0.099	3.483
Freon-12	0.480	143.0	1.12	21.12	0.304	3.266	0.080	3.463
(lead) Glass	0.421	526.4	6.22	46.63	0.061	3.815	0.095	3.074
Kapton, polyimide film	0.513	79.6	1.42	24.59	0.151	2.563	0.160	3.192
Lithium carbonate	0.487	87.9	2.11	29.22	0.055	2.660	0.099	3.542
Methane	0.623	41.7	6.67 $\times 10^{-4}$	0.59	1.626	3.972	0.093	3.626
Methanol	0.562	67.6	0.79	19.21	0.253	2.764	0.090	3.548
Plastic scint., vinyltoluene	0.541	64.7	1.03	21.54	0.146	2.486	0.161	3.239
Polyethylene	0.570	57.4	0.94	21.10	0.137	2.518	0.121	3.429
Propane	0.590	47.1	1.88 $\times 10^{-3}$	0.96	1.433	3.800	0.099	3.592
Lucite	0.539	74.0	1.19	23.09	0.182	2.668	0.114	3.384
Silicon dioxide	0.499	139.2	2.32	31.01	0.139	3.003	0.084	3.506
Tissue, soft (ICRP)	0.551	72.3	1.00	21.39	0.221	2.780	0.089	3.511
Tissue, soft (ICRP four-comp.)	0.550	74.9	1.00	21.37	0.238	2.791	0.096	3.437
Tissue-equiv., gas (methane base)	0.550	61.2	1.06 $\times 10^{-3}$	0.70	1.644	4.140	0.099	3.471
Tissue-equiv., gas (propane base)	0.550	59.5	1.83 $\times 10^{-3}$	0.91	1.514	3.992	0.098	3.516

Data are from [Sternheimer, Berger and Seltzer (1984)]

mic term increases quadratically with $\beta\gamma = (v\gamma)/c$. The reason for such a behavior [see Eq. (2.8)] is that the energy loss depends on $\ln(b_{max}/b_{min})$. Both terms $1/b_{min}$ [Eq. (2.11)] and b_{max} [Eq. (2.9)] increase with $\beta\gamma$. The former refers to the enhancement of the maximum transferable kinetic energy (which goes as $1/b^2$) in a single collision, while the latter refers to the Lorentz dilatation of the maximum impact parameter (conversely, the transverse volume affected by the incoming electric field). Therefore, as the incoming particle energy increases, δ-rays are emitted more energetically along the particle trajectory (see also Sect. 2.1.2). In addition the cylindrical region, surrounding the particle path and experiencing excitation and ionization, is enlarged.

The phenomenon of δ-ray emission is mostly responsible for the difference between the energy lost by a particle traversing a medium (for example, the thin absorbers used in Si or gaseous detectors) and the energy actually deposited inside. As a consequence, although the energy lost continues to increase with increasing energy [Eq. (2.2)], the deposited energy approaches an almost constant value [termed *Fermi Plateau**, see also Eq. (2.29)], which mainly depends on the size and density of the absorber (see discussion in Sect. 2.1.1.4).

By introducing the classical electron radius,

$$r_e = \frac{e^2}{mc^2} \qquad (2.15)$$

(see Appendix A.2) and using Eq. (1.40), we can evaluate the numerical coefficient in Eq. (2.1):

$$\frac{2\pi n z^2 e^4}{mv^2} = \frac{2\pi N mc^2 r_e^2 \rho z^2 Z}{A\,\beta^2}$$

$$= 0.1535 \frac{\rho z^2 Z}{A\,\beta^2} \quad \text{[MeV/cm]}, \qquad (2.16)$$

and, thus, the energy-loss formula becomes

$$-\frac{dE}{dx} = 0.1535 \frac{\rho z^2 Z}{A\beta^2} \left\{ \ln\left[\frac{2mv^2 W_m}{I^2(1-\beta^2)}\right] - 2\beta^2 - \delta - U \right\} \quad \text{[MeV/cm]}, \qquad (2.17)$$

where, to a first approximation, the value of the mean excitation energy is given by [Marmier and Sheldon (1969)]:

$$I \approx 11.5 \times Z \times 10^{-6} \quad \text{[MeV]}$$

or

$$I \approx 9.1 \times Z(1 + 1.9 \times Z^{-2/3})\,10^{-6} \quad \text{[MeV]}.$$

Values of Z/A, ρ and I for several elemental substances, compounds and mixtures are available in literature (e.g., see [Sternheimer, Berger and Seltzer (1984)]; see, also, [Ahlen (1980); Bichsel (1992); ICRUM (1993a, 2005)]) and on the web [NIST (2008)]; some of them are listed in Tables 2.1 and 2.2.

---

*The Fermi Plateau is discussed in Sect. 2.1.1.4.

2.1.1.1   The Barkas–Andersen Effect

The *Barkas effect** denotes the observed difference in ranges of positive and negative particles in emulsion. Barkas and collaborators provided the first indication that π^--ranges were slightly longer than π^+-ranges [Smith, Birnbaum and Barkas (1953)]. Subsequently, Barkas, Dyer and Heckman (1963) obtained a precise result by investigating the process

$$K^- + p \to \Sigma^{\pm} + \pi^{\mp}.$$

For this reaction, range measurements in emulsion[†] allowed them i) to determine that the rest masses[‡] of Σ^+ and Σ^- hyperons were 1189.35 ± 0.15 MeV and 1197.6 ± 0.5 MeV, respectively and ii) to observe the *energy-loss defect* of negative hyperons [i.e., a range larger than expected by about $(3.6 \pm 0.7)\%$]. The energy-loss defect was interpreted as the experimental evidence that a stopping-power theory based on the Born approximation (e.g., see footnote at page 17) must be corrected when the particle velocity becomes comparable to that of atomic electrons in stopping materials [Barkas, Dyer and Heckman (1963); Fano (1963)]. In fact, the next-high-order correction to Born approximation contributes to the stopping power in proportion to the cube of the incident particle's charge $(ze)^3$ [Fano (1963)]. Thus, it results to be of opposite sign for negative and positive particle pairs.

Furthermore, Andersen, Simonsen and Sørensen (1969) carried out high-precision measurements of the stopping powers of aluminum and tantalum for (5–13.5) MeV p and d, and (8–20) MeV ^{3}He and ^{4}He. They determined that the ratio between the stopping powers for the doubly charged and singly charged ions is systematically larger than the factor four predicted by the energy-loss formula. The computed charge-dependent correction to Eq. (2.2) was in agreement to that needed for the range difference between Σ^+ and Σ^- hyperons measured by Barkas, Dyer and Heckman (1963). This effect may better be referred to as the *Barkas–Andersen effect* (see also [ICRUM (2005)]).

For sufficiently slow incoming particles, the collision time becomes long enough for allowing atomic electrons to considerably move during the interaction. When this occurs, the experimental data suggest that i) it results in a smaller effective impact parameter for positively charged particles, thus in an increase of the transferred energy and ii) the reverse occurs for negatively charged particles. Ashley, Ritchie and Brandt (1972) proposed the first theoretical treatment of the Barkas–Andersen effect. Their approach was within the framework of *Bohr's oscillator model* [Bohr (1913)]. However, it was limited to distant collisions, under the assumption that close collisions make a negligible contribution: it showed that the Barkas correc-

---

*It was termed in this way by Lindarhd (1976).

[†]Properties of photographic emulsions are discussed in [Blau (1961)].

[‡]These values are well in agreement, within experimental errors, with those reported in [PDB (2008)].

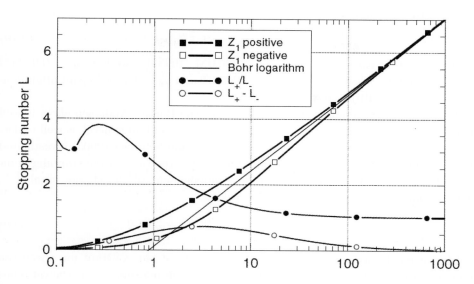

Fig. 2.4 Stopping numbers, ratio and difference of the stopping numbers for positive (L_+) and negative (L_-) incident particles with charge $|Z_1|e$ as a function of ζ' [Eq. (2.20)] (reprinted from *Nucl. Instr. and Meth. in Phys. Res. B* **212**, Sigmund, P. and Schinner, A., Anatomy of the Barkas effect, 110–117, Copyright (2003), with permission from Elsevier). The Bohr logarithm [$= \ln(C\,\zeta')$ with $C = 1.1229$] is also included for comparison.

tion[‡‡] is governed by the parameter (entering Bohr's energy-loss formula)

$$\zeta = \frac{ze^2\omega}{mv^3}, \qquad (2.18)$$

where m and e are the electron rest mass and charge, respectively; ω is the circular velocity of the (isotropically) bound electrons[††]; v and z are the incoming particle velocity and charge number, respectively. Lindhard (1976) determined that close collisions make a significant contribution almost doubling the estimate for the Barkas correction. Equation (2.18) indicates that this latter is relevant i) at small v, ii) large charge number z and iii) large ω, i.e., for inner shells. According to Andersen (1983, 1985) and Bichsel (1990), experimental data[†] are consistent with Barkas corrections[**] estimated by [Lindhard (1976)]. For target elements with high atomic number, Bichsel (1990) extracted the correction to be added to the stopping number from measured stopping powers and found that an accurate expression is

---

[‡‡]As for the Bloch correction, the *Barkas correction* is added to the stopping number [see Eq. (2.3)].

[††]For atoms with mean excitation energy I, this dominant resonance value ω is given by $\omega = I/\hbar$.

[†]These data were based on the determination of the stopping powers for protons, α-particles and lithium ions.

[**]Lindhard's results are considered consistent with experimental data at least for target atomic numbers $Z < 50$ (e.g., see Section 2.3 of [ICRUM (1993a)]).

obtained by the empirical formula

$$\Delta L_{\text{Barkas}} = z \, g_1 \, \beta^{-2 \, g_2}. \tag{2.19}$$

For gold (and for elements with $Z \geq 64$ [ICRUM (1993a)]), we have $g_1 = 0.002833$ and $g_2 = 0.60$; for silver ($Z = 47$), $g_1 = 0.006812$ and $g_2 = 0.45$, while for aluminum ($Z = 13$), $g_1 = 0.001054$ and $g_2 = 0.80$.

Sigmund and Schinner (2001a) showed that the Barkas–Andersen effect is well predicted for the antiproton stopping power in silicon and, in general, is well understood in case of light ions. Furthermore, Sigmund and Schinner (2003) investigated the scaling properties of the effect and found that i) the ratio of the stopping numbers for an ion and its anti-ion is almost independent of the charge number z, increases almost monotonically with decreasing

$$\zeta' = \frac{1}{|\zeta|} = \frac{mv^3}{|z| \, e^2 \omega} \tag{2.20}$$

and has a local maximum at $\zeta' = 0.26$ (Fig. 2.4), i.e., at constant ion speed the relative magnitude of the Barkas effect increases with increasing z, ii) the ratio has a more complex dependence on Z and iii) the difference of these stopping numbers has a maximum at $\zeta' = 3.3$ and becomes negligible for $\zeta' \gtrsim 100$ (Fig. 2.4).

Furthermore, Sigmund (1998) (see also [Sigmund (1997)]) indicates that the ratio of Bloch to Barkas corrections is given by

$$\frac{\text{Bloch correction}}{\text{Barkas correction}} \sim \frac{z \, v}{Z \, v_0},$$

where v_0 is the Bohr velocity.

2.1.1.2   *The Shell Correction Term*

Following the *Pauli exclusion principle* (see for instance textbooks like [Finkelnburg (1964); Eisberg and Resnick (1985)]), the only permitted arrangements of electrons in atoms or molecules are those in which at least one of the four quantum numbers (principal, azimuthal,[§] magnetic and spin) differs. The principal quantum number indicates the shell number. Starting from the inner one, we have 1, 2, 3, 4, ... These shells are indicated with the letters K, L, M, N, ... The orbital quantum numbers (0, 1, 2, 3, ...) are indicated with the letters s, p, d, f, ... (see Appendix A.4 on the Electronic Structure of the Elements).

For an absorber with atomic number Z, the shell correction term

$$U = 2 \left(\frac{C}{Z} \right)$$

in Eq. (2.1) is due to the non-participation of inner (K, L, ...) electrons in the collision loss process (see for instance [Walske (1962); Fano (1963); Marmier and Sheldon (1969); Lindhard (1976); Ahlen (1980)] and references therein), when the

---

[§]It is also called the orbital angular momentum quantum number.

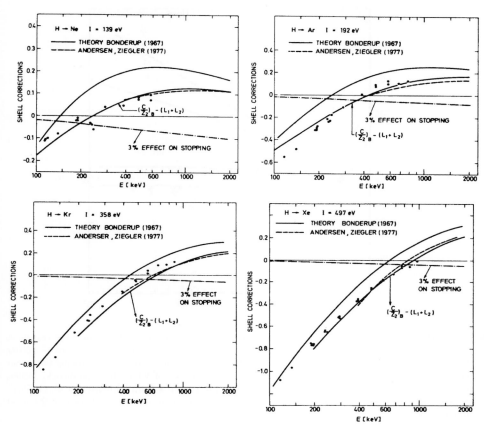

Fig. 2.5 $\left(\frac{C}{Z}\right)$ versus incoming proton kinetic energy in Ne, Ar, Kr and Xe absorbers (adapted and reprinted with permission from [Andersen (1983)]), see text at page 45.

incoming velocity is no longer much larger than that of bound atomic electrons. Detailed calculations of K and L shells were carried out [Walske (1962); Fano (1963); Lindhard (1976)]. The corrections for the M, N, and higher shells of heavy atoms are generally small, except at very low energies (namely for proton kinetic energies ≤ 1 MeV).

The estimated values [Fano (1963)] of $\left(\frac{C}{Z}\right)$ for protons traversing metals are lower than 0.1 when $\beta > 0.05\sqrt{Z}$.

The $\left(\frac{C}{Z}\right)$ values were experimentally determined for protons traversing noble gases and are shown in Fig. 2.5 (see [Andersen (1983)] for explanation on theoretical predictions): the shell correction term cannot be neglected for proton energies lower than a few MeV's. Further treatments and references on this correction term can be found in [ICRUM (1993a)] for protons and α-particles, and in [ICRUM (2005)] for ions heavier than helium. Sigmund (1998) (see also [Sigmund (1997)]) indicates

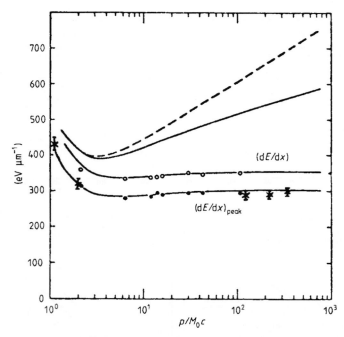

Fig. 2.6   Energy loss in silicon (in units of eV/μm) versus $\beta\gamma$ ($= p/M_0c$, where M_0 is the rest mass of the incoming particle) from [Rancoita (1984)]. From the top the first two curves are: $-dE/dx$ without (broken curve) and with (full curve) the density-effect correction. The following two other curves are compared to experimental data for detector thicknesses of 300 ($\times$ from [Hancock, James, Movchet, Rancoita and Van Rossum (1983)]) and 900 μm ($\circ$ and $\bullet$ from [Esbensen et al. (1978)]): the restricted energy-loss with the density-effect taken into account and the prediction of the most probable energy-loss.

that the ratio of Barkas to shell corrections is given by

$$\frac{\text{Barkas correction}}{\text{shell correction}} \sim \frac{z\,v_0}{2\,Z^{1/3}v},$$

where v_0 is the Bohr velocity; v and z are the incoming particle velocity and charge number, respectively.

A detailed discussion on stopping power, energy straggling and total range for ions with energies[¶] lower than 10 MeV/nucleon is given in [Ziegler, Biersack and Littmark (1985a); Ziegler, J.F. and M.D. and Biersack (2008a)]. However, this goes beyond the field of interest of this book.

2.1.1.3   *The Density Effect and Relativistic Rise*

The energy-loss formula [Eq. (2.1)] shows that, at low β values, the collision energy-loss ($-dE/dx$) decreases rapidly as the velocity increases owing to the $1/\beta^2$ term. At

---

[¶]The reader can find the definition of kinetic energies per nucleon in Sect. 1.4.1.

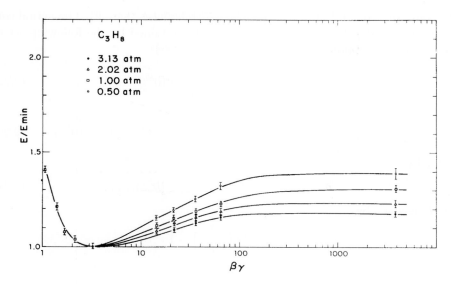

Fig. 2.7 Measured energy-loss (reprinted from *Nucl. Instr. and Meth.* **161**, Walenta, A.H., Fisher, J., Okuno, H. and Wang, C.L., Measurement of the Ionization Loss in the Region of Relativistic Rise for Noble and Molecular Gases, 45–58, Copyright (1979), with permission from Elsevier) in propane normalized to the energy-loss minimum E_{min} versus $\beta\gamma$. The different slopes of the relativistic rise effect depend on the detecting device pressure.

relativistic energies, when the velocity approaches the speed of light and $\beta \approx 1$, the energy-loss curve reaches a minimum, called the *energy-loss minimum*, for

$$\beta_{min}\gamma_{min} \approx 3 \tag{2.21}$$

(corresponding to $\beta_{min} \approx 0.95$). A particle at the energy-loss minimum is called a *minimum ionizing particle (mip)*.

The collision energy-loss will start to grow again beyond its minimum due to the logarithmic term. This latter depends on the increase with the incoming momentum, of both the maximum transferred energy and b_{max}. However, as the range of the distant collision is extending, the atoms close to the path of the particle will produce a polarization, which results in reducing the electric field strength acting on electrons at large distances, as pointed out by Fermi (1939) and others (for instance see [Sternheimer (1961)]).

Bohr ([Ritson (1961)] and references therein) demonstrated that the screening becomes more effective at impact parameters of $\approx \sqrt{(mc^2)/(4\pi ne^2)}$. At distances exceeding this value of the impact parameter, the electric shielding prevents the energy transfer. Furthermore, in the region where the density-effect correction needs to be applied, the shell correction term is negligible.

Systematic studies of this effect were carried out ([Sternheimer (1966); Sternheimer and Peierls (1971); Sternheimer, Berger and Seltzer (1984)]; Section 3.4

in [Groom, Mokhov and Striganov (2001)] and references therein). The actual correction term, δ, due to the density-effect can be obtained by the following set of equations as function of the $\beta\gamma$ value:

$$\delta = \delta_0 \left(\frac{\beta\gamma}{10^{S_0}} \right)^2 \quad \text{(for } \beta\gamma < 10^{S_0}\text{)}, \tag{2.22}$$

$$\delta = 2\ln(\beta\gamma) + C + a \left[\frac{1}{\ln 10} \ln \left(\frac{10^{S_1}}{\beta\gamma} \right) \right]^{md} \quad \text{(for } 10^{S_0} < \beta\gamma < 10^{S_1}\text{)} \tag{2.23}$$

and, finally,

$$\delta = 2\ln(\beta\gamma) + C \quad \text{(for } \beta\gamma > 10^{S_1}\text{)}; \tag{2.24}$$

where C is given by

$$C = -2\ln \left(\frac{I}{h\nu_p} \right) - 1$$

and

$$\nu_p = \sqrt{\frac{ne^2}{\pi m}} \tag{2.25}$$

is the plasma frequency. The parameters S_0, S_1, $h\nu_p$, md and a are given in Tables 2.1 and 2.2. The values of δ_0 (see Table 2.1) are lower than 1, as well as the term $\left(\frac{\beta\gamma}{10^{S_0}} \right)^2$. As a consequence, the correction for the density-effect is small at low energy. In fact, the term in brackets in Eq. (2.1) is not less than ≈ 16 for $\beta\gamma \approx 3$ [namely for particles close to the energy-loss minimum, see Eq. (2.21)], even for high-Z materials for which excitation energies are higher. At larger values of $\beta\gamma$ (where β is ≈ 1 and for $\beta\gamma > 10^{S_1}$), the density-effect correction approaches a linear dependence on $\ln(\beta\gamma)$. The term in brackets in Eq. (2.1) becomes:

$$\ln \left[\frac{2mv^2 W_m}{I^2(1-\beta^2)} \right] - 2 - 2\ln(\beta\gamma) - C = \ln \left[\frac{2m\beta^2\gamma^2 c^2 W_m}{I^2} \left(\frac{1}{\beta^2\gamma^2} \right) \right] - 2 - C$$

$$= \ln \left[\frac{2mc^2 W_m}{(h\nu_p)^2} \right] - 1.$$

The energy-loss formula can be finally rewritten as:

$$-\frac{dE}{dx} = 0.1535 \frac{\rho z^2 Z}{A} \left\{ \ln \left[\frac{2mc^2 W_m}{(h\nu_p)^2} \right] - 1 \right\} \quad \text{[MeV/cm]}. \tag{2.26}$$

In Eq. (2.26), the logarithmic dependence of the energy lost by an incoming massive particle is determined only by the maximum transferred energy W_m in a single collision: the energy-loss increasing beyond its minimum is called *relativistic rise*. This increase [see Eq. (1.29)] depends on $\ln(\beta^2\gamma^2)$, but in the *ultra-relativistic region* it depends on $\ln\gamma$ [see Eq. (1.28)]. In Fig. 2.6 (see [Rancoita (1984)] and references therein), the calculated energy-loss curve (shown by the broken curve, for which the relativistic rise depends on W_m and b_{max}) and the one including the density-effect correction (shown by the full curve, for which the relativistic rise depends on W_m only) are presented for silicon detectors.

In Fig. 2.7 [Walenta, Fisher, Okuno and Wang (1979)], data on energy-loss are presented as a function of the pressure of the detecting gas device. As the pressure increases (i.e., n increases), the value of C, which depends on $\ln\{I/[h\sqrt{(ne^2)/(\pi m)}]\}$, decreases: at fixed values of $\beta\gamma$, the density-effect correction is larger [see at page 48 the expressions for the density-effect correction, e.g., Eqs. (2.23, 2.24)].

2.1.1.4 *Restricted Energy-Loss and Fermi Plateau*

We have seen how the energy loss by collisions continues to rise with the increase of the incoming particle energy. This occurs because the maximum energy transfer W_m in a single collision depends on the Lorentz factor γ (at ultra high energies), or on γ^2 at relativistic energies (see Sect. 1.3.1). This kinetic energy is transferred to electrons (δ-rays) emitted in directions close to the one of the incoming particle and, being energetic, they can travel large distances inside the medium or escape from it. Therefore, the energy transferred by the incoming particle to atomic electrons can result in secondary ionization processes, similar to the primary one, with additional δ-rays traveling far away from the emission point.

Typical radiation or particle detectors are not thick enough to absorb secondary δ-rays fully. Thus, the energy loss suffered by the incoming particle differs from the *deposited energy* inside the detecting medium. Furthermore, we need to define an *effective detectable maximum transferred energy* W_0. This maximum detectable energy is the effective average maximum δ-ray energy that can be absorbed inside the device also taking into account the emission angles.

Table 2.3   Values of the radiation length (X_0), critical energy (ϵ_c) using Eq. (2.121), Molière radius (R_M) using Eq. (2.239) and stopping power for a massive $z = 1$ particle for several materials.

Material	X_0	ϵ_c	R_M	$\left(\frac{dE}{dx}\right)_{mip}$	$\left(\frac{dE}{d\chi}\right)_{mip}$
	cm	MeV	cm	MeV/cm	MeV cm^2/g
Be	35.28	111.47	6.71	2.947	1.595
Al	8.89	40.28	4.68	4.361	1.615
Si	9.36	37.60	5.28	3.877	1.664
Fe	1.76	21.04	1.77	11.419	1.451
Cu	1.44	18.96	1.60	12.571	1.403
W	0.35	8.08	0.92	22.099	1.145
Pb	0.56	7.42	1.60	12.735	1.122
U	0.32	6.77	1.00	20.485	1.081
Water	36.08	70.10	10.91	1.992	1.992
NEMA G10 plate	19.40	52.38	7.85	3.179	1.870
BGO	1.12	10.20	2.33	8.882	1.251

See [Leroy and Rancoita (2000)], Section 6 in [PDB (2008)] and Eq. (2.17).

In order to estimate the deposited energy inside a detector, we have to rewrite the energy-loss formula [Eq. (2.1)], in terms of W_0. Following the treatment by Fano (1963), the electromagnetic interaction for relativistic particles is approximated by the usual static Coulomb interaction (the *longitudinal term*) and with a relativistic *transverse term*. The overall interaction is evaluated by considering three different regions of the transferred energy W

i) $W < W_1$ for which the momentum transfer is $\ll \hbar/a_r$, where a_r is the linear dimension of the target atom:

$$-\left(\frac{dE}{dx}\right) = \frac{2\pi n z^2 e^4}{mv^2}\left\{\ln\left[\frac{2mv^2 W_1}{I^2(1-\beta^2)}\right] - \beta^2\right\},$$

ii) $W_1 < W < W_2$, where the transverse term can be neglected and $W/(mc^2) \ll 1$:

$$-\left(\frac{dE}{dx}\right) = \frac{2\pi n z^2 e^4}{mv^2}\ln\left(\frac{W_2}{W_1}\right),$$

iii) $W_2 < W < W_m$, namely W values large enough to consider the atomic electrons of the target as free:

$$-\left(\frac{dE}{dx}\right) = \frac{2\pi n z^2 e^4}{mv^2}\left\{\ln\left(\frac{W_m}{W_2}\right) - W_m\left(\frac{1-\beta^2}{2mc^2}\right)\right\}.$$

For absorbers thick enough so that the corresponding effective detectable maximum transferred energy (W_0) satisfies the condition of the above-described point (ii) (i.e., $W_2 < W_0 < W_m$), we can rewrite the latter expression as:

$$-\left(\frac{dE}{dx}\right) = \frac{2\pi n z^2 e^4}{mv^2}\left\{\ln\left(\frac{W_0}{W_2}\right) - W_0\left(\frac{1-\beta^2}{2mc^2}\right)\right\}.$$

Summing the three contributions and adding both the density-effect and the shell correction terms, we obtain the so-called *restricted energy-loss formula* valid for $W_m \geq W_0 > W_2$:

$$-\left(\frac{dE}{dx}\right)_{\text{restr}} = \frac{2\pi n z^2 e^4}{mv^2}\left\{\ln\left[\frac{2mv^2 W_0}{I^2(1-\beta^2)}\right] - \beta^2 - W_0\left(\frac{1-\beta^2}{2mc^2}\right) - \delta - U\right\}. \quad (2.27)$$

It has to be noted that when $W_0 \to W_m$ for $W_m \approx 2mc^2\beta^2\gamma^2$, then, Eq. (2.27) reduces to Eq. (2.1).

The term

$$W_0\left(\frac{1-\beta^2}{2mc^2}\right)$$

can be neglected for incoming particle energies beyond the ionization loss minimum or for $W_0 \ll 2mc^2$. When these conditions are satisfied and using Eq. (2.16), for $W_m \geq W_0$ we obtain:

$$-\left(\frac{dE}{dx}\right)_{\text{restr}} = 0.1535\frac{\rho z^2 Z}{A\beta^2}\left\{\ln\left[\frac{2mv^2 W_0}{I^2(1-\beta^2)}\right] - \beta^2 - \delta - U\right\} \quad [\text{MeV/cm}]. \quad (2.28)$$

At high energies (when $\beta \approx 1$ and $\beta\gamma > 10^{S_1}$), we can neglect the shell correction term and express the density correction [Eq. (2.24)] as $\delta = 2\ln(\gamma) + C$, where

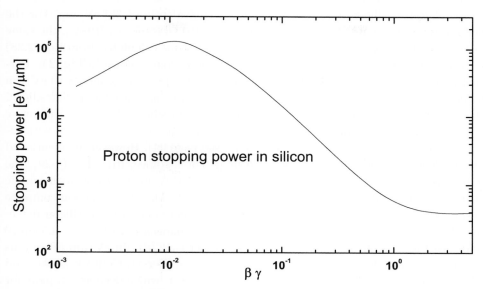

Fig. 2.8   Stopping power (-dE/dx) in eV/μm of protons in silicon for $1.5 \times 10^{-3} \lesssim \beta\gamma \leq 5$ (data from [Berger, Coursey, Zucker and Chang (2005)]). The contribution of the nuclear stopping power is included: it decreases rapidly for $\beta\gamma \gtrsim 3 \times 10^{-3}$; at $\beta\gamma \sim 3 \times 10^{-3}$ it does not exceed 3.4%.

$C = -2\ln\left(\frac{I}{h\nu_p}\right) - 1$ (see page 48) and ν_p (the plasma frequency) is computed using Eq. (2.25). Because W_0 does not depend any more on the incoming particle energy, the collision loss will reach a constant value, called the *Fermi plateau*, computed by rewriting Eq. (2.28) for $W_m \geq W_0$ as:

$$-\left(\frac{dE}{dx}\right)_{\text{plateau}} = 0.1535 \frac{\rho z^2 Z}{A} \left\{ \ln\left[\frac{2mv^2 W_0}{I^2(1-\beta^2)}\right] - 1 - 2\ln(\gamma) - C \right\}$$

$$= 0.1535 \frac{\rho z^2 Z}{A} \ln\left[\frac{2mc^2 W_0}{(h\nu_p)^2}\right] \quad [\text{MeV/cm}]. \qquad (2.29)$$

This equation replaces the formula (2.26) whenever the deposited energy has to be computed instead of the energy-loss formula. In Eq. (2.29) the incoming energy dependent term W_m is substituted by the constant W_0 term. This term has to be evaluated, in turn, for any given detector and detector thickness medium.

In Fig. 2.7 [Walenta, Fisher, Okuno and Wang (1979)], experimental data taken with a gas detector show evidence of the Fermi plateau beyond $\beta\gamma \approx 130$. From the energy-loss minimum to the Fermi plateau, data exhibit the relativistic rise.

In Fig. 2.6, experimental data collected with silicon detectors show the Fermi plateau, as well. The escaped δ-rays result in a lower measured energy-loss deposition (i.e., deposited energy) with respect to the expected one, as derived by the energy-loss formula expressed by Eq. (2.17). For instance at the energy-loss minimum [Eq. (2.21)] and for $z = 1$, from Eq. (2.17) and Table 2.1 the computed value[¶] of $-(dE/dx)$ in silicon is $\approx 3.87\,\mathrm{MeV/cm}$ or equivalently $\approx 387\,\mathrm{eV/\mu m}$. For the $900\,\mu m$ thick silicon detector of [Esbensen et al. (1978); Rancoita (1984)], the value of W_0 is ≈ 0.5 MeV and the restricted energy-loss value calculated from Eq. (2.28) is $\approx 345\,\mathrm{eV/\mu m}$ in agreement with the experimental data shown in Fig. 2.6. Furthermore, the Fermi plateau computed by means of Eq. (2.29) is only $\approx 10\,\mathrm{eV/\mu m}$ higher than the restricted energy-loss at the minimum: in dense media like silicon, the relativistic rise may become almost negligible. In silicon detectors, the active layer thickness is typically $(300\text{--}400)\,\mu m$ and, thus, it can absorb δ-rays with kinetic energies not exceeding $\approx 300\,\mathrm{keV}$ (e.g., see page 94). However, as mentioned above, $900\,\mu m$ thick absorbers were also used. For instance, from Eq. (1.29), the maximum transferred kinetic energy is $\approx 0.3\,(0.5)\,\mathrm{MeV}$ for a proton with kinetic energy of $\approx 128\,(207)\,\mathrm{MeV}$, i.e. with $\beta\gamma \approx 0.54\,(0.70)$. Therefore, the non-complete absorption of emitted δ-rays affects the energy-deposition curve in silicon detectors for incoming protons (or other heavy charged particles) with $\beta\gamma \gtrsim 0.54\text{--}0.70$ (e.g., see Fig. 2.6). In Fig. 2.8, the stopping power $(-dE/dx)$ in $\mathrm{eV/\mu m}$ of protons in silicon is shown for $1.5 \times 10^{-3} \lesssim \beta\gamma \leq 5$ (data from [Berger, Coursey, Zucker and Chang (2005)]). Is included the nuclear stopping power, which decreases rapidly for $\beta\gamma \gtrsim 3 \times 10^{-3}$; at $\beta\gamma \sim 3 \times 10^{-3}$ it does not exceed 3.4% (see also Sect. 2.1.4.1).

For $W_m \geq W_0$, the difference, $\triangle\left(-\frac{dE}{dx}\right)$, between the energy-loss formula [Eq. (2.17)] and the restricted energy-loss equation [Eq. (2.28)] is given by:

$$\triangle\left(-\frac{dE}{dx}\right) = 0.1535\,\frac{\rho z^2 Z}{A\beta^2}\left[\ln\left(\frac{W_m}{W_0}\right) - \beta^2\right] \quad [\mathrm{MeV/cm}]. \qquad (2.30)$$

2.1.1.5 *Energy-Loss Formula for Compound Materials*

Experimental stopping power measurements were mostly carried out on elements rather than compounds. To a first approximation, see for instance [Sternheimer (1961); Fano (1963)], a material containing several atomic species can be treated according to *Bragg's additivity rule*, namely as a combination of atoms contributing to the overall stopping power, separately. The energy-loss formula is valid for compounds and mixtures. But the mean excitation energy I, the ratio Z/A and the density correction term δ appearing in Eq. (2.1) [or for instance in the equivalent expression of Eq. (2.17)] have to be replaced by $\bar{I}$, $\bar{\delta}$, $\langle\frac{Z}{A}\rangle$, which are given by:

$$\ln\bar{I} = \sum_i f_i \ln I_i, \qquad (2.31)$$

$$\bar{\delta} = \sum_i f_i \delta_i, \qquad (2.32)$$

$$\left\langle\frac{Z}{A}\right\rangle = \frac{\sum_i n_i Z_i}{\sum_i n_i A_i}, \qquad (2.33)$$

[¶]This value is in agreement with the one given in Table 2.3, see also Section 6 in [PDB (2008)] and Fig. 2.8.

where n_i is the number of atoms of ith species present in the stopping material, and I_i, A_i, Z_i are the mean excitation energy, the atomic weight and the atomic number of that atomic element, respectively. The *oscillator strength* f_i for the atoms of the ith species is given by

$$f_i \equiv \frac{n_i Z_i}{\sum_i n_i Z_i}. \qquad (2.34)$$

This approach was shown to be usually accurate enough, even though it disregards chemical binding and other materials properties [Fano (1963)]. Equations (2.31–2.33) can also be found expressed using fractions by weight (w_i) of compound or mixture elements

$$w_i = \frac{n_i A_i}{\sum_i n_i A_i}$$

and, thus, with oscillator strengths re-expressed as

$$f_i = \frac{w_i Z_i}{A_i} \bigg/ \left\langle \frac{Z}{A} \right\rangle. \qquad (2.35)$$

Values of Z/A, I and density-effect parameters S_0, S_1, a, and md [see Eqs. (2.22–2.24)] for several compounds and mixtures are available in literature (e.g., see [Sternheimer, Berger and Seltzer (1984)]): some of them are listed in Table 2.2.

At low energy, the application of the additivity rule can introduce errors, because the stopping power contributed by each element is influenced by chemical binding effects. These errors can amount to $\approx$ 15% at energies near the stopping power peak [ICRUM (1993a)]. In addition, the stopping powers also depend on the material phase: they are generally lower for solids than for gases. A detailed discussion of energy-loss in compounds and mixtures is beyond the purpose of this book. The reader can find a literature survey on this subject in Section 3.2 of [ICRUM (1993a)] (see also Section 3.5 of [ICRUM (2005)]).

In [Berger, Coursey, Zucker and Chang (2005)], the databases PSTAR (for protons) and ASTAR (for α-particles) provide the stopping-powers for protons and α-particles in several compounds and mixtures up to 10 and 1 GeV, respectively.

2.1.2 *Energy-Loss Fluctuations*

The energy-loss process experienced by a charged particle in matter is a statistical phenomenon: the collisions responsible for energy losses are (to a first approximation) a series of independent successive events.

Each interaction of the incoming particle with atomic electrons can transfer a different amount of kinetic energy. Equation (2.1) allows one to compute the average energy-loss, but this value undergoes statistical fluctuations. Therefore, the energy lost by a particle, in a given traversed path x, will be characterized by an energy distribution function called the *energy-loss distribution function* or *energy straggling function*. For electrons at high energy, the collision process is not the dominant interaction by which energy is released and, as consequence, is not the main cause of energy-loss fluctuations. On the contrary, this is the dominant process for massive charged particles.

2.1.2.1 δ-Rays, Straggling Function, and Transport Equation

Let us consider the probability $P(\epsilon, E)$ that a particle of energy E traversing a thickness dx of a material loses between ϵ and $\epsilon + d\epsilon$ of its energy. We can write:

$$P(\epsilon, E) = n_A \left[\frac{d\sigma_A(\epsilon)}{d\epsilon} \right] d\epsilon \, dx$$
$$= \omega(\epsilon, E) \, d\epsilon \, dx, \qquad (2.36)$$

where $d\sigma_A(\epsilon)/d\epsilon$ is the atomic differential cross section** for transferring an energy ϵ in the interaction with an atom, n_A is the number of atoms per cm^3. Here

$$\omega(\epsilon, E) = n_A \left[\frac{d\sigma_A(\epsilon)}{d\epsilon} \right]$$

has the meaning of the *differential collision probability* to lose an energy between ϵ and $\epsilon + d\epsilon$, while traversing a thickness dx.

To a first approximation, we can neglect distant collisions which are supposed to give small transferred energies. The expression of the differential collision probability depends on the mass and spin of the incident particle (see Section 2.3 in [Rossi (1964)] and references therein). For a particle with $z = 1$ and mass m_p, kinetic energy E_k (where $E_k = E - m_p c^2$) and energies in units of MeV, we have:
a) with spin 0

$$\omega(\epsilon, E) = 0.1535 \frac{\rho Z}{A \beta^2 \epsilon^2} \left(1 - \beta^2 \frac{\epsilon}{W_m} \right),$$

b) with spin $\frac{1}{2}$

$$\omega(\epsilon, E) = 0.1535 \frac{\rho Z}{A \beta^2 \epsilon^2} \left[1 - \beta^2 \frac{\epsilon}{W_m} + \frac{1}{2} \left(\frac{\epsilon}{E} \right)^2 \right],$$

c) with spin 1

$$\omega(\epsilon, E) = 0.1535 \frac{\rho Z}{A \beta^2 \epsilon^2} \left[\left(1 - \beta^2 \frac{\epsilon}{W_m} \right) \left(1 + \frac{1}{3} \frac{\epsilon}{E_c} \right) + \frac{1}{3} \left(\frac{\epsilon}{E} \right)^2 \left(1 + \frac{1}{2} \frac{\epsilon}{E_c} \right) \right],$$

where

$$E_c = \frac{m_p^2 c^2}{m}$$

and $\omega(\epsilon, E)$ is in (MeV cm)$^{-1}$.

For $\epsilon \ll W_m$ (this implies $\epsilon \ll E_k$ and $\epsilon \ll E$) and $\epsilon \ll E_c$, the differential collision probability is reduced to:

$$\omega(\epsilon, E) = 0.1535 \frac{\rho Z}{A \beta^2 \epsilon^2} \quad [\text{MeV cm}]^{-1}. \qquad (2.37)$$

**The reader can see page 15 for the definition of differential cross section.

Thus, when the transferred energy is very small, the differential collision probability is given by the expression (2.37), which depends on ϵ and β, but is independent of the particle spin. Only for ϵ larger than E_c or when ϵ is no longer much smaller than W_m, the particle spin can play a role (as previously mentioned).

Equations (2.36, 2.37) allow an approximate calculation of the probability of δ-ray emission[*]. The probability $G(\delta, W_\delta)$ to generate a δ-ray with energy larger than W_δ (but $\ll E_k$, E, E_c, W_m) traversing a thickness x is:

$$
\begin{aligned}
G(\delta, W_\delta) &\approx \int_0^x \int_{W_\delta}^{W_m} 0.1535 \, \frac{\rho Z}{A \beta^2 \epsilon^2} \, d\epsilon \, dx' \\
&= x \left[0.1535 \, \frac{\rho Z}{A \beta^2} \int_{W_\delta}^{W_m} \frac{1}{\epsilon^2} \, d\epsilon \right] \\
&= 0.1535 \, x \, \frac{\rho Z}{A \beta^2} \left(\frac{1}{W_\delta} - \frac{1}{W_m} \right) \\
&\approx 0.1535 \, x \frac{\rho Z}{A \beta^2} \frac{1}{W_\delta},
\end{aligned}
\tag{2.38}
$$

where W_δ is in MeV.

For example, if a particle with $\beta = 1$ passes through a $300\,\mu$m thick silicon detector (in which $\frac{Z}{A} \approx 0.5$ and $\rho = 2.33\,\text{g/cm}^3$), the probability to emit a δ-ray with energy larger than 100 keV is $G(\delta, 0.1) \approx 5.4\%$. Furthermore, this formula shows that the emission probability goes as $1/W_\delta$.

In the case of a particle with charge number z, the collision probability increases by a factor z^2. In addition, for $\epsilon \ll E_c$, E_k and E, we can again neglect the spin dependence of the differential collision probability, which reduces to that for a massive particle with spin 0 and charge number z, and becomes the so-called *Rutherford macroscopic differential collision probability* (see [Bichsel and Saxon (1975)] and references therein):

$$
\omega(\epsilon, E)_R = 0.1535 \, \frac{\rho z^2 Z}{A \beta^2 \epsilon^2} \left(1 - \beta^2 \frac{\epsilon}{W_m} \right) \quad [\text{MeV cm}]^{-1}.
\tag{2.39}
$$

Let us consider a large number N_p of particles of energy E traversing a material of thickness x. The probability $f(\epsilon, x) \, d\epsilon$ that the energy loss (traversing a thickness x) is between ϵ and $\epsilon + d\epsilon$ can be determined when an infinitesimal thickness dx is added. By traversing the additional thickness, there is an increase of the number of particles, I_p, losing an energy between ϵ and $\epsilon + d\epsilon$: in fact, particles which have lost less energy, still, can lose energy and, finally, reach the required amount between ϵ and $\epsilon + d\epsilon$. From Eq. (2.36) and indicating with $f(\epsilon - \epsilon', x) \, d\epsilon$ the probability that traversing the thickness x the energy loss differs by ϵ' from the required amount, we have:

$$
I_p = N_p \int_0^\infty [f(\epsilon - \epsilon', x) \, d\epsilon] \, [\omega(\epsilon', E) \, dx] \, d\epsilon'.
$$

---

[*]An example of the emission of δ-rays is shown in Fig. 6.25.

However, there will be also a decrease of particles D_p because some of them, which have already lost the required amount of energy, can lose additional energy. From Eq. (2.36), we have:

$$D_p = N_p f(\epsilon, x)\, d\epsilon \int_0^\infty [\omega(\epsilon', E)\, dx]\, d\epsilon'.$$

Let us write the total variation of the number of particles:

$$N_p\, [f(\epsilon, x + dx) - f(\epsilon, x)]\, d\epsilon = I_p - D_p,$$

from which we obtain

$$f(\epsilon, x + dx) - f(\epsilon, x) = \left[\int_0^\infty f(\epsilon - \epsilon', x)\omega(\epsilon', E)\, d\epsilon'\right] dx - \left[f(\epsilon, x) \int_0^\infty \omega(\epsilon', E)\, d\epsilon'\right] dx,$$

and, finally, we find the so-called *transport equation*

$$\frac{\partial f(\epsilon, x)}{\partial x} = \int_0^\infty f(\epsilon - \epsilon', x)\, \omega(\epsilon', E)\, d\epsilon' - f(\epsilon, x)\, \sigma_t(E), \qquad (2.40)$$

where

$$\sigma_t(E) = \int_0^\infty \omega(\epsilon', E)\, d\epsilon'.$$

In deriving Eq. (2.40), we have assumed that i) the differential collision probability does not vary in a appreciable way because of the energy lost in traversing the thickness x of material and ii) the collisions are statistically independent.

The energy straggling function is the function $f(\epsilon, x)$ which is the solution of the integro-differential Equation (2.40), i.e., the solution of the transport equation.

2.1.2.2 *The Landau–Vavilov Solutions for the Transport Equation*

Solutions of the transport equation were proposed by many authors (see [Landau (1944); Symon (1948); Blunck and Leisegang (1950); Vavilov (1957); Sternheimer (1961); Rossi (1964); Shulek et al. (1966); Bichsel and Saxon (1975)] and references therein).

Landau (1944) solved Eq. (2.40) by a method based on Laplace transforms for *thin absorbers*, namely those for which the condition $\frac{\xi}{W_m} \ll 1$ is satisfied and where ξ is given by:

$$\xi = 0.1535\, x \frac{\rho z^2 Z}{A\beta^2} \quad [\text{MeV}]. \qquad (2.41)$$

In the opposite case $\left(\frac{\xi}{W_m} \gg 1\right)$, i.e., for *thick absorbers*, the function turns into a Gaussian distribution [Landau (1944)]. In his calculations, he used the Rutherford macroscopic collision probability for the limiting case $W_m \to \infty$. Landau's solution of the transport equation is expressed as

$$f(\epsilon, x)_L = \frac{\phi(\lambda)}{\xi} \quad [\text{MeV}^{-1}], \qquad (2.42)$$

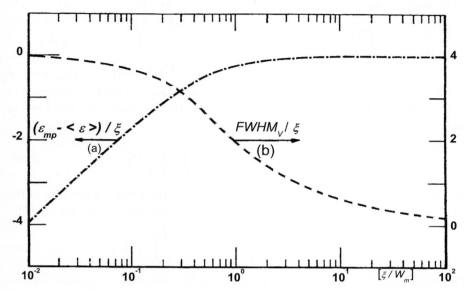

Fig. 2.9 The dashed-dotted curve (a), left hand scale, is the value of $(\epsilon_{mp} - \langle\epsilon\rangle)/\xi$ for the Vavilov solution of the transport equation as function of $\frac{\xi}{W_m}$ (from [Seltzer and Berger (1964)]). The dashed curve (b), right hand scale, is the $FWHM_V$ of the Vavilov distribution in units of ξ (adapted and reprinted with permission from [Seltzer and Berger (1964)]).

where $\phi(\lambda)$ (tabulated in [Boersch-Supan (1961)]) is a function of a universal parameter λ and it is given by:

$$\phi(\lambda) = \frac{1}{2\pi i} \int_{r-i\infty}^{r+i\infty} \exp[u\ln(u) + \lambda u]\, du$$
$$= \frac{1}{\pi} \int_0^{+\infty} e^{(-\pi u/2)} \cos[u(\ln(u) + \lambda)]\, du,$$

where r is an arbitrary real positive constant. $f(\epsilon, x)_L$ has a maximum for

$$\lambda_0 \approx -0.229 \qquad (2.43)$$

and a *Full Width Half Maximum*

$$FWHM_L \simeq 4.02\,\xi \qquad (2.44)$$

(see [Rancoita and Seidman (1982)] and references therein for a more detailed discussion of these parameters). The energy loss corresponding to the maximum of the function $f(\epsilon, x)_L$ is called *the most probable energy-loss* (ϵ_{mp}). This function is characterized by an asymmetric long tail in the region well above ϵ_{mp}, which is called *Landau tail*. This tail is mainly due to fast emitted δ-rays.

Later, Vavilov (1957) solved the equation by introducing the physical limits coming from W_m and using the Rutherford expression provided by Eq. (2.39), i.e., he gave the solution for:

$$\frac{\partial f(\epsilon, x)_V}{\partial x} = \int_0^u f(\epsilon - \epsilon', x)_V \, \omega(\epsilon', E)_R \, d\epsilon' - f(\epsilon, x)_V \int_0^{W_m} \omega(\epsilon', E)_R \, d\epsilon',$$

where $\omega(\epsilon', E)_R = 0$ for $\epsilon' > W_m$ and $u = \epsilon$ for $\epsilon < W_m$ and $u = W_m$ for $\epsilon > W_m$. $f(\epsilon, x)_V$ can be found tabulated in [Seltzer and Berger (1964)]. Furthermore for $\frac{\xi}{W_m} \rightarrow 0$, he demonstrated that $f(\epsilon, x)_V \rightarrow f(\epsilon, x)_L$ and that the Landau parameter λ has the following relationship:

$$\lambda \rightarrow \frac{1}{\xi}(\epsilon - \langle\epsilon\rangle) - \beta^2 - \ln\left(\frac{\xi}{W_m}\right) - 1 + C_E, \tag{2.45}$$

where $C_E = 0.577215$ is the Euler constant and $\langle\epsilon\rangle$ is the mean energy-loss in the material of thickness x. The mean energy-loss can be calculated using Eq. (2.17):

$$\langle\epsilon\rangle = \xi\left\{\ln\left[\frac{2mv^2 W_m}{I^2(1 - \beta^2)}\right] - 2\beta^2 - \delta - U\right\} \quad [\text{MeV}]. \tag{2.46}$$

Experimental data show that the Landau–Vavilov solutions are almost equivalent for $\frac{\xi}{W_m} \leq 0.06$ [Aitken et al. (1969)]. For instance for 300 μm silicon detectors, this limit is already achieved by protons with momenta larger than 550 MeV/c.

In the following, the energy straggling distribution for thin absorbers will be indicated by $f(\epsilon, x)_{L,V}$.

For thick absorbers $\left(\frac{\xi}{W_m} \gg 1\right)$ under the condition that the energy loss is such that the differential collision probability does not vary in an appreciable way along the particle path, the Vavilov energy straggling distribution becomes almost a Gaussian function:

$$f(\epsilon, x)_V \approx \frac{1}{\xi\sqrt{\frac{2\pi W_m}{\xi}\left(1 - \frac{\beta^2}{2}\right)}} \exp\left[-\frac{(\epsilon - \langle\epsilon\rangle)^2}{2W_m\xi\left(1 - \frac{\beta^2}{2}\right)}\right], \tag{2.47}$$

where the standard deviation is

$$\sigma_V \approx \sqrt{W_m\xi\left(1 - \frac{\beta^2}{2}\right)}.$$

In Fig. 2.9 (see [Seltzer and Berger (1964)]), the computed values (curve b) for the $FWHM_V$ of the Vavilov distribution in units of ξ are shown as a function of $\frac{\xi}{W_m}$. It can be seen that the energy-loss distribution narrows as $\frac{\xi}{W_m}$ becomes $\gg 1$. In the same figure (curve a), the difference between the most probable and the mean energy-loss is also shown (see [Seltzer and Berger (1964)]). While the mean energy-loss is larger than the most probable energy-loss for $\frac{\xi}{W_m} \rightarrow 0$, they become almost equal for $\frac{\xi}{W_m} \gg 1$.

2.1.2.3 The Most Probable Collision Energy-Loss for Massive Charged Particles

As previously discussed, for thin absorbers the energy straggling distribution $f(\epsilon, x)_{L,V}$ has a maximum for $\lambda = \lambda_0$ with λ_0 given by Eq. (2.43). The most probable energy-loss is obtained for $\lambda = \lambda_0$ by Eq. (2.45). It can be re-expressed as:

$$
\begin{aligned}
\epsilon_{mp} &= \langle \epsilon \rangle + \xi \left[\lambda_0 + \beta^2 + \ln \left(\frac{\xi}{W_m} \right) + 1 - C_E \right] \\
&= \langle \epsilon \rangle + \xi \left[\beta^2 + \ln \left(\frac{\xi}{W_m} \right) + 0.194 \right] \quad [\text{MeV}].
\end{aligned}
\tag{2.48}
$$

At high energy for $\beta\gamma > 10^{S_1}$ (the parameters S_1 are given in Tables 2.1 and 2.2) and $\beta \approx 1$, the energy-loss formula is approximated by Eq. (2.26). Thus, Eq. (2.48) becomes:

$$
\begin{aligned}
\epsilon_{mp} &= \xi \left\{ \ln \left[\frac{2mc^2 W_m}{(h\nu_p)^2} \right] - 1 \right\} + \xi \left[\ln \left(\frac{\xi}{W_m} \right) + 1.194 \right] \\
&= \xi \left\{ \ln \left[\frac{2mc^2 \xi}{(h\nu_p)^2} \right] + 0.194 \right\} \quad [\text{MeV}].
\end{aligned}
\tag{2.49}
$$

Equation (2.49) shows that, at relativistic energies, *the most probable energy-loss* (ϵ_{mp}) *is independent of the incoming particle energy*. However, it has a *logarithmic dependence on the medium thickness traversed* [see Eq. (2.41)].

As an example (see Appendix A in [Rancoita and Seidman (1982)]), we can evaluate the most probable energy-loss for silicon detectors. Using the $h\nu_p$ value given in Table 2.1, Eq. (2.49) becomes

$$
\epsilon_{mp,\text{Si}} = \xi \left[\ln \xi + 20.97 \right] \quad [\text{MeV}],
\tag{2.50}
$$

where ξ is in MeV. This formula (see Fig. 2.10) is valid for silicon absorbers also at relatively low $\beta\gamma$ values (i.e., > 30). For instance, for $\beta\gamma > 30$ in the expression for the density-effect correction [Eq. (2.23) at page 48] the term $a \left[\frac{1}{\ln 10} \ln \left(\frac{10^{S_1}}{\beta\gamma} \right) \right]^{md}$ is lower than 0.44 and results in the decrease of $\epsilon_{mp,\text{Si}}$ by less than 3%. Thus, the expression given in Eq. (2.50) extends its validity for $\beta\gamma > 30$, i.e., for protons with a momentum larger than $28\,\text{GeV}/c$ or for pions with a momentum larger than $4\,\text{GeV}/c$. From Eq. (2.50), in a $300\,\mu\text{m}$ thick silicon detector relativistic pions and protons will have $\epsilon_{mp,\text{Si}} = 84.2\,\text{keV}$ in agreement with experimental data [Hancock, James, Movchet, Rancoita and Van Rossum (1983)]. Here $\langle \epsilon_{mp,\text{Si}} \rangle$ is about 16% larger than $\epsilon_{mp,\text{Si}}$ (see [Rancoita (1984)]). Furthermore for $\beta\gamma > 100$, in a $300\,\mu\text{m}$ silicon detector (see Fig. 2.6), the observed most probable energy-loss (in eV/μm) is $\approx 5\%$ lower¶ than that in a $900\,\mu\text{m}$ detector, in agreement, within the experimental errors, with the decrease of $\approx 6.5\%$ computed by means of Eq. (2.50). This equation agrees with the most probable energy-losses computed, for large values of

---

¶This value is marginally affected by the shift of the Landau energy-loss peak discussed at page 67.

$\beta\gamma$, in [Bichsel (1988)] and in Section 27 (Figures 27.7 and 27.8) of [PDB (2008)] (for a review of experimental data, see Tables VIII and IX in [Bichsel (1988)]).

As an additional example, we can estimate the most probable energy-loss of spallation protons $\epsilon_{mp,p,\mathrm{Si}}$ in silicon detectors. These protons, generated by hadronic interactions in matter, come out of the target nucleus and typically have momenta between ≈ 500 and $1500\,\mathrm{MeV/c}$, (i.e., kinetic energies between ≈ 125 and $831\,\mathrm{MeV}$; β and γ values between 0.47 and 0.85, and 1.13 and 1.89 respectively). Furthermore, the density-effect factor δ, given by Eq. (2.22), is lower than 0.1 and can be neglected along with the shell correction factor U. By using Eq. (2.17) and approximating* $f(\epsilon, x)_V$ with $f(\epsilon, x)_L$ (Sect. 2.1.2.2), we can rewrite Eq. (2.48) as:

$$
\begin{aligned}
\epsilon_{mp,p,\mathrm{Si}} &= \xi\left\{\lambda_0 + \beta^2 + \ln\left[\frac{\xi}{W_m}\right] + 1 - C_E + \ln\left[\frac{2mv^2 W_m}{I^2(1-\beta^2)}\right] - 2\beta^2 - \delta - U\right\} \\
&= \xi\left[\ln(\xi\beta^2) + \ln\left(\frac{2mc^2}{I^2}\right) - \beta^2 + \ln\gamma^2 + 0.194\right] \\
&= \xi\left[\ln(\xi\beta^2) + 17.54 - \beta^2 + \ln\gamma^2\right] \quad [\mathrm{MeV}],
\end{aligned}
\tag{2.51}
$$

where ξ is in MeV and I is given in Table 2.1. Note that, using a Landau function, the most probable energy-loss[††] is varied by about or less than a percent in the range of incoming proton momenta from 500 up to $1500\,\mathrm{MeV/c}$ for $\xi/W_m \le 0.05$. By averaging the term $(\ln\gamma^2 - \beta^2)$ over this range of incoming proton momenta, Eq. (2.51) can be rewritten, within an additional few % approximation, as:

$$
\epsilon_{mp,p,\mathrm{Si}} = \xi\left[\ln(\xi\beta^2) + 17.84\right] \quad [\mathrm{MeV}].
\tag{2.52}
$$

It has to be noted [see Eq. (2.41)] that the term $\ln(\xi\beta^2)$ does not depend on the proton momentum. $\epsilon_{mp,p,\mathrm{Si}}$ has a $\ln x$ dependence and goes as $1/\beta^2$.

2.1.2.4 *Improved Energy-Loss Distribution and Distant Collisions*

The Landau–Vavilov solutions of the transport equation were derived under the assumption that scatterings occur on quasi-free electrons, and follow the Rutherford collision probability [Eq. (2.39)]. Therefore, as previously discussed, they neglect the electron binding energies. But this latter assumption is a valid approximation only for close collisions.

In general, the distant collisions have also to be taken into account. Modifications[‡‡] were proposed to the Landau–Vavilov energy straggling function ([Blunck and Leisegang (1950); Shulek et al. (1966)], see also [Fano (1963); Bichsel and Yu

---

*The reader can see [Seltzer and Berger (1964)] for a tabulation of $\epsilon_{mp,p}$ in Vavilov's theory and, thus, for estimating its variation with respect to that in Landau's theory.

[††]The energy straggling distribution measured with 37 MeV protons in 10 cm pathlength of argon at a pressure of 1.2 atmospheres indicates that the most probable energy-loss is a few percent larger than that calculated using the Landau–Symon [Gooding and Eisberg (1957)] and Vavilov [Seltzer and Berger (1964)] approaches.

[‡‡]Other approaches related photon absorptions or resonant excitations in silicon to the ionization cross section (e.g., see [Hall (1984); Bak, Burenkov, Petersen, Uggerhøj, Møller and Siffert (1987)].

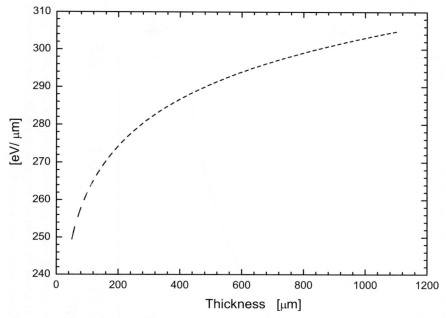

Fig. 2.10   The most probable energy-loss in silicon, calculated by means of Eq. (2.50), divided by the absorber length (in eV/μm) is shown as a function of the absorber thickness (in μm).

(1972); Bichsel and Saxon (1975); Rancoita and Seidman (1982); Hancock, James, Movchet, Rancoita and Van Rossum (1983); Rancoita (1984)]). In these approaches, the Rutherford distribution is replaced by a *realistic differential collision probability* $\omega'(\epsilon, E)$. For close collisions, it has the property that $\omega'(\epsilon, E) \to \omega(\epsilon, E)_R$ but, at low transferred energies, it takes into account that the atomic shell structure affects the interaction. By exploiting the convolution properties of the Laplace transforms (see [Bichsel and Yu (1972); Bichsel and Saxon (1975)] and references therein), it was derived that the transport equation solution [Eq. (2.40)] is given by:

$$f(\epsilon, x)_I = \frac{1}{\sqrt{2\pi\delta_2}} \int_{-\infty}^{+\infty} f(\epsilon - \epsilon', x)_{L,V} \exp\left(-\frac{\epsilon'^2}{2\delta_2}\right) d\epsilon', \qquad (2.53)$$

where $\delta_2 = M_2' - M_{2,R}$ is the square of the standard deviation σ_I of the Gaussian convolving distribution, and M_2' and $M_{2,R}$ are the second moments of the realistic and Rutherford differential collision functions, respectively. The second moment M_2 is defined by:

$$M_2 = \int_0^\infty \epsilon^2 \omega(\epsilon, E) \, d\epsilon.$$

It was shown [Bichsel (1970)] that adding further moments to this correction procedure does not appear to be needed.

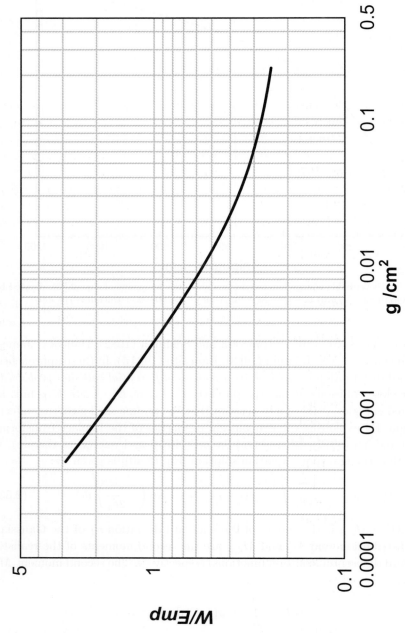

Fig. 2.11 The ratio of $FWHM_I$ over the *most probable energy-loss* (W/E_{mp}) versus the thickness χ in units of g/cm^2 for silicon. $FWHM_I$ was computed as $\approx \sqrt{FWHM_L^2 + FWHM_G^2}$; the most probable energy-loss (for $\beta = 1$) is from Eq. (2.50) and is increased by $\approx 3\%$ to take into account the effect of the Gaussian folding distribution for the improved energy-loss distribution (as discussed at page 67).

The function $f(\epsilon, x)_I$ is called *generalized energy-loss distribution* or *improved energy-loss distribution* (see [Møller (1986); Bichsel (1988)] and references therein for other approaches for the modified straggling function).

To a first approximation, the value of σ_I can be calculated by the *Shulek expression* [Shulek et al. (1966)]:

$$\sigma_I \equiv \sqrt{\delta_2} = \sqrt{\frac{8}{3}\xi\left[\sum_i I_i \frac{Z_i}{Z} \ln\left(\frac{2mc^2\beta^2}{I_i}\right)\right]} \quad [\text{MeV}], \qquad (2.54)$$

where I_i is the excitation energy (i.e., the absorption edges for the various shells (or subshells) of the element) of the ith shell (or subshell), and Z_i is the number of electrons in the ith shell (or subshell). The summation is carried out over those shells for which $I_i < 2mc^2\beta^2$. These excitation energies are given in standard tabulations (see for instance [Carlson (1975)]) and are used to determine the *effective excitation energies* (see [Sternheimer (1966, 1971)]), which allow the calculation of the density-effect correction [Eq. (2.1)]. The value σ_I increases as $\ln\beta^2$ and becomes almost constant for $\beta \to 1$. In addition, σ_I varies as $\sqrt{x}$, where x is the absorber thickness [see Eq. (2.41)]. Comparisons with experimental data are shown in [Hancock, James, Movchet, Rancoita and Van Rossum (1983, 1984); Rancoita (1984)] and reported in Sect. 2.1.2.5.

In Eq. (2.53), the resulting improved energy-loss distribution has an overall value of $FWHM_I$, which is roughly given by $\sqrt{FWHM_L^2 + FWHM_G^2}$, where $FWHM_L \simeq 4.02\,\xi$ varies as x, and $FWHM_G \simeq 2.36\,\sigma_I$ varies as $\sqrt{x}$. $FWHM_L$ and $FWHM_G$ are the $FWHM$'s of the energy straggling distribution [$f(\epsilon - \epsilon', x)_{L,V}$] and of the Gaussian convolving distribution, respectively. As the material thickness decreases, σ_I becomes more and more the dominant term, which determines the overall $FWHM_I$ of the straggling function. Conversely, it is not expected to provide an additional broadening of the distribution at large thicknesses.

For a silicon absorber and using the excitation energies given in [Sternheimer (1966)], we get [Hancock, James, Movchet, Rancoita and Van Rossum (1983, 1984)]:

$$\sigma_{I,\text{Si}} = \sqrt{\frac{8}{3}\xi\left\{\left[\sum_i I_i \frac{Z_i}{Z} \ln\left(\frac{2mc^2}{I_i}\right)\right] + 2\left[\sum_i I_i \frac{Z_i}{Z}\right]\ln\beta\right\}}$$

$$= \sqrt{\frac{8}{3}\xi\,(2.319 + 0.670 \ln\beta)\,10^{-3}} \quad [\text{MeV}], \qquad (2.55)$$

where the excitation energies are those for the K, L, $M_I(3s)$ and $M_{II}(3p)$ electrons. As an example, in $300\,\mu$m thick silicon detector we find $FWHM_L \approx 21.5$ keV, $FWHM_G \approx 13.6$ keV, and an overall $FWHM_I \approx 25.4$ (i.e., 18% larger than $FWHM_L$) for a relativistic $\beta \simeq 1$ and $z = 1$ particle. This is in agreement with the experimental data (see Sect. 2.1.2.5). However, the overall $FWHM_I$ value is almost determined by that of $FWHM_L$ for detectors[‡‡] with a silicon thickness

---

[‡‡]It was observed with plastic scintillators 4 and 10 cm thick [Asano, Hamasaki, Mori, Sasaki, Tsujita and Yusa (1996)].

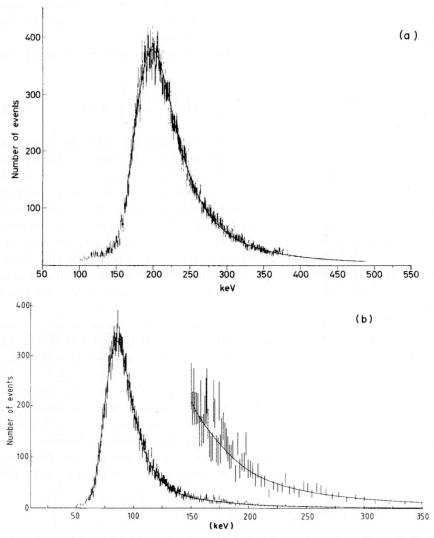

Fig. 2.12   Curves (a) and (b) (adapted and republished with permission from Hancock, S., James, F., Movchet, J., Rancoita, P.G. and Van Rossum, L., *Phys. Rev. A* **28**, 615 (1983); Copyright (1983) by the American Physical Society) show the energy-loss spectra at 0.736 and 115 GeV/c of incoming particle momentum. Continuous curves are the complete fit to experimental data, i.e., the Landau straggling function folded over the Gaussian distribution taking into account distant collisions.

$\gtrsim 1\,\mathrm{mm}$ (i.e., $\gtrsim 0.233\,\mathrm{g/cm^2}$).

In Fig. 2.11, the ratio W/E_{mp} of $FWHM_I$ over the most probable energy-loss (E_{mp}) is shown as a function of $\chi = x\rho$ in units of $\mathrm{g/cm^2}$, for a $z = 1$ and $\beta = 1$ massive particle traversing a silicon medium. The most probable energy-loss E_{mp} was computed by means of Eq. (2.50) and increased by $\approx 3\%$ (see page 67) to take

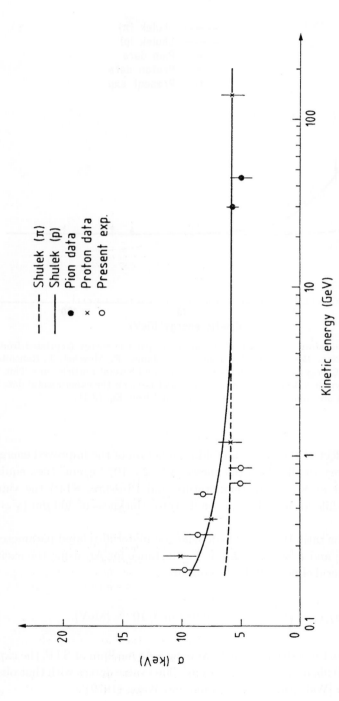

Fig. 2.13 σ_I as function of the incoming particle energy (reprinted from *Nucl. Instr. and Meth. in Phys. Res. Section B1*, Hancock, S., James, F., Movchet, J., Rancoita, P.G. and Van Rossum, L., Energy-Loss Distributions for Single and Several Particles in a Thin Silicon Absorber, 16–22, Copyright (1984), with permission from Elsevier). The experimental data are for pions and protons and are compared with the Shulek expression [Eq. (2.55)].

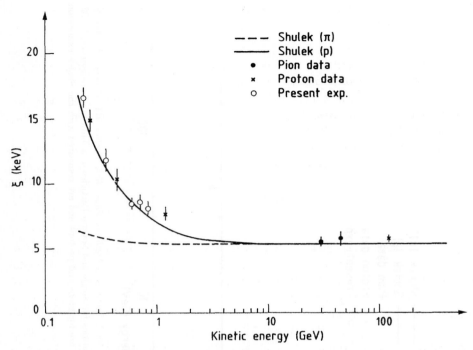

Fig. 2.14 ξ values are shown as function of the incoming particle energy (reprinted from *Nucl. Instr. and Meth. in Phys. Res. Section B***1**, Hancock, S., James, F., Movchet, J., Rancoita, P.G. and Van Rossum, L., Energy-Loss Distributions for Single and Several Particles in a Thin Silicon Absorber, 16–22, Copyright (1984), with permission from Elsevier): the experimental data are for pions and protons and are compared with values calculated from Eq. (2.41).

into account the effect of the Gaussian folding function of the improved energy-loss distribution. For very small silicon thicknesses $\chi \approx 2 \times 10^{-3}$ g/cm^2 (i.e., equivalent to ≈ 1.2 cm of Ar at Standard Temperature and Pressure, STP) the values of W/E_{mp} is ≈ 1.2, while for a typical silicon detector thickness of 300 μm ($\chi = 0.07$) $W/E_{mp} \approx 0.29$.

Furthermore, the ratio W/E_{mp} also depends on material-related parameters like $h\nu_p$ [see Eq. (2.49)] and I [see Eq. (2.54)]. For instance for Ar, using the excitation energy given in [Sternheimer (1971)], we obtain:

$$\sigma_{I,\text{Ar}} = \sqrt{\frac{8}{3}\,\xi\,(3.202 + 0.983\,\ln\beta)\,10^{-3}}\quad[\text{MeV}]. \tag{2.56}$$

For a $z = 1$ and $\beta = 1$ massive particle traversing a Ar medium at STP, the expected ratio W/E_{mp} is ≈ 0.95, at $\chi \approx 2 \times 10^{-3}$ g/cm^2. This value agrees with that observed for Ar gas medium [Walenta, Fisher, Okuno and Wang (1979)].

2.1.2.5 *Distant Collision Contribution to Energy Straggling in Thin Silicon Absorbers*

In thin silicon detectors, deviations from the Landau–Vavilov energy straggling distribution were observed [Esbensen et al. (1978)], systematically studied [Hancock, James, Movchet, Rancoita and Van Rossum (1983, 1984)] up to high energy, and interpreted as due to the distant collisions which were neglected in the Landau–Vavilov theories for thin absorbers (see also [Bak, Burenkov, Petersen, Uggerhøj, Møller and Siffert (1987); Bichsel (1988)]).

These systematic measurements were performed using silicon detectors. In this way, a precise energy determination of the parameters ξ, σ_I, and ϵ_{mp} could be carried out over a large range of proton incoming momenta, from $736\,\mathrm{MeV}/c$ up to $115\,\mathrm{GeV}/c$ [Hancock, James, Movchet, Rancoita and Van Rossum (1983, 1984)]. Data were also taken with pions [Hancock, James, Movchet, Rancoita and Van Rossum (1983, 1984)].

In Fig. 2.12, proton spectra at $736\,\mathrm{MeV}/c$ and $115\,\mathrm{GeV}/c$ incoming momenta are shown. The continuous curves are from Eq. (2.53), namely for a Landau straggling function convolved with a Gaussian distribution. These curves allow the determination of ξ, σ_I, and $\epsilon_{mp,\mathrm{Si}}$ parameters by a fitting procedure. At high energies, the agreement with the data is very accurate. It can be partially seen by the magnification of the Landau tail in Fig. 2.12(b). In Fig. 2.12(a), the data fall below the expectation curve from above $\approx 450\,\mathrm{keV}$, that is, above this threshold value of deposited energy, the Landau and Vavilov functions might exhibit a difference. In fact, fast δ-rays can also escape from the $300\,\mu\mathrm{m}$ thick silicon detector causing a decrease of the deposited energy. This effect is not taken into account by Eq. (2.46) and, consequently, by Eqs. (2.42, 2.45).

The fitted values of ϵ_{mp} are well in agreement with the values expected from Eq. (2.48). But the effective most probable energy-loss of the curve resulting from the convolution, i.e., the improved energy-loss curve, is larger than ϵ_{mp} (i.e., the Landau energy-loss peak) by $\approx 3\%$ [Hancock, James, Movchet, Rancoita and Van Rossum (1983, 1984)]. This is because the Landau function is asymmetric and, when convolved with a symmetric (Gaussian) function, the net result is a light shift of the distribution peak to larger values.

In Fig. 2.13, the $\sigma_{I,\mathrm{Si}}$ values are shown: the experimental data are for pions and protons and are compared with the Shulek expression. Within the experimental errors, it is seen that the data and the predictions from Eq. (2.55) agree. In addition, at high energies $\sigma_{I,\mathrm{Si}}$ becomes constant as expected.

In Fig. 2.14, the ξ values are shown as a function of the incoming energy: the experimental data are for pions and protons, and are compared with values calculated from Eq. (2.41). This general agreement indicates that the energy calibration was correctly performed and that Eq. (2.53) correctly takes into account both the close and distant collision contributions to energy-loss fluctuations in thin absorbers.

2.1.2.6 *Improved Energy-Loss Distribution for Multi-Particles in Silicon*

To a first approximation, by assuming that collision losses are statistically independent, n_p particles with the same velocity βc traversing a thin absorber x will undergo an overall energy-loss equivalent to the one of a single particle with velocity βc traversing a thin absorber (x_p) n_p times thicker, i.e.,

$$x_p = n_p\, x.$$

As previously discussed, the energy-loss distribution for a single particle depends on parameters (i.e., ξ [see Eq. (2.41)], σ_I [see Eq. (2.54)] and ϵ_{mp} [see Eq. (2.48)]), which are functions of the absorber thickness x_p. Similarly, the energy-loss distribution for n_p particles simultaneously crossing the absorber thickness x depends on the corresponding parameters ξ_{np}, $\sigma_{I,np}$, and $\epsilon_{mp,np}$.

For n_p particles traversing an absorber thickness x, we can rewrite Eq. (2.41) as:

$$\xi_{np} = 0.1535\, x\, \frac{\rho z^2 Z}{A\beta^2}\, n_p$$

$$= n_p\, \xi \quad [\text{MeV}]. \tag{2.57}$$

Equation (2.54) is rewritten as:

$$\sigma_{I,np} \equiv \sqrt{\delta_{2,np}} = \sqrt{\frac{8}{3}\xi_{np}\left[\sum_i I_i \frac{Z_i}{Z} \ln\left(\frac{2mc^2\beta^2}{I_i}\right)\right]} \quad [\text{MeV}]. \tag{2.58}$$

For silicon absorbers, Eq. (2.58) becomes

$$\sigma_{I,np,\text{Si}} = \sqrt{\frac{8}{3}\xi_{np}\,(2.319 + 0.670\,\ln\beta)10^{-3}} \quad [\text{MeV}], \tag{2.59}$$

instead of Eq. (2.55) which is valid for a single particle.

Similarly, Eq. (2.48) has be used for calculating the most probable energy-loss of a single particle traversing an absorber n_p times thicker. At high energies for $\beta \approx 1$ and $\beta\gamma > 10^{S_1}$ (the parameters S_1 are given in Tables 2.1 and 2.2), i.e., when Eq. (2.49) replaces Eq. (2.48), we have:

$$\epsilon_{mp,np} = \xi_{np}\left\{\ln\left[\frac{2mc^2\xi_{np}}{(h\nu_p)^2}\right] + 0.194\right\}$$

$$= n_p\,\xi\left\{\ln\left[\frac{2mc^2 n_p\xi}{(h\nu_p)^2}\right] + \ln n_p - \ln n_p + 0.194\right\}$$

$$= n_p\,\xi\left\{\ln\left[\frac{2mc^2\xi}{(h\nu_p)^2}\right] + \ln n_p + 0.194\right\}$$

$$= n_p\,(\epsilon_{mp} + \xi\ln n_p) \quad [\text{MeV}]. \tag{2.60}$$

Equation (2.60) shows that the most probable energy-loss for a set of n_p particles has a $\ln n_p$ dependence.

This set of Eqs. (2.57, 2.58, 2.60) has to be employed for the identification of the multiplicity of relativistic particles by means of energy loss.

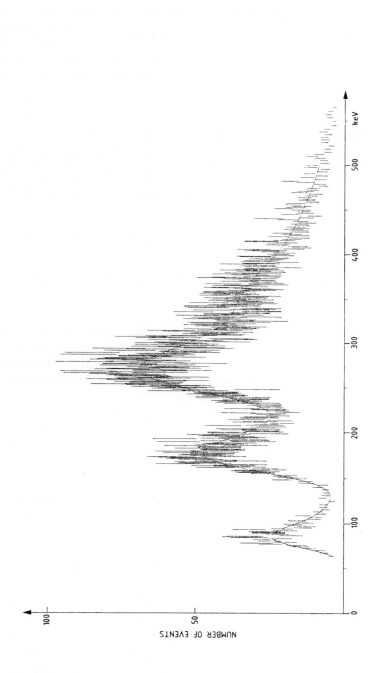

Fig. 2.15 Energy-loss of a relativistic multi-particle production (reprinted from *Nucl. Instr. and Meth. in Phys. Res. Section B1*, Hancock, S., James, F., Movchet, J., Rancoita, P.G. and Van Rossum, L., Energy-Loss Distributions for Single and Several Particles in a Thin Silicon Absorber, 16–22, Copyright (1984), with permission from Elsevier). The fit to experimental data is obtained by assuming Eqs. (2.57, 2.59, 2.60) for the energy-loss distribution.

In Fig. 2.15 [Hancock, James, Movchet, Rancoita and Van Rossum (1984)], it is presented the energy-loss spectrum of 1, 2, 3 and 4 relativistic particles traversing a silicon detector, simultaneously. The production rate depends on the particle multiplicity. The agreement between the fitted (continuous) curve and experimental data indicates that the factors ξ_{np}, $\sigma_{I,np}$ and $\epsilon_{mp,np}$ [see Eqs. (2.57, 2.59, 2.60)] account well for multi-particle collision losses.

2.1.3   *Range of Heavy Charged Particles*

At low energies, but high enough for shell corrections to be neglected, relativistic corrections are not relevant anymore and the stopping power depends on $1/v^2$ [see Eq. (2.2)]. Thus, for non-relativistic particles with charge ze and mass m_p the stopping power becomes, within $\approx 8\%$,

$$-\frac{dE}{dx} \approx 2\pi e^4 \rho N \frac{Z}{A} z^2 \frac{1}{E_k} \frac{m_p}{m} \ln\left(\frac{2mv^2}{I}\right)$$

$$= 0.0784 \,\rho N \frac{Z}{A} z^2 \frac{1}{E_k} \frac{m_p}{m} \ln\left(\frac{2mv^2}{I}\right) \quad [\text{MeV}], \qquad (2.61)$$

where $E_k = \frac{1}{2} m_p v^2$ is the kinetic energy of the particle.

The dependence, confirmed by experimental data, of Eq. (2.61) on both z^2 and $1/v^2$ results in a large multiplicative factor for low energy charged ions. Thus, the *specific ionization*, i.e., the number of charge pairs created per cm by ionization, rises to a maximum when the particle has nearly lost all its energy. This increase of specific ionization is referred to as the *Bragg peak*.

We call the *mass stopping power* of a material the quantity

$$\frac{dE}{d\chi} \equiv -\frac{1}{\rho}\frac{dE}{dx}, \qquad (2.62)$$

where $\chi = \rho x$ is the absorber thickness in units of g/cm^2.

If, for a particle of charge ze and kinetic energy E_k, the mass stopping power is known [Eqs. (2.61, 2.62)] in an absorber a with thickness χ, the *relative stopping power* R_{sp} in a second absorber b with the same thickness χ is given by:

$$R_{sp} = \left[\frac{Z_b}{A_b} z^2 \frac{1}{E_k} \ln\left(\frac{2mv^2}{I_b}\right)\right] \Big/ \left[\frac{Z_a}{A_a} z^2 \frac{1}{E_k} \ln\left(\frac{2mv^2}{I_a}\right)\right]$$

$$= \frac{Z_b A_a}{Z_a A_b} \frac{\ln(2mv^2/I_b)}{\ln(2mv^2/I_a)}. \qquad (2.63)$$

In Fig. 2.16 [PDB (1980)], the stopping powers calculated by means of Eq. (2.1) for various particles (up to α-particles) with kinetic energies above a few MeV's are shown for Pb absorbers and with an indication of the scaling factor up to carbon. Within $\approx 8\%$, Eq. (2.63) can be used for determining the stopping power in other absorbers.

Except for the very final part of the particle range** in matter, the heavy particle path inside an absorber is almost rectilinear, and the mean range†† R_t can be taken equal to the total *path length*, which can be obtained by integrating:

$$R_t = \int_0^{E_{ki}} \frac{dE_k}{(-dE/dx)},$$

where E_{ki} is the particle initial kinetic-energy and $-dE/dx$ is the stopping power (e.g., see Section 5 of [Fano (1963)] and Section 4.9.1 of [Marmier and Sheldon (1969)]). For a non-relativistic particle using Eq. (2.61), we have:

$$R_{t,m_p} = \frac{mm_p}{0.1568\,\rho N \frac{Z}{A} z^2} \int_0^{v_i} \frac{v^3 dv}{\ln\left(\frac{2mv^2}{I}\right)} \tag{2.64}$$

$$\approx \frac{m_p}{z^2} f(v_i), \tag{2.65}$$

where v_i is the particle initial velocity. Once the relationship between the range and the kinetic-energy of a particle with charge $z\,e$ and mass m_p is known for an absorber, one can determine an approximate* relationship for another particle with mass M_p and charge $z_p\,e$ at a shifted kinetic-energy so that both particles have equal v_i^2-values or, equivalently,

$$E_{k,M} = E_{k,m}\left(\frac{M_p}{m_p}\right);$$

thus from Eq. (2.65), for the latter particle we have

$$R_{t,M_p,E_{k,M}} \approx \frac{M_p z^2}{m_p z_p^2} R_{t,m_p,E_{k,m}}.$$

Bragg's measurements show that, at such low energies, the quantity $R_t\,\rho/\sqrt{A}$ is, within 15%, constant for every particle of given energy in different absorbing materials, i.e., the ratio of mean ranges in materials a and b follows the *Bragg–Kleemann rule*:

$$\frac{R_{t,a}}{R_{t,b}} \approx \frac{\rho_b}{\rho_a}\sqrt{\frac{A_a}{A_b}}.$$

In the case of a mixture made of a set of elements with relative mass abundance a_i, the mass value to be used in the Bragg–Kleemann rule is:

$$\sqrt{A_a} = \sum_i a_i \sqrt{A_i}.$$

**In literature (e.g. see discussion in Section 2.9 of [Fano (1963)], the range of a particle traversing a medium can be found defined as the *mean path length* covered before coming to rest or the mean distance traveled along its initial direction (*mean penetration depth*, also termed *mean projected range*). In this book, the range is usually approximated by the mean path length, if not explicitly indicated as the mean penetration depth.

††For ranges determined under the *continuous-slowing-down approximation* (csda), energy-loss fluctuations are neglected and charged particles are assumed to lose energy at a rate given by the stopping power.

*It has been assumed that the I-dependence of the integral in Eq. (2.64) can be neglected.

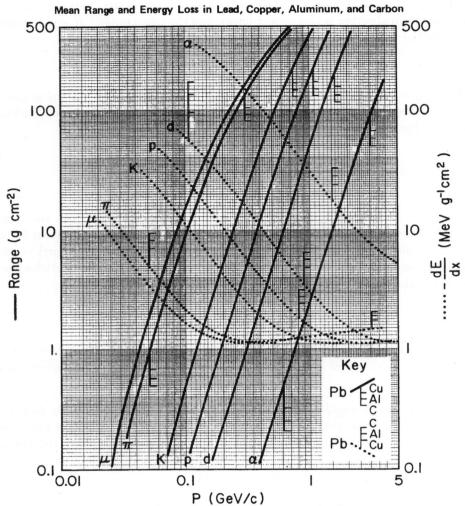

Fig. 2.16   Mean range in $g\,cm^{-2}$ and stopping power in $MeV\,g^{-1}cm^2$ from Eq. (2.1) for various particles in Pb, with scaling to Cu, Al and C (adapted and republished with permission from Kelly, R.L. et al., Particle Data Group, Review of Particle Physics, *Rev. of Mod. Phys.* **52**, S1 (1980); Copyright (1980) by the American Physical Society).

For air, we have $\sqrt{A_{air}} = 3.81$ and $\rho_{air} = 1.21 \times 10^{-3}\,g/cm^3$. A rough estimate of the mean range of protons, R_p, in dry air at STP between a few MeV's and ≈ 200 MeV can be obtained by the *Wilson and Brobeck formula* [Wilson (1947)]:

$$R_{t,air} = 10^2 \times \left(\frac{E_{kin}}{9.3}\right)^{1.8} \quad [cm],$$

where E_{kin} is in MeV. The mean range $R_{t,a}$ in an absorber a can be calculated

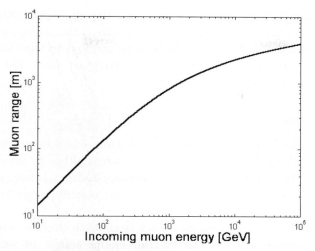

Fig. 2.17   Range of muon, in m, in standard rock computed by means of Eq. (2.66).

from that in air using the relationship:

$$R_{t,a} = 3.2 \times 10^{-4} \times \frac{A_{\mathrm{air}}}{\rho_a} R_{t,\mathrm{air}}.$$

For muons (see Fig. 2.17 and [Wright (1974)]) with incoming energies larger than $10\,\mathrm{GeV}$ in standard rock, with $\rho = 3\,\mathrm{g/cm^3}$, $Z = 11$ and $A = 22$, the range is given by:

$$R_\mu(E_\mu) = \left[\frac{1}{b_\mu} \ln\left(1 + \frac{b_\mu}{a_\mu} E_\mu \right) \right] \left[0.96 \frac{\ln E_\mu - 7.894}{\ln E_\mu - 8.074} \right] \quad [\mathrm{g/cm^2}], \qquad (2.66)$$

where $a_\mu = 2.2\,\mathrm{MeV/(g/cm^2)}$, $b_\mu = 4.4 \times 10^{-6}\,\mathrm{cm^2/g}$, and E_μ is MeV. Muon stopping power and range tables are also found in [Groom, Mokhov and Striganov (2001)].

The ranges of massive particles up to high-energies are reported in [ICRUM (1993a); Grupen (1996); ICRUM (2005)] (see also references therein). The ranges of protons and α-particles are available on web [Berger, Coursey, Zucker and Chang (2005)].

Due to the statistical nature of the energy loss, not all mono-energetic particles will have exactly the same traveled path in an absorber. To a first approximation, this variation, or *range straggling*, follows a normal (Gaussian) distribution with a *straggling parameter*[‡‡] which also depends on i) the massive-particle charge and velocity and ii) the absorber atomic number and ionization energy as discussed for instance in [Marmier and Sheldon (1969)]). However, the path-length distribution is skewed with a tail toward shorter-than-average path lengths, i.e., the most probable

---

[‡‡]The straggling parameter expresses the half-width at $(1/e)$th height of the straggling distribution (see Section 4.9.1 of [Marmier and Sheldon (1969)]).

path length is slightly longer than the average value [ICRUM (1993a)] (see also references therein). Furthermore, useful quantities are i) the projection of the range on the particle-track initial direction, which allows one to determine the penetration depth and ii) the ratio between the *average penetration depth* and *average path length* (this ratio is termed *detour factor*).

2.1.4   *Heavy Ions*

By *heavy ions* is usually meant atoms beyond helium, which do not have a neutral state of charge, i.e., they have a net positive charge, although negative ions can also be generated. In literature, when their atomic number z is not $\gg 1$, they may be also referred to as *light*. In addition, ions with velocity larger than the Bohr velocity[¶], v_0, are commonly termed *swift*.

When atoms or ions having velocities much larger than electron orbital velocities go through a material, they will modify their *charge state*[‡]: at sufficiently high energies their electrons will be stripped and, as bare nuclei, they will proceed to lose energy that results into an *electronic stopping power*. In fact, collision loss processes occur with atomic electrons of the absorber. Only occasionally, will incoming ions interact with the nuclei of the medium. At the beginning, the probability of capture of an electron is very limited. But electrons will be captured as the slowing-down process decreases the incoming particle velocity down to values close to those of orbital electrons. For velocities lower than the orbital velocities, the heavy ions will spend a large fraction of the time in a *neutral charge-state*. In addition, at small velocities the interaction with nuclei in the medium cannot be neglected anymore (Sect. 2.1.4.1).

The ion velocity is usually given in a dimensionless form, like $\beta = v/c$ in units of c, or expressed as

$$\varsigma_0 \equiv \frac{v}{v_0} = \frac{\hbar v}{e^2}$$
$$\simeq 137.036\,\beta,$$

i.e., in units of the Bohr orbital velocity ($v_0 = e^2/\hbar = c\,\alpha$) of the electron in the hydrogen atom (see Appendix A.2); or, finally, in units of $\varsigma = \varsigma_0/z$, i.e., in units of the Bohr orbital velocity of its own K-electron.

At sufficiently high energies, the energy-loss equation can be applied to the collision process (e.g., see Sects. 2.1 and 2.1.1; in addition, see Chapter 3 of [Ziegler, Biersack and Littmark (1985a); Ziegler, J.F. and M.D. and Biersack (2008a)]). Thus, for ion velocities $\varsigma_0 \gg z$, the stopping power increases as $1/\beta^2$ with decreasing velocity; while at intermediate velocities ($\varsigma_0 \approx z^{2/3}$) charge neutralization occurs gradually and begins to dominate the velocity dependence. In fact, as already mentioned, the ion's charge state does not remain constant as it passes through an

---

[¶]The reader can also see Appendix A.2.

[‡]The *charge state* of an ion refers to the instantaneous net charge of an ion.

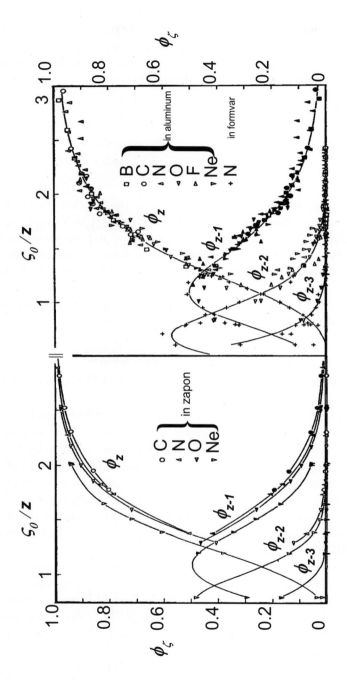

Fig. 2.18 Experimental equilibrium charge distributions for heavy ions at high velocities. The charge fractions ϕ_ζ (i.e., ϕ_z, ϕ_{z-1}, ϕ_{z-2} etc.) versus the velocity expressed as ς_0/z show an approximately universal relationship (adapted and reprinted, with permission, of the Annual Review of Nuclear Science, Volume 13 © 1963 by Annual Reviews *www.annualreviews.org*; [Northcliffe (1963)], see also references therein).

absorber and its velocity is lowered. Empirically, the energy-loss formula must be modified by introducing the parameter $\bar{\varpi}(\beta)$ to weight the charge state. This parameter (see [Northcliffe (1963)] and references therein) can be obtained by energy-loss measurements of protons and ions, with atomic number z and rest mass m_i, in a medium with atomic number Z and rest mass M (m_i and M are in atomic mass units). These measurements allow one to determine

$$-\left(\frac{dE}{dx}\right)_{ion} = -[\bar{\varpi}(\beta)\,z]^2 \left(\frac{dE}{dx}\right)_{proton},$$

with

$$\bar{\varpi}^2(\beta) = \sum_{\zeta=1}^{z} \phi_\zeta(\beta) \left(\frac{\zeta}{z}\right)^2,$$

where ζ is the charge state of the ion and $\phi_\zeta(\beta)$ is the fraction (function of β) of the time spent in that charge state. At energies[*] above $1\,\text{MeV/amu}$, equilibrium charge distributions ϕ_ζ were measured for various ions and are shown in Fig. 2.18 ([Northcliffe (1960)] and references therein). These data indicate that ϕ_ζ (i.e., ϕ_z, ϕ_{z-1}, ϕ_{z-2} etc.) has an approximate universal dependence on the velocity expressed as ς_0/z.

In the velocity region below $\varsigma_0 \approx z^{2/3}$, the electronic stopping power can be expressed as [Lindhard and Sharff (1961)]:

$$-\left(\frac{dE}{dx}\right)_{ion} = 8\,\kappa_e \pi e^2 n_A\, a_0 \frac{zZ}{\sqrt{z^{2/3}+Z^{2/3}}}\,\varsigma_0, \tag{2.67}$$

where κ_e is a numerical constant of order $z^{1/6}$, a_0 is the Bohr radius of the hydrogen atom and n_A is the number of atoms per cm^3 [Eq. (1.39)].

As discussed so far, the ion *effective-charge* varies at low velocity. The effective charge of a hydrogen atom was subject of controversy; but, at present, it is determined that protons do not have a bound electron at any velocity (e.g., see pages 74–79 of [Ziegler, Biersack and Littmark (1985a)] and Section 3.1 of [ICRUM (1993a)]). The charge variation is exhibited by α-particles: for example, Figure 3.1 of [ICRUM (1993a)] shows that this effect decreases the stopping power[‡‡] in water and gold at kinetic energies below $\approx 1.2\,\text{MeV}$. A literature survey on the electronic stopping power of ions with non-constant charge state is given in Section 3.4 of [ICRUM (2005)].

A comparison with experimental data, and a description of computer codes using the Monte-Carlo technique for evaluating both the electronic stopping power and the range of heavy ions are given in [Ziegler, Biersack and Littmark (1985a); Ziegler, J.F. and M.D. and Biersack (2008a)] (see also: [ICRUM (1993a)], for protons and α-particles, [ICRUM (2005)], for heavier ions and Chapter 7 of [Sigmund (2006)]).

---

[*]The reader can find the definition of kinetic energies per amu in Sect. 1.4.1.

[‡‡]Figure 3.1 of [ICRUM (1993a)] also shows that the Barkas effect (see Sect. 2.1.1.1) increases the α-particle stopping power in gold by $\lesssim 4\%$ for kinetic energies above $\approx 2\,\text{MeV}$, but it is negligible in water.

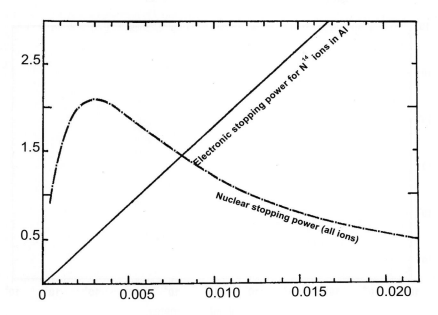

Fig. 2.19 Universal nuclear stopping power (adapted and reprinted, with permission, of the Annual Review of Nuclear Science, Volume 13 © 1963 by Annual Reviews *www.annualreviews.org*; [Northcliffe (1963)]) for heavy ions velocities $\varsigma_0 \leq 1$. The straight line is the electronic stopping power for N^{14} in Al computed by means of Eq. (2.67). The ordinate axis is for the stopping power $-\left(\frac{dE_k}{d\chi_i}\right)_{ion}$ in units of $MeV\,cm^2 mg^{-1}$ multiplied by $\frac{M}{zZm_i}(m_i + M)\sqrt{z^{2/3} + Z^{2/3}}$. The units for the abscissa axis are: $\sqrt{\frac{m_i M/(m_i+M)}{zZ\sqrt{z^{2/3}+Z^{2/3}}}\frac{E_k}{m_i}}$.

2.1.4.1 *Nuclear Stopping Power*

At low velocity ($\varsigma_0 \lesssim 1$), i.e., when the charge neutralization begins to dominate the collision energy-loss process, the ion *electronic stopping power* decreases and the energy loss due to collisions with target nuclei is no longer negligible. This latter process results in the so-called *nuclear stopping power** (e.g., see Sect. 4.2.1.1; in addition, see Chapter 2 of [Ziegler, Biersack and Littmark (1985a); Ziegler, J.F. and M.D. and Biersack (2008a)] and references therein). As the electronic stopping power decreases rapidly, the nuclear stopping power increases approximately as $1/\beta^2$. At a critical velocity

$$\varsigma_{0,c} = \frac{\beta_c}{\alpha} \ll 1,$$

the nuclear stopping power exceeds the electronic stopping power. However, at very low velocities when the screening of the nuclear field of ions and atoms is dominant, the nuclear stopping power reaches a maximum and starts to fall (Fig. 2.19). To a

---

*This process is mainly responsible for displacement damage in semiconductors (e.g., see Sects. 4.2.1, 4.2.1.1 and 4.2.1.3).

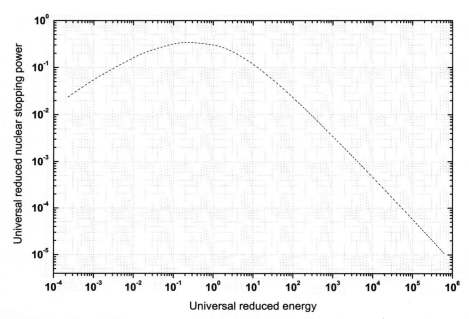

Fig. 2.20 Universal reduced nuclear stopping power $[\mathfrak{R}(\epsilon_{r,U})]$ as a function of the universal reduced energy $(\epsilon_{r,U})$.

first approximation, the nuclear stopping power can be represented by the universal curve given in Fig. 2.19 ([Northcliffe (1963)] and references therein). In this figure, z and m_i are the charge and the mass of the incoming ion, respectively; Z and M are the charge and the mass of the absorber atoms, respectively; finally, E_k is the kinetic energy of the incoming ion in units of MeV and the path length χ_i is in units of mg/cm^2. A detailed comparison of models, a comparison with experimental data, and a description of computer codes using the Monte-Carlo technique for evaluating the nuclear stopping powers for heavy ions are given in [Ziegler, Biersack and Littmark (1985a); Ziegler, J.F. and M.D. and Biersack (2008a)] (see also Section 4 of [ICRUM (1993a)]). In addition, a review of the properties of ion interaction with matter is found in [Balanzat and Bouffard (1993)].

As mentioned above, the energy lost to target nuclei is basically determined by investigating screened Coulomb collisions between two interacting atoms[†] with almost non-relativistic velocities. Using the *Thomas–Fermi model of the atom* (e.g., see Chapters 1 and 2 of [Torrens (1972)]), the scattering of the projectile by target atoms can be treated in terms of an *interatomic Coulomb potential* which is a function of the radial distance r between the two nuclei {e.g., see Equation (3.9)

[†]The reader can find a survey dedicated to interatomic potential treatment in [Torrens (1972)] (see also [Gehlen, Beeler and Jaffee (1972)]).

Table 2.4 Nuclear stopping power fractions in percentage at 1, 10 and 100 keV in Be, Si and Pb absorber for protons (p) and α-particles (α) (data from [Berger, Coursey, Zucker and Chang (2005)]).

Absorber	Particle	1 keV	10 keV	100 keV
Be	p	15.0	1.3	0.2
	α	53.3	12.7	1.0
Si	p	10.2	1.4	0.2
	α	61.0	17.5	1.3
Pb	p	5.1	1.5	0.2
	α	22.9	10.1	1.8

of [Torrens (1972)] or Equation (2-56) of [Ziegler, Biersack and Littmark (1985a)]}

$$V(r) = \frac{zZe^2}{r} \, \Psi_{\mathrm{I}}(r_{\mathrm{r}}), \qquad (2.68)$$

where ez (projectile) and eZ (target) are the charges of the bare nuclei and Ψ_{I} is the *interatomic screening function*. This latter function depends on the *reduced radius* r_{r} given by

$$r_{\mathrm{r}} = \frac{r}{\mathrm{a_I}}, \qquad (2.69)$$

where $\mathrm{a_I}$ is the *screening length* (also termed *screening radius*). Several expressions were suggested for $\mathrm{a_I}$, e.g., by Bohr (1940)

$$\mathrm{a_B} = \frac{\mathrm{a_0}}{\left(z^{2/3} + Z^{2/3}\right)^{1/2}}, \qquad (2.70)$$

by Firsov (1957)

$$\mathrm{a_F} = \frac{C_{\mathrm{TF}} \, \mathrm{a_0}}{\left(z^{1/2} + Z^{1/2}\right)^{2/3}}, \qquad (2.71)$$

and, also, by Lindhard and Sharff (1961)

$$\mathrm{a_L} = \frac{C_{\mathrm{TF}} \, \mathrm{a_0}}{\left(z^{2/3} + Z^{2/3}\right)^{1/2}}, \qquad (2.72)$$

where

$$\mathrm{a_0} = \frac{\hbar^2}{me^2}$$

is the Bohr radius (see Appendix A.2), m is the electron rest mass and

$$C_{\mathrm{TF}} = \left(9\,\pi^2\right)^{1/3} 2^{-7/3} \simeq 0.88534$$

is a constant introduced in the Thomas–Fermi model.

The interatomic potentials [Eq. (2.68)] accounting for the electron clouds of the projectile and target nuclei were calculated by various authors, who provided expressions for the screening functions (e.g., see pages 33–41 of [Ziegler, Biersack and

Littmark (1985a)] or, equivalently, pages 2-16–2-33 of [Ziegler, J.F. and M.D. and Biersack (2008a)] and references therein). Ziegler, Biersack and Littmark (1985a) carried out extensive calculations for a large number of projectile-target combinations and were able to approximate this comprehensive set of numerical results rather accurately[¶] by the so-called *universal screening function* {see also Equation (2-74) in [Ziegler, J.F. and M.D. and Biersack (2008a)]}

$$\Psi_U(r_r^U) \simeq 0.1818 \, \exp(-3.2 \, r_r^U)$$
$$+0.5099 \, \exp(-0.9423 \, r_r^U)$$
$$+0.2802 \, \exp(-0.4028 \, r_r^U)$$
$$+0.02817 \, \exp(-0.2016 \, r_r^U) \,, \tag{2.73}$$

where the reduced radius $r_r^U = r/a_U$ is obtained using the *universal screening length*

$$a_U = \frac{C_{TF} \, a_0}{z^{0.23} + Z^{0.23}}. \tag{2.74}$$

In an elastic scattering of heavy ions, the (angular) differential cross section is determined by the interatomic (central) potential expressed in Eq. (2.68); while the transferred energy T is determined by the scattering angle and maximum transferable energy T_{max} ([Mott and Massey (1965)]; see also Sects. 1.3.1 and 1.5). For a non relativistic scattering, these latter quantities are given in Eq. (1.54), for T, and Eq. (1.53), for T_{max}. Finally, the nuclear stopping power is determined as the average energy transferred in a unit length. Using the universal interatomic potential [Eqs. (2.68, 2.73, 2.74)], Ziegler, Biersack and Littmark (1985a) estimated that, in practical calculations, the *universal stopping power* is approximated[‖] by

$$-\left(\frac{dE}{dx}\right)_{nucl} \simeq 5.1053 \times 10^3 \, \frac{\rho \, z Z \, \mathfrak{R}(\epsilon_{r,U})}{A \, (1 + M/m) \, (z^{0.23} + Z^{0.23})} \quad [\text{MeV/cm}], \tag{2.75}$$

where ρ and A are the density and the atomic weight of the target medium, respectively; m and M are the rest masses of the projectile and target, respectively; $\mathfrak{R}(\epsilon_{r,U})$, shown in Fig. 2.20, is the so-called *(universal) reduced nuclear stopping power* [termed, also, *(universal) scaled nuclear stopping power*] given by {see also Equations (2-89)–(2-90) in [Ziegler, J.F. and M.D. and Biersack (2008a)]}

$$\mathfrak{R}(\epsilon_{r,U}) = \begin{cases} \ln\left(1 + 1.1383 \, \epsilon_{r,U}\right) / \left[2 \left(\epsilon_{r,U} + 0.01321 \, \epsilon_{r,U}^{0.21226} + 0.19593 \, \epsilon_{r,U}^{0.5}\right)\right] \,, \\ \text{for } \epsilon_{r,U} \leq 30, \\ \\ \ln(\epsilon_{r,U})/(2 \, \epsilon_{r,U}), \text{ for } \epsilon_{r,U} > 30, \end{cases}$$

which depends on the dimensionless variable $\epsilon_{r,U}$ [the so-called *(universal) reduced energy*] defined as

$$\epsilon_{r,U} = \frac{a_U}{z Z \, e^2} \left(\frac{M}{m + M}\right) E_k$$
$$= \frac{32.536}{z Z \, (z^{0.23} + Z^{0.23})} \left(\frac{M}{m + M}\right) E_k, \tag{2.76}$$

[¶]These authors discussed the accuracy of calculated electronic and nuclear stopping powers by comparing them to over 13,000 experimental data points taken from over a thousand published papers (e.g., see Chapters 6 and 7 of [Ziegler, Biersack and Littmark (1985a)]).

[‖]The reader can also see Equation 4.15 at page 47 of [ICRUM (1993a)].

where E_k is the kinetic energy of the projectile in the laboratory system; E_k is in keV when the numerical constant in Eq. (2.76) is 32.536 {e.g., see Equation (2-73) of [Ziegler, Biersack and Littmark (1985a)] or Equation (2-88) of [Ziegler, J.F. and M.D. and Biersack (2008a)]}. For instance, from Eq. (2.76) $\epsilon_{r,U}$ is $\approx 6.31 \times 10^5$ for Si-ions with 1 GeV/amu kinetic energy** in a silicon medium. The reduced nuclear stopping powers calculated with other classical atomic models do not differ from the universal for reduced energies $\gtrsim 10$ (e.g., see Figure 2-18 at page 52 of [Ziegler, Biersack and Littmark (1985a)] or Figure 2-18 at page 2-36 of [Ziegler, J.F. and M.D. and Biersack (2008a)]). Furthermore, the universal interatomic potential is used in the SRIM code§ whose latest version is available in [Ziegler, J.F. and M.D. and Biersack (2008b)].

Tabulations of the nuclear stopping powers‡‡ for protons and α-particles are reported in [ICRUM (1993a)] and, also, available in [Berger, Coursey, Zucker and Chang (2005)]. In these tabulations, the nuclear stopping power for α-particles was derived using universal interatomic potentials, screening functions and lengths; while for protons, which are expected to pass through the medium as positive charges*, a screened potential based on the Thomas–Fermi model [Molière (1947)] was used with a screening length

$$a_M = \frac{C_{TF}\, a_0}{Z^{1/3}}$$

and a screening function

$$\Psi_M(r_r^M) \simeq 0.10\, \exp\!\left(-6\, r_r^M\right) + 0.55\, \exp\!\left(-1.2\, r_r^M\right) + 0.35\, \exp\!\left(-0.3\, r_r^U\right),$$

with $r_r^M = r/a_M$ (see Section 4.1 of [ICRUM (1993a)]). For protons and α-particles, these approximated expressions for the scaled nuclear stopping power are accurate to within 1% for $\epsilon_{r,U} < 3$ and to within 5% for $\epsilon_{r,U} > 3$ (Section 4.3 of [ICRUM (1993a)]). In Table 2.4 (data from [Berger, Coursey, Zucker and Chang (2005)]), the fractions (in percentage) of nuclear stopping powers

$$F_{nucl} = -\left(\frac{dE}{dx}\right)_{nucl} \Big/ \left\{\left[-\left(\frac{dE}{dx}\right)_{nucl}\right] + \left[-\left(\frac{dE}{dx}\right)_{ion}\right]\right\}$$

are shown for protons and α-particles in Be, Si and Pb absorbers at kinetic energies of 1, 10 and 100 keV; above 100 keV, F_{nucl} is smaller than $\approx 2\%$.

In Fig. 2.21, the nuclear stopping powers in a silicon medium for α-particles and a few heavy-ions are shown as function of the kinetic energy of incoming ions in

---

** The reader can find a definition of kinetic energies per amu in Sect. 1.4.1.

§ SRIM (Stopping power and Range of Ions in Matter) is a group of programs which calculate the stopping and range of ions (up to 2 GeV/amu) into matter using a quantum mechanical treatment of ion-atom collisions. TRIM (the Transport of Ions in Matter) is the most comprehensive program included and accepts complex targets made of compound materials. SRIM results from the original work by Biersack and Haggmark (1980) and the work by Ziegler on stopping theory reported in: The Stopping and Range of Ions in Matter, volumes 2–6, Pergamon Press (1977–1985). A recently published version is available in [Ziegler, J.F. and M.D. and Biersack (2008a)].

‡‡ The reader can see, in addition, the discussion in [Seltzer, Inokuti, Paul and Bichsel (2001)].

* Only at low velocity protons might have a bound electron, see Sect. 2.1.4.

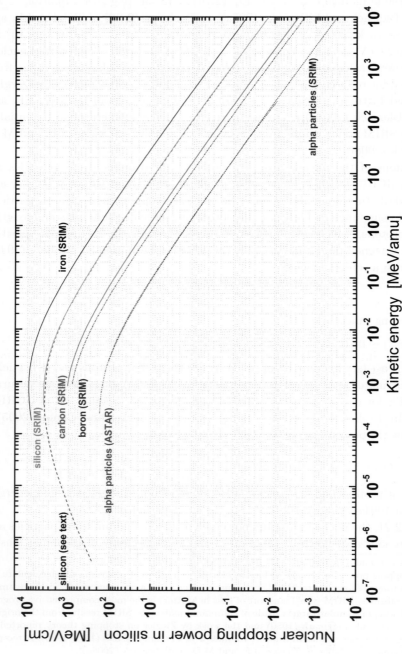

Fig. 2.21 Nuclear stopping powers in a silicon medium for α-particles, boron-, carbon-, silicon- and iron-ions as function of the kinetic energy of incoming ions in units of MeV/amu. The data for α-particles (ASTAR) are from [Berger, Coursey, Zucker and Chang (2005)]; those indicated as "silicon (see text)" are calculated using Eq. (2.75) for incoming ^{28}Si ions; the data for ^{56}Fe ("iron"), ^{28}Si ("silicon"), ^{12}C ("carbon"), ^{11}B ("boron") ions and α-particles (indicated as "SRIM") are those available in [Ziegler, J.F. and M.D. and Biersack (2008b)].

units of MeV/amu[†]. The data for α-particles (ASTAR) are from [Berger, Coursey, Zucker and Chang (2005)], in which the nuclear stopping power is calculated using the relation between the deflection angles and the energy transfers to the recoiling atom in elastic collisions; those indicated as "silicon (see text)" are calculated using Eq. (2.75) for incoming ^{28}Si ions; the data for ^{56}Fe ("iron"), ^{28}Si ("silicon"), ^{12}C ("carbon"), ^{11}B ("boron") ions and α-particles (indicated as "SRIM") are those available in [Ziegler, J.F. and M.D. and Biersack (2008b)]. It has to be remarked that in these calculations, for instance using Eq. (2.75) and in TRIM above 1 MeV/amu, the energy loss into nuclear reactions is not take into account, so that ions may have inelastic energy-losses which are not included in the calculation.

2.1.5 *Ionization Yield in Gas Media*

Along the passage of a particle through a gas medium, a discrete number of ionizing collisions occurs. It is usual to distinguish between *primary* (introduced at page 32) and *total* ionization. The primary ionization is generated directly by the incoming particle, while the *secondary* ionization (introduced at page 49) is generated by fast δ-rays emitted by primary events. The sum of the two contributions is called total ionization. The total number of *ion pairs* (i.e., electrons and positive ions) per unit path is given by:

$$n_T = \frac{\triangle E}{W_i} \text{ [ion pairs/cm]},$$

where $\triangle E$ is the energy deposited in 1 cm of traversed gas, and W_i is the mean energy required to produced an ion pair (see Table 2.5, and [Sauli (1977); Fernow (1986)] and references therein). W_i does not vary appreciably for different gases, as shown in Table 2.5.

The average value of W_i is ≈ 30 eV: it is almost constant at relativistic particle velocities and increases only slightly at low velocities. It has to be noted that the primary ionization is almost linearly dependent on the mean Z value of the gas (with the exception of Xe). For gas mixtures, like for example 70% Ar and 30% isobutane, the primary (n_P) and total (n_T) ionizations are calculated as:

$$n_T = 0.70 \times \frac{2440}{26} + 0.30 \times \frac{4600}{24} = 123 \text{ ion pairs/cm},$$
$$n_P = 0.70 \times 29.4 + 0.30 \times 46 = 34 \text{ ion pairs/cm}.$$

For $n_P \approx 34$ ion pairs per cm the average distance between primary ions is about 300 μm.

The ion pair production is a statistical process. If S pairs are produced on average by a particle, the expected statistical error is $\sqrt{S}$ while the actual error is smaller by a factor $\sqrt{F}$, where F is called the *Fano factor* [Fano (1947)]. In this way, the energy resolution improves by a factor $\sqrt{F}$. Typical Fano factor values

---

[†]The values of the masses in units of amu for ^{56}Fe, ^{28}Si, ^{12}C, ^{11}B and ^{4}He (α-particles) are those used in [Ziegler, J.F. and M.D. and Biersack (2008b)].

Table 2.5   Properties of gases at Standard Pressure and Temperature (STP) for a minimum ionizing particle (*mip*): W_i is the mean energy needed to create a single ion pair, the total ionization and the primary ionization are for 1 cm of path.

Gas	Z	A	ρ g/cm^3	1st Ion. Potent. eV	2nd Ion. Potent. eV	Primary Ion. ion pairs cm^{-1}	Total Ion. ion pairs cm^{-1}	W_i eV	$-dE/dX$ keV/cm
H_2	2	2	8.99×10^{-5}	15.4		5.2	9.2	37	0.34
He	2	4	1.79×10^{-4}	24.6	54.4	5.9	7.8	41	0.32
N_2	14	28	1.25×10^{-3}	15.5		10–19	56	35	1.96
O_2	16	32	1.43×10^{-3}	12.2		22	73	31	2.26
Ne	10	20.2	9.00×10^{-4}	21.6	41.1	12	39	36	1.41
Ar	18	39.9	1.78×10^{-3}	15.8	27.6	29.4	94	26	2.44
Kr	36	83.8	3.74×10^{-3}	14.0	24.4	22	192	24	4.60
Xe	54	131.3	5.89×10^{-3}	12.1	21.2	44	307	22	6.76
CO_2	22	44	1.98×10^{-3}	13.7		34	91	33	3.01
CH_4	10	16	7.17×10^{-4}	13.1		16	53	28	1.48
C_4H_{10}	34	58	2.67×10^{-3}	10.8		46	195	23	4.50

See [Sauli (1977); Fernow (1986)] and references therein.

are between 0.16 and 0.31 depending on both the incoming-particle type and the medium (e.g, see Section 1.1.3 in [Grupen and Shwartz (2008)]).

2.1.6 *Passage of Electrons and Positrons through Matter*

Electrons and positrons lose energy by collisions while traversing an absorber, just as massive charged particles do*. In addition, because of their small mass and depending on their kinetic energy, they will undergo a significant energy-loss by radiative emission[†], i.e., by the so-called *bremsstrahlung emission*. For instance, in a lead absorber the energy loss by radiation becomes the dominant process above a kinetic energy of $\approx 7\,\mathrm{MeV}$. It has to be noted that a detailed energy-loss computation is beyond the scope of this book and can be found, for instance, in [ICRUM (1984b)] (e.g., see also a database available on web in [Berger, Coursey, Zucker and Chang (2005)]).

2.1.6.1 *Collision Losses by Electrons and Positrons*

The treatment of the energy loss by collisions for incoming electrons and positrons follows the same lines as for massive charged particles, i.e., assuming an interaction on quasi-free atomic electrons and neglecting the shell correction term.

The differential cross section for the electron–electron scattering is (at least at low energy) given by the *Mott scattering formula* for two identical particles [Mott (1930)]. At higher energies, the scattering is described by its relativistic extension, namely *Møller's differential cross section* [Møller (1932)]. This differential cross section is used to deal with large energy transfers in the interaction (i.e., in the case of close collisions). Since the outgoing electron of higher energy is considered to be the *primary electron*, the maximum energy transferred is $\frac{1}{2}$ of the incoming electron kinetic energy E_k.

In the case of positrons, the derivation of the energy loss proceeds along exactly the same lines but since positrons and electrons are different particles, the maximum energy transfer allowed is the whole incoming kinetic energy. For collisions with large energy transfers *Bhabha's differential cross section* has to be used [Bhabha (1936)].

However, the differential collision probability for incoming positrons or electrons with kinetic energies $E_k = E - mc^2$ (in units of MeV and where E is the total energy) which are large compared with mc^2 (i.e., $\beta \approx 1$) is reduced to [Rossi (1964)]:

$$\omega(\epsilon, E) = 0.1535 \, \frac{\rho Z}{A \epsilon^2} \ [\mathrm{MeV\ cm}]^{-1},$$

where the transferred energies ϵ are very small with respect to the maximum trans-

---

*In addition, as massive charged particles (e.g., see Sects. 2.1.1 and 2.1.4.1), electrons and positrons can lose energy by displacing atoms of the medium but with lower cross sections compared to massive particles (e.g., see Sects. 4.2.1.4 and 4.2.3.1 for electron–silicon interactions).

[†]The radiative emission occurs also for particles with larger masses (e.g., muons), but at larger energies: for example, for muons it is a relevant energy-loss process above (150–200) GeV. Muon stopping power and range tables are available in [Groom, Mokhov and Striganov (2001)].

ferable energy (and to the incoming kinetic energy). When this latter condition is satisfied, the electron and positron differential cross sections become similar to the one for massive charged particles [see Eq. (2.37)]: only at very large energy transfers, the type of particles and their spins are relevant.

Let us indicate with W_s the transferred energy value above which *Møller's* (for electrons) or *Bhabha's* (for positrons) *differential cross sections* have to used to determine the overall stopping power. For small energy transfers (i.e., for distant collisions) up to $\approx W_s$, the energy loss is related to the oscillator strengths of the atoms and is, to a good approximation, independent of the incoming particle charge. It is given (as for massive charged particles) by [Rohrlich and Carlson (1954)]:

$$-\left(\frac{dE}{dx}\right)_s = \frac{2\pi ne^4}{mv^2}\left\{\ln\left[\frac{2mv^2W_s}{I^2(1-\beta^2)}\right] - \beta^2\right\}, \tag{2.77}$$

where I is the mean excitation energy of the material. Once the correction for the density-effect[‡‡] is added, this latter equation becomes (as shown by Rohrlich and Carlson (1954), see also [Berger and Seltzer (1964)]):

$$-\left(\frac{dE}{dx}\right)^{\pm} = 0.1535\,\frac{\rho Z}{A\beta^2}\left\{\ln\left[\frac{\tau^2(\tau+2)}{2(I^2/mc^2)}\right] + F(\tau)^{\pm} - \delta\right\} \text{[MeV/cm]}, \tag{2.78}$$

where [see Eqs. (1.16, 1.17)]

$$\tau = \frac{E_k}{mc^2},$$

$$\beta = \frac{v}{c} = \frac{\sqrt{\tau(\tau+2)}}{\tau+1},$$

$$\gamma = \tau + 1.$$

The functions $F(\tau)^{\pm}$ (shown in Fig. 2.22) are:
i) for positrons

$$F(\tau)^+ = 2\ln 2 - \frac{\beta^2}{12}\left[23 + \frac{14}{\tau+2} + \frac{10}{(\tau+2)^2} + \frac{4}{(\tau+2)^3}\right],$$

ii) and for electrons:

$$F(\tau)^- = 1 - \beta^2 + \frac{(\tau^2/8) - (2\tau+1)\ln 2}{(\tau+1)^2}.$$

The collision energy-losses for electrons and positrons are expected to differ slightly. In fact [Berger and Seltzer (1964)], the ratio $\left(\frac{dE}{dx}\right)^+ / \left(\frac{dE}{dx}\right)^-$ somewhat depends on the absorbing material: for energies between $(20\text{--}50)\,\text{MeV}$ it is ≈ 1.08 in Al, and becomes ≈ 1.12 in Au. At higher energies ($\approx 1\,\text{GeV}$), it is ≈ 0.98 in Al, and becomes ≈ 0.97 in Au. Further calculations and comparisons are reported in Sects. 11.1–11.3 of [ICRUM (1984b)].

[‡‡]For electrons and positrons, the parameters for the density-effect correction were extensively calculated in the framework of Sternheimer's theory. A database for 278 materials is available in [Seltzer and Berger (1984)] (see also references therein).

For electrons, taking into account that the maximum transferable energy is $E_k/2$, Eq. (2.78) can be rewritten as

$$-\left(\frac{dE}{dx}\right)^- = 0.1535 \frac{\rho Z}{A\beta^2} \left\{\ln\left[\frac{E_k\, mv^2}{2I^2(1-\beta^2)}\right] + F(\tau)^- - \delta\right\}$$

$$= 0.1535 \frac{\rho Z}{A\beta^2} \mathbb{E}(\beta, \gamma, I, W_m, \delta) \quad [\text{MeV/cm}], \qquad (2.79)$$

where the function $\mathbb{E}(\beta, \gamma, I, W_m, \delta)$ is given by:

$$\ln\left[\frac{mv^2 W_m}{I^2(1-\beta^2)}\right] + 1 - \beta^2 + \left(\frac{2\gamma-1}{\gamma^2}\right)\ln 2 - \frac{1}{8}\left(\frac{\gamma-1}{\gamma}\right)^2 - \delta.$$

At high energies ($\gamma \gg 1$ and $\beta \approx 1$), we can estimate the energy-loss difference between electrons and $z = 1$ massive charged particles by means of Eqs. (2.17, 2.79). The energy-loss difference $\triangle_{e-h}$ is given by:

$$\triangle_{e-h} = 0.1535 \frac{\rho Z}{A}\left(\frac{1}{8} - \ln 2 + 2\right)$$

$$= 0.2195 \frac{\rho Z}{A} \quad [\text{MeV/cm}]. \qquad (2.80)$$

This equation shows that the energy loss is larger for electrons than for massive charged particles. For instance, in silicon, the difference $\triangle_{e-h}$ is ≈ 0.255 MeV/cm, i.e., the energy loss for electrons is $\approx 6.6\%$ higher than the one for a massive charged particle at the ionization-loss minimum (see Table 2.3). However, the energy loss increases because of the relativistic rise [see Eq. (2.26)], and the relative energy-loss difference decreases.

To a first approximation [Sternheimer (1961)], Eq. (2.28) can be still used when the effective detectable maximum transferred energy W_0 is much smaller than E_k. In fact, under such conditions, the effective differential collision cross section very slightly depends on the type and the spin of incoming particles, i.e., the collisions are mainly distant ones. Thus, the equation for the restricted energy-loss for electrons and positrons is given by Eq. (2.78), setting $W_s = W_0$ and adding the density-effect correction term:

$$-\left(\frac{dE}{dx}\right)_{restr} = 0.1535 \frac{\rho Z}{A\beta^2}\left\{\ln\left[\frac{2mv^2 W_0}{I^2(1-\beta^2)}\right] - \beta^2 - \delta\right\} \quad [\text{MeV/cm}]. \qquad (2.81)$$

The restricted energy-loss equation for electrons (and positrons) is given by the one [see Eq. (2.28)] for massive and $z = 1$ particles.

2.1.6.2 *Most Probable Energy-Loss of Electrons and Positrons*

As for massive charged particles, the energy-loss process of electrons and positrons undergoes statistical fluctuations. While for the distant collisions the differential probability cross section is almost independent of the type and spin of the incoming particles, these latter characteristics can play a role in close collisions where high-energy transfers occur.

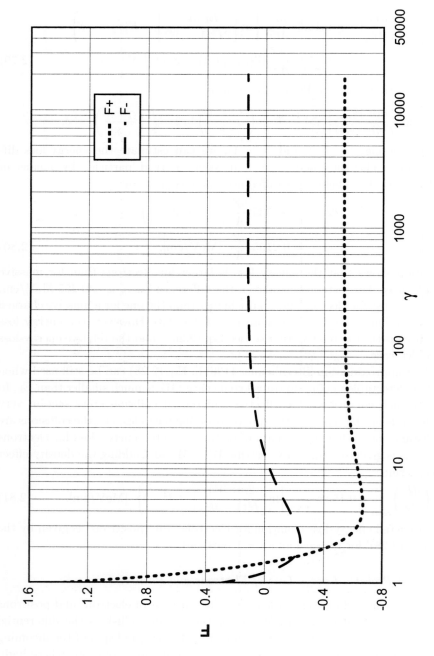

Fig. 2.22 $F(\tau)^+$ and $F(\tau)^-$ as functions of $\gamma = [E_k/(mc^2)] + 1 = \tau + 1$.

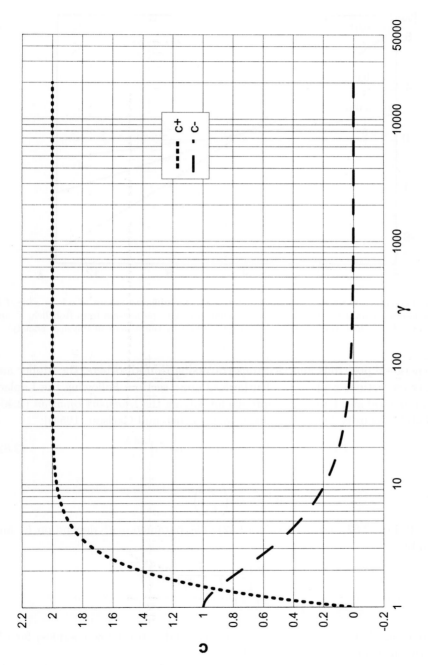

Fig. 2.23 c^+ and c^- as functions of $\gamma = [E_k/(mc^2)] + 1$.

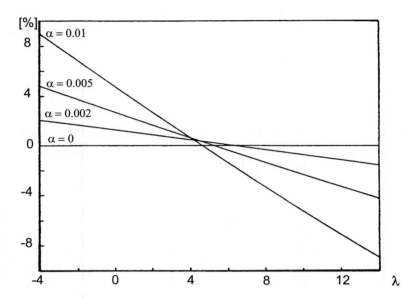

Fig. 2.24　Percentage corrections to the Landau curve for different values of $\alpha^\pm = c^\pm(\xi/E_k)$ versus the Landau parameter λ (adapted and republished with permission from Rohrlich, F. and Carlson, B.C., *Phys. Rev.* **93**, 38 (1954); Copyright (1954) by the American Physical Society).

For energetic positrons and electrons traversing thin absorbers, the Møller and Bhabha cross sections replace the Rutherford cross section. Following the Laplace transform method and Landau's approach for solving the *transport equation* (2.40), the probability density function is given by [Rohrlich and Carlson (1954)]:

$$f(\epsilon, x)_{L,\pm} = \exp\left[-\alpha^\pm(\lambda + \ln \alpha^\pm)\right] \frac{\phi(\lambda)}{\xi}, \qquad (2.82)$$

where

$$\alpha^\pm = c^\pm \left(\frac{\xi}{E_k}\right).$$

For $c^\pm \to 0$, Eq. (2.82) reduces to Eq. (2.42). The parameters $c^\pm$ depend on E_k and are given by:

$$c^+ = \beta^2 \left[2 - \frac{1}{(\gamma + 1)^2}\right],$$

$$c^- = \frac{2\gamma - 1}{\gamma^2}.$$

Their values are between $\simeq 2$ (at high energies) and $\simeq 0$ (at low energies) for c^+, while between 0 and 1 for c^- (Fig. 2.23).

In Fig. 2.24 (see [Rohrlich and Carlson (1954)]), the percentage correction to the Landau curve is given as a function of $\alpha^\pm$: for $\alpha^\pm < 0.001$ corrections to the Landau energy-loss distribution are negligible, i.e., Eq. (2.82) reduces to Eq. (2.42). For instance for a silicon detector $300\,\mu m$ thick and $\beta \approx 1$, ξ is $\approx 5.3\,keV$, namely $\alpha^{\pm} < 0.001$ for electrons or positrons above $10\,MeV$. At lower energies or thicker detectors, corrections can become relevant.

The most probable energy-loss $\epsilon_{mp}^{\pm}$ of the energy straggling distribution [see Eq. (2.82)] occurs at a value $\lambda_0^{\pm}$ given by:

$$\lambda_0^{\pm} \approx \lambda_0 - 2.8\,\alpha^{\pm}, \tag{2.83}$$

where λ_0 is the value of λ at which the Landau curve [Eq. (2.42)] has the peak. Furthermore, the most probable energy-loss is shifted from that one, $\epsilon_{mp,\text{Landau}}$ [Eq. (2.48)], of the Landau distribution by:

$$\Delta\epsilon_{mp}^{\pm} = \epsilon_{mp,\text{Landau}} - \epsilon_{mp}^{\pm} \approx 2.8\,\alpha^{\pm}\,\xi.$$

For instance for a silicon detector $300\,\mu m$ thick and $\beta \approx 1$, this shift is $< 1\,keV$ for electrons above $2\,MeV$: ϵ_{mp} was measured to be $\approx 86\,keV$ for electrons between $(0.8-3)\,MeV$ [Hancock, James, Movchet, Rancoita and Van Rossum (1983)].

There is also a shrinkage of the energy-loss distribution: the $FWHM^{\pm}$ variation with respect to the $FWHM$ of the Landau distribution for massive charged particles (see page 57), $\triangle_{FWHM}^{\pm}$, is given by:

$$\triangle_{FWHM}^{\pm} = FWHM_L - FWHM^{\pm} = 6.6\,\alpha^{\pm}\,\xi. \tag{2.84}$$

For instance, in a thick silicon detector $300\,\mu m$, $\alpha^{\pm} < 0.001$ for electrons or positrons above $10\,MeV$. Thus, the energy-loss curve is narrowed by $\approx 0.2\%$. In thin detectors, the slight shrinkage can be understood by the fact that mostly small (with respect to W_m) energy transfers contribute to the build up of the most probable energy-loss.

2.1.6.3 *Practical Range of Electrons*

Below the critical energy, electrons mainly lose energy by collisions. However, their final path in matter differs from that of heavy particles because they undergo large angular scatterings. At low incoming momentum (for instance $\ll mc$), it is of the order of $\approx m/m_p$ the transferred energy by massive particles of mass m_p [see Eq. (1.26)]. While in a Møller collision, an electron can lose up to half of its kinetic energy. This large amount of energy straggling inside an absorber is directly related to the large deviations in the path of the electrons (and positrons).

A calculation of the mean path length is not meaningful in the case of electrons. Along the incoming direction, the mean penetration depth is much less than the mean path. The absorption curve* for mono-energetic electrons shows typically a long straight portion down to fairly low intensities and large penetration depths,

---

*The absorption curve is the distribution of emerging-electrons (also termed *transmitted electrons*) percentage as a function of absorber thickness (e.g., see [Katz and Penfold (1952)] and Section 1.1.5.4 of [Sternheimer (1961)]) or it can represent a depth-absorbed-dose curve (e.g., see [ICRUM (1984a)] and [Sorcini and Brahme (1994)]).

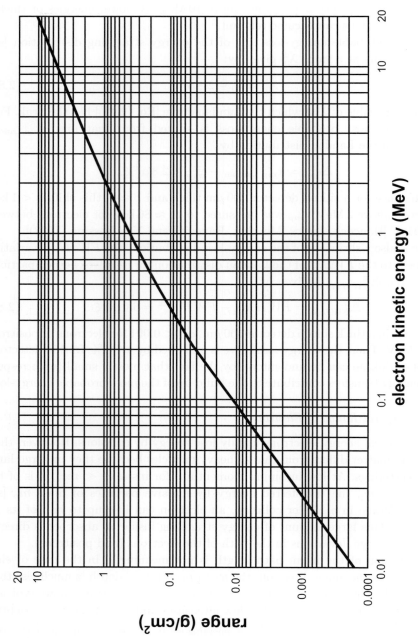

Fig. 2.25   Practical range in g/cm^2 for electrons calculated by means of Eqs. (2.85, 2.86).

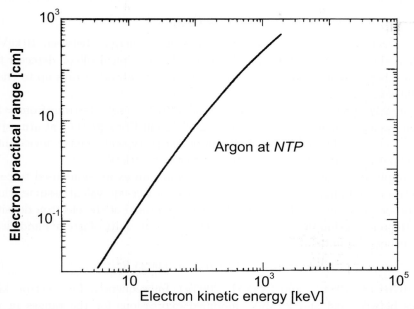

Fig. 2.26 Practical range in cm for soft electrons in Ar at NTP (adapted and reprinted with permission from [Sauli (1977)]).

when plotted in a linear scale. The linear extrapolation of the absorption curve[†] determines the so-called *practical electron range* R_p. Beyond R_p, only a few percent of the incoming electrons are left to be absorbed. Empirical expressions for the practical range, in units of g/cm^2, were proposed for different kinetic-energy intervals in MeV by various authors [Glendenin (1948); Katz and Penfold (1952)]. The following expressions (determined using aluminum absorbers) are from [Katz and Penfold (1952)]:

$$R_p = 0.412\, E^s \qquad (0.01 < E < 3 \text{ MeV}), \qquad (2.85)$$

where E (in MeV) is the electron kinetic energy,

$$s = 1.265 - 0.0954 \ln E;$$

and up to 20 MeV

$$R_p = 0.530\, E - 0.106 \qquad (2.5 < E < 20 \text{ MeV}). \qquad (2.86)$$

For electron energies where the radiation energy-loss is not a significant part of the energy-loss process, the practical ranges of electrons in units of g/cm^2 are

---

[†]In these measurements, the point at which the absorption curve encounters the background is called *range* (e.g., see [Glendenin (1948); Katz and Penfold (1952)] and also Section 1.1.5.4 of [Sternheimer (1961)]).

almost independent of the atomic mass-number of the absorber; thus, to a first approximation {e.g., see Equation (1-21) in Section 1-10 of [Price (1964)]} we have

$$R_{p,Z_1} \approx R_{p,Z_2}. \tag{2.87}$$

The practical range[tt] of electrons with kinetic energies between $10\,\text{keV}$ and $20\,\text{MeV}$ is shown in Fig. 2.25. It has to be noted that typical silicon detector thicknesses are between 300 and $400\,\mu\text{m}$ and, thus, able to absorb δ-rays up to energies of $\approx 300\,\text{keV}$.

Figure 2.26 shows the practical range [Sauli (1977)] of soft electrons from $\approx 3\,\text{keV}$ in argon gas at Normal Temperature (i.e., $20\,^\circ\text{C}$) and Pressure (i.e., at atmospheric pressure), i.e., at NTP condition. As it can be noted, typical counter devices having $1\,\text{cm}$ gas path can absorb δ-rays up to energies of $\approx 30\,\text{keV}$.

For applications of therapeutic electron beams, an expression[‡] used to convert an arbitrary solid phantom-material range into a water-equivalent assumes that, in an absorber, the range is proportional to the reciprocal of the electron density. In water, the relationship between the practical range (in units of g/cm^2) and electron kinetic-energy (in units of MeV) is

$$E_k = 0.22 + 1.98\,R_{p,\text{w}} + 0.0025\,R_{p,\text{w}}^2 \tag{2.88}$$

(from [ICRUM (1984a)] and [ASTM (2003)]). Subsequently, for electron kinetic energies between 0.3 and $25\,\text{MeV}$, modified expressions for the ranges in water ($R_{p,\text{w}}$) and polystyrene ($R_{p,\text{pol}}$) were proposed by Cleland, Lisanti and Galloway (2004):

$$E_k = 0.564 + 1.957\,R_{p,\text{w}} - 0.231\,R_{p,\text{w}}^{-0.130} + 0.0030\,R_{p,\text{w}}^2 \tag{2.89}$$

$$E_k = 0.522 + 1.846\,R_{p,\text{pol}} - 0.189\,R_{p,\text{pol}}^{-0.155} + 0.0045\,R_{p,\text{pol}}^2. \tag{2.90}$$

An expression for practical range can also be obtained using its relationship with the continuous-slowing-down approximation range r_0 (see footnote at page 71) in a material. Scaling laws were proposed by different authors (e.g., [Harder (1970); Andreo, Ito and Tabata (1992); Halbleib, Kensek, Mehlhorn, Valdez, Seltzer and Berger (1992); Zheng-Ming and Brahme (1993); Sorcini and Brahme (1994); Tabata, Andreo and Ito (1994)]). Based on the transport theory for energy deposition of electrons, Harder (1970) suggested that the practical range scales (e.g., see also [Zheng-Ming and Brahme (1993); Sorcini and Brahme (1994)]) as

$$R_p = r_0 \left[0.30 \times \left(\sqrt{\frac{\langle Z \rangle}{\tau}} \right)^{1.31} + 0.83 \right]^{-1}, \tag{2.91}$$

where τ is the electron kinetic energy in units of the electron rest mass and $\langle Z \rangle$ is the mean atomic number of the material. Equation (2.91) is valid for electron energies

---

[tt]It is calculated by means of Eqs. (2.85, 2.86) and it is in units of g/cm^2.

[‡]The reader may see Equation 12 in [Sorcini and Brahme (1994)] (see also Equation 4 in [NACP (1980)] and references therein). In the same reference, other approximate expressions are discussed in view of these applications.

from 5 up to 30 MeV and for materials with atomic number $4 \leq Z \leq 82$. Practical ranges are reported in [Sorcini and Brahme (1994)] for electrons in a kinetic-energy range of $(1–50)$ MeV in Be, C, H_2O, Al, Cu, Ag and U.

2.1.7 *Radiation Energy-Loss by Electrons and Positrons*

When electron (or positron) energies exceed a few tens of MeV, the dominant mechanism of energy loss is by emitting photons, i.e., by *radiative energy-loss*. It is usually called emission by *synchrotron radiation* when occurring for circular acceleration or by *bremsstrahlung* when traversing matter.

The emission of electromagnetic radiation accompanies the acceleration or deceleration of charged particles. Assuming a uniform acceleration during the radiative diffusion, the radiated energy can be easily evaluated in the framework of the classical calculation of an electromagnetic-dipole radiation emission, as for the case of the *Thomson scattering* for *unpolarized radiation* discussed in Sect. 2.3.2.3 (see also Chapter 4 of [Marmier and Sheldon (1969)]). The radiation emitted per unit of time depends quadratically on the velocity variation, i.e.,

$$\frac{dE}{dt} = \frac{2\,e^2}{3\,c^3}\left|\frac{d\vec{v}}{dt}\right|^2,$$

and extends over a continuous range of frequencies. In the case of bremsstrahlung emission by a particle with charge number z and mass m_p in an absorber with atomic number Z, the acceleration $\left|\frac{d\vec{v}}{dt}\right|$ depends on zZe^2/m_p, thus:

$$\left(\frac{dE}{dt}\right)_B \propto \frac{z^2 Z^2}{m_p^2}.$$

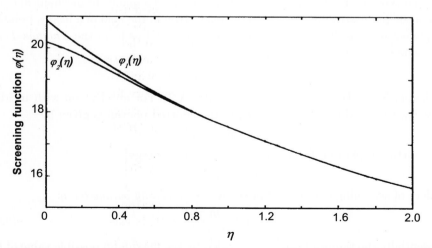

Fig. 2.27 Screening functions $\varphi_1(\eta)$ and $\varphi_2(\eta)$ versus η [Marmier and Sheldon (1969)] (see also [Bethe and Heitler (1934)]).

As a consequence, the bremsstrahlung intensity depends on $z^2 Z^2$, while the collision energy-loss depends on Z (i.e., on the number of electrons per cm^3) as shown in Eq. (2.1), and inversely on the square of the incoming particle mass. For this latter reason, it is much less probable that a massive charged particle (π, K, proton, etc.) radiates photons traversing a medium than an electron or a positron. It has to be noted that the energy emitted by radiation depends on Z^2. However, in the interactions inside a medium, there is an additional contribution due to the Coulomb field of the Z atomic-electrons, each one with charge 1, thus providing an overall Z-dependent contribution.

The probability of radiation depends, in an essential way, on the effective distance between the electron and the nucleus. In a *classical description* of the interaction, when the impact parameter is much larger than the atomic radius, the effective nuclear charge is screened by the atomic electrons and the nuclear field greatly loses its effect on the incoming particle, because the field of outer electrons has to be taken into account. This is referred to as *complete screening*. Conversely, for impact parameters that are small with respect to the atomic radius, we expect *no screening effect*, namely the field acting on the incoming particle can be approximated with the Coulomb field of a point charge Ze at the center of the nucleus. In a quantum-mechanical treatment, the classical description of the interaction is no longer strictly possible. However, one can introduce the concept of the *effective interaction distance*, at which the radiation loss process is affected by the electronic Coulomb field. This distance is of the order of $\hbar/q$, where q is the recoil momentum of the atom after the interaction took place. If $\hbar/q$ is large compared with the atomic radius, the screening effect has to be taken into account.

Under the *Born approximation* (see [Bethe and Heitler (1934); Heitler (1954); Bethe and Ashkin (1953)]), Bethe and Heitler derived a quantum-mechanical calculation of the bremsstrahlung emission by an electron in the field of a heavy, pointlike and spinless nucleus. They determined the atomic radius from the *Thomas–Fermi model*, where it is expressed as

$$a_Z = \frac{a_0}{Z^{1/3}},$$

in which a_0 is the Bohr radius (see Appendix A.2). For an electron with incident energy E_0, the maximum frequency ν_0 of the emitted photon is given by

$$h\nu_0 = E_0 - mc^2.$$

After emitting a photon, for an electron with final energy

$$E = E_0 - h\nu$$

the effect of screening can be evaluated by the *screening parameter* η:

$$\eta = 100 \, \frac{mc^2 h\nu}{E_0 E Z^{1/3}}. \tag{2.92}$$

η is essentially the radius of the atom divided by the maximum possible value of $\hbar/q$ allowed by the energy and momentum conservation in the interaction. For $\eta \gg 1$ the screening effect can be neglected, while for $\eta \simeq 0$ the screening is complete.

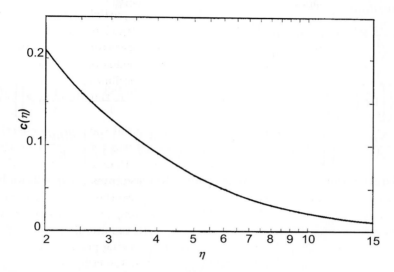

Fig. 2.28 Screening function $c(\eta)$ versus η (adapted and reprinted with permission from [Bethe and Ashkin (1953)], see also [Bethe and Heitler (1934)]).

In the framework of Bethe and Heitler's calculations for the interaction in a nuclear field, the probability that an electron with incoming energy E_0 traversing a thickness x in cm emits a photon with energy between $h\nu$ and $h\nu + d(h\nu)$ is:

$$\Phi(E_0, h\nu)\, dx\, d(h\nu),$$

where $\Phi(E_0, h\nu)\, d(h\nu)$ is in units of cm^{-1}. In this section, all equations referring to $\Phi(E_0, h\nu)\, d(h\nu)$ are expressed in cm^{-1}. $\Phi(E_0, h\nu)$ is called *differential radiation probability*, for which we have

$$\Phi(E_0, h\nu)\, d(h\nu) = 4\,\alpha \frac{N\rho}{A} Z^2 r_e^2 F(E_0, h\nu, \eta) \frac{d(h\nu)}{h\nu} \quad [\text{cm}^{-1}], \qquad (2.93)$$

where, for E_0 large compared with mc^2, the function $F(E_0, h\nu, \eta)$ depends on the screening parameter η:

$$F(E_0, h\nu, \eta) = \left[1 + \left(\frac{E}{E_0}\right)^2\right]\left[\frac{\varphi_1(\eta)}{4} - \frac{1}{3}\ln Z\right] - \frac{2}{3}\frac{E}{E_0}\left[\frac{\varphi_2(\eta)}{4} - \frac{1}{3}\ln Z\right].$$

The functions $\varphi_1(\eta)$ and $\varphi_2(\eta)$ are shown in Fig. 2.27 for $0 < \eta < 2$.

For $\eta \approx 0$ (i.e., for **complete screening**) we have

$$\varphi_1(0) = 4\ln 183 \quad \text{and} \quad \varphi_2(0) = 4\ln 183 - \frac{2}{3},$$

and Eq. (2.93) becomes:

$$\Phi(E_0, h\nu)\, d(h\nu)$$

$$= 4\alpha \frac{N\rho}{A} Z^2 r_e^2 F(E_0, h\nu, 0) \frac{d(h\nu)}{h\nu}$$

$$= 4\alpha \frac{N\rho}{A} Z^2 r_e^2$$

$$\times \left\{ \left[1 + \left(\frac{E}{E_0}\right)^2 \right] \left[\frac{4\ln 183}{4} - \frac{1}{3}\ln Z \right] - \frac{2E}{3E_0} \left[\frac{4\ln 183 - \frac{2}{3}}{4} - \frac{1}{3}\ln Z \right] \right\} \frac{d(h\nu)}{h\nu}$$

$$= 4\alpha \frac{N\rho}{A} Z^2 r_e^2 \left\{ \left[1 + \left(\frac{E}{E_0}\right)^2 - \frac{2}{3}\frac{E}{E_0} \right] \ln\left(\frac{183}{Z^{1/3}}\right) + \frac{1}{9}\frac{E}{E_0} \right\} \frac{d(h\nu)}{h\nu}. \qquad (2.94)$$

For larger η values, the effect of screening decreases and, for $\eta > 2$, we have

$$\varphi_1(\eta) = \varphi_2(\eta) = \varphi(\eta),$$

with

$$\frac{\varphi(\eta)}{4} - \frac{1}{3}\ln Z = \ln\left(\frac{2E_0 E}{mc^2 h\nu}\right) - \frac{1}{2} - c(\eta)$$

and

$$\varphi(\eta) = 4\ln\left(\frac{2E_0 E}{mc^2 h\nu}\right) + \frac{4}{3}\ln Z - 2 - 4\,c(\eta), \qquad (2.95)$$

where the function $c(\eta)$ is shown in Fig. 2.28. The differential radiation probability becomes:

$$\Phi(E_0, h\nu)\, d(h\nu)$$

$$= 4\alpha \frac{N\rho}{A} Z^2 r_e^2 F(E_0, h\nu, \eta > 2) \frac{d(h\nu)}{h\nu}$$

$$= 4\alpha \frac{N\rho}{A} Z^2 r_e^2 \left\{ \left[1 + \left(\frac{E}{E_0}\right)^2 - \frac{2}{3}\frac{E}{E_0} \right] \left[\ln\left(\frac{2E_0 E}{mc^2 h\nu}\right) - \frac{1}{2} - c(\eta) \right] \right\} \frac{d(h\nu)}{h\nu}. \qquad (2.96)$$

For $\eta \gg 1$ (i.e., for **no screening**), $c(\eta)$ becomes negligible and Eq. (2.96) becomes:

$$\Phi(E_0, h\nu)\, d(h\nu)$$

$$= 4\alpha \frac{N\rho}{A} Z^2 r_e^2 \left\{ \left[1 + \left(\frac{E}{E_0}\right)^2 - \frac{2}{3}\frac{E}{E_0} \right] \left[\ln\left(\frac{2E_0 E}{mc^2 h\nu}\right) - \frac{1}{2} \right] \right\} \frac{d(h\nu)}{h\nu}. \qquad (2.97)$$

The equations for the differential radiation probability were derived assuming that the Born approximation can be used. However, the Born approximation gives correct results [Heitler (1954)] only if

$$\frac{2\pi Z e^2}{\hbar\nu} \ll 1 \qquad \text{and} \qquad \frac{2\pi Z e^2}{\hbar\nu_0} \ll 1,$$

where v and v_0 are the electron velocities after and before photon emission, respectively. Because $\alpha = e^2/(\hbar c)$, for electrons moving at speeds v and v_0 close to c, the Born approximation is satisfied for:

$$2\pi Z\alpha \approx 4.6 \times 10^{-2} Z \ll 1.$$

As a consequence, some corrections are needed for high-Z materials, and also for electrons at very high energies (see discussion in [Heitler (1954)]): for the high-Z elements, corrections between $(5\text{--}9)\%$ are expected. Experimental data [Bethe and Ashkin (1953)] indeed indicate a larger radiation emission in a Ta target. The correction is expected to be proportional to $(\alpha Z)^2$.

The above equations were derived taking into account the nuclear field alone. However (see [Bethe and Ashkin (1953)] and references therein), the radiation emission in the field of atomic electrons can contribute to the overall radiation loss process. This contribution is of the order of $1/Z$ of the nuclear contribution. If the energy is large compared with mc^2, but small enough for neglecting the screening effect, Eq. (2.97) is still valid for electrons, just Z has to be replaced by 1 (the atomic electron charge). In this way (i.e., for **no screening**), Eq. (2.97) gives the differential probability of the photon emission for both nuclear and atomic-electron fields, when we replace Z^2 by $Z(Z+1)$:

$$\Phi_{e-n}(E_0, h\nu)\, d(h\nu) = 4\,\alpha \frac{N\rho}{A} Z(Z+1)\, r_e^2$$

$$\times \left[1 + \left(\frac{E}{E_0}\right)^2 - \frac{2}{3}\frac{E}{E_0} \right] \left[\ln\left(\frac{2E_0 E}{mc^2 h\nu}\right) - \frac{1}{2} \right] \frac{d(h\nu)}{h\nu} \quad [\text{cm}^{-1}]. \qquad (2.98)$$

In the limit of **complete screening** ($\eta \simeq 0$), Wheeler and Lamb (1939) derived that the overall electron contribution to the differential radiation probability per unit length is:

$$\Phi_{\text{el}}(E_0, h\nu)\, d(h\nu) = 4\,\alpha \frac{N\rho}{A} Z r_e^2$$

$$\times \left\{ \left[1 + \left(\frac{E}{E_0}\right)^2 - \frac{2}{3}\frac{E}{E_0} \right] \ln\left(\frac{1440}{Z^{2/3}}\right) + \frac{1}{9}\frac{E}{E_0} \right\} \frac{d(h\nu)}{h\nu} \quad [\text{cm}^{-1}]. \qquad (2.99)$$

The term $E/(9E_0)$ is small and can be neglected in calculating the ratio[††] ι:

$$\iota = \frac{Z\Phi_{\text{el}}(E_0, h\nu)}{\Phi(E_0, h\nu)}$$

$$\approx \frac{\ln\left(1440/Z^{2/3}\right)}{\ln\left(183/Z^{1/3}\right)}.$$

The ratio ι is 1.40, 1.29 and 1.14 for $Z = 1$, 10 and 92, respectively. For **complete screening**, the overall differential radiation probability becomes:

$$\Phi_{e-n}(E_0, h\nu)\, d(h\nu) = 4\,\alpha \frac{N\rho}{A} Z(Z+\iota)\, r_e^2$$

$$\times \left\{ \left[1 + \left(\frac{E}{E_0}\right)^2 - \frac{2}{3}\frac{E}{E_0} \right] \ln\left(\frac{183}{Z^{1/3}}\right) + \frac{1}{9}\frac{E}{E_0} \right\} \frac{d(h\nu)}{h\nu} \quad [\text{cm}^{-1}]. \qquad (2.100)$$

[††]This ratio expresses Z times the electron contribution over the nuclear contribution computed by means of Eq. (2.94).

The average radiation loss per cm for an electron with initial energy E_0, the so-called *energy-loss by radiation*, is given by

$$-\left(\frac{dE_0}{dx}\right)_{\text{rad}} = \int_0^{h\nu_0} h\nu\,\Phi_{e-n}(E_0, h\nu)\,d(h\nu)$$

$$= n_A E_0 \Phi_{\text{rad}} \quad [\text{MeV/cm}], \tag{2.101}$$

where E_0 is in units of MeV, $n_A = N\rho/A$ is the number of atoms per cm^3, and Φ_{rad} is:

$$\Phi_{\text{rad}} = \frac{1}{n_A E_0} \int_0^{h\nu_0} h\nu\,\Phi_{e-n}(E_0, h\nu)\,d(h\nu) \quad [\text{cm}^2/\text{nucleus}]. \tag{2.102}$$

In the energy region

$$mc^2 \ll E_0 \ll 137\,mc^2 Z^{-1/3},$$

the screening can be neglected [Heitler (1954); Bethe and Ashkin (1953)]. Using Eq. (2.98) and after integration of Eq. (2.102), for **no screening** we have

$$\Phi_{\text{rad}} = \bar{\Phi}\left[4\ln\left(\frac{2E_0}{mc^2}\right) - \frac{4}{3}\right] \quad [\text{cm}^2/\text{nucleus}], \tag{2.103}$$

where $\bar{\Phi}$ is given by:

$$\bar{\Phi} = \alpha Z(Z+1)\,r_e^2 = 5.8 \times 10^{-28} \times Z(Z+1) \quad [\text{cm}^2]. \tag{2.104}$$

In the energy region

$$E_0 \gg 137\,mc^2 Z^{-1/3},$$

the screening is almost complete [Heitler (1954); Bethe and Ashkin (1953)] except for the production of energetic photons [see Eq. (2.92)]. Using Eq. (2.100) and after integration of Eq. (2.102), for **complete screening** we have

$$\Phi_{\text{rad}} = \bar{\Phi}_c\left[4\ln\left(\frac{183}{Z^{1/3}}\right) + \frac{2}{9}\right] \quad [\text{cm}^2/\text{nucleus}], \tag{2.105}$$

where $\bar{\Phi}_c$ is given by:

$$\bar{\Phi}_c = \alpha Z(Z+\iota)\,r_e^2 = 5.8 \times 10^{-28} \times Z(Z+\iota) \quad [\text{cm}^2]. \tag{2.106}$$

The energy or intensity distribution of the bremsstrahlung is obtained by multiplying the emitted photon energy ($h\nu$) by the differential probability energy distribution per nucleus, i.e.,

$$h\nu\frac{\Phi_{e-n}(E_0, h\nu)}{n_A}\,d(h\nu),$$

where the expression for $\Phi_{e-n}(E_0, h\nu)$ depends on the screening conditions as discussed above. In Fig. 2.29, $[h\nu\,\Phi_{e-n}(E_0, h\nu)]/(n_A\bar{\Phi})$ is shown as a function of $h\nu/(E_0 - mc^2)$. The function

$$\frac{h\nu\,\Phi_{e-n}(E_0, h\nu)}{n_A\bar{\Phi}}$$

is the energy distribution per unit of photon energy interval divided by $\bar{\Phi}$ [see Eq. (2.104)]. The dotted curves are calculated assuming that the screening can be neglected and are valid for all elements, because the Z dependence is contained in the term $\bar{\Phi}$ only. The full curves are for lead, except for the non relativistic energy case whose curve refers to aluminum. As the primary energy increases, the screening is almost complete; the curve marked ∞ is for complete screening. In the region of high-energy photon emission, the screening can be neglected and the curves are those for no screening. At high energies, the $[h\nu \, \Phi_{e-n}(E_0, h\nu)]/(n_A \bar{\Phi})$ distribution shows an approximate flat central region above 0.4 before dropping near to 1: this indicates that $\Phi_{e-n}(E_0, h\nu)$ behaves like $1/h\nu$. Reviews on bremsstrahlung emission are found in [Koch and Motz (1959); Berger and Seltzer (1964); Tsai (1974); ICRUM (1984b)].

In general, the Born approximation meets difficulties to be applied when the atomic number of the target nucleus increases, the initial electron kinetic energy is low and, also, when the photon energy approaches the high-frequency limit ν_0. However, these formulae are in a reasonable agreement with data, except in the high-frequency limit. Exact theoretical calculations are available in the high-frequency limit and agree reasonably well with experimental data [Fano et al. (1959)]. Otherwise, modifications are proposed on the base of experimental data.

For low energy electrons (with kinetic energies lower than 2 MeV), the *Elwert factor* f_E [Elwert (1939)] is applied to multiply the differential probability energy distribution (see Table V in [Koch and Motz (1959)], and also [Berger and Seltzer (1964); Pratt et al. (1977)]):

$$f_E = \frac{\beta_0[1 - \exp(-2\pi\alpha Z/\beta_0)]}{\beta[1 - \exp(-2\pi\alpha Z/\beta)]} ,$$

where β and β_0 are the final and initial electron velocities, respectively, in units of speed of light.

At intermediate kinetic energies (up to ≈ 50 MeV), an empirical *Koch–Motz correction factor* is used (see Table V in [Koch and Motz (1959)] and also [Berger and Seltzer (1964)]). This multiplicative factor is shown in Fig. 2.30 from [Berger and Seltzer (1964)].

Above 50 MeV, the so-called *extreme relativistic region*, cross section calculations with relativistic Coulomb wave functions (Sommerfield–Maue) including screening corrections were derived by Olsen (1955). The computed formula has an additive correction factor to the Born approximation formulae described above. The differential radiation probability for the nuclear field is still given by Eq. (2.93), but the function $F(E_0, h\nu, \eta)$ has to be rewritten as

$$\left[1 + \left(\frac{E}{E_0}\right)^2\right]\left[\frac{\varphi_1(\eta)}{4} - \frac{1}{3}\ln Z - f(Z)\right] - \frac{2}{3}\frac{E}{E_0}\left[\frac{\varphi_2(\eta)}{4} - \frac{1}{3}\ln Z - f(Z)\right],$$

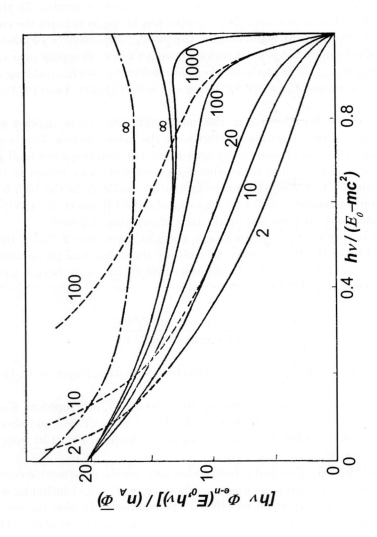

Fig. 2.29 $[h\nu\,\Phi_{e-n}(E_0,h\nu)]/(n_A\bar{\Phi})$ as a function of $h\nu/(E_0-mc^2)$ (adapted and reprinted with permission from [Bethe and Ashkin (1953)]). The numbers labelling the curves indicate the electron energy in units of mc^2. The solid curves are for lead and include the screening effect. The dotted curves are without screening and valid for all Z (see page 101).

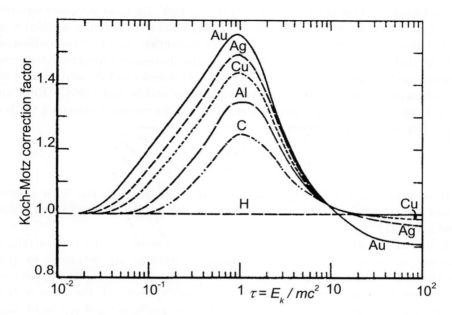

Fig. 2.30   Empirical *Koch–Motz correction factor* as a function of $\tau = E_k/mc^2$ for kinetic electron energies up to 50 MeV (adapted and reprinted with permission from [Berger and Seltzer (1964)]).

where the function $f(Z)$ is given by (see [Davies, Bethe and Maximon (1954)])

$$f(Z) = \begin{cases} 1.2021\,(\alpha\,Z)^2, & \text{for low-}Z, \\[2mm] 0.925\,(\alpha\,Z)^2, & \text{for high-}Z; \end{cases} \tag{2.107}$$

for $0 < \eta < 2$, $\varphi_1(\eta)$ and $\varphi_2(\eta)$ are shown in Fig. 2.27 while, for $\eta > 2$, they are given by Eq. (2.95). Finally, we have

$$\Phi(E_0, h\nu)\,d(h\nu) = 4\,\alpha \frac{N\rho}{A} Z^2 r_e^2\, \mathbb{S}(\eta, Z, E, E_0)\, \frac{d(h\nu)}{h\nu} \quad [\text{cm}^{-1}], \tag{2.108}$$

where $\mathbb{S}(\eta, Z, E, E_0)$ is given by:

$$\left[1 + \left(\frac{E}{E_0}\right)^2\right]\left[\frac{\varphi_1(\eta)}{4} - \frac{1}{3}\ln Z - f(Z)\right] - \frac{2}{3}\frac{E}{E_0}\left[\frac{\varphi_2(\eta)}{4} - \frac{1}{3}\ln Z - f(Z)\right].$$

The corresponding total radiation cross section becomes [Koch and Motz (1959)]:
for **no screening**

$$\Phi_{\text{rad}} = 5.8 \times 10^{-28} \times Z^2 \left[4\ln\left(\frac{2E_0}{mc^2}\right) - \frac{4}{3} - f(Z)\right] \quad [\text{cm}^2/\text{nucleus}] \tag{2.109}$$

and for **complete screening**

$$\Phi_{\text{rad}} = 5.8 \times 10^{-28} \times Z^2 \left[4\ln\left(\frac{183}{Z^{1/3}}\right) + \frac{2}{9} - f(Z)\right] \quad [\text{cm}^2/\text{nucleus}]. \tag{2.110}$$

For kinetic electron energies above $\approx (10\text{--}20)$ MeV, the most adequate formulae are i) Eq. (2.108) for $\eta < 15$, while ii) Born approximation formulae are better for $\eta > 15$ [Berger and Seltzer (1964)]. It has to be noted that, to a first approximation, in the formulae for the differential probability distribution and total radiation cross sections, the term Z^2 has be replaced by $Z(Z+1)$ in order to take into account the interaction on the field of atomic electrons [Berger and Seltzer (1964)]. Koch and Motz (1959) stated that at relativistic energies and for complete screening the best parametrization is $Z(Z+\iota)$, where:

$$\iota = \frac{\ln\left(530/Z^{2/3}\right)}{\ln\left(183/Z^{1/3}\right) + 1/18}.$$

In addition, for the radiative loss these formulae based on the Born approximation are expected to hold for both electrons and positrons [Berger and Seltzer (1964); Tsai (1974)].

It has to be noted that some differences are expected in the bremsstrahlung processes of electrons and positrons, because electrons are attracted by positive charged nuclei and repelled by atomic electrons; whereas the opposite occurs for incoming positrons. However, as previously discussed, the cross sections for electron- and positron-bremsstrahlung exhibit negligible differences at high-energy. At low energy*, the positron–nucleus cross-section is smaller than that for electrons (e.g., see [Feng, Pratt and Tseng (1981); ICRUM (1984b); Kim, Pratt and Seltzer (1984)]); the ratio

$$\mathfrak{R}^{\pm} = \frac{(dE/dx)^{+}_{\text{rad,n}}}{(dE/dx)^{-}_{\text{rad,n}}}$$

of the positron $[(dE/dx)^{+}_{\text{rad,n}}]$ and electron $[(dE/dx)^{-}_{\text{rad,n}}]$ radiative-stopping powers, due to the interactions with the nuclear field, is lower than 0.962 [ICRUM (1984b)] for $E_k/Z^2 \lesssim 2 \times 10^{-2}$ MeV, where E_k is electron and positron kinetic energy in MeV and Z the atomic number of the medium (Fig. 2.31).

2.1.7.1 *Collision and Radiation Stopping Powers*

Calculated collision stopping powers $(dE/d\chi)^{-}$ (where χ is in units of g/cm^2) and radiation stopping powers $(dE/d\chi)^{-}_{\text{rad}}$ for electrons are available in [Berger, Coursey, Zucker and Chang (2005)] and are shown in Figs. 2.32–2.37 for (liquid) water, Al, Si, Fe, W and Pb, respectively.

The uncertainties on these calculated collision stopping powers are estimated to be $(1\text{--}2)\%$ above 100 keV, $(2\text{--}3)\%$ in low-Z materials and $(5\text{--}10)\%$ in high-Z materials between 10 and 100 keV. While in [Berger, Coursey, Zucker and Chang (2005)] the radiative stopping powers are evaluated with a combination of theoretical

---

*At low energy, the screening effect becomes less important [see Eq. (2.92)] for the radiation energy-loss mechanism, which is dominated by photon emission resulting from the interactions in the nuclear field (e.g., see Figure 9.2 of [ICRUM (1984b)]).

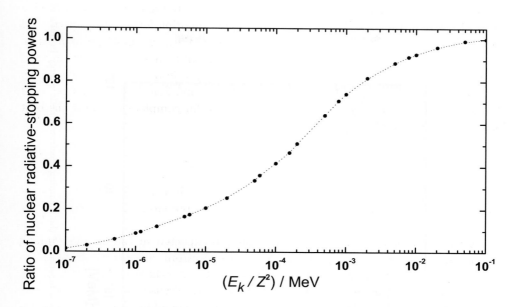

Fig. 2.31 Ratio ($\mathfrak{R}^{\pm}$, see text) of the positron $[(dE/dx)^{+}_{\mathrm{rad,n}}]$ and electron $[(dE/dx)^{-}_{\mathrm{rad,n}}]$ radiative-stopping powers due to the interactions with the nuclear field as a function of $\left(E_k/Z^2\right)$ /MeV, where E_k is the electron and positron kinetic energy in MeV and Z the atomic number of the medium {data point ($\bullet$) from [ICRUM (1984b)]}. The line is to guide the eye.

bremsstrahlung cross sections described in [Seltzer and Berger (1985)], analytical formulae (using a high-energy approximation) are used above 50 MeV, and accurate numerical results are used from [Pratt et al. (1977)] below 2 MeV. The uncertainties on these calculated radiative stopping powers are estimated to be 2% above 50 MeV, (2–5)% between 2 and 50 MeV, and 5% below 2 MeV.

2.1.7.2 *Radiation Yield and Bremsstrahlung Angular Distribution*

The radiation yield, $\Upsilon(E_k)$, is the fraction of the electron energy radiated by an electron with initial kinetic energy E_k for bremsstrahlung emission during the slowing-down process. Under the assumption of a continuous emission, it is given by:

$$\Upsilon(E_k) = \frac{1}{E_k} \int_0^{E_k} \frac{\left(\frac{dE}{dx}\right)^{\pm}_{\mathrm{rad}}}{\left(\frac{dE}{dx}\right)^{\pm}_{\mathrm{tot}}} \, dE', \qquad (2.111)$$

where $(dE/dx)^{\pm}_{\mathrm{tot}}$ is the total energy-loss, which includes both the radiative and the collision losses.

An approximation for the radiation yield was derived by Koch and Motz (1959):

$$\Upsilon(E_k) = \frac{3 \times 10^{-4} \times Z\tau}{1 + 3 \times 10^{-4} \times Z\tau}, \qquad (2.112)$$

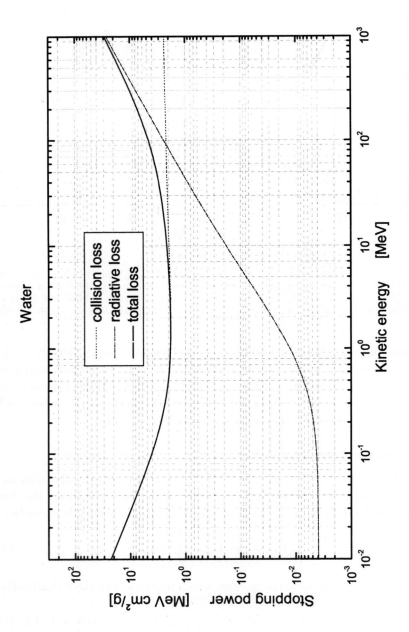

Fig. 2.32   Total, collision and radiation stopping powers in units of MeV cm^2/g as a function of the incoming electron kinetic energy in units of MeV in (liquid) water (data from [Berger, Coursey, Zucker and Chang (2005)]).

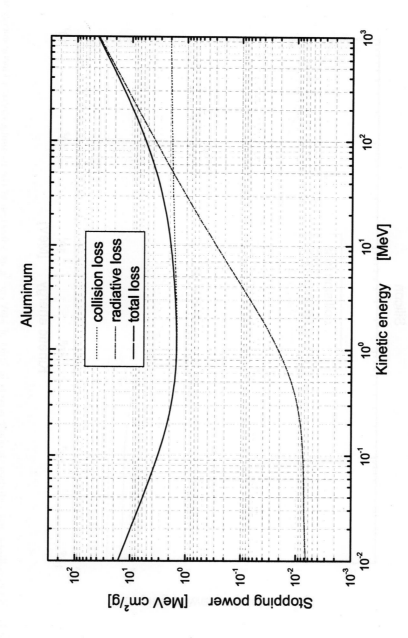

Fig. 2.33 Total, collision and radiation stopping powers in units of $MeV\,cm^2/g$ as a function of the incoming electron kinetic energy in units of MeV in Al (data from [Berger, Coursey, Zucker and Chang (2005)]).

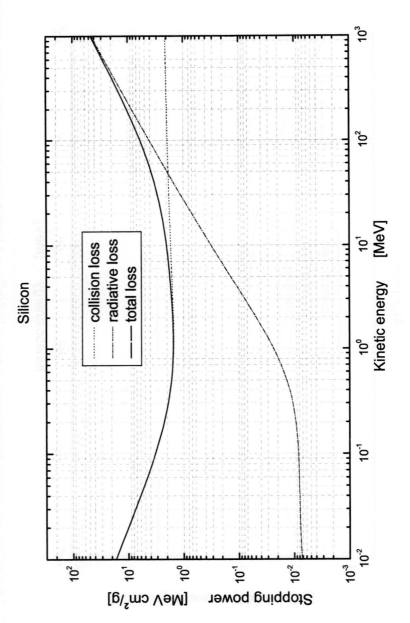

Fig. 2.34 Total, collision and radiation stopping powers in units of MeV cm^2/g as a function of the incoming electron kinetic energy in units of MeV in Si (data from [Berger, Coursey, Zucker and Chang (2005)]).

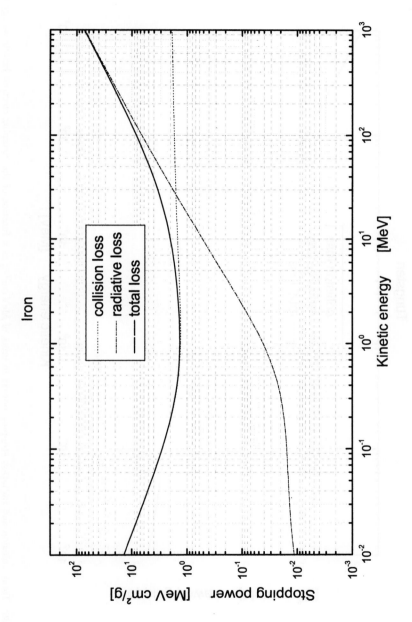

Fig. 2.35 Total, collision and radiation stopping powers in units of MeV cm^2/g as a function of the incoming electron kinetic energy in units of MeV in Fe (data from [Berger, Coursey, Zucker and Chang (2005)]).

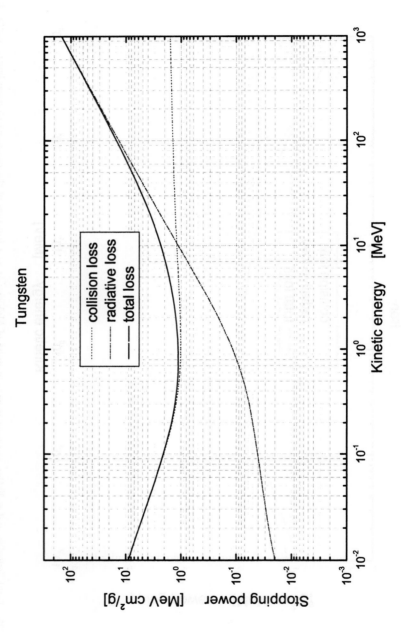

Fig. 2.36 Total, collision and radiation stopping powers in units of MeV cm²/g as a function of the incoming electron kinetic energy, in units of MeV in W (data from [Berger, Coursey, Zucker and Chang (2005)]).

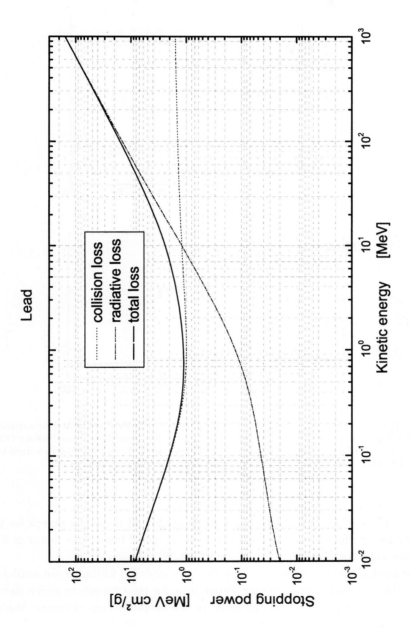

Fig. 2.37 Total, collision and radiation stopping powers in units of MeV cm^2/g as a function of the incoming electron kinetic energy in units of MeV in Pb (data from [Berger, Coursey, Zucker and Chang (2005)]).

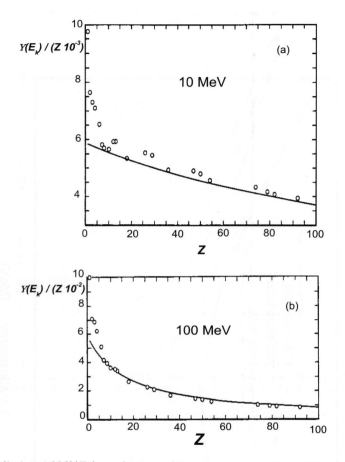

Fig. 2.38   Radiation yield $\Upsilon(E_k)$ as a function of the atomic number Z calculated for a continuous slowing-down process (adapted and reprinted with permission from [Berger and Seltzer (1964)]): (a) $\Upsilon(E_k)/(Z\,10^{-3})$ for electron kinetic energy of 10 MeV, (b) $\Upsilon(E_k)/(Z\,10^{-2})$ for electron kinetic energy of 100 MeV. The curve represents the Koch and Motz formula [Eq. (2.112)].

where $\tau = E_k/mc^2$. This formula seems adequate for all materials except for those with very low Z: for $Z = 1$ it underestimates the yield value by a factor $\simeq 2$, but it becomes valid for Z larger than 6 (see Fig. 2.38).

At non-relativistic energies, no analytical or empirical formulae are available to estimate the bremsstrahlung angular distribution for *thick targets* in which there are additional relevant processes contributing to the overall energy decrease. However, there are some experimental results [Koch and Motz (1959)].

At relativistic energies, estimates of the bremsstrahlung angular distributions were made. These calculations agree fairly well with experimental data as those

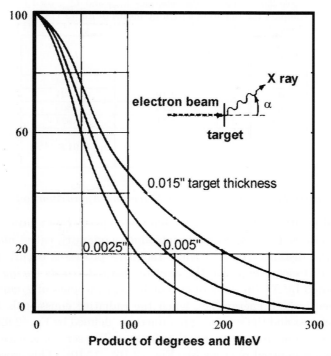

Fig. 2.39   Predictions for the angular distribution for thick target bremsstrahlung of electrons in tungsten with three different thicknesses: 0.0025", 0.005" and 0.015" (adapted and republished with permission from Koch, H.W. and Motz, J.W., *Rev. Mod. Phys.* **31**, 920 (1959); Copyright (1959) by the American Physical Society, see also references therein). The abscissa is the product of the electron kinetic energy in MeV and the angle in degrees: αE_k [degree MeV]. The ordinate is the percentage of the radiated intensity R_α normalized to the radiated intensity at 0°. R_α is defined as the fraction of the total incident electron kinetic energy that is radiated per steradian at angle α.

shown in Fig. 2.39 regarding electrons through tungsten ([Koch and Motz (1959)] and references therein). Analytical expressions for the bremsstrahlung angular distributions at different electron energies and references to experimental data are given in [Koch and Motz (1959)].

Most of the photons coming from a high-energy electron are emitted at relatively small angles. The average emission angle θ_γ is given by:

$$\theta_\gamma = \frac{mc^2}{E_0}, \tag{2.113}$$

where E_0 is the incoming electron energy. The emission cone becomes more and more narrow as the energy increases. In addition, bremsstrahlung photons are, in general, polarized with the polarization vector normal to the plane formed by the photon and incident electron [Segre (1977)].

Table 2.6 Values of the functions $L_{\rm rad}$ and $L'_{\rm rad}$ used to compute the radiation length by means of Eq. (2.118) (see Table B.2 in [Tsai (1974)]).

Element	Z	$L_{\rm rad}$	$L'_{\rm rad}$
H	1	5.31	6.144
He	2	4.79	5.621
Li	2	4.74	5.805
Be	3	4.71	5.924
others	> 4	$\ln\left(184.15\,Z^{-1/3}\right)$	$\ln\left(1194\,Z^{-2/3}\right)$

2.1.7.3 *Radiation Length and Complete Screening Approximation*

Electrons and positrons traversing a medium lose energy by radiation, as described in previous sections. It is convenient to introduce a quantity, called *radiation length*, to measure the distance traveled while radiative processes occur. The radiation length is the distance over which the electron *has reduced its energy* by a *factor e* and it is denoted by X_0 (in units of cm) or by X_{g0} (in units of g/cm^2).

At sufficiently high energy, i.e., when the radiative emission is the dominant energy-loss process and the screening parameter η [defined by Eq. (2.92)] approaches 0, the total radiation cross section is that for complete screening except in the case of high frequency emitted photons [see Eqs. (2.105, 2.110)]. This cross section does not depend on the incoming electron energy E_0. For the case of complete screening in the Born approximation, let us introduce the quantity

$$X_0 = \frac{1}{\left[4 n_A \bar\Phi_c \ln\left(\frac{183}{Z^{1/3}}\right)\right]} \quad [\rm cm], \tag{2.114}$$

where $n_A = N\rho/A$ is the number of atoms per cm^3 and the term[¶] $\bar\Phi_c$ is $\bar\Phi_c = 5.8 \times 10^{-28} \times \alpha Z(Z+\iota)$ [cm^2]. In addition, $\bar\Phi_c$ is proportional to the total radiation cross section $\Phi_{\rm rad}$, i.e., from Eq. (2.105) we have:

$$\bar\Phi_c = \Phi_{\rm rad} \Big/ \left[4\ln\left(\frac{183}{Z^{1/3}}\right) + \frac{2}{9}\right].$$

In this latter expression, the fraction 2/9 can be neglected with respect to the logarithmic term. Thus, Eq. (2.114) for the radiation length X_0 can be finally rewritten as:

$$X_0 \approx \frac{1}{n_A \Phi_{\rm rad}} \quad [\rm cm].$$

In order to understand the physical meaning of the above-defined radiation length, let us introduce the parameter

$$b_0 = \frac{1}{18 \ln\left(\frac{183}{Z^{1/3}}\right)}. \tag{2.115}$$

[¶]The reader can see Eq. (2.106) and the values of ι given in Sect. 2.1.7.

Introducing Eqs. (2.105, 2.114, 2.115), for any given initial energy E in units of MeV, Eq. (2.101) can be rewritten as:

$$-\left(\frac{dE}{dx}\right)_{\text{rad}} = n_A E \Phi_{\text{rad}}$$

$$= E n_A \bar{\Phi}_c \left[4 \ln\left(\frac{183}{Z^{1/3}}\right) + \frac{2}{9}\right]$$

$$= E n_A \bar{\Phi}_c \left[4 \ln\left(\frac{183}{Z^{1/3}}\right) + \frac{4 \ln\left(\frac{183}{Z^{1/3}}\right)}{18 \ln\left(\frac{183}{Z^{1/3}}\right)}\right]$$

$$= 4 n_A \bar{\Phi}_c \ln\left(\frac{183}{Z^{1/3}}\right) E \left[1 + \frac{1}{18 \ln\left(\frac{183}{Z^{1/3}}\right)}\right]$$

$$= \frac{E(1 + b_0)}{X_0} \quad [\text{MeV/cm}].$$

The parameter b_0 is $\simeq 0.012$ for air and $\simeq 0.015$ for Pb ($Z = 82$); as a consequence, it can be neglected. Furthermore, X_0 is independent of the energy E_0. Thus, we can write:

$$-\left(\frac{dE}{dx}\right)_{\text{rad}} \simeq \frac{E}{X_0} \quad [\text{MeV/cm}]. \tag{2.116}$$

After traversing a thickness x of matter, the final energy E_f of an electron of incoming energy E can be calculated by integrating the previous equation. We obtain:

$$\int_E^{E_f} -\frac{dE'}{E'} = \int_0^x \frac{dx'}{X_0}$$

$$\Rightarrow \ln\left(-\frac{E_f}{E}\right) = \frac{x}{X_0},$$

from which, finally, we get

$$E_f = E \, e^{(-x/X_0)}. \tag{2.117}$$

While the meaning of the radiation length is outlined in Eq. (2.117), the actual value depends on the assumptions under which the total radiation cross section is treated. For instance, as described above, Eq. (2.114) is valid when the Born approximation can be employed [Bethe and Ashkin (1953)]. Other authors derived slightly modified expressions for the radiation length (see [Rossi (1964); Dovzhenko and Pomamskii (1964)]).

A comprehensive treatment of the radiation length was derived by Tsai (1974); it includes the effects due to atomic and nuclear form factors for light and heavy elements. The complete expression for the radiation length is given by

$$X_0 = \frac{1}{4 \, n_A \, \alpha r_e^2 \left\{Z^2 \left[L_{\text{rad}} - g(Z)\right] + L'_{\text{rad}}\right\}}$$

$$= 716.405 \, \frac{A}{\rho} \frac{1}{\left\{Z^2 \left[L_{\text{rad}} - g(Z)\right] + L'_{\text{rad}}\right\}} \quad [\text{cm}], \tag{2.118}$$

Table 2.7   Values of the radiation lengths X_{g0} in units of g/cm^2 from Eq. (2.118) (see [Tsai (1974)]), and atomic number Z for elements with Z up to 46.

Element	Z	X_{g0} g/cm^2	Element	Z	X_{g0} g/cm^2
H	1	63.05	Cr	24	14.94
He	2	94.32	Mn	25	14.64
Li	3	82.76	Fe	26	13.84
Be	4	65.19	Co	27	13.62
B	5	52.69	Ni	28	12.68
C	6	42.70	Cu	29	12.86
N	7	37.99	Zn	30	12.43
O	8	34.24	Ga	31	12.47
F	9	32.93	Ge	32	12.25
Ne	10	28.94	As	33	11.94
Na	11	27.74	Se	34	11.91
Mg	12	25.04	Br	35	11.42
Al	13	24.01	Kr	36	11.37
Si	14	21.82	Rb	37	11.03
P	15	21.02	Sr	38	10.76
S	16	19.50	Y	39	10.41
Cl	17	19.28	Zr	40	10.19
Ar	18	19.55	Nb	41	9.92
K	19	17.32	Mo	42	9.80
Ca	20	16.14	Tc	43	9.69
Sc	21	16.55	Ru	44	9.48
Ti	22	16.18	Rh	45	9.27
Va	23	15.84	Pd	46	9.20

where the functions L_{rad} and L'_{rad} are shown in Table 2.6; $g(Z)$ is the *Coulomb correction* approximated by [Tsai (1974)]:

$$g(Z) \approx 1.202\,(\alpha Z)^2 - 1.0369\,(\alpha Z)^4 + 1.008\,\frac{(\alpha Z)^6}{1 + (\alpha Z)^2}.$$

The radiation length values[‡], $X_{go} = X_0 \rho$, are given in Tables 2.7 and 2.8 in units of g/cm^2. The radiation lengths $X_{g0,c}$ of chemical compounds and mixtures of molecules (such as air) can be calculated using the mass fraction F_i of elements and the radiation lengths $X_{go,i}$ of each element shown in Tables 2.7 and 2.8:

$$\frac{1}{X_{g0,c}} = \sum_i \frac{F_i}{X_{g0,i}}. \tag{2.119}$$

The corresponding density ρ_c (in g/cm^3) can be calculated from:

$$\frac{1}{\rho_c} = \sum_i \frac{F_i}{\rho_i}, \tag{2.120}$$

where ρ_i (in g/cm^3) is the density of the ith absorber. For instance, assuming that air consists of 76.9% of nitrogen ($Z = 7$), 21.8% of oxygen ($Z = 8$) and 1.3% of

---

[‡]They were computed by means of Eq. (2.118).

Table 2.8   Values of the radiation lengths X_{g0} in units of g/cm^2 from Eq. (2.118) (see [Tsai (1974)]), and atomic number Z for elements with Z from 47 up to 92.

Element	Z	X_{g0} g/cm^2	Element	Z	X_{g0} g/cm^2
Ag	47	8.97	Yb	70	7.02
Cd	48	8.99	Lu	71	6.92
In	49	8.85	Hf	72	6.89
Sn	50	8.82	Ta	73	6.82
Sb	51	8.72	W	74	6.76
Te	52	8.83	Re	75	6.69
I	53	8.48	Os	76	6.67
Xe	54	8.48	Ir	77	6.59
Cs	55	8.31	Pt	78	6.54
Ba	56	8.31	Au	79	6.46
La	57	8.14	Hg	80	6.44
Ce	58	7.96	Tl	81	6.42
Pr	59	7.76	Pb	82	6.37
Nd	60	7.71	Bi	83	6.29
Pm	61	7.52	Po	84	6.19
Sm	62	7.57	At	85	6.07
Eu	63	7.44	Rn	86	6.29
Gd	64	7.48	Fr	87	6.19
Tb	65	7.37	Ra	88	6.15
Dy	66	7.32	Ac	89	6.06
Ho	67	7.23	Th	90	6.07
Er	68	7.14	Pa	91	5.93
Tm	69	7.03	U	92	6.00

argon ($Z = 18$) by weight, we have

$$\frac{1}{X_{g0,\mathrm{air}}} = \frac{0.769}{X_{g0,\mathrm{N}}} + \frac{0.218}{X_{g0,\mathrm{O}}} + \frac{0.013}{X_{g0,\mathrm{Ar}}},$$

from which we find $X_{g0,\mathrm{air}} = 36.66\,\mathrm{g/cm^2}$.

2.1.7.4   *Critical Energy*

As discussed in previous sections, electrons and positrons undergo both radiative and collision energy-losses. The former is proportional to the particle energy [see Eq. (2.116)], while the latter depends logarithmically on it [see Eq. (2.78)]. Thus, at high energies the dominant energy-loss process is by radiation emission. As the electron energy decreases, the ionization and excitation collisions are more and more important and, finally, becoming the dominant energy-loss process.

The *critical energy* ϵ_c is the energy at which the electron[*] loses an equal amount of energy by radiation and collision. Bethe and Heitler (1934) gave a first approxi-

---

[*]For muons, the values of the critical energy as function of the atomic number are reported in Section 4.5 of [Groom, Mokhov and Striganov (2001)].

mate formula; subsequently another expression was given by Amaldi (1981):

$$\epsilon_c = \frac{550}{Z} \ [\text{MeV}].$$

This expression is valid, within 10%, for absorbers with $Z \geq 13$. An approximate expression was also given by Berger and Seltzer (1964):

$$\epsilon_c = \frac{800}{Z + 1.2} \ [\text{MeV}].$$

A more accurate formula for the critical energy is given by Dovzhenko and Pomamskii (1964):

$$\epsilon_c = B \left(\frac{Z \, X_{0g}}{A} \right)^h \ [\text{MeV}], \tag{2.121}$$

where $B = 2.66$, and $h = 1.11$. This expression is normally employed for calculations in this book. Values of the critical energy for several materials, calculated by means of Eq. (2.121), are shown in Table 2.3.

It has to be noted that Rossi, treating the cascading shower transport inside matter under the so-called "Approximation B" [Rossi (1964)], has indicated another quantity as critical energy, i.e., the value of the electron energy at which the energy loss by collision is given by the electron energy divided by the radiation length (see also the discussion in Section 27.4.3 of [PDB (2008)]). The values of this parameter are, within a few percent, close to the critical energies as defined in this section.

2.2 Multiple and Extended Volume Coulomb Interactions

So far, we have considered phenomena related to the energy released electromagnetically by charged particles traversing a medium. However, there are additional

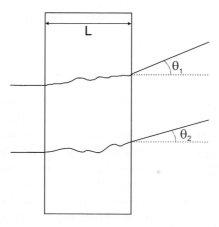

Fig. 2.40   The multiple Coulomb scattering effect on particles traversing a thickness L of material.

effects involving energy transfers small with respect to energy losses by collision or by radiation. They occur at very reduced scale (i.e., for distances of approach smaller than the atomic radius) or at very large scale, so that the medium polarization has to be taken into account. In this section, we describe the effect of the nuclear Coulomb elastic interaction resulting in the so-called *multiple scattering* and relaxation effects of polarized media via the emission of *Čerenkov radiation* and *transition radiation*.

2.2.1 The Multiple Coulomb Scattering

When a charged particle passes in the neighborhood of a nucleus, the most important effect is the deflection of its trajectory. Associated with the deflection, there are photons emitted whose overall energy is usually very small with respect to that of the incoming particle. Cases with large energy emissions are limited statistically. To a first approximation, we treat *elastic Coulomb scatterings*. In an elastic scattering, the total-momentum conservation requires that the incident particle of charge ze acquires an equal, but opposite transverse momentum with respect to the one acquired [see Eq. (2.7)] by the recoil nucleus of charge Ze. This transferred momentum is usually very small in comparison with the incoming particle momentum p. Thus, the incoming particle is scattered at an angle θ given approximately by the ratio of the transverse momentum to the total momentum p, i.e.,

$$\theta \approx \frac{2Zze^2}{bv}p^{-1}$$

$$= \frac{2Zze^2}{bvp}, \tag{2.122}$$

where v is the particle velocity. From Eq. (2.122), the absolute value of the deflection $d\theta$ at an angle θ is related to the impact parameter variation db at b by:

$$d\theta = \frac{2Zze^2}{b^2vp}db$$

$$= \frac{\theta^2vp}{2Zze^2}db. \tag{2.123}$$

The probability of collision dP_{el} of a particle traversing[§] a thickness dx with an impact parameter between b and $b + db$ is

$$dP_{\text{el}} = 2\,n_A\pi b\,db\,dx$$

$$= 2\frac{N\rho\pi}{A}b\,db\,dx,$$

and, by introducing Eq. (2.122) for b and Eq. (2.123) for db, dP_{el} becomes

$$dP_{\text{el}} = 2\frac{N\rho\pi}{A}\left[\left(\frac{2Zze^2}{vp}\right)\theta^{-1}\right]\left[\left(\frac{\theta^2vp}{2Zze^2}\right)^{-1}d\theta\right]dx$$

$$= 2\frac{N\rho\pi}{A}\left(\frac{2Zze^2}{vp}\right)^2\frac{d\theta}{\theta^3}dx,$$

[§]In the traversed material, n_A is the number of atoms with atomic weight A per cm^3 [Eq. (1.39)].

finally, since the solid angle $d\Omega$ can be written as $d\Omega \approx 2\pi\theta d\theta$ for small scattering angles, we obtain:

$$dP_{\text{el}} = \frac{N\rho}{A} \left(\frac{2Zze^2}{vp}\right)^2 \frac{d\Omega}{\theta^4} dx. \tag{2.124}$$

Therefore, the probability of collision can be expressed by

$$dP_{\text{el}} = \Xi(\theta) \, d\Omega \, d\chi,$$

where $d\chi = \rho \, dx$ is the thickness of traversed material in units of g cm^{-2}; $\Xi(\theta)$ shall be referred to as the *differential scattering probability*. From Eq. (2.124) and introducing the classical electron radius $r_e = e^2/mc^2$, the scattering probability is given by

$$\Xi(\theta) \, d\Omega = \frac{N}{A} \left(\frac{2Zze^2}{vp}\right)^2 \frac{d\Omega}{\theta^4}$$
$$= 4N\frac{Z^2}{A}r_e^2 \left(\frac{zmc}{\beta p}\right)^2 \frac{d\Omega}{\theta^4} \quad [\text{g}^{-1}\text{cm}^2]. \tag{2.125}$$

Equation (2.125) is known as the *Rutherford scattering formula* (see Sect. 1.5). The theoretical expression of $\Xi(\theta)$ depends on the spin of the incident particle for large deflections. For small deflections, the spin dependence can be neglected and, to a first approximation, one can use Eq. (2.125) (see [Rossi and Greisen (1941); Rossi (1964)] and references therein). A similar calculation can be done for scattering on atomic electrons. Their contribution is relatively small, being Z times lower.

The finite size of the nucleus and the nuclear field screening by the atomic electrons reduce the validity of Eq. (2.125). By assuming that the electric charge is uniformly distributed over a nuclear sphere with radius

$$r_n \simeq 0.5 \, r_e A^{1/3}$$

(see [Rossi (1964)]), it can be shown [Rossi and Greisen (1941); Rossi (1964)] that the calculated value of $\Xi(\theta)$ is not affected for

$$\theta < \frac{\lambda}{2\pi r_n},$$

where $\lambda = h/p$ is the de Broglie wavelength of the incoming particle, while $\Xi(\theta)$ goes rapidly to 0 at larger values of θ. Thus, we can take into account the finite size of the nucleus considering as maximum deflected angle:

$$\theta_{\max} = \frac{\lambda}{2\pi r_n} \simeq 2\frac{\hbar}{p \, r_e \, A^{1/3}} = 2\frac{mc}{p\,\alpha}A^{-1/3}. \tag{2.126}$$

Furthermore, taking as atomic radius

$$a_Z = \frac{a_0}{Z^{1/3}}$$

(the Thomas–Fermi radius), it can be proven that the nuclear field screening by the outer electrons does not affect the scattering probability for

$$\theta > \frac{\lambda}{2\pi a_Z},$$

while $\Xi(\theta)$ vanishes for lower θ values (see [Rossi (1964)] and references therein). So that, the minimum deflected angle becomes:

$$\theta_{\min} = \frac{\lambda}{2\pi a_Z} = \frac{\hbar}{p\,a_0 Z^{-1/3}} = \frac{mc\,\alpha}{p} Z^{1/3}. \tag{2.127}$$

When a charged particle traverses a thickness $d\chi$ of material, successive small angular deflections can be considered as statistically independent. The mean square of the scattering angle $\langle \theta^2 \rangle$ at a depth $\chi + d\chi$ is given by its value at χ in addition to the mean square of the scattering angle in the thickness $d\chi$:

$$d\langle \theta^2 \rangle = d\chi \int_0^{2\pi} \int_{\theta_{\min}}^{\theta_{\max}} \theta^2 \Xi(\theta)\, d\Omega;$$

thus, by means of Eq. (2.125), we get

$$d\langle \theta^2 \rangle = d\chi \int_0^{2\pi} \int_{\theta_{\min}}^{\theta_{\max}} \theta^2 4N \frac{Z^2}{A} r_e^2 \left(\frac{zmc}{\beta p} \right)^2 \frac{d\Omega}{\theta^4}.$$

Since for small scattered angles $d\Omega \approx \theta\, d\phi\, d\theta$ (where ϕ is the azimuthal angle), one has

$$d\langle \theta^2 \rangle = 4N \frac{Z^2}{A} r_e^2 \left(\frac{zmc}{\beta p} \right)^2 d\chi \int_{\theta_{\min}}^{\theta_{\max}} \frac{2\pi \theta\, d\theta}{\theta^2}$$

and, using Eqs. (2.126, 2.127),

$$d\langle \theta^2 \rangle = 8\pi N \frac{Z^2}{A} r_e^2 \left(\frac{zmc}{\beta p} \right)^2 \ln \left(\frac{\theta_{\max}}{\theta_{\min}} \right) d\chi$$

$$= 8\pi N \frac{Z^2}{A} r_e^2 \left(\frac{zmc}{\beta p} \right)^2 \ln \left(\frac{2 \frac{mc}{p\alpha} A^{-1/3}}{\frac{mc\alpha}{p} Z^{1/3}} \right) d\chi.$$

Furthermore, with the approximation $A \approx 2Z$, we obtain

$$d\langle \theta^2 \rangle = 8\pi N \frac{Z^2}{A} r_e^2 \left(\frac{zmc}{\beta p} \right)^2 \ln \left(\frac{2}{\alpha^2 A^{1/3} Z^{1/3}} \right) d\chi$$

$$\simeq 16\pi N \frac{Z^2}{A} \rho r_e^2 \left(\frac{zmc}{\beta p} \right)^2 \ln \left(\frac{173}{Z^{1/3}} \right) dx,$$

and, finally, since numerically

$$4 \frac{N\rho}{A} \alpha Z^2 r_e^2 \ln \left(\frac{173}{Z^{1/3}} \right) \approx \frac{1}{X_0}$$

where X_0 is the radiation length for the case of complete screening in the Born approximation [see Eqs. (2.106, 2.114)], we have:

$$d\langle \theta^2 \rangle \simeq 4 \frac{\pi}{\alpha} \frac{1}{X_0} \left(\frac{zmc}{\beta p} \right)^2 dx. \tag{2.128}$$

By traversing a thickness L in cm of material (see Fig. 2.40) and assuming that the energy loss can be neglected, we obtain the so-called *Rossi–Greisen equation for the mean square of the scattering angle*:

$$\langle \theta^2 \rangle = \int_0^L 4\frac{\pi}{\alpha}\frac{1}{X_0}\left(\frac{zmc}{\beta p}\right)^2 dx$$

$$= 4\frac{\pi}{\alpha}\frac{L}{X_0}\left(\frac{zmc}{\beta p}\right)^2$$

$$= E_s^2 \frac{L}{X_0}\left(\frac{z}{vp}\right)^2, \qquad (2.129)$$

where

$$E_s^2 = \frac{4\pi (mc^2)^2}{\alpha},$$

i.e., $E_s = 21.2$ MeV. Furthermore, the rms (i.e., the *root mean square*) value of the scattering angle is:

$$\theta^{\mathrm{rms}} = \sqrt{\langle \theta^2 \rangle} = E_s \left(\frac{z}{vp}\right)\sqrt{\frac{L}{X_0}} \ .$$

Instead of considering the total deflection θ, it is often convenient to consider its projection θ_{proj} onto a plane containing the direction of the initial particle trajectory. It can be shown that, under the assumption of small deflections, we have (see, for instance, [Fernow (1986)]):

$$\langle \theta_{\mathrm{proj}}^2 \rangle = \frac{1}{2}\langle \theta^2 \rangle,$$

and

$$\theta_{\mathrm{proj}}^{\mathrm{rms}} = \frac{1}{\sqrt{2}}\theta^{\mathrm{rms}}.$$

The multiple scattering effect is expressed in the Rossi–Greisen equation [Eq. (2.129)] in terms of the radiation length of the material. However, it has to be noted that the radiation length is only used as a numerical approximative factor in Eqs. (2.128, 2.129), since the multiple Coulomb scattering is not a radiative phenomenon.

More detailed calculations were performed [Molière (1947); Bethe (1953)]. In particular in [Molière (1947)], Molière has taken into account the screening effect. The multiple scattering results are roughly Gaussian only for small deflection angles, but with non-Gaussian tails at larger angles. Although Eq. (2.129) can provide an approximate calculation, Highland, Lynch and Dahl [Highland (1975)] proposed a few modifications to the Rossi–Greisen equation in order to reproduce Molière's theory in a more accurate way. For instance, for the 98% of the projected central distribution and to better than 11%, they obtain:

$$\theta_{\mathrm{proj}}^{\mathrm{rms}} = \frac{13.6\,\mathrm{MeV}}{\beta cp}z\sqrt{\frac{L}{X_0}}\left[1 + 0.038\ln\left(\frac{L}{X_0}\right)\right]. \qquad (2.130)$$

Table 2.9 The values of the j and f parameters for calculating the multiple scattering effect on electrons and positrons in Al and Pb absorbers from [Rohrlich and Carlson (1954)].

		j	f
Al	e^+	0.297	0.014
	e^-	0.305	0.034
Pb	e^+	0.311	0.057
	e^-	0.430	0.052

After undergoing multiple Coulomb scattering, particles emerging from a material of thickness L are displaced from their original trajectory (see Fig. 2.40). Let us call X and Y the two coordinate axes in a reference frame on the plane perpendicular to the initial trajectory of the particle and centered on the impinging particle position. The emerging-particle lateral shift was evaluated by Rossi (see [Rossi and Greisen (1941); Rossi (1964)]). For the case in which the energy-loss process can be neglected, the mean square lateral displacement along an axis (for instance the Y axis) is approximately given by

$$\langle Y^2 \rangle = \frac{1}{6} \langle \theta^2 \rangle L^2. \tag{2.131}$$

The cross section (neglecting radiative corrections) for the elastic scattering of electrons and positrons by the Coulomb field of a charge Ze was derived by Mott (1929) and calculated for light and heavy nuclei. Electrons and positrons show sizeable differences with respect to the multiple scattering effect, particularly when they undergo scattering at large angles. For heavy elements, the electron–positron differences are large and can amount up to a factor three. The electron cross section always exceeds the positron cross section. Furthermore, Rohrlich and Carlson (1954) estimated the multiple scattering effect on electrons and positrons taking into account the collision energy-loss. The cosine of the multiple scattering angle averaged over all electrons (or positrons), whose initial kinetic energy $E_0 = \gamma_0 mc^2$ has dropped down to $E = \gamma mc^2$, is given by

$$\langle \cos \theta \rangle = \frac{G(\gamma_0)}{G(\gamma)},$$

where

$$G(\gamma) = \left(\frac{\gamma + 1}{\gamma - 1} \right)^{jZ} \exp \left(f \frac{Z}{\beta} \right).$$

The values of the j and f parameters for electrons and positrons are given in Table 2.9 for Al and Pb absorbers [Rohrlich and Carlson (1954)].

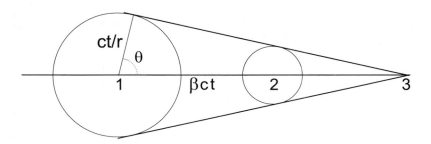

Fig. 2.41   Huyghens construction of the spherical wavefronts at successive times. r is the refraction index of the material and β is the particle velocity in units of the speed of light.

2.2.2   *Emission of Čerenkov Radiation*

The emission of electromagnetic radiation, a remarkable phenomenon discovered by Čerenkov (1937) and explained theoretically by Frank and Tamm (1937), occurs in a medium at the passage of a charged particle with a velocity v larger than the phase velocity of light in that medium, i.e., when

$$v > c_r \equiv \frac{c}{r},$$

where r is the index of refraction. This emitted radiation is referred to as *Čerenkov radiation*.

The classical theory explains this effect by an asymmetric polarization of the medium in front and at the rear of the charged particle, giving rise to a net and time varying electric-dipole momentum. To visualize the effect, let us consider a charged particle traversing a medium. The atoms of the dielectric can be assumed to be approximately spherical in regions far away the particle path, while becoming elongated by interaction with the particle electromagnetic-field, so that the centers of gravity of the positive and negative charge inside atoms do not coincide anymore. Thus, for sufficiently fast[‡] particles, a polarized region is generated following an axial symmetry. In this region, individual atoms act as electric dipoles and create a net overall dipole field. It is this dipole field which is responsible for the emission of the electromagnetic pulses of the Čerenkov radiation.

In general, there is a constructive interference between wavelets propagated from successive areas along the particle path, when the particle velocity is larger than the phase velocity of light in that medium. From an Huygens-type construction in wave optics (see Fig. 2.41), a coherent wavefront is generated and moves with a velocity c_r at angle θ, whose cosine is

$$\cos\theta = \frac{c_r}{\beta c} = \frac{1}{\beta r}. \tag{2.132}$$

[‡]In this case, the particle speed is considered with respect to the phase velocity of light in the medium.

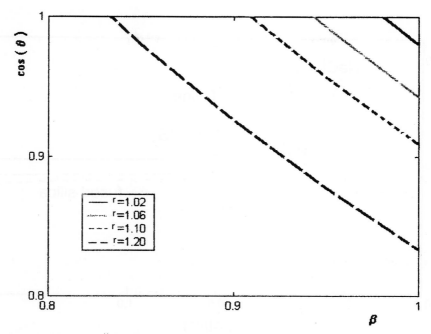

Fig. 2.42 Cosine of the Čerenkov angle θ versus the particle velocity β in units of the speed of light, for various values of the refractive index r.

In Fig. 2.42, the cosine of the Čerenkov angle θ is shown as a function of the particle velocity for various values of the index of refraction. It has to be noted that the particle mass does not play any role upon the emission angle. Furthermore, the emission angle increases with the particle velocity contrary to the radiation emitted in a bremsstrahlung process [see Eq. (2.113)]. In addition, the emission angle depends on the wavelength λ of the Čerenkov radiation, because the index of refraction depends on λ (see for instance Fig. 2.43 and [Physics Handbook (1972)]): the variation $dr(\lambda)/d\lambda$ is referred to as *dispersion* and is largest in the ultraviolet region. Its variation with temperature is generally small. Indices of refraction of liquid and solid media are shown in Fig. 2.43, in Table 2.10 and on the web for $0.041 < \lambda < 41$ nm (e.g., see [Henke, Gullikson and Davis (1993)]). Typical Čerenkov radiations correspond to frequencies between the blue region of the visible and near-visible part of the electromagnetic spectrum. Note that the energy (E_γ) of a photon with a wavelength λ (in vacuum) is given by:

$$E_\gamma = \frac{hc}{\lambda} \approx \frac{1.24}{\lambda} \times 10^3 \ [\text{eV}], \qquad (2.133)$$

where λ is in units of nm.

The Čerenkov radiation starts to be emitted at a *threshold particle velocity*

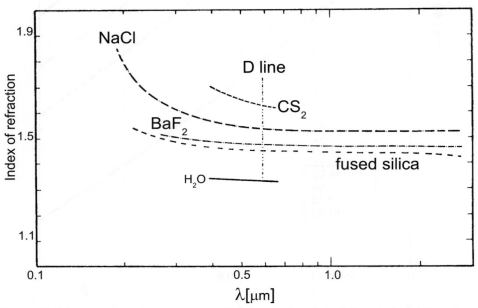

Fig. 2.43 Index of refraction of various materials as a function of the wavelength in units of μm (modified with the permission of Cambridge University Press from [Fernow (1986)], see also [Physics Handbook (1972)]). The vertical dashed line refers to the D line wavelength of Na, which is often used to quote refractive indices.

$\beta_{\text{thres}} c$ and $\theta = 0°$ [see Eq. (2.132)] when

$$\beta_{\text{thres}} = \frac{1}{r},$$
(2.134)

or, rewriting Eq. (2.134) in an equivalent way,

$$\gamma_{\text{thres}} = \frac{1}{\sqrt{1 - \beta_{\text{thres}}^2}} = \frac{r}{\sqrt{r^2 - 1}}.$$
(2.135)

Using Eq. (2.135), we can estimate that for an index of refraction of ≈ 1.58 (i.e., for a plastic scintillator) γ_{thres} is ≈ 1.29, while is ≈ 34.1 for $r \approx 1 + 4.3 \times 10^{-4}$ (i.e., for the CO_2 gas at STP). Conversely, the maximum angle of emission occurs when the particle speed approaches the speed of light:

$$\theta_{\text{max}} = \arccos\left(\frac{1}{r}\right).$$
(2.136)

In practice [see Eq. (2.132)], the condition $\beta r > 1$ is usually satisfied from the ultraviolet to near infrared portion of the electromagnetic spectrum, i.e., for photon wavelengths (*in vacuum*) between $\approx (0.2$–$1.2)$ μm, but does not extend to the X-ray region where r is typically < 1.

Table 2.10   Index of refraction of liquid and solid radiators for sodium light at (20−25) °C (see [Fernow (1986)] and references therein).

Material	r	ρ g/cm^3
Water	1.333	1.00
Carbon tetrachloride	1.459	1.591
Glycerol	1.474	1.26
Toluene	1.494	0.867
Styrene	1.545	0.91
Lucite	1.49	1.16–1.20
Plastic scintillator	1.58	1.03
Crystal quartz	1.54	2.65
Borosilicate glass	1.474	2.23
Lithium fluoride	1.392	2.635
Barium fluoride	1.474	4.89
Sodium iodide	1.775	3.667
Cesium iodide	1.788	4.51
Sodium chloride	1.544	2.165
Silica aerogel	$1+0.25\rho$	0.1–0.3

In a quantum-mechanical treatment of the process [Ginzburg (1940); Marmier and Sheldon (1969)], the classical result is modified to take into account the reaction of the emitted radiation onto the charged-particle motion (see Fig. 2.44), when the incoming velocity is βc. Owing to momentum and energy conservation, the cosine of the Čerenkov angle is (see Chapter 4.9 in [Marmier and Sheldon (1969)]):

$$
\begin{aligned}
\cos\theta_\gamma &= \frac{1}{\beta r} + \left(\frac{\lambda_B}{\lambda}\right)\left(\frac{r^2-1}{2\,r^2}\right) \\
&= \frac{1}{\beta r} + \left(\frac{h}{\lambda}\right)\left(\frac{1}{p}\right)\left(\frac{r^2-1}{2\,r^2}\right),
\end{aligned}
\tag{2.137}
$$

where λ_B and p are the de Broglie wavelength and the momentum of the incoming particle, respectively, and (h/λ) is the momentum of the Čerenkov photon. In Eq. (2.137), a second term is present with respect to Eq. (2.132): this term expresses a small *reaction correction*. For instance, for a 10 GeV/c particle and for an emitted photon with a wavelength[†] (*in vacuum*) of 400 nm, we have

$$
\left(\frac{h}{\lambda}\right)\left(\frac{1}{p}\right) \approx 3\times 10^{-10}
$$

(becoming $\approx 3\times 10^{-9}$ at 1 GeV); thus, this term makes the overall reaction correction negligible.

The intensity* of the Čerenkov radiation for a particle of charge ze traversing a

---

[†]This wavelength corresponds to a photon momentum of $\approx 3\,\text{eV/c}$.

*The intensity is the number of photons emitted per unit length of particle path and per unit of frequency.

thickness dx of an absorber is given by:

$$I_{\check{C}} = \frac{d^2 N_\gamma}{dx \, d\nu}$$
$$= \frac{4\pi^2 z^2 e^2}{hc^2}\left(1 - \frac{1}{\beta^2 r^2}\right)$$
$$= \frac{2\pi z^2 \alpha}{c}\left(1 - \frac{1}{\beta^2 r^2}\right). \tag{2.138}$$

Or equivalently, the intensity per unit of wavelength (*in vacuum*, for which $\lambda\nu = c$) is

$$\frac{d^2 N_\gamma}{dx \, d\lambda} = \frac{2\pi z^2 \alpha}{\lambda^2}\left(1 - \frac{1}{\beta^2 r^2}\right).$$

This equation shows that the number of quanta per wavelength interval is proportional to $1/\lambda^2$ and that the short wavelengths of the spectrum dominate. From Eq. (2.138), the energy loss by Čerenkov radiation becomes

$$-\left(\frac{dE}{dx}\right)_{\check{C}} = \int_{\beta r > 1} \frac{d^2 N_\gamma}{dx \, d\nu} h\nu \, d\nu$$

and, for $\alpha = e^2/(\hbar c)$, it can be expressed by:

$$-\left(\frac{dE}{dx}\right)_{\check{C}} = \frac{4\pi^2 z^2 e^2}{c^2} \int_{\beta r > 1}\left(1 - \frac{1}{\beta^2 r^2}\right)\nu \, d\nu, \tag{2.139}$$

where the integration is extended over all frequencies for which $\beta r > 1$ and ν is the emitted-photon frequency.

To a first approximation, we can assume that the index of refraction is roughly constant (see Fig. 2.43) in the region of wavelengths (*in vacuum*) from ≈ 350 up to ≈ 500 nm, i.e., the region which covers most of the Čerenkov radiation spectrum and overlaps with the highest quantum efficiency region of typical commercial photomultipliers. Equations (2.138, 2.139) can be integrated in this range of wavelengths to obtain both the number of emitted-photons and the particle energy-loss per unit of length. Using Eq. (2.132), we estimate the number of emitted photons per unit of length by means of Eq. (2.138):

$$\frac{dN_\gamma}{dx} = \int_{\nu_1}^{\nu_2} \frac{4\pi^2 z^2 e^2}{hc^2}\left(1 - \frac{1}{\beta^2 r^2}\right) d\nu$$
$$= \int_{\nu_1}^{\nu_2} \frac{2\pi z^2 \alpha}{c} \sin^2\theta \, d\nu$$
$$= \frac{2\pi z^2 \alpha}{c} \sin^2\theta \, (\nu_2 - \nu_1)$$
$$= 2\pi z^2 \alpha \left(\frac{1}{\lambda_2} - \frac{1}{\lambda_1}\right) \sin^2\theta$$
$$\approx 393 \, z^2 \sin^2\theta \; [\text{quanta/cm}], \tag{2.140}$$

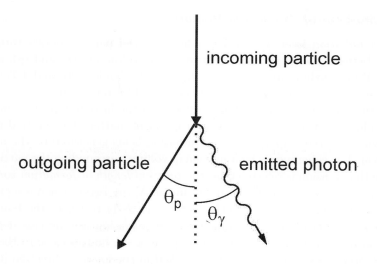

Fig. 2.44 Emission of a photon with energy $h\nu$ at an angle θ_γ from a particle scattered at an angle θ_p.

where λ_2 and λ_1 are 350 and 500 nm, respectively. The energy loss per unit of length can be determined using Eq. (2.139):

$$
\begin{aligned}
-\left(\frac{dE}{dx}\right)_{\check{C},(350-700)\mathrm{nm}}
&= \int_{\nu_1}^{\nu_2} \frac{d^2 N_\gamma}{dx\, d\nu}\, h\nu\, d\nu \\
&= \frac{4\pi^2 z^2 e^2}{c^2} \int_{\nu_1}^{\nu_2} \sin^2\theta\, \nu\, d\nu \\
&= \frac{2\pi^2 z^2 e^2}{c^2} \sin^2\theta \left(\nu_2^2 - \nu_1^2\right) \\
&= 2\pi^2 z^2 e^2 \sin^2\theta \left(\frac{1}{\lambda_2^2} - \frac{1}{\lambda_1^2}\right) \\
&= 2\pi^2 z^2 r_e mc^2 \sin^2\theta \left(\frac{1}{\lambda_2^2} - \frac{1}{\lambda_1^2}\right) \\
&\approx 1.18 \times 10^{-3} \times z^2 \sin^2\theta \;[\mathrm{MeV/cm}].
\end{aligned}
\qquad (2.141)
$$

Even considering larger intervals of integration, the energy loss by Čerenkov radiation is quite small (usually much less than 1%, see Table 2.3) in comparison with the energy loss by collision in solids. In gases with $Z > 7$, the energy loss by Čerenkov radiation can amount to less than 1% of the collision loss of minimum ionizing particles, while for hydrogen and helium it can amount to $\approx 5\%$ (see for instance [Grupen (1996)] and references therein). The energy loss by Čerenkov radiation can be detected because many low energy quanta are emitted in the small solid angle determined by the Čerenkov angle.

2.2.3 *Emission of Transition Radiation*

Transition radiation is emitted when a fast charged particle passes through the boundary between media with different indices of refraction (r_1 and r_2), as shown in Fig. (2.45) ([Ginzburg and Frank (1946)], see also [Goldsmith and Jelley (1959)] for the first experimental observation of the transition radiation).

The reorganization of the field associated with the incoming particle occurs because a sudden change of the dielectric property of matter. The emitted radiation comes from a coherent superposition of radiation fields generated by the molecular polarization. The coherence is kept in a small volume surrounding the particle path, whose length extension is referred to as the *coherent length* or *formation zone*. It can result in an observable amount of energy in the *X-ray* region when a high enough energy particle (i.e., when its Lorentz factor γ is $\gg 1$) traverses the boundary of a macroscopically thick medium. Like Čerenkov emission, the process depends on the particle velocity and is a collective response of the matter close to the particle path. Like bremsstrahlung, it is sharply peaked in the forward direction [Garibyan and Barsukov (1959)]. The intensity of the process, i.e., the overall number of emitted photons, can be enhanced by radiators consisting of several boundaries [Frank (1964)]. The transition-radiation emission depends on a few parameters such as, for instance, the radiator configuration and the Lorentz factor γ of the particle.

In the *X-ray* frequency region, a material behaves as an electron gas (see [Artru, Yodh and Mennessier (1975)]), whose plasma frequency [Eq. (2.25)] is

$$\nu_p = \frac{\omega_p}{2\pi} = \sqrt{\frac{ne^2}{\pi m}},$$

where n is the electron density and m is the electron mass. The corresponding *plasma photon energy* is

$$h\nu_p = \hbar\omega_p = h\sqrt{\frac{Z\rho N e^2}{\pi m A}}$$

$$= \sqrt{4\pi N r_e \hbar^2 c^2}\sqrt{\frac{Z\rho}{A}}$$

$$\approx 28.8\sqrt{\frac{Z\rho}{A}} \text{ [eV]}. \tag{2.142}$$

The plasma photon energies are $\approx 0.7\,\text{eV}$ for normal air, $0.27\,\text{eV}$ for He, $20\,\text{eV}$ for polypropylene and styrene, $13.8\,\text{eV}$ for Li and $24.4\,\text{eV}$ for mylar. The *dielectric constant* of the medium is given by

$$\epsilon(\omega) = 1 - \left(\frac{\omega_p}{\omega}\right)^2 = 1 - \Upsilon^2, \tag{2.143}$$

where Υ is $\ll 1$ in the *X-ray* region. The *formation length* D is of the order of $\approx (\gamma c)/\omega_p$ and represents the largest value of the frequency dependent depth over which the coherent superposition can occur [Jackson (1975)].

The complete expression for the energy radiated is rather complicated. But at large γ, most of the energy is in the forward direction, i.e., for $\theta < \pi/2$, where θ

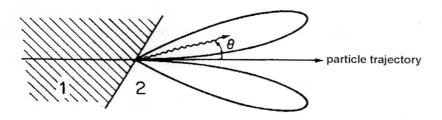

Fig. 2.45   Transition-radiation emission at the boundary between two media with different indices of refraction (adapted and republished with permission from Artru, X., Yodh, G.B. and Mennessier, G., *Phys. Rev. D* **12**, 1289 (1975); Copyright (1975) by the American Physical Society).

is the angle between the direction of the emitted photon and the trajectory of the incoming particle. For the forward direction (see Fig. 2.45), the energy radiated by a particle of charge ze at the boundary between two media, per unit of solid angle and per unit of frequency interval can be approximated by

$$\frac{d^2W}{d\nu\, d\Omega} \simeq z^2 \frac{h\alpha}{\pi^2}\, \theta^2 \left(\frac{1}{\gamma^{-2}+\theta^2+\Upsilon_1^2} - \frac{1}{\gamma^{-2}+\theta^2+\Upsilon_2^2}\right)^2 \qquad (2.144)$$

with $\Upsilon_1 = \frac{\omega_{p,1}}{\omega}$, and $\Upsilon_2 = \frac{\omega_{p,2}}{\omega}$; where $\nu_{p,1} = \omega_{p,1}/(2\pi)$ and $\nu_{p,2} = \omega_{p,2}/(2\pi)$ are the plasma frequencies of the two media (see for instance [Artru, Yodh and Mennessier (1975); Fayard (1988)] and references therein). Equation (2.144) can be rewritten per unit of solid angle and unit of energy as:

$$\frac{d^2W}{d(\hbar\omega)\, d\Omega} = \frac{d^2W}{d(h\nu)\, d\Omega}$$

$$\simeq z^2 \frac{\alpha}{\pi^2}\, \theta^2 \left(\frac{1}{\gamma^{-2}+\theta^2+\Upsilon_1^2} - \frac{1}{\gamma^{-2}+\theta^2+\Upsilon_2^2}\right)^2. \qquad (2.145)$$

The radiation is concentrated in a narrow forward cone, defined by θ^2 and of an order ranging from $\left(\gamma^{-2}+\Upsilon_1^2\right)$ to $\left(\gamma^{-2}+\Upsilon_2^2\right)$. Therefore, the angular distribution is confined to a forward cone for which $\gamma\theta$ is of the order of the unity. As long as the cone is completely contained in the second medium, there is no dependence on the particle incidence angle relative to the boundary. If $\omega_{p,2} \ll \omega_{p,1}$, as it is the case for a dense material and a gas, the most probable emission angle θ_{mp} and the root mean square angle θ_{rms} are given by [Fayard (1988)]:

$$\theta_{mp} \approx \sqrt{\gamma^{-2}+\Upsilon_2^2}$$

and

$$\theta_{rms} \approx \sqrt{\gamma^{-2}+\Upsilon_1^2}.$$

It has to be noted that the distributions given by Eqs. (2.144, 2.145) for an incoming particle with a Lorentz factor γ and two media with $\omega_{p,1}$ and $\omega_{p,2}$ are the same as for an incoming particle with a Lorentz factor γ' and, as media, the vacuum and a material with plasma frequency $\nu'_{p,1} = \omega'_{p,1}/(2\pi)$ where:

$$\omega'_{p,1} = \sqrt{\omega^2_{p,1} - \omega^2_{p,2}},$$

$$\omega'_{p,2} = 0 \quad (\text{i.e., the } vacuum, \text{ for which we have: } \Upsilon'_2 = 0),$$

$$\gamma' = \frac{1}{\sqrt{\gamma^{-2} + \Upsilon^2_2}} \,.$$

We have always $\gamma' < \gamma$ and $\omega'_{p,1} < \omega_{p,1}$. Integrating Eq. (2.145) over Ω, we obtain:

$$\frac{dW}{d(\hbar\omega)} \simeq z^2 \frac{\alpha}{\pi} \left[\left(\frac{\Upsilon^2_1 + \Upsilon^2_2 + 2\gamma^{-2}}{\Upsilon^2_1 - \Upsilon^2_2} \right) \ln\left(\frac{\Upsilon^2_1 + \gamma^{-2}}{\Upsilon^2_2 + \gamma^{-2}} \right) - 2 \right]. \tag{2.146}$$

In Eq. (2.146), $dW/d(\hbar\omega)$ depends on γ/ω only. It can be rewritten introducing γ' and $\Upsilon'_1 = \omega'_{p,1}/\omega$, i.e.,

$$\frac{dW}{d(\hbar\omega)} \simeq z^2 \frac{\alpha}{\pi} \left\{ \left[\frac{(\omega_{p,1}/\omega)^2 + (\omega_{p,2}/\omega)^2 + 2\gamma^{-2}}{(\omega_{p,1}/\omega)^2 - (\omega_{p,2}/\omega)^2} \right] \ln\left[\frac{(\omega_{p,1}/\omega)^2 + \gamma^{-2}}{(\omega_{p,2}/\omega)^2 + \gamma^{-2}} \right] - 2 \right\}$$

$$= z^2 \frac{\alpha}{\pi} \left\{ \left[\frac{(\omega'_{p,1}/\omega)^2 + 2\gamma'^{-2}}{(\omega'_{p,1}/\omega)^2} \right] \ln\left[\frac{\frac{\omega^2_{p,1} - \omega^2_{p,2}}{\omega^2} + \left(\frac{\omega_{p,2}}{\omega}\right)^2 + \gamma^{-2}}{\gamma'^{-2}} \right] - 2 \right\}$$

$$= z^2 \frac{\alpha}{\pi} \left\{ \left[\frac{(\omega'_{p,1}/\omega)^2 + 2\gamma'^{-2}}{(\omega'_{p,1}/\omega)^2} \right] \ln\left[\frac{(\omega'_{p,1}/\omega)^2 + \gamma'^{-2}}{\gamma'^{-2}} \right] - 2 \right\}$$

$$= z^2 \frac{\alpha}{\pi} \left\{ \left[1 + 2\left(\frac{\omega}{\gamma'\omega'_{p,1}} \right)^2 \right] \ln\left[1 + \left(\frac{\omega'_{p,1}\gamma'}{\omega} \right)^2 \right] - 2 \right\}$$

$$= z^2 \frac{\alpha}{\pi} \left[\left(1 + 2\frac{1}{\gamma'^2\Upsilon'^2_1} \right) \ln\left(1 + \gamma'^2\Upsilon'^2_1 \right) - 2 \right]$$

$$= z^2 \frac{\alpha}{\pi} G\left(\frac{1}{\gamma'\Upsilon_1} \right), \tag{2.147}$$

where the function $G\left(\frac{1}{\gamma'\Upsilon_1} \right)$ is given by

$$G\left(\frac{1}{\gamma'\Upsilon_1} \right) = \left[\left(1 + 2\frac{1}{\gamma'^2\Upsilon'^2_1} \right) \ln\left(1 + \gamma'^2\Upsilon'^2_1 \right) - 2 \right] \tag{2.148}$$

and it is shown in Fig. 2.46. For instance, in the case of $\omega_{p,2} \ll \omega_{p,1}$, we can distinguish three regimes as a function of γ
(i) $\gamma \ll \omega/\omega_{p,1} = 1/\Upsilon_1$, i.e., a very low yield for which

$$\frac{dW}{d(\hbar\omega)} \simeq z^2 \frac{\alpha}{6\pi} (\gamma\Upsilon_1)^4 = z^2 \frac{\alpha}{6\pi} \left(\frac{\gamma\omega_{p,1}}{\omega} \right)^4$$

and, in order to have enough yield, it must be noted that there is a frequency cutoff given by $\omega \le \gamma\omega_{p,1}$;

(ii) $1/\Upsilon_1 = \omega/\omega_{p,1} \ll \gamma \ll \omega/\omega_{p,2} = 1/\Upsilon_2$, where there is a logarithmic increase of the yield with γ and we have

$$\frac{dW}{d(\hbar\omega)} \simeq 2z^2\frac{\alpha}{\pi}\left[\ln\left(\gamma\Upsilon_1\right) - 1\right] = 2z^2\frac{\alpha}{\pi}\left[\ln\left(\frac{\gamma\omega_{p,1}}{\omega}\right) - 1\right];$$

(iii) $\gamma \gg \omega/\omega_{p,2} = 1/\Upsilon_2$, in which the yield is almost constant (saturation). The total energy emitted in forward direction by transition radiation per boundary [Fayard (1988)] is calculated by integrating Eq. (2.146):

$$\begin{aligned}
W &= \int_0^\infty \frac{dW}{d(\hbar\omega)}\, d(\hbar\omega) \\
&= z^2\gamma\frac{\alpha\hbar\omega_{p,1}}{3}\frac{\left(1 - \frac{\omega_{p,2}}{\omega_{p,1}}\right)^2}{\left(1 + \frac{\omega_{p,2}}{\omega_{p,1}}\right)} \\
&= z^2\gamma\frac{\alpha\hbar}{3}\frac{\left(\omega_{p,1} - \omega_{p,2}\right)^2}{\omega_{p,1} + \omega_{p,2}}.
\end{aligned} \tag{2.149}$$

It has to be noted that the energy emitted by transition radiation has a linear dependence on the Lorentz factor γ of the incoming particle. This property can be exploited for detector applications [Dolgoshein (1993)]. In the case of medium to vacuum transition for which $\omega_{p,1} = \omega_p$ and $\omega_{p,2} = 0$, Eq. (2.146) reduces to:

$$W = z^2\gamma\alpha\frac{\hbar\omega_p}{3}.$$

Furthermore (see Section 27.7 in [PDB (2008)]), the average number of emitted photons by transition radiation above a minimal threshold $\hbar\omega_0$ is given by:

$$\begin{aligned}
\langle N_\gamma\rangle_{\hbar\omega > \hbar\omega_0} &= \int_{\hbar\omega_0}^\infty \frac{1}{\hbar\omega}\frac{dW}{d(\hbar\omega)}d(\hbar\omega) \\
&= z^2\frac{\alpha}{\pi}\left[\left(\ln\frac{\gamma\hbar\omega_p}{\hbar\omega_0} - 1\right)^2 + \frac{\pi^2}{12}\right].
\end{aligned} \tag{2.150}$$

Thus, the overall number of photons grows as $(\ln\gamma)^2$, but it is constant above a fixed fraction of $\gamma\hbar\omega_p$. For instance, over the typical ionization cluster energy between (2–3) keV in gas detectors, at a lithium to vacuum boundary the number of emitted transition-radiation photons by an electron* of 5 GeV is $\langle N_\gamma\rangle \approx 5.6\,\alpha \approx 2.6 \times 10^{-2}$ with $\hbar\omega > 2\,\text{keV}$. The quantum yield is of the order α, whence the necessity of having a large number of boundary crossings to increase it.

As seen before [i.e., Eq. (2.150)], the photon emission probability from a boundary is quite low, even for low photon detectable energies. Usually, in order to increase the probability of photon emission, many adjacent thin layers (foils) of material referred to as *radiators* are joined together. Thus, the number of surfaces can become large enough to allow a detectable photon-emission. The limiting factor is the photon re-absorption inside the radiator itself. The foil thickness cannot be reduced

---

*The Lorentz factor is $\gamma \approx 10^4$ for an electron of $\approx 5\,\text{GeV}$.

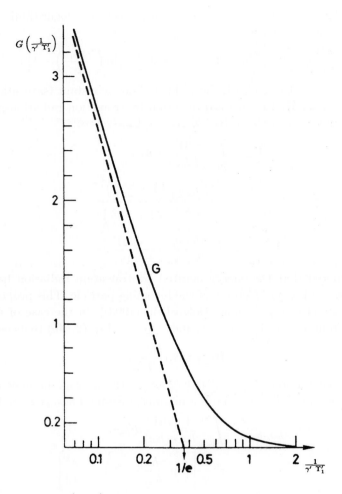

Fig. 2.46   The function $G\left(\frac{1}{\gamma'\Upsilon_1'}\right)$ from Eq. (2.148) as a function of $\frac{1}{\gamma'\Upsilon_1'}$ (adapted and republished with permission from Artru, X., Yodh, G.B. and Mennessier, G., *Phys. Rev.* D **12**, 1289 (1975); Copyright (1975) by the American Physical Society). The dashed line represents $2[\ln(\gamma'\Upsilon_1') - 1]$.

beyond some minimal value, without compromising the creation of the formation zone. It can be shown that [Artru, Yodh and Mennessier (1975); Fayard (1988)] the energy emitted per unit of energy and unit of angle in a foil of thickness l_1 is

$$\left[\frac{d^2W}{d(\hbar\omega)\,d\Omega}\right]_{\text{foil}} = 4\sin^2\left(\frac{\varphi_1}{2}\right)\frac{d^2W}{d(\hbar\omega)\,d\Omega}, \qquad (2.151)$$

where $d^2W/d(\hbar\omega)d\Omega$ is given by Eq. (2.145) for the boundary crossing; $4\sin^2\left(\frac{\varphi_1}{2}\right)$ is the *interference term* and $\varphi_1 = (\gamma^{-2} + \theta^2 + \Upsilon_1^2)\omega l_1/(2c)$. In the relevant region

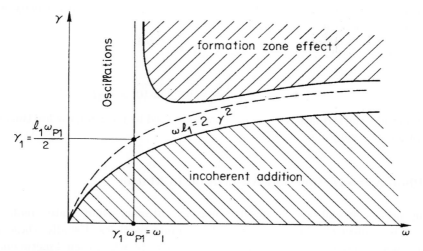

Fig. 2.47   Different regions of the ω–γ plane regarding the single foil yield, in the vacuum case (adapted and republished with permission from Artru, X., Yodh, G.B. and Mennessier, G., *Phys. Rev. D* **12**, 1289 (1975); Copyright (1975) by the American Physical Society). In this figure units are chosen such that $c = \hbar = 1$.

of integration over θ, the values of φ_1 are

$$\varphi_1 \approx \frac{\left(\gamma^{-2} + \Upsilon_1^2\right)\omega l_1}{2c}.$$

Therefore, for thicknesses much lower than Z_1 given by

$$Z_1(\omega) = \frac{2c}{\omega(\gamma^{-2} + \Upsilon_1^2)}, \qquad (2.152)$$

the yield is strongly reduced. For large ω's, we have

$$l_1 \geq Z_1(\omega) \sim \frac{2c\gamma^2}{\omega},$$

where $Z_1(\omega)$ is referred to as the *formation zone* and can be understood as the minimal depth inside a foil required by the electromagnetic field carried by the incoming charged-particle to reach a new equilibrium state inside the medium. The ω–γ plane related to the formation zone is shown in Fig. 2.47 [Artru, Yodh and Mennessier (1975)]. For Lorentz factors larger than $\gamma_1 = \omega_{p,1} l_1/(2\,c)$, the frequency cutoff is no longer $\gamma\omega_{p,1}$ as in the single surface case, but it is rather determined by the formation-zone effect. The condition to have enough yield becomes:

$$\omega < \min(\gamma\omega_{p,1}, \omega_1) \qquad (2.153)$$

where $\omega_1 = \gamma_1\omega_{p,1} = (\omega_{p,1}^2 l_1)/(2c)$ [Artru, Yodh and Mennessier (1975)]. For practical calculations with l_1 in units of μm and the plasma photon energy $h\nu_{p,1}$ in units of eV, we have:

$$\gamma_1 = \frac{\omega_{p,1} l_1}{2\,c} \times 10^{-4} = \frac{h\nu_{p,1} l_1}{2\,c\hbar} \times 10^{-4} \approx 2.5\,h\nu_{p,1} l_1.$$

Furthermore, from Eq. (2.142) assuming $Z_1/A_1 \approx 0.5$ for a foil with density ρ_1 in units of g/cm^3 and l_1 in units of μm, we have:

$$\hbar\omega_1 = \gamma_1(\hbar\omega_{p,1}) = \frac{(\hbar\omega_{p,1})^2 l_1}{2\hbar c} \times 10^{-4}$$

$$\approx \frac{\left(\rho_1 \frac{28.8^2}{2}\right) l_1}{4 \times 10^{-5}} \times 10^{-4} \approx 10^3 \times (\rho_1 l_1) \text{ [eV]}.$$

Experimentally, the formation zone effect was observed by reducing the foil thickness to the order of some μm's.

2.3   Photon Interaction and Absorption in Matter

Ionizing processes embrace fields like nuclear, atomic, solid state, molecular physics. They affect the kinetic energy of incoming particles. Usually, these particles are not removed from the incoming beam, except when their kinetic energy is fully absorbed in matter or (as discussed later) a *shower generation process* is initiated.

In sharp contrast with the behavior of charged particles, any beam of monochromatic photons traversing an absorber exhibits a characteristic *exponential reduction* of the number of its own photons traveling along the original direction. The reason is that, in processes of photon scattering or absorption, each photon is individually removed from the incoming beam by the interaction.

Let us consider a monochromatic photon beam of initial intensity[††] I_0. In addition, let $\sigma_{a,\text{tot}}$ be the total photon atomic cross section for either scattering or absorbing photons with energy equal to the beam energy. In the passage through a thickness dx' of a medium, the number of removed photons[‡‡] per unit of time $-dI$ is proportional to the photon beam intensity I' at depth x' and to the number of target atoms per unit of volume n_A of the traversed material, i.e.,

$$-dI = I' P_{\text{rem}},$$

where $P_{\text{rem}} = n_A \sigma_{a,\text{tot}} dx'$ is the probability for a photon removal in the thickness dx'. In addition, we have

$$-dI = I' n_A \sigma_{a,\text{tot}} dx' = I' \mu_{\text{att},l} dx'.$$

As a consequence, we obtain

$$\frac{dI}{I'} = -\mu_{\text{att},l} dx' \Rightarrow \int_{I_0}^{I} \frac{dI}{I'} = \int_0^x -\mu_{\text{att},l} dx' \Rightarrow \ln \frac{I}{I_0} = -\mu_{\text{att},l} x$$

and, finally,

$$I = I_0 \exp\left[-\left(\mu_{\text{att},l} x\right)\right]. \tag{2.154}$$

[††]The intensity is given by the number of photons per unit of time impinging onto the absorber surface.

[‡‡]These photons can be fully or partially absorbed so that they have no longer their initial energy and initial incoming direction.

The coefficient

$$\mu_{\mathrm{att},l} = n_A\, \sigma_{a,\mathrm{tot}} \quad [\mathrm{cm}^{-1}] \tag{2.155}$$

is the so-called *linear attenuation coefficient* with $\sigma_{a,\mathrm{tot}}$ in $\mathrm{cm}^2/\mathrm{atom}$. In Eq. (2.155), the number of atoms per cm^3 (n_A) of the traversed material is given by $(\rho N)/A$ [Eq. (1.39)], where ρ is the material density in $\mathrm{g/cm}^3$, N is the Avogadro number (see Appendix A.2), A is *atomic weight* (see page 14 and Sect. 1.4.1) of the material. In this section as usual, Z indicates the *atomic number* (Sect. 3.1) of the material. By introducing the absorber density ρ, we get:

$$\mu_{\mathrm{att},m} = \frac{n_A}{\rho}\, \sigma_{a,\mathrm{tot}} \quad [\mathrm{g}^{-1}\mathrm{cm}^2], \tag{2.156}$$

i.e., the so-called *mass attenuation coefficient*. If the absorber is a chemical compound or a mixture, its mass attenuation coefficient $\mu_{\mathrm{att},m}$ can be calculated from the mass attenuation coefficients of its constituent elements $\mu_{\mathrm{att},m,i}$ using the weighted average

$$\mu_{\mathrm{att},m} = \sum_i w_i\, \mu_{\mathrm{att},m,i}\,,$$

where w_i is the proportion by weight of the ith constituent element [Hubbell (1969)].

As mentioned above, the photon interaction on atoms or atomic electrons in matter results in a change of the incoming photon energy and/or of the scattered-photon direction. Atomic electrons can be emitted following the full or partial absorption of the primary photon. Apart resonance effects at frequencies related to

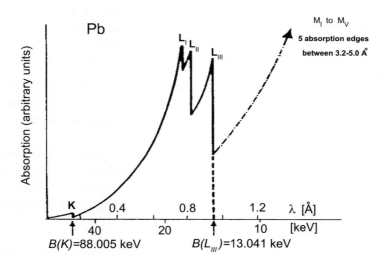

Fig. 2.48   Absorption curve for *X*-rays in Pb as a function of incident photon wavelength and energy (see for instance [Marmier and Sheldon (1969)]), showing the characteristic absorption edges (see Table 2.12).

atomic or nuclear transitions, the main competing and energy dependent processes contributing to the total cross section are:

- the *photoelectric effect*, in which the interaction occurs with the entire atomic electron cloud and results in the complete absorption of the primary photon energy;
- *Thomson* and *Compton scattering* on atomic electrons at photon energies so that the electron binding energies can be neglected and electrons can be treated as *quasi-free*;
- *pair production*, in which the photon incoming energy is high enough to allow the creation of an electron–positron pair in the Coulomb field of an electron or a nucleus.

The photoelectric process dominates at low energies, i.e., below 50 keV for aluminum and 500 keV for lead absorber. As the energy increases, between 0.05 and 15 MeV for aluminum and between 0.5 and 5 MeV for lead, the main contribution to the attenuation coefficient comes from Compton scattering. At larger energies, pair production becomes the dominant mechanism of photon interaction with matter. The photon can be scattered or absorbed by the nucleus. The photonuclear cross section is a measurable effect. However, this kind of process is not easily treated for systematic calculations due to a number of factors. Among these factors, we have both A and Z, and sensitivity to the isotopic abundance. Reviews of the γ-ray interaction processes and practical coefficients tables can be found in Chapter 2 of [Marmier and Sheldon (1969)], and [Hubbell (1969); Messel and Crawford (1970); Hubbell and Seltzer (2004)] and references therein. At present, the tabulations of mass attenuation coefficients are also available on the web (see Sect. 2.3.5).

2.3.1   *The Photoelectric Effect*

When the energy $h\nu$ is larger than the binding energies (B_e) of atomic electrons, photons can be completely absorbed in the interaction with an atom, which, in turn, emits an electron raised into a state of the continuous spectrum. This effect is called *photoelectric effect*.

The interaction involves the entire electron cloud, rather than the individual (*corpuscular*) electron. Furthermore, the atom as a whole takes up the quite small recoil energy to preserve the momentum and energy conservation. Thus, the kinetic energy K_e of the electron after leaving the atom is determined by the equation:

$$K_e = h\nu - B_e. \qquad (2.157)$$

Since a free electron cannot absorb a photon, we should expect that the photoelectric absorption probability is larger for more tightly bound electrons, i.e., for K-shell electrons. In fact, for incoming photon energies larger than K-shell energies, more than about 80% of the photoelectric absorption occurs involving the emission of K-shell electrons (see for instance Chapter V, Section 21 in [Heitler (1954)]). If the photon energy is lower than the binding energy of a shell (see Tables 2.11 and 2.12), an electron cannot be emitted from that shell. Therefore, the absorption curve exhibits the characteristic *absorption edges*, whenever the incoming photon energy coincides with the ionization energy of electrons of K, L, M, ... shells. In addition (see Fig. 2.48), the electron shells (except the K-shell) have substructures with slightly different binding energies, which result in close absorption edges (3 for the L-shell, 5 for the M-shell, etc).

The binding energy depends on the atomic number Z and the electron shell: it decreases, as proceeding towards the outer shells, according to these approximate formulae for the K, L, M shells, respectively:

$$B_e(\mathrm{K}) \approx Ry(Z-1)^2 \ [\mathrm{eV}],$$

$$B_e(\mathrm{L}) \approx \frac{1}{4}Ry(Z-5)^2 \ [\mathrm{eV}],$$

$$B_e(\mathrm{M}) \approx \frac{1}{9}Ry(Z-13)^2 \ [\mathrm{eV}],$$

where $Ry = 13.61$ eV is the Rydberg energy. These formulae are in agreement with the values given in Tables 2.11 and 2.12 within $\pm$ (3–7)% for K-shell absorption edges and within $\pm 10\%$ for L-shell absorption edges.

An exact theoretical calculation of the photoelectric effect presents difficulties and, thus, empirical formulae are used for computing the total (τ_{ph}) and K-shell cross sections per atom. However, an estimate of the K-shell photoelectric cross

Table 2.11 Energies of the absorption edges above 10 keV for elements with Z up to 68 from [Hubbell (1969)].

Element	Z	K-edge keV	Element	Z	K-edge keV
Ga	31	10.368	Sn	50	29.195
Ge	32	11.104	Sb	51	30.486
As	33	11.865	Te	52	31.811
Se	34	12.654	I	53	33.166
Br	35	13.470	Xe	54	34.590
Kr	36	14.324	Cs	55	35.987
Rb	37	15.202	Ba	56	37.452
Sr	38	16.107	La	57	38.934
Y	39	17.038	Ce	58	40.453
Zr	40	17.999	Pr	59	42.002
Nb	41	18.987	Nd	60	43.574
Mo	42	20.004	Pm	61	45.198
Tc	43	21.047	Sm	62	46.849
Ru	44	22.119	Eu	63	48.519
Rh	45	23.219	Gd	64	50.233
Pd	46	24.348	Tb	65	52.002
Ag	47	25.517	Dy	66	53.793
Cd	48	26.716	Ho	67	55.619
In	49	27.942	Er	68	57.487

Table 2.12   Energies of the absorption edges above 10 keV for elements with Z from 69 up to 100 from [Hubbell (1969)].

Element	Z	K-edge keV	L_I-edge keV	L_{II}-edge keV	L_{III}-edge keV
Tm	69	59.380	10.121		
Yb	70	61.300	10.490		
Lu	71	63.310	10.874	10.345	
Hf	72	65.310	11.274	10.736	
Ta	73	67.403	11.682	11.132	
W	74	69.508	12.100	11.538	10.200
Re	75	71.658	12.530	11.954	10.531
Os	76	73.856	12.972	12.381	10.868
Ir	77	76.101	13.423	12.820	11.212
Pt	78	78.381	13.883	13.272	11.562
Au	79	80.720	14.354	13.736	11.921
Hg	80	83.109	14.842	14.212	12.286
Tl	81	85.533	15.343	14.699	12.660
Pb	82	88.005	15.855	15.205	13.041
Bi	83	90.534	16.376	15.719	13.426
Po	84	93.112	16.935	16.244	13.817
At	85	95.740	17.490	16.784	14.215
Rn	86	98.418	18.058	17.337	14.618
Fr	87	101.147	18.638	17.904	15.028
Ra	88	103.927	19.236	18.486	15.444
Ac	89	106.759	19.842	19.078	15.865
Th	90	109.646	20.464	19.683	16.299
Pa	91	112.581	21.102	20.311	16.731
U	92	115.620	21.771	20.945	17.165
Np	93	118.619	21.417	20.596	17.614
Pu	94	121.720	23.109	22.253	18.054
Am	95	124.876	23.793	22.944	18.525
Cm	96	128.088	24.503	23.640	18.990
Bk	97	131.357	25.230	24.352	19.461
Cf	98	134.683	25.971	25.080	19.938
E	99	138.067	26.729	25.824	20.422
Fm	100	141.510	27.503	26.584	20.912

section and the emitted-electron angular distribution can be obtained following Heitler's treatment (Chapter V, Section 21 in [Heitler (1954)]) and using the corrections by Bethe and Ashkin (Section 3 in [Bethe and Ashkin (1953)]).

In the *non-relativistic region**, the *Born approximation* can be used for incoming photon-energies large compared with the ionization energy of the K-shell electrons. The angular distribution of the emitted electrons is expressed by the K-shell differential cross section per atom:

$$\frac{d\tau_{k,B}}{d\Omega} = 4\sqrt{2}\, r_e^2 \frac{Z^5}{137^4} \left(\frac{mc^2}{h\nu}\right)^{7/2} \frac{\sin^2\theta \cos^2\phi}{(1 - \beta\cos\theta)^4}, \tag{2.158}$$

where β is the velocity of the emitted electron in units of the speed of light, θ is the angle between the directions of the incoming photon and the emitted electron, ϕ is

---

*This energy region is for $h\nu \ll mc^2$, where m is the rest mass of the electron.

Table 2.13   Empirical-fit parameters to K-shell photoelectric cross section above 200 keV from [Hubbell (1969)].

n	a_n	b_n	c_n	p_n
1	1.6268×10^{-9}	-2.683×10^{-12}	4.173×10^{-2}	1
2	1.5274×10^{-9}	-5.110×10^{-13}	1.027×10^{-2}	2
3	1.1330×10^{-9}	-2.177×10^{-12}	2.013×10^{-2}	3.5
4	-9.1200×10^{-11}	0	0	4

the angle between the scattering plane[†] and the direction of the incoming radiation polarization (see Chapter V, Section 21 in [Heitler (1954)]). From Eq. (2.158), we note that photoelectrons are mostly emitted along the polarization direction of the incoming radiation (i.e., $\theta = \frac{1}{2}\pi$ and $\phi = 0$). While along the incoming-photon direction (i.e., $\theta = 0$), the differential cross section goes to zero, i.e., no photoelectron is emitted in the very forward direction. Furthermore [see the denominator of Eq. (2.158)], although the photoelectrons can also be emitted in the backward hemisphere (i.e., with emission angles larger than 90°), they will be emitted more and more in the forward hemisphere (i.e., with emission angles lower than 90°) as the photon energy increases. The K-shell total photoelectric cross section can be derived from Eq. (2.158) by neglecting the term $\beta \cos \theta$ in the denominator, then, integrating over the full solid angle and, finally, multiplying the result by a factor 2 to account for two K-shell electrons; thus, we have:

$$
\begin{aligned}
\tau_{k,B} &= 2 \int \frac{d\tau_{k,B}}{d\Omega} d\Omega \\
&\approx 2 \int 4\sqrt{2}\, r_e^2 \frac{Z^5}{137^4} \left(\frac{mc^2}{h\nu} \right)^{7/2} \sin^2 \theta \cos^2 \phi \, d\Omega \\
&= \sigma_{Th} 4\sqrt{2} \frac{Z^5}{137^4} \left(\frac{mc^2}{h\nu} \right)^{7/2},
\end{aligned}
\tag{2.159}
$$

where $\sigma_{Th} = (8/3)\pi r_e^2$ ($\simeq 6.6516 \times 10^{-25}$ cm^2) is the classical *Thomson scattering cross section*. For heavy elements or for incoming photon energies close to those of absorption edges, the *Born approximation* is no longer valid and exact wave functions must be used. Figure 2.49 shows the ratio between the τ_k values, calculated with exact wave functions, and those of $\tau_{k,B}$ [Eq. (2.159)] as a function of $h\nu/(Z^2 Ry)$.

There are a few direct measurements of total and K-shell photoelectric cross sections in the region between 100 keV and 3 MeV. However, most of the theoretical information was derived for the K-shell component of the cross section (see [Hubbell (1969)]). The extensive results by Rakavy and Ron (1965) include almost all the higher shells and cover the range between 1 keV and 2 MeV. In the energy range between 10 and 200 keV, the calculations [Rakavy and Ron (1965)] are in very good agreement with the experimental data. The uncertainties on K-shell cross sections

---

[†]It is the plane determined by the directions of the incoming photon and the emitted electron.

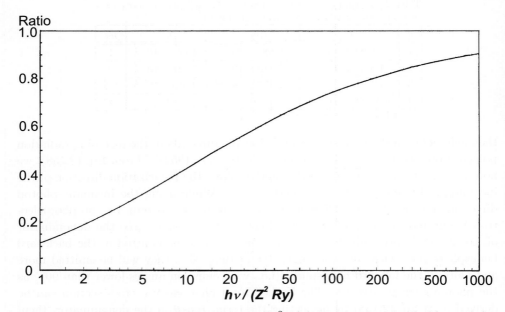

Fig. 2.49   Ratio $\tau_k/\tau_{k,B}$ as a function of $E_\nu = h\nu/(Z^2 Ry)$ (adapted and reprinted with permission from [Bethe and Ashkin (1953)]). $\tau_{k,B}$ is calculated by means of Eq. (2.159).

are estimated to be about 2% in the region (0.2–100) MeV. Over the total photo-electric cross section, the uncertainties are more likely to be about (3–5)% [Hubbell (1969)]. Above 200 keV, an empirical formula for the K-shell photoelectric cross section per atom is:

$$\tau_k \approx Z^5 \sum_{n=1}^{4} \frac{a_n + b_n Z}{1 + c_n Z} E_\gamma^{-p_n} \quad [\text{b/atom}], \tag{2.160}$$

where E_γ is the incoming photon energy in MeV and the parameters a_n, b_n, c_n, and p_n are given in Table 2.13.

To a first approximation, for energies above K-shell binding energies, the total photoelectric cross section $\tau_{\rm ph}$ per atom can be obtained multiplying τ_k by the ratios[§] derived by Kirchner and Davisson [Kirchner et al. (1930)]. These ratios can be computed, within an accuracy of $\pm$ (2–3)%, by the formula [Hubbell (1969)]:

$$\frac{\tau_{\rm ph}}{\tau_k} \approx 1 + 0.01481 \ln^2 Z - 0.000788 \ln^3 Z. \tag{2.161}$$

From Eq. (2.161), we note that, at large atomic numbers, the total photoelectric cross section does not exceed the K-shell cross section by more than $\approx 25\%$ (Fig. 2.50). Other approximate empirical formulae[¶] for photoelectric cross sections can be found and are based on calculations made by Pratt (1960).

[§]These ratios are assumed to be almost energy independent.
[¶]For instance, the reader can see Chapter 3, Section 7 in [Messel and Crawford (1970)].

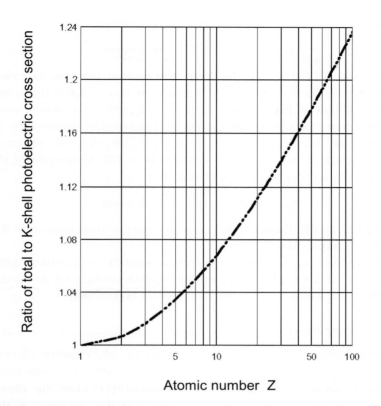

Fig. 2.50 Ratio of the total over K-shell photoelectric cross section as a function of the atomic number Z, from Eq. (2.161).

The contribution to the total linear attenuation coefficient due to photoelectric effect is given by

$$\mu_{att,l,ph} = n_A\, \tau_{ph} = \frac{\rho N}{A}\tau_{ph} \quad [\text{cm}^{-1}] \tag{2.162}$$

with τ_{ph} in cm^2/atom, while the total mass attenuation coefficient due to photoelectric effect is:

$$\mu_{att,m,ph} = \frac{n_A}{\rho}\tau_{ph} = \frac{N}{A}\tau_{ph} \quad [\text{g}^{-1}\text{cm}^2]. \tag{2.163}$$

For instance, let us estimate the photon attenuation coefficients in Al absorber at 20 keV incoming photon energy. At this energy, the photoelectric effect is the dominant absorption process in low-Z media. $\tau_{k,B}$ [computed from Eq. (2.159)] is $\approx 3.11 \times 10^2$ b/atom, while the ratio $\tau_k/\tau_{k,B}$ is ≈ 0.4 for $h\nu/(Z^2 Ry) \approx 8.7$ (Fig. 2.49). Thus, τ_k is about 1.24×10^2 b/atom. Using Eq. (2.161), the ratio of the total to the K-shell photoelectric cross section is estimated to be ≈ 1.09. Therefore, the total photoelectric cross section becomes approximately

$$\tau_{\text{ph}} \approx 1.09\,\tau_k = 1.35 \times 10^2 \text{ b/atom.}$$

Finally, from Eqs. (2.162, 2.163) the linear and mass attenuation coefficients are $\approx 8.1\,\text{cm}^{-1}$ and $3\,\text{g}^{-1}\text{cm}^2$, respectively. Similarly, we can calculate the attenuation coefficients at 300 keV in Pb absorber. At this energy in Pb, the photon absorption is almost due to the photoelectric interaction. τ_k, computed by means of Eq. (2.160), is about 0.86×10^2 b/atom. Therefore, from Eq. (2.161) the ratio of the total to K-shell photoelectric cross section is estimated to be ≈ 1.22. Consequently, the total photoelectric cross section becomes approximately

$$\tau_{\text{ph}} \approx 1.22\,\tau_k = 1.05 \times 10^2 \text{ b/atom.}$$

From Eqs. (2.162, 2.163), the linear and mass attenuation coefficients are $\approx 3.5\,\text{cm}^{-1}$ and $0.3\,\text{g}^{-1}\text{cm}^2$, respectively.

Because the total photoelectric cross section depends on the atomic number Z to a power close to 5, the photon absorption depends strongly on the medium for photon energies for which the photoelectric process is dominant.

2.3.1.1   *The Auger Effect*

As previously discussed, there are processes (like the photoelectric effect) which allow the emission of bound atomic electrons. However, when electrons are ejected from an atomic shell, a vacancy is created in that shell leaving the atom in an excited state. The atom with an electron vacancy in the innermost K-shell can readjust itself to a more stable state by emitting one or more electrons instead of radiating a single X-ray photon. This internal adjustment process is named after the French physicist Pierre-Victor Auger, who discovered it in 1925 [Auger (1925)].

When an electron of the higher L-shell makes a transition to fill a K-shell electron vacancy, the available amount of energy is the difference between the K-shell and L-shell binding energies: $B_e(\text{K}) - B_e(\text{L})$. This energy can be released via a photon (radiative emission) or absorbed by a bound electron of an higher shell, causing its ejection. This soft electron is called *Auger electron*.

The probability of non-radiative transition, with emission of Auger electrons, is larger for low-Z material (see for instance experimental data in [Burhop (1955); Krause (1979)]). The Auger yield decreases with the atomic number Z, and at $Z \approx 30$ the probabilities of X-rays emission from the innermost shell and of the emission of Auger electrons are almost equal. An empirical formula [Burhop (1955)] for the K-fluorescence yield N_K as a function of the atomic number is:

$$\frac{N_K}{1 - N_K} = (-6.4 + 3.4Z - 0.000103Z^3)^4 \times 10^{-8}, \tag{2.164}$$

where N_K (the K-fluorescence yield) is the probability for emitting photons per K-shell vacancy, $1 - N_K$ is the Auger yield (i.e., the probability for ejecting Auger

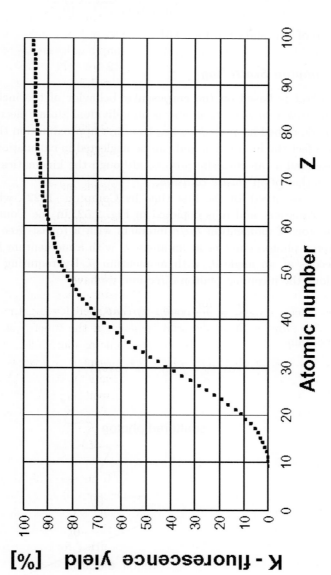

Fig. 2.51   Percentage of the K-fluorescence yield N_K computed by means of Eq. (2.165) as a function of the atomic number Z.

electrons per K-shell vacancy). From Eq. (2.164), we have:

$$N_K = (1 - N_K)(-6.4 + 3.4Z - 0.000103\,Z^3)^4 \times 10^{-8}$$
$$= \frac{(-6.4 + 3.4Z - 0.000103\,Z^3)^4 \times 10^{-8}}{1 + (-6.4 + 3.4Z - 0.000103\,Z^3)^4 \times 10^{-8}}. \tag{2.165}$$

N_K as a function of Z is shown in Fig. 2.51.

2.3.2   *The Compton Scattering*

The *Compton effect* is based on the corpuscular behavior of the incident radiation and it is an *incoherent scattering process* on individual atomic electrons. These electrons can be described as *quasi-free*, i.e., to a first approximation their binding energies do not affect the interaction and can be neglected in calculations. Furthermore, it is considered as an *inelastic process*, although the kinematics description of the reaction is that of an *elastic collision*.

The effect was observed for the first time by Compton (1922), who provided a theoretical explanation, and it is depicted in Fig. 2.52. In the Compton effect, an incoming photon of momentum $h\nu/c$ interacts with a (quasi-)free electron at rest. The scattered photon emerges at an angle θ_ν with a momentum $h\nu'/c$, while the electron recoils at an angle θ_e with momentum $\vec{p}$. By requiring momentum conservation along the incoming photon direction, we have

$$\frac{h\nu}{c} = \frac{h\nu'}{c}\cos\theta_\nu + p\cos\theta_e,$$

from which we obtain

$$p^2c^2\cos^2\theta_e = h^2\,(\nu - \nu'\cos\theta_\nu)^2\,; \tag{2.166}$$

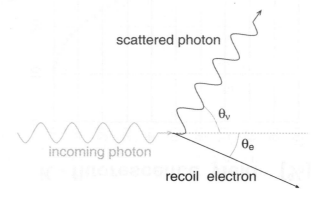

Fig. 2.52   Compton scattering of an incident photon with incoming momentum $h\nu/c$ onto a quasi-free electron which emerges at an angle θ_e. The photon is scattered at an angle θ_ν with a momentum $h\nu'/c$.

while in the perpendicular direction, we get:

$$\frac{h\nu'}{c}\sin\theta_\nu = p\sin\theta_e,$$

from which we find

$$p^2 c^2 \sin^2\theta_e = h^2\nu'^2 \sin^2\theta_\nu. \tag{2.167}$$

Due to energy conservation, we have

$$mc^2 + h\nu = h\nu' + mc^2 + K_e, \tag{2.168}$$

where m and K_e are the mass and kinetic energy of the electron, respectively. Equation (2.168) can be written as:

$$K_e = h\nu - h\nu' = h(\nu - \nu'). \tag{2.169}$$

Furthermore, see Eq. (1.8), the total electron energy is given by:

$$mc^2 + K_e = \sqrt{p^2 c^2 + m^2 c^4}$$
$$\Rightarrow m^2 c^4 + K_e^2 + 2mc^2 K_e = p^2 c^2 + m^2 c^4,$$

from which we have

$$p^2 c^2 = K_e(K_e + 2mc^2). \tag{2.170}$$

By summing Eqs. (2.166, 2.167) and substituting $p^2 c^2$ with the value given by Eq. (2.170), we obtain

$$K_e^2 + 2mc^2 K_e = h^2\left[\left(\nu - \nu'\cos\theta_\nu\right)^2 + \nu'^2 \sin^2\theta_\nu\right]$$
$$= h^2\left(\nu^2 - 2\nu\nu'\cos\theta_\nu + \nu'^2\right),$$

where we can substitute K_e with the value obtained from Eq. (2.169):

$$h^2(\nu - \nu')^2 + 2mc^2 h(\nu - \nu') = h^2\left(\nu^2 - 2\nu\nu'\cos\theta_\nu + \nu'^2\right)$$
$$\Rightarrow h^2\nu^2 - 2h^2\nu\nu' + h^2\nu'^2 + 2mc^2 h(\nu - \nu') = h^2\nu^2 - 2h^2\nu\nu'\cos\theta_\nu + h^2\nu'^2$$

and, finally, we get

$$mc^2 h(\nu - \nu') = h^2\nu\nu'\left(1 - \cos\theta_\nu\right). \tag{2.171}$$

By introducing the *Compton wavelength of the electron*

$$\lambda_e \equiv \frac{h}{mc}, \tag{2.172}$$

considering that the photon wavelength and frequency are related by $\nu\lambda = c$ and, finally, by dividing both terms by $\nu\nu'h^2$, we can rewrite Eq. (2.171) as the so-called *Compton shift formula*:

$$\triangle\lambda \equiv \lambda' - \lambda = \lambda_e\left(1 - \cos\theta_\nu\right). \tag{2.173}$$

The quantity $\triangle\lambda$ is called *wavelength Compton shift*. It increases as the photon scattering angle θ_ν increases. The maximum wavelength shift occurs for $\theta_\nu = 180°$, i.e., for backward scattered photons for which $\triangle\lambda = 2\lambda_e$.

The scattered photon energy depends on the photon scattering angle θ_ν and it is related to the incoming photon energy. The relationship can be derived by rewriting Eq. (2.171) as:

$$hv = hv' + \frac{h^2}{mc^2}\nu\nu'\left(1 - \cos\theta_\nu\right) = hv'\left[1 + \mathcal{E}\left(1 - \cos\theta_\nu\right)\right],$$

where

$$\mathcal{E} \equiv \frac{hv}{mc^2}$$

is the *reduced energy* of the incoming photon. Thus, the fraction of the incoming photon energy carried by the scattered photon is

$$\frac{hv'}{hv} = \frac{1}{1 + \mathcal{E}\left(1 - \cos\theta_\nu\right)}, \tag{2.174}$$

and, conversely, for the cosine of the photon scattering angle:

$$\cos\theta_\nu = 1 - \frac{1}{\mathcal{E}}\left(\frac{hv}{hv'} - 1\right). \tag{2.175}$$

From Eq. (2.174), for photons scattered in the forward direction, i.e., $\theta_\nu \to 0°$, we have $hv' \to hv$ independently of the incoming photon energy hv. In addition, at very low energies for which $\mathcal{E} \ll 1$, the energy of the scattered photon becomes $hv' \approx hv$, independently of the scattering angle θ_ν. Under such circumstances, the electron kinetic energy becomes negligible. From Eqs. (2.169, 2.174), the electron kinetic energy can be rewritten as

$$K_e = hv\frac{\mathcal{E}\left(1 - \cos\theta_\nu\right)}{1 + \mathcal{E}\left(1 - \cos\theta_\nu\right)}, \tag{2.176}$$

or, equivalently, as a function of the recoil electron angle θ_e:

$$K_e = hv\frac{2\,\mathcal{E}\cos^2\theta_e}{\left(1 + \mathcal{E}\right)^2 - \mathcal{E}^2\cos^2\theta_e}. \tag{2.177}$$

The angles θ_e and θ_ν are related by:

$$\tan\theta_e = \frac{1}{1 + \mathcal{E}}\cot\frac{\theta_\nu}{2}. \tag{2.178}$$

In soft collisions, the maximum electron recoil angle is $\theta_e = 90°$ and is reached when the photon is scattered at angle $\theta_\nu = 0°$, while for backward scattered photons, i.e., at the largest Compton shift [see Eq. (2.173)] where $\theta_\nu \to 180°$, the recoiling electron is emitted at angle $\theta_e \to 0°$. From Eqs. (2.176, 2.177), we note that, for backward scattered photons, the electron kinetic energy reaches its maximum value:

$$K_{e,m} = hv\frac{2\,\mathcal{E}}{1 + 2\,\mathcal{E}}.$$

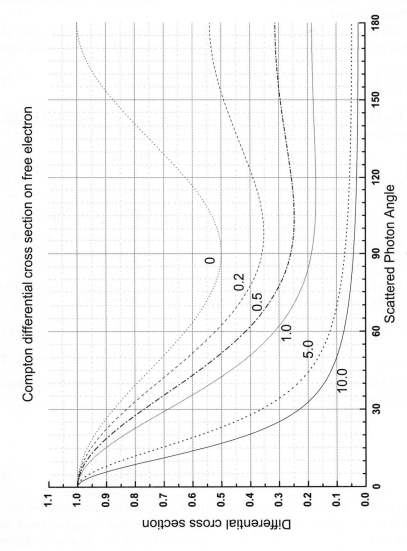

Fig. 2.53 Compton differential cross section on free electrons in units of r_e^2 as a function of the scattered photon angle θ_ν, for reduced energies $\mathcal{E} = 0$, 0.2, 0.5, 1.0, 5.0, and 10. The curves were calculated by means of Eq. (2.180).

2.3.2.1 *The Klein–Nishina Equation for Unpolarized Photons*

The cross section for Compton scattering on quasi-free electrons was first deduced by Klein and Nishina (1929). For present purposes, we will limit ourselves to a summary of the theoretical equations for unpolarized photons. Details of the treatment for both polarized and unpolarized incoming photons can be found in the original paper and in [Heitler (1954)].

The differential cross section for Compton scattering of an unpolarized photon interacting on a quasi-free electron is given by the so-called *Klein–Nishina equation*:

$$\left(\frac{d\sigma}{d\Omega}\right)_{C,e} = \frac{r_e^2}{2}\left(\frac{\nu'}{\nu}\right)^2\left(\frac{\nu'}{\nu} + \frac{\nu}{\nu'} - \sin^2\theta_\nu\right)\ [\text{cm}^2\text{sr}^{-1}/\text{electron}].\qquad(2.179)$$

Since in the Compton scattering the ratio $\frac{h\nu'}{h\nu} = \frac{\nu'}{\nu}$ is given by means of Eq. (2.174), we can rewrite Eq. (2.179) as:

$$\left(\frac{d\sigma}{d\Omega}\right)_{C,e} = \frac{r_e^2}{2}\left[\frac{1}{1 + \mathcal{E}\,(1 - \cos\theta_\nu)}\right]^2$$

$$\times \left\{\left[\frac{1}{1 + \mathcal{E}\,(1 - \cos\theta_\nu)}\right] + 1 + \mathcal{E}\,(1 - \cos\theta_\nu) - \sin^2\theta_\nu\right\}$$

$$= \frac{r_e^2}{2}\left[\frac{1}{1 + \mathcal{E}\,(1 - \cos\theta_\nu)}\right]^3$$

$$\times \left\{1 + [1 + \mathcal{E}(1 - \cos\theta_\nu)]\left(\mathcal{E} - \mathcal{E}\cos\theta_\nu + \cos^2\theta_\nu\right)\right\}$$

$$= \frac{r_e^2}{2}\frac{1}{[1 + \mathcal{E}\,(1 - \cos\theta_\nu)]^3}$$

$$\times \left[-\mathcal{E}\cos^3\theta_\nu + (1 + \cos^2\theta_\nu)(\mathcal{E}^2 + \mathcal{E} + 1) - \mathcal{E}\cos\theta_\nu(2\,\mathcal{E} + 1)\right]\ (2.180)$$

in units of $\text{cm}^2\text{sr}^{-1}/\text{electron}$. Sometimes, Eq. (2.180) can be written equivalently (in units of $\text{cm}^2\text{sr}^{-1}/\text{electron}$) as:

$$\left(\frac{d\sigma}{d\Omega}\right)_{C,e} = \frac{r_e^2}{2}\frac{1 + \cos^2\theta_\nu}{[1 + \mathcal{E}\,(1 - \cos\theta_\nu)]^2}$$

$$\times \left\{1 + \frac{\mathcal{E}^2(1 - \cos\theta_\nu)^2}{(1 + \cos^2\theta_\nu)\,[1 + \mathcal{E}\,(1 - \cos\theta_\nu)]}\right\}.\qquad(2.181)$$

For low energies, i.e., for $\mathcal{E} \to 0$ [when we expect $h\nu' \approx h\nu$ independently of the angle θ_ν of the scattered photon, see Eq. (2.174)], the differential Compton cross section expressed in Eq. (2.181) approaches the so-called *classical Thomson differential cross section* (derived in Sect. 2.3.2.3):

$$\left(\frac{d\sigma}{d\Omega}\right)_{C,e} \to \left(\frac{d\sigma}{d\Omega}\right)_{Th,e},$$

where

$$\left(\frac{d\sigma}{d\Omega}\right)_{Th,e} = \frac{r_e^2}{2}(1 + \cos^2\theta_\nu) \quad [\text{cm}^2\text{sr}^{-1}/\text{electron}]. \tag{2.182}$$

Furthermore, in the forward direction for $\theta_\nu \to 0°$, i.e., when the energy of the scattered photon is $h\nu' \approx h\nu$, independently of the incoming photon energy [see Eq. (2.174)], the differential Compton cross section approaches a constant value:

$$\left(\frac{d\sigma}{d\Omega}\right)_{C,e} \to r_e^2 \quad [\text{cm}^2\text{sr}^{-1}/\text{electron}].$$

For $\theta_\nu \to 180°$, i.e., when photons are scattered backwards, the differential Compton cross section becomes:

$$\left(\frac{d\sigma}{d\Omega}\right)_{C,e} \to r_e^2 \frac{(2\mathcal{E}^2 + 2\mathcal{E} + 1)}{(1 + 2\mathcal{E})^3} \quad [\text{cm}^2\text{sr}^{-1}/\text{electron}].$$

Thus, for $\mathcal{E} \gg 0$, the differential cross section for backward scattered photon decreases as $1/\mathcal{E}$ as $\mathcal{E}$ increases. The Klein–Nishina differential cross section on free electrons is shown in Fig. 2.53, as a function of the reduced photon energy.

The energy distribution of the scattered radiation can be obtained by introducing the expression (2.175) in Eq. (2.179) and integrating over the azimuthal angle ϕ:

$$\int_0^{2\pi} \left(\frac{d\sigma}{d\Omega}\right)_{C,e} d\phi$$

$$= \left(\frac{d\sigma}{d\cos\theta_\nu}\right)_{C,e}$$

$$= \pi r_e^2 \left(\frac{h\nu'}{h\nu}\right)^2 \left\{ \frac{h\nu'}{h\nu} + \frac{h\nu}{h\nu'} - 1 + \left[1 - \frac{1}{\mathcal{E}}\left(\frac{h\nu}{h\nu'} - 1\right)\right]^2 \right\}$$

$$= \pi r_e^2 \left(\frac{h\nu'}{h\nu}\right) \left\{ 1 + \left(\frac{h\nu'}{h\nu}\right)^2 + \frac{1 + 2\mathcal{E}}{\mathcal{E}^2}\left(\frac{h\nu'}{h\nu}\right) + \frac{1}{\mathcal{E}^2}\left(\frac{h\nu}{h\nu'}\right) - 2\frac{(1 + \mathcal{E})}{\mathcal{E}^2} \right\}.$$

Since

$$\left(\frac{d\sigma}{d\cos\theta_\nu}\right)_{C,e} = \mathcal{E}\frac{(h\nu')^2}{h\nu}\left[\frac{d\sigma}{d(h\nu')}\right]_{C,e},$$

we obtain, in units of $\text{cm}^2\text{MeV}^{-1}/\text{electron}$:

$$\left[\frac{d\sigma}{d(h\nu')}\right]_{C,e} = \frac{\pi r_e^2}{\mathcal{E}}\left(\frac{1}{h\nu'}\right)$$

$$\times \left\{ 1 + \left(\frac{h\nu'}{h\nu}\right)^2 + \frac{1 + 2\mathcal{E}}{\mathcal{E}^2}\left(\frac{h\nu'}{h\nu}\right) + \frac{1}{\mathcal{E}^2}\left(\frac{h\nu}{h\nu'}\right) - 2\frac{(1 + \mathcal{E})}{\mathcal{E}^2} \right\}, \tag{2.183}$$

where $h\nu'$ is in units of MeV and varies between [Eq. (2.174)]

$$\frac{h\nu}{1 + 2\mathcal{E}} \leq h\nu' \leq h\nu.$$

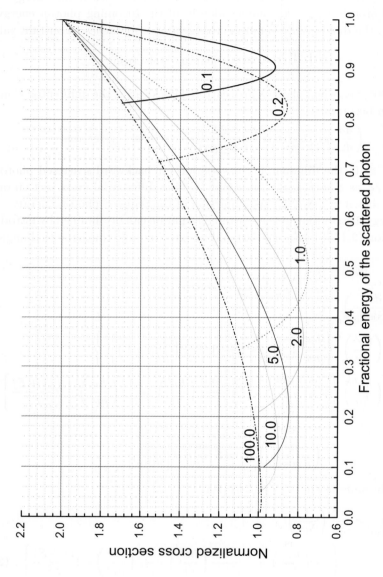

Fig. 2.54 Compton energy distribution of the scattered photon on free electrons as a function of the fractional scattered photon energy $\frac{h\nu'}{h\nu}$, calculated by means of Eq. (2.183) for reduced energies $\mathcal{E} = 0.1$, 0.2, 1.0, 2.0, 5.0, 10.0, and 100.0. In ordinate, the cross section from Eq. (2.183) is shown once divided by $\frac{\pi r_e^2}{\mathcal{E}}\left(\frac{1}{h\nu'}\right)$, i.e., it is the term in brackets in Eq. (2.183).

The term in brackets in Eq. (2.183) is shown in Fig. 2.54. For small values of $\mathcal{E}$, there is a limited spread for the fraction $\frac{h\nu'}{h\nu}$, but a drop of the cross section to a minimum followed by a rise at the lower end of the spectrum. As $\mathcal{E}$ increases, the rise following the minimum gradually decreases.

In the scattering on free electrons, the kinetic energy (K_e) of the recoil electron is obtained by means of Eq. (2.169). Thus, Eq. (2.183) also expresses the differential cross-section for an electron to be scattered with recoil energy $h\nu - h\nu'$. By inspecting the range of $h\nu'$, we get that the kinetic energy of the recoil electron varies between

$$0 \leq K_e \leq 2\,h\nu\frac{\mathcal{E}}{1+2\,\mathcal{E}}, \quad \text{or, equivalently,} \quad 0 \leq \eta_e \leq \frac{2\,\mathcal{E}}{1+2\,\mathcal{E}},$$

where

$$\eta_e = \frac{K_e}{h\nu}. \tag{2.184}$$

Furthermore, Eq. 2.183 can be rewritten in terms of K_e and η_e (in units of $cm^2 MeV^{-1}$/electron) as:

$$\left(\frac{d\sigma}{dK_e}\right)_{C,e} = \frac{\pi r_e^2}{\mathcal{E}}\left(\frac{1}{h\nu - K_e}\right)$$

$$\times \left\{1 + (1-\eta_e)^2 + \frac{1+2\,\mathcal{E}}{\mathcal{E}^2}(1-\eta_e) + \frac{1}{\mathcal{E}^2(1-\eta_e)} - 2\frac{(1+\mathcal{E})}{\mathcal{E}^2}\right\}$$

$$= \frac{\pi r_e^2}{mc^2\mathcal{E}^4}\left(\frac{1}{1-\eta_e}\right)^2 \mathbb{K}(\mathcal{E}, K_e), \tag{2.185}$$

where the function $\mathbb{K}(\mathcal{E}, K_e)$ is given by

$$\mathbb{K}(\mathcal{E}, K_e) = \left[1 + (1-\eta_e)^2\right]\mathcal{E}^2(1-\eta_e) + (1+2\,\mathcal{E})(1-\eta_e)^2$$

$$+ 1 - 2(1+\mathcal{E})(1-\eta_e)$$

$$= \left[\mathcal{E}^2(2 - 2\eta_e + 2\eta_e^2) - \eta_e(1+2\mathcal{E}) - 1\right](1-\eta_e) + 1. \tag{2.186}$$

In Fig. 2.55, it is shown the differential cross-section, calculated by means of Eq. (2.185) and multiplied by $\mathcal{E}$, in units of b/MeV for $\mathcal{E} = 0.1, 0.2, 1, 2, 2.45, 5, 10$ and 100. $\mathcal{E} = 2.45$ corresponds to the average energy of photons emitted from a ^{60}Co source. By an inspection of this latter figure, at the energy value

$$K_{e,edge} = 2\,h\nu\frac{\mathcal{E}}{1+2\,\mathcal{E}}$$

or, equivalently,

$$\eta_{e,edge} = \frac{2\,\mathcal{E}}{1+2\,\mathcal{E}},$$

there is sharp decrease of the differential cross-section, which increases with increasing photon energy. $K_{e,edge}$ is referred to as *Compton-edge*

The average fractional energy $\left\langle\frac{h\nu'}{h\nu}\right\rangle$ decreases as the photon energy increases and it is given by the approximate formula (see Equation (107) in Part II Section 3B of [Bethe and Ashkin (1953)]):

$$\left\langle\frac{h\nu'}{h\nu}\right\rangle \simeq \frac{(4/3) - 3/(2\,\mathcal{E})}{\ln(2\,\mathcal{E} + 1) + 1/2}, \quad \text{for} \quad \mathcal{E} \gg 1.$$

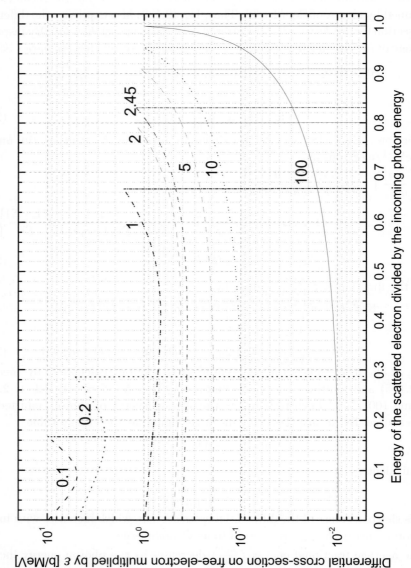

Fig. 2.55   Compton differential cross-sections on a free electron [Eq. (2.185)] multiplied by the incoming photon energy $[\eta_e$, see Eq. (2.184)]. The curves are for $\mathcal{E} = 0.1, 0.2, 1, 2, 2.45$ (i.e., it corresponds to the average energy of photons from a ^{60}Co source), 5.10 and 100.

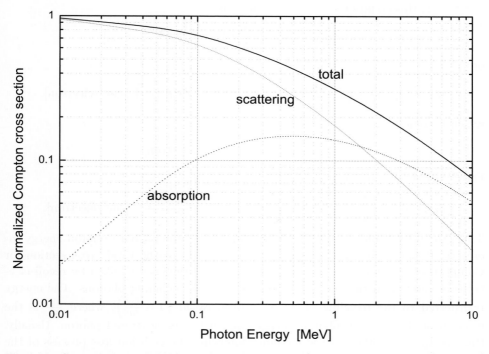

Fig. 2.56 Compton cross sections on a free electron: total [computed by means of Eq. (2.188)], energy absorption [computed by means of Eq. (2.189)] and energy scattering in units of the Thomson cross section σ_{Th} and as functions of the incoming photon energy in MeV.

The total Compton cross section can be derived in units of cm^2/electron by integrating Eq. (2.183) over the scattered photon energy:

$$
\begin{aligned}
\sigma_{C,e} &= \int_{h\nu/(1+2\,\mathcal{E})}^{h\nu} \left[\frac{d\sigma}{d(h\nu')} \right]_{C,e} d(h\nu') \\
&= \frac{\pi r_e^2}{\mathcal{E}} \left\{ \ln\left(1+2\,\mathcal{E}\right) \left[1 - 2\frac{(1+\mathcal{E})}{\mathcal{E}^2} \right] + \frac{1}{2} + \frac{4}{\mathcal{E}} - \frac{1}{2(1+2\,\mathcal{E})^2} \right\} \\
&= \sigma_{Th}\, \frac{3}{8\,\mathcal{E}} \\
&\quad \times \left\{ \ln\left(1+2\,\mathcal{E}\right) \left[1 - 2\frac{(1+\mathcal{E})}{\mathcal{E}^2} \right] + \frac{1}{2} + \frac{4}{\mathcal{E}} - \frac{1}{2(1+2\,\mathcal{E})^2} \right\},
\end{aligned}
\tag{2.187}
$$

where σ_{Th} is the classical Thomson cross section. Equation (2.187) can be rewritten in units of cm^2/electron in an equivalent way as:

$$
\begin{aligned}
\sigma_{C,e} &= 2\,\pi r_e^2 \\
&\quad \times \left\{ \left[\frac{1+\mathcal{E}}{\mathcal{E}^2} \right] \left[2\frac{(1+\mathcal{E})}{1+2\,\mathcal{E}} - \frac{\ln(1+2\,\mathcal{E})}{\mathcal{E}} \right] + \frac{\ln(1+2\,\mathcal{E})}{2\,\mathcal{E}} - \frac{1+3\,\mathcal{E}}{(1+2\,\mathcal{E})^2} \right\}.
\end{aligned}
\tag{2.188}
$$

For $\mathcal{E} \ll 1$, this equation can be expanded* as:

$$\sigma_{C,e} = \sigma_{Th} \left(1 - 2\,\mathcal{E} + \frac{26}{5}\,\mathcal{E}^2 \cdots \right) \ [\text{cm}^2/\text{electron}]$$

or

$$\sigma_{C,e} = \frac{\sigma_{Th}}{(1+\mathcal{E})^2} \left(1 + 2\,\mathcal{E} + \frac{6}{5}\,\mathcal{E}^2 - \frac{1}{2}\,\mathcal{E}^3 + \frac{2}{7}\,\mathcal{E}^4 - \cdots \right) \ [\text{cm}^2/\text{electron}],$$

which become for $\mathcal{E} \to 0$:

$$\sigma_{C,e} \to \sigma_{Th} = \frac{8}{3}\pi r_e^2 \ [\text{cm}^2/\text{electron}].$$

For $\mathcal{E} \gg 1$, Eq. (2.187) can be written as:

$$\sigma_{C,e} \to \sigma_{Th}\,\frac{3}{8\,\mathcal{E}} \left[\ln\left(2\,\mathcal{E}\right) + \frac{1}{2} \right] = \frac{3}{8}\,\sigma_{Th}\,\frac{[2\ln\left(2\,\mathcal{E}\right) + 1]}{\mathcal{E}} \ [\text{cm}^2/\text{electron}].$$

Thus, the total Compton cross section decreases as the primary photon energy increases. It has to be noted that Eq. (2.188) refers to the total cross section for Compton interaction on a free electron. However, the energy taken by recoil electrons is lower than the energy $h\nu$ removed by the incoming photons. The energy transferred to the recoiling electron is given by $h\nu\left(1 - \frac{h\nu'}{h\nu}\right)$, where $\frac{h\nu'}{h\nu}$ is the fraction of the primary photon energy carried by the scattered photon. Usually, this energy is absorbed in the medium, following the collision loss process of the low energy recoiling electron; thus, it is the part of the primary-photon energy deposited in matter after a Compton interaction. The probability, for the recoil kinetic energy K_e [Eq. (2.169)] to be imparted to the electron in a Compton collision, can be evaluated in terms of the so-called *energy absorption cross section*, $\sigma_{C,e,a}$, by integrating the energy distribution of the scattered electron [Eq. (2.183)] weighted by the fractional energy carried by electrons $\left(1 - \frac{h\nu'}{h\nu}\right)$:

$$
\begin{aligned}
\sigma_{C,e,a} &= \int_{h\nu/(1+2\mathcal{E})}^{h\nu} \left(1 - \frac{h\nu'}{h\nu} \right) \left[\frac{d\sigma}{d(h\nu')} \right]_{C,e} d(h\nu') \\
&= 2\,\pi r_e^2 \left[\frac{3 + 11\,\mathcal{E} + 9\,\mathcal{E}^2 - \mathcal{E}^3}{\mathcal{E}^2(1 + 2\,\mathcal{E})^2} - \frac{4\,\mathcal{E}^2}{3(1 + 2\,\mathcal{E})^3} \right] \\
&\quad - 2\pi r_e^2 \left(\frac{3 + 2\,\mathcal{E} - \mathcal{E}^2}{2\,\mathcal{E}^3} \right) \ln(1 + 2\,\mathcal{E}) \ [\text{cm}^2/\text{electron}], \quad (2.189)
\end{aligned}
$$

where $2\,\pi r_e^2 = \frac{3}{4}\,\sigma_{Th}$. The corresponding scattered-energy cross section, $\sigma_{C,e,s}$, i.e., that one which corresponds to the energy carried by scattered photons, is given by

$$\sigma_{C,e,s} = \sigma_{C,e} - \sigma_{C,e,a}.$$

$\sigma_{C,e,s}$ is referred to as the *energy scattering cross section*.

---

*The reader can see i) for the first approximate expression Equation (46), in Section 22 of Chapter V in [Heitler (1954)] and ii) for the second Equation (2.-9) in [Hubbell (1969)], where additional terms of the expansion can also be found.

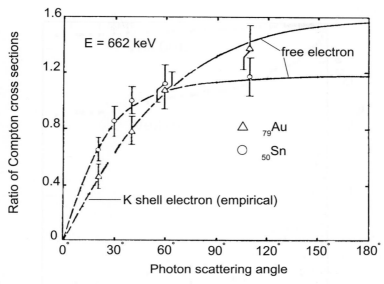

Fig. 2.57 Ratio of the observed Compton scattering from K-shell electrons to Klein–Nishina theory as a function of the angle of the scattered photon (adapted and republished with permission from Motz, J.W. and Missoni, G., *Phys. Rev.* **124**, 1458 (1961); Copyright (1961) by the American Physical Society).

The total [computed by means of Eq. (2.188)], energy scattering and energy absorption [computed by means of Eq. (2.189)] Compton cross sections are shown in Fig. 2.56 for incoming photon energies between 0.01 and 10 MeV. The cross sections are in units of Thomson cross section, i.e., they were divided by σ_{Th}. As the photon energy increases, the energy scattering cross section becomes less and less important with respect to that for energy absorption. On the other hand, the energy scattering part of the total cross section is dominant at lower photon energies, i.e., below ≈ 1 MeV.

2.3.2.2   *Electron Binding Corrections to Compton and Rayleigh Scatterings*

In the treatment of γ-ray transport, electron binding effects were often neglected. The reason being that for low-Z materials K-shell binding energies are low with respect to the photon energies considered, while for high-Z materials (with larger K-shell binding energies) K-shell electrons are a small fraction of the total.

Experimental data indicate a departure from the Klein–Nishina angular distribution [Eq. (2.179)] at low energy photon scattering. Figure 2.57 shows, for Sn and Au, the ratio of the observed differential cross sections on K-shell electrons by 0.662 MeV photons [Motz and Missoni (1961)] to those calculated from Eq. (2.179). At small scattering angles, the effect of binding energies results in the decrease of the distribution, on the contrary at large scattering angles an increase occurs. Nevertheless,

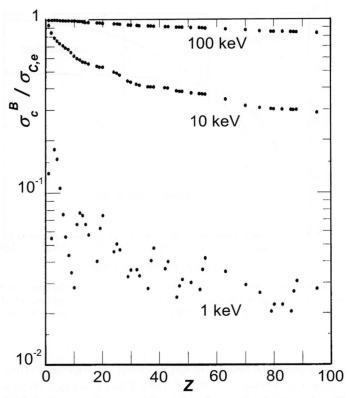

Fig. 2.58   Ratio (calculated by Storm and Israel (1967), see also [Hubbell (1969)]) of the bound electron Compton cross section σ_C^B to Klein–Nishina theory (see page 159) for incoming photon energies of 1, 10 and 100 keV as a function of the atomic number Z.

the total Compton cross section is only slightly affected because these variations almost compensate themselves. The Klein–Nishina differential cross section on free electron, assumed at rest, was generalized to the case of electrons in motion [Jauch and Rohrlich (1955)]: the solid lines in Fig. 2.57 were obtained by averaging on the appropriate distribution of electron velocities. To take into account the electron binding effects, the Klein–Nishina equation [i.e., Eq. (2.179)] is modified by introducing the so-called *incoherent scattering function* $S(q, Z)$:

$$\left(\frac{d\sigma}{d\Omega}\right)_{C,e}^{\text{incoh}} = S(q, Z) \left(\frac{d\sigma}{d\Omega}\right)_{C,e}, \qquad (2.190)$$

where q is the momentum transfer in units of mc^2, with

$$q \approx 2\,\mathcal{E}\sin\left(\frac{\theta}{2}\right)$$

at low momentum transfer, i.e., when binding effects play an important role. The incoherent scattering function gives the probability that an atom be excited or ionized as a result of transferring a recoil momentum q to an atomic electron. For instance, for K-shell electrons only, the observed ratio shown in Fig. 2.57 corresponds to the incoherent scattering function $S(q, Z)$. The total Compton *incoherent cross section* per electron taking into account the binding energy effects can be computed from Eq. (2.190):

$$\sigma_C^B = \int \left(\frac{d\sigma}{d\Omega}\right)_{C,e}^{\text{incoh}} d\Omega. \tag{2.191}$$

Figure 2.58 shows the ratios of $\sigma_C^B/\sigma_{C,e}$ calculated by Storm and Israel (1967) (see also [Hubbell (1969)]), as a function of the atomic number Z for photon energies of 1, 10, 100 keV. The *incoherent cross section* σ_C^B is *almost represented by the Klein–Nishina Compton cross section on free electron* above ≈ 100 keV, *while it decreases for lower photon energies and becomes dependent on the atomic number Z*. When the atom takes part as a whole to the interaction process, there is a phase relation between scattering amplitudes for different atomic electrons. The overall scattering amplitude is a *coherent sum* of individual contributions. The *Rayleigh scattering is a coherent process by which photons interact with atomic bound electrons, leaving the target atom neither excited nor ionized*. This type of interaction occurs at low photon energies and in high-Z materials, i.e., when the Compton process is affected by the binding energies of atomic electrons. Detailed calculations on Rayleigh scattering were carried out (see Section 2.4.4 in [Hubbell (1969)] and references therein). In these calculations, the charge distribution of all Z electrons at once was taken into account by means of an *atomic form factor* $F(q, Z)$ based on atomic models. The form factor square value gives the probability that the recoil momentum q is taken up by the whole Z electrons without absorbing energy. The *Rayleigh differential cross section* is given in terms of the atomic form factor by:

$$\left(\frac{d\sigma}{d\Omega}\right)_R = F^2(q, Z) \frac{r_e^2}{2} \left(1 + \cos^2 \theta_\nu\right) \quad [\text{cm}^2 \text{sr}^{-1}/\text{atom}], \tag{2.192}$$

where $r_e^2 \left(1 + \cos^2 \theta_\nu\right)/2$ is the Thomson differential cross section, i.e., the limit for $\mathcal{E} \to 0$ of the Klein–Nishina differential cross section on free electron. The Rayleigh cross section σ_R is computed by integrating Eq. (2.192):

$$\sigma_R = \int F^2(q, Z) \frac{r_e^2}{2} \left(1 + \cos^2 \theta_\nu\right) d\Omega \quad [\text{cm}^2/\text{atom}]. \tag{2.193}$$

References for the calculations of $S(q, \theta)$ and $F(q, \theta)$ are given in [Hubbell (1969); Berger, Hubbell, Seltzer, Chang, Coursey, Sukumar and Zucker (2005)]. At high energy, the Rayleigh scattering is confined to small angles. For instance more than 50% of scattered photons are within 5° at 1 MeV. At low energy, in particular for high-Z materials, the angular distribution is more enlarged. But in this latter case, the dominant mode of photon interaction is via the photoelectric effect. A practical

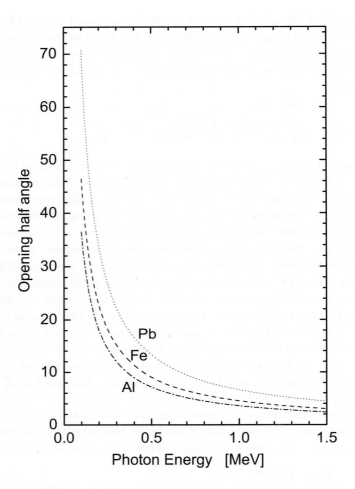

Fig. 2.59   Opening half angle in degrees as computed by means of Eq. (2.194) versus the photon incoming energy in MeV for Al (bottom curve), Fe (mid curve) and Pb (top curve) absorbers.

criterion for determining the Rayleigh angular spread is [Moon (1950)]:

$$\theta_R = 2 \arcsin \left\{ \frac{0.0133\, Z^{1/3}}{E[\text{MeV}]} \right\} \tag{2.194}$$

$$= 2 \arcsin \left[\frac{0.0266\, Z^{1/3}}{\mathcal{E}} \right],$$

where θ_R (in rad) is the opening half angle of a cone containing at least 75% of the Rayleigh scattered photons. Figure 2.59 shows θ_R as a function of the photon incoming energy in Al, Fe and Pb absorbers.

The inclusion of both binding energy corrections to free electron Compton interaction (i.e., the Compton interaction as treated in the Klein–Nishina theory) and

Rayleigh scattering slightly enhances the total photon cross section. The percentage, by which the total cross section is increased, depends on the photon incoming energy and on the atomic number. For instance, at $50\,\mathrm{keV}$ photon energy, it is ≈ 11, 6 and 8% for Al, Fe and Pb, respectively. These values become ≈ 6, 8 and 3% at $100\,\mathrm{keV}$ photon energy, and $\approx 3\%$ for Pb and $< 0.5\%$ for Al and Fe at $1\,\mathrm{MeV}$ photon energy. It has to be noted that, at these photons energies, the photoelectric effect can either become dominant or give an important contribution to the total photon cross section, depending on the atomic number of the material.

2.3.2.3   The Thomson Cross Section

We have seen that at low photon incoming energies, i.e., for $\mathcal{E} \to 0$, the differential Compton cross section* approaches the so-called *classical Thomson differential cross section*. We have also seen at page 156 that, similarly, the total Compton cross section on a free electron (as described in the Klein–Nishina Theory) for $\mathcal{E} \to 0$ approaches the *total Thomson cross section*. Therefore, the Thomson scattering can be regarded as the high-wavelength limit of the Compton interaction.

Using a classical electrodynamic treatment to describe X-rays interaction in matter, Thomson assumed that the incident radiation sets the quasi-free electron into a forced resonant oscillation. The differential scattering cross section can be introduced in classical electrodynamics (see for instance [Jackson (1975)]) as:

$$\left(\frac{d\sigma}{d\Omega}\right)_{Th} = \frac{Radiated\ energy/unit\ of\ time/unit\ of\ solid\ angle}{Incident\ energy\ flux\ in\ energy/unit\ of\ time/unit\ of\ area}.$$

The incident energy flux is the Poynting vector time-average for the incoming plane wave, i.e.,

$$S_{in} = \frac{c}{8\pi} E_i^2 \ \mathrm{erg\,cm^{-2}s^{-1}},$$

where $\vec{E}_i$ is the field amplitude of the plane wave. The energy flux, emitted at an angle θ_ν by an electric dipole set into a constant oscillation with a dipole field strength, is given by $S_o = \frac{c}{4\pi} E_o^2$, where $\vec{E}_o$ is the dipole field strength at a distance r. For an unpolarized incident wave with a field amplitude $\vec{E}_i$, S_o expressed in $\mathrm{erg\,cm^{-2}s^{-1}}$ can be written as:

$$S_o = \frac{e^4}{16\,\pi m^2 c^3 r^2} E_i^2 (1 + \cos^2\theta_\nu)$$

where m is the rest mass of the electron. Moreover, the energy per unit time passing through an area dA normal to the direction of $\vec{r}$ and corresponding to a solid angle $d\Omega = dA/r^2$ is given by:

$$S_o\, dA = \frac{e^4}{16\pi m^2 c^3 r^2} E_i^2 (1 + \cos^2\theta_\nu)(r^2 d\Omega)$$

$$= \frac{e^4}{16\pi m^2 c^3} E_i^2 (1 + \cos^2\theta_\nu)\, d\Omega.$$

---

*The reader can see Eq. (2.181) and the discussion at page 150.

Thus, the Thomson differential cross section [$\left(\frac{d\sigma}{d\Omega}\right)_{Th}$ in cm^2sr^{-1}/electron] becomes:

$$
\left(\frac{d\sigma}{d\Omega}\right)_{Th} d\Omega = \frac{S_o \, dA}{S_{in}}
$$

$$
= \frac{\left[e^4/(16\pi m^2 c^3)\right] E_i^2 (1 + \cos^2 \theta_\nu)}{[c/(8\pi)] \, E_i^2} d\Omega
$$

$$
= \frac{1}{2} r_e^2 (1 + \cos^2 \theta_\nu) \, d\Omega, \tag{2.195}
$$

where $r_e = e^2/(mc^2)$.

The total Thomson cross section can be calculated by integrating the differential cross section [i.e., Eq. (2.195)] over the full solid angle:

$$
\sigma_{Th} = \int \left(\frac{d\sigma}{d\Omega}\right)_{Th} d\Omega
$$

$$
= \frac{1}{2} r_e^2 \int (1 + \cos^2 \theta_\nu) \, d\Omega
$$

$$
= \pi r_e^2 \int_0^\pi (1 + \cos^2 \theta_\nu) \sin \theta_\nu \, d\theta_\nu
$$

$$
= \frac{8}{3} \pi \left(\frac{e^2}{mc^2}\right)^2 \tag{2.196}
$$

$$
= \frac{8}{3} \pi r_e^2 \quad [cm^2/\text{electron}]. \tag{2.197}
$$

In the Thomson scattering, there is no energy dependent term and no change in wavelength for the emitted radiation. The total Thomson cross section per atom ($_a\sigma_{Th}$), assuming that independent interactions occur with each of the Z quasi-free electrons in the atom, is given by:

$$
_a\sigma_{Th} = Z\sigma_{Th} = \frac{8}{3} \pi r_e^2 Z \quad [cm^2/\text{atom}]. \tag{2.198}
$$

It has to be noted that the classical *Thomson formula* is valid only at low frequencies, where quantum-mechanical effects can be neglected.

2.3.2.4 *Radiative Corrections and Double Compton Effect*

The Compton interaction, as described by the Klein–Nishina theory, has to be extended both to the case in which the emission and the reabsorption of virtual photons occurs (i.e., the so-called *radiative correction* of order $\alpha = 1/137$) and to the case in which an additional real photon (usually at low energy) is emitted (i.e., the so-called *double Compton scattering*). Calculations treating both effects can be found for instance in [Mork (1971)]. Once these contributions are both added to the Klein–Nishina cross section, but without taking into account either the electron binding energy effect or the Rayleigh scattering, the total Compton cross section on free electron becomes

$$
\sigma_C = \sigma_{C,e} + \triangle\sigma_C, \tag{2.199}
$$

where $\triangle\sigma_C$ is the cross section variation due to both radiative corrections and the double Compton effect, $\sigma_{C,e}$ is the Klein–Nishina cross section on free electron as given by Eq. (2.187) or equivalently by Eq. (2.188). These corrections are negligible for photons with energies below 100 keV, and amount to $\approx 0.25\%$ and 1% of σ_C at 4 and 100 MeV, respectively. At photon energies of about 1 GeV, $\triangle\sigma_C$ accounts for $\approx 5\%$ of σ_C.

The Compton attenuation coefficients have to be calculated including coherent effects, i.e., those due to the Rayleigh scattering and to electron binding corrections to the free electron cross section, as previously discussed (Sect. 2.3.2.2). However, these effects do not modify the Compton cross section by more than a few percents above ≈ 100 keV. In order to compute the linear and mass Compton attenuation coefficients without coherence effects, we have to use the Compton cross section on atom, which is Z times larger than the cross section σ_C on an atomic electron. The atomic cross section is given by $Z\sigma_C$, where σ_C is calculated using Eq. (2.199) and the linear attenuation coefficient is given by:

$$\mu_{\text{att},l,C}^{\text{nocoh}} = \rho N \frac{Z}{A} \sigma_C \ [\text{cm}^{-1}].$$

The corresponding mass attenuation coefficient becomes

$$\mu_{\text{att},m,C}^{\text{nocoh}} = N \frac{Z}{A} \sigma_C \ [\text{g}^{-1}\text{cm}^2].$$

Furthermore, we can define the linear and mass attenuation coefficients related to the *corrected incoherent Compton cross section* σ_{incoh}, *which accounts for the electron binding corrections of the Z atomic electrons, the radiative corrections and the double Compton effect.* The corrected incoherent Compton cross section per electron is

$$\sigma_{\text{incoh}} = \sigma_C^B + \triangle\sigma_C, \tag{2.200}$$

where σ_C^B is given by Eq. (2.191).

It has to be noted that Z/A is ≈ 0.4–0.5, above $Z = 1$. Thus, the *Compton mass attenuation coefficient is almost independent of the medium* above ≈ 100 keV (see also page 159).

2.3.3 *Pair Production*

As the incoming photon energy exceeds twice the energy corresponding to the electron rest mass, i.e., $2mc^2 \simeq 1.02$ MeV, the production of an electron and positron pair becomes possible (see Fig. 2.60). The process of *pair production* can only occur close to a charged massive object (for instance a nucleus) which takes away the amount of momentum needed to preserve momentum conservation, during the interaction with the Coulomb field of the massive object itself. Furthermore, in the framework of the Dirac theory, this process is intimately related to the bremsstrahlung process.

The photon threshold energy, $E_{\gamma,th}$, depends on the mass of the particle, whose Coulomb field allows the pair production process to happen. The value of $E_{\gamma,th}$ needed to create an electron and positron pair of mass $2m$ can be computed considering that, for a target of mass M at rest in the laboratory system, the minimal final invariant mass is $M_{1,2} = M + 2m$. From Eq. (1.33), we have

$$M_{1,2}^2 = M^2 + \frac{2}{c^4}(E_{\gamma,th}Mc^2) = M^2 + 4mM + 4m^2,$$

from which we obtain:

$$E_{\gamma,th} = 2mc^2 \left(1 + \frac{m}{M}\right). \tag{2.201}$$

Therefore, when the pair production occurs in the proximity of a nucleus the threshold energy is $E_{\gamma,th} \simeq 2m$, while in proximity of an electron $E_{\gamma,th} \simeq 4m$. In addition to the different threshold energy, the pair production in the field of a nucleus exhibits two observable tracks in tracking detectors. Three tracks can be observed for the pair production in the electron field, because of the presence of the fast recoiling electron. The pair production in the field of an electron is sometime referred to as the *triplet production*.

The pair production process becomes the dominant mechanism for photon interactions in matter above $\mathcal{E} \approx 10$ and accounts for almost the whole γ-ray absorption in this energy range.

2.3.3.1 *Pair Production in the Field of a Nucleus*

The quantum-mechanical treatment of pair production in the electron field can be simplified using the Born approximation, as derived by Heitler (1954) (see also references therein), and applies for:

$$\frac{2\pi Z e^2}{\hbar v_\pm} = \frac{2\pi Z \alpha}{\beta_\pm} \ll 1,$$

where $\alpha = e^2/(\hbar c)$ is the fine structure constant and v_+, v_- are the velocities of the positron and electron, respectively.

Similarly to what was discussed in Sect. 2.1.7 regarding the bremsstrahlung emission, the probability of pair creation depends on the effective distance between the incoming photon and the nucleus. The screening can be neglected if energies of both positron and electron are not too high, i.e., if

$$2\alpha Z^{1/3} \frac{E_+ E_-}{E(mc^2)} \ll 1, \tag{2.202}$$

where E_+, E_- are the total energies of the positron and electron, respectively, and E is the incoming photon energy.

The *effective distance* from the nucleus determines the extent to which the electric field of the nucleus is screened by outer electrons. The influence of screening is given by the quantity η, the so-called *screening parameter*:

$$\eta = 100 \, \frac{mc^2}{E} \frac{1}{f_+(1 - f_+)} Z^{-1/3}, \tag{2.203}$$

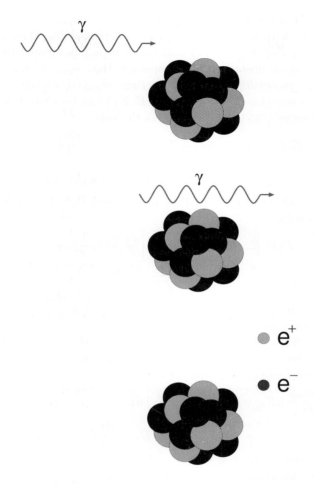

Fig. 2.60   An incoming photon interacts with the Coulomb field of a massive nucleus. Finally, an electron and positron pair emerges.

where f_+ and $(1 - f_+) = f_-$ are the fractional energies carried by the positron and the electron, respectively: $f_+ = E_+/E$, $f_- = E_-/E$. As already mentioned, the nuclear recoil preserves momentum conservation. In addition, to a first approximation, we have $E_+ + E_- = E$. Thus, the positron energy, as well as the electron energy, varies between mc^2 and $E - mc^2$. Let us now rewrite Eq. (2.203) for the parameter η as

$$\frac{1}{\eta} = \frac{E}{mc^2} f_+ (1 - f_+) \frac{Z^{1/3}}{100} = \frac{E_+ E_-}{E(mc^2)} \frac{Z^{1/3}}{100},$$

and, when the screening is negligible [Eq. (2.202)], we obtain:

$$\frac{1}{\eta} = \frac{E_+ E_-}{E(mc^2)} \frac{Z^{1/3}}{100} < 2\,\alpha Z^{1/3} \frac{E_+ E_-}{E(mc^2)} \ll 1.$$

Therefore, η is $\gg 1$ when screening is negligible.

Let us indicate with $\Phi_{\text{pair}}(E, E_+) \, dE_+ \, dx$ the probability that a positron (of energy between E_+ and $E_+ + dE_+$) and electron pair is created by a photon of energy E traversing an absorber of thickness dx. Here $\Phi_{\text{pair}}(E, E_+)$ is referred to as the *differential probability of pair production*: $\Phi_{\text{pair}}(E, E_+) \, dE_+$ is in units of cm^{-1}. For photon energies $E \gg mc^2$, $\Phi_{\text{pair}}(E, E_+)$ can be written as ([Bethe and Heitler (1934); Heitler (1954)] and references therein)

$$\Phi_{\text{pair}}(E, E_+) \, dE_+ = 4\alpha \frac{N\rho}{A} Z^2 r_e^2 P(E, E_+, \eta) \frac{dE_+}{E}$$

$$= 4\bar{\Phi}(Z)_{\text{pair}} \frac{N\rho}{A} P(E, E_+, \eta) \frac{dE_+}{E} \quad [\text{cm}^{-1}], \quad (2.204)$$

where $\bar{\Phi}(Z)_{\text{pair}} = \alpha Z^2 r_e^2 \simeq 5.8 \times 10^{-28} \times Z^2 \, \text{cm}^2$ and $P(E, E_+, \eta)$ depends on the screening parameter η:

$$P(E, E_+, \eta) = \frac{E_+^2 + E_-^2}{E^2} \left[\frac{\varphi_1(\eta)}{4} - \frac{1}{3} \ln Z \right]$$

$$+ \frac{2}{3} \frac{E_+ E_-}{E^2} \left[\frac{\varphi_2(\eta)}{4} - \frac{1}{3} \ln Z \right]. \quad (2.205)$$

The functions $\varphi_1(\eta)$ and $\varphi_2(\eta)$ are shown in Fig. 2.27 for $0 \leq \eta \leq 2$. Therefore, the differential cross section $d\sigma_{\text{pair}}(E, E_+, \eta)/dE_+$ to create a positron (with energy between E_+ and $E_+ + dE_+$) and an electron pair in the nuclear field is obtained dividing the differential probability of pair production [e.g., Eq. (2.204)] by n_A ($= \rho N/A$), i.e., the number of atoms per cm^3:

$$\frac{d\sigma_{\text{pair}}}{dE_+} \, dE_+ = \frac{1}{n_A} \Phi_{\text{pair}}(E, E_+) \, dE_+$$

$$= 4\bar{\Phi}(Z)_{\text{pair}} P(E, E_+, \eta) \frac{dE_+}{E} \quad [\text{cm}^2/\text{atom}], \quad (2.206)$$

where E is the incoming photon energy.

For $\eta \approx 0$, i.e., for **complete screening**, we have

$$\varphi_1(0) = 4 \ln 183 \quad \text{and} \quad \varphi_2(0) = 4 \ln 183 - \frac{2}{3}$$

(see page 97); as a consequence, Eq. (2.204) becomes:

$$\Phi_{\text{pair}}(E, E_+) \, dE_+$$

$$= 4\bar{\Phi}(Z)_{\text{pair}} \frac{N\rho}{A} P(E, E_+, 0) \frac{dE_+}{E}$$

$$= 4\bar{\Phi}(Z)_{\text{pair}} \frac{N\rho}{A}$$

$$\times \left[\left(\frac{E_+^2 + E_-^2}{E^2} + \frac{2}{3} \frac{E_+ E_-}{E^2} \right) \ln \left[183 \, Z^{-1/3} \right] - \frac{1}{9} \frac{E_+ E_-}{E^2} \right] \frac{dE_+}{E}$$

$$= 4\bar{\Phi}(Z)_{\text{pair}} \frac{N\rho}{A}$$

$$\times \left\{ \left[E_+^2 + E_-^2 + \frac{2}{3} E_+ E_- \right] \ln \left[183 \, Z^{-1/3} \right] - \frac{1}{9} E_+ E_- \right\} \frac{dE_+}{E^3}. \quad (2.207)$$

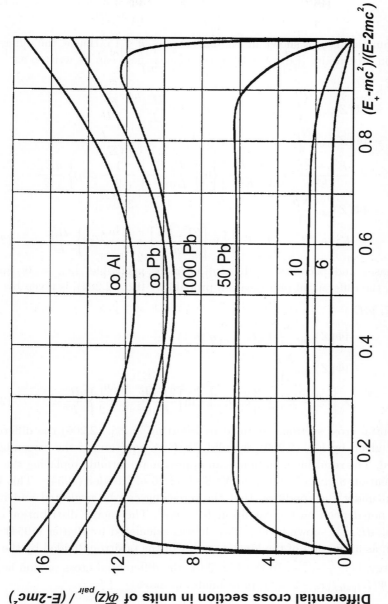

Fig. 2.61   Energy distribution of electron and positron pairs by means of Eq. (2.206), for reduced photon energies $\mathcal{E}$ of 6, 10, 50, 1000 and ∞ [Figure 16 (p. 261) from The Quantum Theory of Radiation 3rd Edition (1954) by Heitler, W., by permission of Oxford University Press], as calculated by Heitler (1954). In ordinate, the differential cross section $\frac{d\sigma_{\text{pair}}}{dE_+}$ for the creation of a positron with energy between E_+ and $E_+ + dE_+$ [Eq. (2.206)] is given in units of $\frac{\bar{\Phi}(Z)_{\text{pair}}}{E-2mc^2}$. In abscissa, the kinetic energy of the positron is divided by the total kinetic energy, i.e., $\frac{E_+-mc^2}{E-2mc^2}$.

For larger values of η (i.e., for $\eta > 2$), the effect of screening decreases. We have $\varphi_1(\eta) \simeq \varphi_2(\eta) \simeq \varphi(\eta)$ (see page 98):

$$\frac{\varphi(\eta)}{4} - \frac{1}{3}\ln Z = \ln\left[\frac{2E_+E_-}{Emc^2}\right] - \frac{1}{2} - c(\eta),$$

where the function $c(\eta)$ is shown in Fig. 2.28 for $2 \leq \eta \leq 15$. Hence, the differential pair creation probability [Eq. (2.204)] in the nuclear field can be written as:

$$\Phi_{\mathrm{pair}}(E, E_+)\, dE_+$$

$$= 4\bar{\Phi}(Z)_{\mathrm{pair}}\frac{N\rho}{A}P(E, E_+, 2 \leq \eta \leq 15)\frac{dE_+}{E}$$

$$= 4\bar{\Phi}(Z)_{\mathrm{pair}}\frac{N\rho}{A}$$

$$\times \left[\frac{E_+^2 + E_-^2}{E^2} + \frac{2}{3}\frac{E_+E_-}{E^2}\right]\left\{\ln\left[\frac{2E_+E_-}{Emc^2}\right] - \frac{1}{2} - c(\eta)\right\}\frac{dE_+}{E}$$

$$= 4\bar{\Phi}(Z)_{\mathrm{pair}}\frac{N\rho}{A}$$

$$\times \left[E_+^2 + E_-^2 + \frac{2}{3}E_+E_-\right]\left\{\ln\left[\frac{2E_+E_-}{Emc^2}\right] - \frac{1}{2} - c(\eta)\right\}\frac{dE_+}{E^3}. \tag{2.208}$$

As η increases and becomes $\gg 1$, $c(\eta)$ becomes negligible, i.e., ≈ 0. For **no screening**, the differential pair creation probability [Eq. (2.204)] is given by:

$$\Phi_{\mathrm{pair}}(E, E_+)dE_+$$

$$= 4\bar{\Phi}(Z)_{\mathrm{pair}}\frac{N\rho}{A}P(E, E_+, \eta \gg 1)\frac{dE_+}{E}$$

$$= 4\bar{\Phi}(Z)_{\mathrm{pair}}\frac{N\rho}{A}$$

$$\times \left[E_+^2 + E_-^2 + \frac{2}{3}E_+E_-\right]\left\{\ln\left[\frac{2E_+E_-}{Emc^2}\right] - \frac{1}{2}\right\}\frac{dE_+}{E^3}. \tag{2.209}$$

The differential cross section, obtained by inserting in Eq. (2.206) the differential probability given by Eq. (2.209), is valid for $E \gg mc^2$ and if the screening can be neglected. The resulting equation is an approximate formula replacing the more general Equation 8 given in Chapter V, Section 26 of [Heitler (1954)]. This latter was derived under the condition that the screening can be neglected and its validity does not depend on the condition $E \gg mc^2$. The energy distribution of the pairs, $d\sigma_{\mathrm{pair}}/dE_+$, is shown in Fig. 2.61. It was calculated by Heitler (1954) using Eq. (2.206), as a function of the kinetic energy of the positron divided by the total kinetic energy, i.e., $\frac{E_+ - mc^2}{E - 2mc^2}$. In the Fig. 2.61, the differential cross section is given in units of $\bar{\Phi}(Z)_{\mathrm{pair}}/(E - 2mc^2)$. In ordinate, we have

$$\left[\frac{d\sigma_{\mathrm{pair}}}{dE_+}\right]\left[\frac{E - 2mc^2}{\bar{\Phi}(Z)_{\mathrm{pair}}}\right].$$

Thus, the area under the curve is again related to the total cross section. The curves are for reduced photon energies $\mathcal{E}$ of 6, 10, 50, 1000 and ∞ [Heitler (1954)]. The calculations for small incoming energies ($\mathcal{E} = 6$ and $\mathcal{E} = 10$) are valid for any element (the screening is neglected). The other curves are calculated for lead, and for $\mathcal{E} = \infty$ in the case of lead and aluminum. It has to be noted that the screening is effective if both the resulting electron and positron are with energies large compared with mc^2. Furthermore, at very high-energy, usually the screening is almost complete, i.e.,

$$2\frac{E_+ E_-}{E(mc^2)} \gg \alpha^{-1} Z^{-1/3}.$$

Therefore, as it can be seen from Eq. (2.207), the energy distribution depends on the atomic number Z. At low photon energies, the energy distribution shows a broad flat maximum. On the contrary at very high energies, the curves show a broad minimum for equal positron and electron energies, and a small maximum when one of the particle receives the whole available energy. However, the nuclear repulsion, if taken into account, should be favoring a slight asymmetry to produce more high-energy positron.

Hough (1948a) provided a convenient approximate formula for the energy distribution without screening for reduced photon energies $2 \leq \mathcal{E} \leq 15$, i.e., in units of cm^2/atom

$$\frac{d\sigma_{\text{pair}}}{dE_+} dE_+ = z_H \left(\frac{d\sigma_{\text{pair}}}{dE_+}\right)_m dE_+$$

$$\times \left\{1 + 0.135 \left[\frac{\left(\frac{d\sigma_{\text{pair}}}{dE_+}\right)_m dE_+}{\bar{\Phi}(Z)_{\text{pair}}} - 0.52\right] z_H (1 - z_H^2)\right\}, \quad (2.210)$$

where

$$z_H = 2\sqrt{\frac{E_+ - mc^2}{E - 2mc^2}\left(1 - \frac{E_+ - mc^2}{E - 2mc^2}\right)},$$

and $\left(\frac{d\sigma_{\text{pair}}}{dE_+}\right)_m dE_+$ is the value of the differential cross section for equal partition of energy ($E_+ = E_- = E/2$), which is obtained from Equation 8 in Chapter V, Section 26 of [Heitler (1954)] and is approximated by [Hough (1948a)]:

$$\left(\frac{d\sigma_{\text{pair}}}{dE_+}\right)_m dE_+ = \bar{\Phi}(Z)_{\text{pair}} (1 - a_H)$$

$$\times \left[\frac{1}{3}\left(4 - a_H^2\right)(b_H - 1) - a_H^2 c_H (c_H - 1) - a_H^4 c_H (b_H - c_H)\right], \quad (2.211)$$

where

$$a_H = \frac{2}{\mathcal{E}}$$

$$b_H = \left[\frac{2}{(1 - a_H^2)}\right] \ln \frac{\mathcal{E}}{2}$$

$$c_H = \frac{1}{\sqrt{1 - a_H^2}} \ln \left[\frac{\mathcal{E}}{2} + \sqrt{\left(\frac{\mathcal{E}}{2}\right)^2 - 1}\right].$$

The second term in the bracket of Eq. (2.210) is to be dropped when it becomes negative (below $\mathcal{E} \approx 4.2$). The differential cross section for equal partition of energy, normalized to $\bar{\Phi}(Z)_{\text{pair}}$ and calculated using Eq. (2.211), is shown in Fig. 2.62 as a function of the reduced photon energy $\mathcal{E}$. While, the values of

$$\left[\frac{d\sigma_{\text{pair}}}{dE_+} dE_+\right] \left[\frac{E - 2mc^2}{\bar{\Phi}(Z)_{\text{pair}}}\right]$$

are shown in Fig. 2.63 as a function of $\frac{E_+ - mc^2}{E - 2mc^2}$; as in Fig. 2.61, dE_+ is replaced by

$$d\left(\frac{E_+ - mc^2}{E - 2mc^2}\right) = \frac{d\left(E_+ - mc^2\right)}{E - 2mc^2}.$$

Thus, the area under the curve is again related to the total cross section. For the energy range $10 < \mathcal{E} < 15$, the differential cross section computed by means of Eq. (2.210) is in overlap with the result obtained using Eq. (2.209). The most suited approximate formulae for selected photon energies can be found in [Hough (1948a)].

For photon energies such that

$$mc^2 \ll E \ll \alpha^{-1}Z^{-1/3}mc^2,$$

the **screening** can be usually **neglected**. The total cross section for the pair production in the nuclear field can be computed by taking into account Eq. (2.209) and integrating under the condition $E \gg mc^2$; one finds:

$$\begin{aligned}
\sigma_{\text{pair},ns} &= \int_{mc^2}^{E-mc^2} \frac{d\sigma_{\text{pair}}}{dE_+} dE_+ \\
&= \int_{mc^2}^{E-mc^2} 4\bar{\Phi}(Z)_{\text{pair}} P(E, E_+, \eta \gg 1) \frac{dE_+}{E} \\
&\simeq 4\bar{\Phi}(Z)_{\text{pair}} \int_{mc^2}^{E} \left[E_+^2 + E_-^2 + \frac{2}{3}E_+ E_-\right] \ln\left[\frac{2E_+ E_-}{Emc^2}\right] \frac{dE_+}{E^3} \\
&\quad - 2\,\bar{\Phi}(Z)_{\text{pair}} \int_0^{E} \left[E_+^2 + E_-^2 + \frac{2}{3}E_+ E_-\right] \frac{dE_+}{E^3} \\
&= \bar{\Phi}(Z)_{\text{pair}} \left[\frac{28}{9} \ln(2\,\mathcal{E}) - \frac{218}{27}\right] \; [\text{cm}^2/\text{atom}].
\end{aligned} \tag{2.212}$$

The result depends on the reduced photon energy $\mathcal{E}$, but the term in bracket does not depend on Z. However, Hough (1948a) has noted that Eq. (2.212) is not sufficiently well approximated and a lower power of $\mathcal{E}$ must be added to the cross section:

$$\sigma_{\text{pair},ns}^{H} = \bar{\Phi}(Z)_{\text{pair}} \left[\frac{28}{9} \ln(2\,\mathcal{E}) - \frac{218}{27} + \frac{6.45}{\mathcal{E}}\right] \; [\text{cm}^2/\text{atom}]. \tag{2.213}$$

For photon energies such that

$$E \gg \alpha^{-1}Z^{-1/3}mc^2 \; (\text{i.e. } \mathcal{E} \gg \alpha^{-1}Z^{-1/3}),$$

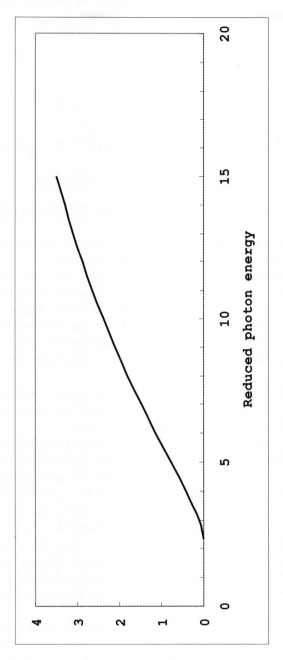

Fig. 2.62   In ordinate $\left[\left(\frac{d\sigma_{\text{pair}}}{dE_+} \right)_m \; dE_+ \right] / \bar{\Phi}(Z)_{\text{pair}}$ calculated using Eq. (2.211) as a function of the reduced photon energy $\mathcal{E}$.

the screening is complete. For **complete screening**, the total cross section for pair production in the nuclear field can be computed by means of Eq. (2.207) and integrating under the condition $E \gg mc^2$; one obtains:

$$
\begin{aligned}
\sigma_{\text{pair},cs} &= \int_{mc^2}^{E-mc^2} \frac{d\sigma_{\text{pair}}}{dE_+} \, dE_+ \\
&= \int_{mc^2}^{E-mc^2} 4\bar{\Phi}(Z)_{\text{pair}} P(E, E_+, 0) \, \frac{dE_+}{E} \\
&\simeq 4\bar{\Phi}(Z)_{\text{pair}} \\
&\quad \times \int_0^E \left\{ \left[E_+^2 + E_-^2 + \frac{2}{3} E_+ E_- \right] \ln\left(183 \, Z^{-1/3} \right) - \frac{1}{9} E_+ E_- \right\} \frac{dE_+}{E^3} \\
&= \bar{\Phi}(Z)_{\text{pair}} \left[\frac{28}{9} \ln\left(183 \, Z^{-1/3} \right) - \frac{2}{27} \right] \quad [\text{cm}^2/\text{atom}]. \quad (2.214)
\end{aligned}
$$

The result is independent of the photon energy, but the term in bracket depends on Z.

The cross sections calculated by means of Eqs. (2.212, 2.213, 2.214) and divided by $\bar{\Phi}(Z)_{\text{pair}}$ are shown in Fig. 2.64 for carbon and aluminum.

In general, deviations from the equations given in this section are expected. These formulae are less accurate for heavy elements, because they involve the Born approximation. For instance, the *Bethe–Maximon theory* [Davies et al. (1954)], which avoids the Born approximation, shows that the cross section for lead is reduced by $\approx 12\%$ at 88 MeV. Tsai (1974) has provided a detailed review on pair production based on QED calculations and with atomic form factors taken into account.

The cross section for pair production rises monotonically and varies approximately linearly with E until an almost constant value is reached above 50 MeV for high-Z materials and at higher energies for low-Z absorbers.

A compilation of cross sections was given by White-Grodstein (1957). In the energy region $E < 5$ MeV, where the screening can be neglected, cross sections were calculated by means of Eq. (2.210) and treated with the inclusion of the *Jaeger–Hulme corrections* [Jaeger and Hulme (1936)]. For $E > 5$ MeV, the corrections involve *Davies–Bethe–Maximon calculations* [Davies, Bethe and Maximon (1954)] and a semi-empirical formula in order to take into account the experimental data. A later compilation made use of empirical corrections to cover the whole range from threshold to high energy for the pair production cross section on the nuclear field $\sigma_{\text{pair},n}$ [Hubbell (1969)] (see also [Hubbell, Gimm and Øverbø (1980)]). This latter is computed taking into account the Coulomb, screening and radiative corrections, described below.

Disregarding screening and radiative corrections, the Coulomb interaction on the nuclear field is described by an expression which gives rise to a perturbation series, the first term of which is the Bethe–Heitler unscreened cross section ($\sigma_{B,ns}$) in the Born approximation. Under Maximon's approximation [Maximon (1968); Hubbell

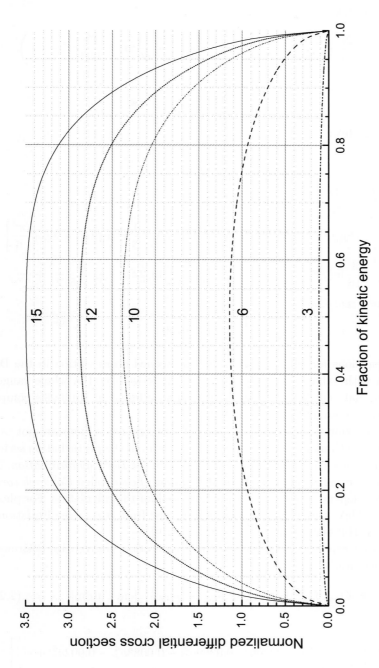

Fraction of kinetic energy

Fig. 2.63 Energy distribution of electron and positron pairs from Eq. (2.210) for the creation of a positron of energy between E_+ and $E_+ + dE_+$ and for reduced photon energies $\mathcal{E} = 3$, 6, 10, 12, and 15: in ordinate $\left[\frac{d\sigma_{\mathrm{pair}}}{dE_+}\, dE_+\right]\left[(E - 2mc^2)/\bar{\Phi}(Z)_{\mathrm{pair}}\right]$ is the normalized differential cross section [i.e., the differential cross section divided by $\bar{\Phi}(Z)_{\mathrm{pair}}/(E - 2mc^2)$], in abscissa we have the kinetic energy of the positron divided by the total kinetic energy, i.e., $\frac{E_+ - mc^2}{E - 2mc^2}$.

(1969)] $\sigma_{B,ns}$ (in units of cm^2/atom) is given by

(a) for $E \leq 2$ MeV:

$$\frac{\sigma_{B,ns}}{\bar{\Phi}(Z)_{\text{pair}}} = \frac{2\pi}{3}\left(\frac{\mathcal{E}-2}{\mathcal{E}}\right)^3\left(1 + \frac{1}{2}\varpi + \frac{23}{40}\varpi^2 + \frac{11}{60}\varpi^3 + \frac{29}{960}\varpi^4 + \cdots\right)$$

with

$$\varpi = \frac{2\mathcal{E}-4}{2+\mathcal{E}+2\sqrt{2\mathcal{E}}},$$

(b) for $E > 2$ MeV:

$$
\begin{aligned}
\frac{\sigma_{B,ns}}{\bar{\Phi}(Z)_{\text{pair}}} = {}& \frac{28}{9}\ln(2\mathcal{E}) - \frac{218}{27} \\
& + \left(\frac{2}{\mathcal{E}}\right)^2\left[6\ln(2\mathcal{E}) - \frac{7}{2} + \frac{2}{3}\ln^3(2\mathcal{E}) - \ln^2(2\mathcal{E}) - \frac{1}{3}\pi^2\ln(2\mathcal{E}) + 2\mathbb{Z} + \frac{\pi^2}{6}\right] \\
& - \left(\frac{2}{\mathcal{E}}\right)^4\left[\frac{3}{16}\ln(2\mathcal{E}) + \frac{1}{8}\right] - \left(\frac{2}{\mathcal{E}}\right)^6\left[\frac{29}{9\times 256}\ln(2\mathcal{E}) - \frac{77}{27\times 512}\right] + \cdots,
\end{aligned}
$$

where *Riemann's zeta-function* is

$$\mathbb{Z} = \sum_n \frac{1}{n^2} = 1.2020569\ldots.$$

These two expansions for $\sigma_{B,ns}$ provide the unscreened cross section in the Born approximation with values accurate within 0.01 percent over the whole range of energies. Additional and detailed formulations can be found in Hubbell's compilation in [Hubbell, Gimm and Øverbø (1980)].

As already mentioned, if both screening and radiative corrections are not taken into account for the Coulomb interaction, the first term of the perturbation series is the Bethe–Heitler *unscreened cross section* ($\sigma_{B,ns}$) *in the Born approximation*. The sum of high order terms of this Born series determines the so-called *Coulomb correction* Δ_B^i, where i depends on the approximation used for the calculation. For photon energies above 5 MeV, this term has to be subtracted from $\sigma_{B,ns}$ (see Equation 42 in Section 2.4 of [Hubbell, Gimm and Øverbø (1980)]).

The *Sörenssen expression* for the Coulomb correction is given by ([Sörenssen (1965)] and references therein):

$$\Delta_B^S = 4\bar{\Phi}(Z)_{\text{pair}} f(Z)\left(\frac{7}{9} - \frac{2}{\mathcal{E}} + \frac{4}{3\mathcal{E}^2} - \frac{8}{9\mathcal{E}^3}\right) \quad [\text{cm}^2/\text{atom}], \tag{2.215}$$

with

$$f(Z) = a_B^2\left[\frac{1}{(1+a_B^2)} + 0.20206 - 0.0369a_B^2 + 0.0083a_B^4 - 0.002a_B^6 + \cdots\right]$$

and $a_B = \alpha Z$.

For large $\mathcal{E}$ values, Eq. (2.215) becomes the expression for the Coulomb correction as calculated by Davies–Bethe–Maximon (1954), i.e.,

$$\Delta_B^{DBM} = \frac{28}{9}\bar{\Phi}(Z)_{\text{pair}}f(Z) \ [\text{cm}^2/\text{atom}].$$

However, in particular for heavy elements, these corrections are not sufficiently accurate for energies much lower than $\approx 100\,\text{MeV}$.

For low energies, exact calculations of the Coulomb correction were carried out by Øverbø, Mork and Olsen (1968), who have provided a low energy unscreened cross section σ^{OMO} well adequate in the energy region $\leq 5\,\text{MeV}$. In general, at energies lower than $\approx 100\,\text{MeV}$, the Coulomb correction was evaluated by Øverbø [Øverbø, Mork and Olsen (1968); Øverbø (1977)] using an expression containing the terms from Eq. (2.215):

$$\Delta_B^O = \Delta_B^S - \frac{\bar{\Phi}(Z)_{\text{pair}}}{\mathcal{E}}\left[0_1 \ln^2 \frac{\mathcal{E}}{2} + 0_2 \ln \frac{\mathcal{E}}{2} + 0_3\left(1 - \frac{2}{\mathcal{E}}\right)\right]$$
$$- \frac{\bar{\Phi}(Z)_{\text{pair}}}{\mathcal{E}^2}\left[0_4 \ln^3 \frac{\mathcal{E}}{2} + 0_5 \ln^2 \frac{\mathcal{E}}{2} + 0_6\left(1 - \frac{2}{\mathcal{E}}\right)\right] \ [\text{cm}^2/\text{atom}], \quad (2.216)$$

where

$$0_1 = a_B^2(-6.366 + 4.14\,a_B^2),$$
$$0_2 = a_B^2(54.039 - 43.126\,a_B^2 + 11.264\,a_B^4),$$
$$0_3 = a_B^2(-52.423 + 49.615\,a_B^2 - 14.082\,a_B^4),$$
$$0_4 = 10.938\,a_B^2\left(1 - \frac{a_B^2}{0.324}\right),$$
$$0_5 = -12.705\,a_B^2\left(1 - \frac{a_B^2}{0.324}\right),$$
$$0_6 = 9.903\,a_B^2\left(1 - \frac{a_B^2}{0.324}\right),$$

in which $a_B = \alpha Z$. Equation (2.216) contains the Davies–Bethe–Maximon correction as leading term. The largest error, occurring within the 5 to 50 MeV energy region, is of the order of a few tenths of a % of the total cross section ([Hubbell, Gimm and Øverbø (1980)] and references therein). However, for reduced energies $3.5 < \mathcal{E} < 10$, the errors are of the order of 0.1%.

An additional correction is that associated with the emission and reabsorption of virtual photons and with the emission of both soft and hard photons. The numerical values of the radiative correction were tabulated by Mork and Olsen (1965). At high reduced energies, $\mathcal{E} > 1000$, and for complete screening we have:

$$\Delta_{\text{rad.corr}} = (0.93 \pm 0.05)\,\% \,.$$

For the cross section, the radiative correction is given by the multiplicative term $(1 + \Delta_{\text{rad.corr}})$ (see Equation 42 in Section 2.4 of [Hubbell, Gimm and Øverbø (1980)]).

The screened cross section ($\sigma_{B,s}$) in the Born approximation was worked out in the high-energy approximation by Bethe and Heitler (1934). We indicate by $\Delta_{B,\mathrm{scr}}$ the difference between the unscreened cross section $\sigma_{B,ns}$ and the screened cross section $\sigma_{B,s}$. However, Tseng and Pratt (1972) demonstrated that the screening correction $\Delta_{B,\mathrm{scr}}$ is not adequately described in the Born approximation at low energy; in addition, they provided a table for the screening correction for $\mathcal{E} < 10$. A more complete set of screening corrections, also valid at higher energies, is given by Øverbø (1979) and it is discussed in detail in [Hubbell, Gimm and Øverbø (1980)].

2.3.3.2   *Pair Production in the Electron Field*

As in the case of bremsstrahlung, the pair production can occur in the Coulomb field of atomic electrons. The recoiling electron can receive a sufficient amount of momentum to emerge as an additional (to the produced positron and electron pair) fast-electron from the interaction. As previously mentioned, this phenomenon is referred to as the *triplet production*, due to three emerging fast-particles.

To a first approximation (see [Bethe and Ashkin (1953)] and references therein), the distribution of recoil momenta, q, of the atomic electron is almost the same as it would be for the nucleus if the pair were produced in the nuclear field. The recoil momentum distribution is given by [Bethe and Ashkin (1953)]

$$P(q)\,dq = \frac{1}{A_q}\frac{mc^2}{mc^2 + q^2}\frac{dq}{q},$$

provided that $q > q_{\min}$, where

$$A_q = \ln\left(\frac{mc}{q_{\min}}\right)$$

and $q_{\min}$ is the largest between

$$q_{\min} = \frac{mc}{\mathcal{E}}\ \text{ or }\ q_{\min} = mc\,\frac{Z^{1/3}}{183}.$$

The momentum distribution is almost uniform in a logarithmic scale. Thus, many recoiling electrons have low energies.

The theory without screening correction is based on *Borsellino–Ghizzetti's calculations* of [Borsellino (1947)] (see also [Hubbell (1969); Hubbell, Gimm and Øverbø (1980)] and references therein). *Ghizzetti's expansion* (here shown up to the first leading term) gives the cross section, σ_e^{BG}, for pair production in the field of an atomic electron **without screening**:

$$\sigma_e^{BG} = \frac{\bar{\Phi}(Z)_{\mathrm{pair}}}{Z^2}\left[\frac{28}{9}\ln(2\,\mathcal{E}) - \frac{218}{27} - \frac{1}{\mathcal{E}}\mathbb{B}(\mathcal{E})\right]\ [\mathrm{cm^2/electron}], \qquad (2.217)$$

where the function $\mathbb{B}(\mathcal{E})$ is given by

$$\mathbb{B}(\mathcal{E}) = \frac{4}{3}\ln^3(2\,\mathcal{E}) - 3\ln^2(2\,\mathcal{E}) + 6.84\ln(2\,\mathcal{E}) - 21.51.$$

For large values of $\mathcal{E}$, the term $1/\mathcal{E}$ can be neglected and the term in brackets of Eq. (2.217) becomes identical to the corresponding term of Eq. (2.212).

For photon energies larger than $\approx 20\,\mathrm{MeV}$, the effect of screening cannot be neglected anymore. Wheeler and Lamb (1939) gave the following expression in the limit of **complete screening** for the differential cross section for pair production in the field of an atomic electron:

$$\frac{d\sigma_{\mathrm{pair}}^{WL}}{dE_+}\, dE_+ = 4\frac{\bar{\Phi}(Z)_{\mathrm{pair}}}{Z^2}\, \mathbb{W}(E_+, E_-, Z)\,\frac{dE_+}{E^3}\quad [\mathrm{cm^2/electron}], \qquad (2.218)$$

where the function $\mathbb{W}(E_+, E_-, Z)$ is given by

$$\mathbb{W}(E_+, E_-, Z) = \left[E_+^2 + E_-^2 + \frac{2}{3}E_+ E_-\right]\ln\left(1440\,Z^{-1/3}\right) - \frac{1}{9}E_+ E_-.$$

This equation, once integrated, provides the total cross section per electron:

$$\sigma_{\mathrm{pair}}^{WL} = \int_{mc^2}^{E-mc^2} \frac{d\sigma_{\mathrm{pair}}^{WL}}{dE_+}\, dE_+$$

$$\simeq 4\frac{\bar{\Phi}(Z)_{\mathrm{pair}}}{Z^2}\int_0^E \left\{\left[E_+^2 + E_-^2 + \frac{2}{3}E_+ E_-\right]\ln\left(1440\,Z^{-1/3}\right) - \frac{1}{9}E_+ E_-\right\}\frac{dE_+}{E^3},$$

from which we obtain, finally,

$$\sigma_{\mathrm{pair}}^{WL} = \frac{\bar{\Phi}(Z)_{\mathrm{pair}}}{Z^2}\left[\frac{28}{9}\ln\left(1440\,Z^{-1/3}\right) - \frac{2}{27}\right]\quad [\mathrm{cm^2/electron}]. \qquad (2.219)$$

The result is independent of the photon energy, but the term in brackets depends on Z, similarly to Eq. (2.214). For instance, Fig. 2.64 shows the cross sections [divided by $\bar{\Phi}(Z)_{\mathrm{pair}}$] for pair production in the field of atomic electrons in carbon and aluminum; these cross sections were computed by means of Eq. (2.217) for no screening and Eq. (2.219) for complete screening, and multiplied by 6 (or 13) to take into account the electrons available in carbon (or in aluminum).

Furthermore, to compute the total cross section for pair production $\sigma_{\mathrm{pair},e}$ in an atomic field, the theoretical treatment must include effects like i) the atomic binding of the target electron, ii) the screening by other atomic electrons and by the nuclear field, iii) the virtual Compton scattering in which the scattered photon (on the atomic electron) generates an electron–positron pair and, also, iv) radiative corrections. In an empirical way and within $\approx 5\%$, $\sigma_{\mathrm{pair},e}$ can be written in terms of the cross section for pair production in the nuclear field $\sigma_{\mathrm{pair},n}$ (see page 172) as [Hubbell (1969)]

$$\sigma_{\mathrm{pair},e} = \frac{\varsigma}{Z}\sigma_{\mathrm{pair},n}\quad [\mathrm{cm^2/atom}],$$

where ς is given by:

$$\varsigma = \frac{3+\alpha Z}{9}\ln\frac{\mathcal{E}}{2} - 0.00635\ln^3\frac{\mathcal{E}}{2}.$$

The total atomic cross section for pair production becomes:

$$\sigma_{\mathrm{pair,tot}} = \sigma_{\mathrm{pair},n} + \sigma_{\mathrm{pair},e} = \left(1 + \frac{\varsigma}{Z}\right)\sigma_{\mathrm{pair},n}\quad [\mathrm{cm^2/atom}]. \qquad (2.220)$$

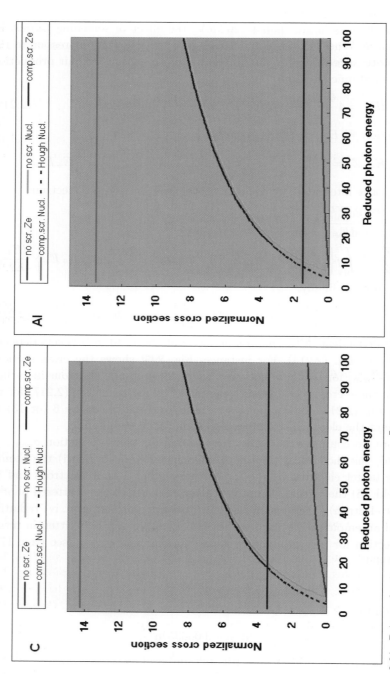

Fig. 2.64 Pair production cross sections divided by $\bar{\Phi}(Z)_{\text{pair}}$ (in ordinate) for carbon(C) and aluminum (Al) as a function of the reduced photon energy $\mathcal{E}$. The asymptotic nuclear pair production cross section without screening is calculated using Eqs. (2.212, 2.213), with screening using Eq. (2.214). For pair production in the atomic electron field, the curves are obtained by means of Eq. (2.217) for no screening and Eq. (2.219) for complete screening [once multiplied by 6 (or 13) to take into account the available electrons in C (or in Al)].

A detailed discussion of triplet production can be found in [Hubbell, Gimm and Øverbø (1980)] (see also references therein).

Furthermore, it has to be noted that the linear and mass attenuation coefficients accounting for the pair production effect can be computed using Eqs. (2.155, 2.156, 2.220).

2.3.3.3 *Angular Distribution of Electron and Positron Pairs*

At high energy, pairs are mostly emitted in the forward direction, similarly to the photon emission in the case of the bremsstrahlung effect. The average angle θ_p between the electron- (positron-) motion direction and the incoming-photon direction is of the order of

$$\theta_p \approx \frac{mc^2}{E}.$$

The angular distribution at small angles is given by [Hough (1948b)]

$$P(\theta_p)\, d\theta_p = \frac{1}{8} \left(\frac{mc^2}{E_1} \right)^2 \frac{E_1^2 + E_2^2}{E_1^2 + \frac{2}{3} E_1 E_2 + E^2} \frac{\cos^2 \frac{\theta_p}{2}}{\sin^4 \frac{\theta_p}{2}} \sin \theta_p\, d\theta_p, \qquad (2.221)$$

where E_1 is the energy of one electron in the pair and E_2 is the energy of the second electron. The probability to produce electrons at angles larger than an angle θ_g can be obtained by integrating Eq. (2.221) over the emission angle:

$$P(\theta_p > \theta_g) = \int_{\theta_g}^{\pi} P(\theta_p)\, d\theta_p = \frac{1}{8} \left(\frac{mc^2}{E_1} \right)^2 \frac{1}{1 - \cos \theta_g}.$$

2.3.4 *Photonuclear Scattering, Absorption and Production*

The photon scattering by the Coulomb nuclear field has more theoretical implications than practical ones. Nuclear Thomson and resonance scatterings, as well as nuclear photoelectric and Delbrück scattering cross sections, are smaller than the cross section of the main photon interaction mechanisms discussed so far. In this section, A is the notation for the mass number (see discussion at page 220 in Sect. 3.1).

The elastic scattering of photons by nuclear Coulomb fields may occur not only via Thomson scattering, whose cross section can be calculated by replacing i) the charge e by Ze (where Z is the atomic number) and ii) the target electron mass m by the nuclear mass in the formula given by Eq. (2.196), but also through the *Delbrück scattering*. The latter scattering occurs via the virtual or real creation of an $e^+ e^-$ pair, which subsequently annihilates.

The photonuclear absorption most likely results in the emission of a neutron, but charged particles, γ-rays or more than a single neutron can also be ejected. For photon energies below the threshold for nucleon removal, i.e., typically 7–8 MeV, only scattering is possible. However, the two low-A nuclei (deuterium and beryllium) have quite low energy thresholds for photoneutron production: 2.226 and 1.666 MeV, respectively. Deuterium is present (although in small amounts) in material containing hydrogen, water and plastics. Thus, it can become a source of neutrons, which has to be taken into account in radiation shielding. In addition, the photoneutron[||] production on deuterium is a mechanism for photoneutron production in heavier nuclei at energies above ≈ 30 MeV. The cross section σ_d is given by

$$\sigma_d = C_d \frac{(E_\gamma - 2.226)^{3/2}}{E_\gamma^3} \quad \text{[mb]}, \tag{2.222}$$

where the incoming photon energy, E_γ, is in MeV and $C_d \approx 61.0$–62.4 [Ericsson (1986)].

For photon energies above the binding energy of the least bound nucleon, the photoneutron emission is possible. Most of the photonuclear reactions up to ≈ 30 MeV occur via dipole excitations of the nucleus known as *giant resonances*. Nonspherical nuclei show a double peak corresponding to vibrations along symmetry axes. The characteristic broad absorption of the photonuclear cross section in the giant resonance region is centered at ≈ 24 MeV for low-A nuclei, but it decreases up to ≈ 12 MeV for the heaviest stable nuclei. The peak width[*], Γ, varies between 3 and 9 MeV, depending on the target nucleus. The threshold energy, for emitting a single neutron (γ,n), varies between ≈ 7 and 8 MeV (for high-A nuclei) and 18 MeV (for low-A nuclei), for proton emission (γ,p) between ≈ 6 and 16 MeV. The emission of other charged particles (for instance α particles) and more than one neutron is also possible. The cross section (σ_{grp}) at the energy resonance peak (E_{grp}) is $\approx 6\%$ of the atomic electron (*electronic*) cross section for low-A nuclei and not larger than $\approx 2\%$ for high-A nuclei.

For photon energies corresponding to the giant resonance maximum, the angular distribution of emitted neutrons behaves as

$$\mathcal{F}(\theta) = A_n + B_n \sin^2 \theta.$$

It shows a symmetrical behavior centered around 90°. As the energy increases, the distribution becomes less symmetric and more forward directed.

For sufficiently massive nuclei ($A > 50$), the shape of the cross section resonance can be approximated by a *Lorenz curve* (see for instance [Fuller and Hayward (1962)]):

$$\sigma_{crs}(E_\gamma) \simeq \sigma_{grp} \frac{E_\gamma^2 \Gamma^2}{(E_{grp}^2 - E_\gamma^2)^2 + E_\gamma^2 \Gamma^2}.$$

For ellipsoidal nuclei, the shape of the cross section is given by a superposition of two Lorenz curves.

[||]It is a neutron emitted by a nuclear target as a result of a photon interaction.
[*]This is the energy difference between points around the peak, for which the cross section decreases by a factor 2.

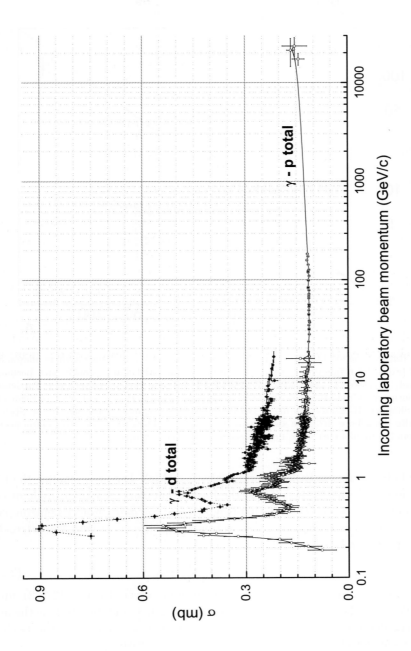

Fig. 2.65 Photon total cross section on proton and deuterium (experimental data and references are available (also via the web) from [PDB (2002)])) as a function of the photon momentum. The lines are to guide the eye.

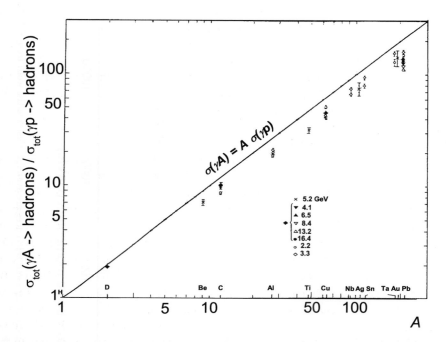

Fig. 2.66   Ratio of the total hadronic photon cross section on nucleus to the hadronic photon cross section on proton as a function of the mass number A (adapted and reprinted with permission from Genzel, H., Joos, P. and Pfeil, W. (1973), Photoproduction of Elementary Particles, in Landolt- Börnstein, Group I: Nuclear and Particle Physics, Vol. 8: Photoproduction of Elementary Particles, Shopper, H. Editor, Springer, Berlin, Chapter 2, Figure 2, page 306; © by Springer-Verlag Berlin Heidelberg 1973; see also [Fassò et al. (1990)]). Data were taken at photon energies ranging from 3.3 to 16.4 GeV. The line represents $\sigma_{\text{tot}}(\gamma A) = A\,\sigma_{\text{tot}}(\gamma p)$.

The photon wavelength for energies above $30\,\mathrm{MeV}$ becomes smaller than the average inter-nucleon distance. Thus, the interaction can occur on a single or on a few nucleons. Most of the reactions happen on a nucleon pair (proton–neutron pair), the so-called *quasi-deuteron*. The cross section σ_{qd} for the *quasi-deuteron interaction theory* proposed by Levinger (1951) is given by

$$\sigma_{qd} = Le\frac{(A - \mathbb{D})\mathbb{D}}{A}\,\sigma_d, \tag{2.223}$$

where σ_d is the cross section on free deuterium, given by Eq. (2.222) up to $\approx 140\,\mathrm{MeV}$; Le is the so-called *Levinger parameter* $\simeq 6.4$; and, finally, $\mathbb{D}$ is the number of quasi-deuteron pairs inside the nucleus. The A-dependence of the Levinger parameter seems to be well represented by [Giannini (1986)]

$$Le = \frac{13.82\,A}{R_n^3},$$

where R_n is the experimental root mean square charge radius in units of fermi [Giannini (1986)]. The angular and energy distributions of emitted nucleons are in principle similar to those on free deuteron, but they are broadened due to the internal motion and re-interactions, before emerging from the nucleus.

As the photon energy increases above the pion production threshold ($\simeq$ 140 MeV), the photon cross section shows peaks corresponding to the excitation of baryonic resonances, as shown in Fig. 2.65 [PDB (2002)] for γ–p and γ–d cross sections. During the photon interaction and the de-excitation of such states, spallation nucleons, nuclear fragments and mesons can be produced.

At high energies, the photon–proton ($\sigma_{\gamma,p}$) and photon–neutron ($\sigma_{\gamma,n}$) cross sections [Bauer, Spital and Yennie (1978)] approach similar values.

High-energy photon–nucleus collisions were studied experimentally and theoretically (see for instance [Lohrmann (1969); Genzel, Joos and Pfeil (1973); Bauer, Spital and Yennie (1978); Engel, Ranft and Roesler (1997)] and references therein). It was pointed out that there are similarities between photon–nucleus and hadron–nucleus features, like for instance the so-called *shadowing effect*, i.e., the decrease of the cross section per nucleon as the mass number increases (see Fig. 2.66 and [Genzel, Joos and Pfeil (1973)]). The photon hadronic-feature was included within the framework of the *Vector Dominance Model* (VDM, see for instance [Bauer, Spital and Yennie (1978); Engel, Ranft and Roesler (1997)] and references therein) or others, like the *Generalized Vector Dominance Model* (GDVDM, see for instance [Engel, Ranft and Roesler (1997)] and references therein). The VDM predicts that, at high energies, a photon behaves like a hadron in strong interactions. In fact, from the quantum-mechanical point of view because of the uncertainty principle, a photon[**] may convert virtually into a vector meson.[††] The lifetime Δt, for such a fluctuation in the laboratory system, is $\Delta t \sim \hbar/(q_L\, c)$, where q_L is the difference between the longitudinal momenta of the photon and of the produced vector meson. In the high-energy limit for which $q_L \rightarrow 0$, Δt might become large enough so that the impinging photon is already in the virtual state of a vector meson. Hence, a strong interaction with the nucleus can occur and can be theoretically treated within the framework of the *Gribov–Glauber formalism* [Glauber (1955)]. As a consequence, shadowing effects, typical of hadron–nucleus interactions, are expected to be also found in the photon–nucleus interaction.

2.3.5 *Attenuation Coefficients, Dosimetric and Radiobiological Quantities*

In Sects. 2.3, 2.3.1, 2.3.2.4, the mass attenuation coefficients were introduced to describe the absortion and/or scatttering of photons in matter. In the present section, the notations are the same of those used in the previous sections.

[**]The spin-parity of a photon is $J^P = 1^-$.

[††]The vector mesons are strongly interacting mesonic particles with $J^P = 1^-$.

The mass attenuation and mass energy-absorption coefficients are extensively used in calculations for γ-ray transport in matter and for photon energy depositions in biological and other materials, e.g., for dosimetric applications [Hubbell (1977)] in the medical and biological context [Hubbell (1999)]. These coefficients were tabulated and discussed in details in literature and, at present, are also available on the web (see [Hubbell (1969); Henke, Gullikson and Davis (1993); Hubbell and Seltzer (2004); Berger, Hubbell, Seltzer, Chang, Coursey, Sukumar and Zucker (2005)] and references therein).

As discussed along this section, the energy and Z-dependence of the mass attenuation coefficients are mainly determined by the total cross sections for photoelectric effect, Compton scattering and pair production. They depend on the incoming photon energy and atomic number. The photonuclear absorption by nuclei mostly results in ejecting nucleons. This interaction contributes by less than $(5–10)\%$ to the total photon cross section in a fairly narrow energy region between a few MeV and a few tens of MeV. This cross section is usually omitted in tabulations due to a) the lack of theoretical models comparable to those available for the description of other photo-interaction processes, b) some incomplete experimental information and, finally, c) its irregular dependence on A (i.e., the relative atomic mass or atomic weight, Sect. 1.4.1) and Z (i.e., the atomic number, Sect. 3.1). The Compton effect dominates at medium reduced-energies, $\mathcal{E}$, and low-Z. At high $\mathcal{E}$ and high-Z, the pair production mechanism is the dominant process of (primary) photon interaction in matter. At low-Z and low $\mathcal{E}$, the photoelectric process becomes the dominant process of photon interaction.

A web database (XCOM) is currently available in [Berger, Hubbell, Seltzer, Chang, Coursey, Sukumar and Zucker (2005)]. The version available on the web gives detailed and updated references to the literature from which numerical approximations were derived. While in previous sections, the emphasis was mainly put on the discussion of principles, on which theoretical models are based; the XCOM database provides photon cross sections for scattering, photoelectric absorption and pair production, as well as total attenuation coefficients, for any element, compound or mixture ($Z \leq 100$), at energies from 1 keV to 100 GeV. At lower energies between 30 eV and 30 keV, mass attenuation coefficients and indeces of refraction were tabulated, based on both measurements and theoretical calculations, and are also available on the web in [Henke, Gullikson and Davis (1993)].

As mentioned above, from the XCOM database we can obtain total cross sections, attenuation coefficients and partial cross sections for the individual following processes: incoherent Compton scattering, coherent Rayleigh scattering, photoelectric absorption, pair production in the field of nuclei and in the field of atomic electrons. The incoherent Compton scattering takes into account the incoherent scattering function $S(q, Z)$, the radiative corrections and the double Compton effect. The values for the coherent Rayleigh scattering were evaluated introducing the atomic form factor $F(q, Z)$. For compounds, the quantities tabulated are par-

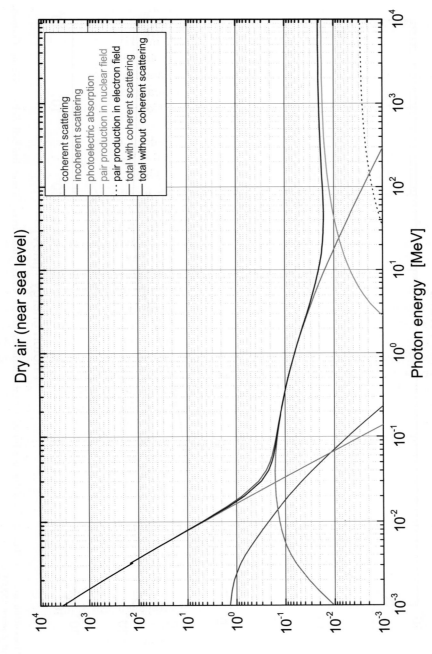

Fig. 2.67 Mass attenuation coefficients for dry air at sea level as a function of the incoming photon energy (data from [Berger, Hubbell, Seltzer, Chang, Coursey, Sukumar and Zucker (2005)]). The air density and Z/A are 1.205×10^{-3} and 0.499, respectively.

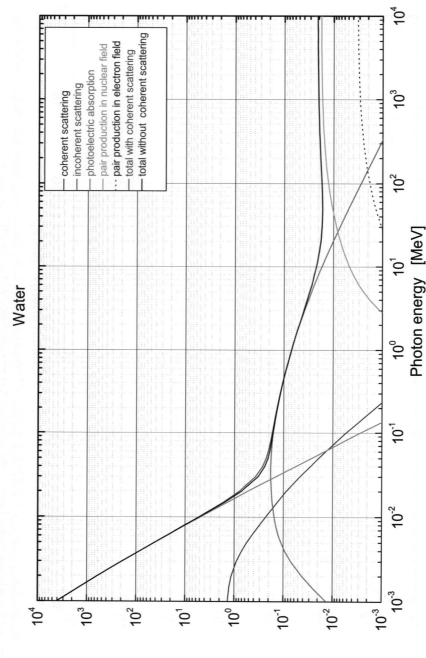

Fig. 2.68   Mass attenuation coefficients for water as a function of the incoming photon energy (data from [Berger, Hubbell, Seltzer, Chang, Coursey, Sukumar and Zucker (2005)]).

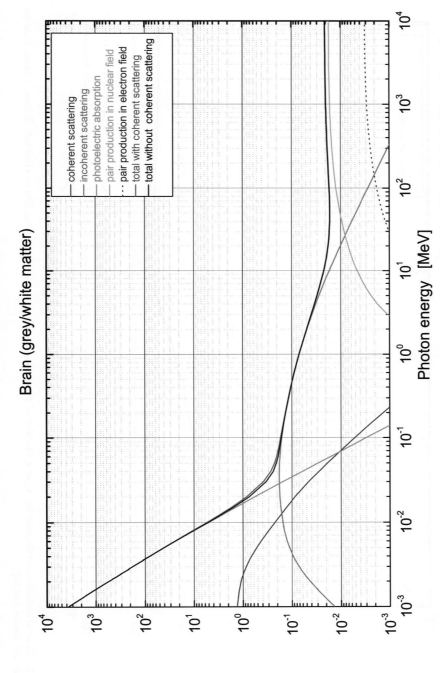

Fig. 2.69 Mass attenuation coefficients for brain as a function of the incoming photon energy (data from [Berger, Hubbell, Seltzer, Chang, Coursey, Sukumar and Zucker (2005)], composition of grey/white matter as from [ICRUM (1989)]). The brain density and Z/A are $1.04\,\text{g/cm}^3$ and 0.552, respectively.

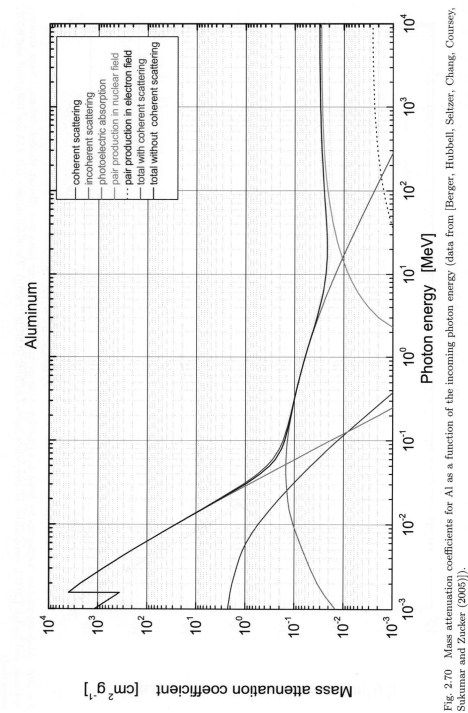

Fig. 2.70 Mass attenuation coefficients for Al as a function of the incoming photon energy (data from [Berger, Hubbell, Seltzer, Chang, Coursey, Sukumar and Zucker (2005)]).

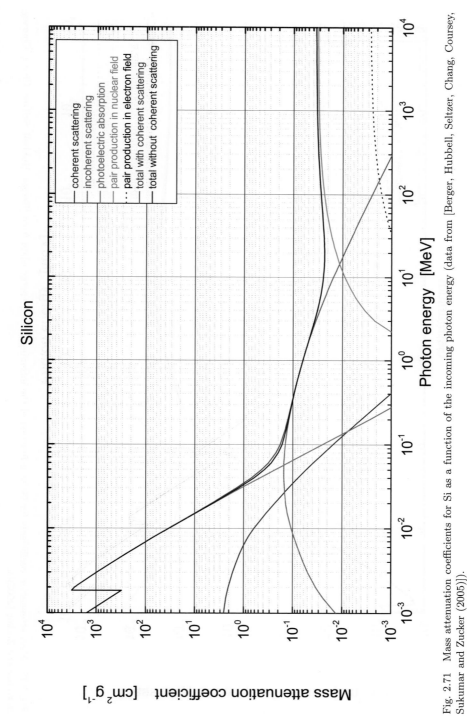

Fig. 2.71 Mass attenuation coefficients for Si as a function of the incoming photon energy (data from [Berger, Hubbell, Seltzer, Chang, Coursey, Sukumar and Zucker (2005)]).

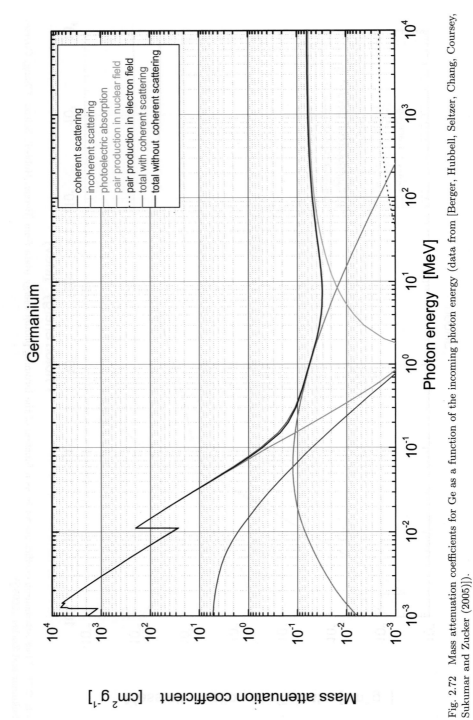

Fig. 2.72 Mass attenuation coefficients for Ge as a function of the incoming photon energy (data from [Berger, Hubbell, Seltzer, Chang, Coursey, Sukumar and Zucker (2005)]).

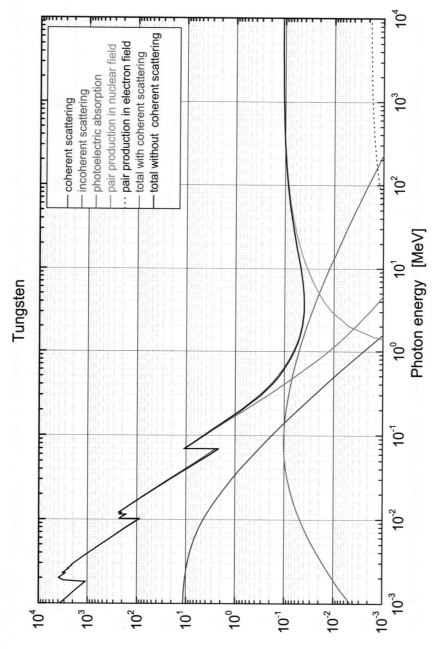

Fig. 2.73 Mass attenuation coefficients for W as a function of the incoming photon energy (data from [Berger, Hubbell, Seltzer, Chang, Coursey, Sukumar and Zucker (2005)]).

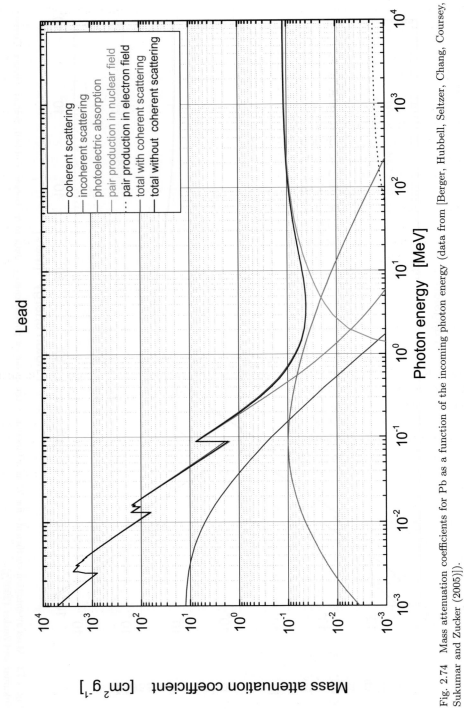

Fig. 2.74 Mass attenuation coefficients for Pb as a function of the incoming photon energy (data from [Berger, Hubbell, Seltzer, Chang, Coursey, Sukumar and Zucker (2005)]).

tial and total mass interaction coefficients. Total attenuation coefficients without the contribution from coherent scattering are also given. Interaction coefficients and total attenuation coefficients for compounds or mixtures are obtained as sums of the corresponding quantities for atomic constituents. For example, in Figs. 2.67–2.74, mass attenuation coefficients from XCOM database [Berger, Hubbell, Seltzer, Chang, Coursey, Sukumar and Zucker (2005)] are shown as a function of the incoming photon energy for air[§], water (H_2O), brain[¶] Al, Si, Ge, W, Pb for photon energies between 1 keV and 10 GeV.

The effect of γ-ray irradiation on matter is mainly indirect, i.e., via charged particles (electrons and positrons) generated in the interaction. It is the dissipation-energy process (mostly by collision energy-losses) of these secondary charged and ionizing particles, which determines the energy deposition of γ-rays in matter. The relationship between the deposited energy and the various physical, chemical and biological effects is usually complex and not fully understood. However, it is commonly assumed that a significant parameter in radiation effects is the *absorbed dose*, which is defined as the *mean energy imparted by ionizing radiation* ($d\bar\varepsilon$) *per unit mass* (dm):

$$\mathcal{D} \equiv \frac{d\bar\varepsilon}{dm}.$$

The mean energy imparted in a volume is given by [ICRUM (1980a)]

$$\bar\varepsilon \equiv R_{in} - R_{out} + \sum Q,$$

where R_{in} is the radiant energy incident on the volume, i.e., the sum of the energies (without taking into account the rest energies) of all (charged and uncharged) ionizing particles entering the volume. R_{out} is the radiant energy emerging from the volume, i.e., the sum of the energies (without taking into account the rest energies) of all (charged and uncharged) ionizing particles leaving the volume. $\sum Q$ is the sum of all changes (which occur in the volume) of the rest mass energy of nuclei and elementary particles in any nuclear transformation. The SI-unit[*] of absorbed dose is designated as the gray[†] (Gy), i.e., in $J\,kg^{-1}$.

At energies larger than the critical energy ϵ_c [Eq. (2.121)] of the medium, the calculation of the energy deposition requires the cascade-development treatment for the transport of electrons and photons. However, at lower energies the energies deposited by γ-rays can be treated in terms of the photon flux and the *mass energy absorption coefficient* $\mu_{att,m,en}$ (or one of its approximation like the *mass absorption coefficient* $\mu_{att,m,ab}$ or the *mass energy transfer coefficient* $\mu_{att,m,tr}$). The mass energy absorption coefficient takes into account all possible modes of energy transfer[‡] to electrons

---

[§]The dry air composition near the sea level is that specified in the web site given for [ICRUM (1989)].

[¶]For brain, the composition of grey/white matter is from [ICRUM (1989)].

[*]The International System of Units is usually indicated as SI.

[†]It is sometimes expressed in rad: 1 rad = 0.01 Gy = 1 cGy.

[‡]The reader can see relatively recent calculations by Seltzer (1993) and, for instance, [ICRUM (1964); Hubbell (1969); Hubbell and Seltzer (2004)].

in the medium, while the mass energy transfer and the mass absorption coefficient neglect some of the modes. The coefficient $\mu_{att,m,en}$ accounts for the escape of fluorescence, Compton scattered, annihilation, and bremsstrahlung photons. The coefficient $\mu_{att,m,tr}$ accounts for the escape of these photons, bremsstrahlung photons excepted. The coefficient $\mu_{att,m,tr}$ accounts for the escape of Compton scattered photons only. The mass energy transfer coefficient is given by:

$$\mu_{att,m,tr} = \frac{N}{A} \left(f_{ph}\, \sigma_{ph} + f_{incoh}\, \sigma_{incoh} + f_{pair,n}\, \sigma_{pair,n} + f_{pair,e}\, \sigma_{pair,e} \right), \quad (2.224)$$

where the cross sections σ_i refer to the processes (previously discussed) of photoelectric absorption, incoherent Compton scattering (see page 163) and pair production in the Coulomb field of nuclei and atomic electrons; the factors f_i represent the average fractions of the photon energy E_γ which were transferred as kinetic energies of charged particles as a result of the interactions. In Eq. (2.224), the coherent scattering cross section was omitted because the associated energy transfer is negligible. The factors f_i in Eq. (2.224) are given by:

$$f_{ph} = 1 - \frac{E_f}{E_\gamma},$$

where E_f is the average energy of the fluorescence radiation emitted per absorbed photon;

$$f_{incoh} = 1 - \frac{\bar{E}_C + E_f}{E_\gamma},$$

where $\bar{E}_C$ is the average energy of the photon scattered by Compton effect;

$$f_{pair,n} = 1 - \frac{2mc^2}{E_\gamma},$$

and

$$f_{pair,e} = 1 - \frac{2mc^2 + E_f}{E_\gamma}.$$

The fluorescence radiation emitted per absorbed photon E_f depends on the distribution of atomic electron vacancies generated in the interaction. Thus, it is differently evaluated for various types of interactions and includes the emission of cascade fluorescence X-rays associated with the atomic relaxation process initiated by the primary vacancy [Carlsson (1971)].

The dosimetric quantity *kerma* ($\mathcal{K}$) *is defined as the sum of the kinetic energies of all those charged particles released by uncharged particles* (in this case photons) *per unit mass*; it can be calculated by multiplying the mass energy transfer coefficient by the photon energy fluence

$$\Psi = \Phi E_\gamma,$$

where Φ is the fluence of photons of energy E_γ. At charged particles equilibrium (i.e., when energy, number and direction are constant throughout the volume of interest) and for negligible bremsstrahlung production, the kerma approaches the absorbed dose. The SI-unit of kerma is the gray.

Further emission of radiation, produced by secondary charged particles, is taken into account by the mass energy absorption coefficient, which is given by

$$
\begin{aligned}
\mu_{att,m,en} = & \frac{N}{A}(1 - g_{ph})f_{ph}\,\sigma_{ph} + \frac{N}{A}(1 - g_{incoh})f_{incoh}\,\sigma_{incoh} \\
& + \frac{N}{A}(1 - g_{pair,n})f_{pair,n}\,\sigma_{pair,n} + \frac{N}{A}(1 - g_{pair,e})f_{pair,e}\,\sigma_{pair,e} \\
= & (1 - g)\,\mu_{att,m,tr}\,,
\end{aligned}
\tag{2.225}
$$

where g_i is the average fraction of the kinetic energy of secondary charged particles produced in a specific type of interaction, that is lost in radiative processes, and g is the weighted average of g_i.

In Fig. 2.75, mass energy absorption coefficients by Hubbell and Seltzer (2004) are shown as function of the incoming photon energy for Al, Si, Pb, brain (grey/white matter), (liquid) water and dry air (near sea level) media** for photon energies between 1 keV and 20 MeV. It has to be noted that at low photon energies, where g is small, $\mu_{att,m,tr} \approx \mu_{att,m,en}$. At 1.5 MeV, their difference increases as the atomic number increases from less than 1% for low-Z media up to $\approx 7\%$ for high-Z media. At 10 MeV and for high-Z absorbers, they differ by larger and energy dependent values; but for air and water, they differ by $\leq 1\%$.

Furthermore, let us introduce the dosimetric quantity *exposure* $\mathcal{X}$, which is related to the mass energy absorption coefficient in air by:

$$
\mathcal{X} = \Psi \mu_{att,m,en}^{air}\,\frac{e}{W_a},
$$

where W_a is the mean energy expended in air per ion-pair and e is the electron charge, while the kerma in air, $\mathcal{K}^{air}$, is related to the energy fluence by

$$
\mathcal{K}^{air} = \mu_{att,m,tr}^{air}\,\Psi.
$$

The SI-unit of exposure is expressed in $C\,kg^{-1}$.

The *equivalent dose*[‡‡] (H_T) in a organ or tissue (T) depends on the absorbed dose (D_T) caused by different radiation types R and averaged over the volume of that organ or tissue (e.g., see Section 4.5.2.1 of [Dietze (2005)] and Section 29 of [PDB (2008)]):

$$
H_T = \sum_R w_R \times D_{T,R},
$$

where the weighting factors w_R are characterizing the biological effectiveness of the specific radiation (R) and can be found, for instance, in Section 29 of [PDB (2008)]

**The dry air and brain composition are the same as for the mass attenuation coefficients from XCOM database (see page 193 and also [ICRUM (1989)]).

[‡‡]In 1977 the the tisse (or organ) dose equivalent was introduced by [ICRP (1977)]. Although the concept was not changed in [ICRP (1991)], important modifications where introduced, e.g., replacing dose equivalent quantities by equivalent dose quantities. The reader can see Section 4.5.2 of [Dietze (2005)].

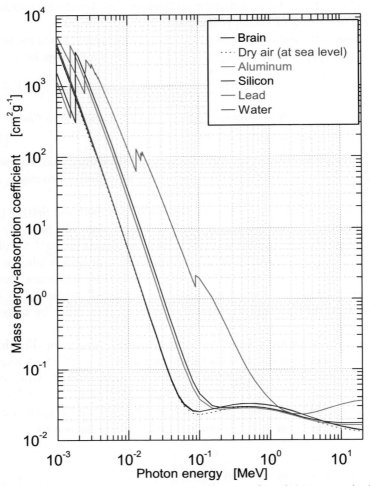

Fig. 2.75 Mass energy-absorption coefficients for brain (grey/white matter), dry air (near sea level), Al, Si, Pb and (liquid) water as function of the incoming photon energy (data from [Hubbell and Seltzer (2004)]). The air and brain densities are 1.205×10^{-3} and $1.04\,\mathrm{g/cm^3}$, respectively. The air and brain Z/A ratios are 0.499 and 0.552, respectively.

(see also references therein). The SI-unit of equivalent dose is designated as the sievert[†] (Sv), i.e., in $\mathrm{J\,kg^{-1}}$.

A treatment of the physical (*stochastic* and *non-stochastic*) quantities and units used to describe radiological and/or radioactive phenomena can be found in [Dietze (2005)]. Stochastic quantities are those used when the phenomena are subject to inherent fluctuations and follow a probability distribution (e.g., the *decay constant* of a radionuclide). For other phenomena, a non-stochastic quantitity (e.g., the

---

[†]It is sometimes expressed in rem: $1\,\mathrm{rem} = 0.01\,\mathrm{Sv} = 1\,\mathrm{cSv}$.

Table 2.14 Measured longitudinal attenuation-lengths of electron-induced cascades (from [Leroy and Rancoita (2000)]).

Material	Incident Energy GeV	Attenuation λ_{att} g/cm^2	X_0	References
Al	6	64.3	2.7	[Bathow et al. (1970)]
Cu	1	41.7	3.2	[Nelson et al. (1966)]
	0.9	34.4	2.7	[Crannell (1967)]
	6	39.0	3.0	[Bathow et al. (1970)]
	1	35.7	2.8	[Yuda et al. (1970)]
	0.6	34.5	2.7	[Yuda et al. (1970)]
Sn	0.9	30.3	3.4	[Crannell (1967)]
Pb	1	22.2	3.5	[Nelson et al. (1966)]
	0.9	23.0	3.6	[Crannell (1967)]
	6	24.7	3.9	[Bathow et al. (1970)]
	1	21.7	3.4	[Yuda et al. (1970)]
	0.6	21.3	3.3	[Yuda et al. (1970)]

fluence or the *absorbed dose*) is defined by averaging in time or over a volume which results in a single value with no inherent fluctuation.

2.4 Electromagnetic Cascades in Matter

Electromagnetic-cascade showers have been one of the most striking phenomena observed in high-energy cosmic rays and were discovered by Blackett and Occhialini (1933) (see also [Blackett, Occhialini and Chadwick (1933)]). An extensive review covering both the electromagnetic (discussed in this section) and hadronic (discussed in the chapter on *Nuclear Interactions in Matter*) cascading shower generation and propagation in matter was provided by Leroy and Rancoita (2000).

The degradation of the incident energy results in a multiplicative process, whose extent is controlled by the magnitude of the incident-particle energy. High energy electrons lose most of their energy by radiation via bremsstrahlung effect and, therefore, produce high energy photons. These high-energy photons, in turn, mainly undergo materialization* in the Coulomb field of a nucleus, or they produce Compton electrons. These electrons and positrons, in turn, radiate new photons which undergo again pair production or Compton scattering, if their energy is sufficiently large. This phenomenon is often referred to as a multiplicative shower or cascade shower.

At high-(and ultra high-) energies the bremsstrahlung and pair production processes are described by the Landau–Pomeranchuk–Migdal (LPM) formulae [Migdal (1956); Landau and Pomeranchuck (1965)], which, at lower energies [Migdal (1956)], reduce to the expressions given by Bethe and Heitler (1934). At very high energy, LPM formulae predict that the production of low-energy photons by high-energy

---

*The pair production process is dominant beyond 100 MeV.

electrons is suppressed in dense media: deviations as large as 30% compared with the Bethe–Heitler formulae were predicted for lead in [Migdal (1956)]. This suppression was confirmed experimentally [Antony et al. (1995)].

In a cascade shower initiated by an electron or a photon, the generated particles are electrons, positrons and photons. This kind of cascade shower is called an *electromagnetic shower*. Soft photons and electrons are generated in the final stage of the multiplication process. These photons mainly interact via the photoelectric process while the soft electrons, including the photoelectric electrons, dissipate their energy through collisions. In this way, the incoming particle energy is absorbed by the medium.

The devices built to measure the energy deposited in their volume from the degradation of the energy carried by incident electrons and/or photons are called *electromagnetic calorimeters*.

2.4.1 *Phenomenology and Natural Units of Electromagnetic Cascades*

Bremsstrahlung and pair production are the dominant interaction processes for high-energy electrons and photons, respectively. Their cross sections become almost energy independent in the case of complete screening, i.e., for incoming electron and photon energies $\gg mc^2/(\alpha Z^{1/3})$, where Z is the atomic number of the absorber. As a consequence, the radiation length[†], X_0, emerges as a natural unit of length and represents the mean-path length of an electron in a material. The absorber depth t is usually given in units of radiation length.

Radiation lengths were calculated and tabulated (see, for instance, [Tsai (1974); PDB (2008); PDG (2008)] and references therein). They are reported in Table 2.3, for various absorbers used for calorimeters, and in Tables 2.7 and 2.8, for elements up to $Z = 92$. In practice, the following approximation is often used [Amaldi (1981)]:

$$X_{g0} = X_0\, \rho \approx 180\frac{A}{Z^2} \ \ [\text{g/cm}^2], \tag{2.226}$$

where X_0 is the value of the radiation length in cm; ρ in g/cm^3 and A are the density and atomic weight (see Sect. 1.4.1 and page 220) of the medium, respectively, and with

$$\frac{\Delta X_0}{X_0} = \frac{\Delta X_{g0}}{X_{g0}} \leq 20\% \text{ for } Z \geq 13.$$

When more than one absorber is present in the showering medium, the overall radiation length can be calculated by means of Eq. (2.119). The corresponding medium density can be calculated from Eq. (2.120).

There is a depth, indicated as $t_{\max}$ and called *shower maximum*, where the largest number of secondary particles created during the multiplication process is reached and, correspondingly, where the largest energy dissipation occurs. Beyond

---

[†]It was discussed in Sect. 2.1.7.3 on *Radiation Length and Complete Screening Approximation*.

that maximum depth, the cascade decays slowly. Each step of the energy degradation is characterized by an increase of secondary particles produced with a decreasing average energy: low energy photons either transfer their energy to low-energy
electrons via Compton scattering or lead to low energy electron production via the
photoelectric effect. This latter phenomenon[§] is relevant or dominant for photon
energies below about a few hundred keV's (Sects. 2.3.1 and 2.3.5). Low energy electrons mostly lose their energy via collisions and, as the cascade evolves along the
calorimeter depth, more and more electrons fall into an energy range where collision
energy-losses dominate radiation energy-losses. In the showering process, the multiplication process is almost stopped when the electron energy finally reaches the
critical energy[¶] ϵ_c. The value of the critical energy, calculated using Eq. (2.121), is
given in Table 2.3 for several materials commonly used as absorber in calorimeters.

2.4.2 *Propagation and Diffusion of Electromagnetic Cascades in Matter*

The electromagnetic processes, generating an electromagnetic cascade, are well described by the quantum-electrodynamics theory (QED), which accounts for the
basic interactions of electrons and photons inside the calorimeter medium (see, for
instance, [Berestetskii, Lifshitz and Pitaevskii (1971); Tsai (1974)] and references
therein). In principle, an analytical description of the cascade behavior is possible. However, the interaction-mechanism complexity[‖] of cascading particles with
the calorimeter medium imposes treatments in which part of the physical processes
is neglected or taken into account in an approximate way. The theory of cascade
showers originates from two independent papers by Bhabha and Heitler (1936) and
by Carlson and Oppenheimer (1936). Afterwards, improvements and amplifications
with respect to the physical and mathematical approximations were proposed by
several authors [Arley (1938); Landau and Rumer (1938); Bhabha and Chakrabarty
(1943); Heitler (1954); Rossi (1964)].

2.4.2.1 *Rossi's Approximation B and Cascade Multiplication of Electro-magnetic Shower*

A simplified analytical model of the cascade development was proposed by
Rossi (1964). In Rossi's formulation, the shower development theory is supposed
to predict the number and the energy distribution of electrons and photons with
energy large compared with a given energy $\eta_0 \approx 5\,\mathrm{MeV}$. Collisions, giving rise to
secondary electrons with energy less than η_0, represent the process in which the

---

[§]The photoelectric effect also depends on the absorber atomic-number (see discussion in
Sect. 2.3.1).

[¶]The critical energy was defined in Sect. 2.1.7.4.

[‖]As it will be seen later, the interaction of cascading particles with low-energy is particularly
complex.

energy, from the flow of the generated shower particles, is lost and eventually deposited in matter via excitation and ionization of atoms. The incoming energies considered are $\gg mc^2/\left(\alpha Z^{1/3}\right)$. Therefore, radiation phenomena and pair production can be described by asymptotic formulae for complete screening, while the Compton effect is neglected. In what is called "Approximation B", collision losses of electrons are taken into account. However, the energy-dependent collision-loss is replaced by a constant collision-loss occurring in a radiation length called critical energy ϵ_c. It is taken, for ϵ_c, the energy lost (by collisions in a radiation length) by electrons of energy equal to the critical energy itself. No difference is considered between electrons and positrons. This definition of ϵ_c differs from that one introduced in Sect. 2.1.7.4, but typically its values are in agreement within a few percent with those ones computed by means of Eq. (2.121) used in the present Sect. 2.4 and in the chapter on *Principles of Particle Energy Determination*. The assumption of constant energy-loss for electrons derives from the fact that in all substances the critical energy is larger than η_0 and for energies above η_0 the collision loss is slowly varying with the incoming electron energy (see Sect. 2.1.7.1 and, for instance, Figs. 2.33–2.37). The assumption of similar energy-loss for electrons and positrons can be justified by considering that electron and positron collision losses do not differ by more than a few % over a wide range of energy (see Sect. 2.1.6.1). It has also to be noted that for $\beta \approx 1$ ($\beta = v/c$, where v is the velocity of the particle and c the speed of light) the collision losses by electrons are $\approx 10\%$ larger than the collision losses by heavy particles, and that the ratio ϵ_c/X_0 is approximately 10% larger than the collision losses for *minimum ionizing particles (mip)*, which occurs at $\beta \approx 0.96$.

The cascade process, leading to the propagation of the shower and energy dissipation, is treated along the longitudinal direction and is described by means of a set of linear integro-differential equations. Because of their formal similarity to equations occurring in the theory of diffusion phenomena, they are often referred to as the *diffusion equations*. "Approximation B" yields identical results for all elements provided thicknesses are measured in units of radiation length and energies in units of critical energy. In "Approximation B", the total track length, T, is the summed length of all individual tracks of charged particles showering and dissipating their energy in the calorimeter medium. It is given by

$$T = \frac{E}{\epsilon_c} \ [X_0], \qquad (2.227)$$

where E is the incoming particle energy. Equation (2.227) can be justified in the following way: at high energies, pair production and bremsstrahlung processes allow the cascade to develop. However, the energy absorption or dissipation in matter mainly proceeds via collision losses at the constant rate per unit of length of $\approx \epsilon_c/X_0$. Therefore, the total path of cascade charged particles in the absorber will be $\approx E \big/ \left(\frac{\epsilon_c}{X_0}\right)$. In reality, there exists a dependence of the total track length on the minimal energy E_c detectable by the calorimetric device [Crawford and

Messel (1962); Nagel (1965); Longo and Sestilli (1975); Amaldi (1981)]. Therefore, the measurable track length $T(\zeta)$ is smaller than T and can be expressed [Amaldi (1981)] as

$$T(\zeta) = F(\zeta) \left(\frac{E}{\epsilon_c} \right) \quad [X_0], \tag{2.228}$$

where

$$F(\zeta) \sim e^\zeta \left[1 + \zeta \ln \left(\frac{\zeta}{1.526} \right) \right],$$

with

$$\zeta = 4.58 \frac{Z}{A} \left(\frac{E_c}{\epsilon_c} \right)$$

and E_c is the cut-off energy, namely the minimum kinetic energy of the electron (positron) that can be detected in the calorimeter.

For a cascade shower, T [or $T(\zeta)$] is proportional to the incident particle energy. Thus, the response of a calorimeter is proportional to the incident energy, provided that it is proportional to the energy deposited by the particles of the shower (i.e., in the absence of saturation effects). Furthermore, the fluctuations in T [or $T(\zeta)$] will determine the intrinsic energy resolution.

The multiplication of electrons and positrons occurs when their energy is much larger than the critical energy and they get absorbed at lower energies, i.e., when collision losses become dominant. The number of charged particles and photons increases rapidly with depth until the maximum, located at the depth $t_{\max}$ of the cascade, is reached, i.e., where the longitudinal development of the shower has a maximum (also called *shower maximum*). This shower behavior can be understood in a simplified way, following crude assumptions. Let E be the energy of an incoming photon, which at a depth $\approx 1\,X_0$ generates a pair of e^+e^- with equal energies. After an additional distance of $\approx 1\,X_0$ both the electron and positron will emit a bremsstrahlung photon. By continuing the process and assuming equal energy sharing among the generated particles, particles will double themselves every radiation length. Thus, the number of particles at depth t is

$$N(t) \approx 2^t,$$

while their energy is

$$E_p(t) \approx \frac{E}{N(t)} = 2^{-t} E.$$

When the particle energy is $E_p \approx \epsilon_c$, the multiplication no longer continues. This occurs at the maximum depth $t_{\max}$ for which

$$E_p(t_{\max}) \approx \epsilon_c \approx 2^{-t_{\max}} E,$$

namely, $t_{\max} \approx \ln(E/\epsilon_c)$. Under "Approximation B", this maximum depth depends on the incoming energy as $\ln\left(\frac{E}{\epsilon_c} \right)$ and is given by

$$t_{\max} = 1.01 \left[\ln \left(\frac{E}{\epsilon_c} \right) - c \right] \quad [X_0], \tag{2.229}$$

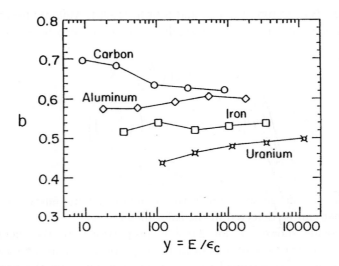

Fig. 2.76 Fitted values of the scale factor b for energy-deposition profiles obtained with the simulation code EGS4 for incident electrons up to ≈ 100 GeV as function of $y = E/\epsilon_c$ (adapted from *Phys. Lett. B* **592**, Eidelman, S. et al., Review of Particle Physics, Edited by Particle Data Group, 1–1109, Copyright (2004), with permission from Elsevier; see also [Leroy and Rancoita (2000)]).

where $c = 1.0$ or 0.5 for incident electrons or photons, respectively. The location of the *shower center-of-gravity* t_{cg}, i.e., the depth at which half of the incident energy was deposited by the cascading particles (also indicated as the *median depth of the shower*), is given by

$$t_{\mathrm{cg}} = 1.01 \left[\ln \left(\frac{E}{\epsilon_c} \right) + d \right] \quad [X_0], \tag{2.230}$$

where $d = 0.4$ or 1.2 for incident electrons or photons, respectively. The number of charged particles at $t_{\max}$ is given by

$$\Pi_e(t_{\max}) = \left[\frac{0.31}{\sqrt{\ln \left(\frac{E}{\epsilon_c} \right)} - e} \right] \frac{E}{\epsilon_c}, \tag{2.231}$$

where $e = 0.37$ or 0.18 for incident electrons or photons, respectively.

2.4.2.2 *Longitudinal Development of the Electromagnetic Shower*

Simulation codes were written to reproduce the behavior of electromagnetic cascades in various media with the best possible accuracy [Nelson, Hirayama and Rogers (1985)]. These codes attempted the inclusion, in a phenomenological way, of the various effects that the analytical approaches failed to describe. An example of such an approach gives the average longitudinal development of cascades, i.e., the energy

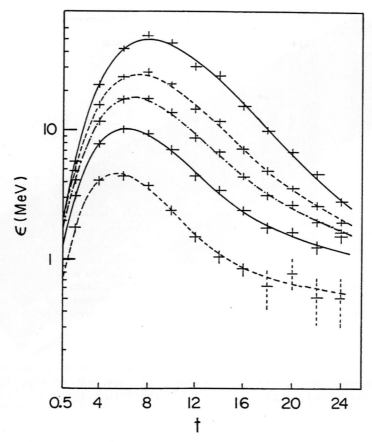

Fig. 2.77 The electromagnetic-cascade longitudinal-development as a function of the calorimeter depth (in units of radiation lengths) at 4, 9, 26, 35, and 49 GeV measured with the silicon/tungsten calorimeter of SICAPO Collaboration shown in Fig. 2.78 (reprinted from *Nucl. Instr. and Meth. in Phys. Res. A* **235**, Barbiellini, G., Cecchet, G., Hemery, J.Y., Lemeilleur, F., Leroy, C., Levman, G., Rancoita, P.G. and Seidman, A., Energy Resolution and Longitudinal Shower Development in a Si/W Electromagnetic Calorimeter, 55–60, Copyright (1985), with permission from Elsevier; see also [Leroy and Rancoita (2000)]). In ordinate with a logarithmic scale, it is shown the energy deposited, in MeV, in each detector.

deposited per unit of radiation length dE/dt in a lead-glass homogeneous calorimeter for primary photons from 0.1 up to 5 GeV [Longo and Sestilli (1975)]. Longo and Sestilli (1975) determined the parametrization of the shower development given by Eq. (2.232) and found t_{max} values for the simulated showers differing only slightly from the values calculated via Eq. (2.229). More recently, the mean longitudinal shower-distribution was studied, in a systematic way, by using simulation codes for a wide range of elements at energies up to about hundred GeV ([Grindhammer et al. (1989); PDB (2008)], see also for instance [Iwata (1979, 1980); Fabjan (1985a,

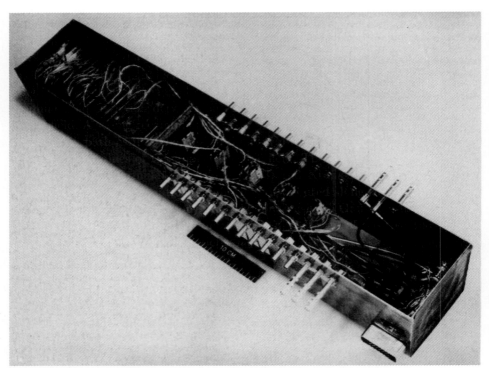

Fig. 2.78   Electromagnetic calorimeter of SICAPO Collaboration: it was the first using silicon detectors as active samplers (reprinted from *Nucl. Instr. and Meth. in Phys. Res. A* **235**, Barbiellini, G., Buksh, P., Cecchet, G., Hemery, J.Y., Lemeilleur, F., Rancoita, P.G., Vismara, G. and Seidman, A., Silicon Detectors and Associated Electronics Oriented to Calorimetry, 216–223, Copyright (1985), with permission from Elsevier). The active area of the detectors was $5 \times 5\,\text{cm}^2$.

1986)]). The longitudinal shower distribution is approximately described by

$$\frac{dE}{dt} = St^{\omega}e^{-bt}, \qquad (2.232)$$

where $b \approx 0.5$, $S = Eb^{\omega+1}/\Gamma(\omega+1)$, E is the incoming particle energy in GeV and Γ is Euler's function. More precise values for b as function of A and E/ϵ_c can be found in [Leroy and Rancoita (2000); PDB (2004)] and are shown in Fig. 2.76. The position of the cascade maximum is located at

$$t_{\text{max}} = \omega/b = \left[\ln\left(\frac{E}{\epsilon_c}\right) + f\right], \qquad (2.233)$$

where $f = -0.5$ or $+0.5$ for incident electrons or photons, respectively.

The slow exponential decay of the cascade beyond its maximum is often expressed as $e^{-t/\lambda_{\text{att}}}$ where λ_{att} is the longitudinal attenuation length. For instance, $\lambda_{\text{att}} = 1/b$ according to Eq. (2.232). Measured values of λ_{att} (in units of g/cm^2 and radiation length X_0) are listed in Table 2.14 for various absorbers; they show very little dependence on the incoming particle energy reflecting the fact that, at

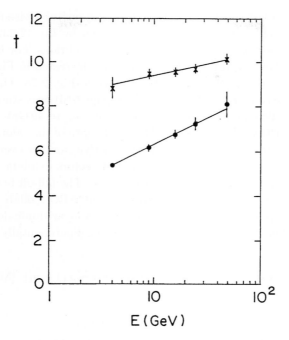

Fig. 2.79 The cascade maximum, t_{max} (lower curve), and the cascade center-of-gravity, t_{cg} (upper curve), in units of radiation lengths as a function of the incoming-electron energy measured with the silicon/tungsten calorimeter of SICAPO Collaboration shown in Fig. 2.78 (reprinted from *Nucl. Instr. and Meth. in Phys. Res. A* **235**, Barbiellini, G., Cecchet, G., Hemery, J.Y., Lemeilleur, F., Leroy, C., Levman, G., Rancoita, P.G. and Seidman, A., Energy Resolution and Longitudinal Shower Development in a Si/W Electromagnetic Calorimeter, 55–60, Copyright (1985), with permission from Elsevier; see also [Leroy and Rancoita (2000)]).

large depths, cascading is mainly due to low-energy photons attenuated with rates corresponding to the respective values of λ_{att}.

The cascade deposits 98% of the incoming-particle energy, i.e., 98% of the shower is longitudinally contained, at an approximate depth given by [Amaldi (1981)]:

$$L(98\%) \approx 3\, t_{\mathrm{cg}} \ [X_0] \tag{2.234}$$

with t_{cg} determined by Eq. (2.230). Other approximate expressions can be found in [Fabjan (1986)]. A further parametrization of the longitudinal shower containment for 95% longitudinal containment, using available experimental data and calculations [Fabjan (1986)], is given by:

$$L(95\%) = (t_{\mathrm{max}} + 0.08Z + 9.6) \ [X_0] \tag{2.235}$$

with t_{max} obtained by Eq. (2.229).

The parametrization of Eq. (2.232) proves to be not fully adequate when applied to the description of data taken with different types of calorimeters, in particular for low values of E_c. Dependences more complex than the one given by Eq. (2.232) were already proposed in the analytical formulations of Heisenberg (1946) and Rossi (1964). Therefore, the parametrization given by Eq. (2.232) has to be improved to account for the modified experimental conditions. For instance, the longitudinal behavior of electromagnetic cascades is illustrated in Fig. 2.77 [SICAPO Collab. (1985a)] for the Si/W calorimeter* shown in Fig. 2.78. The average energy deposition, i.e., the so-called *visible energy* (see page 616), was studied as a function of the depth t for incoming electron energies from 4 up to 49 GeV using that Si/W calorimeter [SICAPO Collab. (1985a)]. The total depth of the calorimeter was 24 X_0 with a silicon detector (25 cm^2 active area) as active sampler every[‡‡] two X_0. The mean deposited energy, ϵ (in MeV), in silicon detectors exhibits a steep rise with the longitudinal depth, followed by a slow decrease. The fall-off beyond the cascade maximum displays a two-component structure [SICAPO Collab. (1985a,b)], each component being approximately exponential with a logarithmic dependence on the incident energy. The fitted curve, interpolating the experimentally measured values of ϵ, is expressd by:

$$\epsilon = \epsilon_0 \left(\frac{t}{2}\right)^{\omega_1} \exp(-b_1 t) + \epsilon_1 \left(\frac{t}{2}\right)^{\omega_2} \exp[-u(t - x_1) - y_1] \quad [\text{MeV}], \qquad (2.236)$$

where

$$\omega_1 = (3.2 \pm 0.5) + (0.3 \pm 0.2) \ln E,$$
$$b_1 = (0.75 \pm 0.10) + (-0.05 \pm 0.04) \ln E,$$
$$\omega_2 = (0.26 \pm 0.10) \ln E,$$
$$u = (0.04 \pm 0.02) \ln E,$$
$$x_1 = (-6.8 \pm 3.5) + (55.9 \pm 20.3) \ln E,$$
$$y_1 = 2.4 \pm 0.4,$$
$$\epsilon_0 = (2.2 \pm 1.2) + (1.5 \pm 0.6) \ln E \quad [\text{MeV}],$$
$$\epsilon_1 = 1 \quad [\text{MeV}] ,$$

and the incoming particle energy, E, is in units of GeV [SICAPO Collab. (1985a)]. In this case, the position of the cascade maximum, determined experimentally, increases with the logarithm of the incident energy (in GeV) according to

$$t_{\max} = (3.97 \pm 0.24) + (1.02 \pm 0.11) \ln E. \qquad (2.237)$$

The values of $t_{\max}$ can be calculated from Eq. (2.229), taking into account that $\epsilon_c \approx 8.08$ MeV for tungsten absorbers. This leads to

$$t_{\max} = 3.87 + 1.01 \ln E$$

in agreement with Eq. (2.237). The location of the center-of-gravity of the cascade, t_{cg}, is determined experimentally and given by

$$t_{cg} = (8.4 \pm 0.5) + (0.45 \pm 0.17) \ln E. \qquad (2.238)$$

---

*This electromagnetic calorimeter was the first one employing silicon detectors as active samplers.
[‡‡]The usage of silicon detectors for high-energy calorimetry was proposed by [Rancoita and Seidman (1984)].

The location of the cascade maximum, $t_{\max}$, and of the cascade center-of-gravity, t_{cg}, are shown in units of radiation length in Fig. 2.79 as a function of the incoming electron energy for the Si/W calorimeter of [SICAPO Collab. (1985a)]. The disagreement between the values of t_{cg}, as predicted from the "Approximation B" [Eq. (2.230)], and the values determined experimentally, i.e., those given in Eq. (2.238), is partially due to the extent of low-energy photon penetration (both lateral and longitudinal) in the absorber. The Compton and photoelectric effects were completely neglected in Rossi's formulation of shower propagation.

2.4.2.3 *Lateral Development of Electromagnetic Showers*

During the cascade development, the energy is degraded into low-energy electrons via ionization, Compton scattering and photoelectric interactions, which generate electrons dissipating their energies mainly (or to a large extent) by collision. The cascade lateral-spread** is caused by several physical processes. The photoelectric and Compton scattering generate secondary electrons which are no longer aligned with incoming-photon directions and can be even emitted in the backward hemisphere in the case of photoelectric electrons. Also, secondary Compton photons* are no longer along primary-photon directions, thus contributing to the widening of the cascade. Furthermore, multiple Coulomb scatterings, of those electrons that cannot radiate but have enough energy to travel away, lead to the spread of electron directions out of the axis defined by the primary particle direction.

The transverse depth unit of a cascade is the Molière radius and is defined as

$$R_M = \left(\frac{E_s}{\epsilon_c}\right) X_0, \qquad (2.239)$$

where $E_s = 21.2$ MeV (see page 122), and ϵ_c is the critical energy [Eq. (2.121)]. Values of R_M for various absorbers are listed in Table 2.3. For rapid estimates of the Molière radius, the following expression can be used [Amaldi (1981)]:

$$R_{gM} \approx 7\frac{A}{Z}, \qquad (2.240)$$

where R_{gM} is the Molière radius in g/cm^2, with

$$\frac{\Delta R_{gM}}{R_{gM}} \leq 10\% \text{ for } Z \geq 13.$$

In a material made of several absorbers, when none of them is negligible in units of X_0, an estimate of the overall Molière radius can be obtained from the expression

$$\frac{1}{R_{gM}} = \frac{1}{E_s} \sum_i \left(f_i \frac{\epsilon_{c,i}}{X_{g0i}}\right), \qquad (2.241)$$

where f_i, $\epsilon_{c,i}$ and X_{g0i} (in g/cm^2) are the weight fraction, critical energy and radiation length of the ith absorber, respectively. The transverse development of

---

**It is the shower development along the direction orthogonal to that of the primary particle.
*These photons may be backward scattered.

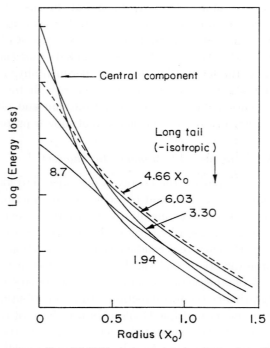

Fig. 2.80   Radial shower profile of 1 GeV electrons in aluminum, i.e., the lateral distribution of energy loss (in arbitrary units) in the shower versus the radial position in radiation lengths (from [Leroy and Rancoita (2000)]; computed curves from [Yuda et al. (1970)]).

electromagnetic cascades in different materials (from low- and medium-Z like Cu to high-Z like Pb) scales fairly accurately with R_M [Nelson et al. (1966); Crannell (1967); Bathow et al. (1970)]. Energy is deposited beyond the cylindrical volume defined in units of the Molière radius. For instance, as much as 10% of the energy is deposited beyond the cylinder with a radius of $\approx 1 R_M$. The 95% radial containment (R_e) for electromagnetic cascades is given by

$$R_e(95\%) = 2R_M. \tag{2.242}$$

The experimental data available in [Yuda (1969); Yuda et al. (1970); Nakamoto et al. (1986); Bormann et al. (1987); De Angelis (1988); SICAPO Collab. (1988a, 1989a, 1991a)] show that the cascade radial profile presents complex features (Figs. 2.80–2.82) requiring an elaborate description [SICAPO Collab. (1988a, 1991a)]. Furthermore, the experimental data indicate (Figs. 2.81–2.84) that the lateral distribution of the shower is mainly determined by the thicker (in units of X_0) shower absorber if there is an additional small (in units of X_0) absorber, as in most cases for readout materials in sampling calorimeters. In Figs. 2.81–2.82, the lateral and longitudinal distributions of the shower for 6 GeV electrons are shown every $2\,X_0$ ($\approx 7\,$mm) of tungsten [SICAPO Collab. (1989a)]. In Figs. 2.83–2.84, the experimental data (at the same energy and with the same readout detectors) were taken inserting 5 mm of G10 absorber ($\approx 2.6\%X_0$) in the middle of the $2\,X_0$ tungsten absorbers [SICAPO Collab. (1989a)]. Although the step length is 7 mm ($2\,X_0$ of tungsten) for the former set of two experimental data and 12 mm for the latter two, the distributions are similar.

The lateral distribution of the shower depends on the calorimeter depth at which it is measured. At least, a two-component structure is needed to describe the transverse profile of the electromagnetic cascade which displays a narrow central and a broad peripheral part [De Angelis (1988); SICAPO Collab. (1988a, 1991a)]. The central part scales as R_M, and is mainly due to multiple-scattering effects produced by the fast electrons responsible for the deposition of most of the incident

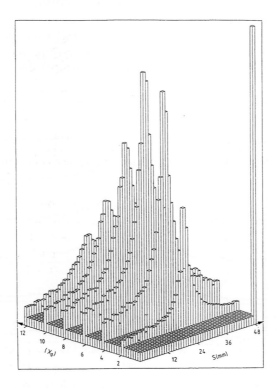

Fig. 2.81 Lateral distribution of the shower for 6 GeV electrons in tungsten absorber up to $12\,X_0$ in steps of $2\,X_0$, S being the 1 mm pitch strip number of the readout silicon detector (adapted from *Nucl. Instr. and Meth. in Phys. Res. A* **279**, Lemeilleur, F., Lamarche, F., Leroy, C., Paludetto, R., Pensotti, S., Rancoita, P.G., Vismara, L. and Seidman, A., Longitudinal and transverse development of electromagnetic showers using silicon detectors, 66–72, Copyright (1989), with permission from Elsevier; see also [Leroy and Rancoita (2000)]). The vertical axis is in arbitrary units.

energy. The peripheral part is mainly due to to the propagation of photons. The spatial distribution of this latter component is determined by the minimum value of an attenuation coefficient which strongly depends on the absorber medium.

Several two-component parameterizations of the transverse shape of an electromagnetic cascade were introduced as a function of the calorimeter depth. Usually, a double exponential form is employed. This form represents the size of the central and peripheral cascade components. A first example is the parameterization of the lateral distribution in one direction perpendicular to the shower axis:

$$Y(y,t) = a_1 \exp\left(-\frac{y}{b_1}\right) + a_2 \exp\left(-\frac{y}{b_2}\right), \qquad (2.243)$$

where y is the distance from the cascade axis and b_1, b_2 are the two lateral attenuation lengths representing the central and peripheral cascade-components, respectively. From lead-glass data, one finds [Bianchi et al. (1989)]: $b_1 = (3.4 \pm 0.1)$ mm, $b_2 = (9.3 \pm 0.3)$ mm, and $a_1/a_2 = 11.9 \pm 0.5$.

The second example is a parametrization adapted from the one used by the ALEPH experiment at LEP to simulate the transverse profile of electromagnetic cascades in a Pb-Scintillating fiber calorimeter [Charlot (1992)]. The radial density

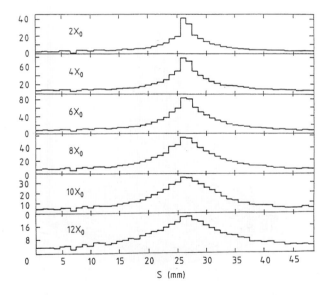

Fig. 2.82 Lateral distribution of the shower for 6 GeV electrons in tungsten absorber up to $12\,X_0$ in steps of $2\,X_0$, S being the 1 mm pitch strip number of the readout silicon detector: the distributions shown are from $2\,X_0$ (top) up to $12\,X_0$ (bottom) (adapted from *Nucl. Instr. and Meth. in Phys. Res. A* **279**, Lemeilleur, F., Lamarche, F., Leroy, C., Paludetto, R., Pensotti, S., Rancoita, P.G., Vismara, L. and Seidman, A., Longitudinal and transverse development of electromagnetic showers using silicon detectors, 66–72, Copyright (1989), with permission from Elsevier; see also [Leroy and Rancoita (2000)]). The vertical axis is in arbitrary units.

distribution, in polar coordinates, on a plane transverse to the shower axis at a shower depth t, is given by

$$F(r, t) = \kappa \left[\exp\left(-\frac{r}{\mu_1}\right) + b(t, E) \exp\left(-\frac{r}{\mu_2}\right) \right], \qquad (2.244)$$

where $\kappa = \mu_1(t, E) + b(t, E)\,\mu_2(t, E)$ and r is the radial distance from the beam axis expressed in units of X_0. The parameters μ_1, μ_2, and b are functions of the energy (E) and the depth (t) according to the empirical formulae:

$$\mu_1(t, E) = 0.025 \exp\left[\sqrt{\frac{t}{p_1(E)}}\right], \qquad (2.245)$$

and

$$b(t, E) = 0.0003 \exp[p_2(E)\,t]; \qquad (2.246)$$

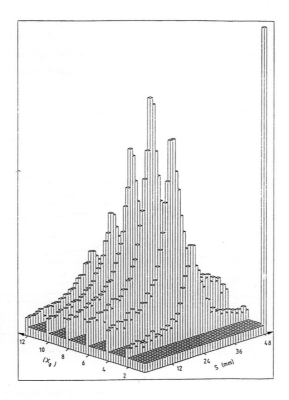

Fig. 2.83 Lateral distribution of the shower for 6 GeV electrons in tungsten and G10 absorbers up to about $12\,X_0$ in steps of about $2\,X_0$, each step consists of $1\,X_0$ tungsten followed by 5 mm G10 ($\approx 2.6\% X_0$) followed again by $1\,X_0$ of tungsten, S being the 1 mm pitch strip number of the readout silicon detector (adapted from *Nucl. Instr. and Meth. in Phys. Res. A* **279**, Lemeilleur, F., Lamarche, F., Leroy, C., Paludetto, R., Pensotti, S., Rancoita, P.G., Vismara, L. and Seidman, A., Longitudinal and transverse development of electromagnetic showers using silicon detectors, 66–72, Copyright (1989), with permission from Elsevier; see also [Leroy and Rancoita (2000)]). The vertical axis is in arbitrary units.

μ_2 is kept constant at all energies and t is in units of radiation length. $p_1(E)$ and $p_2(E)$ are fitted to the data by logarithmic functions of the incoming energy E. This particular parametrization reproduces well the radial profile of cascades up to energy around $20\,\mathrm{GeV}$, but it has difficulty to reproduce the peripheral tail at higher energy [Charlot (1992)].

A third example is provided by the data obtained from a silicon calorimeter. The lateral cascade distributions were measured using a silicon calorimeter with W and U as absorbers for incoming electron energies of 2, 4 and 6 GeV. The experimental data on lateral distributions were fitted to radial energy-distributions using the radial-probability density-function

$$F(r) = \frac{1}{N_1} \frac{\left[\exp\left(-\frac{\sqrt{r}}{\lambda_1}\right) + C_{12}\exp\left(-\frac{r}{\lambda_2}\right)\right]}{r}, \qquad (2.247)$$

where λ_1^2 and λ_2 are attenuation lengths in units of R_M, C_{12} is the relative weight of the two functions and r is the radial distance from the middle of the distribution (assumed to be the shower axis) in units of R_M. N_1 is a normalizing parameter such

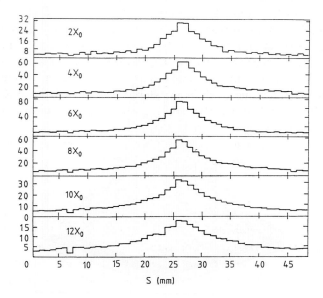

S (mm)

Fig. 2.84 Lateral distribution of the shower for 6 GeV electrons in tungsten and G10 absorbers up to about $12\,X_0$ in steps of about $2\,X_0$, each step consists of $1\,X_0$ tungsten followed by $5\,\mathrm{mm}$ G10 ($\approx 2.6\%X_0$) followed again by $1\,X_0$ of tungsten, S being the $1\,\mathrm{mm}$ pitch strip number of the readout silicon detector: the distributions shown are from $2\,X_0$ (top) up to $12\,X_0$ (bottom) (adapted from *Nucl. Instr. and Meth. in Phys. Res.* A **279**, Lemeilleur, F., Lamarche, F., Leroy, C., Paludetto, R., Pensotti, S., Rancoita, P.G., Vismara, L. and Seidman, A., Longitudinal and transverse development of electromagnetic showers using silicon detectors, 66–72, Copyright (1989), with permission from Elsevier; see also [Leroy and Rancoita (2000)]). The vertical axis is in arbitrary units.

that

$$\int_0^{2\pi} d\phi \int_0^{+\infty} F(r)\, r\, dr = 1, \qquad (2.248)$$

where ϕ is the polar angle. Alternatively, the normalization can be chosen in such a way to reproduce the longitudinal energy profile of the shower. It was observed [SICAPO Collab. (1988a)] that the values of λ_1 are proportional to the calorimeter depth, t, with a slope proportional to $\frac{1}{\sqrt{E}}$ (E is the incoming energy in GeV) and are similar for both U and W absorbers. Furthermore, it was found that the values of λ_2 and C_{12} are almost independent of t and E for both W and U absorbers and scale approximately with R_M. The t-dependence of λ_1 can be explained by the fact that in the electromagnetic-cascade development the energy is mainly carried by relativistic electrons and positrons for small t-values, but mainly by soft electrons, positrons and photons (which are also present at small depths but not dominant) at large t-values. The fast electrons and positrons are able to penetrate more and more inside the calorimeter as the energy increases. This relates the parameter λ_1 to the incoming energy E. The lateral energy distribution of the cascade broadens with increasing absorber depth, although most of the energy is deposited in the central cascade region for depths $t < 6$ [SICAPO Collab. (1988a)]. The two-component structure can take into account the fact that cascades develop i) a narrow central part containing most of the incident energy corresponding to energy deposition by fast electrons and ii) a broad lateral part due to photons and slow electrons scattered away from the cascade axis. It has to be noted that a linear sum of two Bessel functions of different slopes was considered [SICAPO Collab. (1991a)]. The fit to the data is rather good, but the four parameters, thus obtained, are difficult to interpret physically.

The following ansatz was also used [Grindhammer et al. (1990)]: the radial energy profile of the electromagnetic cascade is represented by the function

$$f(r) = \frac{2\, r R^2}{(r^2 + R^2)^2}, \qquad (2.249)$$

where r, the radial distance from the shower axis, and the free parameter R are in units of R_M. This parametrization holds, at least, as long the lateral resolution of the calorimeter is of the order or larger than $\approx 1 R_M$. The parameter R also depends on the calorimeter depth t and on $\ln E$ (E in GeV). A reasonable agreement between this model and experimental data is achieved in the description of the core and halo of cascades.

2.4.2.4 *Energy Deposition in Electromagnetic Cascades*

In electromagnetic cascades, the energy is propagated, both laterally and longitudinally, by an increase in the number of electrons and photons of diminishing energy.

The ionization and excitation processes which precede the energy deposition by collision losses of electrons and positrons were investigated using EGS4 *Monte-Carlo simulations* [Nelson, Hirayama and Rogers (1985)]. The rate of energy deposition depends on the Z-value ($Z_{absorber}$) of the medium, as expected. At the incoming electron energy of 10 GeV [Wigmans (1987)], the simulations indicate that more than 65% of the energy is deposited by collision losses of particles with energy lower than 4 MeV and only $\approx 12\%$ with energy larger than 20 MeV in high-Z materials (like uranium or lead). In low-Z materials (like aluminium), these fractions become ≈ 43 and 32%, respectively. In Fig. 2.85, results of EGS4 simulations from [Wigmans (1987)] are shown for Al, Fe, Sn, Pb and U absorbers. The fraction fr of the energy deposited by particles with energy larger than 20 MeV is approximately represented, in percentage, by

$$fr(> 20\,\text{MeV}) = -9.3\ln(Z_{absorber}) + 54.5, \qquad (2.250)$$

while the fraction of the energy deposited by particles with energy lower than 4 MeV is

$$fr(< 4\,\text{MeV}) = 10.5\ln(Z_{absorber}) + 17.4, \qquad (2.251)$$

which becomes for particles with energy lower than 1 MeV

$$fr(< 1\,\text{MeV}) = 5.0\ln(Z_{absorber}) + 17.0. \qquad (2.252)$$

The sum of the two fractions, for energies larger than 20 MeV and smaller than 4 MeV, amounts to $\approx 76\%$ and is almost independent of the medium.

The electron and positron angular spread was investigated by Monte-Carlo simulations [Fisher (1978); Amaldi (1981)]. The charged particles are expected to spread laterally because of multiple Coulomb, Compton and photoelectric interactions, as previously discussed. The size of this spread depends on R_M/X_0. The mean cosine value of the charged-particle opening angle, θ (in radian), was evaluated by Amaldi (1981) and it is given by

$$\langle \cos\theta \rangle \approx \cos\left(\frac{E_s}{\pi\epsilon_c}\right), \qquad (2.253)$$

where $E_s = 21.2\,\text{MeV}$ and ϵ_c is the critical energy.

2.4.3   *Shower Propagation and Diffusion in Complex Absorbers*

Rossi's analytical formulation of shower transport does not address the problem of shower propagation within a complex medium. By complex medium is meant, in the case of a calorimetric device, a structure subdivided regularly in passive samplers, readout by active media located after each of them. A passive sampler is composed of several absorbers, none of them having its thickness negligible in units of radiation length. Therefore, the shower cannot be considered as generated in a single type of absorber, and its basic characteristics (like the shower maximum depth) cannot be simply related to a well determined critical-energy value.

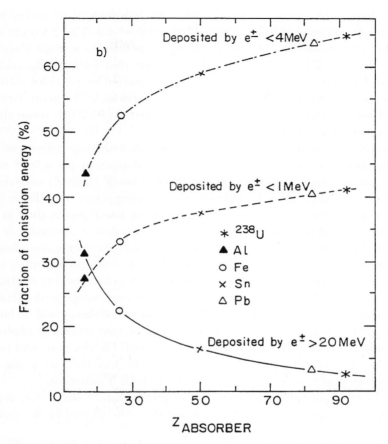

Fig. 2.85 Fraction of the energy of a 10 GeV electron shower that is deposited through ionization by electrons and positrons with energies lower than 1 or 4 MeV or larger than 20 MeV, as a function of the atomic number (Z_{absorber}) of the calorimeter absorber (reprinted from *Nucl. Instr. and Meth. in Phys. Res. A* **259**, Wigmans, R., On the energy resolution of uranium and other hadron calorimeters, 389–429, Copyright (1987), with permission from Elsevier; see also [Leroy and Rancoita (2000)]).

As discussed in previous sections, low-energy photons lead to the generation of low-energy electrons by Compton and photoelectric effects, which have a different Z-dependence, i.e., the Compton cross section is proportional to Z and the photoelectric cross section has a dependence close to Z^5. The Z-dependence largely accounts for the differential attenuation of radiation in absorbers. Most of these generated low-energy electrons are absorbed before they reach the readout medium. In general, thicknesses not small compared with the material radiation-length are needed for absorbing soft electrons and positrons with energy below or close to the critical energy: as previously seen, their dE/dx by collision loss is $\approx \epsilon_c/X_0$. Thus, a significant change in the soft-energy deposition can be expected if materials with large difference among their critical-energy values[‡] are sequentially set in the complex passive-sampler. Results of EGS4 Monte-Carlo simulations (Fig. 2.85) are summarized in Eq. (2.250) and Eq. (2.251) for showers generated in a single absorber. As mentioned in Sect. 2.4.2.4, these equations indicate that for large-$Z_{absorber}$ (like[¶] W, Pb and U) more than 65% of the energy is deposited by particles with energy lower than 4 MeV, while $\approx 12\%$ is deposited by particles with energy larger than 20 MeV. These fractions become lower than 50% and $\approx (30\text{--}20)\%$, respectively, for low-Z absorbers (like[‖] Al and Fe).

When cascading particles are traversing materials with large differences in their critical energies, there is a change in the energy deposition per g/cm^2 resulting from collision and radiative energy-losses. In fact, low-Z (high-Z) materials have large (small) critical-energies. In addition, the stopping power by collision loss and the mass attenuation length per g/cm^2 are larger in low-Z media than in high-Z media. *The energy deposition-rate changes because the shower, composed of charged particles and photons generated in a medium, deposits its energy according to the stopping power and attenuation length of another material.* As an example, when a cascade initiated in a high-Z material enters a low-Z absorber, electrons and positrons with energies between 20 and 7 MeV will lose energy predominantly by collision, while traveling in the high-Z absorber the loss of energy will be largely by radiation. Furthermore, the rate of low-energy photon interactions by photoelectric effect is reduced. If the absorber sequence is reversed, the electrons and positrons with energies between 20 and 7 MeV will reverse the way they are losing energy, and low-energy photons will be more absorbed in high-Z materials.

It can be concluded that *the process of shower transport and energy deposition depends on the complex absorber-structure* and *it is also affected by the sequence of employed absorbers.*

[‡]This corresponds to a large difference in the atomic numbers ($Z_{absorber}$) of absorbers.
[¶]The critical energies of these absorbers are between about 7 and 8 MeV.
[‖]The critical energies of these absorbers are about 40 and 21 MeV, respectively.

Chapter 3

Nuclear Interactions in Matter

In the previous chapter, we discussed the electromagnetic interactions of electrons, positrons, photons and heavy charged-particles with matter. These interactions manifest themselves in many ways, like ionization or excitation of atoms or molecules; they can also result in i) emission of photons or electrons, ii) emission of radiation due to a collective effect in the dielectric medium, iii) electron and positron pair generation and, finally, iv) bremsstrahlung.

On the other hand, there are processes induced by *hadronic particles* or *hadrons*, which are responsible for *nuclear* and *subnuclear interactions*, also called *strong interactions*. Although some hadronic interactions were already considered, like for instance the hadronic feature of photon interactions on nuclei, a general and detailed description of strong interactions is beyond the purpose of this book.

In the present chapter, hadronic interactions will be addressed from a phenomenological point of view, at energies large enough to be outside the field of nuclear physics. In particular, the hadron–nucleus interaction will be discussed in the third section; it is the fundamental phenomenon for the development and propagation of hadronic showers in matter (see, also, the chapter on *Principles of Particle Energy Determination*). Low energy reactions are only mentioned: they are treated, with more details, in specialized nuclear physics textbooks. As an introduction to hadron–nucleus collisions, we will shortly review general properties of nuclei in the first section; while, in the second section, the phenomenology of hadron–nucleus collisions is treated at high energy. Furthermore, complete data sets regarding nuclear and nuclide properties, discussed within this chapter, are available on the web (see, for instance, [NNDC (2008a,b)]).

3.1 General Properties of the Nucleus

The nucleus is the inner constituent of the atom, which is the basic element of matter. A nucleus is a system of M_A *elementary particles*, namely protons and neutrons, held together by attractive nuclear forces which can be treated, to a good approximation, by methods of non-relativistic quantum-mechanics. M_A is called *mass number* (or *atomic mass-number*) and is the sum of the total number of protons, also referred to as *atomic number* Z, and neutrons (N_n) inside the nucleus, i.e.,

$$M_A = Z + N_n. \tag{3.1}$$

Nowadays, we know that the *nuclear force* is not a fundamental force and that it results from *strong interactions* which are gluing quarks together, thus, generating the individual protons and neutrons. These particles are also referred to as *nucleons*, when inside a nucleus. The strong interaction also maintains protons and neutrons together overcoming the repulsive electrostatic force among protons. Therefore, M_A is the number of nucleons inside the nucleus.

Different combinations of Z and N_n are called *nuclides*. Nuclides with the same mass number are called *isobars*. Nuclides with the same Z-value are called *isotopes* (see Appendix A.5). Nuclides with the same value of N_n are called *isotones*. Furthermore, we will follow the traditional nuclear physics notation for nuclides. Thus, for an element whose chemical symbol is X, the corresponding mass, atomic and neutron numbers are indicated as:

$$_Z^{M_A} X_{N_n}.$$

Often, both the atomic and neutron numbers are not explicitly shown. For instance, the carbon isotope with mass number $M_A = 14$ can be written equivalently as: $_6^{14}C_8$, $^{14}C_8$, $_6^{14}C$ or, simply, ^{14}C.

The nucleus can be found in series of quantum states of different energy. The state of lowest energy, the so-called *ground state*, is the state into which the nucleus returns after the emission of one or several photons, after having been excited.

The amount of *nuclear binding energy* indicates the degree of the nucleus stability. This energy, E_b, is related to the mass difference $\triangle M$ between the system (nucleus) mass and the sum of the masses of its constituents. This phenomenon was historically called *mass defect* and, following the mass energy equivalence, it is expressed as:

$$E_b = \triangle M\, c^2. \tag{3.2}$$

The mass of the nucleus is given by:

$$M_{\text{Nuc}} = Z m_p + (M_A - Z)\, m_n - \triangle M, \tag{3.3}$$

where m_p and m_n are the proton and neutron rest-masses, respectively. Since the whole atom is neutral, the number of electrons is equal to the number of protons. The nuclear binding energy is determined from the atomic masses, M_{at}, because they can be measured with a larger precision than nuclear masses [Andi and Wapstra (1993)]. By neglecting the binding energies of the electrons, the atomic mass differs from the nuclear mass (M_{Nuc}) by the the rest-masses of the electrons, i.e.,

$$M_{\text{at}} = Z m + M_{\text{Nuc}}, \tag{3.4}$$

where m is the electron rest-mass. From Eqs. (3.3, 3.4), we have:

$$E_b = [Z M_{H^1} + (M_A - Z) m_n - M_{at}] c^2$$
$$\approx [Z(m_p + m) + (M_A - Z) m_n - Zm - M_{Nuc}] c^2$$
$$= \triangle M c^2, \tag{3.5}$$

where M_{H^1} is the mass of the hydrogen atom, whose binding energy is 13.6 eV and can be neglected. The atomic mass is most often expressed in unified atomic mass units (u). In those units, the mass of the hydrogen atom is $M_{H^1} \simeq 1.00794$ u. As discussed in Sect. 1.4.1, the unified atomic mass unit is equal to 1/12 of the ^{12}C atomic mass in its ground state. The values of the atomic masses of isotopes are periodically reviewed to account for the latest measurements and are also available on the web (e.g., see [Tuli (2000); NNDC (2003); IUPAC (2006); NNDC (2008a)]).

The binding energy per nucleon E_b/M_A is shown in Fig. 3.1 ([Povh, Rith, Scholz and Zetsche (1995)] and references therein; see also [Marmier and Sheldon (1969)]). For most nuclei (all but the very light* ones below $M_A = 11$), the value of the average binding energy per nucleon of an isotope (I), i.e.,

$$B_{nucl}(I, M_A) = \frac{E_b(I, M_A)}{M_A} = \frac{\triangle M(I, M_A) c^2}{M_A}, \tag{3.6}$$

slightly varies around (7.0–8.8) MeV/nucleon.

In practical calculations in which i) the mass difference between proton and neutron can be neglected, i.e.,

$$m_{nucl} \simeq m_p \simeq m_n,$$

where m_{nucl} is the nucleon mass, ii) the dependence of the binding energy on M_A can be neglected, i.e., $B_{nucl}(I, M_A)$ is replaced by an almost constant effective value B_e, and iii) the contribution of electrons to the overall mass is not taken into account, the atomic mass of a nuclide, with atomic number Z and N_n neutrons, is obtained using Eqs. (3.2, 3.4, 3.6), which allow one to approximate Eq. (3.3) as:

$$M_{at} \approx M_{Nuc} \approx M_A m_{nucl} \left(1 - \frac{\triangle M}{M_A m_{nucl}} \right) \tag{3.7}$$

$$= M_A m_{nucl} \left[1 - \frac{B_{nucl}(I, M_A)}{m_{nucl} c^2} \right] \tag{3.8}$$

$$\simeq M_A m_{nucl} \left(1 - \frac{B_e}{m_{nucl} c^2} \right). \tag{3.9}$$

Under these assumptions, from Eq. (3.9) the atomic mass of ^{12}C is:

$$M_{at,^{12}C} = 12 \, u \approx M_{Nuc,^{12}C} \sim 12 \, m_{nucl} \left(1 - \frac{B_e}{m_{nucl} c^2} \right);$$

thus, one finds

$$u \approx m_{nucl} \left(1 - \frac{B_e}{m_{nucl} c^2} \right) \tag{3.10}$$

---

*For ^{12}C, the value of the binding energy per nucleon is $B_{nucl}(^{12}C, M_A) \approx 7.7$ MeV/nucleon; for ^{4}He, ^{7}Li, ^{9}Be and ^{10}B are ≈ 7.1, 5.6, 6.5 and 6.5 MeV/nucleon, respectively.

and, finally, Eq. (3.9) can be re-written as

$$M_{\text{at}} \approx M_A \, \text{u}. \tag{3.11}$$

In summary, *to a first approximation the relative atomic mass, A* (also referred to as atomic weight, see page 15), *of a nuclide is expressed by its mass number*. Furthermore (see Appendix A.2), if we assume $m_{\text{nucl}} \approx 938.92 \, \text{MeV}/\text{c}^2$, i.e., the mean value between the proton and neutron masses, from an inspection of Eqs. (3.8, 3.9) we can note that the variation$^{\parallel}$ of the binding energy per nucleon divided by m_{nucl} slightly affects the resulting value of M_{at} and the expression (3.11) is approximated to about or better than a percent also for light nuclei. From Eq. (3.10) one obtains $B_e \approx 7.42 \, \text{MeV/nucleon}$. Since numerically $A \sim M_A$, in this chapter the symbol A may be found to replace M_A, particularly in theoretical expressions or calculations (e.g., see Sects. 3.2.2-3.2.3) to indicate the number of nucleons in a nucleus with a compact notation.

3.1.1   *Radius of Nuclei and the Liquid Droplet Model*

The size or the *radius of a nucleus* is not completely defined because the nucleus cannot be considered as a rigid sphere. However, under the approximation of a spherical shape, the mean nuclear radius can be determined by scattering with particles such as electrons, neutrons, alpha's, etc. From these measurements, it was concluded that the nuclear radii are proportional to the cubic root of their mass number. From the earliest scattering investigations by Rutherford and Chadwick, it was derived that, except for the lightest nuclei, the nuclear radius, r_n, is given by the relation [Bethe and Ashkin (1953); Povh, Rith, Scholz and Zetsche (1995)]:

$$r_n \simeq r_0 M_A^{1/3} \quad [\text{fm}], \tag{3.12}$$

where $r_0 \simeq 1.2 \, \text{fm} = 1.2 \times 10^{-13}$ cm. For these nuclei, the *mean nuclear density, ρ_0, is approximately constant* and given by:

$$
\begin{aligned}
\rho_0 &\simeq \frac{M_A \, m_{\text{p}}}{\frac{4}{3}\pi r_n^3} \\
&= \frac{M_A \, m_{\text{p}}}{\frac{4}{3}\pi \left(r_0 M_A^{1/3}\right)^3} \\
&= \frac{3 \, m_{\text{p}}}{4 \, \pi r_0^3} \\
&\approx 2 \times 10^{14} \quad [\text{g/cm}^3].
\end{aligned}
\tag{3.13}
$$

The mean density of the nuclear matter is extremely large relatively to ordinary matter.

$^{\parallel}$For all isotopes from ^{3}H up to ^{238}U, the binding energy per nucleon ranges from $\approx$ 2.6 MeV/nucleon up to $\approx$ 8.8 MeV/nucleon; for ^{2}H (deuterium) it is $\approx$ 1.1 MeV/nucleon.

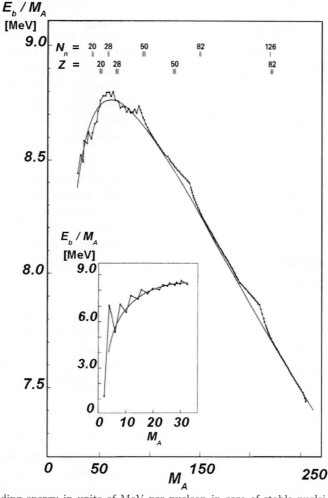

Fig. 3.1 Binding energy in units of MeV per nucleon in case of stable nuclei with even values of mass number M_A as a function of M_A (adapted and reprinted with permission from Povh, B., Rith, K., Scholz, C. and Zetsche, F. (1995), Particles and Nuclei: an Introduction to the Physical Concepts, Figure 2.4, page 18, Springer-Verlag Publ., Berlin Heidelberg New-York; © by Springer-Verlag 1995). The continuous line is determined by the Weizsäcker–Bethe mass formula given in Eq. (3.14) (at page 223).

3.1.1.1   *Droplet Model and Semi-empirical Mass Formula*

In nuclear physics, the analogy with the physics of an incompressible fluid was suggested by the fact that the nuclear density (Sect. 3.1.1) and binding energy per nucleon (Sect. 3.1) are almost constant. Thus, the nucleus can be regarded as a *liquid drop*, in which an almost constant binding energy per nucleon corresponds to a constant heat of vaporization independent of the droplet size.

The liquid drop model was used by von Weizsäcker and Bethe to evaluate nuclear masses ([von Weizsäcker (1936); Blatt and Weisskopf (1952); Bethe and Ashkin (1953); Finkelnburg (1964)]; see also Chapter 2, Section 3 in [Povh, Rith, Scholz and Zetsche (1995)]). In addition, a semi-empirical mass formula was proposed in which the overall nuclear binding energy can be computed on the base of several contributions, whose relative magnitudes are given empirically by adapting a few parameters to measured nuclear masses. In this framework with the parameters a_v, a_s, a_c, a_a and δ_p defined at page 223 (from [Povh, Rith, Scholz and Zetsche (1995)]), five contributions to the binding energy, E_b, are considered:

- Volume Energy
  The major contribution to E_b is given by nucleon interactions mediated by the nuclear forces. As already mentioned, the binding energy per nucleon E_b/M_A slightly varies, except for light (and very heavy) nuclei. Furthermore, the quasi-constancy of the mean nuclear-matter density indicates that the nuclear forces are *short ranged* and involve the nearest nucleon neighbors. The binding volume-energy can be expressed by:
  $$E_b^v \simeq a_v M_A.$$

- Surface Energy
  Nucleons located at the nuclear surface are necessarily less bound than those inside the nucleus, because less nucleons are surrounding them. The number of surface nucleons is proportional to the nucleus surface, which, in turn, is proportional to $M_A^{2/3}$ [Eq. (3.12)]. The effect of the binding surface energy is to decrease the overall binding-energy, thus:
  $$E_b^s \simeq -a_s M_A^{2/3}.$$

- Coulomb Energy
  The electrostatic repulsion among protons has a *long range* characteristic. The resultant Coulomb-energy will decrease the total amount of the nuclear binding-energy. In the electrostatic theory, the energy due to a net charge Ze uniformly distributed over a sphere of radius R is given by $-\frac{3}{5}\frac{(Ze)^2}{R}$. To a first approximation, the Coulomb-energy term can be evaluated as:
  $$E_b^C \simeq -a_c \frac{Z^2}{M_A^{1/3}}.$$

- Asymmetry Energy (neutron excess)
  Light stable nuclei are those for which $N_n \approx Z$. For heavier nuclei, the number of neutrons usually exceed the number of protons. The neutron excess increases as the mass number increases. Therefore, a neutron excess term depending on $N_n - Z$ must be present in the equation. In addition, it has to vanish for $N_n \simeq Z$. The asymmetry term is given by:
  $$E_b^a \simeq -a_a \frac{(N_n - Z)^2}{4M_A}.$$

- Pairing Energy

  Nuclei with an even number of both protons and neutrons show a trend of high stability. Nuclei having an odd number of one type of nucleon and an even number of the other type are less stable. Nuclei with doubly odd numbers of protons and neutrons are even more unstable. This *pairing effect* is taken into account by the additional term:

$$E_b^p \simeq -\frac{\delta_p}{\sqrt{M_A}}.$$

As mentioned above, the *semi-empirical mass formula* was derived by von Weizsäcker and Bethe for the atomic mass $M_{\text{at}}(Z, M_A)$ of an atom constituted by M_A nucleons, Z of them being protons; it can be summarized in the following way:

$$\begin{aligned}
M_{\text{at}}(Z, M_A) &= N_n m_n + Z m_p + Z m - E_b \\
&= N_n m_n + Z m_p + Z m - a_v M_A \\
&\quad + a_s M_A^{2/3} + a_c \frac{Z^2}{M_A^{1/3}} + a_a \frac{(N_n - Z)^2}{4 M_A} + \frac{\delta_p}{\sqrt{M_A}},
\end{aligned} \qquad (3.14)$$

where the parameters are (from [Povh, Rith, Scholz and Zetsche (1995)])

$a_v = 15.67$ MeV/c^2,

$a_s = 17.23$ MeV/c^2,

$a_c = 0.714$ MeV/c^2,

$a_a = 93.15$ MeV/c^2,

$\delta_p = -11.2$ MeV/c^2 for even-even nuclei (even value of M_A),

$\delta_p = 0$ MeV/c^2 for even-odd and odd-even nuclei (odd value of M_A),

$\delta_p = +11.2$ MeV/c^2 for odd-odd nuclei (even value of M_A).

3.1.2 *Form Factor and Charge Density of Nuclei*

The classical interaction of a charged particle, like an electron, with a massive object of charge Ze, like a nucleus, is described by the classical *Rutherford differential cross section* (Sect. 1.5), in which spin dependent effects and target recoil are neglected. The same equation is derived following a non-relativistic quantum-mechanical approach, using the Born approximation, for the case of a point-like target of charge Ze. For a non-point like target, it is possible to demonstrate that the scattering cross section on nucleus can be rewritten as (see for instance Section 5.2 of [Povh, Rith, Scholz and Zetsche (1995)], and also Section 6.3 of [Segre (1977)]):

$$\left(\frac{d\sigma}{d\Omega}\right)_{\text{Rutherford, non point like}} = \left(\frac{d\sigma}{d\Omega}\right)_{\text{Rutherford}} |F(\vec{q})|^2, \qquad (3.15)$$

where $\vec{q}$ is the tri-momentum transfer in the scattering process and $F(\vec{q})$ is the so-called *nuclear form-factor* given by

$$F(\vec{q}) = \int n_c(\vec{r}) \exp\left(\frac{i\vec{q} \cdot \vec{r}}{\hbar}\right) d^3 r,$$

where $\vec{r}$ is the radial position with respect to the scattering center, d^3r is the infinitesimal volume located at $\vec{r}$, and $n_c(\vec{r})$ is the normalized charge-density distribution at $\vec{r}$. The latter is related to the charge-density distribution by:

$$\rho_c(\vec{r}, Z) = Zen_c(\vec{r})$$

with

$$\int \rho_c(\vec{r}) \, d^3r = Ze \int n_c(\vec{r}, Z) \, d^3r = Ze.$$

As a consequence, we have:

$$F(0) = \int n_c(\vec{r}) \, d^3r = 1.$$

At relativistic energies, spin effects cannot be neglected anymore and, for an electron interaction, they are taken into account by the *Mott differential cross section* (as mentioned at pages 85 and 123). Neglecting recoil effects, the Mott cross section can be written as:

$$\left(\frac{d\sigma}{d\Omega} \right)_{\text{Mott}} = \left(\frac{d\sigma}{d\Omega} \right)_{\text{Rutherford}} \left(1 - \beta^2 \sin^2 \frac{\theta}{2} \right), \tag{3.16}$$

where θ is the scattering angle. The meaning of the right-hand multiplicative term in Eq. (3.16) can be understood, for instance, considering that for $\beta \to 1$ it becomes $\sim \cos^2(\theta/2)$. Thus, for $\theta \simeq 0°$, the cross section becomes $\simeq 0$, which accounts for *helicity conservation* in the scattering process.

As seen above, the extended charge-distribution of the nucleus can be described by introducing the nuclear form-factor. Often, for charge distributions which have a spherical symmetry, form factors depend on the value of the tri-momentum transfer ($|\vec{q}|$) and is indicated as $F(q)$, or $F(q^2)$. The form factor can be determined by measuring the experimental differential cross section and taking the ratio to the Mott differential cross section (see for instance [Povh, Rith, Scholz and Zetsche (1995)]), namely:

$$\left(\frac{d\sigma}{d\Omega} \right)_{\text{exp}} = \left(\frac{d\sigma}{d\Omega} \right)_{\text{Mott}} |F(q)|^2.$$

The first measurements of nuclear form-factors were carried out during the years 1950s. In Fig. 3.2, the differential cross section of electrons on ^{12}C nuclei is shown as a function of the electron diffusion angle. It was obtained by electron scattering at 420 MeV [Hofstadter (1957)]: the shape of the differential cross section depends on the nuclear form-factor. The dashed line corresponds to the expected differential cross section calculated using the Born approximation for the electron scattering on the diffused surface of a homogeneous sphere. The differential cross section exhibits the typical diffraction-pattern, in which a minimum is present at $\theta \simeq 51°$, i.e., $|\vec{q}|/\hbar \simeq 1.8$ fm^{-1}.

Systematic experimental measurements have shown that the nucleus has a charge distribution decreasing gradually towards the surface. Therefore, one could conclude

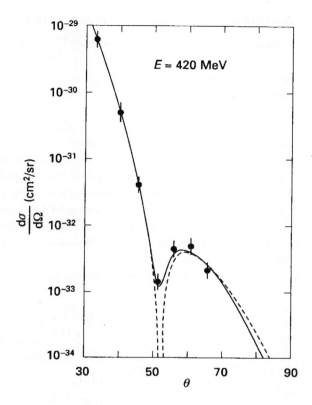

Fig. 3.2   Measurement of the form factor of ^{12}C nucleus obtained by the scattering of 420 MeV electrons (adapted and reprinted, with permission, from the Annual Review of Nuclear Science, Volume 7 © 1957 by Annual Reviews *www.annualreviews.org*; [Hofstadter (1957)]). The differential cross section is shown as a function of the electron scattering angle. The dashed line corresponds to the expected electron scattering on the diffused surface of a homogeneous sphere, following the Born approximation. The continuous line corresponds to a fit to the experimental data.

that Eq. (3.13) is valid only as a simplified expression. Under the approximation of nuclear spherical charge-symmetry and in agreement with the experimental data, the radial charge-density $\rho_c(r, Z)$ is given by

$$\rho_c(r, Z) = \frac{\rho_c(0, Z)}{\exp\left[\frac{r - C_c(M_A)}{Z_0}\right] + 1}, \tag{3.17}$$

where r is the radial distance from the center of the nucleus in fm, $C_c(M_A) \simeq 1.07 \, M_A^{1/3}$ fm for $M_A \geq 30$, $Z_0 \simeq 0.545$ fm and $\rho_c(0, Z)$ is determined from the condition:

$$Ze = 4\pi \int_0^\infty \rho_c(r, Z) \, r^2 dr.$$

### 3.1.3	*Angular and Magnetic Moment, Shape of Nuclei*

Many nuclei have an *intrinsic angular momentum* $\vec{I}_s$, which represents the total angular momentum of the nucleus and is given by the sum of the orbital momentum and the spin of all protons and neutrons inside the nucleus. $\vec{I}_s$ is always an integer or a half integer in units of $\hbar$:

$$|\vec{I}_s| = I_s \hbar. \tag{3.18}$$

Often, I_s is referred to as the *nuclear spin** and varies between 0 and $\frac{9}{2}$ for all known nuclei in their ground state. Nuclei with an even number of nucleons (even mass number) have an integral value of I_s. In most cases, they do not have angular momentum at all, i.e., $I_s = 0$. In particular, even-even nuclei have a ground state with $I_s = 0$. Nuclei with odd mass number always have a half-integral spin.

Nuclei can exist in excited states of higher energy, as will be discussed in the next part of this section. The angular momenta of excited states may differ from those of the corresponding nuclear ground states. It has been shown that there are *selection rules* for transitions between energy states of the nucleus or from neighboring nuclei.

Since transitions between nuclear states with very different angular momenta are strongly forbidden, there exist long-lived excited nuclear states. Nuclei in these long-lived excited states are called *Nuclear isomers*. *Isomeric nuclei*, have the same charge and mass, but are in different energy states, i.e., have different arrangements of their nucleons. An important difference between a normal excited nuclear state and a nuclear isomer is that, for the latter one, the transition probability to a more stable state, particularly to the ground state, is very small. So, like in atomic physics, the isomeric nucleus can be called a *metastable nuclear state*.

As for the case of electron shell, a *magnetic moment* is related to the angular momentum of the electrically charged constituents of the nucleus. These constituents, by their orbital motion, generate electric current densities which contribute to the overall magnetic moment in addition to other effects, like the intrinsic magnetic moment of nucleons. In Appendix A.2, the measured proton and neutron magnetic moments are given in units of the so-called *nuclear magneton*, μ_N, defined as:

$$\mu_N = \frac{e\hbar}{2m_\mathrm{p}}.$$

The positive (negative) sign of the proton (neutron) magnetic moment indicates that it is directed along (opposite) to its angular momentum.

Heavy nuclei show a deviation from a spherical nuclear shape of the order of 1%. An ellipsoidal shape agrees better with the experimental results. Usually, there is a contraction in the direction of the spin axis. This asymmetry is represented by associating an *electric quadrupole moment* with the nucleus. The positive

---

*At page 232, Finkelnburg (1964) noted that this notation is misleading: this designation is correct only for elementary particles, like protons and neutrons. For nuclei consisting of protons and neutrons, $\vec{I}_s$ represents the total angular momentum of the nucleus.

quadrupole sign is assigned to an extension in the direction of the spin axis, while a negative value corresponds to a shortening along this direction.

In Table 3.1, values of the nuclear spin, nuclear magnetic and electrical quadrupole are shown for several nuclides ([Stone (2001)] and references therein).

3.1.4 *Stable and Unstable Nuclei*

An aggregate of nucleons may become a stable and bound nucleus only if the number of one nucleon-type does not largely exceed the other nucleon-type. There is no experimental evidence of a stable nucleus consisting exclusively of neutrons or exclusively of protons.

Unless replenished artificially, unstable nuclides in any given sample decay at a rate $-dN_s(t)/dt$, where the minus sign is introduced because $N_s(t)$ diminishes as the time, t, increases. The *decay rate* is proportional to the number of nuclides in the sample at the time t and is expected to become independent of t, i.e.,

$$\frac{dN_s(t)}{dt} = -\lambda_s N_s(t), \tag{3.19}$$

Table 3.1 Spin, nuclear magnetic moment and electric quadrupole moment for some stable nuclides in their ground state determined by recent measurements (see [Stone (2001)] and references therein).

Element	Z	M_A	Spin	Nuclear Magnetic Moment μ_N	Electric Quadrupole Moment 10^{-24} cm^{-2}
H	1	1	1/2	2.79284734(3)	
		2	1	0.857438228(9)	0.00286(2)
He	2	3	1/2	-2.12749772(3)	
Be	4	9	3/2	-1.1778(9)	0.0529(4)
C	6	13	1/2	0.7024118(14)	
N	7	14	1	0.40376100(6)	0.02001(10)
		15	1/2	-0.28318884(5)	
O	8	17	5/2	-1.89379(9)	-0.26(3)
Al	13	27	5/2	3.6415068(7)	0.1402(10)
Si	14	29	1/2	-0.55529(3)	
K	19	41	3/2	0.21489274(12)	0.060(5)
Ca	20	43	7/2	-1.317643(7)	-0.049(5)
Mn	25	51	5/2	3.5683(13)	0.42(7)
Ga	31	69	3/2	2.01659(5)	0.17(3)
Zr	40	91	5/2	-1.30362(2)	-0.206(10)
Ru	44	99	5/2	-0.641(5)	0.079(4)
		101	5/2	-0.719(6)	0.46(2)
Nd	60	145	7/2	-0.656(4)	-0.314(12)
Ta	73	181	7/2	2.3705(7)	3.17(2)
W	74	183	1/2	0.11778476(9)	
Ir	77	191	3/2	0.1507(6)	0.816(9)
Pb	82	207	1/2	0.58219(2)	

where λ_s is called *decay constant*: it is the inverse of the *mean lifetime*, τ_s:

$$\lambda_s \equiv \frac{1}{\tau_s}.$$

Finally, by integrating Eq. (3.19) between the limits $N_s(t = 0) = N_0$ at $t = 0$ and $N_s(t) = N_s$ at the time t, we have:

$$N_s = N_0 \exp\left(-t\lambda_s\right) = N_0 \exp\left(-\frac{t}{\tau_s}\right).$$

The treatment, extended to the case of more than one nuclide in a substance and to fluctuations in radioactive decays, can be found in [Bethe and Ashkin (1953)].

Among the combinations of nucleons which form bound nuclear aggregates, only a relatively small fraction of them generates stable nuclides. For instance, β-stable nuclides for are the ones whose number of neutrons and protons belong to the narrow band shown* in Fig. 3.3 [Bohr and Mottelson (1969)]. For stable nuclei, an approximated empirical relationship between the mass number M_A and the atomic number Z is given by [Marmier and Sheldon (1969)]:

$$Z = \frac{M_A}{1.98 + 0.0155\, M_A^{2/3}}. \qquad (3.20)$$

This formula shows, for $M_A < 40$, that stable nuclides are those for which $Z \simeq N_n$. For the heaviest nuclei, the number of neutrons can exceed the number of protons up to $\approx 50\%$.

Dynamical instabilities lead to spontaneous nuclear break-up into two or more parts, i.e., α-decay, fission and other related phenomena. The β-*instability* proceeds via a change in the atomic number Z by the emission or absorption of an electron, namely via β-*decay* or *electron capture*.

A large amount of nuclei emits positrons or electrons under reactions, resulting in a net charge change of one unit. The β-radioactive decay originates from fundamental *weak processes* involving nucleons:

$$\mathrm{n} \to \mathrm{p} + \mathrm{e}^- + \bar{\nu}_e \qquad (3.21)$$

$$\mathrm{p} \to \mathrm{n} + \mathrm{e}^+ + \nu_e, \qquad (3.22)$$

where $\bar{\nu}_e$ and ν_e are the electronic antineutrino and neutrino, associated with the electron and the positron, respectively. Also, the process, in which an external orbital electron is absorbed, can occur:

$$\mathrm{p} + \mathrm{e}^- \to \mathrm{n} + \nu_e. \qquad (3.23)$$

The process

$$\mathrm{n} + \mathrm{e}^+ \to \mathrm{p} + \bar{\nu}_e$$

is theoretically possible, but there is no positron available around the nucleus. Processes which proceed following Eqs. (3.21, 3.22) are called electron (β^-) and positron (β^+) β-*decay*, respectively. Equation (3.23) describes the *electron capture process*. The other nucleons in the nucleus can supply to or remove the energy needed for these processes from the nucleon, undergoing the transformation.

---

*The Fig. 3.3 is known as a *Segrè chart*.

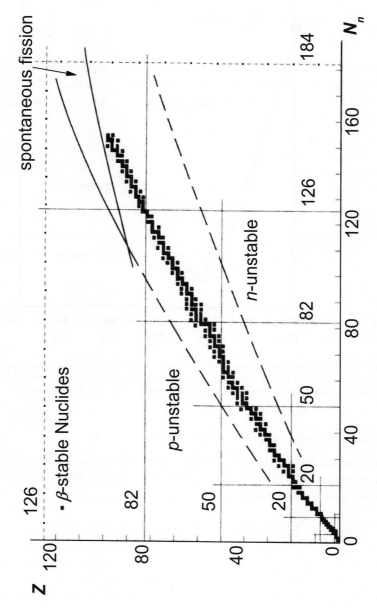

Fig. 3.3 Z versus N_n distribution of β-stable nuclides (adapted with permission from [Bohr and Mottelson (1969)]).

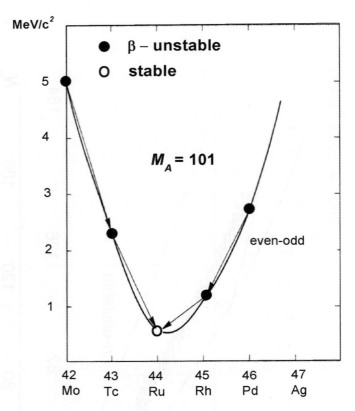

Fig. 3.4   Possible decays of nuclide with the mass number $M_A = 101$. The arrows indicate β-decays (adapted and reprinted with permission from [Segre (1977)]). The origin on the ordinate (mass scale) is chosen arbitrarily. The atomic number Z is shown in abscissa.

3.1.4.1   *The β-Decay and the Nuclear Capture*

Let us consider the *semi-empirical mass formula* derived by von Weizsäcker and Bethe, given in Sect. 3.1.1.1, for isobars with atomic mass $M_{at}(Z, M_A)$. Equation (3.14) can be rewritten as:

$$M_{at}(Z, M_A) = c_1 M_A - c_2 Z + c_3 Z^2 + \frac{\delta_p}{\sqrt{M_A}},$$

where

$$c_1 = a_v + a_s M_A^{-1/3} + \frac{a_a}{4},$$
$$c_2 = (m_n - m_p - m) + a_a,$$
$$c_3 = \frac{a_c}{M_A^{1/3}} + \frac{a_a}{M_A}.$$

The parameters δ_p, a_v, a_s, a_a and a_c are given at page 223. From this equation, we see that the isobar mass depends on Z and Z^2. The isobar mass minimum is found by imposing that

$$\frac{\partial M_{\text{at}}(Z, M_A)}{\partial Z} = 0,$$

i.e., for

$$Z = \frac{c_2}{2c_3} \approx \frac{93.93}{2\left(\frac{0.714}{M_A^{1/3}} + \frac{93.15}{M_A}\right)}. \tag{3.24}$$

For any odd M_A value, the resulting nuclear mass depends on Z following a parabolic curve. For instance, the isobar with $M_A = 101$, in Fig. 3.4, has a minimum corresponding to $Z = 44$ (^{101}Ru), as computed by means of Eq. (3.24). In Fig. 3.4, we see that isobars with neutrons in excess decay with the process given in Eq. (3.21) and emit β^- particles. Isobars, with protons in excess, decay with the process expressed by Eq. (3.22) and emit β^+ particles. However, for nuclides with even M_A value, the decay curves are different for even-even nuclei and odd-odd nuclei. The curves are separated by a term depending twice on $\delta_p/\sqrt{M_A}$. In particular, for heavy nuclei with $M_A > 70$, more than one β-stable nuclide exist, as shown in Fig. 3.5 for isobars with $M_A = 106$. The nuclei $^{106}_{46}$Pd [as predicted by means of Eq. (3.24)] and $^{106}_{48}$Cd are on the lower parabolic curve. The nuclide $^{106}_{48}$Cd is β-stable, because the neighboring odd-odd nuclides have larger masses.

As mentioned at page 228, a nucleus can decay via electron capture. The electron has a finite probability to be inside the nuclide and to allow the proton transformation into a neutron and a neutrino [Eq. (3.23)]. This reaction occurs especially for the case of heavy nuclei, in which K-electrons are close to the nucleus. As a result of a *K-capture*, an outer electron can occupy the inner vacant energy level. Consequently, a characteristic X-ray can be emitted. In Appendix A.6, half-lives and end-point energy emissions are shown for some commonly used radioactive β^+ and β^- sources, from [PDB (2000)].

3.1.4.2   *The α-Decay*

The spontaneous emission of an α-particle by some (heavy) nuclei (see Appendix A.6) is related to the compactness of the He nucleus whose binding energy is $\approx 7\,\text{MeV/nucleon}$ (see Fig. 3.1). Thus, more energy is available for the decay process. However, the α-particle has to penetrate a region of very large potential energy near the nuclear surface. In fact, the α-particle is subjected to a combination of short-range attractive nuclear forces, represented by a square-well potential, and long-range electromagnetic repulsive forces, represented by a Coulomb potential

$$V_c = \frac{2(Z-2)e^2}{r}.$$

For a purely electrostatic *Coulomb potential barrier*, like the one encountered centrally by a positively charged particle, the potential-energy height is determined

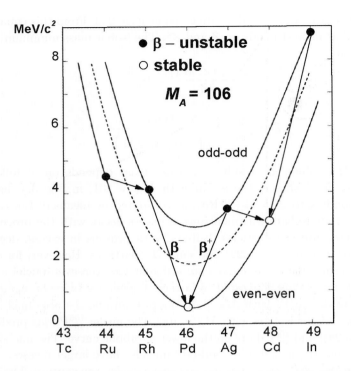

Fig. 3.5   Possible decays of nuclide with the mass number $M_A = 106$. The arrows indicate β-decays (adapted and reprinted with permission from [Segre (1977)]). The origin on the ordinate (mass scale) is chosen arbitrarily. The atomic number Z is shown in abscissa.

by the nuclear radius r_n (Fig. 3.6). For instance, in the case of an α-particle escaping centrally* from an ^{238}U nucleus, the Coulomb barrier is $\approx 24\,\mathrm{MeV}$. In classical mechanics, the potential barrier would prevent such an emission.

The penetration of the potential energy barrier is a quantum-mechanical effect. Its probability depends on the shape and height of the potential-energy barrier, as well as on the kinetic energy of the emitted α-particle. In turn, the probability is related to the lifetime of the decaying nucleus. According to quantum theory, there is a finite probability of *tunneling* through the barrier, the so-called *tunneling effect*, even for α-particle kinetic energies lower than the barrier height. It can be qualitatively understood with the aid of *Heisenberg's uncertainty relation*[‡‡]. For an α-particle of momentum p, we have (e.g., see Equation 4.15 in Chapter IV

---

*An extended discussion on the Coulomb barrier, including non-central cases, is presented in Chapter 7 of [Marmier and Sheldon (1969)] and Chapter 7 of [Segre (1977)].

[‡‡]In Chapter IV (Section 3), Finkelnburg (1964) noted that there are different methods of deducing the uncertainty principle. These methods lead to different terms at the right-hand side of Eq. (3.25), namely h, $\hbar$ or $\hbar/2$, depending on the definition of the uncertainty $\triangle$. In Eq. (3.25), the maximal uncertainty is taken into account with the term h, whereas the smaller values correspond to the mean or the most probable uncertainty, respectively.

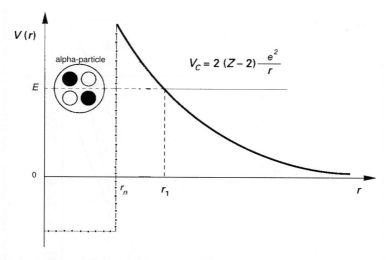

Fig. 3.6 Coulomb potential energy of an α-particle as a function of the distance r from the center of a nucleus of radius r_n (adapted and reprinted with permission from Povh, B., Rith, K., Scholz, C. and Zetsche, F. (1995), Particles and Nuclei: an Introduction to the Physical Concepts, Figure 3.5, page 31, Springer-Verlag Publ., Berlin Heidelberg New-York; © by Springer-Verlag 1995).

of [Finkelnburg (1964)]):

$$\triangle p \triangle x \approx h. \tag{3.25}$$

For a non-relativistic velocity v, because $\triangle p$ cannot be larger than the particle momentum, we get:

$$\triangle x \geq \frac{h}{p} = \frac{h}{m_\alpha v},$$

where $h/(m_\alpha v)$ is the de Broglie wavelength and m_α is the α-particle rest mass. Thus, if the velocity is such that $\triangle x$ exceeds the barrier width at that energy, there is a finite probability that the α-particle finds itself on the other side of the barrier.

The fact, that only a few nuclei with

$$150 \leq M_A \leq 210$$

are *active α-emitters*, can be related to the relative small available decay energies. For $M_A \geq 210$ (Fig. 3.7, see [Rasmussen (1965); Segre (1977)]), decay energies are larger. Hence, heavy nuclei are favored as α-emitters. The resulting half-lives can be long and are given in Appendix A.6. An example of an α-particle emitter is the ^{238}U nuclide which, together with its decay products, contributes to the natural radioactivity background; the ^{222}Rn gas can be assimilated by man and is responsible for about 40% of the natural radioactivity average exposure of human beings.

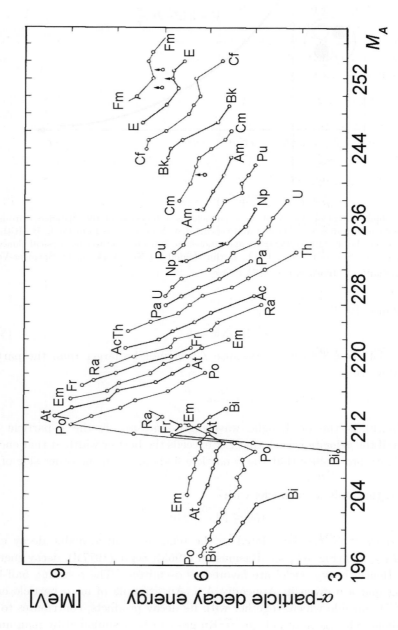

Fig. 3.7 α-particle decay-energy in MeV as a function of the atomic mass number M_A for heavy nuclei (adapted and reprinted with permission from [Rasmussen (1965)]). The lines connect isotopes.

3.1.5 *Fermi Gas Model and Nuclear Shell Model*

The presence of an internal nuclear structure was observed by electron scattering experiments on nuclei, which show a remarkable difference with results obtained by electron–proton elastic scattering.

In *elastic electron scattering on proton*, the emerging electron energy is related to the electron scattering angle. Let us consider the invariant mass of the electron–proton system and indicate with

$$\tilde{p} = (E/c, \vec{p}), \ \tilde{P} = (m_{\mathrm{p}} c, 0), \ \tilde{p}' = (E'/c, \vec{p}'), \ \tilde{P}' = \left(E'_{\mathrm{p}}/c, \vec{P}' \right),$$

the four-momentum of the incoming electron, the target proton at rest, the scattered electron and the recoil proton, respectively, with total energies E, $m_{\mathrm{p}} c^2$, E', E'_{p} and momenta $\vec{p}$, 0, $\vec{p}'$, $\vec{P}'$, respectively, where m_{p} is the proton rest-mass. By considering the invariant mass of the reaction (see page 11), we get:

$$(\tilde{p} + \tilde{P})^2 = (\tilde{p}' + \tilde{P}')^2 \Rightarrow \tilde{p} \cdot \tilde{P} = \tilde{p}' \cdot \tilde{P}'. \tag{3.26}$$

Furthermore, from four-momentum conservation, we derive:

$$\tilde{p} + \tilde{P} = \tilde{p}' + \tilde{P}' \Rightarrow \tilde{P}' = \tilde{p} + \tilde{P} - \tilde{p}'; \tag{3.27}$$

thus, by inserting $\tilde{P}'$ from Eq. (3.27) in Eq. (3.26), we obtain:

$$
\begin{aligned}
E\, m_{\mathrm{p}} &= \tilde{p}' \cdot \tilde{p} + \tilde{p}' \cdot \tilde{P} - p'^2 \\
&= \frac{EE'}{c^2} - \vec{p} \cdot \vec{p}' + E' m_{\mathrm{p}} - p'^2.
\end{aligned} \tag{3.28}
$$

In addition, we have that

$$\tilde{p}' \cdot \tilde{p}' = p'^2 = m^2 c^2$$

(m is the electron rest mass) and, in the case of energetic electrons,

$$E \approx p\,c, \ E' \approx p'c.$$

Since the electron mass can be neglected, Eq. (3.28) becomes:

$$
\begin{aligned}
E\, m_{\mathrm{p}} &= \frac{EE'}{c^2} - pp' \cos\theta + E' m_{\mathrm{p}} - m^2 c^2 \\
&\simeq \frac{EE'}{c^2} - \frac{EE'}{c^2} \cos\theta + E' m_{\mathrm{p}},
\end{aligned}
$$

where θ is the scattering angle of the electron. Finally, we get:

$$E' = \frac{E}{1 + [E/(m_{\mathrm{p}} c^2)]\,(1 - \cos\theta)}, \tag{3.29}$$

which shows that the electron recoil-energy is directly correlated to the scattering angle.

However, in the case of electron scattering on nuclei, which contain many nucleons, the energy spectrum is more complex. For instance, in the experiment of electron scattering on H_2O at incoming energy of $246\,\mathrm{MeV}$ (see Chapter 6 of [Povh, Rith,

Scholz and Zetsche (1995)]), the scattered electron energy spectrum at $\theta = 148.5°$ shows a narrow peak at $\approx 160\,\text{MeV}$ (which corresponds to the elastic electron scattering on the proton of the hydrogen) and a smooth continuous background. The continuous energy spectrum, which has a broad peak slightly lower than the elastic peak, is due to the elastic electron scattering on ^{16}O nucleons. This process is called *quasi-elastic scattering*. The broadening and shift of the quasi-elastic peak can be understood in terms of an elastic interaction on a bound nucleon, which requires a certain amount of energy being emitted from the nucleus, and the presence of an internal nucleon motion inside the nucleus. *Quasi-free nucleons*, inside the nucleus, modify the process kinematics and account for the peak broadening.

The simplest nuclear model, that accounts for momentum distribution of nucleons, is the *free nucleon Fermi gas model*, which was proposed by Fermi 1950 (e.g., see Section H in Chapter VIII therein). In this model, the overall effect of nucleon–nucleon interactions results in a square potential well, as shown in Fig. 3.8. Nucleons with spin $\frac{1}{2}$ do not have a fixed position inside the nucleus, but are free to move and constitute a *degenerate Fermi gas* (see page 810). The nucleons will obey the Fermi–Dirac statistics (Appendix A.7), which in turn accounts for the *Pauli exclusion principle*. The nuclear potential is zero outside the potential-well. At $0\,\text{K}$, since each state cannot contain more than two nucleons of the same kind, we can rewrite Eq. (A.3) of Appendix A.7 for neutrons and protons as:

$$(p_0^{\text{n}})^2 = \hbar^2 \left(3\pi^2 \frac{N_{\text{n}}}{V_{\text{nucl}}} \right)^{2/3} \qquad \Rightarrow \qquad p_0^{\text{n}} = \hbar \left(3\pi^2 \frac{N_{\text{n}}}{V_{\text{nucl}}} \right)^{1/3}, \tag{3.30}$$

$$(p_0^{\text{p}})^2 = \hbar^2 \left(3\pi^2 \frac{Z}{V_{\text{nucl}}} \right)^{2/3} \qquad \Rightarrow \qquad p_0^{\text{p}} = \hbar \left(3\pi^2 \frac{Z}{V_{\text{nucl}}} \right)^{1/3}, \tag{3.31}$$

where V_{nucl} is the nuclear volume, p_0^{n} and p_0^{p} are the *Fermi momenta* of neutrons and protons, respectively. To a first approximation, for a nucleus with

$$N_{\text{n}} = Z = \frac{M_A}{2}$$

and nuclear radius given by Eq. (3.12) (and consequently with $V_{\text{nucl}} = \frac{4}{3}\pi r_0^3 M_A$), the Fermi momentum, p_0, of the nucleons becomes:

$$p_0 = p_0^{\text{n}} = p_0^{\text{p}} = \hbar \left(3\pi^2 \frac{\frac{M_A}{2}}{V_{\text{nucl}}} \right)^{1/3}$$

$$= \hbar \left(3\pi^2 \frac{\frac{M_A}{2}}{\frac{4}{3}\pi r_0^3 M_A} \right)^{1/3}$$

$$= \frac{\hbar}{r_0} \left(\frac{9\pi}{4} \right)^{1/3}$$

$$\approx 250\,\text{MeV}/\text{c}.$$

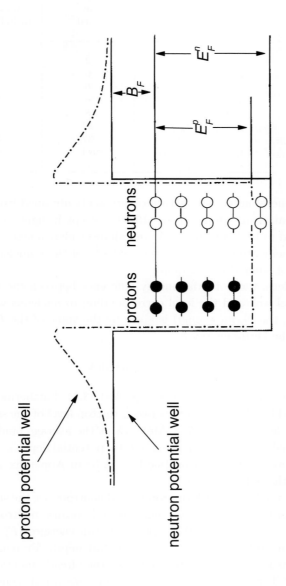

Fig. 3.8   Potential wells and states for neutrons and protons in the framework of the Fermi gas model of the nucleus (adapted and reprinted with permission from Povh, B., Rith, K., Scholz, C. and Zetsche, F. (1995), Particles and Nuclei: an Introduction to the Physical Concepts, Figure 17.1, page 225, Springer-Verlag Publ., Berlin Heidelberg New-York; © by Springer-Verlag 1995).

Table 3.2   Nuclear shells, value of l and j (which appears as subscript), number of particles in the shell and total maximum number of particles including the ones in the shell (see Section 2 in Chapter XIV of [Blatt and Weisskopf (1952)] and references therein).

Nuclear shells	Terms (l_j)	No. of particles in shell	No. of particles up to and including shell
I	$s_{1/2}$	2	2
II	$p_{3/2}, p_{1/2}$	6	8
IIa	$d_{5/2}$	6	14
III	$s_{1/2}, d_{3/2}$	6	20
IIIa	$f_{7/2}$	8	28
IV	$p_{3/2}, f_{5/2}, p_{1/2}, g_{9/2}$	22	50
V	$g_{7/2}, d_{5/2}, d_{3/2}, s_{1/2}, h_{11/2}$	32	82
VI	$h_{9/2}, f_{7/2}, f_{5/2}, p_{3/2}, p_{1/2}, i_{13/2}$	44	126

This result is in general agreement with measurements obtained from quasi-elastic electron nuclear scattering [Moniz et al. (1971)]; except in light nuclei, the Fermi momentum is found to be (240–260) MeV and almost independent of M_A. For light nuclei, the Fermi momentum is lower than expected: for these nuclei, the Fermi gas nuclear model is not sufficiently adequate.

In a stable nucleus, the potential-energy difference between the bottom level of the square potential well and the level corresponding to nucleons with the largest kinetic energy (i.e., the *Fermi energy*) is given by the value of the *Fermi level*. For $p_0 \approx 250$ MeV/c, the Fermi energy is

$$E_F = \frac{p_0^2}{2m_{\mathrm{nucl}}} \approx 33\,\mathrm{MeV}$$

(where m_{nucl} is the mass of the nucleon, see Sect. 3.1). Furthermore, the energy difference, indicated by B_F in Fig. 3.8, between the top level of the square potential well and the Fermi level is about (7–8) MeV. B_F is the average binding energy per nucleon as mentioned at page 219. Hence, the total potential depth is ≈ 40 MeV. The average kinetic energy of the nucleon [see Eq. (A.5) in Appendix A.7] is $\frac{3}{5}E_F$, of the same order as the potential depth.

In stable nuclei, the Fermi levels of protons and neutrons are the same[*]. However, because these nuclei have an excess of neutrons and because neutrons and protons occupy the same nuclear volume, the expected Fermi energies (E_F^n and E_F^p) from Eqs. (3.30, 3.31) are different. Thus, the potential depth for neutrons must be deeper (Fig. 3.8), as proposed by Fermi (e.g., see the already mentioned Section H in Chapter VIII of [Fermi (1950)]). As a result, protons are in general less bound than neutrons.

---

[*]If the Fermi levels of protons and neutrons were different, the nucleus could vary the energetic configuration via β-decay, i.e., the nucleus would be unstable.

Another particular nucleus feature is the existence of the so-called *magic number nuclei*. It has been observed that nuclei, for which Z or $M_A - Z$ or both are equal to 2, 8, 14, 20, 28, 50, 82, 126, have distinctive characteristics, i.e., the largest binding energies (like the nuclides with $N_n = 20$, 28, 50 and 82) and much higher cosmic abundances (as, for instance, the nuclei $^4_2$He, $^{16}_8$O, $^{28}_{14}$Si, $^{40}_{20}$Ca). Furthermore, the stable isotope $^{208}_{82}$Pb is characterized by the magic neutron number 126 and proton number 82. In atomic physics, an analog case is the stability of inert elements, which is attributed to the filling of electron shells. Similarly, the magic number nuclei are considered as an indication of the presence of a nuclear shell structure.

In the *nuclear shell model*[¶], the interaction of any nucleon with remaining nucleons inside the nucleus is represented by a static potential well. The potential well is similar in both extension and shape to the nuclear density distribution. Energy levels in the potential well consist of a series of single particle energy-levels $E(n, l)$, where (as in atomic physics) n is the *principal quantum number* and l is the *orbital angular momentum quantum number*. Their relative spacing is a function of the shape and depth of the potential well. Protons and neutrons fill these levels, but the number of particles in each level is determined by the Pauli exclusion principle. However, in 1948, Mayer and Jensen demonstrated that, because of the spin dependence of nuclear forces, there is a strong spin-orbit potential coupling which results in splitting the (n, l) energy-level in two sublevels (n, l, j) with $j = l \pm \frac{1}{2}$, i.e., there is a spin-orbit coupling which splits the sublevels with $j = l - \frac{1}{2}$ and $j = l + \frac{1}{2}$, so that the latter one[§] is at lower energy. The overall effect is the shell assignment shown in Table 3.2. The shell model also predicts the value of magnetic moments for odd-even and even-odd nuclei. The assignment of definite l values to proton- or neutron-odd nuclei can also be used in the theory of β-decay, in which the decay probability depends on the spin difference between the decaying nucleus and the decay product. A remarkable success of the shell model was its capability to predict that certain β-transitions should exhibit *forbidden*-type spectra. In addition, the model was able to explain why almost all *isomeric states* (see page 226) have long lifetimes.

3.1.5.1 γ Emission by Nuclei

A nucleus can have excited states from which it decays via γ-emission. These nuclear de-excitations provide information on quantum numbers and energy levels of nuclei. Above the ground state, there are many levels with characteristics J^P (spin J and parity P) quantum numbers.

In general, the excitation of an even-even nucleus has the consequence of break-

---

[¶]The reader can see Chapter 17 of [Henley and Garcia (2007)] for a recent review and references on the nuclear shell model.

[§]In this level, the intrinsic spin is parallel to the orbital angular momentum. In addition, this effect is assumed to increase with increasing values of l (e.g., see Section 2 of Chapter IV of [Blatt and Weisskopf (1952)]).

ing a pair of bound nucleons, i.e., it requires $\approx (1\text{--}2)\,\mathrm{MeV}$. Even-even nuclei with $M_A > 20$ may require more than $2\,\mathrm{MeV}$. Even-odd and odd-odd nuclei have many lower excited states of about $100\,\mathrm{keV}$. The electromagnetic radiation emitted by the decay of lower excited energy states can be interpreted as given by a superposition of electric dipole, quadrupole, octopole and other multi-pole transitions (see [Segre (1977); Povh, Rith, Scholz and Zetsche (1995)]), which have different angular distributions and are indicated by E_1, E_2, E_3, etc.. The corresponding multipole magnetic transitions are called M_1, M_2, M_3, etc.. The angular momentum and parity conservations determine *angular momentum* and *parity selection rules*. The treatment of the multipole transition probability is rather complicated and still rather approximative (see, for instance, Chapter 9 in [Marmier and Sheldon (1969)] and references therein).

When γ-decay is so much inhibited by *forbidden transitions* that the mean lifetime of the excited state exceeds $0.1\,\mathrm{s}$, this long-lived nuclear state is called *isomeric state*, as already mentioned at page 226. Isomeric states have typically large spin differences and small energy differences from lower-lying energy levels. They are labeled with a small m following the mass number value. For example, we have the $^{99m}\mathrm{Tc}$ nuclide which has a lifetime of $\approx 6\,\mathrm{h}$ and decays by emitting a photon of $140.5\,\mathrm{keV}$. This nuclide is largely used in nuclear medicine.

Some nuclear γ-ray and intensity standards from [IAEA (1991, 1998)], recommended by the IAEA Coordinate Research Programme (CRP) for calibration of γ-ray measurements, are given in Appendix A.8. Further X- and γ-ray data are available from [IAEA (1991, 1998); ToI (1996, 1998, 1999); Helmer (1999); Helmer and van der Leun (1999, 2000)].

3.2 Phenomenology of Interactions on Nuclei at High Energy

In Sect. 3.1, we treated the general properties and models of nuclei without considering any specific reaction on nuclei. These nuclear properties are due to short-ranged nuclear forces, which, in turn, have their origin in strong interactions among nucleons. Moreover, in addition to the electromagnetic interactions discussed in the chapter on *Electromagnetic Interaction of Radiation in Matter*, hadronic particles like protons (p) and pions (π) also undergo strong interactions on their passage in matter. At high energies, among the resulting effects of strong interactions, there are particle creation and nuclear breakup.

Hadronic interactions on protons were widely studied in *high-energy physics experiments*, during many decades, in order to investigate the fundamental constituents of the so-called *elementary particles*, like the proton itself and the neutron. Because the nucleus is made of nucleons, it has become standard to express the relevant characteristics of the hadronic production on nuclei, in terms of ratios to the corresponding quantities observed in hadronic production on nucleons, i.e., on protons.

Among the salient features of the hadroproduction (and photoproduction) on nuclei from a few GeV up to very high energies, there is the *coherent production* [≈ (8–12)% of the total cross section]. The coherent reactions are interactions where the nucleus remains in its ground state after the collision has occurred. Furthermore, as discussed in Sect. 3.2.2, there is no evidence of a large intranuclear cascade on multiparticle *incoherent production on nuclei* (see for instance [Feinberg (1972); Busza (1977); Otterlund (1977); Halliwell (1978); Otterlund (1980)] and references therein). For many years, experimental data on hadro- and photo-production on nuclei were obtained using photographic *emulsions* exposed to high, very high and ultra-high energy cosmic rays. In the last few decades, accelerator experiments were carried out on both *coherent and incoherent production on nuclei*, as well.

It has to be noted that experiments on relativistic heavy-ion collisions were also performed, but this topic[**] is beyond the purpose of the present book.

3.2.1 *Energy and A-Dependence of Cross Sections*

We already discussed, in the chapter on *Electromagnetic Interaction of Radiation in Matter*, how most of the electromagnetic differential cross sections are strongly peaked in the forward direction and fall off as the energy increases. Away from the forward direction, the high-energy behavior is mostly dominated by the hadronic strong-interaction properties. However, in the forward direction, strong-interaction processes like quasi-elastic and coherent production are relevant at high energies.

The dependence of the total and elastic cross sections of proton (p) and anti-proton (p̄) on proton is shown in Fig. 3.9 versus the incoming laboratory momentum in GeV/c [PDB (2002)]. The main differences on the cross-section behavior appear at incoming particle momenta lower than 2–3 GeV/c. For incoming momenta < 1 GeV/c, the p–p scattering is dominated by the elastic scattering, which is still a relevant fraction, ≈ (15–20)%, of the total cross section at higher energies. At incoming energies of ≈ 200–300 GeV, the total p–p cross section is about 38–40 mb and increases almost logarithmically with s, where s (see page 12) is the square of the total energy in the center-of-mass system divided by c^2. At high-energies, the elastic scattering is a relevant fraction of the total cross section for p–n and p̄–n (see Fig. 3.10), and, also, in $\pi^\pm$–p (see Fig. 3.11) scatterings. Furthermore, the interaction of p and p̄ on the lightest bound nuclide, the deuterium (d), indicates again similar general features (Fig. 3.10).

Total cross sections on nuclei were measured by various experiments (see for instance [Murthy et al. (1975); Busza (1977); Carroll et al. (1979); Roberts et al. (1979)]). The experimental data show that, above 20 GeV, cross sections exhibit a weak energy-dependence [Fernow (1986)]. Complex multi-step processes, associated with an increased nuclear-matter transparency[‡‡] (Sect. 3.2.2) to particles

---

[**]The reader can see Section 16.4 of [Henley and Garcia (2007)], as a short introduction to this topic, and recent reviews like, for instance, [Ludlam (2005); Müller and Nagle (2006)].
[‡‡]The reader can see [Rancoita and Seidman (1982)] and references therein.

produced in the primary interaction on nucleons, can be partially responsible for the weakening of the cross-section energy-dependence. When propagation in matter is considered, the more relevant cross section is the so-called *inelastic cross section*, which is defined as:

$$\sigma_{A(\text{inel})} = \sigma_{A,\text{tot}} - \sigma_{A,\text{el}} - \sigma_{A,\text{q-el}}, \tag{3.32}$$

where $\sigma_{A,\text{tot}}$ refers to the total cross section on the nucleus with atomic weight[¶] A, $\sigma_{A,\text{el}}$ refers to coherent elastic scattering off the whole nucleus (Sect. 3.2.2) and $\sigma_{A,\text{q-el}}$ refers to the quasi elastic process on individual nucleons, in which no additional fast secondary particle is produced but the nucleus might be disrupted. It was determined in neutron–nucleus (n–A) interactions that for $A \geq 9$ [Roberts et al. (1979)]:

$$\sigma_{nA(\text{inel})} \simeq 41.2 A^{0.711} \ [\text{mb}], \tag{3.33}$$

while the total cross section goes as $\sim A^{0.77}$. It has to be noted that, for the nuclear radius value given by Eq. (3.12), we expect that the geometrical cross section is approximatively proportional to r_n^2 (Sect. 1.4), i.e., proportional to $A^{2/3}$ (Sects. 3.1, 3.1.1). Furthermore, the proton–nucleus (p–A) and pion–nucleus (π–A) inelastic cross sections are given by ([Busza (1977)] and references therein):

$$\sigma_{pA(\text{inel})} \simeq 46 A^{0.69} \ [\text{mb}], \tag{3.34}$$

$$\sigma_{\pi A(\text{inel})} \simeq 28.5 A^{0.75} \ [\text{mb}]. \tag{3.35}$$

Finally, in interactions on nuclei, the *absorption cross section* is defined as ([Roberts et al. (1979)] and references therein):

$$\sigma_{A(\text{absorption})} = \sigma_{A(\text{inel})} + \sigma_{A,\text{q-el}} = \sigma_{A,\text{tot}} - \sigma_{A,\text{el}}. \tag{3.36}$$

3.2.1.1 Collision and Inelastic Length

In Sect. 1.4, we introduced the collision length λ_{col} as the mean free path between successive interactions occurring for processes whose total cross section is σ_{tot}. In the case of strong interactions on nuclei, we define the *nuclear inelastic length* (λ_{inel}) as the mean free path between successive inelastic collisions whose cross section is given by Eq. (3.32). By rewriting Eq. (1.41), we have:

$$\lambda_{\text{inel}} = \frac{A}{N \rho \, \sigma_{A(\text{inel})}} \ [\text{cm}] \tag{3.37}$$

where A is the atomic weight, N is the Avogadro constant (see Appendix A.2), $\sigma_{A(\text{inel})}$ is in cm² and ρ in g/cm³.

[¶]As discussed at page 220, to a first approximation, the atomic weight, A (also referred to as relative atomic mass, see page 15), of a nuclide is expressed by its mass number.

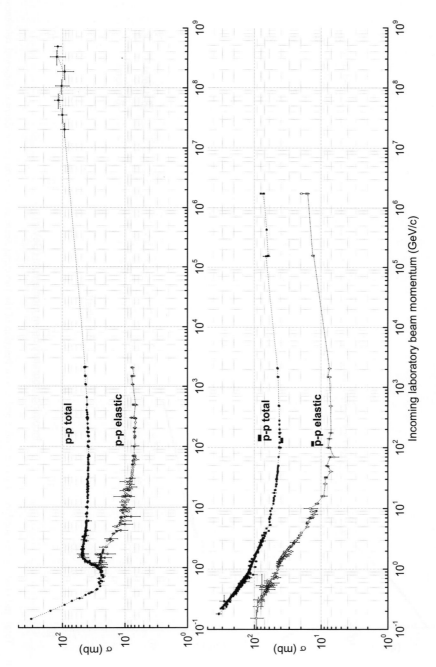

Fig. 3.9 Total and elastic cross sections of p and $\bar{p}$ on p as a function of the incoming beam momentum (experimental data and references are available, also via the web, from [PDB (2002)]). The lines are to guide the eye.

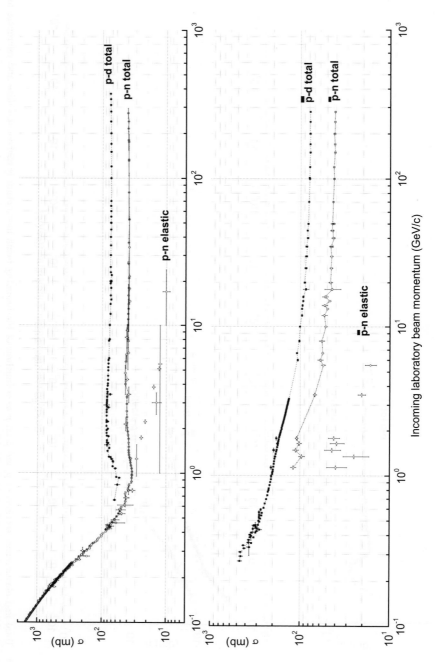

Fig. 3.10 Total and elastic cross sections of p and $\bar{p}$ on d and n as a function of the incoming beam momentum (experimental data and references are available, also via the web, from [PDB (2002)]). The lines are to guide the eye.

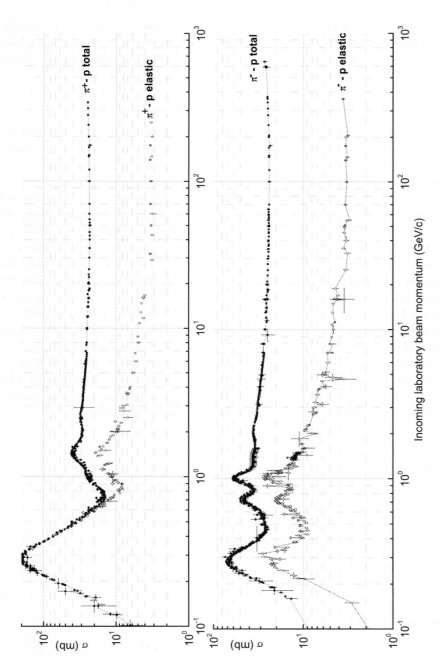

Fig. 3.11   Total and elastic cross sections of $\pi^{\pm}$ on p as a function of the incoming beam momentum (experimental data and references are available, also via the web, from [PDB (2002)]). The lines are to guide the eye.

3.2.2   *Coherent and Incoherent Interactions on Nuclei*

In spite of the (large) number of nucleons involved, the interaction with nuclei is more simple to deal with than interactions with a nucleon, when the nucleus remains in its ground state after the interaction. This kind of interaction is called *coherent interaction on nucleus*, i.e.,

$$h + T_N \rightarrow h^* + T_N.$$

A *coherent nuclear reaction* is an interaction induced by a high energy particle, in which all nucleons, whose amplitudes have strong interference-effects, are participating. The theory of coherent reaction on nuclei was derived by Glauber (see, for instance, [Glauber (1959)] and references therein).

The concept that particle production can occur diffractively, namely coherently, from a nucleus, was first introduced by Feinberg and Pomeranchuk (1956), who pointed out that, at large incoming particle momenta, a minimum momentum transfer, q_m, can be sufficient to produce a final hadronic state with a mass larger than the incoming one, when the wavelength h/q_m is comparable to or larger than the nuclear size. Hence, owing to the uncertainty principle [Eq. (3.25)], the effective dimensions of the region involved can be large compared with the target dimensions. This process was called the *diffraction dissociation* process or *diffractive reaction*. Furthermore, the possibility of nuclear excitation in diffractive collisions can be understood by considering that, if the exchanged momentum is capable of changing the initial particle mass by several hundred MeV/c^2 (or a few GeV/c^2), it should easily allow the target nucleus to acquire a few MeV. Reactions in which the recoil nucleus is a well-defined excited state are called *semi-coherent reactions* (e.g., [Stodolsky (1966)] and references therein), like for instance:

$$h + {}^{12}\text{C} \rightarrow h^* + {}^{12}\text{C}(2^+)$$

and, subsequently,

$${}^{12}\text{C}(2^+) \rightarrow {}^{12}\text{C} + \gamma(4.4\text{MeV}).$$

3.2.2.1   *Kinematics for Coherent Condition*

Let us consider the production reaction:

$$h + T_N \rightarrow h^* + R_N. \tag{3.38}$$

In this reaction,

$$\tilde{h} = \left(E/c, \vec{h}\right), \; \tilde{T}_N = (M_N c, 0), \; \tilde{h}^* = \left(E^*/c, \vec{h}^*\right), \; \tilde{R}_N = (E_R/c, \vec{q})$$

are the four-momenta of the incoming particle, the target nucleus at rest, the diffracted particle and the recoil nucleus, whose rest masses are m_h, M_T, m_{h^*},

M_R and whose momenta are $\vec{h}$, 0, $\vec{h}^*$, $\vec{q}$, respectively. From energy-momentum conservation, the four-momenta transfer squared, t (defined at page 12), between h and h^* and between T_N and R_N must be equal, namely:

$$t = \left(\tilde{h} - \tilde{h}^*\right)^2 = \left(\tilde{T}_N - \tilde{R}_N\right)^2$$
$$= T_N^2 + R_N^2 - 2\tilde{R}_N \cdot \tilde{T}_N$$
$$= M_T^2 c^2 + M_R^2 c^2 - 2M_T E_R. \tag{3.39}$$

For a coherent reaction, i.e., when $M_T = M_R$ and $E_R \simeq q^2/(2\,M_R)$, we have

$$E_R = M_R c^2 + \frac{q^2}{2M_R}$$

and we can rewrite Eq. (3.39) as:

$$-t \simeq q^2 = q_\parallel^2 + q_\perp^2, \tag{3.40}$$

where $|q_\parallel|$ and $|q_\perp|$ are the longitudinal and transverse three-momentum transfer, respectively. Because the incoming momentum is along the longitudinal direction, the transverse three-momentum transfer can be arbitrarily small (at the limit to become zero). Thus, in a coherent reaction, the value of

$$q_\parallel^2 \equiv -t_{\min} \tag{3.41}$$

is kinematically the minimum four-momentum transfer from the target to the recoiling nucleus to produce the particle h^* by an incoming particle h. Furthermore, in coherent reactions the differential cross section, which depends on the four momentum transfer, is usually expressed as a function of the quantity t' defined as:

$$t' \equiv |t - t_{\min}| = q_\perp^2. \tag{3.42}$$

In order to estimate $t_{\min}$, we need first to evaluate the limits on the momentum transfer imposed by the coherent condition.

Let us consider the diffraction scattering in the framework of the *optical model* (e.g. see [Roman (1965); Fisher (1968)]), first developed for the elastic scattering. In this model, the complete interacting system is assumed in complete stationary mode and the time dependence of the wave function is omitted. The incident particle beam moving in the positive z direction is described by a plane wave

$$\Psi_{in} = \exp(ik_p z),$$

where $k_p = p/\hbar$ is the wave vector, i.e., the particle momentum p in units of $\hbar$. At large distances from the scattering center, the scattered wave is an outgoing spherical wave described by

$$\Psi_{scat} = f(\theta)\frac{\exp\left(ik_p r\right)}{r}, \tag{3.43}$$

where k_p is the wave vector of the outgoing particle after elastic scattering, θ is the scattering angle, r is radial distance from the scattering center, $f(\theta)$ is the scattering amplitude, whose related angular differential cross section is given by:

$$\frac{d\sigma}{d\Omega} = |f(\theta)|^2.$$

The solution of *Schrödinger's equation*, when the scattered wave is given by Ψ_{scat}, allows one to establish a relationship (the so-called *optical theorem*) between the total cross section and the imaginary part of the scattering amplitude in the forward direction:

$$\text{Im} f(\theta = 0) = \frac{k_p}{4\pi} \sigma_{\text{tot}}. \tag{3.44}$$

In addition, it can be shown (see, for instance, [Fisher (1968); Bingham (1970)] and references therein) that, in order to produce coherently a diffracted wave on a nuclear target, the phase difference between the outgoing diffracted wave and the incident wave is $\simeq (2r_n q_\parallel)/\hbar$, where r_n is the radius of the nucleus [Eq. (3.12)], and must be lower than π. Then, we can have a constructive interference between contributions from all positions, from the front to the back of the nucleus of diameter $2r_n$. Therefore, the *coherent condition* for the interaction on a nucleus is satisfied when:

$$\frac{2r_n q_\parallel}{\hbar} \leq \pi \quad \Rightarrow \quad q_\parallel \leq \hbar \frac{\pi}{2r_n}. \tag{3.45}$$

By taking a factor ~ 2 lower for the right hand side of Eq. (3.45), the coherent condition is even better satisfied. Since numerically we have

$$\hbar c \simeq 197.3 \, \text{MeV fm},$$

$$m_\pi c^2 = 139.6 \, \text{MeV}$$

(where m_π is the rest mass of the pion), and

$$\frac{\hbar c}{m_\pi c^2} = \frac{\hbar}{m_\pi c} \simeq 1.4 \, \text{fm},$$

we can write the *practical coherent condition*, as:

$$q_\parallel \leq \frac{1}{2} \left(\hbar \frac{\pi}{2r_n} \right) \sim \frac{4.4 \, m_\pi c}{4.8 \, M_A^{1/3}}$$

$$\simeq \frac{m_\pi c}{M_A^{1/3}}$$

$$\sim \frac{m_\pi c}{A^{1/3}} \ [\text{MeV/c}], \tag{3.46}$$

where m_π is in units of MeV/c^2 and the mass number of the target nucleus (M_A) is numerically approximated by the atomic weight (A) of the nuclide (see Sect. 3.1 and, also, the discussion at page 220)

$$M_A \sim A.$$

Let us evaluate the value of $t_{\min}$ for the reaction shown in Eq. (3.38) for a high energy coherent scattering, induced by an incoming particle of momentum p. We have for $q_\perp = 0$:

$$t_{\min} = -q_\parallel^2 = -q^2$$

$$= \left(\tilde{h} - \tilde{h}^* \right)^2$$

$$= m_h^2 c^2 + m_{h*}^2 c^2 - 2 \frac{E}{c} \left[\frac{E}{c} - \frac{1}{c} \left(\frac{q_\parallel^2}{2M_R} \right) \right] + 2p \left(p - q_\parallel \right), \tag{3.47}$$

where $E - (q_\parallel^2/2M_R)$ and $p - q_\parallel$ are the energy and longitudinal momentum of the diffracted hadron h^*, respectively. In a coherent scattering, $q_\parallel$ is small* and, thus, the term $(q_\parallel^2/2M_R)$ can be neglected. Introducing Eq. (1.8), the expression (3.47) becomes

$$-q_\parallel^2 \sim m_h^2 c^2 + m_{h^*}^2 c^2 - 2\left(p^2 + m_h^2 c^2\right) + 2p^2 - 2p\, q_\parallel$$
$$= m_{h^*}^2 c^2 - m_h^2 - 2p\, q_\parallel$$
$$= m_{h^*}^2 c^2 - m_h^2 c^2 - 2p\, q_\parallel,$$

and, finally, by neglecting $q_\parallel^2$ with respect to $2p\, q_\parallel$, we have:

$$\sqrt{|-t_{\min}|} = q_\parallel \sim \frac{m_{h^*}^2 c^2 - m_h^2 c^2}{2p}. \tag{3.48}$$

For instance, from Eq. (3.46), in interactions on C and Pb nuclei, the coherently limited $q_\parallel$ values are ≈ 61 and $24\,\mathrm{MeV/c}$, respectively, within a factor 2. While from Eq. (3.48) at an incoming particle momentum of $25\,\mathrm{GeV/c}$, the largest masses of particles coherently produced are ≈ 1.75 and $1.10\,\mathrm{GeV/c^2}$, respectively, within a factor $\sqrt{2}$.

The coherent condition given in Eq. (3.46) has to be applied also to the transverse momentum transfer $q_\perp$. Thus, the coherent production in a forward cone is limited by:

$$\theta_{\mathrm{coh}} \sim \frac{m_\pi c}{M_A^{1/3}}\left(\frac{1}{p}\right) \sim \frac{m_\pi c}{A^{1/3}}\left(\frac{1}{p}\right).$$

For instance, the coherent production on C and Pb nuclei at an incoming momentum of $25\,\mathrm{GeV/c}$ is almost limited to the angles θ_{coh} of ≈ 2.4 and $1\,\mathrm{mrad}$, respectively.

3.2.2.2 *Coherent and Incoherent Scattering*

In this section, we will employ relativistic units used for calculations, i.e., all quantities are express in units so that

$$\hbar = c = 1.$$

The masses are measured in GeV, i.e., the mass of the proton is $\simeq 0.938\,\mathrm{GeV}$. Because in the International System of Units $\hbar c \simeq 0.1973\,\mathrm{GeV\,fm}$, the conversion factor of the unit of length to cm becomes $1.973 \times 10^{-14}\,\mathrm{cm}$. If the unit of mass were chosen differently, for instance the proton mass, the conversion factor to cm would be modified accordingly. Furthermore, for a compact notation the number of nucleons in a nucleus is indicated with that for the atomic weight[¶] A, instead of the usual symbol M_A.

The coherent elastic scattering on nuclei was first introduced within the framework of *Glauber's multiple scattering theory*. This high-energy approach assumes i)

---

*The largest practical value of $q_\parallel$ can be estimated using Eq. (3.46).

¶As discussed at page 220, to a first approximation the atomic weight (also referred to as relative atomic mass, see page 15) of a nuclide is expressed by its mass number.

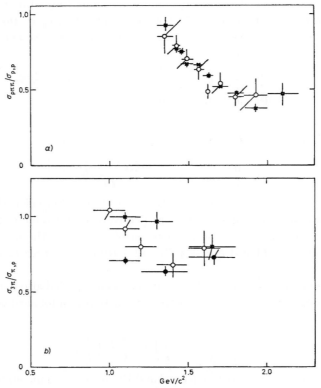

Fig. 3.12   Ratio of the total cross section on nucleon of a coherently produced system over the cross section on a free proton as a function of the produced invariant-mass from [Rancoita and Seidman (1982)] (Copyright of the Società Italiana di Fisica): a) $\sigma_{p\pi\pi}/\sigma_p$ for the reaction $p + A \rightarrow p\pi\pi + A$ (data ∘ at 15.3 GeV from [Mühlemann et al. (1973)] and x at 22.5 GeV from [Rosen (1977)]); b) $\sigma_{3\pi}/\sigma_p$ for the reaction $\pi + A \rightarrow \pi\pi\pi + A$ (data ∘ at 15.0 GeV from [Mühlemann et al. (1973)], data x at 22.5 GeV from [Rosen (1977)] and data • at 40.0 GeV from [Bellini et al. (1981)]).

the *additivity of the phase shifts* from scattering by various nucleons, ii) the *small scattering angles*, which is equivalent to neglecting the longitudinal component of the momentum transfer $q_{\parallel}$, and iii) the so-called *frozen nucleus approximation* whose consequence is to consider stationary target nucleons, so that nuclear wave functions are represented, in the scattering amplitude, by a simple average of scattering amplitudes from fixed centers [Glauber (1959)]. The scattering amplitude is related to a multiple scattering series amplitude, which accounts for single and multiple nucleon interactions. An extension of the theory to large A-nuclei is known as the *optical model*.

The high-energy coherent-dissociation on nuclei was treated, in the framework of the Glauber theory [Kölbig and Margolis (1968)], assuming that a multiple sequence of interactions, initiated in a nucleus, occurs through single interactions with one or more target particles. The incoming particle and the outgoing particle are considered to undergo elastic scatterings, before and after the production took place on a nucleon. In the case of the reaction given in Eq. (3.38), by i) extending the Glauber formalism, ii) ignoring the Coulomb distortion, iii) assuming that the interaction does not depend on the target quantum numbers, and iv) using the optical theorem [Eq. (3.44)], the coherent differential cross section as a function the four-momentum transfer becomes [Kölbig and Margolis (1968)]:

$$\frac{d\sigma_{\text{coh}}}{dt} = \left(\frac{d\sigma_{\text{h-nucl}}}{dt}\right)_{t=0} \mathcal{G}(q) \frac{4}{(\sigma_h - \sigma_{h*})^2 + (\alpha_h \sigma_h - \alpha_{h*} \sigma_{h*})^2}, \qquad (3.49)$$

where $(d\sigma_{\text{h-nucl}}/dt)_{t=0}$ is the differential cross section for the reaction (3.38) on nucleon at $t = 0$; α_h and α_{h*} are the ratio of the real to the imaginary parts of the scattering amplitudes on nucleons; σ_h and σ_{h*} are the total cross sections on nucleon and, finally, $\mathcal{G}(q)$ is given by

$$\left| \int J_0 \left(\vec{q} \cdot \vec{b}\right) \left\{ \exp\left[-\frac{1}{2}(1 - i\alpha_h)\sigma_h T\left(\vec{b}\right)\right] - \exp\left[-\frac{1}{2}(1 - i\alpha_{h*})\sigma_{h*} T\left(\vec{b}\right)\right] \right\} d^2b \right|^2,$$

in which $\vec{b}$ plays the role of the impact parameter; J_0 is the cylindrical Bessel function of the first kind of order zero; and $T\left(\vec{b}\right)$, called the *surface thickness of the target*, is defined in cylindrical coordinates[§] $(\vec{b}, z)$ as

$$T\left(\vec{b}\right) = A \int_{-\infty}^{+\infty} \rho(\vec{b}, z)\, dz,$$

where $\rho(\vec{r})$ describes the single-particle density function[¶] of the nucleus, with

$$\int \rho(\vec{r})\, d\vec{r} = 1.$$

At high energy, the scattering amplitudes on nucleons are mainly imaginary and Eq. (3.49) can be rewritten to a good accuracy as:

$$\frac{d\sigma_{\text{coh}}}{dt} = \left(\frac{d\sigma_{\text{h-nucl}}}{dt}\right)_{t=0} N^2 \left(A; \frac{1}{2}\sigma_h; \frac{1}{2}\sigma_{h*}; q\right), \qquad (3.50)$$

where

$$N\left(A; \frac{1}{2}\sigma_h; \frac{1}{2}\sigma_{h*}; q\right) = \frac{2}{(\sigma_h - \sigma_{h*})^2}$$

$$\times \int J_0 \left(\vec{q} \cdot \vec{b}\right) \left\{ \exp\left[-\frac{1}{2}\sigma_h T\left(\vec{b}\right)\right] - \exp\left[-\frac{1}{2}\sigma_{h*} T\left(\vec{b}\right)\right] \right\} d^2b. \quad (3.51)$$

The coherent production theory was used to determine the values of the total cross section of unstable particles, like meson resonances, whose cross sections are

---

[§]In these coordinates, z is the beam direction, i.e., the longitudinal coordinate.
[¶]It is assumed to be equal for neutrons and protons.

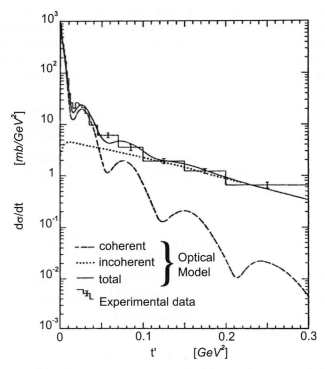

Fig. 3.13   Coherent and incoherent contributions to the differential cross section for 3π production on silver in the A_1 meson-resonance mass-region as a function of t' [defined in Eq. (3.42)] in relativistic units (adapted with permission from [Rosen (1977)]).

represented by the term σ_{h^*} in Eqs. (3.49, 3.50). The A-dependent cross section* can be extracted from measured coherent cross sections on nuclei in both hadro- and photo-production experiments (e.g., see [Mühlemann et al. (1973); Beusch et al. (1975); Rosen (1977); Bellini et al. (1981)]). For instance [Rancoita and Seidman (1982)], in Fig. 3.12 the total cross sections of the $p\pi\pi$ and 3π systems, produced coherently on nuclei, are shown as functions of the final-system invariant-mass (experimental data are from [Mühlemann et al. (1973); Rosen (1977); Bellini et al. (1981)]). In ordinate, the values are normalized to the p–p and π–p cross sections, respectively. The cross sections of final system-states are found to be smaller than the incoming particle cross sections on nucleon and decrease with increasing the produced system invariant-mass. The apparent increase of the final-state cross section was related to the possibility that the production process is more complex

---

*It was experimentally demonstrated that the coherent production cross section depends on the J^P (spin J and parity P) state of the final system (e.g., see [Beusch et al. (1975); Perneger et al. (1978)]).

than a simple one-step reaction (e.g., see [Fäldt (1977)]) or that the produced state is not in its asymptotic condition (see, also, data shown in Sect. 3.2.3). Moreover, in order to produce a well defined final-state and a target to recoil in its ground state, the *exchanged particle* with the interacting one cannot carry arbitrary quantum numbers, i.e., it must respect *selection rules*. A detailed summary of selection mechanisms and selection rules in hadro- and photo-production on nuclei are beyond the scope of this book, and can be found, for instance, in [Bingham (1970)].

Calculations were extended to the case in which the coherent condition of Eq. (3.45) is no longer satisfied, i.e., to the case of *incoherent reactions*. In the limit

$$\sigma_h \to \sigma_{h^*} = \sigma \text{ and } \frac{\sigma_{\rm el}}{\sigma} \ll 1,$$

the differential cross section becomes (see [Kölbig and Margolis (1968); Bingham (1970)] and references therein):

$$\frac{d\sigma_{\rm incoh}}{dt} = \left(\frac{d\sigma_{\rm h-nucl}}{dt}\right) N\left(A; \sigma_h; \sigma_{h^*}; q\right). \tag{3.52}$$

In Fig. 3.13, the differential cross section is shown in relativistic units[*] for the production of the A_1 meson resonance on Ag nuclei by π^- at 23 GeV [Rosen (1977)], with the predicted coherent and incoherent cross sections. The coherent production dominates at low t' values, while, at large values, the incoherent production becomes more and more important. A review, together with a theoretical introduction to this field, can be found in [Gottfried (1972)].

It has to be noted that there is another type of coherent interaction on nuclei, i.e., the electromagnetic interaction of the incoming particle with the nuclear Coulomb-field. In this case, the interaction occurs with a virtual photon of the nuclear Coulomb-field, and the process is known as the *Primakoff effect* [Primakoff (1951)]. The angular differential cross section for such a reaction was derived [Halprin et al. (1966)] for nuclei of spin zero and charge Ze, characterized by a form factor $F(q^2)$ (see definition at page 224), where q is the three-momentum transfer. In a Coulomb coherent production for the process expressed by Eq. (3.38), the particle h^* shall have the decay channel $h + \gamma$ with a radiative decay width $\Gamma_{\gamma,h}$. At high energy[‡] for large nuclear masses, an incoming particle h with spin S_h and a produced particle h^* with spin S_{h^*} (which in turn decays into a multi-particle system X), the Coulomb angular differential cross section is given by [Halprin et al. (1966)]:

$$\frac{d\sigma}{d\Omega} = 8Z^2\alpha|F(q^2)|^2 \frac{\Gamma_{\gamma,h}}{m_{h^*}^3\left(1 - \frac{m_h^2}{m_{h^*}^2}\right)^3} \Xi_{h,h^*} \frac{\theta^2}{(\theta^2 + \delta^2)^2} \frac{\Gamma_{h^* \to X}}{\Gamma_{h^*}}, \tag{3.53}$$

---

[*]In these units, we have $\hbar = c = 1$; see, also, page 5.

[‡]At high-energy, the momentum p of the outgoing system is approximately equal to its energy E and approximately equal to the incoming particle energy.

where θ is the scattering angle,

$$\delta = \frac{m_{h^*}^2 - m_h^2}{2\,E^2},$$
$$q^2 \simeq E^2(\theta^2 + \delta^2);$$

Γ_{h^*} and $\Gamma_{h^* \to X}$ are the decay widths of h^* and $h^* \to X$, respectively; Ξ_{h,h^*} is

$$\Xi_{h,h^*} = \frac{\eta_h'}{\eta_h}\left(\frac{2S_{h^*}+1}{2S_h+1}\right) \tag{3.54}$$

with

$$\eta_h = 1, \text{ for } h \neq \gamma,$$
$$\eta_h = \frac{1}{2}, \text{ for } h = \gamma,$$
$$\eta_h' = 1, \text{ for } m_h \neq 0,$$
$$\eta_h' = \frac{1}{2}(2S_h+1), \text{ for } m_h = 0.$$

For instance, reactions of the type $K_L^0 + \gamma_c \to \bar{K}^{*0}(892) \to K^- + \pi^+$ were investigated by means of the Coulomb coherent production (e.g., see [Carithers (1975)]). The particles K_L^0 and $\bar{K}^{*0}(892)$ have spin 0 and 1, respectively; then, from Eq. (3.54) $\Xi_{K_L^0, \bar{K}^{*0}(892)} = 3$. Furthermore, we have that

$$\frac{\Gamma_{\bar{K}^{*0}(892) \to K^- + \pi^+}}{\Gamma_{\bar{K}^{*0}(892)}} \simeq 1.$$

Because the production is peaked in the forward direction (namely at small scattering angles), we obtain from Eqs. (3.40, 3.42, 3.48):

$$\theta^2 \simeq \left(\frac{q_\perp}{p}\right)^2 = \frac{t'}{p^2},$$
$$\delta \simeq \frac{q_\parallel}{p},$$
$$\theta^2 + \delta^2 \simeq \frac{q_\parallel^2 + q_\perp^2}{p^2} = \frac{-t}{p^2};$$

in addition,

$$d\Omega = \sin\theta\, d\theta\, d\phi \simeq d\left(\frac{1}{2}\theta^2\right)d\phi \simeq \frac{1}{2}d\left(\frac{t'}{p^2}\right)d\phi.$$

Thus, Eq. (3.53), once integrated on the azimuthal angle ϕ, becomes:

$$\frac{d\sigma}{dt'} = 24\pi\alpha Z^2|F(q^2)|^2\Gamma_{K_L^0,\bar{K}^{*0}(892)}\frac{m_{\bar{K}^{*0}(892)}^3}{\left(m_{\bar{K}^{*0}(892)}^2 - m_{K_L^0}^2\right)^3}\frac{t'}{t^2}.$$

3.2.3 *Multiplicity of Charged Particles and Angular-Distribution of Secondaries*

At high and very high energy, the incoherent production on nuclei was first investigated by cosmic rays physicists using *photographic emulsions* [Blau (1961)], mainly. In the study of hadron–nucleus collisions, the nuclear emulsion was used as both target and detector; it consists of hydrogen, a light group of CNO-elements and a heavy group of Ag-Br-elements. The mean atomic weight of the emulsion nuclei is ≈ 60; while for light and heavy nuclei the mean atomic weights are ≈ 14 and 94, respectively.

As already mentioned, one of the salient features in high-energy hadro- and photo-production is the *absence of a strong cascading effect inside the nucleus* (e.g., see [Feinberg (1972); Busza (1977); Otterlund (1977); Halliwell (1978); Otterlund (1980)] and references therein). For instance, existing data from emulsion experiments, both in cosmic rays [Otterlund (1977); Rancoita and Seidman (1982)] and accelerators [Busza (1977); Halliwell (1978)], cover a large energy-range and show that the ratio

$$R_{em} = \frac{\langle n_s \rangle_{em}}{\langle n_s \rangle_p}$$

is less than 2 and, additionally, has a weak (if any) energy dependence on incoming hadron-energy for $E_h > 60\,\text{GeV}$ (Fig. 3.14): $\langle n_s \rangle_{em}$ is the average multiplicity of shower particles** produced in interactions with emulsion nuclei and $\langle n_s \rangle_p$ is the average multiplicity of shower particles produced in hadron–nucleon interactions.

The relationship between the average number of shower particles produced in hadron–nucleon interaction $\langle n_s \rangle_p$ and the corresponding produced on proton $\langle n_{\text{ch}} \rangle$ is given by:

$$\langle n_s \rangle_p = \langle n_{\text{ch}} \rangle - 0.5. \tag{3.55}$$

In a hadron–nucleus interaction, the multiplicity of the produced shower charged particles is compared to $\langle n_s \rangle_p$ in order to take into account, in a better way, the admixture of target nucleons and the number of fast recoiling protons identified as shower particles (a number which is not very well known). In a hadron–proton collision (e.g., [Busza (1977)]), the observed average multiplicity of charged particles is:

$$\langle n_{\text{ch}} \rangle \sim 1.768 \ln s - 2.8,$$

where s (defined at page 12) is the square of the total center-of-mass energy divided by c^2 and in GeV/c^2.

The phenomenon of the weak cascading effect inside the nucleus was systematically studied in accelerator experiments, using other nuclei as targets. The average multiplication of particles inside a nucleus is given by

$$R_A = \frac{\langle n_s \rangle_A}{\langle n_s \rangle_p},$$

**The shower particles are charged relativistic-particles with $\beta > 0.7$.

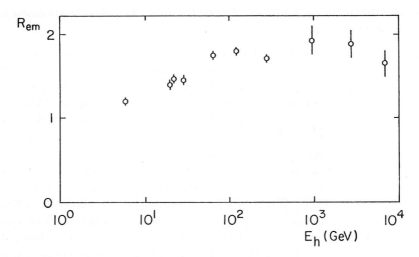

Fig. 3.14   The ratio $R_{em} = \langle n_s \rangle_{em} / \langle n_s \rangle_p$ versus the hadron incoming energy E_h (from [Rancoita and Seidman (1982)], Copyright of the Società Italiana di Fisica; and [Busza (1977)]).

where $\langle n_s \rangle_A$ is the average number of shower particles produced by hadrons interacting on a nucleus of atomic weight A. The observed value of R_A is lower than 2.7, even for massive nuclei.

No naive intranuclear cascade model can account for such a low value of R_A and the energy dependence of R_{em} for incoming hadron-energies up to 10^4 GeV, under the assumption that the cross sections of produced relativistic-particles are of the order of the hadron–proton cross section, i.e., 30–40 mb [Denisov et al. (1973); Busza (1977); Halliwell (1978); Otterlund (1980); Rancoita and Seidman (1982)]. Nevertheless, both the total and the inelastic hadron–nucleus cross sections, and, in addition, the average number of inelastic interactions*, $\bar{\nu}$, inside the nucleus can be derived in the framework of the Glauber theory [Bertocchi and Treleani (1977)]. In fact following the formalism given at page 250, neglecting the correlations in the nuclear wave function and including, between successive scatterings, all the possible intermediate states, the inelastic cross section involving m-nucleons for the production† is written as:

$$\sigma_m = \frac{A!}{(A-m)!m!} \int \left[T\left(\vec{b}\right) \sigma_{h\mathcal{N}(\text{inel})} \right]^m \left[1 - T\left(\vec{b}\right) \sigma_{h\mathcal{N}(\text{inel})} \right]^{A-m} d^2 b,$$

where $\sigma_{h\mathcal{N}(\text{inel})}$ is the inelastic cross-section on nucleon (i.e., $\simeq$ on proton) [Bertocchi and Treleani (1977)] and, for a compact notation, the number of nucleons (M_A) in a nucleus is indicated with the atomic weight¶ A (numerically $A \sim M_A$). The average

---

*The inelastic collision is treated at page 242.

†It is assumed that the remaining $A - m$ nucleons only provide inelastic absorption.

¶As discussed at page 220, to a first approximation the atomic weight, A (also referred to as relative atomic mass, see page 15), of a nuclide is expressed by its mass number.

number of nucleons, participating to the inelastic interaction on nucleus, becomes:

$$\bar{\nu} = \left(\sum_{m=1}^{A} m\sigma_m \right) \left(\sum_{m=1}^{A} \sigma_m \right)^{-1}$$

$$= \frac{1}{\sigma_{hA(\text{inel})}} \int \sum_{m=1}^{A} \frac{mA!}{(A-m)!m!} \left[T\left(\vec{b}\right) \sigma_{h\mathcal{N}(\text{inel})} \right]^m$$

$$\times \left[1 - T\left(\vec{b}\right) \sigma_{h\mathcal{N}(\text{inel})} \right]^{A-m} d^2b$$

$$= \frac{\sigma_{h\mathcal{N}(\text{inel})}}{\sigma_{hA(\text{inel})}} \int AT\left(\vec{b}\right) d^2b$$

$$= \frac{A\sigma_{h\mathcal{N}(\text{inel})}}{\sigma_{hA(\text{inel})}},$$

in which, $\sigma_{hA(\text{inel})}$ is the inelastic cross-section on a nucleus of atomic weight A. Consequently, we have:

$$\bar{\nu} \simeq \frac{A\,\sigma_{hp(\text{inel})}}{\sigma_{hA(\text{inel})}}, \tag{3.56}$$

where $\sigma_{hp(\text{inel})}$ is the inelastic cross-section on proton. $\bar{\nu}$ is the average number of inelastic collisions that the hadron would make with nucleons inside the nucleus if it remained as a single hadron after each collision. It can be used to express the average nuclear thickness in units of the mean free inelastic path of the incident hadron in the nucleus.

For $50 \leq E_h \leq 200\,\text{GeV}$, R_A was observed to be approximatively given by [Busza (1977)]:

$$R_A \approx \frac{1}{2} + \frac{1}{2}\bar{\nu}; \tag{3.57}$$

furthermore, for incoming protons we have $\bar{\nu} \approx 0.70\,A^{0.31}$; while, for incoming pions $\bar{\nu} \approx 0.74\,A^{0.25}$ [Busza (1977); Otterlund (1980)].

To account for the absence of an intranuclear cascade, a space-time development of the hadron–nucleus collision is incorporated in many theoretical models, such as the generalized Glauber model by Gribov (1968), the energy flux model by Gottfried (1974), the two-phase parton model by Nicolaev (1977), the constituent-quark model by Bialas (1978), and the coherent-tube model by Dar [Afek et al. (1977)].

The experimental data on the mean number of shower particles exhibit another characteristic, i.e., the dispersion, D, which is linearly dependent on the mean multiplicity, $\langle n_s \rangle$, of the charged relativistic-particles produced in nuclear interactions (Fig. 3.15):

$$D = 0.56 + 0.58 \langle n_s \rangle,$$

with

$$D = \sqrt{\langle n_s^2 \rangle - \langle n_s \rangle^2}$$

and n_s is multiplicity of the charged relativistic-particles produced in the interaction [Halliwell (1978); Abrasimov et al. (1979)].

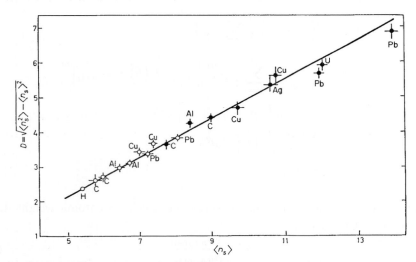

Fig. 3.15 Dispersion D as a function of the mean relativistic charged-particles produced in the interaction (from [Rancoita and Seidman (1982)], Copyright of the Società Italiana di Fisica). Data at 50, 100, 200 GeV are from [Halliwell (1978)], while data at 40 GeV are from [Abrasimov et al. (1979)].

3.2.3.1 *Rapidity and Pseudorapidity Distributions*

Before discussing the experimental data regarding the angular distributions of secondaries produced in interactions on nuclei, let us define a few variables used to describe the p–p interactions (e.g. Chapter 4 in [Perkins (1986)]). At high energies, the longitudinal momentum¶ distribution of the secondaries produced in p–p collision was often discussed in terms of the *rapidity y*, defined as:

$$y = \frac{1}{2} \ln \left(\frac{E + p_\parallel c}{E - p_\parallel c} \right) \tag{3.58}$$

$$= \frac{1}{2} \ln \left(\frac{E + p_\parallel c}{E - p_\parallel c} \frac{E + p_\parallel c}{E + p_\parallel c} \right)$$

$$= \ln \left(\frac{E + p_\parallel c}{\sqrt{p_\perp^2 c^2 + m_s^2 c^4}} \right), \tag{3.59}$$

where

$$E = \sqrt{p_\parallel^2 c^2 + p_\perp^2 c^2 + m_s^2 c^4}, \quad p_\parallel, \ p_\perp, \ m_s$$

are the energy, the *longitudinal* momentum (also, referred to as the *parallel* momentum), the *transverse* momentum (also, referred to as *perpendicular*) and the rest mass of the produced secondary, respectively. In a collision, the maximum

---

¶It is the momentum along the beam direction.

rapidity-value is reached when the incoming particle is elastically scattered in forward direction (i.e., $p_\perp = 0$):

$$y_{\max} = \ln\left(\frac{E_i + p_{i\parallel}c}{m_i c^2}\right) = \ln\left(\gamma_i + \gamma_i \beta_i\right), \tag{3.60}$$

where

$$E_i = \gamma_i m_i c^2, \ \ p_{i\parallel} = \beta_i \gamma_i m_i c, \ \ m_i$$

are the energy, the momentum and the rest mass of the incoming particle, respectively. Particles having rapidity values close to $y_{\max}$ are from the so-called *projectile fragmentation region*. From Eq. (3.58), in the center-of-mass system, a secondary with no longitudinal momentum (i.e., $p_{\mathrm{cm},\parallel} = 0$) has $y_{\mathrm{cm}} = 0$. In addition, the relationship between the rapidity minimum, $y_{\mathrm{cm},\min}$, and maximum, $y_{\mathrm{cm},\max}$, is:

$$y_{\mathrm{cm},\min} = -y_{\mathrm{cm},\max}.$$

In the case of an elastically backward scattered particle, the rapidity is at the minimum. Furthermore, in the center-of-mass system, the elastic scattering in the forward direction produces a backward scattered target-particle having opposite momentum. As a consequence, the forward going particle will be at the rapidity maximum, while the recoil target-particle will be at the rapidity minimum. Thus, particles with rapidity values close to the rapidity minimum are from the so-called *target fragmentation region*.

Let us determine how the rapidity transforms, under Lorentz transformation, to another reference frame moving at velocity βc along the longitudinal direction, i.e., along the incoming particle direction. In the new reference frame, the longitudinal momentum and the energy of the scattered particle are:

$$p'_\parallel c = \gamma \left(p_\parallel c - \beta E\right),$$
$$E' = \gamma \left(E - \beta p_\parallel c\right);$$

while, for the transverse momentum we have

$$p'_\perp = p_\perp.$$

Furthermore, from Eq. (3.59) we get:

$$\begin{aligned}
y' &= \ln\left(\frac{E' + p'_\parallel c}{\sqrt{p'^2_\perp c^2 + m_s^2 c^4}}\right) \\
&= \ln\left[\frac{\gamma\left(E - \beta p_\parallel c\right) + \gamma\left(p_\parallel c - \beta E\right)}{\sqrt{p_\perp^2 c^2 + m_s^2 c^4}}\right] \\
&= \ln\left[\frac{\left(E + p_\parallel c\right)\gamma(1 - \beta)}{\sqrt{p_\perp^2 c^2 + m_s^2 c^4}}\right] \\
&= y + \ln\left[\gamma(1 - \beta)\right] \\
&= y + \frac{1}{2}\ln\left[\frac{(1 - \beta)}{(1 + \beta)}\right].
\end{aligned}$$

Thus, the rapidity y is simply transformed into the rapidity y' (defined in the new reference system) by an additive value, determined by β. The particle rapidity [Eq. (3.59)] can be rewritten as a function of the scattering angle θ and expanded to obtain:

$$y = \frac{1}{2} \ln \left[\frac{\cos^2\left(\frac{\theta}{2}\right) + \frac{m_s^2}{4p^2} + \cdots}{\sin^2\left(\frac{\theta}{2}\right) + \frac{m_s^2}{4p^2} + \cdots} \right]$$

(see for instance Section 38.5.2 of [PDB (2008)]); finally, for $p \gg m_s c$ and $\theta \gg 1/\gamma$, it becomes

$$y \approx -\ln\left[tg\left(\frac{\theta}{2}\right) \right] \equiv \eta, \tag{3.61}$$

where η is the so-called *pseudorapidity*.

One important feature of p–p collisions, which were investigated at ISR machine at CERN, is that the rapidity distribution of secondaries indicates the presence of a plateau at small center-of-mass rapidity values. It is there that the bulk of produced particles is concentrated. At large y_{cm} (close to $y_{\text{cm,max}}$) the distribution falls rapidly.

Experiments on multiparticle production on nuclei do not usually provide the momentum measurement of secondaries, but their production angles, i.e., their peudorapidities, which can be calculated by means of Eq. (3.61). Following the terminology introduced for the rapidity, particles with low η values are from the target fragmentation region, while particles with large η values are from the pro-jectile fragmentation region. The main feature of pseudorapidity distributions for nuclear targets consists of an excess of multiplicity in the target fragmentation region with respect to the pseudorapidity distribution for a proton target (see Fig. 3.16; the experimental data are from [Abdrakhmanov et al. (1974); Furmanska (1977); Abrasimov et al. (1979); Faessler et al. (1979); Lee et al. (1979)]). At large angles, namely in the target fragmentation region, slow secondaries may also come from nuclear rescattering processes (absent in hadron–proton collisions) [Busza (1977); Otterlund (1977); Halliwell (1978); Otterlund (1980)]. The average multiplicity in the forward direction, $3 < \eta < 4$ (i.e., $6° > \theta > 2°$), is approximately the same as in hadron–nucleon collisions [Abdrakhmanov et al. (1974); Furmanska (1977); Abrasi-mov et al. (1979); Faessler et al. (1979); Lee et al. (1979)]. However, in the very forward direction, $\eta > 4$ (i.e., $\theta < 2°$), it becomes lower than the latter one, as it can be seen in Fig. 3.17, where $R(\eta)$ is given by:

$$R(\eta) = \left[\frac{1}{\sigma_{hA(\text{inel})}} \frac{d\sigma_{hA(\text{inel})}}{d\eta} \right] \bigg/ \left[\frac{1}{\sigma_{h\mathcal{N}(\text{inel})}} \frac{d\sigma_{h\mathcal{N}(\text{inel})}}{d\eta} \right] \tag{3.62}$$

[Abdrakhmanov et al. (1974); Furmanska (1977); Azimov et al. (1978); Abrasimov et al. (1979); Faessler et al. (1979); Lee et al. (1979)]. Furthermore, pseudorapidity distributions are only slightly dependent on A for a fixed topology (the topology is the final state particle multiplicity), as it can be observed in the experimental

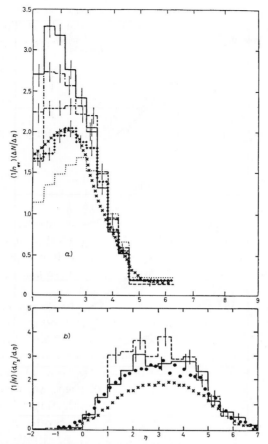

Fig. 3.16   a) Pseudorapidity distribution in π–A interaction at 40 GeV (from [Rancoita and Seidman (1982)], Copyright of the Società Italiana di Fisica): data on Pb (continous line), Cu (dashed-dotted line), Al (broken line), C (black lozenge line) are from [Abrasimov et al. (1979)], data on H (dotted line) are from [Abdrakhmanov et al. (1974)], data on C (crossed line) are from [Faessler et al. (1979)]. b) Pseudorapidity distribution in π–A interaction at 200 GeV: data on Cr (continuous line), W (dashed line), and on H (crossed line) are from [Lee et al. (1979)], data on emulsion are from [Furmanska (1977)]. In the target fragmentation region, the multiplication effect due to the nucleus is visible.

data shown in Fig. 3.18. These data were collected in a multiparticle production experiment on nuclei from C to Pb [Abrasimov et al. (1979); Rancoita and Seidman (1982)]. The global pseudorapidity distribution on a nucleus can be written as:

$$\frac{d\sigma_{hA(\text{inel})}}{d\eta} = \sum_T P_T(A)\frac{d\sigma_T}{d\eta},$$

where $P_T(A)$ is the A-dependent probability for the topology with T-particles in the final state and $d\sigma_T/d\eta$ is the almost A-independent pseudorapidity distribution for this topology.

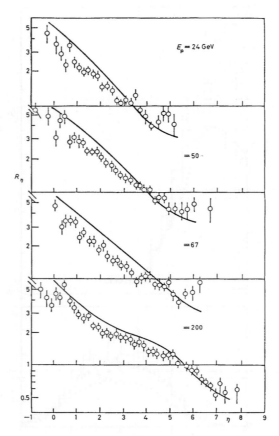

Fig. 3.17 At various energies, the ratio R_η [introduced with the symbol $R(\eta)$ in text, see Eq. (3.62)] is shown as a function of η for p–emulsion interaction (from [Rancoita and Seidman (1982)], Copyright of the Società Italiana di Fisica; see also [Azimov et al. (1978)]). The continuous line was calculated following the additive quark model prediction [Nikolaev (1977)]. It can be seen that, in very forward direction ($\eta > 4$), the multiplicity in emulsion is lower than the corresponding one on p-reaction.

3.2.4 Emission of Heavy Prongs

By extending the phenomenological description of shower secondaries produced in emulsion experiments to other kind of tracks, we observe that shower particles are usually accompanied by recoiling nucleons (among them, there are quasi-direct protons), fragments and boiled-off nucleons emitted by excited nuclear targets. These nuclear products indicate the presence of *spallation processes* (see page 267), which may result in leaving the residual nucleus characterized by different atomic weight and number.

The correlation between the number of relativistic particles (the so-called *shower* multiplicity 2÷4   5÷7   8÷10   11÷13

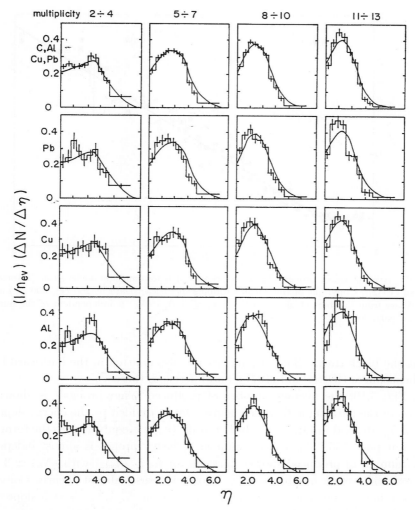

$(1/n_{ev}) \, (\Delta N / \Delta \eta)$

η

Fig. 3.18 Pseudorapidity distributions at fixed topology in the final state for C, Al, Cu, Pb nuclei (from [Rancoita and Seidman (1982)], Copyright of the Società Italiana di Fisica; see also [Abrasimov et al. (1979)]).

particles, n_s, with $\beta > 0.7$) and the charged spallation products, namely the *heavy prongs,* N_h, with $\beta < 0.7$, was widely studied using experimental data collected in emulsion experiments. The heavy prongs are subdivided in *black* and *grey tracks.* The grey (black) tracks, N_g (N_b), are particles with $0.3 \leq \beta \leq 0.7$ ($\beta < 0.3$). This terminology was adopted from emulsion experiments. The number of heavy prongs is interpreted as the number of charged fragments and charged recoiling particles coming from the nuclear target [Otterlund (1977)]. Fast protons are among the grey tracks, while evaporative protons and nuclear fragments are

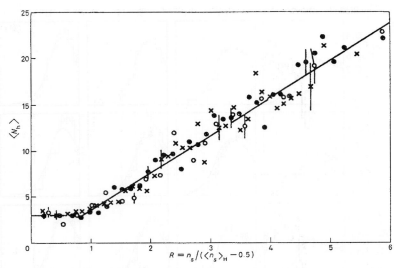

Fig. 3.19 Average heavy prongs $\langle N_h \rangle$ as a function of R for emulsion nuclei (from [Rancoita and Seidman (1982)], Copyright of the Società Italiana di Fisica). Data at 200 ($\bullet$), 300 ($\times$) and 400 ($\circ$) GeV are from [Herbert et al. (1974)]. For $n_s \leq \langle n_s \rangle_{\mathrm{H}}$, $\langle N_h \rangle$ is independent of the number of produced particles.

among the black tracks. Shower particles are also emitted in the backward laboratory hemisphere ([Otterlund (1977)] and references therein).

In Fig. 3.19, the average number of produced heavy-prongs[††] is shown as a function of the parameter R, which is given by n_s divided by the energy dependent term $\langle n_s \rangle_p$, defined in Eq. (3.55). In Fig. 3.19, $\langle n_s \rangle_{\mathrm{H}}$ is equivalent to the term $\langle n_{\mathrm{ch}} \rangle$, defined at page 255. The average number of heavy prongs is almost independent of the number of produced shower particles for $n_s \leq \langle n_s \rangle_p$, with $\langle N_h \rangle \simeq 3$. As R increases, the average number of heavy prongs increases almost linearly. Conversely, there is a linear dependence of the ratio[†] R_{em} on the N_h value, with a slope almost independent of the incoming-hadron energy.

Data from p–W, p–emulsion and p–Cr interactions have allowed the conclusion that the ratio R_A (defined at page 255) as a function of N_h seems to be independent of the nuclear-target mass [Otterlund (1977)]. It was found that $\langle N_h \rangle$ is almost independent of the incoming-hadron energy, but depends on the target mass as $\approx A^{0.7}$. The measured value ([Otterlund (1977)] and references therein) for incoherent π(p)-emulsion interactions is ≈ 6.9 (7.7), where the mean atomic weight of the emulsion nucleus is $\simeq 60$. The average number (e.g., see [Azimov (1977)]) of

---

[††] These experimental data were collected in emulsion experiments at 200, 300 and 400 GeV [Herbert et al. (1974)].

[†] The ratio R_{em}, defined at page 255, is computed without taking into account the number of the heavy prongs (N_h) produced in the interaction. However, it can also be computed as a function of the number of the heavy-prongs produced in the interaction.

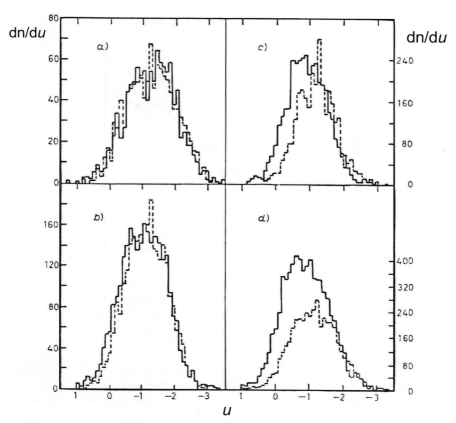

Fig. 3.20 Angular distribution of particles produced in p–emulsion and p–p interactions at 200 GeV as a function of $u = -\eta$, where η is the pseudorapidity defined in Eq. (3.61): (a) $n_s \leq 8$, (b) $9 \leq n_s \leq 16$, (c) $n_s \geq 17$, (d) no n_s selection (from [Rancoita and Seidman (1982)], Copyright of the Società Italiana di Fisica; see also [Alma–Ata–Leningrad–Moscow–Tashkent Collab. (1974)]). The target fragmentation region in p–emulsion and p–p interactions are quite different for $n_s \geq 17$.

grey (black) tracks for incoming-proton energies between 50 and 200 GeV is given by $\simeq A^{0.74}(A^{0.66})$, while the ratio $\langle N_b \rangle / \langle N_g \rangle$ is ≈ 1.7–1.9. Furthermore, pseudorapidity distributions for p–p and p–emulsion interactions show that the A-dependence is marked in events with large n_s, as it can be seen in Fig. 3.20 for $n_s \geq 17$ from [Alma–Ata–Leningrad–Moscow–Tashkent Collab. (1974)]. The A-dependence is mostly observed in the target fragmentation region, as already discussed, whereas a deficit of particles is observed in the projectile fragmentation region.

The frequency of recoil protons with momenta < 1 GeV is rather insensitive to the incident energy and provides a measurement of the thickness of the nuclear matter traversed; also, the excitation energy of the residual nucleus varies very little with energy. The angular distribution of *grey tracks* (most of them being fast protons)

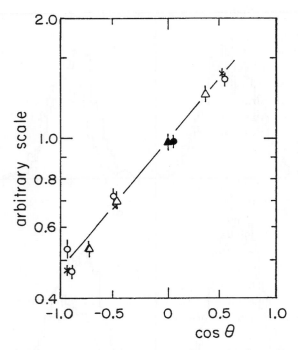

Fig. 3.21 Angular distribution of emitted protons with energies from 60 up to 200 MeV as a function of the emission angle (θ) in the laboratory: (○) π–Pb at 1.6 GeV/c, ($\times$) π–Pb at 6.2 GeV/c, ($\triangle$) p–Ta at 400 GeV/c; the distributions are normalized at 90° (adapted from *Nucl. Phys. A* **335**, Otterlund, I., High energy reactions on nuclei, 507–516, Copyright (1980), with permission from Elsevier; see also [Rancoita and Seidman (1982); Leroy and Rancoita (2000)]) .

was measured for recoil protons with energies between 60 and 200 MeV on π–Pb and p–Ta, and for incoming hadrons with momenta between 1.6 and 400 GeV/c ([Otterlund (1980)] and references therein). It was found (Fig. 3.21) that the distribution has an $\exp(\cos\theta)$ dependence, independently of the incoming-hadron energy within an accuracy of 5%. These data are in agreement with the measured ratio, in emulsion experiments, between the emitted grey tracks in the forward hemisphere and those emitted in the backward hemisphere ([Rancoita and Seidman (1982)] and references therein). These fast particles, also-called *quasi-direct* nucleons, are interpreted as emitted in gentle (*peripheral*) collisions through a fragmentation process.

Although, in this section, we have mostly discussed the emission frequency of charged nucleons (usually protons), the neutron emission probability can be estimated by the proton probability, once the neutron to proton ratio[††], N_n/Z, is taken into account. For evaporation neutrons, the Coulomb barrier will also affect the neutron emission probability (Sect. 3.2.6).

[††]The ratio varies smoothly with A.

3.2.5 *The Nuclear Spallation Process*

The so-called *spallation process* is a nuclear interaction process in which nucleons are *spalled* or *knocked-out*. This process occurs because the primary interaction on a quasi-free nucleon inside the nucleus can be followed by secondary interactions of the struck nucleons with other nucleons. The physical processes inside nuclei seem to be described by *intranuclear cascade models* [Bernardini et al. (1952); Segre (1977)]. Some of the particles taking part in interactions reach the nuclear boundaries and, if sufficiently energetic, can escape. In this step of the spallation process, most of the emitted charged particles are detected as grey tracks in emulsion experiments. Protons and neutrons will be emitted in a ratio, which corresponds, on average, to the N_n/Z ratio in the target nucleus, where N_n and Z are the number of neutrons and protons (i.e., the atomic number), respectively. Other particles remain inside and share their energies with the nucleons inside the nucleus. Physical processes inside nuclei can lead to the production of highly excited hot nuclei with subsequent decay modes, like, multifragmentation, vaporization or conventional fission and evaporation (e.g., [Pochodzalla et al. (1995)]). The further step of the spallation reaction consists of the de-excitation of the resulting intermediate nucleus. The excited nucleus decays by ejecting or evaporating nucleons or light nucleon aggregates, like, deuterons (d), tritium (t), α-particles, etc.; in this step, most of the emitted charged particles are detected as black tracks in emulsion experiments. In the final stage, γ's can also be emitted. In heavy nuclei, the intermediate nucleus can fission, in addition.

Numerous spallation investigations were carried out using many experimental conditions and, then, attempts to represent the cross sections by a suitable approximate analytical function were made. An empirical formula was written by Rudstam (1966) to reproduce the spallation cross section on a target nucleus of mass number A_T and atomic number Z_T, but excluding data with heavy ions and photons as incoming particles. For proton- and neutron-induced reactions and $A_T > 20$, the simplified expression for the emission of a spallation product of mass number A_S and atomic number Z is reduced to [Rudstam (1966)]:

$$\sigma(Z, A_S) \simeq F(A_T)f_2(E) \left\{ \frac{P \exp[-P(A_T - A_S)]}{1 - 0.3/(PA_T)} \right\}$$
$$\times \exp\left(-R\sqrt{|TA_S^2 - SA_S + Z|^3}\right) \quad [\text{mb}], \tag{3.63}$$

where E is the incoming energy, $F(A_T)$ (in mb) is the function shown in Fig. 3.22;

$$f_2(E) \simeq 1$$

for energies $\gtrsim 250\,\text{MeV}$ and increases to ≈ 4 at $\simeq 50\,\text{MeV}$,

$$P \simeq 20 \times \{E[\text{MeV}]\}^{-0.77}, \text{ for } E \leq 2100 \text{ MeV},$$
$$P \simeq 0.056, \text{ for } E > 2100 \text{ MeV}$$

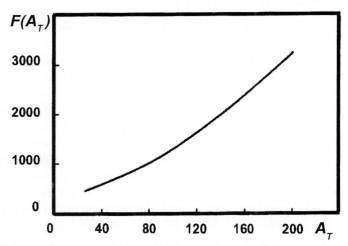

Fig. 3.22　Function $F(A_T)$ versus the target mass number A_T (adapted and reprinted with permission from [Rudstam (1966)]).

and with $PA_T \geq 1$. In addition,

$$R \simeq 11.8 \, A_T^{-0.45},$$
$$S = 0.486,$$
$$T = 0.00038.$$

The parameter P does not depend on the type of projectile used, i.e., protons, neutrons, deuterons, etc. have similar P values.

The total inelastic cross section $\sigma_{A_T,\mathrm{inel}}$ for the target nucleus is given by

$$\sigma_{A_T,\mathrm{inel}} = \sum_{Z,A_S} \sigma(Z, A_S);$$

while, the average mass number of the spallation products, $\langle A_S \rangle$, is

$$\langle A_S \rangle \simeq A_T - \frac{1}{P}.$$

The parameter P is closely related to the first stage in the spallation reaction, namely when intranuclear nucleon cascade interactions are taking place [Rudstam (1966)]. At high energy, i.e., for proton interactions on emulsion nuclei at 22.5 GeV [Winzeler (1965)], it was also found a good agreement between the observed average number of protons emitted and the one expected by means of the parameter P, estimated for the case of charged tracks [Rudstam (1966)]. The parameter R is found to be almost independent of both the incoming-particle type and energy. However, it depends on the mass number of the spallation products. The parameters S and T are related to the peak of the charge distribution.

An inspection of the cross section formula indicates that, for a given combination of target and spallation product, the maximum yield is achieved for:

$$P \simeq \frac{1}{A_T - A_S}.$$

From Eq. (3.63), we get:

$$\sigma(Z, A_S)_{\max} \leq F(A_T) f_2(E) \left\{ \frac{\exp\left(-1 - R\sqrt{|TA_S^2 - SA_S + Z|^3}\right)}{(A_T - A_S)\left[1 - 0.3(A_T - A_S)/A_T\right]} \right\} \quad [\text{mb}],$$

where the inequality sign is valid for $E \geq E_0$.

Nowadays, systematic investigations on spallation reaction are performed in many laboratories and neutron spallation is used to make neutron sources (see, for instance, [ESTP (2003)] and references therein).

3.2.6 *Nuclear Temperature and Evaporation*

During the spallation process, after the primary collision has occurred, fast nucleons with energies much larger than the binding energy are knocked-out, while other nucleons with energies lower or comparable to the binding energy can be trapped inside the nucleus. As a result, the remaining nucleus may be left in an excited state, from which it mainly decays via an *evaporation process*. Hence, additional nucleons or nucleon aggregates (p, n, d, t, ^{3}He, ^{4}He, ...) quit the nucleus; these emitted particles are among the black tracks observed in emulsion experiments. In the nucleus rest frame, all emission directions are equally probable. In fact, it is 1.1–1.2 the measured ratio of black tracks emitted in the forward direction to the ones emitted in the backward direction [Babecki and Nowak (1979)]: the slight forward excess observed is due to the motion of the recoiling residual nucleus.

The energy distribution and the relative abundance of evaporated particles can be estimated by using the *continuum theory for nuclear reaction* [Weisskopf (1937); Blatt and Weisskopf (1952)], in which collisions between a particle (or a photon) and a nucleus are treated by distinguishing two well separated stages. Firstly, we have the formation of a *compound nucleus* in a well defined state, in which the received energy is shared among constituent nucleons. Subsequently, we have the nucleus decay. This latter stage can be treated independently of the first one, i.e., the disintegration mode of the compound system depends on its energy, angular momentum, and parity, but not on the specific way by which it was produced. This two-step process, in which the nuclear interaction is assumed to occur, is referred to as the *Bohr assumption*, whose validity and justification are discussed in [Blatt and Weisskopf (1952)].

To a first approximation, the continuum theory allows one to estimate relative abundances and energy distributions of emitted particles in high energy interactions on nuclei. In calculations based on this theory, there is no *memory* of the way in which the compound nucleus was formed. In the theory, thermodynamical analogies

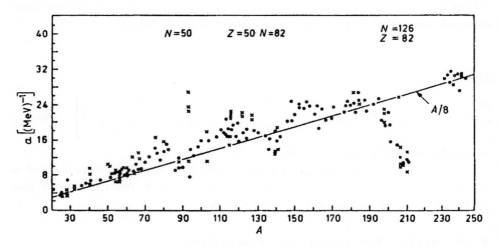

Fig. 3.23   Parameter a as a function of the mass number A. The parameter a appearing in the Fermi-gas level density formula was determined by comparison with the average spacings observed in slow neutron resonances ($\circ$) and evaporation spectra ($\times$) (from [Rancoita and Seidman (1982)], Copyright of the Società Italiana di Fisica; and [Erba, Facchini and Saetta Menichella (1961)]).

were used to describe the general trend of the nuclear process. The excitation energy, stored by a nucleus, can be compared with the heat energy of a solid body or a liquid. The subsequent emission of particles (nucleons or nucleon aggregates) is similar to an evaporation process occurring, for instance, in a liquid. Following such an analogy, a general statistical treatment was first proposed by Weisskopf (1937) for the *evaporation* of particles in highly excited heavy nuclei. The *nuclear evaporation temperature*, T, of the excited nucleus is again defined, in analogy to statistical thermodynamics, as the derivative of the entropy $\mathbb{S}(E)$, which, in turn, depends on the excitation energy E:

$$\frac{1}{T} = \frac{d\mathbb{S}(E)}{dE}, \tag{3.64}$$

where the excitation energy is measured from the nucleus ground state. The temperature, defined in this way, has the dimension of an energy and is equal to the ordinary temperature times the *Boltzmann constant* k (Appendix A.2). For instance, the room temperature of 300 K corresponds to $\simeq 0.026\,\text{eV}$. Referring to nuclear reactions, the temperature is usually expressed in MeV. The entropy of a nucleus, with an excitation energy between E and $E + dE$ [Weisskopf (1937)], is defined as

$$\mathbb{S}(E) = \ln g(E), \tag{3.65}$$

where $g(E)$ is the level density of the excited nucleus. The number of levels between E and $E + dE$ is given by $g(E)\,dE$. Although in statistical thermodynamics the entropy is defined using the logarithm of the number of states available to the system, thermodynamic transitions are characterized by the difference between the entropies of the final and initial states. In the case of nuclear interactions, the entropy difference between two states with excitation energies E_R and E_A is given by the quantity $\ln[g(E_R)/g(E_A)]$, which is dimensionless and independent of the infinitesimal energy-difference*, dE. The energy-level density available for nucleons can be computed by approximating the nucleus to a nuclear Fermi-gas discussed in Sect. 3.1.5. However, it has to be noted that, once a particle is emitted, the overall excitation energy varies so that a lower temperature must be ascribed to the remaining nucleus. A detailed treatment of the evaporation process, based on the continuum theory, is beyond the scope of this book and can be found in [Blatt and Weisskopf (1952)]. In the following, we will present a simplified introduction to this process as given in [Rancoita and Seidman (1982)].

Let us consider the emission of a neutron by an excited nucleus (A) with mass number M_A, excitation energy E and neutron binding energy $B_n < E$. In the continuum theory, the probability $P(W, n)\, dW$, that a neutron is emitted with a kinetic energy between W and $W + dW$, is evaluated by considering the reverse process:

$$R_A + n \rightarrow M_A,$$

where R_A is the mass number of the nucleus after the neutron ejection. The probability is given by:

$$P(W, n)\, dW = k_c m_n (2S + 1)\, W \frac{g(E_R)}{g(E_A)}\, dW, \tag{3.66}$$

where S and m_n are the spin and the rest mass of the neutron, respectively; $g(E_R)$ and $g(E_A)$ are the densities of energy-levels computed for the corresponding energies $E_R = E - B_n - W$ and $E_A = E$; and, finally, k_c is an appropriate constant. For an excited nucleus approximated to a Fermi gas of nucleons with total angular momentum $L = 0$ and equal amount of protons and neutrons, the density of energy levels is:

$$g(E, L = 0) \equiv g(E) \propto \frac{1}{E^2} \exp\left(2\sqrt{aE}\right), \tag{3.67}$$

under the condition that $E < E_F M_A^{1/3}$ and $E > E_F M_A^{-1}$, where E_F is the Fermi energy (see page 238). From Eqs. (3.64, 3.65, 3.67), we have:

$$\frac{1}{T} = \frac{1}{g(E)} \frac{\partial g(E)}{\partial E}$$

$$= E^2 \exp\left(-2\sqrt{aE}\right) \left[\frac{-2\exp\left(2\sqrt{aE}\right)}{E^3} + \frac{\exp\left(2\sqrt{aE}\right)}{E^2} \sqrt{\frac{a}{E}}\right].$$

$$= -\frac{2}{E} + \sqrt{\frac{a}{E}}. \tag{3.68}$$

---

*It was previously introduced to define the number of energy levels.

The values of the constant a (in MeV^{-1}) appearing in Eq. (3.68) were experimentally determined for various nuclei [Erba, Facchini and Saetta Menichella (1961)] and are shown in Fig. 3.23 as a function of the mass number. For temperatures of a few MeV's (easily reached in nuclear interactions) and for a values given in Fig. 3.23, the second term of Eq. (3.68) dominates the first one. The former term can also be neglected for nuclei with low mass number (M_A), such as silicon. Therefore, the excitation energy can be simply related to the nuclear temperature by:

$$E \approx aT^2.$$

Equation (3.66) can be rewritten by introducing the entropy as:

$$P(W, \mathrm{n})\, dW = k_c m_\mathrm{n} \,(2S + 1)\, W \exp[\mathbb{S}\,(E_R) - \mathbb{S}\,(E_A)]\, dW. \qquad (3.69)$$

The entropy $\mathbb{S}\,(E_R)$ can be derived from the entropy at the energy E_A via the Taylor expansion:

$$\mathbb{S}\,(E_R) = \mathbb{S}\,(E_A) - (B_\mathrm{n} + W) \left[\frac{d\mathbb{S}\,(E)}{dE}\right]_{E=E_A} + \cdots$$

$$\simeq \mathbb{S}\,(E_A) - (B_\mathrm{n} + W)\,\frac{1}{T},$$

where T is the temperature of the excited state with excitation energy $E = E_A$. Thus, Eq. (3.69) becomes:

$$P(W, \mathrm{n})\, dW = k_c m_\mathrm{n} \,(2S + 1)\, W \exp\left(-\frac{B_\mathrm{n} + W}{T}\right) dW. \qquad (3.70)$$

In Eq. (3.70), the kinetic energy distribution of emitted neutrons is an exponential function of the emitted particle energy, which goes as $W \exp(-W/T)$ and is peaked at the neutron kinetic energy $W_{n,mp} = T$.

So far, we have considered the evaporation of a neutron. In general, charged particles of mass M can also be ejected in the de-excitation process. The previous considerations are still valid if the Coulomb barrier potential-energy, V_M, is added to the binding energy, B_M, and if no particle is emitted with kinetic energy $W < V_M$. Therefore, by introducing the potential barrier V_M, Eqs. (3.69, 3.70) can be approximated by:

$$P(W, M)\, dW = k_M M \,(2S_M + 1)\, (W - V_M) \exp\left(-\frac{B_M + V_M}{T}\right)$$

$$\times \exp\left(-\frac{W - V_M}{T}\right) dW, \qquad (3.71)$$

where S_M and M are the spin and the rest mass of the emitted particle, respectively, and k_M is an appropriate constant. The Coulomb barrier shifts the most probable kinetic-energy of the emitted charged particles from the nuclear excitation temperature, i.e.,

$$W_{M,mp} = T + V_M.$$

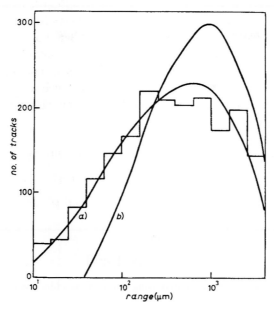

Fig. 3.24   Range distribution in emulsion data for black-tracks in events for which $N_h > 8$ (from [Rancoita and Seidman (1982)], Copyright of the Società Italiana di Fisica; see also [Ciok et al. (1963)]). Both curves are for 50% p, 25% d and 25% He. In a) $T = 7\,\mathrm{MeV}$, $V_\mathrm{p} = V_\mathrm{d} = V_{4\mathrm{He}} = 0$; in b) $T = 7\,\mathrm{MeV}$, $V_\mathrm{p} = 4\,\mathrm{MeV}$, $V_\mathrm{d} = 5\,\mathrm{MeV}$, $V_{4\mathrm{He}} = 8\,\mathrm{MeV}$. The data with many heavy prongs are consistent with high nuclear temperature and negligible Coulomb barrier.

It has to be noted that Eq. (3.71) reduces to Eq. (3.70) in case of neutron emission. We can integrate Eq. (3.71) to obtain the probability of emitting a particle of mass M:

$$
\begin{aligned}
P(M) &= \int_{V_M}^{\infty} P(W, M)\, dW \\
&= k_M M\,(2S_M + 1)\exp\left(-\frac{B_M + V_M}{T}\right) \\
&\quad \times \int_{V_M}^{\infty} (W - V_M)\exp\left(-\frac{W - V_M}{T}\right) dW \\
&= k_M T^2 M\,(2S_M + 1)\exp\left(-\frac{B_M + V_M}{T}\right).
\end{aligned}
\tag{3.72}
$$

Experimentally, the nuclear evaporation was also studied in high energy emulsion experiments, in order to investigate both nuclear temperatures and Coulomb potential-barriers. It was observed that the number of black tracks is related to the level of nuclear excitation. In [Powell, Fowler and Perkins (1959)], the data regarding $Z = 1$ prongs (i.e., p, d and t) with $2 \le N_b \le 12$ and $Z = 2$ prongs (i.e., ^{4}He and ^{3}He) with $2 \le N_b \le 6$ seem to indicate a temperature $T \approx 3\,\mathrm{MeV}$

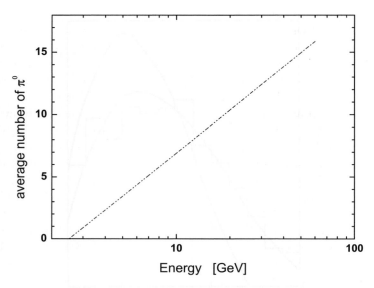

Fig. 3.25 Average number of π°'s computed using Eq.(3.73) as a function of the incoming-particle energy in GeV and $> 2.5\,\mathrm{GeV}$.

and Coulomb barriers $\simeq 4$ (for $Z = 1$) and $\simeq 10\,\mathrm{MeV}$ (for $Z = 2$). While for $Z = 1$ prongs with $N_b \geq 13$ and $Z = 2$ prongs with $N_b \geq 7$, the experimental data seem to indicate a larger temperature and lower Coulomb-barriers. In Fig. 3.24 [Ciok et al. (1963)], the range distribution of black tracks, obtained in high energy interactions on emulsion nuclei, is shown for the case $N_b > 8$. In Fig. 3.24, both continuous curves assume the black tracks are constituted by 50% p, 25% d and 25% He and regard two different sets of values for Coulomb barriers: the curve (a) was calculated for $T = 7\,\mathrm{MeV}$ and negligible Coulomb-barriers for all the particles, while the curve (b) was calculated for $T = 7\,\mathrm{MeV}$ and Coulomb barriers of 4, 5, and 8 MeV for p, d and ^{4}He, respectively. These multiple black prongs data are consistent with a negligible Coulomb-barrier and $T = 7\,\mathrm{MeV}$.

3.3 Hadronic Shower Development and Propagation in Matter

The absorption in matter of high energy hadrons, which undergo strong interactions, develops as a cascade process. The interactions of incoming particles in the absorber volume produce a wide spectrum of secondary particles, which have a variety of subsequent interactions with the nuclei of the medium. This process is similar, in many respects, to the one of electromagnetic showers, which we have examined in Sect. 2.4. However, the production mechanism is more complex. Analytical solutions, even following simplified approaches, are not available [Amaldi (1981); Wigmans (1987); Brückmann et al. (1988); Wigmans (1988); Brau and Gabriel (1989); Wigmans (1991)]. Furthermore, the nucleon binding energy and the fraction of protons among nucleons depend on the mass number (e.g., see [Blatt and Weisskopf (1952); Born (1969)]). Thus, part of the incoming energy will not be deposited (and not be detectable) by collision losses and become *invisible energy*, because of the nuclear break-up. Its fraction, in turn, depends on the mass number.

At present, quantitative approaches to a detailed description of hadronic showers use Monte-Carlo codes, written to simulate both the hadronic cascade development and the performance of the hadron absorbing device, taking into account the physical processes at work during the intra-nuclear cascading and the various many-body final states interactions (e.g., see [Fesefeld (1985); Aarnio et al. (1987); Brückmann et al. (1988); Anders et al. (1989); Brau and Gabriel (1989); Alsmiller et al. (1990); Brun et al. (1992); Giani (1993); Gabriel et al. (1994)] and references therein).

Note that, in Sects. 3.3-3.3.3, the notations are the same as those used in Sects. 2.4-2.4.3; for instance, A, Z, and N are the atomic weight (Sect. 1.4.1 and discussion at page 220), the atomic number (Sect. 3.1) and the Avogadro constant (see Appendix A.2), respectively.

3.3.1 *Phenomenology of the Hadronic Cascade in Matter*

The hadronic cascade is propagated through a succession of various inelastic interactions leading to particles production characterized by a multiplicity of secondaries increasing logarithmically with the available energy, namely increasing as $\approx \ln(s)$, where s is the square of the energy available in the center-of-mass system (see page 12) divided by c^2 (c is the speed of light).

In a hadronic interaction, about half of the incoming energy is carried away by leading particles and the remaining part is absorbed in the production of secondaries. Neutral pions amount, on average, to a third of the produced pions. The nuclear processes[‡], involved in the generation of the hadronic cascade, produce relativistic hadrons (mainly pions), nucleons and nucleon aggregates from spallation (including those ones from the evaporation process), break-up and recoiling nuclear fragments. As discussed in Sect. 3.2.6, the energy distribution and the relative abundance of the evaporated prongs, namely p, n, d, t, ^{4}He, ^{3}He, may be estimated by following the continuum theory for nuclear reactions (e.g., see Chapter 8 in [Blatt and Weisskopf (1952)], and [Weisskopf (1937)]). For instance, an estimate of the nuclear break-up as a result of a high-energy interaction, in particular on silicon, can be found in [Rancoita and Seidman (1982)]. The experimental data indicate large values for nuclear temperatures and a lowering of the Coulomb barrier (e.g., see [Rancoita and Seidman (1982)] and references therein; see, also, [Powell, Fowler and Perkins (1959)]). The spallation processes were widely investigated and

---

[‡]A further discussion on these nuclear processes can be found, for instance, in Sections 2.3.2 and 2.3.3 of [Wigmans (2000)]

Table 3.3   Energy deposition in 5 GeV proton showers neglecting the π^0 component (from [Leroy and Rancoita (2000)], see also [Wigmans (1987)]).

Absorber	U	Pb	Fe
Ionization (fraction due to spallation protons) (%)	38 (0.70)	43 (0.72)	57 (0.74)
Excitation γ's (%)	2	3	3
Neutrons < 20 MeV (%)	15	15	8
Invisible energy, i.e. binding energy and target recoil (%)	45	42	32

are treated in Sect. 3.2.5. An empirical formula, valid for energies > 50 MeV or mass numbers > 20, provides a description of experimental data on spallation reactions and is given in [Rudstam (1966)] (see also Sect. 3.2.5). An important fraction of the secondaries produced in the process is comprised of particles, mainly π^0 mesons, which decay via electromagnetic interaction and generate their own electromagnetic cascade. Thus, hadronic cascades always have an electromagnetic shower component. The number of π^0 fluctuates from event to event, depending on the processes occurring in the various stages of the cascade development.

The average number (n_0) of π^0's (Fig. 3.25) was estimated by Monte-Carlo simulations [Baroncelli (1974)] for E in GeV and > 2.5 GeV (see Equation 3.26 of [Amaldi (1981)] or Equation 88 of [Leroy and Rancoita (2000)] and references therein) as:

$$n_0 \approx 5 \ln(E) - 4.6. \qquad (3.73)$$

The average fraction, f_{em} (Fig. 3.26), of incoming-hadron energy deposited by electromagnetic cascades of secondary particles was also studied using Monte-Carlo simulations [Ranft (1972); Baroncelli (1974)] and estimated by Fabjan and Ludlam (1982) and Wigmans (1987) as a function of the incoming-hadron energy E in GeV for $E \gtrsim 1$ GeV:

$$f_{em} \approx (0.11 - 0.12) \ln(E). \qquad (3.74)$$

To prevent the average electromagnetic fraction from becoming larger than 1 for hadron energies in the ultra high energy range, f_{em} can be parametrized, following another approximate expression [Groom (1990); Gabriel et al. (1994); Groom (2007)], by

$$f_{em} \simeq 1 - E(\text{GeV})^{-n}, \qquad (3.75)$$

where n is 0.15[§]. The two parametrizations give similar results over the energy range between 5 and 200 GeV, where most of the data were collected.

[§] The parameter n is estimated to be ≈ 0.17 in Ref. [Groom (2007)].

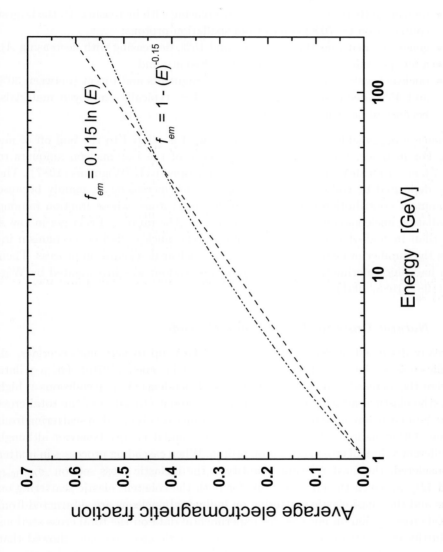

Fig. 3.26 The average fraction, f_{em}, of incoming-hadron energy (in GeV) deposited by electromagnetic cascades of secondary particles as computed using Eqs. (3.74, 3.75).

The purely hadronic-component[¶] of the hadronic cascade deposits its energy in the calorimeter via several mechanisms[‖], which were studied by means of Monte-Carlo simulations and, for instance, summarized in Table 3.3 for 5 GeV proton showers neglecting the π^0 component [Wigmans (1987)]. This energy deposition consists of:

- ionization (between 40% and 60%, decreasing with increasing A), the largest contribution ($\sim 70\%$) comes from spallation protons;
- generated neutrons (between 10% and 15%, decreasing with decreasing A);
- a few percents ($\sim 3\%$) photons (from fission); and,
- nuclear break-up and recoil of nuclear fragments and target (between 30% and 45%, decreasing with decreasing A in typically employed materials, because of nuclear binding energies).

The increase of invisible energy with decreasing A is related to the boil off of nucleons. For instance, the average binding energy of the last nucleon amounts to $\approx 6.4, 7.6$ and 10.5 MeV in U, Pb and Fe nuclei, respectively [Wigmans (1987)]. The energy, deposited by collision losses, decreases with increasing A, mainly because the dominant contribution comes from spallation protons, whose fraction (among the spallation nucleons) depends on the ratio Z/A. The ratio Z/A is larger in low-A nuclei than in high-A nuclei. For ^{238}U nuclei, the nuclear fission mechanism increases the number of neutrons emitted in the nuclear de-excitation process. Their role in hadronic calorimetry with hydrogeneous readout was investigated by Wigmans (1987; 1988; 1991).

3.3.2   *Natural Units in the Hadronic Cascade*

As already discussed in Sect. 3.2.1, above ≈ 5 GeV up to very high energies, all hadronic cross-sections are observed to rise slowly [Giacomelli (1976)]. The p–p data show that the increase is consistent with a $\ln(s)$ dependence. In p–p collisions at high energy, the elastic scattering process amounts to about (15–20)% of the total cross section. Similar behavior and values of cross sections are observed in scattering from neutrons. Other hadrons, for instance π's, have similar energy behavior although with different values of cross-sections on proton. When cascading processes in matter are considered, the most relevant quantity is the inelastic cross section, $\sigma_{hA(\text{inel})}$ [Eq. (3.32)], in which the cross sections for both the coherent elastic scattering on nucleus and the quasi-elastic scattering on individual nucleons are subtracted from the total cross section on nucleus (for experimental data on the total cross sections see [Murthy et al. (1975)]). Measurements of inelastic cross sections showed that these ones are approximately independent of the particle momentum at high energy.

[¶]The hadronic-component is that one in which particles deposit their energy by non-electromagnetic processes.

[‖]The reader can see, for instance, [Amaldi (1981); Wigmans (1987); Brückmann et al. (1988); Wigmans (1988); Brau and Gabriel (1989)].

Table 3.4   Values of atomic number (Z), density (ρ), nuclear collision length [$\lambda_{gA(\text{col})}$], and nuclear interaction length (λ_{gA}, λ_A) for several materials commonly used to absorb hadronic showers (data from [PDB (2008); PDG (2008)]).

Material	Z	ρ [g/cm]^3	$\lambda_{gA(\text{col})}$ [g/cm^2]	λ_{gA} [g/cm^2]	λ_A [cm]
Be	4	1.848	55.3	77.8	42.10
C (amorphous)	6	2.00	59.2	85.8	42.90
Al	13	2.70	69.7	107.2	39.70
Si	14	2.33	70.2	108.4	46.52
Ti	22	4.54	78.8	126.2	27.80
Fe	26	7.87	81.7	132.1	16.77
Cu	29	8.96	84.2	137.3	15.32
Ge	32	5.323	86.9	143.0	26.86
Sn	50	7.31	98.2	166.7	22.80
W	74	19.30	110.4	191.9	9.95
Pb	82	11.35	114.1	199.6	17.59
U	92	18.95	118.6	209.0	11.03

The development of an hadronic cascade along the direction of motion of the incoming particle (referred to as the *longitudinal direction*) is usually described in units of *interaction length* (or *nuclear interaction length*) λ_A. This latter is the nuclear inelastic length for n–A inelastic interaction, i.e.:

$$\lambda_A = \frac{A}{N\rho\,\sigma_{nA(\text{inel})}}, \tag{3.76}$$

where A, N, ρ are the atomic weight, the Avogadro number, the density of the material (in g cm^{-3}), respectively. $\sigma_{nA(\text{inel})}$ (in cm^2) is the inelastic cross section on the nucleus of atomic weight A, measured for incoming neutrons (see Section 6 of [PDB (2004)] and references therein). The weak energy dependence of the inelastic cross-section and, to a first approximation, the similar A-dependence also observed (see Sect. 3.2.1; [Busza (1977)] and references therein) in p–nucleus and π–nucleus justify the choice of that unit to measure the hadronic-cascade depths. The nuclear interaction length λ_{gA} in g cm^{-2} is given by

$$\lambda_{gA} = \rho\lambda_A.$$

The nuclear interaction length is sometimes written using the approximate formula [Amaldi (1981)]

$$\lambda_A \approx 35\frac{A^{1/3}}{\rho} \quad \text{cm.} \tag{3.77}$$

In a similar way, the *nuclear collision length* $\lambda_{A(\text{col})}$ is related to the n–A total cross section. Values of λ_A, $\lambda_{gA} = \rho\lambda_A$ and $\lambda_{gA(\text{col})} = \rho\lambda_{A(\text{col})}$, for commonly employed hadronic-absorbers, are listed in Table 3.4; the nuclear collision and interaction lengths for more than 300 substances are available at [PDG (2008)].

3.3.3 *Longitudinal and Lateral Hadronic Development*

The hadronic cascade, similarly to the electromagnetic cascade, develops along the incoming particle direction (longitudinal direction) as long as the produced secondaries have enough energy to continue the multiplication process. Hadronic cascades were experimentally studied in different materials and for a wide range of incoming-particle energies [Friend et al. (1976); Cheshire et al. (1977); Holder et al. (1978); Sessom et al. (1979); Amaldi (1981); Bock et al. (1981); Della Negra (1981); Muraki et al. (1984); Fabjan (1986); Catanesi et al. (1987); Leroy, Sirois and Wigmans (1986); SICAPO Collab. (1991b, 1993a)].

In the following, we will express the longitudinal shower depth t in units of interaction lengths λ_A [Eq. (3.76)].

To a first approximation, the 95% longitudinal containment of hadronic-shower cascades, $L_{95\%}$, which corresponds to the absorber depth within which 95% of the hadronic-cascade energy will be deposited (on average), is given by [Fabjan (1986)]:

$$L_{95\%} \approx t_{\max} + 2.5\lambda_a, \tag{3.78}$$

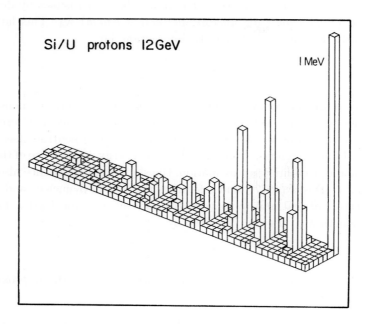

Fig. 3.27   Measured visible-energy for 12 GeV protons as a function of the U calorimeter depth and the strip location (reprinted with permission from Borchi, E. et al., Evidence for compensation and study of lateral shower development in Si/U hadron calorimeters, IEEE Transactions on Nuclear Science, Volume 38, Issue 2, Apr. 1991, pages 403–407, doi: 10.1109/23.289333, © 1991, IEEE, e.g., for the list of the authors see [SICAPO Collab. (1991b)]; see also [Leroy and Rancoita (2000)]). The depleted width of the silicon detectors is $\approx 200\,\mu$m. Each strip is, on average, 36 cm long and 4 cm wide. The total U depth is 50 cm and the lateral coverage is 28 cm.

where

$$t_{\max} \approx 0.2 \ln[E(\text{GeV})] + 0.7$$

is the cascade maximum-depth; λ_a (in units of λ_A) describes the exponential decay of the cascade beyond $t_{\max}$ and varies with the energy as [Fabjan (1986)]

$$\lambda_a = [E(\text{GeV})]^{0.13}.$$

Equation (3.78) (for $L_{95\%}$) describes the data available in the energy range of a few GeV to a few hundred GeV to within $\approx 10\%$ [Fabjan (1986)]. An alternative expression, which agrees within $\approx 10\%$ with Eq. (3.78), is reported by Amaldi (1981) [see also references therein]:

$$L_{95\%} \approx 1 + 1.35 \ln[E(\text{GeV})].$$

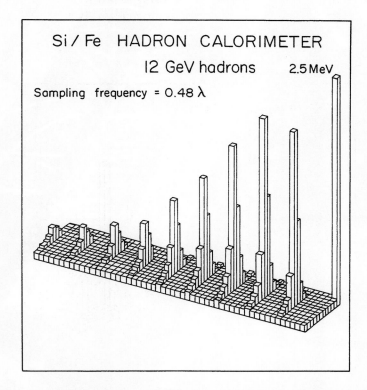

Fig. 3.28   Measured visible-energy for 12 GeV hadrons as a function of the Fe calorimeter depth and the strip location (reprinted with permission from Borchi, E. et al., Evidence for compensation in a Si hadron calorimeter, IEEE Transactions on Nuclear Science, Volume 40 Issue 4, Part 1-2, Aug. 1993, pages 508–515, doi: 10.1109/23.256610, © 1993 IEEE, e.g., for the list of the authors see [SICAPO Collab. (1993a)]; see also [Leroy and Rancoita (2000)]). The depleted width of the combined silicon detectors is $\approx 800\,\mu$m. Each strip is, on average, 40 cm long and 4 cm wide. The total Fe depth is Fe 72 cm and the lateral coverage is 48 cm.

Fig. 3.29   The SICAPO calorimeter using silicon detectors (see Fig. 3.30) as active readout sampler at the PS experimental Hall at CERN.

Data for a larger ($\approx 99\%$) longitudinal containment of hadronic shower cascades are shown in [Bitinger (1990); PDB (1996 and 1998)] and are based on experimental data and shower parametrizations ([Bitinger (1990); PDB (1996 and 1998)] and references therein; see also, e.g., [Bock et al. (1981); Hughes (1986)]).

Normally, after the primary interaction, decays of hadronic resonances created during the energy degradation of the incident hadrons and charge-exchange reactions produce π^0's (mainly) and η's, which will propagate electromagnetically without any further nuclear interactions and, consequently, deposit their energy in the form of electromagnetic cascades [Catanesi et al. (1987)]. As a result, any hadronic cascade has purely hadronic and purely electromagnetic components (Sect. 3.3.1). The size of this electromagnetic component is largely determined by the production of π^0's (and η's) in the first interaction. After the removal of the fluctuations due to the initial location (s_0) of the cascade, the longitudinal profile of a hadronic cascade can be parametrized by two terms describing the electromagnetic and hadronic components of the cascade [Holder et al. (1978); Bock et al.

Fig. 3.30   Kapton-cover for SICAPO silicon active-sampler. The active sampler is an assembled *mosaic* and consists of $2 \times 2 \, \text{cm}^2$ silicon detectors, like the one shown at the bottom right-part of the picture.

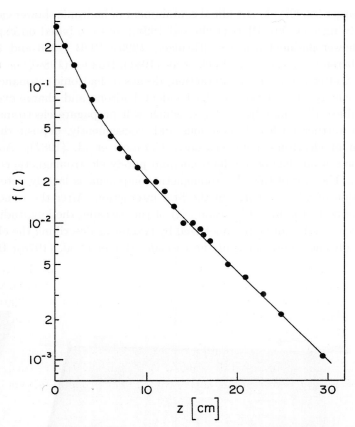

Fig. 3.31   Mean differential lateral profile $f(z)$ as a function of the lateral distance z in cm for incoming antiprotons of 25 GeV (adapted from *Nucl. Instr. and Meth.* **174**, Davidov, V.A. et al., A hodoscope calorimeter for high energy hadrons: IHEP-IISN-LAPP collaboration, 369–377, Copyright (1980), with permission from Elsevier, e.g., for the list of the authors see [Davidov et al. (1980)]; see also [Leroy and Rancoita (2000)]).

(1981); Della Negra (1981)]:

$$dE = E\left\{ f_{em}\left[\frac{x^{\alpha_e-1}}{\Gamma(\alpha_e)}\right]\exp(-x)\,dx + (1-f_{em})\left[\frac{y^{\alpha_h-1}}{\Gamma(\alpha_h)}\right]\exp(-y)\,dy\right\}, \qquad (3.79)$$

where E is the incident energy,

$$x \equiv \beta_e\frac{(s-s_0)}{X_0}$$

is the number of radiation lengths accumulated since the start at s_0 of the cascade multiplied by a dimensionless coefficient β_e, and

$$y \equiv \beta_h\frac{(s-s_0)}{\lambda_A}$$

is the number of interaction lengths accumulated since the start at s_0 of the cascade. α_e, β_e, α_h, β_h are dimensionless coefficients containing a $\ln(E)$ dependence and are given in [Della Negra (1981)] for hadronic showers, generated in Pb and Fe passive absorbers. $\Gamma(\alpha_e)$ and $\Gamma(\alpha_h)$ are the gamma functions ensuring the normalization, i.e.,

$$\Gamma(\alpha_e) = \int_0^\infty x^{\alpha_e - 1} \exp(-x)\, dx \qquad (3.80)$$

and

$$\Gamma(\alpha_h) = \int_0^\infty y^{\alpha_h - 1} \exp(-y)\, dy, \qquad (3.81)$$

in such a way that a fully contained cascade should deposit all its incident energy E. f_{em} represents the fraction of electromagnetic component of the hadronic cascade. The average fraction converted into electromagnetic cascade increases with energy (as discussed at page 276).

Results of measurements (by activation methods) of the longitudinal development of 300 GeV π^- cascades in uranium show a profile that is very similar to electromagnetic cascades, however, the scale is quite different since 300 GeV π^- cascades are contained at the 95% level in 80 cm of uranium, while it only takes 10 cm to achieve the same result for electromagnetic cascades at the same energy [Leroy, Sirois and Wigmans (1986)]. Figure 3.27 (Fig 3.28) shows the average energy deposited longitudinally and laterally [SICAPO Collab. (1991b, 1993a)], i.e., the so-called *visible energy*, in a sampling U (Fe) SICAPO calorimeter (shown in Fig. 3.29) employing silicon active-samplers (Fig. 3.30), as a function of the depth for 12 GeV incoming protons. These experimental data show that the cascade maximum is located at a calorimeter depth of $\approx 1\,\lambda_A$ and, beyond the cascade maximum, the visible energy falls following an approximate exponential dependence on the calorimeter depth [SICAPO Collab. (1991b, 1993a)].

The 95% radial containment ($R_{95\%}$) for hadronic cascades occurs at [Amaldi (1981); Fabjan (1986)]:

$$R_{95\%} \approx 1\lambda_A. \qquad (3.82)$$

The hadronic cascade has a lateral spread due to the production of secondaries at large angles [Friend et al. (1976); Davidov et al. (1980); Muraki et al. (1984)]. Its overall transverse dependence can be approximately described by two components: a main component along the cascade axis which decays fast, and a large and long peripheral component composed mostly of low-energy particles (including neutrons), which carries a relevant fraction of energy away from the cascade axis. It is convenient to describe the transverse profile by the sum of two exponential terms, which represent the central core of the cascade and its halo [Davidov et al. (1980)]:

$$f(z) = A_1 \exp\left(-\frac{|z - z_o|}{b_1}\right) + A_2 \exp\left(-\frac{|z - z_o|}{b_2}\right), \qquad (3.83)$$

where z is the transverse coordinate (in cm) and z_o is the impact point of the cascade. For instance, $A_1/A_2 \approx 2$ (i.e., $A_1 \approx \frac{2}{3}$ and $A_2 \approx \frac{1}{3}$), $b_1 = 2.2$ cm and $b_2 = 7.0$ cm in the case of 25 GeV antiprotons showering in a Scintillator/Fe calorimeter (Fig. 3.31) [Davidov et al. (1980)].

The ansatz [Grindhammer et al. (1989, 1990)], used to describe the lateral energy profile of the electromagnetic cascade [cf. Eq. (2.249)], can be applied to the case of hadronic cascades with r and the free parameter R expressed in units of λ_A and when the lateral resolution of the calorimeter is of the order or larger than $0.1\,\lambda_A$. A good agreement between the model and the data is obtained in the description of the core and halo of the cascades [Grindhammer et al. (1989, 1990)]. The model also reproduces the dependence of the lateral profile on the cascade depth and energy.

Chapter 4

Radiation Environments and Damage in Silicon Semiconductors

Silicon is the active material of radiation detectors and the basic material of electronic devices used in the fabrication and development of electronic circuits. The present technology evolves toward the creation of faster and less power consuming devices of increasingly small sensitive volume and higher density circuits, achieving, now, sub-micron technology with an increase of the number of memory elements. These devices are then used for applications of large to very large scale integration (VLSI) in several fields including particle physics experiments, reactor physics, nuclear medicine and space. Many fields of application present adverse radiation environments that may affect the operation of the devices. These radiation environments are described in Sect. 4.1. These environments are generated by i) the operation of the high-luminosity machines for particles physics experiments, an example is the Large Hadron Collider (LHC), ii) the cosmic rays and trapped particles of various origins in interplanetary space and/or Earth magnetosphere and iii) the operation of nuclear reactors. Their potential threats to the devices operation, in terms of damage by displacement (radiation-induced damage), are also described in that section. A review of the characteristics of space, high-energy and nuclear radiation environments and of physical processes inducing damages in silicon semiconductor operated in radiation environments was provided by Leroy and Rancoita (2007).

The generation and type of damage are linked to processes of energy deposition. The interaction of incoming particles with matter results into two major effects: the collision energy-loss and atomic displacement, which are treated in the chapter on *Electromagnetic Interaction of Radiation in Matter*. Interactions of incoming particles, which result in the excitation or emission of atomic electrons, are referred to as energy-loss by ionization or energy-loss by collisions. The non-ionization energy-loss (NIEL) processes are interactions in which the energy imparted by the incoming particle results in atomic displacements or in collisions, where the knock-on atom does not move from its lattice location and the energy is dissipated in lattice vibrations. The energy deposition by non-ionization processes is much lower (except for neutrons) than that by ionization. However, bulk-damage phenomena result mostly from atomic displacement deposition-mechanisms, the so-called *displacement damage*. Defects, induced by the interaction of radiation with semiconductors, are *primary point-defects*, i.e., *vacancies* and *interstitials*. Clusters of defects are generated when the incident particle, such as fast neutrons, transfers enough energy to the recoil atoms for allowing large cascades of displacements. The change observed in semiconductor conductivity is associated with the formation of defect clusters. Section 4.2 gives a review of the processes of energy deposition and related damage generation including radiation-induced defects. The effect of the processes of energy deposition and induced damage on device parameters evolution is reviewed, in that section, considering the dependence on the type of irradiation particles, their energy and fluence of irradiation. The non-ionizing energy-loss (NIEL) scaling hypothesis is formulated in that section.

The radiation-induced defects have a large impact on bulk properties of silicon and can be investigated by the electrical behavior of semiconductor devices (for instance, radiation detector, diodes and transistors) after irradiation. The degradation rate of minority-carrier lifetimes can be expressed in terms of a damage coefficient. The dependence of this coefficient on the type of substrate, dopant concentration, level of compensation, type, energy and fluence of the irradiating particles are discussed in Sect. 4.3. After irradiation with large fast-neutron fluences at room temperature, centers with energy-levels near the mid-gap (near the *intrinsic Fermi level*) make a significant contribution to carrier generation and determine i) the increase of the leakage current inside the depleted region of silicon devices and ii) modifications to the diode structure and rectification properties. Low resistivity $p - n$ diodes may gradually change their internal structure with increasing fast-neutron fluence. In fact, non-irradiated diodes with an ohmic (n^+) contact can acquire an almost $p - i - n$ structure after irradiation. Section 4.3 reviews the effects of large radiation damages and the formation of $p - i - n$ structure at room temperature, as well as the dependence down to cryogenic temperatures of $I - V$ characteristics on the type and fluence of irradiating particles. The behavior of the junction complex-impedance is additionally discussed down to cryogenic temperatures. The Hall coefficient shows whether the charges are transported by positive or negative carriers in extrinsic semiconductors and, except for the Hall factor or scattering factor, is the reciprocal of the carrier density and electronic charge. The Hall coefficient provides an indication of the type and concentration of dopants in a sample and, when combined with the resistivity, it determines the carrier mobility (Hall mobility). The Hall coefficient is not easily interpreted to determine the type of majority carriers, in the case of materials partly compensated by impurities of the opposite type. Hall coefficient and Hall mobility for various types of materials are reviewed in Sect. 4.3 as functions of the particle fluence and the Frenkel-pairs for irradiations with several types of particles. The case of large displacement damage is also discussed. Finally, Section 4.3 concludes with a brief review of atomic force microscopy (AFM) investigation in irradiated devices.

The basic knowledge, discussed in this chapter, finds applications in understanding the main modifications of silicon radiation-detectors and devices after irradiations. These latter are discussed in Sect. 6.8 and Chapter 7, respectively.

4.1 Radiation Environments

Nowadays, silicon devices are the main components of any electronic system based on very large scale integrated semiconductor technologies. Similarly, since decades, silicon detectors have become the most common semiconductor radiation detector for position and energy measurements. Devices and detectors are often employed in environments degrading their performance (e.g., see [Vavilov and Ukhin (1977); van Lint, Flanahan, Leadon, Naber and Rogers (1980); Messenger and Ash (1992)]). These environments for space payloads and particle experiments differ by the complexity, combination and energy distribution of the radiation and are treated in this section. Other radiation environments, for instance nuclear weapons (e.g., see Chapter 4 in [Messenger and Ash (1992)] and references therein) and nuclear reactors, are beyond the scope of the present book.

4.1.1 *High-Luminosity Machines Environments for Particle Physics Experiments*

Many components of the detection systems of high-energy experiments are exposed to the adverse radiation environment that results from the interactions, with surrounding materials, of particles produced by high rate collisions of incoming beam with target or head-on collisions of particle beams at high luminosity. The origin of this radiation field is well understood. It is known from the time of the CERN intersecting storage rings (ISR) [Johnsen (1973)], the first hadron-collider, that radiation at collider has several sources, namely, particle production at the interaction points, possible local beam losses and beam-gas interactions, although high vacuum condition (up to 10^{-11} Torr) was achieved in the accelerator vacuum chambers. Beams could be accelerated in the ISR up to maximum energy of $31\,\mathrm{GeV}$ with an initial luminosity of about $10^{28}\,\mathrm{cm^{-2}s^{-1}}$. These radiation features were also encountered at various levels at the CERN Sp$\bar{\mathrm{p}}$S-collider and hadron accelerators operated at Fermilab, BNL and elsewhere, until extreme conditions will be reached with the advent of the Large Hadron Collider (LHC) at CERN. Very soon, the LHC will allow head-on collisions of $7\,\mathrm{TeV}$ protons with a rate of about $10^9\,\mathrm{s^{-1}}$ p–p inelastic events per collision, at a (designed) peak luminosity of $1 \times 10^{34}\,\mathrm{cm^{-2}s^{-1}}$, assuming a standard inelastic p–p cross section of $80\,\mathrm{mb}$. With total beam losses not exceeding $10^7\,\mathrm{p\,s^{-1}}$ and beam-gas interactions limited to about $10^2\,\mathrm{m^{-1}s^{-1}}$, the main source of radiation at LHC will be from particles produced in p–p collisions at the interaction points [ATLAS Collab. (1994b)]. High-energy particles from an interaction point begin to cascade when entering the material of surrounding detectors. If the material is thick, the cascade development will continue until most of the charged

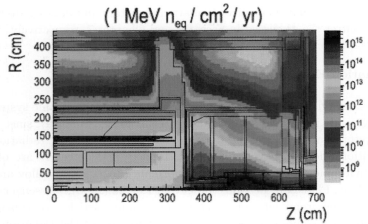

Fig. 4.1   1 MeV neutron equivalent fluence (1 MeV n_{eq}/cm^2/yr) in the ATLAS calorimetry system and vicinity, integrated over a standard high luminosity year; Z (in cm) is the direction along the beam axis and R (in cm) the radius around the Z-direction (from [Leroy and Rancoita (2007)]; see also [ATLAS Collab. (1996)]).

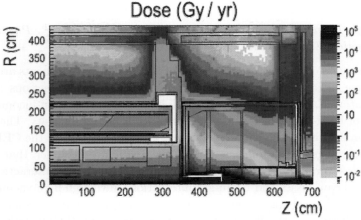

Fig. 4.2   Yearly integrated dose (Gy/yr) for photons above 30 keV in the ATLAS calorimetry system and vicinity for a standard high luminosity year; Z (in cm) is the direction along the beam axis and R (in cm) the radius around the Z-direction (from [Leroy and Rancoita (2007)]; see also [ATLAS Collab. (1996)]).

particles are absorbed. Then, one expects that charged hadrons (mainly protons, pions) will dominate the radiation environment in regions close to the collision point at least for small depth of traversed material (like for the LHC-ATLAS inner detector partly equipped with silicon detectors). Particles backscattered from the calorimeters and cascades leaking out the calorimeters produce neutrons, photons and electrons through their interactions with surrounding material and equipment located in the experimental hall. These particles also affect the region around the collision point, although, neutrons in that region can be moderated by the presence of low-Z material [walls or low-Z planes of inner detectors with Z the atomic number of the material (Sect. 3.1)]. While electromagnetic cascades are rapidly absorbed, neutrons will travel long distance losing their energy gradually. Therefore, the neutron fluence dominates the radiation environment at larger radii. Nuclear capture of thermal neutrons frequently results into production of photons. Large amounts of photons also result from excited-state decay of spallation products and from fast neutron interactions with atomic nuclei. Photons, produced deep inside the material, are rapidly absorbed. Only photons, produced close to the surface of material, have a chance to produce electrons and positrons and are responsible for a low-energy electron component in the environment.

As the showering process takes a relatively long time, this mixed field of neutrons, photons and electrons constitutes a background with no time structure [ATLAS Collab. (1994b)]. The numerous scatterings of neutrons and photons before their capture yield a uniform and isotropic field of low-energy background particles.

The radiation level induced on detector material and equipment will depend on the position of the detector with respect to the collision point. This location is expressed in terms of the *pseudorapidity*, η, defined in Eq. (3.61) and expressed as a function of θ, which is the polar production angle with respect to the beam axis. For experiments to be performed at LHC, the highest radiation levels are expected to occur at high pseudorapidity. As stressed above, the level of radiation also depends on the amount of surrounding materials. In ATLAS example [ATLAS Collab. (1994b, 1996)], the forward calorimeter (FCAL) is covering the range $3.2 \leq \eta \leq 5$. Due to its location, the FCAL is subjected to radiation dose-rates up 10^6 Gy yr^{-1} and a neutron flux (neutron kinetic energy > 100 keV) up to 1×10^9 cm^{-2}s^{-1}. Representing a depth of 9.5 interaction lengths of material, the FCAL contributes to the increase of the radiation level in the hadronic end-cap (HEC) calorimeter due to its location ($1.5 \leq \eta \leq 3.2$) very close to FCAL. The HEC will be subjected to radiation dose up to 0.3 kGy and a neutron fluence of 0.3×10^{14} cm^{-2} over 10 years of operation at the highest luminosity (1×10^{34} cm^{-2}s^{-1}). It is standard to use 1 MeV equivalent neutron fluence to express neutral particle contribution to radiation. The 1 MeV equivalent neutron fluence is the fluence of 1 MeV neutrons producing the same damage* in a detector material as induced by an arbitrary fluence with a specific energy distribution [see Eqs. (4.79, 4.85), Sects. 4.1.3 and 4.2.1]. The displacement damages induced by neutrons in semiconductors can be normalized by displacement damages induced by 1 MeV neutrons through the hardness parameter [see Eqs. (4.79, 4.84), Sects. 4.1.3 and 4.2.1]. The 1 MeV neutron fluence and yearly integrated dose (photons above 30 keV) in the ATLAS calorimetry system

---

*The amount of the induced displacement damage (i.e., the deposited damage energy, see Sect. 4.2.1) can be defined for any kind of incident particle. However, it cannot be assumed to relate a *damage equivalence* until the correlation between measured device degradation and the calculate NIEL deposition is determined (see discussion in Sects. 4.2.1–4.2.1.5 and, also, [Griffin, Vehar, Cooper and King (2007)]).

and vicinity are shown in Figs. 4.1–4.2, respectively, for a standard high luminosity year [ATLAS Collab. (1996)]. As will be discussed later, one has also to consider damages created by ionization processes, which are mainly inflicted to electronics and may alter the good operation of semiconductor devices.

This part of the book addresses the study of semiconductor behavior in radiation environment. However, it is interesting to look briefly at other types of detectors and look at the impact of accelerator generated radiation field on their operation. Again, the example of LHC will be used. Liquid argon is the active medium of FCAL, electromagnetic and HEC calorimeters of the ATLAS experiment. The choice of liquid argon, as active medium of these calorimeters, was largely motivated by the radiation hardness of the technique, compared to scintillator, for instance. The concern about activation of argon through production of ^{41}Ar (half-life of 1.82 hours) via the reaction

$$n +^{40}Ar \rightarrow\ ^{41}Ar + \gamma$$
$$\downarrow$$
$$^{41}K\ + \beta^- + \bar{\nu}_e$$

has been discarded by simulation and measurements performed at SARA [Belymam et al. (1998)]. The short half-life of ^{41}Ar makes dose source negligible during access to the detector for maintenance. Absorbers and structural materials, used in ATLAS liquid argon calorimeters, are radiation hard and stable under irradiation. Under irradiation, oxygen or oxygen-like impurities are released from the surface of materials and equipments immersed in liquid argon. This outgassing leads to pollution of liquid argon and to possible reduction of collected ionization charge and, finally, in turn to the reduction of the calorimeter signal. Then, the operation of the ATLAS liquid argon calorimeters requires limited pollution of liquid argon, below a threshold of (1–2) ppm. This goal can be achieved if the components of the liquid argon calorimeters are certified against release of polluting impurities under irradiation. Therefore, the possible outgassing of components immersed in liquid argon and exposed to high fluences was investigated within the ATLAS Collaboration for several years, in conditions similar to those encountered in the running experiment during LHC operation. The tested materials and equipments were subjected to accelerated irradiation, in order to receive the equivalent of a LHC ten-year dose within a short period of time. Cold test facilities were built at SARA [Belymam et al. (1998)], CERI[§] and Dubna [Leroy et al. (2000a)] for that purpose. In particular, in Dubna, the large beam geometrical acceptance of $800\,cm^2$ at that facility allows the exposure of relatively large area inside a one liter cryostat filled with liquid argon. The Dubna IBR-2 pulsed neutron reactor made possible the irradiation of materials and equipments immersed in liquid argon at high neutron fluences ($\approx 10^{15}\,n\,cm^{-2}\,d^{-1}$ up to a total fluence of $10^{16}\,n\,cm^{-2}$ for a standard measurement period of 11 days) and γ-doses ($\approx 10\,kGy$ per day), simultaneously, as γ's always

---

[§]CERI, Neutron Irradiation Facility, CNRS, Orlans, France.

accompany neutrons in the reactor. Sintered tungsten "slugs" used as absorber in FCAL have been shown to suffer no outgassing [Leroy et al. (2000b)]. The Dubna cold facility also allowed the test of the characteristics integrity of electronic components under the high irradiation level expected at LHC [Leroy et al. (2002c)]. Glues used in calorimeter electronics were selected according to their radiation hardness and mechanical hardness under irradiation [Leroy et al. (2000c)].

The use of scintillator at LHC is mainly constrained by two factors: i) the inter-bunch crossing time of 25 ns which requires a fast scintillator decay time and ii) the high radiation level which requires a radiation hard scintillator. A high density (8.28 g/cm^3) scintillating crystal, PbWO$_4$, which fulfils these constraints is used in CMS [CMS (1997)] for electromagnetic calorimetry. Its short (long) decay time is 5 (15) ns and it is radiation hard at the expense of relatively poor light yield[†].

Semiconductor trackers, consisting of silicon pad or pixel detectors and silicon microstrips detectors, are used in collider experiments [ATLAS Collab. (1994b)]. Silicon pad detectors are also used as active medium of preshower calorimeters [AT-LAS Collab. (1994b); CMS (1998)]. The exposure of silicon detectors or devices to the high level of radiation encountered in high luminosity accelerators lead to degradation of their performances. After intensive investigations, solutions have been found to improve the condition and operation of silicon detectors in such environment. A brief review of these solutions is presented now, more will be discussed in subsequent chapters of this book.

Various types of damage, surface or lattice defects like vacancies (point-like defects) or damaged regions (clusters), are produced when incoming particles deposit their energy in silicon. The importance of the damage created by a specific process depends on its relative cross section compared to those of other possible processes. One can distinguish the ionization process, resulting from the interaction of the incident particle with the atomic electrons (e.g., see Sect. 4.2.3), and non-ionization processes, resulting from the interaction between incoming particles and lattice nuclei (e.g., see Sect. 4.2.1).

4.1.1.1 *Ionization Processes and Collider Environments*

The effect of damage to silicon devices due to ionization process originates from ionizing energy-loss in surface layers. Even at large radii, radiations affect the electronics components located inside and around the LHC detectors through ionization in silicon oxide by photons, protons, pions, ions (*total ionizing dose* - TID). The various aspects of electronics radiation hardness have been extensively discussed in many reports (see for instance [Dentan (1999)] in the framework of the ATLAS experiment). The relevant TID effects for silicon devices employed in LHC environments are summarized in Sect. 4.2.3.2.

[†]The light yield compared to NaI(Tl) is 0.01.

4.1.1.2 *Non-Ionization Processes, NIEL Scaling Hypothesis and Collider Environments*

The non-ionization processes produce bulk defects or displacement damage (e.g., see Sect. 4.2.1). The recoil nuclei are removed from their initial lattice position and displaced into interstitial positions (*interstitials*). However, if the recoil energy of the primary knock-on nucleus is higher than the threshold for atomic displacement, this nucleus can produce displacement of another nucleus. Finally, as long as the recoil energy is above displacement threshold, the cascading-process will continue and a chain of vacancies (V) and interstitials (I) will be produced (e.g., see Sects. 4.2.1 and 4.2.2). The bulk damage effects produced by energetic particles have been shown to be proportional to the displacement damage cross-section equivalent to the *non-ionizing energy-loss* [NIEL, e.g., see Sect. 4.2.1 and Eqs. (4.86, 4.88)]. The proportionality between NIEL-value and resulting damage effects is referred to as the *NIEL scaling hypothesis*. The NIEL value depends on the particle type and energy (e.g., see Sect. 4.2.1.5). However, violations of the NIEL scaling hypothesis were observed. This scaling does not work for calculating N_{eff} in oxygenated silicon [Rose Collab. (2000)] and for low proton energies for standard silicon [Bechevet, Glaser, Houdayer, Lebel, Leroy, Moll and Roy (2002)].

Subjected to increasing fluence, the main effects on the operation of silicon detectors, from a practical point of view, are i) an increase with fluence of the leakage current, ii) an increase of the full-depletion bias voltage to be applied to the detector and iii) a decrease of the charge collection efficiency (CCE). The damaged regions created by radiation dose in the silicon bulk are acting as electrically active defects with deep levels in the forbidden-band gap in silicon. The generation of these additional traps decreases the carrier lifetime and increase the reverse current. For fluences of the order of 10^{14} particles cm^{-2}, i.e., in practice fluences corresponding to one year of operation, in many regions of the LHC, the reverse current can reach values in the range of several tens of microamperes per cm^2 and even more. However, the strong temperature dependence of the reverse current allows it to be minimized by operating the detectors at moderate low temperature (in a range from -5 °C to -10 °C, typically). Re-arrangement between these vacancies and interstitials or their interactions with other defects, impurities and nuclei in the bulk (like oxygen) are responsible of long term effects, e.g., the evolution in time of electrical characteristic of silicon detectors such as annealing and reverse annealing of leakage current and full depletion bias voltage after irradiation. It is observed that the leakage current anneals following irradiation by about 50 % over a period of about two weeks at room temperature. From the measurement of the leakage currents of silicon detectors made of different starting materials irradiated with neutrons, up to fluences of $10^{15}\,\text{n}\,\text{cm}^{-2}$, the leakage current damage parameter is independent of the initial resistivity and impurity concentrations, once normalized to the sensitive volume and to the 1 MeV-neutron equivalent hadron fluence. This parameter is linked to defect clusters which are not affected by the material. The concentration of defects, that are responsible for the leakage current, are in the order of the impurity concentrations and, consequently, migrating interstitials and vacancies produced by the irradiation are responsible for the increase of the leakage current. Annealing of the leakage current is also material independent. The period of annealing is followed by a fast increase of the leakage current. This reverse annealing effect is largely moderated, even suppressed, if the irradiated diode is kept at low temperature, around -5 °C. Therefore, cooling of the detector to moderately low temperatures is a necessity for the reduction of the leakage current so keeping a signal-to-noise ratio required for physics exploitation during colliding beam periods. This cooling is also necessary during machine and detector maintenance periods to avoid reverse-annealing leading to the increase of the reverse leakage current and full depletion bias voltage.

The full depletion bias voltage is the minimal bias voltage to be applied in order to fully deplete the detector[§]. In non-irradiated diodes, the space-charge results from shallow dopants in the silicon. Irradiation brings changes in the doping concentration (donors and/or acceptors) with a build-up of negative space-charge in the depletion region due to creation of deep energy levels. The effective doping concentration being proportional to the full depletion voltage, the value of the full depletion bias voltage is affected and can reach very high and unpractical values after several years of the detector operation in high radiation environment. This has the possible consequence that a fraction of the detectors will have to be operated without being fully depleted. Similarly to the reverse leakage current, an annealing effect is observed for the effective doping concentration (N_{eff}) over a period of about two weeks after irradiation and followed by a reverse annealing, i.e., a fast increase, particularly after several months for diodes kept at room temperature. The amount and rate of reverse annealing depends on the irradiation fluence. For a typical silicon detector $300\,\mu$m thick and $5\,k\Omega$cm resistivity irradiated up to $10^{14}\,p\,cm^{-2}$, the full depletion bias voltage saturates at nearly 350 V. It should be reminded that the operating bias voltage should always be larger than the full depletion bias voltage due to trapping effects. This necessary over-bias depends on the exposure fluence and state of annealing.

The charge collection efficiency (CCE) is degraded down to (10–15) % of the value measured for non-irradiated silicon detector, depending on the irradiation fluence. However, for temperatures below 100 K, temperature range unpractical for LHC, 50 % of the charge is recovered in detectors irradiated up to $2 \times 10^{15}\,n\,cm^{-2}$ (due to the so-called *Lazarus effect*[‡‡] [Palmieri et al. (1998); Bell et al. (1999)]).

In the case of liquid argon, radiation hardness is improved by selection of materials certified against outgassing. For scintillator, one uses a new type of crystal with

---

[§]The silicon detectors are typically $p^+ - n$ junctions (e.g., see Sects. 4.3.2 and 6.8).

[‡‡]It was found out that, at temperatures below 130 K, largely damaged detectors, apparently, exhibited a significant recovery of operating conditions [Palmieri et al. (1998)]. The explanation of this phenomenon, known as the *Lazarus Effect*, is related to the dynamics of the induced defects in the semiconductor bulk.

increased radiation hardness. The technique, called *defect engineering*, is applied to improve the radiation hardness of silicon. It involves the deliberate addition of impurities, oxygen in practice, to the silicon bulk material during the growth process [Rose Collab. (2000)] in order to affect the formation of electrically active defect centers which, in turn, allow control of the macroscopic parameters of silicon devices. The oxygen concentration achieved is $[O] \sim (3\text{--}5) \times 10^{17}\,\mathrm{cm}^{-3}$. It should be noted that high oxygen concentration $\{[O] \sim (4\text{--}20) \times 10^{17}\,\mathrm{cm}^{-3}\}$ can be found in silicon crystals grown by the Czochralski (Cz) method. Cz growth technology was generally characterized by low resistivity ($\leq 900\,\Omega\mathrm{cm}$) not really suited for radiation detector applications. Recently, higher resistivity Cz silicon ($1.2\,\mathrm{k}\Omega\mathrm{cm}$) suitable for radiation detector application became available, after developments in crystal growth technology [Savolainen et al. (2002)]. An explanation of the behavior of oxygen enriched silicon is found by interpreting the role played by interstitial oxygen, which is believed to act as a sink of vacancies. Then, divacancy-oxygen* (V_2O) center defects are identified as main responsible for the radiation-induced negative space-charge. V_2O is produced through the reaction:

$$V + O_i \rightarrow VO, \tag{4.1}$$

followed by

$$VO + V \rightarrow V_2O. \tag{4.2}$$

The addition of oxygen in the bulk material suppresses V_2O formation. This happens through the interaction between interstitial oxygen with vacancies. An increase of interstitial oxygen concentration enhances the ratio of the reactions expressed by Eq. (4.1) and Eq. (4.2), reducing formation of V_2O.

Oxygenation makes no difference to leakage current. Introduction of high concentrations [at the level of $\sim (3\text{--}5) \times 10^{17}\,\mathrm{cm}^{-3}$] of oxygen in silicon detectors has led to improved radiation hardness, when the detectors were exposed to charged hadrons (protons, pions). However, almost no improvement is observed for irradiations of oxygenated silicon materials with neutrons of a few MeV. It should be observed that the proton measurements have been initially done with $24\,\mathrm{GeV/c}$ protons at CERN-PS, while neutron measurements at high fluences have been performed at reactors and CERN-PS for low energy neutrons [(1–2) MeV]. The absence of measurement of high energy neutrons ($\geq 20\,\mathrm{MeV}$) is due to the extreme difficulty of achieving high fluences for neutrons of these energies during short periods of time. For instance, neutrons of $14\,\mathrm{MeV}$ can be produced[†] by $38\,\mathrm{MeV}$ proton on a Be-target with fluences of $10^{14}\,\mathrm{n\,cm}^{-2}\,\mathrm{d}^{-1}$, comparable to those expected at LHC. However, no measurement has been done with oxygenated silicon detectors at this neutron energy. Low energy proton ($\leq 10\,\mathrm{MeV}$) irradiation of silicon, doped with high concentration of oxygen, shows improved radiation hardness of these detectors already observed for

---

*A general presentation of the radiation-induced defects can be found in Sect. 4.2.2.

†For instance at the facility at the Nuclear Physics Institute (Academy of Sciences of the Czech Republic), CZ-25068 Řež near Prague.

24 GeV/c protons [Houdayer et al. (2003)]. The question remains if observed improvement is energy related, i.e., depends on the value of the proton and neutron cross sections. Based on model, simulation [Huhtinen (2002)] and data [Bechevet, Glaser, Houdayer, Lebel, Leroy, Moll and Roy (2002)], radiation hardness improvement of oxygenated silicon is expected when it is exposed to neutrons of at least 20 MeV energy.

The different behavior between protons and neutrons, at least at low particle energy, is of no practical consequences at LHC, since the radiation field in LHC collision areas is largely dominated by charged hadrons, where silicon detectors are operated, neutron acting at larger radii. Therefore the use of diffusion oxygenated standard planar or float zone (DOFZ) silicon will be beneficial. In addition, for proton irradiation, N_{eff} achieved at high fluences is independent of the oxygenated material initial resistivity, but for neutron irradiation, the N_{eff} increase at high fluence is reduced by the use of oxygenated material of low resistivity. Therefore, using oxygenated silicon of low resistivity $[(1-2) \, k\Omega cm]$ would improve their radiation hardness when operated in the mixed field of charged hadrons and neutrons at LHC.

4.1.2 *Space Radiation Environment*

The continuous evolution of mission requirements and their electronic technologies for payloads and spacecrafts (for example, the International Space Station** shown in Fig. 4.3), combined with the need to meet the space environment constraints, particularly radiation, constitute challenges for component engineers and designers. For instance, the increased activities in space for communications, military services and scientific research have required to take into account that the electronics, employed for these purposes, contain advanced devices and electronic systems in VLSI (very large scale integration) and ULSI (ultra large scale integration) technologies, which might have radiation sensitivity.

The space radiation is not homogenously distributed in the Earth magnetosphere and solar cavity (the so-called *heliosphere*) and varies with time. As a consequence, Low- (LEOs), Mid- (MEOs), Geosynchronous/Geostationary-Earth Orbits (GEOs), Geostationary Transfer Orbits (GTOs) and orbits for an Earth Observatory Satellite (EOS) have different requirements for radiation hardness (see, for instance, [Barth (1997)] and references therein). These orbits are distinguished by their degrees of inclination and distances of perigee and apogee. The interplanetary missions are affected by the energetic particles of solar and galactic sources traveling through the interplanetary space (Sect. 4.1.2.4), but are less exposed to radiation effects of the trapped particles in the Earth magnetosphere (Sect. 4.1.2.5). The fluxes of these

---

**Thanks to NASA's courtesy, we show an image (Fig. 4.3) at the final stage of completion. The *Reference Guide to the International Space Station* is available at the web site: *http://www.nasa.gov/mission_pages/station/news/ISS_Reference_Guide.html.*

Fig. 4.3 Computer-generated artist's rendering (Courtesy NASA) of the International Space Station (ISS) when completed. The ISS orbits Earth at an altitude that ranges from 370 to 460 km.

particles are *modulated* by the *solar wind*[‡‡], whose intensity and speed depends on the solar activity.

The presence of this complex environment has drawn attention to the important influence of space radiation on both the computation tools for modeling the space radiation (e.g., see [Barth (1997)], Chapter 1 in [Claeys and Simoen (2002)], [Heynderickx (2002)], Chapter 2 in [Holmes-Siedle and Adams (2002)], [Miroshnichenko (2003)]) and the response to radiation of MOS and bipolar transistors made in VLSI BiCMOS (Bipolar Complementary Metal-Oxide Semiconductor) technologies. These devices are, indeed, the essential part of any circuit used in that environment, thus, the knowledge of irradiation effects on them is critical (e.g., see [Baschirotto et al. (1995b, 1996); Fleetwood et al. (1994); Johnston, Swift and Rax (1994); Baschirotto et al. (1997); Colder et al. (2001, 2002); Codegoni et al. (2004b); Consolandi, D'Angelo, Fallica, Mangoni, Modica, Pensotti and Rancoita (2006)] and references therein; see, also, Sects. 4.2.3.2 and Chapter on *Displacement Damage and Particle Interactions in Silicon Devices*). For instance, bipolar junction transistors (BJT) have important applications in analog or mixed-signal IC's and BiCMOS circuits because of their linearity and excellent matching characteristics, for example. Many of the bipolar integrated circuits in space systems, including operational amplifiers, comparators, voltage regulators, are used to accomplish analog functions. Furthermore, MOS transistors [particularly those of the complementary form (CMOS)] are of wide usage in high-performance and low-power electronics.

Once defined the duration and the orbit of a satellite or a pay-load inside the heliosphere or the Earth magnetosphere, the knowledge of fast charged particle fluences and their time dependence makes possible to determine the expected Frenkel-pairs concentration (*FP*) due to the non-ionizing energy-loss processes (see Sect. 4.2.1) and the absorbed dose mostly determined by collision energy-loss processes (e.g., see the chapters on *Electromagnetic Interaction of Radiation in Matter* and *Nuclear Interactions in Matter*, and [Ziegler, Biersack and Littmark (1985a); Ziegler, J.F. and M.D. and Biersack (2008a)])[*]. Both types of processes result from the interactions of charged particles impinging on semiconductor devices (i.e., on microelectronics).

This knowledge has to be complemented by investigations of possible latch up and single event upset (e.g., see [Messenger and Ash (1997)]) for an effective implementation of any VLSI circuit, whose design can, in turn, depend on the radiation response (e.g., see Sects. 4.2.3.2 and 7.2). Obviously, a non-radiation hard technology cannot be employed in a radiation environment: the space qualification of VLSI technologies with respect to TID is certainly needed for any design consideration and may require the qualification of their basic bipolar devices with regard to space-radiation. As an example, when bipolar integrated circuits are exposed to radiation

---

[‡‡]For instance, the reader can see [White (1970); Encrenaz, Bibring and Blanc (1991); Alurkar (1997); Boella, Gervasi, Potenza, Rancoita and Usoskin (1998); Boella, Gervasi, Mariani, Rancoita and Usoskin (2001); Grieder (2001); Lang (2001); Gosling (2006); Potgieter (2008)].

[*]The reader may also see [Leroy and Rancoita (2004)].

in space, one of the primary effects of degraded operations is the reduction of the current gain.

In the following sections, we present a review about i) the origin of the solar wind and heliospheric (or interplanetary) magnetic-field (Sect. 4.1.2.1), ii) the extension of the heliosphere and the Earth magnetosphere (Sect. 4.1.2.2), iii) how the solar wind affects (modulates) the propagation of galactic cosmic rays in the heliosphere (Sect. 4.1.2.3), iv) fluxes of solar, heliospheric and galactic cosmic rays (Sect. 4.1.2.4) and v) a description of trapped particles in the Earth magnetosphere (Sect. 4.1.2.5).

4.1.2.1 *Solar Wind and Heliospheric Magnetic Field*

Nowadays, we know that the *solar wind* (SW) is a plasma[††] that permeates the interplanetary space and constitutes the *interplanetary medium*. It is, as discussed later, the outer part of the *Sun's corona*[¶] streaming through the solar system and creating the so-called *solar cavity* (also termed *heliosphere*). It is highly variable in both time and space; it consists[†] of protons (about 95% of the positively charged particles), double charged helium ions, a very small number of other positively charged particles and electrons, so that the plasma is electrically neutral. At the orbit of Earth, i.e., at 1 AU from the Sun (see Appendix A.2), experimental observations of SW allowed one to determine that i) the proton density and temperature are $\approx (8\text{--}9)\,\text{proton/cm}^3$ and $\approx 1.2 \times 10^5\,\text{K}$, respectively, ii) the mean wind speed is $\approx 470\,\text{km/s}$, iii) the mean speed of sound is $(50\text{--}63)\,\text{km/s}$ (e.g., see Table 3-1 of [Feynman (1985)], Table 1 of [Gosling (2006)] and references therein) and, thus, iv) the SW speed is *supersonic*[‡‡]. It carries an embedded weak magnetic-field, which results in *modulating* the fluxes of GCRs entering the heliosphere. It has to be noted that until the early 1990s our knowledge of the heliosphere was limited to the *ecliptic plane*. In fact, the investigation of plasma, particles and field in the polar regions of the Sun were among the main goals of the *Ulysses mission*, launched in 1990.

In 1859, the first SW (indirect) observation was made by Carrington, who witnessed what is now called a *solar flare*[*] and noted that a major geomagnetic storm began about 17 hours after the flare. In 1896, Birkeland suggested that the Earth environment is bombarded by "rays of electric corpuscles emitted by the Sun". In the early 1900s, Lindemann suggested that geomagnetic storms may result from an interaction of the geomagnetic-field and plasma clouds ejected by the *solar acti-*

[††]The plasma is an ionized gas, electrically quasi-neutral and exhibiting collective behavior (e.g., see Chapter 2 of [Meyer-Vernet (2007)]).

[¶]The corona with temperatures over 10^6 K is a highly rarefied region above a small region called chromosphere, which extends for ≈ 2000 km above the photosphere. The photosphere, in turn, is a thin layer at the surface of the Sun. The reader may find an introduction to the Sun structure, for instance, in [Aschwanden (2006)].

[†]The reader can find details of solar wind composition, for instance, in [Feynman (1985); Gosling (2006)]).

[‡‡]The *sonic point*, i.e., the position at which the solar wind speed becomes equal to the speed of sound, is located at several solar radii ($R_\odot$).

[*]The solar flare is a giant explosive energy release on the Sun.

vity. In the 1930s, Forbush noted a rapid decrease (the so-called *Forbush decrease*) in the observed GCR intensity following a *coronal mass ejection*. It occurs, as we know now, due to the magnetic field of the SW plasma sweeping some of the galactic cosmic rays away from Earth. In the early 1950s, Biermann concluded that there must be a steady stream of charged particles from the Sun to account for the fact that ionic tails of comets always point away from the Sun. In the mid 1950s, Chapman constructed a *static solar corona model* and calculated the properties of a low-density gas at $\approx 10^6$ K: he determined that this latter is an excellent conductor of heat and must extend way out into space, beyond the orbit of Earth. In other words, the Earth is expected to be inside the solar corona. These results and observations inspired Parker (1958)[§] to formulate a radically new model of the solar corona in which this later is continuously expanding outward, thus solving the discrepancy between the estimated interstellar pressure of $(10^{-12}$–$10^{-13})$ dyn/cm^2 (e.g., see [Parker (1958)], pages 8–11 of [Toptygin (1985)] and Table 1 of [Suess (1990)]) and that of $\approx 0.6 \times 10^{-5}$ dyn/cm^2 [Parker (1958)], computed under the assumption of a static equilibrium in the solar corona (see, also, [Parker (2007)]). Since then, *Parker's model* has demonstrated to provide a general and elegant framework for heliospheric features. However (e.g., see [Fisk (2001); Cranmer (2002)] and references therein), subsequently this model was modified to account additional observations, like for instance those obtained with i) the *Ulysses spacecraft* (launched by the Space Shuttle Discovery in 1990, as already mentioned) designed to reach high solar latitudes, thus out of the ecliptic plane and ii) the *Solar and Heliospheric Observatory* (SOHO, launched in 1995) to study the Sun from its deep core to outer corona and the SW.

As mentioned above, Parker (1958) assumed that i) the solar corona was not in hydrostatic equilibrium, ii) there is a stationary expansion with the *coronal plasma* flowing out, iii) the stationary expansion is spherically symmetric, iv) the kinetic temperature of the gas is maintained (by heating mechanisms) at a uniform value T_0 from the base of the corona at radial distance $r_{co} \approx 10^6$ km to some radius beyond which the heating vanishes and T is negligible, v) electrons and protons are the dominant (and equal in number) particles of the solar coronal plasma, and vi) the stationary expansion satisfies the equation of motion[¶]

$$\rho v \frac{dv}{dr} = -\frac{dP}{dr} - Gm_\odot \frac{\rho}{r^2},\tag{4.3}$$

where r is the radial distance; G and $m_\odot$ are the Newtonian constant and solar mass, respectively (see Appendix A.2); P is the pressure; v is the radial velocity of the corona expansion; the mass density is given by

$$\rho = n_{pl}\left(m_p + m_e\right) = n_{pl}m \approx n_{pl}m_p,\tag{4.4}$$

[§]This article on Parker's theory about the SW was submitted to Astrophysical Journal, but was rejected by the two reviewers. It was saved by then editor Chandrasekhar (1983 Nobel Prize in physics). Parker received the Kyoto and James Clerk Maxwell Prizes in 2003.
[¶]This equation is the *Bernoulli equation* for the solar corona expansion and reduces to the so-called *barometric equation*, when the first term is 0 (i.e., for $dv/dr = 0$). The barometric equation accounts for hydrostatic equilibrium.

in which m_p (m_e) is the proton (electron) rest-mass and, thus,

$$m = m_p + m_e \approx m_p;$$

n_{pl} is the number of protons (or electrons, since the gas is assumed to be neutral and fully ionized) per cm^3 and satisfies the *equation of continuity*

$$\frac{d}{dr}\left(n_{\text{pl}}vr^2\right) = 0 \tag{4.5}$$

for conservation of matter. For an ideal (*mono-atomic*) gas with equal electron and proton temperatures, the pressure is given by

$$P = 2\,n_{\text{pl}}k_BT,$$

where k_B is the Boltzmann constant (Appendix A.2), and the (local) *sound speed*$^{\|}$ is

$$v_{\text{sd}} = \sqrt{\frac{dP}{d\rho}}. \tag{4.6}$$

Equation (4.3) yields a positive expansion velocity, v, as a function of the radial distance, r, for $r \geq r_{\text{co}}$ when [Parker (1958)]

$$\frac{mv^2}{2\,k_BT_0} - \ln\left(\frac{mv^2}{2\,k_BT_0}\right) = -3 - 4\ln\left(\frac{Gmm_\odot}{4\,r_{\text{co}}k_BT_0}\right)$$
$$+4\ln\left(\frac{r}{r_{\text{co}}}\right) + \frac{Gmm_\odot}{rk_BT_0}. \tag{4.7}$$

In Fig. 4.4, it is shown the expansion velocity, $v(r)$, obtained using Eq. (4.7) for an isothermal solar corona with temperatures $T_0 = 10^6$, 1.5×10^6 and 2×10^6 K and $r_{\text{co}} \approx 10^6$ km at the base. Furthermore, Parker extended his treatment to i) the case in which the pressure and temperature are related by the *polytropic law* [Parker (1960, 1963)]

$$P = P_0 \left(\frac{\rho}{\rho_0}\right)^\alpha, \tag{4.8}$$

where α is the polytropic index (for instance, $\alpha = 1$ and $\alpha = 5/3$ for isothermal and adiabatic processes, respectively) and ii) the case of sudden corona-expansion following a large solar flare [Parker (1961a)]. The hydrodynamic expansion of the *coronal plasma* was termed the *solar wind* for the Sun and the *stellar wind* for the other stars by Parker (1960).

For approaching the description of a more complex three-dimensional SW structure as presently observed ([McComas, Elliott, Schwadron, Gosling, Skoug and Goldstein (2003)] and references therein), further discussions, criticisms and, also, improvements to Parker's model can be found, e.g., in [Parker (1961a, 1963);

$^{\|}$Using Eqs. (4.6, 4.8), the local sound speed can be rewritten as $v_{\text{sd}} = \sqrt{\alpha(P/\rho)}$. Thus, the wavespeed depends on the conditions under which the compression takes place. For example, in case of an adiabatic compression of a monoatomic gas we have $\alpha = 5/3$ and, consequently, $v_{\text{sd}} = \sqrt{(5P/3\rho)}$ (e.g, see Sections 17.3 and 21.5 of [Ingard and Kraushaar (1960)]).

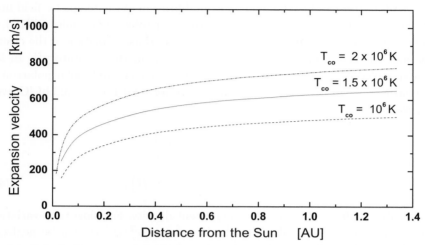

Fig. 4.4 Spherically symmetric hydrodynamic expansion velocity (i.e., the velocity of the solar wind) in km/s as a function of the radial distance (in AU) from the Sun, for an isothermal corona with temperatures T_0 of 10^6, 1.5×10^6 and 2×10^6 K and a radius of $\approx 10^6$ km at the base.

Hundhausen (1972); Feynman (1985)], Chapter 6 of [Cravens (1997)], Chapter 12 of [Gombosi (1998)], Chapter 6 of [Kallenrode (2004)] and Chapter 6 of [Parks (2004)] (see, in addition, references therein). Limitations of this model are related, for instance, to i) observations that the pressure is less isotropic as the plasma moves away from the Sun, ii) the assumption that the flow is considered uniform and radial and iii) proton and electron temperatures, which were assumed equal and isotropic. For example, measurements have determined the presence of a *fast* solar wind component, where electrons have a *core component* with a nearly Maxwellian (kinetic) energy distribution and a high-energy tail (the so-called *halo component*), which has an approximate Maxwellian distribution with a temperature larger by about an order of magnitude with respect to that of the core component. Although not all issues can be treated from the particle point of view, some of these features motivated some authors (e.g., [Lemaire and Scherer (1973); Maksimovic, Pierrard and Lemaire (2001)] and Section 6.7.4 of [Parks (2004)]) to investigate the so-called *kinetic models of the SW*. However, detailed treatments of alternative or extended solar wind models go beyond the purpose and fields of applications of the present book.

As already mentioned, in Parker's model the magnetic-field lines are supposed to be embedded in the streaming particles of the SW. The notion of *frozen-in magnetic-field* originates from the work on electromagnetic hydrodynamic waves by Alfvén* (1942) and plays a relevant role in the description of interplanetary magnetic-field [Parker (1957, 1958)] and planet magnetospheres. Alfvén noted that the motion of matter can be coupled to the modification of the magnetic-field lines, so that these latter follow the matter motion. In magnetohydrodynamics (MHD), we can express the time derivative of the magnetic field $\vec{B}$ as a function of the electric field, $\vec{\mathbb{E}}$, and the center-of-mass plasma-velocity, $\vec{v}_{\rm pl}$, in the laboratory[††]. In fact using the so-called *generalized Ohm's law* and since the Hall, ambipolar polarization and inertial terms can be neglected (e.g., see Sections 4.3.1 and 4.3.2 of [Cravens (1997)]), we have

$$\frac{\partial \vec{B}}{\partial t} = -\nabla \times \left(\frac{\vec{J}}{\wp} - \vec{v}_{\rm pl} \times \vec{B} \right)$$

$$= -\nabla \times \frac{\vec{J}}{\wp} + \nabla \times \left(\vec{v}_{\rm pl} \times \vec{B} \right), \tag{4.9}$$

where $\wp$ is the conductivity and $\vec{J}$ is the current density. For slow time-variations and non-relativistic plasma bulk-speeds[‡], the term $(1/c^2)\,\partial \vec{\mathbb{E}}/\partial t$ can be neglected in Maxwell's equation regarding $\nabla \times \vec{B}$, i.e., we have

$$\nabla \times \vec{B} = \mu_0 \vec{J} + \left(\frac{1}{c^2} \right) \frac{\partial \vec{\mathbb{E}}}{\partial t}$$

$$\approx \mu_0 \vec{J}$$

$$\Rightarrow \vec{J} \approx \frac{1}{\mu_0} \nabla \times \vec{B},$$

where μ_0 is the permeability of the free space (see Appendix A.2). Thus, using a vector calculus identity we can rewrite the first term of the right-hand side of Eq. (4.9) as

$$-\nabla \times \frac{\vec{J}}{\wp} \approx -\frac{1}{\mu_0 \wp} \nabla \times \nabla \times \vec{B}$$

$$= -\frac{1}{\mu_0 \wp} \left[\nabla \left(\nabla \cdot \vec{B} \right) - \nabla^2 \vec{B} \right]$$

$$= \frac{\nabla^2 \vec{B}}{\mu_0 \wp},$$

since

$$\nabla \cdot \vec{B} = 0.$$

---

*Alfvén, who got the Nobel prize in 1970, criticized how the concepts of frozen-in magnetic-field lines and field-line reconnection were used in the theory of magnetosphere (e.g., see [Alfvén (1976)] and references therein).

[††]For the interplanetary magnetic-field, the laboratory reference-frame is that of the solar cavity.

[‡]For example, these conditions are satisfied for a velocity ($v_{\rm pl}$), a time-scale ($\tau_{\rm ch}$) and a length-scale ($L_{\rm ch}$) such that

$$\tau_{\rm ch} \gg \frac{v_{\rm pl} L_{\rm ch}}{c^2};$$

for a discussion, the reader can see, for instance, Section 2.3.3 of [Meyer-Vernet (2007)].

Finally, Eq. (4.9) can be rewritten as

$$\frac{\partial \vec{B}}{\partial t} = \frac{\nabla^2 \vec{B}}{\mu_0 \wp} + \nabla \times \left(\vec{v}_{\text{pl}} \times \vec{B} \right). \tag{4.10}$$

For a plasma composed by i) electrons with rest-mass m_e and concentration n_e and ii) (an equal amount of) single ionized ions with rest-mass $\gg m_e$ and concentration

$$n_{\text{ion}} = n_e = n_{\text{pl}},$$

the electrical conductivity[§] is given by

$$\wp \approx \frac{e^2 n_{\text{pl}}}{m_e \nu_{\text{coll}}}, \tag{4.11}$$

where ν_{coll} is the collision frequency and e is the electron charge. In addition {e.g., see Equation (2.102) of [Meyer-Vernet (2007)]}, $\wp$ is approximated by

$$\wp \approx 6 \times 10^{-4} \times T^{3/2} \ [\Omega\,\text{m}]^{-1}, \tag{4.12}$$

whenever the magnetic-field effect can be neglected, i.e., in the direction i) parallel to $\vec{B}$ and ii) perpendicular to $\vec{B}$, if the particle mean free-path is much smaller than the *radius of gyration*. In general along this latter direction, the opposite condition occurs and, consequently, the electric conductivity results to be strongly reduced perpendicularly to $\vec{B}$. The quantity

$$D_{\text{B}} = \frac{1}{\wp \mu_0} = \frac{m_e \nu_{\text{coll}}}{e^2 n_{\text{pl}} \mu_0} \tag{4.13}$$

is the so-called *magnetic diffusion coefficient*. Equation (4.10) is known as the *magnetic convection-diffusion equation* and accounts for two contributions to the magnetic-field variation: these are due to the a) diffusion, produced by conductive losses (i.e., the first term), and b) convection, produced by the plasma bulk motion (i.e., the second term). Furthermore, the ratio of the magnetic convection and diffusion term

$$R_{\text{m}} = \frac{\text{magnetic convection term}}{\text{magnetic diffusion term}} \tag{4.14}$$

is termed *magnetic Reynolds number* and can be estimated from the expression (e.g., see Section 4.4.2 of [Cravens (1997)] or page 88 of [Meyer-Vernet (2007)])

$$R_{\text{m}} \approx \frac{v_{\text{pl}} L_{\text{ch}}}{D_{\text{B}}}$$
$$= v_{\text{pl}} \wp \mu_0 L_{\text{ch}},$$

where the length-scale L_{ch} is defined in the footnote at page 304. For $R_{\text{m}} \gg 1$ the magnetic diffusion term can be neglected in Eq. (4.10), which reduces to

$$\frac{\partial \vec{B}}{\partial t} \approx \nabla \times \left(\vec{v}_{\text{pl}} \times \vec{B} \right). \tag{4.15}$$

[§]The reader can see, e.g., Section 4.3.2 of [Cravens (1997)], Section 4.3.2 of [Gombosi (1998)], Sections 2.1.3 and 2.3.3 of [Meyer-Vernet (2007)].

When the magnetic convection-diffusion equation is reduced to Eq. (4.15), we are in the condition of a perfectly conducting-fluid moving in a magnetic field, for which it can be proven* that the magnetic flux through any closed contour moving with the plasma is constant. Thus, the fluid can move freely along the *magnetic-field lines* and these latter are *frozen in the fluid* (e.g., see [Parker (1957)] and, also, Chapter 4 of [Cravens (1997)], Section 14.4.2 of [Gombosi (1998)], Section 3.4 of [Kallenrode (2004)], Section 5.5.1 of [Parks (2004)], [Lundin, Yamauchi, Sauvaud and Balogh (2005)] and references therein, Section 2.3.2 of [Meyer-Vernet (2007)]).

For the SW non-relativistic plasma, it can be shown that $\wp$ is large and $R_m \gg 1$ (e.g., see page 291 of [Meyer-Vernet (2007)]) and, thus, the SW can flow with a frozen-in magnetic-field. Furthermore, in the SW at 1 AU from the Sun the mean free path among streaming particles is ≈ 1 AU, as a consequence, particle collisions can hardly occur (e.g., see page 53 of [Meyer-Vernet (2007)]).

The first indication of a solar magnetic-field was obtained in 1905 by Hale. He could demonstrate the presence of strong magnetic-fields in *sunspots* by means of the *Zeeman effect*. These sunspots are supposed to be created deeper in the Sun[†] and, afterwards, moved to the surface by magnetic-buoyancy effects. A large spot is about 2×10^4 km across and may have a central umbra-region with a temperature of $\approx 4.1 \times 10^3$ K (e.g., see Section 1.3 in [Priest (1985)]). Daily observations were started at the Zurich Observatory in 1749. Their number varies and increases with the solar activity following a cycle of ≈ 11 years duration termed the *sunspot cycle*: the cyclic variation of the number of sunspots [also called the (*Schwabe*) *solar cycle* or *Schwabe–Wolf cycle*] was first observed by Heinrich Schwabe between 1826 and 1843 and led Rudolf Wolf to make systematic observations starting in 1848 (Fig. 4.5)). The *relative sunspot number*[‡] (R_s) remains an important index for determining the solar activity level; it is computed using the formula (collected as a daily index of sunspot activity):

$$R_s = k_{ob}(s + 10 \times g), \tag{4.16}$$

where s is the number of individual spots, g is the number of sunspot groups and, finally, k_{ob} is a factor that varies with location and instrumentation (also known as the *observatory factor*). Furthermore, sunspots are used to trace the Sun surface-rotation since Galileo in 1611. Nowadays, it is determined that the *Sun rotation-period* is a function of the *solar latitude* and is called (*sidereal*) *differential rota-*

---

*This is known as *Alfvén's theorem*.

[†]The reader can find a general description of the Sun characteristics, for instance, in [Altrock et al. (1985)] and in [Aschwanden (2006)].

[‡]R_s is also known as the *International sunspot number* or *Zürich number* or *Wolf number*. Note that there are actually at least two "official" sunspot numbers reported. The International Sunspot Number is compiled by the Sunspot Index Data Center in Belgium (*http://sidc.oma.be/index.php*). The NOAA sunspot number is compiled by the US National Oceanic and Atmospheric Administration. The sunspot numbers and smoothed monthly mean sunspot numbers can be found at the web site: *ftp://ftp.ngdc.noaa.gov/ftp.html* or *http://www.ngdc.noaa.gov/stp/SOLAR/ftpsunspotnumber.html#international* (the reader can also see: *http://solarscience.msfc.nasa.gov/SunspotCycle.shtml*).

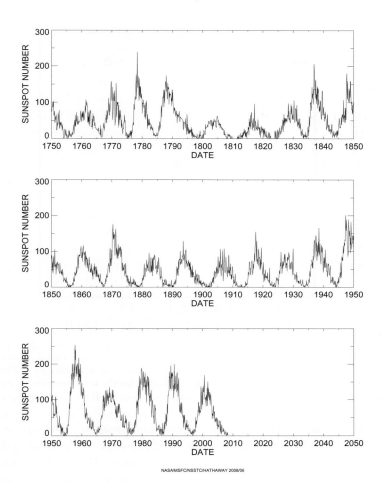

NASA/MSFC/NSSTC/HATHAWAY 2008/06

Fig. 4.5 Monthly averages (updated monthly) of the sunspot numbers show that the number of sunspots visible on the sun waxes and wanes with an approximate 11-year cycle (Courtesy NASA; see the web site *http://solarscience.msfc.nasa.gov/SunspotCycle.shtml*).

tion. The *sidereal rotation* can be approximated by

$$\omega = 14.522 - 2.84 \times \sin^2 b \ [\mathrm{deg/day}], \tag{4.17}$$

where b is the heliographic latitude (see page 77 of [Aschwanden (2006)] and also [Brajša et al. (2001)]). An *effective rotation period* of ≈ 27 days is usually satisfactory to indicate most of the Sun rotation-recurrences. In addition, the *Sun's rotation axis* is tilted by $\approx 7°$ from the axis of the Earth's orbit.

H.D. and H.W. Babcock (1954) observed the Sun magnetic-field over a two years period (1952–1954) using an instrument (called the *solar magnetograph*) utilizing the Zeeman effect and established that the *Sun has a general magnetic dipole*

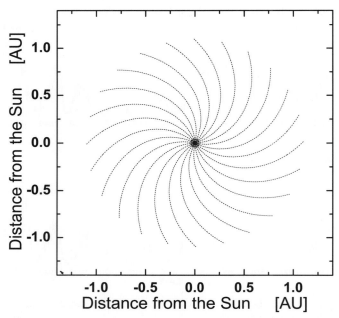

Fig. 4.6 Configuration of the interplanetary magnetic-field (following the the so-called *Archimedean spiral shape*) for the *Parker model* as a function of the radial distance (in AU) from the Sun in the solar equatorial plane for i) a steady solar wind with velocity of 470 km/s from $r_b \approx 10\, R_\odot$ and ii) an angular velocity of 2.9×10^{-6} rad/s for the solar rotation at equator.

field of $\approx 10^{-4}$ T distributed over the visible photosphere*. The magnetic-dipole tilt with respect to the rotation axis varies and depends on the solar cycle: apparently the *magnetic polarity* is *reversed* (e.g., see [Jones, Balogh and Smith (2003)] and references therein) about every 11 years with the evolution of the solar cycle, i.e., a similar magnetic configuration is found about every 22 years (the so-called *Hale cycle* or *solar magnetic cycle*). Parker (1958) (see also [Parker (1957, 1960, 1961a, 1963, 2007)]) suggested that the solar magnetic-field is frozen-in to the flow of the SW non-relativistic plasma, which carries the field with it into *interplanetary space* and generates the so-called *interplanetary magnetic-field* (IMF) or *heliospheric magnetic-field* (HMF). He assumed i) a constant solar rotation with angular velocity ω_c, ii) a simple spherically symmetric emission of the SW (see page 301) and iii) a constant (or approaching an almost constant) wind speed, v_w, at larger radial distances, e.g., for $r > r_b \approx 10\, R_\odot$ §, since beyond r_b the wind speed varies slowly with the distance (see Fig. 4.4). This latter assumption is justified because both the solar gravitation and the outward acceleration by high coronal temperature may be neglected or affect marginally the wind speed for $r \gtrsim r_b$. Thus, in a

---

*The photosphere is the apparent solar surface with a diameter usually considered to be the diameter of the Sun. The photosphere thickness is $\approx 0.014\%$ of $R_\odot$.

§$R_\odot \approx 4.65 \times 10^{-3}$ AU is the solar radius, see Appendix A.2.

spherical system of coordinates (r, θ, ϕ) rotating with the Sun, the outward velocity of a fluid element carrying the magnetic field is approximately[¶] given by [Parker (1958)]

$$\begin{cases} v_r = v_{\mathrm{w}}, \\ v_\theta = 0, \\ v_\phi = -(r - r_b)\,\omega_{\mathrm{c}} \sin\theta. \end{cases} \tag{4.18}$$

Since from Eq. (4.18) and for $r \gg r_b$ we can write

$$\frac{v_\phi}{v_r} = r\frac{d\phi}{dr} = \frac{-r\omega_{\mathrm{c}} \sin\theta}{v_{\mathrm{w}}}, \tag{4.19}$$

the streamline with azimuth ϕ_0 at r_b is obtained after integrating the previous expression, i.e.,

$$\int_{r_b}^{r} dr' = -\int_{\phi_0}^{\phi} \frac{v_{\mathrm{w}}}{\omega_{\mathrm{c}} \sin\theta}\, d\phi'$$
$$\Rightarrow r = r_b - \frac{v_{\mathrm{w}}}{\omega_{\mathrm{c}} \sin\theta}\,(\phi - \phi_0)\,. \tag{4.20}$$

The "spiral angle" (ψ_{sp}) is the angle that the heliospheric magnetic-field line makes to the radial direction and, from Eq. (4.19), is given by

$$\psi_{\mathrm{sp}} = \tan^{-1}\left|\frac{v_\phi}{v_r}\right|\,; \tag{4.21}$$

for instance, in the ecliptic plane it becomes

$$\psi_{\mathrm{sp,e}} \simeq \tan^{-1}\left(\frac{r\omega_{\mathrm{c}}}{v_{\mathrm{w}}}\right).$$

Furthermore, since the plasma velocity is parallel to the magnetic field, the ratios of the corresponding ϕ and r components are same [e.g., see Eqs. (4.18), (4.22)]:

$$\frac{B_\phi(r, \theta, \phi)}{B_r(r, \theta, \phi)} = \frac{v_\phi}{v_r} = \frac{-(r - r_b)\,\omega_{\mathrm{c}} \sin\theta}{v_{\mathrm{w}}}.$$

Thus, for large r the curve describing a magnetic-field line a) can be found using Eq. (4.20) and b) has the shape of an *Archimedean spiral* also termed the *Parker spiral*. In this model, the IMF field-line wraps around a cone, whose surface has an angle θ with respect to the solar rotation axis; i.e., equivalently *the expected spiral pattern consists of field lines on cones of constant heliographic latitude*. In Fig. 4.6, it is shown the Parker spiral for interplanetary magnetic-field lines in the solar equatorial plane $(\theta = 90°)$ as a function of the radial distance (in AU) from the Sun for i) a steady solar wind with velocity of $470\,\mathrm{km/s}$ from $r_b \approx 10\,R_\odot$ and ii) angular velocity (ω_{c}) of $2.9 \times 10^{-6}\,\mathrm{rad/s}$, that is the one of the solar rotation at

---

[¶]A complete derivation of the curve describing a magnetic-field line, i.e., the Archimedean spiral, can be found, for instance, in Section 6.3.3 of [Cravens (1997)], Section 12.3.1 of [Gombosi (1998)] and Section 6.3 of [Kallenrode (2004)].

equator [see Eq. (4.17)]. Finally, since $\nabla \cdot \vec{B} = 0$, the magnetic field at the point (r, θ, ϕ) is [Parker (1958)]

$$\begin{cases} B_r(r,\theta,\phi) = B(r_b,\theta,\phi_0) \, (r_b/r)^2, \\ B_\theta(r,\theta,\phi) = 0, \\ B_\phi(r,\theta,\phi) = -B(r_b,\theta,\phi_0) \, (r_b/r)^2 \, (r-r_b) \, (\omega_c/v_w) \sin\theta. \end{cases} \tag{4.22}$$

$|\vec{B}(r,\theta,\phi)|$ is given by

$$|\vec{B}(r,\theta,\phi)| = B(r_b,\theta,\phi_0) \left(\frac{r_b}{r}\right)^2 \sqrt{1 + \left[(r-r_b)\left(\frac{\omega_c}{v_w}\right)\sin\theta\right]^2}, \tag{4.23}$$

consequently, the magnetic-field strength decreases as $1/r^2$ for $\theta \approx 0°$ or $\approx 180°$ or for radial distances not too large with respect to r_b; whereas for $r \gg r_b$ it varies as $(\sin\theta)/r$. $B(r_b,\theta,\phi_0)$ represents the field at $r = r_b$; under the assumption that only the solar dipole field threads the escaping gas, then

$$B(r_b,\theta,\phi_0) = B_0 \cos\theta, \tag{4.24}$$

where B_0 is the field strength at the poles. For more complicated field structures, $B(r_b,\theta,\phi_0)$ can be approximated in a different way, but in either case, the lines of force follow Eq. (4.22). For instance, if the magnitude of field strength at $r_b \approx 10\,R_\odot$ is $\approx 10^{-6}\,\mathrm{T}$ (e.g., see Table 5.2 at page 180 of [Cravens (1997)]), then from the computation of Eq. (4.23) at Earth orbit (i.e., at $\approx 215\,R_\odot$, see Appendix A.2) we get a field strength of a few nT. This latter value is in agreement with typical observed ones. In general, the Parker model agrees with suitable averages of the IMF over a wide range of heliospheric distances and latitudes. However, for instance, it has to be noted that the field instantaneous-orientation deviates from that predicted.

Experimental observations allowed one to determine that i) while during solar-minimum conditions the solar magnetic-field is approximately dipole-like with a magnetic dipole closely aligned to the solar rotation axis (e.g., see [Smith, Tsurutani and Rosenberg (1978); Smith (1979); Jokipii and Thomas (1981)]), ii) during the declining phase of solar cycle the dipole is more tilted (e.g., see [Hundhausen (1977); Pizzo (1978)]). In addition, as the solar activity approaches its maximum, the large-scale field seems to be not well described by a dipole field alone. Furthermore, in 1976 observations made using the *Pioneer 11 spacecraft* up to about $16°$ above the solar equatorial plane have shown that a) the *magnetic-field sector structure*[tt] (e.g., see Chapter V of [Hundhausen (1972)]) was dependent on latitude, b) this sector structure tends to vanish at higher latitudes, where c) the IMF has the sign of the solar magnetic-field of the appropriate pole. The HMF field-lines consist of those coming i) from the northern solar magnetic hemisphere and directed away from the Sun, and ii) from the southern hemisphere and directed towards the Sun. Field

---

[tt]At 1 AU, close to the ecliptic plane, the magnetic field typically maintains one direction (either towards or away from the Sun) for many days. The region of space in which an orientation is maintained is called *sector*. Usually, the number of sectors varies from about two to four and it is related to the solar activity (e.g., see Section 6.4.3 of [Cravens (1997)]).

lines** from the two hemispheres are separated by a *near-equatorial current sheet* ([Smith, Tsurutani and Rosenberg (1978)] and references therein), also termed the *heliospheric current sheet* (HCS), which is effectively the extension of the solar magnetic equator into the SW. The average position of the HCS is tilted relative to the solar equator and warped‖. Thus, as the Sun rotates, the Earth passes through the current sheet and, consequently, experiences periods of alternating magnetic-field polarities. In practice, the maximum solar latitude of the current sheet is almost equal to the *tilt angle* of the magnetic dipole axis relative to the rotation axis.

Furthermore, the solar corona is frequently dominated by long-lived structures which can be reflected by large-flow patterns of the SW. An example is provided by the so-called *corotating interaction regions* (CIRs) [Pizzo (1978)]. This dynamical process is related to inhomogeneities (in the coronal expansion), which couple with the solar rotation, resulting in significant rearrangement of the material in the interplanetary space: fast streaming material may catch up with slower streams. Thus, Parker-spiral interaction structures may occur at large heliocentric distances (e.g., see [Pizzo (1978)], Section 12.5 of [Gombosi (1998)], Section 6.3 of [Aschwanden (2006)]). This mechanism is responsible for introducing azimuthal gradients and, in addition, meridional gradients transverse to the ecliptic plane. These latter arise because latitude variations stem partly from intrinsic variations in the corona and partly from the latitudinal dependence of the solar rotation [e.g., see Eq. (4.17)].

High-speed SW is known to originate from regions close to centers of the so-called *coronal holes*† [Wang and Sheeley (1993)]. These latter, in turn, are known to exhibit the unusual property of rotating almost rigidly with an angular rotation speed similar to that of the solar equator [Timothy, Krieger and Vaiana (1975); Nash, Sheeley and Wang (1988)], in spite of the differential rotation observed in the photosphere. Large variations of the SW-structure* (Fig. 4.7) during the solar cycle [McComas, Elliott, Schwadron, Gosling, Skoug and Goldstein (2003)] and recurrent energetic particle events (see references in [Fisk (1996)]) were observed for solar latitudes of ±80° by Ulysses spacecraft, which allowed one to observe SW properties a) through the declining phase and solar minimum (first orbit) and b) through the rise to solar maximum, solar maximum and immediately post-maximum (second orbit).

Fisk (1996) pointed out that there were experimental evidences on how some modulation and particle acceleration effects can be induced by mechanisms related

---

**The configurations of field-lines may exhibit the so-called *helmet streamers* [Pneuman and Kopp (1971)] (e.g., see also Section V.6 of [Hundhausen (1972)]), which develop over active regions. The legs of the helmet streamer connect regions of opposite magnetic polarity.

‖ The HCS resembles a *ballerina skirt*.

† Coronal holes are regions where the corona is dark and exhibit low X-ray intensity. Coronal holes are associated with *"open"* *magnetic-field* lines [Pneuman and Kopp (1971); Wang and Sheeley (1993)] and are often found at Sun's poles (in this case, they are also termed *polar holes*).

*Additional results from Ulysses mission can be found in [McComas et al. (2000); Schwadron, McComas, Elliott, Gloeckler, Geiss and von Steige (2005)].

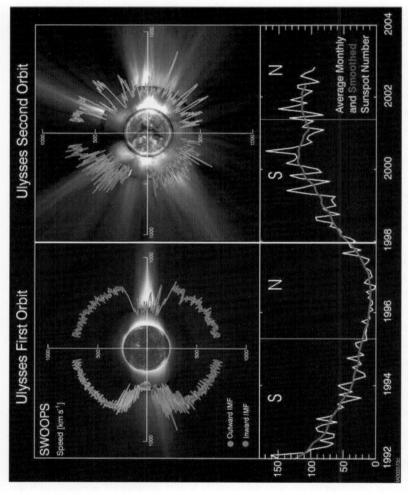

Fig. 4.7 Polar plots of SW speed (top panel) as a function of latitude for the first two orbits of the *Ulysses* spacecraft; sunspot number (bottom panel) shows that the first orbit occurred through the solar cycle declining phase and minimum while the second orbit spanned solar maximum (reprinted with permission from McComas, D.J., Elliott, H.A., Schwadron, N.A., Gosling, J.T., Skoug, R.M. and Goldstein, B.E. (2003), The three-dimensional solar wind around solar maximum, *Geophys. Res. Lett.* **30 (no. 10)**, 1517, doi: 10.1029/2003GL017136, ©2003 American Geophysical Union). Both are plotted over solar images characteristic of solar minimum and maximum. The SW speed is color coded by IMF orientation: red indicates outward pointing IMF, while blue indicates inward pointing.

to CIRs at solar latitudes larger than $\pm 30°$. It has to be noted that, within this range of latitudes, CIRs patterns were observed in the SW by Ulysses. In the *Fisk model*, the open coronal magnetic-field follows the treatment of Wang and Sheeley (1993)[§] and results from the superposition of an *axisymmetric field*, which is related to photospheric material undergoing differential rotation, and a *non-axisymmetric field*, which i) originates from decaying active regions near the solar equator, thus, ii) rotates at the equatorial speed and iii) determines the mechanism by which the boundary of the coronal holes rotates almost rigidly at the equatorial speed. The magnetic field in the inner corona of the polar hole can be represented by a tilted dipole, which has a magnetic axis offset from the solar rotation axis, and which rotates at the equatorial speed. The interplay among a) the differential rotation of photospheric material[‡], b) the SW expansion through the more rigidly rotating polar holes (which are offset from the solar rotation axis) and c) the tilted magnetic dipole results in a SW expansion and, thus, an IMF configuration which can vary from those predicted by Parker [Smith et al. (1995); Fisk (1996); Zurbuchen, Schwadron and Fisk (1997); Fisk (1999); Burger and Hattingh (2001); Fisk (2001); Reinard and Fisk (2004); Schwadron, McComas, Elliott, Gloeckler, Geiss and von Steige (2005); Zurbuchen (2007)]. In addition, field lines may be moved in latitude by the CIR mechanism. As a consequence, field lines are allowed to execute large latitudinal excursions which, for instance, naturally explain 26-day recurrences in cosmic ray events observed at high latitudes. Experimental observations have given indication for the validity of this revised heliospheric magnetic-field ([Zurbuchen, Schwadron and Fisk (1997)] and references therein).

Smith (2004) (see also references therein) has reviewed the characteristics of the radial, azimuthal and North–South components of the magnetic field in the outer heliosphere. He pointed out that recent Ulysses measurements, at both solar minimum and maximum, indicate that i) the radial component is almost independent of solar latitude, ii) the azimuthal component deviates from the Parker values at high latitudes (e.g., see [Banaszkiewicz, Axford and McKenzie (1998)]), iii) a turning of the spiral angle toward the radial direction by tens of degrees is often observed inside the so called *corotating rarefaction regions* (CCR) [Murphy, Smith and Schwadron (2002)] and iv) the North–South component can depart from zero for many days as a result of the tilting of the interface between fast and slow SW streams.

4.1.2.2 *Extension of the Heliosphere and the Earth Magnetosphere*

Parker (1961b) was the first to investigate the extension of the heliosphere. Although since then more work has been done on this topic, details of the SW interaction with the interstellar medium are mostly speculative because we lack ex-

---

[§]The reader can find additional information in [Sheeley, Nash and Wang (1987); Nash, Sheeley and Wang (1988); Wang, Sheeley, Nash and Shampine (1988)] and references therein.

[‡]The heliospheric magnetic-field lines in high speed streams are assumed to be anchored in the differentially rotating photosphere.

tensively direct observations of this space region (e.g., see Chapter IX of [Parker (1963)], Section 1.3 of [Toptygin (1985)], [Suess (1990)], Section 6.61 of [Cravens (1997)], Section 12.7 of [Gombosi (1998)], Section 9 of [Gosling (2006)] and references therein). Parker (1961b,1963) predicted the existence of i) the *heliospheric termination shock* which results because the SW, that is flowing supersonically away from the Sun, must make a transition to subsonic** and ii) the *heliopause*, which is the boundary surface separating the interstellar and SW plasmas and it is located outside the termination shock (Fig. 4.8). A rough estimate of the heliopause location may be obtained from balancing the SW pressure, P_w, and the interstellar medium pressure, P_{IS}, i.e.,

$$P_w = P_{IS}. \tag{4.25}$$

Since the thermal and magnetic pressure†† components of the SW can be neglected {e.g., see Equations (7.17–7.21) at page 349 of [Meyer-Vernet (2007)]}, P_w is practically given by the SW *dynamic pressure* $P_{d,w}$ (also termed the *ram pressure of the SW*), i.e.

$$P_w \sim P_{d,w} \tag{4.26}$$

with (e.g., see Section 7.2.1 of [Meyer-Vernet (2007)])

$$P_{d,w} = \rho v_w^2, \tag{4.27}$$

where ρ and v_w are the SW-plasma mass density and speed, respectively. Under steady-state spherically symmetric condition, the continuity equation [Eq. (4.5)] becomes

$$n_{pl} v_w r^2 = \text{ constant} \sim n_{1AU} \ v_{w,1AU} \times (1 \, \text{AU})^2,$$

**The behavior of gas depends on the speed of sound v_{sd} [Eq. (4.6)], which controls the propagation of disturbances. In the SW, particles are also linked by their embedded magnetic-field, thus, by the *"magnetic pressure"* and, in turn, by the *"Alfvén speed"*, v_A,

$$v_A = \frac{|\vec{B}|}{\sqrt{\mu_0 \rho}},$$

where ρ is the plasma mass-density and μ_0 is the permeability of the free space. The local *fast mode speed* is the characteristic speed with which small amplitude pressure signals propagate in a plasma and it is given by

$$v_f = \sqrt{v_{sd}^2 + v_A^2}.$$

At 1 AU, the mean values of v_{sd} and v_A in units of km/s are 63 and 50, respectively (Table 1 at page 101 of [Gosling (2006)]).

††The *magnetic pressure* is given by:

$$P_M = \frac{|\vec{B}|^2}{2\mu_0},$$

where μ_0 is the permeability of the free space (e.g., see Section 4.5.1 of [Cravens (1997)]). In cgs units it takes the form

$$P_M = \frac{|\vec{B}|^2}{8\pi} \ \text{dyn/cm}^2$$

with $\vec{B}$ in units of Gauss. It can be described as the tendency of neighboring field lines to repulse each other. Note that, in contrast to the gas-dynamic pressure, the magnetic pressure is not isotropic but is always perpendicular to the field (e.g., see Section 3.3.1 of [Kallenrode (2004)]).

where $n_{pl} \approx \rho/m_p$ is the number of protons[‡‡] per cm^3 in the SW plasma [e.g., see Eq. (4.4)]; n_{1AU} and $v_{w,1AU}$ are the SW-plasma density and speed at 1 AU, respectively. Thus, as discussed at page 308, under the assumption that the SW speed is already almost constant beyond the Earth, i.e.,

$$v_w \approx v_{w,1AU} \text{ for } r \gtrsim 1\,\text{AU},$$

we have

$$n_{pl} \sim n_{1AU} \left(\frac{1\,\text{AU}}{r} \right)^2. \tag{4.28}$$

Using Eqs. (4.4, 4.28), the ram pressure can be expressed in terms of the SW characteristics at 1 AU by re-writing Eq. (4.27) as

$$
\begin{aligned}
P_{d,w} &\approx n_{pl} m_p v_w^2 \\
&\approx n_{1AU}\, m_p v_{w,1AU}^2 \left(\frac{1\,\text{AU}}{r} \right)^2 \\
&= P_{1AU} \left(\frac{1\,\text{AU}}{r} \right)^2,
\end{aligned}
\tag{4.29}
$$

where {see Equation (7.18) at page 349 of [Meyer-Vernet (2007)]}

$$P_{1AU} = (2.1\text{–}2.6) \times 10^{-8}\,\text{dyn/cm}^2 \tag{4.30}$$

is the dynamic pressure[¶] at 1 AU. The interstellar medium pressure P_{IS} is given by the sum of i) the magnetic-field pressure, ii) the thermal pressure of the interstellar plasma, iii) the dynamic pressure of the interstellar plasma and iv) the cosmic rays pressure. Suess (1990) estimated that

$$P_{IS} \approx (1.3 \pm 0.2) \times 10^{-12}\,\text{dyn/cm}^2,$$

thus, from Eqs. (4.25, 4.26, 4.29), we obtain that the distance R_{hp} of the heliopause from the Sun is related by

$$P_{IS} = P_{1AU} \left(\frac{1\,\text{AU}}{R_{hp}} \right)^2$$

$$(1.3 \pm 0.2) \times 10^{-12} \approx (2.1\text{–}2.6) \times 10^{-8} \left(\frac{1\,\text{AU}}{R_{hp}} \right)^2$$

and, consequently, we get

$$R_{hp} \sim \sqrt{ \frac{(2.1\text{–}2.6) \times 10^{-8}}{(1.3 \pm 0.2) \times 10^{-12}} }\,\text{AU}$$

$$\approx (1.18\text{–}1.54) \times 10^2\,\text{AU}. \tag{4.31}$$

Since the distance of the heliopause from the Sun is large [Eq. (4.31)], the IMF embedded in the SW cannot be considered to instantaneously propagate through

---

[‡‡]For this approximate calculation the small portion of helium in the SW was neglected.

[¶]Equation (4.30) can be equivalently expressed as: $P_{1AU} = (2.1\text{–}2.6) \times 10^{-9}$ Pa.

the solar cavity. In fact using Eq. (4.31) (see also Appendix A.2), a rough estimate of the *traveling time* (τ_{hp}) *through the heliosphere up to the heliopause* is given by:

$$\tau_{hp} \approx \frac{R_{hp}}{v_{w,1AU}} = \frac{(1.18\text{–}1.54) \times 10^2 \times 1.496 \times 10^8 \,\text{km}}{470 \,\text{km/s}}$$

$$= (3.76\text{–}4.90) \times 10^7 \,\text{s} \sim (1.19\text{–}1.55) \,\text{y}, \qquad (4.32)$$

where $v_{w,1AU} \approx 470\,\text{km/s}$ is the average value of SW speed already almost constant beyond the Earth (e.g., see pages 300, 308). As a consequence, the structure of the IMF depends on the past solar-activity occurred during more than a year. Furthermore, an additional complexity is added to the IMF structure by the delayed propagation of phenomena affecting the SW, for instance those observed at large solar latitudes and discussed at pages 311-313.

Voyager 1 spacecraft has already verified the existence of the termination shock at a heliocentric distance of ≈ 94 AU roughly in the direction of the relative motion of the heliosphere relative[*] to the interstellar medium (e.g., see [Cummings, Stone, McDonald, Heikkila, Lal and Webber (2005)]) and entered the *heliosheath*, which is located between the termination shock and the heliopause (Fig. 4.8).

Outside the heliopause the interstellar flow is deflected around the heliosphere. Furthermore, if the external flow speed is larger than the *fast mode speed* (see footnote at page 314), it is expected the formation of a *bow shock* to deflect the interstellar flow around the heliosphere. The shape of the heliosphere is asymmetric (Fig. 4.8), because of its motion relative[*] to the interstellar medium, and it is largely elongated in the opposite direction, thus, generating the so-called *heliotail*.

Furthermore inside the heliosphere, the *Earth's magnetosphere* is the region of space to which the Earth's magnetic-field[‖] is confined by the SW plasma and, thus, shaped (Fig. 4.9) by its interaction with the SW itself (e.g., see [Knecht and Shuman (1985); Kivelson and Bagenal (2006); Luhmann and Solomon (2006); Schulz (2007)] and references therein). This *cavity* was named *magnetosphere* by T. Gold in 1959 to provide a description of the space region above the *ionosphere*. The first suggestion of such a cavity (sometimes referred to as the *Chapman–Ferrero cavity*) was made by S. Chapman and V. Ferraro in 1930.

The space environment of the Earth is determined by its *almost dipolar internal magnetic-field* that forms an obstacle to the SW with its embedded magnetic-field. Thus, to first approximation, the magnetic field close to the Earth surface has a dipole shape, but moving outwards other contributions become important. Among these we have to mention, for instance, those resulting from the several currents (e.g., the ring current, Birkeland's currents and tail currents) due to charged particles trapped inside the magnetosphere (Sect. 4.1.2.5). Moreover the latitudinal

---

[*]The Sun and heliosphere move at a speed of $\approx 23\,\text{km/s}$ relative to the interstellar medium.

[‖]Note that a number of coordinate systems are employed in the description of geomagnetic phenomena; among the commonly used are geographic, geomagnetic, geocentric solar-ecliptic, geocentric solar-magnetospheric and solar-magnetic coordinate systems. The reader can find the definitions, for instance, in Section 4.1.2 of [Knecht and Shuman (1985)].

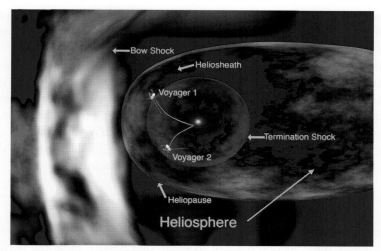

Fig. 4.8   Schematic representation of the heliosphere with Voyager 1 spacecraft crossing into the heliosheath, the region where interstellar gas and solar wind start to mix. (Courtesy NASA/JPL-Caltech).

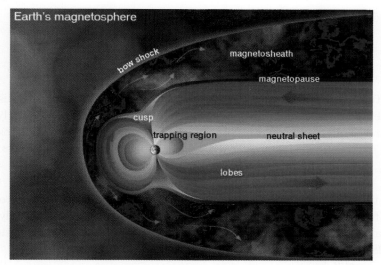

Fig. 4.9   Schematic representation of the magnetosphere to which the Earth's magnetic-field is confined and, thus, shaped by the SW plasma (Courtesy of NASA's Marshall Space Flight Center and Science@NASA).

dependence is not geographically symmetric, because the Earth magnetic dipole is tilted with respect to the Earth rotation axis and shifted from the Earth center. The magnetic field of the Earth magnetosphere can be described (e.g., see Appendix A of [Bobik et al. (2006)] and Section 7.2 of [Schulz (2007)]) using i) the International Geomagnetic Reference Field (IGRF) [Barton (1997)] for representing the main contribution due to the inner Earth and ii) the external magnetic field *model implemented by Tsyganenko and Stern* (e.g., see [Tsyganenko (1995); Tsyganenko and Stern (1996)]) for representing the other contributions.

Similarly to the formation of the heliopause discussed above, the *magnetopause* is the location where the outward magnetic pressure (see footnote at page 314) of the Earth's magnetic-field is counterbalanced by the SW pressure, which is practically given by the SW *dynamic pressure* at 1 AU (P_{1AU}) [see Eq. (4.30)], since the thermal and magnetic combined pressure is less than 1% of the SW ram-pressure [e.g., see Equations (7.17–7.21) at page 349 of [Meyer-Vernet (2007)]]. The *subsolar distance*[‡] (R_{EM}) of the magnetopause can be estimated in SI units from

$$P_{1AU} \approx \frac{\mathbb{F} B_e^2}{\mu_0} \left(\frac{Re}{R_{EM}} \right)^6, \qquad (4.33)$$

where $\mathbb{F} \approx 2$ is a magnetic-field compression factor (e.g., see Section 14.2.2 of [Gombosi (1998)] and, also, Section 8.2.2 of [Kallenrode (2004)]), μ_0 is the permeability of the free space, $B_e \approx 3.11 \times 10^{-5}\,\mathrm{T}$ is the equatorial magnetic-field[§] at the surface of the Earth and Re is the radius of the Earth (Appendix A.2). From Eq. (4.33), we get:

$$R_{EM} \approx \sqrt[6]{\frac{\mathbb{F} B_e^2}{\mu_0\, P_{1AU}}}\, Re \qquad (4.34)$$

$$\approx (9.2\text{–}9.5)\, Re.$$

Thus, on the dayside of the planet, for normal SW conditions the subsolar distance of the magnetopause is $\approx 10\,Re$. However, when the SW pressure is particularly strong, this distance can be reduced to $\approx (6\text{–}7)\,Re$. On the nightside, the magnetosphere is stretched into a long *magnetotail* to a distance, which might extend up to $\approx 1000\,Re$ (Fig. 4.9). The magnetotail has an approximate cylindrical shape of $\approx 40\,Re$ in diameter.

Since the SW is a supersonic flow, a *standing shock wave*, the *bow shock*[¶], develops in front of the magnetopause. Its subsolar distance (R_{BS}) is typically located $(2\text{–}3)\,Re$ ahead of this latter and is approximately given by (e.g., see Equation 14.18 at Page 284 of [Gombosi (1998)])

$$R_{BS} \approx 1.275 \times R_{EM}.$$

At the bow shock the SW plasma is slowed down to subsonic speed and a substantial fraction of its energy is converted into thermal energy. The thermalised SW flow passes through the bow shock and enters the region called *magnetosheath*, but cannot easily penetrate through the magnetopause. However, Dungey (1961) suggested

---

[‡]The subsolar distance is the radial distance towards the Sun to the nose of magnetopause. The *subsolar point* on Earth is where the Sun is perceived to be directly overhead.

[§]The dipolar magnetic-field flux-density falls off with the distance r as $1/r^3$.

[¶]A further discussion about the bow-shock formation and location can be found in [Schwartz (1985)].

that the magnetopause i) is not a complete barrier to the SW and ii) the magneto-sphere structure itself is also controlled by its *reconnection* with the IMF. Further-more, the magnetic field exhibits the *polar cusps*, which separate closed field lines on the dayside of the magnetosphere from field lines swept to its nightside and to which the field lines of the magnetopause converge (Fig. 4.9). The cusp regions are filled with plasma particles from the magnetosheath and, thus, at the cusps these particles can penetrate deep into the Earth's atmosphere. Close to the Earth, in the *inner magnetosphere* the dipolar shape of the magnetic field is almost preserved.

4.1.2.3 *Propagation of Galactic Cosmic Rays through Interplanetary Space*

The Galactic cosmic radiation [e.g., the *Galactic Cosmic Rays* (GCRs), see Sect. 4.1.2.4] is incident on the solar cavity isotropically and constantly and is af-fected by the outwards flowing SW with its embedded magnetic-field and magnetic-field irregularities. For instance, at energies[*] below a few GeV/nucleon, the inten-sity shows a strong dependence on solar activity (e.g., see Sect. 4.1.2.1) with a maximum at the solar minimum and decreases with increasing the solar sunspot number [Smart and Shea (1985)]. This effect is called *modulation of GCRs*. Fur-thermore, differential energy spectra of all high-energy isotopes exhibit a maximum [Fig. 4.12(a) and Fig. 9.37], which shifts towards higher energies and, at the same time, decreases in intensity with increasing solar activity: this effect was observed at 1 AU for protons and other isotopes up to iron [Meyer, Parker and Simpson (1974); Webber and Lezniak (1974); Simpson (1983); Smart and Shea (1985, 1989)].

Parker (1963,1964,1965) proposed a theory for the general properties of the prop-agation of GCRs through the interplanetary space, considering transit time, energy-loss and the outward convection of solar particles. Furthermore, he reached the con-clusion that the SW, and in general the IMF configuration which results, lead to important *cosmic rays modulation-effects* dependant on the solar activity. He based his theory on IMF observations made by the *Explorer XVIII spacecraft*: the IMF recorded-variations were magnetic-field spatial irregularities transported rigidly in the SW (Figure 1 of [Parker (1965)]). These irregularities appear with dimensions of 10^5–10^7 km, which, for instance, are comparable with the radius of gyration of protons with kinetic energies larger than $\approx 100\,\text{MeV}$ and lower than $\approx 10\,\text{GeV}$ in a typical IMF of a few nT. He succeeded in showing that a charged particle, moving in a large-scale field containing small-scale irregularities, is most effectively scat-tered by irregularities which have a scale comparable to the radius of gyration of the particle [Parker (1964)]. When this occurs, the effect of such irregularities is to cause (in the reference frame of the magnetic irregularities) the GCRs to a random walk, which can be considered as a *Markhoff process*[§].

---

[*]The reader can find the definition of kinetic energies per nucleon in Sect. 1.4.1.

[§]A Markhoff process is when the value of a physical parameter at the next measuring time depends only on its value at the present time and not on its value at any previous measuring time, i.e., the system has one time step (or less) "memory".

For describing such a process in the framework of statistical physics [Chandrasekhar (1943); Parker (1965)], one needs to introduce the phase-space[†] distribution function (also termed *distribution function*) $F(\vec{x}, \vec{v}, t)$ for the particle distribution at the time t over position $\vec{x}$ and velocity $\vec{v}$, so that the $d^6 N_p$ number of particles in a $d^3 x$ volume element around the configuration space location $\vec{x}$ with velocity vectors between $\vec{v}$ and $\vec{v} + d^3 v$ is given by:

$$d^6 N_p = F(\vec{x}, \vec{v}, t)\, d^3 x\, d^3 v,$$

with the corresponding normalized phase-space distribution $f(\vec{r}, \vec{v}, t)$ defined as

$$f(\vec{x}, \vec{v}, t) = \frac{F(\vec{x}, \vec{v}, t)}{\iiint F(\vec{x}, \vec{v}, t)\, d^3 v} = \frac{F(\vec{x}, \vec{v}, t)}{n(\vec{x}, t)},$$

where the (*particle*) *number density* (or *particle concentration*) is

$$n(\vec{x}, t) = \iiint F(\vec{x}, \vec{v}, t)\, d^3 v.$$

The *average* (or *bulk*) *velocity* $\vec{v}_{bl}(\vec{x}, t)$ is obtained from

$$\vec{v}_{bl}(\vec{x}, t) = \frac{\iiint \vec{v}\, F(\vec{x}, \vec{v}, t)\, d^3 v}{n(\vec{x}, t)}.$$

The evolution of the (normalized) phase-space distribution function is described by the *Boltzmann equation* (e.g., see Section 2.1.2 of [Cravens (1997)], Section 2.2 of [Gombosi (1998)], Section 5.2.1 of [Kallenrode (2004)]), in which the total time derivative of such a distribution is equal to the rate of change of the phase-space distribution or *collision term*. For a Markhoff process, Chandrasekhar (1943) demonstrated that the collision term can be treated using the so-called *Fokker–Plank approximation* (or *diffusion approximation*) and, consequently, obtaining the so-called *Fokker–Plank equation*.

As already mentioned, within the Parker model (see Sect. 4.1.2.1) magnetic irregularities are carried outwards by the SW with speed $\vec{v}_w$. These irregularities are treated as the elastic scattering centers of GCRs in the reference frame moving with the SW. In the *Parker–Gleeson–Axford–Jokipii treatment* the collision term of the Markhoff process, resulting in the modulation phenomenon of CRs, is firstly evaluated in the SW frame moving with scatterers, then is transformed to a fixed heliocentric frame [Parker (1965); Gleeson and Axford (1967, 1968a,b); Jokipii and Parker (1970); Jokipii (1971)]). In the heliocentric system, in which the spatial coordinates x_i allow one to determine the position $\vec{r}$ at a heliocentric distance r, the derived Fokker–Plank equation[**] is given by

$$\frac{\partial \mathbb{U}}{\partial t} - \frac{\partial}{\partial x_i} \left(\mathcal{K}_{i,j} \frac{\partial \mathbb{U}}{\partial x_J} \right) + \frac{\partial}{\partial T'} \left(\mathbb{U} \frac{dT'}{dt} \right) + \frac{\partial}{\partial x_i} (\mathbb{U}\, v_{w,i}) = 0, \qquad (4.35)$$

[†]In classical non-relativistic statistical mechanics, a particle is represented by a single point in a six-dimensional phase space, i.e., the three spatial (for location $\vec{x}$ in the so-called *configuration space*) and three velocity coordinates (for the location $\vec{v}$ in the so-called *velocity space*). $\vec{x}$ and $\vec{v}$ are the *Eulerian coordinates* of the phase space.

[**]In this section, when an index variable appears twice in a single term, it implies that we are summing over all of its possible values; for instance, the divergence of the SW velocity is expressed by $\partial v_{w,i}/\partial x_i$ (e.g., see [Parker (1965)]). This notation is sometime referred to as the *Einstein summation convention*.

where T' is the kinetic energy of the cosmic-ray particle in the moving frame, $\mathbb{U}(x_i, T', t)$ is the *number density[‡] of CR particles having kinetic energy T' at the position $\vec{r}$ and time t* and, finally, $\mathcal{K}_{i,j}$ is the so-called *diffusion tensor* [Parker (1965); Jokipii and Parker (1970); Jokipii (1971)]. The third term of Eq. (4.35) is derived under the assumption that a possible Fermi acceleration[¶] of CR particles can be neglected [Parker (1965); Jokipii and Parker (1970)]). For a cosmic-ray particle, the largest difference[‡‡] $|\Delta T|$ of its kinetic-energy between the moving and heliocentric frames can be computed from [see Eqs. (1.20, 1.23)]

$$T + m_r c^2 = \gamma_w \left[(T' + m_r c^2) \pm \beta_w \sqrt{(T' + m_r c^2)^2 - m_r^2 c^4} \right]$$

$$= \gamma_w \left[(T' + m_r c^2) \pm \beta_w \sqrt{T'(T' + 2m_r c^2)} \right], \tag{4.36}$$

where T is the kinetic energy in the heliocentric frame, m_r is the rest mass of the cosmic-ray particle and

$$\beta_w = \frac{v_w}{c}.$$

For a typical SW speed of $\approx 470 \, \text{km/s}$ (see page 300), we have $\beta_w \approx 1.57 \times 10^{-3}$ and $\gamma_w \approx 1$ [Eq. (1.4)]. Thus, from Eq. (4.36), we obtain

$$|\Delta T| \approx \beta_w \sqrt{T'(T' + 2m_r c^2)}$$

$$\approx 1.57 \times 10^{-3} \sqrt{T'(T' + 2m_r c^2)}, \tag{4.37}$$

$$\Rightarrow \frac{|\Delta T|}{T'} \approx 1.57 \times 10^{-3} \sqrt{1 + \frac{2m_r c^2}{T'}}. \tag{4.38}$$

From $\approx 100 \, \text{MeV}$ up to $\approx 10 \, \text{GeV}$ per nucleon[*], by means of Eq. (4.38) $|\Delta T|/T'$ is estimated $\approx (0.70\text{--}0.17) \, \%$. Thus, $|\Delta T|$ can be neglected and $T' \approx T$. Furthermore, Jokipii and Parker (1967) (see also [Jokipii and Parker (1970); Jokipii (1971)]) to a first approximation demonstrated that for the number density $[\mathbb{U}(x_i, T, t)]$ of the cosmic-ray particles with kinetic energy T in the heliocentric frame and T' in the moving frame we have

$$\mathbb{U}(x_i, T, t) \approx \mathbb{U}(x_i, T', t).$$

As a consequence, Eq. (4.35) can be rewritten as

$$\frac{\partial \mathbb{U}}{\partial t} - \frac{\partial}{\partial x_i} \left(\mathcal{K}_{i,j} \frac{\partial \mathbb{U}}{\partial x_J} \right) + \frac{\partial}{\partial T} \left(\mathbb{U} \frac{dT}{dt} \right) + \frac{\partial}{\partial x_i} (\mathbb{U} \, v_{w,i}) = 0. \tag{4.39}$$

[‡]The distribution (i.e., the number density) of particles at the position $\vec{r}$ and time t is given by [Parker (1965)]:

$$n(\vec{r}, t) \equiv n(x_i, t) = \int_0^\infty \mathbb{U}(x_i, T', t) \, dT'.$$

[¶]Parker (1965) demonstrated, in Appendix 6 of his article, that Fermi acceleration mechanism, i.e., the stochastic energy gain which may occur in collisions with plasma clouds, provides a negligible energy change.

[‡‡]This occurs when the cosmic-ray particle moves along the direction of the solar wind.

[*]This is the kinetic-energy range of CRs in which the solar modulation i) affects their propagation and ii) can be expressed by means of Eq. (4.35).

For a radial SW with speed v_{w} and an isotropic *diffusion coefficient*[||] $\mathcal{K}$, the appropriate distribution function, $U(r, T, t)$, at heliocentric distance r is a *spherically symmetric modulated number density of CR particles with kinetic energy T*. This function can be obtained as the solution of the Fokker-Plank equation (4.39), by rewriting this latter in spherical coordinates in which the helio-latitude and helio-longitude dependencies vanish, i.e.,

$$\frac{\partial U}{\partial t} = +\frac{1}{r^2} \left[\frac{\partial}{\partial r} \left(r^2 \mathcal{K} \frac{\partial U}{\partial r} \right) \right]$$
$$+ \frac{1}{3r^2} \left[\frac{\partial}{\partial r} \left(r^2 v_{\mathrm{w}} \right) \right] \left[\frac{\partial}{\partial T} \left(\alpha_{\mathrm{rel}} T U \right) \right]$$
$$- \frac{1}{r^2} \left[\frac{\partial}{\partial r} \left(r^2 v_{\mathrm{w}} U \right) \right], \tag{4.40}$$

where α_{rel} is given by

$$\begin{aligned}
\alpha_{\mathrm{rel}} &= \frac{T + 2 m_{\mathrm{r}} c^2}{T + m_{\mathrm{r}} c^2} \\
&= \frac{(\gamma - 1) m_{\mathrm{r}} c^2 + 2 m_{\mathrm{r}} c^2}{(\gamma - 1) m_{\mathrm{r}} c^2 + m_{\mathrm{r}} c^2} \\
&= \frac{\gamma + 1}{\gamma};
\end{aligned} \tag{4.41}$$

m_{r} is the rest mass of the CR particle and γ is its relativistic Lorentz factor (e.g., see Sect. 1.3). Furthermore, we can express $\alpha_{\mathrm{rel}} T$ as:

$$\begin{aligned}
\alpha_{\mathrm{rel}} T &= \left(\frac{T + 2 m_{\mathrm{r}} c^2}{T + m_{\mathrm{r}} c^2} \right) T \\
&= \left(\frac{E_{\mathrm{t}} + m_{\mathrm{r}} c^2}{E_{\mathrm{t}}} \right) \left(E_{\mathrm{t}} - m_{\mathrm{r}} c^2 \right) \\
&= \frac{E_{\mathrm{t}}^2 - m_{\mathrm{r}}^2 c^4}{E_{\mathrm{t}}} \tag{4.42} \\
&= \frac{p_p^2}{E_{\mathrm{t}}} \tag{4.43} \\
&= \beta^2 \gamma m_{\mathrm{r}} c^2, \tag{4.44}
\end{aligned}$$

where E_{t} is the total particle-energy [Eq.(1.12)] and p_p is the particle momentum. Furthermore, $\mathcal{K}$ is a phenomenological *diffusion coefficient*, which is assumed to be isotropic, as already mentioned, and a function of the particle velocity, particle rigidity and heliocentric distance. The first term on the right-hand side of Eq. (4.40) describes the *diffusion* of GCRs by magnetic irregularities; the second[§] accounts for the so-called *adiabatic energy changes* associated with expansions and compressions of the cosmic radiation due to the i) geometrical (radial) divergence proportional

---

[||]This corresponds to the assumption $\mathcal{K}_\perp \approx \mathcal{K}_\parallel$ and to neglect the antisymmetric part in Eq. (4.58).

[§]The cosmic-ray particle is cooled adiabatically and, neglecting Fermi mechanism acceleration

to $2v_{\mathrm{w}}/r$ and ii) acceleration or deceleration (i.e., when $\partial v_{\mathrm{w}}/\partial r \neq 0$) of the SW; the third is related to the resulting *convection effect* [Parker (1965); Gleeson and Axford (1967); Jokipii and Parker (1970)]. Equation (4.40) can be rewritten with the term

$$\frac{\partial U}{\partial t} = 0,$$

if *steady-state modulation conditions* are appropriate, i.e., when the relaxation time of the distribution is short compared with the solar cycle duration (as it is often the case). The presence of any particle source, for instance that for anomalous CRs (Sect. 4.1.2.4), is accounted by adding a term $Q(r,T,t)$ to the right-hand side of Eq. (4.40) [Jokipii (1971); Potgieter (1998)] (see also references therein).

Furthermore, Gleeson and Axford (1967) demonstrated that Eq. (4.40) [originally derived by Parker (1965)] can be obtained by using a *Legendre expansion* of the collision term of the Boltzmann equation and eliminating the so-called *radial current density term* (also termed *streaming term*) $S(r,T,t)$ from the resulting set of Eqs. (4.45, 4.46):

$$\frac{\partial U}{\partial t} = -\frac{1}{r^2}\left[\frac{\partial}{\partial r}\left(r^2 S\right)\right] - \frac{v_{\mathrm{w}}}{3}\left[\frac{\partial^2}{\partial T \partial r}\left(\alpha_{\mathrm{rel}} T U\right)\right], \qquad (4.45)$$

$$S(r,T,t) = v_{\mathrm{w}} U - \frac{v_{\mathrm{w}}}{3}\left[\frac{\partial}{\partial T}\left(\alpha_{\mathrm{rel}} T U\right)\right] - \mathcal{K}\frac{\partial U}{\partial r} \qquad (4.46)$$

$$= C(r,T,t)\, v_{\mathrm{w}} U - \mathcal{K}\frac{\partial U}{\partial r}, \qquad (4.47)$$

where

$$C(r,T,t) = 1 - \frac{1}{3U}\left[\frac{\partial}{\partial T}\left(\alpha_{\mathrm{rel}} T U\right)\right] \qquad (4.48)$$

is the so-called *Compton–Getting factor*[‡]. The streaming term[¶] represents the *radial current density for CR particles with kinetic energy T*. Equation (4.45) can be

---

effects, it can be shown that its kinetic energy T declines as:

$$\frac{dT}{dt} = -\frac{T\alpha_{\mathrm{rel}}}{3}\frac{\partial v_{\mathrm{w},i}}{\partial x_i} = -\frac{T\alpha_{\mathrm{rel}}}{3}\left(\nabla \cdot \vec{v}_{\mathrm{w}}\right)$$

(e.g., see [Parker (1965); Jokipii and Parker (1967, 1970)]). Thus, for a radial SW with speed v_{w} we have

$$\frac{dT}{dt} = -\frac{T\alpha_{\mathrm{rel}}}{3r^2}\frac{\partial}{\partial r}\left(r^2 v_{\mathrm{w}}\right).$$

[‡]This *Compton–Getting effect* deals with the transformation of the differential mean density and current density between reference frames moving relative to each other with constant velocity. Gleeson and Axford (1968a) demonstrated that the Compton–Getting effect results in introducing the *Compton–Getting factor* [Eq. (4.48)] in the streaming term [Eq. (4.47)].

[¶]It is related to the number of particles with speeds between v and $v+dv$ which cross a unit area perpendicular to a unit vector direction in unit time [Gleeson and Axford (1968a)].

rewritten in terms of the Compton–Getting factor using Eqs. (4.47, 4.48), i.e.,

$$
\frac{\partial U}{\partial t} = -\frac{1}{r^2}\left\{\frac{\partial}{\partial r}\left[r^2\left(C\,v_{\mathrm{w}}U - \mathcal{K}\frac{\partial U}{\partial r}\right)\right]\right\} - \frac{v_{\mathrm{w}}}{3}\left[\frac{\partial^2}{\partial T\partial r}(\alpha_{\mathrm{rel}}TU)\right]
$$

$$
= \frac{1}{r^2}\left[\frac{\partial}{\partial r}\left(r^2\mathcal{K}\frac{\partial U}{\partial r}\right)\right] - \frac{1}{r^2}\left[\frac{\partial}{\partial r}(r^2C\,v_{\mathrm{w}}U)\right] - \frac{v_{\mathrm{w}}}{3}\left[\frac{\partial^2}{\partial T\partial r}(\alpha_{\mathrm{rel}}TU)\right]
$$

$$
= \frac{1}{r^2}\left[\frac{\partial}{\partial r}\left(r^2\mathcal{K}\frac{\partial U}{\partial r}\right)\right] - \frac{v_{\mathrm{w}}}{3}\left[\frac{\partial^2}{\partial T\partial r}(\alpha_{\mathrm{rel}}TU)\right]
$$

$$
- \frac{CU}{r^2}\left[\frac{\partial}{\partial r}(r^2\,v_{\mathrm{w}})\right] - v_{\mathrm{w}}\left[\frac{\partial}{\partial r}(CU)\right]
$$

$$
= \frac{1}{r^2}\left[\frac{\partial}{\partial r}\left(r^2\mathcal{K}\frac{\partial U}{\partial r}\right)\right] - \frac{v_{\mathrm{w}}}{3}\left[\frac{\partial^2}{\partial T\partial r}(\alpha_{\mathrm{rel}}TU)\right]
$$

$$
- \frac{CU}{r^2}\left[\frac{\partial}{\partial r}(r^2\,v_{\mathrm{w}})\right] - v_{\mathrm{w}}\left\{\frac{\partial U}{\partial r} - \frac{1}{3}\left[\frac{\partial^2}{\partial T\partial r}(\alpha_{\mathrm{rel}}TU)\right]\right\}
$$

$$
= \frac{1}{r^2}\left[\frac{\partial}{\partial r}\left(r^2\mathcal{K}\frac{\partial U}{\partial r}\right)\right] - v_{\mathrm{w}}\frac{\partial U}{\partial r} - \frac{CU}{r^2}\left[\frac{\partial}{\partial r}(r^2v_{\mathrm{w}})\right]. \tag{4.49}
$$

Furthermore, the modulation of CRs can also be described by means of the so-called omnidirectional distribution function* $f(\vec{r},\vec{p}_p,t)$ of CR particles with momentum $\vec{p}_p$, at the position $\vec{r}$ and time t (e.g., see [Fisk, Forman and Axford (1973); Fisk (1976)]). Forman (1970) derived that the *Compton–Getting factor* is given by

$$
C_p = -\frac{p_p}{3\,f}\frac{\partial f}{\partial p_p} = -\frac{1}{3f}\frac{\partial f}{\partial \ln p_p}, \tag{4.50}
$$

where f is the omnidirectional distribution function averaged over particle directions. Thus, Eq. (4.49) can be rewritten as

$$
\frac{\partial f}{\partial t} = \frac{1}{r^2}\left[\frac{\partial}{\partial r}\left(r^2\mathcal{K}\frac{\partial f}{\partial r}\right)\right] - v_{\mathrm{w}}\frac{\partial f}{\partial r} - \frac{C_p f}{r^2}\left[\frac{\partial}{\partial r}(r^2 v_{\mathrm{w}})\right]
$$

$$
= \frac{1}{r^2}\left[\frac{\partial}{\partial r}\left(r^2\mathcal{K}\frac{\partial f}{\partial r}\right)\right] - v_{\mathrm{w}}\frac{\partial f}{\partial r} + \frac{1}{3r^2}\left[\frac{\partial}{\partial r}(r^2 v_{\mathrm{w}})\right]\left(\frac{\partial f}{\partial \ln p_p}\right). \tag{4.51}
$$

---

*The omnidirectional distribution of particles with momentum $\vec{p}_p$ at the position $\vec{r}$ and time t is related to the differential intensity $\left[J\left(\vec{r},T,\frac{\vec{p}_p}{p_p},t\right)\right]$ of particles with kinetic energy T traveling in the direction $\vec{n} = \vec{p}_p/p_p$ by

$$
J\left(\vec{r},T,\frac{\vec{p}_p}{p_p},t\right) = p_p^2\,f(\vec{r},\vec{p}_p,t)
$$

(see Equation (A3) of [Forman (1970)]), where m_{r} is the rest mass of the particle;

$$
\int f(\vec{r},\vec{p}_p,t)\,dp_{p,x}dp_{p,y}dp_{p,z} = n(\vec{r},t) = \int U(\vec{r},T,t)\,dT
$$

with $n(\vec{r},t)$ the number density of particles; finally, for the kinetic energy (T) the following non-relativistic expression is assumed

$$
dT = \frac{p_p}{m_{\mathrm{r}}}dp_p.
$$

Furthermore, taking into account that

$$
dE_{\mathrm{t}} = dT
$$

where E_{t} is the total particle energy, the reader can also see the equivalence of the above definition of differential intensity with that one expressed in Eq. (4.53).

For a radial SW and an isotropic diffusion coefficient, this equation describes the SW modulation effect by means of the omnidirectional distribution function (f), while Eq. (4.40) by means of the number density (U) of CR particles with kinetic energy T.

When there are no sources or sinks at $r = 0$ and for steady modulation conditions, Gleeson and Axford (1968b) estimated that, to a first approximation, the streaming term, S, is negligible when[¶] $T \geq 400\,\text{MeV/nucleon}$ for protons and $T \geq 200\,\text{MeV/nucleon}$ for α-particles at 1 AU, provided that the radial length characteristic of the radial variation of the diffusion coefficient is $\leq 1\,\text{AU}$. Furthermore, this latter coefficient was assumed to be given by a separable function of r and R^p_{isot}:

$$\mathcal{K}(r,t) = \beta k_1(r,t) k_2(R^p, t),$$

where R^p is the particle rigidity [Eq. (4.71)], $\beta = v/c$, v is the particle velocity and c is the speed of light; $\mathcal{K}(r,t)$ was estimated to be $\approx 10^{22}\,\text{cm}^2/\text{s}$ for particles of a few GeV/nucleon. It has to be noted that above $1\,\text{GV}$ $k_2(R^p, t)$ reduces usually to the value of the particle rigidity and is expressed in units of GV (e.g., see [Gleeson and Axford (1968b); Perko (1987)]). Using Eq (4.42), for $S \approx 0$ we can rewrite Eq. (4.46) in terms of the total particle-energy [E_t, see Eq.(1.12)] and rest mass (m_r) as

$$\mathcal{K}\frac{\partial U}{\partial r} \approx v_w \left[U - \frac{1}{3}\frac{\partial}{\partial T}(\alpha_{\text{rel}} T U) \right]$$

$$\Rightarrow \mathcal{K}\frac{\partial U}{\partial r} = v_w \left\{ U - \frac{1}{3}\frac{\partial}{\partial E_t}\left[\left(\frac{E_t^2 - m_r^2 c^4}{E_t}\right) U \right] \right\}$$

$$= v_w \left\{ U - \frac{1}{3}\frac{\partial}{\partial E_t}\left[\frac{U}{E_t\left(E_t^2 - m_r^2 c^4\right)^{1/2}}\left(E_t^2 - m_r^2 c^4\right)^{3/2} \right] \right\}$$

$$= -\frac{v_w}{3}\left(E_t^2 - m_r^2 c^4\right)^{3/2}\frac{\partial}{\partial E_t}\left[\frac{U}{E_t\left(E_t^2 - m_r^2 c^4\right)^{1/2}} \right], \qquad (4.52)$$

with $U = U(r, E_t)$, in the above equation. Gleeson and Axford (1968b) introduced the *differential intensity*[*], also termed *omnidirectional intensity* (e.g., page 63 of [Jokipii (1971)]),

$$J(r, E_t, t) = \frac{v U(r, E_t, t)}{4\,\pi} = \frac{c\sqrt{E_t^2 - m_r^2 c^4}}{4\,\pi E_t}U(r, E_t, t) \qquad (4.53)$$

and the *modulation parameter* (also termed *modulation strength*)

$$\phi_s(r, t) = \int_r^{r_{\text{tm}}} \frac{v_w(r', t)}{3k_1(r', t)}dr', \qquad (4.54)$$

where it is indicated with r_{tm} the location beyond which the SW terminates. Assuming that v_w and k_1 are almost constant, Eq.(4.54) reduces to

$$\phi_s(r, t) \approx \frac{v_w(t)\,(r_{\text{tm}} - r)}{3k_1(t)} \qquad (4.55)$$

[¶]The reader can find the definition of kinetic energies per nucleon in Sect. 1.4.1.
[*]It expresses the differential CR particle flux per unit of energy and solid angle.

and is usually expressed in units of GV: for instance, a typical weak modulation condition at 1 AU occurs for $\phi_s(1\,\text{AU}) = (0.32\text{–}0.35)\,\text{GV}$. $\phi_s(r,t)$ (as well as k_2) is independent of the species of CR particles. Finally, they derived that the solution of Eq. (4.52) is given by the expression

$$J(r, E_t, t) = J(r_{\text{tm}}, E_t + \Phi_{\text{p}}) \left[\frac{E_t^2 - m_r^2 c^4}{(E_t + \Phi_{\text{p}})^2 - m_r^2 c^4} \right], \tag{4.56}$$

where $J(r_{\text{tm}}, E_t + \Phi_{\text{p}})$ is the undisturbed intensity beyond the SW termination. Φ_{p} is completely determined by $\phi_s(r,t)$, but i) depends on E_t, ii) may be different for each species, iii) accounts for mean energy-losses experienced by particles with kinetic energy T at a location with radial distance r when they come from outside the solar cavity to this location and, thus, iv) may be interpreted as similar to a "potential energy". This approximated solution for determining modulation effects is termed *force-field solution*, while Φ_{p} is called *force-field energy-loss* (e.g., see [Gleeson and Axford (1968b); Gleeson and Urch (1971)]. When modulation is small (i.e., $\Phi_{\text{p}} \ll m_r c^2, T$; see [Gleeson and Axford (1968b); Gleeson and Urch (1971)]), Gleeson and Axford (1968b) determined in addition that

$$\Phi_{\text{p}} = \frac{ZeR^p}{k_2(R^p, t)} \phi_s(r,t),$$

where Ze is the particle charge; thus, using Eqs. (4.44, 4.71) we obtain

$$\begin{aligned}
\Phi_{\text{p}} &= \frac{\beta\gamma m_r c^2}{k_2(R^p, t)} \int_r^{r_{\text{tm}}} \frac{v_{\text{w}}(r', t)}{3k_1(r', t)} dr' \\
&= \frac{\beta^2 \gamma m_r c^2}{3} \int_r^{r_{\text{tm}}} \frac{v_{\text{w}}(r', t)}{\beta k_2(R^p, t)k_1(r', t)} dr' \\
&= \frac{\alpha_{\text{rel}} T}{3} \int_r^{r_{\text{tm}}} \frac{v_{\text{w}}(r', t)}{\mathcal{K}(r', t)} dr'.
\end{aligned} \tag{4.57}$$

For $r = 0$ and assuming $\mathcal{K}$ and v_{w} almost constant, Eq. (4.57) yields

$$\Phi_{\text{p}} \approx \frac{\alpha_{\text{rel}} T v_{\text{w}} r_{\text{tm}}}{3\mathcal{K}},$$

i.e., the mean energy-loss estimated by Parker (1966), in this particular case, for a particle with kinetic energy T after modulation. Numerical solutions of the spherically symmetrical Fokker–Plank equation [Eq. (4.40)] and its force-field approximate solutions were widely investigated (e.g., see [Gleeson and Axford (1968b); Fisk (1971); Perko (1987)] and references therein); at present the modulation effects are also studied using a stochastic simulation approach in order to solve Eq. (4.40) for the transport of the local interstellar spectrum[||] (LIS) to 1 AU in steady-state approximation (e.g., see [Gervasi, Rancoita, Usoskin and Kovaltsov (1999)] and references therein).

In Eq. (4.40), the SW modulation was assumed to affect the differential particle spectrum in a spherically symmetric way resulting in an isotropic *diffusion coefficient* $\mathcal{K}$. However, observed directions of the anisotropies were used as evidence

---

[||] Nowadays, the commonly used LIS is the one from [Burger and Potgieter (1989)].

supporting models for CRs propagation involving diffusion, which is primarily along, and not across, the magnetic-field lines (e.g., see discussion in [Rao et al. (1971)] and references therein): the typical ratio of the *transverse* ($\mathcal{K}_\perp$) to *parallel* ($\mathcal{K}_\parallel$) *diffusion coefficient* is

$$\frac{\mathcal{K}_\perp}{\mathcal{K}_\parallel} \simeq 0.07$$

at 1 GV rigidity [Forman, Jokipii and Owens (1974)] (see also [Jokipii and Parker (1970)] and references therein). The effects of anisotropy in the IMF can be treated by retaining the anisotropic character of the *diffusion coefficient* as a *diffusion tensor* $\mathcal{K}_{i,j}$, which, in a reference system with the 3rd coordinate along the average magnetic-field, takes the simple form*

$$\mathcal{K}_{i,j} = \begin{pmatrix} \mathcal{K}_\perp & -\mathcal{K}_A & 0 \\ \mathcal{K}_A & \mathcal{K}_\perp & 0 \\ 0 & 0 & \mathcal{K}_\parallel \end{pmatrix}; \qquad (4.58)$$

$\mathcal{K}_\perp$ and $\mathcal{K}_\parallel$ are the transverse (or perpendicular) and parallel diffusion coefficients, respectively, and originate the *symmetric part of the diffusion tensor* ($\mathcal{K}^s$). $\mathcal{K}_A$ (also termed *drift diffusion coefficient*) originates the *antisymmetric part of the diffusion tensor* ($\mathcal{K}^a$), contains the effects of gradient and curvature drifts of the particles in the average magnetic-field $\vec{B}$ and, when the scattering mean free path is much greater than the radius of gyration (r_g) of the particle [Parker (1965); Jokipii (1971)], is given by

$$\mathcal{K}_A = \frac{1}{3} v_p \, r_g \, \frac{B_z}{|B_z|},$$

where v_p is the particle velocity: if the magnetic field or the particle charge changes sign, the sign of $\mathcal{K}_A$ changes [Parker (1965); Jokipii and Parker (1970); Jokipii (1971); Jokipii and Levy (1977); Jokipii, Levy and Hubbard (1977); Reinecke and Potgieter (1994); Potgieter (1998)]. Jokipii (1971) remarked how the antisymmetric part of the diffusion tensor does not contribute to the Fokker–Planck transport equation, when the magnetic field is independent of the position (page 46 of [Jokipii (1971)]); sometimes the antisymmetric part is neglected in the literature. Furthermore, it has to be noted that for the IMF pattern described by the Parker model, the radial diffusion coefficient is given by:

$$\mathcal{K}_r = \mathcal{K}_\parallel \cos^2 \psi_{\text{sp}} + \mathcal{K}_\perp \sin^2 \psi_{\text{sp}},$$

where ψ_{sp} is the spiral angle {see Eq. (4.21) and [Jokipii and Parker (1970)]}.

By decomposing the diffusion tensor $\mathcal{K}_{i,j}$ into its symmetric ($\mathcal{K}_{i,j}^s$) and antisymmetric ($\mathcal{K}_{i,j}^a$) parts, Jokipii, Levy and Hubbard (1977) demonstrated that Eq. (4.39) can be rewritten as (see also [Jokipii and Levy (1977)] and footnote at page 323):

---

*The reader can see, for instance, Equation (50) in [Jokipii (1971)], also, [Jokipii and Parker (1970); Jokipii, Levy and Hubbard (1977)] and, in spherical coordinates, [Burger and Hattingh (1998)].

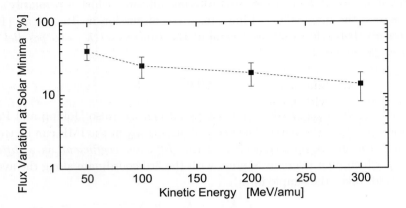

Fig. 4.10   Variation (in percentage) of proton and helium flux between two subsequent solar minima with opposite solar magnetic-field polarities as a function of the particle kinetic-energy (in MeV/amu): data from [Boella, Gervasi, Mariani, Rancoita and Usoskin (2001)]. The dashed line is to guide the eye.

$$
\begin{aligned}
\frac{\partial \mathrm{U}}{\partial t} &= \frac{\partial}{\partial x_i}\left(\mathcal{K}_{i,j}\,\frac{\partial \mathrm{U}}{\partial x_J}\right) - \frac{\partial}{\partial T}\left(\mathrm{U}\,\frac{dT}{dt}\right) - \frac{\partial}{\partial x_i}\left(\mathrm{U}\,v_{\mathrm{w},i}\right)\\
&= \frac{\partial}{\partial x_i}\left(\mathcal{K}_{i,j}\,\frac{\partial \mathrm{U}}{\partial x_J}\right) + \frac{1}{3}\left(\nabla\cdot\vec{v}_{\mathrm{w}}\right)\left[\frac{\partial}{\partial T}\left(\alpha_{\mathrm{rel}}T\mathrm{U}\right)\right] - \frac{\partial}{\partial x_i}\left(\mathrm{U}\,v_{\mathrm{w},i}\right)\\
&= \frac{\partial}{\partial x_i}\left(\mathcal{K}_{i,j}\,\frac{\partial \mathrm{U}}{\partial x_J} - \mathrm{U}\,v_{\mathrm{w},i}\right) + \frac{1}{3}\left(\nabla\cdot\vec{v}_{\mathrm{w}}\right)\left[\frac{\partial}{\partial T}\left(\alpha_{\mathrm{rel}}T\mathrm{U}\right)\right]\\
&= \frac{\partial}{\partial x_i}\left(\mathcal{K}_{i,j}^{s}\,\frac{\partial \mathrm{U}}{\partial x_J} - \mathrm{U}\,v_{\mathrm{w},i}\right) - \langle\vec{v}_{\mathrm{d}}\rangle\cdot\nabla\mathrm{U} + \frac{1}{3}\left(\nabla\cdot\vec{v}_{\mathrm{w}}\right)\left[\frac{\partial}{\partial T}\left(\alpha_{\mathrm{rel}}T\mathrm{U}\right)\right],\quad (4.59)
\end{aligned}
$$

where the *drift velocity* is

$$
\langle v_{\mathrm{d},i}\rangle = \frac{\partial \mathcal{K}_{i,j}^{a}}{\partial x_j}.
$$

The second term on the right-hand side of Eq. (4.59) is the so-called *drift term* and describes the drift effects due to the SW (e.g., [Forman, Jokipii and Owens (1974); Jokipii and Levy (1977); Kota and Jokipii (1983); Le Roux and Potgieter (1990); Potgieter, Le Roux, Burlaga and McDonald (1993); Ferreira, Potgieter, Heber and Fichtner (2003)] and references therein). For an IMF following an Archimedean spiral pattern, the drift velocity can be approximated by

$$
\langle\vec{v}_{\mathrm{d}}\rangle = \frac{p_p\,v_p\,c}{3Ze}\left[\nabla\times\left(\frac{\vec{B}}{B^2}\right)\right],\qquad (4.60)
$$

where Ze is the charge of the CR particle (e.g., see Equation 7.58 at page 274 of [Kallenrode (2004)] and [Potgieter and Moraal (1985)]). The drift velocity has an opposite direction for i) opposite charge states when magnetic-field polarity (A)

of the Sun is the same and ii) the same charge state when the Sun magnetic-field polarity is reversed.

The charge-drift effect can be best observed when the solar activity is at minimum [Potgieter, Le Roux, McDonald and Burlaga (1993)]. For instance, the solar modulation dependence on the sign of CR particles was observed by Garcia-Munoz, Meyer, Pyle, Simpson and Evenson (1986). The charge dependence becomes gradually ineffective with increasing solar activity, as observed using IMP8 satellite data during the period 1973–1995 by Boella, Gervasi, Mariani, Rancoita and Usoskin (2001); the effect exhibited its maximum strength at the solar minima: in Fig. 4.10, it is shown the variation of proton and helium flux (in percentage)

$$1 - R = 1 - \frac{\Phi^-}{\Phi^+} = \frac{\Phi^+ - \Phi^-}{\Phi^+}$$

between two subsequent solar minima as a function of the kinetic energy** (in MeV/amu); Φ^+ and Φ^- are the particles fluxes$^\parallel$ for $A > 0$ and $A < 0$, respectively. The observed particle flux depletion decreases with increasing particle energy.

As discussed above, drift effects are expected to affect the modulation, heliocentric gradient and energy change of CRs‡‡ in the inner solar system [Jokipii and Levy (1977); Garcia-Munoz, Meyer, Pyle, Simpson and Evenson (1986); Reinecke and Potgieter (1994); Bieber and Matthaeus (1997); Boella, Gervasi, Mariani, Rancoita and Usoskin (2001)] (see also references in [Potgieter (1998)]). For instance, when the Sun magnetic-field lines in the northern hemisphere are directed outwards (e.g., for $A > 0$), the inclusion of drift effects [Jokipii and Thomas (1981)] are relevant to account that positively charged particles drift inwards the polar regions and outwards along the current sheet (see page 311); the sense of drift is reversed when the magnetic-field polarity of the Sun is reversed ($A < 0$) (see [Isenberg and Jokipii (1979); Jokipii and Thomas (1981)], Section 7.7.2 of [Kallenrode (2004)] and references therein). As a function of the solar activity, these effects can be accounted and, partially, estimated by means of computer simulations of CRs propagation inside the heliosphere (e.g., see [Potgieter, Le Roux, Burlaga and McDonald (1993); Bobik et al. (2003, 2006a)]).

In addition, the actual requirements of simulating the space radiation environment are addressed by means of empirical or semi-empirical dynamical models of the GCR modulation like, for instance, the *Nymmik–Panasyuk–Pervaja–Suslov model* (e.g., see [Nymmik, Panasyuk, Pervaja and Suslov (1992); Nymmik and Suslov (1996); ISO-15390 (2004)]). In this model at 1 AU beyond Earth's magnetosphere, the effective modulation potential of the heliosphere is the main parameter and is calculated, at the time t, using the expression {see Equation (9) in [Nymmik,

---

**The reader can find a definition of kinetic energies per amu in Sect. 1.4.1.

$^\parallel$ These fluxes are observed at the solar minima and normalized to the same value of neutron monitor counting rate registered at Climax Station (for a discussion about the normalization procedure, the reader can see [Boella, Gervasi, Mariani, Rancoita and Usoskin (2001)]).

‡‡ Effects on modulated spectra can be relevant for particles with rigidities as large as 4 GV [Bieber and Matthaeus (1997)].

Panasyuk and Suslov (1996)] or, equivalently, Equation (2) in [ISO-15390 (2004)]}:

$$\phi_{s,N}(R^p, t) = 0.37 + 3 \times 10^{-4} \times \left[\hat{R}_s(t - \Delta t)\right]^{1.45} \ [\text{GV}], \qquad (4.61)$$

where R^p is the particle rigidity (in GV) and $\hat{R}_s(t-\Delta t)$ is a smoothed mean-monthly average of the Wolf numbers* referred to the time $t-\Delta t$ (e.g., see [Nymmik (2000)]); Δt depends on the particle rigidity, on wether a solar cycle is even or odd, and on the solar cycle phase. Δt expresses the time lag of GCR-flux variation relative to solar activity (see discussion at page 315) and can be calculated using Equations (4–6) of [ISO-15390 (2004)] {or, equivalently, Equations (10-11) in [Nymmik, Panasyuk and Suslov (1996)]}. Outside the magnetosphere, for electrons and nuclei at 1 AU, the modulated differential fluxes [$J(1\,\text{AU}, R^p, i, t)$], i.e., the differential intensities per unit rigidity and solid angle, are obtained from:

$$J(1\,\text{AU}, R^p, i, t) = \frac{C_i\,\beta^{\alpha_i}}{(R^p)^{\gamma_i}} \left[\frac{R^p}{R^p - \phi_{s,N}(R^p, t)}\right]^{\Delta_i(R^p, t)} \ [\text{s}\,\text{m}^2\text{sr}\,\text{GV}]^{-1}, \qquad (4.62)$$

where β [Eq. (4.70)] is the particle velocity divided by the speed of light; C_i, α_i and γ_i are the parameters‖ of the non-modulated rigidity spectrum of the ith particle; $\Delta_i(R^p, t)$ {e.g., see Equation (8) of [ISO-15390 (2004)]} depends on $\phi_{s,N}(R^p, t)$ and on the charge-number (i.e., the atomic number of the nuclide), β and rigidity of the particle. For instance [Grandi (2008)], the measured proton-fluxes could be compared with those expected [using Eq. (4.62)] at 1 AU as function of the proton kinetic-energy expressed in GeV for i) CAPRICE, hydrogen data [Boezio et al. (1999)] at the top of the atmosphere (August 1994) collected when the Sun magnetic-field lines in the northern hemisphere are directed outwards ($A > 0$), ii) AMS-01 data [AMS Collab. (2000a)] at 1 AU (June 1998, $A > 0$) and iii) BESS, proton data [Haino et al. (2004)] at the top of the atmosphere (August 2002) with reversed magnetic-field ($A < 0$). A few systematic departures of the computed curves from the data were observed at low and high energies, but, in general, they do not exceed the uncertainties of the model (e.g., see [ISO-15390 (2004)]). In addition, it has to be remarked that, since the balloon experiments‡ provided their data at the top of the atmosphere, the fluxes may differ from those at 1 AU, due to the propagation of particles through the magnetosphere. Note that a detailed description of this model is beyond the purpose of the present book.

Nowadays, the modulation of CRs is often described by means of the so-called omnidirectional distribution function†† $f(\vec{r}, \vec{p}_p, t)$ of CR particles with momentum $\vec{p}_p$, at the position $\vec{r}$ and time t (see page 324 and, for instance, see [Fisk (1976);

---

*The Wolf number was discussed at page 306.

‖These parameters are listed, for instance, in Tables 1-2 of [ISO-15390 (2004)].

‡The CAPRICE spectrometer was flown from 56.5^0 N and 101.0^0 W to 56.15^0 N and 117.2^0 W [Boezio et al. (1999)]; BESS-TeV spectrometer was launched from 56.5^0 N and 101.0^0 W [Haino et al. (2004)].

††Sometimes, the omnidirectional distribution is introduced as a function of the particle rigidity [Eq. (4.71)].

Potgieter, Le Roux, Burlaga and McDonald (1993); Potgieter (1998); Boella, Gervasi, Mariani, Rancoita and Usoskin (2001); Burger and Sello (2005)] and references therein). Furthermore, it has to be added that different authors attempted to include a latitudinal-dependent SW speed (e.g., see [Burger and Sello (2005)]) and/or Fisk-type heliospheric magnetic-field (see page 313) on estimating the modulation effect (e.g., see [Burger and Hattingh (2001)]). However, even though since then more work has been done, Parker original theory is still important because he introduced the fundamental concepts for dealing with the propagation and modulation of the GCRs in heliosphere (e.g., see [Jokipii (1971); Jokipii, Levy and Hubbard (1977); Jokipii and Thomas (1981)], Chapter V of [Toptygin (1985)], Sections 6.5.3–6.5.4 of [Cravens (1997)], Chapter 13 of [Gombosi (1998)], [Burger and Hattingh (2001)], Chapter 14 of [Schlickeiser (2002)], Section 7.7 of [Kallenrode (2004)]).

4.1.2.4 *Solar, Heliospheric and Galactic Cosmic Rays in the Interplanetary Space*

The cosmic rays[¶] are particles traveling through the interstellar and interplanetary space. Inside the solar cavity and, mainly from the corona, the Sun is a major source of particles and blows off a plasma [i.e., the solar wind (SW), see Sect. 4.1.2.1, [Lang (2001); Gosling (2006)] and references therein], consisting primarily of electrons and protons, but also of alpha particles (approximately 2.8–4 % [Alurkar (1997)]) and other ions (approximately 1%), of which carbon, nitrogen, oxygen, neon, magnesium, silicon and iron are the most abundant (e.g., Chapter 6 in [Grieder (2001)]). The SW transports an embedded magnetic-field and is estimated to extend up to $\approx 100\,$AU (Sect. 4.1.2.2) inside the galactic disk. This *interplanetary magnetic-field* (e.g., see Sect. 4.1.2.1) affects the passage of charged particles [the so-called *Galactic Cosmic Rays* (GCRs)] coming into the solar cavity from our galaxy[‡‡] (see, for instance, Sect. 4.1.2.3 and [Toptygin (1985); Boella, Gervasi, Potenza, Rancoita and Usoskin (1998); Boella, Gervasi, Mariani, Rancoita and Usoskin (2001)] and references therein). The effect is referred to as *solar modulation* and depends on the solar activity. At 1 AU in the ecliptic plane [White (1970); Encrenaz, Bibring and Blanc (1991); Alurkar (1997)], the mean wind velocity is about (400–500) km/s with mean proton (and electron) density of about 5 particles/cm^{-3} which depends on the solar activity (e.g., flares and sunspots).

Furthermore, some evidence has been found for particle acceleration processes, in many cases in association with shock waves, throughout the heliosphere (e.g., see Chapter 3 in [Schlickeiser (2002)]). *Solar flares*, during which a sudden release of

---

[¶]The reader can see, for instance, Section 24 in [PDB (2008)] and, also, [Ginzburg and Syrovatskii (1964); Gaisser (1990)], Chapter 2 of [Schlickeiser (2002)].

[‡‡]The cosmic rays are considered to be originated in our galaxy, although a few of them, in particular at ultra high-energy, can have an extragalactic origin [Allard, Parizot and Olinto (2007)]. A discussion about the local interstellar spectra, i.e., the particle spectra beyond the solar modulation region can be found in [Beliaev, Nymmik, Panasyuk, Peravaya and Suslov (1996)] (see also references therein).

very large amounts of energy occurs in the solar chromosphere, can also proceed by ejecting accelerated particles into space. This variety of processes results in generating particles like i) the *Solar Energetic Particles* (SEPs) with energies[**] up to $> 10\,\text{MeV/nucleon}$, accelerated in the solar corona and lasting a few days (see Chapter 3 in [Toptygin (1985)] and [Smart and Shea (1989); Gabriel (2000); Nymmik (2006)]), ii) the *Energetic Storm Particles* (ESPs) with energies up to $> 500\,\text{MeV}$, accelerated in propagating interplanetary shock and lasting hours [Lario and Decker (2001)], and iii) the *Anomalous Cosmic Rays* (ACRs). The solar wind and solar energetic particles have typically low energies and are mostly prevented to reach the Earth by the magnetosphere or are absorbed in the upper atmosphere. In [ISO-15391 (2004)], a probabilistic model is provided for the $4\text{–}10^4\,\text{MeV}$ SEP-event proton fluences and peak fluxes in the near-Earth space beyond the Earth magnetosphere under varying solar activity

The ACRs (mostly He, N, O, Ne ions) are believed to come from interstellar neutral particles, which have entered into the heliosphere, subsequently ionized by the SW or UV radiation and finally accelerated to energies larger than $10\,\text{MeV/nucleon}$, probably, at the solar wind termination shock (see [Mewaldt et al. (1993); Mewaldt (1996); Tylka et al. (1996)] and references therein). However, collisions in the termination shock region cause some ions to reach multiply charge states [Mewaldt (1996)] and be accelerated to even larger energies. For instance, SAMPEX has observed anomalous cosmic ray oxygen ions at Earth level with energies of about $100\,\text{MeV/nucleon}$ ([Mewaldt (1996)] and Chapter 6 in [Grieder (2001)]).

The GCRs mainly consist of protons and alpha particles (about 98%) [Simpson (1983); Müller (1989); Mewaldt (1994); ISO-15390 (2004); Stanev (2004)] (see also Sect. 9.11.1), in addition there is a lesser amount of low-Z and medium-Z nuclei up to nickel, and an even lesser amount of heavier-Z nuclei[††], where Z is the atomic number (Sect. 3.1). At energies $\gtrsim 10\,\text{GeV/nucleon}$ (e.g., see Sect. 4.1.2.3), the solar modulation marginally affects the energy spectra which exhibit an approximate power law behavior as a function of the kinetic energy per nucleon, E_k, e.g., their intensities being $\propto E_k^{-\gamma}$, where γ is the so-called *differential spectral index*[*] {see Fig. 9.37, [Müller (1989)], and Chapters 5 and 6 in [Grieder (2001)]}[§]. γ is ≈ 2.74 (2.60–2.63) for hydrogen (helium) {see Table 5.1 (5.3) in Section 5.4.1 of [Stanev (2004)]}. The spectral index depends on the element, thus relative nuclear abundances may depend on energy: for instance, observed and calculated abundance-ratios are shown in Figures 7 and 8 of [Engelmann et al. (1990)] for $0.8 < E_k < 35\,\text{GeV/nucleon}$. Above $50\,\text{GeV/nucleon}$, spectral indices of elements

---

[**]The reader can find the definition of kinetic energies per nucleon in Sect. 1.4.1.

[††]Parameters for the description of GCR fluxes and time dependent modulation effects using a semi-empirical model are presented in [Nymmik, Panasyuk, Pervaja and Suslov (1992); Nymmik and Suslov (1996); ISO-15390 (2004)] (see also the discussion at page 329).

[*]It is also referred to as *spectral index*.

[§]In addition, the reader can refer to fluxes and energy distributions of protons, helium, carbon and iron in [Mewaldt (1994)] (see also [Engelmann et al. (1985, 1990)]), Chapter 5 of [Stanev (2004)] and Section 24 of [PDB (2008)] (see also [ISO-15390 (2004)]).

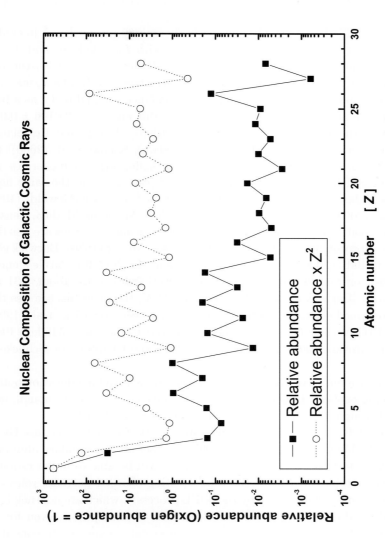

Fig. 4.11 From hydrogen ($Z = 1$) to nickel ($Z = 28$), relative abundances (■) and relative abundances multiplied by Z^2 (○) of GCRs at 16.2 GeV/nucleon normalized to oxygen (i.e., the relative abundance of oxygen is equal to 1) as a function of the atomic number Z (from [Leroy and Rancoita (2007)]). The oxygen flux at this energy is 0.0106 ± 0.0004 m^{-2}s^{-1}sr^{-1}(GeV/nucleon)$^{-1}$ [Engelmann et al. (1990)]. The H and He relatives abundances are averaged from [AMS Collab. (2000a,b); Sanuki et al. (2000)] (see also [AMS Collab. (2002)]). The Li ($Z = 3$) relative abundance is at 5 GeV/nucleon [Mewaldt (1994)].

from hydrogen to iron range from 2.55 to 3.2 (e.g., see Table 3.2 in Chapter 3 of [Schlickeiser (2002)]). For energies from above several GeV to somewhat beyond 100 TeV, the overall intensity of primary nucleons $[I_{\mathrm{nucl}}(E_{\mathrm{tot,isot}})]$ is approximately given by {see Equation (24.2) in [PDB (2008)]}:

$$I_{\mathrm{nucl}}(E_{\mathrm{tot,isot}}) \approx 1.8 \times 10^4\, I_{\mathrm{isot}} \left(\frac{E_{\mathrm{tot,isot}}}{1\,\mathrm{GeV}}\right)^{-\gamma_p} \quad [\mathrm{m}^2\ \mathrm{s\ sr\ GeV/nucleon}]^{-1}, \quad (4.63)$$

where $E_{\mathrm{tot,isot}} = E_{\mathrm{isot}}^{\mathrm{tot}}/I_{\mathrm{isot}}$ in GeV/nucleon is the energy-per-nucleon[¶] [including the rest-mass energy, see Eq. (4.66)] of an isotope with I_{isot} nucleons and, finally, $\gamma_p \approx 2.7$ is the differential spectral index of the CR flux. In Fig. 4.11, the relative abundances of GCRs up to nickel ($Z = 28$) are shown at 16.2 GeV/nucleon[‡], normalized to oxygen (i.e., the oxygen relative abundance is equal to 1), as a function of the atomic number Z. The oxygen flux, at this energy, is 0.0106 ± 0.0004 $\mathrm{m}^{-2}\mathrm{s}^{-1}\mathrm{sr}^{-1}(\mathrm{GeV/nucleon})^{-1}$ [Engelmann et al. (1990)]. The H and He relatives abundances are averaged from [AMS Collab. (2000a,b); Sanuki et al. (2000)] (see also [AMS Collab. (2002)]). Since in matter, at sufficiently high-energies (see Sect. 2.1.4) the collision energy-loss depends on Z^2 [Eq. (2.1)], in the same figure the relative abundances are also shown multiplied by Z^2 (e.g., see [Mewaldt (1994); Codegoni et al. (2004b); Consolandi, D'Angelo, Fallica, Mangoni, Modica, Pensotti and Rancoita (2006)]). In the cosmic radiation, electrons and positrons are less than about 1% of the proton flux at about 10 GeV (e.g., see [Müller (1989); Barwick et al. (1998); AMS Collab. (2000c)] and references therein). In the 5.0 to 50 GeV energy range, the positron and electron differential spectral indexes are about 3.31 and 3.09, respectively [Barwick et al. (1998)]. Above 1 GeV, the positrons are less than 10% of the combined electron–positron flux [Müller (1989); Barwick et al. (1998)], but above 5–10 GeV their fraction has been observed to increase ([Müller (1989); Aharonian and Atoyan (1991); Barwick et al. (1998)] and Chapter 2 of [Schlickeiser (2002)]).

The GCRs are a major radiation concern for missions with long duration outside the Earth magnetosphere. In addition, since they have energies large enough, most of them pass through the magnetosphere and are observed at LEOs.

The flux of GCR's is a possible hazard to spacecraft electronics, because GCR's can penetrate shielding materials [Barth (1997)]. In fact after a standard aluminum shield of 100 mils, fast energetic charged particles[**] will be able to inflict radiation damages. Each isotopic element of these penetrating charged particles releases a dose and loses an amount of energy by the NIEL process[‡‡], which are non-negligible fractions of those due to the proton component. As a consequence, even for not heavy-shielded VLSI components, the proton component is expected to contribute

---

[¶]At high energies per nucleon, we have $E_k \approx E_{\mathrm{tot,isot}}$.

[‡]The Li, $Z = 3$, relative abundance is at 5 GeV/nucleon [Mewaldt (1994)].

[**]These are, for instance, electrons, protons, alpha-particles and iron ions above about 1.2 MeV, 22.5 MeV, 22.5 MeV/nucleon and 90 MeV/nucleon, respectively.

[‡‡]The reader can see Sect. 4.2.1, [Colder et al. (2002); Codegoni et al. (2004b); Consolandi, D'Angelo, Fallica, Mangoni, Modica, Pensotti and Rancoita (2006)] and references therein.

Table 4.1 Dependence (from [Leroy and Rancoita
(2007)]) of the dose rate (Sv/day), due to protons
from the inner radiation belt, on the altitude and
inclination of the orbit {e.g., see [Miroshnichenko
(2003)] and references therein}.

Orbit altitude km	0°	45°	90°
300	0.05	0.05	0.05
500	0.16	0.16	0.16
2500–3000	110.00	26.00	18.00
7500	2.60	3.40	2.40

with an important, but not dominant, fraction of both the dose and the NIEL
deposition.

Furthermore, after introducing the radiation weighting factor to account for
the biological effectiveness, heavier elements like Fe, Si, Mg and O make an even
larger contribution to the *dose equivalent*. Moreover, it has been estimated that,
for $\approx 3\,\mathrm{g/cm}^3$ Al shielding, about 75 % of the dose equivalent is due to GCRs
with kinetic energies lower than 1 GeV/nucleon ([Mewaldt (1994)] and reference
[13] therein). Lockwood and Hapgood (2007) pointed out that, i) while in Earth's
biosphere we typically receive 2 mSv per year from cosmic radiation, ii) the effective
GCR doses in interplanetary space are greater than in the biosphere by factors of
about 90 and 250 at sunspot maximum and minimum, respectively, and that iii)
even in the most favorable time, a manned trip to Mars would let the crew to receive
up to the lifetime radiation allowance for men and more than the double for women.

4.1.2.5 *Trapped Particles and Earth Magnetosphere*

The magnetosphere (see Sect. 4.1.2.2) is the volume of space where the Earth's
magnetic-field extends its effects [Knecht and Shuman (1985); Kivelson and Bagenal
(2006); Schulz (2007)]. Close to the Earth, the magnetic field is approximately
represented by a dipole, slightly offset from the center of the Earth and inclined
by about 11.3° from the axis of rotation. Its magnetic dipole axis intersects the
Earth's surface in two points, the boreal (northern) pole at 78.3° N 69° W and
austral (southern) pole at 78.3° S 111° E. The SW compresses the magnetosphere
side toward the Sun to a distance of (6–10) Earth radii (one Earth radius, Re, is
about 6371 km) as a function of the solar activity and stretches the opposite side
into a tail (the so-called *magnetotail*) to perhaps 1000 Re (Fig. 4.8). The Earth's
radiation belts in the inner magnetosphere*, one of the first major discoveries of
the satellite age [Van Allen, Ludwig and McIlwain (1958); Hess (1968)], are due

---

*For a general description, one can see, for instance, [Spjeldvik and Rothwell (1985)].

to the structure of the magnetic-field lines near the Earth and consist of trapped particles that gyrate around geomagnetic-field lines, travel back and forth along geomagnetic-field lines between conjugate points in opposite hemispheres, and drift around the Earth.

The overall particle flux is due to a combination of *Primary Cosmic Rays*[‡‡] (PCRs), trapped ACRs, trapped solar-wind particles, soft and energetic *Secondary Cosmic Rays* (SCRs), mainly consisting of electrons and protons [Buenerd et al. (2000)] generated by the interactions of PCRs with the atmosphere (e.g., see Section 12 in [Leroy and Rancoita (2000)], Chapter 1 in [Grieder (2001)], [Stanev (2004)], Chapters 6 and 7 in [Grupen (2005)] and references therein). Nevertheless, the Earth magnetosphere prevents the arrival of less energetic cosmic and solar rays. The value of the rigidity, below which the PCRs cannot reach an observation position, is called *local rigidity of geomagnetic cut-off*[‡], R_{cut}^{p} (see, for instance, [Stoermer (1930); Fermi (1950); Shea, Smart and McCracken (1965)] and Section 6.2.3.1 in [Knecht and Shuman (1985)]). The transport of the PCRs, from the Earth magnetosphere boundary at 1 AU through the magnetosphere to the atmosphere, is described by the so-called *transmission function* (TF), which depends on the PCR rigidity, location of the observation point, solid angle and pointing of the detecting instrument (e.g., see [Bobik et al. (2006)] and references therein). The TF becomes 0, when the incoming PCRs cannot reach the observation location any more, thus the particle rigidity is lower than R_{cut}^{p}.

Inside the Earth magnetosphere there are two *Van Allen belt* regions (e.g., see [Hess (1968)]) of trapped fast particles (mostly electrons and protons). In between the two regions, there is the so-called *slot region*. The *inner radiation belt*, discovered by Van Allen [Van Allen, Ludwig and McIlwain (1958)], is relatively compact. In the region above the equator, it extends perhaps up to about 2.4 Re and is centered at about 1.5 Re. It is populated by protons with energies in the (10–100) MeV range and a density of about 15 protons per m^3 [Lang (2001)]. There are also electrons of lower energies, but exceeding 0.5 MeV. The population is fairly stable, but is subject to occasional perturbations due to geomagnetic storms, and varies along a 11 year *solar cycle* (e.g., see Sect. 4.1.2.1). The main sources of particles are solar energetic particles and interactions of cosmic ray in the atmosphere (e.g., see [Hess (1968)]). In the inner radiation belt many satellites, designed to provide telephony (whether for mobile or fixed phones) and internet services for a broad range of users, move on LEOs where they encounter trapped energetic charged particles. These ones represent an hazard primarily for astronauts and also for VLSI electronics. The doses change considerably at small altitudes and depend on the inclination of the orbit plane. For instance, the dose rate for protons of the inner radiation belt is shown in Table 4.1 (e.g., see Sect. 1.1 of [Miroshnichenko (2003)]

[‡‡]Above about 100 MeV/nucleon, the PCRs are mostly GCRs.

[‡]For a practical introduction to the cut-off rigidity of cosmic rays, the reader can see Chapter 1 in [Grieder (2001)]. For practical calculations, the reader can refer to the web sites: *http://hpamsmi2.mib.infn.it/~wwwams/rig.html* and/or *http://pfunc.selfip.biz/riho/index.html*.

and references therein). These dose rates take into account the *radiation weighting factor* for protons[†], but can be adapted to provide rate estimates in other absorbers (e.g., in silicon).

Further out, centered at about (4.5–5.0) Re, the *outer radiation belt* mainly contains electrons with energies up to 10 MeV. This belt is mostly generated by the injection of particles following geomagnetic storms, which make it much more dynamic than the inner belt (it is also subject to day-night variations). It has an equatorial distance of about (2.8–12) Re, with a maximum for electrons above 1 MeV occurring at about 4 Re. Unlike the inner belt, this population fluctuates widely, rising when magnetic storms inject fresh particles from the tail of the magnetosphere, then gradually falling off again. The outer radiation belt coincides with the geostationary orbit of many communications satellites.

A feature of the Van Allen belts is the so-called South Atlantic Anomaly (SAA): the offset of the geomagnetic dipole axis towards the northwest Pacific causes an anomalously weak geomagnetic-field strength in a region which, for an altitude of about 500 km (i.e. at LEO), extends at latitudes between about -55°–0° and longitudes between about -80°–20° [Barth (1997)]. In this region, trapped radiation particles can reach lower altitude before bouncing back to the northern hemisphere. The orbit parameters of a spacecraft (inclination and altitude) determine the number of passes made per day through this region. At an orbit below an altitude of about (500–550) km a considerable part of absorbed radiation dose is caused by passing the SAA.

A few years ago, the analysis of SAMPEX data[§] has confirmed the existence of belts containing trapped ACRs, i.e., heavier nuclei (like N, O and Ne) with energies of the order of 10 MeV/nucleon, already observed by COSMOS satellites [Grigorov et al. (1991)]. These trapped particles have steep energy spectra and are unlikely to be a significant radiation hazard for (lightly-) shielded systems, i.e., with about 50 mils Al equivalent (see for instance [Tylka et al. (1996)]).

PCRs are particles reaching the Earth from the outer space. However the flux of PCRs is only a tiny fraction of the total particle flux observed at the Earth's surface or at LEOs, since a large amount of SCRs can be generated by the interactions of the PCRs with the upper and middle atmosphere. Part of the SCRs can remain trapped by the magnetic field near the Earth. Their flux depends on the flux and energy spectrum of the PCRs and, consequently, on the solar modulation (see Sect. 4.1.2.4). Thus, the observed flux of SCRs is also related to phenomena like the solar activity and magnetic storms. Trapped energetic SCRs are always present and populate a radiation belt at low altitude. The existence of trapped and quasi-trapped high-energy (up to few GeV/c's) particles[‡], generating the so-called *AMS Radiation*

[†]For instance, the radiation weighting factor for protons is 2 (Table 29.1 of [PDB (2008)]). The biological effects of radiation are treated, for instance, in [Stather and Smith (2005)].

[§]One can see, e.g., [Cook et al. (1993); Cummings et al. (1993); Selesnick et al. (1995); Mewaldt, Selesnick and Cummings (1997)].

[‡]These particles are mostly protons, electrons and positrons.

belt located at LEO, was observed by the AMS mission (Space Shuttle flight STS-91, June 1998, see [AMS Collab. (2000a,b,d, 2002)] and references therein). As already mentioned, while the PCRs come from outside the magnetosphere, the SCRs are originated in the Earth atmosphere. They could be distinguished by reconstructing their trajectories[*] from the observation point, i.e., the AMS spectrometer, back to the border of the magnetosphere (e.g., PCRs) or to the border of the atmosphere (e.g., SCRs).

Let us indicate by E_k^{tot} the total kinetic-energy of an isotope with I_{isot} nucleons [i.e., with mass number I_{isot}, Eq. (3.1)] and rest mass m_{isot}

$$m_{\text{isot}} \sim I_{\text{isot}}\, m_{\text{nucl}}$$

where m_{nucl} is the nucleon mass (see page 219). The kinetic energy per nucleon, E_k, is

$$E_k = \frac{E_k^{\text{tot}}}{I_{\text{isot}}}. \tag{4.64}$$

From Eq (1.15), the isotope momentum, p_{isot}, is

$$p_{\text{isot}} = \frac{\sqrt{\left(E_k^{\text{tot}} + m_{\text{isot}}c^2\right)^2 - m_{\text{isot}}^2 c^4}}{c}. \tag{4.65}$$

In addition [see Eq. (1.12)], the isotope total-energy is

$$E_{\text{isot}}^{\text{tot}} = E_k^{\text{tot}} + m_{\text{isot}}c^2. \tag{4.66}$$

Moreover under the assumption that the direction of momentum is perpendicular to the direction of the magnetic field, the momentum p_{isot} and magnetic rigidity R_{isot} of the isotope of nuclear charge Ze (e is the electron charge) can be expressed as a function of the kinetic energy per nucleon by

$$\begin{aligned} R_{\text{isot}} &= \frac{p_{\text{isot}}}{Ze}, \\ &= \frac{\sqrt{\left(E_k^{\text{tot}} + m_{\text{isot}}c^2\right)^2 - m_{\text{isot}}^2 c^4}}{Zec}, \end{aligned} \tag{4.67}$$

from which we get:

$$\begin{aligned} R_{\text{isot}} &= \frac{\sqrt{\left(I_{\text{isot}}E_k + I_{\text{isot}}m_{\text{nucl}}c^2\right)^2 - \left(I_{\text{isot}}m_{\text{nucl}}\right)^2 c^4}}{Zec} \\ &= \left(\frac{I_{\text{isot}}}{Z}\right)\frac{\sqrt{\left(E_k + m_{\text{nucl}}c^2\right)^2 - m_{\text{nucl}}^2 c^4}}{ec} \\ &= \left(\frac{I_{\text{isot}}}{Z}\right)\frac{\sqrt{E_k\left(E_k + 2m_{\text{nucl}}c^2\right)}}{ec}. \end{aligned} \tag{4.68}$$

---

[*]On this subject, the reader can see, for instance, [Bobik et al. (2000, 2001, 2005)] and references therein.

Conversely, by squaring the two sides of Eq. (4.67), we obtain

$$R_{\text{isot}}^2 Z^2 e^2 c^2 + m_{\text{isot}}^2 c^4 = \left(E_k^{\text{tot}} + m_{\text{isot}} c^2\right)^2,$$

and, finally,

$$
\begin{aligned}
E_k &= \frac{1}{I_{\text{isot}}} \left(\sqrt{R_{\text{isot}}^2 Z^2 e^2 c^2 + m_{\text{isot}}^2 c^4} - m_{\text{isot}} c^2 \right) \\
&= c^2 \left[-m_{\text{nucl}} + \sqrt{R_{\text{isot}}^2 \left(\frac{Z}{I_{\text{isot}}}\right)^2 \left(\frac{e}{c}\right)^2 + m_{\text{nucl}}^2} \right].
\end{aligned}
\tag{4.69}
$$

It has to be noted that, using Eqs. (1.6–1.10), from Eqs. (4.65–4.67) we can derive

$$
\begin{aligned}
\beta_{\text{isot}} &= \frac{p_{\text{isot}} c}{E_{\text{isot}}^{\text{tot}}} \\
&= \frac{R_{\text{isot}} Z e c}{\sqrt{(R_{\text{isot}} Z e c)^2 + m_{\text{isot}}^2 c^4}} \\
&= \frac{R_{\text{isot}}}{\sqrt{R_{\text{isot}}^2 + [(m_{\text{isot}} c)/(Z e)]^2}}.
\end{aligned}
\tag{4.70}
$$

Nowadays, it is common to use the quantity (referred to as *particle rigidity*)

$$R_{\text{isot}}^p \equiv \frac{p_{\text{isot}} c}{Z e} \tag{4.71}$$

in units of GV, where p_{isot} is the isotope momentum in GeV/c. Equations (4.68, 4.69) can be rewritten, without any assumption on the momentum and magnetic-field directions, as:

$$R_{\text{isot}}^p = \left(\frac{I_{\text{isot}}}{Z}\right) \frac{\sqrt{E_k \left(E_k + 2 m_{\text{nucl}} c^2\right)}}{e}, \tag{4.72}$$

and

$$E_k = c^2 \left[-m_{\text{nucl}} + \sqrt{\left(\frac{Z}{I_{\text{isot}}}\right)^2 \left(\frac{R_{\text{isot}}^p e}{c^2}\right)^2 + m_{\text{nucl}}^2} \right], \tag{4.73}$$

respectively. Equation (4.70) can be rewritten as

$$\beta_{\text{isot}} = \frac{R_{\text{isot}}^p}{\sqrt{\left(R_{\text{isot}}^p\right)^2 + [(m_{\text{isot}} c^2)/(Z e)]^2}}. \tag{4.74}$$

From Eqs. (1.15, 4.71), the rigidity of a proton with kinetic energy $E_{k,\text{p}}$ is

$$R_{\text{H}}^p = \frac{\sqrt{E_{k,\text{p}} \left(E_{k,\text{p}} + 2 m_{\text{p}} c^2\right)}}{e}, \tag{4.75}$$

where m_{p} is the rest mass of the proton. Since $m_p \approx m_{\text{nucl}}$ (Sect. 3.1), from an inspection of Eqs. (4.72, 4.75) the rigidity of an isotope heavier than proton with kinetic energy per nucleon E_k is

$$R_{\text{isot}}^p \approx \left(\frac{I_{\text{isot}}}{Z}\right) R_{\text{H}}^p, \tag{4.76}$$

where R_H^p is the proton rigidity at the same kinetic energy, i.e., $E_{k,p} = E_k$. In addition, it can be remarked that for the stable** and most abundant isotopes in the heliosphere [i.e., up to nickel (Sect. 4.1.2.4)]

$$\frac{I_{\text{isot}}}{Z} \approx 2.0\text{--}2.3.$$

As a consequence, less energetic (but heavier than proton) PCRs can penetrate deeply the magnetosphere: in any location, the associated geomagnetic cut-off rigidity, R_{cut}^p, requires a kinetic energy of the isotope lower than that of a proton [see Eq. (4.68)]. As discussed in [Bobik et al. (2006)], the rigidity (R_{cut}^p) of geomagnetic cut-off varies as a function of the observation location and increases with decreasing the geomagnetic latitude: the PCRs with rigidities lower than the local value of R_{cut}^p cannot reach that observation region.

This effect is visible in Fig. 4.12(a), which, at the magnetosphere boundary (i.e., at 1 AU), shows the spectra (indicated as "cosmic") of the i) proton and helium differential energy distributions observed using the AMS-01 spectrometer and ii) carbon and iron observed using the HEAO-3-C2 spectrometer (see also [Bobik et al. (2006b); Boella et al. (2008)]). Figure 4.12(a) also shows the proton (helium) spectrum observed with the AMS-01 spectrometer at the Shuttle altitude in the geomagnetic region $0.3 \leq |\Theta_M| \leq 0.4$ ($0 \leq |\Theta_M| \leq 0.4$). In Fig. 4.12(b), the same spectra are shown as function of the particle rigidity expressed in GV. The AMS-01 observations were carried out in June 1998 during the solar cycle 23; HEAO-3 observations from October 1979 up to June 1980, i.e., during the solar cycle 21. In both periods the solar activity was rising from the minimum to the subsequent maximum and the solar magnetic field polarity was positive. The angular acceptance of the detectors, the altitude and the inclination of the orbits were such that the observations were obtained under not too different experimental conditions. For instance, for $0 \leq |\Theta_M| \leq 0.4$ the geomagnetic cut-off rigidity is similar for protons and helium particles [Fig. 4.12(b)], but it affects energetic protons more than helium particles [e.g., see the cut-off effect shown in Fig. 4.12(a)].

At 1 AU and outside the Earth's magnetosphere, the relative abundances to protons for He (He/p), C (C/p) and Fe (Fe/p) nuclei can be calculated using the "cosmic"-data (Fig. 4.12(a)); *above* $0.8\,\text{GeV/nucleon}$, *the relative abundances*‖ are

$$\Re_{\text{He/p}}^{\cos} \approx 0.067, \ \Re_{\text{C/p}}^{\cos} \approx 0.0019 \text{ and } \Re_{\text{fe/p}}^{\cos} \approx 0.00018$$

for He/p, C/p and Fe/p, respectively [Gervasi and Grandi (2008)]. Furthermore at an altitude of 400 km, the relative abundances inside the magnetosphere and

---

**For ^{3}H (tritium) and ^{3}He, this ratio is 3 and 1.5, respectively. Tritium has several different experimentally-determined values of its half-life, the recommended value (e.g., see the web site: *http://nvl.nist.gov/pub/nistpubs/jres/105/4/j54luc2.pdf*) is 4500 ± 8 days (approximately 12.32 years). It decays into ^{3}He [Eq. (3.21)] by the reaction $^3\text{H} \rightarrow^3 \text{He} + e^- + \bar{\nu}_e$.

‖These values are affected by the solar modulation and may differ from those which are derived using abundances at fixed values of kinetic energy per nucleon, large enough so that particles are marginally modulated (e.g., see Fig. 4.11).

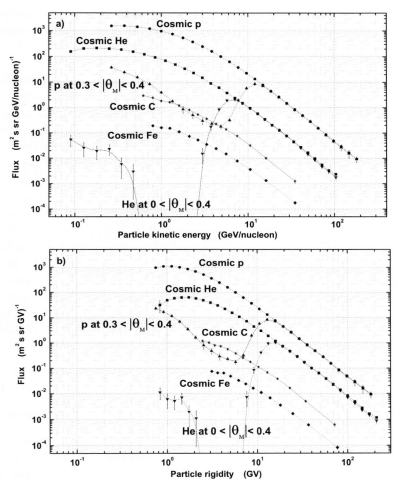

Fig. 4.12 Proton [AMS Collab. (2000a,d)], He [AMS Collab. (2000b)], C [Engelmann et al. (1990)] and Fe [Engelmann et al. (1990)] fluxes as functions of (a) kinetic energy per nucleon in GeV/nucleon and (b) particle rigidity in GV: p (•), He (■), C (★) and Fe (♦) *cosmic* (i.e., these fluxes are observed well above the geomagnetic cut-off and can be considered at about 1 AU), p (▲) in the geomagnetic region $0.3 \leq |\Theta_M| \leq 0.4$ and He (▼) in the geomagnetic region $0 \leq |\Theta_M| \leq 0.4$ (from [Leroy and Rancoita (2007)]). At $|\Theta_M| \approx 0.4$, R_{cut}^p (the associated geomagnetic cut-off proton rigidity) is $\approx$ 7–8 GV.

their ratios with those outside at 1 AU (i.e., the *enhancement factors* E_F's) were computed using a *transmission function* (TF) *approach* (e.g., see [Boella et al. (2008)] and discussion at page 342): for geomagnetic latitudes ranging from $\approx -46°$ up to $\approx 46°$, the values of E_F's depend on the mass number I_{isot} and are[*]

$$2.06 \lesssim E_F \lesssim 3.74.$$

[*]See the discussion in Bobik, P. et al. (2009), Fluxes and Nuclear Abundances of Cosmic Rays inside the Magnetosphere using a TF Approach, *J. Adv. Space Res.*, doi: 10.1016/j.asr.2008.11.020.

Thus, the resulting *magnetospheric isotopic relative-abundances* i) are modified because the proton component is more depleted than the other isotopic components of GCR's and ii) can depend on parameters like the geomagnetic latitude and/or altitude.

It has to be noted that in Fig. 4.12 the proton (for $0.3 \leq |\Theta_M| \leq 0.4$) and He (for $0 \leq |\Theta_M| \leq 0.4$) spectra include both primary and secondary cosmic rays. These latter are responsible for the flux increase below R^p_{cut}. In addition, the same effect is visible in the differential energy spectra for H, He and Fe for one solar cycle from minimum to maximum reported in [Smart and Shea (1989); Gabriel (2000)].

As previously mentioned, in [Bobik et al. (2006)] the passage of the primary protons through the magnetosphere to the AMS-01 spectrometer was treated using the so-called transmission function [Adams et al. (1991); Boberg et al. (1995)], which was determined by back-tracking the proton trajectories to determine those allowed to primaries: the access of a primary proton to the observation location is supposed to be allowed when the back-tracked particle trajectory reaches the magnetopause. The TFs allowed one [Bobik et al. (2006)] to untangle the spectra due to secondary protons from those observed, as function of the geomagnetic latitude, using the primary proton spectrum at 1 AU (the "cosmic"-protons in Fig. 4.12). Some SCRs (mostly re-entrant albedo protons) were also found to populate the rigidity regions above the local rigidity of geomagnetic cut-off [Bobik et al. (2006)].

4.1.3 *Neutron Spectral Fluence in Nuclear Reactor Environment*

In the phenomenon of the neutron-induced nuclear fission, a nucleus captures a neutron and breaks up into two nuclei (e.g., see Sections 5.2 and 5.5 in [Byrne (1994)]). This process is accompanied by a large release of energy, typically about 200 MeV/fission, and the emission of one or more fast neutrons. A controlled self-sustaining chain of fission reactions is called *chain-reacting pile* or *nuclear reactor*. The first reactor was built by Fermi in 1942. The modern nuclear reactors are basically machines that contain and control nuclear fission reactions, while producing electricity. Besides power reactors, there are research reactors (like TRIGA and Godiya types and others, see for instance [Kelly, Luera, Posey and Williams (1988)], Chapter 5 in [Messenger and Ash (1992)] and [Angelescu et al. (1994)]), which can provide sources of fission neutrons, for testing the performance of VLSI transistors (and circuits) and detectors, and allow the simulation of nuclear environments (e.g., nuclear power reactors and weapons or high-luminosity colliding beam accelerators).

The kinetic energies (E) of reactor fission neutrons do not exceed $\approx$ 15 MeV. However, the available fission spectra differ for the energy distribution of both *thermal and fast neutrons* (Table 4.2) and are characterized by a *neutron spectral-fluence* $\phi(E)$ in n/(cm^2 MeV) obtained as

$$\phi(E) = \int \psi(E, t) \, dt, \tag{4.77}$$

where $\psi(E, t)$ is the *neutron spectral-flux* in n/(s cm^2 MeV), and by a neutron fluence

$$\Phi_{\mathrm{n}} = \int_{E_{\min}} \phi(E) \, dE \ \ [\mathrm{n/cm^2}], \tag{4.78}$$

where $E_{\min}$ is the minimum neutron kinetic-energy (typically about 10 keV) to induce radiation damage by atomic displacements (see for instance Section 5.9 in [Messenger and Ash (1992)]). Thus, Φ_{n} accounts for the neutron fluence[‡‡] related to radiation-induced damages, i.e., above 10 keV in n–silicon interactions, the energy released for displacement damages is no longer negligible (see Sect. 4.2.1).

The displacement damages induced by neutrons can be normalized to displacement damages induced by 1 MeV neutrons by means of the *hardness parameter* (κ) also termed *hardness factor* (see Sect. 4.2.1 and, for a further discussion, see for instance [Namenson, Wolicki and Messenger (1982); Kelly, Luera, Posey and Williams (1988); Vasilescu (1997)]). For fission neutrons with spectral fluence $\phi(E)$, the 1 MeV equivalent neutron fluence is given by:

$$\Phi_{\mathrm{eq}}^{1 \ \mathrm{MeV}} = \kappa \int_{E_{\min}} \phi(E) \, dE = \kappa \, \Phi_{\mathrm{n}}. \tag{4.79}$$

$\Phi_{\mathrm{eq}}^{1 \ \mathrm{MeV}}$ is the fluence of 1 MeV neutron needed for generating the same amount of displacement damage compared to the fission fluence Φ_{n} (see, also, footnote at page 291).

4.1.3.1 *Fast Neutron Cross Section on Silicon*

Neutron interactions on nuclei were widely studied and their cross sections determined experimentally: these latter are, for instance, available on the web at the IAEA[‖] - Nuclear Data Service (2007).

In addition, nowadays, for the scattering of a fast neutron on ^{28}Si a systematic evaluation of all important cross sections is reported in [Tagesen, Vonach and Wallner (2002)]: it covers the neutron energy-range (1.75–20) MeV, i.e., for an energy range well above the resonance region (Sect. 4.1.3.2) and of interest for reactions resulting in displacement damage. The evaluation uses experimental data compiled in EXFOR (e.g., see [IAEA - Nuclear Data Service (2007)]) supplemented by very recent ones: the total and elastic cross sections are shown in Fig. 4.13. The elastic cross section was obtained as the difference between the total and non-elastic cross sections. This latter was derived summing the inelastic cross sections and those for reactions like, for example, (n,2n), (n,p), (n,α), (n,d)... (for a complete discussion see [Tagesen, Vonach and Wallner (2002)]).

[‡‡]The total neutron fluence can be computed with $E_{\min} = 0$.

[‖]IAEA is the International Atomic Energy Agency located in Vienna (Austria).

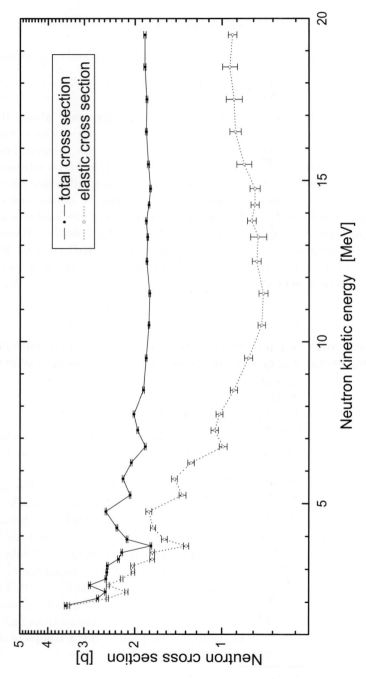

Fig. 4.13   Total and elastic (see text) cross sections (in barn) for neutrons on ^{28}Si with kinetic energies of (1.75–20) MeV. The lines are to guide the eye. Data were obtained from [Tagesen, Vonach and Wallner (2002)].

Table 4.2   Classification of neutrons as a function of kinetic energy.

Energy E_n (eV)	Classification and Subdivision		Subdivision energy-range E_n (eV)
$0 < E_n < 1$	Slow	Ultra cold	$E_n < 2 \times 10^{-7}$
		Very cold	$2 \times 10^{-7} \leq E_n < 5 \times 10^{-5}$
		Cold	$5 \times 10^{-5} \leq E_n \lesssim 0.025$
		Thermal	$E_n \simeq 0.025$
		Epithermal	$0.025 \lesssim E_n \leq 1$
$1 < E_n < 10^6$	Intermediate	including	
		Resonance	$1 < E_n < 100$
$10^6 < E_n < 10^7$	Fast[1]		
$10^7 < E_n < 5 \times 10^7$	Very fast		
$5 \times 10^7 < E_n < 10^{10}$	Ultra fast		
$> 10^{10}$	Relativistic		

[1]: in Section 1.5.5.3 of [Byrne (1994)], a neutron with kinetic energy > 1 MeV is termed fast; in Section 12.2.2 of [Marmier and Sheldon (1970)], a neutron is considered fast for $0.5 < E_n < 10$ MeV.

4.1.3.2   *Energy Distribution of Reactor Neutrons and Classification*

As mentioned in the previous section, reactor-neutron production is one of the most widely used methods of obtaining fast neutrons. The spectral fluence of emerging almost-thermal and fast neutrons depends on the type of reactor and moderator (e.g., see Chapter 12 of [Marmier and Sheldon (1970)], [Kelly, Luera, Posey and Williams (1988)], Chapter 5 in [Messenger and Ash (1992)], Chapter 1 of [Byrne (1994)] and [Angelescu et al. (1994)]). The spectral fluence largely decreases with increasing energy and, usually, it has a high-energy tail which extends up to energies above 10–15 MeV.

Neutrons are classified as a function of their kinetic energy. However, the classification has slightly varied with time. Commonly used terms, their subdivisions and kinetic energy ranges are shown in Table 4.2, following the classification i) for slow (≤ 1 eV), intermediate (1 eV–1 MeV) and fast (> 1 MeV) neutrons from Section 1.5.5.3 of [Byrne (1994)] and ii) above 10 MeV from Section 12.2.2 of [Marmier and Sheldon (1970)]. The energy of 1 eV¶ is chosen to define the energy below which a neutron is termed *slow*. By convention, a *thermal neutron* is taken to have an energy of 0.025 eV. This energy is the mean energy of a neutron at thermal equilibrium with its surrounding at 20 °C; it corresponds to a neutron velocity of $\simeq 2.2$ km/s and a neutron wavelength of $\simeq 1.8$ Å.

In silicon, displacement damage is mostly induced by fast, very fast, ultra fast

---

¶It is the typical energy of the lowest resonance in the reactor moderating material (e.g., see Section 1.5.5.3 of [Byrne (1994)]).

and relativistic neutrons (see Sects. 4.2.1, 4.2.1.2 and 4.2.1.5). In this book, the term *fast neutron* is commonly used for a neutron i) with kinetic energy lower than that corresponding to the ultra fast region and ii) capable to induce a relevant displacement damage (e.g., see Figs. 4.18 and 4.19), thus, typically with kinetic energies $\gtrsim 10\,\mathrm{keV}$ (e.g., see Sects. 4.1.3 and 4.2.1). Furthermore, it has to be noted that the damage induced by a 1 MeV neutron is taken as ASTM-standard reference [ASTM (1985)] (see Sect. 4.2.1).

4.2 Relevant Processes of Energy Deposition and Damage

Particles and photons passing through matter lose energy (or can be absorbed) by means of a variety of interactions and scattering processes, which result in two major effects: the collision energy-loss and atomic displacement. Interactions[§] of incoming particles which result in the excitation or emission of atomic electrons (i.e. ionizing the atom) are referred to as *energy-loss by ionization* or *energy-loss by collisions* (see Sect. 4.2.3). The non-ionization energy-loss (NIEL) processes are interactions in which the energy imparted by the incoming particle results in atomic displacements or in collisions where the knock-on atom does not move from its lattice location and the energy is dissipated in lattice vibrations (phonons), for instance.

The energy deposition by non-ionization processes is much lower (except for neutrons) than that by ionization. However, bulk damage mechanisms result mostly from atomic displacements (the so-called *displacement damages*).

Although the mechanisms of induced damage are somewhat similar in semiconductors and in semiconductor components, we restrict this treatment to what is relevant for silicon radiation-detectors and devices.

4.2.1 *NIEL and Displacement Damage*

Displacements occur when the primary interaction results in the displacement of the recoil atom from its lattice position. The knock-on atoms (PKAs) primarily create *Frenkel defects* (called also *Frenkel-pairs*; see, for instance, [James and Lark-Horovitz (1951); Kraner (1982)]). A Frenkel defect, which differs from dislocations and more complex imperfections, is a pair of *point defects* (i.e., highly localized imperfections arising in a crystal from a one-atom disorder) close enough to exhibit an interaction: a vacancy (V) and an atom in interstitial position (I).

For collisions hard enough to allow large energy transfers, the PKA can collide with other lattice atoms, creating more vacancies and interstitial atoms. At thermal equilibrium, the recoil atoms will be located in interstitial positions, unless some of them recombine with vacancies: some of these point defects are isolated, but for

---

[§]These interactions in matter are reviewed in the chapter on *Electromagnetic Interaction of Radiation in Matter* with an extensive treatment of collision energy-losses in silicon media (see, also, Chapter 2 in [Leroy and Rancoita (2004)]).

recoil energies much larger than the *displacement threshold energy*, E_d, cascading displacements will occur in a closely spaced group of defects (the so-called *cluster* or *cluster of defects*, e.g., see Sect. 4.2.2) within a small spatial region (or even a few small spatial regions). Figure 4.14 shows an example of simulated cascading-displacements by a silicon recoil in bulk silicon after the interaction with an incoming neutron of 50 keV kinetic energy (from [van Lint, Leadon and Colweel (1972)]; see also Chapter 2 of [van Lint, Flanahan, Leadon, Naber and Rogers (1980)]): a few energetic collisions produce other energetic recoils interspaced with many more low-energy transfers. Thus, cascading displacements result in (terminal) *clusters of defects*.

The thermal energy allows some defects to migrate through the crystal and, eventually, be annihilated by the recombination of the $V–I$ pairs or create stable defects in association with other impurities or defects already present (or induced by radiation). The presence of defects and clusters of defects produces changes in the properties of the semiconductor.

In radiation-induced defects, E_d is several times greater than the energy required for adiabatic displacements of atoms from lattice to interstitial positions. E_d depends on the recoil direction and, for silicon, is about (13–33) eV (e.g., see Chapter 1 in [Vavilov and Ukhin (1977)], [Chilingarov, Lipka, Meyer and Sloan (2000)] and references therein). In most of calculations, in particular for incoming neutrons (see Sect. 4.2.1.5), an isotropic value of 25 eV is assumed (e.g., see page 24 of [Dienes and Vineyard (1957)], Section 2.4.2 in [van Lint, Flanahan, Leadon, Naber and Rogers (1980)] and references therein, see also Section V of [Nichols and van Lint (1966)]).

There has been substantial progress in understanding the degradation generated in integrated bipolar transistors[§] and silicon detectors[¶] by fast neutrons, i.e., typically above 10 keV, which induce displacement damages. Neutrons with sufficient large energies, like for instance fast neutrons, can transfer enough kinetic energy to the recoil atom to generate clusters of displacements with a cascading effect.

Most predictions of the neutron energy dependence in semiconductor (i.e. silicon) devices have been based on the amount of non-ionizing energy deposited. In literature (see for instance [Namenson, Wolicki and Messenger (1982); Ougouag, Williams, Danjaji, Yang and Meason (1990); Vasilescu (1997)]), the damage effect due to initial and cascading-displacements induced by neutrons (Sect. 4.2.1.2) is

---

[§]The reader can see, e.g., [Messenger (1966, 1972, 1992); Colder et al. (2001); Claeys and Simoen (2002); Holmes-Siedle and Adams (2002); Codegoni et al. (2004b); Consolandi, D'Angelo, Fallica, Mangoni, Modica, Pensotti and Rancoita (2006)] and references therein.

[¶]One can see, e.g., [Borgeaud, McEwen, Rancoita and Seidman (1983); SICAPO Collab. (1986, 1994b,c, 1995c); Croitoru, Dahan, Rancoita, Rattaggi, Rossi and Seidman (1997); Croitoru, Rancoita, Rattaggi, Rossi and Seidman (1997); Dezillie, Bates, Glaser, Lemeilleur and Leroy (1997); Borchi, Bruzzi, Leroy, Pirollo and Sciortino (1998); Croitoru, David, Rancoita, Rattaggi and Seidman (1998a); Croitoru, Gubbini, Rancoita, Rattaggi and Seidman (1999a); Leroy, Roy, Casse, Glaser, Grigoriev and Lemeilleur (1999a); Mangiagalli, Levalois, Marie, Rancoita and Rattaggi (1999); Golan et al. (2001); Rose Collab. (2001)] and references therein.

expressed by the *damage function* (also referred to as *displacement kerma function*) $D(E)$ in units of MeV cm^2, whose value at 1 MeV is the ASTM standard $D(1\,\text{MeV}) = 95\,\text{MeV}\,\text{mb}$ [ASTM (1985)]. The damage function**, which accounts for both the cross section for neutron–silicon scattering and the energy released in creating displacements, is given by [Codegoni et al. (2004b); Consolandi, D'Angelo, Fallica, Mangoni, Modica, Pensotti and Rancoita (2006)]:

$$D(E) = \sum_k \sigma_k(E) \int f_k(E, E_R) P_k(E_R)\, dE_R, \qquad (4.80)$$

where E is the incoming neutron kinetic-energy, $\sigma_k(E)$ is the cross section for the k-th reaction, $f_k(E, E_R)\, dE_R$ is the probability that a recoil atom is generated with kinetic energy between E_R and $E_R + dE_R$, and, finally, $P_k(E_R)$ is the *partition energy* for the recoil nucleus. This latter is the part of the recoil energy deposited in displacements, calculated in the framework of the Lindhard screened-potential scattering theory [Lindhard, Nielsen, Scharff and Thomsen (1963)], based on the Thomas–Fermi model and further developments [Parkin and Coulter (1977, 1979); Coulter and Parkin (1980)], or, for instance, by means of *Robinson's analytical approximation* [Robinson (1970)] (e.g., for elastic neutron–silicon interactions see Eq. (4.90) for silicon recoil in a silicon medium). The damage function for neutron is reviewed in Sect. 4.2.1.5.

From Equation (4.80), the energy density, E_{dis}, deposited through atomic displacements by neutrons characterized by a neutron spectral-fluence $\phi(E)$ in n cm^{-2} MeV^{-1} (Sect. 4.1.3), is the *damage energy*, E_{de}, (i.e., the *defect producing energy*) imparted per cm^3 and is given by:

$$E_{\text{dis}} = n_{\text{Si}} \int_{E_{\min}} D(E)\, \phi(E)\, dE \quad [\text{MeV/cm}^3], \qquad (4.81)$$

where [see page 14 and Eq. (1.39)]

$$n_{\text{Si}} = \frac{\rho_{\text{Si}}\, N}{A_{\text{Si}}} \qquad (4.82)$$

is the number of atoms per cm^3 in the bulk silicon, ρ_{Si} and A_{Si} are the density and *atomic weight* of the silicon medium, respectively; N is the Avogadro constant (see Appendix A.2) and $E_{\min}$ is the minimum neutron energy for inducing displacement damage. Since the damage function for soft and thermal neutron is much lower than that for fast neutrons (Sect. 4.2.1.5), for spectral fluences of reactor neutrons the integral in Eq. (4.81), thus the value of E_{dis}, marginally depends on $E_{\min}$ up to about 10 keV. For example, in the case of the neutron spectrum of the Triga reactor RC:1 at Casaccia (Rome) the value of E_{dis} varies by no more than 0.5% for $E_{\min} \leq 10\,\text{keV}$ [Codegoni et al. (2004b); Consolandi, D'Angelo, Fallica, Mangoni, Modica, Pensotti and Rancoita (2006)].

**See, for instance, [Ougouag, Williams, Danjaji, Yang and Meason (1990)] and Chapter 5 in [Messenger and Ash (1992)].

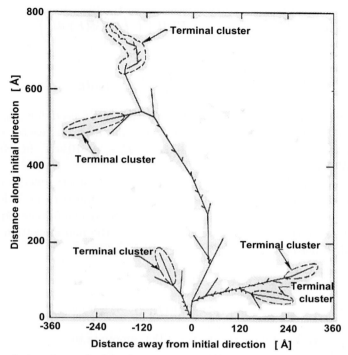

Fig. 4.14 Simulation of cascading-displacements generated by a silicon recoil in bulk silicon after the interaction with an incoming neutron of 50 keV kinetic energy (adapted and reprinted with permission from van Lint, V.A.J., Leadon, R.E., and Colwell, J.F., Energy Dependence of Displacement Effects in Semiconductors, IEEE Transactions on Nuclear Science, Volume 19, Issue 6, Dec. 1972, pages 181–185, doi: 10.1109/TNS.1972.4326830, © 1972 IEEE; see also Chapter 2 of [van Lint, Flanahan, Leadon, Naber and Rogers (1980)]). The distances are in Å.

The concentration of Frenkel-pairs (*FP*), i.e., the number of Frenkel defects per cm^3, can be computed by means of the *modified Kinchin–Pease formula* [Eq. (4.89), see also discussion in Sect. 4.2.1.1] and Eq. (4.81) as:

$$FP \sim \frac{E_{\mathrm{dis}}}{2.5 \, E_d}. \tag{4.83}$$

For the case of the Triga neutron spectrum Eq. (4.83) yields a Frenkel-pairs concentration of about 43.3 per incoming neutron per cm^2 with energy above 10 keV. This value also accounts for the low energy part of the neutron spectrum, i.e., below 10 keV, which contributes by no more than 0.5% to the *FP* overall value.

From Eqs. (4.79, 4.81), we can derive the hardness parameter κ as:

$$n_{\mathrm{Si}} \, D(1\,\mathrm{MeV}) \, \Phi_{\mathrm{eq}}^{1\,\mathrm{MeV}} = \kappa \, n_{\mathrm{Si}} \, D(1\,\mathrm{MeV}) \int_{E_{\mathrm{min}}} \phi(E) \, dE$$

$$= n_{\mathrm{Si}} \int_{E_{\mathrm{min}}} D(E) \, \phi(E) \, dE,$$

from which we get

$$\kappa = \left[\int_{E_{\min}} \frac{D(E)}{D(1\,\text{MeV})}\, \phi(E)\, dE \right] \left(\int_{E_{\min}} \phi(E)\, dE \right)^{-1} \tag{4.84}$$

and, conversely,

$$\Phi_{\text{eq}}^{1\,\text{MeV}} = \int_{E_{\min}} \frac{D(E)}{D(1\,\text{MeV})}\, \phi(E)\, dE. \tag{4.85}$$

As already mentioned, in some calculations the value used for $E_{\min}$ is about $10\,\text{keV}$. Values of the hardness parameter for some neutron reactor spectra can be found in [Angelescu et al. (1994)] and, for $E_{\min} = 10\,\text{keV}$, in [Kelly, Luera, Posey and Williams (1988)].

By replacing the spectral fluence with the neutron kinetic-energy distribution $F(E)$ obtained by integrating the spectral fluence (assumed to be independent of the position) over the exposed area $F(E) = \int \phi(E)\, dS$, we can rewrite Eq. (4.81) and obtain the damage energy lost per unit length, i.e., the non-ionization energy-loss in units of MeV per cm:

$$\frac{dE_{\text{de}}}{dx} = \int E_{\text{dis}}\, dS$$

$$= n_{\text{Si}} \int D(E)\, F(E)\, dE \quad [\text{MeV/cm}]. \tag{4.86}$$

For a neutron with kinetic energy E, the NIEL amounts to

$$\frac{dE_{\text{de}}^{\,\text{n}}}{dx} = n_{\text{Si}}\, D(E). \tag{4.87}$$

Equations (4.86, 4.87) can be used to express the relationship between the damage function and the non-ionizing energy-loss per unit path for any kind of beam particles with a kinetic-energy distribution $F(E)$ or at fixed kinetic-energy E. Equation (4.87) can be rewritten in terms of the differential cross section $d\sigma_k(E)/dE_R$ for any incoming particle with kinetic energy E resulting in a nuclear recoil with kinetic energy E_R:

$$\frac{dE_{\text{de}}}{dx} = n_{\text{Si}} \sum_k \int_{E_d}^{E_R^{\max}} \frac{d\sigma_k(E)}{dE_R}\, P_k(E_R)\, dE_R, \tag{4.88}$$

where $E_R^{\max}$ is the maximum energy transferred to the recoil nucleus and the summation is extended over the possible k reaction channels which allow the generation of a recoil silicon (for example in the elastic scattering) or a residual nucleus.

4.2.1.1 *Knock-on Atoms and Displacement Cascade*

When a semiconductor (as any solid material) is subjected to incident radiation (i.e. neutrons, electrons, protons, ions...), the resulting degree of damage depends on a number of factors. For instance, the type of interaction (strong or electromagnetic), the nature and energy of the incoming particle determine the *initial-damage*, i.e., the number of atoms directly displaced (PKAs) by the incoming particles and their energy spectrum. These PKAs, in turn, can generate a cascade of displacements, i.e., the *cascading damage*. That is, the primary displaced atoms, upon leaving their lattice positions with large enough energies, can interact and displace other atoms. In this way, a displacement cascade is generated upon full dissipation of the primary recoil-energy. In the cascade, the energy dissipation is a combination of elastic interactions (referred to as *nuclear* or *elastic energy-loss*[‡‡]) mostly resulting in displacements and inelastic processes (referred to as *electronic* or *inelastic energy-loss*), in which moving displaced-atoms excite or ionize atomic electrons on their passage[§]. The point defects (see Sect. 4.2.2), created in the silicon lattice at this initial stage, are often referred to as *primary*. *Secondary defects* are created during the diffusion of these primary defects.

The primary recoil energy is deposited by the ionization energy, $E_{\rm loss}$, (including that deposited by recoiling atoms in the cascade when it occurs) and damage energy (also referred to as *partition* or *defect producing energy*), $E_{\rm de}$, accounting for displacements and sub-threshold collisions, which transfer energies lower than E_d. In these latter interactions, the knock-on atom cannot escape from its lattice location and the energy is dissipated in lattice vibrations, as mentioned in Sect. 4.2. The non-ionizing energy-loss (NIEL) is the damage-energy deposited by an incoming particle per unit length (e.g., in units of MeV/cm in this section, if not specified otherwise). As discussed in Sect. 4.2.1, permanent radiation-damage is induced by atoms displaced from their lattice locations and, as a consequence, is related to the amount of damage energy.

As already discussed, the energy and differential cross section of the incoming particle determine the energy spectrum of the primary recoil. However, the secondary interactions can be treated in the framework of atom–atom interactions[*] which, for recoil atoms at low energy, are described by largely screened Coulomb potentials. Kinchin and Pease (1955) estimated the average number of displaced atoms generated by a PKA with recoiling energy above E_d, assuming (i) two-body hard-sphere atomic collisions, (ii) a sharp displacement threshold E_d and (iii) no inelastic energy-losses (i.e. no ionization energy-loss in secondary collisions). Torrens and Robinson (see [Torrens and Robinson (1972); Robinson and Torrens (1974);

[‡‡]For a further discussion see, for instance, Sects. 2.1.4 and 2.1.4.1; see, also, Chapter 2 of [Ziegler, Biersack and Littmark (1985a); Ziegler, J.F. and M.D. and Biersack (2008a)].

[§]For incoming high-energy particles, this type of energy-loss process is commonly referred to as *collision energy-loss process* (e.g., see Sects. 2.1, 2.1.1, 2.1.4 and discussion in the chapter on *Electromagnetic Interaction of Radiation in Matter*; in addition see Chapter 3 of [Ziegler, Biersack and Littmark (1985a); Ziegler, J.F. and M.D. and Biersack (2008a)]).

[*]The reader can see, for instance, Sections 2.2–2.4 of [Dienes and Vineyard (1957)], Section 2.2.8 of [van Lint, Flanahan, Leadon, Naber and Rogers (1980)], [Ziegler, Biersack and Littmark (1985a); Ziegler, J.F. and M.D. and Biersack (2008a)] and references therein.

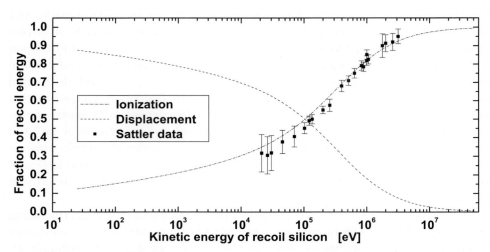

Fig. 4.15 Fractions of kinetic energy of recoil silicon deposited in processes producing defects (dashed line) and ionization-losses (dashed and dotted line) [calculated by means of Eqs. (4.94, 4.95)] (from [Leroy and Rancoita (2007)]). The experimental data (■) are from [Sattler (1965)].

Norgett, Robinson and Torrens (1975)] and references therein) have estimated[¶] the average number of displacements $< N_{FP} >$, i.e., of Frenkel-pairs, by means of extensive computer simulations of collision cascades accounting for inelastic losses and the damage energy left available. The number of Frenkel-pairs generated in a cascade is given by the *modified Kinchin–Pease formula*

$$< N_{FP} > = \begin{cases} 1 \text{ for } E_d < E_{\text{de}} \leq 2\,E_d/\zeta, \\[2mm] \zeta E_{\text{de}}/(2E_d) \text{ for } E_{\text{de}} > 2\,E_d/\zeta, \end{cases} \qquad (4.89)$$

where $\zeta \approx 0.8$ is a constant independent of energy (except for recoil energy close to $2E_d$ [Norgett, Robinson and Torrens (1975)]). The factor ζ is called *displacement efficiency* (e.g., see discussion at pages 425 and 431–435 in [Torrens and Robinson (1972)]). For silicon, as discussed at page 347, the threshold energy to create a displacement is about $25\,\text{eV}$. In Eq. (4.89), the factor 2 was determined in the framework of the above mentioned Kinchin–Pease analytical treatment, in which a sharp displacement threshold energy, E_d, is assumed. When a more realistic interatomic potential is used instead of the hard-sphere model, low-energy transfers are favored, thus resulting in a smaller number of displaced atoms per cascade. The displacement efficiency is the parameter accounting for such a decrease and for the effects of thermal vibrations in the lattice.

The energy depositions E_{loss} and E_{de} can be computed by means of *analytical*

[¶]For Cu, a comparison between $< N_{FP} >$ computed using the modified Kinchin–Pease formula and that created by electrons and ions can be found in [Merkle, King, Baily, Haga and Meshii (1983)].

approximation of Robinson [Robinson (1970)] used, for instance, in the SRIM[‡‡] code (see pages 115–166 in [Ziegler, Biersack and Littmark (1985a)] and also [Ziegler, J.F. and M.D. and Biersack (2008b)]). In a silicon medium, for a recoil silicon nucleus with kinetic energy E_{Si} in eV and atomic number $Z_{Si} = 14$ we have

$$E_{de} = \frac{E_{Si}}{1 + k_d \, g(\epsilon_d)}, \tag{4.90}$$

where:

$$\epsilon_d = 1.014 \times 10^{-2} \, Z_{Si}^{-7/3} E_{Si}$$
$$= 2.147 \times 10^{-5} E_{Si}, \tag{4.91}$$
$$g(\epsilon_d) = \epsilon_d + 0.40244 \, \epsilon_d^{3/4} + 3.4008 \, \epsilon_d^{1/6}, \tag{4.92}$$
$$k_d = 0.1462. \tag{4.93}$$

In the latter expressions ϵ_d, $g(\epsilon_d)$, k_d are dimensionless. Using Eq. (4.90), the fractions of energy deposited by non-ionization (f_{niel}) and by ionization (f_{ion}) can be computed from:

$$f_{niel} = \frac{E_{de}}{E_{Si}}$$
$$= \frac{1}{1 + k_d \, g(\epsilon_d)}, \tag{4.94}$$
$$f_{ion} = \frac{E_{loss}}{E_{Si}}$$
$$= 1 - \frac{E_{de}}{E_{Si}}$$
$$= \frac{k_d \, g(\epsilon_d)}{1 + k_d \, g(\epsilon_d)}. \tag{4.95}$$

In Fig. 4.15, f_{ion} and f_{niel} are shown and compared with the experimental data of Sattler (1965) for neutrons energies between 20 keV and 3.2 MeV (e.g., see Section 2.4.1 in [van Lint, Flanahan, Leadon, Naber and Rogers (1980)] and, also, [Smith, Binder, Compton and Wilbur (1966)]). At these energies, the neutron–silicon total cross section is almost determined by the elastic cross section [Smith, Binder, Compton and Wilbur (1966)], i.e., the interactions result in a recoil silicon. In general, the agreement between f_{ion}, computed by means of Eq. (4.95), and the experimental data is quite satisfactory. As one can see in Fig. 4.15, for recoiling energy up to about 10 keV the fraction f_{niel} is larger than about 70 %, i.e., most of the energy is deposited by NIEL. Above about 100 keV, f_{niel} is lower than 50 %.

Furthermore, since recoil atoms collide with equal mass atoms in the lattice, a large fraction of the incoming recoil energy can be transferred in a collision,

---

[‡‡]Nowadays, an algorithm has been developed for the GEANT4 Monte Carlo package for the computation of screened Coulomb interatomic scattering [Mendenhall and Weller (2005)] for non-relativistic ion–ion interactions. The algorithm allows one to use an arbitrary screening function, for instance the well-established ZBL screening function (see page 48 of [Ziegler, Biersack and Littmark (1985a)]).

but soft collisions can also occur depending on the recoil energy and the effect of interpenetration of orbital electron clouds. Monte-Carlo simulations of displacement cascade generated by a 50 keV recoil silicon (e.g., see Fig. 4.14) show that there are a few collisions with large energy transfers interspersed with many low-energy transfers. These latter are grouped in terminal clusters of displacements.

As mentioned before, Eq. (4.90) expresses the *partition energy* (Sect. 4.2.1) for reactions resulting in a recoil silicon as PKA. In general {see Equations (5–9) in [Norgett, Robinson and Torrens (1975)]}, the partition energy for a recoil nucleus of atomic number $Z_{\rm rec}$ and atomic rest-mass $m_{\rm rec}$ (in atomic mass units) in a medium with atomic number $Z_{\rm m}$ and atomic rest-mass $m_{\rm m}$ can be evaluated by means of the *Norgett–Robinson–Torrens expression*:

$$E_{\rm de} = \frac{E_{\rm rec}}{1 + k'_d\, g(\epsilon'_d)},\tag{4.96}$$

wherein $E_{\rm rec}$ is the kinetic energy of the recoil nucleus and

$$\epsilon'_d = \frac{{\rm a_L}}{Z_{\rm rec} Z_{\rm m}\, e^2}\left(\frac{m_{\rm m}}{m_{\rm rec} + m_{\rm m}}\right) E_{\rm rec},\tag{4.97}$$

$$\rm a_L = \frac{0.88534\, a_0}{\sqrt{Z_{\rm rec}^{2/3} + Z_{\rm m}^{2/3}}} \quad [\text{see Eq. (2.72)}],\tag{4.98}$$

$$g(\epsilon'_d) = \epsilon'_d + 0.40244\, \epsilon'^{3/4}_d + 3.4008\, \epsilon'^{1/6}_d,\tag{4.99}$$

$$k'_d = 0.1337\, Z_{\rm rec}^{2/3}\, m_{\rm rec}^{-1/2},\tag{4.100}$$

where ϵ'_d is the reduced energy (see, for instance, page 80) accounting for the interaction between the atom of the medium and the recoil atomic particle and is calculated using the *Lindhard and Sharff screening length* $\rm a_L$, $E_{\rm rec}$ is the kinetic energy of the recoil nucleus, e is the electronic charge in esu, $\rm a_0 = 0.529177208 \times 10^{-8}$cm is the *Bohr radius* (see Appendix A.2); ϵ'_d, $g(\epsilon'_d)$ and k'_d are dimensionless. Equation (4.97) can be rewritten as:

$$\epsilon'_d = \frac{32.536 \times 10^{-3}}{Z_{\rm rec} Z_{\rm m}\sqrt{Z_{\rm rec}^{2/3} + Z_{\rm m}^{2/3}}}\left(\frac{m_{\rm m}}{m_{\rm rec} + m_{\rm m}}\right) E_{\rm rec}\tag{4.101}$$

with $E_{\rm rec}$ in eV. The damage energy calculated by means of Eq. (4.96) in case of a recoil silicon as PKA differs from that of Eq. (4.90) by less than 4 % for recoil silicon with energies lower than 150 keV (i.e., when more than 45 % of $E_{\rm rec} \equiv E_{\rm Si}$ is deposited as damage energy). The number of Frenkel-pairs generated in a cascade is given by the modified Kinchin–Pease formula, thus, can be calculated by means of Eqs. (4.89, 4.96). Equations (4.96–4.100) were derived according to the method of Lindhard, Nielsen, Scharff and Thomsen (1963) to evaluate the inelastic energy-loss. As discussed by Norgett, Robinson and Torrens (1975), this latter applies to interactions in which $Z_{\rm rec}/Z_{\rm m}$ does not differ too much from unity and for energies lower than $\approx 25\, Z_{\rm rec}^{4/3}\, m_{\rm rec}$ keV.

4.2.1.2 Neutron Interactions

In neutron interactions, the heavy recoil nuclei are of primary importance in determining the amount of damage energy and permanent damages in silicon (as in any other semiconductor). For fast neutrons in the energy range between 50 keV and 14 MeV, relevant reactions are the (n,n) or elastic scattering, (n,n') or inelastic neutron scattering, (n,p) and (n,α) (for a discussion see [Smith, Binder, Compton and Wilbur (1966)]; see, also, Sect. 4.1.3.1).

Since in a nuclear reactor spectrum the high-energy part strongly decreases above (3–5) MeV, the elastic scattering [about (1–10) b] is the most likely process to produce displacements (as already mentioned). The kinetic energy of the recoil silicon depends on the angular differential cross section. For an isotropic angular distribution in the center-of-mass system, all recoil energies in the lattice (i.e. the laboratory-system) from 0 to the maximum transferred energy T^n_{max} are equally probable and the mean energy transferred is $0.5\,T^n_{max}$. At the energy of 0.762 MeV, the angular distribution is almost isotropic, but neutrons with kinetic energies of several MeV tend to scatter in a forward direction (e.g., see Figure 2.21 in [van Lint, Flanahan, Leadon, Naber and Rogers (1980)]) and, therefore, the mean transferred energy is much lower than $0.5\,T^n_{max}$.

In a two body elastic scattering the maximum transferred energy is [see Eq. (1.28) and discussion in Sect. 1.3.1]

$$T_{max} = 2M_A\,c^2(\gamma^2 - 1)\left[1 + \left(\frac{M_A}{m}\right)^2 + 2\gamma\frac{M_A}{m}\right]^{-1}, \qquad (4.102)$$

where M_A is the mass of the target (recoil) atom assumed to be at rest, m is the rest mass of the incoming particle and γ is the *Lorentz factor* of the incoming particle with kinetic energy $E_k = (\gamma - 1)mc^2$. For $\gamma \approx 1$, Eq. (4.102) becomes:

$$T_{max} \approx 4M_A\,c^2(\gamma - 1)\left(1 + \frac{M_A}{m}\right)^{-2}$$

$$= \frac{4\,mM_A}{(m + M_A)^2}\,E_k. \qquad (4.103)$$

In a neutron–silicon elastic scattering, since for fast neutrons $\gamma \approx 1$, from Eq. (4.103) T^n_{max} is given by

$$T^n_{max} = \frac{4\,m_n M_{Si}}{(m_n + M_{Si})^2}\,E^n_k$$

$$\approx 0.133\,E^n_k, \qquad (4.104)$$

where E^n_k is the kinetic energy of the fast neutron, M_{Si} and m_n are the silicon and neutron rest-masses, respectively. For instance as discussed above, on average a neutron with kinetic energy close to 0.762 MeV imparts $\approx 0.5 \times 0.133\,E^n_k = 0.051$ MeV to the recoil silicon, thus, as shown in Fig. 4.15, about 58 % of its recoil energy is dissipated as damage energy. In an elastic scattering of a neutron with a silicon nucleus at rest, the minimum kinetic energy of the neutron to impart 25 eV is about 190 eV [Eq. (4.104)].

4.2.1.3 *Interactions of Protons, α-particles and Heavy-Isotopes*

When protons traverse a material, they can interact through atomic Coulomb scattering (being charged particles) and strong interaction. Furthermore, since the proton rest-mass is $m_p \approx m_n$, from Eq. (4.104) the maximum transferred energy in elastic interactions becomes

$$T_{\max}^p \approx 0.133\, E_k^p, \tag{4.105}$$

where E_k^p is the kinetic energy of the incoming proton. As for neutrons, the minimum kinetic energy of the protons to impart an energy of 25 eV to a recoil silicon at rest is about 190 eV [Eq. (4.105)].

In strong p–silicon interactions, the elastic scattering is a relevant fraction of the total cross section also at high-energies (see for instance [Srour and McGarrity (1988)]), while the non-elastic cross section becomes relevant for protons with kinetic energies above about (5–10) MeV. As shown in [Jun et al. (2003)], at about 10 MeV, the nuclear interactions account for about 15% of the deposited damage-energy, while above 20 MeV the strong (elastic and non-elastic) interactions become the dominant mechanisms (i.e. at 20 MeV the strong interactions account for about 50% of the deposited damage-energy). Below 10 MeV, the atomic Coulomb scattering is the dominant mechanism for producing displacements.

Charged atomic particles passing in a medium undergo energy-loss processes by Coulomb scattering (e.g., see [Northcliffe (1963)], chapter on *Electromagnetic Interaction of Radiation in Matter* and references therein). It is common practice to distinguish between *light* and *heavy ions*. However, it must be considered that this distinction is arbitrary so far the fundamental mechanism of the Coulomb scattering for the energy-loss process is considered. Nonetheless, there is a real practical difference. Light ions (i.e., protons and α-particles) can be regarded as fully ionized and charge invariant over most of the energy region involved in the nuclear and particle physics (e.g., $\gtrsim 1$ MeV see Sect. 2.1.4, [Fano (1963)] and also references therein). Heavier atoms can be fully ionized (by the stripping of their electrons) and proceed as bare nuclei only for velocities largely exceeding the orbital velocities of their electrons. At high energies, light and heavy ions lose energy predominantly by electronic (i.e., collision) energy-loss processes (see [Northcliffe (1963)] and Sect. 2.1.4).

At low energy, the dominant mechanism of energy-loss is by nuclear energy-loss, i.e., the energy of the incoming atomic particle is released in matter by Coulomb interactions occurring with the nuclei of the medium. These interactions are described (as the interactions of the PKA inside the semiconductor medium) by largely screened Coulomb potentials (e.g., see [Ziegler, Biersack and Littmark (1985a); Ziegler, J.F. and M.D. and Biersack (2008a)] and references therein, see also Sect. 2.1.4.1). In the *Ziegler–Biersack–Littmark treatment* (Chapter 2 in [Ziegler, Biersack and Littmark (1985a); Ziegler, J.F. and M.D. and Biersack (2008a)]) it is shown that the nuclear stopping power is properly accounted by the Rutherford scattering with unscreened Coulomb potential for universal reduced energies, $\epsilon_{r,U}$, above 30; whereas, the universal reduced energy is given by [Eq. (2.76)]

$$\epsilon_{r,U} = \frac{a_U}{Z_i Z_t e^2} \left(\frac{m_t}{m_i + m_t} \right) E_k^{tot}, \tag{4.106}$$

in which E_k^{tot} is the kinetic energy of the incoming atomic particle i with rest mass m_i (in atomic mass units) and atomic number Z_i, the stationary PKA has rest mass m_t (in atomic mass units) and atomic number Z_t, e is the electronic charge in esu, the *universal screening length* a_U is

$$a_U = \frac{0.88534\, a_0}{Z_i^{0.23} + Z_t^{0.23}} \tag{4.107}$$

[Eq. (2.74)] and a_0 is the Bohr radius. Equation (4.106) can be rewritten as

$$\epsilon_{r,U} = \frac{32.536 \times 10^{-3}}{Z_i Z_t \left(Z_i^{0.23} + Z_t^{0.23} \right)} \left(\frac{m_t}{m_i + m_t} \right) E_k^{tot}, \tag{4.108}$$

where E_k^{tot} is in units of eV. In a silicon medium, the condition $\epsilon_{r,U} > 30$ is satisfied above[¶] $E_k^{Rut} \simeq$ (75–80) keV/nucleon for nickel isotopes and at lower energies for isotopes[**] with $Z < 28$. To a first approximation, since in the Rutherford scattering the differential cross section is inversely proportional to the square of the transferred energy T [see Eq. (1.56) and discussion in Sect. 1.5], the mean transferred energy $\langle T \rangle$, i.e., the mean kinetic energy of the recoil silicon $\langle E_{Si} \rangle$, above the displacement threshold energy E_d is

$$\langle E_{Si} \rangle \equiv \langle T \rangle$$
$$= \frac{E_d\, T_{max}}{T_{max} - E_d} \ln \left(\frac{T_{max}}{E_d} \right) \quad \text{for} \quad \frac{E_k^{tot}}{I_{isot}} > E_k^{Rut}, \tag{4.109}$$

where I_{isot} is the number of nucleons of the isotope and the maximum transferred energy T_{max} is given for $\gamma \approx 1$ by Eq. (4.103). An approximate expression[*] for the mean number of displaced atoms, induced by the recoil, (i.e. for $\approx \langle N_{FP} \rangle$) is

$$\langle N_{FP} \rangle \approx \frac{T_{max}}{2\,(T_{max} - E_d)} \left[1 + \ln \left(\frac{T_{max}}{2\,E_d} \right) \right] \quad \text{for} \quad \frac{E_k^{tot}}{I_{isot}} > E_k^{Rut} \tag{4.110}$$

(see Equations (29, 30, 35) of [Vavilov and Ukhin (1977)] and Equation (2-31) of [Dienes and Vineyard (1957)]). Furthermore, for $\gamma \approx 1$ and $T_{max} \gg E_d$, we can rewrite Eq. 4.109 as

$$\langle E_{Si} \rangle \approx E_d \ln \left[\frac{4\, m_i\, m_{Si}\, E_k^{tot}}{(m_i + m_{Si})^2\, E_d} \right]. \tag{4.111}$$

Equation (4.111) shows that (for $\gamma \approx 1$) the mean kinetic energy of the recoil silicon increases with the logarithm of the kinetic energy of the incoming isotope. The initial

---

[¶]The reader can find the definition of kinetic energies per nucleon in Sect. 1.4.1.

[**]These isotopes (including nickel) are the most abundant in GCRs; for isotopic abundances, the reader can see Sect. 4.1.2.4 and Fig. 4.11.

[*]This expression was derived assuming $\zeta \simeq 2$ (in the framework of the Kinchin–Pease model) for Eq. (4.89) and $T_{max} > 2E_d$ (see Section 2.4 of [Dienes and Vineyard (1957)]).

displacement is followed by a cascading-process, in which additional displacements are induced by the recoil silicon. From Eqs. (4.89, 4.96, 4.109) and since the fraction of the damage energy is larger than 50% for recoil silicon with energies lower than ≈ 100 keV [e.g., see Eq. (4.90) and Fig. 4.15], we do not expect more than 4–8 Frenkel-pairs created per PKA[§] over most of the range of energies wherein the Coulomb interaction is the dominant scattering mechanism. This energy range is of interest for semiconductor applications. Thus, the developed cascade of Frenkel-pair is typically smaller compared to that induced by strong interacting particles [e.g., see the above discussion about fast neutron and protons with energies larger than (10–20) MeV]. Nonetheless, the dominant mechanism for generating atomic displacements is via silicon-to-silicon cascading-displacement.

As previously said, for protons with kinetic energy larger than about 20 MeV [Jun et al. (2003)], the NIEL deposition results mainly from the elastic and non-elastic strong interactions on silicon. Since isotopes undergo (short-range) strong- and (long-range) electromagnetic-interactions passing through matter (see for instance Section 17.8 in [Eisberg and Resnick (1985)]), let us estimate the energies above which about 50% of the NIEL deposition is expected to be determined by strong isotope–silicon interactions. To a first approximation, the total cross section for strong interactions between two isotopes i and t is proportional to $(R_i + R_t)^2$ (e.g., see Section 11.11 in [Segre (1977)] and Sect. 3.2.1), where R_i and R_t are the nuclear radii of the isotopes in units of r_0 [Eq. (3.12)] computed as:

$$R_{i,t} \simeq M_{i,t}^{1/3} \qquad (4.112)$$

in which M_i and M_t are the nucleon numbers of the isotopes. The total cross section for electromagnetic interactions is proportional to $Z_i^2 Z_t^2$, where Z_i and Z_t are the atomic numbers of the two isotopes. However, because at relativistic energies the region of space at the maximum electric field is relativistically contracted (e.g., see page 37), the collision time[**] is proportional to $1/(v_i \gamma_i)$, where v_i and γ_i are the velocity and Lorentz factor of the incoming isotope, respectively, while the target isotope is supposed to be at rest. Thus, the cross section ratio $C_{C,st}$ of the Coulomb to strong interaction behaves as

$$C_{C,st} \propto \frac{S_{i,t}}{\beta_i \gamma_i}$$

or, equivalently,

$$C_{C,st} \propto \frac{S_{i,t}}{\sqrt{\gamma_i^2 - 1}}, \qquad (4.113)$$

where $\beta_i = v_i/c$, c is the speed of light and

$$S_{i,t} = \left(\frac{Z_i Z_t}{R_i + R_t} \right)^2 .$$

[§]This range of values for $\langle N_{FP} \rangle$ is in agreement with that computed using Eq. (4.110).
[**]This treatment follows that applied in Sect. 2.1.1 to the scattering of a charged particle with atomic electrons resulting in electronic energy-losses.

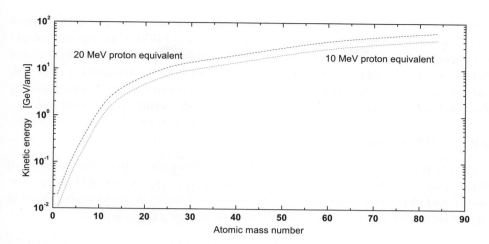

Fig. 4.16 Isotope–silicon interactions: kinetic energies (indicated as "20 MeV proton equivalent") of isotopes in units of GeV/amu corresponding to the values of the Lorentz factor $\gamma_{i,st}$ [Eq. (4.114)] versus the atomic mass-number [Eq. (3.1)] up to ^{84}Kr. The data indicated as "10 MeV proton equivalent" correspond to the kinetic energies of isotopes calculated replacing the Lorentz factor with that corresponding to 10 MeV proton in Eq. (4.114). The curves are the spline fit to the calculated kinetic energies for p, ^{4}He, ^{7}Li, ^{9}Be, ^{11}B, ^{12}C, ^{16}O, ^{20}Ne, ^{24}Mg, ^{28}Si, ^{40}Ar, ^{56}Fe, ^{69}Ga and ^{84}Kr on silicon.

$C_{\mathrm{C,st}}$ decreases as the energy of the incoming isotope increases, i.e., the strong interaction cross section becomes dominant at high-energies. By means of Eq. (4.113), for an isotope–silicon interaction, we can estimate the value of the Lorentz factor $\gamma_{i,st}$ for which the dominant mechanism of the NIEL deposition results from strong interactions, as it occurs for protons at 20 MeV [Jun et al. (2003)], i.e,

$$\frac{S_{\mathrm{i,Si}}}{\sqrt{\gamma_{i,st}^2 - 1}} \approx \frac{S_{\mathrm{H,Si}}}{\sqrt{\gamma_{p,20\mathrm{MeV}}^2 - 1}},$$

where $\gamma_{p,20\mathrm{MeV}}$ is the Lorentz factor of a proton with 20 MeV of kinetic energy. From this latter expression, we obtain

$$\gamma_{i,st} = \sqrt{\left[\gamma_{p,20\mathrm{MeV}}^2 - 1\right]\left(\frac{S_{\mathrm{i,Si}}}{S_{\mathrm{H,Si}}}\right)^2 + 1}. \qquad (4.114)$$

In Fig. 4.16, the kinetic energies in units of GeV/amu[‡], corresponding to the values of the Lorentz factor $\gamma_{i,st}$ of the isotopes [Eq.(4.114)], are shown as function of the mass number [Eq. (3.1)] for isotope–silicon interactions. For instance, the dominant mechanism of the NIEL deposition is expected to result from strong interactions i) for α-particles, ^{7}Li and ^{9}Be isotopes with kinetic energies $\gtrsim 0.12, 0.42$

[‡]The values of the masses in units of amu used in the calculation are those from [Ziegler, J.F. and M.D. and Biersack (2008b)]. The reader can find a definition of kinetic energies per amu in Sect. 1.4.1.

and 0.94 GeV/amu, respectively, and ii) for ^{11}B and heavier isotopes with energies $\gtrsim 1$ GeV/amu. In the same figure, the data indicated as "10 MeV proton equivalent" correspond to the kinetic energies of isotopes calculated replacing the Lorentz factor with that corresponding to 10 MeV proton in Eq. (4.114). As discussed at page 356, up to these energies the NIEL deposition mechanism is expected to be dominated by Coulomb interactions with a non-negligible contribution from strong interactions.

4.2.1.4 *Electron Interactions*

Electrons have a mass much lower than the recoil silicon atom. In an electron–silicon elastic scattering for which $2\gamma\,(m_{\mathrm{Si}}/m_{\mathrm{e}}) \ll (m_{\mathrm{Si}}/m_{\mathrm{e}})^2$, where m_{Si} and m_{e} are the silicon and electron rest masses and γ is the Lorentz factor of the incoming electron of kinetic energy E_k^{e}, the maximum transferred energy[†], $T_{\max}^{\mathrm{e}}$, is

$$T_{\max}^{\mathrm{e}} \simeq 20.02\,\tau(\tau+2) \quad [\mathrm{eV}], \tag{4.115}$$

where $\tau = E_k^{\mathrm{e}}/m_{\mathrm{e}}$, i.e., the kinetic energy of the electron in units of its rest mass. In electron–silicon interactions, the kinetic energy below which $T_{\max}^{\mathrm{e}}$ is lower than 25 eV is about 255 keV.

It has been demonstrated[‖] that the *Mott–McKinley–Feshbach electron cross section* for a Coulomb collision with a point-like nucleus [McKinley and Feshbach (1948); Seitz and Koehler (1956)] is accurate to one percent for nuclei with atomic number up to 40 [Curr (1955)] and, for a nucleus initially at rest, is given by:

$$\frac{d\sigma(E_k)}{dE_{\mathrm{Si}}} = \frac{\pi r_e^2 Z^2}{\gamma^2 \beta^4}\left[1 - \beta\left(\beta - \frac{\pi Z_{\mathrm{Si}}}{137}\right)\frac{E_{\mathrm{Si}}}{T_{\max}^{\mathrm{e}}} - \frac{\pi Z_{\mathrm{Si}}\beta}{137}\sqrt{\frac{E_{\mathrm{Si}}}{T_{\max}^{\mathrm{e}}}}\right]\frac{T_{\max}^{\mathrm{e}}}{E_{\mathrm{Si}}^2}, \tag{4.116}$$

where E_{Si} is the kinetic energy of the recoil silicon nucleus and Z_{Si} is its atomic number; β, γ and r_e^2 are the velocity in units of c (i.e., the speed of light), the Lorentz factor [Eq. (1.4)] and the classical radius of the electron, respectively. Using Eq. (4.116), the expression (for $E_{\mathrm{Si}} < T_{\max}^{\mathrm{e}}$)

$$P(E_{\mathrm{Si}}) = \left(\int_{E_d}^{E_{\mathrm{Si}}} \frac{d\sigma(E_k)}{dE_{\mathrm{Si,r}}}\,dE_{\mathrm{Si,r}}\right)\left(\int_{E_d}^{T_{\max}} \frac{d\sigma(E_k)}{dE_{\mathrm{Si,r}}}\,dE_{\mathrm{Si,r}}\right)^{-1} \tag{4.117}$$

provides an estimate of the probability that the silicon recoils with a kinetic energy, $E_{\mathrm{Si,r}}$, above the displacement threshold-energy (E_d) and $\leq E_{\mathrm{Si}}$. In Fig. 4.17, the normalized integral of the electron cross section above the threshold energy for displacement $E_d = 25$ eV [i.e. $P(E_{\mathrm{Si}})$ calculated by means of Eq. (4.117)] is shown as a function of the recoil silicon energy, E_{Si}, in eV for incoming electrons with kinetic energies of 1, 2, 5, 15, and 50 MeV.

[†]It is calculated by means of Eq. (4.102), assuming that the silicon nucleus is at rest.

[‖]The reader can see [Cahn (1959)], Section 1 of Chapter 1 in [Vavilov and Ukhin (1977)] and Section 2.2.4 in [van Lint, Flanahan, Leadon, Naber and Rogers (1980)].

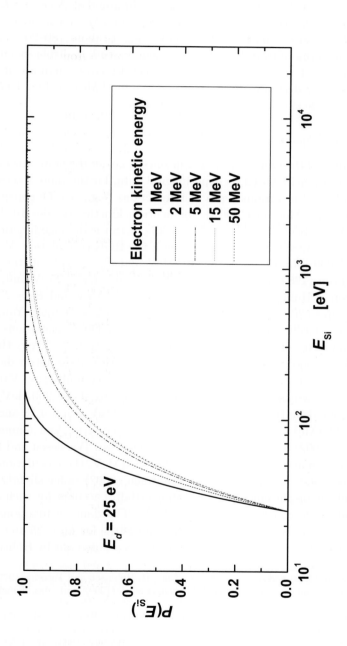

Fig. 4.17 $P(E_{Si})$, normalized integral of the electron cross section calculated from Eq. (4.117), above the threshold energy for displacement $E_d = 25$ eV, as a function of the recoil silicon energy E_{Si} in eV (from [Leroy and Rancoita (2007)]).

In the *modified Kinchin–Pease formula*** the average number of displacements, $< N_{FP} >$ (including the one resulting from the initial knock-on), due to a PKA with damage energy E_{de} is 1 for $E_d < E_{de} < 2.5E_d$ and $< N_{FP} > = E_{de}/(2.5E_d)$ for $E_{de} > 2.5E_d$. In electron–silicon interactions since the recoil energy is usually small (e.g., E_{Si} is lower than about few keV), E_{de} is about $(80\text{–}90)\%$ of E_{Si} (see Fig. 4.15). Furthermore from Eq. (4.115), silicon cascading-displacements may become relevant only for electrons with kinetic energies larger than about 5 MeV. As the electron energy decreases (e.g., see the curve for 1 MeV in Fig. 4.17), the probability for a single silicon displacement increases.

4.2.1.5 *Damage Function*

The evaluation[‡‡] of the damage function in silicon, i.e., of the non-ionization energy-loss, has been primarily based on experimental data. Furthermore a series of calculations*, involving the various types of interactions (e.g., see [Ougouag, Williams, Danjaji, Yang and Meason (1990)]), for neutron kinetic energies $10^{-10} \leq E_k^n \leq 2.4 \times 10^4$ MeV has been also performed . In Fig. 4.18 the damage function (in units of MeV mb) for neutrons is shown including i) the HGK compilation [Vasilescu and Lindstroem (2000)] for $10^{-9} < E_k^n < 10^4$ MeV and $E_d = 25$ eV, ii) the calculation of Van Ginneken (1989) for $10 < E_k^n < 10^3$ MeV and $E_d = 25$ eV and, finally, iii) the calculation of Huhtinen (2002) for $20 < E_k^n < 10^4$ MeV and $E_d = 20$ eV. In the energy range up to 20 MeV, Ougouag [Ougouag, Williams, Danjaji, Yang and Meason (1990)] and Griffin [Vasilescu and Lindstroem (2000)] calculations are almost equivalent and were computed for a displacement energy $E_d = 25$ eV. However (see Fig. 4.19), in the energy region $15 < E_k^n < (18\text{–}20)$ MeV these latter damage functions [Van Ginneken (1989)] are larger than the one computed by Van Ginneken with $E_d = 25$ eV which, in turn, agrees (for E_k^n up to about 200 MeV) with that recently computed by Huhtinen (2002) with $E_d = 20$ eV (Figs. 4.18 and 4.19). For $15 < E_k^n < 200$ MeV, the damage functions computed by Van Ginneken (1989) and Huhtinen (2002) are lower than that of Konobeyev [Vasilescu and Lindstroem (2000)] (Figs. 4.18 and 4.19). For $10^{-3} < E_k^n < 15$ MeV, the calculations of Griffin [Vasilescu and Lindstroem (2000)] and Huhtinen (2002) differ slightly: the different values of the displacement energy can (partially) account for such a difference (e.g., see the discussion in [Huhtinen (2002)]). The damage functions estimated in [Vasilescu and Lindstroem (2000); Huhtinen (2002)] for $E_d = 25$ and 20 eV differ slightly in the energy region $1.5 < E_k^n < 5$ GeV. As pointed out by Huhtinen (2002),

---

**It is given by Eq. (4.89) and see, for instance, [Norgett, Robinson and Torrens (1975)], Chapter 4 in [Ziegler, Biersack and Littmark (1985a)], [Messenger et al. (1999)] and, also, [Kinchin and Pease (1955)].

‡‡One can see, e.g., [Smith, Binder, Compton and Wilbur (1966); Conrad (1971); Namenson, Wolicki and Messenger (1982); Summers et al. (1987); Cheryl, Marshall, Burke, Summers and Wolicki (1988); Summers, Burke, Shapiro, Messenger and Walters (1993)] and references therein.

*The reader can see, for instance, [Van Ginneken (1989); Ougouag, Williams, Danjaji, Yang and Meason (1990); Vasilescu and Lindstroem (2000); Huhtinen (2002)] and references therein.

experimental data are needed to determine which of the computed damage functions is more correct for high-energy neutrons.

The damage function for protons has been computed for $10^{-4} \lesssim E_k^{\mathrm{p}} \lesssim 2.4 \times 10^4\,\mathrm{MeV}$ with $12.9 \leq E_d \leq 25\,\mathrm{eV}$ and compared with experimental data by many authors (e.g., [Summers et al. (1987); Van Ginneken (1989); Summers, Burke, Shapiro, Messenger and Walters (1993); Vasilescu and Lindstroem (2000); Akkerman, Barak, Chadwick, Levinson, Murat and Lifshitz (2001); Huhtinen (2002); Messenger, Burke, Summers and Walters (2002); Jun et al. (2003)]). For protons with kinetic energies above 10 MeV, the damage function was found to be slightly dependent on E_d (see for instance [Summers, Burke, Shapiro, Messenger and Walters (1993)]). Jun et al. (2003) have revised previous calculations up to 200 MeV [Summers, Burke, Shapiro, Messenger and Walters (1993); Akkerman, Barak, Chadwick, Levinson, Murat and Lifshitz (2001)] and extended the proton kinetic energy range up to 1 GeV. These authors have successfully used the Ziegler–Biersack–Littmark (ZBL) screened Coulomb potential [Ziegler, Biersack and Littmark (1985a); Ziegler, J.F. and M.D. and Biersack (2008a)] coupled to the relativistic energy transfer cross section at higher incident energies. In addition, they have employed a charged particle transport code (MCNPX, based on the thin target approximation) to compute the nuclear contribution to non-ionizing energy-loss for protons which accounts for about 15 % (50 %) at 10 (20) MeV (e.g., see [Jun et al. (2003)] and references therein for details of the treatment). The MCNPX code [MCNPX (2006)] uses the cross sections from Barashenkov and Polanski (1994) to determine the probability of elastic and non-elastic interactions. For non-elastic reactions, the kinetic energy of the PKA is computed based on the concept of high-energy intra-nuclear cascade, pre-equilibrium and evaporation physics. In Fig. 4.20, the latest calculations for protons are shown in the kinetic energy ranges i) $2 \times 10^{-4} \leq E_k^{\mathrm{p}} \leq 10^3\,\mathrm{MeV}$ with $E_d = 21\,\mathrm{eV}$ [Jun et al. (2003)], ii) $10^3 \leq E_k^{\mathrm{p}} \leq 2.4 \times 10^4\,\mathrm{MeV}$ with $E_d = 20\,\mathrm{eV}$ [Huhtinen (2002)] and, finally, iii) $15 \leq E_k^{\mathrm{p}} \leq 9 \times 10^3\,\mathrm{MeV}$ with $E_d = 20\,\mathrm{eV}$ [Vasilescu and Lindstroem (2000)]. Furthermore, Huhtinen has also computed the damage function for pions (e.g., see [Huhtinen (2002)] and references therein). In silicon, the damage function for heavy-ions, with a kinetic energy range of $10^{-4} < E_k^{\mathrm{tot}} < 10^3\,\mathrm{MeV}$ and displacement threshold energy $E_d = 21\,\mathrm{eV}$, was calculated using interatomic screened Coulomb potentials (expressed in the form of the ZBL universal potential [Ziegler, Biersack and Littmark (1985a); Ziegler, J.F. and M.D. and Biersack (2008a)]) in the non-relativistic limit by Messenger et al. (2003) and (2004). These calculations are relevant for space missions where spacecrafts and pay-loads are exposed to energetic GCRs in addition to the trapped particles (mostly electrons and protons) populating the radiation belts (Sect. 4.1.2.5). In Fig. 4.21, the damage functions for iron, silicon, boron ions and α-particles (i.e., heavy ions with $Z \leq 28$) are shown as a function of the ion kinetic energy in MeV. However, in order to evaluate properly the displacement damage, they must be extended to about** 1 GeV/nucleon (e.g., for

---

**The reader can find the definition of kinetic energies per nucleon in Sect. 1.4.1.

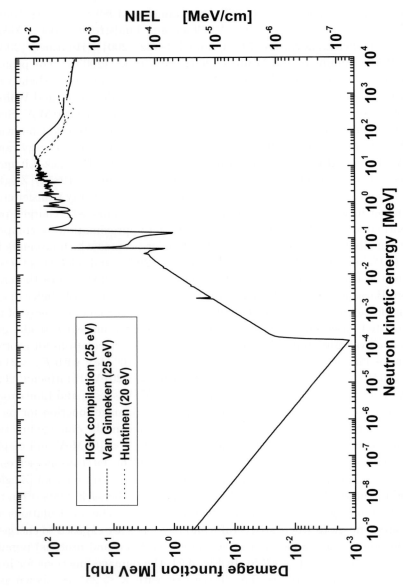

Fig. 4.18 Damage function for neutrons with kinetic energies of $10^{-9} \leq E_k^n \leq 10^4$ MeV (adapted from [Leroy and Rancoita (2007)]). For Van Ginneken (1989) and HGK compilation [Vasilescu and Lindstroem (2000)], the displacement energy is $E_d = 25$ eV; for Huhtinen (2002), $E_d = 20$ eV. The tabulated data are shown interpolated by spline lines.

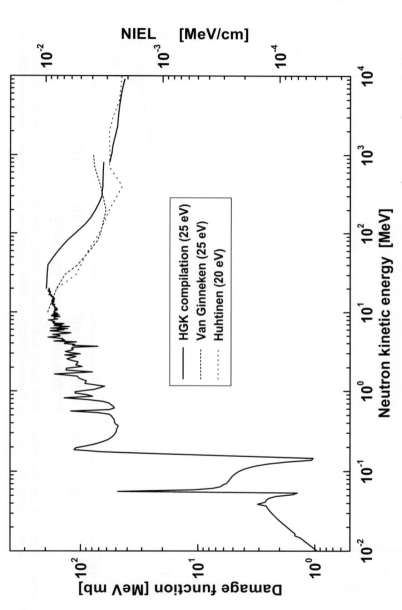

Fig. 4.19 Expanded view of Fig. 4.18: damage function for neutrons with kinetic energies of $10^{-2} \leq E_k^n \leq 10^4$ MeV (adapted from [Leroy and Rancoita (2007)]). For Van Ginneken (1989) and HGK compilation [Vasilescu and Lindstroem (2000)], the displacement energy is $E_d = 25$ eV; for Huhtinen (2002), $E_d = 20$ eV. The tabulated data are shown interpolated by spline lines.

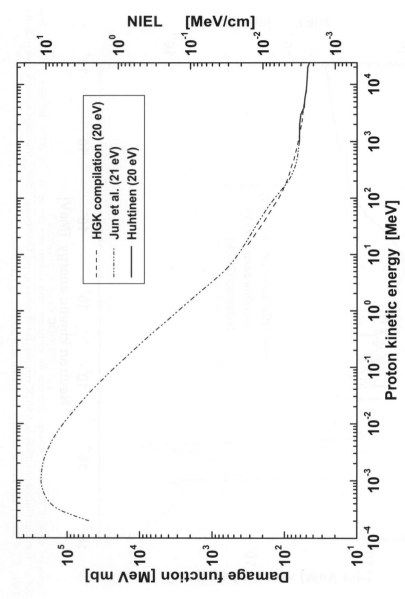

Fig. 4.20 Damage function for protons (adapted from [Leroy and Rancoita (2007)]). The proton kinetic energy range is $2 \times 10^{-4} \le E_k^p \le 2.4 \times 10^4$ MeV. For Jun et al. (2003) $E_d = 21$ eV; for Huhtinen (2002) and HGK compilation [Vasilescu and Lindstroem (2000)] $E_d = 20$ eV. The tabulated data are shown interpolated by spline lines.

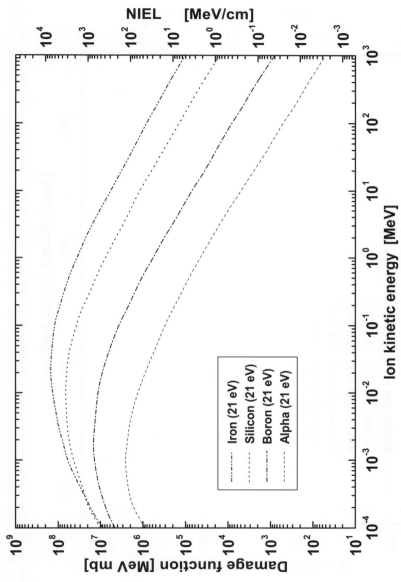

Fig. 4.21 Damage functions in silicon (adapted from [Leroy and Rancoita (2007)]) for iron, boron, silicon ions and α-particles with a kinetic energy range of $10^{-4} \leq E_k^{\text{tot}} \leq 10^3$ MeV and $E_d = 21$ eV [Messenger et al. (2003, 2004)]. The data are shown interpolated by spline lines.

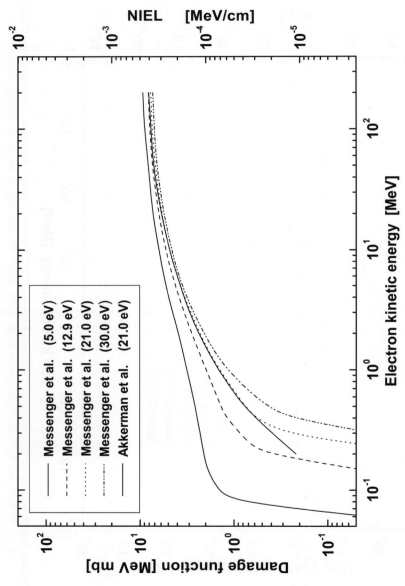

Fig. 4.22 Damage function for electrons with a kinetic energy range of $0.6 \lesssim E_k^e \leq 200$ MeV (adapted from [Leroy and Rancoita (2007)]): for $E_d = 5$, 12.9, 21 and 30 eV from [Messenger et al. (1999)] and $E_d = 21$ eV from [Akkerman, Barak, Chadwick, Levinson, Murat and Lifshitz (2001)]. The tabulated data are shown interpolated by spline lines.

α-particles see Fig. 4.12) and account for strong interactions (see Sect. 4.2.1.3).

Although the damage functions have been computed with different values of E_d, the displacement threshold energy seems not to be the major source of uncertainty. Only for low energy electrons (e.g., $< 2\,\text{MeV}$) and, as a consequence, ^{60}Co γ-ray radioactive-sources, E_d is relevant to estimate the overall non-ionizing energy-loss deposition and concentration of induced displacement damages.

The latest calculations of the damage function for electrons[††] have taken into account displacement energies from 5 to $30\,\text{eV}$[‡‡]. Above a few hundred MeV, the non-ionizing energy-loss is almost independent of the energy of the incoming particle [Van Ginneken (1989)]. However, a slight energy dependence is found in the electron kinetic energy range $50 < E_k^e < 200\,\text{MeV}$ [Van Ginneken (1989); Summers, Burke, Shapiro, Messenger and Walters (1993); Messenger et al. (1999); Akkerman, Barak, Chadwick, Levinson, Murat and Lifshitz (2001)]. In Fig. 4.22, the damage functions for electrons computed by Messenger et al. (1999) and Akkerman, Barak, Chadwick, Levinson, Murat and Lifshitz (2001) with electron kinetic energies in the range $0.6 < E_k^e < 200\,\text{MeV}$ are shown: above $\approx 300\,\text{keV}$, the calculations for $E_d = 21\,\text{eV}$ are in good agreement. In addition, Messenger et al. (1999) have shown (see Fig. 4.22) that the damage function for $E_d = 21$ and $30\,\text{eV}$ does not differ by more than about 2.7% at $5\,\text{MeV}$ and 0.9% above $10\,\text{MeV}$.

^{60}Co γ-ray radioactive-sources are widely used in space qualification procedures and to study radiation-induced damage in silicon devices. In silicon, Compton scattering is the main interaction process of the emitted γ-rays with energies of 1.1732 and $1.3325\,\text{MeV}$ (e.g., see Appendix A.8). The secondary electrons have, usually, enough kinetic energy to displace silicon atoms. Thus, combining the energy spectrum of the slowed-down secondary electrons with the electron damage function (e.g., that given in Fig. 4.22) the damage function for the ^{60}Co γ-ray source can be obtained[†] and sometimes expressed in terms of 1 MeV *electron-equivalent displacement-damage* (e.g., see [Xapsos et al. (1994)]). In addition, the introduction of a shielding material in front of the silicon device can affect the resulting electron spectrum[¶] and, as a consequence, the damage energy deposited. Furthermore, from an inspection of the energy distribution of the scattered Compton electron (e.g., see Fig. 2.55 and Figure 2.16 in [van Lint, Flanahan, Leadon, Naber and

---

[††]The reader can also find NIEL calculations by means of the NEMO code ([Inguimbert and Gigante (2006)] and references therein); they are in agreement with those in [Summers, Burke, Shapiro, Messenger and Walters (1993)] above a few MeV of kinetic energy.

[‡‡]The reader may see [Summers, Burke, Shapiro, Messenger and Walters (1993)] for $E_d = 12.9$ and 21 eV, [Van Ginneken (1989)] for $E_d = 25\,\text{eV}$, [Akkerman, Barak, Chadwick, Levinson, Murat and Lifshitz (2001)] for $E_d = 21\,\text{eV}$ and, finally, [Messenger et al. (1999)] for $E_d = 5, 12.9, 21$ and 30 eV.

[†]One can see, for instance, [Summers, Burke, Shapiro, Messenger and Walters (1993)]; also [Xapsos et al. (1994); Akkerman, Barak, Chadwick, Levinson, Murat and Lifshitz (2001)] for $E_d = 21\,\text{eV}$ and references therein.

[¶]Calculations for shielding thicknesses ranging from $100\,\mu\text{m}$ up to $2\,\text{mm}$ are shown in [Akkerman, Barak, Chadwick, Levinson, Murat and Lifshitz (2001)].

Rogers (1980)]), the NIEL deposition occurs with a limited cascade of displaced silicon atoms (e.g., see Sect. 4.2.1.4).

4.2.2   *Radiation Induced Defects*

The radiation interaction in silicon semiconductors generates *primary point defects*, i.e., vacancies (V) and interstitials (I) as mentioned in Sects. 4.1.1.2 and 4.2.1. Possible structures for the silicon vacancy and interstitial are discussed in [Privitera, Coffa, Priolo and Rimini (1998)], i.e., slightly different configurations may result depending on the charge state of the point defect.

At room temperature, these defects are mobile[§] with low *activation energies* for their motion. For instance, the activation energies for vacancy migration are different for n- and p-type materials, depend on the resistivity and are about (18–45) meV (e.g., see Section 3 in [Claeys and Simoen (2002)]). As a consequence, not all primary defects will result in creating *stable secondary defects* or *defect complexes*, because a non negligible fraction of them will anneal, for example, by an interstitial filling of a vacancy. However, they can also interact with other point defects and impurities[‡‡] (interstitial and substitutional) to form more stable defects (e.g., see [Privitera, Coffa, Priolo and Rimini (1998); Lazanu and Lazanu (2003)]), like for example i) the divacancy referred to as "$G7$-center" (V-V, indicated also as V_2, i.e., two adjacent vacant lattice sites [Watkins and Corbett (1961a, 1965a,b); Cheng, Corelli, Corbett and Watkins (1966)]) in various charge states[††], ii) vacancy-oxygen[**] (V-O) referred to as "A-center" or "$B1$-center" [Watkins and Corbett (1961b); Watkins, Corbett, Chrenko and McDonald (1961)] (see also [Li et al. (1992)] and references therein), iii) vacancy-dopant impurity (for instance, V-P or V-As in n-type silicon) referred to as "E-center" or "$G8$-center" [Watkins and Corbett (1964); Samara (1988); Öğütt and Chelikowsky (2003)] and iv) others (e.g., see Sections 3–5 in Chapter II of Part I of [Vavilov and Ukhin (1977)], Section 7.3 of [van Lint, Flanahan, Leadon, Naber and Rogers (1980)], Sections 2.4 and 3.2.1 of [Claeys and Simoen (2002)], Sections 2.2.1–2.2.4 of [Holmes-Siedle and Adams (2002)] and Sections 3.2.1–3.2.3 of [Kozlovski and Abrosimova (2005)]).

Radiation-induced defects, which are electrically active, are (and have been) extensively studied by means of experimental techniques, for instance the Electron

---

[§]One can see, for instance, Section 24 of [Seitz and Koehler (1956)], Chapters 4 and 5 of [Dienes and Vineyard (1957)], Section 3 in Chapter II of Part I of [Vavilov and Ukhin (1977)] and [Privitera, Coffa, Priolo and Rimini (1998); Hallén, Keskitalo, Josyula and Svensson (1999)].

[‡‡]Impurities (and vacancies) can diffuse (e.g., see Section 2 of [Ravi (1981)]) and, in addition, introduce deep energy-levels in silicon, which can become recombination centers (e.g., see Chapters 2 and 6 of [Milnes (1973)]).

[††]The reader can also see [Kholodar and Vinetskii (1975); Borchi et al. (1989); Bosetti et al. (1995); Bondarenko, Krause-Rehberg Feick and Davia (2004)] and references therein.

[**]EPR investigations have led to the observation of more complex-defects relating vacancies and oxygen(s) (e.g., see [Lee and Corbett (1976)]).

Table 4.3   Energy levels (in eV) of electron (E_i) and hole (H_i) traps determined with the DLTS technique between about 90 and 350 K. E_C and E_V are the energies (with respect to an arbitrary level) of the conduction and valence band edges, respectively (from [Leroy and Rancoita (2007)]).

Level	Energy level from [Bosetti et al. (1995)]	Energy level from [Li, Chen and Kraner (1991); Li et al. (1992)]
E_1	$E_C - (0.16 \pm 0.01)$	$E_C - 0.17$
E_2	$E_C - (0.25 \pm 0.02)$	$E_C - 0.21$
E_3	$E_C - (0.40 \pm 0.02)$	$E_C - 0.41$
E_4	$E_C - (0.54 \pm 0.02)$	$E_C - 0.55$
H_1	$E_V + (0.30 \pm 0.01)$	$E_V + 0.30$
H_2	$E_V + (0.50 \pm 0.02)$	$E_V + 0.54$

Paramagnetic Resonance[†] (EPR), Thermally Stimulated Current[*] (TSC), Photo-Luminescence[||] (PL) and Deep Level Transient Spectroscopy[¶] (DLTS).

Examples of DLTS spectra for $p^+ - n$ silicon junctions with n-type Fz (Float-Zone) and MCz (Magnetic Czochralski) substrates are shown in Figs. 4.23 and 4.24, respectively [Bosetti et al. (1995)] (see also [SICAPO Collab. (1994c)] for details about MCz substrate). These detectors[§] with an active area of $0.5 \times 0.5 \, \text{cm}^2$ were irradiated with a fast neutron[‡] fluence of $3 \times 10^{11} \, \text{n} / \text{cm}^2$. In Figs. 4.23 and 4.24, four energy-levels (E_1, E_2, E_3 and E_4) of electron-traps and two (only one in the case of MCz sample) of hole-traps are observed between about 90 and 350 K. The measured energy-levels of these (electron and hole) traps are in agreement with those determined in [Li, Chen and Kraner (1991); Li et al. (1992)] (see Table 4.3). Furthermore, a complete discussion of the observed defects and their association to determined energy-levels is presented in [Li, Chen and Kraner (1991); Li et al. (1992); Bosetti et al. (1995)] and references therein. From an inspection of Figs. 4.23 and 4.24, we can notice that the ratio (R_trap) of the trap concentrations[‡‡] over donor concentration depends on the type of substrate, as expected. This occurs because the concentrations of impurities in the Fz and MCz substrates are different (see Table 4.4 and [Bosetti et al. (1995)], for electron-traps). The E_1 energy-level corresponds to that of an A-center (vacancy-oxygen). The ratio R_trap for this level is lower in Fz

---

[†]In EPR spectroscopy, the electromagnetic radiation of microwave frequency is absorbed by defects possessing a magnetic dipole moment of unpaired spins.

[*]The TSC technique consists of the measurement of diode currents, generated after filling trap levels by bias pulses, while performing thermal scans [Weisberg and Schade (1968); Schade and Herrick (1969); Bueheler (1972)] (see also Chapter 9 of [Milnes (1973)] and references therein).

[||]A recent review on PL spectroscopy is done in [Davies (1989)].

[¶]The DLTS consists of the measurement of capacitance transients, generated by repetitive filling and emptying of deep levels in the depleted region by bias pulses, while performing a thermal scan [Lang (1974)].

[§]Both types of detectors were manufactured by STMicroelectronics.

[‡]The mean kinetic energy was about 1 MeV [Bosetti et al. (1995)].

[‡‡]The trap concentrations were induced in the Fz and MCz detectors by irradiations with similar fast neutron fluence.

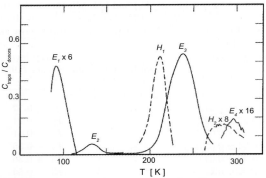

Fig. 4.23   DLTS spectrum as a function of temperature of a silicon detector with an asymmetrical one-side junction and Fz n-type silicon substrate (about $400\,\mu$m thick) cut along the Si crystal direction $< 111 >$ (adapted from *Nucl. Instr. and Meth. in Phys. Res. A* **361**, Bosetti, M. et al., DLTS measurement of energetic levels, generated in silicon detectors, 461–465, Copyright (1995), with permission from Elsevier, e.g., for the list of the authors see [Bosetti et al. (1995)]). The ordinate corresponds to the ratio of concentrations of traps after a fast neutron irradiation of about 3×10^{11} n/cm^2 divided by the donor concentration. The oxygen concentration of the Fz bulk silicon is $(1 \pm 0.04) \times 10^{15}$ cm^{-3} and the resistivity is between 4–6 kΩ cm. Solid and dashed lines are traps for electrons and holes, respectively.

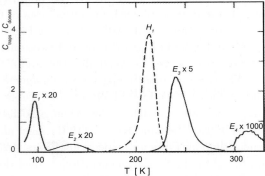

Fig. 4.24   DLTS spectrum as a function of temperature of a silicon detector with an asymmetrical one-side junction and MCz n-type silicon substrate (about $525\,\mu$m thick) cut along the Si crystal direction $< 100 >$ (adapted from *Nucl. Instr. and Meth. in Phys. Res. A* **361**, Bosetti, M. et al., DLTS measurement of energetic levels, generated in silicon detectors, 461–465, Copyright (1995), with permission from Elsevier, e.g., for the list of the authors see [Bosetti et al. (1995)]). The ordinate corresponds to the ratio of concentrations of traps after a fast neutron irradiation of about 3×10^{11} n/cm^2 divided by the donor concentration. The oxygen concentration of the MCz bulk silicon is $(1 \pm 0.04) \times 10^{16}$ cm^{-3} and the resistivity is between 4–6 kΩ cm. Solid and dashed lines are traps for electrons and holes, respectively.

(which contains less oxygen) than in MCz substrate. The E_2 energy-level is that of a double negative charge state of a divacancy (V_2^{--}). While the E_3 energy-level has two components with $E_C - (0.39 \pm 0.01)$ [V_2^-, divacancy in single negative charge state] and $E_C - (0.42 \pm 0.02)$ eV [E-center, i.e., a vacancy-phosphorous complex]. These two latter levels were resolved by an investigation employing the TSC spectroscopy [Borchi et al. (1991)]. The E_3 energy-level has also shown two anneal-ing stages around 150 and 300 °C [Borchi et al. (1989)]. The E_2 and E_3 centers have smaller R_{trap} ratios for MCz compared to Fz detectors, as expected by a *getter ef-fect* [Li et al. (1992); Bosetti et al. (1995)] when oxygen is present at the A-centers, thus avoiding the formation of deeper vacancy centers. In fact, more A-centers were found in the MCz than in the Fz detectors. Furthermore for Fz substrates, R_{trap} for the E_3 level was measured for fast neutron fluences up to $\Phi_n \approx 2 \times 10^{11} \, \text{n/cm}^2$ and found to be linearly dependent on the fluence [Borchi et al. (1989)]:

$$R_{\text{trap}}(E_3) \approx (2.12 \pm 0.06) \times 10^{-12} \times \Phi_n.$$

These data are in agreement with those found in [Bosetti et al. (1995)] for fluences up to $\approx 3 \times 10^{11} \, \text{n/cm}^2$. A linear dependence of $R_{\text{trap}}(E_3)$ as a function of the Kr-ion fluence was also observed up to a maximum fluence of $10^9 \, \text{Kr/cm}^2$ [Croitoru, Rancoita, Rattaggi and Seidman (1999)].

DLTS peaks were also observed in the temperature range (10–90) K for irradia-tions with neutrons, electrons, photons, Ar- and Kr-ions. Both non-irradiated and irradiated samples show a peak related to an oxygen-related donor state, with an energy level $E_l = E_C - E_{od}$ with $E_{od} = (0.11–0.12) \, \text{eV}$ [Walker and Sah (1973); Brotherton and Bradley (1982); Jellison (1982); Kimerling and Benton (1982); Mangiagalli, Levalois, Marie, Rancoita and Rattaggi (1998); Croitoru, Rancoita, Rattaggi and Seidman (1999)]. It has been identified as an oxygen-carbon pairing [Walker and Sah (1973); Jellison (1982)]. A few other (electron trap) levels were also observed at energies $\lesssim 0.1 \, \text{eV}$ below E_C (e.g., see [Walker and Sah (1973); Kimerling and Benton (1982); Croitoru, Rancoita, Rattaggi and Seidman (1999)])

Commonly observed defect levels in Fz, Cz (Czochralski) and MCz silicon inclu-ding those after irradiations can be found in Table 9 at pages 155–157 of [Vavilov and Ukhin (1977)] and also in [Konozenko, Semenyuk and Khivrich (1969); Milnes (1973); Walker and Sah (1973); Borchi et al. (1989); Li et al. (1992); Borchi and Bruzzi (1994); Bosetti et al. (1995); Levinshtein, Rumyantsev and Shur (2000); Lutz (2001); Claeys and Simoen (2002)] (see also references therein); typical experimental errors for energy levels of defects are about (0.5–2.5) meV (e.g., see [Walker and Sah (1973); Jellison (1982); Li et al. (1992); Bosetti et al. (1995)] and Table 4.3).

Defect introduction rates were measured after irradiations with electrons of 1 and 12 MeV (e.g., see [Walker and Sah (1973); Brotherton and Bradley (1982)]). For instance, the introduction rates for the A-center and divacancy were found to be almost constant, although the data indicate a tendency for the rates to decrease with increasing fluence up to a maximum of $5.5 \times 10^{14} \, \text{electrons/cm}^2$ [Brotherton and Bradley (1982)].

Crystals containing defects induced by radiation are unstable, although they tend to stabilize with time, and the concentrations of defects decrease with time if the temperature is typically larger than zero degree Celsius. This mechanism is sometimes referred to as *self-annealing process*. Furthermore, defect concentrations have been studied as a function of the temperature (i.e., by *annealing*) to determine the value at which each type of defect spontaneously dissociates (see Section 5 in Chapter II of Part I of [Vavilov and Ukhin (1977)], Section 3.2.1 of [Claeys and Simoen (2002)], Section 3.2.8 of [Holmes-Siedle and Adams (2002)] and, for instance, [Dannefaer, Mascher and Kerr (1993); Makarenko (2001); Poirier, Avalos, Dannefaer, Schiettekatte, Roorda and Misra (2003)]).

Clusters of defects are generated in semiconductors when the incident particles (e.g., fast neutrons) transfer enough energy to the recoil atoms for allowing large cascades of displacements and, consequently, closely spaced lattice-defects. These clusters introduce large local concentrations of energy-levels in the forbidden gap. For instance (Fig. 4.14), a recoil atom with kinetic energy of 50 keV[§] generates clusters[¶] with subclusters of defects which extend up to about (100–200) Å [van Lint, Leadon and Colweel (1972)]. Typical cluster and subcluster size distributions resulting from proton–, electron– and neutron–silicon interactions are reported in Table 7.2 of [van Lint, Flanahan, Leadon, Naber and Rogers (1980)] (see also [van Lint, Leadon and Colweel (1972)] and references therein). The most direct evidence for the existence of clusters was given by Bertolotti (1968): his electron microscopy observations have provided an excellent physical picture of clusters with a central (disordered) region surrounded by a region attributed to space charge (e.g., see page 311 of [van Lint, Flanahan, Leadon, Naber and Rogers (1980)]). Relevant effects associated to defect clusters are expected in semiconductors: for instance, a change of the type of conductivity was observed in germanium by Cleland *et al*[‖] in the mid of 1950s [Cleland, Crawford and Pigg (1955a,b)]. It has to be noted that a change of the type of conductivity was also observed in silicon (e.g., see Sects. 4.3.5 and 4.3.6, and references therein).

The *Gossick–Gregory–Leadon model* for defect clusters assumes that, for a large local concentration of defects generated by an incoming particle[**], the damage consists of a disordered region (densely populated by defects) surrounded by an almost undisturbed lattice. Thus, a potential well is created because the position of the *Fermi level* relative to the energy bands is different within the disordered region from the outside. As a consequence, the model postulates that the core of the cluster is capable to acquire a net charge by capturing majority-carriers. Furthermore, it predicts a large short-term annealing and the effect of the material resistivity on annealing rate [Gossick (1959); Gregory (1969); Leadon (1970)] (see also Section 6 in Chapter II of Part I of [Vavilov and Ukhin (1977)] and, in addition, [Gregory, Naik and Oldham (1971); Litvinov, Ukhin and Oldham (1989)]). Under the assumption of a spherical cluster, the core region of radius $r_c \approx (10\text{–}25)$ nm (e.g., [Borchi and

---

[§] This is the average energy transferred by a neutron with a kinetic energy of about 0.76 MeV (see Sect. 4.2.1.2).

[¶] A fast neutron is likely to generates $\approx 10^3$ displaced atoms in a volume of approximately 10^{-16} cm^3.

[‖] The experimental results (originally interpreted in the framework of the *James–Lark-Horovitz model* [James and Lark-Horovitz (1951)]) have also motivated Gossick's model for defect clusters [Gossick (1959)].

[**] The model was originally proposed for studying the damage induced by fast neutrons.

Table 4.4 Ratio (R_{trap}) of the trap concentrations over donor concentrations for silicon detectors with Fz and MCz n-type substrate, after irradiation with a fast neutron fluence of about $3 \times 10^{11}\,\text{n}/\text{cm}^2$ (from [Leroy and Rancoita (2007)], see also [Bosetti et al. (1995)]). R_{trap} is given for the 4 energy levels for electron traps shown in Table 4.3. The experimental errors on R_{trap} are about $\pm 1\%$.

Level	Fz substrate	MCz substrate
E_1	0.0750	0.0850
E_2	0.0500	0.0150
E_3	0.5400	0.5000
E_4	0.0180	0.0007

Bruzzi (1994); Gossick (1959); Gregory (1969)] and references therein) is surrounded by a semiconductor volume of radius r_s from which majority-carriers have been partially removed generating a region similar to that depleted in a p-n junction with a potential hill (well) for majority (minority) carriers. The outer boundary must exceed the *Debye–Hückel length* (L_{DH}) in the undisturbed lattice:

$$L_{DH} = \sqrt{\frac{k_B\, T\, \varepsilon}{e^2 N_{\text{imp}}}}, \qquad (4.118)$$

where e is the charge of the electron, k_B is the Boltzmann's constant, T is the absolute temperature, ε is the silicon electric permittivity (e.g., see Appendix A.2) and N_{imp} is the concentration of impurities which contribute current carriers to the undisturbed material. For a high resistivity silicon, L_{DH} is about $4\,\mu\text{m}$ at room temperature.

Since point defects are mobile at room temperature (e.g., see page 370), a re-arrangement process can take place ([Gregory, Naik and Oldham (1971)] and references therein), resulting in a decrease of cluster effects on the semiconductor electronic properties with time. Large annealing factors ($\gtrsim 50$) at early times were experimentally measured [Harrity and Mallon (1970)] and are consistent with the fact that the induced damage depends on the cluster size. In Section 7.4 of [van Lint, Flanahan, Leadon, Naber and Rogers (1980)], one finds a review of experimental data that can be explained by the presence of defect clusters produced by neutron irradiations. These data regard carrier removal, lifetime degradation, mobility changes and short-term thermal annealing. Calculations [Leadon (1970)] based on the cluster model[††] have shown that i) the major features of short-term annealing, ii) the shape of the annealing factor curve with time, iii) the effect of injection level and iv) the material resistivity on the lifetime degradation constants

---

[††]Other cluster models have been proposed with cylindrical clusters [Holmes (1970)] and effective radii of the space-charge region larger than the range of the primary recoil atom [van Lint and Leadon (1966)].

can be all reasonably explained by the cluster nature of the damage induced by fast neutrons.

As discussed in Sects. 4.2.1.1–4.2.1.4 for NIEL processes, the amount of the deposited energy (Sect. 4.2.1.5) and the size of the cascade depend on the type of interaction and the energy of incoming particles. Thus, it is expected that the extension of clusters and subclusters (when generated) can, in turn, depend on the type of interaction between the incoming particle and the atom in the lattice, i.e., the damage effectiveness can be affected by the properties of the defect clusters (e.g., see also [Tsveybak, Bugg, Harvey and Walter (1992)] and references therein). A comparative investigation carried out with Kr-ions, electrons and fast-neutrons has confirmed that the introduction rates of complex-defects normalized to the total displacement cross-section depend on the type of incoming particle: a factor of ≈ 18 for the divacancies and vacancy-doping impurity complexes was found ([Mangiagalli, Levalois and Marie (1998)] and references therein).

4.2.3   *Ionization Energy-Loss and NIEL Processes*

As mentioned at page 346, charged particles heavier than electrons (or positrons) traversing (or being stopped inside) a silicon medium lose energy predominantly by collision energy-losses[¶], i.e., by the excitation and ionization of atoms close to the particle path. For electrons and positrons, the collision energy-loss is the dominant process at energies below the so-called *critical energy* (see Sect. 2.1.7.4) while, above, the main mechanism is the so-called *radiation energy-loss*, i.e., by emitting photons. Photon interactions in matter[†] allow the energy of photons to be partially or fully absorbed in matter, as a consequence of the interaction inside the medium of the scattered (emitted) electrons or of the created electron–positron pairs. As discussed in Sect. 4.2.1.3, light ions (i.e., protons and α-particles) can be regarded as charge invariant over most of the energy region, while heavier isotopes only with velocities largely exceeding the orbital velocities of their electrons can be fully ionized[‡] and lose energy predominantly by electronic energy-loss [Northcliffe (1963); Leroy and Rancoita (2004)], i.e., by the excitation or ionization of atomic electrons of the medium. The energy-loss processes (by collision and radiation) and partial or complete absorption of photons in matter were reviewed in the chapter *on Electromagnetic Interaction of Radiation in Matter*.

Displaced silicon atoms can also undergo both electronic and nuclear energy-losses (see Sect. 4.2.1.1) to dissipate their energy inside the medium. Consequently, neutrons can also deposit energy via the interaction of the recoil silicon with the

---

[¶]The energy-loss formula describing the energy deposited in a medium is that called *restricted energy-loss formula*; see, for instance, Eq. (2.27).

[†]The dominant processes of photon interaction in matter are photoelectric effect, Compton scattering and pair production (e.g., see chapter *on Electromagnetic Interaction of Radiation in Matter*).

[‡]At low energy the isotope becomes partially ionized and the *electronic stopping power* is modified to account for it (see [Northcliffe (1963)]). Furthermore, the collisions with nuclei of the medium cannot be neglected anymore (see Sect. 4.2.1.3).

atoms of the medium through ionization and excitation.

4.2.3.1 *Imparted Dose in Silicon*

When there is no nuclear transformation, the *absorbed dose*[§] in (silicon) devices accounts for the energy deposited by the energy-loss processes of primary particles, those of recoil atoms (i.e. the PKAs generated in the semiconductor medium) and by NIEL processes (discussed in Sect. 4.2.1) resulting in the deposition of damage energy (for instance for creating PKAs, lattice vibrations, ...).

For fast neutrons from nuclear reactors, the contribution (D^{NIEL}) of NIEL processes to the absorbed dose can be estimated using Eq. (4.81), i.e.

$$D^{\mathrm{NIEL}} = \frac{E_{dis}}{6.24 \times 10^9 \times \rho_{\mathrm{Si}}} \tag{4.119}$$

$$\approx \frac{E_{dis}}{1.45 \times 10^{10}} \ [\mathrm{Gy}], \tag{4.120}$$

where E_{dis} is in MeV/cm^3 and $\rho_{\mathrm{Si}} = 2.33\,\mathrm{g/cm^3}$ is the density of the silicon medium. For example in the case of the Triga reactor RC:1 at Casaccia (Rome) [Codegoni et al. (2004b); Consolandi, D'Angelo, Fallica, Mangoni, Modica, Pensotti and Rancoita (2006)], a fluence $2.33 \times 10^{13}\,\mathrm{n/cm^2}$ of fast neutrons with kinetic energies above 10 keV creates a concentration of Frenkel-pairs[*] of about $10^{15}\,\mathrm{cm^{-3}}$ and, by NIEL processes, deposits a dose

$$D^{\mathrm{NIEL}} \approx 4.3\ \mathrm{Gy}. \tag{4.121}$$

The dose deposited by ionization (D^{Ion}) with this fluence of neutrons can be estimated by considering that the elastic scattering is the dominant interaction mechanism for fast reactor neutrons on silicon up to (3–4) MeV[‖]. In addition neutrons with kinetic energy up to a few MeV (see, Chapter 2 in [van Lint, Flanahan, Leadon, Naber and Rogers (1980)]) are isotropically (or almost isotropically) scattered and, finally, the fraction of energy deposited by ionization was also estimated (e.g., see for instance [Smith, Binder, Compton and Wilbur (1966)]) for neutrons with larger kinetic energies, i.e., when the elastic scattering is a relevant, but not dominant, interaction in silicon. D^{Ion} can be calculated by replacing E_{dis} with the energy deposited by ionization (in MeV) per cm^3 in Eq (4.119) [or in Eq (4.120)]. For the Triga reactor neutrons, within 5%, we have

$$D^{\mathrm{Ion}} \approx 4.8\ \mathrm{Gy}. \tag{4.122}$$

Thus, ionization and non-ionization processes result in depositing amounts of energy, which differ by $\approx (10\text{–}11)\,\%$.

[§]The absorbed dose is defined as the mean energy imparted by ionizing radiation per unit mass [ICRUM (1980b, 1993b)] (see also Sect. 2.3.5). It accounts for the net difference of kinetic energies of particle entering and leaving from the medium volume and nuclear transmutations changing the rest masses in the medium volume.

[*]The concentration of Frenkel-pairs can be calculated using Eq. (4.83) with $E_d \approx 25\,\mathrm{eV}$.

[‖]For the Triga reactor RC:1, the neutron fluence above 3.7 MeV is about 2% of the neutron fluence above 10 keV. These high-energy fast neutrons deposit less than about 8% of the damage energy.

Table 4.5 D^{Ion} and particle fluence (from [Leroy and Rancoita (2007)]) for fast reactor neutrons (i.e., for the Triga reactor RC:1 energy spectrum), heavy ions (i.e., ^{12}C, ^{13}C, ^{36}Ar and ^{86}Kr, see page 378), protons and electrons (see page 379). These values refer to similar NIEL energy deposition, i.e., that needed to generate a concentration of Frenkel-pairs of about 10^{15} cm^{-3}.

Particle	Kinetic Energy	Fluence (part./cm^2)	D^{NIEL} (Gy)	D^{Ion} (Gy)
n	> 10 keV	2.33×10^{13}	4.3	≈ 4.8
^{12}C	95.0 MeV/amu	8.10×10^{11}	4.3	$\approx 2.7 \times 10^4$
^{13}C	11.1 MeV/amu	9.30×10^{10}	4.3	$\approx 1.7 \times 10^4$
^{36}Ar	13.6 MeV/amu	1.40×10^{10}	4.3	$\approx 1.8 \times 10^4$
^{86}Kr	60.0 MeV/amu	1.60×10^{10}	4.3	$\approx 2.6 \times 10^4$
p	1 GeV	1.92×10^{13}	4.3	$\approx 5.2 \times 10^3$
e	9.1 MeV	2.58×10^{14}	4.3	$\gtrsim 6.3 \times 10^4$

In case of incoming ions, like for instance in the space radiation environment (see Sect. 4.1.2), the energy deposited by ionization largely exceeds that by NIEL processes. Let us examine, as an example, the energy§ depositions by the four ions used for the investigation reported in [Codegoni et al. (2004b); Consolandi, D'Angelo, Fallica, Mangoni, Modica, Pensotti and Rancoita (2006)]: ^{12}C accelerated at 95 MeV/amu, ^{13}C at 11.1 MeV/amu, ^{36}Ar at 13.6 MeV/amu and ^{86}Kr at 60 MeV/amu. In a silicon bulk depth** down to about 20 μm, a dose $D^{\text{NIEL}} \approx 4.3$ Gy†† is deposited by NIEL-processes [e.g. see Eq. (4.121)] with ion fluences of about 8.1×10^{11}, 9.3×10^{10}, 1.4×10^{10} and 1.6×10^{10} ions/cm^2, respectively. With these ion fluences, the corresponding doses deposited by ionization¶ are about 2.7×10^4, 1.7×10^4, 1.8×10^4 and 2.6×10^4 Gy, respectively. Thus, the ionization doses are larger by almost 4 orders of magnitude with respect to that for NIEL processes, while D^{NIEL} and D^{Ion} differ slightly for neutrons, as discussed above.

Protons are the most abundant particles in cosmic rays and are among the particles populating the radiation belts in the space near the Earth. Protons are commonly used as beam particles in experiments of particle physics. As already mentioned in Sect. 4.2.1.5, for protons with kinetic energies above 10 MeV, the damage function was found to be slightly dependent on E_d (see [Summers, Burke, Shapiro, Messenger and Walters (1993)], for instance). Furthermore, by an inspection of

---

§The reader can find the definition of kinetic energies per amu in Sect. 1.4.1.
**This is the part of the silicon substrate relevant for VLSI components and located in front of the incoming ion beam.
††This damage-energy deposition allows one to generate a concentration of Frenkel-pairs of about 10^{15} cm^{-3} for $E_d \approx 25$ eV.
¶At these energies, the electronic energy-loss is far larger than the nuclear energy-loss and determines the corresponding ionization dose.

Fig. 4.20, we can note that there is a good agreement[‡‡] among the different calcu-lations for the NIEL deposition of protons with kinetic energy of about 1000 MeV ($\beta\gamma \approx 1.81$); the deposition of a damage energy corresponding to $D^{\mathrm{NIEL}} \approx 4.3\,\mathrm{Gy}$ [e.g. see Eq. (4.121)] requires a fluence of about 1.92×10^{13} protons/cm^2. In a sub-strate (500–900) μm thick, the probability of p–Si strong interactions is small[†] and, to a first approximation, does not affect the amount of energy deposited by ioni-zation processes induced by the incoming proton in the silicon absorber. Since fast δ-rays can escape from the medium, the deposited energy has to be estimated by the so-called *restricted energy-loss formula*. For 900 μm thick Si detectors, the restricted energy-loss (based on measured values, see Sect. 2.1.1.3) is about 394 eV/μm and, as a consequence, for such a fluence of protons we have

$$D^{\mathrm{Ion}} \approx 5.2 \times 10^3 \,\, \mathrm{Gy}. \tag{4.123}$$

Electrons with a kinetic energy larger than about (250–260) keV can induce initial displacement damages with a very small (if at all) subsequent cascade, while above a few MeV, displacements result also from a small cascade development. Let us examine, for instance, the energy deposition of electrons with 9.1 MeV kinetic energy used for the investigation reported in [Codegoni et al. (2006)]. In Fig. 4.22, we can see that electrons with a kinetic energy of 9.1 MeV ($\beta\gamma \approx 18.8$) have a NIEL of about 2.42×10^{-4} MeV/cm. Thus, a fluence of about 2.58×10^{14} electrons/cm^2 is required to deposit $D^{\mathrm{NIEL}} \approx 4.3\,\mathrm{Gy}$[*] [e.g. see Eq. (4.121)]. To a first approximation, since the fastest δ-rays can escape from a silicon medium (500–900) μm thick, the collision (i.e. ionization) energy-losses of such electrons are properly described by the restricted energy-loss[‡] deposition, which amounts to about 355 eV/μm. A fluence of about 2.58×10^{14} electrons/cm^2 with kinetic energy of 9.1 MeV deposits an ionization dose

$$D^{\mathrm{Ion}}_{\mathrm{coll}} \approx 6.3 \times 10^4 \,\, \mathrm{Gy}, \tag{4.124}$$

to account for the collision-loss processes. Furthermore, in silicon absorbers, the energy-loss by collisions of these electrons accounts for about 86% of the total energy-loss (i.e., collision and radiative). As a consequence, the overall dose for ionization process, D^{Ion}, is slightly larger because it has also to account for the energy deposition (by ionization) of the secondary electrons emitted by the intera-ctions of radiated photons.

[‡‡]These calculations assume different values [(20–21) eV] for the displacement energy and, as a consequence, the estimated concentrations of Frenkel-pairs may differ up to $\approx 20\,\%$ for $E_d = 25\,\mathrm{eV}$.

[†]The collision length in silicon is about 30.3 cm (see Table 3.4) and the probability of interaction is about 0.3% in a silicon medium 900 μm thick.

[*]The NIEL calculations, reported in Sect. 4.2.1.5, show that the deposited damage-energy slightly depends on the displacement energy. For $E_d = 25\,\mathrm{eV}$ and a fluence of 2.58×10^{14} electrons/cm^2, the concentration of Frenkel-pairs is about 10^{15} cm^{-3}.

[‡]The restricted energy-loss for massive (i.e., protons) particles and electrons are described by similar equations {see Sects. 2.1.1.4 and 2.1.2.3 (or see Sections 2.1.1.3 and 2.1.5.2 of [Leroy and Rancoita (2004)])}.

In summary (see Table 4.5), for particle energies of typical space and high-energy physics environments, the energy deposited by ionization is about 3–4 orders of magnitude larger than that deposited by damage-energy for charged particles and isotopes, while they differ marginally for fast neutrons.

4.2.3.2 *Ionization Damage*

Total ionizing dose (TID) causes the threshold voltage to change due to charge trapping in the SiO_2 gate insulator. Ionization creates electron-hole pairs in MOS gate oxide. In semiconductor oxide layers, such as SiO_2, electrons produced by ionization rapidly move from the oxide layer, under the influence of the electric field created by the bias voltage. The holes, having lower mobility, diffuse slowly near the Si–SiO_2 interface, where they will accumulate as a charged layer, modifying the operating features of the device. The presence of these trapped holes is responsible for an electrical field which induces a negative shift of the threshold voltage, modifying the operating features of the device (NMOS and PMOS transistors). This effect is reduced by the recombination of some of the electrons with holes so reducing, in turn, the amount of holes trapped in the oxide and, consequently, reducing the negative shift of the threshold voltage. This shift is dose-rate dependent as a high dose-rate induces a high instantaneous density of electron-hole pairs, i.e., part of the electrons cannot leave far the oxide and recombine very fast, causing enhancement of the recombination rate and reduction of the shift. In case of low dose-rate, the induced low instantaneous density of electron-hole pairs allows electrons to quickly leave the oxide reducing the recombination effect and the shift. After some time, a fraction of the holes will leave the oxide at the Si–SiO_2 interface and generate defects in Si near the interface reducing the carriers mobility. These defects are traps for electrical charges depending on the bias applied on the gate. In PMOS transistors, the negative bias applied on the gate and the positive charges trapped in the interface states - trapped holes in gate oxide - produce an electrical field which induces a negative shift of the threshold voltage which enhances the threshold voltage shift initially produced by the gate oxide trapped holes. In NMOS transistors, the positive bias applied on the gate and the negative charges trapped at the interface produce an electrical field which induces a positive shift of the threshold voltage which depletes the threshold voltage shift initially produced by the gate oxide trapped holes. This effect in NMOS transistors is also rate dependent since it depends on the density of interface states which depends, in turn, on the holes density accumulated in the gate oxide near the Si–SiO_2 interface. In PMOS transistors, the global threshold-voltage shift is always negative and increases with decreasing dose-rate, while in NMOS transistors the global threshold-voltage shift depends on the ratio between oxide charge density and interface charge density which both increase with decreasing dose-rate. In MOS transistors, TID effects also result in varying the sub-threshold slope (e.g., see Chapter 1 in [Ma and Dressendorfer (1989)], also [Codegoni et al. (2004a)] and references therein). In irradiated bipolar transistors the density of energy states, at the interface between silicon and oxide surrounding the emitter, is high and induces a parasitic base-current from the recombination of minority carriers (injected from the emitter into the base) with majority carriers (from the base). The parasitic current, added to the regular base current decreases the gain of the bipolar transistor. However, at high collector current, saturation of the interface states by carriers from the emitter and recombination with majority carriers from the base brings the parasitic base current to a saturation value, which becomes negligible and therefore the gain is almost unchanged from its value before irradiation. At low collector current, the absence of interface states saturation maintains the parasitic recombination current and degrades the gain compared with its value before irradiation. Again, TID (see page 293) induces higher interface state density at low dose-rate compared with high dose-rate and gain degradation induced by TID is higher at low dose-rate than at high dose-rate. The degradation of bipolar transistor performance by TID that increases when dose-rate decreases is called *low dose-rate effect* (LDRE) and becomes significant with dose-rate $\sim 50\,\mathrm{rad\,s^{-1}}$ and increases continuously with dose-rate decreasing down to $1 \times 10^{-3}\,\mathrm{rad\,s^{-1}}$ or even less. LDRE also occurs for low TID (a few krads). From the above discussion, LDRE depends on the device architecture, the structure and thickness of the oxide layer surrounding the emitter and on bias conditions. LDRE degrades performance of the bipolar circuits, in particular by degrading the gain of amplifiers, of input offset voltages and input bias currents.

The ionizing dose affects other devices, for instance the *field effect transistors*. A treatment of TID effects is beyond the purpose of the present book. However a detailed discussion of these effects can be found in literature (e.g., see [Ma and Dressendorfer (1989)], Chapter 6 of [Messenger and Ash (1992)], Chapter 6 of [Claeys and Simoen (2002)], [Holmes-Siedle and Adams (2002)] and, also, the chapter on *Displacement Damage and Particle Interactions in Silicon Devices* and references therein). A survey of radiation damage in semiconductor devices is given in [ECSS (2005)].

4.3   Radiation Induced Defects and Modification of Silicon Bulk and $p - n$ Junction Properties

Defects and clusters of defects do not behave as donors or acceptors which are intentionally introduced into lattice sites to modify in a controlled-way the intrinsic properties of a semiconductor*.

Radiation-induced defect centers have a major impact on the electrical behavior of semiconductor devices and can deeply affect their properties (e.g., see [Srour, Long, Millward, Fitzwilson and Chadsey (1984); Srour and McGarrity (1988)] and

---

*Studies have been carried out to investigate the improvement of the radiation hardness by the so-called defect-engineering (see for instance [Kozlovski and Abrosimova (2005); Msimanga and McPherson (2006)] and references therein).

references therein). For instance, centers with energy-levels near the mid-gap make a significant contribution to carrier generation. Thermal generation of electron–hole pairs dominates over capture processes, when the free carrier concentrations are much lower than the thermal equilibrium values, like in the depletion regions (e.g., see Chapter 6 in [Grove (1967)]). These centers become the main mechanism for increasing the leakage current in silicon devices after irradiation. Furthermore, the electron-hole recombination occurs when a free carrier of one sign can be captured at defect centers (or recombination centers), followed by capture of a carrier of the opposite sign. Radiation-induced recombination centers is the relevant mechanism to decrease the minority-carrier lifetime which, in turn, is the dominant mechanism for gain degradation in bipolar transistors (e.g., see Sect. 7.1). In addition, donors or acceptors can be compensated by deep-lying radiation-induced centers. That results in the decrease of the concentration of majority carriers. This process (referred to as *carrier removal*) causes the variation of the device properties depending on majority carrier concentration, for example the increase of collector resistance in bipolar transistors. Temporary trapping of carriers can typically occur at a *shallow level*, with no recombination. Both majority- and minority-carriers can be trapped (in separate levels). Moreover, a *tunneling process* can allow the passage through a potential barrier by means of defect levels. For example, there may be a defect-assisted tunneling component of the reverse current in $p-n$ junction diodes.

4.3.1 *Displacement Damage Effect on Minority Carrier Lifetime*

Among the most important semiconductor material-parameters for practical applications in electronic devices, we have the (excess-) carrier lifetime, the equilibrium majority-carrier concentration and the majority-carrier mobility (e.g. see [Schroder (1997)]). To a first approximation[†], the rate at which the electrical properties of semiconductors are degraded by irradiation is often expressed in terms of a *damage coefficient*. For instance, the *minority carrier (recombination) lifetime*, τ, is given by

$$\frac{1}{\tau_{\text{irr}}} - \frac{1}{\tau} \equiv \Delta\left(\frac{1}{\tau}\right)$$

$$= \frac{\Phi_{\text{i}}}{K_{\tau,\text{i}}}, \tag{4.125}$$

where τ_{irr} and τ are the lifetimes after and before the irradiation with a fluence Φ_{i} of particles, respectively; $K_{\tau,\text{i}}$ is the *(recombination) lifetime damage coefficient*[‡], which may depend on a) the type of substrate, b) the dopant concentration, c) the level of compensation and d) the type ("i") and energy of irradiating particles

---

[†]The linear dependence on the particle fluence [see Eq. (4.125)] is expected as long as the steady-state Fermi level is not significantly moved.

[‡]A coefficient that is the reciprocal of that given in Eq. (4.125) is also found and used in literature (e.g., see Section 7.2.1 of [van Lint, Flanahan, Leadon, Naber and Rogers (1980)] and references therein).

(e.g., see Section 5 in Chapter I of Part II of [Vavilov and Ukhin (1977)], Section 3.4 of [Srour, Long, Millward, Fitzwilson and Chadsey (1984)], [Srour and McGarrity (1988)] and references therein). The values of $K_{\tau,i}$ can be usually found in literature. For instance for low-resistivity silicon, experimental results on the lifetime damage constants are presented in [Srour, Othmer and Chiu (1975)] for 0.5, 1.0 and 2.5 MeV electrons with fluences up to $\simeq 3 \times 10^{15}$ e/cm^2 and 10 MeV protons with fluences up to $\simeq 1.2 \times 10^{12}$ p/cm^2 (see also Section 3.4 of [Srour, Long, Millward, Fitzwilson and Chadsey (1984)]). Furthermore, the damage constant $K_{\tau,n}$ was extensively investigated[††] for fast-neutron fluences (typically) up to 10^{11}–10^{12} n/cm^2 and its values[§] are available in literature. For n- and p-type silicon, $K_{\tau,n}$ is almost independent of the silicon resistivity (below a few Ω cm). In addition, the annealing effect on the carrier lifetime was studied at room temperature: the minority-carrier lifetime was found to undergo both short-term and long-term annealing (see pages 32–45 in [Srour, Long, Millward, Fitzwilson and Chadsey (1984)] and [Srour (1973)]).

For a fast-neutron irradiation with a spectral fluence $\phi(E)$ [where E is the kinetic energy of the neutron, see Eq. (4.77)], the ratio $\Phi_n/K_{\tau,n}$ can be re-expressed to account for the creation of recombination centers and the absorption of minority carriers as {see Equation (5.15) of [Messenger and Ash (1992)], and also [Codegoni et al. (2004b); Consolandi, D'Angelo, Fallica, Mangoni, Modica, Pensotti and Rancoita (2006)]}:

$$\frac{\Phi_n}{K_{\tau,n}} = \sigma_m \, v_e n_{\mathrm{Si}} \int_{E_{\min}} \sigma_{c,n}(E) \, \phi(E) \, dE$$
$$= \sigma_m \, v_e n_{\mathrm{Si}} <\sigma_{c,n}> \Phi_n, \tag{4.126}$$

where n_{Si} is the number of atoms per cm^3 in the bulk silicon [Eq. (4.82)], $E_{\min}$ is the minimal threshold of the neutron energy for inducing displacement damage (see page 348), Φ_n is the fast-neutron fluence [Eq. (4.78)], v_e is the average speed of minority carriers, σ_m is the cross section for the absorption of minority carriers by recombination centers; $\sigma_{c,n}(E)$ and $<\sigma_{c,n}>$ are the cross section of neutrons with energy E for creation of recombination centers in silicon and its average value, respectively. $<\sigma_{c,n}>$ is computed according to

$$<\sigma_{c,n}> = \frac{\int_{E_{\min}} \sigma_{c,n}(E) \, \phi(E) \, dE}{\Phi_n}. \tag{4.127}$$

The term

$$C_c = n_{\mathrm{Si}} \int_{E_{\min}} \sigma_{c,n}(E) \, \phi(E) \, dE$$
$$= n_{\mathrm{Si}} <\sigma_{c,n}> \Phi_n \quad [\mathrm{cm}^{-3}] \tag{4.128}$$

[††]The reader can see, e.g., [Messenger (1967b); Srour (1973)], Section 5 in Chapter I of Part II of [Vavilov and Ukhin (1977)], Section 5.9 of [Messenger and Ash (1992)] and references therein.

[§]Below 2–3 Ω cm [Messenger (1967b)], for n-type silicon $K_{\tau,n}$ is $\simeq (1.0$–$1.6) \times 10^5$ s/cm^2 and for p-type silicon $K_{\tau,n} \simeq (1.5$–$3.0) \times 10^5$ s/cm^2.

is the concentration of recombination centers resulting from the displacement processes induced by the fast-neutrons. To a first approximation (e.g., see Sects. 7.1.1 and 7.1.3), in absence of saturation effects[¶], (mostly) for low-resistivity silicon we can assume that the concentration of recombination centers is proportional to the energy deposited by non-ionizing energy-loss (NIEL) processes per unit volume $E_{\rm dis}$ (Sect. 4.2.1) and, consequently [see Eq. (4.83)], to the concentration of Frenkel-pairs (FP) introduced as primary point-defects. For instance, in low-resistivity silicon the concentration of dopants is sufficiently large to keep them participating to the phenomenon of complex-defect formation even at large neutron fluences, although this process reduces the concentration of available impurities. Therefore, Eq. (4.128) can be expressed as

$$E_{\rm dis} \propto \mathcal{C}_{\rm c} = \gamma_{\rm dis}\, FP. \qquad (4.129)$$

For deep defects resulting from primary defects mostly created by cascading-displacement processes, the term $\gamma_{\rm dis}$ may result to be slightly[‖] (if at all) dependent on the type of incoming particles.

By means of Eqs. (4.128, 4.129), Eq. (4.126) can be rewritten as

$$
\begin{aligned}
\frac{\Phi_{\rm n}}{K_{\tau,{\rm n}}} &= \sigma_m\, \nu_e\, \mathcal{C}_{\rm c} \\
&= \sigma_m\, \nu_e \gamma_{\rm dis}\, FP, \qquad (4.130)
\end{aligned}
$$

where σ_m and ν_e are almost independent of the properties of the fast-neutron spectral fluence. These latter terms, similarly to the parameter $\gamma_{\rm dis}$, are expected to be slightly (if at all) dependent on the type of the incoming particle, when deep defects result from primary defects mostly created by cascading-displacement processes. Furthermore, by combining Eqs. (4.125, 4.130), (mostly) for low-resistivity silicon we obtain

$$
\begin{aligned}
\frac{1}{\tau_{\rm irr}} - \frac{1}{\tau} &= \frac{\Phi_{\rm i}}{K_{\tau,{\rm i}}} \\
&\simeq \lambda\, FP, \qquad (4.131)
\end{aligned}
$$

where

$$\lambda = \sigma_m\, \nu_e \gamma_{\rm dis}$$

is almost independent of the type and energy of the incoming particle, but depends on i) the type of substrate, ii) (slightly) the dopant concentration and iii) the level of compensation. Equations (4.129, 4.131) indicate that an *approximate NIEL scaling* (e.g., see Sect. 4.1.1.2) is expected for the variation of the reciprocal of the minority-carrier lifetime in low-resistivity silicon.

Finally, it has to be added that expressions similar to Eq. (4.125) have been formulated for the generation lifetime (e.g., see Section 11.2.4 of [Lutz (2001)]),

---

[¶]For high-resistivity silicon (see discussion in Sect. 4.3.5), it was found that there are secondary defects whose concentrations are not linearly dependent on fluence.

[‖]NIEL scaling violation was observed in high-resistivity silicon (e.g., see Sect. 6.8.3).

minority-carrier diffusion length (e.g., see Section 3.2.7.2 of [Holmes-Siedle and Adams (2002)]), majority-carrier mobility and concentration (e.g., see Section 7.2.2 of [van Lint, Flanahan, Leadon, Naber and Rogers (1980)]).

4.3.2 *Carrier Generation and Leakage Current*

Silicon detectors are usually *one-sided*** $p^+ - n$ junctions[‡] and are referred to as *n-type silicon detectors*. In these detectors, the highly doped p^+-region (usually with a dopant concentration of $\approx 10^{18}\,\mathrm{cm^{-3}}$) extends over $\approx 0.5\,\mu$m at a depth of (2–2.5) μm and is on top of a lowly doped n-substrate (with a dopant concentration of $\approx 6\times 10^{11}$–$2.4\times 10^{12}\,\mathrm{cm^{-3}}$), whereas on the rear-side there is a highly doped n^+-layer of $\approx 0.5\,\mu$m thickness also at a depth of 2–2.5 μm and with a dopant concentration of $\approx 10^{19}\,\mathrm{cm^{-3}}$ (sometimes indicated as n^{++}). One of the purposes of the n^+-region is to allow the n-substrate to make a good external ohmic-connection[††] (e.g., see Section 5.1 of [Sze (1985)] and Appendix B9 of [Ng (2002)]) with, for instance, the external bias supplier. In these devices, the electrical characteristics (for example, $I - V$ and $C - V$) are determined by those of the one-sided $p^+ - n$ junctions.

At room temperature, when a reverse voltage[‡‡], V_{r}, smaller than the breakdown voltage and larger than $(3k_BT)/e \approx 78\,\mathrm{mV}^*$ is applied to a non-irradiated silicon detector, the reverse current density[†] is approximated by the sum of both the diffusion density current in the diffusion regions and the generation current in the depletion region (see, for instance, [Sah, Noyce and Shockley (1957); Moll (1958)], Section 3.4 of [Sze (1985)], Sections 3.7–3.8 of [Messenger and Ash (1992)], Sections 9.1–9.4 of [Bar-Lev (1993)] and Chapter 8 of [Neamen (2002)]):

$$J_{\mathrm{r}} \approx J_{\mathrm{s}} + \frac{e\,n_{\mathrm{int}}W}{\tau_{\mathrm{gn}}}, \qquad (4.132)$$

where τ_{gn} is the *effective generation lifetime*, n_{int} is the intrinsic carrier concentration (e.g., see Chapter 6), W is the depletion layer width and J_{s} is the so-called *saturation current density*. It can be shown that only those generation-recombination centers with an energy level near the intrinsic Fermi level can significantly contribute to the generation rate (see for instance [Sah, Noyce and Shockley (1957)]

**The one-sided $p^+ - n$ junctions [Wolf (1971)] are asymmetrical step-junctions in which the p-side is much more heavily doped than the adjacent n-side.

[‡]For a description of the properties of the $p - n$ junctions, one can see, for instance, Section 6.2 of [Wolf (1971)].

[††]The non-rectifying metal-(n^+-)semiconductor contact results in an almost abrupt junction with the built-in potential determined by the metal work-function, the electron affinity of the semiconductor and the potential difference between the Fermi level and the bottom level of the conduction band; the current transport is mainly due to majority carriers (e.g., see pages 160–171 of [Sze (1985)]).

[‡‡]In this Section, V_{r} indicates the absolute value of the reverse voltage.

*k_B is the Boltzmann constant, T the is temperature in Kelvin and $e = 1.6 \times 10^{-19}\,\mathrm{C}$ is the electronic charge.

[†]The contribution of the surface leakage current (e.g., see Section 10.3 of [Grove (1967)]) has not been taken into account in Eq. (4.132) (see also Sects. 6.1.6).

and Section 3.4 of [Sze (1985)]). For an n-type silicon detector, the saturation density current is expressed by that of the *real diode equation*, i.e.:

$$J_\mathrm{s} \approx e \, \frac{n_\mathrm{int}^2}{N_d} \sqrt{\frac{D_p}{\tau_p}}, \tag{4.133}$$

where N_d is the donor concentration, D_p and τ_p are the diffusion coefficient and the lifetime of holes in the n-region, respectively.

As mentioned above, radiation-induced defect centers, whose energy levels are near the intrinsic Fermi level, can become a relevant source for carrier generation and determine the increase of the volume-generated reverse-bias current inside the depleted region of silicon devices. Since, to a first approximation, the concentration of defects is proportional to the particle fluence, the reverse current after the irradiation ($I_\mathrm{r,irr}$) is expected to increase with increasing particle fluence. Experimental studies* carried out on reverse-biased silicon detectors have confirmed that $I_\mathrm{r,irr}$ (in A) depends, to a first approximation, linearly on the particle fluence:

$$I_\mathrm{r,irr} \simeq I_\mathrm{r} + \alpha_\mathrm{i} \, V_\mathrm{vol} \, \Phi_\mathrm{i}, \tag{4.134}$$

where I_r (in A) is the detector leakage current before the irradiation at a fluence Φ_i (in particles/cm^2), V_vol (in cm^3) is the depleted volume of the detector and α_i (in A/cm) is the so-called radiation-induced *reverse current damage constant*. This latter coefficient depends, in turn, on the type ("i") and energy of the irradiating particles (see Sects. 4.2.1.1–4.2.1.5). The processes resulting in self-annealing at room temperature (see page 373) allow the leakage current to decrease with time.

The time dependence of the leakage current was investigated from 1.5 hours up to 26 months after the irradiations of high-resistivity[†] Fz silicon detectors exposed to 2 MeV neutrons with fluences between 4.8×10^{12} and 1.3×10^{13} n/cm^2 [SICAPO Collab. (1994b)]. The detectors were stored and kept at room temperature. The measurements were performed at a temperature of 20 °C and at full depletion voltage. It has been determined that the reverse current damage constant (α_n) decreases with time as

$$\alpha_\mathrm{n} = 10^a t^{-b} \times 10^{-17} \quad [\mathrm{A/cm}], \tag{4.135}$$

where $a = 1.17 \pm 0.01$, $b = (1.45 \pm 0.86) \times 10^{-1}$ and t (≥ 1.5) is the number of hours after the irradiation [SICAPO Collab. (1994b)]. After a long-term annealing, the reverse current damage constants are $\approx 2 \times 10^{-17}$ [A/cm] and $\approx 3 \times 10^{-17}$ [A/cm] for 1 MeV neutrons and minimum ionizing protons (and pions), respectively (see Section 28.8 of [PDB (2008)] and, also, [Moscatelli et al. (2002)]).

At higher fast-neutron fluences, a stronger rise of $I_\mathrm{r,irr}$ as function of the neutron fluence was observed (see, for instance, Fig. 11.11 in Section 11.2.4 of [Lutz

---

*One can see, for instance, Sect. 6.8.2 and, also, [Fretwurst et al. (1993); Leroy, Glaser, Heijne, Jarron, Lemeilleur, Rioux, Soave and Trigger (1993); SICAPO Collab. (1994b)], [Bechevet, Glaser, Houdayer, Lebel, Leroy, Moll and Roy (2002)], Section 6.4.1 of [Holmes-Siedle and Adams (2002)] and references therein.

†The resistivities were between $(4$–$7)$ kΩ cm.

(2001)]). In fact, at these fluences, the $I - V$ reverse characteristics are largely modified (e.g., see Sect. 4.3.3.1) and determine an increase of the *effective* reverse current damage constant.

4.3.3 *Diode Structure and Rectification Down to Cryogenic Temperature*

The characteristics of semiconductor diodes, like those of other semiconductor devices, are substantially affected by irradiation. The rate and level of radiation changes involve the properties of the bulk region, of the ohmic contact and the surface effects. Thus, at large fluences, both the forward and reverse $I - V$ characteristics are different with respect to those of a non-irradiated diode. After irradiation, these characteristics depend on the type and fluence of particles and on the junction and bulk properties of the silicon. For instance, at large particle fluences, semiconductor diodes lose (largely) the features of being a *rectifying device* to become more similar to linear semiconductor resistors with a low (germanium) or high (silicon) resistivity (see for instance Section 1 in Chapter II of Part II of [Vavilov and Ukhin (1977)], [Croitoru, Gambirasio, Rancoita and Seidman (1996)] and references therein).

4.3.3.1 *Rectification Property Up to Large Fast-Neutron Fluences at Room Temperature*

The rectification property of a $p - n$ diode results from the creation of a depletion region at thermal equilibrium. It is one of the most important characteristics of non-irradiated $p - n$ junctions, that is they allow current to flow easily in one direction. When we apply a *forward bias*, the electrostatic potential across the junction region is reduced from that of the thermal equilibrium [the so-called *built-in potential*, see Eq. (6.12)], thus the diffusion of majority carriers from one to the other side of the junction is enhanced and, as a consequence, the *forward current* increases rapidly with increasing voltage. However, under *reverse bias*, the applied voltage increases the electrostatic potential across the depleted region, as a consequence determines a decrease of the diffusion currents and, finally, results in a small *reverse* (or *leakage*) *current*. As the reverse bias is increased, the reverse current remains very small [Eq. (4.132)], until a critical voltage (the so-called *breakdown voltage*) is reached. Above the breakdown voltage, the reverse current becomes very large[‡].

The forward current-voltage characteristics of a non-irradiated diode accounts for both the diffusion and recombination currents (see for instance [Sah, Noyce and Shockley (1957); Moll (1958)], Section 3.4 of [Sze (1985)], Sections 3.7–3.8 of [Messenger and Ash (1992)], Sections 9.1–9.4 of [Bar-Lev (1993)] and Chapter 8 of [Neamen (2002)]). In fact under low-injection condition[§] and assuming that the

---

[‡]This phenomenon is also referred to as *junction breakdown*.

[§]This condition is satisfied when the injected minority carrier concentrations are small compared

depletion region of the diode has abrupt boundaries, the *ideal diode equation* for the density current due to carrier diffusion ($J_{f,diff}$) is replaced by the *real diode equation*:

$$J_f \simeq J_{f,diff} + \frac{e\, n_{int} W}{2\tau_{rc}} \exp\left(\frac{e\, V_f}{2\, k_B T}\right), \qquad (4.136)$$

where V_f is the applied forward voltage¶, τ_{rc} is the *effective recombination lifetime*, n_{int} is the intrinsic carrier concentration, W is the depletion layer width and $J_{f,diff}$ is given by

$$J_{f,diff} = J_s \left[\exp\left(\frac{e\, V_f}{k_B T}\right) - 1\right]. \qquad (4.137)$$

J_s is the saturation current density and is given by Eq. (4.133) for n-type silicon detectors. The second term of Eq. (4.136) accounts for the so-called *generation current* in the depletion layer width. It can be shown that only those generation-recombination centers with an energy level near the intrinsic Fermi level can contribute significantly to the recombination rate (see for instance [Sah, Noyce and Shockley (1957)] and Section 3.4 of [Sze (1985)]).

In general, the measured forward currents can be represented empirically by:

$$I_f \propto \exp\left(\frac{e\, V_f}{\eta\, k_B T}\right), \qquad (4.138)$$

where η is referred to as *ideality factor*‖. At room temperature, in silicon $p-n$ diodes the recombination current dominates at low-forward voltages, while for $V_f >$ 0.5 V the diffusion current dominates (e.g., see Fig. 6.24 in Section 6.6.b of [Grove (1967)]). At even larger forward currents, the ideality factor becomes again larger than 1. This phenomenon results from the effect of series resistance and from the high-injection of carriers.

In $p-i-n$ diodes, there is an almost-intrinsic (or a weakly doped) layer sandwiched between the heavily doped p^+- and n^+-regions**. In practice, the central region is either weakly doped p-type (indicated as π-) or weakly doped n-type (indicated as ν-) silicon, with typical dopant concentrations of $\approx (10^{12}$–$10^{13})\,\mathrm{cm}^{-3}$ and ranges $\approx (10$–$200)\,\mu m$ (see, for instance, Chapter 2 of [Ng (2002)]). In these diodes, the $I-V$ characteristic is approximately expressed by (e.g., see Section 15.2d of [van der Ziel (1976)], Sections 3.1–3.3.1.1 of [Ghandhi (1977)], Section 11.2 of [Tyagi (1991)], Chapter 2 of [Ng (2002)], see also [Fletcher (1957); Nussbaum (1973)]):

$$I_{f,pin} \propto F_L \exp\left(\frac{e\, V_f}{m\, k_B T}\right), \qquad (4.139)$$

---

with the majority carrier concentrations.

¶In this Section V_f indicates the absolute value of the applied forward voltage.

‖The ideality factor is $\simeq 1$ when the diffusion current dominates, $\simeq 2$ when the recombination current dominates.

**The size of the n^+-region (see for instance the doping profiles of a silicon $p^+ - n - n^+$ power rectifier in Section 18.1.1 of [Tyagi (1991)] and references therein) is such that it does not operate only as an ohmic contact, as it occurs in $p^+ - n$ junction.

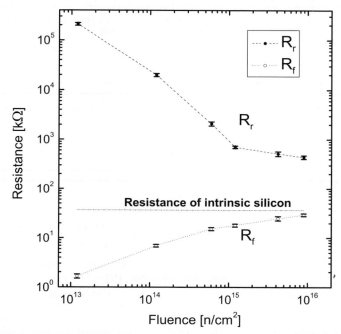

Fig. 4.25   Reverse (R_r) and forward (R_f) resistances in kΩ as functions of the fast neutron fluence in n/cm^2 at room temperature (from [Leroy and Rancoita (2007)]; see also [Croitoru, Gambirasio, Rancoita and Seidman (1996)]). The dotted and dashed lines are to guide the eye. For these detectors the equivalent resistance, in case of an intrinsic silicon medium, was calculated for an intrinsic silicon resistivity of about 2.3×10^5 Ω cm at 300 K (see, for instance, Table 1.20 of [Wolf (1971)]).

where $1 < m < 2$ depends on the dopant concentrations and on the levels of the recombination centers [Nussbaum (1973)]. F_L is a function of the ratio w/L_a, where w is the thickness of the quasi-intrinsic layer and L_a is the ambipolar diffusion length: it reaches its maximum value (≈ 0.3) for $w/L_a \approx 1$ (e.g., see Sections 3.1–3.3.1 of [Ghandhi (1977)]).

At room temperature, the forward (and reverse) $I - V$ characteristics of n-type silicon detectors[††] of 400 μm thickness and resistivity (4–6) kΩ cm have been measured before and after irradiation with fast-neutrons. Before irradiation, for

$$3.8 \lesssim \frac{eV}{k_B T} \lesssim 25$$

these low-doped devices exhibit an ideality factor which is larger than that expected from Eq. (4.138) and becomes ≈ 4 at large V_f [SICAPO Collab. (1995d)][*]. After

---

[††]The dopant concentrations of the layers are close to those indicated at page 385.
[*]The ideality factor was also found to be larger than expected in 300 μm thick devices [Beattie, Chilingarov and Sloan (1998)].

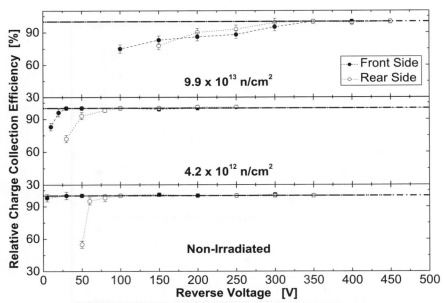

Fig. 4.26   Relative charge-collection efficiency (R_{eff}, see text at page 393) measured with an ^{241}Am α-source located at the front ($\bullet$) and at the rear ($\circ$) of non-irradiated and irradiated silicon detectors with fast-neutron fluences of 4.2×10^{12} n/cm^2 and 9.9×10^{13} n/cm^2 (from [Leroy and Rancoita (2007)]; see also [SICAPO Collab. (1994b,c)]). The dashed line and dotted line are to guide the eye. The dashed and dotted line represents $R_{\text{eff}} = 100\%$.

irradiation[†], the forward $I - V$ characteristic is modified and becomes a function of the fast-neutron fluence [SICAPO Collab. (1995d)]: a similar behavior was experimentally observed in other silicon detectors and photodiodes {e.g., see, for silicon detectors, [Croitoru, Gambirasio, Rancoita and Seidman (1996); Beattie, Chilingarov and Sloan (1998, 2000); Granata et al. (2000)] and, for high-resistivity ($\approx 5\,\text{k}\Omega\,\text{cm}$) n-type photodiodes, [Korde et al. (1989)]}. In addition [see Eq. (4.138)], at the fluence of $\approx 10^{14}$ n/cm^2 $I_{\text{f,irr}}$ depends exponentially on V_{f} (with an ideality factor of ≈ 2 [SICAPO Collab. (1995d)]) only up to $\approx 0.13\,\text{V}$ (i.e., $(eV)/(k_BT) \approx 5$). In practice, for fluences $\gtrsim 1.2 \times 10^{13}$ n/cm^2, the experimental data show that $I_{\text{f,irr}}$ depends linearly on V_{f} [Croitoru, Gambirasio, Rancoita and Seidman (1996)] above a critical forward voltage $V_{\text{f,c}}$ which depends, in turn, on the fluence, i.e.,

$$I_{\text{f,irr}} = \mathcal{C}_{\text{f}}(\Phi_n) + \frac{V_{\text{f}}}{R_{\text{f}}(\Phi_n)}, \text{ for } V_{\text{f}} > V_{\text{f,c}}(\Phi_n). \qquad (4.140)$$

$V_{\text{f,c}}(\Phi_n)$ varies from $\approx 1.2\,\text{V}$ at the fluence of 1.2×10^{13} n/cm^2 up to $\approx 8\,\text{V}$ at 8.9×10^{15} n/cm^2. At the critical voltage[‡] $V_{\text{f,c}}$, $I_{\text{f,irr}}$ is 300–350 μA. The absolute values of $\mathcal{C}_{\text{f}}(\Phi_n)$[§] are small but not negligible compared with $I_{\text{f,irr}}(V_{\text{f,c}})$. Thus,

---

[†]Experimental data are available up to fast-neutron fluences of $\approx 10^{16}$ n/cm^2.

[‡]Before irradiation, for $I_{\text{f}} \approx 350\,\mu$A we have $V_{\text{f}} \lesssim 0.5\,\text{V}$.

[§]$\mathcal{C}_{\text{f}}(\Phi_n)$ is negative for fast-neutron fluences up to $\approx 1.2 \times 10^{14}$ n/cm^2 and positive at larger

Table 4.6   Values of the parameters $m_{\rm eff}$ and $b_{\rm eff}$ in Eq. (4.143) for Fz-detectors with resistivities ranging from 5 to $7\,{\rm k\Omega\,cm}$ [SICAPO Collab. (1994b)]. These parameters have been computed for an α source located at the front and at the rear side of silicon detectors (from [Leroy and Rancoita (2007)]).

	Front Side	Rear Side
$m_{\rm eff}(\times 10^{-2})$	-5.12 ± 0.93	-7.51 ± 0.88
$b_{\rm eff}$	1.64 ± 0.13	1.94 ± 0.12

Eq. (4.140) expresses that the device behaves nearly as an ohmic medium, although a (largely) reduced rectification is maintained even at the largest fluence ([Croitoru, Gambirasio, Rancoita and Seidman (1996)], see also [SICAPO Collab. (1995d); Beattie, Chilingarov and Sloan (1998, 2000); Granata et al. (2000)]). In Fig. 4.25, the values of the forward resistance ($R_{\rm f}$) are shown as a function of the fast-neutron fluence: they approach that of the intrinsic silicon with increasing fluence.

In these detectors irradiated with fast neutron fluences $\gtrsim 1.2 \times 10^{13}\,{\rm n/cm^2}$, the reverse $I - V$ characteristics show that the reverse current $I_{\rm r,irr}$ monotonously increases with $V_{\rm r}$ ([Croitoru, Gambirasio, Rancoita and Seidman (1996)], see also [Granata et al. (2000)]), i.e., the current does not approach a constant value as predicted by Eq. (4.132). At large reverse bias, $I_{\rm r,irr}$ becomes linearly dependent on $V_{\rm r}$:

$$I_{\rm r,irr} = C_{\rm r}(\Phi_{\rm n}) + \frac{V_{\rm r}}{R_{\rm r}(\Phi_{\rm n})}, \quad \text{for } V_{\rm r} > V_{\rm r,c}(\Phi_{\rm n}), \tag{4.141}$$

where $V_{\rm r,c}(\Phi_{\rm n})$ is $\approx 350\,{\rm V}$ at fluences between 1.2×10^{13} and $6 \times 10^{14}\,{\rm n/cm^2}$ and $\approx 90\,{\rm V}$ at $8.9 \times 10^{15}\,{\rm n/cm^2}$. At the critical voltage $V_{\rm r,c}$, $I_{\rm r,irr}$ is $\approx 5.5\,\mu{\rm A}$ ($\approx 300\,\mu{\rm A}$) at $1.2 \times 10^{13}\,{\rm n/cm^2}$ ($8.9 \times 10^{15}\,{\rm n/cm^2}$). In Fig. 4.25, the values of the reverse resistance ($R_{\rm r}$) are shown as a function of the fast-neutron fluence: similarly to those of the forward resistance, they approach that of the intrinsic silicon with increasing fluence. The *rectification ratio*, defined as

$$R_{\rm r/f}(\Phi_{\rm n}) = \frac{R_{\rm r}(\Phi_{\rm n})}{R_{\rm f}(\Phi_{\rm n})}, \tag{4.142}$$

is $\approx 1.27 \times 10^5$ at $1.2 \times 10^{13}\,{\rm n/cm^2}$ and decreases down to ≈ 14.8 at $8.9 \times 10^{15}\,{\rm n/cm^2}$.

In silicon rectifying diodes[‡‡], similarly to low-doped detectors, an increase of the fast-neutron fluence causes a decrease of the forward $I - V$ characteristic. At large

---

fluences [Croitoru, Gambirasio, Rancoita and Seidman (1996)].

[‡‡]One can see, e.g., [Easley (1962); Frank, Poblenz and Howard (1963)], Section 1 in Chapter II of Part II of [Vavilov and Ukhin (1977)] and also Sections 6.5 and 6.8.1 of [Holmes-Siedle and Adams (2002)].

fluences, the $I - V$ characteristic for both $p - n$ and $p - i - n$ rectifying diodes is approximately expressed by Eq. (4.140), where $\mathcal{C}_f(\Phi_n)$, $R_f(\Phi_n)$ and $V_{f,c}(\Phi_n)$ depend on the dopant concentrations. In addition, in $p - n$ diodes $V_f(\Phi_n)$ increases with increasing base thickness; in $p-i-n$ diodes it increases with increasing the thickness of the *quasi-intrinsic i-layer* (Figs. 82–85 at pages 204–207 of [Vavilov and Ukhin (1977)], see also [Shwartz et al. (1966)]). The experimental data on silicon $p - n$ and $p - i - n$ irradiated diodes show a qualitative and quantitative agreement with the predictions based on the *Vavilov–Ukhin model* (see pages 188–208 of [Vavilov and Ukhin (1977)]).

4.3.3.2 *Large Radiation Damage and $p - i - n$ Structure at Room Temperature*

It has been observed that $p - n$ diodes, with resistivities of several $\Omega\,\mathrm{cm}$, may gradually change their internal structure with increasing fast-neutron fluence. In fact, non-irradiated diodes with an ohmic (n^+-) contact can acquire a $p - i - n$ structure after irradiation. This was determined, for instance, by measurements of the potential distribution with respect to the p-region using the method of a moving probe. These measurements show that, in a forward-biased device with a resistivity of $\approx 32\,\Omega\,\mathrm{cm}$, an increased voltage drop occurs in the base-region and in the ohmic-contact with increasing neutron fluence above $\approx 2 \times 10^{13}\,\mathrm{n/cm^2}$ (e.g., see Fig. 86 at page 208 of [Vavilov and Ukhin (1977)]). The $n^+ - n$ ohmic-contact properties are conserved in lower resistivities devices ($\approx 2.1\,\Omega\,\mathrm{cm}$) up to fluence of $\approx 2 \times 10^{14}\,\mathrm{n/cm^2}$.

Similarly to low-resistivity $p - n$ diodes, n-type detectors with an n^+-region as ohmic-contact (e.g., see Sect. 4.3.3.1) gradually change their internal structure. In fact, with increasing particle fluence, the n-doped region becomes almost intrinsic (i, i.e., very weakly doped) and the n^+-layer behaves like a rectifying region. In addition, at fast-neutron fluences of $\approx 10^{14}\,\mathrm{n/cm^2}$, the behavior of the $p^+ - n$ device has been found to be compatible with that of a $p^+ - i - n^+$ diode with an ambipolar diffusion length (e.g., see page 389) of $\approx 200\,\mu\mathrm{m}$ [SICAPO Collab. (1995d)]. This phenomenon has been confirmed [SICAPO Collab. (1994b,c)] by employing fast shaped electronics [i.e., shaped signals of (20–25) ns peaking-time and $\approx 120\,\mathrm{ns}$ base-time] for measuring the charge collection[¶], when an $^{241}\mathrm{Am}$ α-source is located at the front or at the rear of non-irradiated and irradiated silicon detectors[‖]. The total range of an α-particle from an $^{241}\mathrm{Am}$ source is about $25\,\mu\mathrm{m}$ in silicon. As a consequence, the depleted layer width must be $\geq 25\,\mu\mathrm{m}$ in order to obtain the maximum (or full) charge-collection. Before irradiation (see Fig.4.26), full charge-collection is achieved at low reverse bias (a few volts), when the source is located at the front of the de-

---

[¶]A treatment of the charge transport and collection in silicon diodes can be found in the chapter on *Solid State Detectors* (see also Sects. 6.1.4, 6.1.5 and 6.8.2).

[‖]These devices are Fz-silicon detectors ($400\,\mu\mathrm{m}$ thick) with resistivities ranging from 5 to $7\,\mathrm{k}\Omega\,\mathrm{cm}$ [SICAPO Collab. (1994b)].

tector (i.e., of the p^+-region), while the reverse voltage for full depletion ($\approx 100\,\mathrm{V}$) is needed to achieve a full charge-collection when the source is located at the rear side of the detector (i.e., of the n^+-layer for the ohmic-contact). Furthermore, the reverse voltage for maximum charge-collection increases with increasing fluence[**] (e.g., see Fig. 4.26 for $4.2 \times 10^{12}\,\mathrm{n/cm^2}$ and $9.9 \times 10^{13}\,\mathrm{n/cm^2}$). At the fluence of $9.9 \times 10^{13}\,\mathrm{n/cm^2}$ (Fig. 4.26), the maximum charge-collection is obtained for an applied reverse voltage of $\approx 350\,\mathrm{V}$. As expected[*], the applied voltage exceeds largely that needed for full charge-collection before irradiation. In addition, the front and rear sides of the detector seem to behave similarly, i.e., the device operates as a double-sided junction and not as an asymmetrical (one-sided) junction [SICAPO Collab. (1994b)]. In these high resistivity silicon detectors, the charge-collection efficiency (e.g., see Sect. 6.8.2), i.e., the ratio R_{eff} of the charge collected after irradiation with respect to that of before irradiation, was found to have a logarithmic dependence on fluence. The charge collected is roughly that before irradiation up to a fluence of $\approx 3.2 \times 10^{12}\,\mathrm{n/cm^2}$ [SICAPO Collab. (1994b)]. Above this fluence, R_{eff} decreases and is given by

$$R_{\mathrm{eff}} \sim m_{\mathrm{eff}} \ln\left(\Phi_{\mathrm{n}}\right) + b_{\mathrm{eff}}, \text{ for } \Phi_{\mathrm{n}} > 3.2 \times 10^{12}\,\mathrm{n/cm^2}, \quad (4.143)$$

where Φ_{n} is the fast-neutron fluence, and the coefficients m_{eff} and b_{eff}, for Fz-detectors with resistivities ranging from 5 to $7\,\mathrm{k\Omega\,cm}$, are shown in Table 4.6 [SICAPO Collab. (1994b)]. R_{eff} slightly depends on the location of the α-source, i.e., at the front or at the rear side of the silicon detector[††]. In Fig. 4.26, the relative charge-collection efficiency (as a function of the applied reverse voltage) is the fraction (in percentage) of the charge collected with respect to the maximum charge collected after irradiation whose R_{eff} is expressed by means of Eq. (4.143).

After irradiations in the range $(1–10)\times 10^{14}\,\mathrm{n/cm^2}$, high-resistivity silicon detectors are still capable of measuring the energy-loss deposited by minimum-ionizing-particles (*mips*) under both reverse and forward bias [Chilingarov and Sloan (1997); Beattie, Chilingarov and Sloan (2000)]. To reduce the leakage current, these detectors were operated at moderate low-temperature up to about $249\,\mathrm{K}$.

4.3.3.3 *I − V Characteristics Down to Cryogenic Temperature*

The carrier concentration of an extrinsic non-irradiated silicon semiconductor remains essentially constant over a wide range of temperature below $300\,\mathrm{K}$. However, at low temperature the dopant atoms are no longer fully ionized[‡‡] and the electron (hole) concentration depends on the concentration of acceptors (donors) contained in n-type (p-type) silicon. In fact, sophisticated purification techniques can largely

---

[**]The depletion voltage for full charge-collection decreases with decreasing temperature (e.g, see [Granata et al. (2000); Santocchia et al. (2004)]).

[*]One can see, e.g., [Rancoita and Seidman (1982); Borgeaud, McEwen, Rancoita and Seidman (1983); SICAPO Collab. (1994b,c)], see also Section 11.2.4 [Lutz (2001)].

[††]Similar results were also obtained with MCZ detectors [SICAPO Collab. (1994c)].

[‡‡]This temperature range is also referred to as *freeze-out* range of temperature.

reduce "unwelcome" impurities, but not eradicate them completely. In addition with decreasing temperature, while the energy gap (E_g) slightly increases, the Fermi level (E_F) approaches the energy level of the main impurity.

For instance, in an n-type extrinsic semiconductor, the electron concentration (n_0) at low (and very low) temperature ** is given by:

$$n_0 = \sqrt{g_d N_c N_d} \exp\left(\frac{E_d - E_c}{2\,k_B T}\right), \text{ for } N_a \ll n_0 \ll N_d, \tag{4.144}$$

and

$$n_0 = g_d N_c \left[\frac{N_d - N_a}{N_a}\right] \exp\left(\frac{E_d - E_c}{k_B T}\right), \text{ for } n_0 \ll N_a < N_d, \tag{4.145}$$

where N_d and N_a are the concentrations of donor and acceptors, respectively, with $N_d > N_a$, E_d and E_c are the energy level of the principal donor and the energy of the bottom of the conduction band*, respectively; N_c is the *effective density-of-states in the conduction band*† and g_d^{-1} ($= 2$) is the *ground-state degeneracy*‡ of the simple donor (e.g., Section 1.4.3 of [Sze (1981)] and Section 3.1.1 of [Blakemore (1987)]). The expressions regarding the hole concentration in a p-type extrinsic semiconductor at low temperature can be found, for instance, in Section 3.2.2 of [Blakemore (1987)]. It has to be noted that the intrinsic-carrier concentration (n_{int}) also decreases with decreasing temperature. n_{int} is approximately given by (e.g., Section 3.4.1 of [Wolf (1969)])

$$n_{int} \approx 3.73 \times 10^{16} \times T^{3/2} \exp\left(-\frac{7.014}{T} \times 10^3\right) \text{ cm}^{-3}, \tag{4.146}$$

becomes $\sim 10^8 \text{ cm}^{-3}$ at $T \sim 250\,\text{K}$ (e.g., see Fig. 3.4.1 at page 179 of [Wolf (1969)], Fig. 11 at page 19 of [Sze (1981)]) and very small at $T \approx (150\text{--}170)\,\text{K}$. Other expressions and comparisons with experimental data can be found in [Barber (1967)] (see also Sections 1.2.1.3 and 1.2.2.3 of [Gutiérrez, Deen and Claeys (2001)]).

Furthermore {e.g. see Equation (322.11) at page 137 of [Blakemore (1987)]} at very low temperature, the energy of the Fermi level is given by:

$$E_F \sim E_d + k_B T \ln\left[\frac{g_d (N_d - N_a)}{N_a}\right] \text{ for } n_0 \ll N_a < N_d. \tag{4.147}$$

By inspection of Eq. (4.147), we see that the Fermi level moves to that of the principle donor as $T \to 0\,\text{K}$. While, as T increases from absolute zero, it (slightly)

**The reader can see, e.g., Equations (322.7, 322.8) in Section 3.2.2 of [Blakemore (1987)] and, also, Section 1.4.3 of [Sze (1981)].

*$E_c - E_d$ is (0.044–0.045) eV for phosphorus at room temperature in silicon (e.g., see Chapter 1 of [Milnes (1973)]).

†In silicon {e.g., see page 19 of [Sze (1985)] and Equation (220.1) at page 82 of [Blakemore (1987)]}, N_c is $4.829 \times 10^{15} [m_n(T)/m_e]^{3/2} T^{3/2} \text{ cm}^{-3}$, where m_e is the electron rest mass and $m_n(T)$ is the *density-of-states effective mass* for electrons $\approx 1.18\,(1.06)$ at $300\,(4.2)\,\text{K}$ (e.g., see [Barber (1967)] and Section 1.2.1.3 of [Gutiérrez, Deen and Claeys (2001)]; additional data can be found in Section 2.7 of [Adachi (2004)]).

‡For acceptors, the degeneracy factor is 4 {e.g., see Equation (151.1) at page 58 and Equation (220.1) at page 82 of [Blakemore (1987)] and, also, page 19 of [Sze (1985)]}.

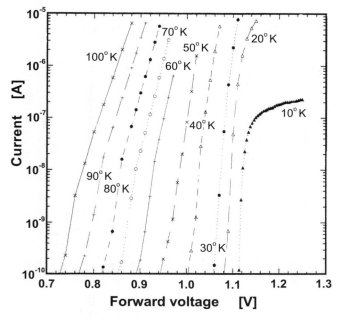

Fig. 4.27 Forward current (in A) as a function of the applied forward voltage (in V) for a non-irradiated n-type silicon detector and temperatures from 10 up to 100 K (from [Leroy and Rancoita (2007)]; see also [Croitoru, Rancoita, Rattaggi, Rossi and Seidman (1996, 1997)]).

rises above E_d in extrinsic semiconductors with a small compensation [i.e. for $g_d (N_d - N_a) > N_a$], before decreasing again at larger temperatures. At 10 K for an acceptor compensation of about 1 %, $E_F - E_d$ is $\approx (2\text{--}4)\,\text{meV}$ {Eq. (4.147), see also Fig. 32.10 at page 137 of [Blakemore (1987)]}.

As mentioned above, the energy gap depends on the temperature: for a weakly doped Fz-material with $N_a = 5 \times 10^{12}\,\text{cm}^{-3}$ and $N_d = 10^{12}\,\text{cm}^{-3}$ [Bludau and Onton (1974)], it is about 1.12 eV at 300 K and about 1.17 eV at $(50\text{--}10)$ K (see also Section 1.2.1.2 of [Gutiérrez, Deen and Claeys (2001)] and, for instance, direct measurements reported in [MacFarlane, McLean, Quarrington and Roberts (1958); Shklee and Nahory (1970); Bludau and Onton (1974)]).

In Fz n-type detectors[§] with resistivity of $\approx (4\text{--}7)\,\text{k}\Omega\,\text{cm}$ (i.e., a dopant concentration of $\approx 1.1 \times 10^{12}\text{--}6.2 \times 10^{11}\,\text{cm}^{-3}$), the acceptor concentration is typically lower than $10^{10}\,\text{cm}^{-3}$. In these n-type extrinsic semiconductors, the freeze-out of dopants cannot be neglected below $\approx (38\text{--}35)$ K {e.g., [Croitoru, Rancoita, Rattaggi, Rossi and Seidman (1996)], see also Eqs. (4.144, 4.145)}; at $(14\text{--}13)$ K the electron concentration is lower than $\approx 10^3\,\text{cm}^{-3}$ and, as a consequence, the medium is practically an insulator.

[§]These detectors are manufactured by STMicroelectronics.

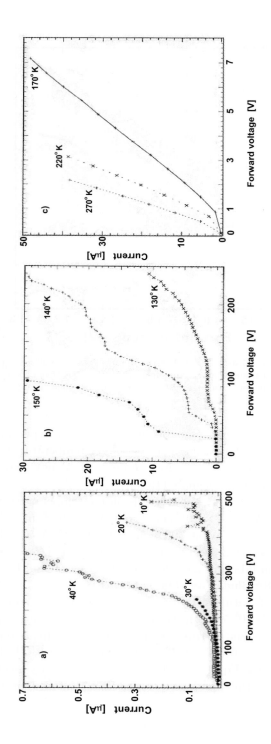

Fig. 4.28 Forward current (in μA) as a function of the applied forward voltage (in V) for an n-type silicon detector irradiated with 5.9×10^{14} n/cm^2 (see also [Croitoru, Rancoita, Rattaggi, Rossi and Seidman (1996, 1997)]): (a) for (10–40) K, (b) for (130–150) K, (c) for (170–270) K {(a) and (b) from [Leroy and Rancoita (2007)]; (c) from [Rancoita (2008)]}.

These devices have a (reverse) leakage current of $\approx 10\,\mathrm{nA}$ at room temperature and lower than $\approx 0.1\,\mathrm{nA}$ at $\approx 100\,\mathrm{K}$. In addition, at room temperature, while full depletion is achieved at a reverse voltage of about $80\,\mathrm{V}$, the device capacitance decreases (see Sect. 6.1.3) as $1/\sqrt{V_r}$ until full depletion is achieved; for temperatures lower than the *critical temperature* (T_c) of about $20\,\mathrm{K}$ (see also Sect. 4.3.4), the capacitance[¶] becomes independent of V_r, i.e., the device behaves like an insulator, where the junction effect is less relevant [Croitoru, Rancoita, Rattaggi, Rossi and Seidman (1997)]. These phenomena are consistent with the large decrease and the freeze-out of free carriers at low and very low temperatures. With decreasing temperature (Fig. 4.27), the forward $I-V$ characteristics exhibit an increasingly sharp dependence on the forward voltage V_f, i.e., a lower forward resistance. In addition [Croitoru, Rancoita, Rattaggi, Rossi and Seidman (1996, 1997)], at $10\,\mathrm{K}$ the *switching voltage* $V_{f,s}$[‖] becomes $\approx 1.12\,\mathrm{V}$, but for V_f slightly above $V_{f,s}$ (at $\approx 1.14\,\mathrm{V}$) the forward current increases less sharply (Fig. 4.27).

As already mentioned, when the temperature is within the freeze-out range, the silicon may be thought of as an insulator. When this occurs, we must consider the phenomena of *single-carrier space-charge-limited current* (e.g., see [Shockley and Prim (1953); Gregory and Jordan (1964)]) and *double-injection conduction*, i.e., the *Lampert–Ashley–Wagener model* (e.g., see [Lampert and Rose (1961); Lampert (1962); Ashley and Milnes (1964); Wagener and Milnes (1964)]), other than the *ohmic conduction*. In these mechanisms, the presence of traps can substantially prevent the conduction upon their filling. The abrupt increase of the forward current at the *junction voltage*[**] ($\approx 1.1\,\mathrm{V}$) was already observed in p^+-n-n^+ junctions at $4.2\,\mathrm{K}$ [Brown and Jordan (1966)] and explained by extending the Lampert–Ashley–Wagener model to account for the junction effects [investigated by Jonscher (1961)].

After irradiation, additional traps are created inside the device and, since the conduction mechanism depends on the concentrations and the energy levels of traps (e.g., [Ashley and Milnes (1964); Brown and Jordan (1966)]), the $I-V$ forward characteristics are largely affected at very low temperatures [Fig. 4.28]. At $10\,\mathrm{K}$ [Croitoru, Rancoita, Rattaggi, Rossi and Seidman (1996)], for a neutron fluence of $1.2\times10^{13}\,\mathrm{n/cm^2}$, $V_{f,s}$ is about that before irradiation, but becomes $>400\,\mathrm{V}$ for $5.9\times 10^{14}\,\mathrm{n/cm^2}$ [Fig. 4.28(a)]. In the framework of double-injection models (e.g., [Ashley and Milnes (1964); Brown and Jordan (1966)]), the switching voltage depends on the concentrations of traps to be filled [Croitoru, Rancoita, Rattaggi, Rossi and Seidman (1996)] and, as a consequence, can become large at high neutron fluence. In addition, the critical temperature[††] increases with the neutron fluence [Croitoru, Rancoita, Rattaggi, Rossi and Seidman (1997)]: for $5.9 \times 10^{14}\,\mathrm{n/cm^2}$ T_c is $\approx 150\,\mathrm{K}$, i.e., is

---

[¶]To determine T_c, the capacitance measurements were performed with a test frequency of $10\,\mathrm{kHz}$.

[‖]$V_{f,s}$ is the average forward voltage for which the forward current increases abruptly from less than $10^{-10}\,\mathrm{A}$ up to more than $10^{-8}\,\mathrm{A}$.

[**]The junction voltage is described by the Sah–Noyce–Shockley theory [Sah, Noyce and Shockley (1957)].

[††]T_c is $\approx 40\,\mathrm{K}$ for $1.2 \times 10^{14}\,\mathrm{n/cm^2}$ [Croitoru, Rancoita, Rattaggi, Rossi and Seidman (1997)].

larger by about an order of magnitude compared to that before irradiation.

Above the critical temperature, the $I-V$ characteristics exhibit an ohmic behavior which can be expressed by Eq. (4.140), where $C_f(\Phi_n)$ and $R_f(\Phi_n)$ also depend on the temperature [see for instance Fig. 4.28(c)]. The forward resistance decreases with increasing temperature: for 5.9×10^{14} n/cm^2, R_f is $\approx 1.3 \times 10^5$, 6.6×10^4, $5.1 \times 10^4 \, \Omega$ at 170, 220, 270 K, respectively [see Fig. 4.28(c)]. At 220 K, for a fluence of 1.2×10^{14} n/cm^2 the forward resistance is $\approx 2.0 \times 10^4 \, \Omega$ [Leroy and Rancoita (2007)].

4.3.4   *Complex Junction Impedance and Cryogenic Temperatures*

The $I-V$ reverse characteristics of the $p^+ - n$ diode, considered in Sects. 4.3.2 and 4.3.3, were derived and also measured under the condition that diode voltages change so slowly that the excess carrier distributions in the two regions of the junction remain in a steady state. When a small sinusoidal voltage is superimposed to the dc bias, the current-voltage behavior cannot be expressed by static characteristics, because the diode shows a capacitive behavior. Thus, at room temperature, the diode impedance will be described by accounting for both the ohmic and capacitive parts of the depleted region (e.g., Section 3.5.2 of [Sze (1985)], Section 9.1 of [Tyagi (1991)] and Section 4.4 of [Shur (1996)]) and, as a consequence, will depend on the frequency (f) of the ac signal. However, the so-called diffusion capacitance[‡‡] can be neglected, when a reverse voltage is applied. In addition, since at low temperature the carriers start to freeze-out (see Sect. 4.3.3.3), the capacitance of the field-free region has been taken into account together with its resistance[*] in the comprehensive model for the *Small-signal ac Impedance of a one-sided $p^+ - n$ Junction Diode* (SIJD) operated under reverse bias [Li (1994a); Croitoru, David, Rancoita, Rattaggi and Seidman (1997a, 1998b)].

The admittance of the equivalent circuit for the SIJD model [Fig. 4.29(a)] is:

$$Y(\omega) = \left[\frac{1}{Y_d(\omega)} + \frac{1}{Y_b(\omega)} \right]^{-1}$$

$$= \frac{Y_d(\omega) \, Y_b(\omega)}{Y_d(\omega) + Y_b(\omega)}, \tag{4.148}$$

where $\omega = 2\pi f$,

$$Y_d(\omega) = \frac{1}{R_d} + j\omega C_d \tag{4.149}$$

and

$$Y_b(\omega) = \frac{1}{R_b} + j\omega C_b \tag{4.150}$$

are the admittances of the depleted and field-free (i.e., the electrical neutral bulk) regions, respectively; C_d (R_d) and C_b (R_b) are the capacitances (resistances) of

---

[‡‡]The diffusion capacitance is related to the minority carriers injected into the neutral region (e.g., Section 9.1 of [Tyagi (1991)] and Section 4.4 of [Shur (1996)]).

[*]One can see, e.g., [Viswanathan, Divakaruni and Kizziar (1991); Divakaruni, Prabhakar and Viswanathan (1994)].

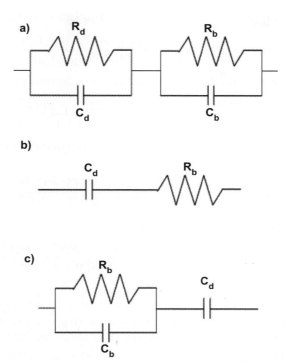

Fig. 4.29 Equivalent circuit for a non-irradiated one-sided junction operated under reverse bias (from [Leroy and Rancoita (2007)]). C_d (R_d) and C_b (R_b) are the capacitances (resistances) of the depleted and field-free regions, respectively: (a) general equivalent circuit, (b) at room temperature for $R_d \gg 1/(\omega C_d)$ and $R_b \ll 1/(\omega C_b)$, (c) at low temperature for $R_d \gg 1/(\omega C_d)$ with $\omega = 2\pi f$, where f is the test frequency.

the depleted and field-free regions, respectively. Introducing Eqs. (4.149, 4.150) in Eq. (4.148), we obtain

$$
\begin{aligned}
Y(\omega) &= \frac{(1 + j\omega C_d R_d)(1 + j\omega C_b R_b)}{(1 + j\omega C_d R_d)R_b + (1 + j\omega C_b R_b)R_d} \\
&= \frac{1 - \omega^2 C_d C_b R_d R_b + j\omega(C_d R_d + C_b R_b)}{(R_d + R_b) + j\omega R_d R_b(C_d + C_b)} \\
&= \frac{1 - \omega^2 C_d C_b R_d R_b + j\omega(C_d R_d + C_b R_b)}{(R_d + R_b)^2 + \omega^2 R_d^2 R_b^2(C_b + C_d)^2} \\
&\quad \times [(R_d + R_b) - j\omega R_d R_b(C_d + C_b)].
\end{aligned}
\tag{4.151}
$$

Finally, using Eq. (4.151), the admittance of the equivalent circuit for the SIJD model [Fig. 4.29(a)] becomes:

$$
\begin{aligned}
Y(\omega) &= G(\omega) + j\, B(\omega) \tag{4.152} \\
&= \frac{1}{R_{\mathrm{r,eff}}(\omega)} + j\,\omega C_{\mathrm{r,eff}}(\omega)
\end{aligned}
$$

with the conductance (i.e., the real part of the admittance) given by

$$G(\omega) = \frac{R_d + R_b + \omega^2 R_d R_b (R_d C_d^2 + R_b C_b^2)}{(R_d + R_b)^2 + \omega^2 R_d^2 R_b^2 (C_b + C_d)^2} \tag{4.153}$$

and the susceptance (i.e., the imaginary part of the admittance) given by

$$B(\omega) = \omega \left[\frac{\omega^2 R_d^2 R_b^2 C_d C_b (C_b + C_d) + R_d^2 C_d + R_b^2 C_b}{(R_d + R_b)^2 + \omega^2 R_d^2 R_b^2 (C_b + C_d)^2} \right]. \tag{4.154}$$

As well known, the conductance is the reciprocal of the equivalent circuit resistance, thus

$$R_{r,\text{eff}}(\omega) = \frac{(R_d + R_b)^2 + \omega^2 R_d^2 R_b^2 (C_b + C_d)^2}{R_d + R_b + \omega^2 R_d R_b (R_d C_d^2 + R_b C_b^2)}; \tag{4.155}$$

while the susceptance expresses the capacitance,

$$C_{r,\text{eff}}(\omega) = \frac{\omega^2 R_d^2 R_b^2 C_d C_b (C_b + C_d) + R_d^2 C_d + R_b^2 C_b}{(R_d + R_b)^2 + \omega^2 R_d^2 R_b^2 (C_b + C_d)^2}, \tag{4.156}$$

of the equivalent circuit multiplied by ω.

At room temperature, for non-irradiated and weakly doped (e.g., see Sect. 4.3.2) devices[*], the reverse current is about or lower than $10\,\text{nA}$ at full depletion voltage and, as a consequence, R_d is typically larger than about $10^4\,\text{M}\Omega$, while C_d is not lower than $1\,\text{pF}$. Thus, even for fully depleted devices, i.e., when C_d is at the minimum, we have that $R_d \gg 1/(\omega C_d)$ at frequencies above a few kHz. In addition, since the time constant of the electrical neutral bulk[†] is the same as the dielectric relaxation time τ_{rel} (e.g., see Sections 7.6 and 7.7 of [Reitz and Milford (1970)] and Section 6.3.4 of [Neamen (2002)])

$$\tau_{\text{rel}} = R_b C_b \tag{4.157}$$

$$= \varepsilon \rho_{\text{bulk}} \tag{4.158}$$

$$\approx (3.2\text{--}7.4) \times 10^{-9}\,\text{s},$$

we have that $R_b \ll 1/(\omega C_b)$ for frequencies $\lesssim 10\,\text{MHz}$. When $R_b \ll 1/(\omega C_b)$ and $R_d \gg 1/(\omega C_d)$, the equivalent circuit for the SIJD model can be approximated by that shown in Fig. 4.29(b) and its admittance becomes:

$$Y_{rT}(\omega) = \frac{\omega^2 R_b C_d^2}{1 + \omega^2 R_b^2 C_d^2} + j\omega \left[\frac{C_d}{1 + \omega^2 R_b^2 C_d^2} \right]. \tag{4.159}$$

With decreasing temperature, the leakage current decreases and the condition $R_d \gg 1/(\omega C_d)$ is satisfied at frequencies lower than a few kHz. Therefore, the equivalent circuit at low and very low temperatures, i.e., when the carriers start to freeze-out (see Sect. 4.3.3.3), can be approximated with that shown in Fig. 4.29(c), whose

---

[*]Typically this type of device is similar to that described at page 397.

[†]ρ_{bulk} is the bulk resistivity [typically $(3\text{--}7)\,\text{k}\Omega\text{cm}$] of the device and ε is the silicon electric permittivity (e.g., see Appendix A.2).

admittance is given by (e.g., see [Viswanathan, Divakaruni and Kizziar (1991); Divakaruni, Prabhakar and Viswanathan (1994); Croitoru, David, Rancoita, Rattaggi and Seidman (1997a, 1998b)]):

$$Y_{lT}(\omega) = \frac{\omega^2 R_b C_d^2}{1 + \omega^2 R_b^2 (C_b + C_d)^2} + j\,\omega\,C_d \left[\frac{1 + \omega^2 R_b^2 C_b (C_b + C_d)}{1 + \omega^2 R_b^2 (C_b + C_d)^2} \right]. \tag{4.160}$$

In the temperature range (indicated as *saturation range* in Fig. 4.30) within which the carrier concentration is almost constant, we can rewrite the conductance term of Eq. (4.160) as (e.g., see [Croitoru, David, Rancoita, Rattaggi and Seidman (1997a)] and also [Viswanathan, Divakaruni and Kizziar (1991); Divakaruni, Prabhakar and Viswanathan (1994); Li (1994a)]):

$$G_{lT}(\omega) = A_d \left[\frac{\omega^2 \varepsilon^2 \rho_{bulk}(d - X)}{X^2 + \omega^2 \varepsilon^2 \rho_{bulk}^2 d^2} \right], \tag{4.161}$$

where A_d is the device active area, ε is the silicon electric permittivity (e.g., see Appendix A.2), X is the depth of the depletion region [e.g., Eq. (6.17)], d is the thickness of the substrate, and ρ_{bulk} is the resistivity[‡] of the extrinsic n-type silicon. It has to be noted that the resistivity depends on the temperature, because the mobility varies (i.e., it increases with decreasing temperature) by almost two orders of magnitude compared to that at 300 K (e.g., [Jonscher (1961); Misiakos and Tsamakis (1994)]). In Eq. (4.161), for a test frequency of 1 MHz the term $\omega\,\varepsilon\,\rho_{bulk}\,d$ is smaller than X even in a partially depleted device. Thus, to a first approximation, the conductance G_{lT} is proportional to the resistivity and then has a similar temperature dependence, i.e., that of $1/\mu_n(T)$. For instance, the measured values of the conductance (Fig. 4.30) for an MCZ-device operated at 5 V of reverse bias indicate that

$$G_{lT}(T) \propto \frac{1}{T^{-\alpha}},$$

where $\alpha \simeq 2.3$ for $170 \lesssim T \lesssim 270$ K and $\alpha \simeq 1.8$ for $70 \lesssim T \lesssim 150$ K [Croitoru, David, Rancoita, Rattaggi and Seidman (1997a,b); David (1997); Croitoru, David, Rancoita, Rattaggi and Seidman (1998b)], in agreement with that expected due to the temperature dependence of the electron mobility found in weakly doped silicon ([Arora, Hauser and Roulston (1982); Misiakos and Tsamakis (1994)], see also page 409 and Section 1.2 of [Müller and Kamins (1977)]).

When the carriers freeze-out, the resistivity increases again because the silicon substrate starts to become similar to an insulator. Furthermore, the function $G_{lT}(\omega, T)/\omega$ has a maximum for

$$\omega_p(T) = \frac{1}{\varepsilon \rho_{bulk}(T)} \frac{X(T)}{d}. \tag{4.162}$$

In the freeze-out range of temperature, the free-carrier concentration depends on both the compensation level and the energy level of the dopant atoms. Thus, it

---

[‡]The resistivity is expressed as $\rho_{bulk} \simeq 1/(e\,\mu_n\,n_0)$, where μ_n and n_0 are the mobility and concentration of electrons, respectively.

is possible to determine the energy level of the dopant atoms by means of the measurement of $\omega_p(T)$ as a function of T ([Viswanathan, Divakaruni and Kizziar (1991); Divakaruni, Prabhakar and Viswanathan (1994)] and references therein). It has to be added that the depth of the depleted layer (X) may depend on the temperature because i) fewer ionized dopants are available to create the space-charge region and ii) the carrier concentrations in the p^+- and n-region exhibit different freeze-out ranges of temperature due to the large difference between their dopant concentrations [see Sect. 4.3.3.3 and references therein]. Furthermore, the resistance of the bulk silicon increases with decreasing temperature so that $\omega R_b(C_b + C_d) \gg 1$ at low frequency. Therefore as experimentally observed[§] (e.g., page 397, see also [Li, Chen and Kraner (1991); Divakaruni, Prabhakar and Viswanathan (1994); Croitoru, David, Rancoita, Rattaggi and Seidman (1997a); Croitoru, Rancoita, Rattaggi, Rossi and Seidman (1997)] and references therein), from Eq. (4.160) the capacitance of the equivalent circuit becomes independent of frequency and approaches

$$C_{r,\text{eff}}(\omega) \simeq \frac{C_b C_d}{C_b + C_d}$$
$$= C_{\text{geo}},$$

where

$$C_{\text{geo}} = \frac{\varepsilon A_d}{d}$$

is the total capacitance of the medium, for R_b so large that it can be ignored in the equivalent circuit shown in Fig. 4.29(c).

The reverse-biased capacitance (C_d) of a $p^+ - n$ junction depends on the depletion depth X [see Eq. (6.35) and discussion in Sect. 6.1.3] which, in turn, depends [Eq. (6.17)] on the applied reverse bias (V_r) and on the donor (i.e., a *shallow impurity*) concentration (N_d), thus, by combining Eqs. (6.17, 6.35) we obtain:

$$C_d = A_d \sqrt{\frac{e\varepsilon N_d}{2(V_0 + V_r)}}, \tag{4.163}$$

where V_0 is the built-in voltage [Eq. (6.12)]. In the substrate, the presence of trapping or deep centers results in a frequency dependence of the capacitance related to the energy levels of the centers because, at the edge of the depletion region, the ionization state of these centers may become unable to follow the variation of the applied voltage over the full spectrum of frequencies. For instance in the *Sah-Reddi model* [Sah and Reddi (1964)], if a junction contains deep impurities, the reverse-biased capacitance $(C_{d,t})$ can be written as

$$C_{d,t} = C_d \left\{ 1 - \frac{\phi_t}{V_0 + V_r} \left[\frac{N_t(\omega)}{N_d} \right] \right\}^{-\frac{1}{2}}, \tag{4.164}$$

where C_d is given by Eq. (4.163), $N_t(\omega)$ is the frequency-dependent concentration of deep impurities, ϕ_t is the energy difference between the Fermi level (of

---

[§]At 18 K in the devices mentioned at page 397, the capacitance is independent of the applied reverse bias at a frequency of 300 Hz.

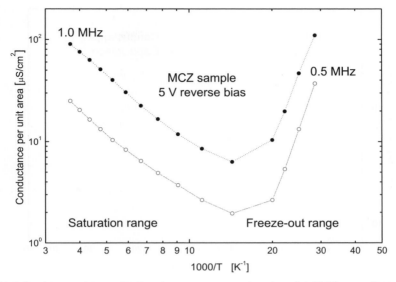

Fig. 4.30 Conductance per unit of active area measured at 1 and 0.5 MHz as a function of the temperature (in 1000/T with T in Kelvin) for a non-irradiated sample with n-type MCZ substrate and operated at a reverse bias of 5 V (adapted from [Leroy and Rancoita (2007)]). The lines are to guide the eye. Within the saturation range of temperature the concentration of carriers is almost constant (i.e., almost all the dopant atoms are ionized), but depends on the temperature in the freeze-out range (see also [Croitoru, David, Rancoita, Rattaggi and Seidman (1997a, 1998b)] and [David (1997)]).

the silicon substrate) and the level of the deep center divided by the electronic charge e. This model has been extended to treat more than one deep center for non-irradiated (e.g., see [Beguwala and Crowell (1974)] and references therein) and irradiated (e.g., see [Li (1994a)] and references therein) devices. Investigations of the frequency and temperature-dependence of the junction capacitance allow the determination of these deep center levels (e.g., see Section 8.3 of [Milnes (1973)], also [Sah and Reddi (1964); Schultz (1971); Beguwala and Crowell (1974)] and references therein). The mechanism of the repetitive filling and emptying of deep levels in the depletion region of a junction is also utilized in the deep-level transient spectroscopy (DLTS, e.g., see Sect. 4.2.2) [Lang (1974)]. The DLTS exploits the thermal discharging of the occupied traps which results from capacitance transients induced by a pulsed bias.

Systematic investigations of the admittance dependence on temperature were carried out for fast-neutron fluences up to $\approx 10^{16}$ n/cm^2 (e.g., [Li (1994a); Croitoru, Gambirasio, Rancoita and Seidman (1996); Croitoru, Rancoita, Rattaggi, Rossi and Seidman (1997); Croitoru, David, Rancoita, Rattaggi and Seidman (1998a)] and references therein) and for temperatures down to 10 K (e.g., [Croitoru, Rancoita, Rattaggi, Rossi and Seidman (1997); Croitoru, David, Rancoita, Rattaggi and Sei-

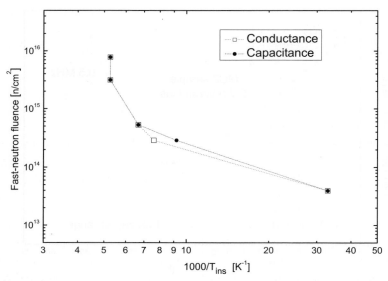

Fig. 4.31   Values of T_{ins} (in K) when the conductance ($\square$) and the capacitance ($\bullet$) of the samples irradiated with fast-neutron fluences up to $\approx 10^{16}$ n/cm^2 become independent of the test frequency (adapted from [Leroy and Rancoita (2007)]; see also [David (1997)]). The lines are to guide the eye.

dman (1998a)] and references therein).

For fast-neutron fluences $\lesssim 10^{13}$ n/cm^2, a *quasi non-irradiated model* by means of Eq. (4.160) can approximately describe the equivalent circuit behavior of the irradiated samples with decreasing temperature. For instance, the conductance exhibits an approximate ω^2 dependence [e.g., see Eq. (4.161)], like the response of the equivalent circuit of non-irradiated devices outside the freeze-out range of temperature [Croitoru, David, Rancoita, Rattaggi and Seidman (1998a)]. However, the *non-freeze-out range* of temperature was found to depend on fluence [Croitoru, David, Rancoita, Rattaggi and Seidman (1998a)]. Moreover, as for non-irradiated devices (Fig. 4.30), the conductance reaches a minimum and then increases again with decreasing temperature. For a fluence of $\approx 10^{11}\,(10^{13})$ n/cm^2 the minimum occurs at $\approx 70\,(150)$ K [David (1997); Croitoru, David, Rancoita, Rattaggi and Seidman (1998a)]. In addition, at room temperature [Li (1994a)], the capacitance behavior of devices irradiated with fast-neutron fluences $\lesssim 3.2 \times 10^{13}$ n/cm^2 is described by Eq. (4.154). In this latter equation, the depletion region capacitance has to account for the trapping centers generated by irradiation, but treated similarly to deep defects [e.g., see Eq. (4.164)] in the framework of the Sah–Reddi model [Sah and Reddi (1964)].

A progressive departure from the *quasi non-irradiated model* description appears for fluences above $\approx 4 \times 10^{13}$ n/cm^2 [Croitoru, David, Rancoita, Rattaggi and Seidman (1998a)]. For instance, the dependence of the conductance on ω goes as ω^m with $m < 2$ [David (1997); Croitoru, David, Rancoita, Rattaggi and Seidman (1998a)]. Furthermore (e.g., see Sect. 4.3.4) for $f = 10\,\text{kHz}$ [Croitoru, Rancoita, Rattaggi, Rossi and Seidman (1997)], the capacitance becomes independent of the applied voltage at a temperature larger than the one corresponding to a non-irradiated device (e.g., see page 397). More generally, it has been demonstrated that, within the experimental errors [David (1997)], the conductance and the capacitance of the irradiated devices become independent of the frequency* and of the applied reverse-bias at a temperature T_{ins} (see Fig. 4.31), which increases with increasing fluence (i.e., with the concentration of the created complex-defects, see Sect. 4.2.2, which are electrically active): T_{ins} is $\approx 190\,\text{K}$ for a fluence of $7.8 \times 10^{15}\,\text{n/cm}^2$. A general model, which has been proposed to account for the behavior of $G(\omega)$ and $C(\omega)$ at these neutron fluences, assumes that i) the irradiation changes the value of the resistivity in localized volumes inside the silicon and that ii) in these latter regions the medium can have a different frequency dependence with respect to that in the undamaged regions (e.g., [Croitoru, David, Rancoita, Rattaggi and Seidman (1998a)], see also [Croitoru, Dahan, Rancoita, Rattaggi, Rossi and Seidman (1997)]).

4.3.5 *Resistivity, Hall Coefficient and Hall Mobility at Large Displacement Damage*

In Sections 4.3.2–4.3.4 some of the main distinctive characteristics of non-irradiated and irradiated silicon semiconductors were apparent from their simple transport properties. Additional investigations were carried out to determine further properties: for example the *Hall coefficient* shows whether the charges are transported by positive or negative carriers in extrinsic semiconductors. The Hall coefficient is of primary importance since, except for a numerical factor (the so-called *Hall factor* or *scattering factor*) of the order of unity, it is the reciprocal of the carrier density and electronic charge§. Thus, it provides an immediate indication of the type and concentration of dopants in the sample and can be combined with the conductivity¶ to determine the carrier mobility (the so-called *Hall mobility*). However, in materials partly compensated by impurities of the opposite type, the simple measurement of the Hall coefficient may not be easily interpreted to determine the type of majority-carriers.

The Hall scattering factor is indicated as $\mathfrak{r}_e$ ($\mathfrak{r}_h$) for electrons (holes) and expresses the effect of the magneto-resistance: it accounts for the energy dependence

---

*The range of employed frequencies were $10 \leq f \leq 10^3\,\text{kHz}$.

§For a general introduction to the Hall effect, the reader may see, for instance, [Putley (1960)], Section 2 of [Blood and Orton (1978)], Chapter 5 of [Smith (1978)], Chapter 3 in [Blood and Orton (1992)] and [Popovic (2004)]; for the Hall factor and its dependence on temperature, the reader may read, for instance, [Norton, Braggins and Levinstein (1973); Kirnas, Kurilo, Litovchenko, Lutsyak and Nitsovich (1974)] and Section 2.6.2 of [Blood and Orton (1978)].

¶The conductivity is the reciprocal of the resistivity.

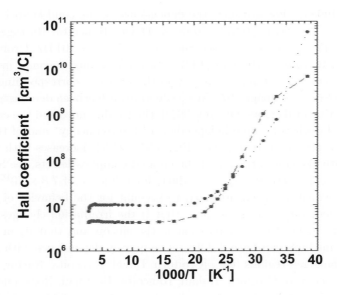

Fig. 4.32   Absolute value of the Hall coefficient, in cm³/C, as a function of the temperature, in
1000/T with T in K, for two non-irradiated n-type samples (from [Leroy and Rancoita (2007)]) with
resistivities of ≈ 2.7 (■) and ≈ 6 (●) kΩcm at room temperature (see [Croitoru, Gubbini, Rancoita,
Rattaggi and Seidman (1999a,b)], respectively). The temperature range is $25 \leq T \leq 350\,\mathrm{K}$. The
lines are to guide the eye.

of the relaxation time with respect to the carrier motion in presence of a magnetic
field (e.g., see Chapter 5 of [Smith (1978)]). The values of the relaxation time‖ for
electrons and holes are not necessarily the same and may depend on the temperature, the energy of the carrier and the shape of the constant-energy surfaces in the
k-vector** space. Experimental determinations and theoretical predictions of the
Hall factor as a function of temperature and dopant concentration under the *low
magnetic-field approximation*†† can be found, for instance, in [Messier and Merlo-
Flores (1963); Norton, Braggins and Levinstein (1973); Rode (1973); Dmitrenko et
al. (1974); Kirnas, Kurilo, Litovchenko, Lutsyak and Nitsovich (1974); Blood and
Orton (1978)]: at room temperature, in weakly doped silicon semiconductors, r_e and
r_h are ≈ 1.15 and 0.80, respectively (e.g., see Chapter 3 of [Blood and Orton (1992)],
[Mangiagalli, Levalois, Marie, Rancoita and Rattaggi (1999)] and, also, [Norton,
Braggins and Levinstein (1973); Rode (1973); Kirnas, Kurilo, Litovchenko, Lutsyak
and Nitsovich (1974)]).

For a partly compensated silicon semiconductor, under low magnetic-field appro-

---

‖ The relaxation time expresses the mean free time between carrier collisions.
** The wave-vector, **k**, is the momentum of the particle divided by $\hbar$.
†† The *low magnetic-field approximation* (e.g., Section 2.2 of [Blood and Orton (1978)] and, also,
Section 5.3.4 of [Smith (1978)]) can be expressed by the condition $\mu_c B \ll 1$ where μ_c and B are
the carrier mobility and the magnetic-field, respectively.

ximation, the Hall coefficient is given by {see Equation (147) at page 128 of [Smith (1978)]}:

$$R_{\mathrm{H}} = -\frac{1}{e}\left[\frac{\mathfrak{r}_e\, b^2 n - \mathfrak{r}_h\, p}{(b\, n + p)^2}\right] \tag{4.165}$$

$$= -\frac{1}{e}\left[\frac{\mathfrak{r}_e\, \mu_e^2 n - \mathfrak{r}_h\, \mu_h^2 p}{(\mu_e\, n + \mu_h\, p)^2}\right]$$

$$= e\rho^2 \mu_h^2 \left(\mathfrak{r}_h\, p - \mathfrak{r}_e\, b^2 n\right), \tag{4.166}$$

where

$$b = \frac{\mu_e}{\mu_h},$$

μ_e (μ_h) is the mobility*, sometimes referred to as *conductivity-mobility*, of electrons (holes); n (p) is the concentration[‡‡] of free electrons (holes) and, finally, the silicon resistivity ρ is

$$\rho = \frac{1}{e(\mu_e\, n + \mu_h\, p)} \tag{4.167}$$

$$= \frac{1}{e\mu_h(b\, n + p)}. \tag{4.168}$$

For instance, at room temperature in an (almost) intrinsic and non-irradiated silicon semiconductor b is ≈ 3 and the Hall coefficient [computed using Eq. (4.165)] is negative, i.e., the current is mainly carried by electrons. For extrinsic silicon under low magnetic-field, Eq. (4.165) can be rewritten as

$$R_{\mathrm{H},e} = -\frac{\mathfrak{r}_e}{n\, e} \quad \text{for } n \gg p \tag{4.169}$$

and as

$$R_{\mathrm{H},h} = \frac{\mathfrak{r}_h}{p\, e} \quad \text{for } p \gg b\, n \text{ and } p \gg b^2 n. \tag{4.170}$$

The Hall coefficient can be experimentally determined by measuring the so-called *Hall voltage* (due to the Hall effect), when a magnetic-field is applied to a semiconductor carrying a current (e.g., see [Putley (1960); Popovic (2004)]).

The *Hall mobility*, μ_{H}, is defined as the product of the Hall coefficient and the conductivity $\wp$:

$$\mu_{\mathrm{H}} = |R_{\mathrm{H}}|\, \wp$$

$$= \frac{|R_{\mathrm{H}}|}{\rho}, \tag{4.171}$$

where

$$\rho = \frac{1}{\wp}$$

---

*The dependence of the mobility on the temperature and concentration is discussed in Sect. 7.1.5 (see [Arora, Hauser and Roulston (1982)] and references therein).
[‡‡]At thermal equilibrium we have $np = n_{\mathrm{int}}^2$ (e.g., see Sect. 6.1).

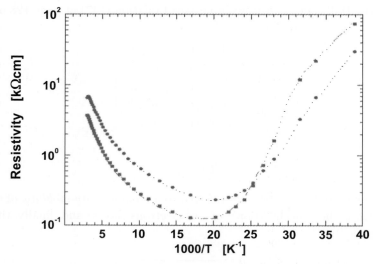

Fig. 4.33   Resistivity, in kΩcm, as a function of the temperature, in 1000/T with T in K, for two non-irradiated samples (from [Leroy and Rancoita (2007)]) with resistivities of ≈ 2.7 (■) and ≈ 6 (●) kΩcm at room temperature (see [Croitoru, Gubbini, Rancoita, Rattaggi and Seidman (1999a,b)], respectively). The temperature range is $25 \leq T \leq 350\,\mathrm{K}$. The lines are to guide the eye.

is the resistivity. It can be determined when combined measurements of the Hall coefficient and resistivity (see later) are performed on the same sample. For an extrinsic silicon semiconductor, under low magnetic-field, from Eqs. (4.169–4.171) we have:

$$\mu_{\mathrm{H},e} = \frac{|R_{\mathrm{H},e}|}{\rho_n}$$
$$= \mathfrak{r}_e\,\mu_e \quad \text{for } n \gg p, \tag{4.172}$$

in which ρ_n is the resistivity of the n-type silicon and

$$\mu_{\mathrm{H},h} = \frac{|R_{\mathrm{H},h}|}{\rho_p}$$
$$= \mathfrak{r}_h\,\mu_h \quad \text{for } p \gg b\,n \text{ and } p \gg b^2 n, \tag{4.173}$$

where ρ_p is the resistivity of the p-type silicon.

The temperature dependence of the Hall coefficient [Eq. (4.169)] of non-irradiated n-type silicon samples with resistivities of ≈ 2.7 [Croitoru, Gubbini, Rancoita, Rattaggi and Seidman (1999b)] and $\approx 6\,\mathrm{k\Omega cm}$ [Croitoru, Gubbini, Rancoita, Rattaggi and Seidman (1999a)] is shown in Fig. 4.32. Above room temperature (i.e., $1000/T \lesssim 3.33$), the rapid decrease of the Hall coefficient with increasing temperature results from the increase of the intrinsic carrier concentration [e.g., see Eq. (4.146)]. $R_{\mathrm{H},e}$ is almost constant up to $1000/T \approx 20$ (18) for the sample with resistivity of $\approx 6\,\mathrm{k\Omega cm}$ ($\approx 2.7\,\mathrm{k\Omega cm}$) resistivity (Fig. 4.32). Below these tempera-tures, the carriers start to freeze-out (e.g., see Sect. 4.3.3.3) and the Hall coefficient increases, as expected.

To make a Hall measurement, the sample is typically prepared in a bridge form (e.g., see Section 5.2 of [Bar-Lev (1993)]) or in a rectangular bar with side arms (e.g., see Section 2.3 of [Blood and Orton (1978)]) or in an arbitrary form by means of van der Paw's method [van der Pauw (1958)]. Thus, the same sample may be used for conductivity measurements. In Fig. 4.33, the resistivity dependence on temperature is shown for non-irradiated samples* with high- ($\approx 6\,\mathrm{k\Omega cm}$ [Croitoru, Gubbini, Rancoita, Rattaggi and Seidman (1999a)]) and low-resistivity ($\approx 2.7\,\mathrm{k\Omega cm}$ [Croitoru, Gubbini, Rancoita, Rattaggi and Seidman (1999b)]). The data exhibit an overall agreement with the predicted resistivity-dependence on temperature, as discussed in Sects. 4.3.3.3 and 4.3.4. For instance, using the data shown in Fig. 4.33 obtained for a sample with a resistivity of $\approx 6\,\mathrm{k\Omega cm}$ [Croitoru, Gubbini, Rancoita, Rattaggi and Seidman (1999a)] in the temperature range $190 \leq T \leq 300\,\mathrm{K}$, the temperature dependence of the electron-mobility is given by:

$$\mu_e = \mu_{0,e}\,T^{-\alpha}\ [\mathrm{cm^2 V^{-1} s^{-1}}], \tag{4.174}$$

where $\mu_{0,e} = (8.5 \pm 0.2) \times 10^8$, T is the temperature (in Kelvin) and the coefficient $\alpha = 2.31 \pm 0.02$ is in agreement with that determined by conductance measurements (e.g., see page 401) and by other authors (e.g., see [Arora, Hauser and Roulston (1982); Misiakos and Tsamakis (1994)]).

It has to be noted that the combined measurements of the Hall coefficient and re-sistivity allow one to determine both the mobility and the free carrier concentration in extrinsic semiconductors [e.g., see Eqs. (4.169, 4.170, 4.172, 4.173)].

Measurements of resistivity, Hall coefficient and Hall mobility were carried out after irradiation** with different types of particles up to large fluences and as a function of temperature. At room temperature the resistivity of the irradiated sam-ples of extrinsic (*p*- and *n*-type) silicon has been observed to increase with respect to that before irradiation§ and to become even larger than that of the intrinsic silicon (e.g., see [Borchi and Bruzzi (1994); Li (1994b); Croitoru, Dahan, Rancoita, Rattaggi, Rossi and Seidman (1997, 1998); Mangiagalli, Levalois, Marie, Rancoita

---

*For these samples, the dependence of Hall coefficients on temperature is that shown in Fig. 4.32.
**The reader can see, e.g., [Konozenko, Semenyuk and Khivrich (1969); Lugakov, Lukashevich and Shusha (1982); Borchi and Bruzzi (1994); Li (1994b); Biggeri, Borchi, Bruzzi and Lazanu (1995); Biggeri, Borchi, Bruzzi, Pirollo, Sciortino, Lazanu and Li (1995); Croitoru, Dahan, Ran-coita, Rattaggi, Rossi and Seidman (1997, 1998); Mangiagalli, Levalois, Marie, Rancoita and Rattaggi (1998); Croitoru, Gubbini, Rancoita, Rattaggi and Seidman (1999a,b); Mangiagalli, Lev-alois, Marie, Rancoita and Rattaggi (1999); Pirollo et al. (1999); Consolandi, Pensotti, Rancoita and Tacconi (2008)].
§One can see, e.g., [Borchi and Bruzzi (1994); Li (1994b); Biggeri, Borchi, Bruzzi and Lazanu (1995); Biggeri, Borchi, Bruzzi, Pirollo, Sciortino, Lazanu and Li (1995); Croitoru, Dahan, Ran-coita, Rattaggi, Rossi and Seidman (1997, 1998); Mangiagalli, Levalois, Marie, Rancoita and Rattaggi (1998); Croitoru, Gubbini, Rancoita, Rattaggi and Seidman (1999a,b); Mangiagalli, Lev-alois, Marie, Rancoita and Rattaggi (1999); Pirollo et al. (1999)], Section 2 in Chapter II of Part II of [Vavilov and Ukhin (1977)], Section 5.13 of [Messenger and Ash (1992)] and references therein.

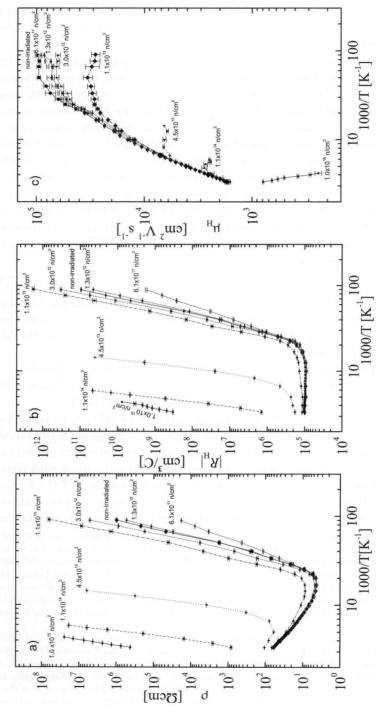

Fig. 4.34 (a) Resistivity (ρ) in Ωcm, (b) absolute value of the Hall coefficient [R_H; see Eqs. (4.166, 4.169)] in cm^3/C and (c) Hall mobility [μ_H, see Eqs. (4.171, 4.172)] in cm^{2}V^{-1}s^{-1} as a function of the temperature from 300 down to 11 K (in 1000/T, where T is in K) for n-type silicon samples with resistivities of $\approx 56\,\Omega$cm before irradiation with fast neutrons: the experimental data are from [Consolandi, Pensotti, Rancoita and Tacconi (2008)]. R_H was observed to be negative before (e.g., for n-type silicon) and after irradiations, i.e., there was no change of the Hall-coefficient sign.

and Rattaggi (1998); Croitoru, Gubbini, Rancoita, Rattaggi and Seidman (1999a,b); Mangiagalli, Levalois, Marie, Rancoita and Rattaggi (1999)]). At very large fluences of fast-neutrons (i.e., above[‡] 10^{15} n/cm^2), Kr-ions (i.e., above 10^{13} Kr/cm^2 [Croitoru, Gubbini, Rancoita, Rattaggi and Seidman (1999a,b)]) and Ar-ions (i.e., above 10^{14} Ar/cm^2 [Mangiagalli, Levalois, Marie, Rancoita and Rattaggi (1998)]) the resistivity is roughly that of the intrinsic silicon.

After irradiation with fast neutrons, the resistivity temperature-dependence [Fig. 4.34(a)] from 300 down to 11 K was investigated for n-type silicon samples with a resistivity of $\approx 56\,\Omega$cm before irradiation [Consolandi, Pensotti, Rancoita and Tacconi (2008)]. From 300 ($1000/T \simeq 3.3$) down to ≈ 60 K ($1000/T \simeq 16.7$) and for a neutron fluence up to 3×10^{12} n/cm^2, the temperature dependence does not exhibit a significant variation with regard to that before irradiation. However, for temperatures lower than ≈ 50 K ($1000/T = 20$) and for a neutron fluence up to 1.3×10^{12} n/cm^2, the resistivity values are lower than the corresponding non-irradiated ones. For neutron fluences above 1.1×10^{13} n/cm^2 a large departure from the behavior of the non-irradiated sample is observed.

For both n- and p-type silicon, the Hall coefficient was measured after irradiations with different type of particles. At large fluences of photons emitted by a ^{60}Co-radioactive source, it was found that, while p-type silicon maintains the same *conductivity type* [i.e., the Hall coefficient remains positive; see Eqs. (4.165, 4.169, 4.170)], the n-type silicon may change conductivity type, i.e., from negative the Hall coefficient may become positive ([Konozenko, Semenyuk and Khivrich (1969)] and references therein). In irradiations with fast-neutrons and 3.8 MeV electrons, the Hall coefficients were observed to show a similar behavior for n-type Fz and p-type silicon [Lugakov, Lukashevich and Shusha (1982)], while no change of conductivity type [e.g., see Figs. 4.34(b)] occurred for the n-type Cz[†] and oxygen enriched silicon samples with $\rho = (56–157)\,\Omega$cm [Lugakov, Lukashevich and Shusha (1982); Consolandi, Pensotti, Rancoita and Tacconi (2008)]. Other experimental data confirmed that n-type silicon with high-resistivity may change type of conductivity with irradiations, for instance, with fast-neutrons and Kr-ions (e.g., see [Borchi and Bruzzi (1994); Biggeri, Borchi, Bruzzi and Lazanu (1995); Biggeri, Borchi, Bruzzi, Pirollo, Sciortino, Lazanu and Li (1995); Croitoru, Dahan, Rancoita, Rattaggi, Rossi and Seidman (1997); Croitoru, Gubbini, Rancoita, Rattaggi and Seidman (1999a,b); Mangiagalli, Levalois, Marie, Rancoita and Rattaggi (1999); Consolandi, Pensotti, Rancoita and Tacconi (2008)] and references therein). In low resistivity samples, no change in conductivity type was observed in irradiations with Ar-ions [Mangiagalli, Levalois, Marie, Rancoita and Rattaggi (1998)].

[‡]The reader can see [Croitoru, Dahan, Rancoita, Rattaggi, Rossi and Seidman (1997); Croitoru, Gubbini, Rancoita, Rattaggi and Seidman (1999a,b)].

[†]These are samples grown by the Czochralski (pulled) method (e.g., see page 296) and contain oxygen and carbon in higher concentrations with respect to those grown by the vacuum float-zone (Fz) method.

As already mentioned, from combined measurements of the resistivity and Hall coefficient we can determine the Hall mobility. The Hall mobility was observed to decrease with increasing particle fluence (e.g., see [Borchi and Bruzzi (1994); Biggeri, Borchi, Bruzzi, Pirollo, Sciortino, Lazanu and Li (1995); Croitoru, Dahan, Rancoita, Rattaggi, Rossi and Seidman (1997, 1998); Mangiagalli, Levalois, Marie, Rancoita and Rattaggi (1998); Croitoru, Gubbini, Rancoita, Rattaggi and Seidman (1999b); Mangiagalli, Levalois, Marie, Rancoita and Rattaggi (1999); Consolandi, Pensotti, Rancoita and Tacconi (2008)] and references therein). For low fluence of fast-neutrons ($\approx 9.9 \times 10^{10}$ n/cm^2), the electron-mobility, which differs from the Hall mobility by the Hall factor [e.g., see Eq. (4.172)], is given by:

$$\mu_{e,\mathrm{irr}} \propto T^{-\alpha},$$

where $\alpha \approx 2.6$ and $200 < T < 300$ K [Croitoru, Gubbini, Rancoita, Rattaggi and Seidman (1999b)]. The coefficient α was about 2.3, before irradiation [e.g., see Eq. (4.174)].

Furthermore since one of the primary effects of irradiation is to cause displacement damage (e.g., see Sect. 4.2.1) and, as a consequence, to generate deep complex-defects (e.g., see Sect. 4.2.2), it is convenient to show the value of Hall mobility as a function of the concentration of Frenkel-pairs (*FP*) created in the samples. In Figs. 4.35 and 4.36, the Hall mobility[‡] [Eq. (4.171)] is shown as a function of the particle fluence and *FP*, respectively, for irradiations with i) fast-neutrons [Croitoru, Gubbini, Rancoita, Rattaggi and Seidman (1999a,b); Consolandi, Pensotti, Rancoita and Tacconi (2008)], ii) 1.5 MeV electrons [Mangiagalli, Levalois, Marie, Rancoita and Rattaggi (1999)], iii) 60 MeV/amu Kr-ions[††] [Croitoru, Gubbini, Rancoita, Rattaggi and Seidman (1999a,b)] and iv) 6.7 MeV/amu Ar-ions [Mangiagalli, Levalois, Marie, Rancoita and Rattaggi (1998)]. The concentrations of Frenkel-pairs for the irradiations with Ar- and Kr-ions were computed using the SRIM-2003 code [Ziegler, Biersack and Littmark (1985a); Ziegler (2006)], for those with electrons by means of the data reported in Sect. 4.2.1.5 and for those with fast-neutrons by combining their spectral-fluence and the data reported in Sect. 4.2.1.5. The experimental data show that i) there is more similarity in the trend of data as a function of *FP* than as a function of the particle fluence [i.e., compare Figs. 4.35 and 4.36], ii) at low *FP* the Hall mobility differs slightly from the electron Hall mobility expected in a non-irradiated and weakly doped n-type silicon [e.g., see Eq. (4.172)], iii) a large decrease of the Hall mobility is observed for $FP \gtrsim 2 \times 10^{14}$ cm^{-3}, iv) in high-resistivity samples, a first change of conductivity type may occur for $FP \gtrsim 10^{15}$ cm^{-3} (i.e., for fast neutrons and Kr-ions) and v) a second change of conductivity type occurs for Kr-ions for $FP \gtrsim 2 \times 10^{17}$ cm^{-3}.

After irradiation with fast neutrons, the temperature-dependence of the Hall coefficient and mobility [Figs. 4.34(b) and (c)] from 300 down to 11 K was investigated

---

[‡]In Figs. 4.35 and 4.36, the Hall mobility is shown maintaining the sign of the Hall coefficient, instead of using its absolute value [see Eqs. (4.171–4.173)], and at room temperature.

[††]The reader can find the definition of kinetic energies per amu in Sect. 1.4.1.

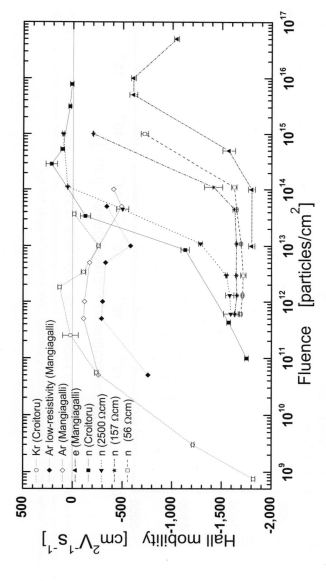

Fig. 4.35 Hall mobility (in $cm^2 V^{-1} s^{-1}$) at room temperature for *n*-type silicon as a function of the particle fluence in particles/cm^2 (adapted from [Leroy and Rancoita (2007)]): 60 MeV/amu Kr-ions (○, high-resistivity) [Croitoru, Gubbini, Rancoita, Rattaggi and Seidman (1999a,b)], 6.7 MeV/amu Ar-ions (◇, for high-resistivity, and ◆, for low-resistivity) [Mangiagalli, Levalois, Marie, Rancoita and Rattaggi (1998)], 1.5 MeV electrons (▲, high-resistivity) [Mangiagalli, Levalois, Marie, Rancoita and Rattaggi (1999)], fast-neutrons (■, high-resistivity) [Croitoru, Gubbini, Rancoita, Rattaggi and Seidman (1999a,b)], fast neutrons (▼, ★ and □ for 2500, 157 and 56 Ωcm before irradiation, respectively) [Consolandi, Pensotti, Rancoita and Tacconi (2008)]. For the above data, before irradiation the resistivities of the high-resistivity samples are in the range $1 \lesssim \rho \lesssim 7 \, k\Omega cm$, while $\lesssim 100 \, \Omega cm$ for low-resistivity samples. The sign of the Hall mobility is the sign of the measured Hall coefficient. The lines are to guide the eye.

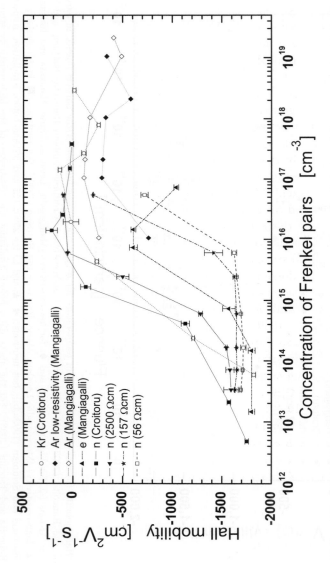

Fig. 4.36 Hall mobility (in $cm^2 V^{-1} s^{-1}$) at room temperature for n-type silicon as a function of the concentration of the Frenkel-pairs in cm^{-3} (adapted from [Leroy and Rancoita (2007)]): the experimental data and corresponding symbols are the same of those shown in Fig. 4.35. The sign of the Hall mobility is the sign of the measured Hall coefficient. The lines are to guide the eye.

for n-type silicon samples with a resistivity of $\approx 56\,\Omega\mathrm{cm}$ before irradiation [Consolandi, Pensotti, Rancoita and Tacconi (2008)]. No change of the Hall-coefficient sign was observed after irradiation, i.e., the Hall-coefficient was negative before (e.g., for n-type silicon) and after irradiation. From 300 ($1000/T \simeq 3.3$) down to $\approx 60\,\mathrm{K}$ ($1000/T \simeq 16.7$) and for a neutron fluence up to $1.1 \times 10^{13}\,\mathrm{n/cm^2}$, it is observed a gradually reduction of the temperature-region extension in which the Hall coefficient is almost constant [Figs. 4.34(b)]. This region is absent for samples irradiated with fluences above $4.5 \times 10^{13}\,\mathrm{n/cm^2}$. For temperatures lower than $\approx 50\,\mathrm{K}$ ($1000/T = 20$) and for a neutron fluence up to $1.3 \times 10^{12}\,\mathrm{n/cm^2}$, the Hall coefficient values are lower than the corresponding non-irradiated ones. In addition, the experimental determinations [Figs. 4.34(c)] indicate i) at all temperatures the Hall mobility of the non-irradiated sample is larger than that of irradiated ones and ii) for these latter samples it gradually decreases with increasing fluence and temperature, except at the largest fluence where it increases with increasing temperature.

As discussed in Sect. 4.2.2, the displacement damage generated by the incoming particles results in the formation of defects with deep levels in the forbidden energy gap. These investigations confirm that large concentrations of deep complex-defects (e.g. see [Konozenko, Semenyuk and Khivrich (1969); Lugakov, Lukashevich and Shusha (1982)] and references therein) are introduced in the silicon and, then, at very large fluences the Fermi level is expected to approach the middle of the forbidden gap[§] irrespective of the incoming particle type and initial kind of conductivity. At the initial stage of irradiation, complex-defects, for instance vacancies with impurity-atoms (donors or acceptors), are introduced that decrease the effective dopant concentration, while, at a later stage, divacancies and other intrinsic defects (e.g, di-interstitial, four-vacancies) become predominant [Konozenko, Semenyuk and Khivrich (1969); Lugakov, Lukashevich and Shusha (1982)]. These defects can be introduced as primary by the cascade mechanism or as secondary by the diffusion of the primary defects. The introduction rate of complex-defects may depend on the incoming particle type (e.g, see Table 9 at pages 155–157 of [Vavilov and Ukhin (1977)] and [Konozenko, Semenyuk and Khivrich (1969); Mangiagalli, Levalois and Marie (1998); Mangiagalli, Levalois, Marie, Rancoita and Rattaggi (1999)] and references therein) and, also, may not be linearly dependent on fluence, as sometimes observed for secondary defects. For example, in irradiations with photons emitted by a ^{60}Co-radioactive source, the divacancy concentration (interpreted as a secondary defect [Konozenko, Semenyuk and Khivrich (1969)]) was determined to be

$$C_{d-v} = 5 \times 10^2 \times \Phi_\gamma^{0.6}\ [\mathrm{cm^{-3}}],$$

where Φ_γ (in $\gamma/\mathrm{cm^2}$) is the photon fluence ([Konozenko, Semenyuk and Khivrich (1969)] and references therein).

[§]Many authors have experimentally determined that the Fermi level approaches the middle of the forbidden gap after irradiation (e.g., see [Konozenko, Semenyuk and Khivrich (1969); Lugakov, Lukashevich and Shusha (1982); Biggeri, Borchi, Bruzzi and Lazanu (1995); Mangiagalli, Levalois, Marie, Rancoita and Rattaggi (1999)] and references therein).

After irradiation with large fluences, the measurement of the Hall coefficient allows one to determine the approximate free-carrier concentration[¶] using Eqs. (4.165, 4.169, 4.170). As a consequence, the resistivity[‖] can be estimated from Eq. (4.168). It has to be noted that, for large fast-neutron fluences, the measured resistivity was found to disagree with that derived by the combined measurement of the Hall coefficient[**] on the same sample [Croitoru, Dahan, Rancoita, Rattaggi, Rossi and Seidman (1997)]. This indicates that a further understanding (as well as investigations) of transport mechanism in irradiated semiconductors is needed.

4.3.6 *AFM Structure Investigation in Irradiated Devices*

Using optical and Atomic Force Microscopy (AFM) techniques along with micro-hardness studies, the characteristics of $p-n$ silicon detectors were widely investigated after fast neutron irradiations for fluences ranging from 10^{11} n/cm^2 up to above 10^{14} n/cm^2 (e.g., see [Golan et al. (1999, 2000, 2001)] and references therein). It has to be noted that, above these fast-neutron fluences, the Hall coefficient was observed to change sign in n-type substrate with resistivity of $(4-6)$ kΩ cm (e.g., see Sect. 4.3.5 and [Croitoru, Dahan, Rancoita, Rattaggi, Rossi and Seidman (1997); Croitoru, Gubbini, Rancoita, Rattaggi and Seidman (1999a); Mangiagalli, Levalois, Marie, Rancoita and Rattaggi (1999)]).

The physical change in the irradiated detector seems to be influenced by the type of defects and their distribution in the bulk material (e.g., [Golan et al. (1999)] and references therein). Investigations have shown morphological changes in silicon detectors. High neutron fluences result in vacancy accumulations and large quantities of interstitial atoms creating dislocation accumulations, which increase the local value of the micro-hardness. For instance, in Fig. 4.37 (left side) the AFM surface topography [Golan et al. (2001)] of a cleaved sample along the $p^+ - n$ doping-distribution and substrate is shown for (I) a non-irradiated device, (II) a device irradiated with a fast neutron fluence of 5.28×10^{14} n/cm^2 and (III) a device irradiated with a fast neutron fluence of 3.12×10^{15} n/cm^2. For a non-irradiated sample [left side of Fig. 4.37 (I)], the defects are homogeneously distributed on the surface of the detector. At large neutron fluences [left side of Figs. 4.37 (II) and 4.37 (III)], accumulations of defects are more observable. An additional set of measurements was carried out on these devices in order to scan the Contact Potential Difference (CPD) along the substrate depth (right side of Fig. 4.37). A sensitivity of less than 10 mV was achieved at an applied bias voltage of 5 V. For the non-irradiated sam-

---

[¶]To determine the free carrier concentration (e.g., see [Konozenko, Semenyuk and Khivrich (1969); Lugakov, Lukashevich and Shusha (1982)]), $\tau_e = \tau_h \simeq 1$ and $b \approx 3$ (e.g., see [Lugakov, Lukashevich and Shusha (1982)]) is assumed in Eq. (4.165), or, the temperature dependence of the Hall factor in Eqs. (4.169, 4.170) is that one given in [Messier and Merlo-Flores (1963)].

[‖]One makes the assumption that μ_h is slightly affected by the irradiation and $b \approx 3$ (e.g., see [Croitoru, Dahan, Rancoita, Rattaggi, Rossi and Seidman (1997)]).

[**]b was assumed to vary in the range $1 < b < 3.3$ (e.g., see [Croitoru, Dahan, Rancoita, Rattaggi, Rossi and Seidman (1997)]).

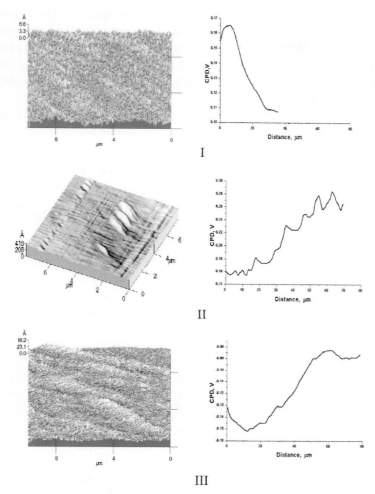

Fig. 4.37 (left side) Atomic Force Microscope (AFM) surface topography of a cleaved sample along the $p^+ - n$ doping-distribution and substrate; (right side) Contact Potential Difference (CPD), in units of V, as a function of the depth in μm of silicon detector junctions with n-type silicon substrate and resistivity of about $6\,\mathrm{k\Omega}\,$cm (before irradiation) (reprinted from *Microelect. Reliab.* **41**, Golan, G. et al., Inversion phenomenon as a result of junction damages in neutron irradiated silicon detectors, 67–72, Copyright (2001), with permission from Elsevier, e.g., for the list of the authors see [Golan et al. (2001)]: (I) for a non-irradiated device, (II) for a device irradiated with a fast-neutron fluence of $5.28 \times 10^{14}\,\mathrm{n/cm^2}$ and (III) for a device irradiated with a fast neutron fluence of $3.12 \times 10^{15}\,\mathrm{n/cm^2}$.

ple [right side of Fig. 4.37 (I)], the behavior of the contact potential follows that expected in the junction region. In this region, the maximum potential of $0.165\,\mathrm{mV}$ is approximately in agreement with that predicted by mean of Eq. (6.12) when the free carrier concentrations of the n-type substrate and p^+-region are taken into account. In addition the width of the depleted region is $\approx 20\,\mu$m, i.e., in agreement with that expected from Eq. (6.26). For samples irradiated with the largest fluence (i.e., 3.12×10^{15} n/cm^2), the contact potential seems to exhibit an inverted behavior [right side of Fig. 4.37 (III)]. Other measurements of CPD as a function of the depth position (Figure 2 in [Golan et al. (2001)]) have provided values lower than those shown in Figs. 4.37 (II) and 4.37 (III) (right side). These later data might indicate a reduced space-charge region.

Chapter 5

Scintillating Media and Scintillator Detectors

The conversion of the energy deposited by incoming particles or radiation into photons can lead to particle detection, energy determination and discrimination between particles of different masses. Therefore, the use of scintillating materials, or the construction of detectors enhancing the Čerenkov or the transition radiation effects, can be of great use in many application fields, such as high-energy physics or medical imaging. The photons emitted in such detectors must be transported, by light guides, to photosensitive devices, e.g., photomultipliers, in order to be collected. In this chapter, the organic and inorganic scintillators will be reviewed, as well as the Čerenkov and transition radiation detectors, with many examples. A section will also be devoted to the description of the light transport and detection techniques.

5.1 Scintillators

Scintillators are materials in which large fractions of incident energy carried by striking particles or radiation are absorbed and transformed into detectable photons (visible or near visible light), later converted into an electric signal. They are used in many applications, e.g., as detecting elements of calorimeters or in medical imaging. This section will put an emphasis on calorimetric applications. Medical applications will be discussed in the chapter devoted to medical physics.

There are two types of light emission: fluorescence and phosphorescence. Fluorescence corresponds to prompt light emission (ns $\rightarrow$ μs) in the visible wavelength range and is temperature independent. Phosphorescence corresponds to light emission over a longer period of time (μs $\rightarrow$ ms, even hours depending on the material) with longer wavelengths compared to fluorescence and is temperature dependent. In practice, only fluorescence, characterized by fast light emission, is useful for the detection of particle or nuclear radiation.

Inorganic and organic scintillators are used for radiation detection (e.g., see [Hofstadter (1974); Knoll (1999)] and references therein). As will be seen later, different mechanisms are at the origin of light production in these two types of scintillators.

There are several parameters characterizing scintillators. The scintillation efficiency, R_s, is the ratio of the average number of emitted photons $\langle N_{\mathrm{ph}} \rangle$ to the energy E_i of the incident radiation absorbed by the scintillator:

$$R_s = \frac{\langle N_{\mathrm{ph}} \rangle}{E_i}. \tag{5.1}$$

The scintillation yield, R_r, is the ratio of the total energy of the emitted light to E_i. It is obtained by multiplying the right-hand side of Eq. (5.1) by the energy, $h\nu$, of the emitted photons:

$$R_r = R_s h\nu. \tag{5.2}$$

The photoelectric efficiency, R_{pe}, of the scintillating material is often quoted in medical physics for absorption of X- and γ-rays. This quantity is the ratio, R_{pe}, of the number of the incident X- and γ-rays that have deposited all their energy in the material, $N_{\mathrm{ph,absorbed}}$, to the total number of photons that have been detected, $N_{\mathrm{ph,detected}}$:

$$R_{\mathrm{pe}} = \frac{N_{\mathrm{ph,absorbed}}}{N_{\mathrm{ph,detected}}}. \tag{5.3}$$

From its definition, R_{pe} features the importance of the absorption power of radiation by the scintillating material. It depends on the incident energy and geometrical size of the scintillating material struck by the radiation.

The scintillation spectrum has to be matched to the spectral sensitivity of the photosensitive device coupled to the scintillator. The wavelength corresponding to the maximum emission of light has to fall within the range of the spectral sensitivity of the photocathode of the photomultiplier (photosensitive device).

As will be seen below, the time response of the scintillator material is controlled by decay time constants.

The growth in time of the scintillation pulse is characterized by a fast rise time, followed by a decay which has two components: a fast exponential decay and a slower decay. The decay time constant of the fast component determines the time response of the scintillating material. This parameter guides the choice of the type of scintillating material for specific applications.

The scintillator material must be transparent to its own scintillation light. It turns out that the emission wavelength is, in general, longer than the absorption wavelength (*Stokes's shift*).

5.1.1 *Organic Scintillators*

Organic scintillators are aromatic hydrocarbon compounds containing a benzenic cycle. In organic scintillators, the mechanism of light emission is a molecular effect. It proceeds through excitation of molecular levels in a primary fluorescent material, which emits bands of ultraviolet (UV) light during de-excitation. This UV light is absorbed in most organic materials with an absorption length of a few mm. The extraction of a light signal becomes possible only by introducing a second fluorescent material in which the UV light is converted into visible light (*wavelength shifter*). This second substance is chosen in such a way that its absorption spectrum is matched to the emission spectrum of the primary fluor, and its emission spectrum, adapted to the spectral dependence of the photocathode quantum efficiency. These two active components of a scintillator are either dissolved in suitable organic liquids or mixed with the monomer of a material capable of polymerization.

Plastic scintillators can be produced in a variety of shapes adapted to the application needs. As an indication, most frequently used are rectangular plates with thicknesses from 0.1 up to 30.0 mm and area from a few mm^2 up to several m^2. They are of current use in a wide range of applications such as beam counters and active medium of sampling calorimeters. In sampling calorimeters (discussed in Sect. 9.2) using scintillators as active medium, the incoming particle energy is measured in a number of plastic scintillating layers interspersed with layers of absorbing material (metal), the latter speeding up the cascade process. A schematic description of a sampling calorimeter is shown in Fig. 9.1 in the chapter on *Principles of Particle Energy Determination*. The scintillation light is coupled from the scintillator layers to the photosensitive device via light guides. However, in most cases, such arrangement is not practical, since it does not favor good transverse and longitudinal segmentations and create large gaps in which energy may be lost. In the case of fixed target experiments with large solid angle coverage or colliding beam experiment, the scintillating light is transported to the photosensitive device via wavelength shifter (WLS) material (see Sect. 5.3 on wavelength shifter).

For organic scintillators, the relation between the emitted light and the energy deposited by an ionizing particle is not linear. The scintillator response depends also on the type of particle and its specific ionization. The relation between emitted light and energy deposited by ionizing particles is expressed by *Birks' Law* [Birks (1951, 1964)]. This law assumes that the response of organic scintillator is linear. Possible deviation from this linearity is explained by quenching interactions between excited molecule created along the path of the ionizing particle, absorbing energy which causes a reduction of the scintillation efficiency. Birks' law correlates the amount of emitted light per unit of length, the differential light output dL/dx, to the stopping power of the ionizing particle dE/dx, namely

$$\frac{dL}{dx} = \left(R_s \frac{dE}{dx} \right) \bigg/ \left(1 + kB \frac{dE}{dx} \right), \tag{5.4}$$

where R_s is the scintillation efficiency [see Eq. (5.1)], B is the *Birks constant* of the medium and k is the *quenching parameter*. Here kB acts as a saturation constant achieving a value ~ 0.01–0.02 g/cm^{2}MeV for a scintillator. In practice, kB is handled as an adjustable parameter to fit experimental data for a specific scintillator with R_s giving the absolute normalization. The total amount of light, L, produced by a particle of energy, E, in a scintillator is obtained by integration of Eq. (5.4).

For small dE/dx (e.g. for fast electrons), Eq. (5.4) reduces to

$$\frac{dL}{dx} \approx R_s \frac{dE}{dx},$$ (5.5)

while for large dE/dx (e.g. for α-particles), Eq. (5.4) becomes:

$$\frac{dL}{dx} \approx \frac{R_s}{kB}.$$ (5.6)

The integration of Eq. (5.6) over x gives the amount of light as a function of the range, $R(E)$, of the particle with energy E

$$L(E) \sim \frac{R_s}{kB} R(E).$$ (5.7)

Practically, kB can be determined from the ratio between Eq. (5.5) and Eq. (5.6).

Organic scintillators can also be liquids. In a liquid scintillator, an organic crystal (solute) is dissolved in a solvent with typical concentrations of ≈ 3 g of solute per liter of solvent. The ionization energy absorbed by the solvent is transferred to the solute which is responsible for fluorescence. This transfer can be achieved as light emission by the solvent, absorption by the solute and re-emission at larger wavelength. The scintillation efficiency increases with the solute concentration [Schram (1963)]. Liquid scintillators have fast response, corresponding to a decay time of a few ns. For instance, the scintillator BIBUQ dissolved in toluene has the best time resolution (≈ 85 ps) of any scintillator currently available [Bengston and Moszynski (1982)]. Liquid scintillators are very sensitive to impurities present in the solvent. This can be turned into an advantage by adding material to increase their efficiency for a specific application. For instance, the efficiency of liquid scintillator for neutron detection can be increased by adding boron, which has a large cross section with neutrons.

Table 5.1 Properties of organic scintillators of current use (see for instance [Anderson (1990); PDB (2002)]).

Properties	naphtalene	anthracene	NE102A	NE110
Density (g/cm^3)	1.15	1.25	1.032	1.032
H/C ratio	0.800	0.714	1.105	1.105
Emission spectrum λ(nm)	348	448	425	437
Decay time constant τ(ns)	11	31	2.5	3.3
Scintillation amplitude (vs anthracene)	11	100	65	60

All organic scintillators have low density and low atomic number and, therefore, have relatively low absorption for charged particles and for γ- and X-rays. Being a mixture of C and a high content of H, the organic scintillators have high absorption for fast neutrons (Table 5.1). The absorption of γ- and X-rays in organic scintillators is dominated by the Compton effect (low-Z material). It is possible to obtain emission spectrum with a maximum wavelength in the range 350–500 nm. For example, anthracene has a maximum wavelength of about 450 nm (Table 5.1). Their scintillation yield for γ and fast electrons is a few percents. The shape of the scintillation pulse is characterized by a fast rise time of the order of 1 ns and a decay time of a few ns (Table 5.1). The decay can be described, as a function of time, by an exponential fast component characterized by a lifetime τ:

$$L(t) \sim e^{-t/\tau}. \tag{5.8}$$

This component is called the main component because it contains the largest part of the emitted light. A delayed (non-exponential) component follows, whose intensity depends on the ionization power of the radiation

$$L(t) \sim \frac{C}{(1 + Dt)^n}, \tag{5.9}$$

where C and D are constants and $n \approx 1$.

5.1.2 *Inorganic Scintillators*

Inorganic scintillators (e.g., see [Swank (1954)]) are ionic crystals doped or not with color centers (activator). Examples of inorganic scintillators without activator are BGO ($Bi_4Ge_3O_{12}$), CeF_3, BaF_2 or $PbWO_4$. Production of luminescence in inorganic scintillator such as NaI(Tl) or CsI(Tl) requires the presence of an activator (Thalium), as explained below. Inorganic scintillators have high density and high atomic number compared to organic scintillators. From these properties, one can immediately expect the inorganic scintillators to have high absorption for γ- and X-rays. They also have high absorption for electrons, alpha, protons and charged heavy particles, in general. Crystals have rather short radiation lengths, between 0.9 and 2.6 cm, and densities from 3.7 up to 8.3 g/cm^3 (Tables 5.2 and 5.3).

Table 5.2 Properties of NaI(Tl), BGO and CeF$_3$ inorganic scintillators (see, for instance, [Leroy and Rancoita (2000)]).

Properties	NaI(Tl)	BGO	CeF$_3$
Density (g/cm^3)	3.67	7.13	6.16
Radiation Length (cm)	2.59	1.12	1.68
Moliere radius (cm)	4.5	2.4	2.6
dE/dx(MeV/cm) [per *mip*]	4.8	9.2	7.9
short decay time (ns)	230	300	~ 5
long decay time (ns)	150[ms]		30
Peak emission(nm): short	415	480	310
Peak emission(nm): long			340
Refractive index at peak emission	1.85	2.20	1.68
light yield (vs NaI(Tl))	1.00	0.15	0.10
light yield γ/MeV	4×10^4	8×10^3	2×10^3

The mechanism of scintillation for these crystals is a lattice effect, rather than a molecular effect as it was the case for organic scintillators. The valence band contains electrons that are bound at the lattice sites. Electrons in the conduction band are free to move throughout the crystal. The passage of a charged particle through the crystal may produce the ionization of the crystal, if the incoming particle transfers sufficient energy to the electrons from the valence band for them to acquire sufficient energy to move into the conduction band, leaving a hole in the valence band. If the incoming energy is not high enough, the electron of the valence band will not reach the conduction band and will form a bound state, called exciton, with the hole. The excitons are located in an exciton band below the conduction band. Finally, there exist activator centers in the scintillator which occupy energy levels in the gap, between the conduction and the valence bands. So, the passage of charged particle through the scintillator medium generates a large number of free electrons, free holes and electron–hole pairs which move around in the crystal lattice until they reach an activator center. They then transform the activation center into an excited state. The subsequent decay of this excited state to the activator center ground state produces emission of light. The decay time of this scintillation light is given by the lifetime of the excited state. Therefore, to help the scintillation efficiency, i.e., to increase the probability of visible light emission, one may introduce a small amount of impurity called activator. Thalium (Tl) is often used as activator. For instance, NaI(Tl) and CsI(Tl) usually contain about 0.1% of Thalium. The light

Table 5.3 Properties of BaF_2, CsI(Tl) and $PbWO_4$ inorganic scintillators (see, for instance, [Leroy and Rancoita (2000)]).

Properties	BaF_2	CsI(Tl)	$PbWO_4$
Density (g/cm^3)	4.89	4.53	8.28
Radiation Length (cm)	2.05	1.85	0.89
Moliere radius (cm)	3.4	3.8	2.2
dE/dx(MeV/cm) [per *mip*]	6.6	5.6	13.0
short decay time (ns)	0.6	> 1000	5
long decay time (ns)	620		15
Peak emission(nm): short	220	550	440
Peak emission(nm): long	310		530
Refractive index at peak emission	1.56	1.80	2.16
light yield (vs NaI(Tl))	0.05(fast)	0.40	0.01
	0.20(slow)		
light yield γ/MeV	10^4	5×10^4	1.5×10^2

emission is characteristic of the activator and it is possible to modify the emission wavelength by varying the activator.

The shape of the scintillation pulse emitted by an inorganic scintillator is characterized by a fast rise time of a few tens of nanoseconds and a decay which has two components: a short (fast or prompt) exponential decay, whose decay time constant ($\tau_{d,s}$) lies between a few nanoseconds and a few hundred of nanoseconds

(Tables 5.2 and 5.3), which governs the main part of the pulse, and a long (slow) delayed component with a decay constant ($\tau_{d,l}$) of few hundred nanoseconds up to microseconds or even milliseconds (Tables 5.2 and 5.3). Then, the number, N, of photons emitted at time t is given by

$$N = A_s\, e^{-t/\tau_{d,s}} + A_l\, e^{-t/\tau_{d,l}}, \tag{5.10}$$

where $\tau_{d,s}$ and $\tau_{d,l}$ are the short and long decay constants, respectively. A_s and A_l are the relative magnitude of the short and long components, respectively. The ratio A_s/A_l varies with the material (see Tables 5.2 and 5.3).

For NaI(Tl), the γ is emitted with $\tau_{d,s} \sim 230$ ns and 80% of the light intensity is emitted in 1μs. Phosphorescence may persist up to $\tau_{d,l} = 150$ ms (Tables 5.2 and 5.3). In BGO, there is a fast component $\tau_{d,s} \sim 60$ ns, accounting for about 10% of the light intensity, and a slow component $\tau_{d,l} \sim 300$ ns, representing 90% of the light intensity (Tables 5.2 and 5.3). BaF_2 is characterized by a very fast component $\tau_{d,s} \sim 0.6$ ns, which represents 20% of the total light intensity and a slow component $\tau_{d,l} \sim 620$ ns.

The decay time of the scintillator should be as short as possible. This becomes an important constraint in experiments at LHC, where the light has to be collected in less than 10 ns since the expected LHC interbunch crossing is 25 ns, requiring the use of new types of crystals such as $PbWO_4$ which has a time decay constant with an average value of 10 ns allowing 85% of the light to be collected during the interbunch crossing time. Radiation hardness of the crystal is also an important issue at LHC where crystals will be exposed to irradiation fluence of 10^{13}–10^{15} hadrons/cm^2 depending on pseudorapidity (see Sect. 3.2.3.1). The choice of $PbWO_4$ crystal, for the electromagnetic calorimeter of CMS experiment at LHC, is a compromise between best possible scintillation performance and best radiation hardness.

In inorganic scintillators, dL/dx also varies with energy but deviations to linearity are weaker in most of the applications and in practice Birk's law is not applied. Non-linearity of the order of 20% can be observed at low energy. This non-linearity of the response to low-energy γ- and X-rays is the consequence of the non-linearity of their response to electrons responsible for the transfer of γ- and X-rays energy to the crystal.

The photoelectric efficiency is maximal for γ- and X-rays of low energy ($<$ 100 keV), for which the interactions with the crystal (high-Z) are largely dominated by the photoelectric effect, $R_{pe} \approx 1$. For higher energies, the value of R_{pe} depends on various factors such as the incident γ energy, the absorption coefficients of the various interaction processes in the crystal, the mechanisms of γ energy deposition in the scintillator, the geometrical size of the detector and the geometry of the experimental setup. Inorganic scintillators such as NaI(Tl) and BGO are standardly used as active elements of *medical imagers*.

Inorganic scintillators generally have two emission bands: a) one is characteristic of the activator; b) the second corresponds to shorter wavelengths and is characteristic of the crystal lattice. Tables 5.2 and 5.3 show that the maximum of the spectral emission takes place in most inorganic scintillators at $220\,\text{nm} \leq \lambda \leq 550\,\text{nm}$ i.e., in overlap with the spectral sensitivity domain of standard photocathodes.

Tables 5.2 and 5.3 summarize the values of decay constants and light yields for various inorganic crystals. The detected signal in these tables is quoted in terms of γ per MeV. One often expresses the detected signal in photoelectrons per MeV (n_{pe}). This conversion into n_{pe} depends on several factors such as the light collection efficiency and the quantum efficiency of the photosensitive device. It has to be noted that the light yields reported in Tables 5.2 and 5.3 have been measured with a photomultiplier with a bialkali photocathode.

Several inorganic scintillators will deteriorate due to water absorption, if exposed to air. This sensitivity to moisture is called hygroscopicity. Scintillators such as NaI(Tl), CsI(Na) are hygroscopic while BGO, CeF$_3$, BaF$_2$, PbWO$_4$ are not. Hygroscopic scintillators have to be placed in an air-tight container to protect them from moisture. CsI(Tl) is only slightly hygroscopic and may be handled without protection.

5.2 The Čerenkov Detectors

As seen in Sect. 2.2.2, a charged particle traveling in a dielectric medium of refractive index $n > 1$ with a speed v greater than the speed of light in the medium ($v > c/n$ or $\beta > 1/n$) emits radiation, the Čerenkov light (Fig. 5.1). It has to be noted that, in Sect. 2.2.2, the refractive index has been indicated by the symbol r and the number of electron per cm^3 by the symbol n, while in the present section the refractive index is indicated by the symbol n. The Čerenkov light is emitted in a forward cone of aperture $\cos\theta = c/(vn) = 1/(\beta n)$. One has

$$\sin^2\theta = 1 - \frac{1}{\beta^2 n^2}. \tag{5.11}$$

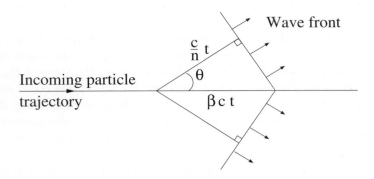

Fig. 5.1 The principle of Čerenkov light emission. The light is emitted in a forward cone of aperture $\cos\theta = [(ct)/n]/(\beta ct) = 1/(\beta n)$.

The maximum emission angle, θ_{max}, corresponds to $\beta = 1$ or $v = c$, i.e., $\cos\theta_{max} = 1/n$. One can express the threshold condition for Čerenkov light emission as

$$\beta_{thr} = v_{thr}/c \geq 1/n \tag{5.12}$$

and the threshold Lorentz factor is

$$\gamma_{thr} = \frac{1}{\sqrt{1 - \beta_{thr}^2}} = \frac{n}{\sqrt{n^2 - 1}}. \tag{5.13}$$

Equations (1.10, 5.13) define a particle threshold-energy, E_{thr}, given by

$$E_{thr} = \gamma_{thr} m_0 c^2, \tag{5.14}$$

where m_0 is the particle rest mass. The threshold Lorentz factor, γ_{thr}, depends on the particle mass for a fixed energy and explains why the measurement of the Čerenkov radiation allows particle identification.

Using Eq. (5.11), the energy emitted by a particle of charge number z and mass m per unit length and unit wavelength moving with a velocity β in a medium of refractive index n is given by [Fernow (1986)]:

$$\frac{d^2 E}{dx\, d\lambda} = 4\pi^2 r_e z^2 mc^2 \frac{1}{\lambda^3}\left(1 - \frac{1}{\beta^2 n^2}\right) \tag{5.15}$$

where $r_e = e^2/m_e c^2$ ($= 2.817938$ fm) is the classical radius of the electron.

The number of photons, N_{ph}, emitted per unit length of the track of the incident particle of charge number z through the medium, in a wavelength interval, is given by Eq. (2.138), which can be re-expressed in terms of $d\lambda$ (see page 128):

$$\frac{dN_{ph}}{dx} = 2\pi\alpha z^2\left(1 - \frac{1}{\beta^2 n^2}\right)\frac{d\lambda}{\lambda^2} \tag{5.16}$$

where $\alpha \sim 1/137$ and n is the refractive index.

In the wavelength interval $d\lambda$, the radiation intensity depends only on the charge number z of the incident particle and on the emission angle, and therefore on the particle velocity. The presence of λ^2 in the denominator of Eq. (5.16) indicates an emission concentrated at shorter wavelength. The emission takes place in the wavelength domains of visible and UV light. If the emission takes place in the range

$$\lambda_{min}(= \lambda_1) - \lambda_{max}(= \lambda_2),$$

after integration, Eq. (5.16) becomes:

$$\frac{dN_{ph}}{dx} = 2\pi\alpha z^2\left(1 - \frac{1}{\beta^2 n^2}\right)\left(\frac{\lambda_2 - \lambda_1}{\lambda_2 \lambda_1}\right). \tag{5.17}$$

It was assumed that n was independent of λ, i.e., the dispersion of the radiator was neglected. In the wavelength domain of the visible and UV light, 400 to 700 nm, which corresponds to the domain of spectral sensitivity of the standard photocathodes (e.g., bialkali photocathodes are sensitive to wavelengths of 400 to 575 nm), Eq. (5.17) becomes

$$\frac{dN_{ph}}{dx} = 491 \text{ cm}^{-1} z^2\left(1 - \frac{1}{\beta^2 n^2}\right). \tag{5.18}$$

The light emitted by a Čerenkov radiator is smaller than that emitted by scintillators of the same dimensions. For a particle with $\beta \sim 1$ in water ($n = 1.33$), $\sin^2\theta \sim 0.43$ [see Eq. (5.11)] and about 210 photons/cm [see Eq. (5.18)] are emitted. Equation (5.15) integrated over the wavelength domain of 400 to 700 nm gives an emitted energy of $\sim 1200 \times 0.43 = 516$ eV/cm corresponding to $\approx 3 \times 10^{-4}$ of the energy lost by a *mip* (~ 2 MeV/cm) in scintillator.

Using the standard relation

$$E = h\nu = \frac{hc}{\lambda} = \frac{2\pi\hbar c}{\lambda}, \tag{5.19}$$

the number of photons emitted per energy interval, dE, is given by:

$$\frac{d^2 N}{dx\,dE} = \frac{d^2 N}{dx\,d\lambda}\frac{d\lambda}{dE} = \frac{\lambda^2}{2\pi\hbar c}\frac{d^2 N}{dx\,d\lambda}. \tag{5.20}$$

Furthermore, using Eq. (5.16), one finds (Frank–Tamm equation)

$$\frac{d^2 N}{dx\,dE} = \frac{\alpha z^2}{\hbar c}\left(1 - \frac{1}{\beta^2 n^2}\right), \tag{5.21}$$

or, equivalently, since $\hbar c = 197.327 10^6$ eV fm and $\alpha \sim 1/137$

$$\frac{d^2 N}{dx\,dE} = 370 \times z^2 \left(1 - \frac{1}{\beta^2 n^2}\right) \quad [\text{eV cm}]^{-1}. \tag{5.22}$$

For a radiator of length L, the number of photoelectrons, N_{pe}, detected in the photosensitive device is proportional to the energy acceptance window

$$\Delta E = E_2 - E_1$$

of the photosensitive device and is given by:

$$N_{\text{pe}} = 370 \text{ eV}^{-1}\text{cm}^{-1} \times Lz^2\left(1 - \frac{1}{\beta^2 n^2}\right)\int_{E_1}^{E_2} \epsilon_{\text{col}}(E)\,\epsilon_{det}(E)\,dE, \tag{5.23}$$

where ϵ_{col} is the efficiency for collecting the Čerenkov light (transmission, reflections) and ϵ_{det} is the quantum efficiency of photo-conversion of the photosensitive device. One defines the factor of merit, N_0, of the detector [Séguinot (1988)] by

$$N_0 = 370 \text{ eV}^{-1}\text{cm}^{-1} \times z^2 \int_{E_1}^{E_2} \epsilon_{\text{col}}(E)\,\epsilon_{det}(E)\,dE \quad [\text{cm}]^{-1}. \tag{5.24}$$

Defining

$$\bar{\epsilon} = \frac{1}{\Delta E}\int_{E_1}^{E_2} \epsilon_{\text{col}}(E)\,\epsilon_{det}(E)\,dE, \tag{5.25}$$

one can write the total number of detected photons as:

$$N_{\text{pe}} = N_0 L \left(1 - \frac{1}{\beta^2 n^2}\right) \tag{5.26}$$

with

$$N_0 = 370 \text{ eV}^{-1}\text{cm}^{-1} \times z^2\bar{\epsilon}\,\Delta E \quad [\text{cm}]^{-1}. \tag{5.27}$$

With a quantum efficiency of photo-conversion of 40% and efficiency for collecting the Čerenkov light of 75%, assuming ϵ_{col} and ϵ_{det} independent of energy, for $z = 1$ one finds

$$N_0 \sim 110 \text{ eV}^{-1}\text{cm}^{-1} \times \Delta E \quad [\text{cm}]^{-1}. \tag{5.28}$$

Čerenkov detectors of various types (threshold, differential and ring-imaging) are used for particle identification.

5.2.1 Threshold Čerenkov Detectors

Threshold Čerenkov detectors can be used to separate particles of same momentum and different masses in a beam. This situation is usually encountered in secondary beams of hadronic accelerator facilities. These secondary beams may contain electrons, pions, kaons and protons of same momentum. The Čerenkov light emitted in a forward cone is reflected back by a spherical mirror into photomultiplier tubes (one or two). Depending on the relative position of the spherical mirror and photomultiplier tubes, a plane mirror can be put at angle with the photomultipier tubes to help focalize the light on the photocathodes (Fig. 5.2).

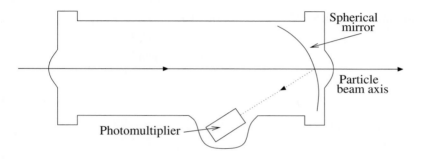

Fig. 5.2 Schematic description of a threshold Čerenkov detector of the type used in CERN-PS secondary hadronic beams (adapted and reprinted, with permission, from the Annual Review of Nuclear Science, Volume 23 © 1973 by Annual Reviews *www.annualreviews.org*; [Litt and Meunier (1973)]). The Čerenkov light is reflected back by a spherical mirror onto the photocathode of the photomultiplier tube (PMT).

Let us assume that one wishes to separate particles "1" and "2" in a beam which also contains particles "3". These particles have the same momentum p and rest masses m_1, m_2 and m_3 ($m_1 < m_2 < m_3$), respectively. Two Čerenkov detectors (C1 and C2) are needed to separate particles "1" and "2". The refractive index of the radiator of C1 has to be adjusted at a value for which particle "2" is just at threshold, i.e., does not radiate: then, $\beta_2{}^2 = 1/n^2$ and $\sin^2 \theta_2 = 0$. From Eq. (5.13), one also has $n^2 = \gamma_2^2/(\gamma_2^2 - 1)$. Then, the production rate of photons from particle "1" is given by Eq. (5.18) with

$$\sin^2 \theta_1 = 1 - \frac{1}{\beta_1{}^2 n^2}. \tag{5.29}$$

The particle energy, E, and its velocity, $\beta = v/c$, are related through

$$\frac{1}{\beta} = \frac{E}{pc} \tag{5.30}$$

and

$$E^2 = m^2 c^4 + p^2 c^2 \tag{5.31}$$

with

$$p = m\beta\gamma c. \tag{5.32}$$

Therefore, using Eq. (5.30), Eq. (5.29) can be written as:

$$\sin^2 \theta_1 = 1 - \frac{E^2}{p^2 c^2 n^2} = \frac{n^2 - 1}{n^2} - \frac{m_1^2 c^2}{p^2 n^2}. \tag{5.33}$$

The particle momentum, p, and its velocity, $\beta = v/c$, are related through (at threshold, $\beta = 1/n$)

$$p = m\beta\gamma c = \frac{m\beta c}{\sqrt{1 - \beta^2}} = \frac{mc}{\sqrt{n^2 - 1}}. \tag{5.34}$$

So, being at the particle "2" threshold, one has

$$\sin^2 \theta_2 = 0 = \frac{n^2 - 1}{n^2} - \frac{m_2^2 c^2}{p^2 n^2}, \tag{5.35}$$

or

$$n^2 - 1 = \frac{m_2^2 c^2}{p^2}. \tag{5.36}$$

Then with the use of Eq. (5.36), Eq. (5.33) becomes

$$\sin^2 \theta_1 = \frac{(m_2^2 - m_1^2) c^2}{p^2 n^2}. \tag{5.37}$$

So, the conditions of Eqs. (5.36, 5.37) have to be fulfilled to achieve separation of particles "1" and "2".

The length of the radiator needed to produce a given number of photoelectrons increases with the square of the momentum. Then, the length L of the radiator can be determined to secure a usable practical counting rate of photons. Taking into account the quantum efficiency ϵ, the number of photoelectrons [Eq. (5.18)] is:

$$N_{\text{pe}} = 491 \text{ cm}^{-1} \times \epsilon L \frac{(m_2^2 - m_1^2) c^2}{p^2 n^2}. \tag{5.38}$$

To obtain N photoelectrons per cm, a radiator length of

$$L = N \frac{p^2 n^2}{491 \times c^2 \epsilon (m_2^2 - m_1^2)} \tag{5.39}$$

is needed with the correct refractive index. If $m_1 = m_\pi = 139.6 \text{ MeV}/c^2$ and $m_2 = m_K = 493.6 \text{ MeV}/c^2$, with $p = 5 \text{ GeV}/c$ and a quantum efficiency of 20%, a radiator length of 17 cm is needed to collect 15 photoelectrons. A radiator with a refractive index $n = 1.005$ [Eq. (5.36)] has to be selected to prevent the radiation of the kaon particles.

To separate particles "2", one has to select a second radiator with refractive index n' set at a value for which particle "3" is at threshold:

$$\sin^2 \theta_3 = 0 = \frac{n'^2 - 1}{n'^2} - \frac{m_3^2}{p^2 n'^2}. \tag{5.40}$$

Consider the case of secondary beams delivered by the CERN Proton-Synchroton (PS) which contains pions, kaons and protons ($m_\pi < m_K < m_p$), one needs to use three Čerenkov detectors to achieve pion, kaon and proton separation. At first, if one uses two aerogel Čerenkov radiators C1 with $n = 1.1$ and C2 with $n' = 1.02$, one can separate pions and kaons [Eq. (5.34)]. Then the threshold momenta (p_{thr}) are:

for pions to emit light in C1:

$$p > p_{\text{thr}} = \frac{m_\pi c}{\sqrt{n^2 - 1}} = 0.3\,\text{GeV}/c \ (n = 1.1); \tag{5.41}$$

for pions to emit light in C1 and C2

$$p > p_{\text{thr}} = \frac{m_\pi c}{\sqrt{n^2 - 1}} = 0.7\,\text{GeV}/c \ (n = 1.02); \tag{5.42}$$

for kaons to emit light in C1

$$p > p_{\text{thr}} = \frac{m_K c}{\sqrt{n^2 - 1}} = 1.1\,\text{GeV}/c \ (n = 1.1); \tag{5.43}$$

for kaons to emit light in C1 and C2

$$p > p_{\text{thr}} = \frac{m_K c}{\sqrt{n^2 - 1}} = 2.5\ \text{GeV}/c \ (n = 1.02). \tag{5.44}$$

Therefore, pion/kaon separation can be achieved in the momentum range below $2.5\,\text{GeV}/c$ with two Čerenkov counters. If one adds a third Čerenkov counter C3 with $n = 1.005$ (not aerogel), one finds threshold momenta to be added to the set of Eqs. (5.41–5.44) such as:

for protons to emit light in C1

$$p > p_{\text{thr}} = \frac{m_p c}{\sqrt{n^2 - 1}} = 2.1\,\text{GeV}/c \ (n = 1.1); \tag{5.45}$$

for protons to emit light in C1 and C2

$$p > p_{\text{thr}} = \frac{m_p c}{\sqrt{n^2 - 1}} = 4.7\,\text{GeV}/c \ (n = 1.02); \tag{5.46}$$

for the three particles to give light in C1, C2 and C3

$$p > p_{\text{thr}} = \frac{m_p c}{\sqrt{n^2 - 1}} = 9.4\,\text{GeV}/c \ (n = 1.005), \tag{5.47}$$

$$p > p_{\text{thr}} = \frac{m_\pi c}{\sqrt{n^2 - 1}} = 1.4\,\text{GeV}/c \ (n = 1.005), \tag{5.48}$$

$$p > p_{\text{thr}} = \frac{m_K c}{\sqrt{n^2 - 1}} = 4.9\,\text{GeV}/c \ (n = 1.005). \tag{5.49}$$

Thus, pion/kaon/proton separation can be achieved in the momentum range below $4.7\,\text{GeV}/c$. For a beam of $2\,\text{GeV}/c$ particles, using this set of three Čerenkov counters with the same values of refractive indices, kaons, protons, and pions would emits light in C1, C2 and C3; kaons and protons would give light only in C1.

Gaseous threshold counters are often used. They offer the possibility to vary the refractive index by pressure adjustment. The refractive index of the radiative medium is given by the *Lorentz–Lorenz law* [Litt and Meunier (1973)]:

$$\frac{n^2 - 1}{n^2 + 2} = \frac{R}{M}\rho, \tag{5.50}$$

where R is the molecular refractivity, M is the molecular weight, and ρ is the gas density. In the case of a gas with $n \sim 1$, Eq. (5.50) becomes (see for instance [Fernow (1986)]):

$$n - 1 = \frac{3}{2}\frac{R}{M}\rho. \tag{5.51}$$

From the *ideal gas law*

$$\frac{P}{RT} = \frac{\rho}{M}, \tag{5.52}$$

where P and T are the pressure and temperature, respectively, Eq. (5.51) becomes:

$$n - 1 = \frac{3}{2}\frac{P}{T}. \tag{5.53}$$

If n_0 is the refractive index of the gas for a given wavelength of light, at temperature T and pressure $P = 1$ atmosphere, one has

$$n_0 - 1 = \frac{3}{2}\frac{1}{T}. \tag{5.54}$$

Combining Eqs. (5.53, 5.54) leads to:

$$n - 1 = (n_0 - 1)P. \tag{5.55}$$

From Eq. (5.13), one obtains ($n \sim 1$)

$$\gamma_{\text{thr}}\beta_{\text{thr}} = \frac{1}{\sqrt{n^2 - 1}} = \frac{1}{\sqrt{2(n - 1)}} = \frac{1}{\sqrt{2(n_0 - 1)P}}. \tag{5.56}$$

Consequently, the momentum threshold is

$$p_{\text{thr}} = m\gamma_{\text{thr}}\beta_{\text{thr}}\, c = \frac{mc}{\sqrt{2(n_0 - 1)P}}. \tag{5.57}$$

Therefore, changing pressure modifies the momentum threshold. In particular, one can decrease the momentum threshold by increasing the gas pressure.

Note that in the case of a gaseous detector, Eqs. (5.36, 5.37) become ($n \sim 1$)

$$n - 1 = \frac{m_2^2 c^2}{2p^2} \tag{5.58}$$

and

$$\theta_1^2 = \frac{(m_2^2 - m_1^2)\, c^2}{p^2}. \tag{5.59}$$

Using Eq. (5.26), one can express Eq. (5.59) as

$$\theta_1^2 = \frac{N_{\text{pe}}}{N_0 L}. \tag{5.60}$$

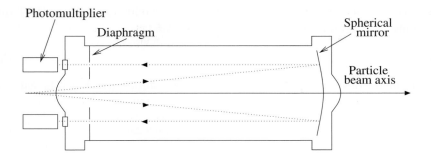

Fig. 5.3 Schematic description of a differential Čerenkov detector close to the type used in the CERN-SPS beams (adapted and reprinted, with permission, from the Annual Review of Nuclear Science, Volume 23 © 1973 by Annual Reviews *www.annualreviews.org*; [Litt and Meunier (1973)]). The Čerenkov light is reflected back by a spherical mirror onto the photocathode of a photomultiplier tube (PMT). The light is filtered by collimators in front of the PMT.

One may also note that:

$$m_2^2 - m_1^2 = \frac{p^2}{c^2} \left[\frac{1}{\gamma_2^2 \beta_2^2} - \frac{1}{\gamma_1^2 \beta_1^2} \right] = \frac{p^2}{c^2 \beta_2^2 \beta_1^2} [\beta_1 - \beta_2] [\beta_1 + \beta_2] . \tag{5.61}$$

The gaseous threshold detector is built with the constraint [Séguinot (1988)]

$$\langle \theta \rangle = \sqrt{\frac{N_{\text{pe}}}{N_0 L}} \leq \sqrt{\frac{(m_2^2 - m_1^2)c^2}{p^2}} = \frac{1}{\beta_1 \beta_2} \sqrt{(\beta_1 - \beta_2)(\beta_1 + \beta_2)} \tag{5.62}$$

and the radiator length

$$L \geq \frac{N_{\text{pe}}}{N_0} \frac{p^2}{(m_2^2 - m_1^2) c^2}, \tag{5.63}$$

or in practice, using the number, n_σ (≥ 3), of standard deviations

$$\frac{(m_2^2 - m_1^2) c^2}{2p^2} \geq n_\sigma \left(\frac{N_{\text{pe}}}{2N_0 L} \right) . \tag{5.64}$$

5.2.2 *Differential Čerenkov Detectors*

The differential Čerenkov detectors allow the tagging of particles in a selected range of velocities. The Čerenkov radiation, produced in the forward cone of opening angle θ in a radiator gas, is reflected by a spherical mirror (Fig. 5.3).

The light is focused to form, in the focal plane, a ring of radius

$$r = f \tan \theta, \tag{5.65}$$

where f is the focal length of the mirror. The radius of curvature of the spherical mirror is $2f$. The selection of the particles velocities is achieved by the presence of a circular diaphragm of radius r and aperture Δr: the width of the diaphragm slit located in front of the photomultiplier tubes. The aperture filters the light. It selects only light produced in the angular range $\Delta\theta$:

$$\Delta r = \frac{f}{\cos^2\theta}\Delta\theta. \tag{5.66}$$

A quality factor of such detectors is the velocity resolution which can be calculated as

$$\frac{\Delta\beta}{\beta} = \frac{\Delta[1/(n\cos\theta)]}{\beta} = \tan\theta\,\Delta\theta. \tag{5.67}$$

Several effects limit the performance of the differential Čerenkov detectors. Multiple scattering in the radiator and beam divergence enlarge the width of the ring of Čerenkov light. Inhomogeneities of the refractive index over the length of the radiator also affect the radius of the ring of light. This can be seen by differentiating $\cos\theta = 1/(\beta n)$ which gives:

$$\Delta\theta = \frac{1}{\tan\theta}\frac{\Delta n}{n}. \tag{5.68}$$

Combined with Eq. (5.66), one finds:

$$\Delta r = \frac{f}{\tan\theta}\frac{\Delta n}{n}\frac{1}{\cos^2\theta}. \tag{5.69}$$

Optical aberrations degrade the angular resolution. In particular, chromatic aberrations enlarge the radius of the Čerenkov light ring [Litt and Meunier (1973)]:

$$\Delta r = f\theta\frac{1}{2\nu}\left(1 + \frac{1}{\theta^2\gamma^2}\right), \tag{5.70}$$

where ν features the gas dispersion. A gas with the highest possible ν should be chosen to minimize Δr. At very high energy, the term $\theta^2\gamma^2$ becomes negligible and Δr is proportional to θ. In summary, the application of differential Čerenkov detectors is based on selecting the particle type by both tuning the radiator index and the diaphragm aperture. It requires the direction of incoming particles to be parallel to the optical axis, which is the case for fixed target experiments or test setups in secondary hadronic beams. CERN users are well acquainted with this type of Čerenkov detector (CEDAR: Čerenkov Differential counter with Achromatic Ring focus), operated in CERN-SPS beams. In the case of colliding beam experiments (head-on collisions), particles are produced over the full solid angle and ring imaging Čerenkov detectors have to be used.

5.2.3 *Ring Imaging Čerenkov (RICH) Detectors*

The RICH detectors [Séguinot and Ypsilantis (1977)] allow the identification of particle by measuring the Čerenkov angle, after a precise measurement of the momentum. A RICH detector consists of two spherical surfaces with a radiator in between (Fig. 5.4).

A spherical mirror (SM), with a radius R_2, has its curvature center located at the point of origin of the particle. The Čerenkov photons emitted within a cone of aperture angle θ are reflected by the SM and focused onto a spherical detecting (SD) surface with radius R_1. Usually, $R_1 = R_2/2$. Therefore, the focal length of SD is $R_2/2$. The photons fall on a circle of radius

$$r = \frac{R_2}{2}\tan\theta \approx \frac{R_2}{2}\theta. \tag{5.71}$$

The radius can be measured, once the center is known, from an associated tracking system. The measurement of r allows the measurement of the Čerenkov angle

$$\theta = \frac{2r}{R_2} = \cos^{-1}\left(\frac{1}{\beta n}\right), \tag{5.72}$$

which in turn allows the determination of the particle velocity:

$$\beta = \frac{1}{n\cos\theta} = \frac{1}{n\cos(2r/R_2)}. \tag{5.73}$$

Once the velocity of the particle is known, the particle mass, m, can be determined via Eq. (5.32), provided the particle momentum, p, is known. This can be seen by rewriting Eq. (5.72) as

$$\theta = \cos^{-1}\left(\frac{E}{pcn}\right), \tag{5.74}$$

where E is the particle energy, or

$$\theta = \cos^{-1}\left(\frac{\sqrt{p^2c^2 + m^2c^4}}{pcn}\right). \tag{5.75}$$

Then, it is possible to distinguish two particles at a number of standard deviation level, n_σ, up to a momentum p whose value is given by [Virdee (1999)]:

$$p = \frac{1}{\sqrt{n_\sigma}}\sqrt{\frac{(m_2^2 - m_1^2)\sqrt{N_{\mathrm{pe}}}}{2\sigma_\theta^{pe}\tan\theta}}, \tag{5.76}$$

where m_2 $(> m_1)$ and m_1 are the particles masses. For aerogel, the maximum number of photons per cm is ranging from 20 to 80 (corresponding to the refractive index range from 1.02 to 1.1). For $N_{\mathrm{pe}} = 50$, $\sigma_\theta = 1\,\mathrm{mrad}$ and $\theta = 38\,\mathrm{mrad}$, pion/kaon separation is achieved up to a momentum of $85\,\mathrm{GeV/c}$ for $n_\sigma = 3$.

5.3 Wavelength Shifters

The light emitted from a plastic scintillator often has to be collected at a location distant from that scintillator. A wavelength shifter (WLS) is used for that purpose. A WLS absorbs the primary scintillation light and re-emits it at a different wavelength to ensure its transport over a relatively long distance towards its collection. The increase of wavelength decreases the absorption of the re-emitted light

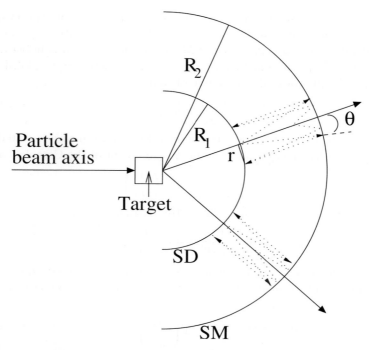

Fig. 5.4 Schematic description of a Ring Imaging Čerenkov (RICH) detector (reprinted from *Nucl. Instr. and Meth.* **142**, Séguinot, J. and Ypsilantis, T., Photo-Ionization and Cherenkov Ring Imaging, 377–391, Copyright (1977), with permission from Elsevier). It consists of a spherical mirror (SM) of radius R_2 and a spherical detecting (SD) surface of radius R_1, $R_1 = R_2/2$.

and favors its transportation over long distances. The WLS material must be insensitive to ionizing particle and Čerenkov light [PDB (2002)]. The choice of the WLS is made with the purpose of matching the emission spectrum to the absorption spectrum of the photocathode of a photomultiplier tube readout. The presence of ultra-violet absorbing additives in the WLS suppresses the response to Čerenkov light.

Bars or plates of wavelength shifters are used in calorimetry to read stacks of scintillator layers for sampling calorimetry applications (Fig. 5.5). The scintillation light, produced by a charged particle in each of the plastic scintillator layers of the sampling calorimeter module, is propagating through each layer until it exits the layer and gets absorbed in a wavelength shifter bar or plate located along the scintillator layers as shown in Fig. 5.5. After absorption, the wavelength shifted (toward a longer wavelength) light is transported via internal reflection to the photosensitive device. The use of WLS material allows the readout of a large amount of scintillator layers over large volume and the transport of the scintillation light over a large distance to a photosensitive device whose area can be kept small (standard PM). It allows one to regroup the readout devices at the back side of the calorimeter module or stack, making possible a compact arrangement of several calorimeter modules or stacks, which can be grouped without dead volumes, and keeping the readout elements such as photomultipliers outside the path of particle beams.

WLS fibers are also used instead of WLS plates (Fig. 5.6). They are very flexible as can be bent to follow distorted paths through and around scintillator plates or tiles and along absorber supports with small loss of signal transmission efficiency, and with the advantage that non-instrumented areas can be reduced to a minimum.

5.4 Transition Radiation Detectors (TRD)

The physics of the phenomenon of transition-radiation (TR) emission has been already treated from a more theoretical point of view in Sect. 2.2.3. TR is the electromagnetic radiation emitted as X-rays, when a charged particle crosses the boundary between two media with different refractive indices. TR emission mostly takes place inside a cone of half-aperture inversely proportional to the Lorentz factor γ of the particle

$$\theta = 1/\gamma. \tag{5.77}$$

TR emission is strongly peaked in the forward direction, i.e., at small angle with respect to the charged particle direction. TR propagation is symmetric, at least regarding the crossing medium. In practice, a TR detector (TRD) consists of a radiator where the X-rays are emitted followed by an X-ray detector.

The radiator is a multilayer stack of foils separated by air gaps. Since the yield of TR photons at each boundary crossing is rather modest ($\approx 1\%$), a large number of radiator foils in the stack is needed to achieve particle detection. However, this number is limited by the absorption of the X-rays by the radiator material. Therefore, the choice of the radiator material has to balance high density to favor high

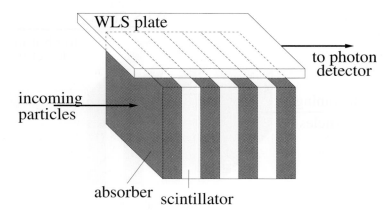

Fig. 5.5 Schematic description of a scintillator stack readout using a wavelength shifter plate.

photon-yield and low X-ray absorption coefficient. The intensity of X-ray emission varies as $\approx Z^{0.5}$. The absorption of X-rays in the radiator material is dominated by the photoelectric effect and the absorption coefficients, then, behave as Z^5. Therefore, the radiator material should have low-Z. The effective number, N_{eff}, of foils in a stack is given by [Fischer, J. et al. (1975)]:

$$N_{\text{eff}} = \frac{1 - e^{-N\sigma}}{1 - e^{-\sigma}}, \tag{5.78}$$

where $\sigma = (\mu\rho t)_f + (\mu\rho t)_g$ with μ, ρ and t being the X-ray absorption coefficient, density and thickness of the material (f stands for foil, and g, for gas in the gap), respectively. N_{eff} varies with the photon energy. For instance, $N_{\text{eff}} \sim 35$ for an X-ray energy of 5 keV and ~ 300 for an X-ray energy of 10 keV in the case of a CH_2 radiator (foil thickness of 20 μm) [Dolgoshein (1993)].

The total energy emitted in the forward direction by transition radiation at a single boundary crossing between two media is given by Eq. (2.149) (see page 133, and also [Garibyan (1960)]). Equation (2.149) shows that the total energy is proportional to the Lorentz factor γ and indicates the possibility to detect X-rays produced by a particle for a large Lorentz factor and, therefore, to identify particles at very high energy, where the operation of Čerenkov detectors becomes very hard and very inefficient.

It has been shown [Artru, Yodh and Mennessier (1975); Fischer, J. et al. (1975)] that large threshold effects occur from constructive interference between radiation emitted by the many boundaries in a periodical arrangements of radiator foils and gaps. These effects produce a threshold behavior of the detector which becomes a particle detector for values of the Lorentz factor $\gamma \gtrsim 1000$. One should note that a saturation value of γ exists (γ_{sat}), which depends on the length, L, of the radiator stack and on the TR wavelength, λ_{tr}, above which destructive interferences occur:

$$\gamma_{\text{sat}} \sim \sqrt{\pi L \lambda_{\text{tr}}} \,. \tag{5.79}$$

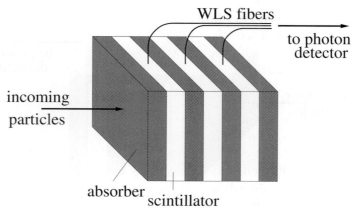

Fig. 5.6 Schematic description of a scintillator calorimeter module readout using wavelength shifter fibers.

Increasing L would enlarge the γ range over which TR can be exploited but, as observed above, this would increase the self-absorption of the radiator medium. Again, one is faced with a compromise, this time between optimization of the useful γ range and the minimization of self-absorption. The threshold factor, $\gamma \sim 1000$, corresponds to a pion energy of 140 GeV ($E_\pi = \gamma m_{pi} c^2$). Therefore, no TR will occur for pions with energy below 140 GeV. It is interesting to observe that $\gamma \sim 19,500$ for an electron of 10 GeV (for instance). In general, pions and electrons in a beam of energy E (in GeV) can be separated on the basis of their Lorentz factor compared to the value of the threshold Lorentz factor: $\gamma_e \sim 19,500\,E(\text{GeV})$, $\gamma_\pi \sim 7.2\,E(\text{GeV})$. If $E < 140$ GeV, electrons will produce TR but pions will not. Additional examples can be found in [Grupen (1996)]. The capability to operate the separation between electrons and pions (hadrons, in general) in a TRD is featured by the rejection factor [Dolgoshein (1993)] defined by:

$$R = \frac{\epsilon_e}{\epsilon_\pi}, \qquad (5.80)$$

where ϵ_e and ϵ_π are the electron and pion detection efficiencies, respectively. The quantity ϵ_π also represents the pion contamination. The overall length, L, of the TRD heavily influences its rejection power. For instance, assuming $\epsilon_e = 90\%$, ϵ_π decreases from 10% for $L = 20$ cm down to 1% for $L = 40$ cm and $\sim 0.1\%$ for $L = 70$ cm [PDB (2002)].

Particles with an energy corresponding to a Lorentz factor above threshold will produce TR which will be detected in the X-ray detector put behind the radiator. This X-ray detector is built with characteristics which are determined by two opposed constraints. On one hand, the X-ray absorption being dominated by the photoelectric effect ($\sim Z^5$), the X-ray detector has to be made of a high-Z material thick enough to maximize the absorption. On the other hand, the X-ray detector has to be thin enough to limit ionization and excitation losses by the charged particle traversing the detector. The X-ray detector is typically a thin wire chamber (10 mm) filled with high-Z gas, such as xenon.

The X-ray detector will detect the sum of ionization loss (dE/dx) of the charged particle traversing the gas chamber and the energy deposition of the X-rays in that chamber. The ionization and excitation signal is created by a large number of low energy transfers to electrons (δ-rays, e.g., see page 55), producing in turn charge clusters proportional to their energy. The absorption of TR will produce few local strong energy depositions. Therefore, energy loss by TR is very localized in contrast to the delocalized ionization and excitation energy which is distributed over the depth of the detector due to fluctuations in the number and energy of the δ-rays. Clearly, delocalization of the ionization energy-loss puts a limit on X-ray detection.

This pattern of energy distribution provides a mean of particle separation which will then require the measurement of the total deposited charge in the wire chamber and its spatial distribution. A complete discussion of the particle identification technique via transition radiation and detector optimization can be found in [Dolgoshein (1993); Egorytchev, Saveliev and Aplin (2000)].

5.5 Scintillating Fibers

Standard scintillating fibers consist of a core having a high refractive index (n_{core}) cladded with a non-scintillating material of lower refractive index (n_{clad}). A scintillating fiber traversed by an ionizing particle or radiation is illustrated in Fig. 5.7. The diameter of the core material depends on the application. For particle tracking, for which the best spatial resolution is needed, the fiber diameter is smaller than 1 mm down to a few tens of μm. For calorimetry applications, for which the measurement of energy deposition is the goal, fibers with diameter of 1 mm and above are used. The thickness of the cladding material is a few microns, typically 2 to 5 μm. The cladding material is surrounded by a thin opaque absorber a few mm thick, called the extra mural absorber (EMA), which captures the untrapped light. The presence of EMA helps to prevent cross-talk between neighbouring fibers, which

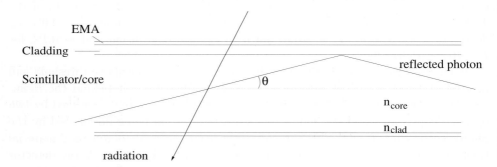

Fig. 5.7 Schematic description of a scintillating fiber. The core (refractive index n_{core}) is cladded with a non-scintillating material (refractive index n_{clad}). The extra mural absorber (EMA) captures the untrapped light.

is essential for particle tracking.

The scintillation light produced in the fiber by the passage of a ionizing particle or radiation is trapped by internal reflections on the fiber core/cladding interface. The phenomenon of internal reflection can be understood from *Snell's law*. Consider a ray of light which propagates from medium "a" of refractive index n_a, into medium "b" of refractive index n_b, with $n_a > n_b$. If θ_a and θ_b are the angles made with the normal for the incident and refracted rays, Snell's law says

$$n_a \sin \theta_a = n_b \sin \theta_b. \tag{5.81}$$

From Eq. (5.81) where one has put $\sin \theta_b = 1$, the critical angle is defined:

$$\theta_c = \sin^{-1} \frac{n_b}{n_a}. \tag{5.82}$$

At the critical angle θ_c, the refracted ray is tangent to the interface at the point of incidence. Most importantly for our purpose, when the angle of incidence is larger than the critical angle, $\theta_a > \theta_c$, no refracted ray appears and the incident ray undergoes total internal reflection from the interface. The operation of scintillating fibers is based on the principle of total internal reflection. In the case of a fiber, Eq. (5.82) becomes:

$$\theta_c = \sin^{-1} \frac{n_{\text{clad}}}{n_{\text{core}}}. \tag{5.83}$$

The emitted scintillation light hitting the core/cladding interface at an angle larger than the critical angle [Eq. (5.83)] will be reflected back into the core. The scintillation light becomes trapped in the fiber and will propagate inside the fiber via total internal reflections on the core/cladding interface until it reaches the fiber end where it will be detected by a photosensitive device.

Typical core materials are polystyrene with $n_{\text{core}} = 1.59$ or glasses such as GS1 (with a few % by weight of Ce_2O_3) with $n_{\text{core}} = 1.69$. Cladding materials are acrylic with $n_{\text{clad}} = 1.49$.

For light emitted at the axis of the fiber or close to it (called *meridional* rays), traveling along the fiber axis, the fraction of light trapped is:

$$f = \frac{1}{2}\left(1 - \frac{n_{\text{clad}}}{n_{\text{core}}}\right). \tag{5.84}$$

In the case of isotropic light propagation inside the fiber, equal amounts are trapped in both directions and the total fraction of trapped light is $f_{\text{tot}} = 2f$. For $n_{\text{core}} = 1.59$ and $n_{\text{clad}} = 1.49$, $\theta_c = 1.21$, and $f_{\text{tot}} = 0.06$. Light rays, off the fiber axis and close to the fiber edge (called *skew* rays), follow spiral paths along the length of the fiber. Adding *meridional* and *skew* rays would modify Eq. (5.84) into [Leutz (1995); White (1988)]:

$$f = \frac{1}{2}\left[1 - \left(\frac{n_{\text{clad}}}{n_{\text{core}}}\right)^2\right], \tag{5.85}$$

which brings $f_{\text{tot}} = 0.12$. Skew rays have larger solid angle acceptance and undergo more reflections at the core/cladding interface than meridional rays. Then, skew rays are very sensible to defects in the core/cladding interface and suffer losses from these defects which reduce their propagation length in the fiber.

The trapped light is transported to the fiber end via a maximum of n_r total internal reflections at the core/cladding interface [Leutz (1995)]:

$$n_r = \frac{l}{d}\cot\theta_c, \tag{5.86}$$

where l and d are the fiber length and diameter, respectively. The critical angle θ_c is given by Eq. (5.83). For a polystyrene core of diameter 0.25 mm and 1 m length and an acrylic cladding, one finds $n_r \sim 1500$.

One also defines the numerical aperture NA as [Leutz (1995)]

$$NA = \sqrt{n_{\text{core}}^2 - n_{\text{clad}}^2}. \tag{5.87}$$

Taking a fiber with a polystyrene core and an acrylic cladding, the numerical aperture is $NA = 0.55$.

For tracking purpose, one wishes to select fibers with diameter as small as possible to achieve good spatial resolution. From Eq. (5.86), one can see that the decrease of the fiber diameter increases the number of reflections and, therefore, additional light losses due to defects in the core/cladding interface are expected. The light intensity I after n_r reflections is related to the initial light intensity I_0 through [Leutz (1995)]:

$$I = I_0 q^{n_r}, \qquad (5.88)$$

where $q \leq 1$ is the reflection coefficient whose value reflects light losses at the core/cladding interface. The combination of Eqs. (5.86, 5.88) gives:

$$\ln \frac{I}{I_0} = -(1-q)\frac{l}{d}\cot\theta_c, \qquad (5.89)$$

where $\ln q$ has been replaced by $(q - 1)$, since q is very close to unity. The experimental value [Leutz (1995)] of $(1 - q)$ is typically $(5 - 6) \times 10^{-5}$, depending on wavelength. The reflection length, Λ_R, is the fiber length after which the initial light intensity is reduced to $1/e$ of its value as the result of reflection losses. It is given by [Leutz (1995)]:

$$\Lambda_R = \frac{1.5}{1-q}\frac{n_{\text{core}}}{NA} d. \qquad (5.90)$$

The total attenuation length, Λ, is the sum [Leutz (1995)] (see discussion in this reference):

$$\Lambda^{-1} = \Lambda_A^{-1} + \Lambda_R^{-1} + \Lambda_{Sc}^{-1}, \qquad (5.91)$$

where Λ_A is defined as the value of the light path length D (in meters) at which the light yield drops to $1/e$ of its initial value at zero D. The attenuation, Λ_{Sc}, quantifies Rayleigh scattering on small density fluctuations in the core material that can deflect a light ray which is no longer totally internally reflected.

5.6 Detection of the Scintillation Light

The ionization energy deposited by a particle traversing a scintillator material is converted into photons. If one considers a plastic scintillator such as NE102A (see Table 5.1), a scintillation photon is produced for about 3 keV of deposited energy, which corresponds to a yield of about 500 photons per centimeter of scintillator material. This scintillation light is of low intensity and has to be optically coupled to a photomultiplier for amplification in order to provide an electrical signal. A schematic view of a photomultiplier is given in Fig. 5.8. The scintillation light has to be transported to the photocathode of the photomultiplier where the impinging photons will be converted into photoelectrons via the photoelectric effect (*Einstein equation*):

$$E = h\nu - \phi, \tag{5.92}$$

where E is the kinetic energy of the emitted electron, ν is the frequency of incident scintillation light and ϕ is the work function. The surface of the photocathode is coated with a photosensitive material that has a low work function to favor electron emission. The probability of this photon to electron conversion is the quantum efficiency QE of the photocathode, i.e., the probability of liberating an electron per photon striking the photocathode:

$$QE(\lambda) = \frac{number\ of\ photoelectrons\ emitted}{number\ of\ photons\ incident\ on\ the\ photocathode}. \tag{5.93}$$

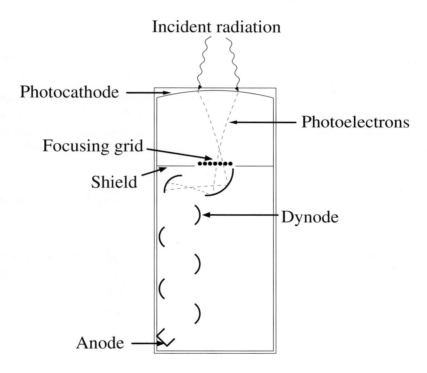

Fig. 5.8   Schematic view of a photomultiplier.

The quantum efficiency depends on the type of photocathode material. The type of photocathode is chosen to have an overlap between its peak spectral response and the characteristic wavelength at maximum emission of the incoming light. The quantum efficiency is typically 20–25%.

A plastic light guide couples the scintillator to the photomultiplier. A "fish tail" shape for the light guide is standard (Fig. 5.9), but other shapes are also used depending on the application. It allows the coupling of photocathode, which are of small area, to a scintillator with a large cross section. The light emitted in the

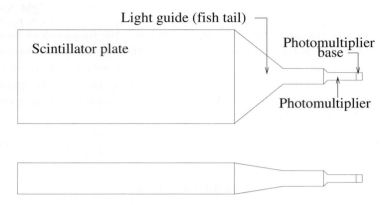

Fig. 5.9   Schematic view of a scintillator coupled to a photomultiplier via a *fish tail* light guide.

scintillator is transported to the photocathode through the scintillator and light guide via total internal reflections. When the emitted light strikes the scintillator surface, total internal reflection occurs when the incident angle is larger than the critical angle [Eq. (5.83)]. In the present situation, Eq. (5.83) becomes[¶]:

$$\theta_c = \sin^{-1} \frac{n_{\text{med}}}{n}. \tag{5.94}$$

The surrounding medium is often air, $n_{\text{med}} = n_{\text{air}} = 1$, and Eq. (5.94) becomes:

$$\theta_c = \sin^{-1} \frac{1}{n}. \tag{5.95}$$

For the example of NE102A above, $n = 1.58$ and therefore Eq. (5.95) gives $\theta_c = 39.2°$. The fraction f of light emitted and transported via total internal reflection in one direction is (e.g., see Equation 7.4 of [Fernow (1986)]):

$$f = \frac{1}{4\pi} \int_0^{2\pi} \int_0^{\theta_c} \sin\theta \, d\theta \, d\phi$$

$$= \frac{1}{2} \left(1 - \cos\theta_c \right)$$

$$= \frac{1}{2} \left(1 - \frac{\sqrt{n^2 - 1}}{n} \right) \simeq 0.11. \tag{5.96}$$

The scintillator and light guide are wrapped with an aluminium foil to prevent the leak, from plastic, of light escaping total internal reflection. The scintillator, light guide and photomultiplier are also wrapped with a layer of black tape to prevent outside light from leaking inside the detector.

[¶]In this case, the cladding is replaced by the surrounding medium of refractive index n_{med} and the core by scintillator of refractive index n.

Geometrical constraints often forces the bending of the light guide (which could be also a bundle of fibers) to achieve the coupling to the photomultiplier. However, there exists a limit for maximum bending beyond which important light losses occur. The maximum bending is calculated from the condition (*Liouville's law*):

$$n^2 - 1 \geq \left(\frac{d}{2r} + 1\right)^2, \tag{5.97}$$

where d is the light guide diameter, r is the bending radius and n is the refractive index relative to air. If the bending radius satisfies condition Eq. (5.97), the light will be transported along the light guide via total internal reflections.

The intensity of light transported in the scintillator and light guide is attenuated by atomic absorption and scattering from imperfections of the reflecting surface (which should be negligible if the surface has been cleanly polished). The reduction of light intensity, I, is a function of distance, l, and light wavelength, λ, can be expressed by:

$$I(l, \lambda) = I(0, \lambda) \exp\left(-\frac{l}{I_{\mathrm{ph}}(\lambda)}\right), \tag{5.98}$$

where $I(l, \lambda)$ is the light intensity at distance l and wavelength λ, $I(0, \lambda)$ is the initial light intensity and $I_{\mathrm{ph}}(\lambda)$ is the photon attenuation length at wavelength λ. As seen in Sect. 5.3, the introduction of a wavelength shifter in the scintillator increases the attenuation length. Taking into account Eqs. (5.93, 5.96, 5.98), the number of photoelectrons produced by the photocathode can be expressed by:

$$n_{\mathrm{pe}} = n_{\mathrm{ph}} \int I(L, \lambda) \, QE(\lambda) \, f(\lambda) \, d\lambda, \tag{5.99}$$

where f, in Eq. (5.96), now depends on λ, and n_{ph} is the number of photons produced in the scintillator of length L. Let us consider the example of a NE102A scintillator counter of 1 cm thickness. Its light output is about 25% of NaI(Tl) (see Table 5.1), i.e., $10^4 \, \gamma/\mathrm{MeV}$ or $1 \, \gamma/100 \, \mathrm{eV}$. A minimum ionizing particle traversing the scintillator thickness loses about $2 \, \mathrm{MeV/cm}$. Therefore, about $2 \times 10^4 \, \gamma$ should be emitted in 1 cm of scintillator. NE102A has an attenuation length of 250 cm which gives an intensity loss of 0.4% in 1 cm according to Eq. (5.98). The fraction of light trapped in scintillator within the critical angle is directly given by Eq. (5.96), since the refractive index of NE102A is 1.58. Assuming a quantum efficiency of 20% for the photocathode and an optical collection efficiency of about 50% (due to imperfections of the reflecting surface and imperfect coupling of the light guide to the photocathode), one finds the number of photoelectrons, n_{pe}, coming off the photocathode:

$$n_{\mathrm{pe}} \approx 220. \tag{5.100}$$

The number of photoelectrons produced at the photocathode follows a Poisson distribution $P(n_{\mathrm{pe}})$ and the probability of production of n_{pe} photoelectrons for a mean number $\bar{n}_{\mathrm{pe}}$ is (Appendix B.1):

$$P(n_{\mathrm{pe}}) = \frac{\bar{n}_{\mathrm{pe}}^{n_{\mathrm{pe}}} e^{-\bar{n}_{\mathrm{pe}}}}{n_{\mathrm{pe}}!}. \tag{5.101}$$

The detector efficiency, ϵ, is the detection probability which is the complement of non-detection probability. Therefore, one has:

$$1 - \epsilon = e^{-\bar{n}_{\mathrm{pe}}}, \tag{5.102}$$

where $e^{-\bar{n}_{\mathrm{pe}}}$ is the inefficiency or the probability of detecting no photoelectron. In this example, Eq. (5.100) gives $e^{-\bar{n}_{\mathrm{pe}}} \simeq 0$ and the efficiency, ϵ, is $\simeq 100\%$.

The photomultiplier is powered by a high voltage power supply. The photocathode, made of alkali metals, is the most negative electrode and put at a large negative voltage, e.g., -1400 V up to -2200 V. The anode is at 0 V.

A set of metallic electrodes, called dynodes, forms the electron multiplier section. The dynodes, set at progressively higher voltage, e.g., steps of $(100\text{--}120)$ V, are located between the photocathode and the anode (see Fig. 5.8). A dynode re-emits electrons when struck by an electron. This emission of secondary electrons by a dynode, via the photoelectric effect [Eq. (5.92)], is similar to photoelectron emission by the photocathode. However, there is a major difference: incoming electrons are responsible for secondary electron emission in the case of the dynode, while incident scintillation photon are responsible for this emission for the photocathode. Like the photocathode, dynodes are made of low work function material to maximize electron emission by photoelectric effect. Voltage to each dynode is delivered through a resistive potential divider consisting of a chain of resistors in series contained in the photomultiplier base (Fig. 5.9). The electrons ejected from the photocathode are accelerated towards the first dynode due to the difference of potential of $(100\text{--}120)$ V between the photocathode and the first dynode, from which they knock out several electrons. These secondary electrons are accelerated, through the potential gradient between the first and second dynodes, towards the second dynode, where each of them ejects more electrons, and so on. The inter-dynode voltage allows the multiplication of secondary electrons via photoelectric effect from dynode to dynode, until they reach the anode where the charge is collected. The multiplication factor (called the gain), M, between the first stage of the n-dynode chain and the anode is

$$M = \delta_1 \delta_2 \delta_3 \ldots \delta_n, \tag{5.103}$$

where δ_i is the multiplication factor at the dynode i of the chain. From Eq. (5.103), a standard multiplication factor of 10^7 with a chain of 12 dynodes can be achieved, if each electron produces about 4 secondary electrons when it strikes a dynode. A chain of 10 dynodes would reduce M to 10^6. The charge collected at the anode for $M = 10^7$ is then

$$Q = e\, 10^7 = 1.6 \times 10^{-19}\mathrm{C} \times 10^7 = 1.6 \times 10^{-12}\mathrm{C} = 1.6 \text{ pC}. \tag{5.104}$$

Since the charge is collected typically within 5 ns, the current at the anode is:

$$i = \frac{dQ}{dt} = \frac{1.6 \times 10^{-12}\mathrm{C}}{5 \times 10^{-9}} = 0.32 \text{ mA}. \tag{5.105}$$

Usually, the signal drives a $50\,\Omega$ load, i.e., the photomultiplier is terminated with a $50\,\Omega$ resistance. Then, a pulse of

$$\delta V = 0.32\,\text{mA} \times 50\,\Omega = 16\,\text{mV} \tag{5.106}$$

is produced. Photomultipliers are also characterized by a rise-time (about 2 ns, typically). The rise-time is usually defined as the time required for the output signal to increase from 10% to 90% of the maximal value. The time elapsed between the illumination of the photocathode and the signal at the anode is the transit time (about 30 ns, typically), which can be understood as the time needed for the electron to travel the length of the photomultiplier. Information about the rise-time and transit time are given by technical specifications released by the manufacturer of the photomultiplier. The output pulse at the anode is subsequently analyzed by an electronics system. A example of such a system is shown in Fig. 5.10.

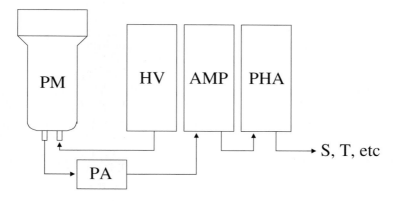

Fig. 5.10 Schematic description of an analyzing electronics system consisting of a preamplifier (PA), a high voltage power source (HV), an amplifier (AMP), a pulse height analyzer (PHA), followed by a spectrometer analyzer (S), a timing unit (T), etc...

The size of the pulse at the anode is proportional to the energy absorbed from the incident radiation by the scintillating material. This is the consequence of the proportionality between the amount of light that reaches the photocathode surface and the amount of photoelectrons ejected. Therefore, the scintillation detector can be used to distinguish between different types of particles or incoming particles of different energy by the shape of the emitted light pulse. The pulse produced by a scintillator detector is controlled by the dominant decay time of the scintillator and the electron transit time through the PM chain. The number of photons emitted after the passage of the ionization radiation follows the exponential law described by Eq. (5.8) with decay constants for various scintillators ranging from a few nanoseconds to a few microseconds (Tables 5.2 and 5.3). The electron transit time is a Gaussian function of time with a standard deviation of about 1 ns. When the scintillator decay constant is 10 ns or larger, the effect of the spread in the electron transit time is negligible.

5.7 Applications in Calorimetry

A variety of scintillating and Čerenkov materials can be used as the active medium of homogeneous calorimeters (see Sect. 9.5) or active samplers of sampling calorimeters. The choice of a particular active medium is made to achieve the best possible energy and position resolutions over an energy range as large as possible (from MeV up to TeV energies), taking into account practical and physical requirements. The energy resolution is determined by the fluctuations in the number of photons or ion pairs produced by the cascade. The light yield, i.e, the number of photons produced per MeV of energy deposition, should be large enough to minimize the contribution of the statistical fluctuations to the energy resolution.

A comprehensive treatment of calorimeter performance can be found in the chapter *Principles of Particle Energy Determination*.

5.8 Application in Time-of-Flight (ToF) Technique

The excellent timing capabilities of scintillator counters make them applicable in time-of-flight (ToF) techniques [Atwood (1981)]. Measurements of particles ToF, combined with the knowledge of their momentum, give information on their mass. Therefore, ToF measurement may allow discrimination between particle of same momentum but different masses.

The speed of a particle of mass m, momentum p, traveling a distance L in a time t is given by

$$v = \frac{L}{t},$$
(5.107)

then,

$$\beta = \frac{v}{c} = \frac{L}{ct}.$$
(5.108)

Using Eq. (5.32) with $\gamma = 1/\sqrt{1 - \beta^2}$, one finds:

$$m = \frac{p}{\beta\gamma c} = \frac{p}{c}\sqrt{\frac{c^2 t^2}{L^2} - 1}.$$
(5.109)

A ToF measurement system consists of two scintillator plastic counters as shown in Fig. 5.11. Plastic scintillators are chosen for their fast response. One of the counters serves as start counter, the other as stop counter. The time information is given by a *time to digital-converter* (TDC) while charge measurement is obtained from *analog to digital converter* (ADC).

If two scintillator counters are separated by a distance L, two particles "1" and "2" of momentum p, with masses m_1 and m_2, velocities v_1 and v_2 and energies E_1 and E_2 will traverse this distance with times of flight

$$t_1 = \frac{L}{v_1} = \frac{L}{c}\frac{1}{\beta_1} = \frac{L}{c}\frac{E_1}{pc} = \frac{L}{c}\frac{\sqrt{m_1^2 c^4 + p^2 c^2}}{pc}$$
(5.110)

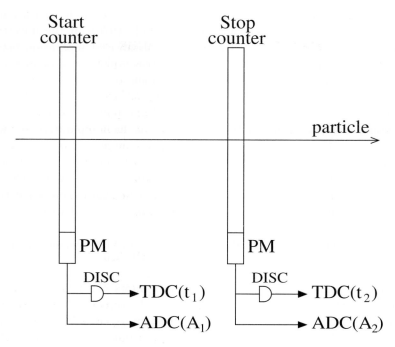

Fig. 5.11 Schematic description of a *ToF* measurement system. The start and stop counters, 1 and 2, respectively, are read each by a photomultiplier (PM). The chain of time measurement for each counter consists of a discriminator (DISC) and a TDC(t_i), while charge measurement is obtained from an ADC(A_i).

and

$$t_2 = \frac{L}{v_2} = \frac{L}{c}\frac{1}{\beta_2} = \frac{L}{c}\frac{E_2}{pc} = \frac{L}{c}\frac{\sqrt{m_2^2 c^4 + p^2 c^2}}{pc}. \tag{5.111}$$

There will be a ToF difference between these two particles of

$$\Delta t = t_2 - t_1 = \frac{L}{pc^2}\left(\sqrt{m_2^2 c^4 + p^2 c^2} - \sqrt{m_1^2 c^4 + p^2 c^2}\right). \tag{5.112}$$

For relativistic particles, $pc \gg mc^2$, Eq. (5.112) may be approximated by:

$$\Delta t = \frac{L}{pc^2}pc\left[\left(1 + \frac{m_2^2 c^2}{2p^2}\right) - \left(1 + \frac{m_1^2 c^2}{2p^2}\right)\right], \tag{5.113}$$

or

$$\Delta t = \frac{Lc}{2p^2}(m_2^2 - m_1^2). \tag{5.114}$$

In practice, the distance L is one or several meters and times of reference such as times of resolution are in picoseconds (ps). Therefore, it is standard to express the ration $\Delta t/L$ in ps/m units:

$$\frac{\Delta t}{L} = \frac{1667}{p^2 c^2}(m_2^2 c^4 - m_1^2 c^4) \ \ [\text{ps/m}]. \tag{5.115}$$

Equation (5.115) gives the maximum momentum to achieve particle separation. For instance [Braunschweig, Koenigs, Sturm and Wallraff (1976)], if the uncertainty on the time-of-flight is $\sigma(\Delta t) = \pm 250\,\text{ps}$ and the distance of separation between two scintillator counters is $L = 5\,\text{m}$, Eq. (5.115) gives a maximum momentum of $\sim 1.9\,\text{GeV/c}$ to achieve pion-kaon separation, for which

$$m_{\text{K}}^2 - m_{\pi}^2 = 0.224\,\text{GeV}^2/\text{c}^4,$$

and a maximum momentum of $\sim 3.3\,\text{GeV/c}$ for proton–kaon separation, for which

$$m_{\text{p}}^2 - m_{\text{K}}^2 = 0.637\,\text{GeV}^2/\text{c}^4.$$

ToF measurement technique does not apply to highly relativistic particles since in that case the ToF difference is close to zero and negligible compared to the standard time resolutions of a few hundred picoseconds achieved with scintillator counters.

In the non-relativistic limit ($\beta \sim 0.1$, $\gamma \sim 1.0$), Eq. (5.112) becomes (p small):

$$\Delta t = \frac{L}{p}(m_2 - m_1) \equiv \frac{L}{p}\Delta m, \tag{5.116}$$

or using Eq. (5.108)

$$\Delta t = \frac{L}{m\beta\gamma c}\Delta m = t\frac{\Delta m}{m}. \tag{5.117}$$

Consequently, for a time resolution of $\Delta t \sim 200\,\text{ps}$ and a flight path $L = 1\,\text{m}$, it is possible to discriminate between low-energy particles to an accuracy better than 1%.

Chapter 6

Solid State Detectors

Solid state detectors are made from semiconductor materials. The semiconductor detectors benefit from a small energy gap between their valence and conduction bands. Therefore, a small energy deposition can move electrons from the valence band to the conduction band, leaving holes behind. When an electric field is applied, the two charge carriers (electron and hole) drift and produce a signal. Therefore, the passage of an ionizing particle can be detected by collecting the charge carriers liberated by energy deposition in the semiconductor. In this chapter, the basic principles of operation and the main features of semiconductor detectors will be explained in details. A specific example, the microstrip[‡] detector, will also be discussed.

The efficiency of silicon detectors at charged and neutral particles detection (note that for neutron detection, the silicon detectors have to be covered by neutron converters (^{6}LiF for slow neutrons and CH_2 for fast neutrons), their small thickness of a few hundreds microns, the possibility to cut silicon detectors to any size and their low power consumption explain their large use, as dosimeters and radiation detectors in particle physics experiments. The silicon devices features and general detection properties of non-irradiated and irradiated silicon detectors based on the standard planar (SP) and MESA technologies are also reviewed. The parameters evolution under irradiation fluence of these types of silicon detectors is also discussed. The study of the peaks evolution with bias voltage can provide precise information on the active medium structure and charge collection performances of the detector. In the case of irradiated detectors, the study of the peak evolution with bias voltage and irradiation fluence allows the measurement of the charge collection efficiency degradation in various regions of the diodes and structure alteration. A review on induced-damage in silicon radiation detectors operated in radiation environments was provided by Leroy and Rancoita (2007).

This chapter includes a discussion of the violation of non-ionizing energy-loss

---

[‡]This detector was firstly designed by Heijne and colleagues at CERN in 1979 for use in high energy particle experiments [Heijne et al. (1980)]. They made segmented Si detectors with narrow strip pattern and matched readout electronics for the small signals. The idea of a linearly segmented silicon diode originated in the Philips Research Laboratories in Amsterdam in around 1963 [Heijne (2003)].

(NIEL) scaling for low energy protons ($\lesssim 10\,\text{MeV}$). Furthermore, it is treated the detection of neutrons with a silicon detector. This detection exploits the interactions of these neutrons with the nuclei of a converter layer adjacent to the detector diode.

6.1 Basic Principles of Operation

The semiconductor materials are characterized by a small gap between the electronic conduction band and the valence band. In the case of silicon, an energy

$$E_g = 1.12\,\text{eV}$$

is needed to excite an electron from the valence band into the conduction band. For comparison, $E_g > 5$ eV for insulators and conductors have their valence and conduction bands in contact. The hole left by an electron in the valence band under some excitation has a positive electric charge.

The jump of electrons from the valence into the conduction band under the excitation generates holes moving in the valence band in a opposite direction to that of electrons in the conduction band. Therefore, a current of holes in the valence band is the counterpart of a current of electrons in the conduction band. Charge carriers, electrons and holes, drifting through a semiconductor under the influence of an electric field, E, have a velocity given by

$$v_e = \mu_e E, \tag{6.1}$$

for electrons, and

$$v_h = \mu_h E, \tag{6.2}$$

for holes.

The quantity μ_e and μ_h are the electron and hole mobility, respectively. The hole mobility ($\mu_h = 450\,\text{cm}^2\,\text{V}^{-1}\,\text{s}^{-1}$ in silicon [Caso (1998)]) is smaller than the electron mobility ($\mu_e = 1350\,\text{cm}^2\,\text{V}^{-1}\,\text{s}^{-1}$ in silicon [Caso (1998)]). The current carried by electrons and holes in a semiconductor is determined by the mobilities. Using Eqs. (6.1, 6.2) (e.g., see Section 1.5.2 of [Sze (1981)]), the current density, J_c, for concentrations n of electron carriers and p of hole carriers, q being the electronic charge, is given by

$$J_c = q\,(n\,\mu_e + p\,\mu_h)\,E. \tag{6.3}$$

There is equilibrium between the generation and recombination of free electrons and holes at a given temperature. Crystals are not perfect and present defects. Various impurities are also present in the crystal. The mean free carrier lifetime is reduced by these defects and various impurities which act as generation or recombination centers. There are n-type and p-type conductivity semiconductors resulting from the introduction of electrically active donor and acceptor impurity atoms, respectively. n-type and p-type semiconductors have an excess of electrons and holes,

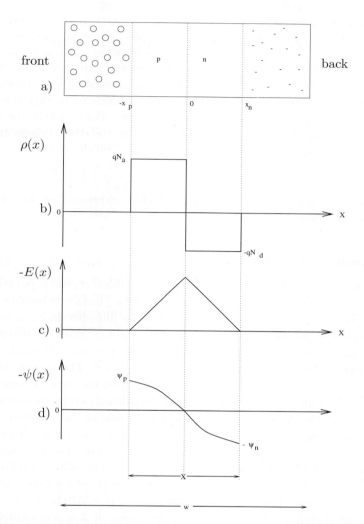

Fig. 6.1 Schematic representation of an unpolarized $p-n$ junction of thickness w. (a) represents the junction, (b) the charge distribution $\rho(x)$, (c) the electric field $-E(x)$ and (d) the electrostatic potential $-\psi(x)$.

respectively. At temperature T, the product of concentrations of electrons n and holes p remains constant and is given by

$$np = n_{int}^2, \tag{6.4}$$

where n_{int} (in cm^{-3}) is the intrinsic carrier concentration. At $T = 300\,\mathrm{K}$ (in silicon), we have $n_{int} \simeq 1.45 \times 10^{10}\,\mathrm{cm}^{-3}$.

For silicon (e.g., see Section 4.1.2 of [Neamen (2002)]), the temperature dependence of n_{int} is given by

$$n_{\text{int}} \simeq 4.1056 \times 10^{21} \times (k_B T)^{3/2} \exp\left(-\frac{E_{\text{g}}}{2k_B T}\right) \ [\text{cm}^{-3}], \tag{6.5}$$

where $E_{\text{g}} = 1.12\,\text{eV}$ (e.g., Sect. 4.3.3.3) is the energy gap in silicon, as already mentioned; $k_B = 8.617 \times 10^{-5}\,\text{eV/K}$ is the Boltzmann constant (see Appendix A.2) and T is the temperature in kelvin. However, one needs to modify this formula to improve the description of the experimental results. Below 700 K, the following phenomenological function gives good agreement with the data ([Morin and Maita (1954)], see also [Lindstroem (1991)]):

$$n_{\text{int}} \simeq 3.873 \times 10^{16} \times T^{3/2} \exp\left(-\frac{1.21}{2k_B T}\right) \ [\text{cm}^{-3}]. \tag{6.6}$$

At low (and very low) temperature, n_{int} is better expressed by Eq. 4.146 (see Sect. 4.3.3.3 and references therein).

6.1.1 *Unpolarized $p-n$ Junction*

A $p-n$ junction is formed when a n-type region in a silicon crystal is put adjacent to a p-type region in the same crystal. In practice, such a junction is built by diffusing acceptor impurities into a n-type silicon crystal or by diffusing donors into a p-type silicon crystal. The junction can be abrupt if the passage from the n-type doping density to p-type doping density is just a step. If the passage is through a gradual change in doping density, the junction is linearly graded. The size of the relative doping densities on each side of the junction dictates the type of junction. For instance, a $p^+ - n$ junction results from an acceptor density on the p-type side being much larger than the donor density on the n-type side of the junction. A charge depleted region occurs at the interface of the n- and p-type regions. This depleted region is created as the result of the diffusion of electrons from n-type material into p-type material and diffusion of holes from p-type to n-type material. This diffusion is the consequence of the motion of carriers from regions of high concentration to regions of low concentration. Therefore, diffusion is responsible for the existence of a space-charge region with two zones: a first zone, of non-zero electric charge, made of filled electron acceptor sites not compensated by holes and a second zone, again of non-zero electric charge made of positively charged empty donor sites not compensated by electrons. If, for instance, the density of acceptors in the p-type region is much lower than the density of donors in the n-type region, the space-charge region extends much deeper into the p-region than into the n^+-region. The net result is the creation of a space-charge region with acceptor centers in the p-region, filled with donor electrons from the n^+-region and not compensated by holes. This space-charge region is called the *depletion region*.

The shape of the electrostatic potential, Ψ, the electric field and the width of the depletion zone of a junction can be obtained by solving the Poisson equation

$$\frac{d^2\Psi}{dx^2} = -\frac{\rho(x)}{\varepsilon}. \tag{6.7}$$

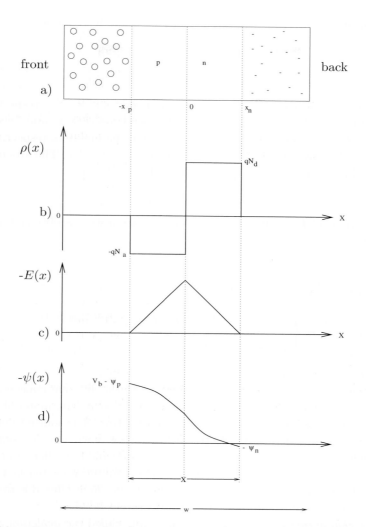

Fig. 6.2 Schematic representation of a $p-n$ junction of thickness w. The applied bias voltage V_b creates a depletion depth X. (a) represents the junction, (b) the charge distribution $\rho(x)$, (c) the electric field $-E(x)$ and (d) the electrostatic potential $-\psi(x)$.

In practice, the lateral spatial extension of the charges represents a few μm^2 which is negligible with respect to the area of the detectors and Eq. (6.7) can be safely solved in one-dimension. In Eq. (6.7) (see Appendix A.2),

$$\varepsilon = \varepsilon_0 \, \varepsilon_{Si} = 1.054 \, pF/cm$$

is the *silicon electric permittivity*, ε_0 and ε_{Si} are the *dielectric constant*[‡‡] *of silicon* and the *permittivity of free space*, respectively. An abrupt junction is assumed with

[‡‡]The dielectric constant is also referred to as *relative permittivity*.

a charge density $\rho(x)$ in the depletion zone approximately given by:

$$\rho(x) = \begin{cases} qN_d & \text{for } 0 \leq x \leq x_n \\ \\ -qN_a & \text{for } -x_p \leq x \leq 0, \end{cases} \tag{6.8}$$

where x_n and x_p are the depletion length on n-side and p-side, respectively, as defined in Fig. 6.1(a). N_d et N_a are the donor (n-type region) and acceptor (p-type region) impurity concentrations on each side of the junction, respectively. The charge density is zero outside the depletion region [Fig. 6.1(b)]. The absence of net total charge in the depletion region is reflected by

$$N_d\, x_n = N_a\, x_p. \tag{6.9}$$

The electric field E resulting from the charges separation is obtained after integration of Eq. (6.7) with

$$E(x_n) = E(-x_p) = 0$$

as boundary condition [Fig. 6.1(c)], i.e.,

$$E(x) = -\frac{d\Psi}{dx} = \begin{cases} E_n(x) = q\,(N_d/\varepsilon)(x - x_n) & \text{for } 0 \leq x \leq x_n \\ \\ E_p(x) = -q\,(N_a/\varepsilon)(x + x_p) & \text{for } -x_p \leq x \leq 0. \end{cases} \tag{6.10}$$

The thicknesses of n^+ and p^+ zones have been neglected.

If no external voltage is applied, E is only due to the different concentrations of electrons and holes at the junction. Diffusion will bring n-type material electrons into the p-type material and holes in opposite direction. The ionized dopants are fixed charges which generate an electric field slowing down the diffusion process since this field pushes electrons back to the n-type side and holes back to the p-type side, until a dynamical equilibrium is reached. This equilibrium is reached when the carrier flux due to the electric field counterbalances the carrier flux due to diffusion, the electron and hole flux summing to zero, separately.

The *free charge carrier depleted region* or simply called the *depletion zone* built above can serve as particle detector. Free charges can be generated in excess of the equilibrium in the depletion region by ionizing particles traversing the diode. An energy of $E_{\text{ion}} = 3.62\,\text{eV}$ is required to produce an electron–hole pair in silicon. The difference between this ionizing energy and the band-gap energy ($3.62\,\text{eV}$–$1.12\,\text{eV}$) is spent on other excitations in the silicon lattice. The charges produced by ionization in the depletion zone are separated and induce an electron and a hole signal. The interest of the junction structure for particle detection is the possibility to polarize it by applying a bias voltage. The voltage drop across the depletion zone, i.e., the electrostatic potential is calculated by integrating Eq. (6.10). Defining the integration constants

$$\Psi(-x_p) = \Psi_p \qquad \text{and} \qquad \Psi(x_n) = \Psi_n,$$

one obtains the electrostatic potential shown in Fig. 6.1(d):

$$\Psi(x) = \begin{cases} \Psi_n(x) = \Psi_n - q\,[N_d/(2\varepsilon)](x - x_n)^2 & \text{for } 0 \le x \le x_n \\[2mm] \Psi_p(x) = \Psi_p + q\,[N_a/(2\varepsilon)](x + x_p)^2 & \text{for } -x_p \le x \le 0. \end{cases} \quad (6.11)$$

The contact potential or built-in voltage, V_0 ($\approx 0.3\text{--}0.6\,\text{V}$ for silicon at $T = 300\,K$), is defined by:

$$\begin{aligned} V_0 &= -\int E(x)\,dx \\ &= \Psi_n - \Psi_p \\ &= \frac{k_B T}{q}\ln\left(\frac{N_a N_d}{n_{\text{int}}^2}\right), \end{aligned} \quad (6.12)$$

with T expressed in kelvin (K).

The depletion depths of the n-type and p-type zones are calculated by imposing the continuity of the potential at $x = 0$ $[\Psi_n(0) = \Psi_p(0)]$ and using Eq. (6.9):

$$x_n = \frac{1}{N_d}\sqrt{\frac{2\varepsilon V_0}{q}\left(\frac{1}{N_a} + \frac{1}{N_d}\right)^{-1}} \quad (6.13)$$

and

$$x_p = \frac{1}{N_a}\sqrt{\frac{2\varepsilon V_0}{q}\left(\frac{1}{N_a} + \frac{1}{N_d}\right)^{-1}}. \quad (6.14)$$

Then, the total depth of the depletion zone is given by:

$$\begin{aligned} X &= x_n + x_p \\ &= \sqrt{\frac{2\varepsilon V_0}{q}\left(\frac{1}{N_a} + \frac{1}{N_d}\right)}. \end{aligned} \quad (6.15)$$

6.1.2 *Polarized $p - n$ Junction*

Without external polarization, the depletion depth is typically a few microns which limits the capability of the junction for particle detection. Therefore, an external voltage is applied between the n- and p-regions: $-V_b < 0$ on the p-side of the junction, the boundary conditions get modified into

$$\Psi(-x_p) = \Psi_p - V_b \qquad \text{and} \qquad \Psi(x_n) = \Psi_n$$

[Fig. 6.2(d)].

As a consequence, V_0 is replaced by $V_0 + V_b$ in Eqs. (6.13, 6.14). Equation (6.15) becomes now

$$\begin{aligned} X &= x_n + x_p \\ &= \sqrt{\frac{2\varepsilon(V_0 + V_b)}{q}\left(\frac{1}{N_a} + \frac{1}{N_d}\right)}. \end{aligned} \quad (6.16)$$

The junction is said to be reverse biased and the depth of the depletion zone grows for increasing V_b until the depletion depth reach the total thickness of the detector (w).

In most applications, n-type detectors and $p^+ - n$ junctions are used $(N_d \ll N_a)$ and from Eq. (6.9) (*neutrality equation*), one finds $x_p \ll x_n$. Therefore, Eq. (6.16) is replaced by:

$$X \approx x_n \approx \sqrt{\frac{2\varepsilon}{qN_d}\,(V_0 + V_b)}. \tag{6.17}$$

The voltage to be applied in order to fully deplete the detector is called the *full depletion voltage*, V_{fd}, i.e., $V_b = V_{fd}$ at $X = w$. From Eq. (6.17) where $X = w$, the full depletion voltage is given by:

$$V_{fd} = \frac{w^2 q N_d}{2\varepsilon} - V_0. \tag{6.18}$$

The silicon detector resistivity is given by

$$\rho = \frac{1}{\mu q\,|N_{\mathrm{eff}}|}, \tag{6.19}$$

where μ is the electron (hole) mobility for a n-type (p-type) detector as given in Eqs. (6.1, 6.2) and N_{eff} is the effective dopant concentration. The use of N_{eff} allows one to write equations valid for both n-type and p-type detectors. It is also intended to reflect the fact that silicon usually contains n-type and p-type impurities. This effective dopant concentration, N_{eff}, is defined as:

$$N_{\mathrm{eff}} = N_a - N_d, \tag{6.20}$$

thus

$$|N_{\mathrm{eff}}| = |N_a - N_d|. \tag{6.21}$$

As will be seen later, N_{eff} can be measured from capacitance–reverse bias (C–V) measurements. Using resistivity and effective doping concentration, Eq. (6.17) becomes:

$$X \approx \sqrt{\frac{2\varepsilon\,(V_0 + V_b)}{q|N_{\mathrm{eff}}|}}, \tag{6.22}$$

or

$$X \approx 0.53\,\sqrt{\rho\,(V_0 + V_b)}\;\;[\mu\mathrm{m}] \tag{6.23}$$

with ρ and $(V_0 + V_b)$ expressed in Ωcm and volts, respectively. Then, the bias voltage at full depletion is given by:

$$V_{fd} = \frac{w^2 q |N_{\mathrm{eff}}|}{2\varepsilon} - V_0, \tag{6.24}$$

or

$$V_{fd} = \frac{w^2}{2\varepsilon\mu\rho} - V_0. \tag{6.25}$$

As shown by Eq. (6.25), the value of V_{fd} depends on the detector resistivity. High resistivity detectors have a low full depletion voltage while low resistivity detectors have a high full depletion voltage.

As already mentioned, in most applications of particle physics, n-type detectors and $p^+ - n$ junction are used. A $p^+ - n$ junction, results from an acceptor density on the p-type side being much larger than the donor density on the n-type side of the junction. Thus, in practical cases the contact potential V_0 can be neglected with respect to the full depletion voltage [see Eq. (6.12)] in Eqs. (6.22, 6.24). Using Eq. (6.19), one can rewrite Eq. (6.22) as

$$X \simeq 0.53 \sqrt{\rho V_b} \quad [\mu\text{m}], \tag{6.26}$$

if ρ is expressed in $\Omega \, \text{cm}$ and V_b in volts. Equation (6.25) can be rewritten as

$$V_{fd} \simeq \frac{w^2}{2\varepsilon\mu\rho}. \tag{6.27}$$

Equation (6.27) shows that high resistivity detectors have low full depletion bias voltage while low resistivity detectors have high full depletion bias voltage.

6.1.3 *Capacitance*

Silicon detectors used in experiments can be considered as parallel plate capacitors. Their capacitance C, in the *parallel plate capacitor approximation*, is given by

$$C = \varepsilon \frac{A}{w}, \tag{6.28}$$

where A is the detector area. This approximation reflects the behavior of the space-charge with voltage in the junction transition region. In fact, in the transition zone of a junction are interfaced two regions of space-charges of equal value but of opposite sign. Using previous notations, one region, located between $-x_p$ et 0, has a total charge of $-AqN_ax_p$ while the other region located between 0 et x_n has a total charge of AqN_dx_n. The depths x_n and x_p depend on the applied voltage via the equations Eqs. (6.13, 6.14) [where V_0 is replaced by $V_0 + V_b$]:

$$x_n = \frac{1}{N_d} \sqrt{\frac{2\varepsilon(V_0 + V_b)}{q} \left(\frac{1}{N_a} + \frac{1}{N_d} \right)^{-1}} \tag{6.29}$$

and

$$x_p = \frac{1}{N_a} \sqrt{\frac{2\varepsilon(V_0 + V_b)}{q} \left(\frac{1}{N_a} + \frac{1}{N_d} \right)^{-1}}. \tag{6.30}$$

Application of an increasing voltage, V (the built-in voltage is included), at the borders of the junction increases the amount of electrons from the n-zone (neutralizing the positive space-charge) and the amount of holes from the p-zone (neutralizing

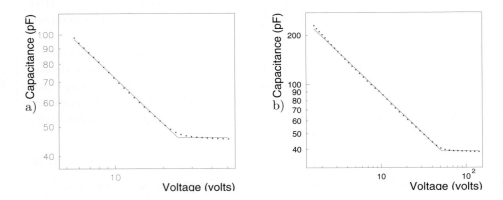

Fig. 6.3 Capacitance (in pF) curves as a function of applied voltage (in V) for two detectors of $1\,\text{cm}^2$ area [Leroy (1998)]: (a) resistivity of $\rho = 8\,\text{k}\Omega\text{cm}$ and thickness $w = 247\,\mu\text{m}$; (b) resistivity of $\rho = 6\,\text{k}\Omega\text{cm}$ and thickness $w = 300\,\mu\text{m}$. The curve represents a fit of Eq. (6.34) to the data points.

the negative space charge). The opposite process is observed when the applied voltage decreases, leading to free charge removal and to an increase of space-charge, Q_t, in the transition region. The variation of Q_t with voltage in the transition zone has the dimension of a capacitance (C_t), i.e.,

$$C_t = \frac{dQ_t}{dV} \tag{6.31}$$

with Q_t given by

$$Q_t = AqN_d x_n = AqN_a x_p. \tag{6.32}$$

From the Eqs. (6.29, 6.30), one obtains

$$\frac{dQ_t}{dV} = \frac{Aq}{2\sqrt{V_0 + V_b}}\sqrt{\frac{2\varepsilon}{q}\left(\frac{1}{N_a} + \frac{1}{N_d}\right)^{-1}}. \tag{6.33}$$

Inserting Eq. (6.15) into Eq. (6.33), one recovers Eq. (6.28) describing the parallel plate capacitor approximation, i.e., two parallel plates separated by a distance X [C_t and C in Eq. (6.28) being obviously the same quantity]:

$$\frac{C}{A} = \frac{\varepsilon}{X} \qquad\qquad \text{for } 0 < X \le w \tag{6.34}$$

and, consequently,

$$C = \varepsilon\frac{A}{X} \qquad\qquad \text{for } 0 < X \le w. \tag{6.35}$$

The detector capacitance is minimal when X is maximal. This occurs when the detector full depletion is achieved, i.e., when $X = w$. For applied voltage larger

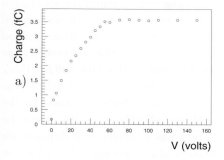

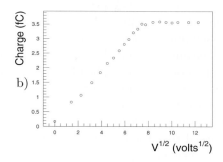

Fig. 6.4   Charge collected with a silicon detector 300 μm thick, area 1 cm^2, exposed to incoming electrons of energy larger than 2 MeV, as a function of the applied bias (a) and as a function of the square root of the applied voltage (b). The value of the full depletion voltage is 60 volts which corresponds to a collected charge of 3.5 fC, as expected from the most probable energy deposited by relativistic electrons in this detector [Leroy (1998)].

than V_{fd}, the capacitance remains constant and a plateau is observed (Fig. 6.3). An estimate of the full depletion voltage can be obtained by taking the value of voltage corresponding to the intersection of the $C-V$ curve with the line of constant geometrical capacitance. The depletion voltage decreases with increasing resistivity, as already observed. The capacitance at full depletion of the 1 cm^2 area detectors shown in Fig. 6.3 is obtained applying Eq. (6.28): one finds 42.5 pF [thickness $w = 247\,\mu$m, Fig. 6.3(a)] and 35.0 pF [thickness $w = 300\,\mu$m, Fig. 6.3(b)].

Furthermore, the measurement of capacitance as a function of the applied bias voltage allows the determination of the full depletion voltage and N_{eff} ($\simeq V_{fd}2\varepsilon/qw^2$), from Eq. (6.24) where V_0 has been neglected. As will be seen in Sect. 6.1.4, the full depletion bias voltage can also be obtained from charge collection measurements. $C-V$ measurements permit the determination of N_{eff} as a function of the depth X of the depletion layer, i.e., the N_{eff} profile. Combining Eqs. (6.24, 6.34) (where $V = V_d + V_0$), one finds:

$$N_{\text{eff}}(X) = \frac{2}{q\varepsilon A^2} \frac{dV}{d(1/C^2)}. \tag{6.36}$$

6.1.4   *Charge Collection Measurements*

Let us turn to the study of the energy deposited by relativistic electrons* in silicon detectors of current use. Since these detectors are thin (a few hundreds microns, typically), the energy deposited by minimum ionizing particles (*mip*) in these silicon detectors follows a *Landau-type spectrum*. The most probable energy deposition by

---

*The energy deposited by relativistic electrons is similar to the energy deposited by minimum ionizing particles (see Sect. 2.1.6.1).

a *mip* is about $80 \, \mathrm{keV}$ per $300 \, \mu\mathrm{m}$ of silicon. If one takes into account the ionization energy $E_{\mathrm{ion}} = 3.62 \, \mathrm{eV}$, that corresponds to about 22,000 electron-hole pairs created per $300 \, \mu\mathrm{m}$ of silicon. Then, a charge of $22,000 \times 1.6 \times 10^{-19} \, \mathrm{C} = 3.5 \, \mathrm{fC}$ is released per $300 \, \mu\mathrm{m}$ of silicon. An example is shown in Fig. 6.4(a) where the collected charge as a function of the applied bias voltage is shown for a silicon detector $300 \, \mu\mathrm{m}$ thick exposed to relativistic electrons (of energy larger than $2 \, \mathrm{MeV}$). Figure 6.4(b) shows the collected charge as a function of $V^{1/2}$ for the same detector exposed to the same incoming electron beam. In agreement with Eqs. (6.22) or (6.26), it can be observed that the detector signal (collected charge) has a square root dependence on the bias voltage below the full depletion voltage and is constant ($Q = 3.5 \, \mathrm{fC}$) as a function of V above it. The measurement of the collected charge as a function of the applied bias voltage is a way to determine the value of the full depletion voltage (V_{fd}), i.e., the value of voltage for which a value of the collected charge constant with voltage begins to be observed. For the example shown is Fig. 6.4, the plateau starts at $V_{fd} \sim 60 \, \mathrm{V}$.

The mean charge collection[†] is proportional to the detector thickness. Figure 6.5 shows the mean charge collection from relativistic electrons ($E_e > 2 \, \mathrm{MeV}$) for fully depleted detectors of various thicknesses [(142–996) $\mu\mathrm{m}$]. The average energy deposition is $\sim (2.70 \pm 0.04) \times 10^2 \, \mathrm{eV} \, \mu\mathrm{m}^{-1}$, in agreement with the expected energy deposition for relativistic electrons.

6.1.5 *Charge Transport in Silicon Diodes*

The movement of the charge carriers generated by an ionizing particle in a detector produces a signal which shape is determined by the charge transport properties of the detector. The charge transport properties of the diode are governed by its electrical characteristics: i) the effective concentration of dopants (N_{eff}), which defines the internal electric field and, thus, the depletion voltage, ii) the electron (μ_e) and hole (μ_h) mobilities, which influence the time needed to collect the charge and iii) the charge trapping lifetime (τ_{th}, τ_{te}), which affects the efficiency of the charge collection. First, let us assume a simplified model where the electron and hole mobilities are constant with the electric field. The electric field inside the depleted region of the detector can be expressed as [Lemeilleur et al. (1994)]:

$$E(x) = -\frac{qN_{\mathrm{eff}}}{\epsilon}(x - w) + \frac{V_b - V_{fd}}{w} = -ax + b, \qquad (6.37)$$

where V_b is the applied bias voltage, V_{fd} is the full depletion voltage,

$$b \equiv \frac{wqN_{\mathrm{eff}}}{\epsilon} + \frac{V_b - V_{fd}}{w} \qquad \text{and} \qquad a = \frac{qN_{\mathrm{eff}}}{\epsilon}.$$

Ramo's theorem [Ramo (1939)] relates the displacement (Δx) of a charge carrier generated by the passage of an incoming particle in the detector to the charge (Δq)

[†]The reader may refer to Sects. 2.1.1, 2.1.6.1 for a further discussion.

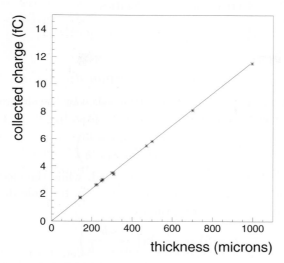

Fig. 6.5   Mean charge collection (fC) for relativistic electrons from a ^{106}Ru source (with $E_e >$ 2 MeV) as a function of the detector thickness (142–996 μm). The line is to guide the eye (reprinted from *Nucl. Instr. and Meth. in Phys. Res. A* **479**, Bechevet, D., Glaser, M., Houdayer, A., Lebel, C., Leroy, C., Moll, M. and Roy, P., Results of irradiation tests on standard planar silicon detectors with 7–10 MeV protons, 487–497, Copyright (2002), with permission from Elsevier).

that it induces on the electrodes:

$$\Delta q = q\frac{\Delta x}{w}. \tag{6.38}$$

The induced current at the electrode is given by

$$i(x) = \frac{dQ}{dt} = \frac{Q}{w}\frac{dx}{dt} = \frac{Q}{w}v(t), \tag{6.39}$$

where

$$v(t) = v_e(t) = \frac{dx_e}{dt} = \mu_e E(x) = \mu_e(-ax + b), \tag{6.40}$$

for electrons, and

$$v(t) = v_h(t) = \frac{dx_h}{dt} = \mu_h E(x) = -\mu_h(-ax + b) \tag{6.41}$$

for holes; μ_e and μ_h are the electron and hole mobility, respectively. Integrating Eqs. (6.40, 6.41) over time for a unit charge generated at $x = x_0$ gives

$$x_e(t) = \frac{b}{a} + \left(x_0 - \frac{b}{a}\right)\exp(-\mu_e at), \tag{6.42}$$

and

$$x_h(t) = \frac{b}{a} + \left(x_0 - \frac{b}{a}\right)\exp(\mu_h at), \tag{6.43}$$

with the corresponding electron and hole velocities:

$$v_e(t) = -\mu_e(ax_0 - b)\,\exp(-\mu_e at) \tag{6.44}$$

and

$$v_h(t) = \mu_h(ax_0 - b)\,\exp(\mu_h at). \tag{6.45}$$

The electron collection time (t_{coll}), i.e., the time taken by the electron generated at $x = x_0$ to reach the electrode located at $x = w$, is given by

$$t_{\mathrm{coll}}(e) = -\frac{1}{\mu_e a}\ln\left(\frac{aw - b}{ax_0 - b}\right). \tag{6.46}$$

In the same way a hole collection time can defined, while the electron generated at $x = x_0$ is drifting towards the electrode at $x = w$, the hole drifts in opposite direction from $x = x_0$ to $x = 0$ during the time:

$$t_{\mathrm{coll}}(h) = \frac{1}{\mu_h a}\ln\left(\frac{-b}{ax_0 - b}\right). \tag{6.47}$$

The current induced by electrons (i_e) and holes (i_h) at the electrodes is given by:

$$i_e(t) = \frac{q\mu_e}{w}(ax_0 - b)\,\exp(-\mu_e at), \tag{6.48}$$

for $0 \leq t \leq t_{\mathrm{coll}}(e)$, and

$$i_h(t) = \frac{q\mu_h}{w}(ax_0 - b)\,\exp(\mu_h at), \tag{6.49}$$

for $0 \leq t \leq t_{\mathrm{coll}}(h)$.

The integration of the currents over time gives the electron and hole collected charge:

$$Q_e(t) = -\frac{q}{aw}(ax_0 - b)\left[\exp(-\mu_e at) - 1\right], \tag{6.50}$$

for $0 \leq t \leq t_{\mathrm{coll}}(e)$, and

$$Q_h(t) = \frac{q}{aw}(ax_0 - b)\left[\exp(\mu_h at) - 1\right], \tag{6.51}$$

for $0 \leq t \leq t_{\mathrm{coll}}(h)$. The total electron collected charge is obtained by integrating Eq. (6.50) over time

$$Q_{\mathrm{tot}}(e) = \frac{q}{w}(w - x_0), \tag{6.52}$$

while the total hole collected charge is obtained from the integration of Eq. (6.51) over time:

$$Q_{\mathrm{tot}}(h) = \frac{q}{w}x_0. \tag{6.53}$$

Summing Eqs. (6.52, 6.53), we get a total collected charge of

$$Q_{\mathrm{tot}} = q, \tag{6.54}$$

i.e., the unit charge initially injected.

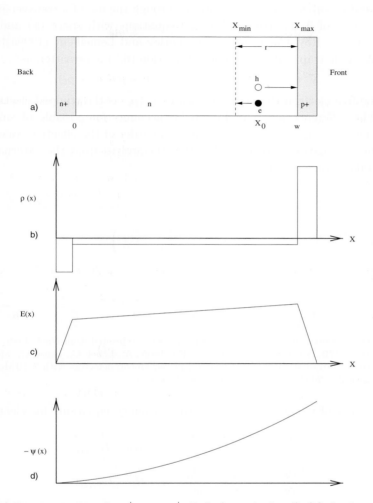

Fig. 6.6   (a) Representation of a $p^+ - n - n^+$ diode (over-depleted), (b) the dopant density profile ρ, (c) the electric field E and (d) the electrostatic potential ψ (reprinted from *Nucl. Instr. and Meth. in Phys. Res. A* **479**, Bechevet, D., Glaser, M., Houdayer, A., Lebel, C., Leroy, C., Moll, M. and Roy, P., Results of irradiation tests on standard planar silicon detectors with 7–10 MeV protons, 487–497, Copyright (2002), with permission from Elsevier).

A more complete treatment of the charge transport problem is presented below. In that case, the mobility will be expressed as a function of the electric field and of other parameters. A planar silicon diode $(p^+ - n - n^+)$ is assumed with the junction and ohmic side located at $x = 0$ et $x = w$, respectively. Figure 6.6 gives a schematic view of the junction under consideration. These electrical characteristics are extracted by solving in one-dimension (through the use of a transverse diffusion term) a system of five partial differential equations with space (x) and time (t) variables ([Leroy, Roy, Casse, Glaser, Grigoriev and Lemeilleur (1999a)] for more details). A single trap state is assumed. Equation (6.7) is rewritten as:

$$\nabla^2 \psi = -\nabla E = -\frac{q}{\varepsilon}(-N_{\text{eff}} - n + p - n_t + p_t), \tag{6.55}$$

n (p) is the free electron (hole) density while n_t (p_t) is the trapped electron (hole) density. The difference $p - n$ in Eq. (6.55) accounts for possible plasma effects, when the electron and hole densities are of the order of the effective concentration of dopants. The carriers are thus shielding themselves from the external electric field and take a longer time to be collected.

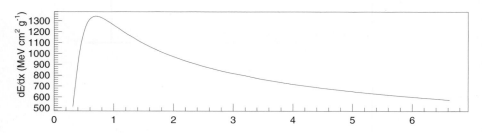

Fig. 6.7 Energy deposition of an alpha-particle in silicon (reprinted from *Nucl. Instr. and Meth. in Phys. Res. A* **479**, Bechevet, D., Glaser, M., Houdayer, A., Lebel, C., Leroy, C., Moll, M. and Roy, P., Results of irradiation tests on standard planar silicon detectors with 7–10 MeV protons, 487–497, Copyright (2002), with permission from Elsevier).

The system of equations involves two continuity equations for electrons and holes:

$$\frac{\partial n}{\partial t} = -\frac{\partial n_t}{\partial t} + \frac{\nabla J_n}{q} + g - Rn - U_{SRH},$$

$$\frac{\partial p}{\partial t} = -\frac{\partial p_t}{\partial t} - \frac{\nabla J_p}{q} + g - Rp - U_{SRH}, \tag{6.56}$$

$$\frac{\partial n_t}{\partial t} = \frac{n}{\tau_{te}} - \frac{p_t}{\tau_{de}} - Rn_t,$$

$$\frac{\partial p_t}{\partial t} = \frac{p}{\tau_{th}} - \frac{p_t}{\tau_{dh}} - Rp_t, \tag{6.57}$$

where g is the electron–hole pair generation function, $\tau_{te/h}$ and $\tau_{de/h}$ are the trapping and de-trapping times, respectively; U_{SRH} is the Shockley–Read–Hall generation-recombination term

$$U_{SRH} = \frac{np - n_{\text{int}}^2}{(p + n_{\text{int}})\tau_{th} + (n + n_{\text{int}})\tau_{te}}. \tag{6.58}$$

The current densities of carriers densities are given by:

$$J_n = qn\mu_e E + \mu_e k_B T \nabla n,$$

$$J_p = qp\mu_h E - \mu_h k_B T \nabla p, \tag{6.59}$$

where k_B and T are the Boltzmann constant and the temperature, respectively; μ_e and μ_h are the electron and hole mobilities, respectively. These mobilities depend on the electric field (E) and the temperature (T). The system of equations reflects a cylindrical symmetry as the induced charges are generated inside a cylinder and are transversely diffused with time. In Eqs. (6.56, 6.57), this is accounted by the term R [Taroni and Zanarini (1969)], which takes into account the transverse diffusion of charge carriers:

$$R = \frac{18D_a}{18D_a t + r_0^2},\tag{6.60}$$

where r_0 is the initial radius of the column of deposited charge, the radius of the column at a time t being

$$r^2(t) = 18\,D_a t + r_0^2;$$

D_a is the ambipolar diffusion coefficient and can be expressed as

$$D_a = \frac{2D_e D_h}{D_e + D_h},\tag{6.61}$$

where

$$D_c = \frac{kT}{q}\mu_c\tag{6.62}$$

are the *Einstein relations* ($c = e,\ h$). Values of the diffusion coefficients, D_c ($c = e,\ h$), are given at page 487. It has to be noted that the minority carrier diffusion length for electrons in p-type material is given by

$$L_e = \sqrt{D_e \tau_e}\tag{6.63}$$

and for holes in n-type material is given by

$$L_h = \sqrt{D_h \tau_h},\tag{6.64}$$

where τ_e and τ_h are the electron and hole carrier lifetime, respectively.

When neglecting the size of the p^+ and n^+ regions, the integration of the one-dimensional Poisson's equation at $t = 0$, for a simple abrupt $p-n$ junction operated in over-depleted mode ($V_b > V_{fd}$), gives

$$E(x,0) = -\frac{qN_{\text{eff}}}{\epsilon}x + \frac{V_b}{w} + \frac{V_{fd}N_{\text{eff}}}{w\,|N_{\text{eff}}|}, \qquad \text{for } 0 < x < w\tag{6.65}$$

with the boundary conditions:

$$E(0,t) = E(w,t) = 0,\tag{6.66}$$

$$\psi(0,t) = V_0 + \psi_p \approx 0 \qquad \text{and} \qquad \psi(w,t) = \psi_p - V_b \approx -V_b,\tag{6.67}$$

where ψ and E are the electrostatic potential and the electric field, respectively; q, ϵ and w are the electrical charge, the silicon electric permittivity and the thickness of the diode, respectively; $V_0 \approx 0.6\,\text{V}$ is the built-in voltage, V_b is the applied bias

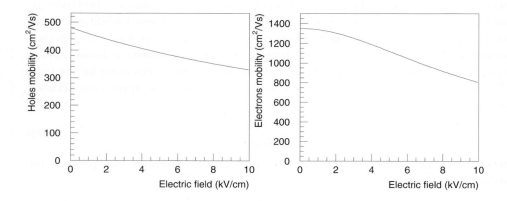

Fig. 6.8 Hole (left) and electron (right) mobility ($\mathrm{cm^2V^{-1}s^{-1}}$) as a function of the electric field (kV/cm), Eq. (6.69), (reprinted from *Nucl. Instr. and Meth. in Phys. Res. A* **479**, Bechevet, D., Glaser, M., Houdayer, A., Lebel, C., Leroy, C., Moll, M. and Roy, P., Results of irradiation tests on standard planar silicon detectors with 7–10 MeV protons, 487–497, Copyright (2002), with permission from Elsevier).

voltage, ψ_p ($-0.6 \leq \psi_p \leq -0.3\,\mathrm{V}$) is the electrostatic potential of the neutral p-type region and V_{fd} is the full depletion bias voltage.

For a relativistic β particle in silicon, the range exceeds the physical size of the detector and a uniform density of electron–hole pairs is created with $n_0 \approx 80\ \mathrm{pairs}/\mu\mathrm{m}$.

The generation function, g in Eq. (6.56, 6.57) at time $t = 0$ is given by:

$$g(x) = \frac{n_0 \times \mathrm{density}(x)\,\delta(x - x_0)}{\sum_{x'} \mathrm{density}(x')} \qquad \text{with } x_\mathrm{min} \leq x_0 \leq x_\mathrm{max}. \qquad (6.68)$$

For an α particle of 5.0 MeV in silicon, the number of electron–hole pairs created is $n_0 \approx 1.4$ million pairs over a range

$$r = x_\mathrm{max} - x_\mathrm{min} \approx 25\,\mu\mathrm{m},$$

while the density of the electron–hole pairs is obtained by interpolation of the data from [Williamson, Boujot and Picard (1966)] shown in Fig. 6.7.

The drift velocity of the charge carrier reaches a saturation value v_s for electric field values around $10^4\,\mathrm{V/cm}$ [Dargys and Kundrotas (1994)]. The empirical equation describing the mobility as a function of the electric field is:

$$\mu(x) = \frac{\mu_0}{[1 + (\mu_0 E(x)/v_s)^m]^{1/m}}, \qquad (6.69)$$

where μ_0 is the zero field mobility, $m = 1$ for holes, $m = 2$ for electrons; $v_s = 1.05 \times 10^7$ and $10^7\,\mathrm{cm/s}$ for electrons and holes, respectively [Dargys and Kundrotas (1994)]. The dependence of the mobilities on the electric field is shown in Fig. 6.8 for holes (left) and electrons (right).

Table 6.1 Examples [Leroy, Roy, Casse, Glaser, Grigoriev and Lemeilleur (1999a); Roy (2000)] of characteristics of the standard planar silicon detectors of current use.

Detector (area)	Current pulse source	thickness (μm)	$|N_{\text{eff}}|$ ($\times 10^{11}$ cm^{-3})	ρ (kΩ cm)
M4 (1 cm^2)	α	317	3.4	12.2
M18 (1 cm^2)	α,β	309	4.1	11.0
M25 (1 cm^2)	α,β	308	2.1	23.0
M35 (1 cm^2)	α	508	1.7	24.0
M49 (1 cm^2)	β	301	4.7	8.9
M50 (1 cm^2)	β	471	1.8	22.8
M53 (1 cm^2)	β	223	5.4	7.7
P88 (0.25 cm^2)	α,β	290	18.0	2.5
P189 (0.25 cm^2)	α,β	294	18.0	2.5
P304 (0.25 cm^2)	α,β	320	7.0	6.0

The mobilities also depend on the temperature and dopant concentrations [Caughey and Thomas (1967)]. The effect of concentration is only appreciable for concentrations over 10^{14} dopants/cm^3. Changing the temperature by 1.5 degree Celsius changes both mobilities by $\approx 1\%$. These features are taken into account via the empirical equation

$$\mu(T, N_{\text{eff}}) = \mu_{\min} + \frac{\mu_0 \, (T/300)^\nu - \mu_{\min}}{1 + (T/300)^\xi \, (N_{\text{eff}}/N_{ref})^\alpha}, \qquad (6.70)$$

where the values used for the electrons (holes) are: $\mu_{\min} = 55.24 \, (49.7)$ cm^2 V^{-1} s^{-1}, $N_{ref} = 1.072 \times 10^{17} \, (1.606 \times 10^{17})$ dopants/cm^3, $\nu = -2.3 \, (-2.2)$, $\xi = -3.8$ (-3.7), $\alpha = 0.73 \, (0.70)$, T is the temperature in kelvin and μ_0 is the mobility at $T = 300$ K. The temperature dependence of mobility is shown in Fig. 6.9 using μ_0(for e) $= 1350$ cm^{2}V^{-1}s^{-1} and μ_0 (for h) $= 480$ cm^2 V^{-1}s^{-1} at $T = 300$ K.

As a consequence of Ramo's theorem [Eq. (6.38)], the observed signal, $V(t)$, is a convolution of the current, $I(t)$, produced by all the individual charge carriers and the response from the system, which is simply an RC circuit. The response of the system is a Gaussian with a characteristic time constant $\sigma = R_a C$, where C is the capacitance of the detector and R_a the input impedance of the amplifier:

$$I(t) = \frac{18 D_a t + r_0^2}{w r_0^2} \int_0^w (\mu_e n + \mu_h p) E \, dx, \qquad (6.71)$$

$$V(t) = \frac{G R_a}{\sigma \sqrt{2\pi}} \sum_{e,h} \int_{-\infty}^{\infty} I(t') \exp\left[-\frac{(t - t')^2}{2\sigma^2}\right] dt', \qquad (6.72)$$

where D_a is the ambipolar diffusion constant [Eq. (6.61)], G the gain of the amplifier and r_0 the initial radius of the column of deposited charge.

As an example, a simple case will be examined where silicon diodes are exposed to electrons from a ^{106}Ru source with an energy > 2 MeV, selected by an external trigger, and to α-particles from an ^{241}Am source with an energy of 5.49 MeV. The setup for measuring the charge induced by the charge carriers generated by ^{106}Ru beta particles is shown in Fig. 6.10. The silicon detector under study is located inside a test box with a ^{106}Ru (β) source fitted in its cover. The source, above the geometrical center of the detector, is collimated to a 5 mm diameter electron beam. The temperature inside the test box is controlled by a water cooling system allowing the selection of temperatures from 6 to 25 °C. Nitrogen is flowing through the box to prevent condensation on the detectors at low temperatures. The current pulse induced by electrons in the detector is detected with a charge pre-amplifier or a current pre-amplifier (according to the goal pursued). The amplification and shaping were ensured by a Ortec 450 research amplifier for the data reported in Figs. 6.4 and 6.5. The shaping integration and differentiation were 100 ns, with a gain of 500. The polarity of the input signals was negative. The output signals were in the −3 volts range. The pulse energy was measured with a peak sensing Lecroy ADC 2259A for a 500 ns gate triggered by the coincidence signal of two photomultipliers detecting photons produced in a scintillator of 1 cm^2 area put behind an iron absorber ($\approx$ 0.5 mm thick) allowing to select minimum ionizing electrons with an energy $\geq$ 2 MeV.

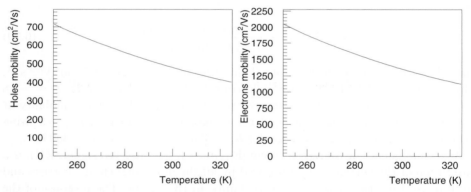

Fig. 6.9   Hole (left) and electron (right) mobility (in cm^{2}V^{-1}s^{-1}) as a function of temperature (in K), Eq. (6.70).

The setup for measuring the charge induced by the charge carriers generated by ^{241}Am α-particles is shown in Fig. 6.11. The charge induced by the collimated α-particles incident on the front side and on the rear side of the detector (the source can be moved on the other side of the detector) can be measured in a way similar to the case of β-particles. However, the setup is somewhat simplified since the charge collection is performed in a self-trigger mode.

A summary of the characteristics of the silicon detectors used for the example is given in Table 6.1. The current pulses induced by particles penetrating the silicon diode are detected by a fast current amplifier with an input impedance $R_a = 50\,\Omega$, providing a gain of $G = 1000$. The pulses are recorded by a LeCroy digital oscillo-

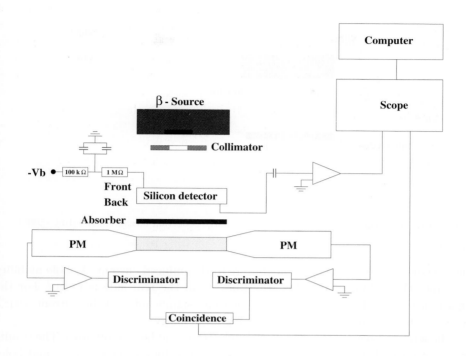

Fig. 6.10   Experimental setup for β-particle charge collection measurements [Roy (2000)].

scope (Fig. 6.12) used in averaging mode, to improve the signal-to-noise ratio [Roy (1994); Leroy et al. (1997b)]. It is necessary to know the concentration of the electrons and holes, as well as the electric field at every space-time coordinate. These quantities can be extracted from the system of partial differential equations introduced above.

In the absence of an analytical solution to the system of the five partial differential equations, considered above, the equations are discretized using Gummel's decoupling scheme [Gummel (1964)] to obtain a numerical solution [Leroy, Roy, Casse, Glaser, Grigoriev and Lemeilleur (1999a); Roy (2000)]. The quantities of interest are extracted by using the code MINUIT [ASG (1992)] to minimize the χ^2 obtained from fitting the numerical solutions of Eq. (6.72) to the experimental data obtained from the measurement of the current pulse response induced by α- and β-particles in the silicon detectors.

Figure 6.13 allows the transport of the charge carriers to be visualized by showing the location of the charge carriers as a function of the collection time, as well as the corresponding signals for incident α- and β-particles. For the α-particles entering the rear side (n^+), the holes (h) drifting in the detector give the main contribution to the induced current. For the α-particles entering the front side (junction side),

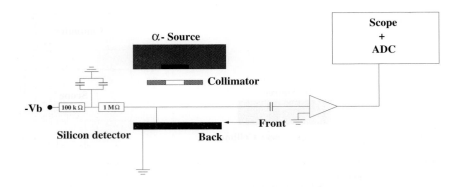

Fig. 6.11   Experimental setup for α-particle charge collection measurements [Roy (2000)].

the electrons (e) are the main contributors to the induced current. As hole mobility is smaller than electron mobility, the pulse from the rear side is longer. For the β-particles, both electrons and holes contribute significantly to the current with a shorter signal for electrons due to their higher mobility.

The model also gives the values of the electrons and holes mobilities. The results are reported in Table 6.2. The average mobilities achieved for electrons and holes for the detectors described in Table 6.1 are: $\mu_h = 491 \pm 10\,\mathrm{cm^2\,V^{-1}\,s^{-1}}$ and $\mu_e = 1267 \pm 20\,\mathrm{cm^2\,V^{-1}\,s^{-1}}$, respectively.

As above discussed, the relevant electrical characteristics of a $p^+ - n$ diode are: first, the effective concentration of dopants which defines the internal electric field and thus determines the depletion voltage; second, the electron and hole mobilities which influence the time needed to collect the charges; and third, the charge trapping lifetimes which affect the efficiency of the charge collection. Also

Table 6.2   Electron and hole mobilities of the detectors, listed in Table 6.1, are extracted from the model fitted to β and α data [Roy (2000)].

Detector	Current pulse source	μ_h $(\mathrm{cm^2\,V^{-1}\,s^{-1}})$	μ_e $(\mathrm{cm^2\,V^{-1}\,s^{-1}})$
M4	α	503.8 ± 2.2	1278 ± 15
M18	$\alpha,\ \beta$	474.4 ± 2.4	1236 ± 15
M25	α	476.0 ± 2	1308 ± 28
M35	α	472.1 ± 3	1272 ± 5
M49	β	546 ± 11	1266 ± 24
M50	β	529 ± 13	1272 ± 20
M53	β	478 ± 12	1350 ± 20
P88	α	459.1 ± 4	1222 ± 20
P189	α	480 ± 20	1340 ± 27
P304	α	495 ± 3	1124 ± 22

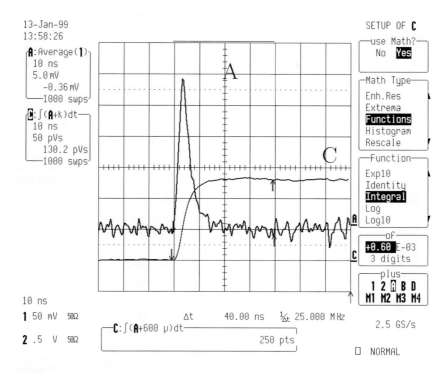

13-Jan-99
13:58:26

A:Average(1)
10 ns
5.0 mV
-0.36 mV
1000 swps

B: ∫(A+k)dt
10 ns
50 pVs
130.2 pVs
1000 swps

A

C

SETUP OF **C**

use Math?
No **Yes**

Math Type
Enh.Res
Extrema
Functions
Histogram
Rescale

Function
Exp10
Identity
Integral
Log
Log10

of
+0.60 E-03
3 digits

plus
1 2 **A** B D
M1 M2 M3 M4

10 ns

1 50 mV 50Ω
C: ∫(A+600 μ)dt

Δt 40.00 ns ¼Δt 25.000 MHz

250 pts

2.5 GS/s

2 .5 V 50Ω

□ NORMAL

Fig. 6.12 The information given by a digital oscilloscope are i) A, representing the average signal produced by the passage of a particle through the detector, as a function of time and ii) C, obtained from the integration of A over time, is a measurement of the collected charge (for instance 170 pVs corresponds to 3.5 fC for a 50 Ω resistance).

after irradiation, these electrical characteristics as a function of the particle fluence (Φ) can be extracted by using the transport model of carriers. The mobility for electrons and holes as a function of fluence can be obtained from the fit of the current pulse response to β-particles incident on detectors exposed to successive levels of fluence. The mobility, after an initial decrease, tends for $\Phi > 5 \times 10^{13}$ particles/cm^2, typically, to a saturation value of $\mu_{sat,e} \approx 1080\,\mathrm{cm^2\,V^{-1}\,s^{-1}}$ and $\mu_{sat,h} \approx 450\,\mathrm{cm^2\,V^{-1}\,s^{-1}}$ for electrons and holes, respectively. The resulting behavior of the mobility as a function of fluence can be represented by [Leroy and Roy (1998)]:

$$\mu_e(\Phi) = \mu_{sat,e} + 272 \times \exp(-5.17 \times 10^{-14} \times \Phi) \tag{6.73}$$

and

$$\mu_h(\Phi) = \mu_{sat,h} + 33.5 \times \exp(-2.09 \times 10^{-14} \times \Phi). \tag{6.74}$$

The effective doping concentration (N_{eff}) is changing with fluence and can be represented as a function of the irradiation fluence, Φ, via

$$|N_{\mathrm{eff}}(\Phi)| = |-N_d \exp(-c\,\Phi) + b\,\Phi|, \tag{6.75}$$

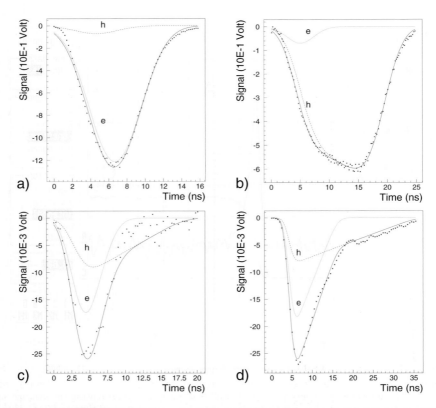

Fig. 6.13   Fits (solid line) of the charge transport model to the current pulse response at $\Phi = 0$ for an α-particle incident on the front side (a), on the rear side (b) of the M25 detector and for a relativistic electron for the M18 (c) and M50 (d) detectors; a bias voltage $V_b = 160$ V is applied in all cases. The individual electron (e) and hole (h) contributions are shown [Leroy, Roy, Casse, Glaser, Grigoriev and Lemeilleur (1999a); Roy (2000)].

where N_a, N_d are the concentration of acceptors and donors at $\Phi = 0$, respectively; c is the donor removal constant and b the acceptor creation rate.

6.1.6   *Leakage or Reverse Current*

For an *ideal unpolarized* $p-n$ *junction*, diffusion of majority carrier cancels the drift of minority carriers and no net current occurs through the junction. The situation changes when the junction gets polarized (reversed mode): a leakage or reversed current (I_r) is observed. The leakage current can be categorized into two types: the bulk and the surface leakage currents. The bulk leakage current has several sources. The electron-hole pairs current (i_g) produced by thermal generation in the depleted region is one of the sources. These pairs are produced by recombination

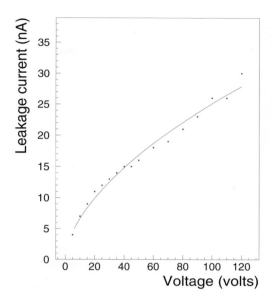

Fig. 6.14   The leakage current as a function of the applied voltage. The curve shows a $\sqrt{V}$ dependence [Roy (2000)].

and trapping centers present in the depleted region. This current is proportional to the active volume of the detector, AX [A is the detector area and X the depletion depth as given by Eq. (6.22)], and to the intrinsic carrier concentration ($n_{\text{int}} = \sqrt{np}$), and inversely proportional to the mean carrier lifetime, τ, which can vary from ~ 100 ns to a few ms depending on the quality or condition of the silicon detector. Then, this current is expressed as :

$$i_g = qAX\frac{n_{\text{int}}}{2\tau}. \tag{6.76}$$

The dependence on X means that this current is depending on the square root of the applied bias voltage (e.g., see Fig. 6.14). Another source of I_r is the current resulting from the motion of electrons from p-region into n-side across the junction and holes moving from n-region into p-side across the junction. In fact, the bulk leakage current is a few nA cm^{-2}, for good quality detectors. This current is increasing up to several tens of μA and even more for irradiated detectors. The bulk leakage current depends on the temperature (Fig. 6.15) as:

$$I_r = T^2 \exp\left(-\frac{E_g}{2k_BT}\right), \tag{6.77}$$

where E_g is defined in Eq. (6.5). This strong dependence on the operating temperature makes necessary to use a renormalization of the measured current. If $T = 20°$C ≈ 293 K is taken as the normalization temperature of reference and T_m the temperature at which the leakage current measurement has been performed, the necessary

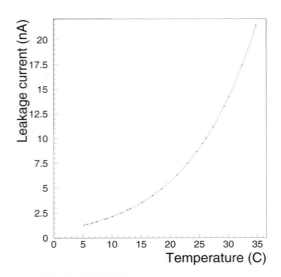

Fig. 6.15 The leakage current as a function of temperature (in °C). The curve represents a fit of Eq. (6.77) to the data points.

temperature correction to the leakage current is done using the equation:

$$I_r(293) = I(T_m) \left(\frac{293}{T_m}\right)^2 \exp\left\{-E_g \bigg/ \left[2k_B\left(\frac{1}{293} - \frac{1}{T_m}\right)\right]\right\}. \qquad (6.78)$$

The surface leakage current is resulting from several sources, such as manufacturing processes and mishandling of detectors (scratches, craters of saliva,...). The surface leakage current does not have dependence on the square root of the applied bias voltage. However, deviations from the expected behavior are extremely difficult to apprehend in practice.

6.1.7 *Noise Characterization of Silicon Detectors*

The noise is a relevant parameter for the detector performance study. The noise is obtained as the root mean square (rms) of the pedestal distribution fitted to a Gaussian function, although this is not always the case since it happens that some noise sources can give non-Gaussian contribution. The pedestal can be measured with a setup configuration where the detector signal is delayed out of the trigger gate. The noise is dominated by the preamplifier noise, controlled by the capacitance at the input of the instrument. Below depletion, the observed noise (ENC_0) versus voltage is proportional to $1/\sqrt{V}$, i.e., it is directly proportional to the diode capacitance. The detection system noise dependence on capacitance can be measured by replacing the silicon detectors by capacitances of known value, up to an external capacitance of 180 pF, which covers adequately the capacitance range of the detectors normally used (thickness $\leq 300\,\mu$m). The noise performance of the charge collection system used in a previous example (Fig. 6.10), expressed in equivalent noise charge ENC, can be represented as a function of capacitance as:

$$ENC_0 = 674 + 3.3 \times C \text{ [pF]}, \tag{6.79}$$

where C is the capacitance seen at the input of the preamplifier. Noise, as a function of the shaping time (θ) analysis, suggests the presence of two components, named as parallel and series noises [Gatti and Manfredi (1986); Leroy et al. (1997a)].

According to this model, the parallel noise ENC_p arises from noise sources in parallel with the detector at the preamplifier input. It depends on the leakage current and, at a given shaping time, it can be calculated (in terms of number of electrons e^-) according to:

$$ENC_p = \sqrt{\left(\frac{I_r\theta}{q}\right) q^2}$$

$$= \left[\sqrt{\left(\frac{I_r\theta}{q}\right)}\right] e^-, \tag{6.80}$$

where I_r is the leakage current, θ is the shaping time and q is the electron charge. ENC_p decreases with temperature as it depends on the leakage current I_r [Eq. (6.77)]. The series component ENC_s derives from the detector impedance at the input of the preamplifier. This series noise component behaves as $1/\theta^{1/2}$ for short shaping times. However, for shaping times $\theta \geq 70$ ns, the series component is practically constant [Leroy et al. (1997a)]. The total noise for a detector is then obtained by summing quadratically the two noise components:

$$ENC^2 = ENC_p{}^2 + ENC_s{}^2. \tag{6.81}$$

The study of the noise of a standard planar detector has shown that the diode capacitance is the main contributor to the series noise. The series noise also represents the total noise of a detector with low leakage current ($I_r \approx 0$). An example of the noise of a standard planar detector (area of $1\,\text{cm}^2$ and thickness of $300\,\mu$m) is shown in Fig. 6.16 as a function of the applied voltage [Leroy (2004a)]. The noise, expressed in ENC, has been measured at $T = 7\,^{\circ}$C. This low temperature minimizes the contribution of ENC_p, which depends on I_r, and hence decreases with T. The data show that the noise follows the diode capacitance dependence on the voltage, in agreement with Eqs. (6.28, 6.79). For applied voltages above the depletion, the series noise reaches a constant value, which corresponds to the minimum level of noise due to the readout electronics: in the example shown in Fig. 6.16, this is about 1000 electrons.

6.2 Charge Collection Efficiency and Hecht Equation

We consider the influence of trapping and recombination on charges produced by ionizing particles in planar (anode A and cathode C are parallel planes as shown in

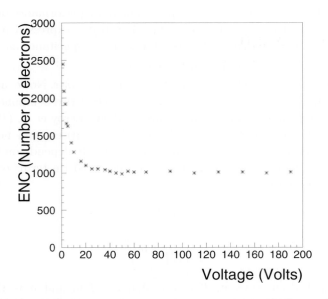

Fig. 6.16   Measured noise, expressed in equivalent noise charge ENC - number of electrons - [Eq. (6.81)], as a function of the applied voltage for a standard planar detector of $1\,\mathrm{cm}^2$ area and $300\,\mu\mathrm{m}$ thickness. The measurement has been performed at $T = 7\,^{\circ}\mathrm{C}$ [Leroy (2004a)].

Fig. 6.17) silicon detectors of the type discussed above. Trapping and recombination are related to defects, such as impurities, or vacancies present in the semiconductor crystal lattice as a result of the manufacturing processes or after exposure of the detectors to high levels of irradiation fluences (as large as 10^{13}–10^{14} neutrons or protons per cm^2). Therefore, trapping and recombination are responsible for loss of charge during the motion of the charge from the point it was created towards its point of collection (anode for electrons and cathode for holes) under a voltage, V, applied across the detector.

The theorem of Ramo for a parallel plate capacitor (the case of a planar silicon detector) states that

$$\frac{dq}{q} = -\frac{x}{w_d},\tag{6.82}$$

w_d is the width of the depleted zone. For a totally depleted detector $w_d = d$, where d is the detector thickness. This implies a current I_i induced by the charge i that is described by Eq. (6.83)

$$I_i = -\frac{q\,v_i}{d}.\tag{6.83}$$

With $v_i = \mu_i\, E$ is the velocity of the charge carrier i ($i = e, h$). E is the electric field taken uniform for our purpose: $E = V/d$, V is the applied bias voltage. In presence of trapping, the current corresponding to the carrier i takes the following form:

$$I_i(t) = I_{0,i}(t) \exp\left(-\frac{t}{\tau_i}\right) = -Q_0 \exp\left(-\frac{t}{\tau_i}\right) \frac{\mu_i\, E}{d},\tag{6.84}$$

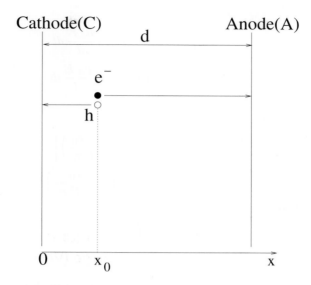

Fig. 6.17   A planar silicon detector with parallel electrodes. The electron and hole created at a point x_0 by an incoming particle are drifting to the anode (A) and cathode (C), respectively, under the field generated by the voltage applied across the detector.

where $I_{0,i}(t)$ is the current induced by charge i in absence of trapping, τ_i is the time the charge carrier i survives before being trapped or recombined $(i = e, h)$. One finds that the charge, Q_i, induced by the charge carrier i on the electrodes is given by

$$Q_i = \int_0^{t_{t_r}} I_i(t)\, dt = -Q_0 \int_0^{t_{t_r}} \exp\left(-\frac{t}{\tau_i}\right) \frac{\mu_i\, E}{d}\, dt, \qquad (6.85)$$

where t_{t_r} is the trapping time, i.e.,

$$t_{t_r} = t_{r,e} = \tau_e = \frac{d - x_0}{\mu_e\, E}$$

for electrons and

$$t_{t_r} = t_{r,h} = \tau_h = \frac{x_0}{\mu_h\, E}$$

for holes. In Eq. (6.85), one takes $x = 0$ for the front electrode ($p^+ - n$ junction) and $x = d$ for the back electrode ($n - n^+$ junction) and $x = x_0$, the position at which an electron-hole pair has been created. The electrons travel toward the back-electrode (anode) over a distance $d - x_0$ and holes travels toward the front-electrode (cathode) over a distance x_0. Taking Eq. (6.85) and adding the contributions of both charge carriers, one finds:

$$Q = Q_e + Q_h \tag{6.86}$$
$$= -Q_0 \int_0^{t_{tr,e}} \exp\left(-t/\tau_e\right) \frac{\mu_e\, E}{d}\, dt$$
$$-Q_0 \int_0^{t_{tr,h}} \exp\left(-t/\tau_h\right) \frac{\mu_h\, E}{d}\, dt$$

or

$$Q = Q_e + Q_h \tag{6.87}$$
$$= \frac{Q_0}{d}\, \mu_e E\, \tau_e \left[1 - \exp\left(-\frac{d - x_0}{\mu_e E\, \tau_e}\right)\right]$$
$$+ \frac{Q_0}{d}\, \mu_h E\, \tau_h \left[1 - \exp\left(-\frac{x_0}{\mu_h E\, \tau_h}\right)\right].$$

Using the simplified notations,

$$\Lambda_e = \mu_e E\, \tau_e \tag{6.88}$$

and

$$\Lambda_h = \mu_h E\, \tau_h, \tag{6.89}$$

one finds

$$Q = Q_0 \left\{ \frac{\Lambda_e}{d}\left[1 - \exp\left(-\frac{d - x_0}{\Lambda_e}\right)\right] + \frac{\Lambda_h}{d}\left[1 - \exp\left(-\frac{x_0}{\Lambda_h}\right)\right] \right\}, \tag{6.90}$$

which is the *Hecht equation* [Hecht (1932)]. Then, the charge collection efficiency, CCE, is the ratio of the collected charge to the initial charge Q_0:

$$\text{CCE} = \frac{Q}{Q_0} \tag{6.91}$$

with $Q_0 = N_0\, q$, where $N_0 = E/E_{\text{ion}}$ is the number of electron-hole pairs created at the point x_0 by an incident particle of energy E and E_{ion} is the energy needed to create an electron-hole pair, $E_{\text{ion}} = 3.62\,\text{eV}$ for silicon. Equation (6.91) can be applied to determine the CCE of detectors illuminated by α-particles.

The range of α-particle in silicon is $25\,\mu\text{m}$ for α-particle from ^{241}Am source and cannot cross the whole depth of the detector which is a few hundreds μm thick in practice. For a source confined under vacuum, the α-particle incident energy is $\sim 5.45\,\text{MeV}$ and $Q_0 = (5.45 \times 10^6/3.62)1.6 \times 10^{-19}\,\text{C} \sim 0.24\,\text{pC}$.

The CCE depends on Λ_e, Λ_h and the point x_0 where the charge has been created. This point is not accessible experimentally. The location of this point being random, the width of the peak in the energy spectrum broadens. The relative broadening is [Iwanczyk, Schnepple and Materson (1992)]:

$$\frac{\sigma(E)}{E} = \frac{2\Lambda_e^2 \Lambda_h^2}{d^3(\Lambda_e - \Lambda_h)} \left(e^{-d/\Lambda_e} - e^{-d/\Lambda_h}\right)$$
$$- \frac{1}{d^4}\left[\Lambda_e^2\left(e^{-d/\Lambda_e} - 1\right) + \Lambda_h^2\left(e^{-d/\Lambda_h} - 1\right)\right]^2$$
$$- \frac{\Lambda_e^3}{2d^3}\left(e^{-2d/\Lambda_e} - 1\right) - \frac{\Lambda_h^3}{2d^3}\left(e^{-2d/\Lambda_h} - 1\right). \tag{6.92}$$

The CCE and width improve with the increase of Λ_e/d and Λ_h/d. Note that

$$\frac{\Lambda_{e,h}}{d} = \frac{\tau_{e,h}\, \mu_{e,h}\, E_f}{d} = \frac{\tau_{e,h}\, \mu_{e,h}\, \rho\, j}{d},$$

where E_f, ρ and $j = i/d$ are the electric field, the semiconductor resistivity and the leakage current density, respectively [Abyzov, Davydov, Kutny, Rybka, Rowland and Smith (1999)]. To optimize the energy resolution, one should maximize $\tau_{e,h}\, \mu_{e,h}\, \rho\, j$ with the constraint, however, that the leakage current, i, contributes to the noise charge

$$ENC_p = \sqrt{\frac{i\theta}{q}},$$

which has to be minimized [θ is the pulse shaping time of the preamplifier, expression already used earlier, see Eq. (6.80)].

For minimum ionizing β-particles, which traverse the whole depth of the detector, calculations similar to those leading to Eq. (6.90) can be done and lead to a ratio of collected charge Q_{coll} to the initial charge Q_0:

$$\mathrm{CCE}_\beta = \frac{Q_{coll}}{Q_0}$$

$$= \frac{\Lambda_e + \Lambda_h}{d} - \left(\frac{\Lambda_e}{d}\right)^2 \left(1 - e^{-d/\Lambda_e}\right) - \left(\frac{\Lambda_h}{d}\right)^2 \left(1 - e^{-d/\Lambda_h}\right). \quad (6.93)$$

In the case of relativistic electrons (selected for instance from β-particles produced by a ^{106}Ru source) traversing a silicon detector $300\,\mu$m thick, the energy loss is about $80\,$keV per $300\,\mu$m and $Q_0 = (80 \times 10^3/3.62)1.6 \times 10^{-19}\,$C $\sim 3.5\,$fC, as seen before.

6.3 Spectroscopic Characteristics of Standard Planar Detectors

Precise spectroscopy technique is a powerful tool for investigating features and functionalities of semiconductor detectors [Casse et al. (1999a); Chren et al. (2001); Houdayer et al. (2002)]. It provides information on the structure and charge collection capabilities of the detectors. The scanning of the diode structure and its sensitive volume by illuminating the front and back sides with heavy charged particles (alpha's and protons) of low energy and well defined range is illustrated in Fig. 6.18. In the case of α-particles, the pulse height is given by:

$$V_{ph} = \eta \frac{q E_\alpha}{E_{ion} C}, \quad (6.94)$$

where E_α is the α-particle energy deposited in silicon, q is the electronic charge, E_{ion} ($= 3.62\,$eV) is the energy necessary to produce a electron-hole pair and, finally, C is the detector capacitance. The factor η is the charge collection efficiency. For high ionizing radiation, such as α-particles, η also reflects possible plasma effects, which

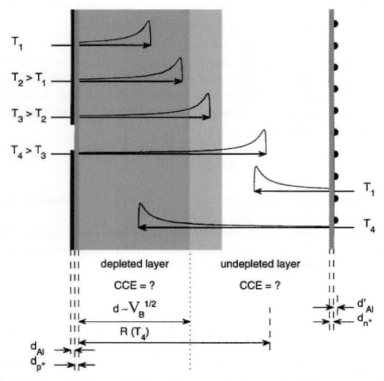

Fig. 6.18   Illustration in the schematic cross-sectional view of a detector of thickness d, of the method for scanning the detector structure and sensitive volume (from [Leroy and Rancoita (2007)]; see also [Casse et al. (1999a)]). The thicknesses of Al contact metallisation and p^+ layer (front side) or n^+ layer (back side) are indicated. The shaded area represents the depleted region (detector sensitive volume). The detector response is measured as a function of the applied bias voltage for protons or alpha particles incident on the front or on the back side at energies $T_4 > T_3 > T_2 > T_1$, corresponding to incoming particle ranges $R(T_4) > R(T_3) > R(T_2) > R(T_1)$ inside the detector. Bragg energy-loss distributions inside the detectors are shown for the various incident energies.

affect the collection efficiency. For instance, in Fig. 6.18 it is shown the dependence of the depth of the depleted (sensitive) layer on the applied bias voltage (V_b) is compared with the Bragg curve distribution of charge produced along the path of the incoming particle.

Pulse-height spectra are obtained by measuring the signal of the diodes with a standard spectroscopy system. The pulse-height spectra are recorded at various applied bias voltages.

The variation of the peak position with the incoming particle energy and the applied bias voltage provides precise information on the spatial efficiency of the charge collection controlled by the electric field configuration inside the diode. This technique requires the use of a reference detector, a silicon surface barrier detector (SSBD) which has a thin entrance window, an excellent energy resolution and 100 %

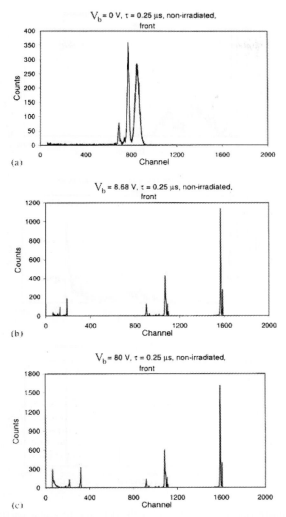

Fig. 6.19 Spectra illustrating the response of a SP detector to alpha particles from Pu, ThC and ThC' sources incident on the front side of the diode for various applied reverse bias voltage: $V_b = 0.0$ (a), 8.7 (b) and 80 V (c) (from [Leroy and Rancoita (2007)]; see also [Casse et al. (1999a)]). The shaping time was $\tau = 0.25\,\mu$s.

charge collection efficiency (CCE). The normalization of the peak position observed with the diodes under study to that observed with the reference detector is used to determine the CCE. Typically, the Full-Width at Half-Maximum (FWHM) at 5.486 MeV (^{241}Am α-line) for the SSBD is 14 keV, giving a relative energy resolution of a fraction of the percent for the whole spectroscopy system (including the reference detector). This very good energy resolution achieved allows the observation of changes in pulse height resulting from changes in CCE and/or influence of the electrode structures. From the splitting and shifts of α-peak, one can determine the

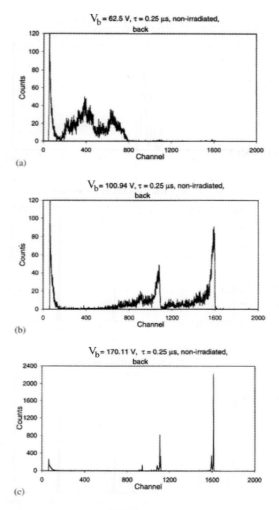

Fig. 6.20   Spectra illustrating the response of a SP detector to alpha particles from Pu, ThC and ThC' sources incident on the back side of the diode for applied reverse bias: $V_b = 62.5$ (a), 100.9 (b) and 170.1 V (c) (from [Leroy and Rancoita (2007)]; see also [Casse et al. (1999a)]). The shaping time was $\tau = 0.25\,\mu s$.

thicknesses of metallic electrodes and implanted layers. The capability to measure shifts at the level of a keV is equivalent to measuring layer thicknesses of about ten nanometers.

Some of the spectra illustrating the response of a Standard Planar (SP) detector to α-particles impinging on the front (back) side of the diode are shown in Fig. 6.19 (Fig. 6.20) for different values of applied reverse bias ($V_b = 0.0,\ 8.7,\ 80\,\mathrm{V}$), with a shaping time $\tau = 0.25\,\mu s$.

The understanding of these peaks behavior is presented in [Casse et al.

(1999a)]. The sensitivity of the spectroscopic method for SP detectors is illustrated in Fig. 6.21. The response of the reference SBBD detector [Fig. 6.21 (a)] is compared with the responses of a SP detector illuminated from the front side [Fig. 6.21 (b)] and from the back side [Fig. 6.21 (c)]. From the splitting and shifts of the α-peak, one can determine the thicknesses of metallic electrodes and of implanted layers as well.

6.3.1 *Energy Resolution of Standard Planar Detectors*

Semiconductor detectors have very good energy resolution. There are several contributions to the resolution such as statistics of electron-hole formation, detector and readout noises. If one considers only the contribution of electron-hole formation statistics, the energy resolution of a planar silicon detector is, in standard notation:

$$\sigma_R = 2.36 \times \sigma(E), \tag{6.95}$$

where $\sigma(E)$ is the standard deviation in the amount of energy deposited in the detector by an incident monoenergetic particle. If N_{e-h} is the average number of electron-hole pairs produced by the incident particle in the material, σ_R can be written as:

$$\sigma_R = 2.36 \times E_{\text{ion}} \sqrt{N_{e-h}}, \tag{6.96}$$

where E_{ion} $(= 3.62\,\text{eV})$ is the average energy to create a electron-hole pair in silicon. If E (expressed in eV) is the energy of the incident particle,

$$N_{e-h} = \frac{E}{E_{\text{ion}}}$$

and one finds

$$\sigma_R = 2.36 \times \sqrt{E E_{\text{ion}}}. \tag{6.97}$$

For an incident particle of 5 MeV on silicon, $\sigma_R = 10\,\text{keV}$. For a 140 keV Compton-photon emitted by a ^{99m}Tc source, $\sigma_R = 1.7\,\text{keV}$.

The other contributions to the energy resolution depend on the particular readout electronic chain used for the measurements. For instance, for an α-particle emitted by an ^{241}Am source, $E = 5.486\,\text{MeV}$, the application of Eq. (6.97) gives $\sigma_R = 10.5\,\text{keV}$, i.e., a relative energy resolution of 0.2%. However, the measurement performed with a standard spectroscopy system composed of a charge-sensitive preamplifier, linear shaping amplifier and multichannel analyzer rather gives for the same alpha line: $\sigma_R = 14\,\text{keV}$ or a relative energy resolution of 0.3%. The difference between the measurement and the value found through Eq. (6.97) is explained by the readout chain noise not accounted in the equation.

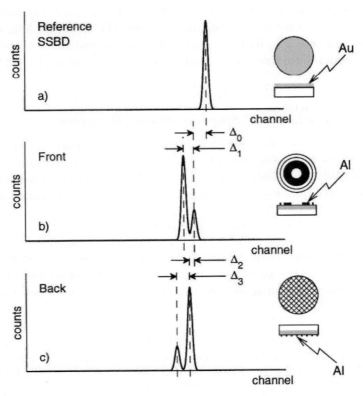

Fig. 6.21　Illustration of the sensitivity of the spectroscopic method (from [Leroy and Rancoita (2007)]; see also [Casse et al. (1999a)]). The response of a reference silicon surface barrier detector (SSBD) (a) is compared to the response of a SP detector illuminated (b) on the front side and (c) on the back side. Shifts among peaks and their splitting are caused by different losses of particles entering through non-sensitive layers on the detector surface. Δ_0: the difference of the thicknesses of layers due to different losses in the Au reference detector electrode and p^+ layer of the measured detector ($\simeq 20\,\mathrm{keV}$), Δ_1: thickness of front Al electrode ($\simeq 86\,\mathrm{keV}$), Δ_2: backside n^+ layer thickness compared to p^+ ($\simeq 33\,\mathrm{keV}$), Δ_3: backside Al grid thickness ($\simeq 122\,\mathrm{keV}$). The thicknesses are expressed in terms of the corresponding energy-losses.

6.4　Microstrip Detectors

The division of the *p-side* of a silicon diode into an array of parallel narrow strips provides a position sensitive detector. The carriers (electrons and holes) created around the incident particle track are basically confined in a tube of $\approx 1\,\mu\mathrm{m}$ of diameter around the track. The charge created drifts under the electric field. In a n-type silicon with p^+-strips, the holes will move to the diode strips and electrons to the backplane (or ground plane), see Fig. 6.22. For instance, for a $300\,\mu\mathrm{m}$ thick detector under a bias voltage of 150 Volts, the electric field is $5.0\,\mathrm{kV/cm}$ and the drift velocities are $6.75 \times 10^6\,\mathrm{cm/s}$ and $2.25 \times 10^6\,\mathrm{cm/s}$ for electron and hole, respectively.

The electrons created closer to the diode strips will take more time to migrate to the backplane. The same is true for the holes created closer to the backplane which will take more time to reach the junction side. During the drift time, the narrow tube of charge carriers broadens as a result of multiple collisions and the carrier distribution obeys a Gaussian law:

$$\frac{dN}{N} = \frac{1}{\sqrt{4\pi D_i t}} \exp\left(-\frac{x^2}{4D_i t}\right), \qquad (6.98)$$

where dN/N is the fraction of charge carriers in the element dx at a distance x from the origin of charge location after a time t. D_i $(i = e, h)$ is the diffusion coefficient for the i-carrier and is given by Eq. (6.62).

At room temperature, Eq. (6.62) gives $D_e = 34.9\,\text{cm}^2/\text{s}$ and $D_h = 11.6\,\text{cm}^2/\text{s}$ using $\mu_e = 1350\,\text{cm}^2\text{V}^{-1}\text{s}^{-1}$ and $\mu_h = 450\,\text{cm}^2\text{V}^{-1}\text{s}^{-1}$, respectively. The root mean square in Eq. (6.98), i.e., $\sqrt{2D_i}$ is the same for electrons and holes since the drift time $t \sim 1/\mu$. The diffusion of charge carriers obviously affects the spatial resolution of the microstrip detectors.

Statistical fluctuations of the energy loss is another effect to be considered with that respect. The statistical nature of the ionization process on the passage of a fast charged particle through matter results in large fluctuations of energy losses in the silicon detector, which are usually very thin compared to the particle range. One has already seen in Sect. 2.1.2 that these fluctuations can be calculated using Landau and Vavilov distributions (Vavilov provides a more accurate solution by introducing the kinematical limit on the maximum transferable energy in a single collision). Binding of electrons in silicon atoms broadens the theoretical Landau distribution of the energy-loss measured in the silicon detector. The correct distribution of energy loss is obtained from the convolution of a Landau distribution and a Gaussian corresponding to a correction for the binding energies of electrons in silicon and taking into account the electronics noise [Shulek et al. (1966); Hancock, James, Movchet, Rancoita and Van Rossum (1983, 1984); Caccia et al. (1987)]. The production of energetic secondary electrons contributes to the energy loss in silicon, in addition to the creation of electron–hole pairs. Then, displacement of the charge distribution is due to secondary electrons ejected with an energy larger than T, which is the kinetic energy imparted to the electrons ejected from the collision of the incoming particle with atomic electrons [Damerell (1984)]. The maximal kinetic energy which can be transferred from a collision of an incoming particle (not an electron) of mass m and energy E with an atomic electron of mass m_e is [Eq. (1.28) on page 10]:

$$T_{\max} = 2m_e c^2 \beta^2 \gamma^2 \left[1 + \left(\frac{m_e}{m}\right)^2 + 2\gamma\frac{m_e}{m}\right]^{-1}$$

$$= \frac{2m_e p^2}{(m^2 + m_e^2 + 2m_e E/c^2)}, \qquad (6.99)$$

where the incoming particle momentum is

$$p = m\gamma\beta c, \quad \text{with} \quad \gamma = \frac{E}{mc^2} \quad \text{and} \quad \beta = \frac{v}{c}$$

(v is the incoming particle velocity). In the case of an incoming electron (see Sect. 2.1.6.1), since the outgoing electron of higher energy is considered to be the *primary electron*, the maximum energy transferred is $\frac{1}{2}$ of the incoming electron kinetic energy.

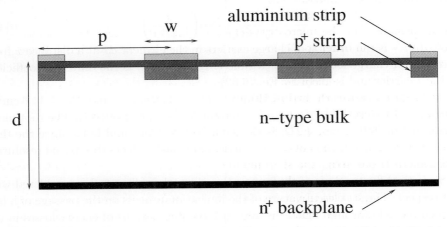

Fig. 6.22 Standard representation of a n-type strip detector; p is the strip pitch, d is the detector thickness and w is the width of the p$^+$ implant.

The probability to eject secondary electrons (δ-rays) of kinetic energy higher or equal to T is decreasing by about 4 orders of magnitude as T increases from 100 eV to 1 MeV (see Sect. 2.1.2.1). However, these high energy electrons, although their number decreases with increasing energy, have an increasing range in silicon. Each ejected electron releases an increasing number of secondaries in silicon with the result [Damerell (1984)] that the mean charge released in silicon increases linearly with $\ln(T)$.

As seen before, a relativistic electron releases about 80 keV in a silicon thickness of 300 μm, which corresponds to 22000 electrons released around its track. If in addition, a δ-ray of 50 keV is produced and drifts perpendicularly to the track, its range will be about 16 μm in silicon and, therefore, 50 keV/3.62 eV $\sim$ 13800 additional electrons will be produced along this ray track and will shift (see [Damerell (1984)]) the centroid of the charge distribution, assuming the secondary charge is produced halfway from the track, by an amount Δ of

$$\Delta = \frac{22000 \times 0 + 13800 \times 8}{22000 + 13800} \sim 3.6\,\mu\text{m}. \tag{6.100}$$

From Eq. (2.38) on page 55 (see also Fig. 10 in [Damerell (1984)]), it is seen that about 4×10^{-4} electrons with an energy greater or equal to 50 keV are ejected per micrometer of track. In the case of a silicon detector 300 μm thick, there is then a 11% probability of producing a δ-ray which shifts the centroid of the charge distribution by about 3 μm. The number of electron–hole pairs produced in a silicon detector will increase with its thickness. However, the probability to produce δ-rays with higher range in silicon will also increase and affect the position resolution. Thinner the detector is, lower will be the probability that such a δ-ray exists and greater will be the spatial resolution of the detector; although there is clearly a practical limit in lowering the thickness of the silicon detector [Manfredi and Ragusa (1985)].

The value of the strip pitch, "p" (see Fig. 6.22), influences largely the design and performances of the microstrip detectors. Intuitively, one would choose it as small as possible in order to have a dense array of strips and the signal spread over several strips. This would allow one to reconstruct the charge distribution and find the impact point of the incoming particle by calculating the center of gravity of the charge distributed over the strips. However, to have the signal really shared by several strips, one should use pitches of the order of a few μm since measured $FWHM$ of charge distribution is of the order of $10 \, \mu$m [Belau et al. (1983)]. This is practically impossible to achieve. In practice, the readout pitch is $20-25 \, \mu$m. The charge will be collected on one strip, the strip number giving the track position. The position can be measured more precisely for tracks generating signals on two strips. In the case of tracks of particles traversing the detector in between two consecutive strips, the position is calculated again by taking the center of gravity of the charge distribution or inferred from the shape of the charge distribution. The precision on the measurement of the position, when several strips share the charge, depends of the relative position of the track with respect to the strips involved since the strip closer to the track will show a larger signal than the strip farther. Then, the precision of the measurement depends on the noise which is expected to be relatively larger for the smaller signal. For a readout pitch of $25 \, \mu$m, the precision of localization is a few microns.

Most of the time, financial costs, geometrical and mechanical constraints may lead to a reduction in the number readout channels. It means that the space between two consecutive active (readout) strips is increased, which means an increase of the effective pitch value, and, consequently, a decrease of the number of events detected by two strips. To remedy this problem, diodes are introduced between the readout strips [England et al. (1981); Kötz et al. (1985)]. The charges collected on those intermediate strips are transferred by capacitive coupling to the readout strips. Such a procedure feeds up the number of events detected on the two readout strips.

More generally, the interplay between the dimension of the pitch, the strip width and the thickness of the diode determines the operational features of the microstrip detector as it influences the noise of the electronics readout. As we already did in this section, it is standard to approximate a silicon microstrip by a planar diode of thickness d and width w. Neglecting the built-in voltage V_0 in Eqs. (6.22, 6.24), the depth, x, of the depleted layer is given as a function of the bias voltage V_b

$$x = \sqrt{\frac{2\epsilon V_b}{q \, |N_{\text{eff}}|}} \, , \tag{6.101}$$

and the bias voltage achieving full depletion is:

$$V_{fd} = \frac{d^2 q |N_{\text{eff}}|}{2\epsilon}. \tag{6.102}$$

The so-called *body capacitance per unit length* for a depletion depth x is given by Eq. (6.28), i.e., in the present case we have

$$c_b = \epsilon \frac{p}{x} = \epsilon \frac{p}{d} \sqrt{\frac{V_{fd}}{V_b}}, \tag{6.103}$$

thus, for $V_b < V_{fd}$, $c_b \propto 1/\sqrt{V_b}$ and for $V_b \geq V_{fd}$, $c_b = \epsilon p/d$.

The relative dimensions of the pitch "p" (p) and strip width "w" (w) ($w/p < 1$ in practice) contribute a term which increases the full depletion voltage [Eq. (6.102)] to the depletion voltage, V'_{fd}, for the strip [Barberis et al. (1994)]

$$V'_{fd} = V_{fd} \left[1 + 2 \frac{p}{d} f\left(\frac{w}{p}\right) \right]$$

$$= \frac{q |N_{\text{eff}}|}{2\epsilon} \left[d^2 + 2 p d f\left(\frac{w}{p}\right) \right], \tag{6.104}$$

and decreases the strip body capacitance, c'_b, per unit length for a depletion depth x [Barberis et al. (1994)]:

$$c'_b = \epsilon \frac{p}{x + pf\left(\frac{w}{p}\right)} = \epsilon \frac{p}{d} \sqrt{\frac{V_{fd}}{V_b + V_{fd} \left[\frac{p}{d} f\left(\frac{w}{p}\right) \right]^2}}. \tag{6.105}$$

At full depletion depth d, we obtain:

$$c'_b = \epsilon \frac{p}{d + pf\left(\frac{w}{p}\right)} = \epsilon \frac{\frac{p}{d}}{1 + \frac{p}{d} f\left(\frac{w}{p}\right)}. \tag{6.106}$$

The function $f\left(\frac{w}{p}\right)$ is given [Barberis et al. (1994)] by:

$$f\left(\frac{w}{p}\right) = -0.00111 \left(\frac{w}{p}\right)^{-2} + 0.0586 \left(\frac{w}{p}\right)^{-1} + 0.240$$

$$- 0.651 \left(\frac{w}{p}\right) + 0.355 \left(\frac{w}{p}\right)^2. \tag{6.107}$$

The total capacitance, C_{tot}, has a dominant contribution to the readout electronic noise. The total capacitance has a linear dependence on the strip width, for a given pitch. In the range $0.1 \leq w/p \leq 0.55$, for a detector $300\,\mu$m thick [Sonnenblick et al. (1991)]:

$$C_{\text{tot}} = \left(0.8 + 1.6 \frac{w}{p} \right) \quad \text{[pF/cm]}, \tag{6.108}$$

for $p = 100\,\mu$m. The interstrip capacitance, C_{int}, can be calculated from:

$$C_{\text{int}} = C_{\text{tot}} - c'_b = \left[0.8 + 1.6 \frac{w}{p} - \epsilon \frac{p}{d + pf\left(\frac{w}{p}\right)} \right] \quad \text{[pF/cm]}, \tag{6.109}$$

where c'_b is given by Eq. (6.106). As an example for $w/p = 0.5$, $d = 300\,\mu$m, $p = 100\,\mu$m, one finds $f\left(\frac{w}{p}\right) = 0.116$, $p/d = 1/3$ and $c'_b = 0.35\,$pF/cm, $C_{\text{tot}} = 1.6\,$pF/cm, and $C_{\text{int}} = 1.25\,$pF/cm. This example illustrates the important contribution of C_{int} to the total capacitance. In fact for $d = 300\,\mu$m and $0.18 \le w/p \le 0.36$ and $50\,\mu$m$\le w/p \le 200\,\mu$m, Eq. (6.109) can be replaced in very good approximation (better than 10%) by the formula [Braibant et al. (2002)]:

$$C_{\text{int}} \sim \left(0.1 + 1.6\frac{w + 20\,\mu\text{m}}{p}\right)\ [\text{pF/cm}]. \tag{6.110}$$

The electronics noise of the charge sensitive amplifier expressed as an equivalent noise charge (ENC) can be parametrized similarly to Eq. (6.79):

$$ENC = a + b\,C_{\text{tot}}\ [\text{pF/cm}], \tag{6.111}$$

where a and b are constants depending on the preamplifier. Typical value of ENC in *peak mode* is [Braibant et al. (2002); Jones (1999)]:

$$ENC = 246 + 36\,C_{\text{tot}}\ [\text{pF/cm}]. \tag{6.112}$$

Using Eq. (6.108) with $w/p = 0.4$, one finds $C_{\text{tot}} = 1.44\,$pF/cm which gives $ENC \approx 300$, comparable to the contribution of 300 electrons reported in [Nygard et al. (1991)] with another preamplifier. One should also consider the contributions of leakage current and bias resistor to the equivalent noise charge. The noise due to the leakage current (typically, 1 nA per strip) is given by Eq. (6.80). For $\theta = 250\,$ns and $I_r = 1\,$nA in Eq. (6.80), one obtains a noise of about 40 electrons. The equivalent noise charge due to bias resistor (ENC_r) is given by:

$$ENC_r = \frac{e}{q}\sqrt{\frac{2\theta kT}{R}}, \tag{6.113}$$

where q is the electronic charge, e is the base of natural logarithm, T is the temperature, k is the Boltzmann constant ($kT = 0.0259\,$eV at $T = 300\,$K) and R is the biasing resistance. For $R = 2\,$MΩ, one finds $ENC_r \sim 550$ electrons.

6.5 Pixel Detector Devices

Most of the pixel detector devices in operation nowadays are based on silicon technology, although such devices based on GaAs, CdTe and CdZnTe are also used. The pixels are small detector elements or cells of a two-dimensional array, organized into rows and columns. One often uses the expression "pixellated silicon detector". The pixels of present generation are usually of squared area with dimensions $\sim 50\,\mu$m $\times$ $50\,\mu$m. Practically, the silicon pixel detector is of the hybrid-type. It is made up of two superimposed layers. The top layer is the detecting or sensor layer, the bottom layer is the readout electronics layer which defines the segmentation of the sensor material and the size of the pixels. Each cell and its corresponding readout chip are connected via conductive bumps (i.e., a drop of solder between two metals pads per pixel) using the so-called bump-bonding and flip-chip technique. This metal bump allows the transfer of the pulse generated in a pixel in the sensor layer to the readout chip. The pixel sensor layer is a $p-n$ diode operated in reverse bias mode. Incident particles on the detector creates electrons and holes in the sensor by ionization. The movement of the separated charges in the applied electric field induces a signal on the pixel electrode and possibly its neighbors. This signal is transmitted to the chip periphery after amplification, discrimination and digitization by the electronic circuitry in the pixel cell of the chip. The amount of ionization charge collected by a series of cells allows the reconstruction of the track of a particle incident on the detector. The track length depends on the incidence angle as particles of large incidence angle (close to or along the normal to the sensor plane create tracks covering a pixel or just a few whereas tracks of particles of low incidence angle cover a larger number of pixels. Obviously, the number of pixel covered by a track also depends on the incident particle type and energy. One should observe that pixels detectors provide large signal-to-noise ratios since the ionization charge Q is collected on a very small area typically $\sim 2500\,\mu m^2$. Therefore, the capacitance at the amplifier input is about $C = 88\,fF$ for a silicon thickness of $300\,\mu m$, leading to very small noise $\sim 100\,e^-$ r.m.s., as this noise is dominated by capacitance. Also, the signal is high as $V = Q/C$. The signal typically consists of the order of $\sim 10,000\,e^-$ charges, and in a segmented matrix with $\sim 50\,\mu m$ size pixels, one can process $\sim 1000\,e^-$ signals with a noise as low as $100\,e^-$ r.m.s [Heijne (2007)]. The small size of a pixel also helps the limitation of the leakage current. A leakage current as high as $1\,mA\,cm^{-2}$ only represents $25\,nA$ for a $50\,\mu m \times 50\,\mu m$ pixel, if the current is evenly distributed [Heijne (2007)]. A further progress in pixel detector technology is to build a pixel detector device which could store the image data in each pixel until readout rather than transmit and process particle signal off-chip. Using a comparator,i.e., setting a threshold, this enables a 1 bit analog-to-digital conversion and accurate analog performance can be achieved by combining linear amplifier and tunable threshold. Examples of such devices are the MediPix-type devices.

6.5.1   *The MediPix-type Detecting Device*

This hybrid silicon pixel device was developed in the framework of the *Medipix Collaboration* [Medipix Collab. (2008)] and originally designed for position sensitive single X-ray photon detection. In a first version, the device called MediPix1, consists of a silicon detector chip (sensor) bonded to a readout chip. The sensor chip is equipped with a single common backside electrode and a front side matrix of 64×64 squared pixels is defined with a pixel size of $170\,\mu m \times 170\,\mu m$ allowed by a $1\,\mu m$ CMOS process. Usually the silicon thickness is $300\,\mu m$ thick, although other thicknesses are possible. Each pixel is connected to its respective readout chain integrated on a readout chip. When a quantum of radiation is absorbed in the sensor layer, the generated charge is transferred through the bump bond to the corresponding electronic chip where after preamplification the charge is fed to a discriminator (comparator) which compares the received signal to a preset threshold. If the signal is above that threshold, the counter of this pixel is incremented by one. This operation is done in each pixel independently. After completion of the data acquisition, the counters of all pixels are read out to obtain an image (a frame). The single lower global threshold for the discriminator allows the rejection of very low energy X-ray photons, so helping the image contrast and provides at the same time suppression against electronic noise or dark current of the active layer, reducing the background in the image. The counter depth of 15-bit in each pixel limits the dynamic range and image contrast. The electronic chip of the *MediPix1 device* only works for positive charges (holes). Then, heavy semiconductors such as GaAs, CdTe and Cd(Zn)Te having low hole mobility are not the best choice for sensor material of MediPix1.

The *MediPix2 hybrid silicon pixel-detector* [Llopart et al. (2002)] is the successor of MediPix1. Most of the content of this chapter will be now about the MediPix2 device as it represents the most powerfull detector of the type, for the time being. The MediPix2 has a front side matrix of 256×256 electrodes. Each pixel has an area of $55\,\mu m \times 55\,\mu m$. The use of $0.25\,\mu m$ CMOS process has allowed this small pixel size. The silicon thickness is $300\,\mu m$, although larger thicknesses ($700\,\mu m$ and 1 mm, for instance) are also possibly utilized. The use of heavy semiconductors like GaAs, CdTe, Cd(Zn)Te is better adapted to MediPix2 compared to MediPix1 as MediPix2 works for electrons and holes. The MediPix2 has additional features compared to MediPix1. The MediPix2 detector has a preamplifier, two discriminators (window discriminator) and a 14-bit counter in each readout pixel cell chain. The presence of a second discriminator allows one to set an energy window with a lower and a upper threshold. The pixel counter is incremented only when the energy of the interacting quantum of radiation falls within this preset energy window. The assembled hybrid pixel device is glued on a printed circuit board (motherboard). There is an interface board between the MediPix2 hybrid pixel device and a Universal Serial Bus (USB), which is presently the most widespread PC interface [Vykydal, Jakubek and Pospisil (2006)]. All necessary detector support, including the detector bias source (up to 100 V), is integrated into one compact system ($64 \times 50 \times 20\,mm^3$). Power supplies are internally derived from the voltage provided by the USB connection, so no power source on the device is required and the bus provides the communication lines (signals, etc.). The interface board has been designed for connection to motherboards carrying one or four MediPix2 pixel devices, permitting, in the latter case, an increased sensitive area. The MediPix2 motherboard connected to the USB interface represents a volume of about $142 \times 50 \times 20\,mm^3$ for the MediPix2-USB device. The MediPix2-USB device is controlled by the pixelman software package using a PC via USB cables [Holy, Jakubek, Pospisil, Uher, Vavrik and Vykydal (2006)]. This software provides a set of plugin's which are used to control different features of the chip or of the experimental environment. The pixelman control can be extented up to 50 Medipix-USB devices, operating simultaneously.

An additional significant advantage of the USB interface is the support of back-side pulse processing. This feature enables the possibility to determine not only the position but also the energy deposited by the interacting particle. Each particle of ionizing radiation interacting in the depleted region of the MediPix2 sensor produces a specific amount of electron-hole pairs proportional to the energy lost by the particle in the sensor. If the total charge is large enough, it can be measured by a charge sensitive preamplifier and, after shaping, sampled by a fast ADC. This feature also allows the generation of a trigger pulse. Particle signals detected by the MediPix2 device can either be integrated in the pixel counters, or the pixel matrix can be read-out at such a frame rate that visualization of single particle events becomes possible. Hence, this circuit can process these signals in various modes either counting or tracking. The counting mode is usable for high detection rates (above $\sim 5 \times 10^3$ events per second per cm^2) when the number of interactions in individual pixels is determined at different thresholds yielding the fraction of radiation with the corresponding ionizing power. By setting thresholds high enough, this mode allows one to differentiate the signal from ionizing particles with poor detection efficiency compared to electrons or low energy X-rays, for instance. Calibration of the devices enables the conversion of the individual counts measured into fluxes of respective types of radiation and dose rates. The tracking mode is applicable at low detection rates. It permits an analysis of individual tracks or traces (with their pattern recognition) of interacting quanta of radiation. In this mode, the threshold is set just above noise level and short time acquisitions are done with the aim to avoid the overlap of characteristic tracks from different particles which then can be separately distinguished. So the tracking mode is based on electronic visualization of tracks and traces of individual quanta of radiation in the sensitive silicon volume.

MediPix2 devices have demonstrated specific capabilities for particle identification relying on pattern recognition techniques. In a single frame, different patterns, that can be associated to different types of particles can be identified with a great level of accuracy. However, the MediPix2 device cannot do a direct study of the energy deposition. This can be achieved with the newly developed TimePix device [Medipix Collab. (2008)]. From its time over threshold (TOT) mode of operation, the TimePix device can count the amount of charge being deposited in a given pixel element, among the 256×256 pixels available, and these counts can be translated into energy with a proper calibration. This added to the pattern recognition capabilities already available from MediPix2, it is possible to measure the amount of a certain type of radiation at a given position.

6.5.2   *Examples of Application: Particle Physics*

Images of MediPix2-USB device responses are characteristic of illuminations by specific radiation, resulting from different charge deposition mechanisms. Then, the MediPix2-USB device makes possible the detection in counting and tracking mode of

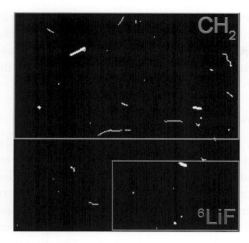

Fig. 6.23 The silicon sensor is covered on the upper half with a CH_2 layer (orange) to detect fast neutrons and part of the lower half with a ^6LiF layer (green) to detect thermal and very slow neutrons [Holy et al. (2006)]. Are seen: the long and thick tracks (Bragg-peak energy deposition) of recoiled protons and big tracks and clusters generated by nuclear products in LiF and via $^{28}Si(n,\alpha)^{25}Mg$ and $^{28}Si(n,p)^{28}Al$ nuclear reactions in silicon. Electrons and photons (always accompanying neutrons) are also seen.

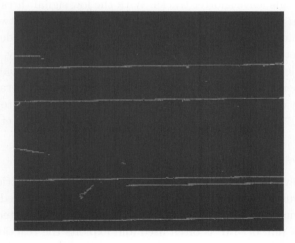

Fig. 6.24 Measurement of muons (mip's) at the CERN SPS with a MediPix2-USB device operated with a 20 V bias voltage [Idarraga (2008)]. The muons are striking the MediPix2 along a direction parallel to the device surface.

a variety of particles produced by collisions in the case of an accelerator [Campbell, Leroy, Pospisil and Suk (2006); Holy et al. (2006)] or subsequent interactions with matter of particles produced by a source. These capabilities can be exploited for real-time measurement of the composition and spectroscopic characteristics of a given radiation field.

MediPix2-USB devices have been intensively and successfully used for detection of single X-ray photons, electrons, minimally ionizing particles (mip's), energetic ions (protons, alpha particles, and other heavy charged ions such as carbon and oxygen) and neutrons. Furthermore, the applications for medical imaging are discussed in Sect. 11.4.

6.5.2.1 *Electrons and Photons*

X-rays or low energy signals generally produce hits in 1 or 2 pixels only (depending on their energy, on the pixel threshold set and other parameters at the base of the charge sharing mechanism (see later)). Beta particles can be identified through their erratically longer, curved tracks. The length of the electron tracks are consistent with the average energy transferred by photons per interaction (dominance of the photoelectric effect for energies lower than 50 keV, while Compton dominates for energies above 100 keV) [Fiederle et al. (2008)]. The direction of a particle beam can be determined from the measurement of impinging angles. Particles along a direction parallel or very close to being parallel to the surface of the MediPix2 device have long tracks inside the sensor material, i.e., tracks covering a number of pixels given by the particle range divided by 55 μm. Particle impinging on the pixels at larger angle leave shorter tracks. With the low signal threshold setting one observes adjacent hits when the particle is near to the border between pixels. This allows one to determine the direction of the particle, and provide coordinates. It is possible to determine the beam direction using the intensity decrease in the MediPix [Fiederle et al. (2008)]. The use of the interaction probability in silicon* for photons of energies from 10 keV up to 160 keV produced by X-ray sources allows the measurement of the photon detection efficiency that is found to range from ~ 1 at 10 keV and 10^{-2} at 160 keV (threshold set at 8 keV) [Fiederle et al. (2008)].

6.5.2.2 *Neutrons*

The MediPix2 device can be used for neutron detection. For detection of thermal or very slow neutrons, the silicon sensor is partially or totally covered with a neutron absorbing material which converts the energy into radiation detectable in silicon (see Sect. 6.7). A film of ^{6}LiF is often used and exploits the reaction

$$\text{n} +^6 \text{Li} \rightarrow \alpha \ (2.05 \, \text{MeV}) +^3 \text{H} \ (2.72 \, \text{MeV})$$

to detect thermal neutrons. The thermal neutron detection efficiency is $\leq 5\%$ for a converter 7 mg/cm^2 thick (see Sect. 6.7). For fast neutrons (or neutrons with kinetic energy higher than hundreds of keV) detection, the silicon sensor is partially or totally covered with a film of hydrogen-rich material, such as polyethylene (CH$_2$, 1.3 mm thick - see Sect. 6.7). The neutrons are detected by recoil protons in the elastic scattering of neutrons on nuclei in CH$_2$. In practice, the sensitive area of the

---

*The probability is equal to $1 - \exp(-\mu x)$, where x is the traversed silicon depth and μ is the sum of photon attenuation coefficients for the photoelectric and Compton effects, mainly.

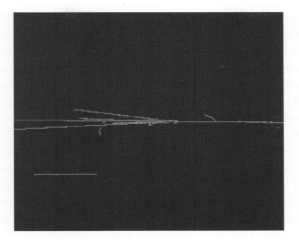

Fig. 6.25 Measurement of pions at the CERN SPS with a MediPix2-USB device operated with a 20 V bias voltage [Idarraga (2008)]. One can see for non-mip pions, the interaction pattern featuring large energy deposition around a vertex with recoil fragments remaining in the sensitive volume of the detector. δ-rays are also visible.

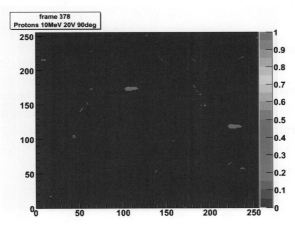

Fig. 6.26 Signature of 10 MeV protons measured at the Van de Graaff tandem of the University of Montreal with a MediPix2-USB device operated with a 20 V bias voltage. The protons after being Rutherford back-scattered on a gold foil are striking the MediPix2 postionned at parallel incidence angle [RBS (2008)]. The proton tracks show an energy deposition characteristic of standard asymmetric feature of a Bragg curve. The threshold was set at 13.3 keV.

MediPix2-USB device is divided in several regions: regions covered with one layer or combination of layers of ^{6}LiF, polyethylene, aluminium materials (aluminium for low energy electrons absorption) and one region is uncovered. A simple example is shown in Fig. 6.23 for a silicon sensor covered on the upper half with a CH_2 layer (orange) to detect fast neutrons and part of the lower half with a ^{6}LiF layer (green) to detect thermal and very slow neutrons. One can see long and rather thick tracks (Bragg peak energy deposition) of recoiled protons and big tracks and clusters generated by nuclear products in LiF and via $^{28}\mathrm{Si}(n,\alpha)^{25}\mathrm{Mg}$ and $^{28}\mathrm{Si}(n,p)^{28}\mathrm{Al}$ nuclear reactions in the body of the silicon detector. Electrons and photons are also seen. However, the huge background of photons which always accompany neutrons can be highly suppressed via threshold setting. Another example can be found with the MediPix2-USB devices used in the ATLAS detector at the CERN-LHC [Campbell, Leroy, Pospisil and Suk (2006); Holy et al. (2006)]. These devices will be exposed, in particular, to a neutron field and accompanying photon field. By comparing the responses from the different regions of a device, uncovered and covered by specific film and foils of given thicknesses, one can determine, in real-time, the spectral composition of the neutron field if one uses the calibration obtained from known sources of neutrons with specific energy, E_n, such as neutron beams from Van de Graaff accelerators ($E_n = 14$–$17\,\mathrm{MeV}$), fission sources ($E_n = 25\,\mathrm{meV}$), $^{252}\mathrm{Cf}$ ($E_n \sim 2\,\mathrm{MeV}$) and AmBe ($E_n \sim 4\,\mathrm{MeV}$) sources.

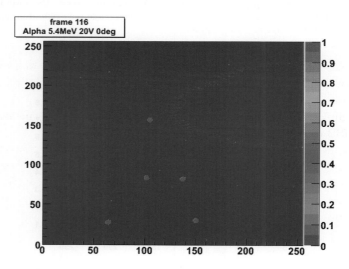

Fig. 6.27   Signature of $5.4\,\mathrm{MeV}$ α-particles from an $^{241}\mathrm{Am}$ source measured with a MediPix2-USB device operated with a $20\,\mathrm{V}$ bias voltage. The α-particles are striking the MediPix2 positioned at perpendicular incidence angle. The threshold was set at $5.6\,\mathrm{keV}$. The α-source was fit into a vacuum chamber of the Van de Graaff Tandem accelerator of the University of Montreal.

6.5.2.3   *Muons, Pions and Protons*

The capability of MediPix2-USB to allow the real-time measurement of microscopic details of muons trajectories, and pions interactions from an hadronic beam has been demonstrated at CERN-SPS [Field and Heijne (2007)] for a MediPix2-USB device placed parallel to the beam [Field and Heijne (2007)]. The muons resulting from the decay of pions produce typical minimum ionizing particle (mip) trails in the sensor. In absence of nuclear interactions the tracks of these mips in MediPix2 are continuous, crossing the rows of pixels of the detector matrix with regular stretches of overlapping rows (Fig. 6.24). Incoming non-mip pions produce interaction patterns characterized by large energy deposition around a vertex with recoil fragments traveling in the detector sensitive volume and leaving long tracks or short tracks depending on recoil angles (Fig. 6.25). δ-rays are also observed. It should be observed that mip pions and muons will leave the same long trails and cannot be distinguished, possibly causing mis-tagging. The same type of measurement can be performed with protons. Dedicated measurements with protons of low energies have been performed with a MediPix2-USB device [RBS (2008)]. An example of 10 MeV protons hitting this MediPix2-USB device along a direction parallel to its surface is shown in Fig. 6.26. Protons of that energy have a total range of up to $\sim 742\,\mu$m (including the longitudinal straggling) in silicon which represents a track length of $742/55 = 14$ pixels in the device.

6.5.2.4 α-Particles and Heavier Ions

Alpha particles and heavier ions deposit large amounts of energy in a very confined space. Therefore, the signature of this type of particle corresponds to a rather symmetric cluster of adjacent pixels.

The example of 5.4 MeV α-particles (from a ^{241}Am source) measured with a MediPix2-USB device is shown in Fig. 6.27.

The measurement with a MediPix2-USB device of carbon and oxygen ions produced from Rutherford back-scattering of 30 MeV carbon and 35 MeV oxygen beams on gold are shown in Fig. 6.28 [RBS (2008)]. Small dots can be observed in these figures. Their origin is explained by the Au X-rays (Kα and Kβ). They are generated via proton induced X-ray emission.

6.5.2.5 Charge Sharing

The shape of a track in a MediPix2-USB device depends on the nature of the interaction in the detector sensitive volume and is influenced by the charge sharing effect. This effect among adjacent pixels takes its origin from the lateral spreading of charges from the interaction of an ionizing particle in the silicon detector. This response to single particle hit is taking the form of a cluster of adjacent pixels. The charge deposited in an adjacent pixel is counted if higher than the threshold set for the comparator. The charge sharing effect can be used as an advantage allowing the improvement of the spatial resolution of the incident particle and exploited for tracking. One can take advantage of heavily ionizing particles (α-particles, carbon, oxygens ions, ...) to understand the effect.

Heavy charged particles produce dense carrier tracks in silicon detectors, as

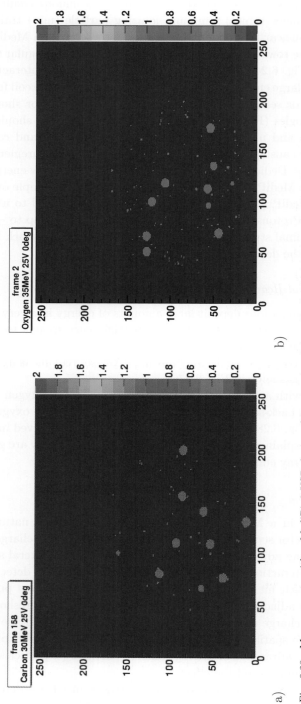

Fig. 6.28 Measurements with a MediPix2-USB device of ions produced from Rutherford back-scattering of ion beams on a gold foil: a) carbon ions (large dots) from a 30 MeV carbon beam and b) oxygen ions (large dots) from a 35 MeV oxygen beam. The ions were striking the MediPix2 perpendicularly. The measurements were performed at the Van de Graaff tandem accelerator of the University of Montreal [RBS (2008)]. The threshold was set at 13.3 keV for both measurements. The bias voltage was 25 V in both cases. Small dots can be observed in these figures. Their origin is explained by the Au X-rays (Kα and Kβ). They are generated via proton induced X-ray emission.

they deposit large amounts of energy in a very confined space. Under the influence of an electric field, the carriers drift towards the corresponding electrode of the device. The lateral spread of the charges are caused by several effects. A plasma effect is caused by the high density of charge carriers released in a column following the particle path of very small radius. When a heavy-charged particle enters a silicon detector, it deposits an amount of energy so large that the conditions to create a plasma are fulfilled. This condition is related to the Debye length (λ_D) which can be calculated using

$$\lambda_D = \sqrt{\frac{\epsilon kT}{q^2 n_c}}, \tag{6.114}$$

where ϵ is the silicon permittivity, k the Boltzman constant (8.617×10^{-5} eV/K), T the temperature, q the elementary charge, and n_c the carrier concentration. The column of carriers is considered to be in a plasma state if the Debye length is small compared with the plasma dimensions (the *plasma condition*). An α-particle of 5.4 MeV (^{241}Am in vacuum) of energy creates 1.5×10^6 pairs in a column 28 μm long with an initial radius of about 1 μm [Tove and Seibt (1967)]. The carrier concentration will therefore be:

$$n_c = \frac{n_{\text{pairs}}}{\pi r^2 l} = \frac{1.5 \times 10^6}{\pi \, (1 \times 10^{-6} \, \text{m})^2 (28 \times 10^{-6} \, \text{m})} = 1.7 \times 10^{22} \, \text{m}^{-3}.$$

The Debye length at room temperature (300 K) for this carrier concentration [Eq. (6.114)] is 0.03 μm, which is much smaller than the initial radius of the column, so fulfilling the *plasma condition*.

The plasma time, t_{p}, is given by the approximate expression [Dearnaley and Northrop (1966); Tove and Seibt (1967)]

$$t_{\text{p}} = n_{\text{lin}}{}^2 \frac{1}{(4\pi\epsilon)^2 E^2 D_a}, \tag{6.115}$$

where n_{lin} is the linear density of the plasma (electrons/cm). Using the dimensions of the initial column of charges, one finds $n_{\text{lin}} = 5.43 \times 10^{10}$ [e/m]. The ambipolar diffusion coefficient D_a can be calculated using:

$$D_a = \frac{2kT}{q} \frac{\mu_e \mu_h}{\mu_h + \mu_e} = 17.48 \, \text{cm}^2/\text{s} \tag{6.116}$$

with $\mu_h (= 450 \, \text{cm}^2 \, \text{V}^{-1} \, \text{s}^{-1})$ and $\mu_e (= 1350 \, \text{cm}^2 \, \text{V}^{-1} \, \text{s}^{-1})$ being the mobilities of holes and electrons in silicon, respectively. The value for the plasma time depends on the applied field, E. For a field strength of 10^5 V/m, one obtains $t_{\text{p}} = 2.5 \, \mu$s. In the case of MediPix2, the silicon sensitive device is 300 μm thick and has a resistivity of ~ 5 kΩcm with a full depletion voltage of ~ 20–25 Volts (value of the voltage to measure all the charges created in this detector by an α-particle of 5.4 MeV). In that case, the typical field strength before full depletion is ~ 10 V/300 μm giving a plasma time $t_{\text{p}} = 22 \, \mu$s. This local high concentration is the source of lateral diffusion which increases initially the cluster size at low field. Present at low field are

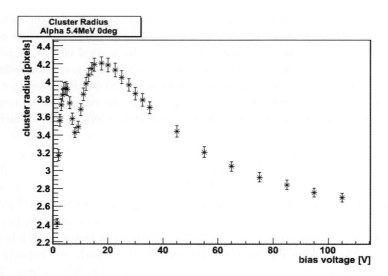

Fig. 6.29   Illustration of the Charge sharing induced by 5.4 MeV α-particles from an ^{241}Am source striking a MediPix2-USB detector perpendicularly. The cluster radius (in pixels) is shown as a function of the applied bias voltage (in volts). The threshold was set at 5.6 keV. The α-source was fit into a vacuum chamber of the Van de Graaff Tandem accelerator of the University of Montreal.

the erosion and funneling effects [Tove and Seibt (1967)]. The erosion effect is the removal of the carriers located at the periphery of the column by the field present in the detector. The longitudinal field forces the outer carriers towards the collecting electrodes thus reducing the concentration of carriers in the plasma. The *funneling effect* [Dearnaley and Northrop (1966); Tove and Seibt (1967)] is also present as it is caused by the alteration of the local field within the plasma. Carriers at the center of the column only feel the electric field from the carriers in the plasma, escape recombination and are funneled to the collecting electrode. Funneling and diffusion reduce the carrier concentration in the plasma. As the electric field near the depleted region is modified, charges inside the column are pushed towards the electrode, decreasing the charge density inside the column hence decreasing the lateral spread. When the depleted region reaches the particle track, more of the charge diffusing laterally is pulled by the detector electric field to the electrode. Therefore, the charge deposited increases at the edges of the cluster and can pass the low threshold of the MediPix2. Finally, when all of the particle track is included in the depleted region, the radius of the cluster decreases because of the increasing strength of the longitudinal electric field. An example of cluster radius (in pixels) as function of the applied bias voltage is shown in Fig. 6.29 for 5.4 MeV α-particles perpendicularly incident on a MediPix2 detector. The shape of the curve shown in this figure is explained along the line developed above: i) first increase in cluster size: the α-particle entering from the back, this shape is mainly due to diffusion;

ii) first decrease in cluster size: the funneling effect sets in. The phenomenon is important when the column approaches the depleted area. The electric field near the depleted region is modified. Charges inside the column are pushed towards the electrode, decreasing the charge density inside the column hence decreasing the lateral spread; iii) second increase in cluster size: the depleted region reaches the alpha particle track. More of the charge diffusing laterally is pulled by the detector's electric field to the electrode. Therefore, the charge deposited increases at the edges of the cluster and can pass the low threshold of the Medipix; iv) second decrease in cluster size: all of the particle track is included in the depleted region. The size of the cluster decreases because of the increasing strength of the longitudinal electric field. A model describing the effects of plasma, diffusion and funneling on the charge collection and charge sharing can be found in [Campbell et al. (2008)].

6.6 Photovoltaic and Solar Cells

A photovoltaic cell is a device which exploits the photovoltaic effect (effect discovered in 1839 by Alexandre-Edmond Becquerel) that converts the electromagnetic energy from a source of light into electrical energy. When the source of light is the sun one speaks about a solar cell. These cells have many applications in current life and activities such as powering electronic calculators, solar modules which generate home electricity, Earth-orbiting satellites, space stations and probes. The principle of operation of a solar cell follows four requirements: i) photogeneration of charge carriers (electrons and holes); ii) separation of charge carriers; iii) transport of electrons to a conductive (metal) contact that will transmit the electricity (wire, for instance); iv) does not require an external voltage source to operate unlike radiation/particles detectors that require a dedicated power supply. The case of solar cells will be explicitly discussed below. In order for a cell to generate power, a voltage and a current need to be generated. Usually, a solar cell is a large area (typically $10 \, \text{cm} \times 10 \, \text{cm}$) silicon $p-n$ junction (or an $n-p$ junction, see below) made by diffusing a n-type dopant into the p-type side of a silicon wafer. Photons (of wavelength less than $1.13 \, \mu\text{m}$ corresponding to a solar energy intensity $\leq 250 \, W \, m^{-2} \, \mu m^{-1}$ at the Earth's surface) from solar radiation falling on the junction generate electron-hole pairs in a light absorbing material contained within the junction structure.

A photon in a solar cell can generate electron-hole pairs if it has an energy equal or larger than the band gap of the semiconductor, here silicon (band gap of $1.12 \, \text{eV}$). However, if the photon has an energy above the band gap, the excess energy above the band gap is lost as heat. Photons with energy smaller than the band gap are not absorbed and are lost. The value of the band gap determines the possible maximum current generated by the cell. The number G of carriers generated depends on the number of incident photons N_{ph}, on the absorption coefficient α of the material, on the incident photon wavelength λ, on the thickness x of the material and on the number of photons N_s at the surface of the material ($N_{\text{ph}} = N_s$ at

$x = 0$). The number of photons is

$$N_{\text{ph}} = N_s\, e^{-\alpha x} \tag{6.117}$$

and G is given by

$$G = -\frac{dN_{\text{ph}}}{dx} = \alpha N_s\, e^{-\alpha x}(1 - R) = \alpha N_{\text{ph}}(1 - R), \tag{6.118}$$

where one has introduced R, the reflectivity at the surface of the cell ($0 \le R \le 1$).

A $p - n$ junction based solar cell consists of a thick n-type crystal covered by a thin p-type layer which will be exposed to sunlight. An external electrical circuit with a load resistance R_{L} through which the electrons will flow is connected across the junction. In order to achieve the so-called forward bias mode condition, the n-side is connected to the positive terminal of the circuit and the p-side is connected to the negative terminal. After passing through the load, the electrons annihilate with holes at the rear contact (p-side), so completing the circuit.

Nowadays due to their greater radiation hardness, solar cells based on $n - p$ junctions are used rather than $p - n$ junctions [van Lint, Flanahan, Leadon, Naber and Rogers (1980)]. A schematic view of such a cell is shown in Fig. 6.30. The cell is covered with an antireflection layer to maximize the incoming light absorption (R as small as possible in Eq. (6.118). fraction of 1 micron) which is an emitter region, an $n - p$ junction layer and a back junction layer (a p-type silicon layer) which is a base region. Two electrical contact layers made of metal are needed to allow the electric current to flow into and out of the cell. The contact layer of the cell, facing incident photons, must present a widely spaced metallic grid pattern to allow light to enter the cell. It also allows reduction of the internal resistance. The back metal contact layer covers the entire back surface of the cell structure. An external load (resistance R_{L}) is connected to the cell. With this structure, the electrons generated in the cell by the conversion of the incoming photons with sufficiently high energy will be freed in the p-type layer and the electric field will send the electrons to the n-side of the junction and the hole to the p-side causing disruption of the electrical neutrality. Then, the electrons cross the now thin depletion region and flowing through a connection wire invade the whole n-type material and reach the negative battery terminal. The external circuit, with the load resistance R_{L}, provides a path through which the electrons can return to the other layer and a current flow is produced that will continue as long as the light strikes the cell. The depth of the junction is chosen so that all the carriers are generated in the base region and none in the emitter region or front face, in general. This is true for light with large wavelength, i.e. typically higher than $0.6\,\mu$m. For light of lower wavelength ($\sim 0.4\,\mu$m, i.e., blue light) corresponding to small absorption coefficients (0.12 (0.8) μm for $\lambda = 0.4$ (0.5) μm), carriers are generated in the front face and diffuse to the surface of the cell where they recombine and do not contribute to the cell current. The product of the voltage V, given by the cell electric field and the current I, caused by the electrons flow is the power W of the cell. The voltage given by the cell electric field is low, typically 0.5–0.6 volt. For instance, a $10 \times 10 \, \text{cm}^2$ cell exposed to direct sunlight with clear skies produces a voltage of ~ 0.5 volt and a current of $\sim 2.4 \, \text{A}$, giving a power of $\sim 1.2 \, \text{W}$. Then, modules composed of several cells have to be used in order to increase the power production. Cells are connected together in series and encapsulated into modules. Usually, a module contains 36 cells to generate an output voltage of 12–18 volts, typically. These modules can be used singly or connected in series and parallel into arrays with larger current and voltage output. The sunlight illumination varies during the day due to the variation of the sunrays incidence on the device and depends on meteorogical conditions. Then, the modules must be integrated with a charge storage system (battery) and with components for power regulation. The battery is used for storing the charge generated during sunny periods and the power regulating components ensure that the power supply is regular and minimize the sensitivity to variations in solar irradiation.

For a large part, we follow now a development that can be found in [van Lint, Flanahan, Leadon, Naber and Rogers (1980); Wahab (1999); Van Zeghbroeck (2007)], for instance. When the light is striking the cell, a photocurrent I_{ph} due to light generated electron-hole pairs is produced opposite to the diode current I_j. One can express the diode current I_j as [Wahab (1999); Van Zeghbroeck (2007)]

$$I_j = I_s \left(e^{qV_0/kT} - 1 \right), \tag{6.119}$$

where I_s (1 pA, typically) and V_0 are the saturation current of the diode and the voltage across the load (or the voltage applied to the junction if the internal resistance is zero), respectively. Therefore, in the presence of light when the load resistance $R_L = 0$, I is the difference

$$I = I_j - I_{\text{ph}}. \tag{6.120}$$

The short-circuit current, I_{sc}, is the current at zero voltage and is equal to I_{ph}. Then, it can be calculated from Eqs. (6.120), (6.119) with

$$I_{\text{ph}} = \frac{q}{h\nu} P_{in} (1 - R) \left(1 - e^{-\alpha x} \right),$$

where P_{in} is the incident optical power and is typically $100 \, \text{mW} \, \text{cm}^{-2}$ for sunlight (average wavelength of about 570 nm). The value $I_{\text{ph}} \sim 25 \, \text{mA/cm}^2$ is found (for $R_L = 0$). This value is consistent with the more precise calculation [van Lint, Flanahan, Leadon, Naber and Rogers (1980)] where a cell of 2 cm^2 area was considered. The exact calculation of this current has been done in [van Lint, Flanahan, Leadon, Naber and Rogers (1980)] taking into account the dependence on the irradiating light wavelength (from 0.5 up to 0.9 μm as the major contribution to the light spectrum). When the load resistance, R_L, is increased an increasing forward bias V_F is applied to the junction and the current is reduced as the potential barrier is lowered. If now $R_L \to \infty$, one has an open circuit. If there is a load resistance R_L connected (in series) with the solar cell, a voltage $V_0 = I \, R_L$ is generated across the load and then

$$I = I_s \, e^{qV_0/kT} - I_s - I_{\text{ph}}. \tag{6.121}$$

For the open circuit, $I = 0$ and $V_0 (I = 0) = V_{oc}$, so that

$$I = I_s \, e^{qV_{oc}/kT} - I_s - I_{ph} = 0. \tag{6.122}$$

Thus,

$$V_{oc} = \frac{kT}{q} \ln \left(\frac{I_s + I_{ph}}{I_s} \right) \sim \frac{kT}{q} \ln \left(\frac{I_{ph}}{I_s} \right). \tag{6.123}$$

From Eq. (6.123) one finds the corresponding value of V_{oc} to be

$$V_{oc} \sim 0.62 \, \text{V} \tag{6.124}$$

($I_{ph} = 25\,\text{mA}$ for an area of $1\,\text{cm}^2$, $I_s = 1.0 \times 10^{-12}\,\text{A}$, $kT/q = 0.0259\,\text{V}$).

The power generated by the solar cell is

$$P = V_0 I = V_0 I_s \left(e^{qV_0/kT} - 1 \right) - V_0 I_{ph}. \tag{6.125}$$

The optimal power release of a solar cell can be calculated from the condition $dP/dV_0 = 0$. If V_M and I_M are the voltage and current corresponding to the optimal power, respectively:

$$\left(\frac{dP}{dV_0} \right)_{V_0 = V_M} = I_s \left(e^{qV_M/kT} - 1 \right) - I_{ph} + I_s \frac{q}{kT} V_M \, e^{qV_M/kT} = 0. \tag{6.126}$$

From which we can extract the voltage corresponding to the maximum power, i.e.,

$$V_M = \frac{kT}{q} \ln \left[\frac{I_{ph} + I_s}{I_s} \frac{1}{1 + qV_M/(kT)} \right]. \tag{6.127}$$

Using Eq. (6.124) and the approximation therein, one finds the transcendental equation which has to be solved by iterations

$$V_M \sim V_{oc} - \frac{kT}{q} \ln \left(1 + \frac{qV_M}{kT} \right). \tag{6.128}$$

With a starting value $V = 0.5\,\text{V}$ in Eq. (6.128), one finds a value of V_M converging to 0.540 ($V_M = 0.50, 0.542, 0.540, 0.540$). The value of I_M is obtained from

$$I_M = I_{ph} \left(1 - \frac{kT}{qV_M} \right) \tag{6.129}$$

or

$$I_M = 24\,\text{mA} \tag{6.130}$$

using $V_M = 0.540\,\text{V}$.

Solar cells are characterized by several efficiency factors.

- The energy conversion efficiency which is defined as

$$\eta = \left| \frac{V_M I_M}{P_{in}} \right| = \left| \frac{P_M}{P_{in}} \right|. \tag{6.131}$$

The ratio η represents the percentage of power converted from absorbed light into electrical energy and collected once the solar cell is connected to an electrical circuit as described above. η is the ratio of the maximum

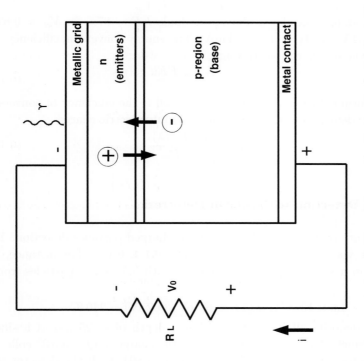

Fig. 6.30 A standard schematic view of an $n - p$ solar cell. The cell is covered with an antireflection layer. The photovoltaic effect occurs in three energy-conversion layers: a top junction layer (a n-type silicon layer), an absorption layer (an $p - n$ junction) and a back junction layer (a p-type silicon layer). Two electrical contact metallic layers allow the electric current to flow into and out of the cell. The front face contact layer present a metallic grid pattern to allow light to enter the cell. The metal back contact layer cover the entire back surface of the cell structure. An external load (load resistivity $R_{\rm L}$) is connected to the cell.

power $P_M = V_M \times I_M$ to the input light irradiance P_{in} expressed in $\mathrm{W\,m^{-2}}$ (in STC conditions, i.e., at a temperature of 25°C and an irradiance of $1\,\mathrm{kW\,m^{-2}}$ with an air mass spectrum 1.5 - technically called AM1.5) [ASTM (2007)] and the surface area of the solar cell expressed in $\mathrm{m^2}$. In the example considered above, one finds (for $P_{in} = 100\,\mathrm{mW\,cm^{-2}}$, $V_M = 0.54\,\mathrm{V}$ and $I_M = 25\,\mathrm{mA}$) $\eta \sim 13\%$.

- The Fill Factor FF which is defined as the ratio of the maximum power of the solar cell to the product of the open-circuit voltage, V_{oc}, and the short circuit current, $I_{sc} \sim I_{\mathrm{ph}}$, i.e.,

$$FF = \frac{V_M I_M}{I_{sc} V_{oc}} = \frac{P_M}{I_{sc} V_{oc}}. \tag{6.132}$$

In the example considered above, one finds $FF \sim 87\%$ for $V_{oc} = 0.62\,\mathrm{V}$, $V_M = 0.54\,\mathrm{V}$ and $I_M \sim I_{sc} \sim 24\,\mathrm{mA}$). The energy conversion efficiency and the Fill Factor are related through:

$$\eta = \frac{I_{sc} V_{oc} FF}{P_{in}}. \tag{6.133}$$

- The quantum efficiency QE of the solar cell is the efficiency of conversion of the incident photons (of wavelength λ) into electric charge:

$$QE = \frac{h\nu}{q} \frac{Isc(\lambda)}{P_{in}}. \tag{6.134}$$

6.7   Neutrons Detection with Silicon Detectors

Silicon detectors are pratically 100% efficient for charged particles detection. This detection is done via direct ionization (see Sects. 6.1.4, 6.1.5). For instance, the charge collected in a silicon detector illuminated with 5.5 MeV α-particles from a ^{241}Am-source is

$$Q = 1.6 \times 10^{-19}\,\mathrm{C} \times (5.5 \times 10^6\,\mathrm{eV}/3.62\,\mathrm{eV}) = 240\,\mathrm{fC},$$

as an α-particle deposits all its energy within a depth of $\sim 25\ \mu\mathrm{m}$ of a silicon detector 300 $\mu\mathrm{m}$ thick. This can be compared to the charge of $Q = 3.5\,\mathrm{fC}$ collected in the same detector for a *mip*[†] which will deposit $\sim 80\,\mathrm{keV}$. If the detector is in good operational condition (not damaged by irradiation or wrong manipulation) this charge will be collected at full depletion voltage. The collection of 100% of the charge will still be possible with a damaged silicon detector provided the bias voltage is increased beyond the value of the full depletion voltage determined for the undamaged detector.

Silicon detectors can detect neutral particles via the measurement of indirect ionization (for photons) or secondary radiation (for neutrons). The detection efficiency is much lower than the one obtained for charged particles in direct ionization (see for instance [Sadrozinski (2000)]). For instance, photons are detected

---

[†]The *mip* is a minimum ionizing particle and was defined at page 47.

Table 6.3   A few examples of neutron reactions with the nuclei of several converters. The probability of these reactions to occur is depending on their energy threshold as compared to the energy of the incoming particle. For ^{10}B(n,α)^{7}Li reaction 93.7% of the reaction leave ^{7}Li in its first excited state, which rapidly de-excites to the ground state (in $\sim$ 100 fs) by emission of a 480 keV γ (I). The remaining 6.3% of the reactions leave ^{7}Li in its ground state (II). The Q-values are from [NNDC (2008c)].

Reaction	Q-value (MeV)	σ (barn)
n+^{6}Li→α(2.05 MeV)+ ^{3}H(2.71 MeV)	4.76	940.[a]
n+^{10}B→α(1.47 MeV)+ ^{7}Li(0.84 MeV)+ γ(0.48 MeV) (I)	2.79	3571.[b]
n+^{10}B→α(1.78 MeV)+ ^{7}Li(1.01 MeV) (II)	2.79	269.[c]
n+^{113}Cd→^{114}Cd + γ(0.56 MeV) + conversion e$^-$	9.0	
n+^{155}Gd→^{156}Gd + γ(0.09, 0.20, 0.30 MeV) + conversion e$^-$	8.5	
n+^{157}Gd→^{158}Gd + γ(0.08, 0.18, 0.28 MeV) + conversion e$^-$	7.9	

[a]: for 0.025 eV (thermal) neutrons.
[b]: 93.7%.
[c]: 6.3%.

through electrons emitted via photoelectric effect (Sect. 2.3.1), Compton scattering (Sect. 2.3.2), and pair production (Sect. 2.3.3). The relative importance of these processes of interaction strongly depends on the material and photon energy. For instance, for silicon, the photoelectric effect is dominant at photon energy, typically lower than 60 keV. Over the energy region from 80 keV up to a few MeV the dominant process is the Compton scattering. At photon energy higher than 1.022 MeV, pair production rapidly increases. The detection efficiency is low as the photon interaction cross sections are themselves low. For instance at 500 keV, assuming a silicon detector $d = 300\,\mu$m thick, fully depleted, the probability, P_C, for Compton interaction is

$$P_C = 1 - \exp\left[-\rho\, d\, \sigma_c(E_\gamma)\right] = 0.5\%$$

($\rho = 4.2 \times 10^{22}$ atoms/cm^3, $\sigma_c(E_\gamma) \sim 4$ barns/atoms). The photoelectric effect (photo-absorption) at that energy is less than 1% of the Compton scattering.

Then, silicon detectors can also be used when low photon detection efficiency is acceptable or when lower energy X-rays (e.g., ≤ 60 keV) have to be measured as it is the case for X-rays imaging, for instance (see Sect. 6.5.1).

6.7.1   *Principles of Neutron Detection with Silicon Detectors*

The detection of neutrons with a silicon detector exploits the interactions of these neutrons with the nuclei of a converter layer adjacent to the detector diode, covering partially or fully the area of the detector active face (Fig. 6.31). Secondary heavy charged particles are produced in the converter material either via nuclear reactions or as energetic recoils from elastic scattering of neutrons on nuclei. These secondary heavy charged particles generated in the converter must have a range larger than the distance between the interaction location in the converter and the converter layer-detector interface, in order to reach and penetrate the silicon layer where they deposit their energy. Charge carriers are then generated in silicon yielding a signal. The particles trajectories must be contained in the solid angle subtending the converter layer-detector interface from the interaction location in the converter. The type of converter used and therefore the mechanism of charged secondary particles generation exploited depend on the neutron energy. Thermal (very slow) neutrons[*] have huge nuclear reaction cross sections and, therefore, converters such as ^{6}LiF, ^{10}B, ^{113}Cd, ^{155}Gd, ^{157}Gd are of standard use for neutrons of these energies. Examples of nuclear reactions of interest are given in Table 6.3.

One should mention also the possibility of using solid-form or bulk detectors [McGregor and Shultis (2004)] for detecting thermal neutrons. These detectors use semiconductor materials partially composed of a neutron reactive material such as boron. In these detectors neutron interactions occur inside the bulk detector itself. These devices consist of a boron compound upon which conductive contacts have been affixed on opposite sides. A voltage is applied across the bulk material. Then, neutrons can be absorbed directly within the detector volume and the charged particle reaction products are released within the detector volume. Any ^{10}B$(n,\alpha)^7$Li reaction in the bulk produces detectable ionization and, since the thermal neutron boron cross section (Table 6.3) dominates over thermal cross sections of other elements, one can define [McGregor and Shultis (2004)] the intrinsic efficiency, $\text{eff}_{\text{th}-\text{n}}$, for detecting thermal neutrons by:

$$\text{eff}_{\text{th}-\text{n}} = 1 - e^{-\Sigma\,l}, \tag{6.135}$$

where Σ is the thermal-averaged macroscopic neutron absorption cross section of boron in the bulk and l is the thickness of the detector. The production of such bulk detectors and their response to thermal or very slow neutrons are discussed in [McGregor and Shultis (2004)].

However, the neutron capture reactions, as the ones reported in Table 6.3, have a cross section dropping significantly for higher energies of neutrons (i.e., fast neutrons such as neutrons of energies from several keV to several MeV and more). Another type of converter has to be chosen and a different mechanism of charged secondary particles generation has to be exploited. Fast neutrons can be detected via energetic recoils produced from their elastic scattering on nuclei of a converter. The energy transferred in elastic scattering from a neutron to a recoil nucleus is the recoil energy given by:

$$E_R = E_n \frac{4 M_n M_R}{(M_n + M_R)^2} \cos^2 \theta, \tag{6.136}$$

where E_n is the incident neutron energy, M_n the neutron mass, M_R the recoil nucleus mass, and θ is the angle in between incident neutron and recoil nucleus directions (laboratory frame). The energy transfer is maximum for $\theta = 0$, i.e, for

---

[*]The reader can find a classification of neutrons in Sect. 4.1.3.2. In addition, the total and elestic corss sections for netron–silicon scattering are shown in Sect. 4.1.3.1.

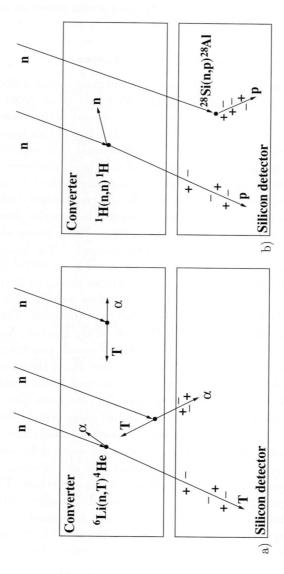

Fig. 6.31 Principle of converter-silicon operation: a) ^{6}LiF converter, α and tritium (T) produced by neutron interactions with the converter layer deposit their energy in silicon. T is dominant (see text); b) CH$_2$ converter, proton recoil produced by the elastic scattering of a neutron on the protons of the CH$_2$ converter. A fraction of incident neutrons may traverse the converter film and interact directly with Si.

neutron–nucleus head-on collisions. Then, the energy of the incident neutron is totally transferred for $M_n = M_R$, i.e., $E_n = E_R$. Therefore, for fast neutrons, it is advantageous to select converters with high hydrogen content and restrain the amount of heavier atoms which reduce E_R and absorb the energy of charged secondaries. Polyethylene (CH_2) appears to be an optimal choice as converter for fast neutrons.

6.7.1.1 *Signal in Silicon Detectors for Thermal Neutrons*

Standard assumptions are made to estimate the signal generated in a silicon detector by thermal neutrons [McGregor et al. (2003); Wielunski et al. (2004)]. A silicon detector, $300\,\mu$m thick with no dead layer, is considered with a converter layer put in front (Fig. 6.31). As reminded above, the efficiency of this type of detector for heavy charged particles is 100%. The flux of thermal neutrons (Sect. 4.1.3.2) is assumed to be independent of the depth d of the (thin) converter. The heavy charged particles produced by the thermal neutron interactions in the converter material have their direction of emission kinematically constrained. Only one of the charged particle reaction products, which are emitted in opposite directions, may cross the converter layer-detector interface into silicon. Being heavy, these charged particles practically travel in straight lines in the converter. The range, $R_i(E)$, of these secondary particles in the converter is calculated from SRIM [Ziegler, J.F. and M.D. and Biersack (2008b)]. Only charged particles produced in the converter at a distance from the converter layer-detector interface smaller than their range $R_i(E)$ in the converter will contribute to the signal in the detector. The probability of signal generation will then depend on $R_i(E)$ and on the number, N_n, of incident neutrons on the converter. It will also depend on the atomic concentration, N_i, of target-isotopes in the converter and on the cross section, σ_i, of the neutron-i-target isotope reaction. Coming back to the example of thermal neutron reactions with ^{6}LiF, as the α-particle and triton travel in opposite directions due to the small kinetic energy of the thermal neutrons, only one of them will be detected from one neutron interaction (Fig. 6.31). In practice, α-particles which have larger losses can hardly leave the converter. Therefore, most of the α-particles are stopped in the converter and do not contribute to the signal (see also [Pospisil (1993)]). In Fig. 6.31 is also illustrated the case where the α-particle and triton (T) are moving in opposite direction parallel to the silicon surface and, therefore, never reach the active volume of the detector and cannot contribute to any signal.

Generally, summing over all particles, $i = \alpha$, T, produced in the neutron reaction, the probability of signal generation can then be expressed as:

$$S = \Sigma_{i=\alpha,\text{T}}\, S_i, \qquad (6.137)$$

where

$$S_i = N_n\, N_i\, \sigma_i\, R_i P. \qquad (6.138)$$

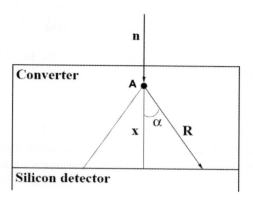

Fig. 6.32   The thermal neutron is produced in A at a distance x from the interface converter-detector. The opening angle of the cone is $\alpha = \arccos(x/R)$.

The factor P in Eq. (6.138) represents the total or integrated geometrical probability that any charge particle generated in the converter reaches the detector and produces a signal. This probability is $P = 0.25$, as calculated below. The point (vertex), A, is the source of isotropic emission of particles produced by the reaction of the thermal neutron with an atom of the converter (Fig. 6.32). These particles are produced in a cone of opening angle α and a particle emitted at angle α stops after the range R in the converter when its energy has gone down to zero (Fig. 6.32). If point A is located at a distance x from the cone basis, i.e., the converter layer-silicon detector interface, one has:

$$\cos \alpha = \frac{x}{R} \tag{6.139}$$

or

$$\alpha = \arccos \left(\frac{x}{R} \right). \tag{6.140}$$

The solid angle, SA, of particle emission is then:

$$SA = \int_0^{2\pi} \int_0^{\alpha} \sin \theta \, d\theta \, d\phi = 2\pi \left(1 - \cos \alpha\right) \tag{6.141}$$

compared to the total sphere solid angle, $SA_{\text{sphere}} = 4\pi$.

Then, the probability, P, that a particle emitted at point, A, with a range R in the converter reaches the converter layer-silicon detector interface at a distance x is

$$\begin{aligned} P(x, R) &= \frac{SA}{SA_{\text{sphere}}} \\ &= \frac{2\pi \left(1 - \cos \alpha\right)}{4\pi} \\ &= \frac{1}{2} \left(1 - \frac{x}{R}\right) \end{aligned} \tag{6.142}$$

or, introducing the variable $\eta = x/R$,

$$P(\eta) = 0.5(1 - \eta). \tag{6.143}$$

Integrating from $\eta = 0$ (origin of the particle at $x = 0$) up to $\eta = 1$ (origin of the particle at $x = R$), one obtains:

$$P \sim 0.5 \int_0^1 P(\eta) \, d\eta = 0.25. \tag{6.144}$$

The factor P in Eq. (6.138) representing the total or integrated geometrical probability that any charge particle generated in the converter goes in a direction opposite to the converter layer-silicon detector interface shown in Fig. 6.32 can be expressed as:

$$P(d - x, R) = \frac{1}{2} \left(1 - \frac{d - x}{R} \right). \tag{6.145}$$

Equation (6.145) is needed when one considers the case of a silicon detector inserted in between two converter layers.

Using Eqs. (6.137, 6.138) the probability of signal generation per incident neutron can be expressed as:

$$\frac{S}{N_n} = \Sigma_i \frac{S_i}{N_n} = \Sigma_i N_i \sigma_i R_i P. \tag{6.146}$$

As an example of estimate one can calculate the signal expected from thermal neutrons (neutron energy of $0.025\,\mathrm{eV}$) using a $^6\mathrm{LiF}$ converter (the Fluorine absorption cross section of $9.6\,\mathrm{mb}$ can be neglected compared to the $^6\mathrm{LiF}$ cross section of $940\,\mathrm{b}$). One can see in Table 6.3, that from neutron interaction with $^6\mathrm{LiF}$, charged particles possibly contributing to the signal are α ($2.05\,\mathrm{MeV}$) and $^3\mathrm{H}$ ($2.71\,\mathrm{MeV}$).

The atomic concentration of target-isotopes in $^6\mathrm{LiF}$ is

$$N_{^6\mathrm{LiF}} = \frac{N\rho}{A} = 6.022 \times 10^{23} \frac{2.635}{25.94} = 6.12 \times 10^{22} \,\mathrm{atoms/cm}^3,$$

where N is the *Avogadro constant* (see Appendix A.2). The ranges of α ($2.05\,\mathrm{MeV}$) and $^3\mathrm{H}$ ($2.71\,\mathrm{MeV}$) in $^6\mathrm{LiF}$, calculated with SRIM [Ziegler, J.F. and M.D. and Biersack (2008b)], are $6.05\,\mu\mathrm{m}$ and $31.67\,\mu\mathrm{m}$, respectively. Then, one finds using Eq. (6.146)

$$\frac{S}{N_n} = \frac{S_{3H}}{N_n} + \frac{S_\alpha}{N_n} = 4.56 \times 10^{-2} + 8.74 \times 10^{-3} = 5.3 \times 10^{-2}. \tag{6.147}$$

The range of the heavy charged particles is limited ($6.05\,\mu\mathrm{m}$ in case of alpha particles and $31.67\,\mu\mathrm{m}$ for tritons). Therefore, if a neutron is captured further away from the converter-silicon boundary, the heavy charged particles do not reach the sensitive volume of the silicon detector. Moreover, as already noted before, some of the heavy charged particles also travel parallel to the silicon surface and do not reach the detector active volume. The detection efficiency increases initially with increasing converter thickness because more neutrons are captured in the converter

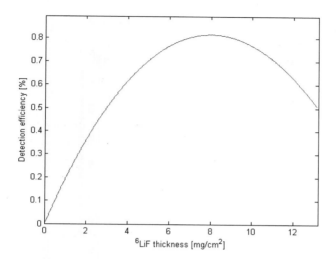

Fig. 6.33 Thermal neutron detection efficiency is shown as a function of the ^{6}LiF converter thickness [Gutierrez (2007)]. The LiF converter optimal thickness occurs at $8\,\text{mg/cm}^2$.

material. However, the detection efficiency starts to decrease at a value, d_{opt}, of LiF thickness. The value of d_{opt} is found by using Eq.(6.145), for instance. The detection efficiency, dP_{det}, of a neutron by the system of a converter layer and an adjacent silicon detector is given by

$$dP_{\text{det}} = \exp(-\mu x)\frac{1}{2}\left(1 - \frac{d-x}{R}\right)\mu\,dx, \tag{6.148}$$

where $\mu\,dx$ is the probability for the interaction of a neutron with a nucleus in the converter material at depth x; $\exp(-\mu x)$ is the probability that a neutron will traverse a path of length x without interaction; $1/2\,[1 - (d - x)/R]$ is the probability that the products of the interaction reaction of the neutron with a nucleus of the converter material will escape into the silicon detector. It can be defined as the ratio between the solid angle defined by the cone whose generatrix equals the path of the secondary particles in the converter material and the total solid angle [Eq. (6.145)]. Integrating Eq. (6.148) over the converter thickness (d) one finds

$$P_{\text{det}} = \frac{1}{2}\left[-\exp(-\mu d) + 1 - \frac{\exp(-\mu d)}{\mu R} - \frac{d}{R} + \frac{1}{\mu R}\right]. \tag{6.149}$$

To find the converter optimal thickness, d_{opt}, one imposes the condition

$$\frac{dP_{\text{det}}}{d(d)} = 0, \tag{6.150}$$

i.e.,

$$d_{\text{opt}} = \frac{1}{\mu}\ln(\mu R + 1). \tag{6.151}$$

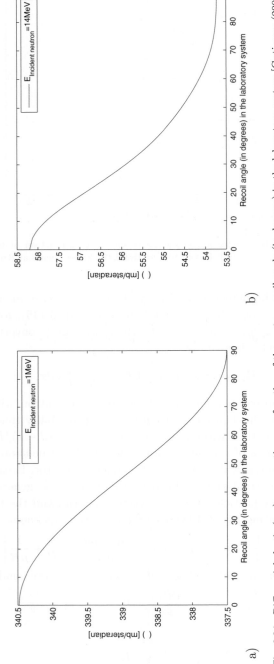

Fig. 6.34 Differential elastic (n,p) cross section as a function of the proton recoil angle (in degrees) in the laboratory system [Gutierrez (2007)]: for an incoming neutron a) of 1 MeV and b) 14 MeV.

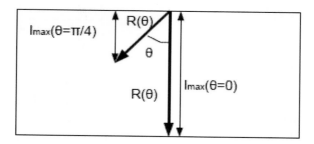

Fig. 6.35   Converter maximal thickness as a function of the proton recoil angle. The incoming neutrons follow a direction along the normal to the converter area [Gutierrez (2007)].

The tritium signal being about 5 times larger than the α signal (Eq. (6.147)), the neutron detection efficiency will be maximum when using the ^{6}LiF optimal thickness for tritium. Equation (6.151) gives for tritium an optimal ^{6}LiF converter thickness

$$d_{\mathrm{opt}} = 30.4\,\mu m = 8.0\,\mathrm{mg/cm}^2. \qquad (6.152)$$

The detection efficiency is shown in Fig. 6.33 as a function of the ^{6}LiF converter thickness [Gutierrez (2007)]. A maximum clearly occurs at the thickness of $8\,\mathrm{mg/cm}^2$ which is the LiF converter optimal thickness. The result is in excellent agreement with the results of Monte Carlo simulations [Uher et al. (2007a)]. For a boron converter, there are four contributions to the signal (Table 6.3) and Eq. (6.138) gives:

$$\frac{S}{N_{\mathrm n}} = \frac{S_{\mathrm{Li,I}}}{N_{\mathrm n}} + \frac{S_{\alpha,\mathrm I}}{N_{\mathrm n}} + \frac{S_{\mathrm{Li,II}}}{N_{\mathrm n}} + \frac{S_{\alpha,\mathrm{II}}}{N_{\mathrm n}}. \qquad (6.153)$$

If one uses

$$N_{^{10}\mathrm B} = \frac{N\,\rho}{A} = 13.03 \times 10^{22}\,\mathrm{atoms/cm}^3$$

and the values of the cross sections for the neutron nuclear reactions in the ^{10}B converter. i.e.,

$$\sigma = 3571\,\mathrm b \text{ for } \mathrm n + {}^{10}\mathrm B \to \alpha\ (1.47\,\mathrm{MeV}) + {}^7\mathrm{Li}\ (0.84\,\mathrm{MeV}) + \gamma\ (0.48\,\mathrm{MeV})$$

and

$$\sigma = 269\,\mathrm b \text{ for } \mathrm n + {}^{10}\mathrm B \to \alpha\ (1.78\,\mathrm{MeV}) + {}^7\mathrm{Li}\ (1.01\,\mathrm{MeV}),$$

one finds:

$$\frac{S}{N_{\mathrm n}} = 2.04 \times 10^{-2} + 4.23 \times 10^{-2} + 1.73 \times 10^{-3} + 3.96 \times 10^{-3} = 6.84 \times 10^{-2}. \quad (6.154)$$

The ranges of ^{7}Li (0.84 MeV), α (1.47 MeV), ^{7}Li (1.01 MeV) and α (1.78 MeV) in ^{10}B, calculated with SRIM [Ziegler, J.F. and M.D. and Biersack (2008b)], are $1.75\,\mu m$, $3.62\,\mu m$, $1.97\,\mu m$ and $4.5\,\mu m$, respectively. Higher neutron reaction cross sections in ^{10}B is compensated by smaller ranges and therefore sensitivities achieved with ^{10}B and ^{6}LiF converters are comparable.

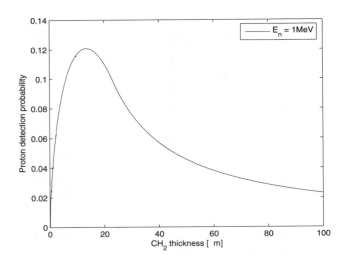

Fig. 6.36   Proton detection probability as a function of the polyethylene thickness (in μm) for incident neutron energy of 1 MeV. This indirectly measures also the fast neutron detection probability [Gutierrez (2007)].

6.7.1.2   *Signals in Silicon Detectors by Fast Neutrons*

These signals are produced by the elastic scattering of fast neutrons on nuclei in the converter. As discussed in Sect. 6.7.1, CH_2 being hydrogen-rich is an optimal choice of converter for the detection of fast neutron with silicon detectors. In that case, the detection efficiency, S_{eff}, of a Si detector as a function of the converter thickness d can be expressed as

$$S_{eff} = \epsilon_p N_0 \left(1 - e^{-\Sigma d}\right) e^{-\mu d} + \epsilon_n N_0 e^{-\Sigma d}, \qquad (6.155)$$

where ϵ_p is the intrinsic efficiency for recoil protons (assumed to be 1 as all the protons reaching the detector generate a signal, 100% charge collection efficiency), ϵ_n is the intrinsic detection for direct neutron detection (a fraction of neutron can cross the converter and produce signals in Si through nuclear reactions, like Si(n,p)Al and Si(n,α)Mg). In Eq. 6.155, the term $\epsilon_p N_0 \left(1 - e^{-\Sigma d}\right) e^{-\mu d}$ represents the flux of recoiling protons produced by neutron interactions with the CH_2 converter nuclei, Σ (cm^{-1}) is the mean macroscopic neutron-converter cross section. In the case of polyethylene,

$$\Sigma \, (\text{cm}^{-1}) = \frac{\rho \, N}{A} \left(n_H \, \sigma_H + n_C \, \sigma_C\right)$$

and μ is the proton absorption coefficient in the converter, which can be calculated from SRIM [Ziegler, J.F. and M.D. and Biersack (2008b)]. Parametrisation of σ_H and σ_C as functions of energy may be found in [Mesquita, Filho and Hamada

(2003)], for instance. The term $\epsilon_n N_0 e^{-\Sigma d}$ represents the fraction of neutrons traversing the converter film and interacting directly with Si. The coefficient N_0 is the rate of neutrons emitted from a source of neutrons (such as an Am-Be source). Equation (6.155) can be fitted to the data to obtain the detection efficiency (count rate) as a function of the converter thickness. For normal incidence, the number of signals per incident fast neutron is given by a formula similar [Wielunski et al. (2004)] to Eqs. (6.137, 6.138) or (6.146)

$$\frac{S}{N_n} = \Sigma_i \frac{S_i}{N_n} = \Sigma_i N_i \, \sigma_i(E_n, \theta) \, R_i[E_R(E_n, \theta)] \, P(E_n), \qquad (6.156)$$

where E_n is the incident neutron energy, N_n the number of incident neutrons on the converter, N_i the atomic concentration of the target-i-isotopes in the converter; $\sigma_i(E_n, \theta)$ is the neutron-i-target isotope elastic scattering cross section and $R_i[E_R(E_n, \theta)]$ is the range of the recoil-i produced by the neutron-i-target isotope elastic scattering. The (n,p) differential elastic cross section for an incident neutron of given energy is [Hopkins and Breit (1971)]

$$\sigma(\theta) = \Sigma_{i=l}^{l} c_i \, P_i \left(1 - 2\cos^2 \theta\right), \qquad (6.157)$$

where P_i is the ith order Legendre polynomial, θ the scattering angle in the laboratory frame. The differential elastic cross section for proton recoil as a function of the recoil angle (laboratory frame) is given in Fig. 6.34 for 1 MeV and 14 MeV incident neutrons. From Fig. 6.34, for both energies, there are more protons forward emitted ($\theta = 0$).

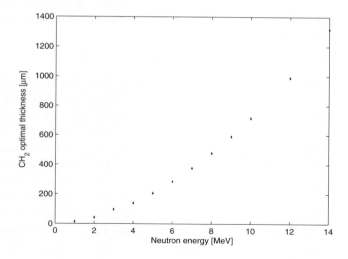

Fig. 6.37 CH$_2$ optimal thickness (μm) as a function of the incident neutron energy (MeV) [Gutierrez (2007)].

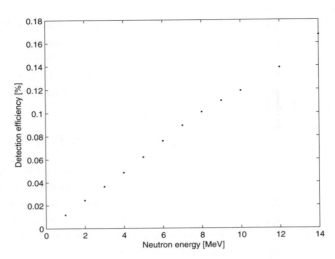

Fig. 6.38 Fast neutron detection efficiency (%) in CH₂ as a function of the incident neutron energy (MeV) [Gutierrez (2007)]. The detection efficiency increases with the incident neutron energy.

The differential elastic cross section is smaller for an incident neutron of 14 MeV compared to 1 MeV as increasing neutron energy opens more inelastic channels, then, decreasing the elastic cross section. The probability that the proton is scattered at a given angle for an elastic scattering can be calculated from the differential cross section. That probability depends on the silicon detector area and the sum of probabilities is normalized to 1. The maximum thickness of the CH_2 converter to be used depends on the proton range, $R(\theta)$. This thickness will change with respect to the recoil energy, $E_R(\theta)$, and is given by

$$l_{\max} = R(\theta) \cos \theta. \tag{6.158}$$

The incident neutrons may follow a direction either along the normal or at angle with the normal to the converter area (Fig. 6.35) [Gutierrez (2007)] In all cases one has to evaluate i) the probability that the proton can come out the converter into the silicon detector, depending on its range in the converter and ii) the probability as a function of the scattering angle. One has also to account for the increase of the collision probability with the converter thickness. Figure 6.36 shows the detection probability of the proton as a function of the converter thickness for neutrons incident along the normal to the converter area [Gutierrez (2007)]. The optimal thickness of the CH_2 converter as a function of the incident neutron energy is shown in Fig. 6.37 [Gutierrez (2007)]. The optimal thickness increases with the incident neutron energy. For incident neutrons along the normal to the converter area, one finds the detection efficiency as a function of the incident neutron energy (Fig 6.38) [Gutierrez (2007)]. The results found in the figures above can be checked with a simple calculation similar to those done for the thermal neutrons (see for

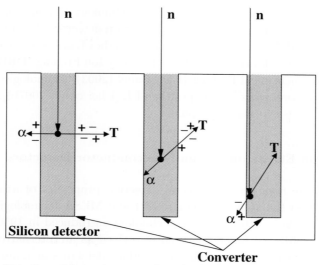

Fig. 6.39   Detection of thermal neutron detection in a detector structure with pores.

instance [Wielunski et al. (2004)]). The signal in silicon covered by a CH_2 converter is essentially given by the recoiling hydrogen, i.e., the proton. Considering the
example of 5 MeV incident neutrons, the maximum energy transfer to hydrogen
$(M_n \sim M_p)$ and carbon are 5 MeV and 1.42 MeV, respectively. These values correspond to a range of 347.8 μm for recoil hydrogen and 3.1 μm for recoil carbon in
polyethylene. At the same time, the elastic scattering cross section for 5 MeV incident neutrons on hydrogen and carbon are 1.63 b and 1.12 b, respectively (the cross
section values can be found in [NNDC (2008a)], for instance). The concentrations of
hydrogen and carbon in polyethylene are 7.986×10^{22} and 3.993×10^{22} atoms/cm^3,
respectively. Therefore, for maximum transfer energy from fast neutrons to hydrogen and carbon in polyethylene, one expects:

$$\frac{S}{N_n P} = 0.45 \times 10^{-2} + 0.14 \times 10^{-4}. \tag{6.159}$$

The signal expected from carbon recoil is negligible compared the signal from hydrogen.

6.7.2   *3-D Neutron Detectors*

The way to improve the limited detection efficiency for neutrons of planar detectors
is to enlarge the surface between the neutron converter and the silicon detector [Uher
et al. (2007b)]. This increases the probability that the heavy charged particles will
reach the detector sensitive volume. Current semiconductor fabrication technologies
allow creation of pores into the silicon detector. The width of these pores can be
from a few micrometers up to hundreds of micrometers. The depth can be hundreds

of micrometers (depending on the pore width, technology and material used). These pore dimensions are well suited for thermal neutron detectors. The principle of neutron detection in a detector structure with pores is shown in Fig. 6.39. Pore depths of $200\,\mu$m or more can be fabricated by Deep Reactive Ion Etching (DRIE) [Wikipedia (2008b)] or Electrochemical Etching (EE) [Badel (2003)] technologies. DRIE offers a wider choice of pores sizes and shapes than EE [Uher et al. (2007b)]. On the other hand DRIE requires expensive equipment.

6.8　Radiation Effects on Silicon Semiconductor Detectors

In this section are reviewed the general detection properties of irradiated silicon detectors based on the standard planar (SP) and MESA technologies. The later, less familiar nowadays, is a rather old technology introduced in 1956 [Tanenbaum and Thomas (1956)] and has been recently revived as an alternative to standard planar technology with the possibility to produce detectors at lower costs [Sopko, Hazdra, Kohout, Mrázek and Pospíšil (1997)]. The section contains a review of the spectroscopic features of SP and MESA detectors and their behavior under particle irradiation. For the comparison, the scanning of non-irradiated and irradiated detectors is done with heavy charged particles (e.g., α-particles and protons). Pulse-height spectra are obtained by measuring the signal of the diodes with a standard spectroscopy system. The study of the peaks evolution with bias voltage can provide precise information on the active medium structure and charge collection performances of the detector. In the case of irradiated detectors, the study of the peak evolution with bias voltage and irradiation fluence allows the measurement of the charge collection efficiency degradation in various regions of the diodes and structure alteration. This section also includes a discussion of the violation of non-ionizing energy-loss (NIEL) scaling. This scaling factor, which expresses the proportionality between NIEL value and resulting damage effects, is modified for low energy protons.

6.8.1　*MESA Radiation Detectors*

The MESA process for building silicon diodes has been known for a long time [Tanenbaum and Thomas (1956)]. It has been recently revived as an alternative to the planar process of silicon detector production that is relatively simpler and cheaper [Sopko, Hazdra, Kohout, Mrázek and Pospíšil (1997)]. This relatively inexpensive high production yield of MESA detectors combined with satisfactory charge collection capabilities and overall performance under irradiation could favor their choice for applications where a large number of detectors is needed such as active medium of trackers (pixels and microstrips) or calorimeters-preshowers (pads) for particle physics experiments. Pad and strip MESA detectors can be produced according either to standard MESA (SM) technology or to planar MESA (PM)

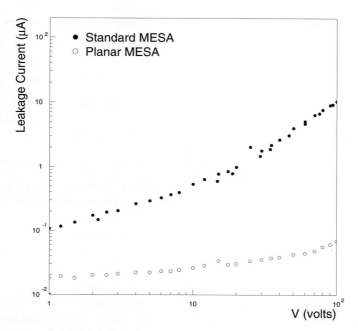

Fig. 6.40 The bulk leakage current (in μA) as a function of the applied bias voltage (in volts) for standard MESA (SM) and planar MESA(PM) detectors, detector area of 25 mm^2 (from [Leroy and Rancoita (2007)]).

technology. The possibility to use SM and PM silicon diodes as radiation detectors has been investigated [Casse et al. (1999b); Chren et al. (2001); Houdayer et al. (2002)]. The SM process has the disadvantage of yielding diodes that have high surface current due to improper passivation of cut edges. The standard MESA technology has been modified in order to decrease the surface current. More generally, one also observes much lower bulk leakage currents for planar MESA (PM) detectors compared to standard MESA (SM) detectors (Fig. 6.40). After PM detectors became available, SM detectors were no longer or rarely used. The leakage current is larger for the SM detector (about $10\,\mu$A for a bias voltage of 100 V, detector area of 25 mm^2) than for the PM detector (around 50 nA for a bias voltage of 100 V, detector area of 25 mm^2).

This improvement observed from SM to PM results from the application of chemical etching for producing a MESA structure with subsequent oxidation of detector edges. As demonstrated next, the PM detectors show electrical features comparable to those of SP detectors.

The leakage currents of the PM and SP detectors used as reference are very low and remain low (a few nA cm^{-2} for SP detectors, a few tens of nA cm^{-2} for PM detectors) with increasing applied bias voltage, even for bias voltage beyond the full

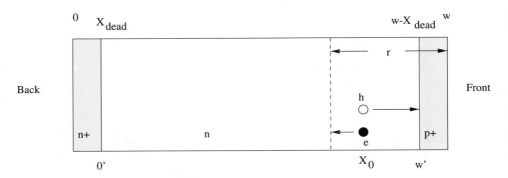

Fig. 6.41 Schematic representation of a MESA detector with the dormant or dead layers on each side; r represents the range of the incident particle (adapted from *Nucl. Instr. and Meth. in Phys. Res. A* **460**, Chren, D., Juneau, M., Kohout, Z., Lebel, C., Leroy, C., Linhart, V., Pospisil, S., Roy, P., Saintonge, A. and Sopko, B., Study of the characteristics of silicon MESA radiation detectors, 146–158, Copyright (2001), with permission from Elsevier).

depletion value.

6.8.1.1 *Electrical features of Planar MESA Detectors*

Several studies were done to compare the performances of planar MESA (PM) detectors to standard planar (SP) detectors (these studies also involved standard MESA (SM) detectors) [Chren et al. (2001); Houdayer et al. (2002)]. The detectors used for these studies were typically PM detectors of average resistivity of $2\,\mathrm{k\Omega\,cm}$, $25\,\mathrm{mm}^2$ area and $280\,\mu\mathrm{m}$ thickness. The SP detectors had a resistivity from $2.5\,\mathrm{k\Omega\,cm}$ to $6.0\,\mathrm{k\Omega\,cm}$, an area from $25\,\mathrm{mm}^2$ to $100\,\mathrm{mm}^2$, and a thickness of $300\,\mu\mathrm{m}$. A charge transport model has demonstrated the existence of dormant or dead layers on each side of the MESA junctions (Fig. 6.41) [Leroy, Roy, Casse, Glaser, Grigoriev and Lemeilleur (1999b)]. A layer of about $5\,\mu\mathrm{m}$ on each side of a PM diode is acting as a dead or dormant layer (this layer was $14\,\mu\mathrm{m}$ deep on each side of a SM diode [Leroy, Roy, Casse, Glaser, Grigoriev and Lemeilleur (1999b)]), consistent with charge collection data. The active thickness of the MESA detector, of which a thickness X_{dead} is considered dormant on each side as shown in Fig. 6.41, can be defined as

$$w' = w - 2\,X_{\mathrm{dead}}.$$

The capacitance-versus-voltage $(C - V)$ curves are shown in Fig. 6.42 for a PM detector and a SP detector of same area $(25\,\mathrm{mm}^2)$. The $C - V$ curves are typical and indicate a full depletion voltage, V_{fd}, in the range of $(120\text{--}150)\,\mathrm{V}$ for both types of detectors (PM and SP). These rather large values of V_{fd} reflect the low resistivity $(2\text{--}2.5)\,\mathrm{k\Omega\,cm}$ of the PM and SP detectors used for the example. The capacitance measured is larger for the PM detector compared to the SP detector since the thickness of the PM detector, $270\,\mu\mathrm{m}$ ($280\,\mu\mathrm{m}$ corrected for the $10\,\mu\mathrm{m}$ dead layer), is smaller than the thickness of the SP detector, of $300\,\mu\mathrm{m}$.

The average charge collection (in fC) from the signal induced by relativistic elec-

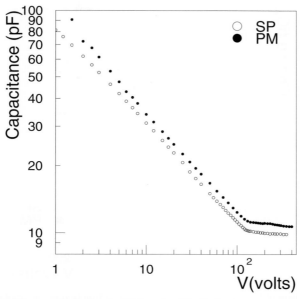

Fig. 6.42 The $C - V$ curve features of a PM and a SP diode are compared. The full depletion voltage, V_{fd}, is found to be in the range of (120–150) V for both type of diodes (reprinted from *Nucl. Instr. and Meth. in Phys. Res. A* **476**, Houdayer, A., Lebel, C., Leroy, C., Roy, P., Linhart, V., Pospisil, S., Sopko, B., Silicon planar MESA diodes as radiation detectors, 588–595, Copyright (2002), with permission from Elsevier).

trons from a ^{106}Ru source is shown in Fig. 6.43 as a function of the applied bias voltage for a PM detector compared to that of a SP detector. Saturation is achieved at and beyond V_{fd} (value consistent with $C - V$ measurements) for the detectors shown in Fig. 6.43. The average charge value of ~ 3.2 fC found at full depletion and beyond for the PM detector is in agreement with an effective thickness of $274\,\mu$m which is (3–4) % less than the thickness of $\sim 283\,\mu$m measured with a micrometer, reflecting the existence of dormant or dead layers which could correspond to the extent of the B or P diffusion zone in the PM process. The total dormant or dead zone is larger for SM detectors ($28\,\mu$m) than for PM detectors ($10\,\mu$m). The average charge value of ~ 3.6 fC measured at full depletion and beyond for a SP detector taken as reference for comparison is consistent with its measured (with a micrometer) thickness of $306\,\mu$m which is about 10% more than the thickness of the PM detector. Note that the ratio of the average charge measured for the PM and SP detectors is equal to the ratio of their thicknesses, as it should be. The resistivity of the particular SP detector used as reference for the measurement was $5.9\,\mathrm{k\Omega\,cm}$, which explains the smaller value of V_{fd} (~ 50 V) observed for this detector in Fig. 6.43.

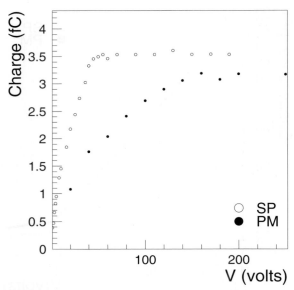

Fig. 6.43   Comparison of the charge collected (in fC) as a function of the applied bias voltage (in volts) for PM and SP detectors illuminated by electrons (*mips*, electron energy larger than 2 MeV from a ^{106}Ru source), (reprinted from *Nucl. Instr. and Meth. in Phys. Res. A* **476**, Houdayer, A., Lebel, C., Leroy, C., Roy, P., Linhart, V., Pospisil, S., Sopko, B., Silicon planar MESA diodes as radiation detectors, 588–595, Copyright (2002), with permission from Elsevier).

6.8.1.2   *Spectroscopic Characteristics of MESA Detectors*

The spectroscopic technique has been applied to the study of PM detectors [Chren et al. (2001); Houdayer et al. (2002)]. Alpha sources (^{239}Pu, ^{241}Am, ^{244}Cm) were used to perform spectroscopic measurements with α-particles in vacuum. The spectra obtained with ^{239}Pu (α-particle energy of 5.157 MeV), ^{241}Am (α-particle energy of 5.486 MeV), and ^{244}Cm (α-particle energy of 5.805 MeV) are shown in Fig. 6.44 for α-particles incident on the front and back sides of a PM detector for two shaping times of 0.25 μs [Fig. 6.44(a)–(c)] and 1.0 μs [Fig. 6.44(d)–(f)]. The comparison of the peaks evolution with voltage for the PM and a silicon surface barrier detector (SSBD) of reference allows the determination of the CCE.

Shifts among peaks and their splitting are caused by different energy losses of particles entering through non-sensitive layers close to or on the detector surface. As already emphasized in the section on standard planar detectors, the capability to measure shifts at the level of a keV is equivalent to measuring layer thicknesses of about ten nanometers. Therefore, an investigation of the detector fine-structure becomes possible [Chren et al. (2001)] as illustrated in Fig. 6.45.

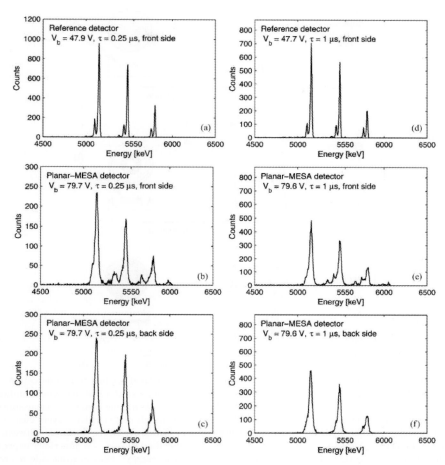

Fig. 6.44 Spectra representing the response of a PM detector to α-particles from Pu, Am, and Cm sources incident on the front side [(b) and (e) for shaping time τ of 0.25 and 1.0 μs, respectively] and back side [(c) and (f) for shaping time τ of 0.25 and 1.0 μs, respectively] of the detector, compared to the response [(a) and (d) for shaping time τ of 0.25 and 1.0 μs, respectively] of a silicon surface barrier detector (SSBD) used as reference. The applied reverse bias voltage was ~ 48 and 80 V for the SSBD and PM detectors, respectively. The fine structure of alpha spectra is clearly resolved by the SSBD. The influence of PM detector electrode structure is explained in Fig. 6.45. The peak positions are determined with a precision better than 5 keV (reprinted from *Nucl. Instr. and Meth. in Phys. Res. A* **460**, Chren, D., Juneau, M., Kohout, Z., Lebel, C., Leroy, C., Linhart, V., Pospisil, S., Roy, P., Saintonge, A. and Sopko, B, Study of the characteristics of silicon MESA radiation detectors, 146–158, Copyright (2001), with permission from Elsevier).

6.8.2 *Results of Irradiation Tests of Planar MESA Detectors*

The characteristics of MESA detectors as a function of fluence were investigated in view of their possible use in high particle fluence environment. The irradiation studies of PM detectors have really been limited up to now to one experiment where

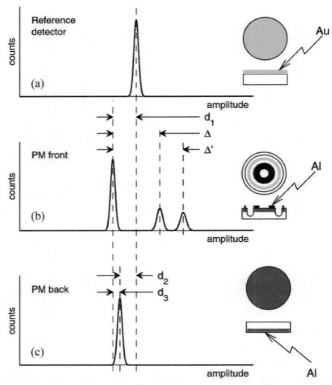

Fig. 6.45   Illustration of the sensitivity of the spectroscopic method applied to PM detectors. The response of a SSBD used as reference (a) is compared to the response of a PM detector illuminated by α-particles (b) on the front side and (c) on the back side. Shifts among peaks and their splitting are caused by different energy-losses of particles entering through non-sensitive layers close to or on the detector surface. d_1 is the difference of front side electrode structure thickness between the reference detector (Au + $p - n$ junction) and the PM detector (Al + $p - n$ junction) expressed in energy losses; d_2 is the difference of thickness expressed in energy losses in front side reference detector structure (Au + $p - n$) and back side PM detector (Al + n^+n). d_3 is the difference expressed in energy losses in front side PM detector structure (Al + $p - n$) thickness and back side PM detector structure (Al + $n^+ - n$) thickness. Also defined are Δ: the energy loss in front side Al of PM detector, and Δ': the energy loss in front side SiO_2 layer in MESA valley of PM detector (reprinted from *Nucl. Instr. and Meth. in Phys. Res. A* **460**, Chren, D., Juneau, M., Kohout, Z., Lebel, C., Leroy, C., Linhart, V., Pospisil, S., Roy, P., Saintonge, A. and Sopko, B, Study of the characteristics of silicon MESA radiation detectors, 146–158, Copyright (2001), with permission from Elsevier).

a set of PM diodes ($25\,mm^2$ area, $\sim 280\,\mu m$ thickness and $\sim 2\,k\Omega\,cm$ resistivity) were exposed to low energy ($10\,MeV$) protons (hardness factor*: theoretical is $\kappa \sim 4.5$, measured is ~ 2.2) [Bechevet, Glaser, Houdayer, Lebel, Leroy, Moll and Roy (2002)]. The effective doping concentration (N_{eff}) of a semiconductor diode can be extracted from measurement of the capacitance (C) as a function of the applied

---

*The reader can see Sect. 4.1.3 (and references therein) for an general treatment of the hardness factor.

Table 6.4   Range of protons in silicon as a function of energy (from [Leroy and Rancoita (2007)]). The ranges were calculated as a function of the proton energy using the SRIM code, which employs Ziegler's stopping tables [Ziegler, Biersack and Littmark (1985a)]

Energy (MeV)	7	8	9	10
Range (μm)	380	476	584	700

bias (V) (see Sect. 6.1.3).

The effective doping concentration (N_{eff}) as a function of fluence (Φ) is given by Eq. (6.75) [Pitzl et al. (1992)]. The values of the parameters of the irradiated PM and SP diodes extracted by fitting Eq. (6.75) to the experimental data (N_{eff} as a function of proton fluence) show that the parameter b is not affected by the MESA process [Bechevet, Glaser, Houdayer, Lebel, Leroy, Moll and Roy (2002)], while the parameter c could be affected since its value should be smaller [Dezillie et al. (1999)], reflecting the higher initial donor concentration (N_d) of the PM diodes compared to the SP diodes.

It is a standard procedure to heat the detectors at 80 C for up to 17 hours to simulate their ageing during the 10 years of operation of the LHC experiments. The irradiation induced change in the effective doping concentration (ΔN_{eff}) as a function of the heating time (t) is given by [Lindstroem, Moll and Fretwurst (1999)]:

$$\Delta N_{\text{eff}}(t) = N_A \exp(-t/\tau_a) + N_C + N_Y \left(1 - \frac{1}{1 + t/\tau_Y}\right), \qquad (6.160)$$

where N_A and τ_a are the short-term annealing constants; N_C represents the so-called *stable damage constant*, which consists of an "incomplete donor removal" and the introduction of negative space-charge proportional to fluence; finally, N_Y and τ_Y are the long-term reverse annealing constants. From results reported in [Bechevet, Glaser, Houdayer, Lebel, Leroy, Moll and Roy (2002)] it is observed that before "inversion"[†] ($\Phi < 10^{13}\,\text{p cm}^{-2}$) the short term annealing (N_A) is about 3 times more in amplitude for SP than for the PM detectors, while after inversion ($\Phi > 10^{13}\,\text{cm}^{-2}$), the difference is about a factor 2. The anti-annealing parameter, N_Y, for PM and SP detectors is comparable before inversion but N_Y is about 25 % larger for the PM detectors after inversion.

The leakage current was recorded (guard ring connected) during the measurements of the capacitance (C) as a function of the applied voltage (V). These measurements were performed at room temperature with the leakage current

---

[†]The reader can also see the discussion in Sects. 4.3.5 and 4.3.6 about the experimental observation that, at large particle fluences, the Hall coefficient changes sign in n-type substrate with high-resistivity (before irradiation).

Table 6.5 Theoretical hardness factors computed from the displacement damage function $D(E)$ normalized to the 1 MeV neutrons value (from [Leroy and Rancoita (2007)]).

Particle	κ_{theo} [D(E)/95 MeV mb]
1 MeV neutrons	1.00
7 MeV protons	7.40
8 MeV protons	5.90
9 MeV protons	5.10
10 MeV protons	4.50
*24 GeV/c protons	0.62

* The value for the 24 GeV/c protons is an experimental value extracted from [Rose Collab. (2001)].

re-normalized to 20 °C [Sze (1981)]. By re-writing Eq. (4.134) [see Sect. 4.3.2], the leakage current per unit volume (I_{vol}) as a function of fluence (Φ) is given by (e.g., [Kraner, Li and Posnecker (1989)]):

$$I_{vol}(\Phi) - I_{vol}(0) = \alpha\,\Phi, \qquad (6.161)$$

where α is the radiation-induced reverse current damage constant [in Eq. (4.134), $I_{vol}(\Phi) = I_{r,irr}/V_{vol}$ and $I_{vol}(0) = I_r/V_{vol}$]. It is found [Bechevet, Glaser, Houdayer, Lebel, Leroy, Moll and Roy (2002)] that the values of the α parameter extracted after heating the detectors for 4 minutes at 80 °C show good agreement between SP and PM detectors.

The irradiation induced change on the α parameter as a function of the heating temperature (T_A) and heating time (t) for detectors (guard ring connected) is given by the phenomenological expression [Moll (1999)]:

$$\alpha_{t/T_A} = \alpha_1 \exp\left(-\frac{t}{\tau_1}\right) + \alpha_0 - \beta \ln\left(\frac{t}{t_0}\right). \qquad (6.162)$$

Equation (6.162) is a convenient parametrization which does not pretend to be based on a physical model [Moll (1999)]. Fitting Eq. (6.162) to the experimental data [Bechevet, Glaser, Houdayer, Lebel, Leroy, Moll and Roy (2002)], one can see that the decrease of the leakage current with annealing is more significant for the SP detectors as compared to the PM detectors.

In summary, the results of PM detectors studies show that detectors produced by planar MESA technology have a satisfactory performance as particle detectors operated in high radiation environment. MESA planar silicon detectors were exposed to low energy protons in order to extract the evolution of the effective doping concentration and leakage current as a function of fluence, up to a fluence of 5×10^{13} p cm^{-2}. The detectors have been heated at 80 °C in order to study the ageing of the detectors. Altogether, the behavior under irradiation of PM and SP detectors is comparable, though the decrease of the leakage current with annealing is more

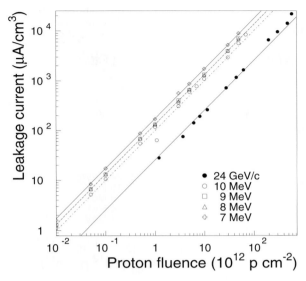

Fig. 6.46 The leakage current (in μA/cm^3) as a function of the proton fluence (Φ in units of $10^{12}\,$p cm^{-2}) after heating for 4 minutes at 80 °C (reprinted from *Nucl. Instr. and Meth. in Phys. Res. A* **479**, Bechevet, D., Glaser, M., Houdayer, A., Lebel, C., Leroy, C., Moll, M. and Roy, P., Results of irradiation tests on standard planar silicon detectors with 7–10 MeV protons, 487–497, Copyright (2002), with permission from Elsevier).

significant for the SP detectors, particularly after inversion. Measurements of the charge collection efficiency (CCE) as a function of fluence were performed [Chren et al. (2001)]. At a fluence $\Phi \approx 10^{14}$ particles cm^{-2}, the charge carrier lifetime degradation due to trapping with increased fluence is responsible for a charge collection deficit of about 12% and about 10% for β-particles incident on MESA and standard planar detectors, respectively. For α-particles incident on the front side, one finds a deficit of 20% for MESA and 25% for standard planar detectors. The deficit increases to 30% and 35% for α-particles incident on the back side of MESA and SP detectors, respectively. These results are in good agreement with existing direct measurements ([Casse et al. (1999a); Leroy et al. (1994)], see also [SICAPO Collab. (1994b,c)]).

6.8.3 *Irradiation with Low-Energy Protons and Violation of NIEL Scaling in High-Resistivity Silicon detectors*

Standard planar silicon detectors were exposed [Bechevet, Glaser, Houdayer, Lebel, Leroy, Moll and Roy (2002)] to low-energy protons in order to extract the evolution of the effective doping concentration and leakage current as a function of fluence, up to a fluence of $7 \times 10^{13}\,$p/cm^2. These measurements were part of a study done for the LHC. The radiation hardness of standard planar silicon detectors exposed

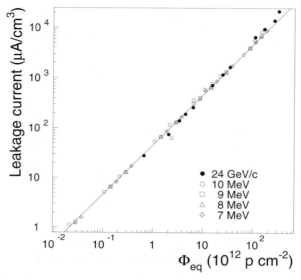

Fig. 6.47 The leakage current (in μA/cm^3) as a function of 1 MeV neutron equivalent fluence ($\Phi_{eq} = \Phi \times \kappa_\alpha$ in units of 10^{12} p cm^{-2}) after heating for 4 minutes at 80 °C (reprinted from *Nucl. Instr. and Meth. in Phys. Res. A* **479**, Bechevet, D., Glaser, M., Houdayer, A., Lebel, C., Leroy, C., Moll, M. and Roy, P., Results of irradiation tests on standard planar silicon detectors with 7–10 MeV protons, 487–497, Copyright (2002), with permission from Elsevier).

to low-energy protons was tested and the results were compared to those obtained from the irradiation of similar material with 24 GeV/c protons. The goal of this study was also to understand the limits of NIEL scaling and therefore the limits for predictions of radiation damage produced by protons with different energies. The resistivity of the detectors used for the study was 2 kΩcm and the proton energies used were 7, 8, 9, 10 MeV. No proton of energy lower than 7 MeV was used to be sure that these protons were not stopped in the 300 μm (295 $\pm$5 μm) thick detectors (Table 6.4 and [Berger, Coursey, Zucker and Chang (2005)]).

Since the LHC experiments will be in operation for at least 10 years, it is necessary to study the long term behavior of silicon detectors to be used in these experiments when exposed to high level of radiations.

The detectors were heated at 80 °C for up to 17 hours in order to extract the damage parameters necessary to simulate the detectors ageing during 10 years of operation of the LHC experiments. A word of explanation is needed to understand the choice of the temperature of 80 °C. Heating detectors after their irradiation has become a standard procedure. This procedure allows the comparison of particle irradiations without the influence of the irradiation time and the samples irradiated at different facilities can be considered as being in the same annealing state. The effective doping concentration, N_{eff}, varies with the diode ageing. At first, N_{eff} (and then V_{fd}) decreases with time after being heated. This period of beneficial annealing

Table 6.6   Value of the parameter α and of the hardness factor κ_α extracted from the leakage current as a function of the proton fluence (Fig. 6.46) after heating the detectors for 4 minutes at 80 °C (from [Leroy and Rancoita (2007)]).

Particle	α [10^{-17} A/cm]	κ_α	$\kappa_\alpha/\kappa_{\text{theo}}$
1 MeV neutrons	4.56	1.0	1.0
7 MeV protons	17.20	3.8 ± 0.1	0.514
8 MeV protons	13.20	2.9 ± 0.1	0.492
9 MeV protons	13.30	2.9 ± 0.1	0.569
10 MeV protons	9.90	2.2 ± 0.1	0.489
24 GeV/c protons	2.54	0.56 ± 0.01	0.903

is followed by a reverse annealing with an increase of N_{eff} (and then of V_{fd}) with time. Heating the detectors at an annealing temperature of 80 °C, is accelerating the room temperature annealing which allows one to reach the intermediate plateau between the beneficial annealing and the reverse annealing after about 4 minutes of heating. The change (ΔN_{eff}) between the value of N_{eff} before irradiation ($N_{\text{eff},0}$), and the value of N_{eff} after irradiation with fluence Φ and after annealing during a period of time t at the temperature T_a is given by:

$$\Delta N_{\text{eff}}(\Phi, t, T_a) = -N_{\text{eff},0} + N_{\text{eff}}(\Phi, t, T_a). \tag{6.163}$$

After heating, N_{eff} has to be time scaled by a factor, Θ, given by [Moll (1999)]:

$$\Theta(T_a) = \exp\left[-\frac{E_a}{k_B}\left(\frac{1}{T_a} - \frac{1}{T_{\text{heat}}}\right)\right], \tag{6.164}$$

where E_a is the activation energy, k_B the Boltzmann constant, T_{heat} the temperature (in K) at which the detector has been heated and T_a the temperature (in K) for which the time has to be scaled. From what is said above, it is understood that an annealing of 4 minutes at 80 °C is applied after each irradiation. After this annealing, ΔN_{eff} is near its minimum, i.e., at the very beginning of the intermediate plateau. This heat treatment of 4 minutes at 80 °C was performed for all detectors in [Bechevet, Glaser, Houdayer, Lebel, Leroy, Moll and Roy (2002)] and corresponds to about 21 days at room temperature (20 °C), as found [Moll (1999)] from Eq. (6.164) with $E_a = 1.33$ eV. Hence, this choice of annealing time and temperature can be viewed as a compromise between i) a temperature high enough to simulate rather realistically the operation scenario of the LHC, in which irradiation (running) periods are interspersed with annealing (maintenance shutdowns) periods and ii) a temperature low enough to excite only the annealing processes occurring at room temperature.

A comparison done with data obtained for 24 GeV/c proton irradiations [Bechevet, Glaser, Houdayer, Lebel, Leroy, Moll and Roy (2002)] shows that higher values of the full depletion voltage are necessary to operate beyond the inversion fluence the detectors irradiated with low-energy protons. As seen before in Sects. 4.1.3 and 4.2.1 for neutrons[§], the concept of radiation hardness factor (κ) arises from the re-normalization of a particle fluence (Φ) of a particular irradiation to the 1 MeV neutron fluence equivalent (Φ_{eq}):

$$\Phi_{eq} = \kappa \, \Phi. \qquad (6.165)$$

The hardness factor can be extracted from various sources. The NIEL scaling hypothesis provides a theoretical value of the hardness factor. This theoretical value is given by the ratio of the displacement damage function[¶] $D_i(E)$ for the particle i and energy E of interest, to the displacement damage function of 1 MeV neutrons {$D_{\text{neutron}}(1\,\text{MeV}) = 95\,\text{MeV mb}$, see page 348 and, for instance, [Namenson, Wolicki and Messenger (1982); ASTM (1985)]}. Theoretical values for $D_i(E)$ (Sects. 4.2.1.3, 4.2.1.5) can be found for low-energy protons using the Rutherford scattering formula along with the Lindhard partition function (see page 348) to separate the displacement damage from the ionization (mathematical expressions available in [Vasilescu (1997)], tabulated values in [Summers, Burke, Shapiro, Messenger and Walters (1993)]). The error on the detector thickness (295 $\pm 5\,\mu$m) corresponds to a 3% error on the displacement damage function for 7 MeV protons (energy at which a maximum variation occurs) [Bechevet, Glaser, Houdayer, Lebel, Leroy, Moll and Roy (2002)]. Finally, the theoretical hardness factor in a finite silicon thickness is found by averaging the damage function over the thickness considered. The theoretical hardness factors extracted that way (Table 6.5) will be compared below with experimental hardness factors extracted from the leakage current of standard planar silicon detectors exposed to (7–10) MeV protons. The measured value of 0.62 [Rose Collab. (2001)] is used as the reference hardness factor for 24 GeV/c protons since a theoretical value cannot be obtained from the method described above for protons of such momentum. This hardness factor is in agreement with the value of $\kappa \approx 0.5$ estimated from [Huhtinen and Aarnio (1993)]. The value of $\kappa_{\text{theo}} \approx 0.93$ extracted from [Van Ginneken (1989)] is widely used. However, this value does not fit the experimental data. The disagreement of experimental data with the values given in [Van Ginneken (1989)], for energies higher than 100 MeV, is mainly due, among other physical features, to the assumption in [Van Ginneken (1989)] that all Lindhard factors are at their plateau values [Bechevet, Glaser, Houdayer, Lebel, Leroy, Moll and Roy (2002)].

Hardness factors have been extracted in [Bechevet, Glaser, Houdayer, Lebel, Leroy, Moll and Roy (2002)] from the leakage current measurements, i.e., from the evolution of leakage current as a function of the irradiation fluence. The measurements of the detector leakage current were done with the guard ring connected. The measurements were performed at room temperature (T_m) and the leakage current was re-normalized to $T = 20\,°$C using Eq. (6.78). The leakage current per unit

---

[§]In case of neutron irradiations, the reader can see Eq. (4.79).
[¶]One can refer to Sects. 4.2.1–4.2.1.5 for a treatment of the diplacement function.

Table 6.7   Value of the parameters in Eq. (6.167) extracted from the effective doping concentration (from [Leroy and Rancoita (2007)]) as a function of the proton fluence (Fig. 6.48). The detectors have been heated 4 minutes at 80 °C. The quoted errors are the errors on the fit. The experimental errors ($\approx$8%) are not included. The parameter $b_{24\,\text{GeV}/c}$ is the slope for 24 GeV/c protons.

Energy	$N_{\text{eff},0}$ $(10^{11}\ \text{cm}^{-3})$	b $10^{-2}\ \text{cm}^{-1}$	c $(10^{-14}\ \text{cm}^2)$	$(b/b_{24\ \text{GeV}/c})$
7 MeV	18.4	4.80 ± 0.12	19.3	8.62 ± 0.25
8 MeV	18.4	4.27 ± 0.12	17.2	7.67 ± 0.24
9 MeV	18.6	4.35 ± 0.12	17.0	7.81 ± 0.24
10 MeV	17.7	3.40 ± 0.10	11.3	6.10 ± 0.20
24 GeV/c	17.9	0.56 ± 0.01	2.12	1.00 ± 0.02

volume (I_{vol}) as a function of fluence (Φ) is given by Eq. (6.161). Figure 6.46 shows the results of fitting Eq. (6.161) to the experimental data for low-energy [(7–10) MeV] protons and high-energy (24 GeV/c) protons after heating for 4 minutes at 80 °C. The values of the α parameter extracted for the various irradiations by this procedure are shown in Table 6.6.

Studies have shown that the evolution of the leakage current as a function of fluence and annealing time is independent of the initial resistivity and impurity concentration [Moll (1999)]. It is thus possible to obtain the hardness factor (κ_α) for the protons by comparing the values of the α parameter obtained from the proton irradiations to the value obtained for 1 MeV neutrons ($\alpha = 4.56 \times 10^{-17}$ A/cm). The α parameter was extracted for the volumic leakage current observed in the silicon detectors heated 4 minutes at 80 °C. The values of the hardness factor (κ_α) extracted by this method are reported in Table 6.6. It can be seen in this table that for low-energy protons, the experimental hardness factors (κ_α) are about half the theoretical hardness factors (κ_{theo}) predicted by the NIEL scaling hypothesis (shown in Table 6.5). The factors (κ_α) and (κ_{theo}) are in agreement for 24 GeV/c protons, with the value $\kappa_\alpha = 0.56 \pm 0.1$ in agreement with the referenced experimental value [Rose Collab. (2001)] of 0.62 and close to the value of ≈ 0.50 estimated from the displacement damage function. For example, when the hardness factors for 10 MeV and 24 GeV/c protons are compared, they differ by a factor of 4 if κ_α is used and by about a factor 7 when using κ_{theo}.

Using this hardness factor (κ_α), the data have been plotted as a function of the equivalent fluence [Φ_{eq}, Eq. (6.165)] in Fig. 6.47.

Comparison between radiation damage inflicted to silicon detectors by low-energy and high-energy protons can also be extracted from the measurement of the slope parameter of the effective doping concentration. The standard method applied to extract the effective doping concentration (N_{eff}) of a semiconductor diode is based on the measurement of the capacitance (C) which is a function of the ap-

Table 6.8   A few examples of proton, neutron and photon reactions with silicon (from [Leroy and Rancoita (2007)]). The probability of these reactions to occur depends on their energy thresholds.

Energy	b $(10^{-2}$ cm$^{-1})$	b/κ_{theo} $(10^{-2}$ cm$^{-1})$	b/κ_{α} $(10^{-2}$ cm$^{-1})$
7 MeV	4.80 ± 0.12	0.649	1.263
8 MeV	4.27 ± 0.12	0.724	1.472
9 MeV	4.35 ± 0.12	0.853	1.500
10 MeV	3.40 ± 0.10	0.756	1.545
24 GeV/c	0.56 ± 0.01	0.903	1.000

plied bias (V) [Eqs. (6.35, 6.22)]. Hence, a plot of the measured capacitance as a function of the applied voltage will yield the value of the voltage (V_{fd}) necessary to fully deplete a detector of thickness w [$X(V_{fd}) = w$]:

$$|N_{\text{eff}}| = \frac{2\,\epsilon\,V_{fd}}{q\,w^2}. \tag{6.166}$$

The effective doping concentration (N_{eff}) as a function of fluence (Φ) can be described by Eq. (6.75) rewritten as:

$$|N_{\text{eff}}| = |-N_{\text{eff},0}\,\exp(-c\,\Phi) + b\,\Phi|, \tag{6.167}$$

where $N_{\text{eff},0}$ is the initial effective doping concentration and c the so-called donor removal constant assuming a complete removal of the initial effective doping concentration after exposure to a high irradiation fluence ($\Phi \gg 1/c$). The parameter b describes the introduction of negative space-charge and depends on the annealing time after irradiation. The value of the parameters extracted by fitting Eq. (6.167) to the experimental data plotted in Fig. 6.48 are shown in Table 6.7.

It is observed by comparing the values of the slope parameter b of N_{eff} (Table 6.7) that the slopes for low-energy protons are larger than those ($b_{24\text{GeV}/c}$) for 24 GeV/c protons by a factor of 6 to ~ 9. This reflects higher radiation damage inflicted to silicon detectors by low-energy protons compared to 24 GeV/c protons. These increase factors are close to the ones predicted by the NIEL scaling hypothesis (factor of 7 to 12, Table 6.5). However, it is not possible to scale the slope parameter b by a single hardness parameter: neither by the theoretical factor (κ_{theo}), nor by the experimental factor (κ_{α}) determined from the α value (Table 6.8).

This investigation of irradiation of Standard planar silicon detectors with low-energy protons demonstrates the limits of applicability of the NIEL scaling hypothesis if used to predict radiation damage produced by different particles with different energies. On one hand, it was demonstrated that the absolute values of the hardness factors determined from the displacement damage function κ_{theo} do not agree with the measured hardness factors κ_{α}. On the other hand, it was shown that individual

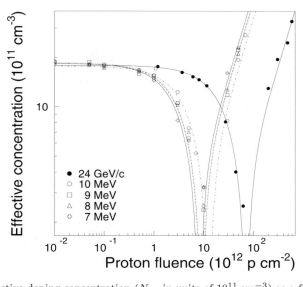

Fig. 6.48 Effective doping concentration (N_{eff} in units of 10^{11} cm^{-3}) as a function of the proton fluence (Φ in units of 10^{12} p cm^{-2}) after heating for 4 minutes at $80\,^\circ$C (reprinted from *Nucl. Instr. and Meth. in Phys. Res. A* **479**, Bechevet, D., Glaser, M., Houdayer, A., Lebel, C., Leroy, C., Moll, M. and Roy, P., Results of irradiation tests on standard planar silicon detectors with 7–10 MeV protons, 487–497, Copyright (2002), with permission from Elsevier).

damage parameters (e.g., leakage current increase and the slope parameter of N_{eff}) have a different scaling with the particle energy.

A simulation model of migration and clustering of the produced primary defects in silicon exposed to various types of hadron irradiation has been developed in [Huhtinen (2002)]. This model predicts violation of NIEL scaling both for the leakage current constant (α) and the b-slope of the effective doping concentration after type-inversion. According to this model the violation is expected in opposite direction and most pronounced in the case of low-energy protons. These simulation results are in very good agreement with the data on low-energy protons reported in [Bechevet, Glaser, Houdayer, Lebel, Leroy, Moll and Roy (2002)], in particular for the 10 MeV proton data.

Chapter 7

Displacement Damage and Particle Interactions in Silicon Devices

The results reviewed in the Chapters on *Radiation Environments and Damage in Silicon Semiconductor* and *Solid State Detectors* and their interpretation show a progressive understanding of the physics phenomena underlying the displacement damage effects resulting from the radiation interaction in silicon semiconductors. Although there is no general model supporting a comprehensive interpretation of the characteristics after irradiation, the generation of complex defects is certainly among the fundamental mechanisms responsible for the degradation of the silicon properties. These deep defects are mostly created by non-ionizing energy-loss (NIEL) processes, which generate primary defects by the initial particle interaction and by the cascade generation due to silicon recoil, and secondary defects caused by the diffusion of the primary point defects (interstitials and vacancies). The different extension of the cascade and concentrations of impurities favor a slight dependence of the damage effects on type and energy of the incoming particle, as well as, on the type and resistivity of the semiconductor. A review of particle interaction and displacement damage in silicon devices operated in radiation environments was provided by Leroy and Rancoita (2007).

Radiation-induced defect centers have a major impact on the electrical behavior of semiconductor devices and deeply affect their properties. For instance, centers with energy-levels near the mid-gap make significant contribution to carrier generation. Thermal generation of electron–hole pairs dominate over capture processes when free carrier concentrations are much lower than the thermal equilibrium values, as in the depletion regions. These centers become the main mechanism for the increase of the leakage current in silicon devices after irradiation. In general, radiation-induced defect centers are the relevant mechanism for the decrease of the minority-carrier lifetime. In addition, donors or acceptors can be compensated by deep-lying radiation-induced centers resulting in the decrease of the concentration of majority carriers. This process causes the variation of the device properties depending on the majority-carrier concentration. Measurements of the bulk and $p-n$ junction properties have shown largely modified behaviors as function of the temperature. Furthermore, there are indications that further understanding of transport mechanisms in irradiated semiconductors is needed by means of additional systematic investigations for low- and high-resistivity silicon.

The particle interactions in the bulk or active volume of silicon devices are responsible for the device degradation resulting from the *absorbed dose*[*] and the so-called *single event phenomena*[††].

As mentioned in Sects. 4.2.3.1, 4.2.3.2, the energy deposited by ionization or collision energy-loss processes accounts for the largest fraction of the total energy deposited in any medium (e.g., in a silicon semiconductor). Silicon based devices are affected by the total (i.e., and by far, ionizing) absorbed dose: for instance in MOS transistors, shifts of the threshold voltage and variations of the sub-threshold slope occur.

Other silicon based devices, like solar cells (Sect. 6.6) and bipolar transistors, are mostly affected by the displacement damage generated by non-ionizing energy-loss processes (e.g., see [Srour, Marshall and Marshall (2003)] and references therein). For instance, at large cumulative irradiations this mechanism was found to be responsible i) for the decrease of the gain of *bipolar transistors*[‡‡] mostly as a result of the *decrease of the minority-carrier lifetime* (e.g., see Sect. 4.3.1) in the transistor base and ii) for the degradation of the series-noise performance of charge-sensitive-preamplifiers with bipolar junction transistors in the input stage mainly because of the *increase of the base spreading-resistance* (e.g., see Sects. 4.3.5, 7.1.5) of these transistors[**]. Furthermore, through systematic measurements (see Sects. 7.1, 7.1.3) it was found that the gain degradation of bipolar transistors manufactured on VLSI technologies depends almost linearly on the amount of displacement damage generated (e.g., the amount of energy deposited by NIEL processes) independently of the type of incoming particle [Colder et al. (2001, 2002); Codegoni et al. (2004a,b, 2006); Consolandi, D'Angelo, Fallica, Mangoni, Modica, Pensotti and Rancoita (2006); D'Angelo, Fallica, Galbiati, Mangoni, Modica, Pensotti and Rancoita (2006)].

A Single Event Effect (SEE) is due to an individual event caused by a single charged particle ($Z \geq 2$) traversing a semiconductor or semiconductor-based device (e.g., integrated circuit, power supply, etc.). A SEE results in a failure of the device, as a consequence of the charge deposited along the path of the incoming particle

---

[*]Both the non-ionizing (e.g., see Sect. 4.2.1) and ionizing (e.g., see Sect. 4.2.3) energy-losses contribute to the *absorbed dose*. The reader can see, for instance, [Vavilov and Ukhin (1977); Srour, Long, Millward, Fitzwilson and Chadsey (1984); Ma and Dressendorfer (1989); Messenger and Ash (1992); Claeys and Simoen (2002); Holmes-Siedle and Adams (2002); ECSS (2005)], Sect. 7.1 and references therein.

[††]One can see, e.g., [Srour, Long, Millward, Fitzwilson and Chadsey (1984); Messenger and Ash (1992, 1997); Claeys and Simoen (2002); Holmes-Siedle and Adams (2002); ECSS (2005)], Sect. 7.2 and references therein.

[‡‡]The reader can refer, e.g., to [Frank and Larin (1965); Messenger (1966); Ramsey and Vail (1970); Messenger (1972); Srour (1973); Vavilov and Ukhin (1977); Srour, Long, Millward, Fitzwilson and Chadsey (1984); Srour and McGarrity (1988); Srour and Hartman (1989); Messenger and Ash (1992); Colder et al. (2001, 2002); Codegoni et al. (2004a,b); ECSS (2005); Codegoni et al. (2006); Consolandi, D'Angelo, Fallica, Mangoni, Modica, Pensotti and Rancoita (2006)] and references therein.

[**]A discussion can be found, e.g., in [Baschirotto et al. (1995a,b, 1996, 1997, 1999)] and references therein.

within the sensitive volume of the device. SEE's are treated in Sects. 7.2–7.2.8.

Substantial progress has been made in the understanding of mechanisms at the origin of SEE and predicting their rate of occurrence for various types of devices and circuits. Monte Carlo tools have been developed which take into account the evolution of modern technology toward faster devices with less power consumption and denser circuits increasingly small. This reduction in size and multiplication of active cells lead to a re-examination of the concepts of critical charge and sensitive volume. The RPP approximation for the sensitive volume, although convenient for rapid estimates, has to be much more refined to account for effects such as funneling, bipolar effects, source-to-drain currents. These effects are the source of sensitivity variations across the devices, not easily reproducible by Monte Carlo. The use of powerful simulation tools such as GEANT4 allows the accurate description of the device geometry, structure, material composition, and at the same time the inclusion of the mechanisms of energy deposition, and cell charge collection efficiency. SEE cross sections for specific species of particles and ions are measured at accelerators and can be convoluted with effective LET spectrum calculated by GEANT4 to infer SEE rates. These predictions are strictly viable for the experimental conditions under which the cross sections have been measured. The ultimate goal is to extend these predicted SEE rates beyond these specific conditions and rely only on GEANT4 fed with refined models of energy deposition and charge collection efficiency in the various regions of the device in order to calculate the charge collected which also depends on the time constant of the process. This also supposes that progress will be made in the determination of the sensitive volume size and shape and of the critical energy which remain a major uncertainty in Monte Carlo simulations. Precise SEE rate predictions may then have a strong impact on device design.

7.1 Displacement Damage in Irradiated Bipolar Transistors

The bipolar transistor* (contraction for *transfer resistor*) consists of two successive junctions ($n - p - n$, or $p - n - p$) made by three semiconductor regions called *emitter* (heavily doped†), *base* and *collector*, respectively. It is one of the most important semiconductor components available, nowadays, on VLSI (Very Large Scale

---

*A general treatment of this topic can be found, e.g., in Chapter 7 of [Grove (1967)], Chapter 6 of [Müller and Kamins (1977)], Chapter 4 of [Sze (1985)], Sections 3.9–3.14 of [Messenger and Ash (1992)], Chapter 10 of [Neamen (2002)] and also [Sze (1981); Goodge (1983); Bar-Lev (1993); Ng (2002); Müller and Kamins - with Chan (2002); Brennan (2005)].

†In VLSI technologies, the emitter- and base-regions are strongly graded with doping concentrations which may exceed 10^{20} and 10^{18} cm^{-3}, respectively; while the collector-region is typically an epitaxial-layer [Sze (1988)] with dopant concentration lower than 10^{16} cm^{-3} (e.g., see Figs. 6.16 and 6.17 at page 304 of [Müller and Kamins (1977)], Fig 11 at page 25 of [Sze (1985)] and Fig 3.27 at page 133 of [Messenger and Ash (1992)] 2nd Edition).

Integration) technologies. It was invented[‡] by a research team at Bell Laboratories in 1947 (see page 133 of [Sze (1981)]). In this device, both the electrons and holes[§] participate in the conduction mechanism.

The *emitter-base junction* is forward-biased and, consequently, allows a large injection into the base-region of minority (for the base) carriers, which are majority carriers inside the emitter-region. The *base-collector junction* is reverse-biased and sufficiently close to the emitter-base junction so that most of these carriers can reach the collector-region, where they are carriers of the majority type. Thus, in the bipolar transistor operation, a large current flows in the reverse-biased base-collector junction due to the large injection of carriers coming from the forward-biased emitter-base junction in its vicinity. Since not all the injected carriers from the emitter will reach the collector-region, two quantities are of great importance in describing the d.c. transistor characteristics[*]:

a) the so-called *base-transport factor* given by

$$\alpha_\mathrm{T} \equiv \frac{I_\mathrm{C,m}}{I_\mathrm{E,m}}, \tag{7.1}$$

where $I_\mathrm{C,m}$ is the current due to the minority (in the base-region, but majority in the emitter-region from which they are injected) carriers reaching the collector and $I_\mathrm{E,m}$ is that due to the minority carriers initially injected from the emitter into the base;

b) the *current gain* given by

$$\beta_0 \equiv \frac{I_\mathrm{C}}{I_\mathrm{B}}, \tag{7.2}$$

where I_C and I_B are the collector and base current, respectively. The collector current accounts for the carriers, which are capable to reach the collector after passing through the base region. The base current accounts for recombination processes in the base, recombination in the emitter-base region and diffusion of carriers from the base into the emitter region. The *minority-carrier flow* [see Eq. (7.1)] provides the basic mechanism of the transistor operation and, as a consequence, of its degradation under irradiation.

In circuit applications, the most often configuration of transistors is the so-called *common-emitter configuration* under *active mode*, that is, the emitter lead is common to the input and output circuits, the emitter-base junction is forward-biased and the base-collector junction is reverse-biased. The collector current can be written as {see Equation (46) at page 128 of [Sze (1985)]}:

$$I_\mathrm{C} = \beta\, I_\mathrm{B} + I_\mathrm{C,o}, \tag{7.3}$$

[‡]In 1948 J. Bardeen and W. Brattain announced the development of the point contact transistor [Bardeen and Brattain (1948)], while in 1949 W. Shockley published his paper on junction diodes and transistors [Shockley (1949)]: all of whom subsequently won the Nobel Prize in Physics in 1956.

[§]This is the reason why is called *bipolar*.

[*]For a general treatment of the bipolar transistor properties, one may refer, for instance, to Chapter 7 of [Grove (1967)], Chapter 6 of [Müller and Kamins (1977)], Chapter 4 of [Sze (1985)] and Chapter 10 of [Neamen (2002)].

where $I_{C,o}$ is the *collector-emitter leakage current* with open base and β is the so-called *common-emitter current gain*[¶]. Under the assumption that the recombination current in the emitter-base depletion region is negligible, β is given by {e.g., see Equations (8) at page 115 and (44) at page 127 of [Sze (1985)]}:

$$\beta \simeq \frac{\gamma \, \alpha_T}{1 - \gamma \, \alpha_T}, \tag{7.4}$$

where $\gamma = I_{E,m}/I_{E,d}$ is the so-called *emitter efficiency*[‖] and $I_{E,d}$ accounts for the electron- and hole-current diffusing across the emitter-base junction, e.g., $I_{E,m}$ and the diffusion current due to carriers injected from the base into the emitter region. γ measures the injected current due to minority carriers with respect to the overall current due to diffusion from the emitter and base regions. For silicon transistors at normal temperature of operation, $I_{C,o}$ is negligible (e.g, see page 250 in [Bar-Lev (1993)]) and, thus, from Eqs. (7.2, 7.3), we have $\beta_0 \approx \beta$. Furthermore, for good transistors we have $\gamma \approx 1$ and $\alpha_T \approx 1$ (e.g., see page 156 of [Lutz (2001)]); as a consequence, since $\gamma \, \alpha_T \lesssim 1$, β_0 is usually large.

In active mode and for low-level injection[**], under the assumptions that the transistor has i) uniform doping in each region, ii) no generation-recombination currents in the depletion regions and iii) neglecting any series resistance in the device, we obtain {e.g., see Equation (36) at page 121 of [Sze (1985)]}:

$$\alpha_T \approx 1 - \frac{W_B^2}{2 \, L_B^2} \quad \text{for} \quad \frac{W_B}{L_B} \ll 1, \tag{7.5}$$

where W_B is the depth of the quasi-neutral base-region[††] and L_B is the diffusion length of the minority-carrier in the base. For the usually well-made (and so-called) *narrow-base transistors*, the condition $W_B/L_B \ll 1$ is satisfied. L_B^2 is given by the minority-carrier life-time (τ_B) multiplied by the diffusion constant (D_B) of these carriers [see Eqs. (6.63, 6.64)]. If the emitter efficiency is close to unity (as it is usual the case), the common-emitter current gain [Eq. (7.4)] can be re-expressed using Eq. (7.5) as:

$$\beta \simeq \frac{\alpha_T}{1 - \alpha_T} \simeq \frac{1}{1 - \alpha_T} = 2 \left(\frac{L_B}{W_B} \right)^2 . \tag{7.6}$$

[¶]This parameter is indicated as h_{fe}, when representing the transistor performance within the framework of small-signal models for low and high frequencies (e.g., see Section 2.2.26 of [Goodge (1983)] and Chapter 15 of [Bar-Lev (1993)]); the static common-emitter current gain is also indicated as h_{FE} in the h-parameter representation (e.g., see Section 3.2.2 of [Sze (1981)] and Section 2.2.17 of [Goodge (1983)]). Furthermore, it is indicated as β_F, i.e., is the so-called forward common-emitter current gain within the framework of the *Ebers–Moll model* (e.g., see the Section 14.3(d) of [Bar-Lev (1993)]).

[‖]The term $\gamma \, \alpha_T$ is the so-called d.c. *common-base current gain* {e.g., see Equation (4.36) at page 86 of [Brennan (2005)]}.

[**]The condition for low-level injection is satisfied when the excess carrier concentration is small in comparison with the doping concentration.

[††]It extends from the edge of the depletion region of the emitter-base junction to that of the base-collector junction.

For graded-base transistors, Eq. (7.5) is replaced by the *Gover–Grinberg–Seidman expression* {Equation (9) of [Gover, Grinberg and Seidman (1972)], see also [Messenger (1973)]}, which is valid for the case of a general impurity distribution:

$$\alpha_T \approx 1 - u_1 \left(\frac{W_B}{L_B}\right)^2 + u_2 \left(\frac{W_B}{L_B}\right)^4 - \cdots, \tag{7.7}$$

where the coefficients u_i depend on the base doping-profile[‡].

For $W_B/L_B \ll 1$, the right side of Eq. (7.7) converges rapidly and the terms beyond the first two can be neglected {e.g., see Equation (31) of [Gover, Grinberg and Seidman (1972)]}. Thus, for $\gamma \approx 1$ the common-emitter current gain can be rewritten using of Eqs. (7.4, 7.7) as:

$$\beta \simeq \frac{\alpha_T}{1 - \alpha_T} \simeq \frac{1}{1 - \alpha_T} = u_1 \left(\frac{L_B}{W_B}\right)^2. \tag{7.8}$$

7.1.1 *Gain Degradation of Bipolar Transistors and Messenger–Spratt Equation*

Since mid-fifties, the effect of fast-neutron irradiations on bipolar transistor was extensively investigated by many authors (e.g., see [Webster (1954); Loferski (1958); Messenger and Spratt (1958); Frank and Larin (1965); Messenger (1967a); Ramsey and Vail (1970); Messenger (1973); Vavilov and Ukhin (1977); Messenger and Ash (1992); Codegoni et al. (2004b); Consolandi, D'Angelo, Fallica, Mangoni, Modica, Pensotti and Rancoita (2006)]). It was shown that the knowledge of the degradation of the minority carrier lifetime allows one to predict changes in the parameters of transistors as a function of the fast-neutron fluence: for instance, Webster, Loferski, Messenger and Spratt could determine that the variation of the reciprocal of the common-emitter current gain (β) linearly depends on the fast-neutron fluence [Webster (1954); Loferski (1958); Messenger and Spratt (1958)]).

As teated in Sect. 7.1, $\beta^\dagger$ [Eq. (7.3)] is related to the lifetime of the minority carriers in the base as shown by the approximate expression (7.17): the decrease of their lifetime with increasing fluence (see Sect. 4.3.1) is the fundamental mechanism which determines the degradation of the transistor gain.

[‡]For instance (see Appendices VII and VIII of [Gover, Grinberg and Seidman (1972)]), for a homogeneous doping-profile in the base we have

$$u_1 = \frac{1}{2};$$

while, for an exponential doping-profile of the type $N = N_0 \exp(-\eta x/W_B)$ we get

$$u_1 = \frac{\eta + \exp(-\eta) - 1}{\eta^2}.$$

[†]As discussed at page 543, in practice the common-emitter current gain does not differ from the current gain β_0 [Eq. (7.2)], thus the two quantities are used with no distinction in Sects. 7.1.1–7.1.5.

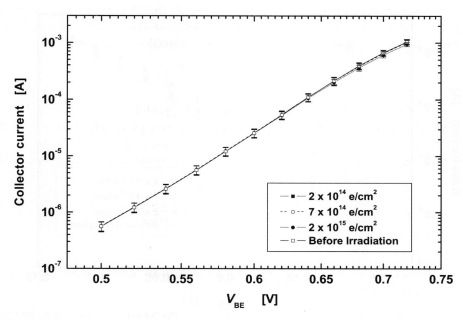

Fig. 7.1 Collector currents (I_C, in A) measured at $25\,^{\circ}$C as a function of the voltage (in V) applied at the emitter-base junction (V_{BE}), before and after irradiation with 9.1 MeV electrons for fluences of 2×10^{14}, 7×10^{14} and 2×10^{15} e/cm^2 (from [Leroy and Rancoita (2007)]; see also [Consolandi, D'Angelo, Fallica, Mangoni, Modica, Pensotti and Rancoita (2006)]).

Let us consider a transistor with narrow-base (i.e., with $W_B/L_B \ll 1$) with a graded doping-profile. As mentioned in Sect. 7.1, when this transistor has a large gain, an emitter efficiency close to 1 and is operated in the common-emitter configuration under active mode, the base transport factor is approximated by Eq. (7.7). Furthermore, it is the dominant factor compared to the common-emitter current gain [see Eqs. (7.4, 7.8)]. Under these assumptions and before irradiation, β can be approximated by the ratio between the minority-carrier lifetime τ_B and the transit time across the base τ_{tr} [Eq. (7.17)]. Since τ_{tr} is the largest contribution to the overall emitter-collector delay time (τ_d), i.e., $\tau_d \approx \tau_{tr}$, the cutoff angular frequency is roughly given by

$$\omega_T \approx \frac{1}{\tau_{tr}}. \tag{7.9}$$

By combining Eqs. (7.17, 7.19), we have, finally,

$$\beta \approx \tau_B \, \omega_T. \tag{7.10}$$

After irradiation with fast neutrons, assuming that the transistor retains a significant common-emitter gain β_{irr} (e.g., $\beta_{irr} \gtrsim 3$ [Messenger (1973)]) and that the

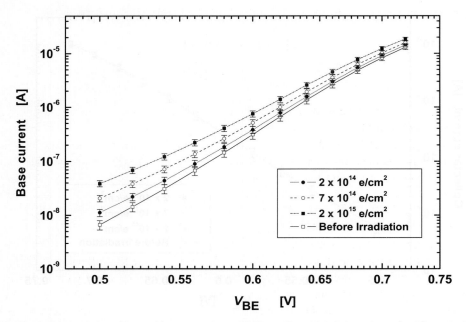

Fig. 7.2   Base currents (I_B, in A) measured at 25 °C as a function of the voltage (in V) applied to the emitter-base junction (V_{BE}), before and after irradiation with 9.1 MeV electrons for fluences of 2×10^{14}, 7×10^{14} and 2×10^{15} e/cm² (from [Leroy and Rancoita (2007)]; see also [Consolandi, D'Angelo, Fallica, Mangoni, Modica, Pensotti and Rancoita (2006)]).

diffusion constant (D_B) of the minority carriers in the base does not vary appreciably, τ_{tr} is almost constant [e.g., see Eq. (7.15)]. Thus, β_{irr} can be written as

$$\beta_{irr} \approx \frac{\tau_{B,irr}}{\tau_{tr}}$$

$$\approx \tau_{B,irr}\,\omega_T, \tag{7.11}$$

where $\tau_{B,irr}$ is the lifetime of the minority carriers after irradiation. Then, the variation of the reciprocal of the common-emitter current gain becomes

$$\Delta\left(\frac{1}{\beta}\right) \equiv \frac{1}{\beta_{irr}} - \frac{1}{\beta}$$

$$\approx \frac{\tau_{tr}}{\tau_{B,irr}} - \frac{\tau_{tr}}{\tau_B}$$

$$= \frac{1}{\omega_T}\left(\frac{1}{\tau_{B,irr}} - \frac{1}{\tau_B}\right). \tag{7.12}$$

As discussed in Sect. 4.3.1, the variation of the reciprocal of the lifetime of the minority-carriers in the base is related to the fast-neutron fluence and the lifetime constant damage; thus, using Eq. (4.125), Eq. (7.12) can be rewritten as:

$$\Delta\left(\frac{1}{\beta}\right) \approx \frac{1}{\omega_T}\Delta\left(\frac{1}{\tau_B}\right)$$

$$= \frac{\Phi_n}{\omega_T\,K_{\tau,n}}, \tag{7.13}$$

where $K_{\tau,n}$ is the lifetime damage coefficient and Φ_n is the fast-neutron fluence (e.g., see Sect. 4.3.1).

Equation (7.13) is the *Messenger–Spratt equation*[‡], which was initially proven for homogeneous base transistors [Messenger and Spratt (1958)], later extended to treat exponentially graded base transistors [Messenger (1967a)] and, finally, using the Gover–Grinberg–Seidman expression [Eq. (7.7)] generalized to cover any base doping-profile [Messenger (1973)]. Messenger (1973) has also noted that i) Ramsey and Vail have shown how the emitter efficiency contribution, resulting from the effect of recombination processes in the emitter-base field-region, is related to the emitter time-constant τ_E [Ramsey and Vail (1970)], ii) Eq. (7.9) is strictly related to the cutoff angular frequency for the base transport, but iii) ω_T can be extended by adding the emitter time constant. Thus, by means of this latter re-expressed value of ω_T, Eq. (7.13) accounts for recombination processes in both the emitter-base field-region and the base (bulk) region.

As discussed in Sect. 4.3.1, the lifetime damage coefficient ($K_{\tau,i}$) and Φ_i depends on the type and energy of the incoming particle. Thus, for the same type of transistor, by inspection of Eqs. (4.125, 7.13) we can see that the variation of the reciprocal of the current gain (or any quantity proportional to it) is expected to exhibit different slopes[§] if irradiated with different types of particle. This behavior was experimentally determined, for instance, by investigations with fast-neutrons for fluences up to $10^{14}\,\mathrm{n/cm^2}$, with 2 MeV electrons up to $2 \times 10^{15}\,\mathrm{e/cm^2}$ and with photons from a ^{60}Co source up to $4.5 \times 10^{17}\,\mathrm{photons/cm^2}$ (e.g., see Figure 42 at page 124 of [Vavilov and Ukhin (1977)], see also other results regarding $\Delta(1/\beta)$ in Figure 5.3 at page 215 of [Holmes-Siedle and Adams (2002)]). Thus, Eq. (7.13) can be rewritten as

$$\Delta\left(\frac{1}{\beta}\right) \approx \frac{\Phi_i}{\omega_T\,K_{\tau,i}}. \tag{7.14}$$

The current gain [see Eqs. (7.2, 7.3)] depends on the ratio between the collector current (I_C) and the base current (I_B). Before and after irradiation with large fluences for which the displacement damage is expected to be dominant, the behavior of the base and collector currents[*] was systematically investigated as a function

---

[‡]A complete derivation of the Messenger–Spratt equation is found, for instance, in Section 5.10 of [Messenger and Ash (1992)].

[§]This is equivalent to exhibit a shift in the double logarithmic plot of $\Delta(1/\beta)$ versus the particle fluence.

[*]The devices were characterized using an HP4142B modular DC source-monitor and the IC-CAP code (e.g., see [IC-CAP (2004)]) controlled by a workstation.

of the voltage applied to the emitter-base junction (V_{BE}). In Fig. 7.1 (Fig. 7.2), the collector (base) currents of $n-p-n$ transistors[¶] with large emitter area region (50 μm $\times$50 μm) are shown before and after irradiation with 9.1 MeV electrons for fluences of 2×10^{14}, 7×10^{14} and 2×10^{15} e/cm² [Leroy and Rancoita (2007)] (see also [Consolandi, D'Angelo, Fallica, Mangoni, Modica, Pensotti and Rancoita (2006)]). Using the electron data given in Table 4.5, these fluences correspond to $D^{NIEL} \approx 3.3$, 11.7 and 33.3 Gy and $D^{Ion} \gtrsim 0.49 \times 10^5$, 1.71×10^5 and 4.88×10^5 Gy, respectively. The data were measured about two years after irradiation and are shown for $0.5 \lesssim V_{BE} \lesssim 0.72$ V, i.e., for $5 \times 10^{-7} \lesssim I_C \lesssim 10^{-3}$ A. For all the investigated fluences, the collector currents are only marginally affected by irradiation (Fig. 7.1). Larger variations are observed in the case of base currents (Fig. 7.2): these currents, as a function of V_{BE}, are systematically larger after irradiation because, in the base, the concentration of recombination centers and the recombination current increase with fluence. These results are in agreement with those found for irradiations with Ar-ions [Codegoni et al. (2004b); Consolandi, D'Angelo, Fallica, Mangoni, Modica, Pensotti and Rancoita (2006)].

The radiation effect on the cutoff frequency f_T of (Si) bipolar transistors (see page 549) has been investigated after irradiation with a fast-neutron fluence of 10^{15} n/cm² and for collector currents $5 \times 10^{-2} < I_C < 5$ mA [Roldan, Ansley, Cressler and Clark (1997)]. After irradiation, f_T shows practically no change with respect to the values before irradiation for I_C up to ≈ 1 mA, but it becomes slightly larger with increasing I_C. These measurements indicate that the overall emitter-to-collector delay time τ_d [Eq. (7.18)] is almost not affected by the fast-neutron irradiation for Si bipolar transistor with large f_T[‖].

The extension of the neutral region in the base and the properties of the minority carriers are related to the transit time of these carriers (τ_{tr}) across the base. For a homogeneous doping-profile of the base, it can be shown {e.g., see Section 7.3(a) of [Grove (1967)]} that τ_{tr} is given by

$$\tau_{tr} = \frac{W_B^2}{2 D_B}, \qquad (7.15)$$

where D_B is the diffusion constant of the minority carriers in the base. For a graded doping-profile of the base and $W_B/L_B \ll 1$, τ_{tr} is given by {Equation (32) of [Gover, Grinberg and Seidman (1972)]}

$$\tau_{tr} = u_1 \frac{W_B^2}{D_B}. \qquad (7.16)$$

It can be remarked that, since $u_1 = 1/2$ for a homogeneous doping-profile in the base (see footnote ‡ at page 544), in such a case Eq. (7.16) reduces to Eq. (7.15). To a first approximation by determining W_B^2 from Eq. (7.15) [Eq. (7.16)] and introducing

---

[¶]These transistors are manufactured according to the HF2BiCMOS technology by STMelectronics [Gola, Pessina and Rancoita (1990)].

[‖]Before irradiation, the peak value of the cutoff frequency is ≈ 28.9 GHz [Roldan, Ansley, Cressler and Clark (1997)].

this parameter in Eq. (7.6) [Eq. (7.8)], we can estimate the common emitter gain for both the homogeneous and graded base doping-profiles, in terms of the transit time across the base and the lifetime of the minority carriers:

$$\beta \simeq \frac{L_B^2}{D_B \tau_{tr}} = \frac{\tau_B}{\tau_{tr}}. \tag{7.17}$$

The transit time of minority carriers across the base may also limit the transistor operation, when the operating frequency ($f = \omega/2\pi$) increases beyond a certain critical frequency. The *cutoff frequency*[*], f_T, is an important figure of merit in transistors. It is defined as the frequency at which the *common-emitter short-circuit current gain* (h_{fe}) is *unity* [Pritchard, Angell, Adler, Early and Webster (1961)]. The frequency response of a transistor can be described in the framework of the *hybrid-π model* (e.g., see Section 4.5 of [Brennan and Brown (2002)]) in which the h-parameter, h_{fe}, is given by $h_{fe} = \partial I_C / \partial I_B$. Furthermore, for $\omega \to 0$ we have that

$$h_{fe} \to \beta$$

{e.g., see Equation (4.5.20) at page 179 of [Brennan and Brown (2002)]}, i.e., h_{fe} reduces to the d.c. common-emitter current gain. The cutoff frequency is inversely proportional to the overall emitter-to-collector delay time τ_d (e.g., see Section 4.3.2 of [Sze (1985)], Sections 4.3 and 4.5 of [Brennan and Brown (2002)] and also [Ramsey and Vail (1970)]):

$$f_T = \frac{1}{2\pi\tau_d}. \tag{7.18}$$

τ_d is the delay resulting from i) the emitter depletion-layer charging time $\tau_E = r_E C_t$, where r_E is the emitter resistance and C_t the electrically measurable total delay capacitance (e.g., see [Ramsey and Vail (1970)], Section 3.3.1 of [Sze (1981)] and Section 5.3.1 of [Sze and Ng (2007)]), ii) the collector depletion-layer transit time and collector charging-time (e.g., see Section 3.3.1 of [Sze (1981)] and Section 5.3.1 of [Sze and Ng (2007)]) and iii) the transit time of the minority carriers across the base region. Generally, this latter term is the most limiting parameter (e.g., see Section 4.3.2 of [Sze (1985)] and Section 4.3 of [Brennan and Brown (2002)]) that influences the transistor frequency-response. From Equation (7.18), we have that the *cutoff angular frequency* is

$$\omega_T = \frac{1}{\tau_d}, \tag{7.19}$$

which (as well as f_T) depends on the collector current I_C [i.e., $\omega_T(I_C)$].

---

[*]This term expresses also the so-called *gain bandwidth product* for the common-emitter gain (e.g., see Section 3.10 of [Messenger and Ash (1992)]).

7.1.2   *Surface and Total Dose Effects on the Gain Degradation of Bipolar Transistors*

Equation (7.13) accounts for the gain degradation of bipolar transistors when the atomic displacement is the dominant damage-mechanism. However, it was found that the gain degradation may result from damages in the silicon lattice and modifications in the *surface properties* (i.e. the so-called *surface effects*) of the crystal. For instance, changes in the recombination properties of surface layers like those close to the emitter-base junction may affect the transistor current-gain. Surface effects may occur even at small radiation doses which, usually, are insufficient to introduce an appreciable bulk damage. Thus, these effects can be investigated with particles of very large $D^{\mathrm{Ion}}/D^{\mathrm{NIEL}}$ ratio between the doses deposited by ionization and atomic displacement (e.g., see Sect. 4.2.3.1), like γ-rays from a ^{60}Co source.

This type of damage was considered to be responsible for the observed non-linear dependence of $\Delta(1/\beta)$ in $n-p-n$ transistors as a function of fluence up to $\approx 10^{15}$ Compton-electron/cm^{2}** (e.g., see Figure 60 at page 165 of [Vavilov and Ukhin (1977)]). These electrons have an average kinetic energy of $\approx 0.53\,\mathrm{MeV}$ and are generated by Compton effect in irradiations with γ-rays from a ^{60}Co source†. The surface component of the damage grows rapidly and approaches saturation for fluences $\gtrsim 3 \times 10^{13}$ Compton-electron/cm^{2}‡‡. Above $\approx 10^{15}$ Compton-electron/cm^{2}, the transistor degradation was found to be dominated by the displacement damage as expected from Eq. (7.12). After irradiations with 2 MeV electrons, a similar behavior is observed for the same type of bipolar transistor (e.g., see Figure 5.3 at page 215 of [Holmes-Siedle and Adams (2002)]). In irradiations of $n-p-n$ transistors with gamma-rays from ^{60}Co source up to a dose of $\approx 0.5\,\mathrm{Mrad}$, the excess base-current was observed to increase non-linearly with increasing dose* [Nowlin et al. (1993)]. Furthermore, this excess base-current was observed to saturate at about 1 Mrad, almost independently of the dose rate [Kosier et al. (1994)]. The linear increase of the inverse of the current gain was observed at large doses (above 5–10 Mrad) after irradiations with gamma-rays from ^{60}Co source and at large fluences ($\gtrsim 10^{14}\,\mathrm{e/cm^{2}}$) with 4.1 MeV electrons [Cheryl, Marshall, Burke, Summers and Wolicki (1988)].

The so-called *lateral* and *vertical* transistors are realized with an architecture which favors the current flow, mainly, parallel to the surface and in the vertical direction, respectively. In irradiations with gamma-rays from ^{60}Co source up to $\approx 0.5\,\mathrm{Mrad}$, a systematic comparison of dose effects on $p-n-p$ transistors has

---

** At this Compton-electron fluence, the concentration of initial Frenkel-pairs is $\approx 4.9 \times 10^{14}\,\mathrm{cm^{-3}}$, for $E_d = (21\text{--}25)\,\mathrm{eV}$.

†† The ratio of the Compton-electron flux to the flux of γ-rays from the ^{60}Co source is ≈ 0.0162 [Summers, Burke, Shapiro, Messenger and Walters (1993)].

‡‡ At this Compton-electron fluence, the concentration of initial Frenkel-pairs is $\approx 1.5 \times 10^{13}\,\mathrm{cm^{-3}}$, for $E_d = (21\text{--}25)\,\mathrm{eV}$.

* To deliver a dose of $\approx 100\,\mathrm{Mrad}$, a fluence of $\approx 1.9 \times 10^{17}\,\gamma/\mathrm{cm^{2}}$ from a ^{60}Co source is needed (e.g., see Figure 5.3 at page 215 of [Holmes-Siedle and Adams (2002)]).

shown that vertical transistors exhibit the least gain degradation [Schimidt et al. (1995)]. Furthermore, it has been observed that lateral $p - n - p$ transistors with lightly-doped emitters degrade more rapidly than devices with heavily-doped emitters [Wu et al. (1997)].

Surface and low dose-rate effects in irradiated transistors are discussed, for instance, in Section 6 of Chapter I of Part II of [Vavilov and Ukhin (1977)], [Cheryl, Marshall, Burke, Summers and Wolicki (1988)], Section 6.14 of [Messenger and Ash (1992)], Chapter 5 of [Claeys and Simoen (2002)] and Section 5.4 of of [Holmes-Siedle and Adams (2002)] (see also references therein).

7.1.3 Generalized Messenger–Spratt Equation for Gain Degradation of Bipolar Transistors

As discussed in Sect. 7.1.1, the Messenger–Spratt equation [Eqs. (7.13, 7.14)] relates the particle fluence to the variation of the reciprocal of the common-emitter current gain for particles fluences sufficiently large to allow the displacement damage in the bulk to be the dominant damage-factor. In Section 4.3.1, we have already discussed that the concentration of Frenkel-pairs (FP) is expected to be proportional to the concentration of deep defects acting as recombination-centers, when these traps are resulting from primary defects mostly created by cascading-displacement processes. As a consequence, the variation of the reciprocal of the minority-carrier lifetime $[\Delta(1/\tau_{\mathrm{B}})]$ is expected to depend linearly on FP almost independently of the type of incoming particle, i.e., to exhibit an *approximate NIEL scaling*.

Under the assumptions regarding the validity of the Messenger–Spratt equation (discussed in Sect. 7.1.1) and these latter about the creation mechanism of recombination centers, we can combine Eqs. (4.131, 7.14) to obtain

$$\Delta\left(\frac{1}{\beta}\right) \approx \frac{1}{\omega_T}\left(\frac{\Phi_i}{K_{\tau,i}}\right)$$
$$= \lambda\,\frac{FP}{\omega_T}, \tag{7.20}$$

where (as discussed in Sect. 4.3.1) λ is almost independent of the type and energy of the incoming particle, but depends on the i) the type of substrate, ii) the dopant concentration (slightly) and iii) the level of compensation. Thus, λ may depend on the type ($n - p - n$ or $p - n - p$) of transistor and VLSI technology (e.g., base width and doping-profile of the graded base). However, the effect on the gain degradation $[\Delta(1/\beta)]$ is expected to depend almost linearly on FP. Equation (7.20) is the *generalized expression for the common-emitter current gain degradation in bipolar transistors* and it is termed *generalized Messenger–Spratt equation*[†] [Consolandi,

---

[†]The treatment described in Sect. 4.3.1 and the *generalized Messenger–Spratt equation* were first derived by Rancoita et al. in [Colder et al. (2001); Codegoni et al. (2004b); Consolandi, D'Angelo, Fallica, Mangoni, Modica, Pensotti and Rancoita (2006)] (see also [Leroy and Rancoita (2007)] and references therein).

D'Angelo, Fallica, Mangoni, Modica, Pensotti and Rancoita (2006)]. This equation extends the Messenger–Spratt equation (7.13) originally derived for fluences of fast neutrons. It relates $\Delta(1/\beta)$ to the concentration of Frenkel-pairs created in the silicon by the displacement damage and accounted for the energy deposited in NIEL processes (treated in Sect. 4.2.1). In addition, it predicts that, similarly to $\Delta(1/\tau_B)$ [see discussion in Sect. 4.3.1], $\Delta(1/\beta)$ is expected to follow an approximate NIEL scaling. Furthermore, Eq. (7.20) can be rewritten as

$$\Delta\left(\frac{1}{\beta}\right) \approx k(I_c)\, FP \tag{7.21}$$

with

$$k(I_c) = \frac{\lambda}{\omega_T}. \tag{7.22}$$

The first experimental evidence of the validity of Eq. (7.20) [Eq. (7.21)] was obtained in irradiations with fast-neutrons with fluences up to $\approx 1.2\times10^{15}\,\mathrm{n/cm^2}$, with ^{12}C-ions[¶] of 95.0 MeV/amu up to $\approx 10^{13}\,\mathrm{C/cm^2}$ and with ^{13}C-ions of 11.1 MeV/amu up to $\approx 10^{13}\,\mathrm{C/cm^2}$ [Colder et al. (2001)] (see also [Colder et al. (2002)]). $\Delta(1/\beta)$ was found to follow an approximate NIEL scaling over the full range of the generated concentration of Frenkel-pairs[‡] [$(6.4 \times 10^{13}$–$1.1 \times 10^{18})\,\mathrm{cm^{-3}}$]. It has to be noted that for same amount of FPs the $D^{\mathrm{Ion}}(\text{Carbon})/D^{\mathrm{Ion}}(\text{neutron})$ dose ratio is larger than 10^3 (see Table 4.5 and [Consolandi, D'Angelo, Fallica, Mangoni, Modica, Pensotti and Rancoita (2006)]). Experimental values of $\Delta(1/\beta)$ as a function of FP are shown in Fig. 7.3 for collector currents of $50\,\mu\mathrm{A}$. In the figure, the data are for $n-p-n$ transistors with (a) small and (b) large emitter area, and (c) vertical and (d) lateral $p-n-p$ transistors[‡‡] manufactured according to the HF2BiCMOS technology [Gola, Pessina and Rancoita (1990)].

As can be seen in Fig. 7.3, these results have been confirmed in irradiations with Ar- and Kr-ions [Codegoni et al. (2004b); Consolandi, D'Angelo, Fallica, Mangoni, Modica, Pensotti and Rancoita (2006)]. Irradiations with 9.1 MeV electrons for fluences up to $2 \times 10^{15}\,\mathrm{e/cm^2}$ have provided a further confirmation of Eq. (7.20) (see Fig. 7.3) [Codegoni et al. (2006); D'Angelo, Fallica, Galbiati, Mangoni, Modica, Pensotti and Rancoita (2006)]. In fact, the ratio $D^{\mathrm{Ion}}(\text{electron})/D^{\mathrm{Ion}}(\text{neutron})$ of the ionization dose of electrons with respect to that of fast-neutrons is larger than four orders of magnitude for the same amount of FP created by displacement damage (see Table 4.5 and [Consolandi, D'Angelo, Fallica, Mangoni, Modica, Pensotti and Rancoita (2006)]). It has to be noted that, as discussed in Sect. 4.2.1.4, the recoil silicon is able to produce a small cascade development for incoming electrons with

---

[¶]The reader can find the definition of kinetic energies per amu in Sect. 1.4.1.

[‡]The concentrations of Frenkel-pairs for C-ions were computed by means of the TRIM simulation code [Ziegler, Biersack and Littmark (1985a); Ziegler (2001)] and for fast-neutrons by the data available in literature for the damage function (e.g., see Sect. 4.2.1.5).

[‡‡]The devices were characterized using an HP4142B modular DC source-monitor and the IC-CAP code (version 2004 [IC-CAP (2004)] or previous versions) controlled by a workstation.

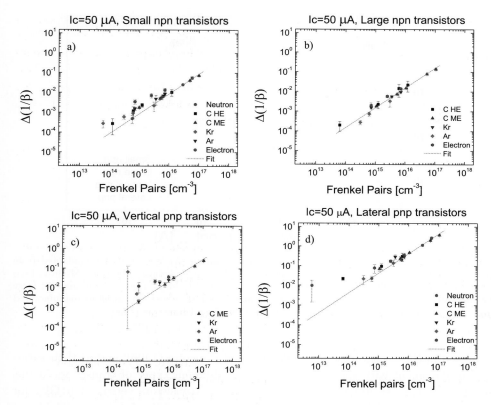

Fig. 7.3 $\Delta(1/\beta)$ as a function of the concentration of Frenkel-pairs (FP) (from [Leroy and Rancoita (2007)]; see also [Consolandi, D'Angelo, Fallica, Mangoni, Modica, Pensotti and Rancoita (2006)]): for $n - p - n$ transistors with (a) small [$5\,\mu$m $\times 5\,\mu$m] and (b) large [$50\,\mu$m $\times 50\,\mu$m] emitter area, and for (c) vertical and (d) lateral $p - n - p$ transistors manufactured according to the HF2BiCMOS technology [Gola, Pessina and Rancoita (1990)]. The collector current is $50\,\mu$A. The incoming particles are fast-neutrons, ^{12}C-ions of 95.0 MeV/amu (high-energy), ^{13}C-ions of 11.1 MeV/amu (medium-energy), ^{36}Ar-ions of 13.6 MeV/amu, ^{86}Kr-ions of 60.0 MeV/amu and electrons of 9.1 MeV [Colder et al. (2001, 2002); Codegoni et al. (2004b, 2006); Consolandi, D'Angelo, Fallica, Mangoni, Modica, Pensotti and Rancoita (2006); D'Angelo, Fallica, Galbiati, Mangoni, Modica, Pensotti and Rancoita (2006)]. The line represents the linear dependence expected from Eq. (7.20).

this kinetic energy. Furthermore, for any value of the collector current, the slope of the fitted straight-line to $\Delta(1/\beta)$ as a function of FP [e.g., see Eqs. (7.21, 7.22)] determines the quantity $k(I_c)$ and, thus, the value of

$$\frac{f_T(I_{\rm C})}{\lambda} = \frac{1}{2\,\pi\,k(I_c)} \tag{7.23}$$

as a function of the collector current [e.g., see Eq. (7.19)]. $f_T(I_{\rm C})/\lambda$ as a function of $I_{\rm C}$ is shown in Fig. 7.4 for the irradiated $n - p - n$ and $p - n - p$ transistors. These curves indicate that i) the cutoff frequencies reach their maxima above $100\,\mu$A as

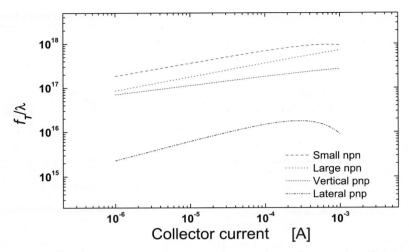

Fig. 7.4 $f_T(I_C)/\lambda$ as a function of the collector current (in A) [see Eq. (7.23)] (adapted from [Leroy and Rancoita (2007)]; see also [Consolandi, D'Angelo, Fallica, Mangoni, Modica, Pensotti and Rancoita (2006)]): for $n - p - n$ transistors with (a) small [$5\,\mu\mathrm{m} \times 5\,\mu\mathrm{m}$] and (b) large [$50\,\mu\mathrm{m} \times 50\,\mu\mathrm{m}$] emitter area, and for (c) vertical and (d) lateral $p - n - p$ transistors manufactured according to the HF2BiCMOS technology [Gola, Pessina and Rancoita (1990)].

expected in this technology [Gola, Pessina and Rancoita (1990)] and ii) their ratios are in agreement with those expected for the different types of transistors in this technology [e.g., $\approx 20\,\mathrm{MHz}$ for lateral $p - n - p$ transistors versus (2–6) GHz for $n - p - n$ transistors]. It has to be noted that below $\approx 100\,\mu\mathrm{A}$ (e.g., see [Colder et al. (2001); Consolandi, D'Angelo, Fallica, Mangoni, Modica, Pensotti and Rancoita (2006)]), $k(I_c)$ can be approximated by

$$k(I_c) = A_{\mathrm{tr}}\, I_c^n, \tag{7.24}$$

where A_{tr} and n depend on the type of transistor. The values found for the coefficient n are $\approx -(0.2\text{–}0.4)$ [Leroy and Rancoita (2007)] (see also [Consolandi, D'Angelo, Fallica, Mangoni, Modica, Pensotti and Rancoita (2006)]) and are in agreement with those available in literature (e.g., see Section 7 of Chapter 1 of Part II in [Vavilov and Ukhin (1977)]).

In summary, these experimental data indicate that, at large values of FP, the dominant processes resulting in gain degradation are well accounted by the processes of non-ionization energy-loss.

7.1.4   *Transistor Gain and Self-Annealing*

The defects induced by radiation, as discussed in Sect. 4.2.2, may be unstable and, thus, their concentration may decrease with time. For instance (e.g., see Sect. 4.3.1), an annealing effect was observed for the lifetime of minority-carriers, thus decreasing the value of $K_{\tau,n}$ (see pages 32–45 in [Srour, Long, Millward, Fitzwilson and Chadsey (1984)], also [Srour (1973)] and references therein). Therefore, the annealing is expected to cause the (slight) increase of the current gain of the bipolar transistors. To a first approximation, this gain increase exponentially approaches a constant value with time.

The annealing effect on $\Delta(1/\beta)$ for large emitter area $n-p-n$ transistors [Gola, Pessina and Rancoita (1990)] has been investigated, at 25 °C, from immediately after the (electron) irradiation up to about 70 months after Argon irradiation [Codegoni et al. (2004b); Rancoita (2005); Consolandi, D'Angelo, Fallica, Mangoni, Modica, Pensotti and Rancoita (2006); Leroy and Rancoita (2007)] and is shown in Fig. 7.5 for three values of the collector currents: $1\,\mu A$, $50\,\mu A$ and $1\,mA$. The irradiations with electrons of 9.1 MeV (2×10^{15} e/cm^2) and ^{36}Ar-ions of 13.6 MeV/amu (10^{11} Ar/cm^2) created initially a concentration $\approx 7.2 \times 10^{15}$ cm^{-3} of Frenkel-pairs. The data show that i) there is evidence for a long-term annealing with a decay time of (10.0 ± 2.98) months and ii) the effect of annealing is similar for devices irradiated with electrons and Ar-ions [Consolandi, D'Angelo, Fallica, Mangoni, Modica, Pensotti and Rancoita (2006)].

7.1.5   *Radiation Effects on Low-Resistivity Base Spreading-Resistance*

The bipolar transistors have a wide range of circuit applications, for example as input device in fast front-end electronics[††]. However, the degradation of transistor properties resulting from radiation damage may impair the expected operation of the whole electronic circuit. In Sections 7.1.1, 7.1.3, we have discussed the decrease of the gain and the increase of the base current of bipolar transistors with increasing fluence. Furthermore from the base lead, a current must flow through the graded-base region up to the recombination centers, the emitter-base junction and, also, the emitter region: the overall resistance, involving different parts of the base region, is referred to as *base spreading-resistance* {e.g., see Section 16.4(d) of [van der Ziel (1976)], Section 7.2 of [Müller and Kamins (1977)] and Section 14.4(c) of [Bar-Lev (1993)]}: the displacement damage may also cause a variation of the effective-doping concentration in the base. In high-resistivity devices (e.g., silicon radiation detectors), the resistivity is largely modified by irradiation (see Sects. 4.3.4, 4.3.5, 6.8.3). As the doping concentrations used in bipolar transistors are several orders of magnitude larger, the radiation-induced change in doping concentration is usually less important, but may not be negligible at large fluences. For instance when these transistors (see Section 5.1.2 of [Gatti and Manfredi (1986)]) are employed as input stage of charge-sensitive-preamplifier (CSP), the increase of transistor base-current and slight variation of the effective doping-

---

[††]The reader may see, e.g., [Gatti and Manfredi (1986); Gola, Pessina and Rancoita (1990); Gola, Pessina, Rancoita, Seidman and Terzi (1992); Baschirotto et al. (1997)] and references therein.

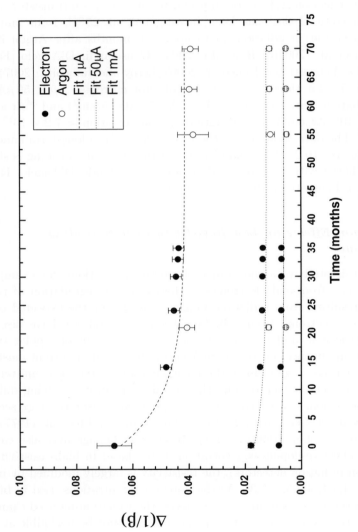

Fig. 7.5   Annealing effect at 25 °C: dependence of $\Delta(1/\beta)$ for large emitter-area transistors [Gola, Pessina and Rancoita (1990)] on the time (in months) after irradiation with electrons (2×10^{15} e/cm^2) and Ar-ions (10^{11} Ar/cm^2) (adapted from [Leroy and Rancoita (2007)]; see also [Rancoita (2005); Consolandi, D'Angelo, Fallica, Mangoni, Modica, Pensotti and Rancoita (2006)]). The initial concentration of Frenkel-pairs is $\approx 7.2 \times 10^{15}$ cm^{-3}. The lines show the fitted curves for collectors currents exponentially decreasing with time. The currents are $1\,\mu A$ (dashed line), $50\,\mu A$ (dotted line) and $1\,mA$ (dashed and dotted line), respectively.

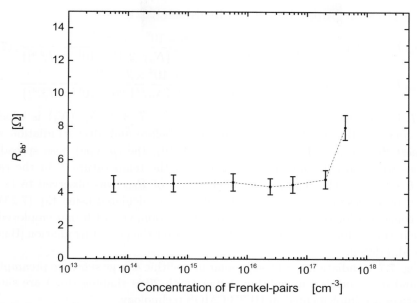

Fig. 7.6 $R_{bb'}$ (in Ω) as a function of the concentration of Frenkel-pairs (in cm^{-3}) (adapted from [Leroy and Rancoita (2007)]; see also [Baschirotto et al. (1997)]). The line is to guide the eye.

concentration of the base spreading-resistance ($R_{bb'}$) are among the relevant causes of degradation of *parallel* and *series noise* of the CSP. This occurs even with very short ($\approx 20\,\text{ns}$) shaping-times[**].

The values of $R_{bb'}$ for $n-p-n$ bipolar transistors, manufactured according to VLSI technology in HF2BiCMOS process [Gola, Pessina and Rancoita (1990)], were determined after irradiation with fast-neutrons up to a fluence of $\approx 9 \times \times 10^{15}\,\text{n}/\text{cm}^2$ by measuring the noise performance of CSPs with these devices employed as input stage [Baschirotto et al. (1997)]: they are shown in Fig. 7.6 as a function of the concentration of the Frenkel-pairs (*FP*) created during irradiation [Leroy and Rancoita (2007)]. $R_{bb'}$ is compatible with the value before irradiation up to $FP \approx 6 \times 10^{16}\,\text{cm}^{-3}$ and is $\approx (70\text{--}75)\%$ larger at the largest fluence, i.e., for $FP \approx 4.3 \times 10^{17}\,\text{cm}^{-3}$, in agreement (see discussion in [Baschirotto et al. (1997)]) with data available in literature (e.g., see Section 5.13 of [Messenger and Ash (1992)]).

Furthermore, $R_{bb'}$ is proportional to the base resistivity, thus to the reciprocal of the carrier mobility [Eq. (6.19)]. For the *p-* (*n-*) type base, the mobility $\mu_p(T)$

[**]The reader can refer, for instance, to [Baschirotto et al. (1993, 1995a,b, 1996, 1997, 1999)] and references therein.

$[\mu_n(T)]$ is given[†] by [Arora, Hauser and Roulston (1982)]:

$$\mu_p(T) = 54.3 \times T_r^{-0.57} + \frac{1.36 \times 10^8 \times T^{-2.23}}{1 + 0.88 \times T_r^{-0.146}\{N_a/[2.35 \times 10^{17} \times T_r^{2.4}]\}}, \quad (7.25)$$

$$\mu_n(T) = 88.0 \times T_r^{-0.57} + \frac{7.4 \times 10^8 \times T^{-2.33}}{1 + 0.88 \times T_r^{-0.146}\{N_d/[1.26 \times 10^{17} \times T_r^{2.4}]\}}, \quad (7.26)$$

where T is the temperature (in kelvin) and $T_r = T/300$; N_a (N_d) is effective dopant concentration in the p- (n-) type base. Before and after irradiation with the largest fluence (see [Baschirotto et al. (1997)]), the (p-type) base spreading-resistance was also determined as a function of the temperature T in the range $283 < T < 343\,\mathrm{K}$. The temperature dependence of $R_{\mathrm{bb'}}$ was observed to be determined by values of $\mu_p(T)$ compatible with those calculated using Eq. (7.25) for dopant concentration of $\approx 10^{18}\,\mathrm{cm}^{-3}$ before irradiation, i.e., with that employed for the base in this technology, and $\approx 5 \times 10^{17}\,\mathrm{cm}^{-3}$ after the largest irradiation [Baschirotto et al. (1997)].

In Fig. 7.7, a radiation hard fast bipolar monolithic charge sensitive preamplifier (CSP) and its associated monolithic shaper (with 20 ns shaping time) are shown: both devices are implemented in HF2BiCMOS technology.

7.2   Single Event Effects

As already mentioned, a *Single Event Effect* (SEE) is due to an individual event caused by a heavy-ion (i.e., a single charged particle with $Z \geq 2$) traversing a semiconductor or semiconductor-based device (integrated circuit, power supply, etc.). A SEE results in a failure of the device. This failure is the consequence of the charge deposition resulting from the electron-hole pairs generated along the path of the incoming particle within the sensitive volume of the device (*direct ionization*). SEE can also be caused by the interactions (elastic, inelastic scattering or spallation) of the incoming particle, neutron or proton, with atoms in the device producing light particles such as protons, deuterons, α-particles, light nuclei such as lithium, beryllium and heavy recoil nuclei such as magnesium. Most of the recoil species are heavier than the original particle and have large ionization capabilities. Then, the recoil particles can also generate tracks of electron-hole pairs along their path (*indirect ionization*). The charge deposition mechanisms are the same as in the case of directly ionizing ions. In both situation, the energy deposition via direct or indirect ionization creates electric charges which, if enough electric charge is created, may modify the functional state of the device, leading to errors. The minimum electric charge that causes the change of state in the device is called the critical charge (Q_c) and a SEE occurs if the collected charge exceeds Q_c. The critical charge is formed

---

[†]These expressions agree (within $\pm 13\%$) with experimental data in the temperature range (250–500) K and up to a dopant concentration of $\approx 10^{20}\,\mathrm{cm}^{-3}$ (for additional data see [Arora, Hauser and Roulston (1982)] and references therein).

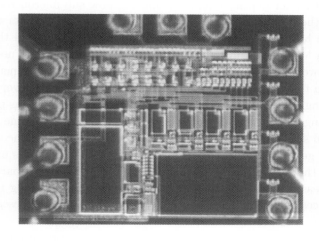

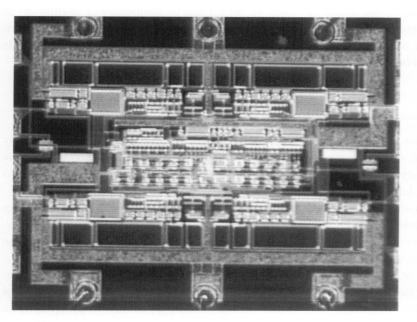

Fig. 7.7 (left) Photograph of a monolithic CSP: four preamplifiers are located on a single die, sharing the biasing network; the chip dimension is $1.62 \times 1.17\,\mathrm{mm}^2$. (right) Photograph of the monolithic shaper associated to the CSP: the circuit area is $1.08 \times 0.93\,\mathrm{mm}^2$. The photographs are reprinted from *Nucl. Instr. and Meth. in Phys. Res. B* **155**, Baschirotto A. et al., A radiation hard bipolar monolithic front-end readout, 120–131, Copyright (1999), with permission from Elsevier, e.g., for the list of the authors see [Baschirotto et al. (1999)].

by energy deposition in a critical volume which depends on the sensitive volume of the device. This sensitive volume, which is the charge collection region, is smaller than the physical volume. The evolution of modern technology goes toward the creation of faster and less power consuming devices of increasingly small sensitive volume and higher density circuits, achieving now sub-micron technology with an increase of the number of memory elements. High density circuits with their smaller feature size have less capacitance, lower operating voltage and the information is stored with less charge or current. The device dimensions are now comparable to and even lower than the range of the striking particles which generate electron-hole pairs within the depth of the device. The occurrence of SEE is very sensitive to the incidence angle of the striking particle upon the circuit device as particles with low incidence angle may deposit more charge in the sensitive volume. This continuous reduction in size and multiplication of cells decreases Q_c and the amount of charge required to generate SEE and opens this possibility to particles of lower energy. It is observed that the relative system soft error rate (SER) for α-particle and neutron effects increases by three orders of magnitude when decreasing the range of sub-micron process technology from $0.25\,\mu$m to $0.05\,\mu m$ [Semico Research Corporation (2002)].

A SEE is inflicted by a single particle, ion (direct ionization) or nucleon (indirect ionization) responsible for damages, which can be temporary or permanent in contrast to the permanent damage in electronics inflicted by an integrated radiation dose, which is usually delivered by a high particle flux, particularly near particle accelerators or in space.

7.2.1 *Classification of SEE*

One classifies SEEs into three categories: *Single Event Upset* (SEU), or simply called *upset*, causing temporary damage, *Single Event Latchup* (SEL) responsible for temporary or permanent damage and *Single Event Burnout* (SEB) leading to permanent damage.

Effects caused by SEEs to particular devices depend on the type and energy of the incident particle and the characteristics of the devices (material, geometry, thickness, etc.). Then, the effect of SEE are device specific and depends on how the resulting corrupted information affects the system including the device.

SEU occurs in memory circuits and logic circuits (DRAM, SRAM, microprocessors, etc.). Semiconductor memory devices store data as the presence or absence of charge carriers in storage wells define the logic state "0" or "1", respectively. The amount of stored charges typically varies from $\sim 0.3 \times 10^6$ up to $\sim 3 \times 10^6$ electrons. The number of electrons which differentiates between "0" or "1" is the critical charge Q_c. In a SEU, the incoming particles release in the proximity of a memory cell an amount of charges exceeding Q_c. This produces a voltage spike, which induces a change to the cell status by a flip $1 \rightarrow 0$ or $0 \rightarrow 1$ generating an error in a bit. It is called a soft error (SE) when the content of the memory device is changed without damaging the device. This occurrence makes the stored data unreliable and the device must be rewritten with correct data. SEU may also change randomly a program of a computer or confuse a processor to the point that it may crash. In general the damage caused by SEU are non-permanent since their effect is canceled by data rewriting or system rebooting.

For high density circuits, charge collection by diffusion or charge sharing from a single ion may lead to multiple-bit upsets. *Multiple bit upsets* (MBU) is a SEU that brings simultaneously corruption of several memory cells (change of logic state) physically adjacent. MBU can be induced by diffusion of charge carriers in the sensitive volume before their collection or by ions striking at low incidence angle [Reed et al. (1997)]. For instance, this is the mechanism responsible for MBU in DRAM [Zoutendyck, Edmonds and Smith (1989a,b); Zoutendyck, Schwartz and Nevill (1989)]. This effect can be removed by a combination of error-correcting codes, that work on a word-by-word basis [Bossen and Hsia (1980)], and lay-out, that prevent physically adjacent bits from belonging to the same word of memory [Crafts (1993)]. Occurrence of MBU is expected to become more serious with the use of increasingly smaller geometries by advanced very large scale integration (VLSI) processes [Zoutendyck, Schwartz and Nevill (1989)].

In a SEL [Bruguier and Palau (1996); Dodd (1996); Pickel (1996)], the energy released by a single particle is injected into parasitic $p - n - p - n$ structures inherent to CMOS technology (proximity of NMOS and PMOS transistors), possibly triggering a short circuit and at the end resulting into the destruction of the device if the feeding voltage is not switched-off immediately. A SEL may be cleared by a power off-on reset.

The SEB is a condition that can cause device destruction due to high current state in a power transistor. In particular, the passage of a particle through the dielectric of a power MOSFET may create a plasma in between silicon and the grid with subsequent diffusion towards the Si-SiO$_2$ interface and increase of the electric field in the gate oxyde leading to destructive burnout. This non-reversible process is called a *single-event gate rupture* (SEGR). For GaAs circuits, SEL and SEB do not occur but GaAs devices are more sensitive to SEU than Si devices [Karp and Gilbert (1993)].

This review will focus on the study of SEU, and soft error (SE), in particular.

SEEs occur randomly since ions, protons or neutrons of various energies and incidence angles hit the circuit randomly. Due to their random nature it is difficult to predict their occurrence simply based on theoretical analysis. However, it is possible to calculate the rate of SEE events from the combination of measurements of their cross sections (the meaning of this cross section will be defined in Sect. 7.2.8) for a specific device exposed to a given radiation environment with models describing the physical mechanisms taking place inside the device. One needs to know the flux or fluence of the irradiating particles. As the sensitivity of a device to SEE depends on its geometry, technology, size, thickness, circuit and memory densities (and speed), rate estimates may have impact on device design.

SEE occur both in space, atmospheric and terrestrial environment.

7.2.2   *SEE in Spatial Radiation Environment*

Anomalies in satellites orbiting around the Earth or in spacecraft operating in deeper space are the result of their exposure to the spatial radiative environment made of several components (a detailed review of these components is presented in Sects. 4.1.2, 4.1.2.4, 4.1.2.5). The primary cosmic rays includes galactic particles entering the solar system. The galactic flux of primary cosmic rays is very low ~ 5 particles cm^{-2} s^{-1}. These cosmic rays are made of very energetic charged particles, dominated by protons ($\approx 87\%$) with an energy range from 100 MeV up to 1 GeV, α-particles ($\approx 12\%$) and heavy-ions representing $\approx 1\%$ of this radiative component with energy ranging from 1 MeV up to 10^5 TeV. The abundance of heavy ions rapidly falls with the atomic number. Iron, $Z = 26$, is the heaviest most significant component, abundances of species heavier than iron are 2–4 orders of magnitude smaller than iron. The large electric charge of these heavy ions gives them an ionization power much higher than that of protons and α-particles. Primary cosmic rays of the highest energy may hit the Earth. The flux of these galactic cosmic rays in the solar system is modulated by the solar activity (Sect. 4.1.2.3). Large solar wind shields the solar system from these particles. The solar wind is made of particles (electrons, protons and α-particles) of energy (< 100 keV) much lower than that of galactic particles. The Sun has an eleven-year cycle in average with four years of low activity and seven years of high activity during which solar flares of variable duration (from a few hours up to several days) occur (Sect. 4.1.2.1). These solar flares produce a flux of particles made of protons (10^{10} p cm^{-2} s^{-1}) of energies from a few MeV up to several hundreds of MeV and heavy ions (10^{10} particles cm^{-2} s^{-1}) with energies ranging from 10 MeV up to 100 GeV. The charged particles trapped in the Van Allen belts are another component of the spatial radiation environment. The belts exist in two regions: the inner zone which extends over a region close to Earth about 1.0 Re up to $\lesssim 2.4$ Re (Re = 6371 km is the radius of the Earth) and the outer zone which extends over a region $\gtrsim 2.4$ Re) up to 9.0 Re. The inner belt is filled by electrons, protons and some heavy ions. Electrons and protons are produced by the β-decay ($n \rightarrow p e^- \bar{\nu}_e$) of neutrons produced by the interaction of cosmic rays with atmospheric particles [Singer (1958); Hess (1959)]. The electrons and protons are trapped by the Earth's magnetic field. The neutrinos, being electrically neutral and with a very small mass, evade the belt into the cosmos. The inner belt reaches a minimum altitude of 250 km, locally, above the Atlantic Ocean off the Brazilian coast, the South Atlantic Anomaly (SAA) as the result of the offset between the Earth's geographical and magnetic axes. The inner zone is dominantly populated with protons of energies in the (10–100) MeV range, their population is affected by the solar wind and flares. The trapped protons flux varies between 1 to 2×10^6 particles cm^{-2} s^{-1} with a maximum around 2.0 Re. In the inner zone, the trapped electrons have energies $\lesssim 5$ MeV. The electron content of the inner zone may vary considerably in time due to phenomena like solar wind or like nuclear explosions in upper atmosphere. A low flux of trapped ions of rigidities corresponding[¶] to a few MeV/nucleon and around (3–4) Re, mainly He and O with traces of C and N, has been observed. These particles are believed to be extracted from the upper layers of the atmosphere during solar storms. SAMPEX data [Cummings et al. (1993)] have also shown the existence of belts included in the inner zone, containing nuclei such as N, O (of ≈ 10 MeV/nucleon) and Ne [of (10–100) MeV/nucleon]. The filling mechanism of these SAMPEX belts is likely due to the interaction of the anomalous cosmic rays with the Earth atmosphere (in anomalous cosmic rays there are more α-particles than protons and much more oxygen than carbon while in the galactic cosmic rays there are more protons than α-particles and equal amounts of oxygen and carbon). Recently, belts below the inner belt zone, at ~ 1.0 Re, have been observed by the AMS experiment [AMS Collab. (2000a,b,d, 2002)] with a relatively large content of electrons, positrons, deuterium and ^{3}He with rigidities of 10^2–10^3 MeV/nucleon. The origin of the AMS belts is believed to result from the interaction of primary cosmic rays with the atmosphere. The outer belt zone is mainly populated with electrons of energies ~ 7 MeV (up to 10 MeV), with a flux of an order of magnitude higher than in the inner zone and peaking at about 5 Re. The trapped electrons flux in the outer zone varies between 10^3 and 2×10^6 particles cm^{-2} s^{-1}. This outer zone is filled by the solar wind.

7.2.3   *SEE in Atmospheric Radiation Environment*

The atmospheric environment originates from the cosmic ray cascades in the Earth's atmosphere and are a concern for avionics. This environment results from the interaction of cosmic rays with atmospheric atoms through ionization or nuclear reactions. The isotopic content of the atmosphere is predominantly made of oxygen (21.8%, $Z = 8$), nitrogen (76.9%, $Z = 7$) and argon (1.3%, $Z = 18$). These nuclear reactions are mostly induced by the primary protons which dominate ($\approx 87\%$) the incident cosmic ray flux and generate secondary particles through elastic and inelastic scatterings, mainly neutrons (90% and more), pions, kaons, muons (from charged pion and kaon decays), photons (from neutral pion decays and photoproduction reactions), muons from pion decays and electrons from muon decays and gamma conversion. The neutron component is of interest for avionics although these neutrons can penetrate the whole atmospheric depth. However, the flux of particles (neutrons, protons, pions and kaons) is decreasing with altitude and latitude. For instance, the flux of neutrons is attenuated by their collisions with atmospheric atoms, thus decreasing the neutron flux at lower altitudes. The Earth's magnetic

---

[¶]The reader can find the definition of kinetic energies per nucleon in Sect. 1.4.1.

field traps the cosmic particles which decreases the probability of their interactions with the atmosphere atoms. As the field lines are closer at the poles, trapping is more effective, and cosmic particles can approach closer to the Earth's surface at polar latitudes compared to equatorial latitudes. Thus, the flux of cosmic particles, in particular neutrons, is larger at polar latitudes compared to equatorial latitudes. The neutron peak flux is about 4 neutron cm^{-2} s^{-1} at 20,000 m. The neutron flux is reduced to about $1/3$ and $1/400$ of the peak flux at 10,000 m and on the ground, respectively [Normand and Baker (1993); Taber and Normand (1993); Sims et al. (1994)]. The SE rate (SER) due to atmospheric neutrons (neutrons with energy higher than 1 MeV) is estimated for a range of sub-micron CMOS SRAM circuits following the formula [Shivakumar, Kistler, Keckler, Burger and Alvisi (2002)]:

$$\text{SER(number)} \sim \phi \times A \times e^{-Q_c/Q_{CE}}, \tag{7.27}$$

where ϕ is the neutron flux with energy > 1 MeV, in particles cm^{-2} s^{-1}, A is the area of the circuit sensitive to particles strikes, in cm^2. The critical charge of the SRAM cell, Q_c (in fC), depends on the circuit characteristics, namely the supply voltage and the effective capacitance of the drain nodes. Q_{CE} (in fC) measures the magnitude of the charge generated by a particle strike, it represents the charge collection efficiency of the circuit. The SER depends on the ratio Q_c/Q_{CE} and on the area of the sensitive region of the device exposed to striking particles and then decreases with the size (length × width) of the device. This model is used in [Hazucha and Svensson (2000)] to evaluate the effect of device scaling on the SER of memory circuits. In [Hazucha and Svensson (2000)] the conclusion is reached that SER/chip of SRAM circuits should increase at most linearly with decreasing feature size.

7.2.4 *SEE in Terrestrial Radiation Environment*

The terrestrial environment is made of the particles that finally hit the Earth. Particles of very high energy (> 1 GeV) can induce cascades which can penetrate the atmosphere down to the sea-level. These particles are usually galactic particles but particles of the highest energy produced during active Sun periods may also reach the Earth level and increase the intensity of cosmic rays at the Earth's surface. However, the additional magnetic field created by the solar wind around the Earth acts as a shield against cosmic rays and reduces their sea-level flux. The most abundant particles are muons. Their large number results from the decay of pions $[\pi^+(\pi^-) \rightarrow \mu^+(\mu^-)\nu_\mu(\bar{\nu}_\mu)]$ generated in the cascading process initiated by cosmic rays collisions with Earth-atmosphere nuclei (Sect. 9.11.2). The muon has a lifetime of 2.2 μs and decays into electron and neutrino-antineutrino $[\mu^-(\mu^+) \rightarrow e^-(e^+)\bar{\nu}_e(\nu_e)\nu_\mu(\bar{\nu}_\mu)]$ creating a sea of electrons and positrons in the atmosphere even close to sea-level. Neutrons are the next most abundant particles of the terrestrial environment. Being electrically neutral, they do not lose energy to the sea of electrons in the atmosphere and may reach the sea-level. Protons equally present to neutrons during the first steps of the cascading process, being electrically charged constantly lose energy to the atmospheric electrons and their number is depleted compared to neutrons at sea-level.

7.2.5 *SEE produced by Radioactive Sources*

Another radiation environment to be faced by integrated circuits (IC) is generated by radioactive sources (α-particles emitters such as Uranium [^{238}U] and Thorium [^{232}Th]) present in their package molding compounds. It has been shown that the α-particles packaging materials can produce upsets [May and Woods (1979)]. ^{238}U has a radioactivity of 6.85×10^{-7} Ci g^{-1}. ^{238}U decays to stable ^{206}Pb with the emission of 8 α-particles with energies ranging from 4.15 to 7.69 MeV. The α-particles having this range of energy can travel from 9.9 to 25.0 μm (average range of 14.3 μm) [Ziegler, Biersack and Littmark (1985b)] in an alumina ceramic with a density of 3.85 g cm^{-3}. Only 25% of the generated α-particles escape from the ceramic. Taking into account this escape factor, assuming that i) at equilibrium all the decay isotopes of ^{238}U will have the same radioactivity as the parent ^{238}U, ii) an average range 14.3 μm of α-particles in ceramic, one can relate the ^{238}U concentration (in ppm) and the α-particle flux in ceramic [Woolley, Lamar, Stradley and Harshbarger (1979)]:

$$1 \text{ ppm } {}^{238}\text{U/g of ceramic} = 0.996 \ \alpha \ \text{cm}^{-2} \text{ hour}^{-1}. \tag{7.28}$$

^{232}Th has a radioactivity of 1.11×10^{-7} Ci g^{-1}. ^{232}Th decays to stable ^{208}Pb with the emission of 6 α-particles with energies ranging from 3.95 to 8.8 MeV. The α-particles having this range of energy can travel from 9.2 to 31.2 μm (average range of 19.1 μm) in an alumina ceramic [Ziegler, Biersack and Littmark (1985b)]. Performing a calculation similar to that done for ^{238}U, one obtains the relation between the ^{232}Th concentration (in ppm) and the α-particle flux in ceramic:

$$1 \text{ ppm } {}^{232}\text{Th/g of ceramic} = 0.162 \ \alpha \ \text{cm}^{-2} \text{ hour}^{-1}. \tag{7.29}$$

One can observe that the ^{238}U/^{232}Th α-particle flux ratio is ~ 6.0. This shows that the α-particles produced by ^{238}U are the dominant cause of soft errors from IC packaging compounds.

Considering the combined presence of ^{238}U and ^{232}Th in the packaging compound, α-particles are emitted at energies ranging from 3.95 up to $\sim$ 8.8 MeV. Alpha-particles with these energies can travel up to 57 μm in silicon substrates (density of 2.33 g cm^{-3}) [Ziegler, Biersack and Littmark (1985b)] and produce up to 2.4×10^6 electron-hole pairs in the silicon component (with a depth $\geq 57 \ \mu$m) corresponding to a deposited charge up to 0.38 pC which may be possibly higher than Q_c for that device. Then, radioactive contamination of the circuit may affect its operation by causing soft errors. However, the number of the emitted α-particles falls with the increase of their energy, decreasing the number of α-particles capable to deposit a charge $\geq Q_c$. In addition α-particles produced near the surface

Table 7.1   Stopping power and projected range in several material used in circuits for incident protons [dE/dx(p) and R(p)] and α-particles [$dE/dx(\alpha)$ and R(α)] (from [Leroy and Rancoita (2007)]). The calculations have been made using data from [Ziegler, Biersack and Littmark (1985b)]. The density of Si, SiO$_2$ and GaAs are 2.33, 2.27 and 5.33 g cm^{-3}, respectively. E_{ion} is the electron-hole pair generation energy.

Circuit Material	Energy (MeV)	dE/dx(p) (keV/μm)	$dE/dx(\alpha)$ (keV/μm)	R(p) (μm)	R(α) (μm)	E_{ion} (eV)
Si	4	16	162	148	18	3.62
SiO2	4	17	178	138	16	17.0
GaAs	4	26	236	95	13	4.8

of the device will escape, contributing only a charge deposition very small compared to Q_c. Then, the soft error rate is decreasing with increasing Q_c. If Q_c is very large, it may be that none of the emitted α-particles will have an energy large enough to create a sufficient amount of electron-hole pairs to achieve a charge $\geq Q_c$ and no soft errors will be generated. If Q_c is very small (let us say smaller than 0.05×10^6) all emitted α-particles will generate soft errors. The number of soft errors per α-particle can be estimated [May and Woods (1979)] from knowing the device cross section (to be discussed later), the depletion region depth, the charge collection areas used to store charge, and the energy spectrum of incident α-particles. The sensitivity factor S is defined to be the fraction of all α-particles hitting the active cell at a specific incidence angle (which depends on the geometry and α-particle fluxes of the various parts of ceramic package, possibly including lid and seal ring). An example of error rate calculation in DRAM's and CCD is given [May and Woods (1979)] (example for 4K test structure):

$$\text{SER} = \text{soft error rate} = A \times \Phi_\alpha \times S, \tag{7.30}$$

where A is device active area, Φ_α the α-particles flux and S the sensitivity factor. For the specific example $A = 0.027 \, \text{cm}^2$, $\Phi_\alpha = 3.8 \, \alpha/(\text{cm}^2 \, \text{h})$ and $S = 0.015$ errors/α giving an error rate of 1.5×10^{-3} errors/h = 150 percent/(1000 h) in good agreement with the observed soft error rate of 200 percent/(1000 h) obtained over 200,000 device hours of testing [May and Woods (1979)]. Purification techniques are needed to minimize the amount of α-particles source in materials. Nowadays, low-α Mold Compounds are used [$0.001 \, \alpha/(\text{cm}^2 \, \text{hour})$] and produce FIT rates at least an order of magnitude lower than FIT (Failures-In-Time, see Sect. 7.2.8) rates produced by standard Mold Compound [$0.04 \, \alpha/(\text{cm}^2 \, \text{hour})$] [Actel (2006)]. In principle, the SER component caused by cosmic rays can be separated from the component due to α-particles emitted by the radioactive sources present in the IC packaging materials and the electronic noise in the measuring device by the measurement of SER at several altitudes above sea level and repeating the measurement underground which should largely eliminate the cosmic rays neutrons and protons fluxes. The

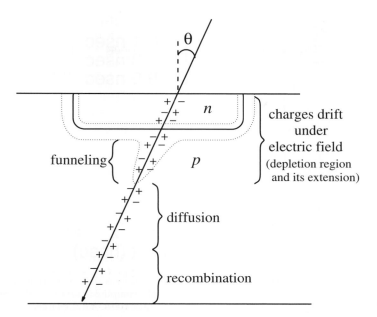

Fig. 7.8 Direct ionization (from [Leroy and Rancoita (2007)]). The striking particle, incident at angle θ, is an heavy ion ($Z \geq 2$).

measurements at several altitudes above sea level allow the investigation of the altitude dependence of the cosmic rays effect on IC. An example is provided by the "Nitetrain" experiment where the presence of rather short-lived ^{210}Po, daughters of radon gas, and long-lived daughters of ^{238}U and ^{232}Th on the chip (DRAM) is taken into account together with cosmic rays striking the chip. The SER is expressed as [O'Gorman (1996); O'Gorman et al. (1996)]

$$\mathrm{SER} = A_0 \exp\left(-t/t_{1/2}\right) + B + C, \tag{7.31}$$

where A_0 is the initial SER due to α-particles emitted by ^{210}Po (rather short-lived, $t_{1/2} = 138\,\mathrm{days}$), B is the SER contribution of the α-particles emitted by ^{238}U and ^{232}Th (long half-lives) and C is the SER contribution from cosmic rays. The SER has been measured at four locations: altitudes of 0.1 km (2 measurements), 1.6 km, 3.1 km and 0.2 km underground [O'Gorman (1996); O'Gorman et al. (1996)]. For the underground measurement, C can be set to zero in good approximation, since cosmic neutrons and protons are stopped by the rock shielding of the underground facility. The insertion of the SER data from the two sea level measurements and the underground measurement into Eq. (7.31) along with the time of each of the measurements provides a system of three equations which can be solved to determine the three unknowns A_0, B and C. That study has shown that the effect of cosmic rays is a strong function of altitude and that cosmic rays are a dominant cause, over α-particle events, of soft errors observed in devices based on NMOS, CMOS and bipolar technologies [O'Gorman (1996); O'Gorman et al. (1996)]. This observation

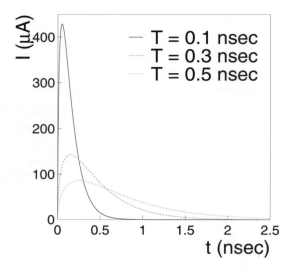

Fig. 7.9   Current pulse shape I(μA) as a function of the collection time, with time constant $T =$ 0.1, 0.3 and 0.5 ns (from [Leroy and Rancoita (2007)]; for the computed curves see also [Srinivasan, Tang and Murley (1994)]). The pulses are normalized to $Q = 100$ fC total charge.

is important since "historically" the α-emitters contamination of package molding compounds was thought to be the main cause of SEU in IC. These results show that memory circuits should be designed with the goal of immunity from upsets by alpha-particles and cosmic rays.

7.2.6   *SEE in Accelerator Radiation Environment*

Another source of SEE in the terrestrial radiation environment comes from the exposure of the experiments electronic devices to the radiation environment created by the operation of particle accelerators. Electronic devices located close to a beam or to an interaction point have shown SEE vulnerability. Observation of SEE occurrences have been reported by the CDF experiment [Tesarek et al. (2005)] at the Tevatron Collider. Failures of power supplies feeding readout electronics and crates, loss of control function, data corruption etc. occurred in systems located on the detector or close to it when beams were in operation. Similar failures are anticipated at the LHC at even higher scale, since the high rate head-on collisions of 7 TeV protons at high luminosity (luminosity peak of $10^{34}\,\mathrm{s}^{-1}cm^{-2}$) will create an unprecedented high radiation field (see Sect. 4.1.1). For instance, leakage out of calorimeters of cascades generated by protons and pions produced by proton-proton interactions will copiously produce neutrons and recoil nuclei through their subsequent hadronic interactions with surrounding materials and equipment. Neutrons and heavy nuclear recoils will interact with silicon components in detector and readout electronics, possibly causing SEE. In particular, heavy ions of rather low energy (less than 10 MeV) and therefore of limited range produced locally in an electronic chip or in an IC may induce a SEE (see Sect. 7.2.7.2). SEE will affect all kinds of semiconductor-based electronics components and devices used by the experiments at LHC such as SRAM, DRAM, FLASH memories, microprocessors, DSP, FPGA and logic programmable state machine and devices. Microelectronics featuring high density of devices and sub-micron technologies increasingly used in these experiments are especially SEE-sensitive.

7.2.7   *SEE Generation Mechanisms*

SEEs are caused by the release of charge by ionizing radiations in a semiconductor device via two mechanisms depending on the striking particle species: direct ioniza-tion if the striking particle is an heavy ion ($Z \geq 2$, such as an α-particle or an iron ion, for instance) and indirect ionization, i.e., ionization by heavy-ions secondaries produced by the interactions of the striking particle, proton, neutron or pion, with atoms of the device.

7.2.7.1   *Direct Ionization*

A charged particle (heavy ion, $Z \geq 2$) traversing a semiconductor (e.g., a reverse-biased $p - n$ device in a memory cell or flip-flop) will loose its energy through interactions with the material. The energy lost is primarily due to the interactions of the ion with bound electrons in the material, causing an ionization of the material and a dense track of electron-hole pairs. It is known that electron-hole pairs gener-ation is an early process occurring within (2–3) μm of an α-particle track (e.g., see Chapter 14 of [Leighton (1959)]) that will have a length of 18 μm in silicon for a 4 MeV α-particle. This α-particle will lose energy at a rate of $dE/dx \approx 162$ keV/μm in silicon. The early process of electron-hole generation and the high rate of energy loss (see Table 7.1) of the incoming ion in the circuit materials (large stopping power) are responsible for direct ionization. The total ionization produced in terms of electric charges depends on the initial particle energy and on the average energy needed to generate electron-hole pairs (3.62 eV for silicon, see Table 7.1). For a 4 MeV α-particle, 1.1×10^6 electron-hole pairs are created in a silicon depth of 18 μm. There are three contributions to the charge collected at the circuit nodes:

(1) when the particle path crosses the depleted region of the junction (Fig. 7.8), it creates electron-hole pairs which will be separated by the junction electric field and the charge carriers (electrons and holes) will be collected at the correspon-ding nodes generating a current pulse of short duration (Fig. 7.9) with a rapid rise time, the *prompt* component of the current pulse.
(2) The excess charge created along the path causes a perturbation of the elec-tric field along the ionizing path (funneling) and the electric field confined to the depletion region extends into silicon beyond the depletion region. Then, the electric field acts on a larger volume and permits the rapid collection of

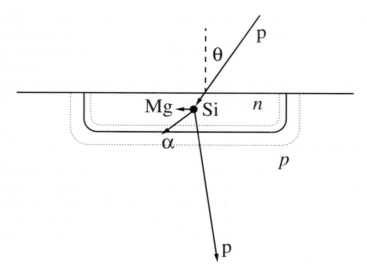

Fig. 7.10   Example of indirect ionization (from [Leroy and Rancoita (2007)]; see also [Petersen (1981)]). The 30 MeV proton incident at an angle θ interacts with silicon producing p, α and a recoiling Mg-ion [the reaction is ^{28}Si(p,pα)^{24}Mg with an energy threshold of 10.34 MeV].

electron-hole pairs generated outside the nominal depleted region which would otherwise recombine or slowly diffuse into the depletion region. These charges from funneling add to the prompt charge and contribute to the *prompt component* of the current pulse. The prompt component consisting of charges collected by electrical drift in the depletion region and through funneling corresponds to a collection time from 100 to $\sim$ 500 ps depending on the time constant of the charge collection process (Fig. 7.9).

(3) Outside the influence of any electric field, the charges are transported by diffusion, some may even recombine (Fig. 7.8), after a range of several microns around the impact point of the incoming particle. These charges separated by a diffusion process takes much longer time to be collected or have to traverse a larger distance through the silicon. They produce a gradual fall time, *delayed component*, following the prompt component. The delayed component is consisting of charges collected by diffusion with a collection time of $\sim$ 600 ps to 1 μs or longer [Dodd, Sexton and Winokur (1994)]. The times associated to the prompt component are in principle short (ps) compared to the devices response times (ns). Therefore, the introduction of sub-micron technologies with response speed allowing the propagation, through the circuit, of short signals ($\sim$ 100 ps) produced by SEE requires the development of adequate techniques to distinguish SEE occurrences from normal circuit signals.

One can represent the shape of the pulse, $I(t)$, as a function of time using a parameter T, which is the time constant for the charge collection process and is the property of the CMOS process used for the device [Shivakumar, Kistler, Keckler, Burger and Alvisi (2002)]:

$$I(t) \approx \frac{Q}{T} \times \sqrt{\frac{t}{T}} \times \exp\left(-\frac{t}{T}\right), \tag{7.32}$$

where Q is the amount of charge collected due to a particle strike. It takes more time for the charge to recombine, if T is large. If T is small, the charge recombines rapidly and the current pulse is of short duration. T decreases as feature size decreases. A method of scaling T, based on feature size, can be found in [Hazucha and Svensson (2000)].

This local current can change the logical state of a cell (SEU). This may occur in digital, analog and optical components. It may also cause loss of device functionality (*Latch up*) or cause device destruction due to a high current state in a power transistor (SEB). It may also induce rupture of oxyde gate (SEGR).

7.2.7.2 *Indirect Ionization*

Direct ionization by high energy protons and neutrons from cosmic rays or produced by the collisions of particles in particle accelerators or the interaction of cosmic rays with atmospheric particles do not create enough charge to cause SEE. In circuits materials (Si, SiO_2, GaAs) a proton, for instance, compared to an α-particle has a stopping power smaller by an order of magnitude and a range larger by an order of magnitude (see Table 7.1). A proton of 30 MeV of the example [Petersen (1981)] in Fig. 7.10 will deposit in the maximum path-length available of $30\,\mu m$ of silicon (the sensitive volume has a transversal size of $20\,\mu m \times 20\,\mu m$ size and a depth of $10\,\mu m$) a direct ionization energy of 0.1 MeV, while the maximum energy deposited by elastic recoils [see Eq. (4.103)] is

$$E_{Si} = 4\,\frac{M_{Si}M_p}{(M_{Si} + M_p)^2}\,E_p \approx 4\,\text{MeV}$$

(with a Si range, in silicon, of $\approx 2.66\,\mu m$) and the maximum energy deposited by evaporated α-particle emission is 5.3 MeV (with a range, in silicon, of $\approx 26.5\,\mu m$). The energy deposited by ^{24}Mg and the evaporated proton are 0.76 MeV (with a range, in silicon, of $\approx 1.22\,\mu m$) and 0.32 MeV (with a range, in silicon, of $\approx 3.28\,\mu m$), respectively. Then, neutrons and protons can produce upsets by indirect ionization through their interactions with atoms in the semiconductor (silicon) producing α-particles and nuclei if their energy is above a given threshold (Fig. 7.10).

As they are much heavier than the original proton or neutron, these α-particles and recoil nuclei can also generate tracks of electron-hole pairs along their paths. These α-particles and recoil nuclei may induce SEE, if they have enough energy, and travel in the right direction after their production to deposit enough charge into the sensitive volume of the circuit. It should be noted that high-energy α-particles produced by these nuclear reactions may have a range large enough to escape from the silicon creating less damage or no damage at all, if emitted very close to the surface of the circuit. The recoiling nuclei deposit energy in the same way as the directly ionizing heavy-ions, but have a very short range. A variety of neutron, proton and gamma reactions with silicon can take place with emission of recoil nuclei, protons, α-particles and γ (Table 7.2). The probability of these reactions is depending on their energy threshold as compared to the energy of the incoming particle. Spallation reactions may also occur, in which the silicon nucleus is broken into oxygen and carbon fragments ($p+Si \rightarrow p+^{12}C+^{16}O$).

Thermal neutrons were shown [Normand (1998)] causing SEE, possibly through the α-particle (1.47 MeV) and lithium (0.84 MeV) nucleus emission after the capture reaction $^{10}B(n,\alpha)^7Li$; boron being present as dopant in semiconductor and passivation layer [Griffin (1997)]. In principle fast neutrons interact with Si-atoms producing relevant recoil energy already at neutron energy around 100 keV, but only neutrons with an incident energy higher than 3 MeV should contribute to SEU cross section [Normand (1998)]. It has been observed [Johansson et al. (1998)] that the measured SEU cross section for SRAM exposed to quasi-monoenergetic neutrons of five different energies, from 22 to 160 MeV, has an energy threshold and is slowly increasing with energy up to a saturation value at about 100 MeV neutrons energy (see Sect. 7.2.8 for a discussion of the cross section shape). The energy dependence of the SEU cross section looks very similar for the variety of SRAM devices tested.

7.2.7.3   *Linear Energy Transfer (LET)*

The capability of an ion ($Z \geq 2$) to interact with a material is a function of its *linear energy transfer* (LET) value. LET is the energy deposited per unit length (dE/dx) along the ion trajectory in the material. The LET is directly proportional to the square of the atomic number of the incident ion and inversely proportional to its energy. The amount of energy deposited (and charge created) in a region of a circuit is proportional to LET as a function of path-length in this region. The LET used in SEU studies is actually the mass stopping power, i.e.,

$$\text{LET}(x) = \frac{1}{\rho} \frac{dE}{dx} \tag{7.33}$$

with ρ being the material density. Then, the LET unit is MeV cm^2 mg^{-1} that becomes the energy loss per *density thickness* ($t_d = \rho x$, where x is the absorber thickness), so that, LET may be quoted roughly independently of the target material of the device. The concept of LET does not apply to striking neutron or proton since SEE in that case are the results of nucleon–nuclear interactions (indirect ionization) and therefore the path-length distribution through the sensitive volume is unknown. Then, one refers to the energy of the incident neutron or proton.

Table 7.2   A few examples of proton, neutron and photon reactions with silicon (from [Leroy and Rancoita (2007)]). The probability of these reactions to occur depends on their energy thresholds.

Reaction	Energy threshold (MeV)	Comment
$n + {}^{28}Si \rightarrow n + {}^{28}Si$	0.00	elastic scattering
$n + {}^{28}Si \rightarrow p + {}^{28}Al$	4.00	
$n + {}^{28}Si \rightarrow \alpha + {}^{25}Mg$	2.75	
$n + {}^{28}Si \rightarrow \alpha + \alpha + {}^{21}Ne$	12.99	
$n + {}^{28}Si \rightarrow p + p + {}^{27}Mg$	13.90	
$n + {}^{28}Si \rightarrow p + \alpha + {}^{24}Na$	15.25	
$n + {}^{28}Si \rightarrow n + {}^{12}C + {}^{16}O$	16.70	
$p + {}^{28}Si \rightarrow p + {}^{28}Si$	0.00	elastic scattering
$p + {}^{28}Si \rightarrow \alpha + {}^{25}Al$	7.99	
$p + {}^{28}Si \rightarrow p + \alpha + {}^{24}Mg$	10.34	see Fig. 7.10
$p + {}^{28}Si \rightarrow p + p + {}^{27}Al$	12.00	
$p + {}^{28}Si \rightarrow p + {}^{12}C + {}^{16}O$	16.70	
$p + {}^{28}Si \rightarrow \alpha + \alpha + {}^{21}Na$	17.48	
$p + {}^{28}Si \rightarrow n + p + {}^{27}Si$	17.80	
$\gamma + {}^{28}Si \rightarrow n + {}^{27}Al$	16.97	
$\gamma + {}^{28}Si \rightarrow n + p + {}^{26}Al$	24.60	
$\gamma + {}^{28}Si \rightarrow n + n + {}^{27}Si$	30.50	
$\gamma + {}^{28}Si \rightarrow \alpha + {}^{24}Mg$	17.10	

7.2.7.4   *Sensitive Volume*

The sensitive volume, often called sensitive node in literature, is often approximated by a rectangular parallelipipeded (RPP approximation) with dimensions length x, width y, and depth d (Fig. 7.11).

As will be seen later (Sect. 7.2.8), the product $x \times y$ (size) can be determined from the saturation SEE cross section. The thickness or depth, d, is often assumed to be of the order of a few microns. Measurement of the SEE sensitivity with the angle of incidence, θ, may help to better define the value of d. This RPP approximation has to be often modified to take into account more complex geometries. However, the RPP approximation is easy for fast calculation. More complex geometries can be treated in simulations using codes such as GEANT4 [Agostinelli et al. (2003)].

7.2.7.5   *Critical Charge*

The RPP model of sensitive volume is assumed in order to estimate the amount of charge to be released by a charged particle striking a logic circuit and that will induce a voltage variation changing the memory logic status (bit flip). One may consider the sensitive volume of the semiconductor-based device, as a parallel plate capacitor (capacitance C), of size $x \times y$ and thickness d; ϵ is the electric permittivity

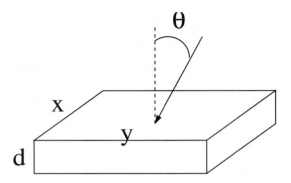

Fig. 7.11 The geometry of the RPP approximation. Usually, the thickness d is a few microns (from [Leroy and Rancoita (2007)]). The incoming particle strikes at an incident angle θ.

of the semiconductor material between the two capacitor plates

$$C = \epsilon \times \frac{S}{d} = \epsilon \times \frac{x \times y}{d}. \tag{7.34}$$

In the case of ions ($Z \geq 2$) the critical charge corresponds to a linear transfer threshold (LET$_{\rm th}$). LET$_{\rm th}$ is defined as the minimum LET needed to cause a single event effect at a particle fluence of 10^7 ions cm^{-2} (according to NASA standards[§]). Then, SEE immunity for a device is defined as LET$_{\rm th} > 100$ MeV cm^2 mg^{-1} [LaBel (1993)].

If a charge Q deposited by a ionizing particle is large enough to induce a voltage change ΔV, a SEU occurs when $Q > Q_c$. The threshold value of LET, LET$_{\rm th}$, needed to produce the voltage change is

$$\mathrm{LET_{th}} \approx \Delta V = \frac{Q}{C}. \tag{7.35}$$

Using Equation (7.34), we obtain

$$Q = \Delta V \times \epsilon \times \frac{x \times y}{d}. \tag{7.36}$$

If $\Delta V = 1.5$ V is the voltage change necessary to flip bit 1 into bit 0 (or vice versa), one has a critical charge of

$$Q_c = 1.58 \times 10^{-16} \times \frac{F}{\mu\mathrm{m}} \frac{x \times y}{\mu\mathrm{m}} \tag{7.37}$$

with a device thickness $d = 1\,\mu\mathrm{m}$ (for silicon, $\epsilon = 1.05 \times 10^{-16}$ F $\mu\mathrm{m}^{-1}$). From Eq. (7.37) or Eq. (7.36), one can derive a scaling law for the critical charge (taking $x = y = L$)

$$Q_c \sim L^2. \tag{7.38}$$

For a sensitive volume size $x \times y = 100\,\mu\mathrm{m}^2$ ($x = y = 10\,\mu\mathrm{m}$) one finds from Eq. (7.37):

$$Q_c = 0.016\,\mathrm{pC}. \tag{7.39}$$

[§]This occurs according to JEDEC: Joint Electron Device Engineering Council, the Semiconductor Engineering Standardization Body of the Electronic Industries Alliance (EIA).

It is interesting to note that the number of electrons, N_e, in the sensitive volume corresponding to this critical charge is:

$$N_e = \frac{0.016 \times 10^{-12}\,\text{C}}{1.6 \times 10^{-19}\,\text{C}} = 10^5, \tag{7.40}$$

where the factor $1.6 \times 10^{-19}\,\text{C}$ is the electronic charge. For a sensitive volume of smaller size, $x = y = 50\,\text{nm}$ and $d = 10\,\text{nm}$, one finds $Q_c = 0.04\,\text{fC}$ and the corresponding number of electrons is $N_e = 250$. This shows that the concepts of critical charge is not necessary applicable to circuit with very small sensitive volume. The critical charge depends on the circuit voltage [Eq. (7.36)] and in principle on the chip temperature. As the voltage drops, the amount of charge needed to induce the bit flip decreases and the circuit becomes more sensitive to upsets. Measurements of upset rates have been performed in various conditions, underground, at ground and above ground levels, varying the *pinch voltage*, which is the voltage reduction from the nominal full operating voltage of the chip. As the pinch voltage is increased, i.e., the chip operating voltage is decreased, the chips are seen more sensitive to ionizing radiation (Figures 6 and 7 in [O'Gorman et al. (1996)]). The dependence of Q_c on temperature is expected to come from the dependence of the charge carrier mobility on temperature [Eq. (6.70)]. The increase of mobility with temperature translates into an increase of the chip gain as the charge are moving faster to the collecting nodes. The increase of Q_c with temperature should mean a better hardness of the chip to soft errors at large temperature. However, experimentally, no significant change in the upset cross section as a function of temperature is observed for a series of bipolar SRAM chips (Table 7 in [Ziegler et al. (1996)]), although the calculated critical charge for the same chips increases by $\sim 40\%$ for a temperature increase from $50\,^{\circ}\text{C}$ to 120–$130\,^{\circ}\text{C}$ (Fig. 11 in [Ziegler et al. (1996)]). This again shows the critical charge as a rather crude parameter, however very useful to compare relative chip upset rates.

From the definition of LET [Eq. (7.33)], the energy deposited, E_{dep}, by a particle along a path of length l through the sensitive volume of the device of density ρ is:

$$E_{\text{dep}} = \text{LET}\,\rho\,l, \tag{7.41}$$

if one assumes that LET and ρ are constant over the distance l. The corresponding deposited charge is then given by

$$Q_{\text{dep}} = \frac{E_{\text{dep}}}{E_{\text{ion}}}\,q, \tag{7.42}$$

where E_{ion} ($= 3.62\,\text{eV}$ in silicon, Table 7.1) is the average energy to create an electron-hole pair, q is the electronic charge.

The LET of a particle can be related to its charge deposition per unit pathlength. For a particle traversing silicon, the conversion factor between charge and energy is

$$\frac{E_{\text{ion}}}{q} = 22.6\,\text{MeV/pC}.$$

Then, one can calculate the equivalent charge transfer rate

$$\frac{\text{LET}\rho}{E_{\text{ion}}/q} = \frac{\text{LET}\rho}{22.60\,\text{MeV/pC}}.$$

For instance, a LET of $97\,\text{MeV cm}^2\text{mg}^{-1}$ corresponds to a charge deposition of $1\,\text{pC}/\mu\text{m}$ in silicon. Using Equation (7.41), the charge deposited in the sensitive volume of the device along the path l can be expressed as a function of the LET of the particle:

$$Q_{\text{dep}} = \frac{\text{LET}\rho\,l\,q}{E_{\text{ion}}}. \tag{7.43}$$

The critical charge, corresponding to a linear transfer threshold LET_{th}, is then given (in silicon) by

$$Q_{\text{dep}} = 0.01 \times \text{LET}\,l \quad [\text{pC}]. \tag{7.44}$$

l is expressed in μm in Eq. (7.44). If one applies the RPP approximation to the sensitive volume with size $x \times y$ and thickness d (Fig. 7.11), one finds that the maximum path-length of the particle inside the volume, l_{max}, is given by:

$$l_{\text{max}} = \sqrt{x^2 + y^2 + d^2}. \tag{7.45}$$

Taking LET at the threshold value, LET_{th}, i.e., the minimum LET required to cause a SEU, one can express the critical charge in silicon as

$$Q_c = 0.01 \times \text{LET}_{\text{th}}\,l_{\text{max}} \quad [\text{pC}]. \tag{7.46}$$

l_{max} and LET_{th} are expressed in μm and $\text{MeV cm}^2\,\text{mg}^{-1}$, respectively, in Eq. (7.46).

Conversely, for experimental purpose one use LET_{th} given by

$$\text{LET}_{\text{th}} = 97 \times \frac{Q_c}{l_{\text{max}}} \quad [\text{MeV cm}^2\,\text{mg}^{-1}] \tag{7.47}$$

with Q_c and l_{max} expressed in pC and μm units, respectively.

A SEU occurs if $Q > Q_c$ which corresponds in the case of indirect ionization to a nuclear recoil energy, E_r, of

$$E_r > 0.226 \times \text{LET}_{\text{th}}\,l \quad [\text{MeV}]. \tag{7.48}$$

A particle of given LET must traverse a minimum path, l_{min}, in order to deposit enough energy to induce an upset. From Eq. (7.47) one finds in silicon:

$$l_{\text{min}} = \frac{Q_c\,E_{\text{ion}}}{q\,\rho\,\text{LET}} = 97 \times Q_c\,\frac{1}{\text{LET}} \quad [\mu\text{m}], \tag{7.49}$$

where one uses the maximum stopping power (in MeV for any particle in the environment. As seen earlier, the galactic cosmic ray environment has a component of very high energy ions with large ionizing power due to their high charge. The flux of heavy ions is dominated by iron ($Z = 26$) for LET $\sim 27\,\text{MeV cm}^2\,\text{mg}^{-1}$. Equation (7.49) becomes for silicon

$$l_{\text{min}} = 3.60 \times Q_c \quad [\mu\text{m/pC}]. \tag{7.50}$$

The dependence of LET_{th} or Q_c on the path length of the particle inside the sensitive volume of the device shows the importance of considering the angle of incidence, θ, of the incoming particle upon the device (Fig. 7.8) [Dodd, Shaneyfelt and Sexton (1997)]. This is particularly important for devices of large size $(x \times y)$ compared to their thickness (d). Deviation from normal results into the increase of the path-length, $l/\cos\theta$, inside the sensitive volume. An effective LET is defined as

$$LET_{eff} = \frac{LET(\theta = 0)}{\cos\theta}. \tag{7.51}$$

The angle at which a SEE may occur for a given LET is the critical angle, θ_c, and is defined as

$$\cos\theta_c = \frac{LET}{LET_c}. \tag{7.52}$$

The particles that generate upset have an angle of incidence between θ_c and 90°. From this definition if $LET > LET_c$, then all incident angles induce SEE. In the case of $LET < LET_c$, there must be a critical angle above which SEE occurs [Binder (1988); Pickel (1996)]. In the case of incident protons and neutrons, the critical charge corresponds to a energy threshold.

7.2.8 SEE Cross-Section

The sensitivity to each type of SEE (SEU, SEL, SEB,...) in a device is estimated from the measurement of the cross section, σ, as a function of LET for ions $(Z \geq 2)$. By counting the number of SEE and knowing how many particles passed through the device, one can calculate the probability of a given particle to cause a SEE. This resulting number, which is the number of upsets divided by the number of particles

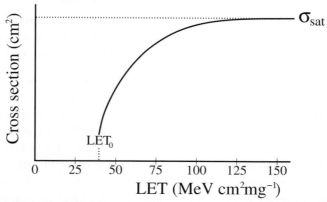

Fig. 7.12   Example of typical upset cross section curve as a function of LET (from [Leroy and Rancoita (2007)]). LET_0 is the LET threshold for the device. σ_{sat} is the saturation cross section and corresponds to the lateral size of the device in the RPP approximation. The shape of the curve is described by a Weibull distribution (see text).

per cm^2 causing the upset, is the cross section of the device. One can express this cross section as:

$$\sigma(\text{LET}) = \frac{N_{\text{events}}}{\Phi} \ [\text{cm}^2], \tag{7.53}$$

where Φ is the total incident ion fluence (ions/cm^2) on the device. For protons or neutrons, the cross section is measured as a function of the energy E of the incident protons or neutrons, i.e.,

$$\sigma(E) = \frac{N_{\text{events}}}{\Phi} \ [\text{cm}^2], \tag{7.54}$$

where Φ is the total incident proton or neutron fluence (protons/cm^2 or neutrons/cm^2) on the device. In both situations N_{events} is the number of SEE events counted during the exposure to the incident flux. The SEE cross section may be expressed per bit or per chip:

$$\sigma(\text{per bit}) = \frac{N_{\text{events}}}{\text{chip}} \frac{1}{\text{bits/chip}} \frac{1}{\Phi}, \tag{7.55}$$

or

$$\sigma(\text{per chip}) = \frac{N_{\text{events}}}{\text{chip}} \frac{1}{\Phi}. \tag{7.56}$$

The *error rate* (SEE) can be expressed as:

$$\text{error rate} = \frac{N_{\text{events}}}{\text{device day}}. \tag{7.57}$$

One preferably uses the FIT (*Failures-in-Time*) rate value to characterize the behavior with time of a device under exposure. The value of the FIT rate is expressed as the rate of error or failure of a device per billion of working hours, i.e, as

$$\text{FIT} = \text{number of errors}/(10^9 \text{ hours device}). \tag{7.58}$$

Then, a *soft error rate* of 1 FIT means that the mean time before an error occurs is a billion device hours. It is frequent to have reported a FIT per Mbit:

$$1\,\text{FIT}/\text{Mb} = 1\,\text{upset}/10^9 \text{ hours}/10^6 \text{ bits}$$

$$= 1 \times 10^{-15}\text{upset}/(\text{bit hours}). \tag{7.59}$$

In the case of a monoenergetic beam of one single type ion incident normal to the surface of the device, σ measures the LET-dependent SEU sensitive area on the chip. For the RPP approximation, the lateral dimensions of the chip are obtained from the measured cross section, assuming a squared surface:

$$x = y = \frac{\sqrt{\sigma_{\text{sat}}}}{\text{number of bits}}. \tag{7.60}$$

It is assumed that there is one sensitive volume per bit, and the shape of the cross section for a given chip (Fig. 7.12) rising from a threshold to a saturation value is understood as the composite response of multiple types of sensitive volume with different thresholds and features. Contrary to the case of direct ionization, for SEEs generated by indirect ionization, it is not possible to interpret the measured cross section in terms of sensitive area of the chip since the cross section convolutes the probability of nucleon–nuclear interaction and the probability that the interaction products recoil will achieve the charge density at various locations in the device needed to induce a SEE.

7.2.8.1 *Calculation of SEU Rate for Ions*

An ion $(Z \geq 2)$ traversing a semiconductor will loose its energy via ionization. The energy deposition along the ion path will generate electron-hole pairs in the semiconductor-based device (direct ionization, Sect. 7.2.7.1). When the ion path crosses a region close to a sensitive node of a circuit (e.g., a $p - n$ junction in a memory cell) the collected charge at the node, if larger than a critical value (Q_c), may induce a SEE. The SEE rate calculation usually assumes a sensitive volume model based on the RPP approximation. In this approximation, the sensitive volume is a rectangular parallelipipeded (Sect. 7.2.7.4). The incident ions are very energetic and have very long ranges compared to the typical dimensions of the sensitive volume. Equations (7.33, 7.42) can be used (in practice with TRIM [Ziegler, Biersack and Littmark (1985b); Ziegler (2001, 2006)]) to estimate the deposited energy by the ions, E_{dep}, across the distance l, assuming a continuous slowing down and a linear trajectory.

The probability to trigger an upset is the probability to have a deposited energy larger than the critical energy E_{th} of the device component and the probability to have an upset is:

$$P_{SEU} = \frac{1}{\Phi_0} \frac{d\,\Phi(\text{LET})}{d\,\text{LET}} P_L[> l(\text{LET})], \qquad (7.61)$$

where $1/\Phi_0 \times (d\,\Phi(\text{LET})/d\,\text{LET})$ is the differential ion flux spectrum as a function of LET, i.e., the probability to encounter an ion with a given LET, Φ_0 is the integrated ion space environment flux and $P_L[> l(\text{LET})]$ is the probability that this particle covers a distance $l(\text{LET})$ such that the deposited energy $E_{dep} > E_{th}$ or the integrated path-length distribution. From Equation (7.33), one finds

$$l(\text{LET}) > \frac{E_{th}}{\rho\,\text{LET}}. \qquad (7.62)$$

Then, the upset rate, N_{SEU}, is expressed as a function of the incident LET spectrum and the path-length distribution [Bradford (1980)]:

$$N_{\text{SEU}} = \frac{A}{4} \int_{\text{LET}_{\min}}^{\text{LET}_{\max}} \frac{d\,\Phi(\text{LET})}{d\,\text{LET}} P_L[> l(\text{LET})] \; d\,\text{LET}, \qquad (7.63)$$

where A is the device total surface area [length: x, width: y (size $S = xy$), thickness: d, $A = 2\,(x\,y + x\,d + d\,y)$], $\text{LET}_{\min}$ and $\text{LET}_{\max}$ are the minimum LET $(= \text{LET}_{th}$ to produce an upset) and the maximum LET $(\sim 10^2\,\text{MeV\,cm}^2\,\text{mg}^{-1}$ value which corresponds to SEE immunity [LaBel (1993)]) of the incident particles distribution, respectively.

The RPP approach can be improved by folding the LET spectrum with the experimental cross section to account for sensitivity variations across the device. The critical LET (LET_c) of the sensitive nodes are not the same, but form a distribution that can be fitted by a *Weibull function* [Eq. (7.68)]. The integrated RPP (IRPP) approach is used to take into account the variation of sensitivity by integrating over a distribution of upset rates corresponding to the variation of cross section as a function of LET. The IRPP approximation uses the differential upset cross section, $d\sigma_{\text{ion}}/d\text{LET}$, the ratio

$$\frac{A}{S} = 2\left(1 + \frac{d}{y} + \frac{d}{x}\right)$$

of the total surface area of the sensitive volume to the surface area, the ratio $d/l_{\max}$ of the total height (d) of the sensitive volume to the maximum length $(l_{\max})$ that can be traveled in the sensitive volume and the differential LET spectrum, $d\phi/d\text{LET}$. The SEU rate is then:

$$N_{\text{SEU}} = \frac{A}{4S} \int_{\text{LET}_{i,\min}}^{\text{LET}_{i,\max}} \frac{d\sigma_{\text{ion}}}{d\text{LET}_i} N_i \; d\text{LET}_i \tag{7.64}$$

with

$$N_i = \int_{(d/l_{\max})\text{LET}_i}^{\text{LET}_{\max}} \frac{d\Phi(\text{LET})}{d\text{LET}} P_L[> l(\text{LET})] \; d\text{LET}. \tag{7.65}$$

$\text{LET}_{i,\min}$ and $\text{LET}_{i,\max}$ are the lower bin limit and the upper bin limit in $d\sigma_{\text{ion}}/d\text{LET}$, respectively.

The path-length distribution used in the RPP approximation can be replaced by an effective LET spectrum, $d\phi/d\text{LET}_{\text{effective}}$ calculated by a Monte-Carlo method. This effective LET spectrum is combined with the ion upset cross section, $\sigma_{\text{SEU}}(LET)$ to produce the upset rate. The effective LET of the incident ion is defined by

$$\text{LET}_{\text{effective}} = \frac{E_{\text{dep}}}{\rho d}, \tag{7.66}$$

where d is the effective thickness of the device component. The upset rate is given by [Inguimbert and Duzelier (2004)]

$$N_{\text{SEU}} = \frac{A}{4\sigma_{\text{sat}}} \int_{\text{LET}_{\min}}^{\text{LET}_{\max}} \frac{d\Phi(\text{LET})}{d\text{LET}}_{\text{effective}} \sigma_{\text{SEU}}(\text{LET}) \; d\text{LET}, \tag{7.67}$$

where σ_{sat} is the ion upset cross section at saturation (Fig. 7.12). N_{SEU} [Eq. (7.67)] was calculated in [Inguimbert and Duzelier (2004)] using GEANT4. The results of [Inguimbert and Duzelier (2004)] are in agreement with CREME96 results [Nymmik, Panasyuk, Pervaja and Suslov (1992)]. However, the use of GEANT4 opens the possibility of accurate modeling of device geometry, structure, material composition and functions of the sensitive volume while including the various mechanisms of energy deposition, toward an accurate estimate of the deposited energy and then upset rate. For accelerator data, where one has the choice to select specific ion beams of well defined energy, it is possible to measure the cross section as a function of LET. Then, the measured cross section is reported in a plot as a function of LET (an example is shown in Fig. 7.12).

This plot is characterized by two parameters: i) the threshold LET (LET_{th}) which is the minimum LET required to cause the specific SEE; ii) the saturation cross section, σ_{sat}, which is approached at very high LET value and corresponds to a limit, at which all the sensitive nodes of the device have been upset. σ_{sat} is associated to the sensitive volume size and can be interpreted as the projection of the sensitive volume in the direction normal to the semiconductor chip. σ_{sat} (cm^2) is given by the product of x and y. Conversely, x and y are determined by the measurement of σ_{sat} [see Eq. (7.60)], taking into account the number of information bits in the chip. In practice, the cross section is measured at several LET values and the fit to the data is done with a Weibull curve, after correction for geometric effects. This function has the form:

$$F(\text{LET}) = 1 - \exp\left[-\left(\frac{\text{LET} - \text{LET}_0}{W}\right)^s\right], \qquad (7.68)$$

where LET$_0$ is the LET threshold for the device, such that $F(x) = 0$ for LET $<$ LET$_0$; W is the width parameter and s is a dimensionless shape parameter. LET$_0$ applies to the overall chip response reflecting the composition of sensitive volumes and their respective thresholds. It differs from LET$_{th}$, defined before, which was related to an individual sensitive volume. Then, the direct ionization (heavy ion) SEE cross section is expressed as

$$\sigma = \sigma_{sat} \times F(\text{LET}). \qquad (7.69)$$

7.2.8.2 *Calculation of SEU Rate for Protons and Neutrons*

Protons and neutrons may generate SEEs via indirect ionization. As observed before, the complexity of the nuclear reactions breaks down the application of the concept of LET and the SEE rate is determined from the proton or neutron fluxes and SEE nucleon cross sections:

$$N_{\text{SEU}} = \int_{E_{min}}^{E_{max}} \frac{d\,\Phi(E)}{d\,E} \, \sigma_{\text{nucleon}}(E) \, d\,E, \qquad (7.70)$$

where $d\,\Phi(E)/d\,E$ is the differential nucleon (proton or neutron) flux spectrum as a function of the proton or neutron energy, E; E_{min} and E_{max} are the minimum and maximum energy of the differential nucleon energy spectrum, respectively; σ_{nucleon} is the nucleon SEE cross section as a function of the nucleon energy. The proton (indirect ionization) SEE cross section is expressed with a 1-parameter or 2-parameter *Bendel function*. The 1-parameter Bendel cross section as a function of the proton energy is practically expressed in units of proton^{-1} cm^{-2} bit^{-1} and given by [Bendel and Petersen (1983)]:

$$\sigma(A, E) = 1 \times 10^{-12} \left(\frac{24}{A}\right)^{14} \left[1 - \exp\left(-0.18\,\sqrt{Y(E)}\right)\right]^4 \qquad (7.71)$$

with

$$Y(E) = (E - A)\left(\frac{18}{A}\right)^{1/2} \quad \text{if } E > A, \qquad (7.72)$$

$$Y(E) = 0 \quad \text{otherwise.} \tag{7.73}$$

σ is the upset cross section depending on E and A which are the proton energy and apparent threshold in MeV, respectively. Cross section data are often better described with the 2-parameter Bendel cross section which is given by [Stapor, Meyers, Langworthy and Petersen (1990)] in units of proton^{-1} cm^{-2} bit^{-1}:

$$\sigma(A, B, E) = \left(\frac{B}{A}\right)^{14} \times 10^{-12} \left\{1 - \exp\left[-0.18\sqrt{Y(E)}\right]\right\}^4 \tag{7.74}$$

with $Y(E)$ defined as in Eq. (7.72). E is the proton energy, the parameters A and B are in MeV. The ratio B/A is the limiting cross section or the fitted cross section at infinite proton energy, see for instance [Oberg et al. (1994)] where $A = 1.5\,\text{MeV}$, $B/A = 0.953$ have been used to fit 148, 200 and 500 MeV proton SEU data for 87C51FC microcontroller at bias voltages of 4.5 and 5.5 V for 148 MeV beam, 4.0 and 5.0 V for 500 MeV beam, and only 4.0 V for 200 MeV beam. There is another two parameter fit parametrization [Shimano et al. (1989)] which has the advantage to separate the energy dependence from σ_{sat} as in Eq. (7.69):

$$\sigma(A, \sigma_{\text{sat}}, E) = \sigma_{\text{sat}} \left\{1 - \exp\left[-0.18\sqrt{Y(E)}\right]\right\}^4 \tag{7.75}$$

with $Y(E)$ given by Eq. (7.72), σ_{sat} corresponds to $(24/A)^{14}$ [Eq. (7.71)] or $(B/A)^{14}$ [Eq. (7.74)] but is determined independently of A. Using one of these parametrizations, Eqs. (7.71–7.75), it is possible to find the SEE rate in any proton environment from the measurement of the SEE cross section at one proton energy.

If the parameter A and σ_{sat} are given, the error $\Delta\sigma(E_i)$ for the proton energy E_i is expressed as [Bendel and Petersen (1983)]

$$\Delta\sigma(E_i) = \sigma(E_i) - \int_0^\infty \sigma(A, \sigma_{\text{sat}}, E)\, f_i(E)\, dE, \tag{7.76}$$

where $\sigma(E_i)$ is a measured cross section, $f_i(E)$ is a energy distribution function for proton of energy E_i. If a Gaussian distribution is assumed for $f_i(E)$, with

$$\int_0^\infty f(E_i)\, dE = 1,$$

A and σ_{sat} can be determined by minimizing the sum of the square of relative errors, i.e., by applying the formula

$$\sum_i \left[\frac{\Delta\sigma(E_i)}{\sigma(E_i)}\right]^2. \tag{7.77}$$

However, one can use heavy-ion data, when available, to predict proton induced upsets. There exist empirical models for proton induced SEU based on heavy-ion data in terms of device cross section as a function of ion LET in order to determine an expression of the proton device cross section as a function of the proton energy. An example of such approach is given with the *PROFIT model* applied to calculate proton induced SEU rates in orbit (due to trapped protons) for various digital circuits parts (silicon-based) [Calvel et al. (1996)]. This model is based on heavy-ion data. Establishing a relationship between upset rates due to heavy ions and upset rates due to nucleons requires several strong assumptions because of the different mechanisms of energy deposition, direct and indirect ionizations. To rely on heavy-ion data, one has to assume that all sensitive cells have the same surface size and depth for both types of data. As observed above they have different threshold LETs (or critical charge). This distribution of threshold LETs can be fitted by a Weibull function. Another strong assumption is to consider proton interacting with the silicon crystal through elastic interaction and ionization produced by the silicon recoil atoms in the silicon crystal, i.e., inelastic interactions are considered as elastic collisions, for the purpose. With this assumption, all proton induced events are due to recoiling silicon atoms, having their LET calculated with TRIM [Ziegler, Biersack and Littmark (1985b); Ziegler (2001, 2006)]. These recoil Si-atoms in Si have a maximum LET value be $15 \, \text{MeV} \, \text{cm}^2 \, \text{mg}^{-1}$, as calculated from TRIM. Therefore, only cells with a threshold $\text{LET} \leq 15 \, \text{MeV} \, \text{cm}^2 \, \text{mg}^{-1}$ will be sensitive to SEU induced by protons. The ions experimental data can be fitted using a Weibull function. The ion device cross section is:

$$\Sigma_{\text{ion}} = \Sigma_0 \left[1 - \exp \left(\frac{\text{LET} - L_0}{W} \right)^S \right], \qquad (7.78)$$

where Σ_0 is the heavy-ions saturated device cross section in cm^2, L_0 is the LET threshold in $\text{MeV} \, \text{cm}^2 \, \text{mg}^{-1}$, W and S are the Weibull parameters obtained from the fit of the curve Σ_{ion} as a function of LET in $\text{MeV} \, \text{cm}^2 \, \text{mg}^{-1}$. Then in the framework of the PROFIT model, the proton device cross section, Σ_p, as a function of the proton energy E_p can be expressed as a function of the ion device cross section:

$$\Sigma_p = \Sigma_0 \left\{ 1 - \exp \left[\frac{\text{LET}[E_{\text{Si}}(E_p)] - L_0}{W} \right]^S \right\} c \, N_{\text{at}} \, \sigma_{\text{nuclear}}(E_p), \qquad (7.79)$$

with $c = 2 \times 10^{-4} \, \text{cm}$ (typical cell depth), $N_{\text{at}} = 5 \times 10^{22}$ Si atoms per cm^3 and

$$E_{\text{Si}}(E_p) = \frac{2 \, M_p M_{\text{Si}}}{(M_p + M_{\text{Si}})^2} \, (1 - \cos \theta) \, E_p, \qquad (7.80)$$

with M_p and M_{Si} are the masses of proton and silicon atom, respectively. The angle θ is the average scattering angle of protons ($\theta = 52^o$, as used in [Calvel et al. (1996)]), $\text{LET}[E_{\text{Si}}(E_p)]$ (calculated with TRIM) is the LET for recoil Si atoms in silicon for E_p proton energy. In Eq. (7.79), $\sigma_{\text{nuclear}}(E_p)$ is the proton-silicon nuclear cross section as a function of the proton energy.

At saturation, the maximum LET of recoil Si-atoms is $15 \, \text{MeV} \, \text{cm}^2 \, \text{mg}^{-1}$ and Eq. (7.79) becomes

$$\Sigma_p^{\text{sat}}(E_p) = \Sigma_{\text{ion}}(\text{LET} = 15 \, \text{MeV} \, \text{cm}^2 \, \text{mg}^{-1}) \, c \, N_{\text{at}} \, \langle \sigma_{\text{nuclear}} \rangle, \qquad (7.81)$$

where Σ_p^{sat} is the saturation proton device cross section and $\langle \sigma_{\text{nuclear}} \rangle$ is the average proton–silicon nuclear cross section in cm^2 ($\langle \sigma_{\text{nuclear}} \rangle = 6 \times 10^{-26} \, \text{cm}^2$ for proton energies between 100 and 200 MeV). Previous works [Rollins (1990); Petersen

(1992)] on SEU rates in orbit induced by trapped protons used the two-parameter Bendel cross section [Eq. (7.74) instead of Eq. (7.81)] with a difficulty to fit the lower energy experimental data. The PROFIT model is used to estimate the proton sensitivity of devices based on heavy-ion measurement but conversely it can also be used to estimate heavy-ion sensitivity based on proton data. Examples of application of the PROFIT model to estimate device sensitivity for heavy ions can be found in [McDonald, Stapor and Henson (1999); Henson, McDonald and Stapor (2006)].

7.2.9   *SEE Mitigation*

SEE mitigation or methods of reducing SEE impacts on data responses of a device and on control of a device or system are developed by device designers. Examples can be found in [LaBel (1996)]. For instance, for mitigation of memories and data-related devices, parity checks may be used [Carlson (1975)]. Parity is a bit added to the end of a data structure, which can state whether an odd or even numbers of "ones" were in that structure. This method allows the detection of an error if an odd number of bits are in error. If an even number of errors occurs, the parity remains correct. This method detects, but does not correct errors. Other methods such as cyclic redundancy check (CRC) coding [Short (1987)] or hamming code [Carlson (1975)] are used for error detection. These mitigation technique may be also applied to some types of control devices, such as microprocessor program memory. In practice one has to add additional hardware and software to the system design in order to perform an effective mitigation. This is an iterative process, since these additions have to be tested by the study of the device response to radioactive sources, particle (pion, proton, neutron) and ion beams.

In the case of the radiation environment created by the operation of a particle accelerator, SEE mitigation of electronic devices may be partially done by suppressing or, at least, by reducing components of the harmful radiation field. Installation of dedicated shieldings may achieve that goal. Tuning of accelerator beam conditions may also help with the attenuation or the removal of the radiation field locally. Components failure rates may be decreased by modifying their operation conditions. For instance, the reduction of the operating voltage of a failing device may restore its normal function without compromising its performances.

Chapter 8

Ionization Chambers

Ionization chambers are detectors filled with a gas or a noble liquid in which the ionizing particle creates ion–electron pairs. The ions and electrons then drift under the applied electric field to the electrodes where they are collected. If the electric field is increased, the electrons have enough energy to ionize the gas themselves. The detectors using this mode of operation are called proportional counters. If the electric field is further increased, for example in a Geiger–Mueller counter, the electrons are so energetic that UV photons are emitted, when they reach the anode. The principle of operation of the ionizing chamber will be explained in detail in this chapter. To demonstrate the usefulness of this type of detector, the study of pollution in liquid argon using an α-cell will be discussed extensively. The chapter will conclude with the principles behind proportional and Geiger–Mueller counter.

8.1 Basic Principle of Operation

This type of detector has been used for many years and its principle of operation is relatively simple. An ionization chamber, in its simplest form, consists of two parallel metallic electrodes (copper for instance) separated by a distance D. High voltages (V), up to several thousand kV, are applied to the anode. This voltage is maintained by an external circuit, characterized by a resistance R and a capacitor C. The gap D between the two plates is filled with a gas or a noble liquid and defines the sensitive volume of the chamber. Ionizing particles traversing the sensitive volume will ionize the gas or noble liquid and produce ion–electron pairs. Positive ions and electrons are the charge carriers. The electric field, $E = V/D$, created by the potential difference across the gap, will cause electrons and positive ions to drift in opposite directions toward the anode and cathode, respectively, where the charge produced by ionizing particles is collected. The number of electrons produced in the gap by n minimum ionizing particles (mip) is:

$$n_- = \frac{n}{W} D \rho \frac{dE}{dx},\qquad(8.1)$$

585

where W is the average energy-loss necessary to create an ion-electron pair in the gas or noble liquid. For example, in argon gas, $W = 26 \, \text{eV}$, $\rho \approx 1.8 \times 10^{-3} \, \text{g/cm}^3$ at STP is the argon gas density, D is the gap thickness, and $dE/dx \approx 1.52 \, \text{MeV}$ is the energy loss per g/cm^2. Using these numbers, one finds for argon gas in the gap:

$$n_- \sim 105 \text{ cm}^{-1} \times nD, \tag{8.2}$$

and, consequently, an electric charge

$$Q \sim (105 \, \text{cm}^{-1} \times nD) \times 1.6 \times 10^{-19} \, \text{C}. \tag{8.3}$$

If one considers the case of an isolated ion-electron pair created at some distance from the anode, the electron drift velocity is

$$v_d(\text{e}) = \frac{dx}{dt} = \mu_e E = \mu_e \frac{V}{D}, \tag{8.4}$$

where μ_e is the electron mobility. In the case of a gas-filled parallel plate chamber, the electron drift velocity also depends on the gas pressure, p, i.e.,

$$v_d(\text{e}) = \frac{\mu_e}{p} E = \frac{V \mu_e}{pD}, \tag{8.5}$$

in which case the mobility is expressed in $\text{bar cm}^2 \, \text{V}^{-1} \, \text{s}^{-1}$.

The collection of these charges produces a drop, ΔV, in voltage across the capacitor of the external circuit. Then, a pulse of height given by

$$\Delta V = \frac{\Delta Q}{C}$$

is produced across the resistor and recorded by the external electronic readout circuit.

In the case of liquid argon, only electrons contribute in practice to the charge collection since the positive ion drift velocity $[v_d(\text{ion})]$ is found to be three to five orders of magnitude smaller than the electron drift velocity $[v_d(\text{e})]$. The electron mobility in liquid argon $(T = 87 \, \text{K})$ is $\mu_e \approx 500 \, \text{cm}^2 \text{V}^{-1}\text{s}^{-1}$ compared to the positive ion mobility $\mu_{\text{ion}} \approx 6 \times 10^{-4} \, \text{cm}^2 \text{V}^{-1}\text{s}^{-1}$ [Gruhn and Edmiston (1978)]. The number of electrons reaching the anode depends on the details of the chamber design, the nature of the sensitive volume and its purity, and on the applied voltage. The ionization electrons are affected by recombination with parent ions (germinate recombination) or with positive ions on their way to the anode (columnar recombination) and by capture by electronegative impurities present in the active medium. The electrons escaping recombination or capture eventually make their way to the anode, where they are collected. The probability for an electron to escape recombination and capture increases with voltage and therefore, the number of electrons collected at the anode increases with voltage up to a saturation value of voltage, V_{sat}, for which the charge created in the sensitive volume by the incoming radiation is all collected. No further increase in the collected charge takes place, when the voltage is increased beyond V_{sat}. V_{sat} is called the saturation voltage. The region, where the charge collected remains roughly constant with voltage, is called the *ionization region*, i.e., the region in which the ionization chambers are operated. We will see in another section that increasing the voltage beyond this saturation voltage leads to the multiplication region, i.e., a region where the electrons are accelerated enough to ionize molecules of the gas and produce secondary electrons.

Taking into account recombination or capture effects, one can write the time decay of the concentration of electrons n_- as:

$$\frac{dn_-}{dt} = -k_n - k_r n_- n_+ - k_s n_- n_s, \tag{8.6}$$

where, in standard notations, k_n is the neutralization factor weighting the neutralization of electrons at the anode, n_+ is the concentration of positive ions, k_r is the recombination rate constant, n_s is the concentration of electronegative impurities and k_s is the capture rate constant (or attachment rate factor). Recombination and attachment phenomena will be discussed in the next section, keeping in mind specific applications using ionization chambers with liquid argon as sensitive volume.

8.2 Recombination Effects

This section treats the various recombination effects, including diffusion, which affect the electron charge collection at the electrodes of the chamber. In absence of impurities in the liquid filling the gap, the charge collection is affected by two processes called *germinate* and *columnar recombinations*.

8.2.1 *Germinate or Initial Recombination*

Germinate or initial recombination takes place when an electron produced by ionization recombines with its parent ion. This recombination happens unless the electron becomes subjected to an electric field larger than the Coulomb field which is maintaining it in the vicinity of the ion. This initial recombination mainly occurs in the case of heavily ionizing particles, such as low energy α particles, and is minimal for *mip*'s. Obviously, the application of large electric fields minimizes the initial recombination.

The ratio of the amount of charge collected after initial recombination, $Q_{\rm rec}$, to the amount of initial charge, Q_0, has been calculated by Onsager (1938):

$$\frac{Q_{\rm rec}}{Q_0} = \exp\left(-\frac{r_{kT}}{r_0}\right)\left[1 + E\left(\frac{e^3}{2\epsilon k^2 T^2}\right)\right], \tag{8.7}$$

where $r_{kT} = e^2/\epsilon kT$ is the Onsager length or radius, r_0 is the thermalization length, k is the Boltzmann constant, T is the temperature, E is the applied electric field, and ϵ is the electric permittivity of the medium.

8.2.2 *Columnar Recombination*

The ionization electrons which escape germinate recombination are free to drift under the influence of the applied electric field. During ionization, specially by heavy particles, ions are created close to the ionizing particles trajectory, within a column along the trajectory. Drifting electrons and positive ions can then recombine. This columnar recombination depends on the particle emission angle with the applied electric field. A strong recombination takes place for large positive ion density along the electrons path as it is the case in liquid. However, this recombination remains small for large values of the applied electric field. The theory of columnar recombination has been formulated by Jaffe (1913) and Kramers (1952), based on the assumptions that the ionization is uniformly distributed along the line of motion of the ionizing particle (z-direction), that the ion density is Gaussian distributed in a column of radius b, and that all ionization is in the form of positive and negative ions. The initial carrier concentrations, $n_\pm(t = 0)$, are:

$$n_\pm(t = 0) = \frac{N_0}{\pi b^2} \exp\left(-\frac{x^2 + y^2}{b^2}\right) \tag{8.8}$$

with N_0 the initial density of ions along the column axis (z-direction). Then, the evolution of the ion, n_+, and electron, n_-, concentrations is described in this model by the equations:

$$\frac{dn_\pm}{dt} = \mp \mu_\pm E \sin\phi \, \frac{\partial n_\pm}{\partial x} + D_\pm \Delta n_\pm - k_r n_- n_+ \tag{8.9}$$

and

$$D_\pm = \frac{\mu_\pm kT}{e}, \tag{8.10}$$

where $\mu_- = \mu_e$ ($\mu_+ = \mu_{\text{ion}}$) is the electron (ion) mobility, ϕ is the emission angle of ionizing particles with respect to the electric field, D_- (D_+) is the diffusion coefficient for electrons (ions) and k_r is the columnar recombination factor. The ratio of the amount of charge collected after columnar recombination, $Q_{\text{rec},c}$, to the amount of initial charge Q_0 has been calculated by Kramers (1952), assuming that the diffusion term is negligible compared to displacement and recombination terms, in an external field ($\zeta = y^2/b^2$):

$$\frac{Q_{\text{rec},c}}{Q_0} = \frac{2f}{\sqrt{\pi}} \int_0^{+\infty} \frac{\sqrt{\zeta}}{f e^\zeta + 1} \, d\zeta \tag{8.11}$$

with

$$f = \frac{Eb \sin\phi}{4\sqrt{\pi} Q_0}. \tag{8.12}$$

8.2.3 *The Box Model*

For electric fields up to a few hundred V/cm in liquid argon, the electron yield, as calculated using Onsager's model, is a linear function of the field and the slope is determined by the temperature T and the electric permittivity ϵ, as used in Eq. (8.7) which can be rewritten as [Buckley et al. (1989)]:

$$Q(E) = Q(E = 0)\,(1 + \beta E), \tag{8.13}$$

where $Q(E = 0)$ is the number of electrons that escape recombination with positive ions in the absence of any external electric field. The slope β is given by (see Eq. (8.7)):

$$\beta = \frac{er_{kT}}{2kT}. \tag{8.14}$$

The comparison of Eq. (8.14) with data shows large discrepancies and indicates that Onsager's model is inappropriate in the case of liquid argon: the measured slope/intercept ratio is about a factor 4 larger than the predicted one [Buckley et al. (1989)]. The explanation for the discrepancy is that even if an electron escape recombination with its parent ion, it may still recombine with another close ion, a possibility not accounted for by Onsager's model which assumes isolated electron–ion pairs [Buckley et al. (1989)].

The defects in the Onsager model has led Thomas and Imel (1987) to propose the *box model*, based on the earlier works of Jaffe (1913) and Kramers (1952). In the box model, the diffusion factor of electrons is negligible since the measured *diffusion factor* is very small, $D_- = (18 \pm 2)\,\mu m/\sqrt{mm}$ [Deiters et al. (1981)] for a drift field of $10\,kV/cm$ and $D_- = (28 \pm 2)\,\mu m/\sqrt{mm}$ [Derenzo et al. (1974)] for a lower drift field of $2.7\,kV/cm$. In addition, the positive ion mobility is set to zero in the box model as a result of the ion drift velocity being three to five orders of magnitude smaller than the electron drift velocity (see above). As a result of these approximations, the evolution of the ion, n_+, and the electron, n_-, concentrations, is described by the equations:

$$\frac{dn_+}{dt} = k_r n_+ n_- \tag{8.15}$$

and

$$\frac{dn_-}{dt} = -\mu_e E \sin\phi\, \frac{dn_-}{dt} - k_r n_+ n_-\,. \tag{8.16}$$

These equations are solved imposing the boundary condition that each electron–ion pair is isolated (box model) and using the initial condition $n_+ = n_-$. Then, the fraction of charge collected, Q/Q_0, is given by:

$$\frac{Q}{Q_0} = \frac{1}{\xi}\,\ln(1 + \xi) \tag{8.17}$$

with

$$\xi = \frac{Nk_r}{4L^2\mu_e E}. \tag{8.18}$$

In Eq. (8.18), E is the applied electric field, k_r is the recombination factor, L is the dimension of the box uniformly populated by the initial distribution of ions and electrons (the ion–electron pairs are isolated), i.e., the box contains N units of each charge at $t = 0$. ξ is the single parameter of the theory, $\xi \to 0$ for perfect charge collection and $\xi \to \infty$ in the case of complete recombination ($\xi E = 470$ kV/cm for α-particles).

A slight modification has been brought to the box model. The quantity ξE was allowed to be a function of E and empirically found to be for α-particles [Andrieux et al. (1999)]:

$$\xi E = a \left(1 - \frac{1}{2} e^{-bE} \right), \tag{8.19}$$

where $a = 416 \pm 1.4$ kV/cm and $b = 0.198 \pm 0.006$ (kV/cm)$^{-1}$.

8.2.4 *Recombination with Impurities*

The electrons which escape recombination (initial and columnar) move under the influence of the applied electric field through the liquid across the electrode spacing to reach the anode. The electron mobility being 10^5 times larger than the positive ion mobility, the total drift current is given by the drift current of electrons towards the anode. The presence of electron attaching impurities in the gap makes it impossible for a fraction of the electrons to reach the anode. This fraction will be converted into negative ions, with a rate constant k_s (Eq 8.6), following:

$$e^- + S \to S^-. \tag{8.20}$$

These negative ions have much smaller mobility compared to electron and they will not contribute to the charge collection at the anode. The number of free electrons remaining at time t is given by:

$$n_-(t) = n_-(t = 0) \, e^{-\eta v_d(e)t}, \tag{8.21}$$

where $n_-(t=0)$ is the number of electrons present initially, $v_d(e)$ is the electron drift velocity in the liquid and η is the attachment coefficient of electrons to an impurity molecule, i.e., the probability of attachment for an electron per centimeter of drift path [Swan (1963)].

The rate constant for the attachment, k_s, is given by [Bakale, Sowada and Schmidt (1976)]:

$$k_s = \int v \, \sigma(v) \, f(v) \, dv, \tag{8.22}$$

where $v \; (= v_d(e))$ is the electron velocity, $f(v)$ is the velocity distribution of the electron, and $\sigma(v)$ is the cross section, as a function of the electron velocity, of the interaction of the electrons with the impurity S. One can express the rate constant, k_s, as a function of the electric field according to:

$$k_s(E) = \int_0^{+\infty} \sigma(E_e) \, f(E_e, E) \, dE_e. \tag{8.23}$$

Table 8.1 Values of the rate constant k_s for different impurity concentrations and electric fields.

Impurity type	E (V/cm)	k_s (mol^{-1}s^{-1})
SF$_6$	30	2×10^{14}
	10^5	10^{13}
N$_2$O	50	3×10^{11}
	10^5	10^{13}
O$_2$	100	10^{11}
	5×10^5	2×10^{10}

Here, E_e is the electron energy, $\sigma(E_e)$ is the capture cross section, i.e., the cross section of the interaction between the electron and the S impurity, and $f(E_e, E)$ is the electron energy distribution for an electron of energy E_e subjected to an electric field E.

The rate constant for the attachment of electrons in liquid argon, k_s, to SF$_6$, N$_2$O, and O$_2$ has been studied and measured [Bakale, Sowada and Schmidt (1976)] as a function of the applied electric field. The different values are found in Table 8.1. As can be seen from that table, k_s decreases with increasing electric field for SF$_6$ and O$_2$ impurities, while it increases for N$_2$O.

The number of electrons captured by impurities during a time dt is given by:

$$\frac{dn_-(t)}{dt} = -k_s\, n_s(t)\, n_-(t). \tag{8.24}$$

Integrating Eq. (8.24) with the assumption of a constant number of impurities, n_s, gives the number of free electrons remaining at time t:

$$n_-(t) = n_-(t = 0)\, e^{-t/\tau_s}, \tag{8.25}$$

where $n_-(t = 0)$ is the number of free electrons initially present in the liquid. One also has the electron lifetime, τ_s, related to the rate constant, k_s, and the number of impurities via

$$\tau_s = \frac{1}{k_s\, n_s}. \tag{8.26}$$

It is useful to introduce the concept of absorption length, λ, of free electron drifting in the liquid under the influence of an electric field E. The absorption length varies with the electron drift velocity and its lifetime as:

$$\lambda = v_d(e)\, \tau_s = \mu_e E \tau_s, \tag{8.27}$$

where the relation between the electron mobility, μ_e, and the drift velocity [$v_d(e) = \mu_e E$] has been used. The absorption length can also be expressed as:

$$\lambda = \mu_e E \frac{1}{k_s\, n_s}. \tag{8.28}$$

The current induced by N electrons collected at the anode as a function of time, assuming a constant drift velocity (v_d), is given by:

$$i(t) = N e \frac{v_d}{D} = \frac{Q_0}{t_D}\, e^{-t/\tau_s}, \tag{8.29}$$

if $0 < t < t_D$, with $t_D = D/v_d$, D being the spacing between the cathode and the anode. If $t > t_D$, the current becomes

$$i(t) = 0. \tag{8.30}$$

The integration of the current over time gives the charge:

$$Q = \frac{Q_0}{t_D} \int_0^t e^{-t'/\tau_s} \, dt', \tag{8.31}$$

or

$$Q = Q_0 \frac{\tau_s}{t_D} \left(1 - e^{-t/\tau_s} \right), \tag{8.32}$$

for $0 < t < t_D$, and

$$Q = Q_0 \frac{\tau_s}{t_D} \left(1 - e^{-t_d/\tau_s} \right), \tag{8.33}$$

for $t \geq t_D$. Using Eq. (8.27), the charge can be expressed as a function of the absorption length:

$$Q = Q_0 \frac{\lambda}{D} \left(1 - e^{-D/\lambda} \right). \tag{8.34}$$

The charge collected at the anode, taking into account the electron–ion recombinations and the capture by impurities of drifting electrons on their way to the anode, is given by combining Eqs. (8.17, 8.34):

$$Q(E) = Q_0 \frac{1}{\xi} \ln(1 + \xi) \frac{\lambda(E)}{D} \left(1 - e^{-D/\lambda(E)} \right). \tag{8.35}$$

One should note the similarity between Eq. (8.35) and Eq. (6.90), the *Hecht equation* (see chapter on *Solid State Detectors*). Indeed, the situations are quite similar. Equation (8.35) expresses the electric charge of electrons depleted by electron recombination with ions and capture by impurities in the sensitive volume of the chamber. Similarly, Eq. (6.90) expresses the depletion of the electric charge by trapping and recombination with impurities and defects present in the semiconductor bulk.

8.3 Example of Ionization Chamber Application: The α-Cell

8.3.1 *The α-Cell*

The α-cell is a very good example of ionization chamber application and we will present it with some details. Such a cell is used to monitor the purity of liquid argon in high radiation environment. Example of such application can be found in the ATLAS experiment [ATLAS Collab. (1994a)] to be performed at the CERN LHC. High-granularity liquid argon calorimetry is a key element of the ATLAS experiment. The liquid argon technique is used for all the ATLAS electromagnetic calorimeters, the electromagnetic barrel and end-caps. It is also applied for the ATLAS hadron calorimeters up to pseudorapidities* $\eta \leq 5$, the forward hadron calorimeter (FCAL) and the hadronic end-cap (HEC). Many components of the ATLAS calorimetry system will be located in the high radiation field of hadrons and gammas resulting from the interactions, with surrounding materials, of hadrons produced by the high rate head-on collisions of 7 TeV protons at an expected peak luminosity of 1×10^{34} cm^{-2}s^{-1} at LHC. The radiation level will depend on the pseudorapidity. In particular, the highest radiation levels are expected to occur at high pseudorapidity ($3 \leq \eta \leq 5$) in the forward calorimeter (FCAL) which will be exposed to a neutron fluence of about 10^{16} n cm^{-2} and a γ-dose of 10^6 Gy over ten years of LHC operation.

The integration of the FCAL with the hadronic end-caps (HEC) also contributes to the radiation level in the hadronic end-caps where a neutron fluence of 10^{12}–10^{14} n cm^{-2} and a γ-dose of 10^3–10^4 Gy are expected over ten years, depending on the pseudorapidity.

The choice of liquid argon for most of ATLAS calorimeters (the hadron calorimeter is based on a sampling technique: plastic scintillating tile embedded in iron absorber due to financial constraints) was largely motivated by the radiation hardness of the technique. However, a large irradiation level can possibly inflict severe damage to the liquid argon calorimeters such as mechanical damage to parts of equipment, breakdown of electronics components or deep modifications of their characteristics and pollution of liquid argon with oxygen or oxygen-like impurities released from the surface of materials and equipments immersed in liquid argon. Therefore, the operation of the ATLAS liquid argon calorimeters requires radiation hard materials and equipment, and a limited pollution of liquid argon, below a threshold of 1–2 ppm. The latter can be achieved, if the components of the liquid argon calorimeters are certified against release of polluting impurities under irradiation. Therefore, the possible outgassing, of components immersed in liquid argon and exposed to high fluences of neutrons and gammas, is being investigated in conditions similar to those encountered in the final detectors during LHC operation. The tested materials and equipments are subjected to accelerated irradiation in order to receive a 10-year dose within a short period of time.

A cold test facility has been built at the Dubna IBR-2 pulsed neutron reactor for that purpose [Leroy et al. (1999a)]. The facility allows the irradiation of materials and equipments immersed in liquid argon at high neutron fluences ($\approx 10^{15}$ n cm^{-2} per day) and accompanying gamma doses (≈ 10 kGy per day). With these fluences and doses, the 10-year dose to be encountered at LHC can be achieved after about ~ 11 days. The large beam geometrical acceptance of 800 cm^2 offered by this facility allows the exposure of relatively large areas.

The argon is liquified at the facility in the argon vessel of the cryostat, using liquid nitrogen, and the level of liquid argon is kept constant during the whole

---

*One can see Eq. (3.61) in Sect. 3.2.3.1 for a definition of the pseudorapidity.

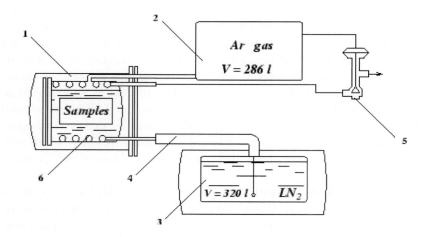

Fig. 8.1   Main cryogenic scheme of the facility: (1) cryostat, (2) receiver of gaseous argon, (3) dewar for liquid nitrogen, (4) liquid nitrogen piping, (5) automatic regulator, (6) argon condenser [Leroy et al. (1999b, 2000a)].

irradiation period. As can be seen on Fig. 8.1, the cryogenic system includes the cryostat (1), the receiver of gaseous argon (2), the dewar filled with liquid nitrogen (3), the piping for liquid nitrogen feeding to the cryostat (4), the automatic system for maintaining constant the level of liquid argon (5), and the argon condenser (6). The samples of materials to be tested are placed inside the argon vessel of the cryostat. The volume of the cryostat is about 1 liter. An ionization chamber (α-cell) installed in the cryostat (Fig. 8.1) is used to check for possible outgassing under irradiation of samples immersed in liquid argon and to monitor the purity of liquid argon. Since α particles have small range and high ionization power in liquid argon, the α source is installed directly in the sensitive volume. An ^{241}Am α-source with an activity of $7.7\,\text{kBq}/4\pi$ is deposited on the cathode of the α-cell and high voltages of 0–$2\,\text{kV}$ are applied to the anode. The gap between anode and cathode is $0.7\,\text{mm}$, which is an order of magnitude larger than the alpha-particle track length ($\approx 0.05\,\text{mm}$) in liquid argon. The α-cell with its readout scheme is illustrated in Fig. 8.2.

8.3.2   *Charge Measurement with the α-Cell*

The electron charge from alpha-ionization of liquid argon in the gap is transported across the cell gap. The collected charge induced by α-particles in liquid argon, in the presence of low impurity concentration ($\leq 10\,\text{ppm}$ of oxygen or oxygen-like impurities) is given by Eq. (8.35). The ratio of the charge signals after and before irradiation is the attenuation factor, $Abs(E)$, given as a function of the electric field,

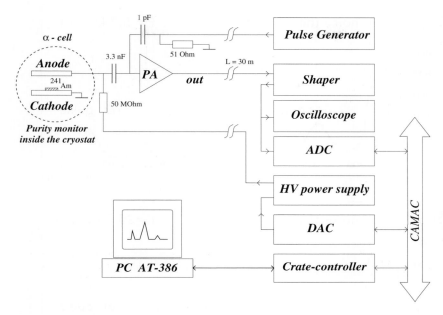

Fig. 8.2 α-cell and its readout scheme. The CAMAC electronics is operated *on-line* by a personal computer [Leroy et al. (1999b, 2000a)].

E (kV/cm), by

$$Abs(E) = \frac{Q(E)}{Q_0 \frac{1}{\xi} \ln(1+\xi)}, \tag{8.36}$$

or

$$Abs(E) = \frac{\lambda(E)}{D} \left(1 - e^{-D/\lambda(E)}\right). \tag{8.37}$$

Here $Q(E)$ and Q_0 are the charge signal after and before irradiation, respectively, where $Q_0 = eE_\alpha/w = 37.1\,\text{fC}$ for an alpha-particle produced in the decay of ^{241}Am (the alpha-particle mean energy, E_α, in that decay is $5.48\,\text{MeV}$ and $W = 23.6\,\text{eV}$ are needed to create an electron–ion pair in pure liquid argon). The function $\frac{1}{\xi}\ln(1+\xi)$ in Eq. (8.36) represents the initial recombination of electron–ion pairs in pure liquid argon [Thomas and Imel (1987)] and is assumed to be unaffected by low impurity concentrations (below $10\,\text{ppm}$). In Thomas and Imel's model [Thomas and Imel (1987)], $\xi E = 470\,\text{kV/cm}$. However, as seen before, discrepancies between their model and data have led to the use of the empirical formula [Andrieux et al. (1999); Leroy et al. (1999a)]:

$$\xi E = a \left(1 - k\,e^{-bE}\right), \tag{8.38}$$

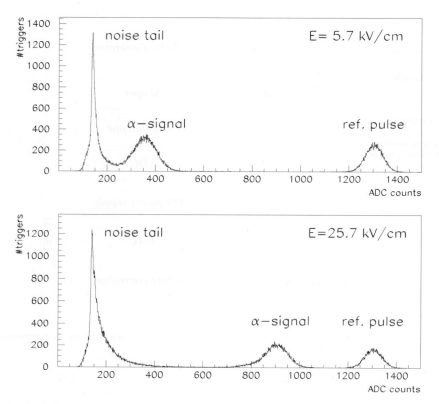

Fig. 8.3 An example of the raw data. Each histogram is an ADC spectrum containing 100k triggers recorded for two different values of the electric field in the liquid argon gap of the α-cell electrodes: 5.7 kV/cm and 25.7 kV/cm. The reference pulse is also shown [Leroy et al. (1999b, 2000a)].

where a, b and k are free parameters to be fitted to the data*. The values of these three parameters depend on the experimental set-up. The values

$$a = (720.6 \pm 5.6)\,\text{kV/cm}, \ b = (0.0833 \pm 0.004)\,(\text{kV/cm})^{-1} \text{ and } k = (0.48 \pm 0.01),$$

used to estimate the liquid argon pollution, were obtained by fitting $Q(E)$ data measured before irradiation.

$\lambda(\text{E})$ in Eq. (8.37) is the charge carrier absorption length and D is the α-cell gap. In the case of oxygen pollution with concentration ρ (ppm), λ is given by [Hofmann et al. (1976)]:

$$\lambda(E) = \frac{\alpha E}{\rho} \tag{8.39}$$

with $\alpha = 14\,\text{mm}^2\text{ppm/kV}$.

---

*Compared to [Andrieux et al. (1999)], an additional parameter k has been introduced in [Leroy et al. (1999a)], this explains the different values of a and b used in Eq. (8.19).

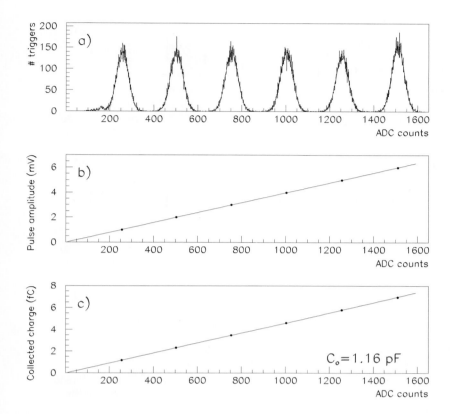

Fig. 8.4 Calibration of the ADC scale: (a) set of calibrated pulses from (1–6) mV range are recorded by the ADC. A Gaussian fit is applied to extract each peak position; (b) generated pulse amplitudes (in mV) are a linear function of ADC counts (the line is a linear fit to the data). The ADC has low intrinsic threshold and pedestal zero value; (c) calibration of the ADC scale for the input capacitor $C_o = 1.16\,\text{pF}$ used in the experiment. The same calibration constants are used for the data processing before and after each irradiation run [Leroy et al. (1999b, 2000a)].

In practice, for comparison between collected charge measurements performed before and after irradiation, a normalization factor, F_N, is introduced to account for possible miscalibration of the readout charge:

$$\frac{Q(E)}{Q_0 \frac{1}{\xi} \ln(1 + \xi)} = F_N \, Abs(E). \tag{8.40}$$

Typical value for F_N is (0.997 ± 0.003) for $Q(E)$ data measured before irradiation.

In the following, most examples of the pollution tests results will be presented in terms of the attenuation factor [the ratio of the charge measured after irradiation to the charge measured before irradiation, Eqs. (8.36, 8.37)] as a function of the electric field. These plots are easy to interpret: significant deviation of the attenuation factor from unity surely indicates the presence of impurities in liquid argon.

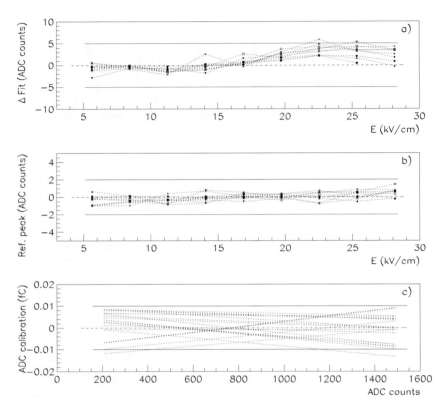

Fig. 8.5 Systematic errors estimated from different runs of measurements: (a) uncertainty of the fit procedure; (b) variation of the reference signal around its average position; (c) uncertainty of the ADC calibration [Leroy et al. (1999b, 2000a)].

The ADC spectra are recorded for various values of the electric field in the range (2–29) kV/cm. Then, one may determine the α-peak position (Fig. 8.3) and obtain the electric field dependence of the charge collected in the liquid argon gap. Examples of ADC-spectra are shown in Fig. 8.3 which represents the raw data collected at two different values of the electric field in the liquid argon gap of the α-cell: 5.7 kV/cm and 25.7 kV/cm. The reference pulse is generated by a pulse generator feeding the test input of the preamplifier (Fig. 8.2) to monitor the gain of the preamplifier (which is shown stable). The non-Gaussian tails of the α-peaks reveal the contribution of pick-up signals. The ADC calibration in terms of collected charge allows the conversion of the ADC counts into charge units (fC). The ADC calibration procedure is shown in Fig. 8.4.

The data processing consists of three main steps: *i*) each recorded ADC spectra have to be fitted by some functional (usually a Gaussian) to find the position of the α-peak, *ii*) the fitted positions have to be corrected assuming the reference

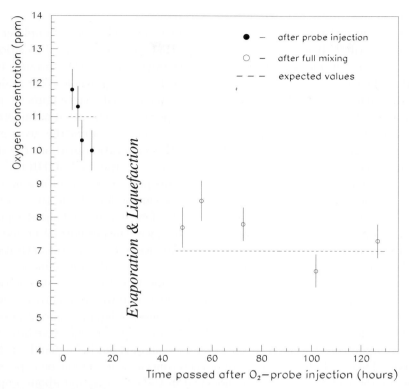

Fig. 8.6 Behavior of the argon purity with time: soon after the oxygen probe injection (black dots) and during a long period ($\approx$ 3 days) after the full mixing of the oxygen probe and the argon gas (open dots). Dashed lines indicate the expected levels of the oxygen concentration, see text [Leroy et al. (1999b, 2000a)].

pulse stability and, finally, *iii*) the ADC counts must be translated into collected charge units (fC) for further analysis. Each step gives a certain contribution to the systematics which has to be summed with the statistical uncertainty from the (Gaussian) fit. These systematics contributions are: uncertainties due to the fitting procedure (± 5 ADC counts), variation of the reference signal around its average position (± 2 ADC counts), and uncertainty of the ADC calibration (± 0.01 fC).

The uncertainties due to the fitting procedure can be estimated from the variations of the peak position as determined from different fitting procedures. It is seen from Fig. 8.3 that, at $E = 5.7\,\text{kV/cm}$, the exponential tail of the noise peak, which is cut by the threshold, should be taken into account. The possible contribution of additional sources of charge at higher E values can be described by additional Gaussian's. On the other hand, each peak of the ADC spectra can be fitted by a single Gaussian within the $\pm 1\,\sigma$ interval around the peak position. The differences among the results obtained from different fitting procedures are presented in Fig. 8.5(a) over an electric field range of $(5.7\text{--}28)\,\text{kV/cm}$. The variations are within the range of ± 5 ADC counts.

The reference pulse position is used to correct the gain of the preamplifier at each electric field value. It is seen from Fig. 8.5(b) that the reference peak positions are varying during the measurements within the range of ± 2 ADC around the average value. This conclusion was checked to be valid over the whole ADC scale.

The calibration of the ADC scale in terms of collected charge units (fC) is obtained from the linear fit of the calibration pulse positions. The fitted parameters are averaged over a set of measurements, and the deviations from this averaged calibration are presented in Fig. 8.5(c). Based on this calibration method, the uncertainty on the collected charge value is expected to be at the level of 0.01 fC.

The detailed analysis of the systematic errors was performed for all irradiation runs. Some significant variations of the systematics can be found due to cryostat activation. The activation of the cryostat is at its lowest after a long stopping period (3–4 months) of the reactor, due to maintenance. During normal period of operation, frequent irradiation every three weeks increases the cryostat activation and thus the systematic errors. In addition, this activation maintains some level of impurities in liquid argon before irradiation. It is important to know the evolution with time of this impurity level. A special test was dedicated to the monitoring over a long period of time of the evolution of the impurity level of liquid argon purposely contaminated with oxygen in the cryostat. About $6\,cm^3$ of oxygen gas were injected into the warm argon vessel of the cryostat which was pumped out beforehand for this test. A maximal value of about 11 ppm of oxygen concentration might be expected in this case for the whole oxygen probe liquefied in the cryostat, and about 7 ppm for the oxygen probe mixed uniformly in the liquid and gaseous states of argon in cryostat and receiver. The oxygen concentration values were determined using Eqs. (8.37–8.39). The results are presented in Fig. 8.6.

The liquid argon purity improves with time after the initial liquefaction of the oxygen probe due to the mixing of the oxygen polluted liquid argon in the cryostat and the argon gas in the receiver. The initial value of the impurity concentration is close to expectation (11 ppm). The "cleaning" could continue until the oxygen concentration in the argon system becomes even. To speed up the process, the oxygen polluted liquid argon was evaporated from the cryostat back into the receiver and liquefied once more. The liquid argon was kept in the cryostat after this full mixing until the end of the test (6 days) and the liquid argon purity remained stable within ± 1 ppm, close to the expected level of ~ 7 ppm. It was found that variations of the monitor signal at the end of the test could be caused by the instability of the high voltage power supply.

8.3.3   *Examples of Pollution Tests Using the α-Cell*

A very first test consisted of a so-called "empty" run. The cryostat filled with liquid argon (with no material for test inside, but everything else being identical to

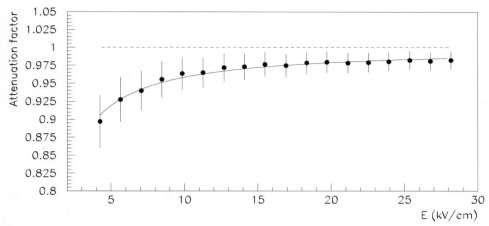

Fig. 8.7 The attenuation factor, as a function of E (in kV/cm), measured after a purposely injection of 2 ppm of oxygen in liquid argon. The curve is the result of fitting Eqs. (8.37–8.39) to the data [Leroy et al. (2002b)].

normal condition of operation) was exposed to a total fast neutron fluence of $(1.0 \pm 0.1) \times 10^{16} \, n/cm^2$ and a γ dose of (96 ± 10) kGy. The collected charge dependence on the electric field was measured before and after irradiation. Assuming that the liquid argon was polluted by oxygen or oxygen-like impurities, Eqs. (8.37–8.39) were fitted to the collected charge measured before and after irradiation to estimate the corresponding impurity concentrations.

It has been found that $\rho = (4.4 \pm 0.3)$ ppm and $\rho = (3.7 \pm 0.2)$ ppm before and after irradiation, respectively. A non-zero value of the initial liquid argon purity concentration was caused by the lack of time for a careful cleaning of the ~ 300 litres receiver, after its preparation (about 20 hours are needed to fill it with pure argon). This non-zero value of ρ in these conditions disappears when the system is cleaned up by multiple flushing with pure argon gas. Then, the liquid argon purity before and after irradiation were found [Leroy et al. (1999b, 2000a)] compatible with $\rho \sim 0$, showing that the cryostat and α-cell system themselves do not pollute argon during irradiation [Leroy et al. (1999a)].

Running periods have been also devoted to various systematic studies, including calibration runs dedicated to check oxygen pollution concentrations, purposely injected in argon by fitting Eqs. (8.37–8.39) to the corresponding electric field curves. The attenuation factor, i.e., in the present case, the ratio between the charge measured after an oxygen injection in liquid argon and the charge measured with pure liquid argon, was analyzed. An example is shown in Fig. 8.7. At the same time, the calibration runs confirmed the amplifier capability to recognize liquid argon pollution at a level better than 1 ppm.

The outgassing under irradiation of PREPREG (epoxy laminate) has been studied, although it is not intended to use this material in the ATLAS liquid argon

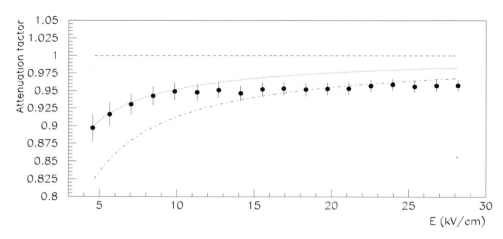

Fig. 8.8   The attenuation factor measured for PREPREG samples as a function of E (in kV/cm) varies from 0.90 at 4.7 kV/cm up to 0.95 for $10 < E < 28.5$ kV/cm, showing the effect of pollution. This measured attenuation factor is compared to the attenuation factor corresponding to an oxygen pollution concentration of 2 ppm (dotted curve) and 4 ppm (dashed lower curve), see text [Leroy et al. (2002b)].

calorimeters. PREPREG being known as a source of pollution in liquid argon, the test was performed to check the α-cell system real capability to detect liquid argon pollution in practical situations. PREPREG samples (total area of $0.22\,\mathrm{m}^2$) were immersed in liquid argon and exposed during a standard period of irradiation of 11 days, for a total fast neutron fluence of $10^{16}\,\mathrm{n\,cm}^{-2}$. The attenuation factor for PREPREG at $10^{16}\,\mathrm{n\,cm}^{-2}$ as a function of E is compared in Fig. 8.8 to the attenuation factor curves corresponding to an oxygen pollution concentration of 2 ppm (dotted curve) and 4 ppm (dashed lower curve) as obtained from calibration runs. This comparison shows a pollution concentration of $\rho \approx 4$ ppm, confirming the known behavior of PREPREG under irradiation as well as the system capability to detect liquid argon pollution.

FCAL resistors, capacitors, and transformers together with capacitors and sintimid disks of the purity monitor have been irradiated in liquid argon to study their possible outgassing at a neutron fluence of $10^{16}\,\mathrm{n\,cm}^{-2}$ at the IBR-2 reactor of JINR, Dubna [Leroy et al. (2002a)]. The comparison of the collected charge measured before and one, two and three days after irradiation shows that no outgassing resulting from irradiation is observable (Fig. 8.9).

The attenuation factor measured three days after irradiation shows that no pollution resulting from irradiation is observable (see Fig. 8.10). These results also demonstrate the good stability of the system response over a period of three days after irradiation.

The examples shown above illustrate the utility of a ionization chamber (α-cell) to certify materials and equipments against release of impurities polluting liquid

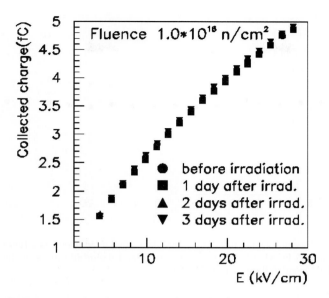

Fig. 8.9 The collected charge as a function of E (in kV/cm) with 72 FCAL resistors, 75 FCAL capacitors and 47 transformers, 5 capacitors and 6 sintimid disks of the purity monitor in liquid argon at a fluence of $1.0 \times 10^{16} \, \mathrm{n \, cm^{-2}}$. The charge measured before irradiation is compared to the charge measured one, two, and three days after irradiation [Leroy et al. (2002b)].

argon in high irradiation environments such as the ones to be met at the LHC.

8.4 Proportional Counters

8.4.1 *Avalanche Multiplication*

Beyond the ionization region, one falls within a region of gas multiplication. When the applied voltage is high enough, the electrons produced by primary ionization are accelerated and gain sufficient energy to ionize molecules of the gas and produce secondary electrons, themselves possibly creating a tertiary ionization. If an electron is produced in a region of sufficiently high uniform electric field, after a free path-length $\lambda = 1/\alpha$ (in standard notation), it will ionize a molecule, i.e., an electron–ion pair will be produced. At this stage, two electrons will derive under the applied field, each producing an electron–ion pair, etc. If n_- is the number of electrons at a given position, after a path-length dx, n_- will increase by an amount:

$$dn_- = n_- \alpha \, dx. \qquad (8.41)$$

If $n_- = n_0$ at $x = 0$, Eq. (8.41) can be easily integrated:

$$n_- = n_0 \, e^{\alpha x}. \qquad (8.42)$$

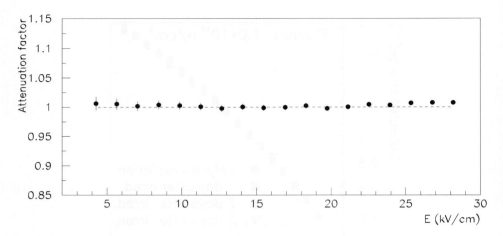

Fig. 8.10   The attenuation factor measured three days after irradiation, as a function of E, for the 72 FCAL resistors, 75 FCAL capacitors, 5 capacitors and 6 sintimid disks of the purity monitor at a fluence of $10^{16} \, \mathrm{n \, cm^{-2}}$ [Leroy et al. (2002a)].

The coefficient α is called the *first Townsend coefficient*. The multiplication factor M is defined from Eq. (8.42) as

$$M = \frac{n_-}{n_0} = e^{\alpha \, x}. \tag{8.43}$$

For non-uniform electric field, α becomes a function of distance $[\alpha(x)]$ and Eq. (8.43) has to be modfied between two points x_1 and x_2:

$$M = \exp \left(\int_{x_1}^{x_2} \alpha(x) \, dx \right). \tag{8.44}$$

No particular geometry for the electric field has been assumed. However, the multiplication factor can be calculated for any geometry, if one knows the *Townsend coefficient* dependence on the electric field. An example of dependence of α on the electric field E often used is [Rose and Korff (1941)]:

$$\frac{\alpha}{p} = A \, e^{-Bp/E}, \tag{8.45}$$

where p is the gas pressure, and A and B are two constants depending on the gas (A and B are expressed in $\mathrm{cm^{-1} \, Torr^{-1}}$ and $\mathrm{V \, cm^{-1} \, Torr^{-1}}$ units, respectively). In the E-region where Eq. (8.45) applies, one can assume a linear dependence of α on the electron energy E_e:

$$\alpha = KNE_e, \tag{8.46}$$

where N is the number of molecules per units volume. The factor K depends on the gas and is expressed in $\mathrm{cm^2 \, V^{-1}}$ units. For example, for argon [Korff (1946)], $A = 14 \, \mathrm{cm^{-1} \, Torr^{-1}}$, $B = 180 \, \mathrm{V \, cm^{-1} Torr^{-1}}$ and $K = 1.81 \times 10^{-17} \, \mathrm{cm^2 V^{-1}}$.

The multiplication factor cannot be increased beyond some limit because beyond some value of the electric field, secondary processes such as emission of photons also

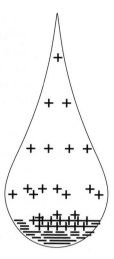

Fig. 8.11 The electrons, with faster drift velocity as compared to the ions, occupy the front of the avalanche while the ions are left behind in the tail, with lower radial extension. The shape of the avalanche is drop-like.

induce avalanches over the whole gas volume, leading to breakdown. The *Raether condition* gives a limit for multiplication before breakdown:

$$\alpha\, x \approx 20, \tag{8.47}$$

which gives $M = 10^8$.

Because of its generation mechanism, the avalanche has a drop-like shape (Fig. 8.11). The electrons drift velocity is much higher as compared to ions. Therefore, the electrons occupy the front of the avalanche, while the ions are left behind in the tail, in decreasing number and radial extension. Since they have been produced in the last average path length, half of the ions are in the front section of the drop, behind the electrons.

8.5 Proportional Counters: Cylindrical Coaxial Wire Chamber

The cylindrical coaxial chamber is a solution to obtain satisfactory gains while benefiting from the proportional mode. Let us consider the example given in [Sauli (1977)] of a thin layer of gas, 1 cm argon in normal conditions of temperature and pressure, in between two parallel planar electrodes. The number of electrons produced by a *mip* in a 1 cm gap is about $n_- \approx 105$, [see Eq. (8.1)]. The resulting pulse height is:

$$V = \frac{Q}{C} = \frac{n_- e}{C} \approx 2\mu\text{V}, \tag{8.48}$$

for $C = 10\,\mathrm{pF}$ (typically). This pulse height of $2\,\mu\mathrm{V}$ is below detection threshold. This signal can be enhanced by avalanche multiplication by increasing the electric field, i.e., the voltage across the electrode gap. In these conditions, one is outside any proportionality regime, since there is no proportionality between the deposited energy and the detected signal, as the signal depends on the avalanche length, and ultimately on the location where the original charge has been produced. Considerable increase of the field can lead very rapidly to voltage breakdown (Raether condition). A solution to this problem is the use of a cylindrical coaxial chamber. The central electrode is a wire of radius r_a, surrounded by a cylindric chamber wall of radius r_b (Figs. 8.12 and 8.13). The polarity is set such that the central wire is the anode and the external cylinder is the cathode. The electric field is of the shape $\sim 1/r$, maximum at the anode and decreasing towards the cathode.

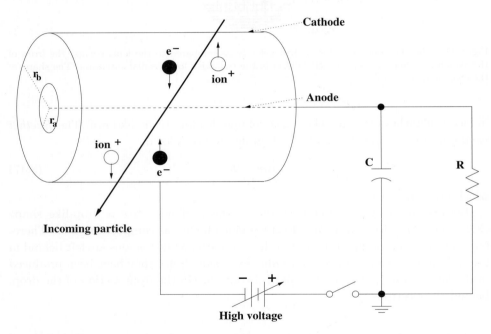

Fig. 8.12 Schematic view of a gas filled cylindric coaxial chamber with a central electrode consisting of a wire of radius r_a, surrounded by a cylindric chamber wall of radius r_b. The electrons and ions created by the passage of an incoming particle are collected at the anode and cathode, respectively.

The electric field at a distance r from the center of the chamber (Fig. 8.14) is, in standard notations,

$$E(r) = \left(\frac{CV_b}{2\pi\epsilon_0} \right) \frac{1}{r} \tag{8.49}$$

and the voltage

$$V(r) = -\frac{CV_b}{2\pi\epsilon_0} \ln\left(\frac{r}{r_a}\right),$$ (8.50)

with the capacitance per units of length:

$$C = \frac{2\pi\epsilon_0}{\ln(r_b/r_a)}.$$ (8.51)

Here $V_b = V(r = r_b)$ is the potential difference between cathode and anode $[V(r = r_a) = 0]$ and ϵ_0 is the electric permittivity of the gas (typically, $\epsilon_0 \sim$ 8–9 pF/m). Under the electric field, the electrons produced by the primary interaction move towards the anode while the positive ions move towards the cathode. For gas-filled cylindrical chambers with electrodes of radii r_a and r_b, the electron drift velocity is expressed as (p is the gas pressure):

$$v_d(\mathrm{e}) = \frac{\mu_e}{p}E = \frac{\mu_e}{p}\frac{V}{r\ln(r_b/r_a)}.$$ (8.52)

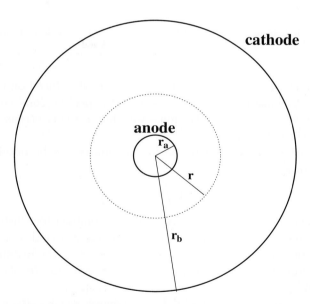

Fig. 8.13 Cross-sectional view of a cylindric coaxial chamber; the central electrode is a wire of radius r_a, surrounded by a cylindric chamber wall of radius r_b.

Close to the anode, the electric field is very strong [Eq. (8.49)] and the multiplication can take place. As seen in the previous section, a drop-like avalanche with electrons at the front and ions behind develops. The lateral diffusion, combined with the small value of r_a, allows the drop-like avalanche to surround the anode. The electrons are rapidly collected (typically in 1 ns) and the cloud of ions is free to

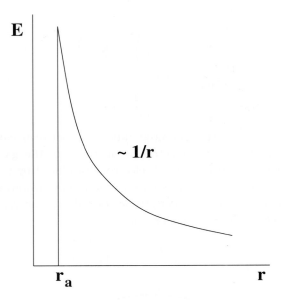

Fig. 8.14 The electric field as a function of the distance r to the anode (central wire of radius r_a).

move towards the cathode. As a consequence, the signal induced on the electrodes in the proportional regime is due to the motion of the positive ions, contrary to the case of ionization chambers where the signal was due to the electrons (very low ion mobility compared to electron).

The multiplication factor for a proportional counter can be obtained from an equation similar to Eq. (8.44):

$$M = \exp\left(\int_{r_a}^{r_c} \alpha(r)\, dr\right), \tag{8.53}$$

where $\alpha(r)$ is calculated from Eqs. (8.46, 8.49) in combination with the electron mean energy obtained from the electric field between two collisions ($E_e = E/\alpha$, remind the definition of α given earlier). The distance r_c is the distance where the multiplication (through avalanche) starts. One can rewrite Eq. (8.53) as a function of the electric field:

$$\ln M = \int_{E(r_a)}^{E(r_c)} \alpha(E)\, \frac{\partial r}{\partial E}\, dE. \tag{8.54}$$

Combining Eqs. (8.49, 8.51, 8.54), one obtains

$$\ln M = \frac{V_b}{\ln(r_b/r_a)} \int_{E(r_a)}^{E(r_c)} \frac{\alpha(E)}{E}\, \frac{dE}{E}. \tag{8.55}$$

If $\alpha(E)$ is a linear function of E, one can find [Diethorn (1956)]:

$$\ln M = \frac{V_b}{\ln(r_b/r_a)} \frac{\ln 2}{\Delta V} \left\{ \ln\left[\frac{V_b}{p r_a\, \ln(r_b/r_a)}\right] - \ln K_d \right\}. \tag{8.56}$$

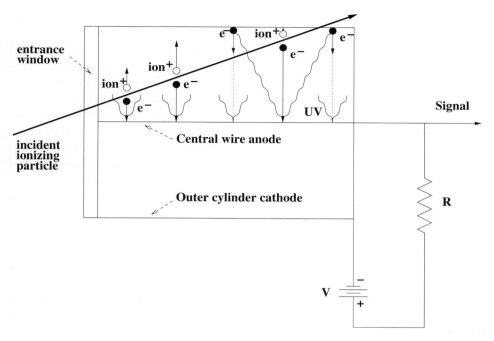

Fig. 8.15   Schematic view of a Geiger-Mueller counter showing its operating principles (see text).

This expression is frequently used. The quantity K_d is the minimal value of E/p (p is the pressure) below which no multiplication occurs. The quantity ΔV corresponds to the voltage gradient to which an electron is subjected during its motion in between two successive ionizing events. The values of ΔV and K_d depend on the gas. For instance, $\Delta V = 36.5$ (29.5) V and $K_d = 6.9$ (10.0) 10^4 V/cm atm for methane (propane) [Fischer, H. et al. (1975)].

8.6   The Geiger–Mueller Counter

The Geiger–Mueller counter [Geiger and Mueller (1928)] is a gas-filled detector built to operate with maximum amplification (Fig. 8.15).

The central wire (anode) is maintained at a very high potential compared to the cylindric wall cathode. When ionization takes place in the gas volume, the electrons are accelerated towards the anode. So, initially, one has an amplification similar to the one encountered in a proportional detector. However, the situation differs very rapidly as electrons strike the central wire so violently that UV photons are emitted. A fraction of these photons reaches the cathode, where they cause emission of supplementary electrons from the cathode wall. Then, these supplementary electrons are accelerated towards the anode where they can also cause UV gamma emission, etc. So, there is propagation of an ionization avalanche through the whole gas volume and along the central wire, with a rapid collection of the electrons.

The positive ions, much slower, escape collection and create a positive cloud around the central wire. The existence of this cloud terminates the avalanche process since the electrons in this region are captured by these positive ions, before reaching the anode. When the cloud of positive ions is near the cathode wall, the electrons extracted from the wall by UV photons are captured. Some of the electrons enter high energy orbits of the ions and can make transitions into lower energy orbits with UV emission, eventually extracting electrons from the cathode wall and triggering another avalanche and, therefore, a secondary pulse. The solution adopted to prevent this secondary pulse is to use a quenching gas. Usually an organic compound, this gas releases electrons, easily. The molecules of this gas can neutralize the positive ions by giving them electrons which allows the conversion of the positive ion-cloud into ionized molecules of the quenching gas.

The Geiger-Mueller counter produces very large pulses since the ionization avalanche releases huge amount of charges. However, as seen above, the number of electrons produced is independent of the applied voltage and independent of the number of electrons produced by the initial ionization. Discrimination between incoming particles is not possible with this type of detector since the signal pulse is independent of the type of radiation.

Chapter 9

Principles of Particle Energy Determination

9.1 Experimental Physics and Calorimetry

Collisions between two particle beams (collider experiments) or by a beam interacting with a fixed-target (fixed-target experiments) or the interaction of cosmic rays with space matter or the Earth's atmosphere (astrophysics, gamma-ray astronomy, cosmic-ray studies) frequently produce a variety of particles through very complicated configurations of events covering a large solid angle.

The geometry of the collisions imposes a defined structure to detectors. For instance, detectors operating in collider experiments have a barrel structure, with endcaps to cover the forward region. The occurrence of a large number of events to investigate and the variety of physics goals pursued impose the construction of a detector composed of many sub-detectors assigned to dedicated tasks. These sub-detectors have complementary capabilities which can be combined in a optimized way to measure the energy, mass, momentum, charge and direction of the particles produced. In particular, the measurement of particle energy is performed with calorimeters. A review of the properties of electromagnetic and hadronic shower propagation in matter and of calorimeters was provided by Leroy and Rancoita (2000).

Calorimeters are instrumented blocks of matter in which the particle to be measured interacts and deposits all its energy in the form of a cascade of particles, whose energy decreases progressively down to the threshold of ionization and excitations that are detectable by the readout media. The deposited energy is detectable in the form of a signal which is proportional to the incoming energy. This proportionality is the base of calorimetry measurement.

Calorimeters contribute also to the measurement of the position and angle of the incident particle. Their different response to electrons, muons, and hadrons can be exploited for particle identification. Neutrinos, which interact only weakly with matter, are detected through the absence of any energy deposit (*missing energy*).

Electromagnetic calorimeters are used to measure the energy deposited by electromagnetic particles (electrons, photons), while the hadron (such as pions, protons) energy measurement is achieved with hadronic calorimeters. Furthermore, calorimeters are classified into two types: homogeneous calorimeters, where the incoming particle energy is measured in a homogeneous block of sensitive absorbing material (lead-glass, sodium iodide (NaI) crystal, bismuth-germanium oxide (BGO) crystal, etc.), and the sampling calorimeters, where the incoming particle energy is measured in a number of sensitive layers interspersed with the layers of absorbing material, the latter speeding up the cascade process. Various active medium (scintillator, silicon*, liquid argon, gas, ...) and absorbers (Fe, Cu, Pb, U, ...) are used.

Calorimeters[†] of all types were used and are currently in operation in many particle physics experiments. They have played a key role in recent experiments leading to fundamental results like the discovery of the vector bosons W and Z^0 (UA1, UA2) [Arnison et al. (1983)] and of the top quark (CDF, D0) [CDF Collab. (1988)]. Major searches and studies, e.g., the search for the quark gluon-plasma in heavy-ion collisions at very high energy (NA-34/HELIOS, NA-35, WA-80) [HE-LIOS Collab. (1988)], the study of e^+e^- collisions at LEP (ALEPH, DELPHI, L3, OPAL) [LEP (1982)], the study of e-p collisions at HERA (H1, ZEUS) [H1 Collab. (1993b)], and the search for possible neutrino oscillations (CHORUS [Eskut et al. (1997)], NOMAD [Altegoer et al. (1998)]), also make large use of calorimeters. This type of detector is also used in cosmic ray (for instance AMS [AMS Collab. (2002)] and PAMELA[‡] [Boezio et al. (2006)]) and gamma-ray experiments (for instance EGRET [Kanbach et al. (1988)], GLAST [GLAST Collab. (1998)]) performed in space and in balloon (such as HEAT [Barwick et al. (1997)], CAPRICE [Bocciolini et al. (1996)]) and in air shower experiments with particle detectors (e.g., KAS-CADE [Klages et al. (1997)]).

Large calorimeters will be also important elements of the central detection systems that are needed to perform the high-luminosity experiments (ATLAS [AT-LAS Collab. (1994a)], CMS [CMS (1994)], ALICE [ALICE (1993)], LHC-B) at the Large Hadron Collider (LHC) in the LEP tunnel at CERN. These experiments provide the opportunity to search for new physics; for instance, the search for heavy leptons, supersymmetric particles, the Higgs particle(s), and the additional vector bosons predicted in many possible extensions of the Standard Model [LHC (1990)]. The radiation environment of LHC-experiments is treated in Sects. 4.1.1–4.1.1.2.

Large and compact calorimeters (like the one based on Si/W) are also needed to be operated at the future International Linear Collider (ILC) (for instance, see [Strom (2008)]).

Independently of its structure, a calorimeter must be of sufficient thickness to allow the particle to deposit all its energy inside the detector volume during the subsequent cascade of particles of lower and lower energy. The total depth of the calorimeter must be large enough to allow a longitudinal containment of the cascade

---

*The usage of silicon detectors as active medium in calorimetry was first proposed by Rancoita and Seidman (1984); silicon electromagnetic and hadronic calorimeters were originally developed by the SICAPO collaboration.

[†]A recent review of calorimeter developments has been provided by Pretzel (2005).

[‡]This pay-load employs a Si/W *imaging calorimeter.*

in order to completely absorb the incoming energy. The necessary longitudinal depth of a calorimeter varies with the incoming energy, E, as $\ln E$ and, therefore, the calorimeter can remain a compact construction even at the highest energies [Fabjan and Ludlam (1982)]. The incoming energy, being distributed among a large number of secondary particles, can be deposited at large angles with respect to the longitudinal axis of the calorimeter. The result is that the transverse containment of the cascade imposes a minimal radial extension of the calorimeter which must, then, reach at least several times the average diameter of a cascade. Therefore, the calorimeter dimensions must be large enough to avoid longitudinal and lateral leakage of cascades.

The granularity is another important requirement for a calorimeter and characterizes the spatial separation between two particles in an event. In electromagnetic calorimeters, it can be increased by adding a so-called *preshower detector*. For example, in the CMS apparatus at LHC a preshower detector[§] is located in front of the *Endcap Electromagnetic Calorimeter* to increase the identification of two closely-spaced photons from π^0 decay. It has to be noted that the granularity is imposed by physics and it is fixed by the minimum angle to be detected between particles, like the detection of a specific particle inside a jet, for instance. Cascades must be separated at least by one cascade diameter in order to be individually recognized. This constraint defines the minimum distance between the detector and the interaction point and requires the use of active layers organized in cells, whose maximum dimensions can allow this cascade separation. In practice, this means that the transverse cell size must be comparable to the lateral cascade dimension. In addition, the detailed measurement of the cascade position during its development in the calorimeter volume also demands the best possible longitudinal granularity and, therefore, requires a longitudinal segmentation of the calorimeter. For instance, tower structures in sampling calorimeters using scintillator as active medium were employed [HELIOS Collab. (1988)] to achieve the needed lateral and longitudinal segmentations.

Calorimeters must have an energy resolution compatible with the experimental goals. In the case of hadron calorimeters, this possibility is mostly dictated by their relative response to the electromagnetic (e) and hadronic (π) cascade, measured by the e/π signal ratio. The equalization between the electromagnetic and hadronic signals ($e/\pi = 1$, i.e., the compensation condition) is the condition for obtaining the linearity of the energy response of the calorimeter to incoming hadronic cascades, and to achieve an energy resolution that improves as the incident energy increases.

However, in large experiments the best performing calorimetry may be in conflict with specific requirements from other sub-detectors owing to the physics goals of the experiments. Therefore, calorimeters in large experiments are not necessarily the most efficient from the point of view of calorimetry principles, but are certainly

---

[§]It consists of two lead radiators each followed by a layer of silicon strip detectors with a strip pitch of 1.9 mm [Topkar et al. (2006)].

optimized towards the achievement of the experiment goals and physics requirements.

Large calorimeters directly apply basic knowledge of the physics of cascade processes in matter and the results of studies, that are often carried out on dedicated
test calorimeters. However, the conditions which prevail during the test measurements have to be remembered, in order to assess the performance of these test
calorimeters with the view of extrapolation to larger calorimeters to be operated
in physics experiments. The physics conditions met in experimental zone by test
calorimeters is different from those encountered in an operating experiment. For instance, the energy resolution of a test calorimeter is assessed from the measurement
in its volume of the energy deposition of particles of a well known dedicated beam,
while a calorimeter operating in a physics experiment will face particles produced
in collisions (collisions between two beams or beam with a target). The energy of
these particles will cover a large range and many will be produced with relatively low
energy, leading to a degradation of the calorimeter energy resolution. The calorimeter performance is also affected by the amount of material in front of the calorimeter
being tested and the effect of the beam conditions, such as for example the momentum spread in the test beam and the precision of beam counters, which vary from
one test to another. Again, the extrapolation of performance to larger calorimeters
in physics experiments has to take into account changing environment and beam
features. Usually, these test environment limitations are quoted in articles and reports in order to separate circumstantial effects from the fundamental properties of
shower propagation and the way the energy is degraded and deposited in matter.

The physics, which governs the electromagnetic (e) cascading and the propagation in complex absorbers, has been reviewed in Sect. 2.4, while the phenomenology
of the hadronic cascading in Sect. 3.3. In this chapter, the instrumental needs due
to shower generation and propagation in matter are pointed out. The chapter is
organized as follows. The electromagnetic calorimeter response, the e/mip ratio
and the local hardening effect (exploited to reduce the e/mip ratio in high-Z absorbers) are discussed in Sect. 9.2. The filtering effect, which can be used to adjust
the e/mip ratio for complex absorbers, is dealt in Sect. 9.3. The energy resolution
of electromagnetic sampling calorimeters is presented in Sect. 9.4. The Sect. 9.5 is
devoted to a review of homogeneous on calorimeters. The use of calorimeters for
position measurement and electron–hadron separation is discussed in Sects. 9.6 and
9.7 respectively. Hadronic calorimetry is introduced in Sect. 9.8, where the intrinsic
properties of hadronic calorimeter such as e/h, h/mip, π/mip and e/π ratios are
dealt. In particular, the relation of the e/π ratio to the compensation condition
is discussed in the same section. Section 9.9 is devoted to the study of the compensation condition and the methods to achieve compensating calorimetry. The
compensation and the hadronic resolution are discussed in Sect. 9.10. In Sect. 9.11,
it is presented a brief review of the study of the cascade behavior in an energy
regime beyond accelerator energies such as the energies characterizing the cosmic

rays, the whole Earth atmosphere serving, then, as radiator volume.

9.1.1 *Natural Units in Shower Propagation*

The comprehensive treatment of the generation and propagation of electromagnetic and hadronic showers is presented and the units employed for their parametrization are discussed in Sects. 2.4-2.4.3 (for electromagnetic cascades) and Sects. 3.3-3.3.3 (for hadronic cascades).

For instance for electromagnetic showers, the radiation length[†], X_0, emerges as a natural unit of length and expresses the mean-path length of an electron in a material. Thus, the longitudinal propagation (referred to as the *longitudinal direction*) and the absorber depth are usually given in units of radiation length. For the lateral development, the unit of the transverse depth of a cascade is the Molière radius (e.g., see Sect. 2.4.2.3).

The development of an hadronic cascade along the direction of motion of the incoming particle is usually described in units of *interaction length* (or *nuclear interaction length*) λ_A (Sect. 3.3.2).

Furthermore, if not otherwise explicitly indicated, in this chapter A, Z and ρ are the atomic weight (Sect. 1.4.1 and discussion at page 220), the atomic number (Sect. 3.1) and the density of the material (in $\mathrm{g\,cm^{-3}}$), respectively.

9.2 Electromagnetic Sampling Calorimetry

9.2.1 *Electromagnetic Calorimeter Response*

Sampling calorimeters are made of layers of passive samplers interleaved with active readout planes (Fig. 9.1). The energy of incoming particles is measured in active layers, usually with a thickness (in units of radiation length) much smaller than the thickness of the corresponding passive samplers, while the cascade process is generated in the passive layers.

For calorimeters having homogeneous sampling the passive samplers have the same thickness in units of radiation length along the longitudinal depth of the calorimeter. For homogeneous sampling calorimeters, the sampling frequency, τ, is defined as the passive sampler absorber thickness between two successive active planes and is expressed in units of *radiation length* (X_0). Therefore, the total number of active readout planes is inversely proportional to τ. Only sampling calorimeters of the homogeneous type will be considered in this chapter, unless explicitly stated otherwise.

The calorimeter response to showering particles is the signal generated by the active readout planes, which sample the incoming particle energy, i.e., it is the measurement of the energy deposited by the electromagnetic cascade in the whole

---

[†]It was discussed in Sect. 2.1.7.3 on *Radiation Length and Complete Screening Approximation*.

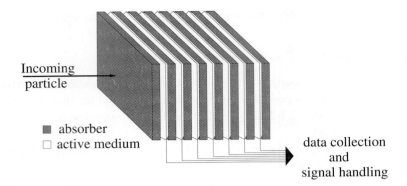

Fig. 9.1   Schematic description of a sampling calorimeter. The passive samplers (absorbers) are interspaced by active readout detector planes (for instance scintillator layer or silicon mosaic) from [Leroy and Rancoita (2000)].

set of the readout planes, after having performed the energy calibration of the device.

This deposited energy is called *visible energy* and, usually, is a tiny fraction of the incoming-particle energy. Therefore, the visible energy, ϵ_{vis}, in sampling calorimeters is the measurement, by the readout planes, of the energy deposited by incident particles in the active layers of the calorimeter (e.g., see [SICAPO Collab. (1985a, 1987)] and references therein).

As seen in Sect. 2.4.2.1, the total track length [Eq. (2.228)] is proportional to the incoming particle energy E. The energy deposited in the readout layers corresponds to the measurement of a fraction of the total track length, i.e., the part related to the overall thickness of the active readout planes, and consequently:

$$\epsilon_{\text{vis}} \propto E. \tag{9.1}$$

Equation (9.1) is the basic principle of calorimetry and has the consequence that the calorimeter response is expected to have a linear dependence on the incident particle energy. The linearity is a fundamental property that is independent of the nature, the thickness of active and absorber media, and of the sampling frequency.

A non-linear calorimeter response might occur, if the readout devices are not providing signals proportional to the deposited energy.

Equation (9.1) holds for electromagnetic cascades, in which almost all incoming particle energy is finally dissipated by processes of atomic ionization and excitation.

In the case of hadronic calorimeters, as will be seen later, an energy-dependent fraction of the incoming energy goes in breaking up nuclei, in low-energy neutrons and in undetectable neutrinos [Gabriel (1978)], thus preventing the linear calorimeter response to the incoming hadron energy. The restoration of Eq. (9.1) for hadron calorimeters is at the core of the problem of compensation, that will be discussed later in this chapter while dealing on hadron calorimetry.

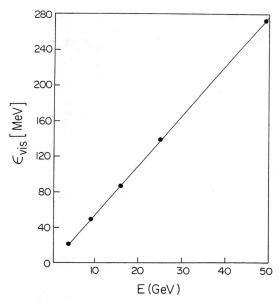

Fig. 9.2 ϵ_{vis} measured with the silicon calorimeter of SICAPO Collaboration shown in Fig. 2.78 for W absorbers as a function of the incoming electron energy (reprinted from *Nucl. Instr. and Meth. in Phys. Res. A* **235**, Barbiellini, G., Cecchet, G., Hemery, J.Y., Lemeilleur, F., Leroy, C., Levman, G., Rancoita, P.G. and Seidman, A., Energy Resolution and Longitudinal Shower Development in a Si/W Electromagnetic Calorimeter, 55–60, Copyright (1985), with permission from Elsevier; see also [Leroy and Rancoita (2000)]). The line represents: $\epsilon_{\text{vis}} = [(5.558 \pm 0.004)\,E[\text{GeV}] + (-1.3 \pm 1.5)]\,[\text{MeV}]$.

Equation (9.1) was found to agree remarkably well with experimental data (see, for example [Bormann et al. (1985); Nakamoto et al. (1986)]). An example [SICAPO Collab. (1985a)] is given in Fig. 9.2, where measured values of ϵ_{vis}, obtained using a Si/W sampling calorimeter, are shown for incoming electron energies between 4 and 49 GeV. The least square fit to the data shown in Fig. 9.2 gives [SICAPO Collab. (1985a)]:

$$\epsilon_{\text{vis}} = [(5.558 \pm 0.004)\,E[\text{GeV}] + (-1.3 \pm 1.5)]\,[\text{MeV}]. \qquad (9.2)$$

The fitted values and their errors indicate the high degree of linearity of the calorimeter response, in agreement with Eq. (9.1).

In sampling calorimeters in which passive samplers are dominant (in terms of thickness expressed in units of radiation length), the total number of active readout planes increases as the sampling frequency decreases and, consequently, the sampled fraction of the track length [Eq. (2.228)] is expected to increase since the longitudinal depth of the calorimeter is fixed. Thus, to a first approximation, Eq. (9.1) becomes:

$$\epsilon_{\text{vis}} \propto \frac{E}{\tau}. \qquad (9.3)$$

Experimental data [SICAPO Collab. (1989c, 1994a)] on ϵ_{vis} for an incoming-electron energy of $4\,\mathrm{GeV}$ are shown in Fig. 9.3 as a function of $1/\tau$. These data were obtained using a Si/Pb sampling calorimeter. The values of τ range between 1.96 and 3.39. The data show a linear dependence of ϵ_{vis} on $1/\tau$. This holds for τ *larger than* ≈ 0.8, as will be discussed at page 626.

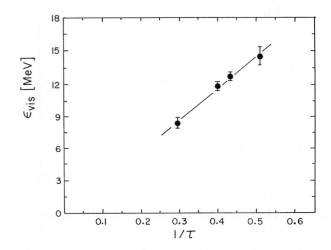

Fig. 9.3　ϵ_{vis} measured with Pb absorbers as a function of $1/\tau$, where τ is the sampling frequency (reprinted from *Nucl. Instr. and Meth. in Phys. Res. A* **345**, Bosetti, M. et al., Systematic investigation of the electromagnetic energy resolution on sampling frequency using silicon calorimeters, 244–249, Copyright (1994), with permission from Elsevier, e.g., for the list of the authors see [SICAPO Collab. (1994a)]; see also [Leroy and Rancoita (2000)]).

9.2.2　*The e/mip Ratio*

In a sampling calorimeter, the visible energy is the result of collision losses in active readout layers by electrons and positrons (generated during the electromagnetic shower development), traversing or ranging out in these layers. It has to be noted that in a homogeneous calorimeter, the full device can be considered as an extended active layer. Because the dominant process of energy dissipation is the process of energy loss by collisions, the calorimeter response to minimum-ionizing and non-showering particles (*mip*), like muons, can be used as a scale or, equivalently, as a *unit of measurement* for the response to electromagnetic showers (*e*) (for instance see [Wigmans (1987, 1988); Gabriel (1989); SICAPO Collab. (1989b,c); Drews et al. (1990); SICAPO Collab. (1990b); Wigmans (1991); SICAPO Collab. (1992a)]).

For a given sampling calorimeter, the ratio of the energy deposited by ionization in active readout planes and passive samplers can be evaluated using the mean dE/dx values for *mip*'s in different materials, as for instance given in [Janni (1982);

PDB (2008)].

Following the energy partition by a minimum ionizing and non showering particle, the shared energy E_s in active readout planes is given by:

$$E_s = E\,F(S) = E\,G(S)\,X_S, \tag{9.4}$$

where E is the incoming particle energy in GeV, $F(S) = G(S)\,X_S$, X_S is the active depth of the readout detectors, and

$$G(S) = \frac{\left(\frac{dE}{dx}\right)_S}{\left(\frac{dE}{dx}\right)_S L_S + \sum_i \left[\left(\frac{dE}{dx}\right)_i L_i\right]}, \tag{9.5}$$

where L_S is the physical thickness of the active readout detectors, L_i is the thickness of the absorber i, $(dE/dx)_S$ is the *mip* average energy-loss per unit of length in active readout planes, and $(dE/dx)_i$ is the *mip* average energy-loss per unit of length in the absorber i (as given in [Janni (1982); PDB (2008)].

It has to be emphasized that *mip*'s are ideal particles. The average energy deposited per unit length by non-showering particles, like muons, in a calorimeter will be always larger than the deposited energy corresponding to the minimum value of the collision energy-loss dE/dx owing to processes like δ-rays production or bremsstrahlung, even if the restricted energy-loss and the density-effect are limiting the relativistic rise. In fact, in dense materials the *Fermi Plateau* (defined on page 51) occurs for mean collision losses close but slightly larger than the dE/dx for $\beta \approx 0.96$, at which the collision energy-loss of a massive particle is minimum, (i.e., it is a *mip*). Nevertheless, the above introduced *mip* values have the advantage of being energy- and thickness-independent and allows a comparison among different calorimeter structures.

In Eq. (9.4), $F(S)$ represents the fraction of the energy lost by a *mip* particle depositing its energy by collision losses in similar (with respect to the sampling electromagnetic calorimeter structure) but indefinitely extended calorimeter, so that the *mip* particle is completely absorbed. The value of $F(S)$ can be computed by considering a single passive sampler, which in turn can be made of several absorbers, and a single readout active plane. For instance for a 10 GeV incoming-electron energy, the calculated values of E_s for a calorimeter with one X_0 of Pb passive samplers are ≈ 210 and 770 MeV for 400 μm thick silicon detectors and 3 mm thick plastic scintillators, respectively. Thus only about 2.1 and 7.7%, respectively, of the incoming electron energy is shared by active samplers in these two calorimeters.

The *e/mip* ratio is defined as:

$$e/mip \equiv \frac{\epsilon_{\text{vis}}}{E_s}, \tag{9.6}$$

where ϵ_{vis} is the energy deposited in the sensitive part of the calorimeter by the electromagnetic cascade initiated by a particle of energy E; E_s is defined in Eq. (9.4). It provides a way to determine how much the energy deposition processes for showering particles and *mip*'s differ each other.

From Eqs. (9.4, 9.6), we can write

$$\epsilon_{\text{vis}} = (e/mip)E_s = (e/mip)\, F(S)\, E. \tag{9.7}$$

Equation (9.7) is the way Equation (9.1) is rewritten in terms of the e/mip ratio.

In the previous section we have seen how the experimental data show a remarkable linear dependence of the measured ϵ_{vis} values on the incoming particle energy. These data and Eq. (9.7) indicate that the e/mip ratio is *energy independent*. Furthermore, both experimental data and cascade simulations (see page 623) support the fact that the e/mip ratio is almost *independent of the thickness of readout layers* for practical sampling calorimeters. The e/mip ratio is a fundamental characteristic of the structure of sampling calorimeters, essentially related to the difference between the readout and absorber Z values (see Sect. 9.2.2.1).

If the sampling calorimeter has a dominant (in units of radiation length) passive absorber a, with thickness L_a and radiation length X_{0a}, and if collision energy-losses in readout active detectors are small compared with those in passive absorbers (these conditions are usually fulfilled in most sampling calorimeters), from Eq. (9.4) one finds:

$$G(S) = \frac{\left(\frac{dE}{dx}\right)_S}{\left(\frac{dE}{dx}\right)_S L_S + \sum_i \left[\left(\frac{dE}{dx}\right)_i L_i\right]} \approx \frac{\left(\frac{dE}{dx}\right)_S}{\left(\frac{dE}{dx}\right)_a X_{0a}\tau}, \tag{9.8}$$

where $\tau = L_a/X_{0a}$. Considering that $F(S) = G(S)\, X_S$ and using Eq. (9.8), Eq. (9.7) can be rewritten as:

$$\epsilon_{\text{vis}} = \frac{(e/mip)}{\tau}\left[\frac{\left(\frac{dE}{dx}\right)_S X_S}{\left(\frac{dE}{dx}\right)_a X_{0a}}\right] E. \tag{9.9}$$

This equation [see Eq. (9.3)] shows the visible energy is inversely proportional to the sampling frequency τ, as long as the e/mip ratio is *independent of τ*. This occurs *for τ larger than ≈ 0.8*, as discussed at page 626. Since $\left(\frac{dE}{dx}\right)_a \approx \epsilon_{ca}/X_{0a}$ to better than 10%, Eq. (9.9) can be approximated by

$$\epsilon_{\text{vis}} \approx \frac{(e/mip)}{\tau \epsilon_{ca}}\left[\left(\frac{dE}{dx}\right)_S X_S\right] E, \tag{9.10}$$

where ϵ_{ca} is the critical energy (see Sect. 2.1.7.4) of the passive absorber a. By knowing the e/mip ratio from Eq. (9.7) [or Eq. (9.9)], it is possible to determine the expected visible-energy for any electromagnetic calorimeter.

In this section, we have seen how the e/mip ratio can be almost considered a characteristic or intrinsic property of the calorimeter structure, namely of the type of readout detectors and the kind of passive absorbers used, and is almost independent of the sampling frequency τ. In addition and within $\approx 10\%$ as shown in Table 9.1, since readout active detectors (like scintillators, liquid argon or silicon detectors, etc.) usually have low-Z values, e/mip values (see [Fabjan (1985a); Wigmans (1987); SICAPO Collab. (1989c)] and references therein) are ≈ 0.85 for medium-Z passive absorbers (like Fe or Cu) and ≈ 0.65 for high-Z passive absorbers (like W, Pb or U).

Table 9.1 Selection of measured e/mip ratios from [Leroy and Rancoita (2000)]. L_a and S are the passive sampler and active readout thicknesses. Errors are quoted when available in the original reference.

Passive sampler	Readout type	L_a (mm)	S (mm)	e/mip	Reference
Al	Liquid Ar	1.0	3.0	1.00	[Cerri et al. (1983)]
Fe	Liquid Ar	1.5	2.0	0.90	[Fabjan (1985a)]
Fe	Si det.+G10 plates	35.0	0.2	0.78±0.02	[SICAPO Collab. (1989c)]
Fe	Si det.	35.0	0.2	0.82±0.02	[SICAPO Collab. (1989c)]
Cu	Si det.	15.0	0.3	0.85	[SICAPO Collab. (1989c); Lindstroem et al. (1987)]
Cu	Scintillator	5.0	2.5	0.84±0.05	[Botner et al. (1981)]
Pb	Liquid Ar	3.0	2.4	0.54	[Dubois et al. (1985)]
Pb	Si det.+G10 plates	11.0	0.2	0.69±0.04	[SICAPO Collab. (1989c)]
Pb	Si det.	11.0	0.2	0.76±0.04	[SICAPO Collab. (1989c)]
Pb	Si det.	6.0	0.3	0.71	[SICAPO Collab. (1989c); Lindstroem et al. (1987)]
Pb	Scintillator	2.1	6.3	0.51	[Stone et al. (1978)]
Pb	Scintillator	5.0	5.0	0.65±0.10	[Bernardi (1987)]
Pb	Scintillator	6.4	6.0	0.77	[Duffy et al. (1984)]
Pb	Scintillator	10.0	2.5	0.67±0.03	[Bernardi et al. (1987)]
U	Gas	4.5	6.0	0.65±0.04	[Arefiev et al. (1989)]
U	Liquid Ar	1.6	4.4	0.51	[Dubois et al. (1985)]
U	Liquid Ar	1.7	2.0	0.61	[Fabjan (1985a)]
U	Liquid Ar	4.0	3.2	0.50±0.05	[Aronson et al. (1988)]
U	Scintillator	3.0	2.5	0.70±0.05	[HELIOS Collab. (1987)]
U	Scintillator	3.0	2.5	0.65±0.03	[Anders et al. (1988)]
U	Scintillator	3.2	3.0	0.74±0.03	[Bernardi (1987)]
U	Scintillator	3.2	5.0	0.68±0.04	[Bernardi (1987)]
U	Scintillator	10.0	5.0	0.61	[Catanesi et al. (1987)]

Finally, it has to be noted [Eq. (9.7)] that the visible energy in a calorimeter is usually only a few percent of the overall incoming particle energy even for high-Z passive samplers. It may be reduced to a fraction of a percent when low-Z passive samplers are used (or for large sampling frequencies).

9.2.2.1   *e/mip Dependence on Z-Values of Readout and Passive Absorbers*

As discussed in Sect. 2.4, the development of the electromagnetic cascade inside a medium can be understood within the well established and complete framework of the QED theory.

Initial analytical treatments and interpretations have dealt with the shower development and propagation in a single medium. They have disregarded physical processes of signal generation in additional media, which are present in sampling calorimeters. For instance basic properties, like the energy dependence of the depth of the shower maximum and the longitudinal profile, were predicted in "Approximation B" (discussed in Sect. 2.4.2.1) considering the fast cascade multiplication and neglecting low-energy processes (namely those induced by soft electrons, positrons and photons), by which almost all the energy is deposited in matter. The main reason for neglecting these soft processes was that they were not important for the determination of the properties of shower transport inside the absorber: very soft electrons and positrons are ranging out locally in few tens (hundreds) of microns. Only the more energetic charged particles undergo approximately constant collision energy-losses.

It has already been noted (see page 207) that some cascade characteristics depend on the soft-component of the shower: soft photons, electrons and positrons are mainly responsible for the transverse development of the shower. Furthermore, low-energy photons below $\approx 1\,\mathrm{MeV}$ undergo differential absorption in matter.

In Fig. 9.4, mass attenuation coefficients from [Leroy and Rancoita (2000)] (see also [Hubbell (1969); Hubbell, Gimm and Øverbø (1980)]) in units of cm^2/g are shown for materials typically used in electromagnetic calorimeters as readout detectors or passive samplers. Above an energy between $\simeq 2$ and $3\,\mathrm{MeV}$, the pair-production effect dominates and the multiplication process is still important. For photon energies between $\simeq 1$ and 2 MeV, the Compton scattering dominates and, because of its dependence on the electron density of the medium, the mass attenuation coefficients depend on the ratio Z/A, which varies only slightly from one material to another. As the photon energy decreases below $\simeq 1\,\mathrm{MeV}$ for high-Z materials and even lower energy for lower-Z materials ($\approx 100\,\mathrm{keV}$ in Al), the photoelectric effect becomes more and more important. As a consequence, the attenuation mass coefficient is strongly dependent on the atomic number Z of the material. Furthermore in the photoelectric effect, the interaction essentially occurs with the entire atomic electron cloud rather than with individual *corpuscular electrons*. At low photon energies (for $E_\gamma <$ a few hundreds keV), the emerging photoelectrons are very markedly emitted in the transverse direction with respect to the direction of the incoming photon and enlarge the shower size.

As discussed above, the low-energy phenomena, which control the energy deposition via soft photon interactions in matter during the development of electromagnetic cascades, depend on the difference of atomic numbers between the absorber and the active medium. For a fixed amount of matter in g/cm^2, the flux of low energy photons is less attenuated (see Fig. 9.4) in low-Z than in high-Z media. Consequently, less soft photoelectrons are generated and locally absorbed in low-Z materials (like detector active readouts as for instance scintillators, liquid argon or silicon detectors, etc.) compared with medium-Z materials (like Fe and Cu passive samplers), and much less compared with high-Z materials (like W, Pb and U passive samplers). Thus, the e/mip value is expected to depend on the Z difference among detector active readout and passive samplers.

Many experimental results have confirmed this predicted behavior of the e/mip ratio (see [Fabjan (1985a); Wigmans (1987); SICAPO Collab. (1989c)] and references therein). Wigmans (1987), using the EGS4 code [Nelson, Hirayama and Rogers (1985)], has systematically simulated electromagnetic shower cascades in sampling calorimeters with liquid argon and scintillator readout detectors as a function of the Z-value of passive samplers. His results are in agreement with experimental data and other simulations [Flauger (1985); del Peso and Ro (1990)] and have confirmed [Wigmans (1987)] that the e/mip ratio mainly depends on the differential absorption of low energy photons.

The e/mip ratios are < 1 for $Z_{absorber} > Z_{readout}$, ≈ 1 for $Z_{absorber} \approx Z_{readout}$, and > 1 for $Z_{absorber} < Z_{readout}$. Simulations have shown that the energy deposition is quite uniform in thin (in units of radiation length) readout layers [Flauger (1985)]. These simulation results and experimental data indicate that the value of e/mip is *almost independent of the readout layer thickness* for practical calorimeters.

Several e/mip values, obtained experimentally, are shown in Table 9.1: the data are mainly given for liquid argon, scintillator, and silicon as readout detectors and for Al, Fe, Cu, Pb and U as passive samplers. As pointed out before and within $\approx 10\%$ (see also [Fabjan (1985a); Wigmans (1987); SICAPO Collab. (1989c)] and references therein), e/mip values are ≈ 0.85 for medium-Z passive absorbers (like Fe or Cu) and ≈ 0.65 for high-Z passive absorbers (like W, Pb or U). Because in typical sampling calorimeters e/mip ratios are < 1, the phenomenon was also referred to as *electromagnetic sampling inefficiencies* [Gabriel (1989)].

Sometimes phenomena leading to "sampling inefficiencies", i.e., to the e/mip ratio dependence on $Z_{absorber}$ and $Z_{readout}$, were interpreted as being due to the so-called *transition effects*.

Although the calorimeter response is based on the energy deposited in the readout detectors after the shower transits from passive samplers to active planes, it has to be recalled that the term transition effect was introduced by Pinkau (1965) in the framework of "Approximation B". In this formulation, not all soft-photon

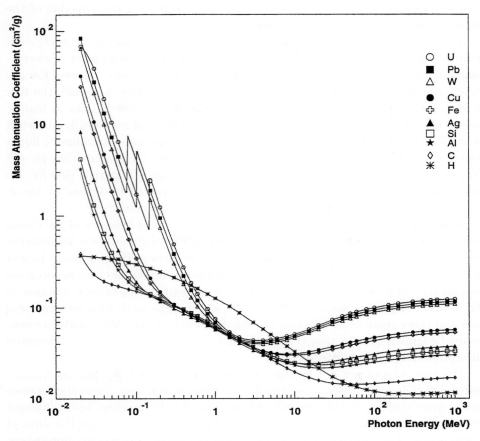

Fig. 9.4 Mass attenuation coefficients in units of cm^2/g for photons between $10\,keV$ and $1\,GeV$ (from [Leroy and Rancoita (2000)], see also [Hubbell (1969); Hubbell, Gimm and Øverbø (1980)]).

interactions with the medium are taken into account. But, as it has been seen, these interactions play a major role in the process of energy deposition in matter. Pinkau argued that, when a cascade propagates through matter, the low-energy electrons absorbed by ionization loss are replaced by new ones from the cloud of γ rays. However, if the cascade moves into a medium of higher critical energy the absorption process is stopped, while the replenishment by γ-rays takes place at the old rate since it is determined by the radiation length and not by the critical energy. In his argument, soft γ-ray interactions in matter are neglected. These interactions, when occurring via the photoelectric process (disregarded in the framework of "Approximation B"), have a dependence on the atomic numbers of materials stronger than their dependence on radiation lengths. The latter one occurs at higher photon energies, when pair-production processes dominate. In addition, soft secondary

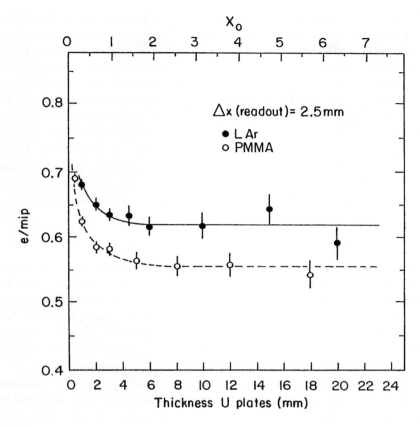

Fig. 9.5 The *e/mip* ratio as a function of the thickness of passive samplers for U/LAr and U/scintillator calorimeters (reprinted from *Nucl. Instr. and Meth. in Phys. Res. A* **259**, Wigmans, R., On the energy resolution of uranium and other hadron calorimeters, 389–429, Copyright (1987), with permission from Elsevier; see also [Leroy and Rancoita (2000)]).

electrons are not absorbed instantaneously by collision losses. In "Approximation B" (see Sect. 2.4.2.1), because the analytical treatment is oriented to provide a description of energetic electrons and positrons propagating in matter, soft secondary electrons are taken out of the propagating stream of particles. In other words, these charged particles are supposed to dissipate their energy immediately.

In reality as seen in Sect. 2.4.2.1, the collision energy-loss per unit of length by soft electrons and positrons, having energies close to the critical energy, is $\approx \epsilon_c/X_0$. They are absorbed in thicknesses which are fractions, not necessarily small, of radiation length. The readjustment of shower characteristics is expected to take place gradually. Pinkau's predictions were found to be in disagreement with the experimental data [Crannell et al. (1969)].

Furthermore, it has to be noted that the presence of thin, in units of radiation length, G10 plates (or other low-Z layers, having large ϵ_c values) in the calorimeter configuration does not introduce any sizable modification to the transversal and longitudinal cascade development and more generally leaves the containment properties of electromagnetic cascades unchanged, as was experimentally observed [SICAPO Collab. (1989a, 1991a)] (see page 207 and also Figs. 2.81 and 2.83). This point is demonstrated by the values of fitted radial and longitudinal parameters of the cascade energy distribution, which are almost the same independently of the fact that G10 plates are used in between or next to readout detectors or not at all. Only the normalization coefficient is modified because of the e/mip variation. Also, Wigmans (1987) has drawn the conclusion from his simulations that this transition effect in crossing the boundary between two layers with different Z plays a very minor role: the shower will adapt itself "adiabatically" to the new situation.

9.2.2.2   *e/mip Dependence on Absorber Thickness*

Low energy photons can interact via Compton or photoelectric effects in the passive sampler close to the readout layer. Thus, soft electrons generated in the passive absorber may enter the readout detector and contribute to the visible energy. The electromagnetic response is expected to increase slightly, when the passive sampler is thin. In turn, because this effect is related to electrons generated close to the readout detectors, the e/mip ratio will not vary, when the absorber thickness is increased.

Extensive simulations [Wigmans (1987)] have investigated the e/mip ratio dependence on the thickness of the passive sampler. Medium-Z (like iron) and high-Z (like uranium) absorbers and commonly employed readout detectors (like liquid argon and scintillator plates), and dependence on the energy of incoming particles, were considered. It is observed that the e/mip ratio slightly increases for very thin absorber plates with a thickness smaller than $\approx 0.8 X_0$. For larger passive samplers and within the simulation uncertainties, the e/mip ratio does not depend on the absorber thickness. These results cover well the range of sampler thicknesses commonly employed in sampling calorimeters.

An example of such simulations is given in Fig. 9.5, where the e/mip ratio is shown as a function of the thickness of the uranium passive sampler for liquid argon and plastic scintillator readouts [Wigmans (1987)]. The effect of the e/mip ratio dependence on the absorber thickness is visible, but for plates thicker than $\approx (2.5\text{–}3.0)\,\mathrm{mm}$, i.e., above $\approx 0.8\,X_0$, this effect apparently does not play any further role.

In summary, the e/mip ratio can be considered to be *independent of the passive sampler thickness for thicknesses above* $\approx 0.8\,X_0$, i.e., for most practical sampling calorimeters.

9.2.3   *e/mip Reduction in High-Z Sampling Calorimeters:*
The Local Hardening Effect

A considerable fraction of the energy deposition in the calorimeter, which increases logarithmically with the absorber Z-value, as discussed at page 214, occurs through low-energy particles. In the case of uranium (similarly in other high-Z materials like tungsten and lead), particles with an energy below 1 MeV account for about 40 % of the total energy deposition (Fig. 2.85) and are related to the low-energy photons generated during the cascade development. In these media, the interaction between generated soft-photons and matter is dominated by the Compton scattering and the photoelectric effect for energies below ≈ 4 MeV and, mainly, by the photoelectric effect below ≈ 1 MeV.

The contribution of the Compton cross section to the total cross section for photon interactions in a given material is proportional to its Z-number (in fact there are Z electrons per atom). The dependence of the photoelectric cross section on the atomic number of the material varies somewhat with the energy of the photon. However, it approximately goes between Z^4 and Z^5 at these energies. Therefore, the interaction cross section of soft photons is much larger in high-Z absorbers than in low-Z media.

In high-Z absorbers, the transverse shower size, for which $\approx 95\%$ of the energy is contained within $2R_M$ [Eq. (2.242)], is larger in units of X_0 (see Table 2.3) than in low-Z media because of the way electrons and positrons are generated and scattered, as discussed in Sect. 2.4.2. The mean cosine of charged-particle angles formed with the shower axis varies as $1/\epsilon_c$, as given in Eq. (2.253). Consequently, large particle angles are expected for high-Z materials, since these materials have low ϵ_c values (see Table 2.3).

The soft electron component was studied by simulations [Fisher (1978)] for the case of a Pb proportional chamber quantameter. The angular distribution of shower electrons is peaked along the shower axis. However, there is a large angle spread. About 12% of all electrons are backscattered in the sampling gap. For low energy electrons with energy < 3 MeV, the forward peak vanishes and the angular distribution (although not fully forward-backward symmetric) behaves like $\approx \cos\theta$, where θ is the angle with respect the shower axis [Fisher (1978); Amaldi (1981)].

According to Amaldi's estimates on the Z-dependence of the soft electron component [Amaldi (1981)], there is a sizeable fraction of large angle low energy electrons, a part of them moving backwards. The fraction of large angle soft electrons varies as $1/\epsilon_c^2$ and is about 40% in Pb. Soft electrons are locally absorbed: their mean range is about $2.0\,\mathrm{g/cm^2}$ for an energy of ≈ 4 MeV, and about $0.4\,\mathrm{g/cm^2}$ for an energy of ≈ 1 MeV. However in sampling calorimeters, low-energy photons from the electromagnetic cascade may convert into electrons sufficiently close to the surface of the absorber plate to escape and contribute to the measured signal. Although for thin readout layers these electrons can give a substantial contribution to the overall signal, the e/mip ratio is < 1 for high-Z passive samplers in sampling calorimeters. To further decrease the e/mip ratio, low-Z passive absorbers have to be added in between high-Z and readout layers, bringing to three the number of media of the sampling calorimeter.

The additional low-Z absorbers have a lower photoelectric cross section than high-Z samplers. Thus photons, whose energy distribution has been determined by the high-Z samplers, will propagate in a medium with a relative lower photoelectric cross section. These additional absorbers have to be thin in units of radiation lengths in order to absorb soft electrons generated close to the surface in the high-Z samplers, while at the same time these additional thin passive samplers are not (or marginally) becoming part of the shower generation process.

Taking into account the mean absorption range of 1 MeV electrons (about $0.4\,\mathrm{g/cm^2}$) and Eq. (2.253), thicknesses (along the shower axis) of $\approx 0.3\,\mathrm{g/cm^2}$ are needed to prevent a large number of soft electrons from entering the readout detectors and consequently to reduce the e/mip ratio. This effect was referred to as the *local hardening effect* [SICAPO Collab. (1989b)], because softer electrons and positrons are locally taken out of the charged-particle stream, thus hardening the shower while traversing readout layers.

The degree of hardening and consequently the size of the reduction of the e/mip ratio depend both on the thickness of low-Z absorbers and, because of the photoelectric Z-dependence on the photon cross section, on the difference between the Z_{absorber} and the low-Z of added thin absorber plates. This allows the fine tuning of the calorimeter response to electromagnetic cascades. In addition, as it will be seen in following sections, the exploitation of the local hardening effect is a way to achieve the compensation condition in hadronic calorimetry.

A reduction of the energy detected in electromagnetic showers, due to the presence of low-Z absorbers in front of silicon mosaics in a Si/U calorimeter, was observed in [Pensotti, Rancoita, Simeone, Vismara, Barbiellini and Seidman (1988); Pensotti, Rancoita, Seidman and Vismara (1988)]. This reduction can be exploited to equalize the response of a Si/U calorimeter to incoming electrons and hadrons. The evidence of the local hardening effect was found and systematically studied for Si/U and Si/W calorimeters [SICAPO Collab. (1989b)]. This effect was also investigated by means of EGS4 simulations [Brückmann et al. (1988); Wigmans (1988); Lindstroem et al. (1990); Hirayama (1992)]. In this way, it was possible to evaluate the reduction of the e/mip ratio in the case of U/scintillator calorimeter with uranium plates wrapped in stainless steel.

Experimentally, the local hardening effect is found to be almost *independent* of *the incoming electron energy* [SICAPO Collab. (1989b,d)], and of the *sampling frequency* [SICAPO Collab. (1992b)]. In other words, the reduction of visible energy $\Delta\epsilon/\epsilon$ depends on the thickness of the low-Z absorber, but is independent of the thickness of the passive high-Z absorber, where we define:

$$\frac{\Delta\epsilon}{\epsilon} = \frac{\epsilon_{\text{vis}}(\text{no low}-Z) - \epsilon_{\text{vis}}(\text{low}-Z)}{\epsilon_{\text{vis}}(\text{no low}-Z)}$$

$$= 1 - \frac{\epsilon_{\text{vis}}(\text{low}-Z)}{\epsilon_{\text{vis}}(\text{no low}-Z)}. \tag{9.11}$$

Moreover, the insertion of low-Z material downstream, behind the silicon detectors, is almost twice as effective as inserting the same thickness in front of the readout detectors [SICAPO Collab. (1988b)]. This occurs because the back-scattered electrons are softer than the forward electrons.

The downstream insertion (behind the silicon detectors) of G10 plates with thicknesses varying from 0.5 up to 5.0 mm in a Si/U calorimeter configuration is at the origin of the local hardening effect. Thus, it is responsible for the reduction of the visible energy

$$\frac{\Delta\epsilon}{\epsilon} = \frac{\epsilon_{\text{vis}}(\text{no G10}) - \epsilon_{\text{vis}}(\text{G10})}{\epsilon_{\text{vis}}(\text{no G10})} = 1 - \frac{\epsilon_{\text{vis}}(\text{G10})}{\epsilon_{\text{vis}}(\text{no G10})}$$

shown in Fig. 9.6 from [SICAPO Collab. (1989b); Leroy and Rancoita (2000)]. The value of $\Delta\epsilon/\epsilon$ is $\approx 28\%$ for $\epsilon_{\text{vis}}(\text{G10} = 5\text{mm})$. By means of Eq. (9.4) the change in the shared energy, due to a 5 mm G10 absorber added to the passive sampler, is given by

$$\frac{\Delta E_s}{E_s}(\text{G10} = 5\text{mm}) = \frac{E_s(\text{no G10}) - E_s(\text{G10} = 5\text{mm})}{E_s(\text{no G10})}$$

$$= 1 - \frac{E_s(\text{G10} = 5\text{mm})}{E_s(\text{no G10})} \approx 10\%.$$

The corresponding e/mip ratio reduction is

$$e/mip(\text{G10} = 5\text{mm}) = \frac{\epsilon_{\text{vis}}(\text{G10} = 5\text{mm})}{E_s(\text{G10} = 5\text{mm})}$$

$$= \frac{\epsilon_{\text{vis}}(\text{no G10})\left[1 - \frac{\Delta\epsilon}{\epsilon}(\text{G10} = 5\text{mm})\right]}{E_s(\text{no G10})\left[1 - \frac{\Delta E_s}{E_s}(\text{G10} = 5\text{mm})\right]}, \tag{9.12}$$

therefore,

$$e/mip(\text{G10} = 5\text{mm}) \approx 0.8 \, \frac{\epsilon_{\text{vis}}(\text{no G10})}{E_s(\text{no G10})} = 0.8 \, e/mip(\text{no G10}). \tag{9.13}$$

In addition, the e/mip ratio is not modified by inserting low-Z layers between the uranium plates. The visible energy was measured to be reduced by about 9.5%, instead of $\approx 28\%$, when 5 mm G10 plates were located inside the uranium samplers and not adjacent to the silicon readout detectors [SICAPO Collab. (1989b)]. This reduction corresponds to the decrease of E_s, as discussed above, and is explained by the increase of the overall passive sampler (low- and high-Z plates). Thus, the e/mip ratio is almost constant when no local hardening effect occurs, as expected.

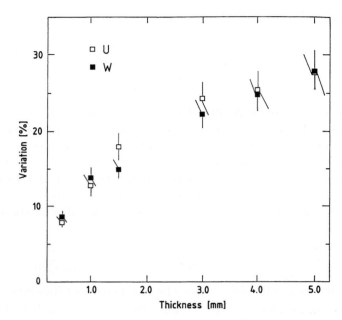

Fig. 9.6   Reduction in percentage of the electromagnetic visible-energy, measured with Si/U and Si/W calorimeters, expressed as $\Delta\,\epsilon/\epsilon[\%] = 100[\epsilon_{\mathrm{vis}}(\text{no G10}) - \epsilon_{\mathrm{vis}}(\text{G10})]/\epsilon_{\mathrm{vis}}(\text{no G10})$ versus the G10 plate thickness inserted at the rear of the silicon readout detectors (from [Leroy and Rancoita (2000)], see also [SICAPO Collab. (1989b)]).

The local hardening effect requires thin (in units of radiation length) low-Z absorbers. If their thickness is increased so much that they start to become part of the shower generation process, the e/mip ratio will increase again as indicated by EGS4 simulations [Wigmans (1988)] and the experimental data of Fig. 9.6. In fact, the data show an approximate logarithmic increase of $\Delta\epsilon/\epsilon$ with the G10 thickness, while [Eq. (9.4)] the value of $1/E_s$ is linearly increasing with the G10 thickness.

A further reduction of the e/mip ratio is achievable by inserting low-Z plates both behind and in front of readout detectors. The effect was also observed, when different low-Z absorbers (like polyethylene and aluminum) [SICAPO Collab. (1992b)] are employed.

The extent to which the reduction in the e/mip ratio can be achieved depends on the readout layer thickness in g/cm^2. As the readout layers thickness increases, the maximum reduction achievable decreases, for instance we have e/mip(0.5 mm Fe) $\approx 0.92\ e/mip$(noFe) using scintillator readout plates 2.6 mm thick in a uranium (uranium plates wrapped in stainless steel) sampling calorimeter [Wigmans (1988)].

Although the local hardening effect largely reduces the e/mip ratio in calorimeters employing high-Z media, it was also observed when medium-Z passive samplers (like Fe) were used [SICAPO Collab. (1993b)]. However, as predicted, the effect is smaller. In an iron calorimeter (with 30 mm-thick passive samplers), the visible energy is reduced by $\approx 8\%$, while E_s is reduced by $\approx 1.2\%$ when plexiglas layers (1.6 mm in front of and 0.8 mm behind the silicon readout detectors) are introduced.

9.3 Principles of Calorimetry with Complex Absorbers

For complex absorbers we mean a set of media with large differences in their critical energies and with thicknesses, which are non-negligible for the process of shower generation. These absorbers can be used as passive samplers in sampling calorimetry.

We have seen how, in the generation and propagation process of a shower in complex absorbers, low-energy photons lead to the generation of low-energy electrons by Compton and photoelectric effects, which have a different Z dependence. This Z-dependence explains the differential attenuation of radiation in absorbers. Most of these generated low-energy electrons are generally absorbed before they reach the readout medium. Thus, a significant change in the soft-energy deposition can be expected if materials with large differences among their critical-energy values, i.e., with large difference in Z_{absorber} values, are set in sequence in the complex absorber sampler.

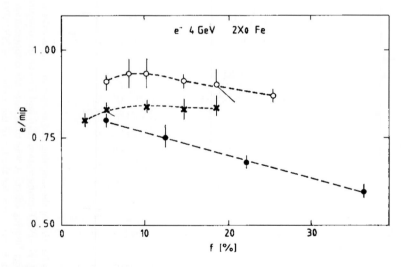

Fig. 9.7 e/mip as a function of the Pb fraction, f, in the passive sampler in various calorimeter configurations measured using 4 GeV electrons: Pb+Fe+Pb [sandwich configuration (○)], Fe+Pb [F-configuration (×)] and Pb+Fe [R-configuration (●)] (reprinted from *Phys. Lett. B* **222**, Borchi, E. et al., Electromagnetic shower energy filtering effect. A way to achieve the compensation condition (e/π=1) in hadronic calorimetry, 525–532, Copyright (1989), with permission from Elsevier, e.g, for the list of the authors see [SICAPO Collab. (1989c)]; see also [Leroy and Rancoita (2000)]). In these configurations, the thickness of the Fe absorber in each passive sampler was corresponding to 1.99 X_0. The lines are to guide the eye.

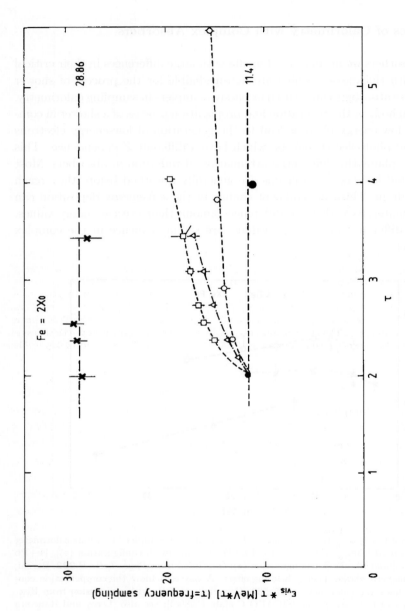

Fig. 9.8 $\epsilon_{vis}\tau$ in MeV τ as a function of the sampling frequency τ in various calorimeter configurations measured using 4 GeV electrons: Pb+Fe+Pb [sandwich configuration ($\square$)], Fe+Pb [F-configuration ($\triangle$)] and Pb+Fe [R-configuration ($\circ$)] (reprinted from *Phys. Lett. B* **222**, Borchi, E. et al., Electromagnetic shower energy filtering effect. A way to achieve the compensation condition ($e/\pi=1$) in hadronic calorimetry, 525–532, Copyright (1989), with permission from Elsevier; e.g, for the list of the authors see [SICAPO Collab. (1989c)]; see also [Leroy and Rancoita (2000)]). In the above configurations, the thickness of the Fe absorber in each passive sampler was corresponding to 1.99 X_0. The configuration with Pb ($\times$) and Fe ($\bullet$) absorbers alone are also shown. The lines are to guide the eye.

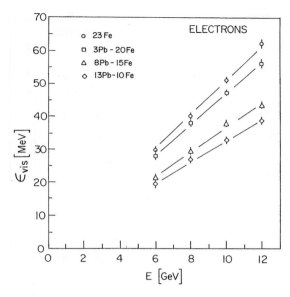

Fig. 9.9 Mean energy deposited by electrons in the Si/(Fe,Pb) calorimeter as a function of the incoming energy (reprinted from *Nucl. Instr. and Meth. in Phys. Res. A* **332**, Borchi, E. et al., Systematic investigation of the electromagnetic filtering effect as a tool for achieving the compensation condition in silicon hadron calorimetry, 85–90, Copyright (1993), with permission from Elsevier, e.g, for the list of the authors see [SICAPO Collab. (1993b)]; see also [Leroy and Rancoita (2000)]). The sampling thickness is 23 mm. The top points correspond to the case where there is no Pb in the absorber. The other points correspond to configurations where the thickness of Pb in the PbFe–Si–PbFe absorber configuration is (from top to bottom) 3.0, 8.0, and 13.0 mm. The lines are to guide the eye.

In general, when complex absorbers are employed *the energy deposition rate will change because the shower composed of charged particles and photons generated in a medium has to deposit its energy according to the stopping power and the attenuation length of another material.* It can be concluded, as seen on page 214, that the process of shower energy deposition is affected by the sequence of absorbers. As a consequence, the *e/mip* ratio also varies depending on the sequence of the absorbers, when a complex absorber is employed. However, an additional slight variation may be expected as a function of the type and thickness of the readout detector. This phenomenon was experimentally observed and referred to as *the electromagnetic shower energy filtering effect* [SICAPO Collab. (1989c)].

It has to be emphasized that low-energy phenomena, i.e., the energy loss of electrons and positrons close to the critical energy and soft-photon interactions by Compton or photoelectric effects, are the main processes of shower energy deposition. While interactions by soft electrons and positrons were considered in an approximate way, the full complexity of low-energy photon interactions was not taken into account at all in the framework of "Approximation B". Consequently, as

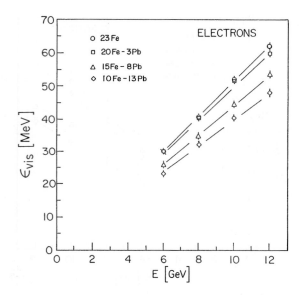

Fig. 9.10   Mean energy deposited by electrons in the Si/(Fe,Pb) calorimeter as a function of the incoming energy (reprinted from *Nucl. Instr. and Meth. in Phys. Res. A* **332**, Borchi, E. et al., Systematic investigation of the electromagnetic filtering effect as a tool for achieving the compensation condition in silicon hadron calorimetry, 85–90, Copyright (1993), with permission from Elsevier, e.g, for the list of the authors see [SICAPO Collab. (1993b)]; see also [Leroy and Rancoita (2000)]). The sampling thickness is 23 mm. The top points correspond to the case where there is no Pb in the absorber. The other points correspond to configurations where the thickness of Pb in the FePb–Si–FePb absorber configuration is (from top to bottom) 3.0, 8.0, and 13.0 mm. The lines are to guide the eye.

remarked on page 623, the *transition effect* introduced by Pinkau does not play any effective role in explaining the behavior of the visible energy and the *e/mip* ratio for a sampling calorimeter with complex absorbers.

9.3.1   *The Filtering Effect and how to tune the e/mip Ratio*

The reduction of the *e/mip* ratio as the result of the reduction of the electromagnetic visible-energy in a sampling calorimeter was experimentally observed using both Fe and Pb plates as passive samplers and silicon detectors as active read-out media [SICAPO Collab. (1989c)]. The *e/mip* ratios of three passive sampler configurations* along the shower axis were investigated: Fe+Pb (F-configuration), Pb+Fe (R-configuration) and Pb+Fe+Pb (sandwich configuration). They are shown

---

*In these configurations, the thickness of the Fe absorber in each passive sampler was corresponding to $1.99 \, X_0$.

in Fig. 9.7 as function of the Pb absorber fraction (f) in the passive sampler, with

$$f = \frac{L_{Pb}}{L_{Pb} + L_{Fe}}, \tag{9.14}$$

where L_{Fe} and L_{Pb} are the thicknesses of Fe and Pb absorbers, respectively. Data were also taken with Fe and Pb plates alone (as passive samplers): in Fig. 9.8 the corresponding values of the visible energies (ϵ_{vis}) multiplied by the sampling frequency (τ), measured using 4 GeV electrons, are shown as function of τ. For $f \gtrsim 5\%$, the e/mip ratios resulted to be

$$e/mip \ (\mathrm{Pb+Fe+Pb}) > e/mip \ (\mathrm{Fe+Pb}) > e/mip \ (\mathrm{Pb+Fe}).$$

Furthermore, a strong evidence of *anomalous* energy deposition during the shower propagation was obtained by the measurements of e/mip ratios for Fe alone, Pb alone and a R-configuration, in which the Pb absorber was a fraction

$$f \approx 36\%$$

of the total Pb+Fe length ($= L_{Pb} + L_{Fe}$); the e/mip ratios found were 0.78±0.02, 0.69±0.04, and 0.60±0.02, respectively. These measurements indicate that, using Pb and Fe absorbers together (Fe+Pb), the e/mip value is lower than the corresponding values obtained, when Fe and Pb absorbers are used alone [SICAPO Collab. (1989c)]. Furthermore for the R-configuration, the measured e/mip ratio was found to decrease linearly with the increase of the Pb fraction f (Fig. 9.7).

The visible-energy variation was also experimentally observed, when scintillators were used as the readout medium. The SDC calorimeter group has carried out measurements using the so-called *Hanging-File Calorimeter*, in which scintillator plates were read out by wavelength shifting fibers [Beretvas et al. (1993)]. Two of the tested configurations [Byon-Wagner (1992); Beretvas et al. (1993); Job et al. (1994)] had the passive sampler made of 25.4 mm Fe and 3.2 mm Pb absorbers: along the shower axis, one had the aluminium cladded scintillator sandwiched by lead upstream and iron downstream (as for the F-configuration), the other one by iron upstream and lead downstream (as for the R-configuration). An analysis of experimental data (see [Byon-Wagner (1992); Job et al. (1994)] and Sect. 9.9.2) shows that the e/mip ratio is larger in the former configuration than in the latter one, again because of the *filtering effect* (experimentally observed for the first time by SICAPO Collaboration in [SICAPO Collab. (1989c)]).

A systematic study of the filtering effect was performed ([SICAPO Collab. (1993b)] and references therein) using silicon detectors as readout medium. Data were taken for Fe+Pb (F-configuration) and Pb+Fe (R-configuration) with an overall fixed length of 23 mm for the passive sampler, as a function of the Pb thickness (no Pb, and 3, 8, 13 mm of Pb). For a given Pb thickness and configuration, the thickness of lead was kept constant, as well as the respective positions of the absorbers plates inside each passive sampler. Data were also taken with Fe as the only absorber for comparison. The mean energies deposited in the calorimeter for incoming electron energies between 6 and 12 GeV are shown in Figs. 9.9 (for R-configuration) and 9.10 (for F-configuration). In all cases, ϵ_{vis} is found to be a linear function of the incident energy. The value of E_s does not depend on the absorber sequence order, i.e., on the configuration, as can be seen from Eq. (9.4) for fixed Fe and Pb thicknesses. Then, Eq. (9.6) shows that the ratio of the measured ϵ_{vis} for the two configurations gives the ratio of their e/mip values:

$$\frac{\epsilon_{\text{vis}}(\text{F})}{\epsilon_{\text{vis}}(\text{R})} = \frac{e/mip(\text{F})}{e/mip(\text{R})}. \tag{9.15}$$

The ratio (averaged over the incoming energies) of the visible electromagnetic energies, i.e., the ratio of the e/mip values measured in the F- to the R-configurations, is observed to increase with the thickness of Pb: 1.08 ± 0.02 (3 mm of Pb), 1.20 ± 0.02 (8 mm of Pb), and 1.23 ± 0.03 (13 mm of Pb). Data were also taken with 30 mm deep samplers, as a function of the Pb thickness (no Pb, and 5, 10, 15 mm of Pb). The ratio of the electromagnetic visible-energy measured in the F- to the R-configurations, for the 30 mm thick passive samplers and averaged over the incoming energies, varies from 1.17 ± 0.02 (5 mm of Pb) up to 1.26 ± 0.02 (15 mm of Pb). Thus, the general behavior of the e/mip ratios as a function of the Pb thickness, observed in the data obtained with 23 mm thick samplers, is confirmed. These determined ratios of e/mip values are shown in Fig. 9.11 as a function of the Pb thickness: the behavior indicates a value increasing almost linearly with the Pb thickness. All above reported results can be understood in terms of the filtering effect.

The combination of low-Z (Fe) and high-Z (Pb) materials in passive samplers of a silicon calorimeter leads to an electromagnetic cascade energy transformation effect, generated by changing the medium generating the shower when the critical energies differ greatly ($\epsilon_c = 7.4$ MeV and 21.0 MeV for Pb and Fe, respectively). The value of ϵ_c is the value of the electron energy below which the energy loss by ionization begins to dominate the energy loss by bremsstrahlung. Consequently, the increase of the critical energy of the absorber favors the energy loss by ionization with respect to the energy loss by radiation. The energy spectra of incident electrons (and positrons) are softer in a high-Z absorber (Pb) than in a low-Z (Fe) absorber. The stopping power by collision is larger and dominates the stopping power by radiation up to higher electron energies, namely up to about 21 MeV, in Fe absorber compared with lead. Then, for R-configurations, in which the electromagnetic cascade develops at first in lead and subsequently in iron, the forward-going electrons are absorbed faster. These experimental data were reproduced by EGS4 Monte-Carlo simulations [Hirayama (1992)], which were also predicting that the filtering effect can be applied to sampling calorimeters using scintillators as the active media.

The effect depends on the Pb fraction, f, since the number of low energy electrons generated in lead increases by increasing f, and is also related to the Fe sampling thickness, because this thickness determines the maximum absorbable electron energy.

The reduction of the electromagnetic visible-energy for the R-configuration with respect to the F-configuration is the result of a larger energy deposition in the Fe absorber, before the shower traverses the active readout detectors. This reduction increases with the thickness of Pb in the absorber. As a consequence, the e/mip ratio of the R-configuration decreases compared with the one of the F-configuration, as shown in Fig. 9.11. The largest observed reduction [SICAPO Collab. (1993b)] takes place for the case where 15 mm Fe and 15 mm Pb absorbers are used as passive samplers:

$$\frac{e/mip(\text{F-configuration})}{e/mip(\text{R-configuration})} \approx 1.26,$$

namely,

$$e/mip(15\text{mmPb} + 15\text{mmFe}) \approx 0.79\, e/mip(15\text{mmFe} + 15\text{mmPb}). \qquad (9.16)$$

In summary, the filtering effect is produced by the combination of low- and high-Z materials in the absorber, which leads to a transformation of the development of the electromagnetic cascade energy. Here, two absorbers are used to transform the distribution of the cascade energy before it reaches the readout, which has to be properly located in the calorimeter configuration (high-Z)-(low-Z)-readout in order to obtain the required electromagnetic visible-energy reduction and consequently a decreased e/mip value.

In the case of the filtering effect, the e/mip ratio is controlled by the modified intrinsic absorption of soft particles, mainly in the passive samplers during the shower generation and propagation. This is a feature unique to the filtering effect. In fact, it should be stressed that, in the case of the local hardening effect, the properties of soft electrons were also exploited, but the additional low-Z absorbers had to be thin in units of radiation length in order not to modify the shower generation and propagation characteristics. Indeed, the filtering effect uses the combination of low-Z and high-Z absorber materials, as cascade generators, in order to produce a considerable transformation of the initial incoming cascade.

9.3.2 *e/mip Reduction by Combining Local Hardening and Filtering Effects*

The e/mip ratio can be further reduced by combining the local hardening and filtering effects [SICAPO Collab. (1989c, 1992b)], when medium- (or low-) and high-Z materials are jointly used as shower generators: the high-Z material will generate soft-charged particles also going backwards with respect to the shower axis (see Sect. 9.2.3). These electrons can be absorbed by introducing thin, in units of radiation length (in order not to take part in the shower-generation process), low-Z additional absorbers between the high-Z material and the readout detector. In this way, part of the soft shower charged component can be absorbed before entering the readout detector, thus providing a further reduction of the visible energy.

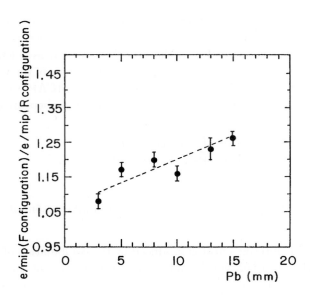

Fig. 9.11   Ratio of e/mip values of F- to R-configuration for passive samplers made of Fe and Pb absobers, as a function of Pb absorber thickness (from [Leroy and Rancoita (2000)]). The line is to guide the eye.

Results obtained with 30 mm thick passive samplers ([SICAPO Collab. (1993b)] and Sect. 9.3.1) can be compared with results obtained, when plexiglas layers (1.6 mm in front and 0.8 mm behind every active plane of silicon detectors) are inserted in R- and F-configurations. The ratio between e/mip values, of the electromagnetic visible-energy measured in F- to R-configurations modified by the insertion of plexiglas plates, increases from 1.17 ± 0.02 up to 1.29 ± 0.02 and is similar to the ratio shown in Fig. 9.10 when no plexiglas layer was inserted in similar configurations.

However, the overall change in the electron visible-energy is best observed (and shown in Table 9.2) when the energy $\epsilon_{\text{vis}}(\text{R} + \text{P})$ deposited in R-configurations, for the 3 mm thick passive sampler in which plexiglas layers $[(1.6 + 0.8)\,\text{mm}]$ are inserted, is compared with the energy $\epsilon_{\text{vis}}(\text{F})$ deposited in corresponding F-configurations (with no plexiglas layer). The value of

$$\mathcal{D} \equiv \frac{\epsilon_{\text{vis}}(\text{F}) - \epsilon_{\text{vis}}(\text{R} + \text{P})}{\epsilon_{\text{vis}}(\text{R} + \text{P})} \qquad (9.17)$$

is as large as $(41.0 \pm 1.0)\%$ for 15 mm of Fe and 15 mm of Pb in the passive sampler. Taking into account that from Eq. (9.17) $\epsilon_{\text{vis}}(\text{R} + \text{P}) = \epsilon_{\text{vis}}(\text{F})/(1 + \mathcal{D})$, the ratio $e/mip(\text{R} + \text{P})$ with plexiglas layers and 15 mm Pb + 15 mm Fe absorbers (R+P configuration) is given by

$$e/mip(\text{R} + \text{P}) = \frac{\epsilon_{\text{vis}}(\text{R} + \text{P})}{E_s(\text{R} + \text{P})} = \frac{\epsilon_{\text{vis}}(\text{F})}{(1 + \mathcal{D})} \left[\frac{1}{E_s(\text{R} + \text{P})} \right]. \qquad (9.18)$$

Table 9.2 Pb thickness versus the ratio $\mathcal{D} \equiv [\epsilon_{\text{vis}}(\text{F}) - \epsilon_{\text{vis}}(\text{R}+\text{P})]/\epsilon_{\text{vis}}(\text{R}+\text{P})$ in percentage, where $\epsilon_{\text{vis}}(\text{R}+\text{P})$ is the energy deposited in the R-configurations for the 30 thick samplers and plexiglas layers $[(1.6 + 0.8)\,\text{mm}]$, $\epsilon_{\text{vis}}(\text{F})$ is the corresponding energy deposited in F-configurations (with no Plexiglas) from [Leroy and Rancoita (2000)].

Pb thickness (mm)	Ratio $\mathcal{D}$ (%)
5	32.9±1.0
10	34.1±1.0
15	41.0±1.0

Since $\mathcal{D}$ is $\approx 41\%$, and $E_s(\text{R}+\text{P})$ only $\approx 1.2\%$ lower than $E_s(\text{R}) = E_s(\text{F})$ (namely the E_s value for the F- and R-configuration for 15 mm Fe and 15 mm Pb absorbers), the final result is:

$$e/mip(\text{R}+\text{P}) \approx \frac{\epsilon_{\text{vis}}(\text{F})}{E_s(\text{F})}\left[\frac{1}{(1+\mathcal{D})}\right] \approx 0.71\, e/mip(\text{F}). \qquad (9.19)$$

As discussed in Sect. 9.2.3, the insertion of low-Z plates (like polyethylene, plexiglas, aluminium, G10 plates or others) leads to similar effects when these absorbers are inserted in between the thick passive sampler and the readout detector.

A large e/mip ratio reduction, of about 18%, was obtained by combining the two effects in a calorimeter with plastic scintillator active samplers, and Pb and Fe passive absorbers [Beretvas et al. (1993)]. In this case the local hardening effect was achieved by inserting low-Z plates in front of and behind the active media, i.e., by *cladding* the scintillator with 1.6 mm thick aluminum sheets [Beretvas et al. (1993)].

The possible use of polyethylene, a hydrogen-rich material, to achieve the tuning of the e/mip ratio presents the advantage that it plays the role of a neutron moderator, as indicated by activation studies in [SICAPO Collab. (1991d)], and may then contribute to reinforcing the radiation hardness of calorimeters, whose readout detectors are not too sensitive to very slow neutrons.

A further increase of the calorimeter radiation hardness is obtained, when low (or medium) Z absorbers, like Fe, are employed as passive samplers. In fact (see Sect. 3.3 on *Hadronic Shower Development and Propagation in Matter*), during the hadronic cascading process, fewer neutrons are generated in low-A (which are also low-Z) materials than in high-A (which are also high-Z) materials.

Indeed experiments at the LHC collider, for instance, will face an environment of high-level radiation challenging the survival of the calorimeters. The neutrons with fluences estimated to reach up between 10^{12} and $10^{13}\,\text{cm}^{-2}$ year^{-1} are at the core of the problem. The large cross section of neutrons on hydrogen nuclei suggests introducing hydrogeneous material in calorimeters, in order to moderate the effect of neutrons and reduce the damage inflicted by neutrons on the calorimeter materials and equipment, especially the detectors and their readout electronics. Preliminary results were reported on experiments involving the exposure to very intense beams at the CERN PS of Na, In, and Rh foils, with and without polyethylene foils in a dump calorimeter [SICAPO Collab. (1991d)]. These results indicate that the low-energy neutron flux should be reduced by as much as an order of magnitude with polyethylene foils with a thickness of the order of 1 cm. Studies, carried out by Monte-Carlo simulations, have confirmed this possible use of low-A materials as neutron moderators [Russ (1990); Russ et al. (1991)].

9.4 Energy Resolution in Sampling Electromagnetic Calorimetry

9.4.1 *Visible Energy Fluctuations*

The energy deposited in a sampling calorimeter is sampled and measured in a number of readout layers, which interspace the passive samplers. The energy deposition is dominated by collision losses of positrons and electrons generated in the process of energy degradation during the cascade development.

The energy deposited in active readout detectors fluctuates on an event-to-event basis and is affected by the number and energy distribution of charged particles, which do not usually cross the same amount of active samplers. At any depth in a calorimeter, the fluctuation of the number of electrons and positrons is intrinsically related to the statistical nature of the cascade process. Therefore, the average number of electrons and positrons traversing active samplers will undergo statistical fluctuations. Moreover, the collision loss stopping power of soft-energy electrons (and positrons) is energy dependent. These particles deposit more energy than fast electrons (or positrons) moving closer to the shower axis and are also more spread laterally. Thus, the direction of average motion of soft charged particles makes large angles with respect to the shower axis. Furthermore, because of the statistical nature of the shower process, the cascade energy and spatial distributions are characterized by fluctuations, i.e., deviations from their average behavior.

In sampling calorimeters, these fluctuations result in fluctuations of the sampled visible-energy ϵ_{vis}, i.e., in widening the distribution of the measured ϵ_{vis}. They are usually called *energy sampling fluctuations* [Amaldi (1981); Fabjan (1985a, 1986)]. As discussed above, this is related to the fluctuations in the number of particles and in the energy deposited by electrons and positrons traversing (or being absorbed in) the active layers of the calorimeter.

The overall physical process of energy deposition in a medium is determined by the mean energy deposition per charged particle traversing the active readout layers multiplied by the average numbers of particles N_{cp}. To a first approximation, the value of N_{cp} can be estimated as:

$$N_{cp} = \frac{\epsilon_{\mathrm{vis}}}{E_{\mathrm{el-loss}}}, \tag{9.20}$$

where $E_{\mathrm{el-loss}}$ is the collision energy-loss by electrons (or positrons, whose difference can be neglected within our approximation) in a single active readout layer.

As discussed at page 200 and in Sect. 2.1.6, the electron collision energy-losses are $\approx 10\%$ larger than the collision energy-losses of a massive *mip*. In addition, in a

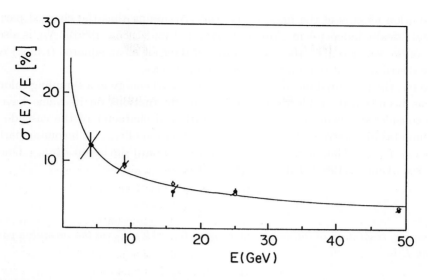

Fig. 9.12 Energy resolution $\sigma(E)/E$ in (%) versus E, data from the Si/W electromagnetic sampling calorimeter of SICAPO Collaboration (Fig. 2.78) with $\tau = 2$ (reprinted from *Nucl. Instr. and Meth. in Phys. Res. A* **235**, Barbiellini, G., Cecchet, G., Hemery, J.Y., Lemeilleur, F., Leroy, C., Levman, G., Rancoita, P.G. and Seidman, A., Energy Resolution and Longitudinal Shower Development in a Si/W Electromagnetic Calorimeter, 55–60, Copyright (1985), with permission from Elsevier; see also [Leroy and Rancoita (2000)]).

shower generator medium, namely the passive sampler of the calorimeter, the mean cosine value of the charged-particle opening-angle $\langle\cos\theta\rangle$ is given by Eq. (2.253). In the active readout layer of thickness X_S, the average value $E_{\text{el}-\text{loss}}$ is given by

$$E_{\text{el}-\text{loss}} \approx 1.1 \left(\frac{dE}{dx}\right)_S X_S \left(\frac{1}{\langle\cos\theta\rangle}\right). \qquad (9.21)$$

In a sampling calorimeter with a dominant passive sampler of thickness L_a and radiation length X_{0a}, ϵ_{vis} is given by Eq. (9.9) and using Eq. (9.21) one obtains

$$N_{cp} = (e/mip) \left[\frac{\left(\frac{dE}{dx}\right)_S X_S}{\left(\frac{dE}{dx}\right)_a X_{0a}}\right] \left[\frac{\langle\cos\theta\rangle}{1.1\left(\frac{dE}{dx}\right)_S X_S}\right] \frac{E}{\tau}. \qquad (9.22)$$

As seen on page 200, the ratio ϵ_{ca}/X_{0a} is $\approx 1.1\,(dE/dx)_a$, where ϵ_{ca} is the critical energy of the passive sampler. The previous equation can be finally written as:

$$N_{cp} = (e/mip)\frac{E}{\tau}\left[\frac{\langle\cos\theta\rangle}{1.1\left(\frac{dE}{dx}\right)_a X_{0a}}\right]$$
$$= (e/mip)\left(\frac{\langle\cos\theta\rangle}{10^{-3}\epsilon_{ca}}\right)\frac{E}{\tau}, \qquad (9.23)$$

where the incoming particle energy E is in units of GeV, the critical energy of the passive sampler, ϵ_{ca}, is units of MeV and τ is the sampling frequency in units of radiation lengths of the passive material. Assuming that the charged particles are statistically independent [Amaldi (1981); Fabjan (1985a, 1986)], N_{cp} is also the value of the variance, i.e., the square value of the root mean squared (r.m.s.) error, of the distribution of the number of charged particles.

So far, the measured mean value of the deposited energy in a sampling calorimeter has been called visible energy (ϵ_{vis}), but fluctuations on its value have not been considered. Owing to the intrinsic statistical character of the cascade phenomenon (as discussed above), the energy distribution of ϵ_{vis}, for incoming particles of energy E, is a Gaussian distribution with standard deviation $\sigma(\epsilon_{\mathrm{vis}})$. One has (e.g., see [Amaldi (1981); Fabjan (1985a, 1986)]):

$$\frac{\sigma(\epsilon_{\mathrm{vis}})}{\epsilon_{\mathrm{vis}}} \approx \frac{1}{\sqrt{N_{cp}}}.$$

The energy resolution of the sampling calorimeter, $\sigma(E)/E$, on the incoming particle energy E is

$$\frac{\sigma(E)}{E} \equiv \frac{\sigma(\epsilon_{\mathrm{vis}})}{\epsilon_{\mathrm{vis}}}, \tag{9.24}$$

and using Eq. (9.23), one obtains:

$$\frac{\sigma(E)}{E} = \frac{\sigma(\epsilon_{\mathrm{vis}})}{\epsilon_{\mathrm{vis}}} \approx \frac{1}{\sqrt{N_{cp}}} = K\sqrt{\frac{\tau}{E}}. \tag{9.25}$$

As in Eq. (9.23), τ is the sampling frequency in units of radiation length of the passive material, the incoming particle energy, E, is in units of GeV and K is given by

$$K = \sqrt{\frac{10^{-3}\epsilon_{ca}}{(e/mip)\langle\cos\theta\rangle}} \approx (3.16\%)\sqrt{\frac{\epsilon_{ca}}{(e/mip)\langle\cos\theta\rangle}}, \tag{9.26}$$

where the passive sampler critical energy ϵ_{ca} is in units of MeV.

Although Eq. (9.26) is similar to Equation 8 given by Amaldi (1981), the e/mip ratio replaces the function of the minimum detectable energy $F(\zeta)$ (defined on page 201) in Eq. (9.26). In this equation, the e/mip ratio takes properly into account the difference of the energy deposition per g/cm^2 between active readout layers and passive samplers.

9.4.1.1 *Calorimeter Energy Resolution for Dense Readout Detectors*

Equation (9.26) was obtained under the assumption that electrons (and positrons) depositing their energy by collision losses have similar statistical weights. This assumption is acceptable for sufficiently thick dense active readout detectors, e.g., a few cm of scintillator, so that the energy deposited by collision losses (for instance, $\approx 2\,\mathrm{MeV/cm}$ in scintillator) by fast charged particles moving forwards and close to the shower axis are not negligible compared with the energy deposition by collision losses of laterally spread soft electrons. These soft particles have their energies not exceeding a few MeV's and can be fully absorbed.

However in thin (although dense) detectors, e.g. a few mm of liquid argon or a few mm of scintillator or a few hundreds of microns of silicon detectors, particles passing at large angle deposit much larger energy than those passing close to the shower axis. As a consequence, N_{cp} can be overestimated even by a factor 2, corresponding to particles traversing (without being fully absorbed) at an angle of about 60^0 to the shower axis. This factor is reduced, when the mean traversing angle increases. To a rough approximation, for a dense but thin readout detector, we can assume that the N_{cp} value given in Eq. (9.23) has to be multiplied by a reducing factor $1/(1 + \langle \cos \theta \rangle)$. As already mentioned in the previous section, the mean cosine value of the charged particle opening angle $\langle \cos \theta \rangle$ is given by Eq. (2.253). Under these approximations, the calorimeter energy resolution is again given by Eq. (9.25), but the parameter K becomes:

$$
K = \sqrt{\frac{10^{-3}\epsilon_{ca}[1 + P_d \langle \cos \theta \rangle]}{(e/mip) \langle \cos \theta \rangle}}
$$

$$
\approx (3.16\%)\sqrt{\frac{\epsilon_{ca}[1 + P_d \langle \cos \theta \rangle]}{(e/mip) \langle \cos \theta \rangle}}, \tag{9.27}
$$

where the passive sampler critical energy ϵ_{ca} is in units of MeV, and the parameter P_d is 0 for thick and dense readout layers [in this case, Eq. (9.27) transforms into Eq. (9.26)] and 1 for thin and dense readout layers. Within the approximations used, Eq. (9.27) accounts for effects of differential collision energy-losses by charged particles, impinging at different angles, traversing (or being absorbed in) dense readout detectors. These effects are often referred to as *path-length fluctuations* [Amaldi (1981); Fabjan (1985a, 1986)], and *Landau fluctuations* [Amaldi (1981); Fabjan (1985a, 1986); Grupen (1996)]. These effects were introduced in the sampling calorimetry to take into account the degradation of the energy resolution because of the different path of large-angle particles (*path-length fluctuations*) and large energy transfers in ionization process (*Landau fluctuations*). However, the latter fluctuations add a minor contribution to the deterioration of the calorimeter energy resolution [Fabjan (1985a)].

In Eq. (9.27), the e/mip ratio depends on the Z values of the readout and absorber materials (see Sect. 9.2.2). Furthermore, the parameter K (as it is the case for the energy and angular distributions of soft electrons) depends on the critical energy of the passive sampler. The calorimeter resolution $\sigma(E)/E$, for a fixed sampling frequency and incoming particle energy, improves by increasing the visible-energy sampled as long as sampling fluctuations are the major contribution to the overall calorimeter resolution and both Landau and path length fluctuations can be taken into account simply by the multiplicative factor discussed for Eq. (9.27). Experimental data are in agreement with Eq. (9.25) predicting the energy resolution dependence on $1/\sqrt{E}$. In Fig. 9.12, the energy resolution [SICAPO Collab.

Table 9.3 Selection of measured energy resolutions for sampling calorimeters with dense readout detectors from [Leroy and Rancoita (2000)]. Errors are quoted when available in the original reference.

passive sampler	readout	τ (X_0)	energy resolution at 1 GeV (%)	K (%)	Reference
Al [see (1)] (marble)	Scintillator (3 cm) [see (1)]	≈ 1	20.0	20.0	[Amaldi (1981); Fabjan (1985a)]
Fe	Scintillator	1.4	23.0	19.4	[Diddens et al. (1980)]
Fe	liquid Ar	8.5×10^{-2}	7.5	25.7	[Abramowicz et al. (1981)]
Fe	Si det.	2.0	32.2 ± 1.1	22.8 ± 0.8	[Fabjan (1985a); Fabjan et al. (1977)]
W	Si det.	2.0	24.6 ± 0.4	17.6 ± 0.3	[SICAPO Collab. (1994a)]
Pb	Scintillator	1.8×10^{-1}	7.5	17.7	[SICAPO Collab. (1985a)]
Pb	Scintillator	2.1×10^{-1}	9.0	19.4	[Hofmann et al. (1979)]
Pb	Scintillator	1.8	23.5 ± 0.5	17.6 ± 0.4	[Schneegans et al. (1982)]
Pb	liquid Ar	3.1×10^{-1}	[see(2)] >20 GeV	14.4	[Drews et al. (1990)]
Pb	Si det.	0.9	16.5 ± 0.5	17.4 ± 0.5	[Buckardt et al. (1988)]
Pb	liquid Ar	3.2×10^{-1} Accordion	$9.6 - 10.1$	16.9-17.8	[Nakamoto et al. (1985)]
U	Scintillator	0.5	11.0	15.6	[Aubert et al. (1991)]
U	Scintillator	1	16.5 ± 0.5	16.5 ± 0.5	[Carosi et al. (1984)] [Drews et al. (1990)]

(1): it means that the marble used as passive sampler is equivalent to Al. The scintillator used as readout is 3 cm thick.

(2): the energy resolution is given by: $\frac{\sigma(E)}{E} = \sqrt{\left(\frac{0.1}{E}\right)^2 + \left(\frac{0.075}{\sqrt{E}}\right)^2 + (0.005)^2}$.

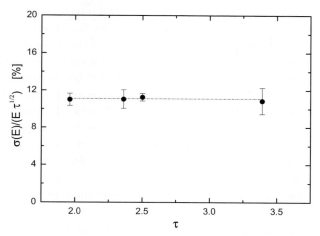

Fig. 9.13 $\sigma(E)/(E\tau^{1/2})$ (in %) measured for Pb absorbers versus τ at the incoming energy of 4 GeV (data from [SICAPO Collab. (1994a)]). The line is the weighted mean of the data values.

(1985a)] of a sampling calorimeter with tungsten sampler thicknesses of $2\,X_0$ and silicon readout detectors is shown for incoming electron energies between 4 and 49 GeV. The energy resolution is well represented by the fitted curve following Eq. (9.25) [SICAPO Collab. (1985a)].

At fixed energy, Eq. (9.25) predicts that the energy resolution depends on $\sqrt{\tau}$. In Fig. 9.13, the energy resolution divided by $\sqrt{\tau}$, i.e., $\sigma(E)/(E\tau^{1/2})$ in %, is shown as a function of τ, for electrons of 4 GeV energy and passive samplers made of Pb plates [SICAPO Collab. (1994a)]. The values of $\sigma(E)/(E\tau^{1/2})$ do not depend on τ, as expected from Eq. (9.25).

A summary of experimental data on electromagnetic sampling calorimeters using dense readout detectors (namely silicon detectors, scintillators and liquid argon ionization chambers) is presented in Table 9.3. Reported data include the type of readout detector, the sampling frequency τ, the measured energy resolution fitted by Eq. (9.25) and evaluated at 1 GeV of incoming electron energy, and the parameter K. The experimental energy resolutions are measured by employing low- (Al), medium- (Fe), and high-Z (W, Pb, U) passive samplers, which are the most frequently used absorbers. The sampling frequency varies between $\approx 9 \times 10^{-2}$ and 2. However, in spite of experimental uncertainties and different calorimeter parameters, for a given type of passive sampler the values of K do not differ by more than $\approx 10\%$ from their mean values. These averages of experimental K values, for calorimeters using the same passive samplers, are shown in Table 9.4 together with the K values calculated by means of Eq. (9.27). In Table 9.4, the e/mip ratios were estimated by averaging the experimental data¶ shown in Table 9.1. The agreement among estimated and averaged experimental K values (shown in Table 9.4) is sati-

---

¶The same value of Pb sampler was assumed for W absorber.

sfactory and indicates that Eqs. (9.25, 9.27) take well into account the fluctuations of the visible energy for dense readout detectors. So far, it was assumed, in the comparison of experimental data with predictions, that the intrinsic statistical fluctuations are the limiting constraints on the calorimeter energy resolution. In a real detector, many instrumental effects add to these statistical fluctuations and worsen the energy resolution. They depend on the type of device (sampling or homogeneous calorimeter) and readout medium. The instrumental noise, detection statistics and intercalibration errors affect the overall energy resolution and, particularly, limit the detector performance at high energies. In [Engler (1985)] and references therein, their influence on calorimeter resolution is treated in a general way. Furthermore in Sect. 9.4.2, the degradation of the calorimeter resolution induced by visible-energy losses, due to the finite dimensions of the device, is evaluated.

9.4.1.2 *Calorimeter Energy Resolution for Gas Readout Detectors*

Sampling calorimeters were also built with gas detectors as active readout devices. As discussed in the previous section, the wide spread of electron angles is among the causes of degradation of the calorimeter energy resolution. These path-length fluctuations depend on the thickness of readout detectors, namely they increase as the thickness decreases (as discussed above). In the same way, Landau fluctuations [as shown by Amaldi (1981) in Equation 21] are larger for calorimeters in which the deposited energy is measured. However, in calorimeters using streamer-tubes as readout detectors, tracks are counted at least as long as the particles are close to the shower axis. For each ionization track one streamer is formed and, consequently, Landau fluctuations affect less the energy resolution.

In general, these fluctuations cannot be neglected, when gas detectors are used as active readout samplers. The energy resolutions of multiwire proportional quantameters were investigated by means of Monte-Carlo simulations [Fisher (1978)] and compared with experimental data. Fisher (1978) has shown, see Fig. 9.14, that *path length* (also termed *track length*) and *Landau fluctuations* worsen the *sampling fluctuations* in these quantameters [Fisher (1978)] by a factor ≈ 2. The energy resolutions obtained using Monte-Carlo simulations were found to agree with those measured with Pb [Nordberg (1971)] and Fe [Anderson et al. (1976)] gas quantameters [Fisher (1978)]. Measured energy resolutions of calorimeters employing gas readout detectors are given for Fe and Pb passive samplers in Table 9.5. The average measured K values for Fe and Pb absorbers are 32% and 26%, respectively. Although the gas readout detectors are operated in different way, the measured K values do not differ by more than $\approx 15\%$ from their respective average values. These average values, in turn, are in agreement with K-parameters derived from Eq. (9.27) (see Table 9.4) once multiplied by a factor $\approx \sqrt{2}$. This factor can be taken into account empirically in Eqs. (9.26, 9.27). For a gas calorimeter (within

Table 9.4 Comparison of K values from measured energy resolutions and calculated by means of Eq. (9.27) from [Leroy and Rancoita (2000)].

passive sampler	readout	(e/mip) [see(1)]	$\langle \cos\theta \rangle$ from Eq. (2.253)	K from Eq. (9.27) (%)	K experim. [see(2)] (%)
Al [see(3)] (marble)	thick	1.00	0.986	20.2	20.0
Fe	thin	0.83	0.949	22.8	22.6
W	thin	0.66	0.671	17.4	17.6
Pb	thin	0.66	0.614	17.2	17.3
U	thin	0.63	0.543	17.5	16.1

(1): average values from data of Table 9.1; it was assumed that W and Pb have equal (e/mip) ratios.
(2): average values from data of Table 9.3.
(3): it means that the marble used as passive sampler is equivalent to Al.

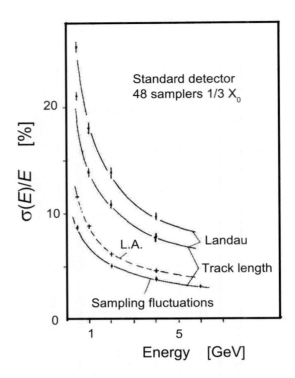

Fig. 9.14 Contributions of sampling, track length (also termed path length) and Landau fluctuations to the energy resolution of a Pb gas quantameter as computed by Fisher (1978) (see also [Amaldi (1981); Leroy and Rancoita (2000)]), using Monte-Carlo simulations (adapted from *Nucl. Instr. and Meth.* **156**, Fisher, H.G., Multiwire Proportional Quantameters, 81–85, Copyright (1978), with permission from Elsevier). The curve labeled "L.A." regards the expected resolution for a liquid argon calorimeter (for details see [Fisher (1978)]).

an approximation of $\approx 15\%$), one has

$$K \approx (3.16\%)\sqrt{\frac{\epsilon_{ca}(1 + P_d \langle \cos\theta \rangle)(1 + P_g)}{(e/mip)\langle \cos\theta \rangle}}, \qquad (9.28)$$

where ϵ_{ca} is the critical energy of the passive sampler in units of MeV, $\langle \cos\theta \rangle$ is given by Eq. (2.253); the parameter P_d is 0 for thick and dense readout layers [in this case, Eq. (9.28) transforms into Eq. (9.26)] and 1 for thin and dense readout layers [in this case Eq. (9.28) transforms into Eq. (9.27)]. In both cases, the parameter P_g is 0. For sampling calorimeters with gas readout detectors, the parameters P_g and P_d are both equal to 1. Then, the energy resolution of an electromagnetic calorimeter can be finally written as

$$\frac{\sigma(E)}{E} \approx (3.16\%)\sqrt{\frac{\epsilon_{ca}(1 + P_d \langle \cos\theta \rangle)(1 + P_g)}{(e/mip)\langle \cos\theta \rangle}}\sqrt{\frac{\tau}{E}}. \qquad (9.29)$$

Table 9.5 Selection of measured energy resolutions for sampling calorimeters with gas readout detectors from [Leroy and Rancoita (2000)]. Errors are quoted when available in the original reference.

passive sampler	readout	τ (X_0)	energy resolution at 1 GeV (%)	K (%)	Reference
Fe	MWPC	1.6	≈ 47.2	≈ 37.4	[Fisher (1978)],[Anderson et al. (1976)]
Fe	streamer mode	1.2	28.8	26.4	[Baumgart (1987); Baumgart et al. (1988a)]
Fe	streamer mode	2.8	55.0	32.6	[Catanesi et al. (1986)]
Pb	MWPC	5.6×10^{-1}	≈ 20.6	≈ 27.6	[Fisher (1978); Nordberg (1971)]
Pb	saturated prop.-mode	0.5	16.2	22.9	[Atac et al. (1983)]
Pb	saturated prop.-mode	3.6×10^{-1}	16.5	27.6	[Videau et al. (1984)]

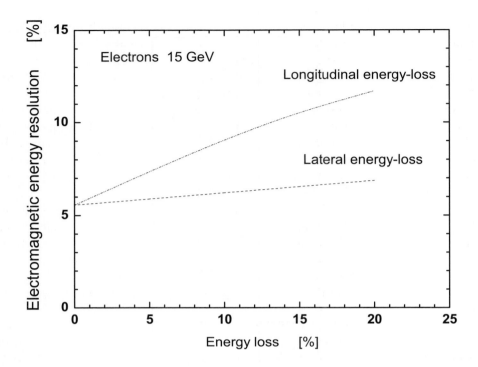

Fig. 9.15   Energy resolution, $\sigma(E)/E$ in %, measured as a function of the fraction in % of lateral and longitudinal deposited-energy losses (curves from [Amaldi (1981)], see also [Leroy and Rancoita (2000)]) for the marble calorimeter of CHARM Collaboration [Diddens et al. (1980)].

9.4.2   *Effect of Limited Containment on Energy Resolution*

So far in the previous sections, the energy resolution of the calorimeter was considered for negligible lateral and longitudinal visible-energy losses, namely for the case where the calorimeter longitudinal depth and lateral width are large enough to contain the shower cascade completely. However, the energy resolution expressed by Eq. (9.24) is degraded by the effects of the finite dimensions of the calorimeter.

The energy resolution is increased by leakage fluctuations caused by longitudinal, E_L, and lateral, E_l, energy-losses. The size of these losses depends on the dimensions of the calorimeter. The longitudinal energy-losses display a weak dependence on the incoming energy and an even weaker dependence on the material. The longitudinal shower development can be described, as seen for instance in Eq. (2.236), by a function, whose parameters depend on the incoming energy E as $\ln E$. Thus, finite dimensions of a calorimeter result in losses of the average deposited energy depending on $\ln E$ [Engler (1985); Grupen (1996)]. Moreover, because of the statistical nature of the cascade process (including the location of the first interaction of the primary particle along the shower axis), the finite longitudinal dimension will affect the energy resolution on an event-by-event basis.

The lateral energy-losses do not critically depend on the incoming energy, on the type of incident particles (electrons or photons), or on the absorber material at a depth where the cascade is extended over a small transverse distance in units of the Molière radius. For large transverse distances in units of Molière radius, a slight dependence on the above parameters is related to soft interactions contributing to the energy deposition process. Fluctuations coming from the transverse leakage come out to be less effective compared to those caused by the longitudinal leakage [Amaldi (1981)]. Thus, the longitudinal energy-loss contribution to the degradation of the energy resolution is expected to be larger than the contribution depending on lateral energy-losses (see the discussion on the lateral shower development on page 207).

The longitudinal energy leakage can be calculated by integrating one of the equations given in Sect. 2.4.2.2. Similarly, the lateral energy leakage can be calculated by integrating one of the equations given in Sect. 2.4.2.3. However, these equations give the average behavior of the shower development only approximately. In a sampling calorimeter, because the visible energy is proportional to the incoming energy [see Eq. (9.1)], the fraction of visible-energy lost laterally or longitudinally is given by the fraction of the lateral or longitudinal leakage of the incoming energy, respectively.

As a rule of thumb [Engler (1985)], the deterioration of the energy resolution is expected to increase proportionally to the longitudinal leakage. These effects were experimentally investigated. The marble fine-grained calorimeter of the CHARM Collaboration [Diddens et al. (1980)] was used to determine the effect of lateral and longitudinal losses. The experimental data allowed Amaldi (1981) to determine that the longitudinal losses are affecting more the energy resolution than the lateral losses, as expected (Fig. 9.15). For energy losses lower than 15%, the calorimeter energy-resolution is almost linearly related to the average fraction in % of the lateral energy lost ($lt = 100 \times [E_l/E]$) and of the longitudinal energy lost ($Lg = 100 \times [E_L/E]$), according to

$$\frac{\sigma(E)}{E} \approx \left[\frac{\sigma(E)}{E}\right]_0 (1 + 6.6 \times 10^{-2} Lg), \tag{9.30}$$

and

$$\frac{\sigma(E)}{E} \approx \left[\frac{\sigma(E)}{E}\right]_0 (1 + 1.4 \times 10^{-2} lt), \tag{9.31}$$

where (from Fig. 9.15) $\left[\frac{\sigma(E)}{E}\right]_0 \approx 5.58\%$ is the measured calorimeter resolution in the absence of energy losses.

For highly segmented readout planes such as mosaic planes made of silicon detectors, dead zones between the detectors in the mosaics result in losses of visible energy, indicated by E_D. These dead zones can affect the energy resolution [SICAPO Collab. (1994a)]. In the following, the fraction in % of visible-energy lost by the mosaic dead zones is indicated by De (i.e., $De = 100 \times [E_D/E]$). These losses are subjected to event-by-event fluctuations which are more similar to the longitudinal energy-losses than to the lateral energy-losses. The dead area energy-losses start much closer to the impinging point than the lateral energy-losses which appear after some transverse distance from the impinging point and are controlled by the Molière radius.

Their effect on the calorimeter energy resolution was investigated as a function of the incoming electron energy with Fe and Pb passive samplers and employing the same readout detectors (namely silicon mosaic planes) [SICAPO Collab. (1994a)]. These passive materials have similar Molière radii ($\approx$ 1.8 and 1.6 cm respectively), and consequently similar values of De. The data were collected for sampling frequencies $\tau = 1.99$ and 3.98 for Fe absorbers, and $\tau = 1.96$, 2.36, 2.50 and 3.39 for Pb absorbers [SICAPO Collab. (1994a); Furetta et al. (1995)]. In Figs. 9.16 and 9.17, the measured energy-resolutions for the two calorimeters with Fe (and $\tau = 1.99$) and Pb (and $\tau = 3.39$) absorbers are shown as a function of $1/\sqrt{E}$. The experimental data show (in both cases) a linear dependence on $1/\sqrt{E}$ and are in agreement with

$$\frac{\sigma(E)}{E} = K_{le}\sqrt{\frac{\tau}{E}}. \tag{9.32}$$

Similarly to the cases of longitudinal and lateral energy-losses, these experimentally measured energy-resolutions [SICAPO Collab. (1994a)] are consistent with

$$\frac{\sigma(E)}{E} = \left[\frac{\sigma(E)}{E}\right]_0 [1 + (6.8 \pm 1.7) \times 10^{-2} De], \tag{9.33}$$

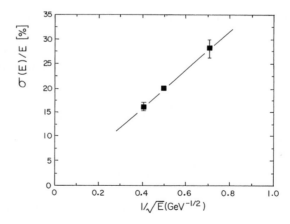

Fig. 9.16   Energy resolution $\sigma(E)/E$ in % measured as a function of $1/\sqrt{E}$ (E is the incoming electron energy) for fixed visible-energy losses due to the mosaic structure of the active readout silicon plane and Fe passive sampler (reprinted from *Nucl. Instr. and Meth. in Phys. Res. A* **345**, Bosetti, M. et al., Systematic investigation of the electromagnetic energy resolution on sampling frequency using silicon calorimeters, 244–249, Copyright (1994), with permission from Elsevier, e.g., for the list of the authors see [SICAPO Collab. (1994a)]; see also [Leroy and Rancoita (2000)]).

where $\left[\frac{\sigma(E)}{E}\right]_0$ is the calorimeter resolution in the absence of energy-losses [e.g., see Eq. (9.25)], and, thus, can be expressed in a general form by

$$\frac{\sigma(E)}{E} = K[1 + (6.8 \pm 1.7) \times 10^{-2}De]\sqrt{\frac{\tau}{E}}. \qquad (9.34)$$

As a consequence, from Eqs. (9.32–9.34) we have that

$$K_{le} = K[1 + (6.8 \pm 1.7) \times 10^{-2}De],$$

where K is the parameter of the energy resolution, in the absence of dead-area energy-losses [see Eqs. (9.25, 9.27)]. As expected from Eq. (9.34), the ratio R_K of the measured K_{le} value for the Fe and Pb calorimeters is $R_K = 28.3/22.2 \approx 1.27$ and is consistent with the value calculated from Table 9.4, i.e., ≈ 1.31, without energy losses.

The visible-energy losses, due to the leakage of the incoming particle energy, degrade the calorimeter resolution. In fact, the K_{le} values are larger than those discussed in previous sections on energy resolutions for no dead-area energy-losses.

The multiplicative factor De, appearing in Eqs. (9.33, 9.34), is similar (within the experimental errors) to the one for longitudinal energy-loss (Lg) given in Eq. (9.30) since, in both cases, energy losses are subjected to event-to-event fluctuations.

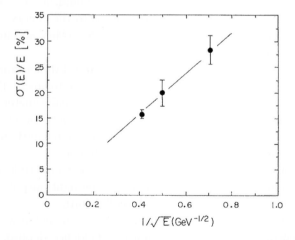

Fig. 9.17 Energy resolution $\sigma(E)/E$ in % measured as a function of $1/\sqrt{E}$ (E is the incoming electron energy) for fixed visible-energy losses due to the mosaic structure of the active readout silicon plane and Pb passive sampler (reprinted from *Nucl. Instr. and Meth. in Phys. Res. A* **345**, Bosetti, M. et al., Systematic investigation of the electromagnetic energy resolution on sampling frequency using silicon calorimeters, 244–249, Copyright (1994), with permission from Elsevier, e.g., for the list of the authors see [SICAPO Collab. (1994a)]; see also [Leroy and Rancoita (2000)]).

9.5   Homogeneous Calorimeters

9.5.1   *General Considerations*

A homogeneous calorimeter is a block or a matrix of blocks of sensitive absorbing material where the energy of an incoming particle is degraded and measured via the collection of visible photons. The absorbing material or active medium of the calorimeter is either scintillator, where fluorescence is produced by the passage of an ionizing particle or material with a high refractive index transparent to the Čerenkov photons produced by charged particles, which move with a velocity

$$v = c/n$$

larger than that of light through the medium of refractive index $n > 1$ (for a treatment of scintillator detectors see the chapter on *Scintillating Media and Scintillator Detectors*, while for a treatment of the Čerenkov effect see Sect. 2.2.2).

A variety of scintillating and Čerenkov materials (see the chapter on *Scintillating Media and Scintillator Detectors*) can be used as the active medium of homogeneous calorimeters. The choice of a particular active medium is made with the goal to achieve the best possible energy and position resolution over an energy range as large as possible (from MeV up to TeV energies) taking into account practical and physical requirements.

The energy resolution is determined by the fluctuations in the number of photons or ions pairs produced by the cascade. The light yield, i.e, the number of photons produced per MeV of energy deposition, should be large enough to minimize the contribution of the statistical fluctuations to the energy resolution. Typically, the amount of collected light should exceed $100\,\gamma/\text{MeV}$ and in fact can exceed $4\times10^4\,\gamma/\text{MeV}$ for NaI(Tl) (Table 5.2).

The size of the sensitive material length and lateral dimensions are important constraints. The total length should be sufficiently large to ensure that energy leakage of the electromagnetic cascades at the back of the calorimeter is small enough not to degrade the energy resolution, even for very energetic electrons and photons. High density sensitive materials have short radiation length and small Molière radius, so favoring good compactness of the calorimeter.

The length of the sensitive material has also to be compatible with light transmission properties and light attenuation length. The light output along the length of the crystal should not vary (uniform light collection).

The lateral size of the sensitive material is chosen in such a way that incoming particles fired centrally into a block at normal incidence deposit a large fraction between 70 and 80% of their energy in this particular block, while energy deposited in the neighboring blocks is large enough to measure the coordinates of the center of gravity and, consequently, of the impact point. This can be achieved by selecting a material with a small Molière radius (typically $\leq$ a few cm's, see Table 5.2) resulting in narrow cascades and a matrix of sensitive material blocks with the lateral size of a cell of the order of the Molière radius. However, transverse dimensions of a sensitive block in a matrix result most generally from practical considerations, as a compromise between the spatial resolution and the number of readout channels whose number increases fast with the segmentation.

The decay time of scintillation should be as short as possible. This becomes an important constraint in experiments at the LHC, where the light has to be collected in less than $\sim 10\,\mathrm{ns}$ since the expected LHC interbunch crossing time is $25\,\mathrm{ns}$, requiring the use of new types of crystals such as $PbWO_4$.

The readout of the homogeneous calorimeters is realized through photomultiplier tubes or photodiodes of various types collecting the produced light. Factors such as the noise, the quantum efficiency and the gain of the photosensitive device contribute to the energy resolution of the calorimeter.

In Sect. 9.9, the mechanisms for achieving the compensation in (hadronic) calorimeters will be described. These mechanisms do not apply to homogeneous calorimeters. These mechanisms are based on the fact that only a small fraction of the cascade energy is deposited in the active medium of the calorimeter. Depending on the nature of the active and passive medium, there are ways to equalize the calorimeter response to electromagnetic and non-electromagnetic components of the cascade. In the case of homogeneous calorimeters, there is no way to compensate for binding energy-losses occurring in the non-electromagnetic component of the cascade. This is supported by the results of measurements performed with homogeneous hadronic calorimeters of e/h ratio not equal to 1. Then, the non-linearity of the calorimeter response and the deviation from a Gaussian response are observed (Sect. 9.8.1).

The properties of scintillators, mechanisms of light emission and readout techniques are reviewed in [Suffert (1988); Cushman (1992); Bourgeois (1994)] and are treated in Sect. 5.1. These elements are necessary in order to understand the characteristics of both homogeneous calorimeters and hydrogeneous readouts in hadronic sampling calorimetry (Sect. 9.9.1).

The inorganic scintillators were used in small-scale homogeneous calorimeters, since their relative high cost makes their use prohibitive in large-scale calorimeters. However, BGO was used in the large electromagnetic calorimeter of the L3 experiment at LEP machine (at CERN). BaF_2 was also used in smaller arrays [Wisshak et al. (1989)]. The decay constants of BGO, NaI(Tl) and CsI(Tl) are in the hundreds of ns and more. It precludes the use of these crystals for high-rate applications such as those to be encountered in experiments at the LHC machine.

Organic scintillators are aromatic compounds produced from a benzenic cycle. In organic scintillators, the mechanism of light emission is not a lattice effect, but rather a molecular effect. It proceeds through excitation of molecular levels in a primary fluorescent material, which emits bands of ultraviolet (UV) light during de-excitation. This UV light is readily absorbed in most organic materials transparent in the visible wavelength region, with an absorption length of a few mm. The extraction of a light signal becomes possible only by introducing a second fluorescent material in which the UV light is converted into visible light (*wavelength shifter*). This second substance is chosen in such a way that its absorption spectrum is matched to the emission spectrum of the primary flour, and its emission should be adapted to the spectral dependence of the quantum efficiency of the photocathode. These two active components of a scintillator are either dissolved in suitable organic liquids or mixed with the monomer of a material capable of polymerization. The polymer can then be cast into any shape for practical applications. Most frequently used are rectangular plates with thicknesses from 0.2 up to 5.0 mm and area from a few mm^2 up to several m^2 and are of current use as active medium of sampling calorimeters. Organic scintillators are very rarely utilized in homogeneous calorimeters with the exception of liquid scintillators. In a liquid scintillator, an organic crystal (solute) is dissolved in a solvent (with typical concentrations of ≈ 3 g/l). The scintillation efficiency depends on the solvent used. Xylene and toluene give good relative pulse height with a PPO solute [Schram (1963)]. The scintillator BIBUQ dissolved in toluene has the best time resolution (≈ 85 ps) of any scintillator currently available [Bengston and Moszynski (1982)].

Čerenkov detectors, used as calorimeters, are composed of liquids (with refractive index $n \approx$ between 1.2 and 1.5) or solids ($n \approx$ between 1.4 and 1.8). Čerenkov material used in practice are represented by lead glass of various types such as SF-5, SF-6, PbF$_2$, and their properties are shown in Table 9.6. The Čerenkov light is emitted in a forward cone with aperture $\cos\theta = c/(vn) = 1/(\beta n)$. The number of photons, N_{ph}, emitted per unit length of track of the incident particle of charge number z through the medium in the $d\lambda$ is given by Eq. (5.16) [see also Eq. (2.138) on page 128]. In the wavelength interval, $d\lambda$, the radiation intensity depends only on the charge number z of the incident particle and on the emission angle, and therefore on the particle velocity. The presence of λ^2 in the denominator of Eq. (5.16) [see also Eq. (2.140) on page 128] indicates an emission concentrated at shorter wavelength. The emission is taking place in the wavelength domains of visible and UV light. This is one reason why low light yields are often obtained, since only part

Table 9.6 Properties of Čerenkov materials of current use (see for instance [Leroy and Rancoita (2000)]).

Properties	SF-5	SF-6	PbF$_2$	Water
Density (g/cm^3)	4.08	5.20	7.66	1.00
Radiation Length (cm)	2.54	1.69	0.95	36.40
λ_{cut} (nm)	350	350	260	$\sim$300
Refractive index	1.67	1.81	1.82	1.33
Light yield (pe/GeV)	600	900	2000	
Light yield [vs NaI(Tl)]	1.5×10^{-4}	2.3×10^{-4}	10^{-3}	

of the spectrum is sampled. In the wavelength domain of the visible and UV light which corresponds to the domain of spectral sensitivity of the standard photocathodes (300 to 600 nm for bialkali photocathodes), the light emitted by a Čerenkov radiator is smaller than that emitted by scintillators of the same dimensions. The Čerenkov light is produced very fast and essentially with no time spread (with the exception of light propagation effects in crystals of large dimensions). The timing properties are not governed by the specific Čerenkov material itself but rather by the dimensions, geometry of the block or container (for liquid Čerenkov), and by the rise time of the photosensitive device.

The threshold effect linked to Čerenkov light production ($v > c/n$) is also another reason for low light yields since the cutoff energy (E_{cut}) is high [Sowerby (1971)]. As an example, in the case of SF-5 ($n = 1.7$), $E_{\text{cut}} \approx 0.15$ and ≈ 0.3 MeV for electrons and photons, respectively. Table 9.6 shows the properties of Čerenkov materials of standard use.

9.5.2 *Energy Measurement*

Ideally, all the incoming particle energy is deposited in the volume (which is the active medium) of the homogeneous calorimeters [Nakamoto et al. (1986); Bormann et al. (1987)], and the number of photons or ions pairs n_p produced by the cascade is given by a formula similar to Eq. (9.25):

$$n_p = \frac{E}{\bar{E}}, \qquad (9.35)$$

where $\bar{E}$ is the mean energy needed to create a photon or an ion pair.

Since this creation process is a random process, the intrinsic energy resolution is dominated by the statistical fluctuations of the number of detected primary processes, and is given by

$$\frac{\sigma(E)}{E} = \frac{\sqrt{n_p}}{n_p} = \frac{\sqrt{E}}{E} = \frac{\text{constant}}{\sqrt{E}}. \qquad (9.36)$$

The threshold of detection is usually small enough that the number of primary processes, that can be detected, is large and gives good intrinsic energy resolution. However, as already discussed in Sect. 9.2 for the case of sampling calorimeters, Eq. (9.36) only applies to cascades fully contained in the calorimeter volume. Ideally, the calorimeter must be of sufficient thickness to allow the particle to deposit all its energy inside the detector volume in the subsequent cascade of increasingly lower energy particles. The total depth of the calorimeter must be large enough to allow a longitudinal containment of the cascade in order to completely absorb the incoming energy. The necessary longitudinal depth of a calorimeter varies with the incoming energy E, as $\ln E$.

The incoming energy being distributed among a large number of secondary particles, energy can be deposited at large angles with respect to the longitudinal axis of the calorimeter. Thus, the requirement of transverse containment of the cascade imposes a minimal radial extension of the calorimeter which must reach, at least, several times the average diameter of a cascade.

In practice, the dimensions of the homogeneous calorimeters are limited by the cost of crystals. Thus, some fraction of the cascades escapes through the sides and the rear of any realistic homogeneous calorimeter. This leakage of energy causes the energy resolution to depend on the dimensions of the calorimeter. The measured value of the constant in Eq. (9.36) accounts for the leakage fluctuations caused by longitudinal and lateral energy-losses. The increase of the value of this constant with the energy leakage reflects the deterioration of the energy resolution.

The readout of the homogeneous calorimeters is mostly achieved through photomultiplier tubes or photodiodes of various types collecting the produced light, and instrumental effects, such as the gain of the photosensitive device, contribute to the energy resolution. The number of photoelectrons, n_{pe}, produced by an electromagnetic cascade is relatively small and can present large statistical fluctuations. The contribution to the overall energy resolution from these statistical fluctuations is [Akrawy et al. (1990); Cushman (1992)]:

$$\left[\frac{\sigma(E)}{E}\right]_{pe} = \sqrt{\frac{(1 + 1/g)}{\bar{n}_{pe}}}, \tag{9.37}$$

where g is the gain of the photosensitive device, $\bar{n}_{pe}$ is the mean number of photoelectrons, and E is the energy of the shower in GeV units.

For phototubes with a gain of 10^6, the second term in Eq. (9.37) is negligible. However, for photodiodes the gain is small compared with phototubes. With a photodiode gain of about 1, the gain term in Eq. (9.37) will be large and leads to a degradation of the energy resolution. As an example, the following results are obtained for the OPAL end cap electromagnetic calorimeter (EMEC), made of a large array of lead glass (total of 2264 lead glass blocks) [Akrawy et al. (1990)], for triode (gain ~ 10) readout (in a magnetic field of 0.4 T):

$$\bar{n}_{pe} = 1.8E \times 10^3$$

and

$$\left[\frac{\sigma(E)}{E}\right]_{pe} = \frac{0.027}{\sqrt{E}}.$$

The importance of electronic (stochastic) noise in the amplifier readout circuit grows for low-gain photosensitive devices. The contribution of this noise to the energy resolution can be parameterized as [Akrawy et al. (1990)]:

$$\sqrt{m} \, \frac{\sigma_n}{\bar{n}_{pe}g} \times 100\%, \tag{9.38}$$

where σ_n is the noise equivalent electrons rms of each amplifier contribution [Gatti and Manfredi (1986)], and m is the total number of channels collecting the energy deposited by the electromagnetic shower. In the case of low-energy incident particles, this contribution [Eq. (9.38)] can largely affect the energy resolution even for $\sigma_n \approx$ several hundreds of electrons [Akrawy et al. (1990); Cushman (1992)].

The resolution achievable with lead-glass detectors is determined by the fluctuations of the electromagnetic shower development, by the Čerenkov light absorption in a radiator $\sigma_0(E)$, and by the photoelectron statistics ($\sim \sqrt{\bar{n}_{pe}}$). One can write [Prokoshkin (1980)]:

$$\frac{\sigma(E)}{E} = \frac{1}{E}\left[\sigma_0{}^2(E) + \frac{E}{g\xi_r}\right]^{1/2}, \tag{9.39}$$

where ξ_r is the ratio of the photocathode area to the radiator (counter) exit area. Here $\bar{n}_{pe} = g\xi_r E$ and g is about 1000 photoelectrons per GeV. At $\xi = 0.5$, the energy resolution can be parameterized as:

$$\frac{\sigma(E)}{E} = 0.0064 + 0.042\frac{1}{\sqrt{E}}, \tag{9.40}$$

where $\sigma_0(E)$ can be obtained from Eqs. (9.39, 9.40). The value of $\sigma_0(E)/E$ decreases more slowly than the law $E^{-1/2}$ when the incoming energy increases, as a consequence of large fluctuations of Čerenkov insertion of a light filter between the radiator and the photosensitive device, which absorbs the short-wave fraction of the spectrum [Prokoshkin (1980)]. The contribution from the photoelectron statistics to the resolution [Eq. (9.40)] decreases with the increase of the incoming energy.

The energy resolution of the OPAL EMEC end cap [Akrawy et al. (1990)] has an energy dependence well described by $\sim 5\%/\sqrt{E}$ at 6 GeV for a array of nine lead glass blocks with no material in front. The addition of $1.5\,X_0$ of lead in front of the lead glass $20.5\,X_0$ deep reduces the cascade leakage from the back and enlarges the lateral spread since the cascade starts earlier. The energy deposited in the lead glass is lower, since a fraction of the incoming energy is lost in the lead layer. The fluctuations of this energy loss degrades the energy resolution. However, at 50 GeV the improved containment almost exactly compensates for the increased fluctuations in energy caused by the lead layer, and the energy resolution is found to be $\sim 1.8\%$ at that energy. The energy resolution is $\sim 3.2\%$ at 6 GeV [Akrawy et al. (1990)].

Deviations from the $E^{-1/2}$ law exist for crystal calorimeters. These are attributed to energy leakage [Hughes et al. (1972)] and other instrumental effects such as the number of photons absorbed or refracted, or the shape of the spectral overlap between photocathodes and photon energy. For instance, the energy resolution of NaI(Tl) is rather found to obey a law $E^{-1/4}$. The energy resolution of the Crystal Ball detector is $\frac{\sigma(E)}{E} = (2.7 \pm 0.2)\%\, E^{-1/4}$ for $16\,X_0$ of NaI(Tl) [Oreglia et al. (1982)]. The energy resolution of CUSB detector at CESR is $\frac{\sigma(E)}{E} = 3.9\%\, E^{-1/4}$ for $8.8\,X_0$ of NaI(Tl). Tables 9.7 and 9.8 report the energy resolution found for various homogeneous calorimeters. From Table 9.7, NaI(Tl) shows an energy resolution of $\frac{\sigma(E)}{E} = 2.8\%\, E^{-1/4}$ and $\frac{\sigma(E)}{E} = 0.9\%\, E^{-1/4}$ for a calorimeter depth of $16\,X_0$ and $24\,X_0$ [Hughes et al. (1972)], respectively. The improved energy resolution in the second case is consistent with a larger longitudinal containment.

Table 9.7   Energy resolution of homogeneous calorimeters (from [Leroy and Rancoita (2000)]).

Readout	depth X_0	Energy resolution $\sigma(E)/E$	Reference	Comments
Lead glass	20.5	$5\%/\sqrt{E}$	[Akrawy et al. (1990)]	CEREN25 for lead glass
	22.0	3.2% (at 6 GeV)	[Akrawy et al. (1990)]	$1.5X_0$ of Pb in front of the lead glass array
	22.0	1.8% (at 50 GeV)	[Akrawy et al. (1990)]	$1.5X_0$ of Pb in front of the lead glass array
NaI(Tl)	8.8	$3.9\% \, E^{-1/4}$	[Klopfenstein (1983)]	
	16.0	$2.7\% \, E^{-1/4}$	[Oreglia et al. (1982)]	
	16.0	$2.8\% \, E^{-1/4}$	[Partridge (1980)]	
	24.0	$0.9\% \, E^{-1/4}$	[Hughes et al. (1972)]	
CsI(Tl)	16.2	3.8% (at 0.18 GeV)	[Blucher et al. (1986)]	
	16.2	1.6% (at 5 GeV)	[Bebek (1988)]	
	21.6	$\sim 2\%$ (4 to 20 GeV)	[Grassman et al. (1985)]	leakage fluctuations: 1.3% and 0.7% at 4 and 20 GeV
BaF$_2$	19.5	$\leq 1\%$ (4 to 40 GeV)	[Lorenz, E., Mageras, G. and Vogel, H. (1986)]	

Table 9.8 Energy resolution of other homogeneous calorimeters (from [Leroy and Rancoita (2000)]).

Readout	Depth X_0	Energy resolution $\sigma(E)/E$	Reference	Comments
BGO	21.5	0.5% ($E \geq 10$ GeV)	[Sumner (1988)]	
	21.5	2.0% ($E \sim 1$ GeV)	[Sumner (1988)]	
	21.5	5% ($E \sim 0.1$ GeV)	[Sumner (1988)]	
	22.3	$\leq 1\%$ ($E > 4$ GeV)	[Kampert et al. (1994)]	
SCG1-C	20.5	$1.46\%/\sqrt{E} + 1.6\%$	[Wagoner et al. (1985)]	
	20.5	$3.43\%/\sqrt{E} + 0.5\%$	[Wagoner et al. (1985)]	$3.5X_0$ converter
				$0.2X_0$ hodoscope in front
CeF$_3$	25.0	0.5% ($E \geq 50$ GeV)	[Auffray et al. (1996)]	
PbWO$_4$	24.0	$3.5\%/\sqrt{E} + 0.35\%$ ($10 \leq E \leq 150$ GeV)	[Peigneux et al. (1996)]	photomultiplier readout
	24.0	$6\%/\sqrt{E} + 0.5\%$ ($10 \leq E \leq 150$ GeV)	[Peigneux et al. (1996)]	avalanche photodiode readout

CsI(Tl) is obtained by doping CsI with thalium at a level varying from 150 to 2000 ppm. It has a rather short radiation length (1.85 cm) allowing compact calorimeters and a peak emission of 550 nm (Table 5.2), which, in turn, permits a readout by silicon photodiodes [Blucher et al. (1986); Bebek (1988); Fukushim (1992)]. CsI(Tl) was used in homogeneous electromagnetic calorimeters. An example is provided by CLEO [Blucher et al. (1986); Bebek (1988)] where the electromagnetic cascade detector consists of 8000 CsI(Tl) crystals readout with Si photodiodes. A prototype of this detector made of an array of 465 crystals providing a depth of $16.2\,X_0$ was tested in a 180 MeV positron beam giving a measured energy resolution of 3.8%. This array was later installed in CLEO and the energy spectrum of electrons from Bhabha events at 5 GeV were measured with an energy resolution of 1.6%. Earlier data with $21.6\,X_0$ of CsI(Tl) [Grassman et al. (1985)] gave an energy resolution of about 2% for incoming electron energy ranging between 4 and 20 GeV. Leakage fluctuations contributed 1.3% and 0.7% at 4 GeV and 20 GeV, respectively.

From the point of view of calorimetry applications, BaF_2 suffers from several handicaps, namely a long radiation length (2.05 cm, Table 5.2), a large Molière radius (3.4 cm, Table 5.2) and a relatively small ratio of hadronic interaction length (see Sect. 3.3) to radiation length, hampering electron–hadron separation [Majewski and Zorn (1992)]. However, BaF_2 has a high density $(4.88\,g/cm^3)$, which favors detector compactness, a very short decay time of its fast scintillation component $\tau_{d,1} \sim 0.6$ ns (Table 5.2), which favors applications where light collection has to be achieved fast, and excellent radiation hardness. These features had motivated the choice of this crystal for SSC-GEM (see references in [Majewski and Zorn (1992)]). An energy resolution of $\leq 1\%$ for incident electron energies between 4 and 40 GeV is obtained [Lorenz, E., Mageras, G. and Vogel, H. (1986)] (corrected for beam momentum spread and leakage) by using a readout scheme involving fluorescent flux concentrators and silicon photodiodes.

BGO gives 7% of the light output of NaI(Tl) at room temperature. The light output varies with temperature with a decrease of 1% per degree at 20 °C [Suffert (1988)] and therefore can be improved by cooling of the crystal. The relatively high output and green spectral response allow the replacement of photomultipliers by photodiodes for the readout due to the availability of photodiodes of large area and low noise, nowadays [Sumner (1988); Kampert et al. (1994)]. Energy resolution of $\frac{\sigma(E)}{E} \sim 0.5\%$, $\leq 2.0\%$ and $\sim 5\%$ are obtained for incoming energy $E \geq 10$ GeV, ~ 1 GeV and at 0.1 GeV for the L3 electromagnetic calorimeter consisting of 10734 BGO crystals corresponding to a total depth of $\sim 21.5 X_0$ [Sumner (1988)]. These results are compatible with the energy resolution $\frac{\sigma(E)}{E}$ better than 1% measured for incident energies above 4 GeV for a total length of $22.3\,X_0$ [Kampert et al. (1994)].

The scintillating glass (SCG1-C) is a crystal permitting the achievement of an energy resolution comparable to that of NaI(Tl) and BGO and produced at much lower costs. The SCG1-C is composed of BaO (43.4%), SiO_2 (42.5%) and other high-Z materials LiO_2 (4.0%), MgO (3.3%), K_2O (3.3%), Al_2O_3 (2.0%), Ce_2O_3 (1.5%) with a density of $3.36\,g/cm^3$ and a radiation length of $4.35\,cm$. The Ce_2O_3 acts as a scintillating component as well as wavelength shifter for short wavelengths, i.e., shifting the UV photons into the sensitive blue region of the photocathode. The light produced by SCG1-C has two components with a fast Čerenkov component which represents about 15% of the total light output.

The energy resolution (and the position resolution, as will be discussed in Sect. 9.6) is often represented by an expression, where the account for different effects (like the readout electronic noise, momentum spread, the amount of material in front of the calorimeter being tested) modifies Eq. (9.36) into (see also [Engler (1985)])

$$\frac{\sigma(E)}{E} = \frac{1}{\sqrt{E}}\left(a_i + \frac{b_i}{\sqrt{E}} + c_i\sqrt{E}\right), \qquad (9.41)$$

where a_i is the intrinsic resolution term, b_i the noise contribution and c_i a constant term (E is expressed in GeV).

For SCG1-C, the energy resolution is measured to be

$$\frac{\sigma(E)}{E} = \frac{(1.46 \pm 0.1)\%}{\sqrt{E}} + (1.63 \pm 0.05)\%$$

for $20.5\,X_0$ of scintillating glass for positrons in the energy range between 1 and $25\,GeV$ [Wagoner et al. (1985)]. When adding an active converter $3.5\,X_0$ deep, and a $0.2\,X_0$ shower position hodoscope, the constant term in the energy resolution was reduced and the resolution found to be [Wagoner et al. (1985)]:

$$\frac{\sigma(E)}{E} = \frac{(3.43 \pm 0.18)\%}{\sqrt{E}} + (0.50 \pm 0.08)\%.$$

The energy resolution with the active converter was better since a better cascade containment was then achieved and also allowed to correct for conversion point fluctuations within the glass. These fluctuations were degrading the energy resolution because of the differential absorption of light by glass. The photon yield was measured [Wagoner et al. (1985)] to be $1.6 \times 10^4\,\gamma/GeV$.

The "Crystal Clear" Collaboration [Lecoq et al. (2001)] has conducted studies in order to produce CeF_3 crystal with optimal properties with respect to their possible utilization in experiments at LHC. This collaboration has tested a CeF_3 matrix of 9 crystals readout via silicon photodiodes, $25\,X_0$ deep, with a lateral segmentation of $3 \times 3\,cm^2$ and $2 \times 2\,cm^2$ (at the 4 corners) exposed at CERN-SPS to electron, muon and pion beams of momenta ranging from 10 to $150\,GeV/c$. An energy resolution of 0.5% for energies above $50\,GeV$ was achieved [Auffray et al. (1996)].

Lead tungstate $PbWO_4$ crystals were finally selected as active material of electromagnetic calorimeters for CMS [CMS (1994)] and ALICE [ALICE (1993)] at LHC. $PbWO_4$ is grown from a 50%-50% mixture of lead oxide (PbO) and tungsten oxide (WO_3). Its high refractive index (Table 5.2) helps the light propagation along the crystal but limits the efficiency of the light extraction. The decay time constant has a mean value of 10 ns and about 85% of the light is collected in 25 ns. The emission spectrum has two broad bands at 440 nm and 530 nm. The light yield is about $150 \, \gamma/\text{MeV}$ for small samples and about between 50 and $80 \, \gamma/\text{MeV}$ for 20 cm long crystals. The light output varies with temperature with a decrease of 1.98% per degree at 20 °C [CMS (1994)] and therefore can be improved by cooling of the crystal. For instance, the light yield at $-20\,°\text{C}$ is 2.3 times larger than at $+20\,°\text{C}$. Beam tests of electromagnetic calorimeters made of $PbWO_4$ crystals were performed at several energies and with various types of photosensitive devices. Namely, a comparative study between photomultiplier and Si avalanche photodiodes was performed [Peigneux et al. (1996)]. The energy resolution measured using photomultipliers (without unfolding the beam momentum spread) is

$$\frac{\sigma(E)}{E} = \frac{3.5\%}{\sqrt{E(\text{GeV})}} + 0.35\%$$

for a depth of $24 \, X_0$ and for incoming electron energies ranging from 35 to 150 GeV. The energy resolution was found to be

$$\frac{\sigma(E)}{E} = \frac{6.0\%}{\sqrt{E(\text{GeV})}} + 0.5\%$$

(electrons of energy ranging from 20 up to 150 GeV) when the measurements were carried out with avalanche photodiodes, showing the necessity of further development for crystal and avalanche photodiodes. The effect of putting a preshower detector with 16 mm of lead (or $2.9 \, X_0$) in front of the crystals was tested. The preshower detector increases the effective length of the calorimeter, with the consequence that the flux of particle leaking out of the calorimeter is decreased and the energy resolution is improved [Peigneux et al. (1996)].

9.6 Position Measurement

The impact point of a γ-ray or an electron can be measured by exploiting the longitudinal segmentation and transverse granularity of the calorimeter, which define cells. If the granularity of the crystals is chosen to be equal or smaller than the Molière radius, the lateral spread of the electromagnetic cascade over several crystal modules allows the reconstruction of the impact point of the incident particle. Then, the particle impact coordinates are defined by measuring the energy deposit of the cascade in transverse directions. The position of the impact point is usually measured by using the center of gravity of the energies, E_i, deposited in the modules:

$$\bar{x} = \frac{\sum_i x_i E_i}{\sum_i E_i}, \tag{9.42}$$

where x_i are the coordinates of each module hit with respect to the central one, and E_i is the energy deposited in the cell i. However, it is known that such methods calculates an impact point systematically shifted towards the center of the module hit by the particle. There exists several ways to correct this effect; see [Anzivino et al. (1993)], for instance.

In order to locate the position of a shower within the calorimeter, it is necessary for the energy to be shared between a number of calorimeter cells. The precision of this measurement increases with the number of calorimeter cells hit by the cascade particles and decreases with the crystal cell size. However, the intrinsic resolution of such a measurement is small, since the spatial distribution of the incoming showers is rather limited. This is a consequence of the dependence on the incoming energy of the processes contributing to the cascade transverse development: *bremsstrahlung* and *pair creation* create electrons and positrons at angle

$$\theta_{b,pc} \sim \frac{m_e c^2 \ln \left[E / \left(m_e c^2 \right) \right]}{E},$$

where $E(\text{GeV})$ is the particle energy (e or γ). Angular spread of the cascade is also generated by multiple scattering which, for relativistic particles traversing a thickness x, results in an angular divergence

$$\theta_{ms} \sim 0.02 \frac{\sqrt{x/X_0}}{E}.$$

As E increases, more higher energy electrons, positrons and photons are generated in the cascade. As a consequence of their $1/E$ dependence, $\theta_{b,pc}$ and θ_{ms} decrease with the increase of the energy, and the particles in the cascade do not spread out forming the central core of the cascade. The electrons and positrons, with energies close to the critical energy, are responsible for the dominant part of the energy loss and do not modify the pattern of energy loss, since the critical energy is a characteristic of the material and not linked to the incoming energy [Kondo and Niwa (1984)].

For homogeneous calorimeters, a tower structure with a lateral side comparable to the width of incoming electromagnetic cascades can be used to measure the position of a photon with a spatial resolution much smaller than the cell size. This structure favors two-photon separation and a spatial resolution improving close to the cell edges [Amendolia et al. (1980)]. In the end, the choice of the cell dimensions is a compromise between good position resolution, cascade containment in a tower consisting of a moderate number of lateral cells and the total number of readout channels. Good position resolution as well as a good knowledge of the transverse cascade shape, which is important (as it will be shown below) to achieve good electron–hadron separation, favor a small cell size while cascade containment in a few cells favors large cell size. A cell dimension in the neighborhood of about one Molière radius is usually taken, corresponding to about 75% of the cascade energy deposited in the center cell.

In practice, the detector matrix is made of a large number of total absorption counters each with a lateral size comparable with the electromagnetic cascade width. These absorption counters are made of scintillating crystal (Table 5.2) or lead glass (Table 9.6). In the case of lead glass (SF-5) used in [Prokoshkin (1980)], the dependence of the coordinate accuracy σ_y on the cell size d at the incoming energy of 25 GeV is parametrized as

$$\sigma_y(d) = \sigma_y(d = 0) \, e^{d/d_0}. \tag{9.43}$$

For $d \leq 3\,\mathrm{cm}$, σ_y is constant while for $d > 3\,\mathrm{cm}$, $\sigma_y(d = 0) = 0.8\,\mathrm{mm}$ and $d_0 = 6.5\,\mathrm{cm}$ [Prokoshkin (1980)] (the size of a lead glass cell in [Prokoshkin (1980)] was $4.5 \times 4.5 \times 65\,\mathrm{cm}^3$).

As discussed above, when the incoming energy increases, the lateral size of the electromagnetic cascade remains almost unchanged. Since the number of cascade particles increases with the incoming energy, the coordinates accuracy will scale with energy as

$$\sigma_y(E) = \frac{\mathrm{constant}}{\sqrt{E}}, \tag{9.44}$$

if the lateral correlations of the number of particles in the cascade is negligible. The scaling law, given by Eq. (9.44), was verified in the energy range between 2 and 40 GeV [Akapdjanov et al. (1977)]. The two-photon separation is $\sim 5\,\mathrm{cm}$.

The position resolution as a function of the energy for a calorimeter organized in pointing cascade can also be parameterized as [Zhu (1993)]:

$$\sigma_x(E) = \frac{3\,e^{0.4D}}{\sqrt{E}} \tag{9.45}$$

where D is the cell size in radiation lengths. Again, this illustrates the importance of the cell dimension.

The improvement of the spatial resolution with increasing energy is also observed in the EMEC-OPAL calorimeter large lead glass array (see above) [Akrawy et al. (1990)]. For electrons fired centrally into a lead glass block at normal incidence, the position resolution is (in mm) 9.9, 6.3, 4.4, 3.1 and 2.5 mm at 5, 10, 20, 35, and 50 GeV, respectively. The multiple scattering contributions which amount to $\sim 25\,\mathrm{mm}/E$ has not been unfolded from these numbers [Akrawy et al. (1990)].

BGO data on position measurement are also found to follow the scaling law as given in Eq. (9.44). The measured position resolution can be parameterized as

$$\sigma_x(E) \sim \frac{3.3}{\sqrt{E(\mathrm{GeV})}} \,\mathrm{mm}$$

for electron momenta ranging from between 0.5 and 6 GeV/c [Kampert et al. (1994)]. This result is not in disagreement with the measured spatial resolution found better than 2 mm above 2 GeV for the BGO L3 electromagnetic calorimeter [Sumner (1988)]. The position resolution achieved with the CsI(Tl) CLEO II detector at CESR is $\sim 0.3\,\mathrm{cm}$ and 1.2 cm at 5 GeV and 180 MeV, respectively.

The position resolution achieved with the matrix of CeF$_3$ crystals of [Auffray et al. (1996)] is $\sigma_x = 0.67\,\mathrm{mm}$ and $\sigma_y = 0.7\,\mathrm{mm}$ for 50 GeV electrons. Then, the angular resolution is 15 mrad for the 50 GeV electrons corresponding to

$$\sigma_\theta \sim \frac{100}{\sqrt{E}} \,\mathrm{mrad}$$

(E in GeV) [Auffray et al. (1996)].

The small Molière radius of PbWO$_4$ results in narrow cascades. However the energy deposited in the cells around the impact point is large enough to enable the measurement of impact coordinates. The combination of preshower and crystal matrix may achieve relatively good position and angular resolutions [Peigneux et al. (1996)]. The position resolution on the center of gravity x_{cg} as a function of the incoming energy, is

$$\sigma(x_{cg}) = \left[\frac{(2.02 \pm 0.48)}{\sqrt{E}} + (0.29 \pm 0.06) \right] \text{ mm},$$

E in GeV for the PbWO$_4$ calorimeter of [Peigneux et al. (1996)]. The transverse spread of the cascade after $\sim 3\,X_0$ of lead is a few mm. This gives a spatial resolution of the preshower detector

$$\sigma(x_{cg}) = \left[\frac{(1.58 \pm 0.46)}{\sqrt{E}} + (0.36 \pm 0.06) \right] \text{ mm},$$

with a stochastic term smaller than the one found for the crystals but with a larger constant term, possibly because of noise and cross talk. The angular resolution of the preshower-crystal system is found to be

$$\sigma_\theta = \left[\frac{(36.5 \pm 6.5)}{\sqrt{E}} + (4.1 \pm 0.8) \right] \text{ mrad}.$$

The position resolution is degraded for sampling calorimeters. The shower spreads transversely with an exponential fall-off that becomes flatter for larger longitudinal depths. This lateral decrease projected onto a plane is given as [Engler (1985)]

$$E_x \approx \exp\left(-\frac{4x}{R_M} \right), \tag{9.46}$$

where x is the coordinate orthogonal to the shower axis. R_M is smaller for the absorber compared to the active medium (see Table 2.3) and therefore the cascade in the absorber is wider transversely than in the active medium. As a consequence, interspersing active medium planes with absorber layers leads to a larger overlap of contiguous showers. The position resolution, found for a fully projective lead and scintillating fiber calorimeter, is [Anzivino et al. (1993)]

$$\sigma_x = \frac{(6.21 \pm 0.17)}{\sqrt{E}} + (0.32 \pm 0.03) \text{ [mm]}. \tag{9.47}$$

The space resolution of the large-scale prototype of the ATLAS Pb/LAr accordion electromagnetic calorimeter was determined by comparing the shower center-of-gravity, reconstructed in the calorimeter, with the extrapolated impact point provided by a beam chamber system (three multiwire proportional chambers used to extrapolate the particle trajectory to the calorimeter front face). By fitting the data to the expected $1/\sqrt{E}$ law, the space resolution was found to be (the incoming energy E is in GeV) [Gingrich et al. (1995)]:

$$\sigma_\phi = (0.186 \pm 0.021) + \frac{3.87 \pm 0.05}{\sqrt{E}} \ [\text{mm}] \tag{9.48}$$

in the direction perpendicular to the accordion waves, and

$$\sigma_\theta = (0.210 \pm 0.015) + \frac{4.70 \pm 0.05}{\sqrt{E}} \ [\text{mm}] \tag{9.49}$$

in the other direction.

9.7 Electron Hadron Separation

In calorimetry, the electron hadron separation is based on the difference in the cascade profiles (e.g., see [Baumgart et al. (1988b)]). This difference is enhanced in materials with very different radiation and interaction lengths. Their ratio, using the approximated expressions given in Eqs. (2.226, 3.77), and for $Z/A \approx 0.5$ becomes:

$$\frac{\lambda_{gA}(\text{g/cm}^2)}{X_{g0}(\text{g/cm}^2)} \sim 0.12 \times Z^{4/3}. \tag{9.50}$$

Therefore, an electromagnetic calorimeter which has a depth of about one interaction length already provides by itself a discrimination between hadrons and electrons or photons due to the low probability for hadronic interaction in this calorimeter volume (e.g., see [Heijne et al. (1983)]).

A modular calorimeter can be used to detect electrons and photons against strongly interacting particles. The discrimination capability between electrons, photons and hadrons of the calorimeter is accomplished through the analysis of the lateral dispersion and the longitudinal distribution of the cascade. The hadronic cascades present larger lateral dispersions than electromagnetic cascades. The dispersion depends on the point of incidence relative to the module boundaries. A cut on the lateral dispersion allows the rejection of a large fraction of hadrons at the price of a rather modest loss in electrons.

Obviously, the lateral dispersion analysis becomes inefficient at low energy since then a lower number of calorimeter modules is involved in the cascade. Therefore, the cut on lateral dispersion has to be combined with a cut on energy deposition in the calorimeter modules. In general, the energy deposited by hadronic cascades will be larger in the rear part compared to the front part of the calorimeter. The electromagnetic cascades show the opposite behavior. The measurement of the ratio E_{front}/E of the energy deposited in the front part defined in the calorimeter E_{front} to the total energy E allows the analysis of the longitudinal dispersion of the cascade. A rectangular cut is then performed in the scatter plot of the particle lateral dispersion versus E_{front}/E.

It turns out that the longitudinal dispersion analysis proves to be more efficient at low energy, where the lateral dispersion analysis is less efficient [Kampert et al.

(1994); Auffray et al. (1996)]. The procedure combining lateral and longitudinal information allows to achieve a good electron–pion separation with efficiency in pion rejection between 90 and 99%, so giving a level of $\approx 10^{-3}$ for the pion contamination for BGO ($22.3 X_0$ and $0.5 \leq p \leq 6\,\mathrm{GeV/c}$, where p is the particle momentum) [Kampert et al. (1994)], CeF_3 ($25 X_0$ and $10 \leq p \leq 150\,\mathrm{GeV/c}$) [Auffray et al. (1996)], $PbWO_4$ ($24 X_0$ and $10 \leq p \leq 150\,\mathrm{GeV/c}$) [Peigneux et al. (1996)]. The dispersion analysis can be generalized by using a method using the momenta of the cascade. This method can be applied, when the incident particles hit the module with its momentum non-parallel to the longitudinal module axis, a case in which the dispersion is not well defined [Gomez, Velasco and Maestro (1987)].

The separation between electrons, photons and hadrons is achieved in sampling calorimeters by installing an electromagnetic calorimeter in front followed by an hadron calorimeter. This takes advantage of the difference in the lateral and longitudinal shower profile of hadrons and electrons, as discussed above. In practice, this is accomplished by dividing the sampling calorimeter modules into an electromagnetic section several radiation lengths deep and a hadronic section with a depth of several interaction lengths.

The calorimeter modules can be read out via wavelength shifting (WLS) plates or fibers. The fiber of each module or tower is split into two parts: an electromagnetic fiber read through the front of the calorimeter and an hadronic fiber read through the back. Using this method, one may expect more than 99% of pions rejected for more than 99% electron efficiency [Bagdasarovand and Goulianos (1993)]. However, in many cases electrons and pions are efficiently separated when using the tagging information provided by devices such as Čerenkov differential counter put ahead of the calorimeter modules [HELIOS Collab. (1987)]. In the case of calorimeters with scintillating fiber layers as active medium, the calorimeter towers are made of two longitudinal sections defined by short and long scintillating fibers in lead achieving a separation between electrons and charged pions. Pion rejection factors of the order of 100 were obtained by using this longitudinal segmentation within one lead block with short and long fibers [Dagoret-Campagne (1993)].

9.8 Hadronic Calorimetry

9.8.1 *Intrinsic Properties of the Hadronic Calorimeter*

Hadronic calorimeters are mostly employed in High Energy Physics experiments, where hadrons can have energies above tens or hundreds GeV. We will limit ourselves by considering only *primary hadrons with energies above a few* GeV's. At lower energies, the energy loss by ionization alone becomes increasingly important and the calorimeter performance may change considerably. Furthermore to compare experimental (intrinsic) properties of calorimeters, the *hadron reference energy* is usually taken at 10 GeV.

In dealing with intrinsic hadronic properties, we will consider calorimeter in which cascades are fully contained. In fact, we have seen in Sect. 9.4.2 how longitudinal losses of electromagnetic visible-energy, amounting to a few percent, can affect the calorimeter energy resolution. Later (Sect. 9.10.3), we will discuss similar effects for hadronic calorimeters. From Eq. (3.78), it can be seen that about between 4.5 and 5.6 λ_A are needed in order to contain longitudinally 95% of hadronic cascades for incoming hadrons with energy between 10 and 50 GeV. For instance, the corresponding absorber thicknesses (see values of λ_A in Table 3.4) are about between 77 and 95 cm for iron or lead, and about between 48 and 60 cm for uranium, respectively.

The large absorber depth needed is a main reason why homogeneous calorimeters can hardly be employed in hadronic calorimetry at high energies. Homogeneous (or quasi-homogeneous calorimeters, based on liquid argon [Cerri et al. (1989); Fabjan (1995b)]) were developed (e.g., see [Hughes et al. (1969); Benvenuti et al. (1975)]). However, the hadronic energy resolution of homogeneous calorimeters does not improve below $\approx 10\%$, even at energies as high as 140 GeV [Benvenuti et al. (1975)]. As it will be discussed in following sections, only in sampling calorimeters the energy resolution can be improved up to values, where instrumental effects start to dominate.

Hadronic calorimeters are usually sampling devices made of (as for electromagnetic calorimeters, see Sect. 9.2) layers of passive samplers interleaved with active readout planes.

The incoming particle energy is measured by means of the energy ϵ_{vis} deposited in the active layers with a thickness, usually, much smaller than the thickness of the corresponding passive samplers.

Hadronic cascades are mostly generated in passive absorbers. Similarly to the case of electromagnetic calorimeters, the sampling frequency, τ, is defined as the passive sampler absorber thickness between two successive active planes, expressed in units of interaction length (λ_A). Only sampling calorimeters with constant τ along the calorimeter depth will be considered in this book, unless explicitly stated otherwise.

9.8.1.1   *The e/h, e/π, h/mip and π/mip Ratios*

While the energy (ϵ_{vis}) deposited in active layers mainly comes from processes of energy loss by collisions, interactions leading to the generation of charged secondary particles in electromagnetic and hadronic cascades (see Sects 2.4 and 3.3, respectively) are different. In particular, for hadronic showers, there is a large A-dependent fraction of the incoming hadron energy spent in releasing nucleons and nucleon aggregates (like α-particles), that are bound in atomic nuclei. As discussed by many authors (see [Fabjan et al. (1977); Fabjan (1986); HELIOS Collab. (1987); Wigmans (1987); Brückmann et al. (1988); Fesefelt (1988); Wigmans (1988); Brau and Gabriel (1989); SICAPO Collab. (1989b,c, 1991b); Wigmans (1991, 2000); Byon-

Wagner (1992); Beretvas et al. (1993); SICAPO Collab. (1993b); Gabriel et al. (1994); SICAPO Collab. (1995a,b, 1996); Groom (2007)] and references therein), this process has no equivalent in electromagnetic cascades and largely contributes to the so-called *hadronic invisible-energy*. Therefore, at the same incoming energy, we expect that the signal for electrons (or photons), e, differs (being usually larger) from the signal for *ideal hadrons*, h.

These *ideal hadrons* are hadrons not accompanied by electromagnetic cascading processes, i.e., for which $f_{em} \approx 0$, where f_{em} is the average fraction (see page 276) of incoming hadron energy deposited by electromagnetic cascades of secondary particles and has been introduced in Sect. 3.3.1. We have:

$$e/h \equiv \frac{\epsilon_{\text{vis}}(e)}{\epsilon_{\text{vis}}(h)}, \tag{9.51}$$

where $\epsilon_{\text{vis}}(e)$ and $\epsilon_{\text{vis}}(h)$ are the visible energies, at the same incoming particle energy, for electromagnetic and ideal-hadron showers, respectively. The e/h ratio is usually *larger than* 1.

The process of nuclear interactions leading to hadronic shower multiplication quickly degrades the incoming hadron energy. Thus, weakly energy dependent nuclear processes (see Sect. 3.3) occur (on average) at incoming secondary hadron energies much lower than the incoming primary hadron energy. To a first approximation, we expect that, for an ideal hadron, the fraction of invisible energy is independent of the incoming energy and is A-dependent. As in the case of electromagnetic showers for which $\epsilon_{\text{vis}}(e)$ *is proportional to* E_e [where E_e is the incoming electron, or photon energy, see Eq. (9.1)], we have that

$$\epsilon_{\text{vis}}(h) \propto E_h, \tag{9.52}$$

where E_h is the incoming energy of an ideal hadron. Thus, the e/h ratio can be considered *as an intrinsic energy independent property of the calorimeter* [HELIOS Collab. (1987); Wigmans (1987); Brückmann et al. (1988); Fesefelt (1988); Wigmans (1988); Brau and Gabriel (1989); SICAPO Collab. (1989c); Acosta et al. (1991); SICAPO Collab. (1991b); Wigmans (1991); Gabriel et al. (1994); SICAPO Collab. (1995a,b, 1996)]. We expect that the e/h ratio may depend on the *type* and the *thickness* of both the *passive samplers* and *active layers*.

The hadronic cascade of *real hadrons* is made of a pure electromagnetic component and a pure hadronic component; the respective fractions of the total incoming energy are f_{em} and

$$f_h = 1 - f_{em}, \tag{9.53}$$

respectively.

The visible energy $\epsilon_{\text{vis}}(\pi)$, corresponding to a real hadron signal (π), can be written in terms of pure electromagnetic and ideal hadronic signals (at the same energy):

$$\epsilon_{\text{vis}}(\pi) = f_{em}\epsilon_{\text{vis}}(e) + (1 - f_{em})\,\epsilon_{\text{vis}}(h). \tag{9.54}$$

In Eq. (9.54), both $\epsilon_{\text{vis}}(e)$ and $\epsilon_{\text{vis}}(h)$ are expected to be proportional to the incoming particle energy. However, the term f_{em} has an approximate *logarithmic energy dependence* [see Eq. (3.74)]. Then, in general, we expect that $\epsilon_{\text{vis}}(\pi)$ *is not linearly dependent on the incoming hadron energy*, contrary to what occurs for the visible energy of electromagnetic showers, as experimentally observed ([Abramowicz et al. (1981); HELIOS Collab. (1987); Wigmans (1988); Acosta et al. (1991); SICAPO Collab. (1996)] and references therein).

From Eqs. (9.51, 9.54), the e/π signal ratio is:

$$
\begin{aligned}
e/\pi &\equiv \frac{\epsilon_{\text{vis}}(e)}{\epsilon_{\text{vis}}(\pi)} \\
&= \frac{\epsilon_{\text{vis}}(e)}{f_{em}\epsilon_{\text{vis}}(e) + (1 - f_{em})\,\epsilon_{\text{vis}}(h)} \\
&= \frac{[\epsilon_{\text{vis}}(e)/\epsilon_{\text{vis}}(h)]}{f_{em}\,[\epsilon_{\text{vis}}(e)/\epsilon_{\text{vis}}(h)] + (1 - f_{em})} \\
&= \frac{e/h}{1 - f_{em}(1 - e/h)}.
\end{aligned}
\tag{9.55}
$$

The e/π ratio contains the energy dependent term f_{em} and, therefore, depends on *the energy of the incoming hadron*. In addition, it depends on the *atomic weight of the absorber* and on *the Z-values of the passive samplers and active layers*. This latter Z-dependence is mostly caused by the electromagnetic shower component.

From Eq. (9.55), we can obtain the e/h ratio as a function of the e/π ratio, i.e.,

$$
\begin{aligned}
(e/\pi)\,[1 - f_{em}(1 - e/h)] &= e/h \\
\Rightarrow (e/\pi)\,(1 - f_{em}) &= e/h\,[1 - f_{em}(e/\pi)]\,,
\end{aligned}
$$

and, finally, we have:

$$
e/h = \frac{(e/\pi)(1 - f_{em})}{1 - f_{em}(e/\pi)}.
\tag{9.56}
$$

The e/h ratio of a calorimeter can be evaluated experimentally by measuring the e/π ratio at the same (and known) incoming energy for electrons and hadrons. For instance, the e/h ratio of a lead / scintillating-fiber calorimeter was determined by a set of experimental data taken from 5 up to $150\,\text{GeV}$ [Acosta et al. (1990, 1991)]. The measured ratios of calorimeter responses to electrons and hadrons are shown in Fig. 9.18 as a function of the incoming particle energy. The two curves overimposed represent the energy dependent e/π ratios obtained by using Eq. (9.55), for $f_{em} = 0.11\ln[E(\text{GeV})]$ [as given by Eq. (3.74)] and $e/h = 1.15$, and for $f_{em} = [1 - E(\text{GeV})^{-0.15}]$ {e.g., see Eq. (3.75) and [Groom (1990)]} and $e/h = 1.16$.

In a hadronic sampling calorimeter, the visible energy is the result of collision losses in the active readout layers by charged particles produced during the multiplication process. As was the case for electromagnetic calorimeters [see Sect. 9.2.2

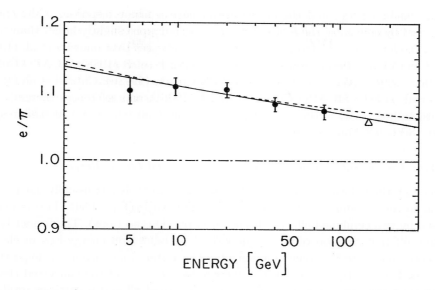

Fig. 9.18 The e/π ratio (adapted from *Nucl. Instr. and Meth. in Phys. Res. A* **308**, Acosta, D. et al., Electron, pion and multiparticle detection with a lead/scintillating-fiber calorimeter, 481–508, Copyright (1991), with permission from Elsevier, e.g., for the list of the authors see [Acosta et al. (1991)]; see also [Leroy and Rancoita (2000)]) measured in a Lead/Scintillating-fiber calorimeter. The continuous line corresponds to $e/h = 1.15$ and f_{em} is calculated by means of Eq. (3.74). The dashed line corresponds to $e/h = 1.16$ and $f_{em} = [1 - E(\text{GeV})^{-0.15}]$ {e.g., see Eq. (3.75) and [Groom (1990)]}.

and Eq. (9.6)], the hadronic calorimeter response to minimum-ionizing and non-showering particles[||] (*mip*'s), like muons, can be used as a scale or, equivalently, as a unit of measurement of the response to hadronic showers.

The h/mip and π/mip ratios, for ideal and real hadrons, are defined by

$$h/mip \equiv \frac{\epsilon_{\text{vis}}(h)}{E_s} \quad \text{and} \quad \pi/mip \equiv \frac{\epsilon_{\text{vis}}(\pi)}{E_s}, \qquad (9.57)$$

where E_s is the energy shared in the active readout planes following the partition of a *mip* particle, given by Eq. (9.4). The value of $\epsilon_{\text{vis}}(h)$ and E_s are expected to be proportional to the incoming (ideal) hadron energy. As a consequence, the h/mip *ratio is an intrinsic energy independent property of the calorimeter*, as is the e/h ratio. The π/mip ratio depends on the hadron incoming energy [see Eq. (9.54)], owing to the energy dependent term f_{em}.

From Table 3.3, we see that the fraction of incoming energy lost through collisions, mostly due to spallation protons, varies from $\approx 40\%$ (high-A nuclei) to $\approx 60\%$ (low-A nuclei). In calorimeters, where neutrons contribute in a minimal way to the

---

[||]One can see, for instance, [HELIOS Collab. (1987); Wigmans (1987); Brückmann et al. (1988); Wigmans (1988); SICAPO Collab. (1989b,c, 1990b, 1991b); Wigmans (1991); SICAPO Collab. (1992a, 1993b); Gabriel et al. (1994); SICAPO Collab. (1995a,b, 1996)]

overall visible-energy and the active media have a linear response to the energy deposited by collisions, the h/mip ratio is expected to be slightly larger than ≈ 0.4 and ≈ 0.6 for high-A and low-A nuclei, respectively (see [Abramowicz et al. (1981); HELIOS Collab. (1987); Brückmann et al. (1988); Fesefelt (1988); SICAPO Collab. (1989b,c, 1991b); Wigmans (1991); SICAPO Collab. (1993b); Gabriel et al. (1994); Job et al. (1994); SICAPO Collab. (1995a,b, 1996)], and references therein). The corresponding values of the e/mip ratios are about 0.50–0.65 and 0.70–0.85, respectively (Sect. 9.2.2 and Table 9.1).

9.8.1.2 *Compensating Condition $e/h = e/\pi = 1$ and Linear Response*

In general, the calorimeter response to hadrons $[\epsilon_{\mathrm{vis}}(\pi)]$ is intrinsically not proportional to the incoming particle energy, as expressed by Eq. (9.54). This is mainly caused by two combined effects (discussed in previous sections). The former is due to the different fraction of energy deposited by ionization energy-loss in electromagnetic and pure hadronic components. The latter comes from the logarithmic [e.g., see Eq. (3.74)] increase of the average fraction f_{em} of the converted electromagnetic energy resulting from the photon decays of π^0 and η particles produced during the hadronic cascade (Sect. 3.3.1).

From Eqs. (9.6, 9.54, 9.57), the overall visible-energy (electromagnetic and purely hadronic) deposited by hadrons in the calorimeter can be written as:

$$
\begin{aligned}
\epsilon_{\mathrm{vis}}(\pi) &= (\pi/mip)E_s \\
&= E_s\left[(e/mip)f_{em} + (h/mip)(1 - f_{em})\right] \\
&= EF(S)\left\{(h/mip) + \left[(e/mip) - (h/mip)\right]f_{em}\right\} \\
&= EF(S)(h/mip)\left\{1 + \left[(e/h) - 1\right]f_{em}\right\}, \qquad (9.58)
\end{aligned}
$$

where (for mip's) the energy shared E_s and the fraction of energy lost $F(S)$ are given by Eq. (9.4); E is the incoming particle (hadron) energy. Equivalently, once Eq. (9.7) is taken into account, Eq. (9.58) can be rewritten as:

$$
\begin{aligned}
\epsilon_{\mathrm{vis}}(\pi) &= EF(S)(h/mip)\left\{1 + \left[(e/h) - 1\right]f_{em}\right\}\frac{e/mip}{e/mip} \\
&= EF(S)(e/mip)\frac{h}{e}\left\{1 + \left[(e/h) - 1\right]f_{em}\right\} \\
&= \epsilon_{\mathrm{vis}}(e)\left\{(h/e) + \left[1 - (h/e)\right]f_{em}\right\} \\
&= \epsilon_{\mathrm{vis}}(e)\left\{1 - f_h\left[1 - (h/e)\right]\right\}, \qquad (9.59)
\end{aligned}
$$

where f_h [Eq. (9.53)] is the fraction of energy deposited by the purely hadronic shower component and $\epsilon_{\mathrm{vis}}(e)$ is the visible energy for an electron/photon of the same energy of the hadron entering in the calorimeter.

When no absolute conversion energy scale is determined for measuring the deposited energy $[\epsilon_{\mathrm{vis}}(\pi)]$, the calorimeter response to hadrons can be calibrated by using electron beams of known energies [Groom (1992); ATLAS Collab. (1997)]. Thus, the calorimeter (linear) response to electrons is taken as equal to

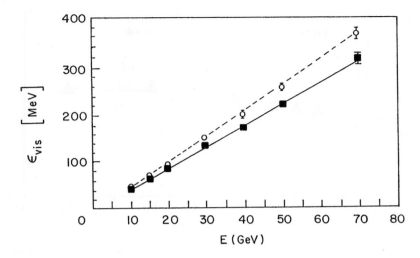

Fig. 9.19 Measured visible energies $\epsilon_{\rm vis}$ for the Si/Fe (○) and the almost compensating Si/Pb+Fe (dark squares) calorimeters (reprinted from *Nucl. Instr. and Meth. in Phys. Res. A* **368**, Furetta, C., Gambirasio, A., Lamarche, F., Leroy, C., Pensotti, S., Penzo, A., Rattaggi, M. and Rancoita, P.G., Determination of intrinsic fluctuations and energy response of Si/Fe and Si/Fe + Pb calorimeters up to 70 GeV of incoming hadron energy, 378–384, Copyright (1996), with permission from Elsevier; see also [Leroy and Rancoita (2000)]).

the incoming beam energy $E_{\rm beam}$ and used as an energy scale for the calorimeter response E_π to hadrons of identical energy. The response to hadrons (scaled from the electron response), which is sometimes called the *mean visible hadron energy* (e.g., [Groom (1992); ATLAS Collab. (1997)]), can be written as:

$$E_\pi = E_{\rm beam}[1 - (1 - f_{em})(1 - h/e)]$$
$$= E_{\rm beam}[1 - f_h(1 - h/e)]. \tag{9.60}$$

This equation follows from Eq. (9.59) once $\epsilon_{\rm vis}(e)$ is replaced by $E_{\rm beam}$ and, as a consequence, $\epsilon_{\rm vis}(\pi)$ by E_π. Although E_π is introduced in a way similar to $\epsilon_{\rm vis}(\pi)$, it does not indicate a similar quantity, the former being a relative scaled energy, and the latter an absolute measurement of the deposited energy in the active samplers. In the following, when we need to make use of the scaled hadronic energy, we will refer to it by explicitly indicating its notation E_π.

In the previous section, we have seen that both e/h and h/mip ratios are energy independent intrinsic properties of a sampling calorimeter. We must note [from Eqs. (9.55, 9.56)] that if $e/h = 1$, we have $e/\pi = 1$ and vice-versa, i.e., the ratio e/π becomes energy-independent. The condition $e/h = e/\pi = 1$ is called the *compensating condition* or *compensation condition* (see [Wigmans (1987); Fesefeldt (1988); Wigmans (1988); SICAPO Collab. (1989b,c, 1991b); Wigmans (1991); SICAPO Collab. (1993b); Job et al. (1994); SICAPO Collab. (1995a,b, 1996)] and references

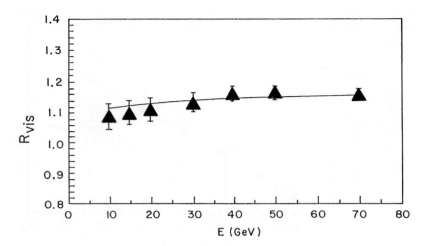

Fig. 9.20   Ratio $R_{\mathrm{vis}} = \epsilon_{\mathrm{vis}}(Fe)/\epsilon_{\mathrm{vis}}(Fe + Pb)$ versus the incoming hadron energy (reprinted from *Nucl. Instr. and Meth. in Phys. Res. A* **368**, Furetta, C., Gambirasio, A., Lamarche, F., Leroy, C., Pensotti, S., Penzo, A., Rattaggi, M. and Rancoita, P.G., Determination of intrinsic fluctuations and energy response of Si/Fe and Si/Fe + Pb calorimeters up to 70 GeV of incoming hadron energy, 378–384, Copyright (1996), with permission from Elsevier; see also [Leroy and Rancoita (2000)]).

therein). Furthermore from Eqs. (9.6, 9.57), and from Eqs. (9.51, 9.55), we have

$$e/h = \frac{e/mip}{h/mip} \quad \text{and} \quad e/\pi = \frac{e/mip}{\pi/mip} ;$$

furthermore, when the compensating condition is satisfied, we have also that

$$h/mip = \pi/mip = e/mip$$

independently of the incoming particle energy. At very high energy, we obtain

$$e/\pi \to 1, \quad \text{since} \quad f_{em} \to 1$$

(see discussion on f_{em} on page 276). As a consequence, at very high energies, the calorimeter becomes *almost compensating*.

Equation (9.58) shows that it is only when $e/mip = h/mip$ (i.e., for $e/h = 1$, the compensation condition) that the calorimeter response is proportional to the incoming hadron energy, E, independently of the energy dependent term f_{em}. Thus, we have

$$\epsilon_{\mathrm{vis}}(\pi) = (h/mip)E_s = (h/mip)\, F(S)\, E, \qquad (9.61)$$

in a similar way to Eq. (9.7) for the electromagnetic cascades. It means that for a hadronic calorimeter the most important requirement, i.e., its linear response, can be fulfilled only under a very particular operating condition, namely the compensation condition.

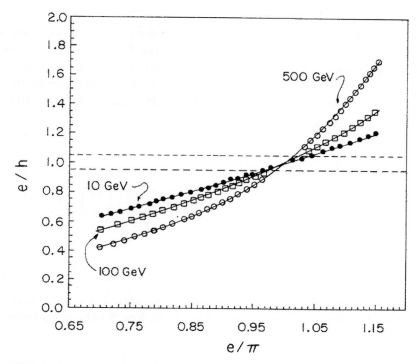

Fig. 9.21 e/h versus e/π [from Eqs. (3.74, 9.56)] at 10, 100, 500 GeV from [Leroy and Rancoita (2000)]. The two dotted lines represent $e/h = 0.95$ and 1.05, respectively.

The restoration of the linear calorimeter response to the incoming hadron energy, when the compensation condition is satisfied, was experimentally observed, see e.g., [HELIOS Collab. (1987); SICAPO Collab. (1996)]. In Fig. 9.19, the measured visible hadroniv energies, ϵ_{vis}, in MeV, for Si/Fe and Si/Fe+Pb sampling calorimeters are shown as a function of the incoming hadron energy [SICAPO Collab. (1996)]. The Si/Fe+Pb calorimeter is operated under an almost compensating condition, i.e., $e/h \approx 1$. The overimposed curve to the data (dark squares) is well represented by a straight line crossing the origin of the axes. A slight deviation from linearity, smaller than 3.5% over the full energy range, is observed for the experimental data from the Si/Fe calorimeter (open circles), for which $e/h \approx 1.16$. Figure 9.20 shows the ratio $R_{\mathrm{vis}} = \epsilon_{\mathrm{vis}}(Fe)/\epsilon_{\mathrm{vis}}(Fe + Pb)$ versus the energy. The overimposed line is calculated using Eq. (9.54), with f_{em} as given by Eq. (3.74), and taking into account the e/h values of the two calorimeters [SICAPO Collab. (1996)]. The R_{vis} dependence on the incoming hadron energy is due to the non-linear energy response of the Si/Fe calorimeter ($\approx 3.5\%$ over the entire energy range, as mentioned above).

A linear energy response was also observed in compensating U/plastic-scintillator calorimeters, like the one from [HELIOS Collab. (1987)] for hadrons of energy between 8 and 200 GeV.

The extent to which the lack of compensation affects the non-linear calorimeter response to incoming-hadron energy depends on the e/h ratio. We can describe the non-linear behavior, N_l, of a hadronic calorimeter between the reference energy of 10 GeV and any incoming-hadron energy E (in GeV), by considering its linear response to electromagnetic cascades for which $\epsilon_{\mathrm{vis}}(e, E) = (E/10)\epsilon_{\mathrm{vis}}(e, 10\ \mathrm{GeV})$:

$$
\begin{aligned}
N_l &= \frac{\left|\epsilon_{\mathrm{vis}}(\pi, E)\left(\frac{10}{E}\right) - \epsilon_{\mathrm{vis}}(\pi, 10\ \mathrm{GeV})\right|}{\epsilon_{\mathrm{vis}}(\pi, 10\ \mathrm{GeV})} \\
&= \left|\frac{\epsilon_{\mathrm{vis}}(\pi, E)}{\epsilon_{\mathrm{vis}}(\pi, 10\ \mathrm{GeV})}\left(\frac{10}{E}\right) - 1\right| \\
&= \frac{\epsilon_{\mathrm{vis}}(e, 10\ \mathrm{GeV})}{\epsilon_{\mathrm{vis}}(e, E)}\left(\frac{E}{10}\right)\left|\frac{\epsilon_{\mathrm{vis}}(\pi, E)}{\epsilon_{\mathrm{vis}}(\pi, 10\ \mathrm{GeV})}\left(\frac{10}{E}\right) - 1\right| \\
&= \left|\frac{e/\pi(10\ \mathrm{GeV})}{e/\pi(E)} - 1\right| \\
&= \frac{|e/\pi(10\ \mathrm{GeV}) - e/\pi(E)|}{e/\pi(E)},
\end{aligned}
\tag{9.62}
$$

where the ratio e/π [see Eq. (9.55)] depends on the intrinsic ratio e/h and on the energy dependent term f_{em}. Thus, using Eq. (9.55), Eq. (9.62) can equivalently be rewritten as

$$
\begin{aligned}
N_l &= \left|\left[\frac{e/h}{1 - f_{em}(10\ \mathrm{GeV})(1 - e/h)} - \frac{e/h}{1 - f_{em}(E)(1 - e/h)}\right]\left[\frac{e/h}{1 - f_{em}(E)(1 - e/h)}\right]^{-1}\right| \\
&= \left|\frac{f_{em}(10\ \mathrm{GeV})(1 - e/h) - f_{em}(E)(1 - e/h)}{1 - f_{em}(10\ \mathrm{GeV})(1 - e/h)}\right| \\
&= \left|[f_{em}(10\ \mathrm{GeV}) - f_{em}(E)]\left[\frac{1 - e/h}{1 - f_{em}(10\ \mathrm{GeV})(1 - e/h)}\right]\right|.
\end{aligned}
$$

In Fig. 9.21, the curves, relating the two ratios e/h and e/π, are calculated by means of Eqs. 3.74, (9.56), for e/h differing by 5% from the compensating value. From Fig. 9.21 we can estimate that N_l [see Eq. (9.62)] is $\approx 2\%$ at 500 GeV. N_l is $\approx 6.7\%$ (3.7%) at 500 GeV (100 GeV), for $e/h \approx 1.15$.

9.9 Methods to Achieve the Compensation Condition

The e/π signal ratio is determined by processes that occur at the nuclear and atomic levels during the cascade development. A large amount of the deposited energy in a hadron calorimeter goes into breaking nuclei (binding energy) or into low-energy neutrons and, therefore, is partially not visible. In addition, charged π and μ decays produce secondary particles like ν, which are undetected. As discussed in Sect. 3.3, the nuclear effects, which dominate pure hadronic cascades, have no counterpart in electromagnetic cascades. Therefore, the calorimeter response to the electromagnetic and non-electromagnetic part of the hadronic cascade is different, with the consequence that, in general, the electromagnetic (e) to hadronic (π) signal ratio $e/\pi \neq 1$.

The energy independent equalization between the electromagnetic and the hadronic signals ($e/\pi = 1$, i.e., the compensation condition, see page 675) is the condition to restore the linear response of the calorimeter to hadronic cascades, and, as will be discussed in the next section, to obtain an energy resolution that improves as $1/\sqrt{E}$, where E is the incoming hadron energy.

Calorimeter compensation physics, see e.g., [Wigmans (1987); Brückmann et al. (1988)], deals with the flux, the deposition and the fluctuation of the energy coming from the cascade process. The energy deposition in both passive and active samplers mainly occurs via collision losses of i) relativistic hadrons produced in the interactions on target nuclei, ii) quasi-direct spallation protons coming from the target nuclei and iii) electrons and positrons generated by the electromagnetic component of the hadronic cascade. Additionally, prompt gamma radiation may be emitted from excited fission products, or after particle evaporation from residual nuclei. Moreover, sources of delayed gamma photons are neutron capture processes, when they occur. These photons can subsequently interact via the Compton or the photoelectric effects and, in turn, generate a fast electron. Neutrons may transfer part of their energy in neutron–nucleus collisions, particularly on low-A nuclei. Furthermore, since most of the protons contributing to the hadronic signal are highly non-relativistic, the saturation properties of the detecting medium for densely ionizing particles (for instance, the scintillator light output) are of crucial importance. This effect was intensively investigated, when the first compensation calorimeters were operated (e.g., see [Brau and Gabriel (1985); HELIOS Collab. (1987); Wigmans (1987); Brückmann et al. (1988); Brau, Gabriel and Rancoita (1989)] and references therein).

Various approaches to the realization of the compensation condition exist. One method mainly depends on the quantity of hydrogen contained in the active medium, and on the A-value of the passive samplers. The increase of the hadronic signal, i.e., of the h/mip ratio, can be obtained by detecting part of the energy carried by neutrons, which are generated in the cascade process. In fact, when hydrogen rich materials (for instance plastic scintillators or scintillating fibers) are employed as active media, neutron–proton scattering processes generate fast recoiling and ionizing protons in active samplers (see, e.g., [Fabjan et al. (1977); Abramowicz et al. (1981); De Vincenzi et al. (1986); Bernardi et al. (1987); HELIOS Collab. (1987); Drews et al. (1990); Ros (1991)]).

The e/mip ratio can also be tuned by an appropriate choice of both passive samplers and active media (see Sects. 9.2 and 9.3 and, e.g., [SICAPO Collab. (1990a, 1991b); Byon-Wagner (1992); SICAPO Collab. (1992a); Beretvas et al. (1993)] and references therein).

In the following, we will call *almost compensating calorimeter* a calorimeter for which the measured e/π ratio is $\approx$ between 0.95 and 1.05, within $\approx 5\%$ experimental errors, for incoming hadron energies ≥ 10 GeV.

9.9.1   *Compensation Condition by Detecting Neutron Energy*

The role of neutron energy detection in compensating or almost compensating calorimetry and, in particular, the role of the nuclear fission in the building of the signal of uranium calorimeters (see e.g., [Leroy, Sirois and Wigmans (1986)]) have intensively investigated experimentally and, also, by using Monte-Carlo simulations (e.g., see [Fabjan et al. (1977); Abramowicz et al. (1981); De Vincenzi et al. (1986); Bernardi et al. (1987); HELIOS Collab. (1987); Drews et al. (1990); Ros (1991)], and also [Wigmans (1987); Brückmann et al. (1988)], Section 3.3.3.2 in [Wigmans (2000)]).

In hadronic cascades, neutrons (n) are produced either by spallation¶ or, mostly, are evaporated from highly excited nuclear fragments (see [Fraser et al. (1965); Patterson and Thomas (1981); Brau and Gabriel (1985); Leroy, Sirois and Wigmans (1986); Wigmans (1987); Brückmann et al. (1988)] and references therein). The energy spectrum has a peak below 10 MeV, corresponding to evaporation neutrons. For heavy elements, the evaporation peak is slightly shifted to lower energies. For fissionable nuclei (^{238}U), a large amount of fast neutrons are generated owing to subsequent fission processes. Those neutrons with energies larger than about 1 MeV are likely to dissipate their energy through inelastic (n,γ) reactions on heavy nuclei. At lower energies (below 100 keV), neutrons are captured by nuclei. Neutrons emerging with energies above 10 MeV are mainly produced by spallation processes and have an energy spectrum almost independent of the atomic number of the absorber element. The few very highly energetic neutrons, that are generated, travel some distance inside the calorimeter before interacting. They behave similarly to highly energetic charged hadrons. Furthermore, the longitudinal profile of neutron generated in hadronic cascades was measured in iron and lead absorbers for incoming hadrons of 24 and 200 GeV ([Russ et al. (1991)] and references therein).

The proton-induced neutron yield, $Y(E, A)$, was measured from light to heavy nuclei with incoming protons above ≈ 0.5 GeV [Fraser et al. (1965)]. To a first approximation, it is given by [Patterson and Thomas (1981)]:

$$Y(E, A) = 0.1 \times (E - 0.12)(A + 20), \text{ for non-fissionable nuclei} \qquad (9.63)$$

and

$$Y(E, A) = 50 \times (E - 0.12), \text{ for fissionable nuclei (like } {}^{238}\text{U}), \qquad (9.64)$$

where E (in GeV) is the incoming hadron energy and A is the target mass.

From Eqs. (9.63, 9.64), we expect ≈ 45 n/GeV per incoming proton on depleted uranium absorbers and ≈ 20 n/GeV on lead absorber. These values agree with the

---

¶The relative probability of neutron-to-proton spallation increases as A increases, because the Z/A ratio decreases.

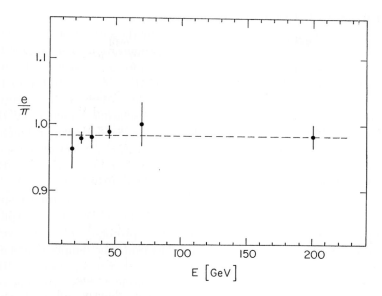

Fig. 9.22   The measured e/π ratio with the compensating uranium/scintillator calorimeter real-ized by the HELIOS Collaboration (reprinted from *Nucl. Instr. and Meth. in Phys. Res. A* **262**, Åkesson, T. et al., Performance of the uranium/plastic scintillator calorimeter for the HELIOS experiment at CERN, 243–263, Copyright (1987), with permission from Elsevier, e.g., for the list of the authors see [HELIOS Collab. (1987)]; see also [Leroy and Rancoita (2000)]).

experimental data from [Fraser et al. (1965); Leroy, Sirois and Wigmans (1986)] (see also [Russ et al. (1991)] and references therein). Therefore, when uranium absorbers are employed, the neutron yield is more than doubled with respect to a high-A nucleus, such as lead.

In hadron calorimeters employing hydrogenous material as active devices, like scintillators or scintillating fibers, the compensation condition can be achieved by increasing the relative contribution of the pure hadronic signal, namely by increasing the h/mip ratio. Indeed, the neutrons, generated in the last stages of the hadronic cascade development where processes at the nuclear level occur, transfer part of their energy (E_n) to the protons of the scintillator. The maximum recoil energy ($E_{R,\mathrm{max}}$) transferred from an incoming non-relativistic neutron to a recoiling nucleus of atomic weight A is:

$$E_{R,\mathrm{max}} = \frac{4A}{(1+A)^2}\, E_n. \tag{9.65}$$

For neutron–hydrogen collisions, the full energy of the neutron can be transferred to the recoiling proton ($E_{p,R,\mathrm{max}}$), namely $E_{p,R,\mathrm{max}} = E_n$. This results in an increase of the hadronic signal, whose amount can be varied by modifying either the sampling fraction or the ratio of the thicknesses of the passive sampler and the active medium, i.e., by increasing the number of protons available as targets for neutrons [Leroy,

Sirois and Wigmans (1986)].

The increase of the hadronic signal is limited by the saturation of the scintillator response, which occurs in the presence of dense ionization loss, i.e., for fast protons. In scintillators, this phenomenon was studied for many years and is described to a first approximation‖ by *Birks' law* [Birks (1951)] expressed in Eq. (5.4), where dL/dx is the light yield per unit path length, dE/dx is the specific energy-loss by collisions for the charged particle, S is the normal scintillation efficiency, the parameter kB (see [Knoll (1999)]) describes the quenching which occurs for high-density ionization. Extended calculations regarding the saturation effect of the scintillator response in hadronic calorimetry can be found in the literature, e.g., see [Brau and Gabriel (1985); Brau et al. (1985); Brau and Gabriel (1989)].

The calorimeter performance also depends on the relative time response for the various components of the calorimeter signal. The energy loss by collisions of both charged relativistic particles and spallation protons occurs at a very short time after the arrival of the incoming primary hadron. The temporal dependence of fission and neutron capture was studied for uranium samplers (see e.g. [Brau and Gabriel (1989)]). It lasts a few hundreds ns. The majority of the hadronic responses of uranium calorimeters occurs in the first (50–100) ns. As consequence, the h/mip ratio will depend on the integration time of the signal. The typical integration time is about 100 ns.

A compensating uranium/plastic scintillator calorimeter was built by the HE-LIOS Collaboration [HELIOS Collab. (1987)]. In Fig. 9.22, the e/π ratio, measured up to 200 GeV, is shown. Their calorimeter modules consisted of uranium and scintillator layers 3.0 mm and 2.5 mm thick, respectively. This calorimeter had a ratio $R_{p/a}$ of ≈ 1.2, for the thicknesses of the passive relative to active samplers.

The compensation condition or an almost compensating condition were also achieved by the ZEUS Collaboration (see [Klanner et al. (1988); d'Agostini et al. (1989); Tiecke (1989); Behrens (1990); Drews et al. (1990); Ros (1991)] and references therein), with devices for which the $R_{p/a}$ ratios were between 1.07 and 1.26. In Fig. 9.23, the e/π ratio, measured up to 100 GeV with the ZEUS prototype calorimeter, is shown [Behrens (1990); Ros (1991)]. This calorimeter consisted in uranium and scintillator layers 3.3 mm and 2.6 mm thick, respectively. The ZEUS Collaboration measurements show the calorimeter is compensating or almost compensating above (2–3) GeV, i.e., the e/π value is close to 1 independently of the incoming hadron energy. However, as discussed in Sect. 9.8.1 and in [Wigmans (1987)], below ≈ 2 GeV the calorimeter properties, like the e/π value, differ from the intrinsic ones at higher energies (see Fig. 9.23).

For $R_{p/a}$ values largely differing from ≈ 1.2, the uranium/plastic scintillator calorimeter is no longer a compensating calorimeter. The measured e/π ratio, at 10 GeV, was ≈ 1.12 for $R_{p/a} \approx 0.8$ [Åkesson et al. (1985)] and ≈ 0.81 for $R_{p/a} \approx 2.0$ [Catanesi et al. (1987)]. Furthermore ([Brückmann and Kowalski (1986); Ros

---

‖Other more refined approximations are given in [Knoll (1999)] and in references therein.

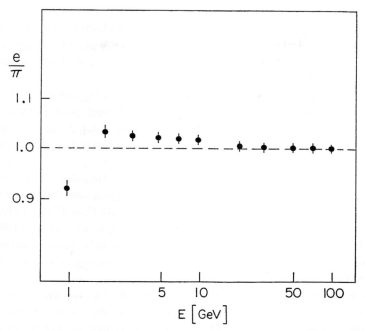

Fig. 9.23 The measured e/π ratio with the compensating uranium/scintillator calorimeter real-ized by the ZEUS Collaboration (reprinted from *Nucl. Phys. B (Proc. Supp.)* **23 (Issue 1)**, Ros, E., Influence of instrumental effects on the performance of the ZEUS calorimeter, 51–61, Copyright (1991), with permission from Elsevier; see also [Leroy and Rancoita (2000)] and [Behrens (1990)]).

(1991)] and references therein), the experimental data show a decrease (as expected) of $\approx 7\%$ in the e/π ratio when the integration times are increased from ≈ 100 ns to $\approx 1\,\mu s$.

An almost compensating lead/plastic scintillator calorimeter was built by the ZEUS Collaboration [Bernardi et al. (1987)]. Their calorimeter modules consisted of lead and scintillator layers 10.0 mm and 2.5 mm thick, respectively. This calorime-ter had a ratio $R_{p/a}$ of ≈ 4 between the thicknesses of the passive to active sam-plers. The measured e/π ratio was ≈ 1.05, above 10 GeV.

A consequence of these results is that the use of uranium is not an absolute requirement to achieve compensation, when hydrogeneus detectors are employed as readout media. However, it is fair to say that uranium still possesses the advantage of easy calibration with the noise of its natural radioactivity. Furthermore, *an almost non-variable $R_{p/a}$ ratio of about 1.2 for passive uranium samplers and of about 4 for passive lead samplers is needed to achieve compensation with hydrogeneus active media.*

Lead and scintillating fibres (or strips) were used in calorimetry, for instance by the SPACAL Collaboration ([Acosta et al. (1991); DeSalvo (1995)] and references therein) and the CHORUS ([Buontempo et al. (1995)] and references therein). They have achieved an e/π ratio ≈ 1.1 (or larger) at $10\,\text{GeV}$.

Data from Fe/scintillator calorimeter with a longitudinal tile configuration [Ariztizabal et al. (1994)] show that such a calorimeter was not compensating (see also [Botner (1981)]. The e/π ratio was larger than ≈ 1.2 at an incoming particle energy of $20\,\text{GeV}$.

Other readout media using hydrogeneous liquids were investigated ([Pripstein (1991)] and references therein) for low-A (e.g., see [Aubert et al. (1993)]) and high-A passive sampler. A detailed study of their almost compensating properties (as a result of Monte-Carlo simulation predictions) is presented in [Brau and Gabriel (1989)].

Hadronic calorimeters employing liquid argon** as the active medium were and are presently used in high-energy physics experiments (e.g., see [Fabjan et al. (1977); H1 Collab. (1988); D0 Collab. (1989); SSC Detector R&D at BNL (1990); D0 Collab. (1993); Axen et al. (1993); H1 Collab. (1993a); Gingrich et al. (1994); Guida (1995); ATLAS Collab. (1997)] and references therein). These calorimeters are also non-compensating, when uranium absorbers constitute the passive samplers (e.g., see [Kondo et al. (1984)]).

The underlying phenomena that are responsible for the significant difference with respect to the plastic scintillator calorimeters, are mainly explained taking into account the neutron cross sections and the saturation effects of the active medium (e.g., see [Brau and Gabriel (1985); Wigmans (1987); Brau and Gabriel (1989)]). The neutron cross section on hydrogen continues to rise below $(1\text{--}2)\,\text{MeV}$, while in argon falls. In this way, the neutron cross section is larger on hydrogen than on argon nucleus. In addition, the maximum recoil energy transferred to an argon nucleus is [from Eq. (9.65)] $\approx 9.5\%$ of the incoming neutron energy, while it is the entire neutron energy for an interaction on hydrogen. Therefore, on average, the recoil nucleus will emerge with a lower velocity for the same incoming neutron energy.

Following the suggestion of Brau and Gabriel (1985), saturation effects in liquid argon can be expressed in the form of Birks' law [Eq. (5.4)]. In scintillators, the value of the parameter kB is $\approx (0.01\text{--}0.02)\,\text{g\,cm}^2\text{MeV}^{-1}$, while in liquid argon it is $0.0045\,\text{g\,cm}^2\text{MeV}^{-1}$ [Fabjan et al. (1977); Babaev et al. (1979); Brau and Gabriel (1985)]. For electrons at all energies, it is assumed that $kB = 0$. Although liquid argon is less saturating than plastic scintillator, the kinematic constraint on the energy transferred in neutron–nucleus collision leads (as mentioned above) to a greater suppression of the signal, because the highly ionizing recoil nuclei receive only one-tenth the relative energy of the less highly ionizing recoil protons in scintillator. The e/mip ratio is also slightly larger in the case of liquid argon, whose Z-value is larger than the one of plastic scintillator (see Sect. 9.2.2).

In uranium/liquid argon calorimetry, the e/π ratio measured at $10\,\text{GeV}$ is ≈ 1.09 ([D0 Collab. (1993)], e.g., see also [D0 Collab. (1989); Guida (1995)]), or even larger

---

**Reviews on liquid-argon calorimetry are given in [Gordon (1991); Fabjan (1995b)].

(e.g., see [SSC Detector R&D at BNL (1990)]), depending on sampler thicknesses and integration times.

Iron cladding of high-Z passive absorbers was investigated for reducing the e/π ratio (e.g., see [Gordon (1991); Mockett and Boulware (1991)]), but the compensation is hardly achievable. It was also suggested (e.g. see [D0 Collab. (1989); Cennini et al. (1990); Fabjan (1995b)]) that, by the addition of dopants containing hydrogen (like methane) to liquid argon, the e/π ratio might be reduced. The experimental results [D0 Collab. (1989)] do not confirm that the compensation condition can be achieved when methane is added to liquid argon.

In large liquid-argon calorimeters, different passive samplers have often been used inside the same device (e.g., [H1 Collab. (1988, 1993a); Gingrich et al. (1994); ATLAS Collab. (1997)]). The choice of the types and the sequence of absorbers is related to constraints coming from the experimental apparatus and the physics of the experiment. In particular, lead absorbers were employed in the first part of the device. This part is able to longitudinally contain almost all the energy deposited by electromagnetic showers and is called the *electromagnetic section*. In the second part, called the *hadronic section*, iron or copper absorbers were used. These calorimeters are intrinsically non-compensating, e.g., the measured e/π ratio is ≈ 1.17 at 20 GeV for the Pb (passive samplers of the electromagnetic section) and Cu (passive samplers of the hadronic section) calorimeter of the H1 Collaboration [H1 Collab. (1988)]. However, methods for restoring the calorimeter response almost to linearity were developed taking into account the response of each active sampler (e.g., [H1 Collab. (1988, 1993a); Gingrich et al. (1994); ATLAS Collab. (1997)]) or the intrinsic e/h ratios of the two sections of the calorimeter (e.g., [ATLAS Collab. (1997)]).

The method of *weighted energies*, developed by the H1 Collaboration in [H1 Collab. (1988, 1993a)] (see also [Abramowicz et al. (1981); Beretvas et al. (1993)]), takes into account the fluctuations of the electromagnetic energy deposition in the electromagnetic and hadronic sections of the calorimeter and requires to determine an energy-dependent weight for the response of each active sampler.

The method of the *benchmark approach*, developed by the ATLAS Collaboration [ATLAS Collab. (1997)], consists in a two-step procedure, in which the fractions of the scaled hadronic energies E_π (see the definition on page 675) deposited in the two sections of the calorimeter are determined.

A comparison of these two methods is given in [ATLAS Collab. (1997)]. It is shown that the calorimeter response is readjusted to linearity, within $\pm 2\%$ for incoming hadrons of energy between 20 and 300 GeV, by using both methods. However, it must be remembered that in hadronic jets (e.g., see [Perkins (1986)]) with the same energy, the number of particles and their energies vary from event to event. As a consequence, these procedures have to be carefully studied and tuned when used in a high energy physics experiments [H1 Collab. (1988); Wigmans (1988, 1991)].

9.9.2   *Compensation Condition by Tuning the e/mip Ratio*

The compensating (or almost compensating) condition can be achieved by tuning the electromagnetic response of the calorimeter. In Sects. 9.2.3, 9.3, it was discussed how an intrinsic calorimetric property, namely the e/mip ratio, can be largely decreased by choosing and locating the set of passive samplers in the suited sequence.

Mainly two effects, or their combination ([SICAPO Collab. (1993c)] and references therein), were exploited to provide such a tuning. The former is the local hardening effect*, realized by inserting thin low-Z absorbers next to the readout detectors when high-Z absorbers constitute the passive samplers. The latter is the filtering effect[†], obtained by using a combination of low-Z and high-Z materials as absorbers, both of them non-negligible in units of radiation length. The local hardening effect exploits the way the energy deposition occurs in the final stage of the electromagnetic cascade, when soft electrons and photons interact. The filtering effect makes use of the different electron (and positron) energy distributions generated by electromagnetic showers in passive media, namely combining passive media whose critical energies have greatly different values. The property of these absorbers is such that the radiation losses by electrons (and positrons) dominate at different values of electron (and positron) energy in the subsequent passive media. Moreover, these effects were studied by using Monte-Carlo simulations (e.g., see [Wigmans (1988); Brau and Gabriel (1989); Brau, Gabriel and Rancoita (1989); Mockett and Boulware (1991); Job et al. (1994)]).

In hadronic cascades, the visible energy of the pure hadronic component (Sect. 3.3.1) is mainly generated by collision losses by charged particles (spallation protons and relativistic secondary particles).

The pure hadronic visible-energy is expected to be affected neither by the location of passive absorbers nor by effects related to the limited path needed to fully absorb very soft charged particles. In fact, both secondaries and recoil protons (involved in collision loss processes) undergo negligible (if any) radiation losses and, being fast, can travel long distances inside the calorimeter before being absorbed.

The overall calorimeter response to the energy deposited by collisions may depend on saturation effects (even in the case of fast protons) occurring in active samplers, by the passage of densely ionizing particles. Also the neutron contribution to the hadronic visible-energy depends on the property of the active medium. As discussed in the previous section, the maximum recoiling energy [Eq. (9.65)] transferred from an incoming non-relativistic neutron is E_n (i.e., the kinetic energy of the incoming neutron) onto a hydrogen target, like in scintillators, but only 9.5% of E_n on an argon nucleus and $\approx 13.3\%$ of E_n on silicon nucleus. For instance, the saturation affects detectors based more on scintillation media than on liquid argon

---

*The experimental evidence for this effect was given by the SICAPO Collaboration, see [SICAPO Collab. (1989b)] and references therein, and also Sect. 9.2.3.

[†]The experimental evidence for this effect was given by the SICAPO Collaboration, see [SICAPO Collab. (1989c)] and Sect. 9.3.

[see Sect. 9.9.1 and the values of kB coefficient in Eq. (5.4)], while saturation is very small in other devices, like silicon detectors ([Hancock, James, Movchet, Rancoita and Van Rossum (1983, 1984); Rancoita (1984); Brau and Gabriel (1989); Brau, Gabriel and Rancoita (1989)] and references therein). In particular, silicon detectors demonstrated to have an extremely linear response for very high densities of ionization, as for heavily ionizing ions above $\approx$ a few MeV [Sattler (1965); Wilkins et al. (1971)]. Some saturation effects are observed for recoils of silicon atoms below ≈ 3 MeV [Sattler (1965)]. However, even for silicon ions of kinetic energy as low as 100 keV nearly a half of the deposited energy is detected as an observable output signal relatively to an electron of the same energy [Sattler (1965)]. For recoiling nuclei, the saturation properties of the active medium determine the fraction of the recoil energy, which can contribute to the overall hadronic visible-energy: this fraction increases as the detector response approaches linearity. We can conclude that the purely hadronic visible-energy is expected, to a first approximation, to be marginally affected when tuning the e/mip value (e.g., see [SICAPO Collab. (1991c)] and references therein). A systematic investigation of the local hardening effect on hadronic cascades was carried out for incoming electrons and protons of 8, 10, and 12 GeV [SICAPO Collab. (1990a, 1991b)], employing a calorimeter with silicon detectors as active media and passive uranium samplers (each 1.5 cm thick). Additional low-Z G10 plates were located on the front and at the rear of the active readout detectors in order to tune the electromagnetic response (Sect. 9.2.3). The G10 plates (of equal thickness on the front and at the rear of silicon readout planes) were 1, 3, and 5 mm thick. The silicon planes had a support structure made by two G10 sheets 0.2 and 1.0 mm thick, respectively. The visible-energy reduction, $\frac{\Delta\epsilon}{\epsilon}$(G10), measured (from an average of measurements at the three incoming-particle energies) for protons and electrons is shown in Fig. 9.24 ([Leroy and Rancoita (2000)], see also [SICAPO Collab. (1990a, 1991b)]), being

$$\frac{\Delta\epsilon}{\epsilon}(\text{G10}) = \left[\frac{\epsilon_{\text{vis}} - \epsilon_{\text{vis}}(\text{G10})}{\epsilon_{\text{vis}}}\right] = \left[1 - \frac{\epsilon_{\text{vis}}(\text{G10})}{\epsilon_{\text{vis}}}\right], \qquad (9.66)$$

where ϵ_{vis} and ϵ_{vis}(G10) are the visible energy without and with additional G10 plates, respectively. In the following, we will indicate with $\frac{\Delta\epsilon}{\epsilon}(e, \text{G10})$ the visible-energy reduction for electromagnetic showers, i.e., for incoming electrons, and with $\frac{\Delta\epsilon}{\epsilon}(\pi, \text{G10})$ the visible-energy reduction for incoming hadrons. The data show that $\frac{\Delta\epsilon}{\epsilon}$(G10) increases as the G10 plate thickness increases, as expected. These results, for incoming electrons, are in agreement with previous measurements performed using electromagnetic calorimeters (see Sect. 9.2.3, and [SICAPO Collab. (1988b, 1989b,d, 1992b)]). Equation (9.54) gives the hadronic visible-energy, $\epsilon_{\text{vis}}(\pi)$, in terms of the visible energies of the purely electromagnetic, $\epsilon_{\text{vis}}(e)$, and purely hadronic, $\epsilon_{\rm vis}(h)$, components and, consequently [see also Eq. (9.55)],

$$
\begin{aligned}
\frac{\Delta\epsilon}{\epsilon}(\pi, \mathrm{G10}) &= \frac{\epsilon_{\rm vis}(\pi) - \epsilon_{\rm vis}(\pi, \mathrm{G10})}{\epsilon_{\rm vis}(\pi)} \\
&= \frac{f_{em}\left[\epsilon_{\rm vis}(e) - \epsilon_{\rm vis}(e, \mathrm{G10})\right]}{\epsilon_{\rm vis}(\pi)} \\
&\quad + \frac{(1 - f_{em})\left[\epsilon_{\rm vis}(h) - \epsilon_{\rm vis}(h, \mathrm{G10})\right]}{\epsilon_{\rm vis}(\pi)} \\
&= \frac{f_{em}\Delta\epsilon(e, \mathrm{G10}) + (1 - f_{em})\Delta\epsilon(h, \mathrm{G10})}{\epsilon_{\rm vis}(\pi)} \\
&= f_{em}\frac{\Delta\epsilon(e, \mathrm{G10})}{\epsilon_{\rm vis}(\pi)}\left[\frac{\epsilon_{\rm vis}(\pi)}{\epsilon_{\rm vis}(e)}\right](e/\pi) + (1 - f_{em})\frac{\Delta\epsilon(h, \mathrm{G10})}{\epsilon_{\rm vis}(\pi)} \\
&= (e/\pi)f_{em}\frac{\Delta\epsilon}{\epsilon}(e, \mathrm{G10}) + (1 - f_{em})\frac{\epsilon_{\rm vis}(h) - \epsilon_{\rm vis}(h, \mathrm{G10})}{\epsilon_{\rm vis}(\pi)} \\
&= (e/\pi)f_{em}\frac{\Delta\epsilon}{\epsilon}(e, \mathrm{G10}) \\
&\quad + [1 - f_{em}]\left[\frac{\epsilon_{\rm vis}(h)}{\epsilon_{\rm vis}(\pi)}\right]\left[1 - \frac{\epsilon_{\rm vis}(h, \mathrm{G10})}{\epsilon_{\rm vis}(h)}\right] \\
&= (e/\pi)f_{em}\frac{\Delta\epsilon}{\epsilon}(e, \mathrm{G10}) \\
&\quad + \left[\frac{\epsilon_{\rm vis}(\pi) - f_{em}\epsilon_{\rm vis}(e)}{\epsilon_{\rm vis}(\pi)}\right]\left[1 - \frac{\epsilon_{\rm vis}(h, \mathrm{G10})}{\epsilon_{\rm vis}(h)}\right] \\
&= (e/\pi)f_{em}\frac{\Delta\epsilon}{\epsilon}(e, \mathrm{G10}) \\
&\quad + [1 - (e/\pi)f_{em}]\left[1 - \frac{h/mip(\mathrm{G10})}{h/mip}\frac{E_s(\mathrm{G10})}{E_s}\right],
\end{aligned} \tag{9.67}
$$

where the e/π ratio is the one obtained without additional G10 plates.

For instance (Fig. 9.24) the measured energy reduction, $\frac{\Delta\epsilon}{\epsilon}(e, \mathrm{G10} = 5\ \mathrm{mm})$, for incoming electrons is $\approx 40\%$ with the insertion of $5\,\mathrm{mm}$ G10 plates, while for incoming protons the measured value of $\frac{\Delta\epsilon}{\epsilon}(\pi, \mathrm{G10} = 5\ \mathrm{mm})$ is $\approx 12\%$. The lower reduction of visible energy for incoming protons is consistent with the value predicted by Eq. (9.67). In fact at these energies, f_{em} is $\approx (25\text{–}27)\%$ on average [Eq. (3.74)]. In addition, the measured e/π ratio (Fig. 9.25), without additional G10 plates, is ≈ 1.22.

In order to estimate $\frac{\Delta\epsilon}{\epsilon}(\pi, \mathrm{G10} = 5\ \mathrm{mm})$, we have to take into account the fact that the local hardening effect acts on the electromagnetic component of the hadronic shower and that, to a first approximation even for the case of 5 mm thick G10 low-Z additional absorbers, we have $\epsilon_{\rm vis}(h) \approx \epsilon_{\rm vis}(h, \mathrm{G10} = 5\ \mathrm{mm})$. In fact, the slight decrease of the purely hadronic visible-energy, due to the decrease of the shared energy (E_s) when the G10 plates are inserted, is mostly compensated by the slight increase of the h/mip ratio, since the low-Z (i.e., low-A) materials have a larger A/Z ratio than the uranium absorbers (see discussion in [SICAPO Collab.

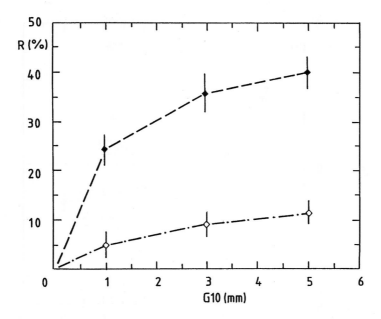

Fig. 9.24   Percentage (R) of the visible-energy reduction for protons (bottom data) and electrons (top data) versus the thickness of the added G10 plates (reprinted from *Phys. Lett. B* **242**, Angelis, A.L.S. et al., Evidence for the compensation condition in Si/U hadronic calorimetry by the local hardening effect, 293–298, Copyright (1990), with permission from Elsevier, e.g., for the list of the authors see [SICAPO Collab. (1990a)]; see also [Leroy and Rancoita (2000); SICAPO Collab. (1991b)]). $R(\%)$ is defined as: $R[\%] = 100\frac{\Delta\epsilon}{\epsilon}(\mathrm{G}10) = 100\left[\frac{\epsilon_{\mathrm{vis}}-\epsilon_{\mathrm{vis}}(\mathrm{G}10)}{\epsilon_{\mathrm{vis}}}\right]$, where ϵ_{vis} is the visible energy without additional G10 plates and $\epsilon_{\mathrm{vis}}(\mathrm{G}10)$ is the visible energy with additional G10 plates.

(1990a, 1991b)], and Sect. 3.3.1). Consequently, the second term in Eq. (9.67) can be neglected, namely we have:

$$\frac{\Delta\epsilon}{\epsilon}(\pi, \mathrm{G}10) \approx (e/\pi)f_{em}\frac{\Delta\epsilon}{\epsilon}(e, \mathrm{G}10). \qquad (9.68)$$

Thus, from Eq. (9.68) we get $\frac{\Delta\epsilon}{\epsilon}(\pi, \mathrm{G}10 = 5\,\mathrm{mm}) \approx 1.22 \times 0.26 \times 40\% \approx 12.6\%$ in agreement with experimental data shown in Fig. 9.24.

The local hardening effect allows the tuning of the e/mip ratio in such a way that the compensation condition is achieved. In Fig. 9.25, the e/π ratio [SICAPO Collab. (1990a, 1991b, 1993c)] measured with the Si/U calorimeter, described above, is shown as a function of the thickness of G10 plates inserted. The experimental data show that the e/π ratio decreases from ≈ 1.22 up to ≈ 0.87, when 5 mm G10 plates are inserted. The compensation condition is obtained for a G10 additional thickness of $1.2 \pm 0.2\,\mathrm{mm}$. The calorimeter properties which determine the value of the e/π ratio are mostly due to the added low-Z material, which causes the occurrence of the local hardening effect.

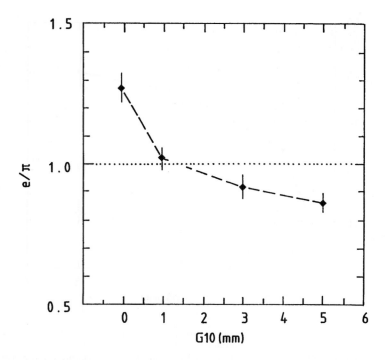

Fig. 9.25 Measured values of the e/π ratio for a Si/U calorimeter (see text) versus the thickness of additional G10 plates (reprinted from *Phys. Lett. B* **242**, Angelis, A.L.S. et al., Evidence for the compensation condition in Si/U hadronic calorimetry by the local hardening effect, 293–298, Copyright (1990), with permission from Elsevier, e.g., for the list of the authors see [SICAPO Collab. (1990a)]; see also [Leroy and Rancoita (2000); SICAPO Collab. (1991b)]). The line is to guide the eye.

The local hardening effect was achieved and studied for low-Z materials such as G10, Al and polyethylene. The reduction of the electromagnetic visible-energy depends, to some extent, on the thickness and type of the readout layers. In fact, for typical detector thicknesses of $400\,\mu$m silicon, $2.5\,$mm scintillator and $2\,$mm liquid argon, *mip*'s (which can be considered as scale units for energy deposition of fast particles) deposit ≈ 140, 500, and $440\,$keV, respectively, while the low-Z absorbers (for the local hardening effect) act mainly on soft electrons, generated in regions of the high-Z passive samplers very close to surfaces adjacent to the active readout media. This generated visible-energy reduction, $\frac{\Delta\epsilon}{\epsilon}(\text{low}-Z)$, is almost independent of the sampling fraction (Sect. 9.2.3), i.e., of the thickness of the high-Z passive absorber.

From Eq. (9.66), we introduce the fraction $R_{lh}(\text{low}-Z)$ defined as:

$$R_{lh}(\text{low}-Z) = 1 - \frac{\Delta\epsilon}{\epsilon}(\text{low}-Z) = \frac{\epsilon_{\text{vis}}(\text{low}-Z)}{\epsilon_{\text{vis}}}. \qquad (9.69)$$

$R_{lh}(\text{low}-Z)$ is independent of the thickness of the passive sampler, but depends on the thickness (and the location, i.e., in front or at the rear, see Sect. 9.2.3) of the low-Z material added for achieving the local hardening effect.

Let us consider incoming electrons and protons on calorimeters which have different high-Z thicknesses U and U', respectively. The ratio of the two e/π calorimeter ratios can be written as:

$$R(e/\pi) \equiv \frac{e/\pi(U, \text{low}-Z)}{e/\pi(U', \text{low}-Z)} = \frac{\epsilon_{\text{vis}}(e, U, \text{G10})}{\epsilon_{\text{vis}}(e, U', \text{G10})} \frac{\epsilon_{\text{vis}}(\pi, U', \text{G10})}{\epsilon_{\text{vis}}(\pi, U, \text{G10})}. \tag{9.70}$$

From Eqs. (9.6, 9.69), the electromagnetic visible-energy becomes:

$$\epsilon_{\text{vis}}(e, U, \text{G10}) = \epsilon_{\text{vis}}(e, U) R_{lh}(\text{G10}) = R_{lh}(\text{G10}) (e/mip) E_s(U), \tag{9.71}$$

where $R_{lh}(\text{G10})$ is almost independent of the sampling frequency (i.e., independent of the thickness of the high-Z passive sampler) and, as discussed in Sect. 9.2.2, e/mip becomes independent of the sampling frequency, τ, for values of τ larger than ≈ 0.8. Thus we have:

$$\frac{\epsilon_{\text{vis}}(e, U, \text{G10})}{\epsilon_{\text{vis}}(e, U', \text{G10})} \simeq \frac{E_s(U)}{E_s(U')}.$$

This way, the ratio $R(e/\pi)$ of Eq. (9.70) can be written as:

$$R(e/\pi) \simeq \frac{E_s(U)}{E_s(U')} \frac{\epsilon_{\text{vis}}(\pi, U', \text{G10})}{\epsilon_{\text{vis}}(\pi, U, \text{G10})}. \tag{9.72}$$

Figure 9.26 shows the ratio $R(e/\pi)$ for silicon calorimeters with U sampler 2.5 cm (data from [SICAPO Collab. (1995a)]) and 1.5 cm thick (data from [SICAPO Collab. (1990a)]) as function of the added G10 plate thickness. The ratio $E_s(2.5\,\text{cm U})/E_s(1.5\,\text{cm U})$ was calculated by means of Eq. (9.4). The ratios $\epsilon_{\text{vis}}(\pi, 1.5\,\text{cm U}, \text{G10})/\epsilon_{\text{vis}}(\pi, 2.5\,\text{cm U}, \text{G10})$ were evaluated from an average (over the incoming hadron energies of 8, 10, and 12 GeV) of the measured visible energies in the two experiments mentioned above. Experimental data indicate that the $R(e/\pi)$ ratio is quite consistent with 1, namely the e/π ratio does not depend on the thickness of the uranium passive samplers, but on the thickness of the added low-Z absorbers.

Once the compensating condition or an almost compensating condition (as a result of the local hardening effect) is achieved in calorimeters marginally sensitive to the energy carried by neutrons, it becomes possible to vary the calorimeter sampling frequency, while keeping the e/h ratio almost consistent with 1. Thus unlike the case of active detectors with hydrogeneous materials, we can operate compensating calorimeters in which *the ratio of the thicknesses of passive samplers to active readout media is variable*. In this way, it is possible to vary the contribution of sampling fluctuations to the overall calorimeter energy resolution (see next section) by modifying the sampling frequency, but keeping the compensation condition.

The achievement of the compensation condition, in the case of an absorber made of Pb and Fe, results from the exploitation of the filtering effect. This was demonstrated and systematically studied by an experiment performed by the SICAPO

Collaboration [SICAPO Collab. (1992a, 1993a)], using silicon detectors as readout medium. The sampling hadron calorimeter consisted of sets of absorbers made of Fe or Fe and Pb plates, interspaced with silicon readout mosaics. In each sampling, the thickness of Pb and Fe, as well as the respective positions of the absorbers plates, are kept constant. Each set of absorbers is 23 mm thick. The mosaic supporting structure is equivalent to ≈ 2.5 mm thick Fe absorber and this thickness has to be added to the overall thickness of the sampling absorbers. The incoming particle energies are 6, 8, 10, and 12 GeV. The energy deposited by the incoming electrons and pions was measured for the F-configuration FePb–Si–FePb [$\epsilon_{\rm vis}({\rm F})$] and for the R-configuration PbFe–Si–PbFe [$\epsilon_{\rm vis}({\rm R})$], where a reduction of the visible energy is expected, as a consequence of the filtering effect [SICAPO Collab. (1993a,c)] (see Sect. 9.3). In both configurations, the thickness of Pb was varied from 3 mm to 13 mm. The thickness of Pb was kept constant inside each sampling in each configuration. Data were also taken with Fe as the only absorbers for comparison.

Let us define the ratio

$$R_{\rm FR} = \frac{\epsilon_{\rm vis}({\rm F})}{\epsilon_{\rm vis}({\rm R})}.$$

For incoming electrons, we have that the value of $R_{\rm FR}$ corresponds to the ratio of the e/mip values for the F- and R-configurations [see Eq. (9.15) and Sect. 9.3.1], because the shared energy, E_s, does not depend on the sequence of absorbers [see Eqs. (9.4, 9.5)]:

$$R_{\rm FR}(e) = \frac{\epsilon_{\rm vis}(e, {\rm F})}{\epsilon_{\rm vis}(e, {\rm R})} = \frac{e/mip({\rm F})}{e/mip({\rm R})}. \tag{9.73}$$

$R_{\rm FR}(e)$ was found to increase almost linearly (Fig. 9.11) as a function of the Pb thickness, above a few mm of Pb absorber, when the filtering effect is acting.

For hadronic cascades we have to remember that the filtering effect acts on the electromagnetic component of the hadronic shower, and that $\epsilon_{\rm vis}(h, {\rm F}) \approx \epsilon_{\rm vis}(h, {\rm R})$ and $E_s({\rm F}) = E_s({\rm R})$, since there is no change of the thickness of samplers (passive and active) in the two configurations, but only an interchange of the passive sampler locations. As a consequence, h/mip values are unaffected. The second term of Eq. (9.67) can be neglected, when the equation expresses the visible-energy reduction for the case of the filtering effect; namely the visible-energy reduction between the F- and the R-configuration is expressed by Eq. (9.68). This equation can be rewritten as a function of electromagnetic visible energies as:

$$\frac{\Delta\epsilon}{\epsilon}(\pi, {\rm FR}) \equiv \left[\frac{\epsilon_{\rm vis}(\pi, {\rm R}) - \epsilon_{\rm vis}(\pi, {\rm F})}{\epsilon_{\rm vis}(\pi, {\rm R})} \right] = \left[1 - \frac{\epsilon_{\rm vis}(\pi, {\rm F})}{\epsilon_{\rm vis}(\pi, {\rm R})} \right]$$

$$= 1 - R_{\rm FR}(\pi) \approx e/\pi({\rm R})\, f_{em}\, \frac{\Delta\epsilon}{\epsilon}(e, {\rm FR})$$

$$= e/\pi({\rm R})\, f_{em} \left[1 - \frac{\epsilon_{\rm vis}(e, {\rm F})}{\epsilon_{\rm vis}(e, {\rm R})} \right]$$

$$= e/\pi({\rm R})\, f_{em}\, [1 - R_{\rm FR}(e)], \tag{9.74}$$

where

$$R_{FR}(\pi) = \frac{\epsilon_{vis}(\pi, F)}{\epsilon_{vis}(\pi, R)}$$

and $e/\pi(R)$ is the e/π ratio of the R-configuration; finally, we have:

$$R_{FR}(\pi) \approx 1 - e/\pi(R) \, f_{em} \, [1 - R_{FR}(e)]. \tag{9.75}$$

The mean energy deposited in the calorimeter $\epsilon_{vis}(e)$ (found to be linear as a function of the incident energy, e.g., see Figs. 9.9 and 9.10) decreases with the thickness of Pb in the absorber. Using identical passive absorber thicknesses, the values of the visible-energy, measured with the PbFe–Si–PbFe R-configuration [$\epsilon_{vis}(R)$], are lower than those measured with the FePb–Si–FePb F-configuration [$\epsilon_{vis}(F)$] for incoming both electrons and pions, as expected because of the filtering effect.

For instance, the ratio $R_{FR}(e)$ for incoming electrons is 1.23 ± 0.05 for a Pb thickness of 13 mm. The $e/\pi(R)$ ratio (Fig. 9.27), measured for the PbFe–Si–PbFe configuration with 13 mm of Pb plates, is 0.89 ± 0.01 [SICAPO Collab. (1992a,

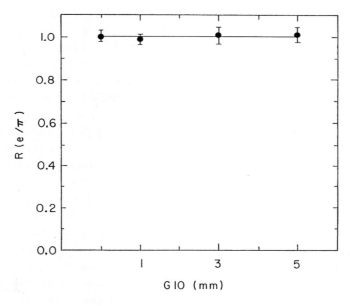

Fig. 9.26   The ratio $R(e/\pi)$ (reprinted from *Nucl. Instr. and Meth. in Phys. Res. A* **361**, Furetta, C., Leroy, C., Pensotti, S., Penzo, A. and Rancoita, P.G., Experimental determination of the intrinsic fluctuations from binding energy losses in Si/U hadron calorimeters, 149–156, Copyright (1995), with permission from Elsevier; see also [Leroy and Rancoita (2000)]) for a U sampler 2.5 cm (data from [SICAPO Collab. (1990a)]) and 1.5 cm (data from [SICAPO Collab. (1995a)]]) thick versus the added G10 plates thickness (on the front and on the rear the silicon mosaics). The silicon mosaics had supporting structure made by two G10 sheets of 0.2 and 1 mm thick, respectively. The line represents $R(e/\pi) = 1$.

1993a)]. Furthermore, for instance at 6 [12] GeV, the expected average fraction of the converted electromagnetic energy in a pion cascade is $f_{em} \approx (20\text{--}22)\%$ [(27--30)%], with an average value among 6, 8, 10 and 12 GeV of $\approx (24\text{--}26)\%$. Using Eq. (9.75), and measured $e/\pi(\mathrm{R})$, $R_{\mathrm{FR}}(e)$ and calculated f_{em} ratios, the estimated value of $R_{\mathrm{FR}}(\pi)$ for incoming pions becomes:

$$R_{\mathrm{FR}}(\pi) \approx 1 - 0.89 \times 0.25 \times (1 - 1.23) \approx 1.05.$$

This latter value is quite consistent within with the measured ratio:

$$R_{\mathrm{FR,measured}}(\pi) = 1.06 \pm 0.03.$$

The different rate of reduction of the visible energy for electrons and pions allows to tune the e/π ratio and, therefore, to achieve the compensation condition in a Si/Pb+Fe hadron calorimeter. A fundamental characteristic of the filtering effect, in PbFe–Si–PbFe calorimeters, is the linear decrease of the e/mip ratio as the Pb fraction, f [Eq. (9.14)], increases (Sect. 9.3). This dependence on f was experimentally observed [SICAPO Collab. (1989c)] for the entire f-range investigated, namely for f values between $\approx (5\text{--}36)\%$.

The e/π values for the PbFe–Si–PbFe R-configuration (from an average of measurements performed at incoming particle energies of 6, 8, 10, and 12 GeV) are shown in Fig. 9.27 (from [SICAPO Collab. (1992a)]) as a function of the thickness of Pb in the Pb+Fe absorber. It is observed that the e/π ratio decreases from the value 1.11 ± 0.02, when Fe is the only Fe absorber, to the value 0.89 ± 0.01 for a Pb thickness of 13 mm. From these experimental data, the thickness of Pb to be inserted in the absorber in order to achieve the compensation condition ($e/\pi = 1$) is estimated to be (5.4 ± 1.0) mm. The Pb thickness of (5.4 ± 1.0) mm corresponds to a fraction f of Pb present in the absorber of about (22--24)% (in calculating this fraction, the thickness of Fe takes into account the amount of material serving as support of the silicon mosaics). This experimental result is in agreement with the prediction, based on Monte-Carlo simulations, for which the compensation has to be achieved for $f \approx 25\%$ in a PbFe–Si–PbFe calorimeter with Fe absorber thickness of about 35 mm [SICAPO Collab. (1989c)] (see also [Brau and Gabriel (1989)]).

Other quasi compensating calorimeters using values of f between $\approx (20\text{--}25)\%$ were operated [SICAPO Collab. (1995b, 1996)]. Among these compensating (or almost compensating) calorimeters, the minimal thickness of Fe absorber sampler used is ≈ 15 mm (i.e., $\approx 0.85\,X_0$). In general, the ratios of the thicknesses of Pb absorbers to Fe absorbers (for which the compensation condition was achieved) are ≈ 0.94, when the ratio is calculated in units of radiation lengths, and ≈ 0.29, when the ratio is calculated in units of interaction lengths.

Once the appropriate value of f is maintained (as well as the configuration) for keeping the compensation, the overall passive sampler thickness can be readjusted, i.e., *the thicknesses of passive samplers can be varied* to modify the sampling fluctuations contribution to the overall hadronic energy resolution (see next section).

The filtering effect can also act, when other low-Z and high-Z materials are combined as passive samplers (Sect. 9.3). The value of f, required to achieve the compensation condition, depends on the h/mip ratio of the calorimetric structure. In fact, this value is related to the calorimeter response to the purely hadronic energy, i.e., also to the type of active detectors. As a consequence, the ratio (of ≈ 1) between the radiation lengths of Fe and Pb absorbers needs to be optimized, when replacing silicon readout with other active detectors.

The e/π ratios measured in the FePb–Si–FePb F-configuration (see [SICAPO Collab. (1993a,c)]) never achieved values close to compensation, except when the Pb thickness inserted in the absorber represents a fraction (f) larger than 50%. The decrease of the e/π ratio value, in the case of the FePb–Si–FePb F-configuration, is expected because the e/π ratio value for the Pb absorber is lower than the one for the Fe absorber. Thus, as the Pb fraction increases the overall e/π value has to decrease slightly. However, the compensation condition is not reached without a further reduction of the calorimeter response to the electromagnetic component. This reduction is achievable by exploiting the propagation properties of the electromagnetic shower in complex absorbers (as discussed above and in Sect. 9.3).

These experimental results were reproduced by simulations [Giani (1993)] using the GEANT program code [Brun et al. (1992)] and the FLUKA package [Aarnio

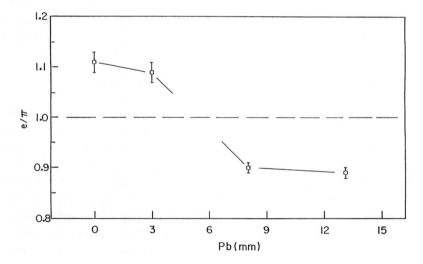

Fig. 9.27 The e/π ratio for PbFe–Si–PbFe R-configuration as a function of the thickness of Pb absorber in the passive sampler (the overall thickness, including the Fe plate, of the passive sampler is 23 mm). The data are from an average at incoming particle energies of 6, 8,10, and 12 GeV (reprinted from *Phys. Lett. B* **280**, Borchi, E. et al., Evidence for compensation in a Si/(Fe, Pb) hadron calorimeter by the filtering effect, 169–174, Copyright (1992), with permission from Elsevier, e.g. for the list of the authors see [SICAPO Collab. (1992a)]; see also [Leroy and Rancoita (2000)]). The line is to guide the eye.

et al. (1987)] for the hadronic interactions (see also [Hirayama (1992)]). These simulations, using the geometry and the exact composition of the sampling material (at the 1% level) of the experimental setup, have yielded results which are in good agreement with the measured e/π ratio values for both configurations [SICAPO Collab. (1993a)]. The simulations indicate that the precise knowledge of the composition of the supporting structure is a very important element for determining the agreement between experimental and Monte-Carlo data.

The local hardening and filtering effects can be combined. Soft electrons survived to the filtering effect in Pb-Fe samplers can be absorbed by additional thin (in units of radiation lengths) low-Z plates inserted in the calorimeter configuration next to the active detectors.

As already discussed (e.g., see Sect. 9.2.3), the local hardening effect modifies the e/mip ratio, also when other active media, like plastic scintillators, are employed. The maximum achievable e/mip ratio reduction was estimated [Wigmans (1988)] to be $\approx 8\%$ by EGS4 Monte-Carlo calculations. For instance, the Reconfigurable-Stack calorimeter (see [Byon-Wagner (1992); Beretvas et al. (1993); Job et al. (1994)], and references therein) used plastic scintillators as readout media and a combination of Pb (3.2 mm thick) and Fe (25.4 mm thick) absorbers as passive samplers, with a value $f \approx 11\%$. The plastic scintillators were cladded with 1.6 mm Al plates, in order to vary the mip ratio by exploiting the local hardening effect. The measurements [Job et al. (1994)] using F- and R-configuration have shown that the e/π ratio at 10 GeV was decreased by ≈ 0.17 (with respect to the one measured with the F-configuration), reaching the value of 1.10 ± 0.03 with the R-configuration.

9.10   Compensation and Hadronic Energy Resolution

The particle multiplication in calorimeters is based on statistical processes (although due to different types of interactions in matter) for both electromagnetic, already discussed in Sect. 9.4.1, and hadronic cascades. In sampling devices the hadronic visible-energy depends (Sects. 9.8, 9.9) on the number of secondary ionizing particles and on particles (like fast neutrons) able to produce ionizing recoil protons in hydrogeneous active media or ionizing recoil nuclei in any other detectors. Statistical fluctuations (e.g., see [Amaldi (1981); Fabjan (1985a, 1986)]) of the number of created and recoil particles cause fluctuations of the visible-energy deposited in the readout detectors and, consequently, limit the obtainable *energy resolution* which, for an incoming hadron with energy E, is given by

$$\frac{\sigma(E)}{E} \equiv \frac{\sigma(\epsilon_{\mathrm{vis}})}{\epsilon_{\mathrm{vis}}(\pi)}, \tag{9.76}$$

where $\sigma(\epsilon_{\mathrm{vis}})$ is the standard deviation of the distribution of the hadronic visible-energy $\epsilon_{\mathrm{vis}}(\pi)$.

Because of the statistical nature of the cascade phenomenon, the energy resolution is expected to improve, that is $\sigma(E)/E$ decreases, with the increase of the energy. In electromagnetic sampling calorimetry (Sect. 9.4), we have seen that the linear relationship between the incoming energy and the visible energy determines both the $1/\sqrt{E}$ dependence of the energy resolution and the Gaussian-like distribution of $\epsilon_{\text{vis}}(e)$.

For hadronic fully contained showers, the resulting visible-energy distribution is almost Gaussian (deviations may occur in the tail of the distribution [Fabjan et al. (1977)]) with a peak located at the position of the mean visible-energy. However, generally the energy resolution does not scale as $1/\sqrt{E}$. Therefore in hadronic calorimetry, the energy resolution is worsened.

In sampling calorimeters, only a small fraction of the visible energy is deposited in readout detectors. As in sampling electromagnetic devices, fluctuations in the number of crossed active samplers by ionizing particles and in the deposited energy by collision loss processes mostly contribute to broaden the visible-energy distribution and to build up the so-called *sampling fluctuations* (e.g., [Amaldi (1981); Wigmans (1987)]). Furthermore, a non-negligible fraction of the incoming energy is invisible (see Sect. 3.3.1), since it is spent in processes like the nuclear break-up. Fluctuations in the amount of this invisible energy result in the enlargement of the visible-energy distribution and constitute the *intrinsic resolution* (e.g., see [Amaldi (1981); Fabjan (1986); Wigmans (1987); Brückmann et al. (1988); Fesefelt (1988)]). These nuclear processes (as discussed in Sect. 3.3) are not present in electromagnetic showers. In hadronic cascades, the visible energy, $\epsilon_{\text{vis}}(\pi)$, is usually no longer proportional to the incoming hadron energy, since there is always an electromagnetic component, whose fraction varies with the energy (Sect. 3.3). This fraction, f_{em}, undergoes fluctuations on an event-to-event basis (see Sect. 3.3.1 and, e.g., [Amaldi (1981); Fabjan (1985a, 1986); Wigmans (1987); Fesefelt (1988); Wigmans (1988)]). As the e/π ratio becomes quite different from 1 (i.e., as the calorimeter becomes less and less compensating), the contribution, of fluctuations of a non-Gaussian nature, to f_{em} is more and more important and affects the $1/\sqrt{E}$ behavior of the energy resolution. This latter one, in turn, becomes related to the e/π ratio [Amaldi (1981); Wigmans (1987)]. In Figs. 9.28 and 9.29, measured hadronic energy resolutions are shown for a calorimeter employing hydrogeneous readout [Catanesi et al. (1987); HELIOS Collab. (1987); Acosta et al. (1991)] and for silicon readout [SICAPO Collab. (1996)]. The data indicate[||] that the calorimeter energy resolution scales as $1/\sqrt{E}$, but an additive term is present for non compensating calorimeters. This term increases, as expected, as the calorimeter becomes less and less compensating.

The *energy resolution* $\sigma(E)/E$ of a *sampling hadron calorimeter* (when instru-

---

[||] These data were confirmed by other measurements, e.g., see [Bleichert et al. (1987); Baumgart et al. (1988a); Ros (1991); Wigmans (1991); Fabjan (1995b); SICAPO Collab. (1995a,b)] and references therein.

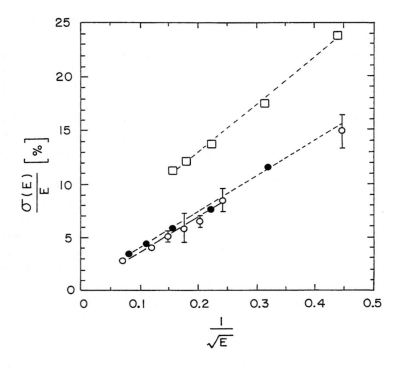

Fig. 9.28   Energy resolution $\sigma(E)/E$ in [%] measured by calorimeters with hydrogeneous detectors as a function of $1/\sqrt{E}$, where E is the incoming hadron energy in GeV (from [Leroy and Rancoita (2000)]): uranium/scintillator ($\circ$) compensating calorimeter [HELIOS Collab. (1987)]; data corrected for effects of light attenuation from the lead/scintillating fiber ($\bullet$) calorimeter [Acosta et al. (1991)] (with $e/\pi \sim 1.1$ at 10 GeV), uranium/scintillator ($\square$) calorimeter [Catanesi et al. (1987)] (with $e/\pi \sim 0.8$ at 10 GeV). The lines are to guide the eyes.

mental contributions can be neglected) is expressed as [Wigmans (1987, 1988, 1991)]:

$$\frac{\sigma(E)}{E} = \frac{C}{\sqrt{E}} = \frac{C_0}{\sqrt{E}} + \phi(e/\pi), \tag{9.77}$$

where E is the incoming hadron energy in GeV;

$$C_0 = \sqrt{\sigma_{\text{intr}}^2 + \sigma_{\text{samp}}^2}$$

contains the contributions from the intrinsic resolution (σ_{intr}), mostly due to nuclear binding energy-losses, and from the sampling fluctuations (σ_{samp}), added in quadrature; the function $\phi(e/\pi)$ takes into account effects related to non compensation; and, finally,

$$C = C_0 + \phi(e/\pi)\sqrt{E}.$$

In Eq. (9.77), the function $\phi(e/\pi)$ is called the *constant term* and dominates the energy resolution at high energy, since C_0 is almost independent of energy. Experimental data show that $\phi(e/\pi)$ vanishes, when the compensation condition is

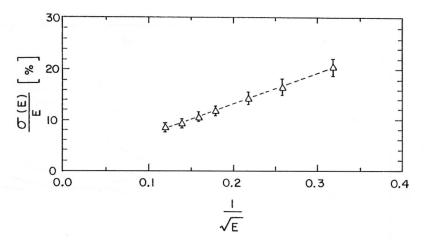

Fig. 9.29   Energy resolution, $\frac{\sigma(E)}{E}$ in [%], measured by an almost compensating PbFe–Si–PbFe calorimeter and corrected for the effects of visible-energy losses as a function of $1/\sqrt{E}$, where E is the incoming hadron energy in GeV (from [Leroy and Rancoita (2000)]; data from Table 1 of [SICAPO Collab. (1996)]). The line is to guide the eyes.

achieved, i.e., $\phi(e/\pi \sim 1) \sim 0$. One can also fit the experimental data to an expression that adds the scaling and the constant term in quadrature (see e.g., [Catanesi et al. (1987); Young et al. (1989); Acosta et al. (1991); Fabjan (1995b)] and references therein). However, it is interesting to note that WA80 Collaboration obtained results, favoring Eq. (9.77) over a very wide range and up to high energies [Young et al. (1989)].

As mentioned before, the small fraction of the energy sampled in the active samplers is at the origin of sampling fluctuations σ_{samp}. Their contribution to the overall energy resolution [Eq. (9.77)] was measured [Fabjan et al. (1977)] for low-A and high-A passive samplers and found to depend on the amount of energy lost in a passive sampler (e.g., see [Fabjan and Ludlam (1982)]). Further measurements [Tiecke (1989); Drews et al. (1990)] have confirmed such a dependence (e.g., [SICAPO Collab. (1995a,b, 1996)]). The contribution to the energy resolution from sampling fluctuations can be written as [Drews et al. (1990)]:

$$\left(\frac{\sigma}{E}\right)_{\text{samp}} = \frac{\sigma_{\text{samp}}}{\sqrt{E}} \approx 0.115\sqrt{\frac{\Delta\epsilon_{mip}[\text{MeV}]}{E}}, \tag{9.78}$$

where $\Delta\epsilon_{mip}$ is the energy deposited by a minimum ionizing particle (*mip*) in a passive sampler, whose thickness is supposed to be kept constant in the whole calorimeter. As an example, for hadron calorimeters with U samplers of 4 mm thickness, the contribution of sampling fluctuations to the overall energy resolution is:

$$\left(\frac{\sigma}{E}\right)_{\text{samp}} \approx \frac{0.33}{\sqrt{E}}.$$

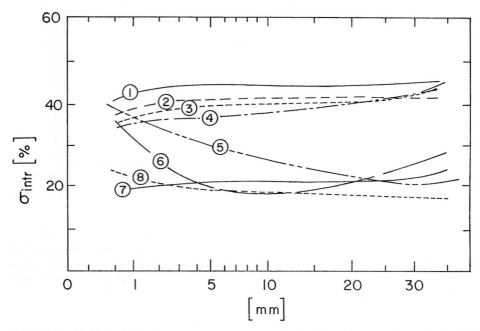

Fig. 9.30 Intrinsic resolutions (σ_{intr}) in [%] calculated by means of Monte-Carlo simulations for different passive and active samplers (namely liquid argon [LAr], scintillator [PMMA] and silicon detector [Si]) as a function of the passive sampler thickness in mm (from [Leroy and Rancoita (2000)], see also [Wigmans (1987)] for details). The curves are for: (1) Pb/0.4 mm Si $\sim$ Pb/2.5 mm LAr and no neutron capture, (2) Pb/0.4 mm Si $\sim$ Pb/2.5 mm LAr for 20% neutron capture, (3) U/0.4 mm Si for 20% neutron capture, (4) U/2.5 mm LAr for 20% neutron capture, (5) Pb/2.5 mm PMMA, (6) U/2.5 mm PMMA, (7) Fe/0.4 mm Si $\sim$ Fe/2.5 mm LAr for 20% neutron capture, (8) Fe/5 mm PMMA.

For an Fe (Pb) sampler with a thickness of 6 mm, one finds $\approx \frac{0.30}{\sqrt{E}} \left(\frac{0.32}{\sqrt{E}} \right)$.

The intrinsic resolution was also studied with Monte-Carlo simulations[††] and measured, usually, with compensating or quasi-compensating calorimeters[‡‡].

In Fig. 9.30, intrinsic resolutions (σ_{intr}) from [Leroy and Rancoita (2000)] (see also [Wigmans (1987)]) are shown for low-A (Fe) and high-A (Pb and U) passive samplers, some active media, a small percentage of neutron capture and no neutron capture (see [Wigmans (1987); Leroy and Rancoita (2000)] for details). These results show that, for non-hydrogeneous active media, the intrinsic resolution is larger than for hydrogeneous detectors, since the amount of invisible energy is decreased

---

[††]One can see [Amaldi (1981); Brau and Gabriel (1985); Wigmans (1987); Brückmann et al. (1988); Fesefelt (1988); Wigmans (1988); Brau and Gabriel (1989); Brau, Gabriel and Rancoita (1989); Wigmans (1991)] and references therein.

[‡‡]The reader can see, for instance, [Fabjan et al. (1977); HELIOS Collab. (1987); Tiecke (1989); Drews et al. (1990); SICAPO Collab. (1995a,b, 1996)] and references therein.

when its fluctuation is reduced. As a consequence, *compensating calorimeters with hydrogeneous readout are expected to minimize the contribution from the intrinsic resolution.* However these latter ones (Sect. 9.9.1) require an almost fixed ratio between passive and active sampler thicknesses.

For non-hydrogeneous readout, the neutron contribution to the overall visible-energy is small and the estimated intrinsic energy resolutions are

$$\left(\frac{\sigma}{E}\right)_{\text{intr}} = \frac{\sigma_{\text{intr}}}{\sqrt{E}} \approx \begin{cases} \frac{0.2}{\sqrt{E}} \text{ for low-}A \text{ (like Fe) absorbers,} \\ \\ \frac{0.4}{\sqrt{E}} \text{ high-}A \text{ (like Pb or U) absorbers.} \end{cases}$$

These values depend weakly on the passive sampler thickness. The intrinsic resolutions, σ_{intr}, were determined for low-A and high-A passive samplers (Table 9.9) and are in agreement with calculated ones shown in Fig. 9.30.

For example, in compensating calorimeters with hydrogeneous readout, i.e., with a ratio $R_{p/a}$ of ≈ 1.2 for U samplers (Sect. 9.9.1), the intrinsic energy resolution is $\approx 0.20/\sqrt{E}$. An overall energy resolution [HELIOS Collab. (1987)] of $\approx 0.34/\sqrt{E}$ (Fig. 9.28) was achieved with U samplers, each with a thickness of 2.9% of an interaction length. Also the tuning of the *mip* ratio allows one to reach the compensating condition in Pb+Fe calorimeters (Sect. 9.9.2). In this case, the thickness ratio of the Fe absorber* to the Pb absorber is ≈ 3. The measured intrinsic energy resolution is $\approx 0.25/\sqrt{E}$. For an almost-compensating Pb+Fe calorimeter (see Table 1 of [SICAPO Collab. (1996)]), the overall energy resolution achieved is $\approx 0.66/\sqrt{E}$ (Fig. 9.29) with passive samplers, each with a thickness of 14.3% of an interaction length. For this type of calorimeter the energy resolution can be improved by decreasing the passive sampler thickness (while keeping the Fe to Pb sampler thickness ratio fixed, see Sect. 9.9.2) to reduce the contribution of the sampling fluctuations to the overall energy resolution [Eq. (9.78)]. As already discussed, for these compensating (or almost compensating) calorimeters the $\phi(e/\pi)$ term is negligible and the calorimeter energy resolution scales as $1/\sqrt{E}$ (see Figs. 9.28 and 9.29).

9.10.1 *Non-Compensation Effects and the $\phi(e/\pi)$ Term*

The function $\phi(e/\pi)$ in Eq. (9.77) takes into account effects of non-compensation. It should be emphasized that there exists no analytical expression for the function $\phi(e/\pi)$, since it depends on several complex processes at the nuclear level, which are at the basis of the e/π ratio value. This ratio (Sect. 9.8) depends on both the active and passive samplers. Measured e/π ratios, mostly at the reference energy of 10 GeV, were given in Sect. 9.9 for different types of calorimeters (for estimated values based on Monte-Carlo simulations see references cited on page 670). Their values cover a wide range: $\approx (0.8\text{--}1.3)$.

---

*The contribution of Fe absorbers to the overall intrinsic resolution is lower than the one due to Pb absorbers.

Table 9.9 Intrinsic resolutions σ_{intr} in [%], measured with hydrogeneous media (plastic scintillators) and non-hydrogeneous media (silicon detectors) from [Leroy and Rancoita (2000)].

Passive sampler	Hydrogeneous readout [see(1)]	Non-hydrogeneous readout	Reference
Fe		(16.2–17.5)	[SICAPO Collab. (1995b)]
Fe		19.8±2.4	[SICAPO Collab. (1996)]
Fe+Pb [see(2)]		24.5	[SICAPO Collab. (1995b)]
Fe+Pb [see(2)]		25.9±2.1	[SICAPO Collab. (1996)]
Pb		(44.7–48.5)	[SICAPO Collab. (1995b)]
Pb		49.9±14.1	[SICAPO Collab. (1996)]
Pb	13.4±4.7		[Drews et al. (1990)]
Pb	11±5		[Tiecke (1989)]
U	20.4±2.4	47.8±1.9	[SICAPO Collab. (1995a)]
U	19±2		[Drews et al. (1990)]
U	22		[Tiecke (1989)]
U			[Fabjan et al. (1977); Fabjan (1985a)]

(1): measured with compensating or almost compensating calorimeters (Sect. 9.9.1).
(2): the ratio of thicknesses between Fe and Pb absorbers makes the calorimeters almost compensating (Sect. 9.9.2).

The function $\phi(e/\pi)$ vanishes for $e/\pi \sim 1$, i.e., for compensating (or almost compensating) calorimeters when the equalization of the calorimeter response to electromagnetic and hadronic cascades is achieved (Sects. 9.8, 9.9). Furthermore, as discussed so far, $\phi(e/\pi)$ mainly depends on the e/π ratio, but weakly (if at all) on both the sampling fluctuations and intrinsic resolution of the calorimeter.

The additional term $\phi(e/\pi)$ was estimated using Monte-Carlo simulations by Wigmans (1987) (see references given in Sect. 9.8, and also, e.g., Section 27.9.2 on *Hadronic calorimeters* in [PDB (2002)]) as a function of the e/π ratio and of the incoming hadron energy. *These simulations show that $\phi(e/\pi)$ depends slightly on the incoming hadron energy.* As an example, the value $\phi(e/\pi)$, when e/π ratio is between $\approx (1.11-1.26)$ at the reference energy of 10 GeV, increases by no more than ≈ 0.003 in the energy range (10–70) GeV ([SICAPO Collab. (1996)] and also e.g., [Wigmans (1987)]). From [SICAPO Collab. (1996)] for e/π ratio ≈ 1.11 at 10 GeV, we have $\phi(e/\pi) \approx 0.0094, 0.0101, 0.0104, 0.0109, 0.0115, 0.0122$ and 0.0128 at 10, 15, 20, 30, 40, 50 and 70 GeV respectively.

Experimental data on $\phi(e/\pi)$ obtained with calorimeters providing a ratio $e/\pi \sim 1.11$ at 10 GeV ([Acosta et al. (1991)], see also [SICAPO Collab. (1995a,b, 1996)] and references therein) are found to be in agreement with the simulations of [Wigmans (1987)].

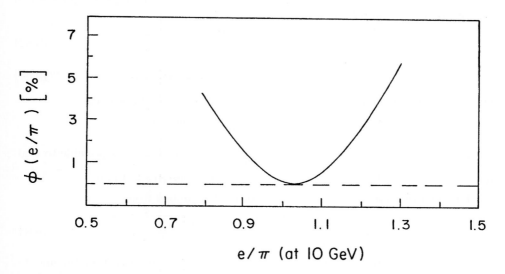

Fig. 9.31 $\phi(e/\pi)$ in [%] is shown as a function of the e/π ratio at the reference energy of 10 GeV (from [Leroy and Rancoita (2000)]).

In Fig. 9.31, $\phi(e/\pi)$ is shown in percentage (namely multiplied by a factor 100) as a function of the e/π ratio at the reference energy of 10 GeV. This curve was obtained by interpolating the simulation results of [Wigmans (1987)]. The function $\phi(e/\pi)$ is given, within a $\approx 10\%$ approximation, by:

$$\phi(e/\pi) \approx -0.207 \ln(e/\pi) - 4.5 \times 10^{-3} \quad \text{for } e/\pi < 0.955, \tag{9.79}$$

$$\phi(e/\pi) \approx 0.281 \ln(e/\pi) - 2.1 \times 10^{-2} \quad \text{for } e/\pi > 1.115. \tag{9.80}$$

For $0.95 < e/\pi < 1.10$, the function $\phi(e/\pi)$ is lower than 0.007 (see Fig. 9.31).

9.10.2 *Determination of Effective Intrinsic Resolutions*

The energy resolution, $\sigma(E)/E$, in non-compensating calorimeters does not scale as $1/\sqrt{E}$, owing to the non-vanishing term $\phi(e/\pi)$ (Sect. 9.10.1). The overall effect is to make the parameter C of Eq. (9.77) dependent on the incoming energy. We can rewrite Eq. (9.77) ([SICAPO Collab. (1995a,b, 1996)] and references therein) as:

$$\frac{\sigma(E)}{E} = \frac{C(E)}{\sqrt{E}} = \frac{\sqrt{\sigma_{\text{eff}}^2 + \sigma_{\text{samp}}^2}}{\sqrt{E}}, \tag{9.81}$$

where $\sigma_{\text{eff}} = \sqrt{C(E)^2 - \sigma_{\text{samp}}^2}$ and σ_{samp} is given by Eq. (9.78).

The *effective intrinsic resolution* σ_{eff} includes both the effect of the intrinsic resolution and the effect of non-compensation on the calorimeter energy resolution, the latter being accounted for by the constant term $\phi(e/\pi)$. Using Eqs. (9.78, 9.77), the effective intrinsic resolution can be expressed as:

$$\sigma_{\text{eff}} = \sqrt{C(E)^2 - \sigma_{\text{samp}}^2} \tag{9.82}$$

$$= \sqrt{\sigma_{\text{intr}}^2 + \phi^2(e/\pi)\, E + 2\phi(e/\pi)\, C_0 \sqrt{E}}. \tag{9.83}$$

For a compensating calorimeter [i.e., $e/\pi \sim 1$ and $\phi(e/\pi) \sim 0$], we get

$$\sigma_{\text{eff}} \equiv \sigma_{\text{intr}}.$$

Once the calorimeter energy resolution $\left[\frac{\sigma(E)}{E}\right]$ was measured, it is possible to evaluate $C(E) = \left[\frac{\sigma(E)}{E}\right]\sqrt{E}$ and the value of σ_{eff} from Eqs. (9.83, 9.77).

The intrinsic resolution [Eq. (9.83)] can be determined from

$$\sigma_{\text{intr}} = \sqrt{\left[C(E) - \phi(e/\pi)\sqrt{E}\right]^2 - \sigma_{\text{samp}}^2}, \tag{9.84}$$

where $\phi(e/\pi)$ can be estimated [see Sect. 9.10.1 and references therein, and for instance Eqs. (9.79, 9.80)], once the e/π ratio of the calorimeter was measured.

In Figs. 9.32, and 9.33, for instance, both σ_{intr} [σ_{samp} is obtained from Eq. (9.78)] and σ_{eff} [from Eq. (9.83)] are shown for silicon sampling calorimeters using U (Fig. 9.32) and Fe (Fig. 9.33) passive samplers. For the uranium calorimeter (Fig. 9.32), σ_{eff} was determined as a function of the G10 plate thickness, which are interleaved in order to achieve the local hardening effect ([SICAPO Collab. (1995a)]

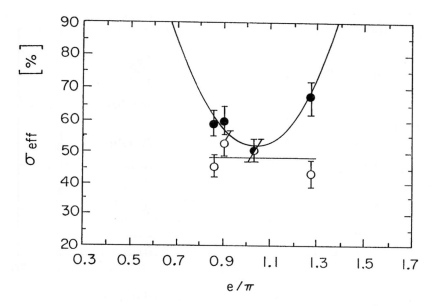

Fig. 9.32 σ_{eff} (•) in [%] as a function of the e/π ratio in a Si/U sampling calorimeter (reprinted from *Nucl. Instr. and Meth. in Phys. Res. A* **361**, Furetta, C., Leroy, C., Pensotti, S., Penzo, A. and Rancoita, P.G., Experimental determination of the intrinsic fluctuations from binding energy losses in Si/U hadron calorimeters, 149–156, Copyright (1995), with permission from Elsevier; see also [Leroy and Rancoita (2000)]). The intrinsic energy resolution (○) are also shown. The e/π ratio depends on the amount of G10 plates used for the local hardening effect (see also Sect. 9.9.2).

and Sect. 9.9.2). When the e/mip ratio is tuned to achieve the compensation condition by the local hardening effect, we can observe the effective intrinsic resolution approaching the intrinsic resolution. For the iron calorimeter (Fig. 9.33), the effective intrinsic resolution is given as a function of the incoming hadron energy [SICAPO Collab. (1996)]. σ_{eff} increases [Eq. (9.83)] because the constant term $\phi(e/\pi)$ is non vanishing for non-compensating calorimeters.

From the experimental data, we can observe that *the intrinsic resolution is independent of the incoming energy,* as expected. Measured values of intrinsic resolutions are given in Table 9.9 (data are from [Fabjan (1985a); HELIOS Collab. (1987); Tiecke (1989); Drews et al. (1990); SICAPO Collab. (1995a,b, 1996)]).

9.10.3 *Effect of Visible-Energy Losses on Calorimeter Energy Resolution*

In Sects. 9.10.1, 9.10.2, the energy resolution of the calorimeter was investigated for negligible visible-energy losses, i.e., the calorimeter physical sizes are large enough to almost fully contain the hadronic cascades. However, the energy resolution (as discussed in Sect. 9.4.2 for electromagnetic showers) is affected by the finite dimen-

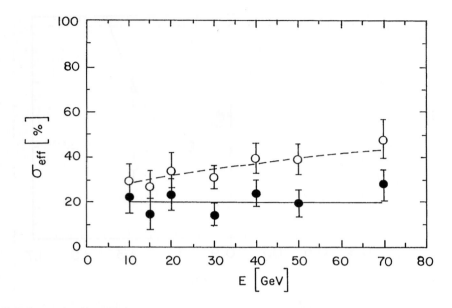

Fig. 9.33 σ_{eff} (o) in [%] in a Si/Fe sampling calorimeter (reprinted from *Nucl. Instr. and Meth. in Phys. Res. A* **368**, Furetta, C., Gambirasio, A., Lamarche, F., Leroy, C., Pensotti, S., Penzo, A., Rattaggi, M. and Rancoita, P.G., Determination of intrinsic fluctuations and energy response of Si/Fe and Si/Fe + Pb calorimeters up to 70 GeV of incoming hadron energy, 378–384, Copyright (1996), with permission from Elsevier; see also [Leroy and Rancoita (2000)]) as a function of the incoming hadron energy. The intrinsic energy resolution (•) are also shown.

sions of calorimeters, which can cause leakage of the deposited energy of hadronic showers. Furthermore in Sect. 9.4.2, we have seen that the limited shower containment degrades the energy resolution almost linearly with increasing energy-loss.

In a sampling hadron calorimeter, the measured energy resolution $\left[\frac{\sigma(E)}{E}\right]_{\text{exp}}$ is given by:

$$\left[\frac{\sigma(E)}{E}\right]_{\text{exp}} \equiv \left[\frac{\sigma(\epsilon_{\text{vis}})}{\epsilon_{\text{vis}}(\pi)}\right], \qquad (9.85)$$

where $\sigma(\epsilon_{\text{vis}})$ is the standard deviation value of the Gaussian-like distribution of the hadronic visible-energy, $\epsilon_{\text{vis}}(\pi)$, measured in the calorimeter when no correction for energy losses are included. These losses can be longitudinal, lateral, and dead-area visible-energy losses (as discussed in Sect. 9.4.2 for electromagnetic calorimeters). The latter ones have particularly to be considered for calorimeters with high granularity, as is the case for silicon readout.

The effects of energy losses on the energy resolution were experimentally measured. In particular, for hadronic cascades, the linear dependence of the hadronic resolution on lateral and longitudinal energy-losses was observed up to energy losses of about 15% [Amaldi (1981)] (see Fig. 9.34). To a first approximation (for energy

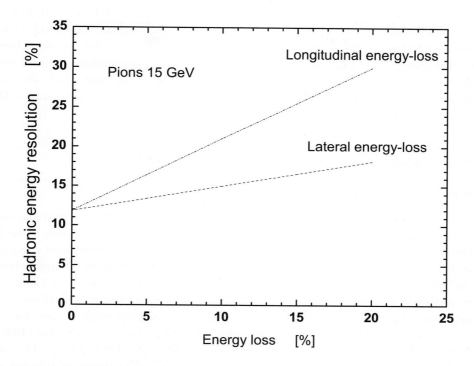

Fig. 9.34 Hadronic energy resolution, $\sigma(E)/E$ in [%], measured as a function of the fraction in [%] of lateral and longitudinal deposited-energy losses (curves from [Amaldi (1981)], see also [Leroy and Rancoita (2000)]) for the marble calorimeter of CHARM Collaboration [Diddens et al. (1980)].

losses up to 15%), the combined effect of the overall energy lost on the measured energy resolution can be expressed, following the discussion in [Amaldi (1981); SICAPO Collab. (1995b)] (and references therein), according to

$$\left[\frac{\sigma(E)}{E}\right]_{\exp} \approx \frac{\sigma(E)}{E}\left[1 + \sum_P (\lambda_P P)\right]$$

$$= \frac{C}{\sqrt{E}}\left[1 + \sum_P (\lambda_P P)\right], \qquad (9.86)$$

where P is the fraction in % of lateral (E_l), longitudinal (E_L) and dead area (E_D) energy losses to the total visible-energy; λ_P is a coefficient depending on the kind of energy lost (namely, lateral, longitudinal or dead area energy-loss); and $\frac{\sigma(E)}{E} = \frac{C}{\sqrt{E}}$ is the expected calorimeter energy resolution in absence of visible-energy losses, i.e., the one referred to in Eq. (9.77).

In Fig. 9.34 (see discussion in [Amaldi (1981); Leroy and Rancoita (2000)]), we can observe that fluctuations produced by lateral energy-losses are less effective, by a factor $\approx (2$–$3)$, than longitudinal energy-losses, as in the case of electromagnetic showers (Sect. 9.4.2 and references therein), i.e., $\lambda(E_l) < \lambda(E_L)$. Furthermore, it was also shown [SICAPO Collab. (1994a)] that the longitudinal energy-losses are as effective as the dead area energy-losses, i.e., $\lambda(E_L) \approx \lambda(E_D)$, since both of them are related to event-to-event fluctuations. The experimental data allowed Amaldi (1981) to determine the curves shown in Fig. 9.34, from which one can derive that, for hadronic cascades, the values of λ_l and $\lambda(E_L) \approx \lambda(E_D)$ are 0.027 and 0.077, respectively.

9.11 Calorimetry at Very High Energy

9.11.1 *General Considerations*

The concept of cascade and its propagation in matter were studied in previous sections at the energy range obtainable at particle accelerators. This section turns to the study of the cascade behavior in an energy regime beyond accelerator energies, such as the energies characterizing cosmic rays in a band from about 10^8 up to about 10^{20} eV, with a flux of particles decreasing rapidly with the increase of the energy. A considerable number of excellent articles and books have been published over the years covering the topic of cosmic ray physics and related subjects (see for instance [Greisen (1960); Berezinskii, Bulanov, Dogiel, Ginzburg and Ptuskin (1990); Knapp et al. (2003); Allard, Parizot and Olinto (2007); Matthiae (2008)]). The aim of this section is rather to discuss the physics of cascading processes relevant to cosmic rays[††].

From our point of view, cosmic rays are subatomic particles and nuclei (helium, carbon, nitrogen, oxygen up to iron), which hit the Earth's atmosphere. Primary cosmic rays, which are particles (such as electrons, protons, helium nuclei and nuclei synthesized in stars) accelerated at astrophysical sources, are distinguished from secondary cosmic rays produced by the interaction of primaries with the interstellar matter (the same cosmic rays can be considered as a part of the interstellar medium, see [McCray and Snow (1981); Berezinskii, Bulanov, Dogiel, Ginzburg and Ptuskin (1990)] and references therein). This distinction is not strict. While most of the observed electrons and protons are believed to be of primary origin, positrons and antiprotons are generally considered to be secondaries. However, there exists the possibility that a fraction of the observed positrons and antiprotons is of primary origin. For instance, the measurement of the energy spectrum of cosmic antiprotons provides important information on cosmic ray propagation models and might be sensitive to possible novel processes such as the annihilation of supersymmetric weakly interacting massive neutral particles, called neutralinos, or the evaporation of primordial black holes (e.g., see [AMS Collab. (1999)]).

Antiprotons, which are produced by collisions of cosmic rays with the interstellar

---

[††]For a general introduction to cosmic rays and trapped particle in the Earth magnetosphere, the reader may see Sects. 4.1.2.4 and 4.1.2.5.

medium, suffer a kinematical suppression and will show a characteristic energy spectrum at kinetic energies below 1 GeV, which falls off towards lower energies. Therefore, the antiprotons primarily produced by novel processes are predicted to emerge at these low energies (see, for example, [Jungman and Kamionkowski (1994)] for antiproton spectra from neutralino masses of 30 and 60 GeV/c^2). In Fig. 9.35, Local Interstellar energy Spectra (LIS) of antiprotons from different production models from [Webber and Potgieter (1989); Simon and Heinbach (1996); Gaisser and Schaefer (1997)] are shown. These models differ by the choice of main parameters and their estimated values (see, for example, [Boella et al. (1998)] for a discussion on these production models).

However for observations close to the Earth, it has to be considered that the Sun emits a plasma wind with an embedded magnetic field (Sect. 4.1.2.1), which may prevent the propagation of low energy cosmic ray particles inside the heliosphere. Furthermore in the heliosphere, interactions with particles coming from the Sun generate an adiabatic process of energy loss for the incoming cosmic rays. Thus, the LIS spectra are largely modified, particularly for energies up to a few GeV's. The effect of solar modulation depends on solar activity (Sect. 4.1.2.3) and results in a time dependence of the interstellar energy spectra at the Earth orbit (Fig. 9.36, see also [Boella et al. (1998)]).

The exact origin of cosmic rays is presently unknown. Charged particles which dominate (> 99.9%) hadronic cosmic radiation cannot be tracked back to their origin, since they are deflected by the weak galactic magnetic field and, then, the reach Earth uniformly. Therefore, high energy photons serve an important role, when one tries to find the origin of cosmic rays. Since they are uncharged and reach the Earth undeflected by the galactic magnetic field, their possible detection can lead to the identification of their source.

At present, data exist (see, for example, [Werber et al. (1991); Simpson (1983)]) on the energy spectra of various particles and nuclei from hydrogen to iron with kinetic energies of up to hundreds of GeV per nucleon (Fig. 9.37). The fraction of secondary nuclei in the cosmic rays decreases with the increase of the energy. Direct measurements of the intensity of protons and helium nuclei give energies[¶] of the order of 3×10^5 GeV/nucleon. For energies beyond 10^5 GeV, the spectra consist almost exclusively of data from the measurements of extended air showers. Between 10 and 10^6 GeV, the full cosmic ray spectrum is described by a power law as a function of the particle (kinetic-) energy [Yodh (1987)]:

$$\frac{dN_f}{dE} \propto E^{-\gamma_p}, \tag{9.87}$$

where $\gamma_p \approx 2.7$ (see, for instance, Sect. 4.1.2.4; [Berezinskii, Bulanov, Dogiel, Ginzburg and Ptuskin (1990); Gaisser (1990)] and references therein). Possible differential galactic spectra (i.e., with slightly different γ_p values[‡‡]) were observed for protons and helium nuclei [Randall and Van Allen (1986); AMS Collab. (2002)].

[¶] The reader can find the definition of kinetic energies per nucleon in Sect. 1.4.1.
[‡‡] γ_p is the differential spectral index see page 332.

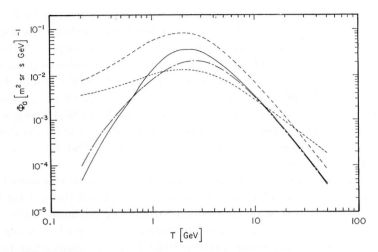

Fig. 9.35 Estimates (from [Leroy and Rancoita (2000)]) of interstellar antiproton fluxes (Φ_a) from production models as a function of the kinetic energy T in GeV: Webber and Potgieter (dashed line), leaky box [Webber and Potgieter (1989)]; Simon and Heinbach (dotted line), diffusive reacceleration [Simon and Heinbach (1996)]; Simon and Heinbach (dot-dashed line), leaky box [Simon and Heinbach (1996)]; Gaisser and Shaefer (solid line), leaky box [Gaisser and Schaefer (1997)].

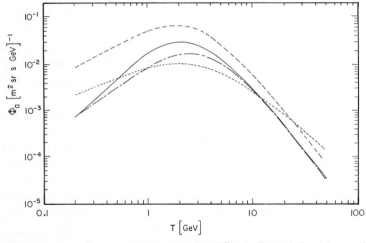

Fig. 9.36 Estimates (from [Leroy and Rancoita (2000)]) of effect of the solar modulation as a function of the kinetic energy T in GeV at minimum activity (Sect. 4.1.2.3) and at the Earth orbit for the interstellar antiproton fluxes (Φ_a) shown in Fig. 9.35.

A knee is observed in the spectrum at $E \approx 3 \times 10^6$ GeV. The knee is usually attributed to a change in the nature of the propagation of cosmic rays and/or a decrease in the efficiency of confining particles at very high energies in the Galaxy [Greisen (1966); Berezinskii, Bulanov, Dogiel, Ginzburg and Ptuskin

(1990)]. The particle spectrum presents an ankle between 10^9 and 10^{10} GeV. This ankle is usually interpreted as the result of the injection of a high energy flux of extragalactic origin [Greisen (1966)].

At low energies, where intensities are high, it is possible to study cosmic rays through direct measurements via detectors operated in balloons, satellites, space Shuttles and on board of the International Space Station Alpha. At energies beyond 10^{14} eV, the intensities are very low and the study of the cosmic rays is performed through cascade showers resulting from the degradation of the incident cosmic ray energy in the atmosphere. The extent of these showers is determined by the size of the incident particle (electrons, positrons, photons or hadrons) energy. The radiator volume is now very large since it consists of the whole Earth atmosphere from the ground level up to about an altitude of 30 km. Since the air shower is spread over a large area with a low flux of particles, detectors with large collection areas have to be used. Their array sensitive area cover up to several 10^4 m^2 [Bott-Bodenhausen et al. (1992)] and effective detection over a volume of up to 100 km^3 [Cassiday (1985)].

9.11.2   *Air Showers (AS) and Extensive Air Showers (EAS)*

The cosmic rays interact with the terrestrial atmosphere creating *air showers* (AS). These interactions produce secondaries through showering processes over long distances in the atmosphere. If the cosmic ray energy is high enough, the secondaries will reach eventually the ground level where they are detected. The atmosphere, which acts as a very thick and low density radiator, forces the spread of secondaries over a large area into so-called *extensive air showers* (EAS). This spread disperses the secondaries far enough apart to resolve and count them. In principle, these secondaries bring information about incident time, direction and energy.

The ratio of extragalactic cosmic rays to local cosmic ray fluxes is thought to be 10^{-4}–10^{-5}. If the universe were baryon symmetric, about half of the extragalactic cosmic rays should be antimatter. Fragmentation processes in the galaxy of origin and modulation in our galaxy will reduce the flux of antimatter. Therefore, experiments with an antiproton flux sensitivity at the level of (or better than) 10^{-3} m^{-2}sr^{-1}s^{-1}GeV^{-1} with an antiproton–proton separation capability at the 10^{-7} level or better are required to study primordial antimatter.

A component from solar origin is present in primary cosmic rays. Variations in the magnitude of this component are clearly correlated with solar activity [Berezinskii, Bulanov, Dogiel, Ginzburg and Ptuskin (1990)]. As mentioned above, the significance of this solar component is small for kinetic energies above 1 GeV and consequently these particles are not generally catalogued as cosmic rays. Thus, the main part of the primary cosmic rays reach the Earth's atmosphere from interstellar space, produced in our Galaxy with the exception perhaps of particles with ultra high energies $(10^8$–$10^{10})$ GeV which are presumably of extragalactic origin, where they were accelerated by very large magnetic fields or produced by large size objects or regions.

There are hadron showers initiated by hadrons, and electromagnetic showers generated by electrons, positrons and photons. The EAS's generated by hadrons develop an electromagnetic component, since an important fraction of the secondaries produced in the process is made of π^0 and η mesons, which decay via electromagnetic interaction and generate their own electromagnetic showers (as explained in Sect. 3.3). Electromagnetic showers or electromagnetic components contain photons, electrons and positrons. The mechanism of energy loss for these particles is pair production for photons and bremsstrahlung for electrons and positrons. Thus, the number of photons, electrons and positrons rapidly increases with the atmosphere depth until the electron and positron energy is down to the critical energy ($\epsilon_c = 81\,\mathrm{MeV}$ in air), afterwards which they will lose the remaining energy through ionization and the number of particles will decrease.

Muons are also involved in the generation of hadron and electromagnetic showers. The charged mesons ($\pi^\pm$, $K^\pm$) produced in the hadron showers decay into muons and are responsible for the presence of a large muon component in hadron initiated showers. This component has a broad distribution since the muons are produced high in the atmosphere at the start of the hadron shower. Muons can also be produced in electromagnetic induced showers through the decay of photoproduced $\pi^\pm$ and $K^\pm$, but less copiously with respect to hadronic induced shower, since $\sigma(\gamma\text{-air})/\sigma(\text{nucleon-air}) \approx 1.4\,\mathrm{mb}/300\,\mathrm{mb}$. Another source of muons in an electromagnetic induced shower is hadroproduction by electrons, but this contributes even less than photoproduction. Thus, the lateral spread of hadron initiated showers is much wider than that of electromagnetic showers, especially at low altitude. As was the case for calorimeters operated with accelerators (Sect. 9.8), a large fraction of the hadron shower energy $(20\text{--}30)\%$ goes into nuclear excitation or is carried away by neutrinos. These observations have consequences for the choice of techniques used in the study of high energy cosmic rays.

The high energy cascade, produced by the interaction of ultra high energy particles entering the top of the atmosphere, can be detected by either the observation of the electromagnetic radiation emitted in the atmosphere by the shower particles via Čerenkov radiation or visible nitrogen fluorescence, or by the direct observation of the particles in the cascade. The combination of several techniques of measurement allows the identification of particles necessary for the search of point sources of high energy cosmic radiation. The separation between γ's and hadrons is essential to this search since γ (like ν), as already stressed, can possibly be traced back to their sources. In contrast, the sources of charged particles escape retracing since they reach the Earth surface uniformly after being deflected by the galactic magnetic field. The γ-hadron separation requires high angular resolution and the possibility of measuring the muon component in the electromagnetic induced showers.

Air Čerenkov telescopes are used for the observation of the fast Čerenkov light flash emitted by a shower generated in the atmosphere [Weekes (1988)]. The charged

Table 9.10   The value of the refractive index, Čerenkov angle (θ_c) and Čerenkov threshold momenta for electrons $[p_c(e)]$, pions $[p_c(\pi)]$ and protons $[p_c(p)]$, as a function of altitude from [Leroy and Rancoita (2000)].

Altitude (km)	Refractive index	θ_c (degree)	$p_c(e)$ (MeV)	$p(\pi)$ (GeV)	$p_c(p)$ (GeV)
30	1.00000424	0.17	175	48	322
20	1.00001734	0.34	87	24	159
15	1.00003506	0.48	61	17	112
10	1.00007091	0.68	43	12	79
8	1.00009398	0.78	37	10	68
6	1.0001245	0.90	32	9	59
4	1.0001651	1.04	28	8	52
2	1.0002188	1.20	24	7	45

particles traveling at velocities greater than the speed of light in a medium emit electromagnetic radiation via the Čerenkov effect (Sects. 2.2.2, 9.5). EAS's contain relativistic charged particles with momenta above the Čerenkov threshold. The Čerenkov angle of the emitted radiation and the Čerenkov threshold particle velocity are given by Eqs. (2.132, 2.134), respectively. Since the Čerenkov effect is related to the refractive index of the traversed medium, the altitude will be a factor affecting the detection of EAS's by their Čerenkov light. At very high altitude i.e., in the upper layers of the Earth atmosphere, the atmospheric pressure is low and the refractive index is close to unity. As the altitude decreases, the refractive index and the Čerenkov angle increase, while the particle threshold momentum decreases (Table 9.10).

The observation of Čerenkov light, produced in the atmosphere by highly relativistic particles in EAS's, provides a tool for investigating the longitudinal structure of the EAS. The intensity of the Čerenkov light is proportional to the total energy dissipated in the atmosphere. This method of measurement provides a good angular resolution. When the shower size is very large, the effective detecting area will be extended by observing the scintillation light, produced by EAS particles at high altitudes or the scattering of electromagnetic waves by an ionized column produced by the EAS. The application of such methods allows the study of EAS's of very large size that can be due to primary cosmic rays of energies greater than 10^{20} eV.

The main shower development at high altitudes can be observed by its radial Čerenkov light pattern at the Earth ground level. At high altitudes, the lateral dispersions of electromagnetic and hadron induced showers are quite different and the measurement of the Čerenkov light provides a way to separate photons from hadrons. The interactions of the charged particles radiating Čerenkov light with the atmosphere modify their trajectories and, in particular, the multiple Coulomb scattering spreads the electron paths. The lateral spread of the shower and the fact that many particles in the shower have a momentum close to the threshold momentum i) create a situation of overlap for photons generated at different altitudes and ii) complicates in practice the measurement of the γ/hadron ratio. Therefore, there is a need to combine Čerenkov measurements with scintillators or tracking chamber arrays that sample the shower tail, when it reaches the Earth ground level. Significant progress in the understanding of the cascading mechanisms, γ-hadron separation, and improved sensitivity in the search for point-like sources or diffuse γ radiation is expected from the combination of Earth based scintillator and muon detector arrays with a matrix of air Čerenkov counters recording the shower parameters from ground level up to the higher layers of the atmosphere [Bott-Bodenhausen et al. (1992)].

The passage of EAS's through the atmosphere can also be detected via the measurement of the nitrogen fluorescence light given off by relativistic charged particles in the shower as performed in the Fly's Eye detector [Cassiday (1985)]. The main difference between fluorescence and Čerenkov light lies in the angular distribution as the Čerenkov light is distributed along the shower direction while the fluorescence light has an isotropic distribution. The distribution of the number of photons of fluorescence is approximated by [Chiavassa and Ghia (1996)]:

$$\frac{d^2 N_f}{dl\, d\Omega} = \frac{y_l N_e}{4\pi}, \tag{9.88}$$

where N_e is the number of electrons in the EAS (see Sect. 9.11.3.1 below) and y_l is the fluorescent yield $\approx 4\,\gamma$/electron/m. The fact that fluorescence light is emitted isotropically from the EAS permits experiments such as the Fly's Eye detector [Cassiday (1985)] to detect at large distances. The experiments carried out with this detector include: i) a direct measurement of the proton–air cross section (at $\sqrt{s} = 30\,\text{TeV}$), ii) an analysis of the primary cosmic ray spectrum in the energy range $(10^{16}$–$10^{20})\,\text{eV}$, iii) an extraction of the composition of the high energy cosmic ray primaries, iv) a search for anisotropies in arrival directions, v) a search for deeply penetrating showers indicative of primary neutrinos, possible heavy-lepton production and quark matter in the primary flux and, finally, vi) a search for sources of γ rays near $10^{15}\,\text{eV}$.

9.11.3 *Electromagnetic Air Showers*

9.11.3.1 *Longitudinal Development*

The longitudinal development of the electromagnetic cascades is parameterized as a function of the *age* parameter s. For each sub-shower induced by photon (from π^0 and η decays) or by electrons of energy E, the age s of this sub-shower at depth t is given by

$$s = \frac{3t}{t + 2y}, \tag{9.89}$$

where t is the depth measured in radiation length units ($\sim 37.0\,\text{g/cm}^2$ in air) and $y = \ln(E_0/\epsilon_c)$; E_0 is the energy of the primary photon or electron, and ϵ_c is the critical energy (as given in Sect. 2.1.7.4, $\epsilon_c = 81\,\text{MeV}$ for air).

The age parameter evolves from $s = 0$ at the point of the first interaction to $s = 1$ at the shower maximum, and continues to increase $(s > 1)$ beyond the shower maximum. The longitudinal development of an electromagnetic cascade can be described by a parametrization of the number of charged particles (electrons and positrons), N_e, in the shower as a function of the depth [Hillas (1982)]:

$$N_e(E_0, E, t) = \frac{0.31}{\sqrt{y}} \exp\{t[1.0 - 1.5\ln(s)]\}. \tag{9.90}$$

Here E is the threshold energy of electrons. N_e is often called the *size of the shower*. As an example of an application, Eq. (9.90) was used in [Fenyves et al. (1988)] with a modified form in order to describe the longitudinal development of the electromagnetic component of EAS's generated by 10^{14}–10^{16} eV proton and iron nuclei:

$$N_e{}^f(E_0, E, t) = \alpha_p\, A(E)\, N_e(E_0, E, t_1), \tag{9.91}$$

where α_p is the number of primary particles (photons or electrons) generating the shower; $A(E)$ is the fraction of electrons having energies larger than E compared with the total number of electrons. The modified age parameter, s_1, is calculated as a function of the modified depth, t_1, according to:

$$s_1 = \frac{3t_1}{t_1 + 2y}. \tag{9.92}$$

The modified depth, t_1, is given by

$$t_1 = t + a_{\pi,\gamma}(E), \tag{9.93}$$

where $a_\pi(E)$ and $a_\gamma(E)$ account for the different development and different $t_{\max}$ values of electron- and photon-induced showers, respectively, with different electron threshold energies ($t_{\max}$ is the depth where the shower reaches its maximum). Examples of values for $A(E)$, $a_\pi(E)$ and $a_\gamma(E)$ obtained from fits to the data [Fenyves et al. (1988)] are given in Table 9.11.

9.11.3.2 *Lateral Development*

The lateral distribution of particles in an electromagnetic shower is usually described by a parametrization suggested by Nishimura and Kamata (1958):

$$f(r/R_M, s, E) = C(s) \left(\frac{r}{R_M}\right)^{s-2} \left(1 - \frac{r}{R_M}\right)^{s-4.5}, \tag{9.94}$$

Table 9.11 Values of the parameters for the longitudinal and lateral development of electromagnetic air showers from [Leroy and Rancoita (2000)].

E	5 MeV	10 MeV	15 MeV	20 MeV
$A(E)$	0.67	0.59	0.52	0.48
$a_\pi(E)$	0.60	0.80	0.92	1.0
$a_\gamma(E)$	0.00	0.20	0.32	0.40
$b_\pi(E)$	0.20	0.40	0.52	0.60
$b_\gamma(E)$	-0.40	-0.20	-0.08	0.00

where s is given by Eq. (9.89), E is the threshold energy of the charged particle, r is the perpendicular distance from the shower axis, and R_M is the Molière radius at the level of observation (80 m at sea level and 100 m at 2 km); furthermore, one imposes the normalization condition [Chiavassa and Ghia (1996)]

$$2\pi \int_0^\infty (r/R_M) f(r/R_M, s, E) \, d(r/R_M) = 1. \tag{9.95}$$

The function $f(r/R_M, s, E)$ represents the probability that a charged particle falls at a distance r from the shower axis with an area of R_M^2. It also means that the factor $C(s)$ is constrained to be:

$$C(s) = \frac{1}{2\pi} \frac{\Gamma(4.5 - s)}{\Gamma(s)\,\Gamma(4.5 - 2s)}. \tag{9.96}$$

If N_e is the size of the shower, the charged particle density, ρ_e, as a function of the distance r from the shower axis is given by

$$\rho_e(r) = \frac{N_e}{R_M^2} f(r/R_M, s, E). \tag{9.97}$$

Equation (9.94) was applied in [Fenyves et al. (1988)] to the description of the lateral distribution of the electromagnetic component of an EAS generated by 10^{14}–10^{16} eV protons and iron nuclei. Equation (9.94) was modified for the purpose by replacing R_M by $R'_M = 0.5\,R_M$. The best fit was obtained for $R'_M = 45$ m at 850 g/cm^2 and $R'_M = 37.5$ m at sea level. The age parameter was given by

$$s = \frac{3t_2}{t_2 + 2y} \tag{9.98}$$

with

$$t_2 = t + b_{\pi,\gamma}(E). \tag{9.99}$$

Here $b_\pi(E)$ and $b_\gamma(E)$ account for the different development and aging of electron- and photon-induced showers, respectively, with different electron threshold energies. Examples of values for $b_\pi(E)$ and $b_\gamma(E)$ are also given in Table 9.11.

9.11.4 *Hadronic Extensive Air Showers*

A hadron-induced EAS has a development following closely the steps described in Sect. 3.3. Basically, the interaction of a incident nucleon with a nucleus in the upper atmosphere produces many hadrons. Each of these secondary hadrons will further interact with atmospheric nuclei or decay into other hadrons (π, K), leptons ($e^\pm$, ν's) and γ's. The occurrence of a decay or of an interaction depends on the atmosphere density. Pions and kaons generated in the upper atmosphere, which also corresponds to the earlier stage of the shower development, have a higher probability to decay. In the lower atmosphere, the probability to have an interaction instead of a decay is significantly higher. The interaction length $\lambda_{g,\text{air}}$ [Eq. (3.76)] in air

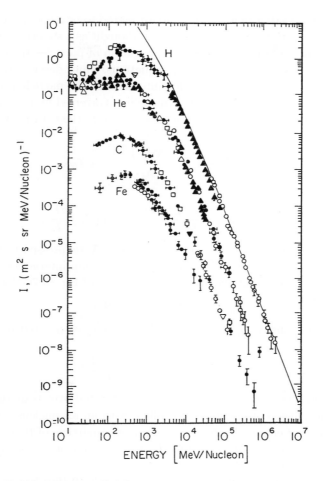

Fig. 9.37   Energy spectra of various nuclei from hydrogen to iron up to energies of hundreds of GeV per nucleon on the Earth as a function of the kinetic energy per nucleon (from [Simpson (1983)], reprinted, with permission, from the Annual Review of Nuclear and Particle Science, Volume 33 ©1983 by Annual Reviews www.annualreviews.org; see also [Leroy and Rancoita (2000)]).

is $80 \, \text{g/cm}^2$. The depth of the atmosphere is commonly put at about $13 \, \lambda_{g,\text{air}}$ and allows for a complete development of the hadronic shower.

   As described in Sect. 3.3, there is a multiplication of particles accompanied with a decrease of the average energy of the secondaries along the shower development. The number of particles in the shower will reach a maximum at a depth, which depends on the primary energy (E_0), on the type of primary particle, and on the history of interaction of secondary particles. The interaction of a nucleus with the atmosphere can be viewed as the interaction of A independent nucleons each with the energy E_0/A. The interaction length of a nucleus in the atmosphere is somewhat reduced compared to that of a nucleon and estimated to be a few g/cm^2. The interaction of a nucleus with the atmosphere can be thought as a superposition of nucleon interactions with the atmosphere. The propagation of particles through the atmosphere is described by a system of cascade equations, which describe the transport of particles through the atmosphere, taking into account the particle properties, their interactions, and the properties of the atmosphere traversed by the particles. Using the equations from Gaisser's textbook [Gaisser (1990)], the transport of nucleons in the atmosphere is described by

$$\frac{dN_{E_0}(E, X)}{dX} = -\frac{N_{E_0}(E, X)}{\lambda_{gN(E)}} + I_{NN}, \tag{9.100}$$

where

$$I_{NN} = \int_E^\infty \frac{N_{E_0}(E', X)}{\lambda_{gN(E')}} F_{NN}(E, E') \frac{dE'}{E}, \tag{9.101}$$

and where $N_{E_0}(E, X)$ is the nucleon flux at depth X (expressed in g/cm^2) in the atmosphere, and E_0 is the primary energy. The nucleon interaction length in air, $\lambda_{gN(E)}$ in g/cm^2, is given at the energy E by means of Eq. (3.76) for A = air, i.e.,

$$\lambda_{gN(E)} \approx \lambda_{g,\text{air}} \approx 80 \text{ g/cm}^2.$$

The atmospheric density, ρ, depends on the altitude, as pointed out earlier. The function

$$F_{NN}(E, E') = E_c \frac{dn_c(E_c, E_a)}{dE_c} \tag{9.102}$$

is the (dimensionless) inclusive cross section, integrated over transverse momenta, for an incident nucleon of energy E' to collide with an air nucleus and produce an outgoing nucleon of energy E; dn_c is the number of particle of type c produced on average in the energy bin, dE_c, around E_c per collision of an incident particle of type a [Gaisser (1990)].

If one takes into account the secondary pions, Eq. (9.100) has to be associated with the following equations [Sokolsky (1989); Gaisser (1990)]:

$$\frac{d\Pi_{E_0}(E, X)}{dx} = -\Pi_{E_0}(E, X) \left[\frac{1}{\lambda_{g\pi}(E)} + \frac{\epsilon_\pi}{EX \cos\theta} \right] + I_{N\pi^c} + I_{\pi^c N\pi^c}, \tag{9.103}$$

where

$$\int_E^\infty \frac{N_{E_0}(E', X)}{\lambda_{gN(E')}} F_{N\pi^c}(E, E') \frac{dE'}{E} \tag{9.104}$$

and

$$I_{\pi^c N\pi^c} = \int_E^\infty \frac{\Pi_{E_0}(E', X)}{\lambda_{g\pi(E')}} F_{\pi^c N\pi^c}(E, E') \frac{dE'}{E}, \tag{9.105}$$

in which Π_{E_0} is the average number of pions and $\lambda_{g\pi}(E)$ is the pion interaction length in air at the energy E; the term

$$\frac{\epsilon_\pi}{EX \cos\theta}$$

accounts for pion decay at the air shower zenith angle θ;

$$\epsilon_\pi = \frac{m_\pi c^2 h_0}{c\tau_\pi}$$

(τ_π is the pion lifetime) and h_0 varies with the altitude: $h_0 = 8400$ m at sea level and, for a vertical atmospheric depth $X_v < 200\,\mathrm{g/cm^2}$, $h_0 = 6400$ m [Gaisser (1990)]. π^0's decay ($\pi^0 \to \gamma\gamma$) before they have the chance to interact and, consequently, do not feed-back the hadronic cascade. The functions $F_{N\pi^c}$ and $F_{\pi^c\pi^c}$ are defined analogously to Eq. (9.102) for the processes

$$N + \mathrm{air} \to \pi^c + \mathrm{anything}$$

and

$$\pi^c + \mathrm{air} \to \pi^c + \mathrm{anything}.$$

The function representing the process

$$a + \mathrm{air} \to b + \mathrm{anything}$$

is [Sokolsky (1989)]:

$$F_{ab}(E, E_0) = \frac{\pi}{\sigma_{\mathrm{inel}}} \int E \frac{d\sigma_{ab}}{d^3 p}\, dp_T^2. \tag{9.106}$$

Applying *Feynman scaling* [Feynman (1969); Gaisser and Yodh (1980)], F_{ab} depends only on one variable and can be rewritten as:

$$F_{ab}\left(\frac{E}{E_0}\right) = F_{ab}(X_F),$$

where

$$X_F = \frac{2p_L}{\sqrt{s}}$$

is the *Feynman variable*. p_L is the particle momentum component parallel to the incident particle direction. As a consequence of Feynman scaling hypothesis, the multiplicity, n, of secondary particles (mostly pions) produced in hadron (nucleon or π) nucleon collisions depends logarithmically on energy

$$n = a \ln(E). \tag{9.107}$$

The shower generation is then a process in which a particle almost interacts once every interaction length, producing a number of secondaries that is roughly constant per interaction. The maximum of the shower depth increases logarithmically with E_0. In the case of Feynman scaling violation, the multiplicity increases faster with energy, from $E^{1/4}$ to $E^{1/2}$. Then, the shower develops faster and earlier in the upper atmosphere. Indeed, violations of the Feynman scaling are increasingly present as the energy increases due to the increased hard QCD scattering contribution. As pointed out in [Battistoni and Grillo (1996)], the scaling hypothesis remains a useful tool for the understanding of hadronic shower behavior in the atmosphere.

9.11.5 *The Muon Component of Extensive Air Showers*

The muon component of the air shower results from the decay of secondary pions and kaons produced by the interaction of nucleons in the atmosphere. The muon cascade grows towards a maximum and then slowly decays due to the $\mu^+(\mu^-) \rightarrow e^+(e^-)\,\bar{\nu}_\mu(\nu_\mu)\,\nu_e(\bar{\nu}_e)$ decays and its small cross section for radiation and pair production. The behavior of the muon component differs from that of the electromagnetic component, which rapidly increases with the atmospheric depth traversed and gets rapidly absorbed after the maximum. In fact, the electron–photon component is about one or two order of magnitude larger than the muon component, in the initial stage of the cascade development. This smaller (muon) component grows to a maximum and remains almost constant because of the long interaction length of muons, while the electromagnetic cascade, which initially represents the highest number of particles in the shower, has a relatively small range compared to the thickness of the atmosphere. Therefore, the electron–photon component of the shower becomes completely absorbed, after having reached a maximum, leaving muons as the dominant population in the shower as sea level is reached. The lateral distribution of muons, as a function of the perpendicular distance from the shower axis, is given in units of muons/m^2 by [Greisen (1960)]:

$$\rho_\mu(r) = \frac{\Gamma(2.5)}{2\pi\,\Gamma(1.5)\,\Gamma(1.5)\,(R/\text{meter})}\,N_\mu\,r^{-0.75}\left(1 + \frac{r}{R}\right)^{-2.5}, \tag{9.108}$$

where N_μ is the number of muons and R is a few hundred meters (455 m and 320 m in [Aglietta et al. (1997); Greisen (1960)], respectively). The different behavior displayed by the muon and electromagnetic components offers the possibility of studying the shower development by a measurement of the ratio of the number of muons to electrons. The relation between the number of muons, N_μ, and the number of electrons, N_e, is:

$$N_\mu = K N_e{}^n \tag{9.109}$$

with $n = 0.75$ [Aglietta et al. (1997)] and K a normalization factor.

Chapter 10

Superheated Droplet (Bubble) Detectors and CDM Search

A superheated droplet (Bubble) detector consists of a suspension of liquified gas droplets (active medium) dispersed in a polymerized gel [Ing, Noulty and McLean (1997)]. Presently, these droplet detectors consist of an emulsion of metastable superheated freon-like $C_x F_y$ (like CF_3Br, CCl_2F_2, C_3F_8, C_4F_{10}) droplets of 5 to $100\,\mu m$ diameter, dispersed in an aqueous solution. An appropriate concentration of heavy salt (e.g., CsCl) is added to the solution in order to obtain the same density as the droplets $[(1.3$–$1.6)\,g/cm^3]$, thus preventing the downward migration of the droplets. The solution is then polymerized to prevent the upward migration of the bubbles (in the sensitive mode) created, when enough energy is deposited in a droplet by an incoming particle. By applying an adequate pressure, the boiling temperature can be raised, allowing the emulsion to be kept in a liquid state. Under this external pressure, the detectors are insensitive to radiation. By removing the external pressure, the liquid becomes superheated and sensitive to radiation. The interactions, between the incoming radiation and the nuclei of the active medium, can lead to sufficient energy deposition to trigger a liquid-to-vapor phase transition. Another technique exists to prepare droplet detectors. This technique, developed by Apfel (1979), consists in dispersing droplets of a liquified gas in a viscous gel. The droplet size (typical diameter of $100\,\mu m$) is uniform and the droplets are maintained in suspension by viscosity. When they burst, the resulting bubbles migrate to the surface of the gel. Based on our own experience, this chapter will only treat detectors prepared according the BTI technique [Ing, Noulty and McLean (1997)].

The need to achieve a minimal energy deposition to induce the phase transition features droplet detectors as threshold detectors. Their sensitivity to various types of radiation strongly depends on the temperature and pressure of operation. The liquid-to-vapor transition is explosive in nature and accompanied by an acoustic shock wave, which can be detected by piezoelectric sensors. These detectors are re-usable by re-compressing the bubbles back to droplets.

Droplet detectors of small volume, typically $10\,ml$ [representing $(0.1$–$0.2)\,g$ of active material], are currently used in several applications such as portable neutron dosimeters for personal dosimetry or for measuring the radiation fields in irradi-

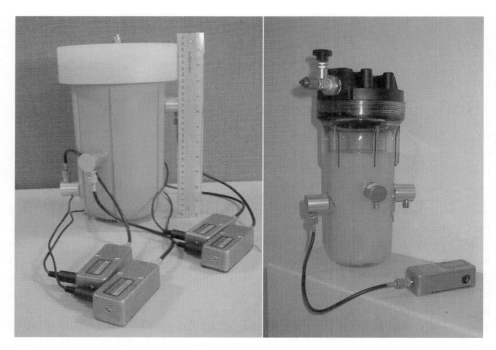

Fig. 10.1   A 1.5 l detector module equipped with piezoelectric sensors [Barnabé-Heider et al. (2002)].

ation zones near particle accelerators or reactors. These detectors can also serve for detection of radon or be used for the measurement of the transit neutron dose received in-flight by aviation crews, an important component of the total radiation exposure. More details and a description of other applications can be found in [Ing, Noulty and McLean (1997)]. For such small volumes (low gas loading), the counting of bubbles accumulated for a period of time can be visually performed.

More recently, the PICASSO group [Barnabé-Heider et al. (2002)] has developed large volume detectors of the type shown in Fig. 10.1 with the aim to perform a direct measurement of neutralinos, predicted by minimal supersymmetric models as Cold Dark Matter (CDM) particles. The very low interaction cross section between CDM and the detector active medium nuclei requires the use of very massive detectors to achieve a sensitivity level, allowing the detection of CDM particles in the galactic environment [Boukhira et al. (2000)]. Detectors of large volumes (1.5 and 3 litres) are shown in Fig. 10.1 in containers capable to hold pressures up to 10 bars. Piezo-sensors are glued on the container surface for signal detection. Typical C_xF_y gas loading presently achieved for this type of detector is in the (5–10) g/litre range.

Droplet detectors have high efficiency for detecting neutrons while being insensitive to minimum ionizing particles (*mips*) and to nearly all sources of background, when operated at proper temperature and pressure. One can observe, from kinematical considerations [see Eq. (10.5), below], that nuclear recoil thresholds in droplet detectors can be obtained in the same range for neutrons of low energy (e.g., from 10 keV up to a few MeV) and massive neutralinos (mass of 60 GeV/c^2 up to 1 TeV) with no sensitivity to *mips* and γ-radiation. Therefore, for CDM searches, the droplet detector response to neutrons has to be fully investigated. The heavy salt, present in the gel at production stage, contains α-emitters (U/Th and daugthers), which are the ultimate background at normal temperature of operation. Other backgrounds only contribute to the detector signal for higher temperatures [Boukhira et al. (2000)]. Purification techniques are applied to remove these α-emitters [Di Marco (2004)]. Presently, contamination levels of 10^{-11} g/g for U and 10^{-10} g/g for Th are obtained, toward a final goal around 10^{-14} g/g. Regardless of the level of the purity achieved, the response of droplet detectors to α-particles has to be fully understood. Recently*, the PICASSO collaboration observed for the first time a significant difference between acoustic signals induced by neutrons and α-particles in a detector based on superheated liquids. This observation brings the possibility of improved background suppression in CDM searches based on superheated liquids.

Sections in this chapter are devoted to neutron and α-particle response measurements. These data provide an understanding of the physics mechanisms at the base of droplet detector operation. Finally, a section is dedicated to the search for CDM particles and, in particular, to the spin dependent part of their cross section.

10.1 The Superheated Droplet Detectors and their Operation

The response of a droplet detector to incoming particles or radiation is determined by the thermodynamics properties of the active gas, such as operating temperature and pressure. The detector operation can be understood in the framework of *Seitz's theory* [Seitz (1958)], in which bubble formation is triggered by a heat spike in the superheated medium produced, when a particle deposits energy within a droplet.

The droplet should normally make a transition from the liquid phase (high potential energy) to the gaseous phase (lower potential energy). However, undisturbed, the droplet is in a metastable state, since it must overcome a potential barrier to make the transition from the liquid to the gas phase. This transition can be achieved if the droplet receives an extra amount of energy, such as the heat due to the energy deposited by incoming particles. The potential barrier is given by *Gibbs equation*:

$$E_c = \frac{16\pi}{3} \frac{\sigma(T)^3}{(p_i - p_0)^2}, \tag{10.1}$$

where the externally applied pressure, p_0, and the vapor pressure in the bubble, p_i, are functions of the temperature T. The difference between these two pressures

---

*See Aubin, F. et al. (2008), Discrimination of nuclear recoils from alpha particles with superheated liquids, *arXiv:0807.1536v1* [physics.ins-det].

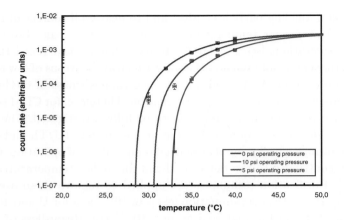

Fig. 10.2   Detector response to 200 keV neutrons as a function of temperature at various pressures. The 8 ml detector is loaded with a 100% C_4F_{10} gas. The 200 keV neutrons used for these measurements were obtained from ^{7}Li(p,n)^{7}Be reactions at the tandem facility of the Université de Montréal [Barnabé-Heider et al. (2002)].

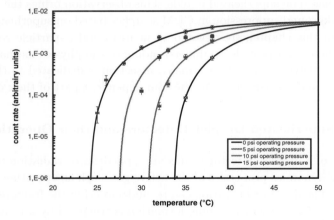

Fig. 10.3   Detector response to 400 keV neutrons as a function of temperature at various pressures. The 8 ml detector is loaded with a 100% C_4F_{10} gas. The 400 keV neutrons used for these measurements were obtained from ^{7}Li(p,n)^{7}Be reactions at the tandem facility of the Université de Montréal [Barnabé-Heider et al. (2002)].

determines the degree of *superheat*. The surface tension of the liquid-vapor interface at a temperature T is given by

$$\sigma(T) = \frac{\sigma_0(T_c - T)}{(T_c - T_0)},$$

where T_c is the critical temperature of the gas (defined as the temperature at which the surface tension is zero), σ_0 is the surface tension at a reference temperature T_0 (usually the boiling temperature T_b). For a combination of two gases, T_b and T_c can be adjusted by choosing the mixture ratio. For instance $T_b = -19.2\,°C$, $T_c = 92.6\,°C$

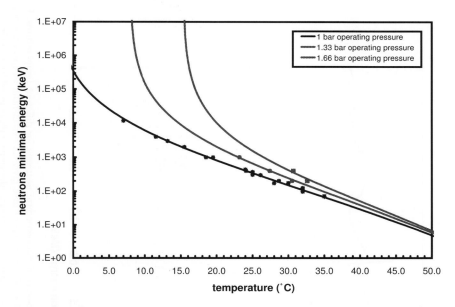

Fig. 10.4   Minimal neutron-energy ($E_{R,\text{th}}$) as a function of temperature for various pressures of operation [Barnabé-Heider et al. (2002)].

for a detector loaded with a gas mixture; $T_b = 1.7\,^\circ\text{C}$, $T_c = 113.0\,^\circ\text{C}$ for a detector loaded with 100% C_4F_{10} gas [Boukhira (2002)]. Various types of droplet detectors responses, corresponding to different gas mixtures, can be unified via their amount of so-called reduced superheat, s, introduced in [d'Errico (1999)] and defined as

$$s = \frac{T - T_b}{T_c - T_b}. \tag{10.2}$$

Bubble formation will occur, when a minimum energy $E_{R,\text{th}}$ deposited, for example, by the nuclear recoils induced by neutrons, exceeds the threshold value, E_c, within a distance $l_c = aR_c$, where the critical radius R_c is given by:

$$R_c = \frac{2\sigma(T)}{(p_i - p_0)}. \tag{10.3}$$

A value $a \approx 2$ is suggested from Apfel, Roy and Lo (1985). However, larger values of a (up to 13) can be found in literature [Bell, Oberle, Rohsenow, Todreas and Tso (1973)] and from recent simulation studies [Barnabé-Heider et al. (2004)], as will be seen below. If dE/dx is the effective energy deposition per unit distance, the energy deposited along l_c is

$$E_{\text{dep}} = dE/dx \times l_c.$$

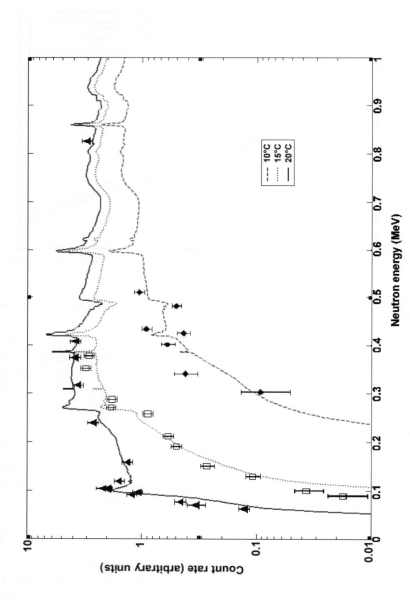

Fig. 10.5 Response of a droplet detector (loaded with a gas mixture) to beams of monoenergetic neutrons for different temperatures. Using the known tabulated neutron cross section on ^{12}C and ^{19}F, the fit to the data for different temperatures gives an exponential temperature dependence for $E^i_{th}(T)$, and the efficiency $\epsilon^i(E_n, T)$ is obtained with $\alpha = 1.0 \pm 0.1$ [Genest (2004)]. In ordinate, the count rate is in arbitrary units.

Note that dE/dx is a function of the energy of the nuclei recoiling after their collision with the incident particle. The condition to trigger a liquid-to-vapor transition is $E_{\text{dep}} \geq E_{R,\text{th}}$. Since it is not the total energy deposited that will trigger a liquid-to-vapor transition, but the fraction of this energy transformed into heat, the actual minimum or threshold energy $E_{R,\text{th}}$ for recoil detection is related to E_c by an efficiency factor

$$\eta = \frac{E_c}{E_{R,\text{th}}}$$

$(2 < \eta < 6\%)$ [Apfel (1979); Harper and Rich (1993)]. Since the threshold energy value depends on the temperature and pressure of operation, the detector can be set into a regime where it mainly responds to nuclear recoils, allowing discrimination against background radiations such as mip's and γ-rays.

10.1.1 *Neutron Response Measurement*

As an example, we will consider the case of nuclear recoils induced by neutrons of low energy ($E_n \leq 500$ keV). It is possible to determine precisely the threshold energy as a function of temperature and pressure by exposing the detector to monoenergetic neutrons at various temperatures and pressures of operation. The detector responses (count rates) to monoenergetic neutrons of 200 and 400 keV, measured as a function of temperature for various pressures of operation, are shown in Figs. 10.2 and 10.3, respectively, for a 8 ml detector.

From such curves (e.g., Figs. 10.2 and 10.3), one can extract the threshold temperature T_{th} for a given neutron energy by extrapolating the curves to their lowest point (a few degrees below the measured lowest point). Then, it is possible to represent the neutron threshold energy as a function of temperature for various pressures (Fig. 10.4).

As can be seen from Fig. 10.4, for a practical range of temperature of operation, $E_{R,\text{th}}$ follows a temperature dependence expressed as:

$$E_{R,\text{th}} = E_b \, e^{-K(T-T_b)}, \tag{10.4}$$

where K is a constant to be determined experimentally and E_b is the threshold energy at the boiling temperature T_b.

To understand the response of this type of detector to neutrons, one has to study the interaction of neutrons with the nuclei (e.g., ^{19}F and ^{12}C) of the active material. The interactions of neutrons with the freon-like droplets ($C_x F_y$) lead to recoils of ^{19}F and ^{12}C nuclei, inducing the phase transition. Inelastic collisions are possible if the center-of-mass kinetic energy of the neutron–nucleus system is higher than the first excitation level of the nucleus (1.5 and 4.3 MeV for ^{19}F and ^{12}C, respectively). Absorption of a neutron by the nucleus followed by an ion, proton or alpha-particle emission requires a neutron threshold energy of 2.05 MeV. The absorption of a neutron by the nucleus may also lead to the emission of γ, but the

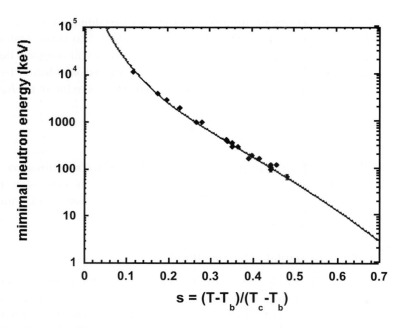

Fig. 10.6 Minimal neutron-energy as a function of the reduced superheat parameter s [Barnabé-Heider et al. (2002)].

droplet detectors are only sensitive to γ at high temperature of operation. In our example, the neutron energy is lower than $500\,\text{keV}$. Therefore, it mainly interacts through elastic scattering on fluor and carbon nuclei. Assuming neutron elastic scattering on nucleus, the recoil energy $E^i{}_R$ of the nucleus i is given by:

$$E^i{}_R = \frac{2m_n m_{N_i} E_n (1 - \cos\theta)}{(m_n + m_{N_i})^2}, \tag{10.5}$$

where E_n is the incident neutron energy and θ is the neutron scattering angle in the center-of-mass system; m_n and m_{N_i} are the masses of the neutron and the nucleus i, respectively. The nucleus recoil energy is zero if the nucleus i recoils along the neutron incident direction. The recoil energy of the nucleus i is maximum if $\theta = 180°$:

$$E^i{}_{R,\max} = \frac{4m_n m_{N_i} E_n}{(m_n + m_{N_i})^2} = f_i E_n. \tag{10.6}$$

The f_i factor is the maximal fraction of the energy of the incident neutron that can be transmitted to the nucleus i: $f_i = 0.19$ and 0.28 for ${}^{19}\text{F}$ and ${}^{12}\text{C}$, respectively. The ranges for ${}^{19}\text{F}$ and ${}^{12}\text{C}$ depend on the value of $E_{R,\max}$ [Eq. (10.6)] and on their specific energy-losses dE/dx, which can be calculated using TRIM, a code calculating the transport of ions in matter [Ziegler and Biersack (2000)].

At a given neutron energy E_n, the recoiling nuclei i ($i = {}^{19}\text{F}$, ${}^{12}\text{C}$) are emitted with an angular distribution: every angle is associated with a specific recoil energy,

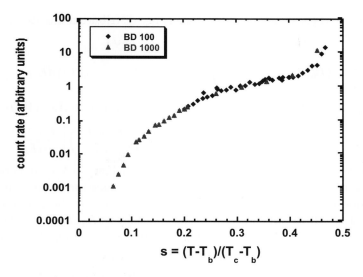

Fig. 10.7   The response of two detectors one called BD100 (loaded with a gas mixture) and the other called BD1000 (loaded with 100% C_4F_{10}) to neutrons from an AcBe source as a function of reduced superheat s. The use of the reduced superheat parameter allows the unification of the response of the two detectors. With parameter s, the response for different gases is computed with $T_{c,\mathrm{eff}} = 0.9\,T_c$ [Barnabé-Heider et al. (2002)].

ranging between $0\,\mathrm{keV}$ up to a maximum energy $E^i{}_{R,\mathrm{max}}$. Therefore, the nucleus recoil energy distribution, $dn_i/dE^i{}_R$, is determined by the ^{19}F and ^{12}C recoil angular distributions. Not all recoil energy depositions are detectable, since there exists a threshold recoil energy, $E^i{}_{R,\mathrm{th}}$, below which no phase transition is triggered. $E^i{}_{R,\mathrm{th}}$ depends on the temperature and pressure of operation and is related to the neutron energy threshold, $E^i{}_{\mathrm{th}}$, by the relation $E^i{}_{R,\mathrm{th}} = f^i E^i{}_{\mathrm{th}}$. The threshold energy has an exponential dependence on temperature [Eq. (10.4) and Fig. 10.4]. The probability, $\mathrm{P}(E^i{}_R, E^i{}_{R,\mathrm{th}}(T))$, that a recoiling nucleus i at an energy near threshold will generate an explosive droplet-to-bubble transition is zero for $E^i{}_R < E^i{}_{R,\mathrm{th}}$ and will increase gradually up to 1 for $E^i{}_R > E^i{}_{R,\mathrm{th}}$. This probability can be expressed as:

$$P(E^i{}_R, E^i{}_{R,\mathrm{th}}(T)) = 1 - \exp\left[-\alpha \frac{E^i{}_R - E^i{}_{R,\mathrm{th}}(T)}{E^i{}_{R,\mathrm{th}}(T)}\right], \qquad (10.7)$$

where α is a parameter to be determined experimentally. Therefore, the efficiency $\epsilon^i(E_n, T)$, that a i-type recoil nucleus triggers a droplet-to-bubble phase transition at a temperature T after being hit by a neutron of energy E_n, is given by comparing the integrated recoil spectrum with and without threshold:

$$\epsilon^i(E_n, T) = \int_{E^i{}_{R,\mathrm{min}}}^{E^i{}_{R,\mathrm{max}}} \frac{dn_i}{dE^i{}_R} P(E^i{}_R, E^i{}_{R,\mathrm{th}}(T))\, dE^i{}_R \left/ \int_0^{E^i{}_{R,\mathrm{max}}} \frac{dn_i}{dE^i{}_R}\, dE^i{}_R \right. . \quad (10.8)$$

For neutrons of energy lower than $500\,\mathrm{keV}$, collisions with ^{19}F and ^{12}C are elastic and isotropic and consequently, the recoil energy distribution $dn_i/dE^i{}_R = 1$. Equation (10.8) becomes:

$$\epsilon^i(E_n, T) = \frac{1}{E^i{}_{R,\max}} \int_{E^i{}_{R,\min}}^{E^i{}_{R,\max}} P(E^i{}_R, E^i{}_{R,\mathrm{th}}(T)) \, dE^i{}_R. \tag{10.9}$$

Using $E^i{}_{R,\max} = f^i E_n$ and $E^i{}_{R,\min} = f^i E^i{}_{\mathrm{th}}$, we can rewrite Eq. (10.9) as:

$$\epsilon^i(E_n, T) = 1 - \frac{E^i{}_{\mathrm{th}}(T)}{E_n} - \left[1 - \exp\left(-\alpha \frac{E_n - E^i{}_{\mathrm{th}}(T)}{E^i{}_{\mathrm{th}}(T)}\right)\right] \frac{E^i{}_{\mathrm{th}}(T)}{\alpha E_n}. \tag{10.10}$$

A detailed description of the neutron data has to include the energy dependence of the neutron cross section on ^{12}C and ^{19}F, which contains many resonances in the energy region relevant to our example. One can determine the parameters α and $E^i{}_{\mathrm{th}}(T)$ in Eq. (10.10) and, therefore, the efficiency $\epsilon^i(E_n, T)$, from the measured count rate (per second), $R(E_n, T)$, of liquid-to-vapor transitions for monoenergetic neutrons of energy E_n at temperature T:

$$R(E_n, T) = \Psi(E_n) \, V_l \, \Sigma_i \, \epsilon^i(E_n, T) \, N^i \sigma^i{}_n(E_n), \tag{10.11}$$

where $\Psi(E_n)$ is the flux of monoenergetic neutrons of energy E_n, V_l is the volume of the superheated liquid, N^i and $\sigma^i{}_n(E_n)$ are the atomic number density of species i in the liquid and the corresponding neutron cross section, respectively. The fit of Eq. (10.11) to the data, giving the count rate as a function of the neutron energy, is shown in Fig. 10.5 for different temperatures. It gives an exponential temperature dependence for $E^i{}_{\mathrm{th}}(T)$. The efficiency $\epsilon^i(E_n, T)$ is obtained from the fit, as well as $\alpha = 1.0 \pm 0.1$, found to be temperature independent.

In general, the cross section, σ, of neutron–nucleus interaction is the sum of two terms:

$$\sigma = \sigma_{SD} + \sigma_C, \tag{10.12}$$

where σ_{SD} and σ_C are the spin-dependent and coherent cross sections, respectively. The coherent cross section goes like A^2 (A being the atomic mass number of the nucleus, see Sect. 3.1). In the case of the CDM search experiment PICASSO, the data are analyzed by exploiting the feature that the interaction of neutrons, with the superheated carbo-fluorates, is dominated by the spin dependent cross section on ^{19}F, due to a dominant magnetic term in ^{19}F ($\sigma_{SD} \gg \sigma_C$). One should note that the interaction of neutrons with superheated active gas containing bromine ($A \sim 80$) would be dominated by the coherent cross section. The possibility to fabricate superheated droplet detectors with an active gas containing bromine has to be investigated. Returning to our present case, the minimum detectable recoil energy for ^{19}F nuclei is extracted from $E^i{}_{\mathrm{th}}(T)$ and is given by [Boukhira et al. (2000)]

$$E^F{}_{R,\min}(T) = 0.19 E^F{}_{\mathrm{th}} = 9 \, (\mathrm{keV}) \times \exp\left(-\frac{T - 20}{6.45}\right). \tag{10.13}$$

Combining Eqs. (10.7, 10.13), one finds the droplet phase transition probability as a function of the recoil energy deposited by a ^{19}F nucleus when it is struck by a neutron. For instance, at $T = 20\,°C$, $E^F{}_{R,\text{min}} = 9\,\text{keV}$ and

$$P(E_R, E^F{}_{R,\text{min}}) = 1 - \exp\left[\frac{-1.0\,(E_R - 9\,\text{keV})}{9\,\text{keV}}\right].$$

For $E_R = 30\,\text{keV}$ and $15\,\text{keV}$, $P(30\,\text{keV}, 9\,\text{keV}) = 0.9$ and $P(15\,\text{keV}, 9\,\text{keV}) = 0.49$. The sensitivity curve shows that the detectors are 80% efficient at $30\,°C$ for $E_R \geq 5\text{keV}$ recoils and at $20\,°C$ for $E_R \geq 25\,\text{keV}$ recoils. This sensitivity is exploited in cold dark matter search [Boukhira (2002)]. Knowing the ^{19}F nuclei recoil spectrum expected from neutralino interactions and knowing the detector response as function of temperature [Eqs. (10.7, 10.13)], it is possible to determine the detector efficiency for a given neutralino mass and a given operating temperature. The temperature dependence of the detection efficiency provides a way to discriminate against background contributions.

In the case of a polyenergetic beam, such as a radioactive source, one has to integrate Eq. (10.11) over the neutron energy spectrum:

$$R(T) = \int_0^{E_{n,\text{max}}} \psi(E_n)\,\epsilon^i(E_n, T)\,\sigma^i{}_n(E_n)\,dE_n, \tag{10.14}$$

where $\psi(E_n)$ is the *neutron spectral-flux* (Sect. 4.1.3); $\sigma^i{}_n(E_n)$ is known from tables of neutron cross section on ^{12}C and ^{19}F and $\epsilon^i(E_n, T)$ has been obtained above. For a given neutron energy spectrum at low temperature only the high energy neutrons take part in the process of liquid-to-vapor transition. Since the threshold energy is decreasing as the temperature increases (Fig. 10.4), low energy neutrons, in addition to high energy neutrons, are also detected. So, for a polyenergetic neutron source, $\epsilon^i(E_n, T)$ should increase with temperature. At high enough temperature, all the neutrons in the spectrum are contributing to phase transitions and $\epsilon^i(E_n, T)$ becomes constant with temperature since no other neutrons will add their contribution to phase transitions. For a monoenergetic neutron source, there is only one strong increase of $\epsilon^i(E_n, T)$ at the operation temperature corresponding to the energy of phase transition and $\epsilon^i(E_n, T)$ is constant for higher temperatures. The use of droplet detectors as neutron spectrometers is based on the possibility to make the detector sensitive to different ranges of neutron energies by varying the temperature of operation.

In order to unify the presentation of count rates and minimal neutron energy for detectors with different neutron threshold energies, it is convenient to use the reduced superheat parameter, s, defined in Eq. (10.2). As can be seen in Fig. 10.6, the minimal neutron energy, or threshold energy, is an exponential function of s:

$$E_{\text{th}} = E_0 e^{-\alpha_s(s - s_0)}, \tag{10.15}$$

where α_s and E_0 are two parameters to be fitted from the data. E_0 is the threshold

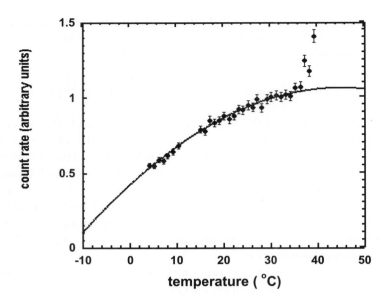

Fig. 10.8　Response of a 1.5 l detector (loaded with 100% C_4F_{10}) to α-particles emitted from the gel, doped with 20 Bq ^{232}U. Above 38 °C, the detector becomes sensitive to background gamma-rays [Barnabé-Heider et al. (2002)].

energy at a value of reference $s = s_0$. With parameter s, the response* for different gases is computed with $T_{c,\mathrm{eff}} = 0.9\,T_c$.

Figure 10.7 shows the response of two detectors (1 and 2 have different active gases and hence have different neutron energy thresholds) to neutrons from an AcBe source as a function of reduced superheat.

10.1.2　*Alpha-Particle Response Measurement*

The α-particles produced outside the superheated droplet detectors cannot be detected. Due to their short range in matter, α-particles are stopped in the detector wall. Only α-particles produced within the volume of the detector can be detected. The heavy salt used to equalize densities of droplets and solution and other ingredients mixed in the gel at the present stage of detector fabrication contains α-emitters, such as U/Th and daughters. This α-background is the ultimate background at normal temperatures of superheated droplet detectors operation in CDM search experiments since other backgrounds, such as γ-rays and mip's, contribute only to the detector signal at higher temperature. The α-response is studied with detectors doped with a source (^{232}U or ^{241}Am) of a known α-activity, introduced as a soluble salt, and uniformly distributed in the polymerized gel, but not present in

---

*Vaporization for organic liquids takes place at an absolute temperature which is about 90% of the critical temperature at atmospheric pressure [Eberhart, Kremsner and Blander (1975)].

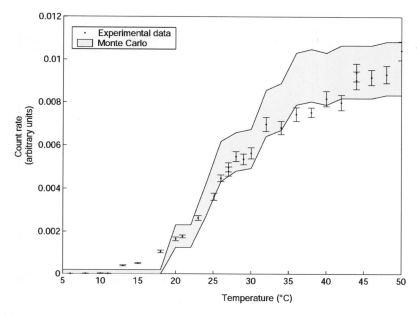

Fig. 10.9   Detector response (count rate) of a 1 l detector loaded with a 100% C_4F_{10} gas (loading of 0.7%), whose gel is doped with 20 Bqs of ^{241}Am as a function of temperature. A critical length of $L = 18\,R_c$ is necessary to fit the data [Barnabé-Heider et al. (2004)].

the droplets themselves, since the source compounds used in these spiked detectors is hydrophilic and the freon droplets are hydrophobic.

Figure 10.8 shows the measured response of a 0.7% droplet loading detector to α-particles emitted from the gel doped with 20 Bq ^{232}U. Above 38 °C, the detector becomes sensitive to background γ-rays. Previous studies were performed with a detector of the same loading [Boukhira et al. (2000)] doped with 10 Bq ^{241}Am. It was observed that the count rate increases above 0 °C and stays nearly constant from 20 to 40 °C at a level of about 3 counts per minute. From the droplet distribution, the range of α-particles and the activity per unit volume, 8 counts per minutes is expected from geometrical considerations. Thus, 0.5% of the α-particles are detected in a plateau region. Above 40 °C, the count rate increases fastly. Monte-Carlo studies [Barnabé-Heider et al. (2004)] have been performed, where α-particles were randomly generated in the gel with an energy spectrum corresponding to the ^{241}Am decay. These Monte-Carlo studies indicate that the experimental efficiency is too high for the phase transition to be caused by elastic collisions between α-particles and nuclei in the droplets. This leads to the suggestion that the phase transition is triggered by the α-particles ionization loss in the droplets. The fabrication process leads to no diffusion of ^{241}Am in the droplets and the experimental efficiency is low enough, to discard surfactant effect as described in [Pan and Wang (1999)]. Under those assumptions, the contribution of the recoil short-range ^{237}Np

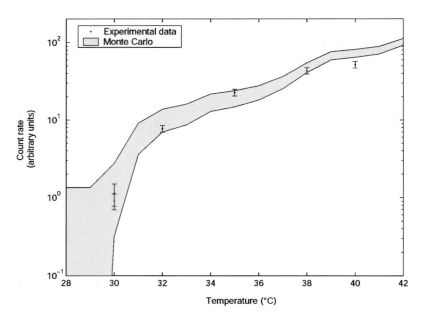

Fig. 10.10   SBD-1000 response (count rate) to 200 keV neutrons as a function of temperature. The volume of the detector is 8 ml. The simulated response gives a loading of $0.61 \pm 0.06\%$ [Barnabé-Heider et al. (2004)].

can be neglected. Furthermore, under the assumption that the recoil nucleus triggering vaporization at neutron threshold is fluorine, the dE/dx required to trigger a phase transition is too high to explain the efficiencies seen in the α case. This is not completely understood. However, it suggests that the minimal energy deposited at neutron threshold must be defined by the carbon recoil, as assumed in calculating the efficiency η (see above). Taking the probability function, Eq. (10.7), and this minimal energy requirement, the critical length as a function of temperature and the value of α were deduced from the fit to the data (Fig. 10.9).

A critical length of $L = 18\,R_c$ and $\alpha = 1$ are necessary to fit these data. This value of the critical length obtained is not inconsistent with the value found in [Bell, Oberle, Rohsenow, Todreas and Tso (1973)]. The values $L = 18\,R_c$ and $\alpha = 1$ are validated from the simulation of the neutron response, using the same energy deposition and critical length requirements. As an example, the result of the fit to the data for 200 keV neutrons is shown in Fig. 10.10 for 200 keV neutrons, where the data at 1 atmosphere are from Fig. 10.2. The simulated response gives a loading $0.61 \pm 0.06\%$, in agreement with the 0.7% loading of the detector used for the α-particle measurement (Fig. 10.9).

Neutrons, with energy above threshold, produced from radioactivity in the environment material or spallation induced by cosmic rays muons, are a possible source of background for α-particles data. In particular, neutrons are produced by the

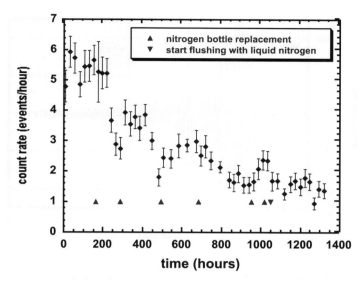

Fig. 10.11   Background counting rate of 1.5 l detector (loaded with 100% C_4F_{10}) as a function of time. The background count rate depends on the quality-purity of nitrogen used for flushing. Fluctuation stabilized and minimized with liquid nitrogen boil off [Barnabé-Heider et al. (2002)].

interaction of cosmic-ray muons with the detector environment, such as shielding, rocks or concrete walls. In principle, shielding against these background neutrons can be achieved by their moderation with paraffin or water, bringing their energies below threshold (< 50 keV at room temperature).

10.1.3   *Radon Detection*

The superheated droplet detectors can serve for the detection of radon, ^{222}Rn, from ^{238}U-decay chain. Radon is a radioactive gas created by the decay of ^{226}Ra. It decays further with $T_{1/2} = 3.85$ days, $E_\alpha = 5.49$ MeV, to isotopes of ^{218}Po ($T_{1/2} = 3.05$ min, $E_\alpha = 6.00$ MeV), ^{214}Pb ($T_{1/2} = 26.8$ min, β emitter), ^{214}Bi ($T_{1/2} = 19.7$ min, β emitter) and ^{214}Po ($T_{1/2} = 0.164$ ms, $E_\alpha = 7.69$ MeV). Thus, radon is an α-emitter. When present in the detector environment, it can diffuse into the detector, where it induces an α-signal which can be measured. In the case of an experiment such as CDM search, radon is a source of background. Flushing the detector container with pure nitrogen largely reduces radon, achieving a typical radon count rate of 1–2 events per hour, in the example shown in Fig. 10.11 at room temperature and at a pressure of 1 atm. Isolation from radon must be secured during detector fabrication, storage, and operation. Hermetic wall and lid for the detector container and use of pure compression gas should achieve that goal.

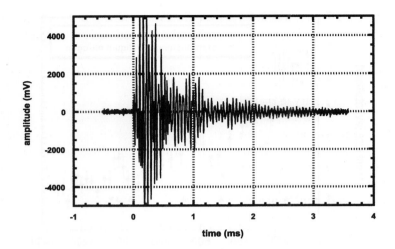

Fig. 10.12   Example of an electrical signal produced by the passage of droplet-to-bubble transition sound wave through a piezoelectric sensor.

10.1.4   *Spontaneous Nucleation*

In CDM search experiments, a possible source of background events is spontaneous phase transition due to homogeneous nucleation in the superheated liquid. The metastability limit of a superheated liquid is described in [Eberhart, Kremsner and Blander (1975)] and references therein. Vaporization for organic liquids takes place at an absolute temperature which is about 90% of the critical temperature (T_c), at atmospheric pressure. A very rapid decrease of the spontaneous nucleation flux, P (bubbles s^{-1}cm^{-3}) with decreasing temperature is expected [Eberhart, Kremsner and Blander (1975)] and follows an exponential form [Boukhira et al. (2000)]:

$$P(\text{bubbles s}^{-1} \text{ cm}^{-3}) \propto \exp\left[-\frac{E_c(T)}{k_B T}\right], \tag{10.16}$$

where E_c is the critical energy, and k_B is the Boltzmann constant (see Appendix A.2). Detectors have been tested in a shielded environment [Boukhira et al. (2000)]. The temperature dependence of the signal was measured and shown to decrease by several order of magnitude over a temperature interval of 1 °C near the temperature region $T \sim 0.9\,T_c$, where such nucleation should dominate. Therefore, spontaneous nucleation is not playing any significant role, when the detector is used at normal operation temperature (often at room temperature).

10.1.5   *Signal Measurement with Piezoelectric Sensors*

The explosive droplet-to-bubble transition generates an acoustic signal which can be detected by piezoelectric sensors adapted to the acoustic emission spectrum. These sensors, two or more, are glued to the surface of the detector container (Fig. 10.1) and coupled to high gain, low noise preamplifiers whose frequency response is optimized to suppress lower frequency acoustic noise. An example of an acoustic signal transformed into an electronic signal is shown in Fig. 10.12.

The signal produced by the explosive droplet-to-bubble transition is transmitted through the gel as a pressure front, then through the container wall to the piezo-sensor. The sound velocity in the gel has been measured to be 1600 ± 100 m/s, close to the sound velocity in the plastic materials used in the container fabrication [Gornea (2002)]. The signal shapes and frequency responses are dependent on i) the energy released in the liquid-to-vapor phase transition, ii) the distance traveled by the sound in the gel leading to signal attenuation as a function of event-sensor distance, iii) the temperature and pressure of operation and iv) the recording history of the detector, mainly the number of events that have occurred before the measured signal since the last compression of the detector. The dependence of the signal amplitude as a function of the number of events counted after a recompression and pressure release cycle, for various temperatures of operation, shows a decrease in the counting and mean maximum amplitude. This reflects the detector depletion, starting with the largest droplets, fewer in number but containing a larger fraction of the active volume. The amplitude attenuation for various sensor-event distances indicates that signals can be obtained with adequate efficiency up to 20 cm from the source. The signal amplitude increases with increasing temperature and decreasing pressure of operation. This observation follows the expectation that the energy released in a droplet explosion increases with temperature and decreases with pressure. It allows one to set well-defined limits on the temperature and pressure ranges of operation. The piezo-sensors are set to discriminate against low frequency noise, while favoring higher frequencies useful for timing purpose [Gornea (2002)]. Fast Fourier transform analysis of pulses within specific frequency windows selected by the sensor response allows acoustic noise rejection, yielding a clean radiation-induced signal at a cost of a loss ($< 10\%$) of efficiency.

10.2 Search of Cold Dark Matter (CDM)

In many models of supersymmetry, the *neutralino*, χ, is assumed to be the *Lightest Supersymmetric Particle* (LSP). Particularly, in models of supersymmetry with R-parity conserved. The R-parity is defined as

$$R = (-1)^{3B+L+2S}$$

for a particle of spin S, baryon number B and lepton number L: $R = 1$ for *Standard Model* particles and $R = -1$ for the *superpartners*. The neutralino is stable and regarded by many as the most promising candidate for *Cold Dark Matter* (CDM). Experiments doing a direct search of cold dark matter, like PICASSO [Barnabé-Heider et al. (2005a)], SIMPLE [Girard R et al. (2005)], and others, are based on the technique of detection of nuclear recoils induced in the elastic scattering of neutralino on a nucleus N [of atomic mass* m_N and atomic number Z]:

$$\chi + N \to \chi + N,$$

as the energy, E_χ, of neutralinos is very small, for instance $E_\chi \approx 40\,\mathrm{keV}$ for a neutralino mass of $100\,\mathrm{GeV}$. These experiments measure the neutralino–nucleus cross section as a function of the neutralino mass (this mass is presently unknown but expected to be very heavy, $100\,\mathrm{GeV}$ or more). The cross sections of these neutralino–nucleus interaction processes can be divided in two separate types. The coherent part or spin-independent part, described by an effective scalar coupling between the neutralino and the nucleus, is proportional to the number of nucleons in the nucleus [i.e., proportional to the square of the mass number (see page 218) of the nucleus (A^2)]. The coherent part receives contribution from the scattering of quark,

$$\chi + q \to \chi + q,$$

described by the Lagrangian $\mathrm{L} \sim \chi\chi\bar{q}q$. The second part of the neutralino–nucleus cross section is the incoherent part or spin-dependent part and results from an axial current interaction of a neutralino with constituent quarks. This interaction, which couples the spin of a neutralino to the total spin of the nucleus, depends on the spin of the nucleus and is described by the Lagrangian $\mathrm{L} \sim (\chi\gamma^\mu\gamma^5\chi)(\bar{q}\gamma_\mu\gamma_5 q)$. Therefore, depending on the target nucleus atomic mass the experiment can be sensitive either to spin-dependent or to spin-independent neutralino–nucleus interactions. In particular, one considers the example of the PICASSO experiment [Barnabé-Heider et al. (2005a)], which is measuring the neutralino–^{19}F cross section. This cross section is largely dominated by the spin-dependent contribution. The measurement of the spin-dependent neutralino–^{19}F cross section, as a function of the neutralino mass, provides neutralino–^{19}F and neutralino–nucleon exclusion limits. From the data, one determines: i) the neutralino–proton and the neutralino–neutron cross sections as functions of the neutralino mass and, ii) the relation between neutralino–proton and neutralino–neutron spin-dependent coupling constants. One can also obtain limits on spin-independent cross section for PICASSO by extracting the neutralino–proton and neutralino–neutron spin-independent coupling constants from spin-independent experiments (with large atomic mass m_N).

10.2.1 *Calculation of the Neutralino–Nucleon Exclusion Limits*

For experiments based on the technique of nuclear recoils detection, one can express the differential interaction rate per unit detector mass as (e.g., see [Bottino et al. (1997); Genest and Leroy (2004); Giuliani (2005)]):

$$\frac{dR}{dE_R} = N_T \frac{\rho_\chi}{m_\chi} \int_{v_{\min}(E_R)}^{v_{\max}(E_R)} f(v) \frac{d\sigma(v, E_R)}{dE_R}\, dv, \tag{10.17}$$

---

*The difference between the nuclear mass and the atomic mass is negligible.

Table 10.1 value of atomic number Z, mass number A (see discussion at page 220) and the ratio Z/A for nuclei typically used in direct searches for dark matter particle.

Nucleus	Z	A	Z/A
^{19}F	9	19	0.47
^{23}Na	11	23	0.48
^{40}Ca	20	40	0.50
^{73}Ge	32	73	0.44
^{127}I	53	127	0.42
^{131}Xe	54	131	0.41

where E_R is the recoil energy, N_T is the number[||] of target nuclei per unit mass; ρ_χ is the local density of neutralino($\rho_\chi \sim 0.3\,\mathrm{GeV/cm^3}$) of mass m_χ (in units of GeV/c^2 in Sects. 10.2.1-10.2.2), v is the neutralino velocity in the detector rest frame, with $v_{\min}$ and $v_{\max}$ the minimum and maximum velocities, respectively. The velocities will be converted into E_R for the calculations. $f(v)$ is the neutralino velocity distribution in the detector rest frame and is given by:

$$f(v)\,d^3v = \frac{\exp\left(-v^2/v_0^2\right)}{\pi^{3/2}\,v_0^3}\,d^3v \tag{10.18}$$

with the velocity dispersion given by

$$\bar{v} = \langle v^2 \rangle^{1/2} = \left(\frac{3}{2}\right)^{1/2} v_0 \approx 270\text{--}280\,\mathrm{km\,s^{-1}} \tag{10.19}$$

($v_0 \approx 220\text{--}230\,\mathrm{km\,s^{-1}}$). $d\sigma(v, E_R)/dE_R$ is the differential cross section for neutralino–nucleus scattering and is the sum of two contributions:

$$\frac{d\sigma}{dE_R} = \left(\frac{d\sigma}{dE_R}\right)_{\mathrm{SI}} + \left(\frac{d\sigma}{dE_R}\right)_{\mathrm{SD}} \tag{10.20}$$

where $(d\sigma/dE_R)_{\mathrm{SI}}$ is the spin-independent (SI) or coherent part of the cross section and $(d\sigma/dE_R)_{\mathrm{SD}}$ is the spin-dependent (SD) part of the cross section.

10.2.1.1 *Spin-Independent or Coherent Cross Section*

The spin-independent (SI) or coherent cross section is given by ([Bottino et al. (1997)], for instance)

$$\left(\frac{d\sigma}{dE_R}\right)_{\mathrm{SI}} = \frac{2G_{\mathrm{F}}^2 m_N}{\pi v^2}\left[Zg_{\mathrm{p}} + (A - Z)g_{\mathrm{n}}\right]^2 \left[F_{\mathrm{SI}}(E_R)\right]^2 \tag{10.21}$$

where: G_{F} is the Fermi coupling constant, $G_{\mathrm{F}} \approx 1.16610^{-5}\,\mathrm{GeV^{-2}}$ (Appendix A.2); m_N is the nucleus mass, v is the neutralino velocity in the laboratory frame; Z and A are the nuclear charge and the mass number of the target nucleus, respectively; g_{p}

[||]The number of target nuclei per gram of material is $N_T = N_a/A_a$, where A_a is the relative atomic mass (or atomic weight, i.e., Sect. 1.4.1 and discussion at page 220) of the material and N_a is the Avogadro constant. For ^{19}F, we have $A_a = 19$.

and g_n are the effective neutralino-proton and neutralino-neutron coupling constants for SI interactions; F_{SI} is the spin-independent form factor [Eq. (10.25)]. The recoil energy, E_R, is given by

$$E_R = \frac{\mu_{\chi N}^2 v^2 (1 - \cos\theta^*)}{m_N}, \qquad (10.22)$$

where $\mu_{\chi N}$ is the neutralino–nucleus (N) *reduced mass*

$$\mu_{\chi N} = \frac{m_\chi m_N}{m_\chi + m_N}$$

and θ^* is the scattering angle in the neutralino–nucleus center-of-mass frame. From Eq. (10.22), one obtains:

$$E_{R,max} = E_R(\theta^* = \pi) = \frac{2\mu_{\chi N}^2 v^2}{m_N} \qquad (10.23)$$

and, therefore,

$$v^2 = \frac{m_N}{2\mu_{\chi N}^2} E_{R,max}. \qquad (10.24)$$

As above mentioned, $F_{SI}(E_R)$ is the spin-independent form factor and following [Engel (1991)]; it is written as

$$F_{SI}(E_R) = 3\frac{J_1(qR_0)}{qR_0} e^{-\frac{1}{2}s^2 q^2}, \qquad (10.25)$$

where $q^2 = 2m_N E_R$ is the squared three-momentum transfer and J_1 is the spherical Bessel function of index 1. The parameter s is the thickness parameter of the nuclear surface, $s \sim 0.9\,\text{fm}$. The parameter R_0 is related to the nuclear radius and thickness,

$$R_0 = \sqrt{R^2 - 5s^2}$$

with $R = 1.2\,A^{1/3}\,\text{fm}$. In the case of ^{19}F, $R = 3.2\,\text{fm}$, $R_0 = 2.5\,\text{fm}$ with $s = 0.9\,\text{fm}$. The form factor is normalized to unity for $q^2 = 0$. Indeed, we have

$$J_1(qR_0) = \frac{\sin(qR_0)}{(qR_0)^2} - \frac{\cos(qR_0)}{qR_0}. \qquad (10.26)$$

In the limit $qR_0 \to 0$, Eq. (10.26) becomes:

$$\frac{J_1(qR_0)}{qR_0} = \frac{1}{(qR_0)^3}\left[qR_0 - \frac{(qR_0)^3}{6} + \cdots\right] - \frac{1}{(qR_0)^2}\left[1 - \frac{(qR_0)^2}{2} + \cdots\right], \qquad (10.27)$$

i.e.,

$$J_1(qR_0) = \frac{1}{3}. \qquad (10.28)$$

Therefore, for $qR_0 \to 0$, one obtains

$$F_{SI}(0) = 1. \qquad (10.29)$$

Equation (10.25) is obtained from a density [Helm (1956)] in the form $\int \rho_0(r')\rho_1(r-r')\,d^3r'$, where ρ_0 is constant inside the radius R_0 and

$$\rho_1(r) = \exp\left(-\frac{1}{2}s^2q^2\right).$$

The Fourier transform of the function ρ, which represents a nearly constant interior density and a surface of thickness s, gives Eq. (10.25).

Integrating Eq. (10.21) over E_R, one finds

$$\sigma_{\rm SI}^{\chi N}(v) = \int_0^{E_{R,\max}} \left(\frac{d\sigma}{dE_R}\right)_{\rm SI} dE_R; \tag{10.30}$$

using Eq. (10.24),

$$\sigma_{\rm SI}^{\chi N}(v) = \frac{4G_{\rm F}^2\mu_{\chi N}^2}{\pi}\,[Zg_{\rm p} + (A-Z)g_{\rm n}]^2\,\frac{1}{E_{R,\max}}\int_0^{E_{R,\max}}[F_{\rm SI}(E_R)]^2\,dE_R \tag{10.31}$$

or

$$\sigma_{\rm SI}^{\chi N}(v) = \frac{4G_{\rm F}^2\mu_{\chi N}^2}{\pi}\,[Zg_{\rm p} + (A-Z)g_{\rm n}]^2\,G_{\rm SI}(v), \tag{10.32}$$

where one has defined

$$G_{\rm SI}(v) = \frac{1}{E_{R,\max}}\int_0^{E_{R,\max}}[F_{\rm SI}(E_R)]^2\,dE_R. \tag{10.33}$$

The quantity $G_{\rm SI}(v)$ was calculated for several nuclei (^{19}F, ^{73}Ge, ^{127}I) and the results are shown in Fig. 10.13.

One can establish the following identity:

$$\left[1 - \frac{g_{\rm p}-g_{\rm n}}{g_{\rm p}+g_{\rm n}}\left(1-\frac{2Z}{A}\right)\right]^2 A^2\left[\frac{g_{\rm p}+g_{\rm n}}{2}\right]^2 = \left[g_{\rm p}+g_{\rm n} - (g_{\rm p}-g_{\rm n})\left(1-\frac{2Z}{A}\right)\right]^2$$

$$\times\left[\frac{1}{g_{\rm p}+g_{\rm n}}\right]^2 A^2\left[\frac{g_{\rm p}+g_{\rm n}}{2}\right]^2$$

$$= [Zg_{\rm p}+(A-Z)g_{\rm n}]^2. \tag{10.34}$$

Therefore, Eq. (10.32) can be rewritten as:

$$\sigma_{\rm SI}^{\chi N}(v) = \frac{4G_{\rm F}^2\mu_{\chi N}^2}{\pi}\left[\frac{g_{\rm p}+g_{\rm n}}{2}\right]^2\left[1 - \frac{g_{\rm p}-g_{\rm n}}{g_{\rm p}+g_{\rm n}}\left(1-\frac{2Z}{A}\right)\right]^2 A^2 G_{\rm SI}(v) \tag{10.35}$$

or

$$\sigma_{\rm SI}^{\chi N}(v) = \frac{4G_{\rm F}^2\mu_{\chi N}^2}{\pi}g^2A^2G_{\rm SI}(v), \tag{10.36}$$

where one has defined

$$g = \frac{g_{\rm p}+g_{\rm n}}{2}\left[1 - \frac{g_{\rm p}-g_{\rm n}}{g_{\rm p}+g_{\rm n}}\left(1-\frac{2Z}{A}\right)\right]. \tag{10.37}$$

In the limit $v \to 0$, $G_{\rm SI}(0) \to 1$ and Eq. (10.36) becomes

$$\sigma_{\rm SI}^{\chi N}(0) = \frac{4G_{\rm F}^2\mu_{\chi N}^2}{\pi}g^2A^2. \tag{10.38}$$

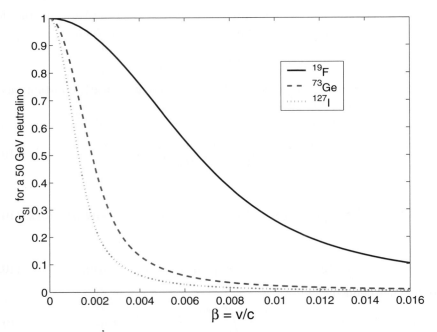

Fig. 10.13 G_{SI} [Eq. (10.33)] as a function of $\beta = v/c$ for ^{19}F, ^{73}Ge, ^{127}I.

Since Z/A is nearly constant for the nuclei typically used in direct searches for dark matter particle (Table 10.1), the coupling constant g is approximately independent of the target nucleus used in the experiment. Therefore, one defines a so-called generalized spin-independent neutralino–nucleon(p) cross section

$$\sigma_{\mathrm{SI}}^{\chi\mathrm{P}} = \frac{4{G_{\mathrm{F}}}^2 \mu_{\chi\mathrm{P}}^2}{\pi} g^2, \qquad (10.39)$$

which is related to the spin-independent neutralino–nucleus(N) cross section via

$$\sigma_{\mathrm{SI}}^{\chi\mathrm{P}} = \frac{\mu_{\chi\mathrm{P}}^2}{\mu_{\chi N}^2} \frac{\sigma_{\mathrm{SI}}^{\chi N}}{A^2}. \qquad (10.40)$$

Furthermore, one has the relation between g_p and g_n:

$$g_\mathrm{p} = \frac{1}{Z} \frac{\sqrt{\sigma_{\mathrm{SI}}^{\chi N}(0)\pi}}{2 G_{\mathrm{F}} \mu_{\chi N}} + \left(1 - \frac{A}{Z}\right) g_\mathrm{n}. \qquad (10.41)$$

Then, in principle one can estimate the coupling g_p and g_n from the experimental limits on spin-independent neutralino cross sections on two different nuclei.

10.2.1.2   *Spin-Dependent or Incoherent Cross Section*

The spin-dependent (SD) cross section is given by (see for instance [Bottino et al. (1997); Tovey et al. (2000)] and references therein)

$$\sigma_{\chi N}^{\mathrm{SD}}(v) = \int_0^{E_{R,\mathrm{max}}} \left(\frac{d\sigma}{dE_R}\right)_{\mathrm{SD}} dE_R, \qquad (10.42)$$

where

$$\left(\frac{d\sigma}{dE_R}\right)_{\mathrm{SD}} = \frac{2G_{\mathrm{F}}^2 m_N}{\pi v^2} 8\Lambda^2 J(J+1)\left[F_{\mathrm{SD}}(E_R)\right]^2, \qquad (10.43)$$

$F_{\mathrm{SD}}(E_R)$ is the spin-dependent form factor and

$$\Lambda = \frac{1}{J}\left[a_{\mathrm{p}}\langle S_{\mathrm{p}}\rangle + a_{\mathrm{n}}\langle S_{\mathrm{n}}\rangle\right]. \qquad (10.44)$$

$\langle S_{\mathrm{p}}\rangle$ and $\langle S_{\mathrm{n}}\rangle$ are the expectation values of the proton and neutron spin within the nucleus N, respectively; a_{p} and a_{n} are the neutralino–proton and neutralino–neutron coupling coefficients, respectively. We have:

$$\langle S_{\mathrm{p}}\rangle = \langle N|S_{\mathrm{p}}|N\rangle = \frac{1}{2}$$

if $N = 1\,\mathrm{proton}$ and $\langle S_{\mathrm{p}}\rangle = 0.441$ if $N = {}^{19}\mathrm{F}$;

$$\langle S_{\mathrm{n}}\rangle = \langle N|S_{\mathrm{n}}|N\rangle = \frac{1}{2}$$

if $N = 1\,\mathrm{neutron}$ and $\langle S_{\mathrm{n}}\rangle = -0.109$ if $N = {}^{19}\mathrm{F}$.

Using Eq. (10.24), we obtain

$$\sigma_{\chi N}^{\mathrm{SD}}(v) = \frac{4G_{\mathrm{F}}^2 \mu_{\chi N}^2}{\pi}\left[8\Lambda^2 J(J+1)\right]\frac{1}{E_{R,\max}}\int_0^{E_{R,\max}}]F_{\mathrm{SD}}(E_R)]^2\,dE_R \qquad (10.45)$$

or

$$\sigma_{\chi N}^{\mathrm{SD}}(v) = \frac{4G_{\mathrm{F}}^2 \mu_{\chi N}^2}{\pi}\left[8\Lambda^2 J(J+1)\right]G_{\mathrm{SD}}(v) \qquad (10.46)$$

with

$$G_{\mathrm{SD}}(v) = \frac{1}{E_{R,\max}}\int_0^{E_{R,\max}} F_{\mathrm{SD}}{}^2(E_R)dE_R. \qquad (10.47)$$

If one notes that $v \to 0$ ($\beta = v/c \sim 220/300{,}000 \sim 7\times 10^{-4}$) and $G_{\mathrm{SD}}(v) \to 1$, therefore, one can write Eq. (10.46) as

$$\sigma_{\chi N}^{\mathrm{SD}}(0) = \frac{4G_{\mathrm{F}}^2 \mu_{\chi N}^2}{\pi}\frac{8(J+1)}{J}\left[a_{\mathrm{p}}\langle S_{\mathrm{p}}\rangle + a_{\mathrm{n}}\langle S_{\mathrm{n}}\rangle\right]^2. \qquad (10.48)$$

One usually rewrites Eq. (10.48) as

$$\sigma_{\chi N}^{\mathrm{SD}}(0) = 4G_{\mathrm{F}}^2 \mu_{\chi N}^2 C_N^{\mathrm{SD}}, \qquad (10.49)$$

where the factor

$$C_N^{\mathrm{SD}} = \frac{8(J+1)}{J\pi}\left[a_{\mathrm{p}}\langle S_{\mathrm{p}}\rangle + a_{\mathrm{n}}\langle S_{\mathrm{n}}\rangle\right]^2$$

is called the *total enhancement factor* in the spin-dependent neutralino–nucleus elastic cross section. If one defines the nucleon contribution to the total enhancement factor as

$$C_{\mathrm{nucleon}(N)}^{\mathrm{SD}} = \frac{8}{\pi}\frac{(J+1)}{J}a_{\mathrm{nucleon}}^2\langle S_{\mathrm{nucleon}}\rangle^2 = F\,a_{\mathrm{nucleon}}^2\langle S_{\mathrm{nucleon}}\rangle^2 \qquad (10.50)$$

with

$$F = \frac{8(J+1)}{\pi J},$$

one has

$$a_{\text{nucleon}} \langle S_{\text{nucleon}} \rangle = \pm \sqrt{\frac{C_{\text{nucleon}(N)}^{\text{SD}}}{F}}. \tag{10.51}$$

Introducing Eq. (10.51) in Eqs. (10.48, 10.49), one finds:

$$\sigma_{\chi N}^{\text{SD}}(0) = 4G_{\text{F}}^2 \mu_{\chi N}^2 \left[\sqrt{C_{\text{p}(N)}^{\text{SD}}} \pm \sqrt{C_{\text{n}(N)}^{\text{SD}}} \right]^2. \tag{10.52}$$

One can express the total neutralino cross section on nucleus from the cross section on nucleons as

$$\sigma_{\chi N}^{\text{SD}}(0) = \left(\sqrt{4G_{\text{F}}^2 \mu_{\chi N}^2 C_{\text{p}(N)}^{\text{SD}}} \pm \sqrt{4G_{\text{F}}^2 \mu_{\chi N}^2 C_{\text{n}(N)}^{\text{SD}}} \right)^2. \tag{10.53}$$

The calculation of $\langle S_{\text{p}} \rangle$ and $\langle S_{\text{n}} \rangle$ for nuclear shells models shows non-vanishing proton and neutron contributions to the nuclear spin [Tovey et al. (2000)]. The supersymmetric and nuclear degrees of freedom are generally not decoupled. This prevents the direct extraction of a universal χ–nucleon cross section contrary to the spin independent case [Bottino et al. (1997)]. In order to compare experiments performed on different nuclei, one introduces neutralino-proton, neutralino-neutron cross sections instead of using neutralino–nucleus cross sections. Such a formalism can be found in [Tovey et al. (2000); Genest and Leroy (2004); Giuliani (2005)], for instance.

Equation (10.52) can be rewritten as

$$\sigma_{\chi N}^{\text{SD}}(0) = \left(\sqrt{\sigma_{\text{p}(N)}^{\text{SD}}} \pm \sqrt{\sigma_{\text{n}(N)}^{\text{SD}}} \right)^2, \tag{10.54}$$

where one has defined the proton and neutron contributions to the total cross section:

$$\sigma_{\text{nucleon}(N)}^{\text{SD}} = 4G_{\text{F}}^2 \mu_{\chi N}^2 C_{\text{nucleon}(N)}^{\text{SD}}. \tag{10.55}$$

One has also the cross sections for isolated nucleon

$$\sigma_{\text{nucleon}}^{\text{SD}} = 4G_{\text{F}}^2 \mu_{\chi p}^2 C_{\text{nucleon}}^{\text{SD}}. \tag{10.56}$$

From Eqs. (10.55, 10.56), one finds

$$\frac{\sigma_{\text{nucleon}}^{\text{SD}}}{\sigma_{\text{nucleon}(N)}^{\text{SD}}} = \frac{4G_{\text{F}}^2 \mu_{\chi\text{nucleon}}^2 C_{\text{nucleon}}^{\text{SD}}}{4G_{\text{F}}^2 \mu_{\chi N}^2 C_{\text{nucleon}(N)}^{\text{SD}}}. \tag{10.57}$$

One has a relation between the cross sections obtained for protons (neutrons) in a nucleus and cross sections for free protons (free neutrons):

$$\sigma_{\text{p}}^{\text{SD}} = \sigma_{\text{p}(N)}^{\text{SD}} \frac{\mu_{\chi p}^2}{\mu_{\chi N}^2} \frac{C_{\text{p}}^{\text{SD}}}{C_{\text{p}(N)}^{\text{SD}}}. \tag{10.58}$$

and

$$\sigma_n^{SD} = \sigma_{n(N)}^{SD} \frac{\mu_{\chi n}^2}{\mu_{\chi N}^2} \frac{C_n^{SD}}{C_{n(N)}^{SD}}. \tag{10.59}$$

When one achieves the limit in the number of events in a direct detection experiment, one assumes that the total neutralino–nucleus (N) cross section is dominated by the proton (or neutron) contribution only , i.e.,

$$\sigma_{\chi N}^{SD} \sim \sigma_{p(N)}^{SD} \tag{10.60}$$

and

$$\sigma_{\chi N}^{SD} \sim \sigma_{n(N)}^{SD}. \tag{10.61}$$

Equations (10.58, 10.59) become

$$\sigma_p^{SD} = \sigma_{\chi(N)}^{SD} \frac{\mu_{\chi p}^2}{\mu_{\chi N}^2} \frac{C_p^{SD}}{C_{p(N)}^{SD}} \tag{10.62}$$

and

$$\sigma_n^{SD} = \sigma_{\chi(N)}^{SD} \frac{\mu_{\chi n}^2}{\mu_{\chi N}^2} \frac{C_n^{SD}}{C_{n(N)}^{SD}}. \tag{10.63}$$

The use of the ratios $C_p^{SD}/C_{p(N)}^{SD}$ and $C_n^{SD}/C_{n(N)}^{SD}$ ensures the cancellation of the a_p and a_n terms which makes the calculation model-independent.

If $\sigma_{\chi N}^{SDlim}$ is the upper limit of neutralino-target cross section from the data, one has the inequality

$$\sigma_{\chi N}^{SD} \leq \sigma_{\chi N}^{SDlim}. \tag{10.64}$$

So, one can define a limit on the neutralino-proton cross section, σ_p^{SDlim} corresponding to the neutralino–nucleus N cross section limit $\sigma_{\chi N}^{SDlim}$ as

$$\sigma_p^{SDlim} = \sigma_{\chi N}^{SDlim} \frac{\mu_{\chi p}^2}{\mu_{\chi N}^2} \frac{C_p^{SD}}{C_{p(N)}^{SD}} \tag{10.65}$$

and, analogously, for neutrons

$$\sigma_n^{SDlim} = \sigma_{\chi N}^{SDlim} \frac{\mu_{\chi n}^2}{\mu_{\chi N}^2} \frac{C_n^{SD}}{C_{n(N)}^{SD}} \tag{10.66}$$

with (one assumes the same mass for proton and neutron):

$$\frac{\mu_{\chi N}^2}{\mu_{\chi p}^2} = \frac{\mu_{\chi N}^2}{\mu_{\chi n}^2} \left[\frac{1 + m_\chi/m_p}{1 + m_\chi/m_N} \right]^2. \tag{10.67}$$

Equations (10.65) and (10.66) allows the calculation of the exclusion limits on the neutralino-proton and neutralino-neutron cross sections as a function of the neutralino mass.

Coming back to Eq. (10.54), $\sigma_{\chi N}^{SD}(0)$ can be expressed as

$$\sigma_{\chi N}^{SD}(0) = \left(\sqrt{\sigma_p^{SD} \kappa_p} \pm \sqrt{\sigma_n^{SD} \kappa_n} \right)^2, \tag{10.68}$$

where one has defined

$$\kappa_{\rm p} = \frac{\mu_{\chi N}^2}{\mu_{\chi p}^2} \frac{C_{{\rm p}(N)}^{\rm SD}}{C_{\rm p}^{\rm SD}} \tag{10.69}$$

and

$$\kappa_{\rm n} = \frac{\mu_{\chi N}^2}{\mu_{\chi p}^2} \frac{C_{{\rm n}(N)}^{\rm SD}}{C_{\rm n}^{\rm SD}}. \tag{10.70}$$

Then, it follows

$$1 = \left(\sqrt{\frac{\sigma_{\rm p}^{\rm SD} \kappa_{\rm p}}{\sigma_{\chi N}^{\rm SD}}} \pm \sqrt{\frac{\sigma_{\rm n}^{\rm SD} \kappa_{\rm n}}{\sigma_{\chi N}^{\rm SD}}} \right)^2. \tag{10.71}$$

Using the inequality of Eq. (10.64), one has

$$1 \geq \left(\sqrt{\frac{\sigma_{\rm p}^{\rm SD} \kappa_{\rm p}}{\sigma_{\chi N}^{\rm SDlim}}} \pm \sqrt{\frac{\sigma_{\rm n}^{\rm SD} \kappa_{\rm n}}{\sigma_{\chi N}^{\rm SDlim}}} \right)^2. \tag{10.72}$$

Since

$$\frac{\kappa_{\rm p}}{\sigma_{\chi N}^{\rm SDlim}} = \frac{1}{\sigma_{\rm p}^{\rm SDlim}} \tag{10.73}$$

and

$$\frac{\kappa_{\rm n}}{\sigma_{\chi N}^{\rm SDlim}} = \frac{1}{\sigma_{\rm n}^{\rm SDlim}}, \tag{10.74}$$

then Eq. (10.72) becomes

$$1 \geq \left(\sqrt{\frac{\sigma_{\rm p}^{\rm SD}}{\sigma_{\rm p}^{\rm SDlim}}} \pm \sqrt{\frac{\sigma_{\rm n}^{\rm SD}}{\sigma_{\rm n}^{\rm SDlim}}} \right)^2. \tag{10.75}$$

If one considers one isolated proton ($J = 1/2$, $\langle S_{\rm p} \rangle = 1/2$), one finds

$$C_{\rm p}^{\rm SD} = \frac{8}{\pi} \frac{(J+1)}{J} a_{\rm p}^2 \langle S_{\rm p} \rangle^2 = \frac{6}{\pi} a_{\rm p}^2 \tag{10.76}$$

and [via Eq. (10.56)]

$$\sigma_{\rm p}^{\rm SD} = 4 G_{\rm F}^2 \mu_{\chi p}^2 \frac{6}{\pi} a_{\rm p}^2 = \frac{24}{\pi} G_{\rm F}^2 \mu_{\chi p}^2 a_{\rm p}^2. \tag{10.77}$$

For an isolated neutron ($J = 1/2$, $\langle S_{\rm n} \rangle = 1/2$), we obtain

$$C_{\rm n}^{\rm SD} = \frac{8}{\pi} \frac{(J+1)}{J} a_{\rm n}^2 \langle S_{\rm n} \rangle^2 = \frac{6}{\pi} a_{\rm n}^2 \tag{10.78}$$

and [via Eq. (10.56)]

$$\sigma_{\rm n}^{\rm SD} = 4 G_{\rm F}^2 \mu_{\chi p}^2 \frac{6}{\pi} a_{\rm p}^2 = \frac{24}{\pi} G_{\rm F}^2 \mu_{\chi p}^2 a_{\rm n}^2. \tag{10.79}$$

Introducing Eqs. (10.77, 10.79) in Eq. (10.75), one finds

$$\frac{\pi}{24 G_F{}^2 \mu_{\chi p}{}^2} \geq \left(\frac{a_p}{\sqrt{\sigma_p{}^{SDlim}}} \pm \frac{a_n}{\sqrt{\sigma_n{}^{SDlim}}} \right)^2 . \tag{10.80}$$

The limit represents an allowed band in the a_p/a_n plane. The sign of the slope is determined by the sign appearing in parentheses in Eq. (10.80). This latter sign is determined by the sign of the ratio $\langle S_p \rangle / \langle S_n \rangle$. The optimal limit on the a_p and a_n couplings could be obtained by using two target nuclei having $\langle S_p \rangle / \langle S_n \rangle$ ratios of opposite signs. The allowed region should then be comprised within the intersection of the two bands with slopes of opposite signs [Tovey et al. (2000)]. An example of limits calculation in the a_p/a_n plane will be given later for the PICASSO experiment (Sect. 10.2.2).

10.2.1.3 *Calculation of $\langle S_p \rangle$ and $\langle S_n \rangle$ in Nuclei*

The total enhancement factor, C_N^{SD}, in the spin-dependent neutralino–nucleus elastic cross section has been defined previously as

$$C_N^{SD} = \frac{8}{\pi} \left(a_p \langle S_p \rangle + a_n \langle S_n \rangle \right)^2 \frac{J+1}{J}, \tag{10.81}$$

where a_p and a_n are effective χ–proton and χ–neutron couplings; J is the total nuclear spin and $\langle S_p \rangle = \langle N | S_p | N \rangle$ and $\langle S_n \rangle = \langle N | S_n | N \rangle$ are the expectation values of the proton and neutron spins within the nucleus.

In the case of free nucleons $\langle S_p \rangle = 0.5$ and $\langle S_n \rangle = 0.5$, obviously, and the corresponding enhancement factors are $C_p = (6/\pi) a_p^2$ and $C_n = (6/\pi) a_n^2$.

In the case of nuclei, the calculation of $\langle S_p \rangle$ and $\langle S_n \rangle$ depends on the nuclear model. The calculation performed in three models are presented: the extreme single-particle model (ESPM)([Engel and Vogel (1989)]), the odd-group model (OGM)([Engel and Vogel (1989)]) and the extended odd-group model (EOGM) ([Engel and Vogel (1989)]). The odd-group model uses the measured nuclear magnetic moment to obtain $\langle S_p \rangle$ and $\langle S_n \rangle$. The extended odd-group model uses the measured nuclear magnetic moment and the ft-value for the Gamow-Teller β-decay, leading to more realistic predictions for $\langle S_p \rangle$ and $\langle S_n \rangle$. Finally, the nuclear angular moment and the nuclear magnetic moment can be calculated from the values of $\langle S_p \rangle$ and $\langle S_n \rangle$ obtained from shell models. Comparison is done with the experimental nuclear magnetic moment for several nuclei. One turns now to the detailed calculation of $\langle S_p \rangle$ and $\langle S_n \rangle$.

Assuming charge symmetry of nuclear forces, the magnetic moments of the odd-proton (μ_p) and odd-neutron (μ_n) members of a mirror pair are [Buck and Perez (1983)]:

$$\mu_p = g_p L_o + G_p S_o + g_n L_e + G_n S_e \tag{10.82}$$

and

$$\mu_n = g_n L_o + G_n S_o + g_p L_e + G_p S_e, \tag{10.83}$$

Table 10.2 Test of $\langle S_{\rm p} \rangle$ and $\langle S_{\rm n} \rangle$ calculated from shell models; "small" and "large" refer to the size of model spaces which adequately describe the configuration mixings in the nuclei [Ressell et al. (1993)].

Nucleus	J	$S_{\rm p}$	$S_{\rm n}$	$L_{\rm p}$	$L_{\rm n}$	J [Eq. (10.105)]	Ref.	μ [Eq. (10.106)]	$\mu_{\rm exp}$
^{19}F	1/2	0.4751	-0.0087	0.2235	-0.1899	0.50	[Divari et al. (2000)]	2.911	2.629
^{29}Si	1/2	-0.0019	0.1334	0.0183	0.3498	0.499(6)	[Divari et al. (2000)]	-0.503	-0.555
^{23}Na	3/2	0.2477	0.0199	0.9115	0.3207	1.499(8)	[Divari et al. (2000)]	2.219	2.218
^{73}Ge	9/2	0.030	0.378	0.361	3.732	4.501	[Dimitrov, Engel and Pittel (1995)]	-0.918	-0.879
^{73}Ge	9/2	0.005	0.496	0.40	3.596	4.497	[Ressell et al. (1993)]–"small"	-1.470	-0.879
^{73}Ge	9/2	0.011	0.468	0.491	3.529	4.499	[Ressell et al. (1993)]–"large"	-1.238	-0.879

where $g_p = 1$, $G_p = 5.586$, $g_n = 0$ and $G_n = -3.826$ (all in nuclear magnetons units) are the orbital g and spin G factors for proton and neutron. The odd- and even-nucleon contributions to the angular moment are labeled by the subscripts "o" and "e", respectively. For both nuclei, the total angular moment J is equal to:

$$J = L_o + S_o + L_e + S_e \tag{10.84}$$

and

$$J_e = L_e + S_e, \tag{10.85}$$

where S_i and L_i $(i = o, e)$ are the nuclear spin and orbital moment components, respectively. Equations (10.84) and (10.85) allow one to rewrite Eqs. (10.82, 10.83) as

$$\begin{aligned} \mu_p &= g_p L_o + G_p S_o + g_n (J - L_o - S_o - S_e) + G_n S_e \\ &= g_p J + (G_p - g_p)(S_o - S_e) - (g_p - g_n) J_e \\ &\quad + (G_p - g_p + G_n - g_n) S_e \end{aligned} \tag{10.86}$$

and

$$\begin{aligned} \mu_n &= g_n L_o + G_n S_o + g_p (J - L_o - S_o - S_e) + G_p S_e \\ &= g_n J + (G_n - g_n)(S_o - S_e) + (g_p - g_n) J_e \\ &\quad + (G_p - g_p + G_n - g_n) S_e. \end{aligned} \tag{10.87}$$

If even-nucleon contributions are neglected in Eqs. (10.86, 10.87), one finds:

$$\mu_p \approx g_p J + (G_p - g_p) S_o, \tag{10.88}$$

giving

$$S_o = \frac{\mu_p - g_p J}{(G_p - g_p)}, \tag{10.89}$$

and

$$\mu_n \approx g_n J + (G_n - g_n) S_o, \tag{10.90}$$

giving

$$S_o = \frac{\mu_n - g_n J}{(G_n - g_n)}. \tag{10.91}$$

Equations (10.89, 10.91) are the odd-group model (OGM) expressions in which the odd-nucleon spin is related to the nuclear magnetic moment μ and the gyromagnetic factors $G_{p,n}$ and $g_{p,n}$. They allow the calculation of S_o using the experimental magnetic moment. For instance, ^{19}F is an odd-proton nucleus with $J = 1/2$, a $S_{1/2}$ state, and with $\mu_{exp} = 2.629$. Therefore, Eq. (10.89) gives the value:

$$\begin{cases} \langle S_p \rangle = S_o = [2.629 - 1.0 \times (1/2)] / (5.586 - 1.0) = 0.46, \\ \\ \langle S_n \rangle = 0 \end{cases} \tag{10.92}$$

Table 10.3   Values of $\langle S_\mathrm{p}\rangle$ and $\langle S_\mathrm{n}\rangle$ for $^{19}\mathrm{F}$ calculated from various models. The corresponding ratios $R_\mathrm{p} \equiv C_{\mathrm{p}(N)}{}^{\mathrm{SD}}/C_\mathrm{p}{}^{\mathrm{SD}}$ and $R_\mathrm{n} \equiv C_{\mathrm{n}(N)}{}^{\mathrm{SD}}/C_\mathrm{n}{}^{\mathrm{SD}}$ are also listed.

S_p	S_n	R_p	R_n	Reference
0.4751	-0.0087	0.903	0.0003	[Divari et al. (2000)]
0.368	-0.001	0.542	1×10^{-6}	[Leroy (2004b)], $g_\mathrm{A}/g_\mathrm{V} = 1.25$
0.415	-0.047	0.689	0.0088	[Leroy (2004b)], $g_\mathrm{A}/g_\mathrm{V} = 1.00$
0.441	-0.109	0.778	0.0475	[Pacheco and Strottman (1989)]

and, obviously, $\langle L_\mathrm{p}\rangle = 0.04$, $\langle L_\mathrm{n}\rangle = 0$. For instance, $^{73}\mathrm{Ge}$ is an odd-neutron nucleus with $J = 9/2$, a $G_{9/2}$ state, and with $\mu_\mathrm{exp} = -0.879$. Therefore, Eq. (10.91) gives the value:

$$\begin{cases} \langle S_\mathrm{n}\rangle = S_\mathrm{o} = \left[-0.879 - 0.0 \times (9/2)\right]/(-3.826 - 0.0) = 0.23, \\ \langle S_\mathrm{p}\rangle = 0 \end{cases} \qquad (10.93)$$

with $\langle L_\mathrm{n}\rangle = 4.27$ and $\langle L_\mathrm{p}\rangle = 0$

In the extreme single-particle model(ESPM), it is assumed that the entire spin of the nucleus comes from the single last unpaired proton or neutron. Thus, one finds for a nucleus with an unpaired proton

$$\begin{cases} \langle S_\mathrm{p}\rangle = S_\mathrm{o} = 1/2 \left\{[J(J+1) + 3/4 - l(l+1)]/(J+1)\right\}, \\ \langle S_\mathrm{n}\rangle = 0 \end{cases} \qquad (10.94)$$

and, vice versa, for a nucleus with an unpaired neutron. It gives $\langle S_\mathrm{p}\rangle = 0.50$, $\langle S_\mathrm{n}\rangle = 0$, and $\langle L_\mathrm{p}\rangle = \langle L_\mathrm{n}\rangle = 0$ for $^{19}\mathrm{F}$, and $\langle S_\mathrm{n}\rangle = 0.50$, $\langle S_\mathrm{p}\rangle = 0$, $\langle L_\mathrm{p}\rangle = 0$ and $\langle L_\mathrm{n}\rangle = 4.0$ for $^{73}\mathrm{Ge}$.

Returning to Eqs. (10.86, 10.87), one considers the effects of meson currents in the nucleus. It is believed that the dominant one-pion exchange mechanism does not modify the isoscalar moment

$$\mu_\mathrm{IS} = \mu_\mathrm{p} + \mu_\mathrm{n}.$$

The mean effect of these currents is to introduce additional terms $-\mu_\mathrm{M}$ and $+\mu_\mathrm{M}$ into Eqs. (10.86, 10.87), respectively. Heavy-vector-meson exchange currents contribute a term μ_x in Eq. (10.86). Model calculations suggest a form:

$$\mu_x = -x\left(S_\mathrm{o} - S_\mathrm{e}\right), \qquad (10.95)$$

with a factor x to be determined later. One can rewrite Eqs. (10.86, 10.87) as

$$\begin{aligned} \mu_\mathrm{p} = {}& g_\mathrm{p}\, J + (G_\mathrm{p} - g_\mathrm{p} - x)\left(S_\mathrm{o} - S_\mathrm{e}\right) - (g_\mathrm{p} - g_\mathrm{n})\, J_\mathrm{e} - \mu_\mathrm{M} \\ & + (G_\mathrm{p} - g_\mathrm{p} + G_\mathrm{n} - g_\mathrm{n})\, S_\mathrm{e} \end{aligned} \qquad (10.96)$$

and

$$\begin{aligned} \mu_\mathrm{n} = {}& g_\mathrm{n}\, J + (G_\mathrm{n} - g_\mathrm{n})\left(S_\mathrm{o} - S_\mathrm{e}\right) + (g_\mathrm{p} - g_\mathrm{n})\, J_\mathrm{e} \\ & + \mu_\mathrm{M} + (G_\mathrm{p} - g_\mathrm{p} + G_\mathrm{n} - g_\mathrm{n})\, S_\mathrm{e}. \end{aligned} \qquad (10.97)$$

Then, the isoscalar moment is [summing Eqs. (10.96, 10.97)]:

$$\mu_{IS} = (g_p + g_n) J + (G_p - g_p + G_n - g_n - x) (S_o - S_e)$$
$$+2 (G_p - g_p + G_n - g_n) S_e \qquad (10.98)$$

or, using $g_p = 1$, $G_p = 5.586$, $g_n = 0$ and $G_n = -3.826$,

$$\mu_{IS} = J + (0.76 - x) (S_o - S_e) + 1.52 S_e \qquad (10.99)$$

or

$$\frac{\mu_{IS}}{J} = 1 + (0.76 - x) \left(\frac{S_o - S_e}{J}\right) + 1.52 \frac{S_e}{J}. \qquad (10.100)$$

The difference $(S_o - S_e)$ and μ_{IS} are obtained from a global fit to magnetic moments in each of the two nuclei and the *ft*-value for the Gamow-Teller β-decay from one to the other:

$$R^2 (S_o - S_e)^2 = \left(\frac{6170}{ft} - 1\right) \frac{J}{J+1}, \qquad (10.101)$$

where $R = g_A/g_V$ is the ratio of the axial-vector to vector weak-interaction coupling coefficients.

For ^{19}F, one obtains [Buck and Perez (1983)] from a fit to magnetic moment and *ft*-value data: $\mu_p/J = 5.2576$, $\mu_n/J = -3.7708$ and, therefore, $\mu_{IS}/J = 1.487$. The quantity $R(S_o - S_e)/J = 0.9226$ is also found. From the original fit, a quenched value for R was found:

$$R = 1.00 \pm 0.02.$$

If one uses this value, one obtains

$$\frac{S_o - S_e}{J} = 0.9226. \qquad (10.102)$$

The quantity x in Eq. (10.100) can be found from a table [Raman, Houser, Walkiewicz and Towner (1978)]. For ^{19}F, the table gives $\mu_x = -0.018$. However, to adapt the present normalization with that of the table, one needs to multiply the values found in the table by a factor 2. So, one finds $\mu_x = -0.036$ for ^{19}F. The value of x can be extracted using Eqs. (10.95, 10.102) to find:

$$x = \frac{0.036}{0.461} = 0.078. \qquad (10.103)$$

Combining Eqs. (10.100, 10.102, 10.103), one finds:

$$\begin{cases} S_e = -0.047, \\ \\ S_o = 0.415, \end{cases} \qquad (10.104)$$

which are reported in [Engel and Vogel (1989)]. If one uses the unquenched value

$$R = 1.249 \pm 0.006 \approx 1.25$$

obtained in neutron decay experiments, one finds

$$\frac{S_o - S_e}{J} = \frac{0.9226}{1.25} = 0.738$$

and $x = 0.036/0.369 = 0.098$ and therefore, $S_e = -0.001$ and $S_o = 0.368$.

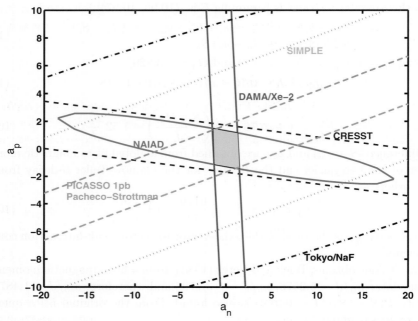

Fig. 10.14   Exclusion plot a_p versus a_n for $\sigma_{\chi p} = 1$ pb and $m_\chi = 50\,\mathrm{GeV\,c^{-2}}$ [Genest and Leroy (2004)], [Leroy (2004b)] using $\langle S_p \rangle$, $\langle S_n \rangle$ calculated from model [Pacheco and Strottman (1989)]. The PICASSO result [Barnabé-Heider et al. (2005b)] is compared to the results of several experiments: CRESST [Seide et al. (2002)], DAMA-Xe-2 [Bernabei et al. (1998)], SIMPLE [Girard R et al. (2005)], Tokyo/NaF [Takeda et al. (2003)]. Other updated and new experimental results can be found in [Genest (2007)].

10.2.1.4   *Shell Models Calculation and Validation of $\langle S_p \rangle$ and $\langle S_n \rangle$*

The nuclear angular moment and the nuclear magnetic moment can be calculated from the value of $\langle S_p \rangle$ and $\langle S_n \rangle$, as calculated from shell models, and comparison with the experimental nuclear magnetic moment can be done (Table 10.2) as one test of the shell model. Using again notations familiar to shell models, the angular momentum is:

$$J = \langle L_p \rangle + \langle S_p \rangle + \langle L_n \rangle + \langle S_n \rangle \tag{10.105}$$

and the magnetic moment is given by

$$\mu = g_p \langle L_p \rangle + G_p \langle S_p \rangle + g_n \langle L_n \rangle + G_n \langle S_n \rangle . \tag{10.106}$$

The calculated magnetic moment of ^{73}Ge with free particle g-spin factors show significant improvement in agreement with measured values if quenched g-spin factors are used (consistent with isovector quenching found in the sd [Brown and Wildenthal (1987)] and fp [Richter et al. (1991)] shells[‡‡] when fitting to magnetic moment data). Using quenched isovector component of the spin, one finds effective

---

[‡‡]The reader can find an introduction to nuclear shells in Sect.3.1.5.

g-spin factors: $G_p = 4.80$ and $G_n = -3.04$. Then, the calculated magnetic moments of ^{73}Ge are $\mu = -1.084$ and -0.879 for [Ressell et al. (1993)]-"small" and [Ressell et al. (1993)]-"large", respectively.

10.2.2 *The PICASSO Experiment, an Example*

As said previously the PICASSO Experiment [Barnabé-Heider et al. (2005a)] is measuring the neutralino–^{19}F cross section. This cross section is largely dominated by the spin-dependent contribution. The neutralino induced recoil (R) spectrum of ^{19}F is given by [Lewin and Smith (1996)]:

$$\frac{dR}{E_R} \approx c_1 \frac{R_{0,P}}{E_R} [F_{SD}(E_R)]^2 \exp\left(-c_2 \frac{E_R}{\langle E_R \rangle}\right), \qquad (10.107)$$

where $F_{SD}(E_R)$ is the form factor for spin-dependent intereactions: $F_{SD}(E_R) < 1$, it is due to the finite size of the nucleus and dependent mainly on nuclear radius and recoil energy;

$$R_{0,P} = \frac{403}{A_T \, m_\chi} [\sigma_{SD}/(1\,\text{pb})] [\rho_\chi/(0.3\,\text{GeV cm}^{-3})] [\langle v\chi \rangle / (230\,\text{km s}^{-1})] \quad (10.108)$$

is the total rate of neutrino–nucleus interaction (assuming zero momentum transfer), $R_{0,P}$ is expressed in counts per kg and per day; A_T is the atomic mass number of the target atoms (^{19}F in the present case), ρ_χ in GeV c^{-2} the mass density of the neutralino (the local neutralino mass density at the position of the solar system is assumed to be $0.3\,\text{GeV cm}^{-3}$), $\langle v\chi \rangle$ is the relative average neutralino velocity ($230\,\text{km s}^{-1}$ is the velocity dispersion of the dark matter halo);

$$E_R = 2 \frac{m_A \, m_\chi^2}{(m_A + m_\chi)^2} \langle v\chi^2 \rangle \qquad (10.109)$$

is the mean recoil energy with m_A the nucler mass of the recoil nucleus, $F^2(E_R) \sim 1$ for ^{19}F and small momentum transfer; $c_1 = c_2 = 1$ for $v_E = 0$ (v_E is the velocity of the earth relative to the dark matter distribution) [Lewin and Smith (1996)]; and $c_1 = 0.75$ and $c_2 = 0.56$ for $v_E = 244\,\text{km s}^{-1}$) [Lewin and Smith (1996)]. Combining the ^{19}F recoil spectra expected from neutralino interactions [Eq. (10.107)] and the measured detector threshold for ^{19}F recoil energy at a given operating temperature (T), one can determine the neutralino detection efficiency, $\epsilon(m_\chi, T)$) as a function of the neutralino mass and operating temperature. Then, the observable neutralino count rate, R_{obs}, as a function of temperature, neutralino mass and cross section is given by:

$$R_{\text{obs}}(m_\chi, \sigma_{SD}, T) = \frac{c_1}{c_2} R_{0,P}(m_\chi, \sigma_{SD}) \epsilon(m_\chi, T))$$
$$= 1.34 \, R_{0,P}(m_\chi, \sigma_{SD}) \epsilon(m_\chi, T)). \qquad (10.110)$$

The cross section, σ_{SD}, is given by Eq. (10.62). A combined fit of alpha background and neutralino response to the PICASSO data brought an upper limit of 1.31 pb on $\sigma_{\chi p}$, 21.5 pb on $\sigma_{\chi n}$ for a mass $m_\chi = 29\,\text{GeV c}^{-2}$ [Barnabé-Heider et al. (2005b)].

One can also calculate the $a_\mathrm{p}/a_\mathrm{n}$ limits. If one considers the example of a cross section $\sigma_{\chi p} = \sigma_\mathrm{p}^{\mathrm{SDlim}} = 1\,\mathrm{pb}$ for $m_\chi = 50\,\mathrm{GeV\,c^{-2}}$, one finds:

$$\frac{\pi}{24 G_\mathrm{F}^2 \mu_{\chi p}^2} = 2.92\,\mathrm{pb^{-1}}. \tag{10.111}$$

$\sigma_\mathrm{p}^{\mathrm{SDlim}}$ being fixed, from the ratio of Eqs. (10.65, 10.66) assuming the same mass for proton and neutron, one finds

$$\frac{\sigma_\mathrm{p}^{\mathrm{SDlim}}}{\sigma_\mathrm{n}^{\mathrm{SDlim}}} = \frac{C_{p(N)}^{\mathrm{SD}}/C_\mathrm{p}^{\mathrm{SD}}}{C_{n(N)}^{\mathrm{SD}}/C_\mathrm{p}^{\mathrm{SD}}} \tag{10.112}$$

with

$$\frac{C_{p(N)}^{\mathrm{SD}}}{C_\mathrm{p}^{\mathrm{SD}}} = \frac{(8/\pi)a_\mathrm{p}^2\,\langle S_\mathrm{p}\rangle^2\,[(J+1)/J]}{(6/\pi)a_\mathrm{p}^2} = \frac{4}{3}\left(\frac{J+1}{J}\right)\langle S_\mathrm{p}\rangle^2 = 0.778 \tag{10.113}$$

and

$$\frac{C_{n(N)}^{\mathrm{SD}}}{C_\mathrm{p}^{\mathrm{SD}}} = \frac{(8/\pi)a_\mathrm{n}^2\,\langle S_\mathrm{n}\rangle^2\,[(J+1)/J]}{(6/\pi)a_\mathrm{n}^2} = \frac{4}{3}\left(\frac{J+1}{J}\right)\langle S_\mathrm{n}\rangle^2 = 0.0475. \tag{10.114}$$

One has used the values

$$\begin{cases} \langle S_\mathrm{p}\rangle = 0.441, \\[2mm] \langle S_\mathrm{n}\rangle = -0.109, \end{cases} \tag{10.115}$$

given in [Pacheco and Strottman (1989)] (Table 10.3).

One can determine the neutralino mass independent ratio:

$$\frac{\sigma_\mathrm{p}^{\mathrm{SDlim}}}{\sigma_\mathrm{n}^{\mathrm{SDlim}}} = \frac{1/0.778}{1/4.75 \times 10^{-2}} = 0.061. \tag{10.116}$$

Therefore, fixing $\sigma_{\chi p} = \sigma_\mathrm{p}^{\mathrm{SDlim}} = 1\,\mathrm{pb}$ and choosing $m_\chi = 50\,\mathrm{GeV\,c^{-2}}$, one obtains from Eq. (10.116), the value $\sigma_{\chi n} = \sigma_\mathrm{n}^{\mathrm{SDlim}} = 16.37\,\mathrm{pb}$. Then, the value $\sigma_{\chi p} = \sigma_{p(N)}^{\mathrm{SDlim}} = 1\,\mathrm{pb}$ corresponds to $\sigma_{\chi F}^{\mathrm{SDlim}} = 158.78\,\mathrm{pb}$.

Equation (10.80) for $m_\chi = 50\,\mathrm{GeV\,c^{-2}}$ reads

$$\frac{2.92}{\mathrm{pb}} \geq \left(\frac{a_\mathrm{p}}{\sqrt{1.0\,\mathrm{pb}}} \pm \frac{a_\mathrm{n}}{\sqrt{16.37\,\mathrm{pb}}}\right)^2, \tag{10.117}$$

giving the two exclusion boundary limits [Genest and Leroy (2004)], [Leroy (2004b)]:

$$a_\mathrm{p} = 1.708 + 0.247 a_\mathrm{n} \tag{10.118}$$

and

$$a_\mathrm{p} = -1.708 + 0.247 a_\mathrm{n}. \tag{10.119}$$

The lines corresponding to Eqs. (10.118,10.119) represents the PICASSO experiment exclusion limits (Fig. 10.14) for a neutralino of $50\,\mathrm{GeV\,c^{-2}}$ mass and $\sigma_{\chi p} = 1\,\mathrm{pb}$.

Chapter 11

Medical Physics Applications

The knowledge about the physics governing the interactions of particles with matter and particle detection finds applications in the field of nuclear medicine imaging technique. This technique uses the injection into the patient of radionuclides directly emitting photons, or of radiopharmaceuticals, labeled with a positron emitting isotope. Photons directly produced by radionuclides or produced by the annihilation of positrons emitted by the radiopharmaceutical with body electrons are detected by radiation detectors. This allows one to reconstruct three dimensional images representing the distribution of radioactivity inside the patient's body and to measure metabolic, biochemical and functional activities in tissue. Magnetic Resonance Imaging (MRI) is another imaging technique, which does not require the use of any radioactive material and uses instead the non-zero nuclear spin, an intrinsic property found in nuclei (see page 226). MRI uses magnetic fields* varying from 0.2 to 2 T and radio-frequency (RF) waves to observe the magnetization change of the non-zero spin nuclei. The hydrogen isotope ^{1}H, which has a nuclear spin of $\frac{1}{2}$, is a major component of the human body and is used as the main source of information.

Two techniques exploit the interaction of the produced photons with the active material of radiation detectors (imager or scanner): *Single Photon Emission Computed Tomography* (SPECT) and *Positron Emission Tomography* (PET). We have seen, in Sect. 2.3, that photons interact with matter in several ways. These are Compton scattering, photoelectric effect, and pair production. The two other possible interactions are discarded for medical applications: Rayleigh or coherent scattering (see Sect. 2.3.2.2) and photonuclear absorption (see Sect. 2.3.4). Rayleigh scattering is a process predominant in the forward direction, i.e., in which photons scatter from atomic electrons without exciting or ionizing the atom and, therefore, no energy is absorbed in that process. Also, the incoming photon beam (usually from a source) is hardly altered by this process. The Rayleigh cross section, σ_{coh} (see Sect. 2.3.2.2), may be large for low photon energies (around 1 keV or less) and rapidly decreases with the photon energy. However, for practical energies faced in medical applications, σ_{coh} is much smaller than the Compton (incoherent) cross

---

*1 T $= 10^4$ G. The Earth magnetic field is ≈ 0.5 G.

section. In particular, at energies where coherent and Compton cross sections compete (the coherent cross section could even dominates the Compton cross section), they are both largely dominated by the photoelectric cross section.

The photonuclear absorption is a nuclear interaction, in which the photon is absorbed by the nucleus. It becomes relevant for photons with energy beyond a few MeV's, although smaller than the pair production cross section, but these energies are not encountered in medical applications. The energies encountered in medical physics range from a few keV up to a few MeV.

The photoelectric effect is dominant for high-Z materials, while Compton scattering dominates for low-Z materials at the photon energies used in medical physics (see Fig. 2.75 and consider water as the material closest to tissue). The pair production process does not contribute, since the photon energy of the sources used in SPECT and the photon energy in PET (0.511 MeV) are lower than the threshold energy ($2\,mc^2 = 1.022$ MeV, where m is the electron rest-mass) for creating an electron–positron pair.

The detection probability of a photon emitted by a source and experiencing Compton scattering in the body depends on the amount of energy lost as a result of that scattering. The initial energy (E_0) of a photon and its energy after Compton scattering (E_A) are related through

$$E_A = E_0 \left/ \left[1 + \frac{E_0(1 - cos\theta)}{mc^2} \right] \right. , \qquad (11.1)$$

where θ is the angle between the initial and final direction (after Compton scattering) of the photon. Basically, the probability of Compton interaction has a weak dependence on the atomic number and decreases with the photon energy. The photon does not disappear in the Compton interaction and is available for further interaction (with a decreased energy) in another detector, giving directional information. This will be exploited in Compton camera, as we will see below.

The photoelectric effect (Sect. 2.3.1) is an interaction of a photon with a tight-bound atomic electron: K-electron, L-electron, M-electron, ... The photon of energy, $h\nu$, is completely absorbed and a photoelectron is ejected with a kinetic energy K_e given by

$$K_e = h\nu - B_e, \qquad (11.2)$$

where B_e is the binding energy of the electron. The photoelectric effect is inversely related to the power of ≈ 3.5 of the photon energy and directly related to $\approx$ the fifth power of the atomic number. The photoelectric effect cross section also depends on the electron shell. The photoelectric cross section as a function of the photon energy presents several discontinuities at low energy. These discontinuities are called absorption edges and correspond to energies below which it is impossible to eject certain electrons from the atom. Below the K-edge, the photon cannot eject a K-electron but still can eject a L-electron or a M-electron (Sect. 2.3.1). For instance, the vacancy left in the K-shell after interaction is immediately filled by the transition of one electron of an outer shell accompanied by the emission of X-rays or an Auger electron. Therefore, contrary to the Compton scattering, the photon is totally absorbed in the radiation detector and there is no directional information provided by the gamma or X-rays in the photoelectric process.

11.1 Single Photon Emission Computed Tomography (SPECT)

Single Photon Emission Computed Tomography (SPECT) has become a routine technique in medical applications [Brooks and DiChiro (1976)]. This gamma-ray imaging technique proceeds through the injection into the patient of a radioactive substance, which emits photons of well-defined energy. The distribution of radionuclides, position and concentration inside patient's body is externally monitored through the emitted radiation deposited in a photon detector array rotating around the body. This rotation allows the acquisition of data from multiple angles. This procedure allows the study of organs behaviors, bringing the possibility to reveal signs of malfunctioning as early as possible.

Organ imaging requires a radiation of sufficient energy to penetrate the body tissues. However, the radiation energy must remain low enough to allow its absorption in the detecting device. Therefore, photons with an energy ranging between 50 keV and several hundreds of keV can be used for imaging. Photons in this energy range are produced by specific radionuclides. A widely-used radionuclide is ^{99m}Tc, an isomer of technetium with a half-life of 6.02 hours, which decays emitting 140.5 keV (89%) photons (see Fig. 11.1).

Other sources like ^{201}Tl, ^{178}Ta and ^{133}Xe, emitting lower energy photons, are also used. For instance, ^{201}Tl emits 135 keV (2%) and 167 keV (8%) photons and (69–83) keV mercury K X-rays (90%).

Photons, produced after injection of ^{99m}Tc in the patient's body, will eventually reach a detector where their energy deposition is measured. The organ structure and its evolution are then visualized from the resulting photon absorption patterns. The images are the projection of a three-dimensional distribution onto a two-dimensional plane. This can be achieved by rotating the detector around the patient. Series of two-dimensional projections are taken from different directions.

To create the two-dimensional projections, Anger cameras are often used. First, the photon emitted from within the patient crosses a collimator. Then, it reaches a scintillator. The point of scintillation corresponds exactly to the plane coordinates of the point of emission. Once the initial photon has reached the scintillator, it excites all the photomultipliers. Analyzing the intensity of the signal coming from every photomultiplier allows the determination of the plane coordinates. Finally, the intensity of every photomultiplier's signal is added. If it equals the energy of the photon emitted, the information will be kept and helps the formation of the image. If the energy is inferior, it means that the photon was scattered. Therefore, the wrong coordinates were found and the information is rejected.

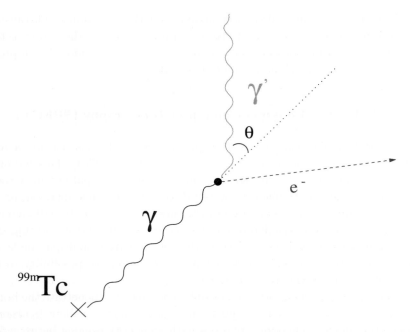

Fig. 11.1   Photons of 140.5 keV energy emitted by a ^{99m}Tc source of standard use in SPECT, decaying to its isomer ^{99m}Tc, followed by a Compton scattering on a nucleus producing a recoiling electron. θ is the Compton scattering angle.

These cameras need to be operated with a collimator in front of the detecting material. The collimator is usually made of a thick high-Z metal plate, drilled with a huge number of small holes. The collimator holes allow only incident radiation perpendicular to the detector surface and eliminate the obliquely-incident photons i.e., the secondary photons produced by the interaction of the primary photons (from the source) with biological matter (tissue, bones, etc...). These obliquely-incident photons, by activating several detector readout cells, may prevent the image formation or, at best, degrade the image focusing. Their removal is necessary to obtain a source image of high quality. Compton scattered photons (see Fig. 11.1) have a lower energy than primary photons and can be rejected by energy discrimination. For a given energy window, only Compton photons scattered at sufficiently small angle can be detected [Eq. (11.1)]. If one assumes that the detector has perfect energy resolution and that a rectangular energy window is applied, centered at E_0 with a width ΔE (in keV), the maximum allowable scattering angle at the first order, $\theta_{\max}$, is given by:

$$\theta_{\max} = \cos^{-1}\left\{1 - mc^2\Delta E\left[E_0(2E_0 - \Delta E)\right]^{-1}\right\}. \tag{11.3}$$

As can be seen from Eq. (11.3), the probability to detect a 140.5 keV photon from a ^{99m}Tc source in a 15% wide rectangular energy window is zero if the scattering

Table 11.1   Characteristics of semiconductor materials for use in medical imaging.

Characteristics	Si	GaAs	CdTe	HgI$_2$
Atomic number (Z)	14	31,33	48,52	80,53
Density (g/cm^3)	2.33	5.32	5.80	6.30
Band gap (eV)	1.12	1.42	1.50	2.10
Energy per e-h pairs (eV)	3.62	4.20	4.43	4.15
Electron mobility (cm^2 V^{-1} s^{-1})	1350	8500	1000	100
Hole mobility (cm^2 V^{-1} s^{-1})	450	400	70	4

angle $\theta > 45.2°$. If the energy window is increased to 20%, the corresponding angle becomes 53.5°. For a single scattered photon to be detected at a position x of the collimator, the photon must have been emitted and come out of the body along a path confined to a cone with aperture $\theta \leq \theta_{max}$.

The necessity to use a collimator has the adverse consequence to decrease the detection efficiency and this is a limitation to SPECT. The collimator reduces the number of normal impinging photons to about 10^{-4} of their original number, after passing the collimator. As a consequence, a higher dose has to be given to the patient in order to provide sufficient statistics for elaborating an accurate image.

Semiconductors have been also considered as detecting medium of gamma scanner. The properties of several semiconductors for this purpose are listed in Table 11.1. A priori, the low atomic number ($Z = 14$) of silicon is seen as a handicap for this type of application, although silicon detectors naturally offer excellent spatial and energy resolutions. However, it is possible to use silicon as active medium devices of gamma camera by combining silicon with high-Z material (W or Pb, for instance) to form a sampling calorimeter of high effective atomic number which will serve as SPECT detector. The Silicon Collimated Photon Imaging Calorimeter, (SiCPICal) [D'Angelo, Leroy, Pensotti and Rancoita (1995)], consists of a superposition of 200 active silicon layers interspersed with 120 μm tungsten layers. A silicon active layer is made of 145 strips, each strip being 400 μm thick. Each silicon layer is organized in high spatial resolution readout cells ($550 \times 550 \, \mu$m^2). A detector like SiCPICal has a high-Z detection volume with high photon conversion, exploiting at the same time the excellent spatial and energy resolution provided by silicon. The single readout pixel can be operated at about 10 MHz, thus no electronic dead time affects the overall detector performances. This detector can be operated in two different modes. It can select events in which the incoming photons have interacted either by both Compton and photoelectric effects or by photoelectric effect only. The operation mode depends on the current discriminator threshold setting.

The formation of a focused image is made possible by the presence of a collimator with a variable structure, located in front of the detector. This allows one to keep the minimal image size, of about 0.3 mm^2 for a point source, independently of the distance between the collimator and the point source.

As stressed above, the detector can be operated selecting, at the same time, photoelectric and Compton photon interactions. In this way, up to 43% of the impinging photons provide events for the image formation. However, in order to reduce the background of low and medium energy photons, generated by Compton interaction in the patient's body, a metallic Sn filter can be utilized. It reduces to a negligible amount the γ's with energy lower than 100 keV.

The detector can also be operated by selecting photoelectric interacting photons alone, which are about 4% of the total number of 140.5 keV incoming photons. It achieves a very high energy resolution of about 1%. In this way, a strong reduction of background, due to photons interacting in the patient's body, is expected. Thus, highly resolved images are expected, even if the number of counted events is reduced.

The photoelectric operational mode seems very attractive for extending the usage of SiCPICal to lower energy photons, for which the photoelectric cross section increases. In the (60–80) keV photon energy range, the photoelectric interaction probability in SiCPICal is (43–23)%, and the energy resolution is (2.3–1.7)%.

The possibility of using high-Z semiconductor materials like cadmium telluride (CdTe and CdZnTe) and mercuric iodide (HgI_2) has been envisaged. These materials, widely available, present however drawbacks for their use in SPECT detectors. Indeed, hole transport is poor in these materials and charges are heavily trapped and cannot be collected in a practical amount of time. Due to incomplete charge collection, the size of the output pulse becomes dependent on the exact position of interaction of the radiation in the detector volume, and spoils the energy resolution. The accumulation of uncombined trapped charges leads to polarization, which further inhibits the charge collection. Overall, the performances of the detector can change over a period of time. However, it is possible to build CdZnTe imaging devices that only relies on the collection of electrons. A device of this type (the coplanar orthogonal anode detector) has been successfully tested [Tousignant et al. (1999)] and is able to measure the position of interaction in 3-dimensions with a spatial resolution of 300 μm and an energy resolution of 1% and 2.6% *FWHM* at 662 keV and 122 keV, respectively. Bridgman CdTe and CdZnTe crystal growth, with cadmium vapor pressure control, can produce crystals that are highly donor doped and highly electrically conducting. After annealing in tellurium vapors, they are transformed into highly compensated state of high resistivity and high sensitivity to photons [Lachish (1999)]. These detectors, after proper equipment with ohmic contacts and a grounded guard-ring around the positive contact, have fast electron collection time: for a detector $d = 1$ mm thick operated at a bias of 150 volts (electron and hole mobility, $\mu_e \approx 1000$ cm^{2}V^{-1}s^{-1}, and $\mu_h \approx 70$ cm^{2}V^{-1}s^{-1}, respectively), the transit time of an electron from contact to contact is

$$t = d^2/(\mu_e V) = 66\,\text{ns},\qquad(11.4)$$

while the transit time for a hole (should there be no trapping) under the same conditions is:

$$t \approx 1\,\mu\text{s}.\qquad(11.5)$$

Adjusting the shaping time of the charge collection system [(50–120) ns], such a detector is not sensitive to hole trapping and only collects the electron contribution to the signal.

The Compton camera concept is emerging for SPECT applications. However,to our knowledge, no practical application has been achieved yet. The real prospect in that direction is represented by the Medipix detector device (Sect. 6.5.1). The Compton camera is based on a method which allows the reconstruction of the direction of the primary photon coming from the object to be imaged by iterative back projection methods [Brechner and Singh (1990)]. Therefore, a Compton camera can be operated without collimator, offering a great advantage over Anger camera. This absence of collimator translates into higher detection efficiency. The Compton camera should permit the acquisition of data representing multiple angular views of the source distribution from a single position consequently reducing the need of camera motion. This increased sensitivity allows also a reduction of the dose delivered to the patient (lower activity level and shorter half-lives) and a reduction of the time spent by the patient in front of the scanner (complete immobility of the patient which could be painful just after surgery or during a suffering period). Reduced angular motion of the camera, meaning less time spent between angular stops, also help this latter aspect. The absence of a collimator has another important implication as this piece of metal is heavy and its insertion in the detecting system could alter the precision and mechanical stability of the system.

In principle, the Compton camera consists of a scatter detector and a absorption detector (Fig. 11.2).

The scatter detector generates and detects Compton interactions. The Compton scattered photons are depositing their energy in the absorption detector. The material of the scattering detector must have sizable Compton scattering cross-sections for the energy of photon released by the source (standard energy range from 100 up to 600 keV). Semiconductors, such as Si and GaAs of large thicknesses (1 mm and more), are usually envisaged for active medium of the scatter plane. The material composing the absorption detector should have large photo-absorption cross-sections in this energy range. Usually NaI(Tl), CsI(Tl) and BGO are used for absorption detector. The use of heavy semiconductors such as CdZnTe (CZT) is also possible.

The energy of the photons emitted by the source is known exactly. Therefore, the sum of energies deposited in the scattering and absorption detectors can be used to reject photons produced by Compton scattering in the patient.

The photon energy after Compton scattering in the scatter plane, E_A, is related to the incident photon energy, E_0, via Eq. (11.1). E_A is also the energy deposited in the absorption detector. E_S is the energy lost by the photon in the scatter plane and this energy is converted into kinetic energy of the recoil electron. Applying the conservation of energy, one finds

$$E_S = E_0 - E_A. \tag{11.6}$$

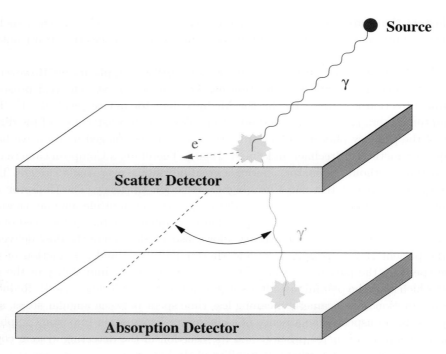

Fig. 11.2   The Compton camera with a scatter plane and an absorption plane. The location of photon impact in the scatter plane is obtained from the measurement of the recoiling electron. The Compton scattered photon is absorbed in the absorption plane.

The locations of the two interactions (one in the scattering detector and the second in the absorption detector), coupled with the scattering angle, limits the photon possible source location to a cone whose axis is in line with the positions of the two interactions and an aperture defined by the scattering angle. The reconstruction of the source distribution is going through the reconstruction of the direction of the primary photon coming from the object to be imaged. This reconstruction is performed by iterative back projection methods (see for instance [Brechner and Singh (1990)]).

As an illustration of the scatter plane role in a Compton camera principle, let us calculate the number of Compton events generated in a scatter plane made of silicon by a gamma emitting source that is injected into a patient. The source is ^{99m}Tc with a typical dose injection of $2\,\mu$Ci. ^{99m}Tc emits photons of 140.5 keV energy. The probability of having a photon of a specific energy emitted in a decay is represented by the factor p. For the example of ^{99m}Tc, $p = 0.889$ for a photon energy of 140.5 keV.

The activity of the radioisotope is measured in transformation per second. The special unit of activity is the becquerel (Bq, $1\,$Bq $= 1$ transformation/sec) that is expressed in curies (Ci)

$$1\,\text{Ci} = 3.7 \times 10^{10}\,\text{Bq}. \tag{11.7}$$

The number of atoms, N_A, present in the source and the activity, A, of the source are related by the transformation constant λ as:

$$A = \lambda N_A, \tag{11.8}$$

where

$$\lambda = \frac{\ln 2}{t_{1/2}} = \frac{0.693}{t_{1/2}}$$

and $t_{1/2}$ is the half-life of the radioisotope. One should note that the mean-life of the radioisotope is

$$\tau = \frac{1}{\lambda} = 1.443 \times t_{1/2}.$$

The activity decays with time, t, according to an exponential law, as:

$$A = \lambda N_A = \lambda N_0\, e^{-\lambda t} = A_0\, e^{-0.693t/t_{1/2}}. \tag{11.9}$$

In the equation above, $A_0 = \lambda N_0$ with N_0 is the number of atoms present at $t = 0$.

The source inside the body is eliminated by natural means. Some radiopharmaceuticals are being eliminated faster, some slower but it usually takes a few hours. The effective half-life, $t_{1/2}(\text{eff})$, is calculated by adding physics λ and biological λ_{bio}, i.e.,

$$\lambda(\text{eff}) = \frac{\ln 2}{t_{1/2}(\text{eff})} = \lambda + \lambda_{\text{bio}}.$$

The half-life of ^{99m}Tc is $t_{1/2} = 6.02$ hours and $\tau = 8.69$ hours. The total number of photons, N_{ph}, of energy 140.5 keV emitted by the ^{99m}Tc source during a period of time t is:

$$N_{\text{ph}} = p\, N_0 \left(1 - e^{-0.693t/t_{1/2}}\right). \tag{11.10}$$

Therefore, the number of 140.5 keV photons emitted by the source after one hour ($N_0 = 7.4 \times 10^4$ for a ^{99m}Tc injection dose of $2\,\mu$Ci which corresponds to 7.4×10^4 transformations/sec) is

$$N_{\text{ph}} \sim 0.7 \times 10^4. \tag{11.11}$$

Assuming no attenuation between the source and the detector and an isotropic emission of the source, the number of 140.5 keV incident photons reaching the detector can be defined as:

$$N_{\text{incident}} = f N_{\text{ph}}$$

with

$$f = \frac{\text{area of the detector}}{4\pi\,(\text{distance to the source})^2}. \tag{11.12}$$

We consider a scatter plane made of pads or pixels of silicon detectors representing a total area of $1.4\,\text{cm} \times 1.4\,\text{cm}$ (such as a MediPix device, Sect. 6.5.1) and a thickness of $1\,\text{mm}$, typical for medical applications (Fig. 11.3). Then, the fraction f of photons reaching the sensitive layer, located at $1\,\text{cm}$ from the source, is [using Eq. (11.12)]

$$f \sim 0.2,$$

giving the number of photons reaching the sensitive layer:

$$N_{\text{incident}} \sim 1400. \tag{11.13}$$

The number of photons, N_{int}, interacting in the detector of thickness Δy is given by:

$$N_{\text{int}} = \mu N_{\text{incident}} \Delta y, \tag{11.14}$$

where μ is the total attenuation coefficient. It is the sum of four components (see Sect. 2.3.5): the photoelectric (τ_{pe}), the coherent scattering (σ_{coh}), the Compton or incoherent scattering (σ_C) and the pair production (κ_{pair}) attenuation coefficients, i.e.,

$$\mu = \tau_{\text{pe}} + \sigma_{\text{coh}} + \sigma_C + \kappa_{\text{pair}}.$$

Equation 11.14 is an approximation which is valid because the detector width is small. A more precise calculation would use

$$N_{\text{int}} = f N_{\text{ph}} \left(1 - e^{-\mu \Delta y} \right). \tag{11.15}$$

The percentage of the Compton interactions is:

$$100 \sigma_C / (\tau_{\text{pe}} + \sigma_{\text{coh}} + \sigma_C + \kappa_{\text{pair}}).$$

The number, N_C, of Compton scattered photons in the detector of thickness Δy in the scatter plane is:

$$N_C = N_{\text{int}} \times \sigma_C / \mu = \mu \, N_{\text{incident}} \, \Delta y \, \sigma_C / \mu, \tag{11.16}$$

using Eq. (11.14). For a ^{99m}Tc source, at the photon energy of $140.5\,\text{keV}$, we have

$$\frac{\sigma_C}{\rho} = 1.33 \times 10^{-1}\,\text{cm}^2/\text{g}.$$

The silicon density being $2.33\,\text{g/cm}^3$, one has a linear Compton scattering attenuation coefficient $\sigma_C = 0.30989\,\text{cm}^{-1}$. Then, the average number of Compton scattered photons in the scatter plane per second for one hour of source decay for a distance detector-source of $1\,\text{cm}$ and an injected dose of $2\,\mu\text{Ci}$ [using Eqs. (11.13, 11.16) with $\Delta y = 0.1\,\text{cm}$]:

$$N_C \sim 44 \ s^{-1}.$$

The previous calculation was obviously approximate. If the distance between the detector and the source is not so small compared with the detector's dimensions, one should use an exact formula to calculate $f \Delta y$:

$$f \Delta y = 4 \times \int_0^{l/2} dx \int_0^{l/2} dz \frac{\Delta y}{4\pi (x^2 + z^2 + d^2)}.$$

Because of the finite size of the detector, the area $l/2 \times l/2$ has to be divided (as indicated in Fig. 11.4) in order to calculate Δy correctly:

$$\epsilon = \frac{wl}{2(d+w)}.$$

In region 1, we have

$$\Delta y = w\sqrt{1 + tg^2\alpha + tg^2\beta} = \frac{w}{d}\sqrt{x^2 + z^2 + d^2}.$$

In regions 2 and 3, we obtain

$$\Delta y = \left(\frac{l}{2} - x\right)\sqrt{1 + \frac{1}{tg^2\alpha} + \frac{tg^2\beta}{tg^2\alpha}} = \left(\frac{l}{2x} - 1\right)\sqrt{x^2 + z^2 + d^2}.$$

We now find:

$$f\Delta y = 4 \times \int_0^{l/2-\epsilon} dx \int_0^{l/2-\epsilon} dz \frac{\frac{w}{d}\sqrt{x^2 + z^2 + d^2}}{4\pi(x^2 + z^2 + d^2)}$$

$$+ 8 \times \int_{l/2-\epsilon}^{l/2} dx \int_0^{l/2-\epsilon} dz \left(\frac{l}{2x} - 1\right)\frac{\frac{w}{d}\sqrt{x^2 + z^2 + d^2}}{4\pi(x^2 + z^2 + d^2)}$$

$$+ 8 \times \int_{l/2-\epsilon}^{l/2} dx \int_{l/2-\epsilon}^{x} dz \left(\frac{l}{2x} - 1\right)\frac{\frac{w}{d}\sqrt{x^2 + z^2 + d^2}}{4\pi(x^2 + z^2 + d^2)}.$$

Finally, we obtain

$$f\Delta y = \frac{w}{\pi d} \int_0^{l/2-\epsilon} dx \int_0^{l/2-\epsilon} dz \frac{1}{\sqrt{x^2 + z^2 + d^2}}$$

$$+ \frac{2}{\pi} \int_{l/2-\epsilon}^{l/2} dx \int_0^{l/2-\epsilon} dz \left(\frac{l}{2x} - 1\right)\frac{1}{\sqrt{x^2 + z^2 + d^2}}$$

$$+ \frac{2}{\pi} \int_{l/2-\epsilon}^{l/2} dx \int_{l/2-\epsilon}^{x} dz \left(\frac{l}{2x} - 1\right)\frac{1}{\sqrt{x^2 + z^2 + d^2}}.$$

This last integral can only be evaluated numerically.

From Eq. (11.17), it is easy to see that the minimum energy of the scattered photon will be: $E_A = 90.7\,\text{keV}$, where

$$E_A = \frac{E_0}{1 + E_0(1 - \cos\theta)/mc^2}, \tag{11.17}$$

where θ is the scattering angle, $E_0 = 140.5\,\text{keV}$ and $mc^2 = 0.511\,\text{keV}$ is the electron rest mass. The average energy of the scattered photon is $E_A \approx 114\,\text{keV}$. This was taken from the graph found in [Johns and Cunningham (1983)] representing the fraction of the photon's energy transferred to the electron as a function of the energy of the incident photon.

The energy of the electron recoil is about 26 keV. The signal in the scatter silicon detector is in electron equivalent:

$$26\,\text{keV}/3.62\,\text{eV} = 7,200\ e^-, \tag{11.18}$$

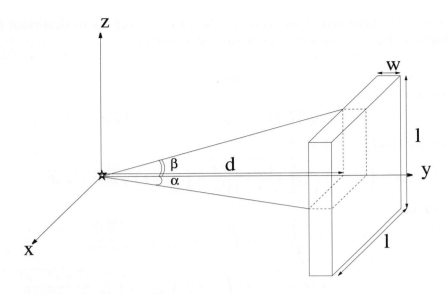

Fig. 11.3   Idealized representation of a source-detector system.

which correspond to a collected charge of $7,200 \times 1.610^{-19}\,\mathrm{C} \approx 1.2\,\mathrm{fC}$. There are two sources of noise: the noise from the preamplifier and the noise from leakage current. Typical preamplifier noise for a shaping time of $2\,\mu\mathrm{s}$ is

$$\mathrm{ENC[preamp]} = 60e^- + \left(11e^-/\mathrm{pF}\right) C_{\mathrm{det}}.$$

The silicon detector used for the estimate has a capacitance of

$$C_{\mathrm{det}} = 1.05\mathrm{pF/cm} \times 2.0\,\mathrm{cm}^2/0.1\,\mathrm{cm} \;\approx 21\,\mathrm{pF}$$

and therefore $\mathrm{ENC[preamp]} = 290e^-$. The leakage current of silicon detectors strongly depends on temperature [see Eq. (6.77)] and contributes to a large extent to the detector noise, spoiling energy and spatial resolution. Therefore, it is needed to operated silicon detectors at temperature as low as possible (low means, in practical cases, room temperature and lower). However, standard planar (float-zone) silicon detector can be safely operated at room temperature with a leakage current at the level of a few $\mathrm{nA/cm}^2$. Typical noise from leakage current for a shaping time of $2\,\mu\mathrm{s}$ is $\mathrm{ENC[leakage]} = 150\sqrt{I_r\,(\mathrm{nA})}$. For a detector of the type and size used in the calculation, $I_r \approx 2\,\mathrm{nA}$ giving $\mathrm{ENC[leakage]} = 212\,e^-$. The total noise is then

$$\mathrm{ENC} = \sqrt{\mathrm{ENC[preamp]}^2 + \mathrm{ENC[leakage]}^2} \approx 360\,e^-$$

and a signal to noise ratio

$$S/N \sim 20.$$

First feasibility experimental studies of using silicon pad detectors as scatter plane of a Compton camera has been reported in [Weilhammer et al. (1995)]. These pads were consisting of $\sim 345\mu$m thick detectors with size as low as $150\,\mu$m $\times 150\,\mu$m covering an area of 2.4 mm segmented electrodes on one side which were connected to their individual signal processing circuits through metal lines on top of the detector.

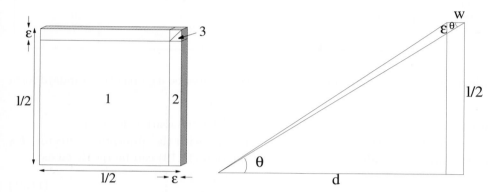

Fig. 11.4   Illustration of the divisions necessary to calculate Δy precisely.

The next step is to shrink the size of the silicon pads, down to the micrometer scale, and to finally achieve the stage of pixel detectors, each pixel detector being on very small electrode. Each detecting units is composed of one single diode attached to its own readout electronics, the diode being finely segmented, providing high two-dimensional spatial resolution. The concept of active pixel detectors rely on electronics cells being equally dimensioned and close to the corresponding pixel, keeping the collection electrode capacitance very small and a large signal-to-noise ratio. The application of the concept of silicon (and other type of semiconductors) pixel arrays to practical systems for medical imaging is under way [Mikulec (2000)].

As discussed in Sect. 6.3.1, semiconductor detectors may achieve very good energy resolution (σ_R) because a large number of electron-holes is created inside the semiconductor. However, other contributions can limit the overall energy resolution, for instance those depending on the particular readout electronic chain used for the measurements.

The dependence of angular uncertainties on the energy resolution has to be taken into account. Several models to calculate these angular uncertainties exist and differ, according to the way the energy discrimination is done [Ordonez, Bolozdynya and Chang (1997a)]. If the energy discrimination is done with the scatter detector, E_S is measured in the scatter detector. Combining Eqs. (11.1, 11.6), the scattering angle is:

$$\cos\theta = 1 + mc^2 \frac{E_S}{E_0(E_S - E_0)}. \tag{11.19}$$

The angular uncertainties are:

$$\sigma_\theta = \frac{mc^2 \sigma_{E_S}}{(E_0 - \langle E_S \rangle)^2 \sin\theta}, \tag{11.20}$$

Table 11.2   Positron emitting isotopes currently used in PET.

e^+ emitting isotope	half life (minutes)	production reaction	e^+ effective range in body (mm)	end point energy (MeV)
^{18}F	110.	^{18}O(p,n)^{18}F	1.4	0.635
^{15}O	2.	^{15}N(p,n)^{15}O, ^{14}N(d,n)^{15}O	4.5	1.7
^{13}N	10.	^{16}O(p,α)^{13}N	3.0	1.20
^{11}C	20.	^{14}N(p,α)^{11}C	2.1	0.97

where $\langle E_S \rangle$ is the mean value of the E_S measurements; σ_{E_S} has the standard form

$$\sigma_{E_S} = k\sqrt{a + bE_S},$$

where k, a and b depend on the material used for the scatter detector.

If the energy discrimination is done, instead, with the absorption detector, E_A is measured in the absorption detector. Equation (11.19) can be rewritten as

$$\cos\theta = 1 + mc^2\frac{(E_A - E_0)}{E_0 E_A}. \tag{11.21}$$

The angular uncertainties are:

$$\sigma_\theta = \frac{mc^2 \sigma_{E_A}}{\langle E_A \rangle^2 \sin\theta}, \tag{11.22}$$

where $\langle E_A \rangle$ is the mean value of the E_A measurements; σ_{E_A} can also be parameterized as

$$\sigma_{E_A} = k'\sqrt{a' + b'E_A},$$

where k', a' and b' depend on the material used for the absorption detector.

There exist other ways to discriminate energy. The scattering angle can be calculated with both measured energies E_S and E_A:

$$\cos\theta = 1 - mc^2\frac{E_S}{E_0 E_A}. \tag{11.23}$$

The resulting angular uncertainties are:

$$\sigma_\theta = \frac{mc^2}{E_0 \langle E_A \rangle^2 \sin\theta}\sqrt{(\langle E_A \rangle\,\sigma_{E_S})^2 + (\langle E_S \rangle\,\sigma_{E_A})^2}. \tag{11.24}$$

If one takes into account that the photon emitted by the source can interact with moving electrons bound to nuclei, one has to use the concept of electron pre-collision momentum [Ordonez, Bolozdynya and Chang (1997b)]. One defines

$$p_z = -mc\frac{E_0 - E_A - E_0 E_A\left(1 - \cos\theta\right)/mc^2}{\sqrt{E_0^2 + E_A^2 - 2E_0 E_A \cos\theta}}. \tag{11.25}$$

Here p_z is the projection of the electron's pre-collision momentum on the momentum transfer vector of the photon. The effects of electron motion on the angular uncertainty are described by [Ordonez, Bolozdynya and Chang (1997b)]

$$\sigma_\theta = \frac{1}{A_\theta^2}\left[(A_S\,\sigma_{E_S})^2 + (A_A\,\sigma_{E_A})^2 + (A_{p_z}\,\sigma_{p_z})^2\right], \tag{11.26}$$

where

$$A_\theta = \left(\frac{1}{mc^2} - \frac{\langle p_z \rangle}{mcw} \right) \langle E_A \rangle E_0 \sin \theta, \tag{11.27}$$

$$A_S = 1 - \frac{\langle E_A \rangle}{mc^2} (1 - \cos \theta) + \frac{\langle p_z \rangle}{mcw} (E_0 - \langle E_A \rangle \cos \theta), \tag{11.28}$$

$$A_A = \left(\frac{1}{mc^2} - \frac{\langle p_z \rangle}{mcw} \right) (\langle E_A \rangle + E_0)(1 - \cos \theta), \tag{11.29}$$

$$A_{p_z} = \frac{w}{mc}, \tag{11.30}$$

with

$$w = \sqrt{E_0^2 + \langle E_A \rangle^2 - 2E_0 \langle E_A \rangle \cos \theta}. \tag{11.31}$$

Here $\langle E_S \rangle$, $\langle E_A \rangle$, and $\langle p_z \rangle$ represent the mean value of E_S, E_A, and p_z, respectively. The parameter σ_{p_z}, appearing in the Doppler broadening term $(A_{p_z} \sigma_{p_z})$, can be estimated from the width of the total Compton profile, $J_n(p_z)$, of the target nucleus [Biggs, Mendelsohn and Mann (1975)].

Heavier semiconductors, i.e., semiconductors with larger atomic number such as GaAs or CdZnTe, can be used as absorption plane of Compton camera. For heavier semiconductors, the photoelectric effect has the same magnitude or dominates over Compton scattering and represents the largest contribution to the total attenuation of photons in the material. In the case of GaAs, for 140.5 keV photons, the photoelectric absorption is 0.0139 cm^2/g compared to the Compton scattering contribution of 0.0115 cm^2/g. For CdZnTe, the photoelectric absorption is 0.0108 cm^2/g and the Compton scattering contribution is 0.0457 cm^2/g.

11.2 Positron Emission Tomography (PET)

Positron emission tomography (PET) is a nuclear medical imaging technique, which relies on the measurement of the distribution of a radioactive tracer or radiopharmaceutical labeled with a positron emitting isotope injected into a patient.

Several positron emitters are used for the purpose of PET. The most common emitters are ^{11}C, ^{13}N, ^{15}O and ^{18}F. The radiopharmaceutical, labeled with a positron emitting isotope, is a form of glucose which is injected into the patient. The PET scan will reveal areas where the glucose is consumed in excess of the normal body needs, such as in a growing tumor.

The emitter lifetime is obviously very important, since it must be large enough to allow the transportation of practical doses from the production facility to the location, where the patient is treated. At the same time, the radioisotope has to be short lived enough to reduce the amount of activation left into the patient. The half-life of several emitters are listed in Table 11.2.

The radioactive tracer 2-[^{18}F]fluro-2-deoxy-D-glucose (^{18}FDG) is a substance widely used in nuclear medicine for nuclear imaging. The radioactive element is the isotope ^{18}F, which has a half-life of 110 min (Table 11.2) and decays via positron emission. ^{18}FDG is a sugar analogue, where one or several of the hydrogen atoms are substituted by a ^{18}F atom. FDG accumulates in organs where glucose is used, as the primary source of energy and therefore FDG is used for instance in studies of the glucose metabolism of the brain and heart. The traditional method of producing ^{18}FDG consists of using a proton beam of about 10 MeV on a target of enriched water $H_2^{18}O$ via the reaction $^{18}O(p,n)^{18}F$. Such proton beams are available at many Van der Graaff tandem accelerators and cyclotrons located close to where the patient is treated. This reaction cross-section has a threshold of around 2.57 MeV and resonance around 5.13 MeV with a maximum cross-section of 697 mb. The rate of production of fluorine-18 from 100% ^{18}O-enriched water targets can be calculated as a function of the proton energy. For instance, for a 10 MeV proton beam, the proton range in water is 119 mg/cm^2 and the production rate of fluorine-18 is 39.1 mCi/Ah. Taken into account the energy dissipation in the front target foils (estimated to 0.5 MeV), one has about 70 mCi/h fluorine-18 yield, for 2 μA beam, corresponding to a ^{18}FDG yield of 20 mCi/h.

The positron emitted by the radioactive tracer or radiopharmaceutical, via the decay $p \rightarrow n \nu_e e^+ ([Z, A] \rightarrow [Z - 1, A] \nu_e e^+)$, annihilates very close to the emission point (≤ 1 mm) with an electron of the body to produce a pair of 511 keV photons emitted back-to-back (Fig. 11.5). The effective range of positrons and end point energy for most used isotopes in PET are shown in Table 11.2. These photons traverse the body and enter the active medium of the PET detector placed on a ring.

The PET camera is detecting two photons emitted back-to-back in coincidence. The observed pair of back-to-back photons defines an axis along which the disintegration of the radioactive element has taken place. The line connecting two detected photons is called a *chord*. The time correlation between detected photons permits the selection of pairs in coincidence and their association to a chord. The positron emitters can then be traced back as they participate in biological processes. The simultaneous detection of several pairs of photons indicates the rate of disintegration along different axes and enables one to determine the distribution of the compound in the body and to draw conclusions as to the proper functioning of tissues and organs. The time interval between the detection of these two photons is a few ns *FWHM*, typically (2–5) ns. A good timing resolution minimizes the accidental coincidence rate and permits the use of the arrival time difference to determine the radioisotope position along the chord.

An excellent spatial resolution (< 5 mm *FWHM* [Moses, Derenzo and Budinger (1994)]) is needed along the two directions corresponding to the axial and trans-axial directions of the tomographic devices. The detector spatial resolution helps to achieve the quality of spatial resolution in the reconstructed image.

The photon detector must combine an angular coverage, large enough to intercept the photons of interest, and a high spatial resolution, in order to account for

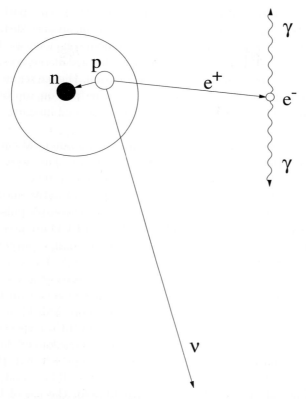

Fig. 11.5 The positron emitted by the radioactive tracer or radiopharmaceutical annihilates with an electron of the body to produce a pair of 511 keV photons emitted back-to-back.

the details of the body area under investigation, as resolved as possible.

Optimized spatial, timing and energy resolutions are factors determining the sensitivity of the detector and permit the reduction of the dose, injected into the patient.

The use of readout electronic chain with low noise is of great importance in order to optimize the signal-to-noise ratio. Good energy resolution typically means < 100 keV [Moses, Derenzo and Budinger (1994)]. Good energy resolution helps the rejection of secondary photons produced by Compton scattering of primary photons from the source (the ones of interest) on biological matter (tissues, bones, ...).

Strict mechanical constraints apply to the building of imagers. The ensemble of active elements has to be well mechanically adjusted, avoiding cracks. The absence of collimator, an heavy metal layer, in PET imagers helps the mechanical stability of the system.

The detectors used in *PET* scanners are composed of scintillation blocks (or other active materials) and systems placed on a ring. The diameter of the ring defines the distance between detectors. This distance influences the spatial resolution of the imaging system. Better resolution is obtained for smaller distance. However, smaller diameter also favors higher number of random coincidences. The granularity of the scintillating crystal assembly, defined by the size of the crystal elements, is the primary factor determining the spatial resolution. Higher segmentation lead to better spatial resolution. About fifteen rings can be put on top of each other to form an array of detectors that permits three-dimensional imaging. A resolution of about 6 mm in each direction is achieved.

The selection of a crystal is made according to its potential energy resolution, which depends on the mean number of photoelectrons. The noise of the readout electronic chain is a limiting factor [Leroy and Rancoita (2000)].

The use of high atomic number (high-Z) material as active medium of the detector gives high photon detection sensitivity and an acceptable pulse-height resolution. Sensitive materials such as NaI(Tl), CsI(Tl) and BGO are used. The assembly of these crystals along a ring structure allows the building of large active medium volumes providing the necessary large angular coverage. The crystals must have large light yield and high detection efficiency for photons of 511 keV of energy for PET and of 140 keV or less for SPECT. The light yields of NaI(Tl) and CsI(Tl) are large compared with BGO. It is standard to choose NaI(Tl) as reference. The relative light yield of CsI(Tl) and BGO are 0.40 and 0.15, respectively. The crystals must have large stopping power for photon energy ranges faced in SPECT and PET. High-Z and high density materials have to be selected. BGO has higher density ($7.13 \, \text{g/cm}^3$) than NaI(Tl) ($3.67 \, \text{g/cm}^3$) and CsI(Tl) ($4.53 \, \text{g/cm}^3$). K-edge location has to be taken into account and would favor the use of BGO. The high counting rate and, in some cases, the relatively short lifetime of the radioisotope require minimal dead time, of the order of a few μs. The decay times of BGO (300 ns) and NaI(Tl) (250 ns) are comparable and much lower than the CsI(Tl) decay time (1000 ns).

The medical imagers represent a large volume of active material and therefore the cost of the active material is an issue. Most of the time, the imagers in operation in the medical field are purchased from commercial company (not universities) following market prices. The price of crystals ranges from a few dollars to (10–15) dollars per cm^3.

The radiation hardness of the detecting material is also an element of consideration. Although the detector is exposed to doses much smaller than those encountered in other fields (space, accelerator and reactor environments) radiation degradation can possibly be observed with time and lead to detecting material replacement. Good radiation hardness extends the lifetime of the detecting devices, avoiding frequent replacements.

Therefore, from the point of view of best performance for a crystal to be used as the active medium of a medical scanner, one is looking for a crystal having a light yield comparable to that of NaI(Tl), a density comparable to that of BGO, but with a decay time much smaller than BGO, while remaining affordable. Several new types of scintillating materials have been or are being developed for a new generation of medical scanners. Among these, the yttrium aluminium perovskite (YAP:Ce) [Baccaro et al. (1995)] has a light efficiency of about 40% relative to NaI(Tl), a density of $5.37\,g/cm^3$, lower than BGO but higher than NaI(Tl) and CsI(Tl). YAP:Ce has a rather high-Z value ($Z = 39$) which guarantees good photon absorption. YAP:CE has a decay time of 25 ns which is another advantage over BGO (decay time of 300 ns). The detection of two 511 keV photons by coincidence by two YAP:Ce crystal bundles (5×5 pillars of $0.2 \times 0.2 \times 3.0\,cm^3$) coupled to position sensitive photomultiplier tubes have given a spatial resolution of 1.2 mm $FWHM$, a time resolution of 2.0 ns $FWHM$ and a large efficiency of 70% with a threshold of 150 keV [Del Guerra (1997)].

Following the development of YAP, the Crystal Clear Collaboration [Lecoq (2000)] has developed Luthetium Aluminium Perovskite (LuAP) crystals. LuAP has a high light yield and a very short decay time of 18 ns. The density of LuAP is also high ($8.34\,g/cm^3$). The peak emission of 380 nm is well adapted to avalanche photodiode readout, allowing in turn compact detecting system.

PET detector modules can also be built from photon converter, readout by Multiwire proportional chambers (MWPC). The converter consists of either a high-Z metal such as lead or tungsten or a combination of crystal such as $BaF_2/TMAE$ gas. The 511 keV photon are converted into photoelectrons, which are collected by a MWPC, generating a timing pulse and identifying the interaction position. This type of scanner offers the advantage to be of moderate costs, no photomultiplier being used. However they also present several disadvantages like the photon converter lower efficiency to detect single 511 keV photons {(10–30)% as opposed to 90% for BGO [Moses, Derenzo and Budinger (1994)]}, decreasing the coincident even detection efficiency. The number of photoelectrons per 511 keV interaction is very small, causing a poor energy resolution. Poor limited spatial resolution {(5–11) mm $FWHM$ [Moses, Derenzo and Budinger (1994)]} for BaF_2 and poor timing resolution (88 ns [Moses, Derenzo and Budinger (1994)]) for high-Z metal converters are also disadvantages.

11.3 Magnetic Resonance Imaging (MRI)

This imaging technique has an advantage compared to SPECT and PET since it does not require the use of any radioactive material. Instead, it uses an intrinsic property found in some nuclei: the non-zero nuclear spin. MRI uses magnetic fields varying from 0.2 to 2 T and radiofrequency (RF) waves, to observe the magnetization change of the non-zero spin nuclei. The hydrogen isotope, ^1H, which has a nuclear spin of $\frac{1}{2}$, is a major component of the human body and will be used as the main source of information.

11.3.1   *Physical Basis of MRI*

Let us consider the behavior of the nucleus of ^{1}H under the influence of a magnetic field [Desgrez, Bittoun and Idy-Peretti (1989)]. The proton has a spin of $\frac{1}{2}$ and, therefore, has two observable states $S_z = +\frac{1}{2}$ or $S_z = -\frac{1}{2}$. The energy difference between the two states is:

$$\Delta E = h\gamma B, \tag{11.32}$$

where γ is the gyro-magnetic ratio which is characteristic of each atom. In the case of hydrogen, $\gamma = 42.58\,\mathrm{MHz/T}$. For a 2 T magnetic field, this gives an energy of $35.1\,\mu\mathrm{eV}$. The related frequency is 85.16 MHz, in the RF range, since the resonance or Larmor frequency is given by:

$$\nu = \gamma B. \tag{11.33}$$

As for the individual magnetization held within the nucleus, it is expressed by:

$$\vec{\mu} = 2\pi\gamma\vec{S}. \tag{11.34}$$

We have $\vec{M} = \sum \vec{\mu}$, where $\vec{M}$ is the net magnetization of the system. When a magnetic field is applied, the majority of the nuclei will align in the same direction, giving $\vec{M} \propto \vec{B}_0$, according to a Boltzmann distribution:

$$\frac{N_-}{N_+} = \exp\left(-\frac{\Delta E}{kT}\right). \tag{11.35}$$

To simplify, let us choose $\vec{B}_0 = B_0\vec{z}$. As previously seen, it is possible to change the magnetization of a single nucleus, if it is reached by a photon of energy $E = h\gamma B$. A RF wave that equalizes the populations $N_- = N_+$, giving a net magnetization of $M_z = 0$, is called a saturation pulse or 90° impulsion. After that impulsion, the system will return to its equilibrium according to [Hornak (2002)]:

$$M_z = M_0\left(1 - e^{-t/T_1}\right). \tag{11.36}$$

T_1 is called the spin-lattice relaxation time. If it is a 180° impulsion (complete inversion of populations), then the equilibrium will be recovered like:

$$M_z = M_0\left(1 - 2e^{-t/T_1}\right). \tag{11.37}$$

A 90° impulsion brings the net magnetization in the XY plane. It then starts to precess around the z-axis at the Larmor frequency. To reach equilibrium, it will decrease as:

$$M_{xy} = M_{xy_\tau}\left(e^{-t/T_2}\right), \tag{11.38}$$

where T_2 is the spin-spin relaxation time and τ, the time marking the end of the impulsion. It decreases as the spins of the individual nuclei dephase. There is a dephasing because each nucleus has its own magnetic field affecting the surrounding nuclei. Therefore, the precession around the z-axis is done at several slightly different

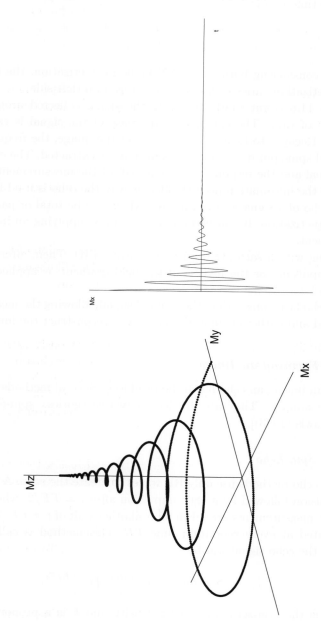

Fig. 11.6   Evolution of the net magnetization's orientation and evolution of M_x as a function of time after a saturation pulse.

resonance frequencies. There is an other source of dephasing: the non-uniformity of the magnetic field. These two effects combine to change the observed spin-spin relaxation time to T_2^*:

$$\frac{1}{T_2^*} = \frac{1}{T_2} + \frac{1}{T_{2,\text{inhomo}}}. \qquad (11.39)$$

When considering both z and XY plane magnetization, the orientation of the net magnetization varies as illustrated in Fig. 11.6 (left side).

Figure 11.6 (right side) represents the signal collected around the x-axis as a function of time. The exponential decrease of the signal is called a FID (Free Induction Decay). To be able to reconstruct the image, the frequency ω, the spin-lattice and spin-spin relaxation times have to be extracted. The exponential nature of the signal and the response time required for the measurements make it hard to gather all the information needed. That is why the echo is used instead.

The echo of a signal is a re-phasing, which can be total or partial, of the transverse magnetization. It can be done in two ways: applying an inversion pulse or a field gradient.

Starting with a saturation pulse, we have a FID. Then, after a waiting period, a 180° impulsion, or the inverse of the field gradient, is applied as illustrated in Fig. 11.7.

The echo lasts longer than the original signal, allowing the measuring equipment to respond and gather enough information to reconstruct the image.

11.3.2　*Forming an Image*

To form an image, an echo has to be produced. Several methods can be used. Here are four examples. They are divided in two categories: spin-echo and gradient-echo [Sprawls (1993)].

11.3.2.1　*Spin-Echo*

The spin-echo methods use only RF waves to create the echo. A sequence used is a 90° impulsion followed by a 180° impulsion after $t = TE/2$, where TE is the echo time. The measurements are taken after another wait of $t = TE/2$. The manoeuvres are repeated at every repetition time TR. This method is called spin-echo. The height of the echo signal will be:

$$S = k\rho \left(1 - e^{-TR/T_1}\right) e^{-TE/T_2}, \qquad (11.40)$$

where ρ is the non-zero spin nuclei density and k is a proportionality constant, which depends on the measuring equipment.

The inversion-recovery method uses the same idea but inverses the sequence. After waiting TR, an inversion pulse is applied. After $t = TI$, where TI is the inversion

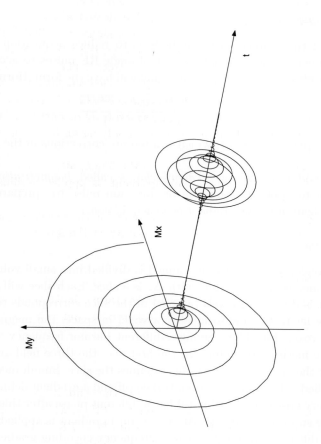

Fig. 11.7 Representation of the transverse magnetization orientation during a FID and an echo as a function of time.

time, a 90° impulsion is given to the system and the measurement is done immediately after. In this case, we have [Hornak (2002)]:

$$S = k\rho \left(1 - 2e^{-TI/T_1} + e^{-TR/T_1}\right). \tag{11.41}$$

The advantage of the spin-echo methods is that it is independent of T_2^* and, therefore, of the inhomogeneities of the magnetic field. Unfortunately, these methods require longer acquisition time.

11.3.2.2 *Gradient-Echo*

The main goal of the gradient-echo methods is to reduce acquisition time. The Small Angle Gradient Echo (SAGE) uses small angle RF pulses to accelerate the *longitudinal magnetization's recovery*. The signal will have the form [Hornak (2002)]:

$$S = k\rho \frac{\left(1 - e^{-TR/T_1}\right) \sin\theta \, e^{-TE/T_2^*}}{\left(1 - \cos\theta \, e^{-TR/T_1}\right)}. \tag{11.42}$$

Here, there is a dependence on T_2^*, which will require corrections in the data treatment.

Another method that uses a *gradient-echo* is called *magnetization preparation*. The idea is to apply a saturation or inversion pulse to "prepare" the longitudinal magnetization to the gradient-echo acquisition.

11.3.2.3 *Space Positioning*

First of all, the region that has to be scanned is divided into small volumes called *voxels*. Each volume enclosed in the width Δz is a slice. Each slice will correspond to an image that is divided in pixels (a voxel in the slice corresponds to a pixel in the image). In order to be able to localize a voxel in space, the magnetic field is different in each voxel (Fig. 11.8), giving a different Larmor frequency.

There are two methods of coding. In the first one, the three field gradients are applied one after the other. First, a gradient defines the slice. Simultaneously, a 90° impulsion is applied. The gradient is then turned off and a gradient defining x (or y) is turned on. Every abscissa (or ordinate) has a different phase after this gradient is turned off (phase encoding gradient). At last, a third gradient is applied. It gives to every y (or x) a different Larmor frequency (frequency encoding gradient). During the application of this gradient, the FID occurs. In the second method, there is also a slice selection accompanied by a saturation pulse. It is followed by the simultaneous application of the other 2 gradients accompanied by the FID.

The information gathered by the measuring instruments is a function of time. Using a 2-dimensional Fourier transform, the information is translated in frequencies, which are finally translated in spatial coordinates. Several techniques can be used in MRI. Research is still underway to discover the most efficient technique time- and quality-wise.

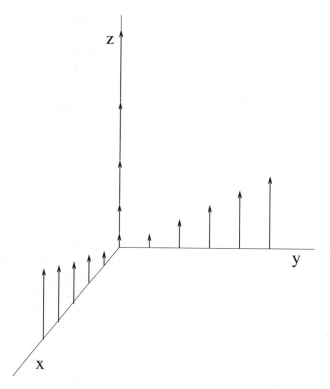

Fig. 11.8   The spatial positioning magnetic field gradients applied to encode each voxel with a specific Larmor frequency. All the magnetic fields are in the z-direction. The specific magnetic field applied to a point in space is the addition of the X-, Y- and Z-gradients.

11.3.2.4   *Flows*

Like all the other imaging techniques, MRI requires the immobility of the patient. However, the movement inside the patient cannot be controlled. With MRI, this movement can be used to have an other type of imaging: flow imaging. This technique is used for angiographies. Three types of sequences can be used: time-of-flight, phase contrast and contrast enhanced angiographies [Hornak (2002)].

On a regular MRI image, blood vessels seem empty because the atoms that receive the 90° impulsion do not receive the 180° and, therefore, no echo is created. The idea behind the time-of-flight angiography technique is to follow the atoms that received the saturation pulse, i.e., to do a second slice selection with the inversion pulse in order to have an echo. The sequence used for this technique is a spin-echo sequence with a 90° and a 180° impulsion with different frequencies.

The phase contrast angiography technique introduces a bipolar gradient in the slice selecting gradients, i.e., between the saturation and inversion pulses in a spin-echo sequence. This gradient is constituted of two inverse gradient lobes applied one after the other. Of course, for the immobile atoms, this gradient has no effect since each lobe will nullify the effect of the other one. But, a moving nucleus will be affected. Therefore, to see the blood movement, one has to take the image twice (once with and once without the bipolar gradient) and subtract one from the other. The immobile matter will disappear, while the moving matter will have different intensities depending on its velocity.

The last technique is the most used: the contrast enhanced angiography. It is based on the principle that the relaxation time, T_1, changes when a paramagnetic contrast agent is injected into the blood. In brief, T_1 changes with the blood's surroundings. And as it changes, contrasts can be seen. A rapid data acquiring sequence is used.

11.3.2.5 *Functional MRI*

Using a sequence of impulsions and gradients called the echo-planar imaging, it is possible to gather the information needed for a whole image in the short time period TR. This allows one to take several images pictures per second just like a video. This technique opens new avenues to study the human body. For example, it is possible to track the blood flows in the brain allowing an elaborate study of the brain's reaction to stimuli.

The idea behind echo-planar imaging is to sweep all sections of a k-space (the Fourier transform of the image) [Hornak (2002)]. First, a saturation pulse and a slice selective gradient are applied to the system. Then, simultaneously a phase and a frequency encoding gradients are injected, bringing the initial data to a "corner" of the k-space. A bit after, an inversion pulse is applied. Finally, the following sequence is used to sweep the k-space: a phase gradient is applied, immediately followed by a frequency gradient, during which data is gathered; then, another phase encoding gradient is applied after which another echo is produced; and the previous steps are repeated, until the entire k-space is swept.

11.4 *X*-Ray Medical Imaging with MediPix Devices

The MediPix device, discussed in Sects. 6.5.1-6.5.2.5, can be exploited for medical imaging. Medical imaging is typically in the 5–80 keV X-ray energy range. The photoelectric effect is the dominant interaction between X-ray photons and detecting material in that energy range. In this section, we look at the application of MediPix-type devices in X-ray medical imaging. The MediPix-type devices, when exposed to X-ray beams, produce real-time digital images that can be stored and analyzed. These devices can improve the quality of image as the X-ray detection is improved while the noise is reduced. This improvement of the image quality allows the reduction of the radiation exposure to the patient. The image quality is expressed in terms of several parameters. Some of the most important parameters are briefly reviewed below. However, this list is non-exhaustive and readers who wants to complete their information are invited to consult [Mikulec (2000); Pfeiffer (2004); Stoehr (2005); Norlin (2007)], for instance.

11.4.1 *The Contrast*

In the 5–80 keV energy range, tissue, bone and calcifications have very distinct absorption coefficients and high-contrast images can be formed. The contrast, C, is describing the difference in intensities I_1 and I_2 between two adjacent regions (several pixels) which present different absorptions of X-rays. If one uses a photon counting system, the contrast can be expressed in terms of counts between the two adjacent regions. If N_1 and N_2 are the photons counted in the two regions (normalized to the same area), the contrast is defined as [Mikulec (2000); Anton et al. (2006)]:

$$C = \frac{I_1 - I_2}{I_1 + I_2} = \frac{N_1 - N_2}{N_1 + N_2}. \tag{11.43}$$

Since the numbers of counted photons have a Poisson noise, one defines a signal-difference-to-noise ratio ($SDNR$) which measures the visibility of a given contrast [Anton et al. (2006)]:

$$SDNR = \frac{N_1 - N_2}{\sigma_{N_1 - N_2}} = C\sqrt{N_1 + N_2}, \tag{11.44}$$

where $\sigma_{N_1 - N_2}$ is the variance of the counts distribution. For a contrast C, a particular value of $SDNR$ can be achieved by either long exposure to photon counting (large incident photon fluence) or a large pixel-area [Anton et al. (2006)].

11.4.2 *The Modulation Transfer Function*

One defines the modulation transfer function (MTF) as the spatial frequency response of an imaging system or a component. The spatial frequency is measured in units of cycles per mm (c/mm) or also in units of line pairs per millimeter (lp/mm). The MTF is the contrast at a given spatial frequency relative to low frequencies and therefore measures the quality of transmission (in to out) of the contrast of an object through an imaging chain for a given spatial frequency. High spatial frequencies correspond to fine image detail. Signals which change often over a given distance have a high spatial frequency. Signals with low spatial frequency, i.e., which change slowly over the same distance are easier to detect than those of high spatial frequency. The MTF at a spatial frequency ν is defined as the ratio between the modulation of a sinusoidal pattern M_{in} and the modulation of image M_{out} obtained after transmission [Pfeiffer (2004)]:

$$MTF(\nu) = \frac{M_{out}(\nu)}{M_{in}(\nu)}. \tag{11.45}$$

The $MTF(\nu)$ ranges between 0 and 1 for all frequencies. The MTF can be described by analytical functions. In the case of square pixel (such is the case of MediPix1 and MediPix2) of size l^2 with uniform sensitivity over the whole pixel area, the theoretical limit for MTF is [Pfeiffer (2004)]:

$$MTF(\nu) = \frac{\sin(\pi\nu l)}{\pi\,\nu l}. \tag{11.46}$$

This parametrization comes from the possibility to decompose the information for a given image into a set of sinusoidal functions of different amplitude. Therefore, one can say that the MTF gives the spatial response of a detector to a sinusoidal input stimulus. There are several factors affecting the MTF: a) focal spot blur: because the X-ray source is not point-like and causes blur in the X-ray image. Standard procedures to minimize this effect is to reduce the focal spot size, optimizing (increasing) the distance between the source and the subject of imaging and decreasing the distance between the subject of imaging and the X-ray detector (basically what one would do with a camera); b) the pixel size: the pixel size has to be as small as possible ($55\,\mu$m $\times$ $55\,\mu$m for MediPix2) possibly smaller than the details of the image; c) scattering of photons produced from the conversion of absorbed X-rays: this effect can be minimized by the choice of an adequate conversion material.

11.4.3   *The Detective Quantum Efficiency*

The detective quantum efficiency (DQE) is the signal-to-noise ratio (SNR) transfer function. It measures how the SNR at the input ($SNR_{\rm in}$, SNR of the incoming X-ray flux) of an imaging system is transferred to the output ($SNR_{\rm out}$, SNR of the image) and measures its possible degradation by the imaging system. Then, the DQE is expressed as:

$$DQE(\nu) = \frac{SNR_{\rm out}^2(\nu)}{SNR_{\rm in}^2(\nu)}. \tag{11.47}$$

The noise is from various origins: the electronic, fixed pattern and quantum noises. The electronic noise is produced by the read-out components of the imaging system. The fixed pattern noise originates from gain and off-set value variations among pixels. The so-called quantum noise is the result of the random nature of the X-ray photons: the number $n_{\rm in}$ of X-ray photons produced by the source incident per unit area and unit time on the detector has fluctuations that follow a Poisson distribution with a variance $\sqrt{n_{\rm in}}$. Then, with $SNR_{\rm in} = \sqrt{n_{\rm in}}$, Eq. (11.47) can be written as

$$DQE(\nu) = \frac{SNR_{\rm out}^2(\nu)}{n_{\rm in}}. \tag{11.48}$$

$SNR_{\rm out}$ is determined by using the noise power spectrum (NPS). The NPS describes the noise transfer properties of the imaging system. The NPS is a quantity which accounts for the distribution of noise variations with spatial frequency. Technically, the calculation of the NPS uses the Fourier transform of noise image to determine the variance of noise power present at each spatial frequency. The shape of the NPS shows where the noise power is concentrated in frequency space [Riederer, Pelc and Chesler (1978); Boedeker, Cooper and McNitt-Gray (2007)]. In simpler terms, the NPS gives the variance of a noise process, but distributes it as a function of the spatial frequency. Hence, using the normalized NPS (NNPS) and the MTF, one has [Pfeiffer (2004)]:

$$SNR_{\text{out}}^2 = \frac{MTF^2(\nu)}{NNPS(\nu)}. \tag{11.49}$$

Combining Eqs. (11.48), (11.49), one finds:

$$DQE(\nu) = \frac{MTF^2(\nu)}{NNPS(\nu) \times n_{\text{in}}}. \tag{11.50}$$

One can consult, for instance, [Mikulec (2000); Pfeiffer (2004); Stoehr (2005); Norlin (2007)] for a review of results of the measurements of MTF, NPS and DQE done with MediPix1 and MediPix2. A final remark for this section is the phenomenon of charge sharing and its implication for X-ray imaging. As discussed in Sect. 6.5.2.5, charge sharing results from several mechanisms with the consequence that deposited charges are shared among adjacent pixels. Particle physics may use the charge sharing effect for improved tracking. For medical physics charge tracking is viewed as an adverse phenomenon as it decreases the photon detection efficiency. Depending on the threshold set for the MediPix2-type device, incoming photons can be counted a single time or several times. There are several ways envisaged to reduce the charge sharing effect and even to suppress it. It has been proposed to develop a readout electronics which can sum the charge of several pixels [Llopart et al. (2002)]. This solution has however a drawback as summing charges collected in several pixels yields additional noise. The problem of charge summing and adequate noise control will be hopefully solved with MediPix3, the successor of MediPix2 [Medipix Collab. (2008)]. Another solution advanced for suppression of the charge sharing effect is to develop 3-D structure detectors in place of the standard planar pixel detectors. It is advocated that the transverse electric field in a 3-D structure (perpendicular to the incoming photon) will force the electron to drift to the correct pixel [Norlin (2007)]. This development has to be accompanied with an increase of the silicon thickness for maintaining a practical quantum efficiency.

Appendix A

General Properties and Constants

A.1   Conversion Factors

The conversion factors are from Appendix B.9 of [Taylor (1995)].

Factors in **boldface** are exact

To convert from	to	Multiply by	
ACCELERATION			
acceleration of free fall, standard (g_n)	meter per second squared (m/s^2)	**9.806 65**	**E+00**
foot per second squared (ft/s^2)	meter per second squared (m/s^2)	**3.048**	**E−01**
gal (Gal)	meter per second squared (m/s^2)	**1.0**	**E−02**
inch per second squared (in/s^2)	meter per second squared (m/s^2)	**2.54**	**E−02**
ANGLE			
degree (°)	radian (rad)	1.745 329	E−02
gon (also called grade) (gon)	radian (rad)	1.570 796	E−02
gon (also called grade) (gon)	degree (°)	**9.0**	**E−01**
mil	radian (rad)	9.817 477	E−04
mil	degree (°)	**5.625**	**E−02**
minute (')	radian (rad)	2.908 882	E−04
revolution (r)	radian (rad)	6.283 185	E+00
second (")	radian (rad)	4.848 137	E−06
AREA AND SECOND MOMENT OF AREA			
acre (based on U.S. survey foot)[9]	square meter (m^2)	4.046 873	E+03
are (a)	square meter (m^2)	**1.0**	**E+02**
barn (b)	square meter (m^2)	**1.0**	**E−28**
circular mil	square meter (m^2)	5.067 075	E−10
circular mil	square millimeter (mm^2)	5.067 075	E−04
foot to the fourth power (ft^4)[17]	meter to the fourth power (m^4)	8.630 975	E−03
hectare (ha)	square meter (m^2)	**1.0**	**E+04**
inch to the fourth power (in^4)[17]	meter to the fourth power (m^4)	4.162 314	E−07
square foot (ft^2)	square meter (m^2)	**9.290 304**	**E−02**
square inch (in^2)	square meter (m^2)	**6.4516**	**E−04**
square inch (in^2)	square centimeter (cm^2)	**6.4516**	**E+00**
square mile (mi^2)	square meter (m^2)	2.589 988	E+06
square mile (mi^2)	square kilometer (km^2)	2.589 988	E+00
square mile (based on U.S. survey foot) (mi^2)[9]	square meter (m^2)	2.589 998	E+06
square mile (based on U.S. survey foot) (mi^2)[9]	square kilometer (km^2)	2.589 998	E+00
square yard (yd^2)	square meter (m^2)	8.361 274	E−01

CAPACITY (see VOLUME)

DENSITY (that is, MASS DENSITY — see MASS DIVIDED BY VOLUME)

ELECTRICITY and MAGNETISM			
abampere	ampere (A)	**1.0**	**E+01**
abcoulomb	coulomb (C)	**1.0**	**E+01**
abfarad	farad (F)	**1.0**	**E+09**
abhenry	henry (H)	**1.0**	**E−09**
abmho	siemens (S)	**1.0**	**E+09**
abohm	ohm (Ω)	**1.0**	**E−09**
abvolt	volt (V)	**1.0**	**E−08**
ampere hour (A · h)	coulomb (C)	**3.6**	**E+03**

To convert from	to		Multiply by
biot (Bi)	ampere (A)	**1.0**	**E+01**
EMU of capacitance (abfarad)	farad (F)	**1.0**	**E+09**
EMU of current (abampere)	ampere (A)	**1.0**	**E+01**
EMU of electric potential (abvolt)	volt (V)	**1.0**	**E−08**
EMU of inductance (abhenry)	henry (H)	**1.0**	**E−09**
EMU of resistance (abohm)	ohm (Ω)	**1.0**	**E−09**
ESU of capacitance (statfarad)	farad (F)	1.112 650	E−12
ESU of current (statampere)	ampere (A)	3.335 641	E−10
ESU of electric potential (statvolt)	volt (V)	2.997 925	E+02
ESU of inductance (stathenry)	henry (H)	8.987 552	E+11
ESU of resistance (statohm)	ohm (Ω)	8.987 552	E+11
faraday (based on carbon 12)	coulomb (C)	9.648 531	E+04
franklin (Fr)	coulomb (C)	3.335 641	E−10
gamma (γ)	tesla (T)	**1.0**	**E−09**
gauss (Gs, G)	tesla (T)	**1.0**	**E−04**
gilbert (Gi)	ampere (A)	7.957 747	E−01
maxwell (Mx)	weber (Wb)	**1.0**	**E−08**
mho	siemens (S)	**1.0**	**E+00**
oersted (Oe)	ampere per meter (A/m)	7.957 747	E+01
ohm centimeter (Ω · cm)	ohm meter (Ω · m)	**1.0**	**E−02**
ohm circular-mil per foot	ohm meter (Ω · m)	1.662 426	E−09
ohm circular-mil per foot	ohm square millimeter per meter (Ω · mm²/m)	1.662 426	E−03
statampere	ampere (A)	3.335 641	E−10
statcoulomb	coulomb (C)	3.335 641	E−10
statfarad	farad (F)	1.112 650	E−12
stathenry	henry (H)	8.987 552	E+11
statmho	siemens (S)	1.112 650	E−12
statohm	ohm (Ω)	8.987 552	E+11
statvolt	volt (V)	2.997 925	E+02
unit pole	weber (Wb)	1.256 637	E−07

ENERGY (includes WORK)

To convert from	to		Multiply by
British thermal unit$_{\text{IT}}$ (Btu$_{\text{IT}}$)[11]	joule (J)	1.055 056	E+03
British thermal unit$_{\text{th}}$ (Btu$_{\text{th}}$)[11]	joule (J)	1.054 350	E+03
British thermal unit (mean) (Btu)	joule (J)	1.055 87	E+03
British thermal unit (39 °F) (Btu)	joule (J)	1.059 67	E+03
British thermal unit (59 °F) (Btu)	joule (J)	1.054 80	E+03
British thermal unit (60 °F) (Btu)	joule (J)	1.054 68	E+03
calorie$_{\text{IT}}$ (cal$_{\text{IT}}$)[11]	joule (J)	**4.1868**	**E+00**
calorie$_{\text{th}}$ (cal$_{\text{th}}$)[11]	joule (J)	**4.184**	**E+00**
calorie (mean) (cal)	joule (J)	4.190 02	E+00
calorie (15 °C) (cal$_{15}$)	joule (J)	4.185 80	E+00
calorie (20 °C) (cal$_{20}$)	joule (J)	4.181 90	E+00
calorie$_{\text{IT}}$, kilogram (nutrition)[12]	joule (J)	**4.1868**	**E+03**
calorie$_{\text{th}}$, kilogram (nutrition)[12]	joule (J)	**4.184**	**E+03**
calorie (mean), kilogram (nutrition)[12]	joule (J)	4.190 02	E+03
electronvolt (eV)	joule (J)	1.602 177	E−19
erg (erg)	joule (J)	**1.0**	**E−07**
foot poundal	joule (J)	4.214 011	E−02
foot pound-force (ft · lbf)	joule (J)	1.355 818	E+00
kilocalorie$_{\text{IT}}$ (kcal$_{\text{IT}}$)	joule (J)	**4.1868**	**E+03**
kilocalorie$_{\text{th}}$ (kcal$_{\text{th}}$)	joule (J)	**4.184**	**E+03**
kilocalorie (mean) (kcal)	joule (J)	4.190 02	E+03

To convert from	to		Multiply by
kilowatt hour (kW · h)	joule (J)	**3.6**	**E+06**
kilowatt hour (kW · h)	megajoule (MJ)	**3.6**	**E+00**
quad (10^{15} Btu$_{IT}$)[11]	joule (J)	1.055 056	E+18
therm (EC)[25]	joule (J)	**1.055 06**	**E+08**
therm (U.S.)[25]	joule (J)	**1.054 804**	**E+08**
ton of TNT (energy equivalent)[26]	joule (J)	**4.184**	**E+09**
watt hour (W · h)	joule (J)	**3.6**	**E+03**
watt second (W · s)	joule (J)	**1.0**	**E+00**

ENERGY DIVIDED BY AREA TIME

erg per square centimeter second

[1obrktδ1rul/(cm^2 · s)]	watt per square meter (W/m^2)	**1.0**	**E−03**
watt per square centimeter (W/cm^2)	watt per square meter (W/m^2)	**1.0**	**E+04**
watt per square inch (W/in^2)	watt per square meter (W/m^2)	1.550 003	E+03

FLOW (see MASS DIVIDED BY TIME or VOLUME DIVIDED BY TIME)

FORCE

dyne (dyn)	newton (N)	**1.0**	**E−05**
kilogram-force (kgf)	newton (N)	**9.806 65**	**E+00**
kilopond (kilogram-force) (kp)	newton (N)	**9.806 65**	**E+00**
kip (1 kip=1000 lbf)	newton (N)	4.448 222	E+03
kip (1 kip=1000 lbf)	kilonewton (kN)	4.448 222	E+00
ounce (avoirdupois)-force (ozf)	newton (N)	2.780 139	E−01
poundal	newton (N)	1.382 550	E−01
pound-force (lbf)[24]	newton (N)	4.448 222	E+00
pound-force per pound (lbf/lb) (thrust to mass ratio)	newton per kilogram (N/kg)	**9.806 65**	**E+00**
ton-force (2000 lbf)	newton (N)	8.896 443	E+03
ton-force (2000 lbf)	kilonewton (kN)	8.896 443	E+00

FORCE DIVIDED BY AREA (see PRESSURE)

FORCE DIVIDED BY LENGTH

pound-force per foot (lbf/ft)	newton per meter (N/m)	1.459 390	E+01
pound-force per inch (lbf/in)	newton per meter (N/m)	1.751 268	E+02

HEAT
Available Energy

British thermal unit$_{IT}$ per cubic foot (Btu$_{IT}$/ft^3)	joule per cubic meter (J/m^3)	3.725 895	E+04
British thermal unit$_{th}$ per cubic foot (Btu$_{th}$/ft^3)	joule per cubic meter (J/m^3)	3.723 403	E+04
British thermal unit$_{IT}$ per pound (Btu$_{IT}$/lb)	joule per kilogram (J/kg)	**2.326**	**E+03**
British thermal unit$_{th}$ per pound (Btu$_{th}$/lb)	joule per kilogram (J/kg)	2.324 444	E+03
calorie$_{IT}$ per gram (cal$_{IT}$/g)	joule per kilogram (J/kg)	**4.1868**	**E+03**
calorie$_{th}$ per gram (cal$_{th}$/g)	joule per kilogram (J/kg)	**4.184**	**E+03**

Coefficient of Heat Transfer

British thermal unit$_{IT}$ per hour square foot degree Fahrenheit [Btu$_{IT}$/(h · ft^2 · °F)]	watt per square meter kelvin [W/(m^2 · K)]	5.678 263	E+00
British thermal unit$_{th}$ per hour square foot degree Fahrenheit [Btu$_{th}$/(h · ft^2 · °F)]	watt per square meter kelvin [W/(m^2 · K)]	5.674 466	E+00
British thermal unit$_{IT}$ per second square foot degree Fahrenheit [Btu$_{IT}$/(s · ft^2 · °F)]	watt per square meter kelvin [W/(m^2 · K)]	2.044 175	E+04
British thermal unit$_{th}$ per second square foot degree Fahrenheit [Btu$_{th}$/(s · ft^2 · °F)]	watt per square meter kelvin [W/(m^2 · K)]	2.042 808	E+04

To convert from	to	Multiply by

Density of Heat

British thermal unit$_{IT}$ per square foot
(Btu$_{IT}$/ft^2) .... joule per square meter (J/m^2) .... 1.135 653   E+04

British thermal unit$_{th}$ per square foot
(Btu$_{th}$/ft^2).. joule per square meter (J/m^2) .... 1.134 893   E+04

calorie$_{th}$ per square centimeter (cal$_{th}$/cm^2) .... joule per square meter (J/m^2) .... **4.184**   **E+04**

langley (cal$_{th}$/cm^2) .... joule per square meter (J/m^2) .... **4.184**   **E+04**

Density of Heat Flow Rate

British thermal unit$_{IT}$ per square foot hour
[Btu$_{IT}$/(ft^2 · h)] .... watt per square meter (W/m^2) .... 3.154 591   E+00

British thermal unit$_{th}$ per square foot hour
[Btu$_{th}$/(ft^2 · h)].. watt per square meter (W/m^2) .... 3.152 481   E+00

British thermal unit$_{th}$ per square foot minute
[Btu$_{th}$/(ft^2 · min)] .... watt per square meter (W/m^2) .... 1.891 489   E+02

British thermal unit$_{IT}$ per square foot second
[Btu$_{IT}$/(ft^2 · s)]... watt per square meter (W/m^2) .... 1.135 653   E+04

British thermal unit$_{th}$ per square foot second
[Btu$_{th}$/(ft^2 · s)]... watt per square meter (W/m^2) .... 1.134 893   E+04

British thermal unit$_{th}$ per square inch second
[Btu$_{th}$/(in^2 · s)] .... watt per square meter (W/m^2) .... 1.634 246   E+06

calorie$_{th}$ per square centimeter minute
[cal$_{th}$/(cm^2 · min)] .... watt per square meter (W/m^2) .... 6.973 333   E+02

calorie$_{th}$ per square centimeter second
[cal$_{th}$/(cm^2 · s)] .... watt per square meter (W/m^2) .... **4.184**   **E+04**

Fuel Consumption

gallon (U.S.) per horsepower hour
[gal/(hp · h)] .... cubic meter per joule (m^3/J) .... 1.410 089   E−09

gallon (U.S.) per horsepower hour
[gal/(hp · h)] .... liter per joule (L/J) .... 1.410 089   E−06

mile per gallon (U.S.) (mpg) (mi/gal).............. meter per cubic meter (m/m^3)................. 4.251 437   E+05

mile per gallon (U.S.) (mpg) (mi/gal).............. kilometer per liter (km/L) .... 4.251 437   E−01

mile per gallon (U.S.) (mpg) (mi/gal)^{22} .... liter per 100 kilometer (L/100 km) ....divide 235.215 by number
                                                         of miles per gallon

pound per horsepower hour [lb/(hp · h)] .... kilogram per joule (kg/J) .... 1.689 659   E−07

Heat Capacity and Entropy

British thermal unit$_{IT}$ per degree Fahrenheit
(Btu$_{IT}$/°F)... joule per kelvin (J/k) .... 1.899 101   E+03

British thermal unit$_{th}$ per degree Fahrenheit
(Btu$_{th}$/°F)... joule per kelvin (J/k) .... 1.897 830   E+03

British thermal unit$_{IT}$ per degree Rankine
(Btu$_{IT}$/°R) .... joule per kelvin (J/k) .... 1.899 101   E+03

British thermal unit$_{th}$ per degree Rankine
(Btu$_{th}$/°R)... joule per kelvin (J/k) .... 1.897 830   E+03

Heat Flow Rate

British thermal unit$_{IT}$ per hour (Btu$_{IT}$/h)........... watt (W)...................................... 2.930 711   E−01

British thermal unit$_{th}$ per hour (Btu$_{th}$/h) .... watt (W)...................................... 2.928 751   E−01

British thermal unit$_{th}$ per minute (Btu$_{th}$/min) .... watt (W)...................................... 1.757 250   E+01

British thermal unit$_{IT}$ per second (Btu$_{IT}$/s)......... watt (W)...................................... 1.055 056   E+03

British thermal unit$_{th}$ per second (Btu$_{th}$/s) .... watt (W)...................................... 1.054 350   E+03

calorie$_{th}$ per minute (cal$_{th}$/min) .... watt (W)...................................... 6.973 333   E−02

calorie$_{th}$ per second (cal$_{th}$/s) .... watt (W)...................................... **4.184**   **E+00**

kilocalorie$_{th}$ per minute (kcal$_{th}$/min) .... watt (W)...................................... 6.973 333   E+01

kilocalorie$_{th}$ per second (kcal$_{th}$/s) .... watt (W)...................................... **4.184**   **E+03**

ton of refrigeration (12 000 Btu$_{IT}$/h)................ watt (W)...................................... 3.516 853   E+03

To convert from	to	Multiply by

Specific Heat Capacity and Specific Entropy

British thermal unit$_{IT}$ per pound degree Fahrenheit
[Btu$_{IT}$/(lb · °F)]...................................joule per kilogram kelvin [J/(kg · K)]..........**4.1868**   E+03

British thermal unit$_{th}$ per pound degree Fahrenheit
[Btu$_{th}$/(lb · °F)]..................................joule per kilogram kelvin [J/(kg · K)]..........**4.184**   E+03

British thermal unit$_{IT}$ per pound degree Rankine
[Btu$_{IT}$/(lb · °R)] ....joule per kilogram kelvin [J/(kg · K)]..........**4.1868**   E+03

British thermal unit$_{th}$ per pound degree Rankine
[Btu$_{th}$/(lb · °R)]..................................joule per kilogram kelvin [J/(kg · K)]..........**4.184**   E+03

calorie$_{IT}$ per gram degree Celsius
[cal$_{IT}$/(g · °C)]...................................joule per kilogram kelvin [J/(kg · K)]..........**4.1868**   E+03

calorie$_{th}$ per gram degree Celsius
[cal$_{th}$/(g · °C)]...................................joule per kilogram kelvin [J/(kg · K)]..........**4.184**   E+03

calorie$_{IT}$ per gram kelvin [cal$_{IT}$/(g · K)] ....joule per kilogram kelvin [J/(kg · K)]..........**4.1868**   E+03

calorie$_{th}$ per gram kelvin [cal$_{th}$/(g · K)] ....joule per kilogram kelvin [J/(kg · K)]..........**4.184**   E+03

Thermal Conductivity

Britsh thermal unit$_{IT}$ foot per hour square foot degree Fahrenheit
[Btu$_{IT}$ · ft/(h · ft^2 · °F)]...........................watt per meter kelvin [W/(m · K)].............1.730 735   E+00

Britsh thermal unit$_{th}$ foot per hour square foot degree Fahrenheit
[Btu$_{th}$ · ft/(h · ft^2 · °F)] ....watt per meter kelvin [W/(m · K)].............1.729 577   E+00

Britsh thermal unit$_{IT}$ inch per hour square foot degree Fahrenheit
[Btu$_{IT}$ · in/(h · ft^2 · °F)]...........................watt per meter kelvin [W/(m · K)].............1.442 279   E−01

Britsh thermal unit$_{th}$ inch per hour square foot degree Fahrenheit
[Btu$_{th}$ · in/(h · ft^2 · °F)]. ....watt per meter kelvin [W/(m · K)].............1.441 314   E−01

Britsh thermal unit$_{IT}$ inch per second square foot degree Fahrenheit
[Btu$_{IT}$ · in/(s · ft^2 · °F)]. ....watt per meter kelvin [W/(m · K)].............5.192 204   E+02

Britsh thermal unit$_{th}$ inch per second square foot degree Fahrenheit
[Btu$_{th}$ · in/(s · ft^2 · °F)] ....watt per meter kelvin [W/(m · K)].............5.188 732   E+02

calorie$_{th}$ per centimeter second degree Celsius
[cal$_{th}$/(cm · s · °C)] ....watt per meter kelvin [W/(m · K)].............**4.184**   E+02

Thermal Diffusivity

square foot per hour (ft^2/h) ....square meter per second (m^2/s)...............**2.580 64**   E−05

Thermal Insulance

clo ....square meter kelvin per watt (m^2 · K/W) ......1.55   E−01

degree Fahrenheit hour square foot per British thermal unit$_{IT}$
(°F · h · ft^2/Btu$_{IT}$)..................................square meter kelvin per watt (m^2 · K/W) ......1.761 102   E−01

degree Fahrenheit hour square foot per British thermal unit$_{th}$
(°F · h · ft^2/Btu$_{th}$).................................square meter kelvin per watt (m^2 · K/W) ......1.762 280   E−01

Thermal Resistance

degree Fahrenheit hour per British thermal unit$_{IT}$
(°F · h/Btu$_{IT}$).....................................kelvin per watt (K/W)........................1.895 634   E+00

degree Fahrenheit hour per British thermal unit$_{th}$
(°F · h/Btu$_{th}$) ....kelvin per watt (K/W)........................1.896 903   E+00

degree Fahrenheit second per British thermal unit$_{IT}$
(°F · s/Btu$_{IT}$) ....kelvin per watt (K/W)........................5.265 651   E−04

degree Fahrenheit second per British thermal unit$_{th}$
(°F · s/Btu$_{th}$).....................................kelvin per watt (K/W)........................5.269 175   E−04

Thermal Resistivity

degree Fahrenheit hour square foot per British thermal unit$_{IT}$ inch
[°F · h · ft^2/(Btu$_{IT}$ · in)] ....meter kelvin per watt (m · K/W) ....6.933 472   E+00

degree Fahrenheit hour square foot per British thermal unit$_{th}$ inch
[°F · h · ft^2/(Btu$_{th}$ · in)]...........................meter kelvin per watt (m · K/W) ....6.938 112   E+04

To convert from	to	Multiply by	
LENGTH			
ångström (Å)	meter (m)	**1.0**	**E−10**
ångström (Å)	nanometer (nm)	**1.0**	**E−01**
astronomical unit (AU)	meter (m)	1.495 979	E+11
chain (based on U.S. survey foot) (ch)[9]	meter (m)	2.011 684	E+01
fathom (based on U.S. survey foot)[9]	meter (m)	1.828 804	E+00
fermi	meter (m)	**1.0**	**E−15**
fermi	femtometer (fm)	**1.0**	**E+00**
foot (ft)	meter (m)	**3.048**	**E−01**
foot (U.S. survey) (ft)[9]	meter (m)	3.048 006	E−01
inch (in)	meter (m)	**2.54**	**E−02**
inch (in)	centimeter (cm)	**2.54**	**E+00**
kayser (K)	reciprocal meter (m^{-1})	1	E+02
light year (l.y.)[19]	meter (m)	9.460 73	E+15
microinch	meter (m)	**2.54**	**E−08**
microinch	micrometer (μm)	**2.54**	**E−02**
micron (μ)	meter (m)	**1.0**	**E−06**
micron (μ)	micrometer (μm)	**1.0**	**E+00**
mil (0.001 in)	meter (m)	**2.54**	**E−05**
mil (0.001 in)	millimeter (mm)	**2.54**	**E−02**
mile (mi)	meter (m)	**1.609 344**	**E+03**
mile (mi)	kilometer (km)	**1.609 344**	**E+00**
mile (based on U.S. survey foot) (mi)[9]	meter (m)	1.609 347	E+03
mile (based on U.S. survey foot) (mi)[9]	kilometer (km)	1.609 347	E+00
mile, nautical[21]	meter (m)	**1.852**	**E+03**
parsec (pc)	meter (m)	3.085 678	E+16
pica (computer) (1/6 in)	meter (m)	4.233 333	E−03
pica (computer) (1/6 in)	millimeter (mm)	4.233 333	E+00
pica (printer's)	meter (m)	4.217 518	E−03
pica (printer's)	millimeter (mm)	4.217 518	E+00
point (computer) (1/72 in)	meter (m)	3.527 778	E−04
point (computer) (1/72 in)	millimeter (mm)	3.527 778	E−01
point (printer's)	meter (m)	3.514 598	E−04
point (printer's)	millimeter (mm)	3.514 598	E−01
rod (based on U.S. survey foot) (rd)[9]	meter (m)	5.029 210	E+00
yard (yd)	meter (m)	**9.144**	**E−01**
LIGHT			
candela per square inch (cd/in^2)	candela per square meter (cd/m^2)	1.550 003	E+03
footcandle	lux (lx)	1.076 391	E+01
footlambert	candela per square meter (cd/m^2)	3.426 259	E+00
lambert[18]	candela per square meter (cd/m^2)	3.183 099	E+03
lumen per square foot (lm/ft^2)	lux (lx)	1.076 391	E+01
phot (ph)	lux (lx)	**1.0**	**E+04**
stilb (sb)	candela per square meter (cd/m^2)	**1.0**	**E+04**
MASS and MOMENT OF INERTIA			
carat, metric	kilogram (kg)	**2.0**	**E−04**
carat, metric	gram (g)	**2.0**	**E−01**
grain (gr)	kilogram (kg)	**6.479 891**	**E−05**
grain (gr)	milligram (mg)	**6.479 891**	**E+01**
hundredweight (long, 112 lb)	kilogram (kg)	5.080 235	E+01
hundredweight (short, 100 lb)	kilogram (kg)	4.535 924	E+01

To convert from	to		Multiply by	
kilogram-force second squared per meter (kgf · s²/m)	kilogram (kg)		**9.806 65**	**E+00**
ounce (avoirdupois) (oz)	kilogram (kg)		2.834 952	E−02
ounce (avoirdupois) (oz)	gram (g)		2.834 952	E+01
ounce (troy or apothecary) (oz)	kilogram (kg)		3.110 348	E−02
ounce (troy or apothecary) (oz)	gram (g)		3.110 348	E+01
pennyweight (dwt)	kilogram (kg)		1.555 174	E−03
pennyweight (dwt)	gram (g)		1.555 174	E+00
pound (avoirdupois) (lb)[23]	kilogram (kg)		4.535 924	E−01
pound (troy or apothecary) (lb)	kilogram (kg)		3.732 417	E−01
pound foot squared (lb · ft²)	kilogram meter squared (kg · m²)		4.214 011	E−02
pound inch squared (lb · in²)	kilogram meter squared (kg · m²)		2.926 397	E−04
slug (slug)	kilogram (kg)		1.459 390	E+01
ton, assay (AT)	kilogram (kg)		2.916 667	E−02
ton, assay (AT)	gram (g)		2.916 667	E+01
ton, long (2240 lb)	kilogram (kg)		1.016 047	E+03
ton, metric (t)	kilogram (kg)		**1.0**	**E+03**
tonne (called "metric ton" in U.S.) (t)	kilogram (kg)		**1.0**	**E+03**
ton, short (2000 lb)	kilogram (kg)		9.071 847	E+02

MASS DENSITY (see MASS DIVIDED BY VOLUME)

MASS DIVIDED BY AREA

ounce (avoirdupois) per square foot (oz/ft²)	kilogram per square meter (kg/m²)		3.051 517	E−01
ounce (avoirdupois) per square inch (oz/in²)	kilogram per square meter (kg/m²)		4.394 185	E+01
ounce (avoirdupois) per square yard (oz/yd²)	kilogram per square meter (kg/m²)		3.390 575	E−02
pound per square foot (lb/ft²)	kilogram per square meter (kg/m²)		4.882 428	E+00
pound per square inch (*not* pound force) (lb/in²)	kilogram per square meter (kg/m²)		7.030 696	E+02

MASS DIVIDED BY CAPACITY (see MASS DIVIDED BY VOLUME)

MASS DIVIDED BY LENGTH

denier	kilogram per meter (kg/m)		1.111 111	E−07
denier	gram per meter (g/m)		1.111 111	E−04
pound per foot (lb/ft)	kilogram per meter (kg/m)		1.488 164	E+00
pound per inch (lb/in)	kilogram per meter (kg/m)		1.785 797	E+01
pound per yard (lb/yd)	kilogram per meter (kg/m)		4.960 546	E−01
tex	kilogram per meter (kg/m)		**1.0**	**E−06**

MASS DIVIDED BY TIME (includes FLOW)

pound per hour (lb/h)	kilogram per second (kg/s)		1.259 979	E−04
pound per minute (lb/min)	kilogram per second (kg/s)		7.559 873	E−03
pound per second (lb/s)	kilogram per second (kg/s)		4.535 924	E−01
ton, short, per hour	kilogram per second (kg/s)		2.519 958	E−01

MASS DIVIDED BY VOLUME (includes MASS DENSITY and MASS CONCENTRATION)

grain per gallon (U.S.) (gr/gal)	kilogram per cubic meter (kg/m³)		1.711 806	E−02
grain per gallon (U.S.) (gr/gal)	milligram per liter (mg/L)		1.711 806	E+01
gram per cubic centimeter (g/cm³)	kilogram per cubic meter (kg/m³)		**1.0**	**E+03**
ounce (avoirdupois) per cubic inch (oz/in³)	kilogram per cubic meter (kg/m³)		1.729 994	E+03
ounce (avoirdupois) per gallon [Canadian and U.K. (Imperial)] (oz/gal)	kilogram per cubic meter (kg/m³)		6.236 023	E+00
ounce (avoirdupois) per gallon [Canadian and U.K. (Imperial)] (oz/gal)	gram per liter (g/L)		6.236 023	E+00
ounce (avoirdupois) per gallon (U.S.) (oz/gal)	kilogram per cubic meter (kg/m³)		7.489 152	E+00
ounce (avoirdupois) per gallon (U.S.) (oz/gal)	gram per liter (g/L)		7.489 152	E+00

To convert from	to		Multiply by
pound per cubic foot (lb/ft³)	kilogram per cubic meter (kg/m³)	1.601 846	E+01
pound per cubic inch (lb/in³)	kilogram per cubic meter (kg/m³)	2.767 990	E+04
pound per cubic yard (lb/yd³)......................	kilogram per cubic meter (kg/m³)	5.932 764	E−01
pound per gallon [Canadian and U.K. (Imperial)] (lb/gal)	kilogram per cubic meter (kg/m³)	9.977 637	E+01
pound per gallon [Canadian and U.K. (Imperial)] (lb/gal)	kilogram per liter (kg/L)	9.977 637	E−02
pound per gallon (U.S.) (lb/gal)....................	kilogram per cubic meter (kg/m³)	1.198 264	E+02
pound per gallon (U.S.) (lb/gal)....................	kilogram per liter (kg/L)	1.198 264	E−01
slug per cubic foot (slug/ft³)......................	kilogram per cubic meter (kg/m³)	5.153 788	E+02
ton, long, per cubic yard	kilogram per cubic meter (kg/m³)	1.328 939	E+03
ton, short, per cubic yard	kilogram per cubic meter (kg/m³)	1.186 553	E+03

MOMENT OF FORCE or TORQUE

dyne centimeter (dyn · cm)........................	newton meter (N · m)........................	**1.0**	**E−07**
kilogram-force meter (kgf · m).....................	newton meter (N · m)........................	**9.806 65**	**E+00**
ounce (avoirdupois)-force inch (ozf · in)	newton meter (N · m)........................	7.061 552	E−03
ounce (avoirdupois)-force inch (ozf · in)	millinewton meter (mN · m)..................	7.061 552	E+00
pound-force foot (lbf · ft)	newton meter (N · m)........................	1.355 818	E+00
pound-force inch (lbf · in).........................	newton meter (N · m)........................	1.129 848	E−01

MOMENT OF FORCE or TORQUE, DIVIDED BY LENGTH

pound-force foot per inch (lbf · ft/in)	newton meter per meter (N · m/m)...........	5.337 866	E+01
pound-force inch per inch (lbf · in/in)	newton meter per meter (N · m/m)...........	4.448 222	E+00

PERMEABILITY

darcy[15]..	meter squared (m²).........................	9.869 233	E−13
perm (0 °C)	kilogram per pascal second square meter [kg/(Pa · s · m²)]...........................	5.721 35	E−11
perm (23 °C).....................................	kilogram per pascal second square meter [kg/(Pa · s · m²)]...........................	5.745 25	E−11
perm inch (0 °C).................................	kilogram per pascal second meter [kg/(Pa · s · m)]............................	1.453 22	E−12
perm inch (23 °C)	kilogram per pascal second meter [kg/(Pa · s · m)]............................	1.459 29	E−12

POWER

erg per second (erg/s).............................	watt (W)...................................	**1.0**	**E−07**
foot pound-force per hour (ft · lbf/h)...............	watt (W)...................................	3.766 161	E−04
foot pound-force per minute (ft · lbf/min)	watt (W)...................................	2.259 697	E−02
foot pound-force per second (ft · lbf/s)	watt (W)...................................	1.355 818	E+00
horsepower (550 ft · lbf/s)	watt (W)...................................	7.456 999	E+02
horsepower (boiler)	watt (W)...................................	9.809 50	E+03
horsepower (electric)..............................	watt (W)...................................	**7.46**	**E+02**
horsepower (metric)...............................	watt (W)...................................	7.354 988	E+02
horsepower (U.K.).................................	watt (W)...................................	7.4570	E+02
horsepower (water)................................	watt (W)...................................	7.460 43	E+02

PRESSURE or STRESS (FORCE DIVIDED BY AREA)

atmosphere, standard (atm)........................	pascal (Pa).................................	**1.013 25**	**E+05**
atmosphere, standard (atm)........................	kilopascal (kPa)............................	**1.013 25**	**E+02**
atmosphere, technical (at)[10].......................	pascal (Pa).................................	**9.806 65**	**E+04**
atmosphere, technical (at)[10].......................	kilopascal (kPa)............................	**9.806 65**	**E+01**
bar (bar)...	pascal (Pa).................................	**1.0**	**E+05**
bar (bar)...	kilopascal (kPa)............................	**1.0**	**E+02**

To convert from	to	Multiply by	
centimeter of mercury (0 °C)[13]	pascal (Pa)	1.333 22	E+03
centimeter of mercury (0 °C)[13]	kilopascal (kPa)	1.333 22	E+00
centimeter of mercury, conventional (cmHg)[13]	pascal (Pa)	1.333 224	E+03
centimeter of mercury, conventional (cmHg)[13]	kilopascal (kPa)	1.333 224	E+00
centimeter of water (4 °C)[13]	pascal (Pa)	9.806 38	E+01
centimeter of water, conventional (cmH$_2$O)[13]	pascal (Pa)	**9.806 65**	**E+01**
dyne per square centimeter (dyn/cm^2)	pascal (Pa)	**1.0**	**E−01**
foot of mercury, conventional (ftHg)[13]	pascal (Pa)	4.063 666	E+04
foot of mercury, conventional (ftHg)[13]	kilopascal (kPa)	4.063 666	E+01
foot of water (39.2 °F)[13]	pascal (Pa)	2.988 98	E+03
foot of water (39.2 °F)[13]	kilopascal (kPa)	2.988 98	E+00
foot of water, conventional (ftH$_2$O)[13]	pascal (Pa)	2.989 067	E+03
foot of water, conventional (ftH$_2$O)[13]	kilopascal (kPa)	2.989 067	E+00
gram-force per square centimeter (gf/cm^2)	pascal (Pa)	**9.806 65**	**E+01**
inch of mercury (32 °F)[13]	pascal (Pa)	3.386 38	E+03
inch of mercury (32 °F)[13]	kilopascal (kPa)	3.386 38	E+00
inch of mercury (60 °F)[13]	pascal (Pa)	3.376 85	E+03
inch of mercury (60 °F)[13]	kilopascal (kPa)	3.376 85	E+00
inch of mercury, conventional (inHg)[13]	pascal (Pa)	3.386 389	E+03
inch of mercury, conventional (inHg)[13]	kilopascal (kPa)	3.386 389	E+00
inch of water (39.2 °F)[13]	pascal (Pa)	2.490 82	E+02
inch of water (60 °F)[13]	pascal (Pa)	2.4884	E+02
inch of water, conventional (inH$_2$O)[13]	pascal (Pa)	2.490 889	E+02
kilogram-force per square centimeter (kgf/cm^2)	pascal (Pa)	**9.806 65**	**E+04**
kilogram-force per square centimeter (kgf/cm^2)	kilopascal (kPa)	**9.806 65**	**E+01**
kilogram-force per square meter (kgf/m^2)	pascal (Pa)	**9.806 65**	**E+00**
kilogram-force per square millimeter (kgf/mm^2)	pascal (Pa)	**9.806 65**	**E+06**
kilogram-force per square millimeter (kgf/mm^2)	megapascal (MPa)	**9.806 65**	**E+00**
kip per square inch (ksi) (kip/in^2)	pascal (Pa)	6.894 757	E+06
kip per square inch (ksi) (kip/in^2)	kilopascal (kPa)	6.894 757	E+03
millibar (mbar)	pascal (Pa)	**1.0**	**E+02**
millibar (mbar)	kilopascal (kPa)	**1.0**	**E−01**
millimeter of mercury, conventional (mmHg)[13]	pascal (Pa)	1.333 224	E+02
millimeter of water, conventional (mmH$_2$O)[13]	pascal (Pa)	**9.806 65**	**E+00**
poundal per square foot	pascal (Pa)	1.488 164	E+00
pound-force per square foot (lbf/ft^2)	pascal (Pa)	4.788 026	E+01
pound-force per square inch (psi) (lbf/in^2)	pascal (Pa)	6.894 757	E+03
pound-force per square inch (psi) (lbf/in^2)	kilopascal (kPa)	6.894 757	E+00
psi (pound-force per square inch) (lbf/in^2)	pascal (Pa)	6.894 757	E+03
psi (pound-force per square inch) (lbf/in^2)	kilopascal (kPa)	6.894 757	E+00
torr (Torr)	pascal (Pa)	1.333 224	E+02

RADIOLOGY

curie (Ci)	becquerel (Bq)	**3.7**	**E+10**
rad (absorbed dose) (rad)	gray (Gy)	**1.0**	**E−02**
rem (rem)	sievert (Sv)	**1.0**	**E−02**
roentgen (R)	coulomb per kilogram (C/kg)	**2.58**	**E−04**

SPEED (see VELOCITY)

STRESS (see PRESSURE)

To convert from	to	Multiply by

TEMPERATURE

degree Celsius (°C)	kelvin (K)	$T/\text{K} = t/\text{°C} + \mathbf{273.15}$
degree centigrade[16]	degree Celsius (°C)	$t/\text{°C} \approx t/\text{deg. cent.}$
degree Fahrenheit (°F)	degree Celsius (°C)	$t/\text{°C} = (t/\text{°F} - \mathbf{32})/\mathbf{1.8}$
degree Fahrenheit (°F)	kelvin (K)	$T/\text{K} = (t/\text{°F} + \mathbf{459.67})/\mathbf{1.8}$
degree Rankine (°R)	kelvin (K)	$T/\text{K} = (T/\text{°R})/\mathbf{1.8}$
kelvin (K)	degree Celsius (°C)	$t/\text{°C} = T/\text{K} - \mathbf{273.15}$

TEMPERATURE INTERVAL

degree Celsius (°C)	kelvin (K)	**1.0**	E+00
degree centigrade[16]	degree Celsius (°C)	1.0	E+00
degree Fahrenheit (°F)	degree Celsius (°C)	5.555 556	E−01
degree Fahrenheit (°F)	kelvin (K)	5.555 556	E−01
degree Rankine (°R)	kelvin (K)	5.555 556	E−01

TIME

day (d)	second (s)	**8.64**	**E+04**
day (sidereal)	second (s)	8.616 409	E+04
hour (h)	second (s)	**3.6**	**E+03**
hour (sidereal)	second (s)	3.590 170	E+03
minute (min)	second (s)	**6.0**	**E+01**
minute (sidereal)	second (s)	5.983 617	E+01
second (sidereal)	second (s)	9.972 696	E−01
shake	second (s)	**1.0**	**E−08**
shake	nanosecond (ns)	**1.0**	**E+01**
year (365 days)	second (s)	**3.1536**	**E+07**
year (sidereal)	second (s)	3.155 815	E+07
year (tropical)	second (s)	3.155 693	E+07

TORQUE (see MOMENT OF FORCE)

VELOCITY (includes SPEED)

foot per hour (ft/h)	meter per second (m/s)	8.466 667	E−05
foot per minute (ft/min)	meter per second (m/s)	**5.08**	**E−03**
foot per second (ft/s)	meter per second (m/s)	**3.048**	**E−01**
inch per second (in/s)	meter per second (m/s)	**2.54**	**E−02**
kilometer per hour (km/h)	meter per second (m/s)	2.777 778	E−01
knot (nautical mile per hour)	meter per second (m/s)	5.144 444	E−01
mile per hour (mi/h)	meter per second (m/s)	**4.4704**	**E−01**
mile per hour (mi/h)	kilometer per hour (km/h)	**1.609 344**	**E+00**
mile per minute (mi/min)	meter per second (m/s)	**2.682 24**	**E+01**
mile per second (mi/s)	meter per second (m/s)	**1.609 344**	**E+03**
revolution per minute (rpm) (r/min)	radian per second (rad/s)	1.047 198	E−01
rpm (revolution per minute) (r/min)	radian per second (rad/s)	1.047 198	E−01

VISCOSITY, DYNAMIC

centipoise (cP)	pascal second (Pa · s)	**1.0**	**E−03**
poise (P)	pascal second (Pa · s)	**1.0**	**E−01**
poundal second per square foot	pascal second (Pa · s)	1.488 164	E+00
pound-force second per square foot (lbf · s/ft²)	pascal second (Pa · s)	4.788 026	E+01
pound-force second per square inch (lbf · s/in²)	pascal second (Pa · s)	6.894 757	E+03
pound per foot hour [lb/(ft · h)]	pascal second (Pa · s)	4.133 789	E−04
pound per foot second [lb/(ft · s)]	pascal second (Pa · s)	1.488 164	E+00
rhe	reciprocal pascal second [(Pa · s)⁻¹]	**1.0**	**E+01**
slug per foot second [slug/(ft · s)]	pascal second (Pa · s)	4.788 026	E+01

To convert from	to		Multiply by

VISCOSITY, KINEMATIC

centistokes (cSt)	meter squared per second (m^2/s)	**1.0**	**E−06**
square foot per second (ft^2/s)	meter squared per second (m^2/s)	**9.290 304**	**E−02**
stokes (St)	meter squared per second (m^2/s)	**1.0**	**E−04**

VOLUME (includes CAPACITY)

acre-foot (based on U.S. survey foot)[9]	cubic meter (m^3)	1.233 489	E+03
barrel [for petroleum, 42 gallons (U.S.)](bbl)	cubic meter (m^3)	1.589 873	E−01
barrel [for petroleum, 42 gallons (U.S.)](bbl)	liter (L)	1.589 873	E+02
bushel (U.S.) (bu)	cubic meter (m^3)	3.523 907	E−02
bushel (U.S.) (bu)	liter (L)	3.523 907	E+01
cord (128 ft^3)	cubic meter (m^3)	3.624 556	E+00
cubic foot (ft^3)	cubic meter (m^3)	2.831 685	E−02
cubic inch (in^3)[14]	cubic meter (m^3)	1.638 706	E−05
cubic mile (mi^3)	cubic meter (m^3)	4.168 182	E+09
cubic yard (yd^3)	cubic meter (m^3)	7.645 549	E−01
cup (U.S.)	cubic meter (m^3)	2.365 882	E−04
cup (U.S.)	liter (L)	2.365 882	E−01
cup (U.S.)	milliliter (mL)	2.365 882	E+02
fluid ounce (U.S.) (fl oz)	cubic meter (m^3)	2.957 353	E−05
fluid ounce (U.S.) (fl oz)	milliliter (mL)	2.957 353	E+01
gallon [Canadian and U.K. (Imperial)] (gal)	cubic meter (m^3)	**4.546 09**	**E−03**
gallon [Canadian and U.K. (Imperial)] (gal)	liter (L)	**4.546 09**	**E+00**
gallon (U.S.) (gal)	cubic meter (m^3)	3.785 412	E−03
gallon (U.S.) (gal)	liter (L)	3.785 412	E+00
gill [Canadian and U.K. (Imperial)] (gi)	cubic meter (m^3)	1.420 653	E−04
gill [Canadian and U.K. (Imperial)] (gi)	liter (L)	1.420 653	E−01
gill (U.S.) (gi)	cubic meter (m^3)	1.182 941	E−04
gill (U.S.) (gi)	liter (L)	1.182 941	E−01
liter (L)[20]	cubic meter (m^3)	**1.0**	**E−03**
ounce [Canadian and U.K. fluid (Imperial)] (fl oz)	cubic meter (m^3)	2.841 306	E−05
ounce [Canadian and U.K. fluid (Imperial)] (fl oz)	milliliter (mL)	2.841 306	E+01
ounce (U.S. fluid) (fl oz)	cubic meter (m^3)	2.957 353	E−05
ounce (U.S. fluid) (fl oz)	milliliter (mL)	2.957 353	E+01
peck (U.S.) (pk)	cubic meter (m^3)	8.809 768	E−03
peck (U.S.) (pk)	liter (L)	8.809 768	E+00
pint (U.S. dry) (dry pt)	cubic meter (m^3)	5.506 105	E−04
pint (U.S. dry) (dry pt)	liter (L)	5.506 105	E−01
pint (U.S. liquid) (liq pt)	cubic meter (m^3)	4.731 765	E−04
pint (U.S. liquid) (liq pt)	liter (L)	4.731 765	E−01
quart (U.S. dry) (dry qt)	cubic meter (m^3)	1.101 221	E−03
quart (U.S. dry) (dry qt)	liter (L)	1.101 221	E+00
quart (U.S. liquid) (liq qt)	cubic meter (m^3)	9.463 529	E−04
quart (U.S. liquid) (liq qt)	liter (L)	9.463 529	E−01
stere (st)	cubic meter (m^3)	**1.0**	**E+00**
tablespoon	cubic meter (m^3)	1.478 676	E−05
tablespoon	milliliter (mL)	1.478 676	E+01
teaspoon	cubic meter (m^3)	4.928 922	E−06
teaspoon	milliliter (mL)	4.928 922	E+00
ton, register	cubic meter (m^3)	2.831 685	E+00

To convert from	to	Multiply by

VOLUME DIVIDED BY TIME (includes FLOW)

cubic foot per minute (ft³/min)	cubic meter per second (m³/s)................	4.719 474	E−04
cubic foot per minute (ft³/min)	liter per second (L/s).........................	4.719 474	E−01
cubic foot per second (ft³/s)	cubic meter per second (m³/s)................	2.831 685	E−02
cubic inch per minute (in³/min).....................	cubic meter per second (m³/s)................	2.731 177	E−07
cubic yard per minute (yd³/min).....................	cubic meter per second (m³/s)................	1.274 258	E−02
gallon (U.S.) per day (gal/d)	cubic meter per second (m³/s)................	4.381 264	E−08
gallon (U.S.) per day (gal/d)	liter per second (L/s).........................	4.381 264	E−05
gallon (U.S.) per minute (gpm) (gal/min)..........	cubic meter per second (m³/s)................	6.309 020	E−05
gallon (U.S.) per minute (gpm) (gal/min)..........	liter per second (L/s).........................	6.309 020	E−02

WORK (see ENERGY)

A.2   Physical Constants

Most values of physical constants and conversion factors are available on the web (e.g., see [PDB (2008); Mohr and Taylor (2005)]).

Quantity	Symbol	Value
speed of light in vacuum	c	$299\ 792\ 458\,\mathrm{m\,s^{-1}}$
Planck constant	$h = 2\pi\hbar$	$6.6260693 \times 10^{-34}\,\mathrm{J\,s}$
conversion constant	$\hbar\, c$	$197.326968\,\mathrm{MeV\,fm}$
electron charge magnitude	e	$1.60217653 \times 10^{-19}\,\mathrm{C}$ $4.803204 \times 10^{-10}\,\mathrm{esu}$
electron mass	m_e	$0.510998918\,\mathrm{MeV}/c^2$ $9.1093826 \times 10^{-28}\,\mathrm{g}$
proton mass	m_p	$938.272029\,\mathrm{MeV}/c^2$ $1.67262171 \times 10^{-24}\,\mathrm{g}$
neutron mass	m_n	$939.565360\,\mathrm{MeV}/c^2$ $1.67492728 \times 10^{-24}\,\mathrm{g}$
deuteron mass	m_d	$1875.61282\,\mathrm{MeV}/c^2$ $3.34358335 \times 10^{-24}\,\mathrm{g}$
unified atomic unit mass	u	$931.4940\,\mathrm{MeV}/c^2$ $1.66053886 \times 10^{-24}\,\mathrm{g}$
Bohr magneton	$\mu_B = \frac{e\hbar}{2m_e}$	$5.788381804 \times 10^{-11}\,\mathrm{MeV/T}$
nuclear magneton	$\mu_N = \frac{e\hbar}{2m_p}$	$3.152451259 \times 10^{-14}\,\mathrm{MeV/T}$
electron magnetic moment	μ_e	$-1.0011596521859\,\mu_B$ $-1.83828197107 \times 10^3\,\mu_N$
proton magnetic moment	μ_p	$1.521032206 \times 10^{-3}\,\mu_B$ $2.792847351\,\mu_N$
neutron magnetic moment	μ_n	$-1.04187563 \times 10^{-3}\,\mu_B$ $-1.91304273\,\mu_N$
Avogadro constant	N	$6.0221415 \times 10^{23}\,\mathrm{mol^{-1}}$

continued on next page

continued from previous page

Quantity	Symbol	Value
Boltzmann constant[†]	k_B	$1.3806504 \times 10^{-23}\,\mathrm{J\,K^{-1}}$ $8.617343 \times 10^{-5}\mathrm{eV\,K^{-1}}$
permittivity of free space	ε_0	$8.854187817 \times 10^{-12}\mathrm{F\,m^{-1}}$
Si dielectric constant[*]	$\varepsilon_{\mathrm{Si}}$	11.9
Si electric permittivity[*]	$\varepsilon = \varepsilon_{\mathrm{Si}}\,\varepsilon_0$	$1.054 \times 10^{-10}\,\mathrm{F\,m^{-1}}$ $1.054\,\mathrm{pF\,cm^{-1}}$
permeability of free space	μ_0	$4\pi \times 10^{-7}\mathrm{N\,A^{-2}}$ $12.566370614 \times 10^{-7}\mathrm{N\,A^{-2}}$
fine-structure constant	$\alpha = e^2/(4\pi\epsilon_0\hbar c)$ α $= e^2/(\hbar c)$, e in esu	$1/137.03599911$
classical electron radius	$r_e = e^2/(4\pi\epsilon_0 m_e c^2)$ r_e $= \frac{e^2}{(m_e c^2)}$, e in esu	$2.817940325\,\mathrm{fm}$
Compton wavelength of the electron	$\lambda_e \equiv \frac{h}{m_e c}$	$2.426310238 \times 10^{-10}\mathrm{cm}$
Bohr radius of the Hydrogen atom	a_0 $= (4\pi\epsilon_0\hbar^2)/(m_e c^2)$ $a_0 = r_e/\alpha^2$ $= \hbar^2/(m_e e^2)$	$0.5291772108 \times 10^{-8}\mathrm{cm}$
Bohr velocity	$v_0 = c\,\alpha$ $v_0 = e^2/\hbar$, e in esu	$c/137.03599911$

continued on next page

[†]The symbol used in this book is k_B to avoid confusion with other symbols, while that universally adopted is k.
[*]At room temperature.

continued from previous page

Quantity	Symbol	Value
Rydberg constant for infinitely large nuclear mass	$R_\infty = \frac{m_e c^2 \alpha^2}{2hc}$ $= \frac{2\pi^2 e^4 m_e}{(4\pi\epsilon_0)^2 h^3 c}$ R_∞ $= \frac{2\pi^2 e^4 m_e}{h^3 c}$, e in esu	$1.0973731568525 \times 10^5 \mathrm{cm}^{-1}$
Rydberg constant for nuclear mass M	R $= \frac{R_\infty}{(1 + m_e/M)}$	
Rydberg energy	$Ry = R_\infty hc$ $= (m_e c^2 \alpha^2)/2$	$13.6056923\,\mathrm{eV}$
classical Thomson cross section	$\sigma_{Th} = (8/3)\pi r_e^2$	$6.65245873 \times 10^{-25}\mathrm{cm}^2$ $0.665245873\,\mathrm{b}$
atomic radius in the *Thomas–Fermi* model	$a_Z = a_0/Z^{1/3}$	
Astronomical Unit[*]	AU	$1.4959787 \times 10^8\,\mathrm{km}$
Parsec	pc	$3.08567758066631 \times 10^{13}\,\mathrm{km}$ $2.06265 \times 10^5\,\mathrm{AU}$ $3.262\,\mathrm{ly}$ (light years)
volumetric radius of the Earth	Re	$6.370998685023 \times 10^3\,\mathrm{km}$ $6.370998685023 \times 10^8\,\mathrm{cm}$
solar mass[*]	$m_\odot$	$1.989 \times 10^{33}\,\mathrm{g}$
solar radius[*]	$R_\odot$	$6.955 \times 10^5\,\mathrm{km}$
solar mean density[*]	$\rho_\odot$	$1.409\,\mathrm{g/cm}^3$
Newtonian constant of gravitation	G	$6.67428 \times 10^{-11}\,\mathrm{m}^3/(\mathrm{kg\,s}^2)$ $6.70881 \times 10^{-39}\,\hbar c/(\mathrm{GeV}/c^2)^2$

continued on next page

[*]Value from page 72 of [Aschwanden (2006)].

continued from previous page

Quantity	Symbol	Value
Fermi coupling constant	$G_{\mathrm{F}}/(\hbar c)^3$	$1.16637(1)\times 10^{-5}\ \mathrm{GeV}^{-2}$
Stefan–Boltzmann constant	σ $= \pi^2 k_B^4/(60\hbar^3 c^2)$	$5.670400\ \times 10^{-8}\ \mathrm{W/(m^2\,K^4)}$
air density at NTP i.e., (20 °C, 1 atm)		$0.001205\ \mathrm{g\,cm^{-3}}$
air density at STP i.e., (0 °C, 1 atm)		$0.0012931\ \mathrm{g\,cm^{-3}}$
unit of absorbed dose for deposited energy	Gy 1 Gy = 100 rad	$10^4\mathrm{erg\,g^{-1}}$ $6.24 \times 10^{12}\mathrm{MeV\,kg^{-1}}$
unit of activity	Bq $1\ \mathrm{Bq} = 1\ \mathrm{dis.\,s^{-1}}$	$1/(3.7 \times 10^{10})\,\mathrm{Ci}$

A.3   Periodic Table of Elements

This version of the Periodic Table is based on that recommended by the Commission on the Nomenclature of Inorganic Chemistry and published in IUPAC Nomenclature of Inorganic Chemistry, Recommendations (1990). The definition of (standard) atomic weight for an element is given in Sect. 1.4.1. For more precise values of atomic weights see the table of 1997 recommended values [*Pure Appl. Chem.* **71**, 1593-1607 (1999)]. For elements with no stable nuclides the mass of the longest-lived isotope has been quoted in brackets. The names for elements 110 to 118 are temporary and are based on the 1978 recommendations [*Pure Appl. Chem.* **51**, 381-384 (1979)]. Elements marked with a ‡ have recently been reported (see, for instance, [Novov et al. (1999); Oganessian et al. (1999a,b)]).

Periodic Table of Elements

1	2	3	4	5	6	7	8	9	10	11	12	13	14	15	16	17	18
1 H 1.0079																	2 He 4.0026
3 Li 6.941	4 Be 9.0122											5 B 10.811	6 C 12.011	7 N 14.007	8 O 15.999	9 F 18.998	10 Ne 20.180
11 Na 22.990	12 Mg 24.305											13 Al 26.982	14 Si 28.086	15 P 30.974	16 S 32.066	17 Cl 35.453	18 Ar 39.948
19 K 39.098	20 Ca 40.078	21 Sc 44.956	22 Ti 47.867	23 V 50.942	24 Cr 51.996	25 Mn 54.938	26 Fe 55.845	27 Co 58.933	28 Ni 58.693	29 Cu 63.546	30 Zn 65.39	31 Ga 69.723	32 Ge 72.61	33 As 74.922	34 Se 78.96	35 Br 79.904	36 Kr 83.80
37 Rb 85.468	38 Sr 87.62	39 Y 88.906	40 Zr 91.224	41 Nb 92.906	42 Mo 95.94	43 Tc (98)	44 Ru 101.07	45 Rh 102.91	46 Pd 106.42	47 Ag 107.87	48 Cd 112.41	49 In 114.82	50 Sn 118.71	51 Sb 121.76	52 Te 127.60	53 I 126.90	54 Xe 131.29
55 Cs 132.91	56 Ba 137.33	57-71 *	72 Hf 178.49	73 Ta 180.95	74 W 183.84	75 Re 186.21	76 Os 190.23	77 Ir 192.22	78 Pt 195.08	79 Au 196.97	80 Hg 200.59	81 Tl 204.38	82 Pb 207.2	83 Bi 208.98	84 Po (209)	85 At (210)	86 Rn (222)
87 Fr (223)	88 Ra (226)	89-103 #	104 Rf (261)	105 Db (262)	106 Sg (263)	107 Bh (264)	108 Hs (265)	109 Mt (268)	110 Uun (269)	111 Uuu (272)	112 Uub (269)		114 Uuq +		116 Uuh +		118 Uuo +

* Lanthanide series

57 La 138.91	58 Ce 140.12	59 Pr 140.91	60 Nd 144.24	61 Pm (145)	62 Sm 150.36	63 Eu 151.96	64 Gd 157.25	65 Tb 158.93	66 Dy 162.50	67 Ho 164.93	68 Er 167.26	69 Tm 168.93	70 Yb 173.04	71 Lu 174.97

\# Actinide series

89 Ac (227)	90 Th 232.04	91 Pa 231.04	92 U 238.03	93 Np (237)	94 Pu (244)	95 Am (243)	96 Cm (247)	97 Bk (247)	98 Cf (251)	99 Es (252)	100 Fm (257)	101 Md (258)	102 No (259)	103 Lr (262)

A.4   Electronic Structure of the Elements

For this version of the Electronic Structure of the Elements (reproduced with the permission from Groom, D.E. et al. (2000), Table 5.1, pages 78-79, Review of Particle Physics, Particle Data Group, *The Eur. Phys. Jou. C* **15**, 1; © by SIF, Springer-Verlag 2000), the electronic configuration and ionization energies are mainly taken (except those marked by *) from [Martin and Lise (1995)]. For instance, the electron configuration for silicon indicates a neon electronic core (see neon) plus two 3s electrons and two 3p electrons. The ionization energy is the least necessary energy to remove to infinity one electron from an atomic element.

Electronic configurations of the elements are available on the web [Kotochigova, Levine, Shirley, Stiles and Clark (1996)]. Furthermore in this Reference, data for atomic electronic structure calculations have been generated to provide a standard reference for results of specified accuracy under commonly used approximations. Results are presented there for total energies and orbital energy eigenvalues for all atoms from H to U, at microHartree accuracy in the total energy, as computed in the local-density approximation (LDA), the local-spin-density approximation (LSD), the relativistic local-density approximation (RLDA), and scalar-relativistic local-density approximation (ScRLDA).

	Element		Electron configuration $(3d^5 = $ five $3d$ electrons, *etc.*$)$				Ground state $^{2S+1}L_J$	Ionization energy (eV)
1	H	Hydrogen	$1s$				$^2S_{1/2}$	13.5984
2	He	Helium	$1s^2$				$^1S_0$	24.5874
3	Li	Lithium	$(\text{He})\,2s$				$^2S_{1/2}$	5.3917
4	Be	Beryllium	$(\text{He})\,2s^2$				$^1S_0$	9.3227
5	B	Boron	$(\text{He})\,2s^2\ 2p$				$^2P_{1/2}$	8.2980
6	C	Carbon	$(\text{He})\,2s^2\ 2p^2$				$^3P_0$	11.2603
7	N	Nitrogen	$(\text{He})\,2s^2\ 2p^3$				$^4S_{3/2}$	14.5341
8	O	Oxygen	$(\text{He})\,2s^2\ 2p^4$				$^3P_2$	13.6181
9	F	Fluorine	$(\text{He})\,2s^2\ 2p^5$				$^2P_{3/2}$	17.4228
10	Ne	Neon	$(\text{He})\,2s^2\ 2p^6$				$^1S_0$	21.5646
11	Na	Sodium	$(\text{Ne})\,3s$				$^2S_{1/2}$	5.1391
12	Mg	Magnesium	$(\text{Ne})\,3s^2$				$^1S_0$	7.6462
13	Al	Aluminum	$(\text{Ne})\,3s^2\ 3p$				$^2P_{1/2}$	5.9858
14	Si	Silicon	$(\text{Ne})\,3s^2\ 3p^2$				$^3P_0$	8.1517
15	P	Phosphorus	$(\text{Ne})\,3s^2\ 3p^3$				$^4S_{3/2}$	10.4867
16	S	Sulfur	$(\text{Ne})\,3s^2\ 3p^4$				$^3P_2$	10.3600
17	Cl	Chlorine	$(\text{Ne})\,3s^2\ 3p^5$				$^2P_{3/2}$	12.9676
18	Ar	Argon	$(\text{Ne})\,3s^2\ 3p^6$				$^1S_0$	15.7596
19	K	Potassium	$(\text{Ar})\quad 4s$				$^2S_{1/2}$	4.3407
20	Ca	Calcium	$(\text{Ar})\quad 4s^2$				$^1S_0$	6.1132
21	Sc	Scandium	$(\text{Ar})\,3d\ 4s^2$	T			$^2D_{3/2}$	6.5615
22	Ti	Titanium	$(\text{Ar})\,3d^2\ 4s^2$	r	e		$^3F_2$	6.8281
23	V	Vanadium	$(\text{Ar})\,3d^3\ 4s^2$	a	l		$^4F_{3/2}$	6.7463
24	Cr	Chromium	$(\text{Ar})\,3d^5\ 4s$	n	e		$^7S_3$	6.7665
25	Mn	Manganese	$(\text{Ar})\,3d^5\ 4s^2$	s	m		$^6S_{5/2}$	7.4340
26	Fe	Iron	$(\text{Ar})\,3d^6\ 4s^2$	i	e		$^5D_4$	7.9024
27	Co	Cobalt	$(\text{Ar})\,3d^7\ 4s^2$	t	n		$^4F_{9/2}$	7.8810
28	Ni	Nickel	$(\text{Ar})\,3d^8\ 4s^2$	i	t		$^3F_4$	7.6398
29	Cu	Copper	$(\text{Ar})\,3d^{10}4s$	o	s		$^2S_{1/2}$	7.7264
30	Zn	Zinc	$(\text{Ar})\,3d^{10}4s^2$	n			$^1S_0$	9.3942
31	Ga	Gallium	$(\text{Ar})\,3d^{10}4s^2\ 4p$				$^2P_{1/2}$	5.9993
32	Ge	Germanium	$(\text{Ar})\,3d^{10}4s^2\ 4p^2$				$^3P_0$	7.8994
33	As	Arsenic	$(\text{Ar})\,3d^{10}4s^2\ 4p^3$				$^4S_{3/2}$	9.7886
34	Se	Selenium	$(\text{Ar})\,3d^{10}4s^2\ 4p^4$				$^3P_2$	9.7524
35	Br	Bromine	$(\text{Ar})\,3d^{10}4s^2\ 4p^5$				$^2P_{3/2}$	11.8138
36	Kr	Krypton	$(\text{Ar})\,3d^{10}4s^2\ 4p^6$				$^1S_0$	13.9996
37	Rb	Rubidium	$(\text{Kr})\quad 5s$				$^2S_{1/2}$	4.1771
38	Sr	Strontium	$(\text{Kr})\quad 5s^2$				$^1S_0$	5.6949
39	Y	Yttrium	$(\text{Kr})\,4d\ 5s^2$	T			$^2D_{3/2}$	6.2171
40	Zr	Zirconium	$(\text{Kr})\,4d^2\ 5s^2$	r	e		$^3F_2$	6.6339
41	Nb	Niobium	$(\text{Kr})\,4d^4\ 5s$	a	l		$^6D_{1/2}$	6.7589
42	Mo	Molybdenum	$(\text{Kr})\,4d^5\ 5s$	n	e		$^7S_3$	7.0924
43	Tc	Technetium	$(\text{Kr})\,4d^5\ 5s^2$	s	m		$^6S_{5/2}$	7.28
44	Ru	Ruthenium	$(\text{Kr})\,4d^7\ 5s$	i	e		$^5F_5$	7.3605
45	Rh	Rhodium	$(\text{Kr})\,4d^8\ 5s$	t	n		$^4F_{9/2}$	7.4589
46	Pd	Palladium	$(\text{Kr})\,4d^{10}$	i	t		$^1S_0$	8.3369
47	Ag	Silver	$(\text{Kr})\,4d^{10}5s$	o	s		$^2S_{1/2}$	7.5762*
48	Cd	Cadmium	$(\text{Kr})\,4d^{10}5s^2$	n			$^1S_0$	8.9938

49	In	Indium	$(\mathrm{Kr})\,4d^{10}\,5s^2\;5p$		$^2P_{1/2}$	5.7864
50	Sn	Tin	$(\mathrm{Kr})\,4d^{10}\,5s^2\;5p^2$		$^3P_0$	7.3439
51	Sb	Antimony	$(\mathrm{Kr})\,4d^{10}\,5s^2\;5p^3$		$^4S_{3/2}$	8.6084
52	Te	Tellurium	$(\mathrm{Kr})\,4d^{10}\,5s^2\;5p^4$		$^3P_2$	9.0096
53	I	Iodine	$(\mathrm{Kr})\,4d^{10}\,5s^2\;5p^5$		$^2P_{3/2}$	10.4513
54	Xe	Xenon	$(\mathrm{Kr})\,4d^{10}\,5s^2\;5p^6$		$^1S_0$	12.1298
55	Cs	Cesium	$(\mathrm{Xe})\qquad 6s$		$^2S_{1/2}$	3.8939
56	Ba	Barium	$(\mathrm{Xe})\qquad 6s^2$		$^1S_0$	5.2117
57	La	Lanthanum	$(\mathrm{Xe})\qquad 5d\;\;6s^2$	L	$^2D_{3/2}$	5.5770
58	Ce	Cerium	$(\mathrm{Xe})\,4f\;\;5d\;\;6s^2$	a	$^1G_4$	5.5387
59	Pr	Praseodymium	$(\mathrm{Xe})\,4f^3\qquad 6s^2$	n	$^4I_{9/2}$	5.464
60	Nd	Neodymium	$(\mathrm{Xe})\,4f^4\qquad 6s^2$	t	$^5I_4$	5.5250
61	Pm	Promethium	$(\mathrm{Xe})\,4f^5\qquad 6s^2$	h	$^6H_{5/2}$	5.58
62	Sm	Samarium	$(\mathrm{Xe})\,4f^6\qquad 6s^2$	a	$^7F_0$	5.6436
63	Eu	Europium	$(\mathrm{Xe})\,4f^7\qquad 6s^2$	n	$^8S_{7/2}$	5.6704
64	Gd	Gadolinium	$(\mathrm{Xe})\,4f^7\;\;5d\;\;6s^2$	i	$^9D_2$	6.1498*
65	Tb	Terbium	$(\mathrm{Xe})\,4f^9\qquad 6s^2$	d	$^6H_{15/2}$	5.8638
66	Dy	Dysprosium	$(\mathrm{Xe})\,4f^{10}\qquad 6s^2$	e	$^5I_8$	5.9389
67	Ho	Holmium	$(\mathrm{Xe})\,4f^{11}\qquad 6s^2$	s	$^4I_{15/2}$	6.0215
68	Er	Erbium	$(\mathrm{Xe})\,4f^{12}\qquad 6s^2$		$^3H_6$	6.1077
69	Tm	Thulium	$(\mathrm{Xe})\,4f^{13}\qquad 6s^2$		$^2F_{7/2}$	6.1843
70	Yb	Ytterbium	$(\mathrm{Xe})\,4f^{14}\qquad 6s^2$		$^1S_0$	6.2542
71	Lu	Lutetium	$(\mathrm{Xe})\,4f^{14}5d\;\;6s^2$		$^2D_{3/2}$	5.4259
72	Hf	Hafnium	$(\mathrm{Xe})\,4f^{14}5d^2\;\,6s^2$	T	$^3F_2$	6.8251
73	Ta	Tantalum	$(\mathrm{Xe})\,4f^{14}5d^3\;\,6s^2$	r e	$^4F_{3/2}$	7.5496
74	W	Tungsten	$(\mathrm{Xe})\,4f^{14}5d^4\;\,6s^2$	a l	$^5D_0$	7.8640
75	Re	Rhenium	$(\mathrm{Xe})\,4f^{14}5d^5\;\,6s^2$	n e	$^6S_{5/2}$	7.8335
76	Os	Osmium	$(\mathrm{Xe})\,4f^{14}5d^6\;\,6s^2$	s m	$^5D_4$	8.4382*
77	Ir	Iridium	$(\mathrm{Xe})\,4f^{14}5d^7\;\,6s^2$	i e	$^4F_{9/2}$	8.9670*
78	Pt	Platinum	$(\mathrm{Xe})\,4f^{14}5d^9\;\;6s$	t n	$^3D_3$	8.9587
79	Au	Gold	$(\mathrm{Xe})\,4f^{14}5d^{10}6s$	i t	$2S_{1/2}$	9.2255
80	Hg	Mercury	$(\mathrm{Xe})\,4f^{14}5d^{10}6s^2$	o s	$^1S_0$	10.4375
81	Tl	Thallium	$(\mathrm{Xe})\,4f^{14}5d^{10}6s^2\;6p$	n	$^2P_{1/2}$	6.1082
82	Pb	Lead	$(\mathrm{Xe})\,4f^{14}5d^{10}6s^2\;6p^2$		$^3P_0$	7.4167
83	Bi	Bismuth	$(\mathrm{Xe})\,4f^{14}5d^{10}6s^2\;6p^3$		$^4S_{3/2}$	7.2855*
84	Po	Polonium	$(\mathrm{Xe})\,4f^{14}5d^{10}6s^2\;6p^4$		$^3P_2$	8.4167
85	At	Astatine	$(\mathrm{Xe})\,4f^{14}5d^{10}6s^2\;6p^5$		$^2P_{3/2}$	
86	Rn	Radon	$(\mathrm{Xe})\,4f^{14}5d^{10}6s^2\;6p^6$		$^1S_0$	10.7485
87	Fr	Francium	$(\mathrm{Rn})\qquad 7s$		$^2S_{1/2}$	4.0727
88	Ra	Radium	$(\mathrm{Rn})\qquad 7s^2$		$^1S_0$	5.2784
89	Ac	Actinium	$(\mathrm{Rn})\qquad 6d\;\;7s^2$		$^2D_{3/2}$	5.17
90	Th	Thorium	$(\mathrm{Rn})\qquad 6d^2\;7s^2$		$^3F_2$	6.3067
91	Pa	Protactinium	$(\mathrm{Rn})\,5f^2\;6d\;\;7s^2$	A	$^4K_{11/2}$	5.89
92	U	Uranium	$(\mathrm{Rn})\,5f^3\;6d\;\;7s^2$	c	$^5L_6$	6.1941
93	Np	Neptunium	$(\mathrm{Rn})\,5f^4\;6d\;\;7s^2$	t	$^6L_{11/2}$	6.2657
94	Pu	Plutonium	$(\mathrm{Rn})\,5f^6\qquad 7s^2$	i	$^7F_0$	6.0262
95	Am	Americium	$(\mathrm{Rn})\,5f^7\qquad 7s^2$	n	$^8S_{7/2}$	5.9738
96	Cm	Curium	$(\mathrm{Rn})\,5f^7\;6d\;\;7s^2$	i	$^9D_2$	5.9915*
97	Bk	Berkelium	$(\mathrm{Rn})\,5f^9\qquad 7s^2$	d	$^6H_{15/2}$	6.1979*
98	Cf	Californium	$(\mathrm{Rn})\,5f^{10}\qquad 7s^2$	e	$^5I_8$	6.2817*
99	Es	Einsteinium	$(\mathrm{Rn})\,5f^{11}\qquad 7s^2$	s	$^4I_{15/2}$	6.42
100	Fm	Fermium	$(\mathrm{Rn})\,5f^{12}\qquad 7s^2$		$^3H_6$	6.50
101	Md	Mendelevium	$(\mathrm{Rn})\,5f^{13}\qquad 7s^2$		$^2F_{7/2}$	6.58
102	No	Nobelium	$(\mathrm{Rn})\,5f^{14}\qquad 7s^2$		$^1S_0$	6.65
103	Lr	Lawrencium	$(\mathrm{Rn})\,5f^{14}\qquad 7s^2\;7p?$		$^2P_{1/2}?$	
104	Rf	Rutherfordium	$(\mathrm{Rn})\,5f^{14}6d^2\;7s^2?$		$^3F_2?$	6.0?

A.5 Isotopic Abundances

IUPAC recommended isotopic compositions in percentage: uncertainties are shown by the last decimals in italic (reprinted from *Int. J. of Mass Spectrom., formerly Int. J. of Mass Spectrom. and Ion Proc.* **123**, De Bièvre, P. and Taylor, P.D.P., Table of the Isotopic Compositions of the Elements, 149-166, Copyright (1993), with permission from Elsevier).

Isot.	Comp.%	Isot.	Comp.%	Isot.	Comp.%	Isot.	Comp.%	Isot.	Comp.%
^{1}H	99.9851	^{54}Fe	5.81	^{96}Ru	5.526	^{136}Ce	0.191	^{180}W	0.134
^{2}H	0.0151	^{56}Fe	91.7230	^{98}Ru	1.886	^{138}Ce	0.251	^{182}W	26.32
		^{57}Fe	2.21	^{99}Ru	12.71	^{140}Ce	88.4810	^{183}W	14.31
^{3}He	0.0001373	^{58}Fe	0.281	^{100}Ru	12.61	^{142}Ce	11.0810	^{184}W	30.6715
^{4}He	99.9998633			^{101}Ru	17.01			^{186}W	28.62
		^{55}Mn	100	^{102}Ru	31.62	^{138}La	0.09022		
^{6}Li	7.52			^{104}Ru	18.72	^{139}La	99.90982	^{184}Os	0.021
^{7}Li	92.52	^{58}Ni	68.0779					^{186}Os	1.5830
		^{60}Ni	26.2238	^{102}Pd	1.021	^{141}Pr	100	^{187}Os	1.63
^{9}Be	100	^{61}Ni	1.1401	^{104}Pd	11.148			^{188}Os	13.37
		^{62}Ni	3.6342	^{105}Pd	22.338	^{142}Nd	27.1312	^{189}Os	16.18
^{10}B	19.92	^{64}Ni	0.9261	^{106}Pd	27.333	^{143}Nd	12.186	^{190}Os	26.412
^{11}B	80.12			^{108}Pd	26.469	^{144}Nd	23.8012	^{192}Os	41.08
		^{59}Co	100	^{110}Pd	11.729	^{145}Nd	8.306		
^{12}C	98.903					^{146}Nd	17.199	^{185}Re	37.402
^{13}C	1.103	^{63}Cu	69.173	^{103}Rh	100	^{148}Nd	5.763	^{187}Re	62.602
		^{65}Cu	30.833			^{150}Nd	5.643		
^{14}N	99.6349			^{106}Cd	1.254			^{190}Pt	0.011
^{15}N	0.3669	^{64}Zn	48.63	^{108}Cd	0.892	^{144}Sm	3.11	^{192}Pt	0.796
		^{66}Zn	27.92	^{110}Cd	12.4912	^{147}Sm	15.02	^{194}Pt	32.96
^{16}O	99.76215	^{67}Zn	4.11	^{111}Cd	12.808	^{148}Sm	11.31	^{195}Pt	33.86
^{17}O	0.0383	^{68}Zn	18.84	^{112}Cd	24.1314	^{149}Sm	13.81	^{196}Pt	25.36
^{18}O	0.20012	^{70}Zn	0.61	^{113}Cd	12.228	^{150}Sm	7.41	^{198}Pt	7.22
				^{114}Cd	28.7328	^{152}Sm	26.72		
^{19}F	100	^{69}Ga	60.1089	^{116}Cd	7.4912	^{154}Sm	22.72	^{191}Ir	37.35
		^{71}Ga	39.8929					^{193}Ir	62.75
^{20}Ne	90.483			^{107}Ag	51.8397	^{151}Eu	47.815		
^{21}Ne	0.271	^{70}Ge	21.234	^{109}Ag	48.1617	^{153}Eu	52.215	^{196}Hg	0.151
^{22}Ne	9.253	^{72}Ge	27.663					^{198}Hg	9.978
		^{73}Ge	7.731	^{112}Sn	0.971	^{152}Gd	0.201	^{199}Hg	16.8710
^{23}Na	100	^{74}Ge	35.942	^{114}Sn	0.651	^{154}Gd	2.183	^{200}Hg	23.1016
		^{76}Ge	7.442	^{115}Sn	0.341	^{155}Gd	14.805	^{201}Hg	13.188
^{24}Mg	78.993			^{116}Sn	14.531	^{156}Gd	20.474	^{202}Hg	29.8620
^{25}Mg	10.001	^{74}Se	0.892	^{117}Sn	7.687	^{157}Gd	15.653	^{204}Hg	6.874
^{26}Mg	11.012	^{76}Se	9.3611	^{118}Sn	24.2311	^{158}Gd	24.8412		
		^{77}Se	7.636	^{119}Sn	8.594	^{160}Gd	21.864	^{197}Au	100
^{27}Al	100	^{78}Se	23.789	^{120}Sn	32.5910				
		^{80}Se	49.6110	^{122}Sn	4.633	^{156}Dy	0.061	^{203}Tl	29.52414
		^{82}Se	8.736	^{124}Sn	5.795	^{158}Dy	0.101	^{205}Tl	70.47614
^{28}Si	92.231					^{160}Dy	2.346		
^{29}Si	4.671	^{75}As	100	^{113}In	4.32	^{161}Dy	18.92	^{204}Pb	1.41
^{30}Si	3.101			^{115}In	95.72	^{162}Dy	25.52	^{206}Pb	24.11
		^{78}Kr	0.352			^{163}Dy	24.92	^{207}Pb	22.11
^{31}P	100	^{80}Kr	2.252	^{120}Te	0.0962	^{164}Dy	28.22	^{208}Pb	52.41

continued on next page

continued from previous page

Isot.	Comp.%	Isot.	Comp.%	Isot.	Comp.%	Isot.	Comp.%	Isot.	Comp.%
^{32}S	95.029	^{82}Kr	11.61	^{122}Te	2.6034				
^{33}S	0.754	^{83}Kr	11.51	^{123}Te	0.9082	^{159}Tb	100	^{209}Bi	100
^{34}S	4.218	^{84}Kr	57.03	^{124}Te	4.8166				
^{36}S	0.021	^{86}Kr	17.32	^{125}Te	7.1396	^{162}Er	0.141	^{232}Th	100
				^{126}Te	18.951	^{164}Er	1.612		
^{35}Cl	75.777	^{79}Br	50.697	^{128}Te	31.691	^{166}Er	33.62	^{234}U	0.00555
^{37}Cl	24.237	^{81}Br	49.317	^{130}Te	33.801	^{167}Er	22.9515	^{235}U	0.720012
						^{168}Er	26.82	^{238}U	99.274560
						^{170}Er	14.92		
^{36}Ar	0.3373	^{84}Sr	0.561	^{121}Sb	57.368				
^{38}Ar	0.0631	^{86}Sr	9.861	^{123}Sb	42.648				
^{40}Ar	99.6003	^{87}Sr	7.001			^{165}Ho	100		
		^{88}Sr	82.581	^{124}Xe	0.101				
^{39}K	93.258144			^{126}Xe	0.091	^{168}Yb	0.131		
^{40}K	0.011711	^{85}Rb	72.165 20	^{128}Xe	1.913	^{170}Yb	3.056		
^{41}K	6.730244	^{87}Rb	27.835 20	^{129}Xe	26.46	^{171}Yb	14.32		
				^{130}Xe	4.11	^{172}Yb	21.93		
^{40}Ca	96.94118	^{89}Y	100	^{131}Xe	21.24	^{173}Yb	16.1221		
^{42}Ca	0.6479			^{132}Xe	26.95	^{174}Yb	31.84		
^{43}Ca	0.1356	^{90}Zr	51.453	^{134}Xe	10.42	^{176}Yb	12.72		
^{44}Ca	2.08612	^{91}Zr	11.224	^{136}Xe	8.91				
^{46}Ca	0.0043	^{92}Zr	17.152			^{169}Tm	100		
^{48}Ca	0.1874	^{94}Zr	17.384	^{127}I	100				
		^{96}Zr	2.802			^{174}Hf	0.1623		
^{45}Sc	100			^{130}Ba	0.1062	^{176}Hf	5.2065		
		^{92}Mo	14.844	^{132}Ba	0.1012	^{177}Hf	18.6064		
^{46}Ti	8.01	^{94}Mo	9.253	^{134}Ba	2.41727	^{178}Hf	27.2974		
^{47}Ti	7.31	^{95}Mo	15.925	^{135}Ba	6.59218	^{179}Hf	13.6296		
^{48}Ti	73.81	^{96}Mo	16.685	^{136}Ba	7.85436	^{180}Hf	35.1007		
^{49}Ti	5.51	^{97}Mo	9.553	^{137}Ba	11.234				
^{50}Ti	5.41	^{98}Mo	24.137	^{138}Ba	71.707	^{175}Lu	97.412		
		^{100}Mo	9.633			^{176}Lu	2.592		
^{50}V	0.2502			^{133}Cs	100				
^{51}V	99.7502	^{93}Nb	100			^{180}Ta	0.0122		
						^{181}Ta	99.9882		
^{50}Cr	4.34513								
^{52}Cr	83.78918								
^{53}Cr	9.50117								
^{54}Cr	2.3657								

A.6 Commonly Used Radioactive Sources

In this table (reproduced with the permission from Groom, D.E. et al. (2000), Table 26.1, page 190, Review of Particle Physics, Particle Data Group, *The Eur. Phys. Jou. C* **15**, 1; © by SIF, Springer-Verlag 2000), half-lives and energy (or end-point energy) emissions are shown for some commonly used radioactive β^+, β^-, α and γ sources.

Revised November 1993 by E. Browne (LBNL).

Nuclide	Half-life	Type of decay	Particle Energy Emission (MeV)	prob.	Photon Energy Emission (MeV)	prob.
$^{22}_{11}$Na	2.603 y	β^+, EC	0.545	90%	0.511	Annih.
					1.275	100%
$^{54}_{25}$Mn	0.855 y	EC			0.835	100%
					Cr K x rays 26%	
$^{55}_{26}$Fe	2.73 y	EC			Mn K x rays:	
					0.00590	24.4%
					0.00649	2.86%
$^{57}_{27}$Co	0.744 y	EC			0.014	9%
					0.122	86%
					0.136	11%
					Fe K x rays 58%	
$^{60}_{27}$Co	5.271 y	β^-	0.316	100%	1.173	100%
					1.333	100%
$^{68}_{32}$Ge	0.742 y	EC			Ga K x rays 44%	
$\to ^{68}_{31}$Ga		β^+, EC	1.899	90%	0.511	Annih.
					1.077	3%
$^{90}_{38}$Sr	28.5 y	β^-	0.546	100%		
$\to ^{90}_{39}$Y		β^-	2.283	100%		
$^{106}_{44}$Ru	1.020 y	β^-	0.039	100%		
$\to ^{106}_{45}$Rh		β^-	3.541	79%	0.512	21%
					0.622	10%
$^{109}_{48}$Cd	1.267 y	EC	0.063 e^-	41%	0.088	3.6%
			0.084 e^-	45%	Ag K x rays 100%	
			0.087 e^-	9%		
$^{113}_{50}$Sn	0.315 y	EC	0.364 e^-	29%	0.392	65%
			0.388 e^-	6%	In K x rays 97%	
$^{137}_{55}$Cs	30.2 y	β^-	0.514 e^-	94%	0.662	85%
			1.176 e^-	6%		
$^{133}_{56}$Ba	10.54 y	EC	0.045 e^-	50%	0.081	34%
			0.075 e^-	6%	0.356	62%
					Cs K x rays 121%	
$^{207}_{83}$Bi	31.8 y	EC	0.481 e^-	2%	0.569	98%
			0.975 e^-	7%	1.063	75%
			1.047 e^-	2%	1.770	7%
					Pb K x rays 78%	
$^{228}_{90}$Th	1.912 y	6α:	5.341 to 8.785		0.239	44%
		$3\beta^-$:	0.334 to 2.246		0.583	31%
					2.614	36%
($\to ^{224}_{88}$Ra	$\to ^{220}_{86}$Rn	$\to ^{216}_{84}$Po	$\to ^{212}_{82}$Pb	$\to ^{212}_{83}$Bi	$\to ^{212}_{84}$Po)	
$^{241}_{95}$Am	432.7 y	α	5.443	13%	0.060	36%
			5.486	85%	Np L x rays 38%	
$^{241}_{95}$Am/Be	432.2 y	6×10^{-5} neutrons (4–8 MeV) and				
		$4 \times 10^{-5}\gamma$'s (4.43 MeV) per Am decay				
$^{244}_{96}$Cm	18.11 y	α	5.763	24%	Pu L x rays $\sim$ 9%	
			5.805	76%		
$^{252}_{98}$Cf	2.645 y	α (97%)	6.076	15%		
			6.118	82%		
		Fission (3.1%)				
		≈ 20 γ's/fission; 80% < 1 MeV				
		≈ 4 neutrons/fission; $\langle E_n \rangle = 2.14$ MeV				

"Emission probability" is the probability per decay of a given emission; because of cascades these may total more than 100%. Only principal emissions are listed. EC means electron capture, and e^- means monoenergetic internal conversion (Auger) electron. The intensity of 0.511 MeV e^+e^- annihilation photons depends upon the number of stopped positrons. Endpoint $\beta^\pm$ energies are listed. In some cases when energies are closely spaced, the γ-ray values are approximate weighted averages. Radiation from short-lived daughter isotopes is included where relevant.

Half-lives, energies, and intensities are from E. Browne and R.B. Firestone, *Table of Radioactive Isotopes* (John Wiley & Sons, New York, 1986), recent *Nuclear Data Sheets*, and *X-ray and Gamma-ray Standards for Detector Calibration*, IAEA-TECDOC-619 (1991).

Neutron data are from *Neutron Sources for Basic Physics and Applications* (Pergamon Press, 1983).

A.7 Free Electron Fermi Gas

A number of physical properties in metals can be understood in terms of the free gas electron model, which was put forward by Drude in 1900 to explain the metallic conductivity. According to this model, some (conduction) electrons are free to move around the whole conductor volume. These electrons behave as molecules of a perfect gas. Forces among conduction electrons and ions are neglected. This classical theory accounts, among others, for the derivation of Ohm's law and, furthermore, for the relation between electrical and thermal conductivity. While it fails to explain other phenomena like, for instance, the heat capacity and the paramagnetic susceptibility of conduction electrons.

In free classical electron model, the electron kinetic energies can have any value and, as the temperature decreases, the average kinetic energy decreases linearly with the temperature becoming zero at $0\,\mathrm{K}$ (K is the absolute temperature in units of kelvin). In fact at thermal equilibrium, the average kinetic electron energy is $\frac{3}{2}kT$ (k is the Boltzmann constant) and is derived assuming that electrons obey to the classical Maxwell–Boltzmann statistics.

In quantum-mechanics not all the energy levels are permitted and the continuous energy distribution is replaced by a discrete set of energies. In addition, electrons have an intrinsic spin angular momentum of $\frac{1}{2}\hbar$ and, at $0\,\mathrm{K}$, they must occupy energy levels consistent with the Pauli exclusion principle, i.e., their mean energy is far from zero. The statistics taking into account the Pauli exclusion principle is the Fermi–Dirac statistics. A gas is called *degenerate* when deviations from classical properties occur. In 1927, Pauli and Sommerfield pointed out that the electron gas within a metal must be treated as a *degenerate gas*, whose properties are essentially different from those of an ordinary gas. It turns out that, at ordinary temperatures, the energy distribution differs very little from the one at $0\,\mathrm{K}$.

Following the treatment in Section 4.2 of [Bleaney, B.I. and Bleaney, B. (1965)], let us consider the *momentum space*, where the coordinates are the components (p_x, p_y, p_z) of the momentum instead of the components of the position (x, y, z). In this space the particle momentum is represented by a point. The magnitude and direction of the momentum are the length and the direction of the radius vector from the origin of the coordinate system and to the momentum point. From *Heisenberg's uncertainty relation* (see page 232) we have that momentum components cannot be determined more precisely* that:

$$\triangle x\,\triangle y\,\triangle z\triangle p_x\,\triangle p_y\,\triangle p_z = h^3.$$

Therefore, if the electron is constrained to be inside a volume V, we have:

$$P = \frac{h^3}{V}, \tag{A.1}$$

---

*In statistical mechanics, for a phase-space of one degree of freedom one quantum state occupies a *volume* $\triangle x\,\triangle p_x = h$ (for the generalization to more degrees of freedom, e.g., see discussion at page 247 of [Morse (1969)]).

where P is the size of the corresponding volume-element in the momentum space. The sphere momentum-volume, for which the momentum is between p and $p + dp$, is $4\pi p^2 dp$; once divided by the volume-element in the momentum-space [Eq. A.1], it becomes:

$$v_{el,p} = 4\pi p^2 \frac{dp}{P} = 4\pi p^2 \frac{V}{h^3}\, dp. \qquad (A.2)$$

Due to the Pauli exclusion principle, only two electrons (with opposite spins) can have assigned the same volume-element position in the momentum-space.

Furthermore, the total kinetic energy is the sum of the individual electron kinetic energies $W_i = p_i^2/(2m)$, where m is the electron rest-mass. This energy will be the smallest when $\sum_i [p_i^2/(2m)]$ has its minimum value.

The minimum energy value is represented by a sphere of radius p_0 and volume $\frac{4}{3}\pi p_0^3$ in the momentum-space. p_0 is called the *Fermi momentum*. The total number of electron, N_{el}, can be derived considering that in each momentum-volume element, P, two electrons of opposite spins can be found. We have:

$$N_{el} = 2\frac{\frac{4}{3}\pi p_0^3}{P} = \frac{8}{3}\pi p_0^3 \frac{V}{h^3}.$$

From which, we get:

$$E_F = \frac{p_0^2}{2m} = \frac{\hbar^2}{2m}\left(3\pi^2 n\right)^{2/3}, \qquad (A.3)$$

where $n = N_{el}/V$ is the electron density in the metal (e.g., for further treatments and discussions, the reader can see Section 4.2 of [Bleaney, B.I. and Bleaney, B. (1965)] and Section 11-11 of [Eisberg and Resnick (1985)]; E_F is the so-called *Fermi energy*, namely the energy of the highest level occupied at $0\,K$. The *Fermi temperature* is given by:

$$T_F = \frac{E_F}{k}.$$

Equation (A.3) shows that E_F depends on the metal, since n varies with the material. From Eqs. (A.2, A.3), the number of electron states per unit volume with electron energies between W and $W + dW$, $g(W)\,dW$, is:

$$g(W)\,dW = 2\frac{v_{el,p}}{V}$$

$$= 8\pi p^2 \frac{dp}{h^3}$$

$$= 16\pi W m^2 \frac{dW}{h^3 p}$$

$$= 16\pi \sqrt{W} m^2 \frac{dW}{h^3 \sqrt{2m}}$$

$$= \frac{1}{2\pi^2}\sqrt{W}\left(\frac{2m}{\hbar^2}\right)^{3/2} dW$$

$$= \frac{3n}{2}\frac{\sqrt{W}}{E_F^{3/2}}\,dW, \qquad (A.4)$$

where $g(W)$ is called the *density of states*. The mean electron energy $\langle W \rangle$ can be calculated as:

$$\langle W \rangle = \frac{1}{n} \int_0^{E_F} W g(W) \, dW$$

$$= \frac{3}{2E_F^{3/2}} \int_0^{E_F} W^{3/2} \, dW$$

$$= \frac{3}{2E_F^{3/2}} \frac{2E_F^{5/2}}{5} = \frac{3}{5} E_F. \tag{A.5}$$

The Fermi energy varies from ≈ 3 to $7\,\mathrm{eV}$ in usual metals and is $\gg kT$ at all ordinary temperatures. Thus, the classical gas model cannot be applied for the case of a free electron gas.

The energy distribution at a finite temperature, T, is derived by the Fermi–Dirac statistics as mentioned above. Thus, the number of electron per unit volume with an energy between W and $W + dW$ is:

$$dn = f(W,T) \, g(W) \, dW = f(W,T) \frac{1}{2\pi^2} \sqrt{W} \left(\frac{2m}{\hbar^2}\right)^{3/2} dW, \tag{A.6}$$

where $f(W,T)$, the probability that an electron occupies a given state, is given by

$$f(W,T) = \frac{1}{\exp\left(\frac{W - E_F}{kT}\right) + 1}. \tag{A.7}$$

The function $f(W,T)$ is called the *Fermi factor*. The total number of electron per unit volume n is:

$$n = \int_0^\infty f(W,T) \, g(W) \, dW$$

$$= \frac{1}{2\pi^2} \left(\frac{2m}{\hbar^2}\right)^{3/2} \int_0^\infty \frac{\sqrt{W}}{\exp\left(\frac{W - E_F}{kT}\right) + 1} \, dW.$$

From Eq. (A.7), we can see that E_F is the energy at which the probability of a state of being occupied by an electron is $f(W,T) = \frac{1}{2}$.

A.8 Gamma-Ray Energy and Intensity Standards

The table lists some γ-ray energies and intensity standards, recommended by the IAEA Co-ordinated Research Programme for calibration of γ-ray measurements [IAEA (1991, 1998)]. The γ-ray energy is in keV, the half-life in days and P is the emission probability. Uncertainties in the data are presented in the format $123(x)$, where x is the uncertainty in the last figure or figures quoted in the prime number, expressed at the 1σ confidence level; thus, $12.132(17)$ means 12.132 ± 0.017, and $0.0425(8)$ means 0.0425 ± 0.0008. X-ray sources can be found in [IAEA (1991, 1998)]. The data are the result of the work of an IAEA Coordinated Research Project 1986 to 1990. Further X- and γ-ray data are available from [ToI (1996, 1998, 1999); Helmer (1999); Helmer and van der Leun (1999, 2000)].

Source	Half-life (days)	E_γ (keV)	P
^{22}Na	950.8 ± 0.9	1274.542(7)	0.99935(15)
^{24}Na	0.62356 ± 0.00017	1368.633(6)	0.999936(15)
		2754.030(14)	0.99855(5)
^{46}Sc	83.79 ± 0.04	889.277(3)	0.999844(16)
		1120.545(4)	0.999874(11)
^{51}Cr	27.706 ± 0.007	320.0842(9)	0.0986(5)
^{54}Mn	312.3 ± 0.4	834.843(6)	0.999758(24)
^{56}Co	77.31 ± 0.19	846.764(6)	0.99933(7)
		1037.844(4)	0.1413(5)
		1175.099(8)	0.02239(11)
		1238.287(6)	0.6607(19)
		1360.206(6)	0.04256(15)
		1771.350(15)	0.1549(5)
		2015.179(11)	0.03029(13)
		2034.759(11)	0.07771(27)
		2598.460(10)	0.1696(6)
		3201.954(14)	0.0313(9)
		3253.417(14)	0.0762(24)
		3272.998(14)	0.0178(6)
		3451.154(13)	0.0093(4)
		3548.27(10)	0.00178(9)
^{57}Co	271.79 ± 0.09	14.4127(4)	0.0916(15)
		122.0614(3)	0.8560(17)
		136.4743(5)	0.1068(8)
^{58}Co	70.86 ± 0.07	810.775(9)	0.9945(1)
^{60}Co	1925.5 ± 0.5	1173.238(4)	0.99857(22)
		1332.502(5)	0.99983(6)

continued on next page

continued from previous page

Source	Half-life (days)	E_γ (keV)	P
^{65}Zn	244.26 ± 0.26	1115.546(4)	0.5060(24)
^{75}Se	119.64 ± 0.24	96.7344(10)	0.0341(4)
		121.1171(14)	0.171(1)
		136.0008(6)	0.588(3)
		264.6580(17)	0.590(2)
		279.5431(22)	0.250(1)
		400.6593(13)	0.115(1)
^{85}Sr	64.849 ± 0.004	514.0076(22)	0.984(4)
^{88}Y	106.630 ± 0.025	898.042(4)	0.940(3)
		1836.063(13)	0.9936(3)
^{94}Nb	$7.3 \pm 0.9 \times 10^6$	702.645(6)	0.9979(5)
		871.119(4)	0.9986(5)
^{95}Nb	34.975 ± 0.007	765.807(6)	0.9981(3)
^{109}Cd	462.6 ± 0.7	88.0341(11)	0.0363(2)
^{111}In	2.8047 ± 0.0005	171.28(3)	0.9078(10)
		245.35(4)	0.9416(6)
^{113}Sn	115.09 ± 0.04	391.702(4)	0.6489(13)
^{125}Sb	1007.7 ± 0.6	176.313(1)	0.0685(7)
		380.452(8)	0.01518(16)
		427.875(6)	0.297(3)
		463.365(5)	0.1048(11)
		600.600(4)	0.1773(18)
		606.718(3)	0.0500(5)
		635.954(5)	0.1121(12)
^{125}I	59.43 ± 0.06	35.4919(5)	0.0658(8)
^{134}Cs	754.28 ± 0.22	475.364(3)	0.0149(2)
		563.240(4)	0.0836(3)
		569.328(3)	0.1539(6)
		604.720(3)	0.9763(6)
		795.859(5)	0.854(3)
		801.948(5)	0.0869(3)
		1038.610(7)	0.00990(5)
		1167.968(5)	0.01792(7)
		1365.185(7)	0.03016(11)
^{137}Cs	$1.102 \pm 0.006 \times 10^4$	661.660(3)	0.851(2)
^{133}Ba	3862 ± 15	80.998(5)	0.3411(28)
		276.398(1)	0.07147(30)
		302.853(1)	0.1830(6)
		356.017(2)	0.6194(14)
		383.851(3)	0.08905(29)

continued on next page

continued from previous page

Source	Half-life (days)	E_γ (keV)	P
^{139}Ce	137.640 ± 0.023	165.857(6)	0.7987(6)
^{152}Eu	4933 ± 11	121.7824(4)	0.2837(13)
		244.6989(10)	0.0753(4)
		344.2811(19)	0.2657(11)
		411.126(3)	0.02238(10)
		443.965(4)	0.03125(14)
		778.903(6)	0.1297(6)
		867.390(6)	0.04214(25)
		964.055(4)	0.1463(6)
		1085.842(4)	0.1013(5)
		1089.767(14)	0.01731(9)
		1112.087(6)	0.1354(6)
		1212.970(13)	0.01412(8)
		1299.152(9)	0.01626(11)
		1408.022(4)	0.2085(9)
^{154}Eu	3136.8 ± 2.9	123.071(1)	0.412(5)
		247.930(1)	0.0695(9)
		591.762(5)	0.0499(6)
		692.425(4)	0.0180(3)
		723.305(5)	0.202(2)
		756.804(5)	0.0458(6)
		873.190(5)	0.1224(15)
		996.262(6)	0.1048(13)
		1004.725(7)	0.182(2)
		1274.436(6)	0.350(4)
		1494.048(9)	0.0071(2)
		1596.495(18)	0.0181(2)
^{198}Au	2.6943 ± 0.0008	411.8044(11)	0.9557(47)
^{203}Hg	46.595 ± 0.013	279.1967(12)	0.8148(8)
^{207}Bi	$1.16 \pm 0.07 \times 10^4$	569.702(2)	0.9774(3)
		1063.662(4)	0.745(2)
		1770.237(9)	0.0687(4)
^{228}Th decay chain	698.2 ± 0.6	84.373(3)	0.0122(2)
		238.632(2)	0.435(4)
		240.987(6)	0.0410(5)
		277.358(10)	0.0230(3)
		300.094(10)	0.0325(3)
		510.77(10)	0.0818(10)
		583.191(2)	0.306(2)
		727.330(9)	0.0669(9)

continued on next page

continued from previous page

Source	Half-life (days)	E_γ (keV)	P
^{228}Th decay chain	698.2 ± 0.6	860.564(5)	0.0450(4)
		1620.735(10)	0.0149(5)
		2614.533(13)	0.3586(6)
^{239}Np	2.35 ± 0.004	106.123(2)	0.267(4)
		228.183(1)	0.1112(15)
		277.599(2)	0.1431(20)
^{241}Am	$1.5785 \pm 0.0024 \times 10^5$	26.345(1)	0.024(1)
		59.537(1)	0.360(4)
^{243}Am	$2.690 \pm 0.008 \times 10^6$	43.53(1)	0.0594(11)
		74.66(1)	0.674(10)

Appendix B
Mathematics and Statistics

B.1 Probability and Statistics for Detection Systems

This Appendix reproduces, with the permission, the Sections 27 and 28, pages 191–201, from Groom, D.E. et al. (2000), Review of Particle Physics, Particle Data Group, *The Eur. Phys. Jou. C* **15**, 1; ⓒ by SIF, Springer-Verlag 2000. Complementary information on elements from the theory of statistics, errors and their propagation can be found in Chapter 10 of [Melissinos and Napolitano (2003)] (see also Chapter 10 of [Melissinos (1966)]).

27. PROBABILITY

Revised May 1996 by D.E. Groom (LBNL) and F. James (CERN). Updated September 1999 by R. Cousins (UCLA).

27.1. General [1–6]

Let x be a possible outcome of an observation. The probability of x is the relative frequency with which that outcome occurs out of a (possibly hypothetical) large set of similar observations. If x can take any value from a *continuous* range, we write $f(x;\theta)\,dx$ as the probability of observing x between x and $x+dx$. The function $f(x;\theta)$ is the *probability density function* (p.d.f.) for the *random variable* x, which may depend upon one or more parameters θ. If x can take on only *discrete* values (*e.g.*, the non-negative integers), then $f(x;\theta)$ is itself a probability, but we shall still call it a p.d.f. The p.d.f. is always normalized to unit area (unit sum, if discrete). Both x and θ may have multiple components and are then often written as column vectors. If θ is unknown and we wish to estimate its value from a given set of data measuring x, we may use statistics (see Sec. 28).

The *cumulative distribution function* $F(a)$ is the probability that $x \leq a$:

$$F(a) = \int_{-\infty}^{a} f(x)\,dx \ . \tag{27.1}$$

Here and below, if x is discrete-valued, the integral is replaced by a sum. The endpoint a is expressly included in the integral or sum. Then $0 \leq F(x) \leq 1$, $F(x)$ is nondecreasing, and $\mathrm{Prob}(a < x \leq b) = F(b) - F(a)$. If x is discrete, $F(x)$ is flat except at allowed values of x, where it has discontinuous jumps equal to $f(x)$.

Any function of random variables is itself a random variable, with (in general) a different p.d.f. The *expectation value* of any function $u(x)$ is

$$E[u(x)] = \int_{-\infty}^{\infty} u(x)\,f(x)\,dx \ , \tag{27.2}$$

assuming the integral is finite. For $u(x)$ and $v(x)$ any two functions of x, $E(u+v) = E(u) + E(v)$. For c and k constants, $E(cu+k) = cE(u) + k$.

The nth moment of a distribution is

$$\alpha_n \equiv E(x^n) = \int_{-\infty}^{\infty} x^n f(x)\,dx \ , \tag{27.3a}$$

and the nth moment about the mean of x, α_1, is

$$m_n \equiv E[(x - \alpha_1)^n] = \int_{-\infty}^{\infty} (x - \alpha_1)^n f(x)\,dx \ . \tag{27.3b}$$

The most commonly used moments are the mean μ and variance σ^2:

$$\mu \equiv \alpha_1 \tag{27.4a}$$
$$\sigma^2 \equiv \mathrm{Var}(x) = m_2 = \alpha_2 - \mu^2 \ . \tag{27.4b}$$

The mean is the location of the "center of mass" of the probability density function, and the variance is a measure of the square of its width. Note that $\mathrm{Var}(cx + k) = c^2 \mathrm{Var}(x)$.

Any odd moment about the mean is a measure of the skewness of the p.d.f. The simplest of these is the dimensionless coefficient of skewness $\gamma_1 \equiv m_3/\sigma^3$.

Besides the mean, another useful indicator of the "middle" of the probability distribution is the *median* x_{med}, defined by $F(x_{\mathrm{med}}) = 1/2$; *i.e.*, half the probability lies above and half lies below x_{med}. For a given *sample* of events, x_{med} is the value such that half the events have larger x and half have smaller x (not counting any that have the same x as the median). If the sample median lies between two observed x values, it is set by convention halfway between them. If the p.d.f. for x has the form $f(x - \mu)$ and μ is both mean and median, then for a large number of events N, the variance of the median approaches $1/[4Nf^2(0)]$, provided $f(0) > 0$.

Let x and y be two random variables with a joint p.d.f. $f(x,y)$. The *marginal* p.d.f. of x (the distribution of x with y unobserved) is

$$f_1(x) = \int_{-\infty}^{\infty} f(x,y)\,dy \ , \tag{27.5}$$

and similarly for the marginal p.d.f. $f_2(y)$. We define $f_3(y|x)$, the *conditional* p.d.f. of y given fixed x, by

$$f_3(y|x)\,f_1(x) = f(x,y) \ . \tag{27.6a}$$

Similarly, $f_4(x|y)$, the conditional p.d.f. of x given fixed y, is

$$f_4(x|y)\,f_2(y) = f(x,y) \ . \tag{27.6b}$$

From these definitions we immediately obtain Bayes' theorem [2]:

$$f_4(x|y) = \frac{f_3(y|x)\,f_1(x)}{f_2(y)} = \frac{f_3(y|x)\,f_1(x)}{\int f_3(y|x)\,f_1(x)\,dx} \ . \tag{27.7}$$

The mean of x is

$$\mu_x = \int_{-\infty}^{\infty}\int_{-\infty}^{\infty} x\,f(x,y)\,dx\,dy = \int_{-\infty}^{\infty} x\,f_1(x)\,dx \ , \tag{27.8}$$

and similarly for y. The *correlation* between x and y is

$$\rho_{xy} = E\left[(x - \mu_x)(y - \mu_y)\right]/\sigma_x\,\sigma_y = \mathrm{Cov}(x,y)/\sigma_x\,\sigma_y \ , \tag{27.9}$$

where σ_x and σ_y are defined in analogy with Eq. (27.4b). It can be shown that $-1 \leq \rho_{xy} \leq 1$. Here "Cov" is the covariance of x and y, a 2-dimensional generalization of the variance.

Two random variables are *independent* if and only if

$$f(x,y) = f_1(x)\,f_2(y) \ . \tag{27.10}$$

If x and y are independent then $\rho_{xy} = 0$; the converse is not necessarily true except for Gaussian-distributed x and y. If x and y are independent, $E[u(x)\,v(y)] = E[u(x)]\,E[v(y)]$, and $\mathrm{Var}(x + y) = \mathrm{Var}(x) + \mathrm{Var}(y)$; otherwise, $\mathrm{Var}(x + y) = \mathrm{Var}(x) + \mathrm{Var}(y) + 2\mathrm{Cov}(x,y)$, and $E(u\,v)$ does not factor.

In a *change of continuous random variables* from $\boldsymbol{x} = (x_1, \ldots, x_n)$, with p.d.f. $f(\boldsymbol{x}) = f(x_1, \ldots, x_n)$, to $\boldsymbol{y} \equiv (y_1, \ldots, y_n)$, a one-to-one function of the x_i's, the p.d.f. $g(\boldsymbol{y}) = g(y_1, \ldots, y_n)$ is found by substitution for $(x_1, \ldots, x_n)$ in f followed by multiplication by the absolute value of the Jacobian of the transformation; that is,

$$g(\boldsymbol{y}) = f\left[w_1(\boldsymbol{y}), \ldots, w_n(\boldsymbol{y})\right] |J| \ . \tag{27.11}$$

The functions w_i express the *inverse* transformation, $x_i = w_i(\boldsymbol{y})$ for $i = 1, \ldots, n$, and $|J|$ is the absolute value of the determinant of the square matrix $J_{ij} = \partial x_i/\partial y_j$. If the transformation from $\boldsymbol{x}$ to $\boldsymbol{y}$ is not one-to-one, the situation is more complex and a unique solution may not exist. For example, if the change is to $m < n$ variables, then a given y may correspond to more than one x, leading to multiple integrals over the contributions [1].

To change variables for discrete random variables simply substitute; no Jacobian is necessary because now f is a probability rather than a probability density.

If f depends upon a parameter set $\boldsymbol{\alpha}$, a change to a different parameter set $\phi_i = \phi_i(\boldsymbol{\alpha})$ is made by simple substitution; no Jacobian is used.

27.2. Characteristic functions

The characteristic function $\phi(u)$ associated with the p.d.f. $f(x)$ is essentially its (inverse) Fourier transform, or the expectation value of $\exp(iux)$:

$$\phi(u) = E(e^{iux}) = \int_{-\infty}^{\infty} e^{iux} f(x)\,dx \ . \tag{27.12}$$

It is often useful, and several of its properties follow [1].

It follows from Eqs. (27.3a) and (27.12) that the nth moment of the distribution $f(x)$ is given by

$$i^{-n}\frac{d^n\phi}{du^n}\bigg|_{u=0} = \int_{-\infty}^{\infty} x^n f(x)\,dx = \alpha_n \ . \tag{27.13}$$

Thus it is often easy to calculate all the moments of a distribution defined by $\phi(u)$, even when $f(x)$ is difficult to obtain.

If $f_1(x)$ and $f_2(y)$ have characteristic functions $\phi_1(u)$ and $\phi_2(u)$, then the characteristic function of the weighted sum $ax + by$ is $\phi_1(au)\phi_2(bu)$. The addition rules for common distributions (*e.g.*, that the sum of two numbers from Gaussian distributions also has a Gaussian distribution) easily follow from this observation.

Let the (partial) characteristic function corresponding to the conditional p.d.f. $f_2(x|z)$ be $\phi_2(u|z)$, and the p.d.f. of z be $f_1(z)$. The characteristic function after integration over the conditional value is

$$\phi(u) = \int \phi_2(u|z)\, f_1(z) dz \ . \tag{27.14}$$

Suppose we can write ϕ_2 in the form

$$\phi_2(u|z) = A(u)e^{ig(u)z} \ . \tag{27.15}$$

Then

$$\phi(u) = A(u)\phi_1(g(u)) \ . \tag{27.16}$$

The semi-invariants κ_n are defined by

$$\phi(u) = \exp\left(\sum_1^\infty \frac{\kappa_n}{n!}(iu)^n\right) = \exp\left(i\kappa_1 u - \tfrac{1}{2}\kappa_2 u^2 + \ldots\right) \ . \tag{27.17}$$

The κ_n's are related to the moments α_n and m_n. The first few relations are

$$\kappa_1 = \alpha_1 \ (= \mu, \text{ the mean})$$
$$\kappa_2 = m_2 = \alpha_2 - \alpha_1^2 \ (= \sigma^2, \text{ the variance})$$
$$\kappa_3 = m_3 = \alpha_3 - 3\alpha_1\alpha_2 + 2\alpha_1^2 \ . \tag{27.18}$$

27.3. Some probability distributions

Table 27.1 gives a number of common probability density functions and corresponding characteristic functions, means, and variances. Further information may be found in Refs. 1–7; Ref. 7 has particularly detailed tables. Monte Carlo techniques for generating each of them may be found in our Sec. 29.4. We comment below on all except the trivial uniform distribution.

27.3.1. *Binomial distribution*: A random process with exactly two possible outcomes is called a *Bernoulli* process. If the probability of obtaining a certain outcome (a "success") in each trial is p, then the probability of obtaining exactly r successes ($r = 0, 1, 2, \ldots, n$) in n trials, without regard to the order of the successes and failures, is given by the binomial distribution $f(r; n, p)$ in Table 27.1. If r successes are observed in n_r trials with probability p of a success, and if s successes are observed in n_s similar trials, then $t = r + s$ is also binomial with $n_t = n_r + n_s$.

27.3.2. *Poisson distribution*: The Poisson distribution $f(r; \mu)$ gives the probability of finding exactly r events in a given interval of x (*e.g.*, space and time) when the events occur independently of one another and of x at an average rate of μ per the given interval. The variance σ^2 equals μ. It is the limiting case $p \to 0$, $n \to \infty$, $np = \mu$ of the binomial distribution. The Poisson distribution approaches the Gaussian distribution for large μ.

Two or more Poisson processes (*e.g.*, *signal* + *background*, with parameters μ_s and μ_b) that independently contribute amounts n_s and n_b to a given measurement will produce an observed number $n = n_s + n_b$, which is distributed according to a new Poisson distribution with parameter $\mu = \mu_s + \mu_b$.

27.3.3. *Normal or Gaussian distribution*: The normal (or Gaussian) probability density function $f(x; \mu, \sigma^2)$ given in Table 27.1 has mean $\bar{x} = \mu$ and variance σ^2. Comparison of the characteristic function $\phi(u)$ given in Table 27.1 with Eq. (27.17) shows that all semi-invariants κ_n beyond κ_2 vanish; this is a unique property of the Gaussian distribution. Some properties of the distribution are:

rms deviation = σ

probability x in the range $\mu \pm \sigma = 0.6827$

probability x in the range $\mu \pm 0.6745\sigma = 0.5$

expection value of $|x - \mu|$, $E(|x - \mu|) = (2/\pi)^{1/2}\sigma = 0.7979\sigma$

half-width at half maximum = $(2\ln 2)^{1/2}\sigma = 1.177\sigma$

The cumulative distribution, Eq. (27.1), for a Gaussian with $\mu = 0$ and $\sigma^2 = 1$ is related to the error function erf(y) by

$$F(x; 0, 1) = \tfrac{1}{2}\left[1 + \text{erf}(x/\sqrt{2})\right] \ . \tag{27.19}$$

The error function is tabulated in Ref. 7 and is available in computer math libraries and personel computer spreadsheets. For a mean μ and variance σ^2, replace x by $(x - \mu)/\sigma$. The probability of x in a given range can be calculated with Eq. (28.36).

For x and y independent and normally distributed, $z = ax + by$ obeys $f(z; a\mu_x + b\mu_y, a^2\sigma_x^2 + b^2\sigma_y^2)$; that is, the weighted means and variances add.

The Gaussian gets its importance in large part from the *central limit theorem*: If a continuous random variable x is distributed according to *any* p.d.f. with finite mean and variance, then the sample mean, $\bar{x}_n$, of n observations of x will have a p.d.f. that approaches a Gaussian as n increases. Therefore the end result $\sum^n x_i \equiv n\bar{x}_n$ of a large number of small fluctuations x_i will be distributed as a Gaussian, even if the x_i themselves are not.

(Note that the *product* of a large number of random variables is not Gaussian, but its logarithm is. The p.d.f. of the product is *lognormal*. See Ref. 6 for details.)

For a set of n Gaussian random variables x with means μ and corresponding Fourier variables u, the characteristic function for a one-dimensional Gaussian is generalized to

$$\phi(x; \mu, S) = \exp\left[i\mu \cdot u - \tfrac{1}{2}u^T S u\right] \ . \tag{27.20}$$

From Eq. (27.13), the covariance about the mean is

$$E\left[(x_j - \mu_j)(x_k - \mu_k)\right] = S_{jk} \ . \tag{27.21}$$

If the x are independent, then $S_{jk} = \delta_{jk}\sigma_j^2$, and Eq. (27.20) is the product of the c.f.'s of n Gaussians.

The covariance matrix S can be related to the correlation matrix defined by Eq. (27.9) (a sort of normalized covariance matrix). With the definition $\sigma_k^2 \equiv S_{kk}$, we have $\rho_{jk} = S_{jk}/\sigma_j\sigma_k$.

The characteristic function may be inverted to find the corresponding p.d.f.

$$f(x; \mu, S) = \frac{1}{(2\pi)^{n/2}\sqrt{|S|}} \ \exp\left[-\tfrac{1}{2}(x - \mu)^T S^{-1}(x - \mu)\right] \tag{27.22}$$

where the determinant $|S|$ must be greater than 0. For diagonal S (independent variables), $f(x; \mu, S)$ is the product of the p.d.f.'s of n Gaussian distributions.

Table 27.1. Some common probability density functions, with corresponding characteristic functions and means and variances. In the Table, $\Gamma(k)$ is the gamma function, equal to $(k-1)!$ when k is an integer.

Distribution	Probability density function f (variable; parameters)	Characteristic function $\phi(u)$	Mean	Variance σ^2
Uniform	$f(x;a,b) = \begin{cases} 1/(b-a) & a \leq x \leq b \\ 0 & \text{otherwise} \end{cases}$	$\dfrac{e^{ibu} - e^{iau}}{(b-a)iu}$	$\bar{x} = \dfrac{a+b}{2}$	$\dfrac{(b-a)^2}{12}$
Binomial	$f(r;n,p) = \dfrac{n!}{r!(n-r)!}\, p^r q^{n-r}$ $r = 0,1,2,\ldots,n\; ;\quad 0 \leq p \leq 1\; ;\quad q = 1-p$	$(q + pe^{iu})^n$	$\bar{r} = np$	npq
Poisson	$f(r;\mu) = \dfrac{\mu^r e^{-\mu}}{r!}\; ;\quad r = 0,1,2,\ldots\; ;\quad \mu > 0$	$\exp[\mu(e^{iu} - 1)]$	$\bar{r} = \mu$	μ
Normal (Gaussian)	$f(x;\mu,\sigma^2) = \dfrac{1}{\sigma\sqrt{2\pi}}\, \exp(-(x-\mu)^2/2\sigma^2)$ $-\infty < x < \infty\; ;\quad -\infty < \mu < \infty\; ;\quad \sigma > 0$	$\exp(i\mu u - \tfrac{1}{2}\sigma^2 u^2)$	$\bar{x} = \mu$	σ^2
Multivariate Gaussian	$f(\boldsymbol{x};\boldsymbol{\mu}, S) = \dfrac{1}{(2\pi)^{n/2}\sqrt{\lvert S \rvert}}$ $\times \exp\left[-\tfrac{1}{2}(\boldsymbol{x}-\boldsymbol{\mu})^T S^{-1}(\boldsymbol{x}-\boldsymbol{\mu})\right]$ $-\infty < x_j < \infty;\quad -\infty < \mu_j < \infty;\quad \det S > 0$	$\exp\left[i\boldsymbol{\mu}\cdot\boldsymbol{u} - \tfrac{1}{2}\boldsymbol{u}^T S \boldsymbol{u}\right]$	$\boldsymbol{\mu}$	S_{jk}
χ^2	$f(z;n) = \dfrac{z^{n/2-1}e^{-z/2}}{2^{n/2}\Gamma(n/2)}\; ;\quad z \geq 0$	$(1 - 2iu)^{-n/2}$	$\bar{z} = n$	$2n$
Student's t	$f(t;n) = \dfrac{1}{\sqrt{n\pi}}\dfrac{\Gamma[(n+1)/2]}{\Gamma(n/2)}\left(1 + \dfrac{t^2}{n}\right)^{-(n+1)/2}$ $-\infty < t < \infty\; ;\quad n \text{ not required to be integer}$	—	$\bar{t} = 0$ for $n \geq 2$	$n/(n-2)$ for $n \geq 3$
Gamma	$f(x;\lambda,k) = \dfrac{x^{k-1}\lambda^k e^{-\lambda x}}{\Gamma(k)}\; ;\quad 0 < x < \infty\; ;$ $k \text{ not required to be integer}$	$(1 - iu/\lambda)^{-k}$	$\bar{x} = k/\lambda$	k/λ^2

For $n = 2$, $f(\boldsymbol{x};\boldsymbol{\mu}, S)$ is

$$f(x_1, x_2;\; \mu_1, \mu_2, \sigma_1, \sigma_2, \rho) = \frac{1}{2\pi\sigma_1\sigma_2\sqrt{1-\rho^2}}$$

$$\times \exp\left\{ \frac{-1}{2(1-\rho^2)}\left[\frac{(x_1-\mu_1)^2}{\sigma_1^2} - \frac{2\rho(x_1-\mu_1)(x_2-\mu_2)}{\sigma_1\sigma_2} \right.\right.$$

$$\left.\left. + \frac{(x_2-\mu_2)^2}{\sigma_2^2} \right] \right\} . \tag{27.23}$$

The marginal distribution of any x_i is a Gaussian with mean μ_i and variance S_{ii}. S is $n \times n$, symmetric, and positive definite. Therefore for any vector $\boldsymbol{X}$, the quadratic form $\boldsymbol{X}^T S^{-1} \boldsymbol{X} = C$, where C is any positive number, traces an n-dimensional ellipsoid as $\boldsymbol{X}$ varies. If $X_i = (x_i - \mu_i)/\sigma_i$, then C is a random variable obeying the $\chi^2(n)$ distribution, discussed in the following section. The probability that $\boldsymbol{X}$ corresponding to a set of Gaussian random variables x_i lies *outside* the ellipsoid characterized by a given value of C ($= \chi^2$) is given by Eq. (27.24) and may be read from Fig. 27.1. For example, the "s-standard-deviation ellipsoid" occurs at $C = s^2$. For the two-variable case ($n = 2$), the point $\boldsymbol{X}$ lies outside the one-standard-deviation ellipsoid with 61% probability. (This assumes that μ_i and σ_i are correct.) For $X_i = x_i/\sigma_i$, the ellipsoids of constant χ^2 have the same size and orientation but are centered at $\boldsymbol{\mu}$. The use of these ellipsoids as indicators of probable error is described in Sec. 28.6.2.

27.3.4. χ^2 distribution: If $x_1,\ldots,x_n$ are independent Gaussian distributed random variables, the sum $z = \sum^n (x_i - \mu_i)^2/\sigma_i^2$ is distributed as a χ^2 with n *degrees of freedom*, $\chi^2(n)$. Under a linear transformation to n dependent Gaussian variables x_i', the χ^2 at each transformed point retains its value; then $z = \boldsymbol{X}'^T V^{-1} \boldsymbol{X}'$ as in the previous section. For a set of z_i, each of which is $\chi^2(n_i)$, $\sum z_i$ is a new random variable which is $\chi^2\left(\sum n_i\right)$.

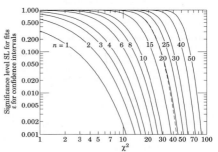

Figure 27.1: The significance level versus χ^2 for n degrees of freedom, as defined in Eq. (27.24). The curve for a given n gives the probability that a value at least as large as χ^2 will be obtained in an experiment; *e.g.*, for $n = 10$, a value $\chi^2 \gtrsim 18$ will occur in 5% of a large number of experiments. For a fit, the SL is a measure of goodness-of-fit, in that a good fit to a correct model is expected to yield a low χ^2 (see Sec. 28.5.0). For a confidence interval, ε measures the probability that the interval *does not* cover the true value of the quantity being estimated (see Sec. 28.6). The dashed curve for $n = 20$ is calculated using the approximation of Eq. (27.25).

Fig. 27.1 shows the signficance level (SL) obtained by integrating the tail of $f(z; n)$:

$$\mathrm{SL}(\chi^2) = \int_{\chi^2}^{\infty} f(z; n) \, dz \; . \tag{27.24}$$

This is shown for a special case in Fig. 27.2, and is equal to 1.0 minus the cumulative distribution function $F(z = \chi^2; n)$. It is useful in evaluating the consistency of data with a model (see Sec. 28): The SL is the probability that a random repeat of the given experiment would observe a greater χ^2, assuming the model is correct. It is also useful for confidence intervals for statistical estimators (see Sec. 28.6), in which case one is interested in the unshaded area of Fig. 27.2.

Since the mean of the χ^2 distribution is equal to n, one expects in a "reasonable" experiment to obtain $\chi^2 \approx n$. Hence the "reduced χ^{2}" $\equiv \chi^2/n$ is sometimes reported. Since the p.d.f. of χ^2/n depends on n, one must report n as well in order to make a meaningful statement. Figure 27.3 shows χ^2/n for useful SL's as a function of n.

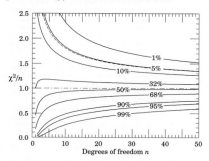

Figure 27.3: Significance levels as a function of the "reduced χ^{2}" $\equiv \chi^2/n$ and the number of degrees of freedom n. Curves are labeled by the probability that a measurement will give a value of χ^2/n greater than that given on the y axis; e.g., for $n = 10$, a value $\chi^2/n \gtrsim 1.8$ can be expected 5% of the time.

For large n, the SL is approximately given by [1,8]

$$\mathrm{SL}(\chi^2) \approx \frac{1}{\sqrt{2\pi}} \int_y^{\infty} e^{-x^2/2} \, dx \; , \tag{27.25}$$

where $y = \sqrt{2\chi^2} - \sqrt{2n-1}$. This approximation was used to draw the dashed curves in Fig. 27.1 (for $n = 20$) and Fig. 27.3 (for SL = 5%). Since all the functions and their inverses are now readily

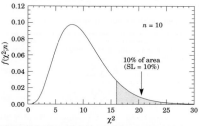

Figure 27.2: Illustration of the significance level integral given in Eq. (27.24). This particlar example is for $n = 10$, where the area above 15.99 is 0.1.

available in standard mathematical libraries (such as IMSL, used to generate these figures, and personal computer spreadsheets, such as Microsoft ℝ Excel [9]), the approximation (and even figures and tables) are seldom needed.

27.3.5. *Student's t distribution*: Suppose that x and $x_1, \ldots, x_n$ are independent and Gaussian distributed with mean 0 and variance 1. We then define

$$z = \sum_1^n x_i^2 \; , \quad \text{and} \quad t = \frac{x}{\sqrt{z/n}} \; . \tag{27.26}$$

The variable z thus belongs to a $\chi^2(n)$ distribution. Then t is distributed according to a Student's t distribution with n degrees of freedom, $f(t; n)$, given in Table 27.1.

The Student's t distribution resembles a Gaussian distribution with wide tails. As $n \to \infty$, the distribution approaches a Gaussian. If $n = 1$, the distribution is a *Cauchy* or *Breit-Wigner* distribution. The mean is finite only for $n > 1$ and the variance is finite only for $n > 2$, so for $n = 1$ or $n = 2$, the central limit theorem is not applicable to t.

As an example, consider the *sample mean* $\bar{x} = \sum x_i/n$ and the *sample variance* $s^2 = \sum (x_i - \bar{x})^2/(n-1)$ for normally distributed random variables x_i with unknown mean μ and variance σ^2. The sample mean has a Gaussian distribution with a variance σ^2/n, so the variable $(\bar{x} - \mu)/\sqrt{\sigma^2/n}$ is normal with mean 0 and variance 1. Similarly, $(n-1)\,s^2/\sigma^2$ is independent of this and is χ^2 distributed with $n-1$ degrees of freedom. The ratio

$$t = \frac{(\bar{x} - \mu)/\sqrt{\sigma^2/n}}{\sqrt{(n-1)\,s^2/\sigma^2\,(n-1)}} = \frac{\bar{x} - \mu}{\sqrt{s^2/n}} \tag{27.27}$$

is distributed as $f(t; n-1)$. The unknown true variance σ^2 cancels, and t can be used to test the probability that the true mean is some particular value μ.

In Table 27.1, n in $f(t; n)$ is not required to be an integer. A Student's t distribution with nonintegral $n > 0$ is useful in certain applications.

27.3.6. *Gamma distribution*: For a process that generates events as a function of x (*e.g.*, space or time) according to a Poisson distribution, the distance in x from an arbitrary starting point (which may be some particular event) to the kth event belongs to a *gamma* distribution, $f(x; \lambda, k)$. The Poisson parameter μ is λ per unit x. The special case $k = 1$ (*i.e.*, $f(x; \lambda, 1) = \lambda e^{-\lambda x}$) is called the *exponential* distribution. A sum of k' exponential random variables x_i is distributed as $f(\sum x_i; \lambda, k')$.

The parameter k is not required to be an integer. For $\lambda = 1/2$ and $k = n/2$, the gamma distribution reduces to the $\chi^2(n)$ distribution.

References:

1. H. Cramér, *Mathematical Methods of Statistics,* Princeton Univ. Press, New Jersey (1958).

2. A. Stuart and A.K. Ord, *Kendall's Advanced Theory of Statistics*, Vol. 1 *Distribution Theory* 5th Ed., (Oxford Univ. Press, New York, 1987), and earlier editions by Kendall and Stuart.

3. W.T. Eadie, D. Drijard, F.E. James, M. Roos, and B. Sadoulet, *Statistical Methods in Experimental Physics* (North Holland, Amsterdam and London, 1971).

4. L. Lyons, *Statistics for Nuclear and Particle Physicists* (Cambridge University Press, New York, 1986).

5. B.R. Roe, *Probability and Statistics in Experimental Physics,* (Springer-Verlag, New York, 1992).

6. G. Cowan, *Statistical Data Analysis* (Oxford University Press, Oxford, 1998).

7. M. Abramowitz and I. Stegun, eds., *Handbook of Mathematical Functions* (Dover, New York, 1972).

8. R.A. Fisher, *Statistical Methods for Research Workers,* 8th edition, Edinburgh and London (1941).

9. Microsoft ℝ is a registered trademark of Microsoft corporation.

28. STATISTICS

Revised April 1998 by F. James (CERN). Updated February 2000 by R. Cousins (UCLA).

A probability density function $f(x; \alpha)$ (p.d.f.) with known parameters α enables us to predict the frequency with which random data x will take on a particular value (if discrete) or lie in a given range (if continuous). Here we are concerned with the inverse problem, that of making inferences about α from a set of actual observations. Such inferences are part of a larger subject variously known as statistics, statistical inference, or inverse probability.

There are two different approaches to statistical inference, which we may call Frequentist and Bayesian. In the former, the frequency definition of probability (Sec. 27.1) is used, and it is usually meaningless to define a p.d.f. in α (for example, a parameter which is a constant of nature has a value which is fixed). In Frequentist statistics, one can compute confidence intervals as a function of the observed data, and they will contain ("cover") the unknown true value of α a specified fraction of the time in the long run, as defined in Sec. 28.6.

In Bayesian statistics, the concept of probability is not based on limiting frequencies, but is more general and includes *degree of belief*. With this definition, one may define p.d.f.'s in α, and then inverse probability simply obeys the general rules of probability. Bayesian methods allow for a natural way to input additional information such as physical boundaries and subjective information; in fact they *require* as input the *prior* p.d.f. for any parameter to be estimated. Using Bayes' Theorem (Eq. (27.7)), the prior degree of belief is updated by incoming data.

For many inference problems, the Frequentist and Bayesian approaches give the same numerical answers, even though they are based on fundamentally different assumptions. However, for exact results for small samples and for measurements near a physical boundary, the different approaches may yield very different confidence limits, so we are forced to make a choice. There is an enormous amount of literature devoted to the question of Bayesian vs non-Bayesian methods, much of it written by people who are fervent advocates of one or the other methodology, which often leads to exaggerated conclusions. For a reasonably balanced discussion, we recommend the following articles: by a statistician [1], and by a physicist [2]. A more advanced comparison is offered in Ref. 3.

In high energy physics, where experiments are repeatable (at least in principle) the frequentist definition of probability is normally used. However, Bayesian equations are often used to treat uncertainties on luminosity, background, *etc.* If the result has poor properties from a Frequentist point of view, one should note that the result is not a classical confidence interval.

Frequentist methods cannot provide the probability that a theory is true, or that a parameter has a particular value. (Such probabilities require input of prior belief.) Rather, Frequentist methods calculate probabilities that various data sets are obtained given specified theories or parameters; these frequencies are often calculated by Monte Carlo methods. As described below, confidence intervals are constructed from such frequencies, and therefore do not represent degree of belief.

The Bayesian methodology is particularly well-adapted to *decision-making*, which requires subjective input not only for prior belief, but also for risk tolerance, etc. Even primarily Frequentist texts such as Ref. 4 outline Bayesian decision theory. However, the usefulness of Bayesian methods as a means for the communication of experimental measurements is controversial.

Recently, the first Workshop on Confidence Limits [5] was held at CERN, where proponents of various statistical methods presented and discussed the issues. One sees that there was not a consensus on the best way to report confidence limits. We recommend the web site and eventual proceedings as a starting point for discussion of these issues. The methods described below use the Frequentist definition of probability, except where noted.

28.1. Parameter estimation [3, 4, 6–9]

Here we review *parametric* statistics in which one desires estimates of the parameters α from a set of actual observations.

A *statistic* is any function of the data, plus known constants, which does not depend upon any of the unknown parameters. A statistic is a random variable if the data have random errors. An *estimator* is any statistic whose value (the *estimate* $\widehat{\alpha}$) is intended as a meaningful guess for the value of the parameter α, or the vector $\boldsymbol{\alpha}$ if there is more than one parameter.

Since we are free to choose any function of the data as an estimator of the parameter α, we will try to choose that estimator which has the best properties. The most important properties are (a) *consistency*, (b) *bias*, (c) *efficiency*, and (d) *robustness*.

(a) An estimator is said to be *consistent* if the estimate $\widehat{\alpha}$ converges to the true value α as the amount of data increases. This property is so important that it is possessed by all commonly used estimators.

(b) The *bias*, $b = E(\widehat{\alpha}) - \alpha$, is the difference between the true value and the expectation of the estimates, where the expectation value is taken over a hypothetical set of similar experiments in which $\widehat{\alpha}$ is constructed the same way. When $b = 0$ the estimator is said to be unbiased. The bias depends on the chosen metric, i.e., if $\widehat{\alpha}$ is an unbiased estimator of α, then $(\widehat{\alpha})^2$ is generally not an unbiased estimator of α^2. The bias may be due to statistical properties of the estimator or to *systematic* errors in the experiment. If we can estimate the b we can subtract it from $\widehat{\alpha}$ to obtain a new $\widehat{\alpha}' \equiv \widehat{\alpha} - b$. However, b may depend upon α or other unknowns, in which case we usually try to choose an estimator which minimizes its average size.

(c) *Efficiency* is the inverse of the ratio between the *variance of the estimates* $\mathrm{Var}(\widehat{\alpha})$ and the minimum possible value of the variance. Under rather general conditions, the minimum variance is given by the Rao-Cramér-Frechet bound:

$$\mathrm{Var}_{\min} = [1 + \partial b/\partial \alpha]^2 / I(\alpha) \; ; \tag{28.1}$$

$$I(\alpha) = E\left\{ \left[\frac{\partial}{\partial \alpha} \sum_i \ln f(x_i; \alpha) \right]^2 \right\} \; .$$

(Compare with Eq. (28.6) below.) The sum is over all data and b is the bias, if any; the x_i are assumed independent and distributed as $f(x_i; \alpha)$, and the allowed range of x must not depend on α. *Mean-squared error*, mse $= E[(\widehat{\alpha} - \alpha)^2] = V(\widehat{\alpha}) + b^2$ is a convenient quantity which combines in the appropriate way the errors due to bias and efficiency.

(d) *Robustness*; is the property of being insensitive to departures from assumptions in the p.d.f. due to such factors as noise.

For some common estimators the above properties are known exactly. More generally, it is always possible to evaluate them by Monte Carlo simulation. Note that they will often depend on the unknown α.

28.2. Data with a common mean

Suppose we have a set of N independent measurements y_i assumed to be unbiased measurements of the same unknown quantity μ with a common, but unknown, variance σ^2 resulting from measurement error. Then

$$\widehat{\mu} = \frac{1}{N} \sum_{i=1}^{N} y_i \tag{28.2}$$

$$\widehat{\sigma}^2 = \frac{1}{N-1} \sum_{i=1}^{N} (y_i - \widehat{\mu})^2 \tag{28.3}$$

are unbiased estimators of μ and σ^2. The variance of $\widehat{\mu}$ is σ^2/N. If the common p.d.f. of the y_i is Gaussian, these estimates are uncorrelated. Then, for large N, the standard deviation of $\widehat{\sigma}$ (the "error of the error") is $\sigma/\sqrt{2N}$. Again if the y_i are Gaussian, $\widehat{\mu}$ is an efficient estimator for μ. Otherwise the mean is in general not the most efficient estimator. For example, if the y follow a double-exponential distribution [$\sim \exp(-\sqrt{2}|y - \mu|/\sigma)$], the most efficient estimator of the mean is the sample median (the value for which half the y_i lie above and half below). This is discussed in more detail in Ref. 4, Sec. 8.7.

If σ^2 is known, it does not improve the estimate $\hat{\mu}$, as can be seen from Eq. (28.2); however, if μ is known, substitute it for $\hat{\mu}$ in Eq. (28.3) and replace $N - 1$ by N, to obtain a somewhat better estimator of σ^2.

If the y_i have different, known, variances σ_i^2, then the weighted average

$$\hat{\mu} = \frac{1}{w} \sum_{i=1}^{N} w_i \, y_i \; , \qquad (28.4)$$

is an unbiased estimator for μ with smaller variance than an unweighted average; here $w_i = 1/\sigma_i^2$ and $w = \sum w_i$. The standard deviation of $\hat{\mu}$ is $1/\sqrt{w}$.

28.3. The method of maximum likelihood

28.3.1. *Parameter estimation by maximum likelihood*:

"From a theoretical point of view, the most important general method of estimation so far known is the *method of maximum likelihood*" [6]. We suppose that a set of independently measured quantities x_i came from a p.d.f. $f(x; \boldsymbol{\alpha})$, where $\boldsymbol{\alpha}$ is an unknown set of parameters. The method of maximum likelihood consists of finding the set of values, $\hat{\boldsymbol{\alpha}}$, which maximizes the joint probability density for all the data, given by

$$\mathscr{L}(\boldsymbol{\alpha}) = \prod_i f(x_i; \boldsymbol{\alpha}) \; , \qquad (28.5)$$

where $\mathscr{L}$ is called the likelihood. It is usually easier to work with $\ln \mathscr{L}$, and since both are maximized for the same set of $\boldsymbol{\alpha}$, it is sufficient to solve the *likelihood equation*

$$\frac{\partial \ln \mathscr{L}}{\partial \alpha_n} = 0 \; . \qquad (28.6)$$

When the solution to Eq. (28.6) is a maximum, it is called the *maximum likelihood estimate* of $\boldsymbol{\alpha}$. The importance of the approach is shown by the following proposition, proved in Ref. 3:

If an efficient estimate $\hat{\boldsymbol{\alpha}}$ of $\boldsymbol{\alpha}$ exists, the likelihood equation will have a unique solution equal to $\hat{\boldsymbol{\alpha}}$.

In evaluating $\mathscr{L}$, it is important that any normalization factors in the f's which involve $\boldsymbol{\alpha}$ be included. However, we will only be interested in the maximum of $\mathscr{L}$ and in ratios of $\mathscr{L}$ at different $\boldsymbol{\alpha}$'s; hence any multiplicative factors which do not involve the parameters we want to estimate may be dropped; this includes factors which depend on the data but not on $\boldsymbol{\alpha}$. The results of two or more independent experiments may be combined by forming the product of the $\mathscr{L}$'s, or the sum of the $\ln \mathscr{L}$'s.

Most commonly the solution to Eq. (28.6) will be found using a general numerical minimization program such as the CERN program MINUIT [10], which contains considerable code to take account of the many special cases and problems which can arise.

Under a one-to-one change of parameters from $\boldsymbol{\alpha}$ to $\boldsymbol{\beta} = \boldsymbol{\beta}(\boldsymbol{\alpha})$, the maximum likelihood estimate $\hat{\boldsymbol{\alpha}}$ transforms to $\boldsymbol{\beta}(\hat{\boldsymbol{\alpha}})$. That is, the maximum likelihood solution is invariant under change of parameter. However, many properties of $\hat{\boldsymbol{\alpha}}$, in particular the bias, are not invariant under change of parameter.

28.3.2. *Uses of $\mathscr{L}$: $\mathscr{L}(\alpha)$ is not a p.d.f. for α*:

Recall the definition of a probability *density* function: a function $p(\alpha)$ is a p.d.f. for α if $p(\alpha)d\alpha$ is the *probability* for α to be within α and $\alpha + d\alpha$. The likelihood function $\mathscr{L}(\alpha)$ is *not* a p.d.f. for α, so in general it is nonsensical to integrate the likelihood function with respect to its parameter(s).

Consider, for example, the Poisson probability for obtaining n when sampling from a distribution with mean α: $f(n; \alpha) = \alpha^n \exp(-\alpha)/n!$. If one obtains $n = 3$ in a particular experiment, then

$\mathscr{L}(\alpha) = \alpha^3 \exp(-\alpha)/6$. Nothing in the construction of $\mathscr{L}$ makes it a probability *density*, *i.e.*, a function which one can multiply by $d\alpha$ in order to obtain a probability.

In Bayesian theory, one applies Bayes' Theorem to construct the posterior p.d.f. for α by multiplying the prior p.d.f. for α by $\mathscr{L}$:

$$p_{\text{posterior}}(\alpha) \propto \mathscr{L}(\alpha) \times p_{\text{prior}}(\alpha).$$

If the prior p.d.f. is uniform, integrating the posterior p.d.f. may give the appearance of integrating $\mathscr{L}$. But note that the prior p.d.f. crucially provides the *density* which makes it sensible to multiply by $d\alpha$ to obtain a probability. In non-Bayesian applications, such as those considered in the following subsections, only likelihood *ratios* are used (or equivalently, differences in $\ln \mathscr{L}$).

Because $\mathscr{L}$ is so useful, we strongly encourage publishing it (or enough information to allow the reader to reconstruct it), when practical.

28.3.3. *Confidence intervals from the likelihood function*:

The covariance matrix V may be estimated from

$$V_{nm} = \left(E\left[-\frac{\partial^2 \ln \mathscr{L}}{\partial \alpha_n \, \partial \alpha_m} \Big|_{\hat{\alpha}} \right] \right)^{-1} \; . \qquad (28.7)$$

(Here and below, the superscript -1 indicates matrix inversion, followed by application of the subscripts.)

In the large sample case (or a linear model with Gaussian errors), $\mathscr{L}$ is Gaussian, $\ln \mathscr{L}$ is a (multidimensional) parabola, and the second derivative in Eq. (28.7) is constant, so the "expectation" operation has no effect. This leads to the usual approximation of calculating the error matrix of the parameters by inverting the second derivative matrix of $\ln \mathscr{L}$. In this asymptotic case, it can be seen that a numerically equivalent way of determining s-standard-deviation errors is from the contour given by the $\boldsymbol{\alpha}'$ such that

$$\ln \mathscr{L}(\boldsymbol{\alpha}') = \ln \mathscr{L}_{\max} - s^2/2 \; , \qquad (28.8)$$

where $\ln \mathscr{L}_{\max}$ is the value of $\ln \mathscr{L}$ at the solution point (compare with Eq. (28.32), below). The extreme limits of this contour parallel to the α_n axis give an approximate s-standard-deviation confidence interval in α_n. These intervals may not be symmetric and in pathological cases they may even consist of two or more disjoint intervals.

Although asymptotically Eq. (28.7) is equivalent to Eq. (28.8) with $s = 1$, the latter is a better approximation when the model deviates from linearity. This is because Eq. (28.8) is invariant with respect to even a non-linear transformation of parameters $\boldsymbol{\alpha}$, whereas Eq. (28.7) is not. Still, when the model is non-linear or errors are not Gaussian, confidence intervals obtained with both these formulas are only approximate. The true coverage of these confidence intervals can always be determined by a Monte Carlo simulation, or exact confidence intervals can be determined as in Sec. 28.6.1.

28.3.4. *Application to Poisson-distributed data*:

In the case of Poisson-distributed data in a counting experiment, the unbinned maximum likelihood method (where the index i in Eq. (28.5) labels events) is preferred if the total number of events is very small. (Sometimes it is "extended" to include the total number of events as a Poisson-distributed observable.) If there are enough events to justify binning them in a histogram, then one may alternatively maximize the likelihood function for the contents of the bins (so i labels bins). This is equivalent to minimizing [11]

$$\chi^2 = \sum_i \left[2(N_i^{\text{th}} - N_i^{\text{obs}}) + 2N_i^{\text{obs}} \ln(N_i^{\text{obs}}/N_i^{\text{th}}) \right] . \qquad (28.9)$$

where N_i^{obs} and N_i^{th} are the observed and theoretical (from f) contents of the ith bin. In bins where $N_i^{\text{obs}} = 0$, the second term is zero. This function asymptotically behaves like a classical χ^2 for purposes of point estimation, interval estimation, *and goodness-of-fit*. It also guarantees that the area under the fitted function f is equal to the sum of the histogram contents (as long as the overall normalization of f is effectively left unconstrained during the fit), which is not the case for χ^2 statistics based on a least-squares procedure with traditional weights.

28.4. Propagation of errors

Suppose that $F(x; \alpha)$ is some function of variable(s) x and the fitted parameters α, with a value $\hat{F}$ at $\hat{\alpha}$. The variance matrix of the parameters is V_{mn}. To first order in $\alpha_m - \hat{\alpha}_m$, F is given by

$$F = \hat{F} + \sum_m \frac{\partial F}{\partial \alpha_m} (\alpha_m - \hat{\alpha}_m) , \qquad (28.10)$$

and the variance of F about its estimator is given by

$$(\Delta F)^2 = E[(F - \hat{F})^2] = \sum_{mn} \frac{\partial F}{\partial \alpha_m} \frac{\partial F}{\partial \alpha_n} V_{mn} , \qquad (28.11)$$

evaluated at the x of interest. For different functions F_j and F_k, the covariance is

$$E[(F_j - \hat{F}_j)(F_k - \hat{F}_k)] = \sum_{mn} \frac{\partial F_j}{\partial \alpha_m} \frac{\partial F_k}{\partial \alpha_n} V_{mn} . \qquad (28.12)$$

If the first-order approximation is in serious error, the above results may be very approximate. $\hat{F}$ may be a biased estimator of F even if the $\hat{\alpha}$ are unbiased estimators of α. Inclusion of higher-order terms or direct evaluation of F in the vicinity of $\hat{\alpha}$ will help to reduce the bias.

28.5. Method of least squares

The *method of least squares* can be derived from the maximum likelihood theorem. We suppose a set of N measurements at points x_i. The ith measurement y_i is assumed to be chosen from a Gaussian distribution with mean $F(x_i; \alpha)$ and variance σ_i^2. Then

$$\chi^2 = -2 \ln \mathscr{L} + \text{constant} = \sum_i \frac{[y_i - F(x_i; \alpha)]^2}{\sigma_i^2} . \qquad (28.13)$$

Finding the set of parameters α which maximizes $\mathscr{L}$ is the same as finding the set which minimizes χ^2.

In many practical cases one further restricts the problem to the situation in which $F(x_i; \alpha)$ is a linear function of the α_m's,

$$F(x_i; \alpha) = \sum_n \alpha_n f_n(x_i) , \qquad (28.14)$$

where the f_n are k linearly independent functions (e.g., 1, x, x^2, ..., or Legendre polynomials) which are single-valued over the allowed range of x. We require $k \leq N$, and at least k of the x_i must be distinct. We wish to estimate the linear coefficients α_n. Later we will discuss the nonlinear case.

If the point errors $\epsilon_i = y_i - F(x_i; \alpha)$ are Gaussian, then the minimum χ^2 will be distributed as a χ^2 random variable with $n = N - k$ degrees of freedom. We can then evaluate the goodness-of-fit (significance level) from Figs. 27.1 or 27.3, as per the earlier discussion. The significance level expresses the probability that a *worse* fit would be obtained in a large number of similar experiments under the assumptions that: (a) the model $y = \sum \alpha_n f_n$ is correct and (b) the ϵ_i are Gaussian and unbiased with variance σ_i^2. If this probability is larger than an agreed-upon value (0.001, 0.01, or 0.05 are common choices), the data are *consistent* with the assumptions; otherwise we may want to find improved assumptions. As for the converse, most people do not regard a model as being truly *inconsistent* unless the probability is as low as that corresponding to four or five standard deviations for a Gaussian (6×10^{-3} or 6×10^{-5}; see Sec. 28.6.2). If the ϵ_i are not Gaussian, the method of least squares still gives an answer, but the goodness-of-fit test would have to be done using the correct distribution of the random variable which is still called "χ^2."

Minimizing χ^2 in the linear case is straightforward:

$$-\frac{1}{2} \frac{\partial \chi^2}{\partial \alpha_m} = \sum_i f_m(x_i) \left(\frac{y_i - \sum_n \alpha_n f_n(x_i)}{\sigma_i^2} \right)$$

$$= \sum_i \frac{y_i f_m(x_i)}{\sigma_i^2} - \sum_n \alpha_n \sum_i \frac{f_n(x_i) f_m(x_i)}{\sigma_i^2} . \qquad (28.15)$$

With the definitions

$$g_m = \sum_i y_i f_m(x_i) / \sigma_i^2 \qquad (28.16)$$

and

$$V_{mn}^{-1} = \sum_i f_n(x_i) f_m(x_i) / \sigma_i^2 , \qquad (28.17)$$

the k-element column vector of solutions $\hat{\alpha}$, for which $\partial \chi^2 / \partial \alpha_m = 0$ for all m, is given by

$$\hat{\alpha} = V \, g . \qquad (28.18)$$

With this notation, χ^2 for the special case of a linear fitting function (Eq. (28.14)) can be rewritten in the compact form

$$\chi^2 = \chi^2_{\min} + (\alpha - \hat{\alpha})^T V^{-1} (\alpha - \hat{\alpha}) . \qquad (28.19)$$

Nonindependent y_i's

Eq. (28.13) is based on the assumption that the likelihood function is the product of independent Gaussian distributions. More generally, the measured y_i's are not independent, and we must consider them as coming from a multivariate distribution with nondiagonal covariance matrix S, as described in Sec. 27.3.3. The generalization of Eq. (28.13) is

$$\chi^2 = \sum_{jk} [y_j - F(x_j; \alpha)] S_{jk}^{-1} [y_k - F(x_k; \alpha)] . \qquad (28.20)$$

In the case of a fitting function that is linear in the parameters, one may differentiate χ^2 to find the generalization of Eq. (28.15), and with the extended definitions

$$g_m = \sum_{jk} y_j f_m(x_k) S_{jk}^{-1}$$

$$V_{mn}^{-1} = \sum_{jk} f_n(x_j) f_m(x_k) S_{jk}^{-1} \qquad (28.21)$$

solve Eq. (28.18) for the estimators $\hat{\alpha}$.

The problem of constructing the covariance matrix S is simplified by the fact that contributions to S (not to its inverse) are additive. For example, suppose that we have three variables, all of which have independent statistical errors. The first two also have a common error resulting in a positive correlation, perhaps because a common baseline with its own statistical error (variance s^2) was subtracted from each. In addition, the second two have a common error (variance a^2), but this time the values are anticorrelated. This might happen, for example, if the sum of the two variables is a constant. Then

$$S = \begin{pmatrix} \sigma_1^2 & 0 & 0 \\ 0 & \sigma_2^2 & 0 \\ 0 & 0 & \sigma_3^2 \end{pmatrix}$$

$$+ \begin{pmatrix} s^2 & s^2 & 0 \\ s^2 & s^2 & 0 \\ 0 & 0 & 0 \end{pmatrix} + \begin{pmatrix} 0 & 0 & 0 \\ 0 & a^2 & -a^2 \\ 0 & -a^2 & a^2 \end{pmatrix} . \qquad (28.22)$$

If unequal amounts of the common baseline were subtracted from variables 1, 2, and 3—e.g., fractions f_1, f_2, and f_3, then we would have

$$S = \begin{pmatrix} \sigma_1^2 & 0 & 0 \\ 0 & \sigma_2^2 & 0 \\ 0 & 0 & \sigma_3^2 \end{pmatrix}$$

$$+ \begin{pmatrix} f_1^2 s^2 & f_1 f_2 s^2 & f_1 f_3 s^2 \\ f_1 f_2 s^2 & f_2^2 s^2 & f_2 f_3 s^2 \\ f_1 f_3 s^2 & f_2 f_3 s^2 & f_3^2 s^2 \end{pmatrix} . \qquad (28.23)$$

While in general this "two-vector" representation is not possible, it underscores the procedure: Add zero-determinant correlation matrices to the matrix expressing the independent variation.

Care must be taken when fitting to correlated data, since off-diagonal contributions to χ^2 are not necessarily positive. It is even possible for all of the residuals to have the same sign.

Example: straight-line fit

For the case of a straight-line fit, $y(x) = \alpha_1 + \alpha_2\,x$, one obtains, for independent measurements y_i, the following estimates of α_1 and α_2,

$$\widehat{\alpha}_1 = (g_1\,\Lambda_{22} - g_2\,\Lambda_{12})/D \; , \tag{28.24}$$

$$\widehat{\alpha}_2 = (g_2\,\Lambda_{11} - g_1\,\Lambda_{12})/D \; , \tag{28.25}$$

where

$$(\Lambda_{11}, \Lambda_{12}, \Lambda_{22}) = \sum (1, x_i, x_i^2)/\sigma_i^2 \; , \tag{28.26a}$$

$$(g_1, g_2) = \sum (1, x_i) y_i/\sigma_i^2 \; . \tag{28.26b}$$

respectively, and

$$D = \Lambda_{11}\,\Lambda_{22} - (\Lambda_{12})^2 \; . \tag{28.27}$$

The covariance matrix of the fitted parameters is:

$$\begin{pmatrix} V_{11} & V_{12} \\ V_{12} & V_{22} \end{pmatrix} = \frac{1}{D} \begin{pmatrix} \Lambda_{22} & -\Lambda_{12} \\ -\Lambda_{12} & \Lambda_{11} \end{pmatrix} \; . \tag{28.28}$$

The estimated variance of an interpolated or extrapolated value of y at point x is:

$$(\widehat{y} - y_{\text{true}})^2 \Big|_{\text{est}} = \frac{1}{\Lambda_{11}} + \frac{\Lambda_{11}}{D} \left(x - \frac{\Lambda_{12}}{\Lambda_{11}} \right)^2 \; . \tag{28.29}$$

28.5.1. *Confidence intervals from the chisquare function:*

If y is not linear in the fitting parameters $\boldsymbol{\alpha}$, the solution vector may have to be found by iteration. If we have a first guess $\boldsymbol{\alpha}_0$, then we may expand to obtain

$$\frac{\partial \chi^2}{\partial \boldsymbol{\alpha}} \Big|_{\alpha} = \frac{\partial \chi^2}{\partial \boldsymbol{\alpha}} \Big|_{\alpha_0} + V_{\alpha_0}^{-1} \cdot (\boldsymbol{\alpha} - \boldsymbol{\alpha}_0) + \dots \; , \tag{28.30}$$

where $\partial \chi^2/\partial \boldsymbol{\alpha}$ is a vector whose mth component is $\partial \chi^2/\partial \alpha_m$, and $(V_{mn}^{-1}) = \frac{1}{2}\partial^2 \chi^2/\partial\alpha_m\partial\alpha_n$. (See Eqns. 28.7 and 28.17. When evaluated at $\widehat{\alpha}$, V^{-1} is the inverse of the covariance matrix.) The next iteration toward $\widehat{\alpha}$ can be obtained by setting $\partial \chi^2/\partial\alpha_m|_{\alpha} = 0$ and neglecting higher-order terms:

$$\boldsymbol{\alpha} = \boldsymbol{\alpha}_0 - V_{\alpha_0} \cdot \partial \chi^2/\partial\boldsymbol{\alpha}|_{\alpha_0} \; . \tag{28.31}$$

If V is constant in the vicinity of the minimum, as it is when the model function is linear in the parameters, then χ^2 is parabolic as a function of $\boldsymbol{\alpha}$ and Eq. (28.31) gives the solution immediately. Otherwise, further iteration is necessary. If the problem is highly nonlinear, considerable difficulty may be encountered. There may be secondary minima, and χ^2 may be decreasing at physical boundaries. Numerical methods have been devised to find such solutions without divergence [9,10]. In particular, the CERN program MINUIT [10] offers several iteration schemes for solving such problems.

Note that minimizing any function proportional to χ^2 (or maximizing any function proportional to $\ln \mathscr{L}$) will result in the same parameter set $\widehat{\alpha}$. Hence, for example, if the variances σ_j^2 are known only up to a common constant, one can still solve for $\widehat{\alpha}$. One cannot, however, evaluate goodness-of-fit, and the covariance matrix is known only to within the constant multiplier. The scale can be estimated at least roughly from the value of χ^2 compared to its expected value.

Additional information can be extracted from the behavior of the normalized residuals (known as "pulls"), $r_j = (y_j - F(x_j; \boldsymbol{\alpha})/\sigma_j$, which should themselves distribute normally with mean 0 and rms deviation 1.

If the data covariance matrix S has been correctly evaluated (or, equivalently, the σ_j's, if the data are independent), the s-standard deviation limits on each of the parameters are given by a set $\boldsymbol{\alpha}'$ such that

$$\chi^2(\boldsymbol{\alpha}') = \chi^2_{\min} + s^2 \; . \tag{28.32}$$

This equation gives confidence intervals in the same sense as 28.8, and all the discussion of Sec. 28.3.3 applies as well here, substituting $-\chi^2/2$ for $\ln \mathscr{L}$.

28.6. Exact confidence intervals

The unqualified phrase "confidence intervals" refers to frequentist (also called classical) intervals obtained with a construction due to Neyman [12], described below. Approximate confidence intervals are obtained in classical statistics from likelihood *ratios* as described in the preceding subsections. The validity of the approximation (in terms of coverage; see below) should be checked (typically by the Monte Carlo method) when in doubt, as is usually the case with small numbers of events.

Intervals in Bayesian statistics, usually called credible intervals or Bayesian confidence intervals, are obtained by integrating the posterior p.d.f. (based on a non-frequency definition of probability), and in many cases do not obey the defining properties of confidence intervals. Correspondingly, confidence intervals do not in general behave like credible intervals.

In the Bayesian framework, all uncertainty including systematic and theoretical uncertainties can be treated in a straightforward manner: one includes in the p.d.f. one's degree of belief about background estimates, luminosity, etc. Then one integrates out such "nuisance parameters." In the Frequentist approach, one should have exact coverage no matter what the value of the nuisance parameters, and this is not in general possible. If one performs a Bayesian-style integration over nuisance parameters while constructing nominally Frequentist intervals, then coverage must be checked.

28.6.1. *Neyman's Construction of Confidence intervals:*

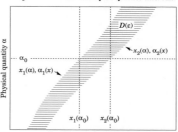

Figure 28.1: Confidence intervals for a single unknown parameter α. One might think of the p.d.f. $f(x; \alpha)$ as being plotted out of the paper as a function of x along each horizontal line of constant α. The domain $D(\varepsilon)$ contains a fraction $1 - \varepsilon$ of the area under each of these functions.

We consider the parameter α whose true value is fixed but unknown. The properties of our experimental apparatus are expressed in the function $f(x; \alpha)$ which gives the probability of observing data x if the true value of the parameter is α. This function must be known in order to interpret the results of an experiment. For a large complex experiment, f is usually determined numerically using Monte Carlo simulation.

Given $f(x; \alpha)$, we can find for every value of α, two values $x_1(\alpha, \varepsilon)$ and $x_2(\alpha, \varepsilon)$ such that

$$P(x_1 < x < x_2; \alpha) = 1 - \varepsilon = \int_{x_1}^{x_2} f(x; \alpha)\,dx \; . \tag{28.33}$$

This is shown graphically in Fig. 28.1: a horizontal line segment $[x_1(\alpha, \varepsilon), x_2(\alpha, \varepsilon)]$ is drawn for representative values of α. The union of all intervals $[x_1(\alpha, \varepsilon), x_2(\alpha, \varepsilon)]$, designated in the figure as the domain $D(\varepsilon)$, is known as the *confidence belt*. Typically the curves $x_1(\alpha, \varepsilon)$ and $x_2(\alpha, \varepsilon)$ are monotonic functions of α, which we assume for this discussion.

Upon performing an experiment to measure x and obtaining the value x_0, one draws a vertical line through x_0 on the horizontal axis.

The confidence interval for α is the union of all values of α for which the corresponding line segment $[x_1(\alpha, \varepsilon), x_2(\alpha, \varepsilon)]$ is intercepted by this vertical line. The confidence interval is an interval $[\alpha_1(x_0), \alpha_2(x_0)]$, where $\alpha_1(x_0)$ and $\alpha_2(x_0)$ are on the boundary of $D(\varepsilon)$. Thus, the boundaries of $D(\varepsilon)$ can be considered to be functions $x(\alpha)$ when constructing D, and then to be functions $\alpha(x)$ when reading off confidence intervals.

Such confidence intervals are said to have Confidence Level (CL) equal to $1-\varepsilon$.

Now suppose that some unknown particular value of α, say α_0 (indicated in the figure), is the true value of α. We see from the figure that α_0 lies between $\alpha_1(x)$ and $\alpha_2(x)$ if and only if x lies between $x_1(\alpha_0)$ and $x_2(\alpha_0)$. Thus we can write:

$$P[x_1(\alpha_0) < x < x_2(\alpha_0)] = 1 - \varepsilon = P[\alpha_2(x) < \alpha_0 < \alpha_1(x)] \,. \tag{28.34}$$

And since, by construction, this is true for any value α_0, we can drop the subscript 0 and obtain the relationship we wanted to establish for the probability that the confidence limits will contain the true value of α:

$$P[\alpha_2(x) < \alpha < \alpha_1(x)] = 1 - \varepsilon \,. \tag{28.35}$$

In this probability statement, α_1 and α_2 are the random variables (not α), and we can verify that the statement is true, as a limiting ratio of frequencies in random experiments, for any assumed value of α. In a particular real experiment, the numerical values α_1 and α_2 are determined by applying the algorithm to the real data, and the probability statement is (all too frequently) misinterpreted to be a statement about the true value α since this is the only unknown remaining in the equation. It should however be interpreted as the probability of obtaining values α_1 and α_2 which include the true value of α, in an ensemble of identical experiments. Any method which gives confidence intervals that contain the true value with probability $1 - \varepsilon$ (no matter what the true value of α is) is said to have the correct *coverage*. The frequentist intervals as constructed above have the correct *coverage* by construction. Coverage is a critical property of confidence intervals [2]. (Power to exclude false values of α, related to the length of the intervals in a relevant measure, is also important.)

The condition of coverage Eq. (28.33) does not determine x_1 and x_2 uniquely, since any range which gives the desired value of the integral would give the same coverage. Additional criteria are thus needed. The most common criterion is to choose *central intervals* such that the area of the excluded tail on either side is $\varepsilon/2$. This criterion is sufficient in most cases, but there is a more general *ordering principle* which reduces to centrality in the usual cases and produces confidence intervals with better properties when in the neighborhood of a physical limit. This ordering, which consists of taking the interval which includes the largest values of a likelihood ratio, is briefly outlined in Ref. 3 and has been applied to prototypical problems by Feldman and Cousins [13].

For the problem of a counting rate experiment in the presence of background, Roe and Woodroofe [14] have proposed a modification to Ref. 13 incorporating *conditioning*, i.e., conditional probabilities computed using constraints on the number of background events actually observed. This and other prescriptions giving frequentist intervals have not yet been fully explored [5].

28.6.2. *Gaussian errors:*

If the data are such that the distribution of the estimator(s) satisfies the central limit theorem discussed in Sec. 27.3.3, the function $f(x; \alpha)$ is the Gaussian distribution. If there is more than one parameter being estimated, the multivariate Gaussian is used. For the univariate case with known σ,

$$1 - \varepsilon = \int_{\mu-\delta}^{\mu+\delta} e^{\frac{-(x-\mu)^2}{2\sigma^2}} \, dx = \mathrm{erf}\left(\frac{\delta}{\sqrt{2}\,\sigma}\right) \tag{28.36}$$

is the probability that the measured value x will fall within $\pm\delta$ of the true value μ. From the symmetry of the Gaussian with respect to x and μ, this is also the probability that the true value will be within

Table 28.1: Area of the tails ε outside $\pm\delta$ from the mean of a Gaussian distribution.

ε (%)	δ	ε (%)	δ
31.73	1σ	20	1.28σ
4.55	2σ	10	1.64σ
0.27	3σ	5	1.96σ
6.3×10^{-3}	4σ	1	2.58σ
5.7×10^{-5}	5σ	0.1	3.29σ
2.0×10^{-7}	6σ	0.01	3.89σ

$\pm\delta$ of the measured value. Fig. 28.2 shows a $\delta = 1.64\sigma$ confidence interval unshaded. The choice $\delta = \sqrt{\mathrm{Var}(\mu)} \equiv \sigma$ gives an interval called the *standard error* which has $1 - \varepsilon = 68.27\%$ if σ is known. Confidence coefficients ε for other frequently used choices of δ are given in Table 28.1.

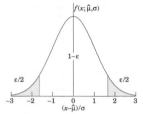

Figure 28.2: Illustration of a symmetric 90% confidence interval (unshaded) for a measurement of a single quantity with Gaussian errors. Integrated probabilities, defined by ε, are as shown.

For other δ, find ε as the ordinate of Fig. 27.1 on the $n = 1$ curve at $\chi^2 = (\delta/\sigma)^2$. We can set a one-sided (upper or lower) limit by excluding above $\mu + \delta$ (or below $\mu - \delta$); ε's for such limits are $1/2$ the values in Table 28.1.

For multivariate $\boldsymbol{\alpha}$ the scalar $\mathrm{Var}(\mu)$ becomes a full variance-covariance matrix. Assuming a multivariate Gaussian, Eq. (27.22), and subsequent discussion the standard error ellipse for the pair $(\hat{\alpha}_m, \hat{\alpha}_n)$ may be drawn as in Fig. 28.3.

The minimum χ^2 or maximum likelihood solution is at $(\hat{\alpha}_m, \hat{\alpha}_n)$. The standard errors σ_m and σ_n are defined as shown, where the ellipse is at a constant value of $\chi^2 = \chi_{\min}^2 + 1$ or $\ln \mathscr{L} = \ln \mathscr{L}_{\max} - 1/2$. The angle of the major axis of the ellipse is given by

$$\tan 2\phi = \frac{2\rho_{mn}\,\sigma_m\,\sigma_n}{\sigma_m^2 - \sigma_n^2} \,. \tag{28.37}$$

For non-Gaussian or nonlinear cases, one may construct an analogous contour from the same χ^2 or $\ln \mathscr{L}$ relations. Any other parameters $\hat{\alpha}_\ell, \ell \neq m, n$ must be allowed freely to find their optimum values for every trial point.

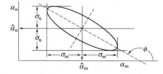

Figure 28.3: Standard error ellipse for the estimators $\hat{\alpha}_m$ and $\hat{\alpha}_n$. In this case the correlation is negative.

Table 28.2: $\Delta\chi^2$ corresponding to $(1 - \varepsilon)$, for joint estimation of k parameters.

$(1 - \varepsilon)$ (%)	$k = 1$	$k = 2$	$k = 3$
68.27	1.00	2.30	3.53
90.	2.71	4.61	6.25
95.45	4.00	6.18	8.03
99.	6.63	9.21	11.34
99.73	9.00	11.83	14.16

For any unbiased procedure (*e.g.*, least squares or maximum likelihood) used to estimate k parameters α_i, $i = 1, \ldots, k$, the probability $1 - \varepsilon$ that the true values of all k parameters lie within an ellipsoid bounded by a fixed value of $\Delta\chi^2 = \chi^2 - \chi^2_{\min}$ may be found from Fig. 27.1. This is because the *difference*, $\Delta\chi^2 = \chi^2 - \chi^2_{\min}$, obeys the "$\chi^2$" p.d.f. given in Table 27.1, if the parameter n in the formula is taken to be k (rather than degrees-of-freedom in the fit). In Fig. 27.1, read the ordinate as ε and the abscissa as $\Delta\chi^2$. The correct values of ε are on the $n = k$ curve. For $k > 1$, the values of ε for given $\Delta\chi^2$ are much greater than for $k = 1$. Hence, using $\Delta\chi^2 = s^2$, which gives s-standard-deviation errors on a single parameter (*irrespective of the other parameters*), is not appropriate for a multi-dimensional ellipsoid. For example, for $k = 2$, the probability $(1 - \varepsilon)$ that the true values of α_1 and α_2 simultaneously lie within the one-standard-deviation error ellipse ($s = 1$), centered on $\hat{\alpha}_1$ and $\hat{\alpha}_2$, is only 39%.

Values of $\Delta\chi^2$ corresponding to commonly used values of ε and k are given in Table 28.2. These probabilities assume Gaussian errors, unbiased estimators, and that the model describing the data in terms of the α_i is correct. When these assumptions are not satisfied, a Monte Carlo simulation is typically performed to determine the relation between $\Delta\chi^2$ and ε.

28.6.3. *Upper limits and two-sided intervals*:

When a measured value is close to a physical boundary, it is natural to report a one-sided confidence interval (often an upper limit). It is straightforward to force the procedure of Sec. 28.6.1 to produce only an upper limit, by setting $x_2 = \infty$ in Eq. (28.33). Then x_1 is uniquely determined. Clearly this procedure will have the desired coverage, but *only if we always choose to set an upper limit*. In practice one might decide after seeing the data whether to set an upper limit or a two-sided limit. In this case the upper limits calculated by Eq. (28.33) will not give exact coverage, as has been noted in Ref. 13.

In order to correct this problem and assure coverage in all circumstances, it is necessary to adopt a *unified procedure*, that is, a single ordering principle which will provide coverage globally. Then it is the *ordering principle* which decides whether a one-sided or two-sided interval will be reported for any given set of data. The unified procedure and ordering principle which follows from the theory of likelihood-ratio tests [3] is described in Ref. 13. We reproduce below the main results.

28.6.4. *Gaussian data close to a boundary*:

One of the most controversial statistical questions in physics is how to report a measurement which is close to the edge or even outside of the allowed physical region. This is because there are several admissible possibilities depending on how the result is to be used or interpreted. Normally one or more of the following should be reported:

(a) The actual measurement should be reported, even if it is outside the physical region. As with any other measurement, it is best to report the value of a quantity which is nearly Gaussian distributed if possible. Thus one may choose to report mass squared rather than mass, or $\cos\theta$ rather than θ. For a complex quantity z close to zero, report $\mathrm{Re}(z)$ and $\mathrm{Im}(z)$ rather than amplitude and phase of z. Data carefully reported in this way can be unbiased, objective, easily interpreted and combined (averaged) with other data in a straightforward way, even if they lie partly or wholly outside the physical region. The reported error is a direct measure of the intrinsic accuracy of the result, which cannot always be inferred from the upper limits proposed below.

(b) If the data are to be used to make a decision, for example to determine the dimensions of a new experimental apparatus for an improved measurement, it may be appropriate to report a Bayesian upper limit, which must necessarily contain subjective belief about the possible values of the parameter, as well as containing information about the physical boundary. Its interpretation requires knowledge of the prior distribution which was necessarily used to obtain it.

(c) If it is desired to report an upper limit that has a well-defined meaning in terms of a limiting frequency, then report the Frequentist confidence bound(s) as given by the unified approach [3], [13]. This algorithm always gives a non-null interval (that is, the confidence limits are always inside the physical region, even for a measurement well outside the physical region), and still has correct global coverage. These confidence limits for a Gaussian measurement close to a non-physical boundary are summarized in Fig. 28.4. Additional tables are given in Ref. 13.

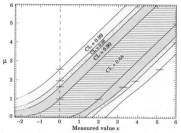

Figure 28.4: Plot of 99%, 95%, 90%, and 68.27% ("one σ") confidence intervals (using the unified approach as in Ref. 13) for a physical quantity μ based on a Gaussian measurement x (in units of standard deviations), for the case where the true value of μ cannot be negative. The curves become straight lines above the horizontal tick marks. The probability of obtaining an experimental value at least as negative as the left edge of the graph ($x = -2.33$) is less than 1%. Values of x more negative than -1.64 (dotted segments) are less than 5% probable, no matter what the true value of μ.

28.6.5. *Poisson data for small samples*:

When the observable is restricted to integer values (as in the case of Poisson and binomial distributions), it is not generally possible to construct confidence intervals with exact coverage for all values of α. In these cases the integral in Eq. (28.33) becomes a sum of finite contributions and it is no longer possible (in general) to find consecutive terms which add up exactly to the required confidence level $1 - \varepsilon$ for all values of α. Thus one constructs intervals which happen to have exact coverage for a few values of α, and unavoidable over-coverage for all other values.

In addition to the problem posed by the discreteness of the data, we usually have to contend with possible background whose expectation must be evaluated separately and may not be known precisely. For these reasons, the reporting of this kind of data is even more controversial than the Gaussian data near a boundary as discussed above. This is especially true when the number of observed counts is greater than the expected background. As for the Gaussian case, there are at least three possibilities for reporting such results depending on how the result is to be used:

(a) The actual measurements should be reported, which means (1) the number of recorded counts, (2) the expected background, possibly with its error, and (3) normalization factor which turns the number of counts into a cross section, decay rate, *etc.* As with Gaussian data, these data can be combined with that of other experiments, to make improved upper limits for example.

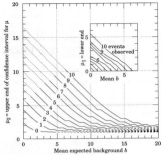

Figure 28.5: 90% confidence intervals $[\mu_1, \mu_2]$ on the number of signal events as a function of the expected number of background events b. For example, if the expected background is 8 events and 5 events are observed, then the signal is 2.60 or less with 90% confidence. Dotted portions of the μ_2 curves on the upper left indicate regions where μ_1 is non-zero (as shown by the inset). Dashed portions in the lower right indicate regions where the probability of obtaining the number of events observed or fewer is less than 1%, even if $\mu = 0$. Horizontal curve sections occur because of discrete number statistics. Tables showing these data as well as the CL = 68.27%, 95%, and 99% results are given in Ref. 13. There is considerable discussion about the behavior of the intervals when the number of observed events is less than the expected background; see Ref. 5

(b) A Bayesian upper limit may be reported. This has the advantages and disadvantages of any Bayesian result as discussed above. The noninformative priors (based on invariance principles rather than subjective degree of belief) recommended in the statistics literature for Poisson mean are rarely, if at all, used in high energy physics; they diverge for the case of zero events observed, and they give upper limits which undercover when evaluated by the Frequentist criterion of coverage. Rather, priors uniform in the counting rate have been used by convention; care must be used in interpreting such results either as "degree of belief" or as a limiting frequency.

(c) An upper limit (or confidence region) with optimal coverage can be reported using the unified approach of Ref. 13. At the moment these confidence limits have been calculated only for the case of exactly known background expectation. The main results can be read from Fig. 28.5 or from Table 28.3; more extensive tables can be found in Ref. 13.

Table 28.3: Poisson limits $[\mu_1, \mu_2]$ for n_0 observed events in the absence of background.

n_0	CI = 90%		CI = 95%	
	μ_1	μ_2	μ_1	μ_2
0	0.00	2.44	0.00	3.09
1	0.11	4.36	0.05	5.14
2	0.53	5.91	0.36	6.72
3	1.10	7.42	0.82	8.25
4	1.47	8.60	1.37	9.76
5	1.84	9.99	1.84	11.26
6	2.21	11.47	2.21	12.75
7	3.56	12.53	2.58	13.81
8	3.96	13.99	2.94	15.29
9	4.36	15.30	4.36	16.77
10	5.50	16.50	4.75	17.82

None of the above gives a single number which quantifies the quality or sensitivity of the experiment. This is a serious shortcoming of most upper limits including those of method (c), since it is impossible to distinguish, from the upper limit alone, between a clean experiment with no background and a lucky experiment with fewer observed counts than expected background. For this reason, we suggest that in addition to (a) and (c) above, a measure of the sensitivity should be reported whenever expected background is larger or comparable to the number of observed counts. The best such measure we know of is that proposed and tabulated in Ref. 13, defined as the average upper limit that would be attained by an ensemble of experiments with the expected background and no true signal.

References:

1. B. Efron, Am. Stat. **40**, 11 (1986).

2. R.D. Cousins, Am. J. Phys. **63**, 398 (1995).

3. A. Stuart and A. K. Ord, *Kendall's Advanced Theory of Statistics*, Vol. 2 *Classical Inference and Relationship* 5th Ed., (Oxford Univ. Press, 1991), and earlier editions by Kendall and Stuart. The likelihood-ratio ordering principle is described at the beginning of Ch. 23. Chapter 31 compares different schools of statistical inference.

4. W.T. Eadie, D. Drijard, F.E. James, M. Roos, and B. Sadoulet, *Statistical Methods in Experimental Physics* (North Holland, Amsterdam and London, 1971).

5. Workshop on Confidence Limits, CERN, 17-18 Jan. 2000, www.cern.ch/CERN/Divisions/EP/Events/CLW/. See also the later Fermilab workshop linked to the CERN web page.

6. H. Cramér, *Mathematical Methods of Statistics*, Princeton Univ. Press, New Jersey (1958).

7. B.P. Roe, *Probability and Statistics in Experimental Physics*, (Springer-Verlag, New York, 208 pp., 1992).

8. G. Cowan, *Statistical Data Analysis* (Oxford University Press, Oxford, 1998).

9. W.H. Press *et al.*, *Numerical Recipes* (Cambridge University Press, New York, 1986).

10. F. James and M. Roos, "MINUIT, Function Minimization and Error Analysis," CERN D506 (Long Writeup). Available from the CERN Program Library Office, CERN-IT Division, CERN, CH-1211, Geneva 21, Switzerland.

11. For a review, see S. Baker and R. Cousins, Nucl. Instrum. Methods **221**, 437 (1984).

12. J. Neyman, Phil. Trans. Royal Soc. London, Series A, **236**, 333 (1937), reprinted in *A Selection of Early Statistical Papers on J. Neyman* (University of California Press, Berkeley, 1967).

13. G.J. Feldman and R.D. Cousins, Phys. Rev. **D57**, 3873 (1998). This paper does not specify what to do if the ordering principle gives equal rank to some values of x. Eq. 23.6 of Ref. 3 gives the rule: all such points are included in the acceptance region (the domain $D(\varepsilon)$). Some authors have assumed the contrary, and shown that one can then obtain null intervals..

14. B.P. Roe and M.B. Woodroofe, Phys. Rev. **D60**, 053009 (1999).

B.2   Table of Integrals

In this Appendix, the most commonly used integrals for the purposes of this book are listed. More detailed treatments can be found in specialized Mathematical Handbooks.

$$\int (ax+b)^r \, dx = \frac{1}{a \, (r+1)} (ax+b)^{r+1} + C, \qquad\qquad r \neq -1$$

$$\int \frac{1}{\sqrt{(a^2 - x^2)}} \, dx = \arcsin \frac{x}{a} + C$$

$$\int \frac{1}{\sqrt{(a^2 + x^2)}} \, dx = \operatorname{arcsinh} \frac{x}{a} + C$$

$$\int \sqrt{(a^2 - x^2)} \, dx$$
$$= \frac{x}{2}\sqrt{(a^2 - x^2)} + \frac{a^2}{2} \arcsin \frac{x}{a} + C$$

$$\int \sqrt{(a^2 + x^2)} \, dx$$
$$= \frac{x}{2}\sqrt{(a^2 + x^2)} + \frac{a^2}{2} \operatorname{arcsinh} \frac{x}{a} + C$$

$$\int \frac{1}{ax+b} \, dx = \frac{1}{a} \ln|(ax+b)| + C$$

$$\int xe^x \, dx = e^x(x-1) + C$$

$$\int e^{ax} \, dx = \frac{1}{a} e^{ax} + C, \qquad\qquad a \neq 0$$

$$\int a^x \, dx = \frac{1}{\ln a} a^x + C$$

$$\int \ln x \, dx = x(\ln x - 1) + C, \qquad\qquad (x > 0)$$

$$\int x \ln x \, dx = \frac{1}{2} x^2 \left(\ln x - \frac{1}{2}\right) + C, \qquad\qquad (x > 0)$$

$$\int \frac{1}{x \ln x} \, dx = \ln|\ln x| + C, \qquad\qquad (x > 0)$$

$$\int x^m (\ln x)^n \, dx \qquad\qquad\qquad m \,,\, n \text{ integers, and } \neq -1$$
$$= -\frac{n}{m+1} \int x^m (\ln x)^{n-1} \, dx, \qquad\qquad \text{and } (x > 0)$$

$$\int \sin x \, dx = -\cos x + C$$

$$\int \cos x \, dx = \sin x + C$$

$$\int x \sin x \, dx = \sin x - x \cos x + C$$

$$\int \cos^2 x \, dx = \frac{1}{2}(\sin x \cos x + x) + C$$

$$\int x \sin x^2 \, dx = -\frac{1}{2} \cos x^2 + C$$

$$\int \frac{1}{1+\sin ax} \, dx = -\frac{1}{a} \tan\left(\frac{ax}{2} - \frac{\pi}{4}\right) + C$$

continued on next page

continued from previous page

$\int \frac{1}{1+\cos ax} \, dx = -\frac{1}{a} \tan \frac{ax}{2} + C$

$\int x \sinh ax \, dx = \frac{1}{a} \cosh ax + C$

$\int x \cosh ax \, dx = \frac{1}{a} \sinh ax + C$

$\int \frac{\sin x}{\sin^4(x/2)} \, dx = \int \frac{4\cos(x/2)}{\sin^3(x/2)} \, d(x/2) = -\frac{2}{\sin^2(x/2)} + C$

$\int \frac{\sin x}{\sin^2(x/2)} \, dx = \int \frac{4\cos(x/2)}{\sin(x/2)} \, d(x/2) = 4 \ln |\sin(x/2)| + C$

Bibliography

Aarnio, P.A. et al. (1987). FLUKA86 User's Guide and enhancements to the FLUKA86 program (FLUKA87), CERN TIS-RP/168 and TIS-RP/190.
See also:
Fassò, A. et al. (1993). FLUKA: present status and future developments, presented at the IV International Conference on Calorimetry in High Energy Physics, 21–26 September, La Biodola, Italy.
See also the web site:
http://fluka.web.cern.ch/fluka/

Abdrakhmanov, E.O. et al. (1974). *Nucl. Phys. B* **72**, 189.

Abramowicz, H. et al. (1981). *Nucl. Instr. and Meth.* **180**, 429.

Abrasimov, A.T. et al. (1979). *Nucl. Phys. B* **158**, 11.

Abyzov, A.S., Davydov, L.N., Kutny, V.E., Rybka, A,V., Rowland, M.S. and Smith, C.F. (1999). 11th International Workshop on Room Temperature Semiconductor X- and Gamma-Ray Detectors and Associated Electronics, Vienna, Reference Number P1.

Acosta, D. et al. (1990). *Nucl. Instr. and Meth. in Phys. Res. A* **294**, 193.

Acosta, D., Buontempo, S., Caloba, L., Caria, M., DeSalvo, R., Ereditato, A., Ferrari, R., Fumagalli, G., Goggi, G., Hao, W., Hartjes, F.G., Henriques, A., Jenni, P., Linssen, L., Livan, M., Maio, A., Mapelli, L., Mondardini, M.R., Ong, B., Paar, H.P., Pastore, F., Poggioli, L., Polesello, G., Riccardi, F., Rimoldi, A., Scheel, C.V., Schmitz, J., Seixas, J.M., Simon, A., Sivertz, M., Sonderegger, P., Souza, M.N., Thome, Z.D., Vercesi, V., Wang, Y., Wigmans, R., Xu, C. and You, K. (1991). Electron, pion and multiparticle detection with a lead/scintillating-fiber calorimeter, *Nucl. Instr. and Meth. in Phys. Res. A* **308**, 481–508, doi: 10.1016/0168-9002(91)90062-U.

Actel (2006). Single-Event Effects in FPGAs: Ground Level and Atmospheric Background Radiation Effects in FPGAs, available on the web site:
http://www.actel.com/documents/FirmErrorPIB.pdf

Adachi, S. (2004). Handbook on Physical Properties of Semiconductors (Volume 1), Kluwer Academic Publisher, Boston.

Adams, J.R. Jr et al. (1991). *Astrophys. J.*, **377**, 292.

Afek, Y. et al. (1977). The collective tube model for high energy particle–nucleus and nucleus–nucleus collision, Proc. of the Topical meeting on Multiparticle Production on Nuclei at Very High Energies, edited by Bellini, G., Bertocchi, L. and Rancoita, P.G., IAEA-SMR-21, 591.

Aglietta, M. et al. (EAS–TOP Collab.) (1997). Comparison of the electron and muon data in extensive air showers with the expectations from a cosmic ray composition and hadron interaction model, Proceedings of the Vulcano Workshop on Frontier Objects

in Astrophysics and Particle Physics, Vulcano, Italy, 27 May-1 June 1996, Giovannelli, F. and Mannocchi, G. Editors, IPS, Bologna IPS Conference Proceedings No
57, 395.

Agostinelli, S. et al. (2003). *Nucl. Instr. and Meth. in Phys. Res. A* **506**, 250.
See also the web site:
http://geant4.web.cern.ch/geant4/

Aharonian, F.A. and Atoyan, A.M. (1991). *J. Phys. G: Nucl. Part. Phys.* **17**, 1769, doi:
10.1088/0954-3899/17/11/021.

Ahlen, S. (1980). *Rev. Mod. Phys.* **52**, 121.

Aitken, D.W. et al. (1969). *Phys. Rev.* **179**, 393.

Akapdjanov, G.A. et al. (1977). *Nucl. Instr. and Meth.* **140**, 441.

Åkesson, T. et al. (1985). *Nucl. Instr. and Meth. in Phys. Res. A* **241**, 17.

Akkerman, A., Barak, J., Chadwick, M.B., Levinson, J., Murat, M. and Lifshitz, Y. (2001).
Radiat. Phys. Chem. **62**, 301.

Akrawy, M. et al. (1990). *Nucl. Instr. and Meth. in Phys. Res. A* **290**, 76.

Alfvén, H. (1942). *Nature* **14**, 405.

Alfvén, H. (1976). *J. Geophys. Res.* **81**, 4019.

ALICE (1993). Letter of Intent, CERN/LHCC/93-16.

Allard, D., Parizot, E. and Olinto, A.V. (2007). *Astropart. Phys.* **27**, 61.

Alma–Ata–Leningrad–Moscow–Tashkent Collab. (1974). *Sov. J. Nucl. Phys.* **19**, 536.

Alsmiller, R.G. et al. (1990). *Nucl. Instr. and Meth. in Phys. Res. A* **295**, 337.

Altegoer, J. et al. (NOMAD Collab.) (1998). *Nucl. Instr. and Meth. in Phys. Res. A* **404**,
96.

Altrock, R.C. et al. (1985). The Sun, Chapter 1 of Handbook of Geophysics and the Space
Environment, Editor Jursa, A.S., Air Force Geophysics Lab, Springfield.

Alurkar, S.K. (1997). Solar and Interplanetary Disturbances (Chapter 4), World Scientific,
Singapore.

Amaldi, U. (1981). Fluctuations in Calorimetry Measurements, *Physica Scripta* **23**, 409–
424, doi: 10.1088/0031-8949/23/4A/012; published from 2006 by Institute of Physics
on behalf of the Royal Swedish Academy of Sciences for the Science Academies and
the Physical Societies of the Nordic Countries.

Amendolia, S.R. et al. (1980). Report Pisa 80-4.

AMS Collab.: Alcaraz, J. et al. (1999). *Phys. Lett. B* **461**, 387.

AMS Collab.: Alcaraz, J. et al. (2000) (a). *Phys. Lett. B* **472**, 215.

AMS Collab.: Alcaraz, J. et al. (2000) (b). *Phys. Lett. B* **494**, 193.

AMS Collab.: Alcaraz, J. et al. (2000) (c). *Phys. Lett. B* **484**, 10.

AMS Collab.: Alcaraz, J. et al. (2000) (d). *Phys. Lett. B* **472**, 215.

AMS Collab.: Aguilar, M. et al. (2002). *Phys. Reports* **366**, 331.

Anders, B. et al. (1988). *Nucl. Instr. and Meth. in Phys. Res. A* **270**, 140.

Anders, B. et al. (1989). *Nucl. Instr. and Meth. in Phys. Res. A* **277**, 56.

Andersen, H.H., Simonsen, H. and Sørensen, H. (1969). *Nucl. Phys. A* **125**, 171.

Andersen, H.H. (1983). *Phys. Scripta* **28**, 268.

Andersen, H.H. (1985). The Barkas Effect and Other Higher-Order Z_1 Contributions to
the Stopping Power, in Semiclassical Descriptions of Atomic and Nuclear Collisions,
Proc. of the Niels Bohr Centennial Conf., Copenhagen (25–28 March 1985), North-
Holland, Amsterdam.

Andersen, H.H. and Ziegler, J.F. (1977). Hydrogen Stopping powers and Ranges in all
Elements, in The Stopping and Range of Ions in Matter, Vol. 2, Ziegler, J.F. Editor,
Pergamon Press, New York.

Andersen,H.H., Simonsen, H. and Sørensen, H. (1969). *Nucl. Phys. A* **125**, 171.

Anderson, D.F. (1990). *Nucl. Instr. and Meth. in Phys. Res. A* **287**, 606.

Anderson, R.L. et al. (1976). Proposal PEP-6 SLAC.

Andi, G. and Wapstra, A.H. (1993). *Nucl. Phys. A* **565**, 1.

Andreo, P., Ito, R. and Tabata, T. (1992). Tables of Charge- and Energy-Deposition Distributions in elemental Materials irradiated by plane-parallel Electron Beams with Energies between 0.1 and 100 MeV, University of Osaka Technical Report 1.

Andrieux, M.L. et al. (1999). *Nucl. Instr. and Meth. in Phys. Res. A* **427**, 568.

Angelescu, T. et al. (1994). *Nucl. Instr. and Meth. in Phys. Res. A* **345**, 303.

Anton, G. et al. (2006). *Nucl. Instr. and Meth. in Phys. Res. A* **563**, 116.

Antony, P.L. et al. (1995). *Phys. Rev. Lett.* **75**, 1949.

Anzivino, G. et al. (1993). Proceedings of the 4th International Conference on Calorimetry in High Energy Physics, La Biodola, Isola d'Elba, Italy, 19–25 September, Editors Menzione, A. and Scribano, A., World Scientific 1993, 401.

Apfel, R.E. (1979). *Nucl. Instr. and Meth.* **162**, 603.

Apfel, R.E., Roy, S.C. and Lo, Y.C. (1985). *Phys. Rev. A* **31**, 3194.

Arley, N. (1938). *Proc. Roy. Soc. A* **168**, 519.

Arefiev, S. et al. (1989). *Nucl. Instr. and Meth. in Phys. Res. A* **285**, 403.

Arista, N.R. and Lifschitz, A.F. (1999). *Phys. Rev. A* **59**, 2719.

Ariztizabal, F. et al. (1994). *Nucl. Instr. and Meth. in Phys. Res. A* **349**, 384.

Arnison, G. et al. (UA1 Collab.) (1983). *Phys. Lett. B* **122**, 103; *Phys. Lett. B* **122**, 103; *Phys. Lett. B* **126**, 398; *Phys. Lett. B* **129**, 273; *Phys. Lett. B* **134**, 469 (1984); *Phys. Lett. B* **135**, 250 (1984).
See also:
Banner, M. et al. (UA2 Collab.) (1983). *Phys. Lett. B* **122**, 476.
Bagnaia P. et al. (UA2 Collab.) (1983). *Phys. Lett. B* **129**, 130; *Zeit. Phys. C* **24**, 1 (1984).

Aronson, S. et al. (1988). *Nucl. Instr. and Meth. in Phys. Res. A* **269**, 492.

Arora, N.D., Hauser, J.R. and Roulston, D.J. (1982). *IEEE Trans. on Electron Devices* **29 (no. 2)**, 292, doi: 10.1109/16.121685.

Artru, X., Yodh, G.B. and Mennessier, G. (1975). *Phys. Rev. D* **12**, 1289.

Asano, Y., Hamasaki, H., Mori, S., Sasaki, S., Tsujita, Y. and Yusa, K. (1996). *J. Radioanlyt. Nucl. Chem.* **210 (no. 1)**, 79.

Aschwanden, M.J. (2006). The Sun, Encyclopedia of the Solar System - 2nd Edition, McFadden, L.-A., Weissman, P.R. and Johnson, T.V. Editors, Academic Press, San Diego.

ASG (Application Software Group) (1992). CERN Program Library D506, MINUIT Function Minimization and Error Analysis.

Ashley, K.L. and Milnes, A.G. (1964). *J. of App. Phys.* **35**, 369.

Ashley, J.C., Ritchie, R.H. and Brandt, W. (1972). *Phys. Rev. B* **5**, 2393.
See also:
Jackson, J.D. and McCarty, R.L. (1972). *Phys. Rev. B* **6**, 4131.

ASTM (1985). Standard Practice for characterizing neutron energy fluence spectra in terms of equivalent monoenergetic neutron fluence for radiation-hardness testing electronics *E72285*
See also:
Revision 04 (2005).

ASTM (2003). Standard Practice for dosimetry in an Electron Beam Facility for Radiation Processing for Energies between 300 keV and 25 MeV *ISO/ASTM 51649:2002E*.

ASTM (2007). Terrestrial Reference Spectra for Photovoltaic Performance Evaluation, *ASTM G-173-03*.

Atac, M. et al. (1983). *Nucl. Instr. and Meth. in Phys. Res.* **205**, 113.

ATLAS Collab. (1994) (a). Advanced Toroidal LHC Apparatus, CERN/LHCC/94-43, LHCC/P2 15 December.

ATLAS Collab. (1994) (b). Technical Proposal for a General-Purpose pp Experiment at the Large Hadron Collider at CERN, LHCC/P2, CERN/LHCC/94-43.

ATLAS Collab. (1996). Liquid Argon Calorimeter Technical Design Report, CERN/LHCC/96-41.

ATLAS Collab.: Ajaltouni, Z. et al. (1997). *Nucl. Instr. and Meth. in Phys. Res. A* **387**, 333.

Atwood, W.B. (1981). Time-of-Flight Measurements, 8th SLAC Summer Institute on Particle Physics, Stanford, CA, USA, Proceedings Ed. Mosher, A., SLAC Stanford.

Aubert, B. et al. (1991). *Nucl. Instr. and Meth. in Phys. Res. A* **309**, 438; *Nucl. Instr. and Meth. in Phys. Res. A* **321**, 467 (1992).

Aubert, B. et al. (1993). *Nucl. Instr. and Meth. in Phys. Res. A* **334**, 383.

Auffray, E. et al.(1996). *Nucl. Instr. and Meth. in Phys. Res. A* **378**, 171.

Auger, P. (1925). *Copt. Rend.* **180**, 65.

Axen, D. et al. (SLD Collab.) (1993). *Nucl. Instr. and Meth. in Phys. Res. A* **328**, 472.

Azimov, S.A. (1977). Inelastic Hadron-Nucleus interactions in the energy range from 20 up to 200 GeV, Proc. of the Topical meeting on Multiparticle Production on Nuclei at Very High Energies, edited by Bellini G., Bertocchi L. and Rancoita P.G., IAEA-SMR-21, 83.

Azimov, S.A. et al. (1978). *Phys. Lett. B* **73**, 500.

Babaev, A. et al. (1979). *Nucl. Instr. and Meth.* **160**, 427.

Babcock, H.D. and H.W. (1954). *Astrophys. J.* **121**, 349.

Babecki, J. and Nowak, G. (1979). On Differences between some Characteristics of slow Particles produced in Iteractions of Pions and Protons with Nuclei at High Energies, report no. 1050-PH (Krakow).

Baccaro, S. et al. (1995). *Nucl. Instr. and Meth. in Phys. Res. A* **361**, 209.

Badel, X. (2003). Development of macropore arrays in silicon and related technologies for X-ray imaging applications, Ph.D. Thesis in *Materials and Semiconductor Physics* (*1st part*), Stockholm, KTH, Royal Institute of Technology, integrated part of 3D Radiation Imaging Detectors (3D-RID) European Project.

Bagdasarovand, S. and Goulianos, K. (1993). Proceedings of the 4th International Conference on Calorimetry in High Energy Physics, La Biodola, Isola d'Elba, Italy, 19–25 September 1993, Menzione, A. and Scribano, A. Editors, World Scientific, Singapore, 524.

Bak, J.F., Burenkov, A., Petersen, J.B.B., Uggerhøj, E., Møller, S.P. and Siffert, P. (1987). *Nucl. Phys. B* **288**, 682.

Bakale, G., Sowada, U. and Schmidt, W.F. (1976). *Journ. Phys. Chem.* **80**, 2556.

Balanzat, E. and Bouffard, S. (1993). *Solid State Phenomena* **30** & **31**, 7.

Banaszkiewicz, M., Axford, W.I. and McKenzie, J.F. (1998). *Astron. Astrophys.* **337**, 940.

Barashenkov, V.S. and Polanski, A. (1994). Electronic Guide for Nuclear Cross-Sections, *JINR E2-94-417*, Dubna.

Barate, R., Bonamy, P., Borgeaud, P., Burchell, M., David, M., Lemoigne, Y., Magneville, C., Movchet, J., Noon, P., Poinsignon, J., Primout, M., Rancoita, P.G. and Villet, G. (1985). *Nucl. Instr. and Meth. in Phys. Res.* **235**, 235, doi: 10.1016/0168-9002(85)90558-3.

Barber, H.D. (1967). *Solid–State Elec.* **10**, 1039.

Barberis, E. et al. (1994). *Nucl. Instr. and Meth. in Phys. Res. A* **342**, 90.

Bardeen, J. and Brattain, W.H. (1948). *Phys. Rev.* **74**, 230.

Barkas, W.H., Dyer, J.N. and Heckman, H.H. (1963). *Phys. Rev. Lett.* **11 (no. 1)**, 26; **11 (no. 3)**, 138.

Bar-Lev, A. (1993). Semiconductors and Electronic Devices - 3rd Edition, Prentice Hall, Hemel Hempstead.

Barnabé-Heider, M. et al. (2002). Proceedings of the 7th International Conference on Advanced Technology and Particle Physics, October 15–19 2001, Villa Olmo, Como, Italy, Barone, M., Borchi, E., Huston, J., Leroy, C., Rancoita, P.G., Riboni, P.L. and Ruchti, R. Editors, World Scientific, Singapore, 85.

Barnabé-Heider, M. et al. (2004). Proceedings of the 8th International Conference on Advanced Technology and Particle Physics, October 6–11 2003, Villa Erba, Como, Italy, Barone, M., Borchi, E., Leroy, C., Rancoita, P.G., Riboni, P.L. and Ruchti, R. Editors, to be published by World Scientific, Singapore.

Barnabé-Heider, M. et al. (2005) (a). *Nucl. Instr. and Meth. in Phys. Res. A* **555**, 184.

Barnabé-Heider, M. et al. (2005) (b). *Phys. Lett. B* **624**, 186.

Baroncelli, A. (1974). *Nucl. Instr. and Meth.* **118**, 445.

Barth, J. (1997). Short Course on Applying Computer Simulation Tools to Radiation Effects Problems, 1997 IEEE Nuclear and Space Radiations Effects Conferences, Snowmass Village, 21–25 July.

Barton, C.E. (1997). International Geomagnetic Reference Field: The Seventh Generation, *J. Geomag. Geoelectr.* **49**, 123.

Barwick, S.W. et al. (HEAT Collab.) (1997). *Nucl. Instr. and Meth. in Phys. Res. A* **400**, 34.

Barwick, S.W. et al. (1998). *Astrophys. J.* **498**, 779.

Baschirotto A., Boella G., Cappelluti I., Castello R., Cermesoni, M., Gola, A., Pessina, G., Pistolesi, E., Rancoita, P.G. and Seidman, A. (1999). A radiation hard bipolar monolithic front-end readout, *Nucl. Instr. and Meth. in Phys. Res. B* **155**, 120–131, doi: 10.1016/S0168-583X(99)00234-7.

Baschirotto, A., Bosetti, M., Castello, R., Cermesoni, M., Gola, A., Rancoita, P.G., Rattaggi, M., Redaelli, M., Seidman, A. and Terzi, G. (1993). *Nucl. Phys.B (Proc. Suppl.)* **32**, 535, doi: 10.1016/0920-5632(93)90068-H.

Baschirotto, A., Castello, R., Gola, A., Onado, C., Pessina, G., Rancoita, P.G., Redaelli, M. and Seidman, A. (1996). *Nucl. Instr. and Meth. in Phys. Res. B* **114**, 327, doi: 10.1016/0168-583X(96)00228-5.

Baschirotto, A., Castello, R., Gola, A., Pessina, G., Rancoita, P.G., Rattaggi, M., Redaelli, M. and Seidman, A. (1995) (a). *Nucl. Phys.B (Proc. Suppl.)* **44**, 621, http://dx.doi: 10.1016/S0920-5632(95)80096-4.

Baschirotto, A., Castello, R., Gola, A., Pessina, G., Rancoita, P.G., Rattaggi, M., Redaelli, M. and Seidman, A. (1995) (b). *Nucl. Instr. and Meth. in Phys. Res. A* **362**, 466, doi: 10.1016/0168-9002(95)00287-1.

Baschirotto, A., Castello, R., Onado, C., Pessina, G., Rancoita, P.G. and Seidman, A. (1997). *Nucl. Instr. and Meth. in Phys. Res. B* **122**, 73, doi: 10.1016/S0168-583X(96)00640-4.

Bathow, G. et al. (1970). *Nucl. Phys. B* **20**, 592.

Battiston, R. (2002). The Alpha Magnetic Spectrometer, a Particle Physics Experiment in Space, Proc. of the 7th ICATPP (15–19 October 2001, Como, Italy), ISBN 981-238-180-5, Editors Barone, M., Borchi, E., Huston, J., Leroy, C., Rancoita, P.G., Riboni, P. and Ruchti, R., World Scientific, Singapore, 3.

Battistoni, G. and Grillo, A. F. (1996). Introduction to high energy cosmic ray physics, the Fourth Trieste School on Non Accelerator Particle Astrophysics, Bellotti, E., Carrigan, R.A., Giacomelli, G. and Paver, N. Editors, World Scientific, Singapore,

341.

Bauer, T.H., Spital, R.D. and Yennie, D.R. (1978). *Rev. Mod. Phys.* **50**, 261.

Baumgart, R. (1987). Ph.D. Thesis, Siegen University.

Baumgart, R. et al. (1988) (a). *Nucl. Instr. and Meth. in Phys. Res. A* **268**, 105.

Baumgart, R. et al. (1988) (b). *Nucl. Instr. and Meth. in Phys. Res. A* **272**, 722.

Beattie, L.J., Chilingarov, A. and Sloan, T. (1998). Forward Bias I–V Characteristics for Heavily Irradiated Silicon Diodes, *ROSE/TN/98-1*.

Beattie, L.J., Chilingarov, A. and Sloan, T. (2000). *Nucl. Instr. and Meth. in Phys. Res. A* **439**, 293.

Bebek, C. (1988). *Nucl. Instr. and Meth. in Phys. Res. A* **265**, 258.

Bechevet, D., Glaser, M., Houdayer, A., Lebel, C., Leroy, C., Moll, M. and Roy, P. (2002). *Nucl. Instr. and Meth. in Phys. Res. A* **479**, 487, doi: 10.1016/S0168-9002(01)00925-1.

Beguwala, M. and Crowell, C.R. (1974). *Solid–State Elec.* **17**, 203.

Behrens, U. et al. (1990). *Nucl. Instr. and Meth. in Phys. Res. A* **289**, 115.

Belau, E. et al. (1983). *Nucl. Instr. and Meth. in Phys. Res.* **214**, 253.

Belau, E., Klanner, R., Lutz, G., Neugebauer, E. and Wylie, A. (1983). *Nucl. Instr. and Meth. in Phys. Res.* **205**, 99, doi: 10.1016/0167-5087(83)90177-1.

Beliaev, A.A., Nymmik, R.A., Panasyuk, M.I., Peravaya, T.I. and Suslov, A.A. (1996). *Radiat. Meas.* **26 (no. 3)**, 481.

Bell, C.R., Oberle, N.P., Rohsenow, W., Todreas, N. and Tso, C. (1973). *Nucl. Sci. Eng.* **53**, 458.

Bell, W.H. et al. (1999). *Nucl. Instr. and Meth. in Phys. Res. A* **435**, 187.

Bellini, G. et al. (1981). The 3π–nucleon cross section in coherent production on nuclei at 40 GeV/c, CERN-EP/81-40.

Belymam, A. et al. (1998). *Nucl. Instr. and Meth. in Phys. Res. B* **134**, 217.

Bendel, W.L. and Petersen, E.L. (1983). *IEEE Trans. on Nucl. Sci.* **30**, 4481.

Bengston, B. and Moszynski, M. (1982). *Nucl. Instr. and Meth. in Phys. Res.* **204**. 129.

Benvenuti, A. et al. (1975). *Nucl. Instr. and Meth.* **125**, 447.

Berestetskii, V.B., Lifshitz, E.M. and Pitaevskii, L.P. (1971). Relativistic Quantum Theory (translated from Russian by Sykes, J.B. and Bell, J.S.), Pergamon Press, London.

Beretvas, A. et al. (1993). *Nucl. Instr. and Meth. in Phys. Res. A* **329**, 50.

Berezinskii, V.S., Bulanov, S.V., Dogiel, V.A., Ginzburg, V.L. and Ptuskin, V.S. (1990). Astrophysics of Cosmic Rays, North Holland, Amsterdam.

Berger, M.J. and Seltzer, S.M. (1964). Tables of Energy Losses and Ranges of Electrons and Positrons from: Studies in Penetration of charged particles in matter, Publication 1133, National Academy of Sciences - National Research Council, Washington D.C., 205.

Berger, M.J., Coursey, J.S., Zucker, M.A. and Chang, J. (2005). ESTAR, PSTAR and ASTAR: Computer Programs for Calculating Stopping-Power and Range Tables for Electrons, Protons and Helium Ions (version 1.2.3), National Institute of Standards and Technology, Gaithersburg, MD; available on the web site (2007): *http://physics.nist.gov/Star* originally published as: Berger, M.J. (1993). NISTIR 4999, National Institute of Standards and Technology, Gaithersburg, MD. The databases ESTAR (for electrons), PSTAR (for protons) and ASTAR (for α-particles) provide stopping-power and range tables according to methods described in [ICRUM (1984b)] and [ICRUM (1993a)]; stopping-power and range tables can be calculated for electrons in any user-specified material and for protons and α-particles in 74 materials.

Berger, M.J., Hubbell, J.H., Seltzer, S.M., Chang, J., Coursey, J.S., Sukumar, R. and Zucker, D.S. (2005). XCOM: Photon Cross Section Database (version 1.3), National Institute of Standards and Technology, Gaithersburg, MD; available on the web site (2007):

http://physics.nist.gov/xcom

Originally published as:

Berger, M.J. and Hubbell, J.H. (1987). XCOM: Photon Cross Sections on a Personal Computer, NBSIR 87-3597, National Bureau of Standards (former name of NIST), Gaithersburg, MD; and as:

Berger, M.J. and Hubbell, J.H. (1990). NIST X-ray and Gamma-ray Attenuation Coefficients and Cross Sections Database, NIST Standard Reference Database 8, Version 2.0, National Institute of Standards and Technology, Gaithersburg, MD.

Bernabei, R. et al. (1998). *Phys. Lett. B* **436**, 379.

Bernardi, E. (1987). DESY Internal report F1-87-01.

Bernardi, E. et al. (1987). *Nucl. Instr. and Meth. in Phys. Res. A* **262**, 229.

Bernardini, G. et al. (1952). *Phys. Rev.* **88**, 1017.

See also:

Goldberger, M.L. (1948). *Phys. Rev.* **74**, 1268.

Golubeva, Ye.S., Iljinov, A.S., Botvina, A.S. and Sobolevsky, N.M. (1988). *Nucl.Phys. A* **483**, 539.

Cugnon, J., Mizutani, T. and Vandermeulen, J. (1981). *Nucl. Phys. A* **352**, 505.

Bertini, H.W. et al. (1963). *Phys.Rev.* **134**, 1801.

Yariv, Y. et al. (1979). *Phys.Rev. C* **20**, 2227.

Cloth, P. et al. (1988). HERMES, A Monte Carlo Program System for Beam materials interaction studies, Report Juel 2203, ISSN 0366-0885.

Bertocchi, L. and Treleani, D. (1977). *J. Phys. G* **3**, 147.

Bertolotti, M. (1968). Radiation Effects in Semiconductors, Proc. of the Santa Fe Conference, Vook, F.L. Ed., Plenum Press, New York, 311.

Bethe, H.A. (1930). *Ann. Physik* **5**, 325.

Bethe, H.A. (1953). *Phys. Rev.* **89**, 1256.

See also:

Scott, W.T. (1963). *Rev. Mod. Phys.* **35**, 231.

Marion, J.B. and Zimmermann, B.A. (1967). *Nucl. Instr. and Meth.* **51**, 93.

Bethe, H.A. and Ashkin, J. (1953). Passage of Radiation through Matter, in Segre, E. (Editor), Experimental Nuclear Physics, Vol. 1, Part 2, John Wiley & Sons, New York.

Bethe, H.A. and Heitler, W. (1934). *Proc. Roy. Soc. A* **146**, 83; *Proc. Cambridge Phil. Soc.* **30**, 524.

Beusch, W., Polgar, E., Websdale, D., Freudenreich, K., Gentit, F.X., Muhlemann, P., Pernegr, J., Wetzel, W., Astbury, P., Lee, J.G., Letheren, M., Bellini, G., Di Corato, M., Palombo, F., Rancoita, P.G. and Vegni, G. (1975). *Phys. Lett. B* **55**, 97, doi: 10.1016/0370-2693(75)90196-3.

See also:

Ting, S.C.C. (1970). Interactions of Photons with Matter, in Methods in Subnuclear Physics Vol. 5, Nikolić, M. Ed., New York, 307.

Bauer, T.H. et al. (1978). *Rev. of Mod. Phys.* **50**, 261.

Bhabha, H.J. (1936). *Proc. Roy. Soc A* **154**, 195.

Bhabha, H.J. and Chakrabarty, S.K. (1943). *Proc. Roy. Soc. A* **181**, 267.

See also:

Tamm, I. and Belenky, S. (1946). *Phys. Rev.* **70**, 660.

Bhabha, H.J. and Heitler, W. (1936). *Proc. Roy. Soc. A* **159**, 432.

Bialas, A. (1978). Particle production from nuclear targets and the structure of the hadrons, Fermilab-Pub-78/75-THY.

Bianchi, F. et al. (1989). *Nucl. Instr. and Meth. in Phys. Res. A* **279**, 473, doi: 10.1016/0168-9002(89)91294-1.

Bichsel, H. (1970). *Phys. Rev. B* **1**, 2854.

Bichsel, H. (1988). *Rev. of Mod. Phys.* **60**, 663.

Bichsel, H. (1990). *Phys. Rev. A* **41**, 3642.

Bichsel, H. (1992). *Phys. Rev. A* **46**, 5761.

Bichsel, H. and Saxon, R.P. (1975). *Phys. Rev. A* **11**, 1266.

Bichsel, H. and Yu, S. (1972). *IEEE Trans. Nucl. Science* **19**, 172.

Bieber, J.W. and Matthaeus, W.H. (1997). *Astrophys. J.* **485**, 655.

Biersack, J.P. and Haggmark, L. (1980). *Nucl. Instr. and Meth.* **174**, 257.

Biggeri, U., Borchi, E., Bruzzi, M. and Lazanu, S. (1995). *Nucl. Instr. and Meth. in Phys. Res. A* **360**, 131.

Biggeri, U., Borchi, E., Bruzzi, M., Pirollo, S., Sciortino, S., Lazanu, S. and Li, Z. (1995). *Nucl. Instr. and Meth. in Phys. Res. A* **400**, 113.

Biggs, F., Mendelsohn, L.B. and Mann, J.B. (1975). *Atomic Data and Nuclear Data Tables* **16**, 201.

Binder, D. (1988). *IEEE Trans. on Nucl. Sci.* **35**, 1570.

Bingham, H.H. (1970). Review of Coherent Multiparticle Production Reactions from Nuclei, Proceedings of the Topical Seminar on Interactions of Elementary Particles with Nuclei, Bellini, G., Bertocchi, L. and Bonetti, S. Editors, Trieste September 15–17, Publ. by Ist. Naz. di Fis. Nucl sez. di Trieste, 38

Birks, J. (1951). *Proc. Phys. Soc. A* **64**, 874.

Birks, J. (1964). The Theory and Practice of Scintillation Counting, New York, Macmillan.

Bitinger, D. (1990). Depth of Calorimetry for SSC Experiments, Proc. of the Workshop on Calorimetry for SSC, Tuscaloosa (USA), 13–17 March 1989, Donaldson, R. and Gilchriese, M.G.D. Editors, World Scientific, Singapore, 91.

Blackett, P.M.S. and Occhialini, G.P.S. (1933). *Proc. Roy. Soc. A* **139**, 699.

Blackett, P.M.S., Occhialini, G.P.S and Chadwick, J. (1933). *Proc. Roy. Soc. A* **144**, 235.

Blakemore, J.S. (1987). Semiconductor Statistics, Dover Publications, Mineola, New York.

Blatt, J.M. and Weisskopf, V.F. (1952). Theoretical Nuclear Physics, John Wiley & Sons Publ., New York.

Blau, M. (1961). Photographic Emulsions, Methods of Experimental Physics Vol. 5 part A, Yuan, L.C.L. and Wu, C.-S. Editors, Academic Press, New York, 268.

Bleaney, B.I. and Bleaney, B. (1965). Electricity and Magnetism, Oxford University Press, London.

Bleichert, B. et al. (1987). *Nucl. Instr. and Meth. in Phys. Res. A* **254**, 529; *Nucl. Instr. and Meth. in Phys. Res. A* **241**, 43 (1985).

Bloch, F. (1933). *Z. Physik* **81**, 363 and *Ann. Phys.* **16**, 285.

Blood, P. and Orton, J.W. (1978). *Rep. Prog. in Phys.* **41**, 157.

Blood, P. and Orton, J.W. (1992). The Electrical Characterization of Semiconductors: Majority Carriers and Electron States, Academic Press, London.

Blucher, E. et al. (1986). *Nucl. Instr. and Meth. in Phys. Res. A* **249**, 201.

Bludau, W. and Onton,A. (1974). *J. of App. Phys.* **45**, 1846.

Blunck, O. and Leisegang, S (1950). *Z. Phys.* **128**, 500.
    See also:
    Symon, K.R. (1948). Flutuations in Energy Loss by High Energy Charged Particles

in Passing through Matter (Thesis), Harward Univ.

Boberg, P.R. et al. (1995). *Geophys. Res. Lett.*, **22**, 1133.

Bobik, P., Boella, G., Boschini, M.J., Gervasi, M., Grandi, D., Kudela, K., Pensotti, S. and Rancoita, P.G. (2006). Magnetospheric Transmission Function Approach to disentangle Primary from Secondary Cosmic Ray Fluxes in the Penumbra Region, *J. Geophys. Res.* **111**, A05205, doi: 10.1029/2005JA011235.
See also:

Bobik, P., Boschini, M.J., Gervasi, M., Grandi, D., Micelotta, E. and Rancoita, P.G. (2005). *Int. J. of Modern Phys. A* **20**, 6678, doi: 10.1142/S0217751X05029782.

Bobik, P., Boschini, M.J., Gervasi, M., Grandi, D., Kudela, K., Potenza, M. and Rancoita, P.G. (2000). Particle tracing to study the properties of the Earth magnetosphere, Proc. of the IX GIFCO Conference, Lecce (24–26/5/2000), Editors Aiello, S. and Bianco, A., SIF, Bologna, *Conference Proc.* **Vol 68**, 311.

Bobik, P., Boschini, M.J., Gervasi, M., Grandi, D., Kudela, K., Potenza, M. and Rancoita, P.G. (2001). Study of cosmic ray access to a Space detector by particle tracing in the Earth magnetosphere, Proceedings of the Vulcano Workshop 2000 (Frontier Objects in Astrophysics and Particle Physics), Vulcano (21–27/5, 2000), Editors Giovannelli, F. and Mannocchi, G., SIF, Bologna, *Conference Proc.* **Vol 73**, p. 379.

Bobik, P., Boschini, M.J., Gervasi, M., Grandi, D., Micelotta, E. and Rancoita, P.G. (2005). A Back-Tracing Code To Study The Evolution of the Magnetosphere Transmission Function for Primary Cosmic Rays in The Inner Magnetosphere: Physics and Modeling, Proc. of the CHAPMAN Conference (25–29 August 2003, Helsinki, Finland), Editors Pulkkinen, T.I., Tsyganenko, N.A. and Friedel, R.H.W., American Geophysical Union (AGU), Washington D.C., *Geophysical Monograph Series*, Volume **155**, ISBN 0-87590-420-3, 301.

Bobik, P., Boschini, M.J., Gervasi, M., Grandi, D. and Rancoita, P.G. (2006) (a). A 2D Stochastic Montecarlo for the Solar Modulation of GCR: a Procedure to fit Interplanetary Parameters comparing to the Experimental Data, Proc. of the 9th ICATPP (17–21 October 2005, Como, Italy), ISBN 981-256-798-4, Editors Barone, M., Borchi, E., Gaddi, A., Leroy, C., Price, L., Rancoita, P.G. and Ruchti, R., World Scientific, Singapore, 206.

Bobik, P., Boschini, M.J., Gervasi, M., Grandi, D. and Rancoita, P.G. (2006) (b). Ions Abundance close to the Earth Surface: the Role of the Magnetosphere, Proc. of the 9th ICATPP (17–21 October 2005, Como, Italy), ISBN 981-256-798-4, Editors Barone, M., Borchi, E., Gaddi, A., Leroy, C., Price, L., Rancoita, P.G. and Ruchti, R., World Scientific, Singapore, 928.
See also:

Bobik, P., Boella, G., Boschini, M.J., Gervasi, M., Grandi. D., Kudela, K., Pensotti, S. and Rancoita, P.G. (2006). The Transmission Function of Helium, Carbon and Iron and their flux inside the magnetosphere, presented at the 20th European Cosmic Rays Symposium (5–8 September 2006, Lisbon, Portugal).

Bobik, P., Gervasi, M., Grandi, D., Kudela, K., Micelotta, E., Rancoita, P.G. and Usoskin, I. (2003). 2D Stochastic Simulation Model of Cosmic Ray Modulation: Comparison with Experimental Data, Proceedings of the Solar Variability as an input to the Earth's Environment Conference (23–28 June 2003, Tatranska Lomnica, Slovakia), ESA SP-535, 637.

Bocciolini, M. et al. (CAPRICE Collab.) (1996). *Nucl. Instr. and Meth. in Phys. Res. A* **370**, 403.

Bock, R.K. et al. (1981). *Nucl. Instr. and Meth. in Phys. Res.* **186**, 533.

Boedeker, K.L., Cooper, V.N. and McNitt-Gray, M.F. (2007). *Phys. Med. Biol.* **52**, 4027.

Boella, G. et al. (1998). *Astropart. Phys.* **9**, 261.
See also:
Effect of the solar modulation at minimum activity on expected ratio of antiproton to proton flux, Proc. of COSMO-97 International Workshop on Particle Physics and Early Universe, Ambleside (UK), World Scientific, Singapore, 273 (1998).

Boella, G, Boschini, M.J., Gervasi, M., Grandi, D., Pensotti, S. and Rancoita, P.G. (2008). Evaluation of the flux of CR nuclei inside the magnetosphere, Proc. of the 10th ICATPP (8–12 October 2007, Como, Italy), Editors Barone, M., Gaddi, A., Leroy, C., Price, L., Rancoita, P.G. and Ruchti, R., World Scientific, Singapore, 875.

Boella, G., Gervasi, M., Mariani, S., Rancoita, P.G. and Usoskin, I. (2001). *J. Geophys. Res.* **106**, 355, doi: 10.1029/2001JA900075.

Boella, G., Gervasi, M., Potenza, M.A.C., Rancoita, P.G. and Usoskin, I. (1998). *Astropart. Phys.* **9**, 261, doi: 10.1016/S0927-6505(98)00022-X.

Boersch-Supan, W. (1961). *Res. Natl Bur. Stand.* **65B**, 245.

Boezio, M. et al. (1999). *Astrophys. J.* **518**, 457.

Boezio, M. et al. (2006). *Astropart. Phys.* **26**, 111.

Bohr, A. and Mottelson, B.R. (1969). Nuclear Structure, Volume I Nuclear Deformations, W.A. Benjamin Inc., New York; 2nd Edition, World Scientific, Singapore (1998).

Bohr, N. (1913). *Phil. Mag.* **25**, 10; *Phil. Mag.* **30**, 581 (1915).

Bohr, N. (1940). *Phys. Rev.* **58**, 654; *Phys. Rev.* **59**, 270 (1941).

Bondarenko, V., Krause-Rehberg, R., Feick, H. and Davia, C. (2004). *J of Mat. Scie.* **39**, 919.

Borchi, E. and Bruzzi, M. (1994). *La Riv. Del Nuovo Cimento* **17 (no. 11)**, 1.

Borchi, E., Bertrand, C., Leroy, C., Bruzzi, M., Furetta, C., Paludetto, R., Rancoita, P.G., Vismara, L. and Giubellino, P. (1989). *Nucl. Instr. and Meth. in Phys. Res. A* **279**, 277, doi: 10.1016/0168-9002(89)91093-0.

Borchi, E., Bruzzi, M., Leroy, C., Pirollo, S. and Sciortino, S. (1998). *Nucl. Phys. B (Proc. Suppl.)* **61B**, 481, doi: 10.1016/S0920-5632(97)00606-3.

Borchi, E., Furetta, C., Leroy, C., Macii, R., Manoukian-Bertrand, C., Paludetto, R., Pensotti, S., Rancoita, P.G., Rattaggi, M., Seidman, A. and Vismara, L. (1991). *Nucl. Instr. and Meth. in Phys. Res. A* **301**, 215, doi: 10.1016/0168-9002(91)90461-X.

Borgeaud, P., McEwen, J.C., Rancoita, P.G. and Seidman, A. (1983). *Nucl. Instr. and Meth. in Phys. Res.* **211**, 363, doi: 10.1016/0167-5087(83)90260-0.

Bormann, M. et al. (1985). *Nucl. Instr. and Meth. in Phys. Res. A* **240**, 63.

Bormann, M. et al. (1987). *Nucl. Instr. and Meth. in Phys. Res. A* **257**, 479.

Born, M. (1969). Atomic Physics, Blackie & Son Ltd., Glasgow.

Borsellino, A. (1947). *Nuovo Cim.* **4**, 112; *Helv. Physica Acta* **20**, 136..
See also:
Ghizzetti, A. (1947). *Revista de Matematica y Fisica Teorica* (Univ. Nacianal de Tucuman, Argentina) **6**, 37.

Bosetti, M., Croitoru, N., Furetta, C., Leroy, C., Pensotti, S., Rancoita, P.G., Rattaggi, M., Redaelli, M., Rizzatti, M. and Seidman, A. (1995). DLTS measurement of energetic levels, generated in silicon detectors, *Nucl. Instr. and Meth. in Phys. Res. A* **361**, 461–465, doi: 10.1016/0168-9002(95)00277-4.

Bossen, D.C. and Hsia, M.Y. (1980). *IBM J. Res. Develop.* **24 (no. 3)**, 390.

Botner, O. (1981). *Physica Scripta* **23**, 556.

Botner, O. et al. (1981). *Nucl. Instr. and Meth.* **179**, 45.

Bott-Bodenhausen, M. et al. (1992). *Nucl. Instr. and Meth. in Phys. Res. A* **315**, 236.

Bottino, A. et al. (1997). *Phys. Lett. B* **402**, 113.

Boukhira, N. (2002). Etalonnage aux neutrons d'un détecteur à gouttelettes surchauffées pour la recherche de la matière sombre, MSc. Thesis, University of Montreal.

Boukhira, N. et al. (2000). *Astropart. Phys.* **14**, 227.

Bourgeois, Ch. (1994). Des Processus de Base aux Détecteurs, cours à l'École Internationale Joliot-Curie de Physique Nucléaire 1994, Maubuisson, France, 12–17 septembre, 1.

Bradford, J.T. (1980). *IEEE Trans. on Nucl. Sci.* **27**, 942.

Braibant, S. et al. (2002). *Nucl. Instr. and Meth. in Phys. Res. A* **485**, 343.

Brajša, R. et al. (2001). An Analysis of the Solar Rotation Velocity by Tracing Coronal Features, Recent Insights into the Physics of the Sun and Heliosphere: Highlights from SOHO and Other Space Missions, Proceedings of IAU Symposium 203, Brekke, P., Fleck, B. and Gurman, J.B. Editors, Published by Astronomical Society of the Pacific, San Francisco, Vol. 203, ISBN: 1-58381-069-2, 2001, p. 377.

Brau, J.E. and Gabriel, T.A. (1985). *Nucl. Instr. and Meth. in Phys. Res. A* **238**, 489, doi: 10.1016/0168-9002(85)90490-5.

Brau, J.E. and Gabriel, T.A. (1989). *Nucl. Instr. and Meth. in Phys. Res. A* **279**, 40, doi: 10.1016/0168-9002(89)91061-9.

Brau, J.E. et al. (1985). *IEEE Trans. Nucl. Sci.* **32**, 1.

Brau, J.E., Gabriel, T.A. and Rancoita, P.G. (1989). Prospects and tests of hadron calorimetry with silicon detectors, Proc. of the Summer Study on High Energy in the 1990's (June 27 – July 15 1988, Snowmass, USA), Jensen, S. Editor, World Scientific, Singapore, 824.

Braunschweig, W., Koenigs, E., Sturm, W. and Wallraff, W. (1976). *Nucl. Instr. and Meth.* **134**, 261.

Brechner, R.R. and Singh, M. (1990). *IEEE Trans Nucl. Sci.* **37**, 1328.

Brennan, K.F. (2005). Introduction to Semiconductor Devices, Cambridge University Press, Cambridge.

Brennan, K.F. and Brown, A.S. (2002). Theory of Modern Electronic Semiconductor Devices, John Wiley & Sons, New York.

Brooks, R.A. and DiChiro, G. (1976). *Principles in Medicine and Biology* **32(5)**, 689.

Brotherton, S.D. and Bradley, P. (1982). *J. of Appl. Phys.* **53**, 5720.

Brown, B.A. and Wildenthal, B.H. (1987). *Nucl. Phys. A* **474**, 290.

Brown, J.M. and Jordan, A.G. (1966). *J. of App. Phys.* **37**, 337.

Brückmann, H. and Kowalski, H. (1986). The dependence of calorimeter responses on the gate width, ZEUS Inter. Note 86/13.

Brückmann, H. et al. (1988). *Nucl. Instr. and Meth. in Phys. Res. A* **263**, 136.

Brun, R. et al. (1992). GEANT3 CERN Program Library, Long Write-up. See also Ref. [Agostinelli et al. (2003)] and the web site: *http://geant4.web.cern.ch/geant4/*

Bruguier, G. and Palau, J.M. (1996). *IEEE Trans. on Nucl. Sci.* **43**, 522.

Buck, R. and Perez, S.M. (1983). *Phys. Rev. Lett.* **50**, 1975.

Buckardt, H. et al. (1988). *Nucl. Instr. and Meth. in Phys. Res. A* **268**, 116.

Buckley, E. et al. (1989). *Nucl. Instr. and Meth. in Phys. Res. A* **275**, 364.

Bueheler, M.G. (1972). *Solid–State Elec.* **15**, 69.

Buenerd, M. et al. (2000). *Phys. Lett. B* **489**, 1.

Buontempo, S. et al. (1995). *Nucl. Phys. (Proc. Supp.) B* **44**, 45; *Nucl. Instr. and Meth. in Phys. Res. A* **349**, 70 (1994).

Burger, R.A. and Hattingh, M. (1998). *Astrophys. J.* **505**, 244.

Burger, R.A. and Hattingh, M. (2001). Effect of Fisk-type Heliospheric Magnetic Field on the Latitudinal Transport of Cosmic Rays, Proc. of the ICRC 2001, 07–15 August,

Hamburg, Germany, p. 3698.

Burger, R.A. and Potgieter, M.S. (1989). *Astrophys. J.* **339**, 501.

Burger, R.A. and Sello, P.C. (2005). *Adv. Space Res.* **35**, 643.

Burhop, E.H.S. (1955). *J. Phys. Radium* **16**, 625.
See also:
Hagedorn, H.L and Wapstra A.H. (1960). *Nucl. Phys.* **15**,146.

Busza, W. (1977). *Acta Phys. Pol. B* **8**, 333.

Byon-Wagner, A. (1992). Beam test of reconfigurable-stack calorimeter (Hanging File Calorimeter), talk given at the Third International Conference on Advanced Technology and Particle Physics, Villa Olmo, Como, Italy, June 22–26.

Byrne, J. (1994). Neutrons, Nuclei and Matter, Institute of Physics Publishing, London.

Caccia, M. et al. (1987). *Nucl. Instr. and Meth. in Phys. Res. A* **260**, 124.

Cahn, J.H. (1959). *J. of Appl. Phys.* **30**, 1310.

Calvel, P. et al. (1996). *IEEE Trans. on Nucl. Sci.* **43**, 2827.

Campbell, M. et al. (2008). *Nucl. Instr. and Meth. in Phys. Res. A* **591**, 38.

Campbell, M., Leroy, C., Pospisil, S. and Suk, M. (2006). MPX-USB-ATLAS-2006, ATL-R-MA-0001, CERN.

Carithers, W.C. (1975). Neutral Beam Studies of Coherent Production on Nuclei from 6 – 16 GeV/c, Proc. of the Topical Meeting on Hugh-Energy Collisions involving Nuclei, Bellini, G., Bertocchi, L. and Rancoita, P.G. Editors, Ed. Compositori, Bologna, 307.

Carlson, A.B. (1975). Communication Systems, McGraw Hill, New York.

Carlson, J.F. and Oppenheimer, J.R. (1936). *Phys. Rev.* **51**, 220.

Carlson, T.A. (1975). Photoelectron and Auger Spectroscopy, Plenum Press, New York.
See also:
Carlson, T.A. (1969). Handbook of Physics and Chemistry, Chemical Rubber Publishing Co., Cleveland.

Carlsson, G.A. (1971). *Health Phys.* **20**, 653.

Carosi, R. et al. (1984). *Nucl. Instr. and Meth. in Phys. Res. A* **219**, 361.

Carroll, A. et al. (1979). *Phys. Lett. B* **80**, 319.

Caso, C. et al. (1998). *The Europ. Phys. Jou. C* **3**, 1.

Casse, G. et al. (1999) (a). *Nucl. Instr. and Meth. in Phys. Res. A* **434**, 118.

Casse, G. et al. (1999) (b). *Il Nuovo Cimento A* **112**, 1.

Cassiday, G.L. (1985). *Ann. Rev. Nucl. Part. Sci.* **35**, 321.

Catanesi, M.G. et al. (1986). *Nucl. Instr. and Meth. in Phys. Res. A* **247**, 438.

Catanesi, M.G. et al. (1987). *Nucl. Instr. and Meth. in Phys. Res. A* **260**, 43.

Caughey, D.M. and Thomas, R.E. (1967). *Proc. IEEE* **55**, 2192.

CDF Collab.: Ascoli, G. et al. (1988). *Nucl. Instr. and Meth. in Phys. Res. A* **268**, 33.
See also:
Bedeschi, F. et al. (1988). *Nucl. Instr. and Meth. in Phys. Res. A* **268**, 50.
Abe, F. et al. (1988). *Nucl. Instr. and Meth. in Phys. Res. A* **271**, 387; *Phys. Rev. D* **50**, 2966 (1994); *Phys. Rev. Lett.* **73**, 225 (1994).
Abachi, S. et al. (D0 Collab.) (1994). *Nucl. Instr. and Meth. in Phys. Res. A* **338**, 185; *Phys. Rev. Lett.* **72**, 2138; *Phys. Rev. Lett.* **73**, 225; *Phys. Rev. D* **52**, 4877 (1995); *Phys. Rev. Lett.* **74**, 2362 (1995); *Phys. Rev. Lett.* **75**, 3997 (1995).

Cennini, P. et al. (1990). Study of Liquid Argon Dopants for LHC Hadron Calorimetry, CERN / DRDC/90-34 DRDC/P6.

Čerenkov, P.A. (1937). *Phys. Rev.* **52**, 378.
See also:
Jelley, J. (1958). Čerenkov Radiation and its Applications, London, Pergamon.

Cerri, C. et al. (1983). *Nucl. Instr. and Meth. in Phys. Res.* **214**, 217.

Cerri, C. et al. (1989). *Nucl. Instr. and Meth. in Phys. Res. A* **227**, 227.

Chandrasekhar, S. (1943). *Rev. Mod. Phys.* **15**, 1.

Charlot, C. (1992). Proceedings of the Second International Conference on Calorimetry in High Energy Physics, Capri, Italy, 1991, Ereditato, A. Editor, World Scientific, Singapore, 382.

Cheng, L.J., Corelli, J.C., Corbett, J.W. and Watkins, G.D. (1966). *Phys. Rev.* **152**, 761.

Cheryl, J.D., Marshall, P.W., Burke, E.A., Summers, G.P. and Wolicki, E.A. (1988). *IEEE Trans. on Nucl. Sci.* **35 (no. 6)**, 1208.

Cheshire, D.L. et al. (1977). *Nucl. Instr. and Meth.* **141**, 219.

Chiavassa, A. and Ghia, P.L. (1996). Extensive Air Showers: the primary spectrum and VHE-UHE γ-ray astronomy, Lectures given at the Fourth Trieste School on Non Accelerator Particle Astrophysics, Bellotti, E., Carrigan, R.A., Giacomelli, G. and Paver, N. Editors, World Scientific, Singapore, 401.

Chilingarov, A. and Sloan, T. (1997). *Nucl. Instr. and Meth. in Phys. Res. A* **399**, 35.

Chilingarov, A., Lipka, D., Meyer, J.S. and Sloan, T. (2000). *Nucl. Instr. and Meth. in Phys. Res. A* **449**, 277.

Chren, D., Juneau, M., Kohout, Z., Lebel, C., Leroy, C., Linhart, V., Pospisil, S., Roy, P., Saintonge, A. and Sopko, B. (2001). *Nucl. Instr. and Meth. in Phys. Res. A* **460**, 146.

Ciok, C. et al. (1963). *Nucl. Phys.* **40,** 260.

Claeys, C. and Simoen, E. (2002). Radiation Effects in Advanced Semiconductor Materials and Devices, Springer-Verlag, Berlin.

Cleland, J., Crawford, J.H. and Pigg, J.C. (1955) (a). *Phys. Rep.* **98**, 1742.

Cleland, J., Crawford, J.H. and Pigg, J.C. (1955) (b). *Phys. Rep.* **99**, 1170.

Cleland, M.R., Lisanti, T.F. and Galloway, R.A. (2004). *Rad. Phys. and Chem.* **71**, 583.

CMS (1994). Compact Muon Solenoid, CERN/LHCC/94-38, LHCC/P1 Geneva.

CMS (1997). ECAL Technical Design Report, CERN/LHCC 97-33.

CMS (1998). Tracker Technical Design Report, CERN/LHCC 98-6, April.

Codegoni, D. et al. (2004) (a). Radiation Effects on Bipolar and MOS Transistors Made in BiCMOS Technology, Proc. of the 8th ICATPP (6–10 October 2003, Como, Italy), ISBN 981-238-860-5, World Scientific, Singapore, 622.

Codegoni, D. et al. (2004) (b). *Nucl. Instr. and Meth. in Phys. Res. B* **217**, 65, doi: 10.1016/j.nimb.2003.08.044.

Codegoni, D. et al. (2006). Behavior of Irradiated BICMOS Components for Space Applications, Recent Advances in Multidisciplinary Applied Physics, Reprinted 2006, Proc. of the Aphys 2003, First International Meeting on Applied Physics, Badajoz 13–18/10/2003, ISBN 0-08-044648-5, Elsevier, Amsterdam, 587.

Colder, A, et al. (2001). *Nucl. Instr. and Meth. in Phys. Res. B* **179**, 397, doi: 10.1016/S0168-583X(01)00582-1.

Colder, A. et al. (2002). Effects of ionizing radiation on BiCMOS components for space application, Proc. of the European Space Component Conference, Toulose 24–27 September 2002, ESA SP-507, p. 377.

Conrad, E.E. (1971). *IEEE Trans. on Nucl. Sci.* **18**, 200.
See also:
  Rogers, V.C. et al. (1975). *IEEE Trans. on Nucl. Sci.* **22**, 2326.
  Rogers, V.C. et al. (1976). *IEEE Trans. on Nucl. Sci.* **23**, 875.
  Burke, E.A. et al. (1986). *IEEE Trans. on Nucl. Sci.* **33**, 1276.

Compton, A.H. (1922). *Bull. Natl. Res. Council (US)* 4, Part 2, No. **20**, 10; *Phys. Rev.* **18**, 96 (1921); *Phys. Rev.* **21**, 207 and 483 (1923); *Phys. Rev.* **24**, 168 (1924).

Consolandi, C., D'Angelo, P., Fallica, G., Mangoni, R., Modica, R., Pensotti, S. and

Rancoita, P.G. (2006). *Nucl. Instr. and Meth. in Phys. Res. B* **252**, 276, doi: 10.1016/j.nimb.2006.08.018.

Consolandi, C., Pensotti, S., Rancoita, P.G. and Tacconi, M. (2008). Determination of Hall coefficient for netron irradiated samples down to cryogenic temperatures and its dependendance on resistivity at room temperature, Proc. of the 10th ICATPP (8–12 October 2007, Como, Italy), Editors Barone, M., Gaddi, A., Leroy, C., Price, L., Rancoita, P.G. and Ruchti, R., World Scientific, Singapore, 547.

Cook, W.R. et al. (1993). *IEEE Trans. Geoscie. Remote Sensing* **31**, 557.

Coulter, C.A. and Parkin, D.M. (1980). *J. Nucl. Mat.* **88**, 249.

Crafts, D. (Intel Corp.) (1993). Private communication in:
    Cossett, C.A. et al. (1993). *IEEE Trans. on Nucl. Sci.* **40**, 1845.

Cranmer, S.R. (2002). Solar Wind Acceleration in Coronal Holes, Proceedings of the SOHO-11 Symposium: from Solar Minimum to Solar Maximum (Davos 11–15 March 2002), Noordwijk, ESA Publications Division, ESA SP-508, p. 361.

Crannell, C.J. (1967). *Phys. Rev.* **161**, 310.

Crannell, C.J. et al. (1969). *Phys. Rev.* **182**, 1435.

Cravens, T. (1997). Physics of Solar System Plasmas, Cambridge University Press, Cambridge.

Crawford, D.F. and Messel, H. (1962). *Phys. Rev.* **128**, 2352.

Croitoru, N., Dahan, R., Rancoita, P.G., Rattaggi, M., Rossi, G. and Seidman, A. (1997). *Nucl. Instr. and Meth. in Phys. Res. B* **124**, 542, doi: 10.1016/S0168-583X(97)00055-4.

Croitoru, N., Dahan, R., Rancoita, P.G., Rattaggi, M., Rossi, G. and Seidman, A. (1998). *Nucl. Phys. B (Proc. Suppl.)* **61B**, 456, doi: 10.1016/S0920-5632(97)00602-6.

Croitoru, N., David, G., Rancoita, P.G., Rattaggi, M. and Seidman, A. (1997) (a). *Nucl. Instr. and Meth. in Phys. Res. B* **132**, 199, doi: 10.1016/S0168-583X(97)00388-1.

Croitoru, N., David, G., Rancoita, P.G., Rattaggi, M. and Seidman, A. (1997) (b). *Nucl. Instr. and Meth. in Phys. Res. A* **386**, 156, doi: 10.1016/S0168-9002(96)01111-4.

Croitoru, N., David, G., Rancoita, P.G., Rattaggi, M. and Seidman, A. (1998) (a). *Nucl. Instr. and Meth. in Phys. Res. B* **134**, 209, doi: 10.1016/S0168-583X(98)00555-2.

Croitoru, N., David, G., Rancoita, P.G., Rattaggi, M. and Seidman, A. (1998) (b). *Nucl. Phys. B (Proc. Suppl.)* **61B**, 470, doi: 10.1016/S0920-5632(97)00604-X.

Croitoru, N., Gambirasio, A., Rancoita, P.G. and Seidman, A. (1996). *Nucl. Instr. and Meth. in Phys. Res. B* **111**, 297, doi: 10.1016/0168-583X(95)01383-0.

Croitoru, N., Gubbini, E., Rancoita, P.G., Rattaggi, M. and Seidman, A. (1999) (a). *Nucl. Phys.B (Proc. Suppl.)* **78**, 657, doi: 10.1016/S0920-5632(99)00620-9.

Croitoru, N., Gubbini, E., Rancoita, P.G., Rattaggi, M. and Seidman, A. (1999) (b). *Nucl. Instr. and Meth. in Phys. Res. A* **426**, 477, doi: 10.1016/S0168-9002(98)01421-1.

Croitoru, N., Rancoita, P.G., Rattaggi, M. and Seidman, A. (1999). DLTS Measurements of energetic levels in the temperature range $10 < T < 350$ K, Proceedings (summary) of the 5th European Conference RADECS 99, Abbaye de Fontevraud (France 14–17/08/1999), C–26.

Croitoru, N., Rancoita, P.G., Rattaggi, M., Rossi, G. and Seidman, A. (1996). *Nucl. Instr. and Meth. in Phys. Res. B* **114**, 120, doi: 10.1016/0168-583X(96)00204-2.

Croitoru, N., Rancoita, P.G., Rattaggi, M., Rossi, G. and Seidman, A. (1997). *Nucl. Instr. and Meth. in Phys. Res. A* **388**, 340, doi: 10.1016/S0168-9002(97)01255-2.

Cummings, A.C., Stone, E.C., McDonald, F.B., Heikkila, B.C., Lal, N. and Webber, W.R. (2005). Characteristics of the Solar Wind Termination Shock Region from Voyager 1 Observations, Proceedings of the 29th ICRS Conference, Pune (India 3–10/08/2005), **vol. 2**, 17; available at the web site:

http://icrc2005.tifr.res.in/htm/conf_proceedings.htm
See also:
Ness, N.F., Burlaga, L.F., Acuna, M.H., Lepping, R.P. and Connerney, J.E.P. (2005). Studies of the Termination Shock and Heliosheath at > 92 AU: Voyager 1 Magnetic Field Measurements, Proceedings of the 29th ICRS Conference, Pune (India 3–10/08/2005), **vol. 2**, 39.

Cummings, J.R. et al. (1993). *IEEE Trans. Nucl. Sci.* **40** 1459; *Geophys. Res. Lett.* **20**, 2003.

Curr, R.M. (1955). *Proc. Phys. Soc.* (London) **A68**, 156.

Cushman, P. (1992). Electromagnetic and hadronic calorimeters, Instrumentation in high energy physics, V 9, Advanced series on directions in high energy physics, Sauli, F. Editor, World Scientific, Singapore, 281.

D0 Collab.: Abolins, M. et al. (1989). *Nucl. Instr. and Meth. in Phys. Res. A* **280**, 36.

D0 Collab.: Abachi, S. et al. (1993). *Nucl. Instr. and Meth. in Phys. Res. A* **324**, 53.

Dagoret-Campagne, S. (1993). Proceedings of the 4th International Conference on Calorimetry in High Energy Physics, La Biodola, Isola d'Elba, Italy, 19–25 September 1993, Menzione, A. and Scribano, A. Editors, World Scientific, Singapore, 176.

d'Agostini, G. et al. (1989). *Nucl. Instr. and Meth. in Phys. Res. A* **274**, 134.

Damerell, C.J.S. (1984). Developments in Solid State Vertex Detectors, Proceedings of the Twelfth SLAC Summer Institute on Particle Slac Summer Institute, 43.

D'Angelo, P., Fallica, G., Galbiati, A., Mangoni, R., Modica, R., Pensotti, S. and Rancoita, P.G. (2006). Investigation of VLSI Bipolar Transistors Irradiated with Electrons, Ions and Neutrons for Space Application, Proc. of the 9th ICATPP (17–21 October 2005, Como, Italy), ISBN 981-256-798-4, Editors Barone, M., Borchi, E., Gaddi, A., Leroy, C., Price, L., Rancoita, P.G. and Ruchti, R., World Scientific, Singapore, 832.

D'Angelo, P., Leroy, C., Pensotti, S. and Rancoita, P.G. (1995). *Nucl. Phys. B (Proc. Suppl.)* **44**, 719; *Nucl. Sci. Jou.* **32, No. 5**, 413.

Dannefaer, S., Mascher, P. and Kerr, D. (1993). *J. of Appl. Phys.* **73**, 3740.

Dargys, A. and Kundrotas, J. (1994). Handbook on Physical Properties of Ge, Si, GaAs and InP, Science and Encyclopedia Publishers, Vilnius, Lithuania.

David, G. (1997). Radiation Damage Effects on the Dependence of the Frequency and Temperature with respect to the Admittance of Silicon Detectors for Experiments in High-Energy Physics, *Thesis University of Milan.*

Davidov, V.A., Donskov, S.V., Inyakin, A.V., Kachanov, V.A., Kakauridze, B.D., Khaustov, G.V., Kulik, A.V., Lednev, A.A., Mikhailov, Yu.K., Prokoshkin, Yu.D., Rodnov, Yu.V., Sadovsky, S.A., Starzev, A.V., Lagnaux, J.P., Binon, F., Bricman, C., Stroot, J.P., Roosen, R., Dufournaud, J., Peigneux, J.P., Gouanerec, M., Sillouc, D. and Duteil, P. (1980). A hodoscope calorimeter for high energy hadrons: IHEP-IISN-LAPP collaboration, *Nucl. Instr. and Meth.* **174**, 369–377.

Davies, G. (1989). *Phys. Rep.* **176**, 83.

Davies, H., Bethe, A.H. and Maximon, L.C. (1954). *Phys. Rev.* **93**, 788.

Davies, H. et al. (1954). *Phys. Rev.* **93**, 788 and references therein.

De Angelis, A. (1988). *Nucl. Instr. and Meth. in Phys. Res. A* **271**, 455.

Dearnaley, G. and Northrop, D.C. (1966). Semiconductor counters for nuclear radiations - 2nd Edition, John Wiley & Sons, New York; see also:
Northrop, D.C. and Simpson O. (1962). *Proc. Pys. Soc.* **80**, 262.

De Bièvre, P. and Taylor, P.D.P. (1993). *Int. J. of Mass Spectrom.*, formerly *Int. J. of Mass Spectrom. and Ion Proc.* **123**, 149; updated IUPAC (International Union of Pure and Applied Chemistry) recommendations on atomic weights of the elements and isotopic composition are available on the web site:

http://www.iupac.org/divisions/II/II.1/

de Ferraiis, L. and Arista, N.R. (1984). *Phys. Rev. A* **29**, 2145.

Deiters, K. et al. (1981). *Nucl. Instr. and Meth.* **180**, 45.

Del Guerra, A. (1997). Review of Detectors for Medical application, Proceedings of the First International Four Seas Conference, Trieste, Italy 26 June – 1 July 1995, Gougas, A.K., Lemoigne, Y., Pepe-Altarelli, M., Petroff, P., Wultz, C.E. Editors, CERN 97-06, 217.

Della Negra, M. (1981). *Phys. Scripta* **23**, 469.

del Peso, J. and Ro, E. (1990). *Nucl. Instr. and Meth. in Phys. Res. A* **295**, 330.

Denisov, S.P. et al. (1973). *Nucl. Phys. B* **61**, 62.

Dentan, M. (1999). ATLAS Technical Coordination Radiation Hardness Assurance, see web page:
http://www-hep2.fzu.cz/pixpage/IRR/RHAWG1.pdf

Derenzo, S. et al. (1974). *Nucl. Instr. and Meth.* **122**, 319.

d'Errico, F. (1999) *Radiation Protection Dosimetry* **84**, 55.

DeSalvo, R. (1995). *Nucl. Phys. (Proc. Supp.) B* **44**, 123.

Desgrez, A., Bittoun, J. and Idy-Peretti, I. (1989). 1. Bases physiques de l'IRM, Masson, Paris, 102.

De Vincenzi, M. et al. (1986). *Nucl. Instr. and Meth. in Phys. Res. A* **243**, 348.

Dezillie, B., Bates, S., Glaser, M., Lemeilleur, F. and Leroy, C. (1997). *Nucl. Instr. and Meth. in Phys. Res. A* **388**, 314, doi: 10.1016/S0168-9002(97)00005-3.

Dezillie, B. et al. (1999). *IEEE Trans. on Nucl. Sci.* **46**, 221.

Diddens, A.N. et al. (1980). *Nucl. Instr. and Meth.* **178**, 27.

Dienes, G.J. and Vineyard, G.H. (1957). Radiation Effects in Solids, in Monographs in Physics and Astronomy, Vol. II, Interscience Publishers Inc., New York.

Diethorn, W. (1956). NYO-6628.

Dietze, G. (2005). Radiological Quantities and Units, in Landolt–Börnstein, Group VIII: Advanced Materials and Technologies, Vol. 4: Radiological Protection, Martienssen, W. Editor in Chief, Springer-Verlag, Berlin.

Dimitrov, V.I., Engel J. and Pittel, S. (1995). *Phys. Rev. D* **51**, 291.

Di Marco, M. (2004). Development and characterization of PICASSO superheated droplet detectors aiming at dark matter for a new limit on spin-dependent neutralino cross section, PhD Thesis, University of Montreal.

Divakaruni, R., Prabhakar, V. and Viswanathan, C.R. (1994). *IEEE Trans. on Electron Devices* **ED-41**, 1405.

Divari, P.C. et al. (2000). *Phys. Rev. C* **61**, 054612-1.

Dmitrenko, N.N. et al. (1974). *Phys. Status Solidi A* **26**, K156.

Dodd, P.E. (1996). *IEEE Trans. on Nucl. Sci.* **43**, 561.

Dodd, P.E., Sexton, F.W. and Winokur, P.S. (1994). *IEEE Trans. on Nucl. Sci.* **41**, 2005.

Dodd, P.E., Shaneyfelt, M.R. and Sexton, F.W. (1997). *IEEE Trans. on Nucl. Sci.* **44**, 2256.

Dolgoshein, B. (1993). *Nucl. Instr. and Meth. in Phys. Res. A* **326**, 434.

Dovzhenko, O.I. and Pomamskii, A.A. (1964). *Trans. Sov. Phys. JETP* **18**, 187.

Drews, D. et al. (1990). *Nucl. Instr. and Meth. in Phys. Res. A* **290**, 335.

Dubois, R. et al. (1985). SLAC-PUB-3813.

Duffy, M.E. et al. (1984). *Nucl. Instr. and Meth. in Phys. Res. A* **228**, 37.

Dungey, J.W. (1961). *Phys. Rev. Lett.* **6**, 47.

Easley, J.W. (1962). *Nucleonics* **20** (no. **7**), 51.

Eberhart, J.G., Kremsner, W. and Blander, M. (1975). *Jou. of Colloid and Interface Sci.* **50, No. 2**, 369.

ECSS (2005). Space Engineering - Methods for the calculation of radiation received and its effects and a policy for design margins, European Cooperation for Space Standardization, *ECSS-E-10-12*, ESA Publications Division, Noordwijk.

Eisberg, R. and Resnick, R. (1985). Quantum Physics of Atoms, Molecules, Solids, Nuclei and Particles - 2nd Edition, John Wiley & Sons, New York.

Egorytchev, V., Saveliev, V. and Aplin, S.J. (2000). *Nucl. Instr. and Meth. in Phys. Res. A* **453**, 346.

Elwert, G. (1939). *Ann. Physik* **34**, 178.

Encrenaz, T., Bibring, J.P. and Blanc, M. (1991). The Solar System (Chapter 4), Springer-Verlag, Berlin.

Engel, J. (1991). *Phys. Lett. B* **264**, 114.

Engel, J. and Vogel, P. (1989). *Phys. Rev. D* **40**, 3132.

Engel, R., Ranft, J. and Roesler, S. (1997). *Phys. Rev D* **55**, 6957.

Engelmann, J.J. et al. (1985). *Astron. Astrophys.* **148**, 12.

Engelmann, J.J. et al. (1990). *Astron. Astrophys.* **223**, 96.

England, J.B.A. et al. (1981). *Nucl. Instr. and Meth. in Phys. Res.* **185**, 43.

Engler, J. (1985). *Nucl. Instr. and Meth. in Phys. Res. A* **235**, 301.

Erba, E., Facchini, U. and Saetta Menichella, E. (1961). *Nuovo Cimento* **22**, 1237.

Ericsson, M. (1986). Photonuclear reactions and dispersion relations, Proc. of 5th Course of the International School of Intermediate Energy Nuclear Physics, Bergere, R., Costa, S. and Shaerf, C. Editors, World Scientic, Singapore.
See also:
Wu, R.J. and Chang, C.C. (1977). *Phys. Rev. C* **16**, 1812.

Esbensen, H. et al. (1978). *Phys. Rev. B* **18**, 1039.

Eskut, E. et al. (CHORUS Collab.) (1997). *Nucl. Instr. and Meth. in Phys. Res. A* **401**, 7.

ESTP (2003). The European Spallation Target Project, on the web site
http://ik332.ikp.kfa-juelich.de/nessi/nessi.html
See also:
The European Spallation Neutron Source, on the web site
http://www.kfa-juelich.de/ess/
Spallation neutron Source, on the web site
http://www.sns.gov/
IFMIF (International Fusion Materials Irradiation Facility), on the web site
http://insdell.tokai.jaeri.go.jp/IFMIFHOME/ifmif_home_e.html

Fabjan, C.W. (1985) (a). Calorimetry in high-energy Physics, CERN-EP/85-54.

Fabjan, C.W. (1995) (b). *Nucl. Instr. and Meth. in Phys. Res. A* **360**, 228.

Fabjan, C.W. (1986). *Nucl. Instr. and Meth. in Phys. Res. A* **252**, 145.
See also:
Fabjan, C.W. (1985). Concepts and Techniques in High Energy Physics III, Ferbel, T. Editor, Plenum Press, New York, 281.

Fabjan, C.W. and Ludlam, T. (1982). *Ann. Rev. Nucl. Part. Sci.* **32**, 335.

Fabjan, C.W. et al. (1977). *Nucl. Instr. and Meth.* **141**, 61.

Faessler, M.A. et al. (1979). *Nucl. Phys. B* **157**, 1.

Fäldt, G. (1977). Theory of Coherent Production on Nuclei, Proceedings of the Topical Meeting on Multiparticle Production on Nuclei at Very High Energies, Bellini, G., Bertocchi, L. and Rancoita, P.G. Editors. IAEA-SMR-21, 387.

Fano, U. (1947). *Phys. Rev.* **72**, 26.

Fano, U. (1963). *Ann. Rev. of Nucl. Scie.* **13**, 1.

Fano, U., Koch, H.W. and Motz, J.W. (1959). *Phys. Rev.* **112**, 1679.

Fayard, L. (1988). Transition Radiation, Cours donné a l'Ecole JoliotCurie, Maubuisson, Gironde (26–30 Sept.).

Fassò, A., Goebel, K., Höfert, M., Ranft, J. and Stevenson, G. (1990). Shielding Against High Energy Radiation, in Landolt–Börnstein, Group I: Nuclear and Particle Physics, Vol. 11: Photoproduction of Elementary Particles, Shopper, H. Editor, Springer-Verlag, Berlin.

Feinberg, E.L. (1972). *Phys. Rep. C* **5**, 237.

Feinberg, E.L. and Pomeranchuk, I. (1956). *Supp. Nuovo Cimento* **3**, 652.

Feng, I.J., Pratt, R.H. and Tseng, H.K. (1981). *Phys. Rev. A* **24**, 1358.

Fenyves, E.J. et al. (1988). *Phys. Rev. D* **37**, 649.

Fermi, E. (1939). *Phys. Rev.* **56**, 1242 and *Phys. Rev.* **57**, 485 (1940).

Fermi, E. (1950). Nuclear Physics, Univ. of Chicago Press, Chicago.

Fernow, R. (1986). Introduction to experimental particle physics, Cambridge University Press.

Ferreira, S.E.S., Potgieter, M.S., Heber, B. and Fichtner, H. (2003). *Annales Geophys.* **21**, 1359.

Fesefeld, H.C. (1985). PITHA Report 85/02, RWTH Aachen.

Fesefelt, H. (1988). *Nucl. Instr. and Meth. in Phys. Res. A* **263**, 114.

Feynman, J. (1985). Solar wind, Chapter 3 of Handbook of Geophysics and the Space Environment, Editor Jursa, A.S., Air Force Geophysics Lab, Springfield.

Feynman, R.P. (1969). *Phys. Rev. Lett.* **23**, 1415.

Fiederle, M. et al. (2008). *Nucl. Instr. and Meth. in Phys. Res. A* **591**, 75.

Field, R. and Heijne, E. (2007). *Nucl. Instr. and Meth. in Phys. Res. A* **577**, 595.

Finkelnburg, W. (1964). Structure of Matter, Springer-Verlag, Berlin and Academic Press Inc. Publishers, New York.

Firsov, O.B. (1957). *Zh. Eksp. Teor. Fiz.* **32**, 1464; *Zh. Eksp. Teor. Fiz.* **33**, 696; *Zh. Eksp. Teor. Fiz.* **34**, 447, (1958).

Fischer, H. et al. (1975). Proceedings of the International Meeting on Proportional and Drift Chambers, angle measurement and space resolution in proportional chambers, Dubna.

Fischer, J. et al. (1975). *Nucl. Instr. and Meth.* **127**, 525.

Fisher, C.M. (1968). Coherent Production Processes, in Methods in Subnuclear Physics Vol. 1, Intern. School of Elementary Particle Physics, Herceg-Novi Yu, Nikolić, M. Editor, Gordon and Breach Science Publ. Inc., New York, 235.

Fisher, H.G. (1978). Multiwire Proportional Quantameters, *Nucl. Instr. and Meth.* **156**, 81–85.

Fisk, L.A. (1971). *J. Geophys. Res.* **76 (no. 1)**, 221.

Fisk, L.A. (1976). *J. Geophys. Res.* **81 (no. 25)**, 4646.

Fisk, L.A. (1996). *J. Geophys. Res.* **101 (no. 47)**, 15547.

Fisk, L.A. (1999). *Adv. Space Res.* **23 (no. 3)**, 415.

Fisk, L.A., Forman, M.A. and Axford, W.I. (1973). *J. Geophys. Res.* **78 (no. 7)**, 995.

Fisk, L.A. and Schwadron, N.A. (2001). *Space sci. Rev.* **97**, 21.

Flauger, W. (1985). *Nucl. Instr. and Meth. in Phys. Res. A* **241**, 72.

Fleetwood, D.M. et al. (1994). *IEEE Trans. Nucl. Sci.* **41**, 1871.

Fletcher, N. (1957). *Proc. of the IRE* **45**, 862.

Forman, M.A. (1970). *Planet. Space Sci.* **18**, 25.

Forman, M.A., Jokipii, J.R. and Owens, A.J. (1974). *Astrophys. J.* **192**, 535.

Frank, I.M. (1964). Nobel Lecture 1958, Nobel Lecture, Elsevier.

Frank, I. and Tamm, I. (1937). *Dokl. Akad. Nauk. SSSR* **14**, 109.

Frank, M. and Larin, F. (1965). *IEEE Trans. on Nucl. Sci.* **12 (no. 5)**, 126.

Frank, M., Poblenz, F.W. and Howard, D.W. (1963). *IEEE Trans. on Nucl. Sci.* **10 (no. 5)**, 93.

Fraser, J.S. et al. (1965). *Phys. Can.* **21**, 17.
See also:
Barthtolomew, G.A. and Tunnicliffe, P.R. Editors (1966). CRNL Rep. AECL-2600 Vol. VII, 12.

Fretwurst E. et al. (1993). *Nucl. Instr. and Meth. in Phys. Res. A* **326**, 357.

Friend, B. et al. (1976). *Nucl. Instr. and Meth.* **136**, 505.

Fukushim, M. (1992). Electromagnetic Calorimeter for KEK B Factory, Proceedings of International Workshop on B-Factories: Accelerators and Experiments Kikutani, E. and Matsuda, T. Editors, KEK, Tsukuba, Japan November 17–20, 368.

Fuller, E.G. and Hayward, E. (1962). Nuclear Reactions, Endt, P.M. and Smith, P.B. Editors, North Holland Publ., Amsterdam, Vol. 2 Chapter 3.

Furetta, C. et al. (1995). *Nucl. Sci. Journ.* **32, no. 2**, 85.

Furetta, C., Leroy, C., Pensotti, S., Rancoita, P.G. and Rattaggi, M. (1994). *Nucl. Sci. Jou.* **31, no. 4**, 292.

Furmanska, B. et al. (1977). *Acta Phys. Pol.* **8**, 973.

Gabriel, S.B (2000). Cosmic Rays and Solar Protons in the near Earth Environment and their entry into the Magnetosphere, talk given at the *Workshop on The Utilization of a Future European Space Weather Service* 12 December 2000 ESTEC, Noordwijk, The Netherlands.

Gabriel, T.A. (1978). *Nucl. Instr. and Meth.* **150**, 145.

Gabriel, T.A. (1989). Detectors for the Superconducting Collider Design Concepts and Simulation, Proc. of the Workshop on calorimetry for the Supercollider, Tuscalosa, Al, March 13–17, Donaldson, R. and Gilchriese, M.G.D. Editors, World Scientific, Teaneck, NJ, 121.

Gabriel, T.A. et al. (1994). *Nucl. Instr. and Meth. in Phys. Res. A* **338**, 336.

Gaisser, T.K. (1990). Cosmic Rays and Particle Physics, Cambridge University Press, Cambridge.

Gaisser, T.K. and Schaefer, R.K (1997). *Adv. Space Res.* **19/5**, 174.
See also:
Astrophys. J. **394**, 174 (1992).

Gaisser, T.K. and Yodh, G.B. (1980). *Ann. Rev. of Nucl. Part. Sci.* **30**, 475.

Garcia-Munoz, M., Meyer, P., Pyle, K.R., Simpson, J.A. and Evenson, P. (1986). *J. Geophys. Res.* **91**, 2858.

Garibyan, G.M. (1959). *Zh. Exp. Teor. Fiz.* **37**, 527.
See also:
Barsukov, K.A. (1959). *Zh. Exp. Teor. Fiz.* **37**, 1106.

Garibyan, G.M. (1960). *Soviet Phys. JETP* **10**, 372; **33**, 23 (1971); *Yerevan Report ЕФН-27* (1973).

Gatti, E. and Manfredi, P.F. (1986). *La Riv. del Nuovo Cimento* **Vol. 9, N. 1**.

Gehlen, P.C., Beeler, J.R. Jr. and Jaffee, R.I. Editors (1972). Interatomic Potentials and Simulations of Lattice Defects, Plenum Press, New York.

Geiger, H. and Marsden, E. (1913). *Phil. Mag. [6]* **25**, 604.

Geiger, H. and Mueller, W. (1928). *Physik Zeits* **29**, 839.

Genest, M.-H. (2004). Private communication.

Genest, M.-H. (2007). Recherche du neutralino avec les détecteurs ATLAS et PICASSO, Doctoral Thesis, University of Montreal, June.

Genest, M.-H. and Leroy, C. (2004). Calculation of the neutralino–nucleon exclusion limits, *PICASSO Scientific / Technical Report*, **PSTR-04-016**, 20 novembre.

Genzel, H., Joos, P. and Pfeil, W. (1973). Photoproduction of Elementary Particles, in Landolt - Börnstein, Group I: Nuclear and Particle Physics, Vol. 8: Photoproduction of Elementary Particles, Shopper, H. Editor, Springer-Verlag, Berlin.

Gervasi, M. and Grandi, D. (2008). Private communication.

Gervasi, M., Rancoita, P.G., Usoskin, I.G. and Kovaltsov, G.A. (1999). *Nucl. Phys. B (Proc. Supp.)* **78**, 26; Gervasi, M., Rancoita, P.G. and Usoskin, I.G.. Transport of Galactic Cosmic Rays in the Heliosphere: Stochastic Simulation Approach, Proceedings of the 26th ICRS Conference, Salt Lake City 17–25/08/1999, **vol. 7**, 69.

Ghandhi, S.K. (1977). Semiconductor Power Devices, John Wiley & Sons, New York.

Giacomelli, G. (1976). *Phys. Rep.* **23**, 123.

Giani, S. (1993). Proceedings of the Third International Conference on Advanced Technology and Particle Physics, Villa Olmo, Como, Italy, 1992, Borchi, E., Ferbel, T., Nygren, D., Penzo, A. and Rancoita, P.G. Editors, *Nucl. Phys. B (Proc. Suppl.)* **32**, 361.

Giannini, M.M. (1986). Photonuclear Reaction above the Giant Dipole Resonance, Proc. of 5th Course of the International School of Intermediate Energy Nuclear Physics, Bergere, R., Costa, S. and Shaerf, C. Editors, World Scientic, Singapore, 97.

Gilmore, R.S. (1992). Single Particle Detection And Measurement, Taylor & Francis, London.

Gingrich, D.M. et al. (1995). *Nucl. Instr. and Meth. in Phys. Res. A* **364**, 290.

Gingrich, G. et al. (RD3 Collab.) (1994). *Nucl. Instr. and Meth. in Phys. Res. A* **344**, 39; *Nucl. Instr. and Meth. in Phys. Res. A* **355**, 295 (1995).

Ginzburg, V.L. (1940). *J. Phys. (USSR)* **2**, 441; *Zh. Exp. Teor. Fiz.* **10**, 589.

Ginzburg, V. and Frank, I. (1946). *Zh. Exp. Teor. Fiz.* **16**, 15.

Ginzburg, V.L. and Syrovatskii, S.I. (1964). The Origin of Cosmic Rays, The Macmillan Company, New York.

Girard, R. et al. (2005). *Phys. Lett. B* **621**, 233.

Giuliani, F. (2005). *Phys. Rev. Lett.* **95**, 101301.

GLAST Collab. (1998). Gamma-ray Large Area Space Telescope, A Summary by the GLAST Facility Science Team and NASA/Goddard Space Flight Center; GLAST, Report of the Gamma Ray Astronomy Program Working Group, April (1997).

Glauber, R.J. (1955). *Phys. Rev.* **100**, 242.
    See also:
    Gribov, V.N. (1970). *Sov. Phys. JETP* **30**, 709.
    Bertocchi, L. (1972). *Nuovo Cimento A* **11**, 45.

Glauber, R.J. (1959). Lectures in Theretical Physics, Vol. 1, Brittin, W.E. and Dunham, L.G. Editors, Interscience, New York.
    See also:
    Yennie, D.R. (1971). Hadronic Interactions of Electrons and Photons, Cummings, J. and Osborn, H. Eds., Academic Press, London.
    Gottfried, K. (1972). Photon and Hadron Interactions in Nuclei, TH. 1564 - CERN.

Gleeson, L.J. and Axford, W.I. (1967). *Astrophys. J.* **149**, L115.

Gleeson, L.J. and Axford, W.I. (1968) (a). *Astrophys. Space Sci.* **2**, 431.

Gleeson, L.J. and Axford, W.I. (1968) (b). *Astrophys. J.* **154**, 1011.

Gleeson, L.J. and Urch, I.H. (1971). *Astrophys. Space Scie.* **11**, 288; **25**, 387 (1973).

Glendenin, L.E. (1948). *Nucleonics* **2**, 12.
    See also:
    Kobetich, E.J. and Katz, R. (1968). *Phys. Rev.* **170**, 391.

Gola, A., Pessina, G. and Rancoita, P.G. (1990). *Nucl. Instr. and Meth. in Phys. Res. A* **292**, 648, doi: 10.1016/0168-9002(90)90183-7.

Gola, A., Pessina, G., Rancoita, P.G., Seidman, A. and Terzi, G. (1992). *Nucl. Instr. and Meth. in Phys. Res. A* **320**, 317, doi: 10.1016/0168-9002(92)90792-3.

Golan, G. et al. (1999). *Microelect. Reliab.* **39**, 1497, doi: 10.1016/S0026-2714(99)00089-X.

Golan, G., Rabinovich, E., Inberg, A., Oksman, M., Rancoita, P.G., Rattaggi, M., Gartsman, K., Seidman, A. and Croitoru, N. (2000). *Microelect. J.* **31**, 937, doi: 10.1016/S0026-2692(00)00093-8.

Golan, G., Rabinovich, E., Inberg, A., Axelevitch, A., Lubarsky, G., Rancoita, P.G., Demarchi, M., Seidman, A. and Croitoru, N. (2001). Inversion phenomenon as a result of junction damages in neutron irradiated silicon detectors, *Microelect. Reliab.* **41**, 67–72, doi: 10.1016/S0026-2714(00)00212-2.

Goldsmith, P. and Jelley, J.V. (1959). *Phil. Mag.* **4**, 836.

Gombosi, T.I. (1998). Physics of the Space Environment, Cambridge University Press, Cambridge.

Gomez, J.J., Velasco, J. and Maestro, E. (1987). *Nucl. Instr. and Meth. in Phys. Res.* **205**, 284.

Goodge, M.E. (1983). Semiconductor Device Technology, Howard W. Sams & Co., Inc., Indianapolis.

Gooding, T.J. and Eisberg, R.M. (1957). *Phys. Rev.* **105**, 357.

Gordon, H.A. (1991). Liquid Argon Calorimetry for the SSC, Proc. of the Symposium on Detector Research and Development for the Superconducting Supercollider, Fort Worth, Texas, 15–18 October 1990, Dombeck, T., Kelley, V. and Yost, G.P. Editors, World Scientific, Singapore, 100.

Gornea, R. (2002). Système d'acquisition des données et de contrôle du détecteur à gouttelettes surchauffées dans le cadre du projet PICASSO, MSc. Thesis, University of Montreal.

Gosling, J.T. (2006). The Solar Wind, Encyclopedia of the Solar System - 2nd Edition, McFadden, L.-A., Weissman, P.R. and Johnson, T.V. Editors, Academic Press, San Diego.

Gossick, B.R. (1959). *J. of Appl. Phys.* **30**, 1214.

Gottfried, K. (1972). Photon and Hadron Interactions on Nuclei, Proceedings of the 1972 CERN School of Physics, CERN 72–12, 55.

Gottfried, K. (1974). *Phys. Rev. Lett.* **32**, 957; Coherent and incoherent multiple production on nuclei, Proc. of the V International Conference on High Energy Physics and Nuclear Structure, Tibell, G. Editor, 79.

Gover, A., Grinberg, J. and Seidman, A. (1972). *IEEE Trans. on Electron Devices* **ED-19 (no. 8)**, 967.

Granata, V. et al. (2000). *Philos. Mag.* **B 80**, 811.

Grandi, D. (2008). Private communication.

Grassman, H. et al. (1985). *Nucl. Instr. and Meth. in Phys. Res. A* **235**, 319.

Green, D. (2000). The Physics of Particle Detectors, Cambridge University Press, Cambridge.

Gregory, B.L. (1969). *IEEE Trans. on Nucl. Sci.* **NS-16**, 53.

Gregory, B.L. and Jordan, A.G. (1964). *Phys. Rev.* **134**, A1378.

Gregory, B.L., Naik, S.S. and Oldham, W.G. (1971). *IEEE Trans. on Nucl. Sci.* **NS-18**, 181.

Greisen, K. (1960). *Ann. Rev. Nucl. Sci.* **10**, 63.

Greisen, K. (1966). *Phys. Rev. Lett.* **16**, 748.

Gribov, V. N. (1968). *Ž. Èksp. Teor. Fiz.* **88**, 392.

Grieder, P.K.F. (2001). Cosmic Rays at Earth, Elsevier, Amsterdam.

Griffin, P.J. (1997). *IEEE Trans. on Nucl. Sci.* **44**, 2079.

Griffin, P.J., Vehar, D.W., Cooper, P.J. and King, D.B. (2007). *IEEE Trans. on Nucl. Sci.* **54**, 2288.

Grigorov, N.L. et al. (1991). *Geophys. Res. Lett.* **18**, 1959.

Grindhammer, G. et al. (1989). The fast simulation of electromagnetic and hadronic showers, Proc. of the Workshop on calorimetry for the Supercollider, Tuscaloosa, A1, March 13–17, Donaldson, R. and Gilchriese, M.G.D. Editors, World Scientific, Teaneck, NJ, 151.

Grindhammer, G. et al. (1990). *Nucl. Instr. and Meth. in Phys. Res. A* **290**, 469.

Groom, D.E. (1990). Proceedings of the Workshop on Calorimetry for the Superconducting Super Collider, The University of Alabama, Tuscaloosa, Al, USA, March 13–17th 1989, Donaldson, R. and Gilchriese, M.G.D. Editors, World Scientific, Singapore, 59.

Groom, D.E. (1992). Proc. of the II International Conference on Calorimetry in High Energy Physics, Capri 1991, Ereditato, A. Editor, World Scientific, Singapore, 376.

Groom, D.E. (2007). *Nucl. Instr. and Meth. in Phys. Res. A* **572**, 633, doi: 10.1016/j.nima.2006.11.070.

Groom, D.E., Mokhov, N.V. and Striganov, S.I. (2001). Muon stopping-power and range tables, 10 MeV–100 TeV, *Atomic Data and Nuclear Data Tables* **78**, 183.

Grove, A.S. (1967). Physics and Technology of Semiconductor Devices, Wiley, New York.

Gruhn, C.R. and Edmiston, M.D. (1978). *Phys. Rev. Lett.* **40**, 407.

Grupen, C. (1996). Particle Detectors, Cambridge University Press, Cambridge; see also [Grupen and Shwartz (2008)].

Grupen, C. (2005). Astroparticle Physics Springer, Berlin.

Grupen, C. and Shwartz, B. (2008). Particle Detectors - 2nd Edition, Cambridge University Press, Cambridge, ISBN-13: 9780521840064.

Guida, J. (D0 Collab.) (1995). *Nucl. Phys. B (Proc. Supp.)* **44**, 158.
    See also:
    Wimpenny, S. (D0 Collab.) (1989). *Nucl. Instr. and Meth. in Phys. Res. A* **279**, 107.

Gummel, H. K. (1964). *IEEE Trans. Electron Devices* **11 (no. 10)**, 455.

Gutierrez, A. (2007). "Étude de la sensibilité des détecteurs au silicium aux neutrons lents et neutrons rapides", *Rapport PHY3030 Université de Montréal*.

Gutiérrez, E.A., Deen, M.J. and Claeys, C. (2001). Low Temperature Electronics: Physics, Devices, Circuits and Applications, Academic Press, San Diego.

H1 Collab.: Braunschweig, W. et al. (1988). *Nucl. Instr. and Meth. in Phys. Res. A* **265**, 419.

H1 Collab.: Braunschweig, W. et al. (1993) (a). *Nucl. Instr. and Meth. in Phys. Res. A* **336**, 460; *Nucl. Instr. and Meth. in Phys. Res. A* **275**, 246 (1989); Results of a Pb - Fe Liquid Argon Calorimeter, DESY 89-022 (1989).

H1 Collab.: Abt, I. et al. (1993) (b). DESY preprint 93-103; DESY Internal Report H1-96-01 (1996).
    See also:
    H1 Collab. (1997). *Nucl. Instr. and Meth. in Phys. Res. A* **386** 310.
    ZEUS Collab. (1993). The ZEUS Detector Status, Report 1993 DESY.
    Derrick, M. et al. (1992). *Phys. Lett. B* **293** 465; *Z. Phys. C* **63**, 391 (1994).

Hagedorn, R. (1964). Relativstic Kinematics, New York, Benjamin.

Haino, S. et al. (2004). *Phys. Lett. B* **594**, 35.

Halbleib, J.A., Kensek, R.P., Mehlhorn, T.A., Valdez, G., Seltzer, S.M. and Berger, M.J. (1992). ITS version 3.0: The integrated TIGER series of coupled electron/photon Monte Carlo transport codes, Sandia Nat. Labs. Report SAND91-1634

Hall, G. (1984). *Nucl. Instr. and Meth. in Phys. Res.* **220**, 356.

Hallén, A., Keskitalo, N., Josyula, L. and Svensson, B.G. (1999). *J. of App. Phys.* **86 (no. 1)**, 214.

Halliwell, C. (1978). A review of h–nucleus interactions at high energies, Proc. of the VIII International Symposium on Multiparticle Dynamics, Centre de Recherches Nucleaires (Strasbourg) Editor, D-1.

Halprin, A. et al. (1966). *Phys. Rev.* **152**, 1295.

Hamacher, K., Coenen, H.H. and Stoecklin, G. (1986). *Jou. Nucl. Med.* **27**, 235. See also:
Hamacher, K., Blessing, G. and Nebeling, B. (1990). *Appl. Radiat. Isot.* **41**, 49.

Hancock, S., James, F., Movchet, J., Rancoita, P.G. and Van Rossum, L. (1983). *Phys. Rev. A* **28**, 615, doi: 10.1103/PhysRevA.28.615.

Hancock, S., James, F., Movchet, J., Rancoita, P.G. and Van Rossum, L. (1984). *Nucl. Instr. and Meth. in Phys. Res. B* **1**, 16, doi: 10.1016/0168-583X(84)90472-5.

Harder, D. (1970). Some General Results from the Transport Theory of Electron Absorption, Proceedings of the 2nd Symposium on Microdosimetry, Ebert, H.G. Editor, Report EUR 4452 d-f-e, Brüssel, 567.

Harper, M.J. and Rich, J.C. (1993). *Nucl. Instr. and Meth. in Phys. Res. A* **336**, 220.

Harrity, J.W. and Mallon, C.E. (1970). *IEEE Trans. on Nucl. Sci.* **NS-17**, 100.

Hazucha, P. and Svensson, C. (2000). *IEEE Trans. on Nucl. Sci.* **47**, 2586.

Hecht, K. (1932). *Zeit. Phys.* **77**, 235.

Heijne, E.H.M. (2003). *Nucl. Instr. and Meth. in Phys. Res. A* **509**, 1, doi: 10.1016/S0168-9002(03)01541-9.

Heijne, E.H.M. (2007). *Nucl. Instr. and Meth. in Phys. Res. A* **571**, 7.

Heijne, E.H.M. et al. (1983). *Nucl. Instr. and Meth. in Phys. Res. A* **262**, 437.

Heijne, E.H.M., Hubbeling, L., Hyams, B.D., Jarron, P., Lazeyras, P., Piuz, F., Vermeulen, J.C. and Wylie A. (1980). *Nucl. Instr. and Meth. in Phys. Res. A* **178**, 331.

Heisenberg, W. (1946). Cosmic Radiation, Dover Publication, New York.

Heitler, W. (1954). The Quantum Theory of Radiation - 3rd Edition, Clarendon Press, Oxford; reproduced as unaltered replication in agreement with Oxford University Press by Dover Publications, Inc., New York (2000).

HELIOS Collab.: Åkesson, T., Angelis, A.L.S., Corriveau, F., Devenish, R.C.E., Di Tore, G., Fabjan, C.W., Lamarche, F., Leroy, C., McCubbin, M.L., McCubbin, N.A., Olsen, L.H., Seman, M., Sirois, Y., Wigmans, R., and Willis, W.J. (1987). Performance of the uranium/plastic scintillator calorimeter for the HELIOS experiment at CERN, *Nucl. Instr. and Meth. in Phys. Res. A* **262**, 243–263.

HELIOS Collab.: Åkesson, T. et al. (1988). *Zeit. für Physik C* **38**, 383; *Phys. Lett. B* **214**, 295. See also:
Bamberger, A. et al. (NA-35 Collab.) (1987). *Phys. Lett. B* **184**, 271.
Albrecht, R. et al. (WA-80 Collab.) (1987). *Phys. Lett. B* **199**, 297.

Helm, R.H. (1956). *Phys. Rev.* **104**, 1466.

Helmer, R.G. (1999). *Nucl. Instr. and Meth. in Phys. Res. A* **422**, 518.

Helmer, R.G. and van der Leun, C. (1999). *Nucl. Instr. and Meth. in Phys. Res. A* **422**, 525.

Helmer, R.G. and van der Leun, C. (2000). *Nucl. Instr. and Meth. in Phys. Res. A* **450**, 35.

Henke, B.L., Gullikson, E.M. and Davis, J.C. (1993). *At. Data Nucl. Tables* **54**, 181. For updated information see the web site (2008):
http://henke.lbl.gov/optical_constants/

Henley, E.M. and Garcia, A. (2007). Subatomic Physics - 3rd Edition, World Scientific,

Singapore.

Henson, B.G., McDonald, P.T. and Stapor, W.J. (2006). 16Mb SDRAM Proton Radiation Effects Measurements and Analysis, available at the web site: *http://www.radiation-effects.com/pdf/16MbSDRAM.pdf*

Herbert, J. et al. (1974). *Phys. Lett. B* **48**, 467; *Phys. Rev. D* **15**, 1867 (1977). See also: Otterlund, I. et al. (1978). Nuclear Interactions of 400 GeV protons in emulsions, LUIP 7804.

Hess, W.N. (1959). *Phys. Rev. Lett.* **3**, 11.

Hess, W.N. (1968). The Radiation Belt and Magnetosphere, Blaisdell Publishing Company, Waltham.

Heynderickx, D. (2002). *Intern. J. of Modern Physics A* **17**, 1675.

Highland, V.L. (1975). *Nucl. Instr. and Meth.* **129**, 497. See also: Lynch, G.R. and Dahl, O.I. (1991). *Nucl. Instr. and Meth. in Phys. Res. B* **58**, 6.

Hillas, A.M. (1982). *Jou. Phys. G: Nucl. Phys.* **8**, 1461.

Hirayama, H. (1992). *IEEE Trans. Nucl. Sci.* **40**, 503.

Hofmann, W. et al. (1976). *Nucl. Instr. and Meth.* **135**, 151.

Hoffman, W. et al. (1979). *Nucl. Instr. and Meth.* **163**, 77.

Hofstadter, R. (1957). *Ann. Rev. Nucl. Sci.* **7**, 231.

Hofstadter, R. (1974). Twenty-Five years of Scintillation counting, HEPL Report No. 749, Stanford University.

Holder, M. et al. (1978). *Nucl. Instr. and Meth.* **151**, 317.

Holmes, R.R. (1970). Bell Telephone Laboratories Report to Advanced Ballistic Missile Defence Agency, October 1.

Holmes-Siedle, A. and Adams, L. (2002). Handbook of Radiation Effects - 2nd Edition, Oxford University Press, Oxford.

Holy, T. et al. (2006). See web site: *https://edms.cern.ch/document/815615/1*

Holy, T., Jakubek, J., Pospisil, S., Uher, J., Vavrik, D. and Vykydal, Z. (2006). *Nucl. Instr. and Meth. in Phys. Res. A* **563**, 254.

Hopkins, J.C. and Breit, G. (1971). The ^{1}H(n,n)^{1}H scattering observables for high-precision fast-neutron measurement, *Nuclear data Tables* **A.9**, 137.

Hornak, J.P. Ph.D. (2002). The basics of MRI, available on the web site: *http://www.cis.rit.edu/htbooks/mri/*

Houdayer, A., Lebel, C., Leroy, C., Roy, P., Linhart, V., Pospisil, S., Sopko, B., Courtemanche, S. and Stafford, M.C. (2002). *Nucl. Instr. and Meth. in Phys. Res. A* **476**, 588.

Houdayer, A. et al. (2003). *Nucl. Instr. and Meth. in Phys. Res. A* **512**, 92.

Hough, P.V.C. (1948) (a). *Phys. Rev.* **73**, 266.

Hough, P.V.C. (1948) (b). *Phys. Rev.* **74**, 80.

Hubbell, J.H. (1969). Photon Cross Sections, Attenuation Coefficients and Energy Absorption Coefficients from 10 kev to 100 GeV, NSRDS-NBS 29, U.S. Department of Commerce, National Bureau of Standards.

Hubbell, J.H. (1977). *Rad. Res.* **70**, 58; *Int. Jou. App. Radiat. Iso.* **33**, 1269 (1982). See also: Pletachy, E.F. et al. (1978). Tables and Graphs of Photon-Interaction Cross sections from 0.1 to 100 MeV derived from LLNL Evaluated-Nuclear-Data Library, LLNL Report UCRL-5400, Vol.6, Rev. 2.

Hubbell, J.H. (1999). *Phys. Med. Biol.* **R1**.

Hubbell, J.H. and Seltzer, S.M. (2004). Tables of X-Ray Mass Attenuation Coefficients and Mass Energy-Absorption Coefficients (version 1.4); available on the web site at National Institute of Standards and Technology, Gaithersburg, MD (2007): *http://physics.nist.gov/PhysRefData/XrayMassCoef/cover.html* *http://physics.nist.gov/xaamdi* Originally published as: NISTIR 5632, National Institute of Standards and Technology, Gaithersburg, MD (1995).

Hubbell, J.H., Gimm, H.A. and Øverbø, I. (1980). *Phys. Chem. Ref. Data* **9**, 1023.

Hughes, E. (1986). Measurements of Hadronic and Electromagnetic Shower development between 10 and 140 GeV by an Iron-Scintillator Calorimeter, CDHS Internal report.

Hughes, E.B. et al. (1969). *Nucl. Instr. and Meth.* **75**, 130.

Hughes, E.B. et al. (1972). Stanford University Report No. 627.

Huhtinen, M. (2002). *Nucl. Instr. and Meth. in Phys. Res. A* **491**, 194.

Huhtinen, M. and Aarnio, P.A. (1993). *Nucl. Instr. and Meth. in Phys. Res. A* **335**, 580.

Hundhausen, A.J. (1972). Coronal Expansion and Solar Wind, Springer-Verlag, New York.

Hundhausen, A.J. (1977). An Interplanetary View of Coronal Holes, Coronal Expansion and High Speed Wind Streams, Zirker, J. Editor, Colorado Associated University Press, Boulder, Colorado.

IAEA (1991). IAEA Coordinated Research Project 1986 to 1990 with the following participants: Bambynek, W., Barta, T., Christmas, P., Coursol, N., Debertin, K., Helmer, R.G., Jedlovszky, R., Nichols, A.L., Schima, F.J., Yoshizawa, Y. and with Lorenz, A. and Lemmel, H.D. as IAEA Scientific Secretaries, X-Ray and Gamma-Ray Standards for Detector Calibration, IAEA-TECDOC-619. Data are available on the web site: *http://www.nucleartraining.co.uk/iaea/index.htm*

IAEA (1998). IAEA Coordinated Research Project 1986 to 1990 with the following participants: Bambynek, W., Barta, T., Christmas, P., Coursol, N., Debertin, K., Helmer, R.G., Jedlovszky, R., Nichols, A.L., Schima, F.J., Yoshizawa, Y. and with Lorenz, A. and Lemmel, H.D. as IAEA Scientific Secretaries, ZZ XG, Radionuclide Decay Parameters for Gamma and X-Ray Detector Calibration, IAEA1257 ZZ-XG.

IAEA - Nuclear Data Service (2007). Access to evaluated nuclear reaction cross section libraries (ENDF) and experimental nuclear reaction data (EXFOR) is available online from IAEA at the web site: *http://www-nds.iaea.org/*

IC-CAP (2004). IC-CAP version 2004 (Integrated Circuit Characterization and Analysis Program), Agilent Technologies Inc. Headquarters, 5301 Stevens Creek Blvd., Santa Clara, CA 95051 (USA). See also the web site: *http://eesof.tm.agilent.com/products/iccap_main.html*

ICRP - International Commission on Radialogical Protection (1977). ICRP Publication 26, *Ann. ICRP* **1** (3).

ICRP - International Commission on Radialogical Protection (1991). 1990 Recomendation of the International Commission on Radialogical Protection, ICRP Publication 60, *Ann. ICRP* **21** (1–3).

ICRUM - International Commission on Radiation Units and Measurements (1964). Appendix I of Physical Aspects of Irradiation, Recomendation of the ICRU (ICRU report 10b), NBS Handbook 85, Washington D.C., U.S. Govt. Printing Off. See also: Evans, R.D. (1963). Am. Inst. of Phys. Handbook, 2nd Ed., McGraw-Hill, New York.

Berger, R.T. (1961). *Rad. Res.* **15**, 1.

ICRUM - International Commission on Radiation Units and Measurements (1980) (a). Radiation and Quantities Units, ICRU Report 33, Bethesda MD; Quantities and Units in Radiation Protection Dosimetry, ICRU Report 51, Bethesda MD (1993).

ICRUM - International Commission on Radiation Units and Measurements (1980) (b). Radiation and Quantities Units, ICRU Report 33, Bethesda MD.

ICRUM - International Commission on Radiation Units and Measurements (1984) (a). Electrons Beams Energies between 1 and 50 MeV, ICRU Report 35, Bethesda MD.

ICRUM - International Commission on Radiation Units and Measurements (1984) (b). Stopping Powers for Electrons and Positrons, ICRU Report 37, Bethesda MD.

ICRUM - International Commission on Radiation Units and Measurements (1989). Tissues Substitutes in Radiation Dosimetry and Measurements, ICRU Report 44, Bethesda MD.
    See also data on compositions of various human tissues from [ICRUM (1989)] available on web (2007):
    http://physics.nist.gov/PhysRefData/XrayMassCoef/tab2.html

ICRUM - International Commission on Radiation Units and Measurements (1993) (a). Stopping Powers and Ranges for Protons and Alpha Particles, ICRU Report 49, Bethesda, MD.

ICRUM - International Commission on Radiation Units and Measurements (1993) (b). Quantities and Units in Radiation Protection Dosimetry, ICRU Report 51, Bethesda, MD.

ICRUM - International Commission on Radiation Units and Measurements (2005). Stopping of Ions Heavier than Helium, ICRU Report 73, *J. of ICRU* **Vol. 5 No.1**.

Idarraga, J. (2008). Private communication.

Ido, T. et al. (1978). *Jou. Labelled Compd. Radiopharm.* **14**, 175.

Ing, H., Noulty R.A. and McLean, T.D. (1997). *Radiation Measurements* **27**, 1.

Ingard, U. and Kraushaar, W.L. (1960). Introduction to Mechanics, Matter and Waves, Addison-Wesley, Reading (MA).

Inguimbert, C. and Duzelier, S. (2004). *IEEE Trans. on Nucl. Sci.* **51**, 2805.

Inguimbert, C. and Gigante, R. (2006). *IEEE Trans. on Nucl. Sci.* **53**, 1967.

Isenberg, P.A. and Jokipii, J.R. (1979). *Astrophys. J.* **242**, 746.

ISO-15390 (2004). Space Environment (Natural and Artificial) - Galactic Cosmic Ray Model, Ref. no. ISO 15390:2004(E).

ISO-15391 (2004). Space Environment (Natural and Artificial) - Probabilistic Model for Fluences and Peak Fluences of Solar Energetic Particles, Part I Protons, Version October 2004, Ref. no. ISO TS-15391.
    See also the web site:
    http://srd.sinp.msu.ru/nymmik/models/sep.php

IUPAC: Wieser, M.E. (2006). Atomic weights of the elements 2005 (IUPAC Technical Report), *Pure Appl. Chem.* **Vol. 78 (No. 11)**, 2051 doi: 10.1351/pac200678112051.

Iwanczyk, J.S., Schnepple, W.F. and Materson, M.J. (1992).*Nucl. Instr. and Meth. in Phys. Res. A* **322**, 421.

Iwata, S. (1979). Calorimeters (Total Absorption Detectors) for high energy experiments, Nagoya University preprint DPNU-3-79.

Iwata, S. (1980). Calorimeters Nagoya University preprint DPNU 13-80.

Jackson, J.D. (1975). Classical Electrodynamics, 2nd ed., John Wiley, New York.

Jaeger, C.J. and Hulme, H.R. (1936). *Proc. Roy. Soc. (London) A* **153**, 443.
    See also:
    Jaeger, C.J. (1936). *Nature* **137**, 781; **148**, 86 (1941).

Jaffe, G. (1913). *Ann. Phys.* **42**, 303.

James, H.M. and Lark-Horovitz, K. (1951). *Z. Phys. Chem. (Leipzig)* **198**, 107.

Janni, J.F. (1982). *Atomic Data and Nuclear Data Tables* **27**, 147.

Jauch, J.M. and Rohrlich, F. (1955). The Theory of Photons and Electrons [Equations (11)–(17)], Addison-Wesley, Cambridge.

Jellison, G.E. Jr. (1982). *J. of Appl. Phys.* **53**, 5715.

Job, P.K. et al. (1994). *Nucl. Instr. and Meth. in Phys. Res. A* **340**, 283.

Johansson, K. et al. (1998). *IEEE Trans. on Nucl. Sci.* **45**, 2519.

Johns, H.E. and Cunningham, J.R. (1983). The Physics of Radiology - 4th Edition, Charles, C. Thomas Publisher, Fig. 5.8, 153.

Johnsen, K. (1973). *Proc. Nat. Acad. Sci. USA* **70 No. 2**, 619.

Johnston, A.H., Swift, G.M. and Rax, B.G. (1994). *IEEE Trans. Nucl. Sci.* **41**, 2427.

Jokipii, J.R. (1971). *Rev. Geophys. Space Phys.* **9 (no. 1)**, 27.

Jokipii, J.R. and Levy, E.H. (1977). *Astrophys. J.* **213**, L85.

Jokipii, J.R., Levy, E.H. and Hubbard, W.B. (1977). *Astrophys. J.* **213**, 861.

Jokipii, J.R. and Parker, N.E. (1967). *Planet. Space Sci.* **15**, 1375.

Jokipii, J.R. and Parker, N.E. (1970). *Astrophys. J.* **160**, 735.

Jokipii, J.R. and Thomas, B. (1981). *Astrophys. J.* **243**, 1115.

Jones, G.H., Balogh, A. and Smith, I.J. (2003). *Geophys. Res. Lett.* **30 (no.19)**, 8028, doi: 10.1029/2003GL017204.

Jones, L.L. (1999). The APV25 deep submicron readout chip for CMS detectors, Proceedings of the Fifth Workshop on Electronics for LHC experiments, CERN/LHCC 99-33, 162.

Jonscher, A.K. (1961). *Brit. J. of App. Phys.* **12**, 363.

Jun, I. et al. (2003). *IEEE Trans. on Nucl. Sci.* **50**, 1924.

Jungman, G. and Kamionkowski, M. (1994). *Phys. Lett. D* **49**. 2316.

Kallenrode, M.-B. (2004). Space Physics - 3rd Edition, Springer, Berlin.

Kamata, K. and Jishimura, N. (1958). *Progr. Theoret. Phys.* (Kyoto), *Suppl.* **6**, 93.

Kampert, K.H. et al. (1994). A High Resolution BGO Calorimeter with Longitudinal Segmentation, University of Munster preprint, IKS-MS-94/0301.

Kanbach, G. et al. (1988). EGRET Collaboration (Meeting on Problems of Gamma-Ray Astronomy and Planned Experiments, Odessa, Ukrainian SSR, May 4–8, 1987), Space Science Reviews (ISSN 0038-6308), Vol. 49, no. 1–2, 69.

Karp, S. and Gilbert, B.K. (1993). *IEEE Trans. on Aerospace and Elect. Syst.* **29**, 310.

Katz, L. and Penfold, A.S. (1952). *Rev. Mod. Phys.* **24**, 28.

Kelly, J.G., Luera, T.F., Posey, L.D. and Williams, J.G. (1988). *IEEE Trans. Nucl. Sci.* **35**, 1242.

Kholodar, G.A. and Vinetskii, V.L. (1975). *Phys. Status Solidi A* **30**, 47.

Kim, L., Pratt, R.H. and Seltzer, S.M. (1984). *Bull. Am. Phys. Soc.* **29**, 806.

Kirchner, F. et al. Editors (1930). Handbuch der Experimentalphysk, Vol. 1, Leipzig, Akademische Verlagsgesellschaft, 256.
See also:
Davisson, C.M. (1965). Alpha, Beta and Gamma Ray Spectroscopy, Vol. 1, Chapter 2 and Appendix 1, Siegbhan, K. Editor, North-Holland Pub., Amsterdam.

Kimerling, L.C. and Benton, J.L. (1982). *Appl. Phys. Lett.* **39**, 410.

Kinchin, G.H. and Pease, R.S. (1955). *Rep. Prog. in Phys.* **XVIII**, 1.

Kirnas, I.G., Kurilo, P.M., Litovchenko, P.G., Lutsyak, V.S. and Nitsovich, V.M. (1974). *Phys. Status Solidi A* **23**, K123.

Kivelson, M.G. and Bagenal, F. (2006). Planetary Magnetospheres, Encyclopedia of the Solar System - 2nd Edition, McFadden, L.-A., Weissman, P.R. and Johnson, T.V.

Editors, Academic Press, San Diego.

Klages, H.O. et al. (KASCADE Collaboration) (1997). The Kascade Experiment, *Nucl. Phys. B (Proc. Suppl.)* **52**, 92.

Milke, J. et al. (1997). Proc. 25th ICRC, Durban, HE 1.2.28.

Klanner, R. et al. (1988). *Nucl. Instr. and Meth. in Phys. Res. A* **265**, 200.

Klein, O. and Nishina, Y. (1929). *Z. Physik* **52**, 853.

Kleinknecht, K. (1998). Detectors for Particle Radiation - 2nd Edition, Cambridge University Press, Cambridge.

Klopfenstein, C. (1983). *Phys. Lett. B* **130**, 444.
See also:
Schamberger, R.D. et al. (1984). *Phys. Lett. B* **138**, 225.
Böhringer, T. et al. (1980). *Phys. Rev. Lett.* **44**, 1111.

Knapp, J., Heck, D., Sciutto, S., Dova, M.T. and Risse, M. (2003). *Astropart. Phys.* **19**, 77.

Knecht, D.J. and Shuman, B.M. (1985). The Geomagneic Field, Chapter 4 of Handbook of Geophysics and the Space Environment, Editor Jursa, A.S., Air Force Geophysics Lab, Springfield.

Knoll, G.F. (1999). Radiation Detection and Measurement - 3rd Edition, John Wiley & Sons, New York.

Koch, H.W. and Motz, J.W. (1959). *Rev. Mod. Phys.* **31**, 920.

Kölbig, K.S. and Margolis, B. (1968). *Nucl. Phys. B* **6**, 85.
See also:
Trefil, J.S. (1969). *Phys. Rev.* **180**, 1366.

Kondo, T. and Niwa, K. (1984). Electromagnetic shower size and containement at high energies, Proceedings of the 1984 Summer Study on the Design and Utilization of the Superconducting Super Collider, June 23 – July 13, Snowmass, Colorado, Donaldson, R. and Morfin, J.G. Editors, 559

Kondo, T. et al. (1984). A simulation of electromagnetic showers on iron, lead and uranium liquid argon calorimeters usung EGS and its implications to e/h ratios in hadron calorimetry, Proceedings of the 1984 Summer Study on the Design and Utilization of the Superconducting Super Collider, June 23 – July 13 1984, Snowmass, Colorado, Donaldson, R. and Morfin, J.G. Editors, 556.

Konozenko, D., Semenyuk, A.K. and Khivrich, V.I. (1969). *Phys. Status Solidi* **35**, 1043.

Korde, R. et al. (1989). *IEEE Trans. on Nucl. Sci.* **NS-36**, 2169.

Korff, S. A. (1946). Electrons and nuclear counters, Van Nostrand, New York.

Kosier, S.L. et al. (1994). *IEEE Trans. on Nucl. Sci.* **41 (no. 6)**, 1964.

Kota, J. and Jokipii, J.R. (1983). *Astrophys. J.* **265**, 573.

Kotochigova, S., Levine, Z.H., Shirley, E.L., Stiles, M.D. and Clark, C.W. (1996). Atomic Reference Data for Electronic Structure Calculations (version 1.2), National Institute of Standards and Technology, Gaithersburg, MD; available on the web site:
http://physics.nist.gov/PhysRefData/DFTdata/contents.html
See also:
Kotochigova, S., Levine, Z.H., Shirley, E.L., Stiles, M.D. and Clark, C.W. (1997). *Phys. Rev. A* **55**, 191; erratum **56**, 5191.
In particular, the ground electronic configurations of the elements H through U and their first cations are available on the web site:
http://physics.nist.gov/PhysRefData/DFTdata/configuration.html

Kötz, U. et al. (1985). *Nucl. Instr. and Meth. in Phys. Res. A* **235**, 481.

Kozlovski, V. and Abrosimova, V. (2005). Radiation Defect Engineering, World Scientific, Singapore.

Kramers, H.A. (1952). *Physica* **18**, 665.

Kraner, H.W. (1982). *IEEE Trans. on Nucl. Sci.* **29**, 1088.

Kraner, W., Li, Z. and Posnecker, K.U. (1989). *Nucl. Instr. and Meth. in Phys. Res. A* **279**, 266.

Krause, M.O. (1979). *Jou. Phys. Chem. Ref. Data* **8**, 307.
See also:
Krause, M.O. and Oliver J.H. *Jou. Phys. Chem. Ref. Data* **8**, 329.

LaBel, K. (1993). Single event effects specification, see web site:
http://radhome.gsfc.nasa.gov/radhome/papers/seespec.htm

LaBel, K. (1996). SEE mitigation: Methods of reducing SEE impacts, see web site:
http://radhome.gsfc.nasa.gov/radhome/papers/seeca6.htm

Lachish, U. (1999). *Nucl. Instr. and Meth. in Phys. Res. A* **436**, 146.

Lampert, M.A. (1962). *Phys. Rev.* **125**, 126.

Lampert, M.A. and Rose, A. (1961). *Phys. Rev.* **121**, 26.

Landau, L. (1944). *Jou. of Phys. (USSR)* **8**, 201.

Landau. L.D. and Pomeranchuck, I.J. (1965). The collected papers of Landau, Pergamon Press London.

Landau, L.D. and Rumer, G. (1938). *Proc. Roy. Soc. A* **166**, 213.

Lang, D.V. (1974). *J. of Appl. Phys.* **45**, 3023.

Lang, K.R. (2001). The Cambridge Encyclopedia of the Sun, Cambridge University Press, Cambridge.

Lario, D. and Decker, R.B. (2001). Re-examination of the October 20, 1989 ESP event, Proc. of the ICRC 2001, 07–15 August, Hamburg, Germany, p. 3485.

Lazanu, I. and Lazanu, S. (2003). *Physica Scripta* **67**, 388.

Leadon, R.E. (1970) *IEEE Trans. on Nucl. Sci.* **NS-17**, 110.

Lecoq, P. (2000). New Scintillating Crystals for Medical Imaging, Proceedings of VIIIth International Conference on High Energy Physics, CALOR99, 13–19 June 1999, Lisbon, Portugal, Barreira, G., Gomes, A., Maio, A., Tomé, B., Varanda, M.J. Editors, World Scientific, Singapore, 667.

Lecoq, P. et al. (2001). The Crystal Clear Collaboration, R&D Proposal for the study of new, fast and radiation hard scontillators for calorimetry at LHC, CERN/DRDC P27/91-1, Project RD18.

Lee, M.Y. et al. (1979). *Phys. Rev. D* **19**, 55.

Lee, Y.-H. and Corbett, J.W. (1976). *Phys. Rev. B* **13 (no. 6)**, 2653.

Leighton, R.B. (1959). Principles of Modern Physics, McGraw-Hill, New York.

Lemaire, J. and Scherer, M. (1973). *Rev. Geophys. Space Phys.* **11**, 427.

Lemeilleur, F. et al. (1994). Charge Transport in Silicon Detectors, Conference Proceedings, Vol. 46, Large Scale Application and Radiation Hardness of Semi-Conductors, Baldini, A. and Focardi, E. Editors, IPS, Bologna, 167.

Leo, W.R. (1994). Techniques for Nuclear and Particle Physics Experiments - 2nd Edition, Springer-Verlag, Berlin.

LEP (1982). ALEPH Apparatus for LEP PHysics LEPC/82-3/I 1; DELPHI Detector with Lepton Photon and Hadron Identification LEPC/82-8/I 6; L3 LEPC/82-5/I 3; OPAL an Omni Purpose Apparatus for Lep LEPC/82-4/I 2.

Le Roux, J.A. and Potgieter, M.S. (1990). *Astrophys. J.* **361**, 275.

Leroy, C. (1998). Personal logbook.

Leroy, C. (2004) (a). Private communication.

Leroy, C. (2004) (b). Calculation of the expectation values of proton and neutron groups spin in nuclei, *PICASSO Scientific / Technical Report* **PSTR-04-010**, 1 June.

Leroy, C. and Rancoita, P.G. (2000). Physics of cascading shower generation and propagation in matter: principles of high-energy, ultrahigh-energy and compensating calorimetry, *Rep. Prog. in Phys.* (April) **63, no. 4**, Institute of Physics Publishing, 505–606, doi: 10.1088/0034-4885/63/4/202.

Leroy, C. and Rancoita, P.G. (2004). Principles of Radiation Interaction in Matter and Detection - 1st Edition, World Scientific, Singapore, ISBN 981-238-909-1.

Leroy, C. and Rancoita, P.G. (2007). Particle Interaction and Displacement Damage in Silicon Devices operated in Radiation Environments, *Rep. Prog. in Phys.* (April) **70, no. 4**, 403–625, Institute of Physics Publishing, doi: 10.1088/0034-4885/70/4/R01.

Leroy, C. and Roy, P. (1998). Current pulse response expected for silicon detectors at LHC, *UdeM HEP preprint* **98-004**.

Leroy, C. et al. (1994). Study of electrical properties and charge collection of silicon detectors under neutron, proton and gamma irradiations, Proc. IVth Int. Conf. on Calorimetry in High Energy Physics, Editors Menzione, A. and Scribano, A., World Scientific, Singapore, 627.

Leroy, C. et al. (1997) (a). *Nucl. Instr. and Meth. in Phys. Res. A* **388 3**, 289.

Leroy, C. et al. (1997) (b). *Nucl. Inst. and Meth. in Phys. Res. A* **388**, 289.

Leroy, C. et al. (1997) (c). *Nucl. Instr. and Meth. in Phys. Res. A* **388**, 289, doi: 10.1016/S0168-9002(97)00004-1.

Leroy, C. et al. (1999) (a). Radiation Hardness Studies of Components of the ATLAS Forward and Hadronic End Cap Calorimeters at Dubna, Proceedings of the VIII International Conference on Calorimetry in High Energy Physics, CALOR99, June 13–19 1999, Lisbon, Portugal.

Leroy, C. et al. (1999) (b). Liquid argon pollution tests of the ATLAS detector materials at IBR-2 reactor in Dubna, ATLAS-LARG internal note, ATL-LARG-99-010.

Leroy, C. et al. (2000) (a). *Particles and Nuclei, Letters* **51 [102]**, 5.

Leroy, C. et al. (2000) (b). *Particles and Nuclei, Letters* **51 [102]**, 25.

Leroy, C. et al. (2000) (c). *Particles and Nuclei, Letters* **51 [102]**, 20.

Leroy, C. et al. (2002) (a). ATLAS Note, ATL-LARG-2002-003.

Leroy, C. et al. (2002) (b). Proceedings of the 7th International Conference on Advanced Technology and Particle Physics, October 15–19 2001, Villa Olmo, Como, Italy, Barone, M., Borchi, E., Huston, J., Leroy, C., Rancoita, P.G., Riboni, P.L. and Ruchti, R. Editors, World Scientific, Singapore, 800.

Leroy, C. et al. (2002) (c). Irradiation test of ATLAS liquid argon forward calorimeter (FCAL) electronics components, ATL-LARG-2002-003.

Leroy, C., Glaser, M., Heijne, E.H.M., Jarron P., Lemeilleur, F., Rioux, J., Soave, C. and Trigger, I. (1993). *Technical Report*, **CERN/ecp 93-12**.

Leroy, C., Roy, P., Casse, G., Glaser, M., Grigoriev, E. and Lemeilleur, F. (1999) (a). *Nucl. Instr. and Meth. in Phys. Res. A* **426**, 99, doi: 10.1016/S0168-9002(98)01478-8.

Leroy, C., Roy, P., Casse, G., Glaser, M., Grigoriev, E. and Lemeilleur, F. (1999) (b). *Nucl. Instr. and Meth. in Phys. Res. A* **426**, 99, doi: 10.1016/S0168-9002(98)01478-8.

Leroy, C., Sirois, Y. and Wigmans, R. (1986). *Nucl. Instr. and Meth. in Phys. Res. A* **252**, 4.

Leutz, H. (1995). *Nucl. Instr. and Meth. in Phys. Res. A* **364**, 422.

Levinger, J.S. (1951). *Phys. Rev.* **84**, 43.

Levinshtein, M., Rumyantsev, S. and Shur, M. (2000) (first published 1996). Handbook series on Semiconductor Parameters (vol 1): Si, Ge, C (Diamond), GaAs, GaP, GaSb, InAs, InP, InSb, World Scientific, Singapore.

Lewin, J.D. and Smith, P.F. (1996). *Astropart. Phys.* **6**, 87.

LHC (1990). See for instance articles in the Proceedings of the Large Hadron Collider Workshop Vol. I–III Aachen 4–9 October 1990, Jarlskog, G. and Rein, D. Editors,

CERN 90-10 ECFA 90-133 (1990).

See also:

Kazakov, D. (1991). Beyond the Standard Model, CERN-JINR School of Physics Egmond-aan-Zee 25 June–8 July 1989; CERN 91-07.

Bagger, J.A. (1991). Physics Beyond the Standard Model, Lecture given at the 1991 Theoretical Advanced Studies Institute Boulder Colorado JHU-TIPAC-910038.

Li, Z. (1994) (a). *IEEE Trans. on Nucl. Sci.* **41 (no. 4)**, 948.

Li, Z. (1994) (b). *Nucl. Instr. and Meth. in Phys. Res. A* **342**, 105.

Li, Z., Chen, W. and Kraner, H.W. (1991). *Nucl. Instr. and Meth. in Phys. Res. A* **308**, 585.

Li, Z., Kraner, H.W., Verbitskaya, E., Eremin, V., Ivanov, A., Rancoita, P.G., Rattaggi, M., Fonash, S.J., Rubinelli, F.A., Dale, C. and Marshall, P. (1992). *IEEE Trans. on Nucl. Sci.* **39 (no. 6)**, 1730, doi: 10.1109/23.211360.

Lindhard, J. (1976). *Nucl. Instr. and Meth.* **132**, 1 and references therein.

Lindhard, J. and Scharff, M. (1961). *Phys. Rev.* **124**, 128.

Lindhard, J. and Sørensen, A.H. (1996). *Phys. Rev. A* **53**, 2443.

Lindhard, J., Nielsen, V., Scharff, M. and Thomsen, P.V. (1963). *Kgl. Danske Vidensk. Selsk. Mat.-Fys. Medd.* **33**, 10.

Lindstroem, G. (1991). *SITP-Internal note Feb. 10.*

Lindstroem, G. et al. (1987). SICAPO Collab. meeting (February 1987), private communication.

Lindstroem, G. et al. (1990). MC simulations with EGS4 for calorimeters with thin silicon, Proceedings of the Workshop on Calorimetry for Supercollider, Tuscaloosa March 13–17 1989, Donaldson, R. and Gilchriese, M.G.D. Editors, World Scientific Singapore, 215.

Lindstroem, G., Moll, M. and Fretwurst, E. (1999). *Nucl. Instr. and Meth. in Phys. Res. A* **426**, 1.

Lingren, C.L. and Butler, J.F. (1998). *IEEE, Trans. on Nucl. Sci.* **45**, 1723.

Litt, J. and Meunier, R. (1973). *Ann. Rev. Nucl. Sci.* **23**, 1.

Litvinov, V.L., Ukhin, N.A. and Oldham, W.G. (1989). *Phys. Status Solidi B* **154**, 507.

Llopart, X. et al. (2002). *IEEE Trans. on Nucl. Sci.* **49**, 2279.

Lockwood, M. and Hapgood, M. (2007). *News and Reviews in Astron. & Astrophys.* **48 (no. 6)**, 6.11.

Loferski, J.T. (1958). *J. of Appl. Phys.* **29**, 35.

Lohrmann, E. (1969). Electromagnetic Interactions and Photoproduction, Proc. of the Lund Int. Conf. on Elementary Particles, Lund June 25 – July 1, 11.

See also:

Gottfried, K. (1972). Neutrino Gamma and Hadron Interactions in Nuclear Matter, Proc. of the 1972 CERN School of Physics, CERN 72-12, 55.

Longo, E. and Sestilli, I. (1975). *Nucl. Instr. and Meth.* **128**, 283.

Lorenz, E., Mageras, G. and Vogel, H. (1986). *Nucl. Instr. and Meth. in Phys. Res. A* **249**, 235.

Ludlam, T. (2005). *Nucl. Phys. A* **750**, 9.

Lugakov, P.F., Lukashevich, T.A. and Shusha, V.V. (1982). *Phys. Status Solidi A* **74**, 445.

Luhmann, J.G. and Solomon, S.C. (2006). The Sun-Earth Connection, Encyclopedia of the Solar System - 2nd Edition, McFadden, L.-A., Weissman, P.R. and Johnson, T.V. Editors, Academic Press, San Diego.

Lundin, R., Yamauchi, M., Sauvaud, J.-A. and Balogh, A. (2005). *Annales Geophysicae* **23**, 2565.

Lutz, G. (2001). Semiconductor Radiation Detectors, Springer-Verlag, Berlin.

Ma, T.P. and Dressendorfer, P.V, (Editors) (1989). Ionizing Radiation Effects in MOS Devices and Circuits, John Wiley & Sons, New York.

MacFarlane, G.G., McLean, T.P., Quarrington, J.E. and Roberts, V. (1958). *Phys. Rev.* **111**, 1245.

Majewski, S. and Zorn, C. (1992). Fast scintillators for high radiation levels, Instrumentation in high energy physics, V 9, Advanced series on directions in high energy physics, Sauli, F. Editor, World Scientific, Singapore, 157.

Makarenko, L.F. (2001). *Physica B* **308–310**, 465.

Maksimovic, M., Pierrard, V. and Lemaire, J. (2001). *Astrophys. Space Sci.* **277**, 181.

Manfredi, P.F. and Ragusa, F. (1985). *Nucl. Instr. and Meth. in Phys. Res.* A **235**, 345.

Mangiagalli, P., Levalois, M. and Marie, P. (1998). *Nucl. Instr. and Meth. in Phys. Res.* B **146**, 321.

Mangiagalli, P., Levalois, M., Marie, P., Rancoita, P.G. and Rattaggi, M. (1998). *Nucl. Phys. B (Proc. Suppl.)* **61B**, 464, doi: 10.1016/S0920-5632(97)00603-8.

Mangiagalli, P., Levalois, M., Marie, P., Rancoita, P.G. and Rattaggi, M. (1999). *Eur. Phys. J. AP* **6**, 121, doi: 10.1051/epjap:1999161.

Marmier, P. and Sheldon, E. (1969). Physics of Nuclei and Particles, Vol. I, Academic Press, New York.

Marmier, P. and Sheldon, E. (1970). Physics of Nuclei and Particles, Vol. II, Academic Press, New York.

Martin, W.C. and Lise, W.L. (1995). Atomic Spectroscopy in Atomic, Molecular and Optical Physics References Book, AIP Press, New York.

Matthiae, G. (2008). The Auger Experiment Status and Results, Proc. of the 10th ICATPP (8–12 October 2007, Como, Italy), Editors Barone, M., Gaddi, A., Leroy, C., Price, L., Rancoita, P.G. and Ruchti, R., World Scientific, Singapore, 229.

Maximon, C.L. (1968). *Res. Nat. Bur. Stds.* **72** B, 79.

May, T.C. and Woods, M.H. (1979). *IEEE Trans. on Electron Devices* **ED-26**, 2.

McComas, D.J., Barraclough, B.L., Funsten, H.O., Gosling, J.T., Santiago-Muõz, E., Skoug, R.M., Goldstein, B.E., Neugebauer, M., Riley, P. and Balogh, A. (2000). *J. Geophys. Res.* **105 (no. A5)**, 10419.

McComas, D.J., Elliott, H.A., Schwadron, N.A., Gosling, J.T., Skoug, R.M. and Goldstein, B.E. (2003). The three-dimensional solar wind around solar maximum, *Geophys. Res. Lett.* **30 (no. 10)**, 1517, doi: 10.1029/2003GL017136.

McCray, R. and Snow, T.P. (1981). *Astrophys. J.* **248**, 1166.
    See also:
    Cowie, L.L. and Somgaila, A. (1986). *An. Rev. Astron. Astrophys.* **24**, 499.

McDonald, P.T., Stapor, W.J. and Henson, B.G. (1999). PC603 32-Bit RISC μP Radiation Effects Study, *GOMAC Digest of Papers* **26**.

McGregor, D.S. and Shultis, J.K. (2004). *Nucl. Instr. and Meth. in Phys. Res.* A **517**, 180.

McGregor, D.S. et al. (2003). *Nucl. Instr. and Meth. in Phys. Res.* A **500**, 272.

McKinley, A. Jr. and Feshbach, H. (1948). *Phys. Rev.* **74**, 1759.

MCNPX (2006). See web site:
    http://mcnpx.lanl.gov/

Medipix Collab. (2008). Home page at the web site:
    http://medipix.web.cern.ch/MediPix2/

Melissinos, A.C. (1966). Experiments in Modern Physics, Academic Press, San Diego.

Melissinos, A.C. and Napolitano, J. (2003). Experiments in Modern Physics - 2nd Edition, Academic Press, Burlington.

Mendenhall, M.H. and Weller, R.A. (2005). *Nucl. Instr. and Meth. in Phys. Res.* B **227**, 420.

Merkle, K.L., King, W.E, Baily, A.C., Haga, K. and Meshii, M. (1983). *J. Nucl. Mater.* **117**, 4.

Mesquita, C.H., Filho, T.M. and Hamada, M.M. (2003). *IEEE Trans. on Nucl. Sci.* **50**, 1170.

Messel, H. and Crawford, D.F. (1970). Electron-Photon Shower Distribution, Tables for Lead, Copper and Air Absorbers, Pergamon Press, London.

Messenger, G.C. (1966). *IEEE Trans. on Nucl. Sci.* **13 (no. 6)**, 141.

Messenger, G.C. (1967) (a). *Proc. of the IEEE* **55**, 414.

Messenger, G.C. (1967) (b). *IEEE Trans. on Nucl. Sci.* **14 (no. 6)**, 88.

Messenger, G.C. (1972). *IEEE Trans. on Nucl. Sci.* **55**, 160.

Messenger, G.C. (1973). *IEEE Trans. on Nucl. Sci.* **20**, 809.

Messenger, G.C. (1992). *IEEE Trans. on Nucl. Sci.* **39**, 468.

Messenger, G.C. and Ash, M.S. (1992). The Effects of Radiation on Electronic Systems - 2nd Edition, Van Nostrand Reinhold Company, New York; 1st Edition (1986).

Messenger, G.C. and Ash, M.S. (1997). Single Event Phenomena, Chapman & Hall, New York.

Messenger, G.C. and Spratt, J.P. (1958). *Proc. IRE* **46**, 1038.

Messenger, S.R., Burke, E.A., Summers, G.P. and Walters, R.J. (2002). *IEEE Trans. on Nucl. Sci.* **49**, 2690.

Messenger, S.R. et al. (1999). *IEEE Trans. on Nucl. Sci.* **46**, 1595.

Messenger, S.R. et al. (2003). *IEEE Trans. on Nucl. Sci.* **50**, 1919.

Messenger, S.R. et al. (2004). *IEEE Trans. on Nucl. Sci.* **51**, 2846.

Messier, J. and Merlo-Flores, J.G. (1963). *J. Phys. Chem. Solids* **24**, 1539.

Mewaldt, R.A. (1994). *Adv. Space Res.* **14**, 737.

Mewaldt, R.A. (1996). *Astrophys. J.* **466**. L43.

Mewaldt, R.A. et al. (1993). *Geophys. Res. Lett.* **20**, 2263.

Mewaldt, R.A., Selesnick, R.S. and Cummings, J.R. (1997). Anomalous Cosmic Rays: The Principal Source of High Energy Heavy Ions in the Radiation Belts, *Radiation Belts Models and Standards* (AGU Geophysical Monograph) **97**, 35.

Meyer, P., Parker, E.N. and Simpson, J.A. (1974). *Physics Today* **27**, 10.

Meyer-Vernet, N. (2007). Basics of the Solar Wind, Cambridge University Press, Cambridge.

Migdal, A.B. (1956). *Phys. Rev.* **103**, 1811.

Mikulec, B. (2000). Single Photon Detection with Semiconductor Pixel Arrays for Medical Imaging Applications, Ph. Thesis, University of Vienna, Austria, June, CERN-THESIS-2000-021.

Milnes, A.G. (1973). Deep Impurities in Semiconductors, John Wiley & Sons, New York.

Miroshnichenko, L.I. (2003). Radiation Hazard in Space, Kluwer Academic Press, Dordrecht.

Misiakos, K. and Tsamakis, D. (1994). *App. Phys. Lett.* **64 (15)**, 2007.

Mockett, P.M. and Boulware, M. (1991). Electromagnetic shower sampling inefficiency in liquid argon sampling calorimeters and the sampling efficiency dependence on low-Z cladding of high-Z aborber plates, Symposium on Detector Research and Development for the Superconducting Supercollider, Fort Worth, Texas, 15–18 October 1990, Dombeck, T., Kelley, V. and Yost, G.P. Editors, World Scientific, Singapore, 382.

Mohr, P.J. and Taylor, B.N. (2005). The 2002 CODATA Recommended Values of the Fundamental Physical Constants, web Version 4.2, National Institute of Standards and Technology, Gaithersburg, MD 20899, 24 May 2005; available on the web site (2007):

http://physics.nist.gov/cuu/Constants/index.html
See also:
Mohr, P.J. and Taylor, B.N. (2005). *Rev. of Mod. Phys.* **77**, 1.
Molière, G. (1947). *Z. Naturforsh.* **A2**, 133; **A3**, 78 (1948).
Moll, J.L. (1958). *Proc. of the IRE* **46**, 1076.
Moll, M. (1999). Radiation Damage in Silicon Particle Detectors - microscopic defects and macroscopic properties, *Ph.D. thesis University of Hamburg*, DESY-THESIS-1999-040, ISSN: 1435-8085.
Møller, C. (1932). *Ann. Physik* **14**, 531.
Møller, S.P. (1986). Experimental Investigation of Energy loss and Straggling together with Inner Shell-Excitation of Relativistic Projectiles in solids, Thesis, Institute of Physics, University of Aarhus (Denmark).
Other approaches (see also references therein) for evaluating the modified straggling function by distant collsions are found for instance in:
Talman, R. (1979). *Nucl. Instr. and Meth.* **159**, 189.
Hall, G. (1984). *Nucl. Instr. and Meth. in Phys. Res. B* **1**, 16.
Lindhard, J. (1985). *Physica Scripta* **32**, 72.
Moniz, E.J. et al. (1971). *Phys. Rev. Lett.* **26**, 445.
See also:
Whitney, R.R. et al. (1974). *Phys. Rev. C* **9**, 2230.
Moon, P.B. (1950). *Proc. Pys. Soc. (London) A* **63**, 1189.
See also:
Evans, R.D. (1958). Encyclopedia of Physics Vol. XXXIV, Springer Pub., Berlin/Göttingen/Heidelberg, 218.
Morin, F.J. and Maita, J.P. (1954). *Phys. Rev.* **96**, 28.
Mork, J. (1971). *Phys. Rev. A* **4**, 917.
Mork, J.K. and Olsen, H. (1965). *Phys. Rev.* **140**, 1661.
Morse, P.M. (1969). Thermal Physics, Benjamin inch., New York.
Moscatelli, F. et al. (2002). *Nucl. Instr. and Meth. in Phys. Res. B* **186**, 171.
Moses, W.W., Derenzo, S.E. and Budinger, T.F. (1994). *Nucl. Instr. and Meth. in Phys. Res. A* **353**, 189.
Mott, N.F. (1929). *Proc. Roy. Soc. A* **124**, 426; *A* **135**, 429 (1932).
See also:
McKinley, W.A. and Feshbach, H. (1948). *Phys. Rev.* **74**, 1759.
Dalitz, R.H. (1951). *Proc. Roy. Soc. A* **206**, 509.
Barlett, J.H. and Watson, R.E. (1940). *Proc. Am. Acad. Arts. Sci.* **74**, 53.
Massey, H.S.W. (1942). *Proc. Roy. Soc. A* **181**, 14.
Mott, N.F. (1930). *Proc. Roy. Soc. A* **126**, 259.
Mott, N.F. and Massey, H.S.W. (1965). The Theory of Atomic Collisions - 3rd Edition, Oxford University Press, London.
Motz, J.W. and Missoni, G. (1961). *Phys. Rev.* **124**, 1458.
Msimanga, M. and McPherson, M. (2006). *Mat. Scie. and Engin. B* **127**, 46.
Mühlemann, P. et al. (1973). *Nucl. Phys. B* **59**, 106.
Müller, D. (1989). *Adv. Space Res.* **9** (12), 31.
Müller, R.S. and Kamins, T.I. (1977). Device Electronics for Integrated Circuits - 2nd Edition, John Wiley & Sons, New York.
Müller, R.S. and Kamins, T.I. (with Chan, M.) (2002). Device Electronics for Integrated Circuits - 3rd Edition, John Wiley & Sons, New York.
Müller, B. and Nagle, J.L. (2006). *Ann. Rev. Nucl. Part. Sci.* **56**, 93.
Muraki, Y. et al. (1984). Radial and longitudinal behaviour of nuclear-electromagnetic

cascade showers induced by 300 GeV protons in lead and iron absorbers, ICR-report 117-84-6, Tokyo.

Murphy, N., Smith, E.J. and Schwadron, N.A. (2002). *Geophys. Res. Lett.* **29 (22)**, 23-1.

Murthy, P.V.R. et al. (1975). *Nucl. Phys. B* **92**, 269.

NACP - Nordic Association of Clinical Physics (1980). *Acta Radiol. Oncology* **19 (no. 1)**, 55.

Nagel, H.H. (1965). *Z. Physik* **186**, 319.

Nakamoto, A. et al. (1985). *Nucl. Instr. and Meth. in Phys. Res. A* **238**, 53.

Nakamoto, A. et al. (1986). *Nucl. Instr. and Meth. in Phys. Res. A* **251**, 275.

Namenson, A.I., Wolicki, E.A. and Messenger, G.C. (1982). *IEEE Trans. on Nucl. Sci.* **29**, 1018.

Nash, A.G., Sheeley, N.R. Jr. and Wang, Y.-M. (1988). *Solar Phys.* **117**, 359.

Neamen, D.A. (2002). Semiconductor Physics and Devices - 3rd Edition, Tata McGraw–Hill Science, New Delhi.

Nelson, W.R. et al. (1966). *Phys. Rev.* **149**, 201.

Nelson, W. R., Hirayama, H. and Rogers, D.W.O. (1985). The EGS4 Code System SLAC Report 165.

Ng, K.K. (2002). Complete Guide to Semiconductor Devices - 2nd Edition, John Wiley & Sons, New York.

Nichols, D.K. and van Lint V.A.J. (1966). Solid State Physics **vol. 18**, edited by Seitz, F. and Turnbull, D., Academic Press Inc., New York, 1.

Nikolaev, N.N. (1977). Particle–nucleus interactions at high energies, Proc. of the Topical meeting on Multiparticle Production on Nuclei at Very High Energies, Bellini, G., Bertocchi, L. and Rancoita, P.G. Editors, IAEA-SMR-21, 159.

NIST (2008). A partial list of the mean excitation energy is avaialble at:
http://physics.nist.gov/PhysRefData/XrayMassCoef/tab1.html
http://physics.nist.gov/PhysRefData/XrayMassCoef/tab2.html

NNDC (2003). 2003 Atomic Mass Evaluation, National Nuclear Data Center, BNL, Upton NY 11973-5000; see the web site:
http://www.nndc.bnl.gov/masses/#atomic
See also:
Wapstra, A.H., Audi, G. and Thibault, C. (2003). Nucl. Phys. A **729**, 129.
Audi, G., Wapstra, A.H. and Thibault, C. (2003). Nucl. Phys. A **729**, 337.
Coursey, J.S., Schwab, D.J. and Dragoset, R.A. (2005). Atomic Weights and Isotopic Compositions (version 2.4.1), National Institute of Standards and Technology, Gaithersburg, MD; available at the web site:
http://physics.nist.gov/Comp
originally published as:
Coplen, T.B. (2001). Atomic Weights of the Elements 1999, *Pure Appl. Chem.* **73(4)**, 667.
Rosman, K.J.R. and Taylor, P.D.P. (1998). Isotopic Compositions of the Elements 1997, *J. Phys. Chem. Ref. Data* **27(6)**, 1275.
Audi, G. and Wapstra, A.H. (1995). The 1995 Update To The Atomic Mass Evaluation, *Nucl. Phys. A* **595(4)**, 409.

NNDC (2008) (a). The National Nuclear Data Center (NNDC) and other memnbers of the IAEA-sponsored International Nuclear Structure and Decay Data (NSDD) and Nuclear Reaction Data (NRDC) Networks and the U.S. Nuclear Data Program (US-NDP) provide access to many of the bibliographic and numeric databases mantained by members of these groups. Access can be obtained by FTP, TELNET and WEB; see the web site:

http://www.nndc.bnl.gov/index.jsp

NNDC (2008) (b). National Nuclear Data Center, BNL, Upton NY 11973-5000; see the web site:

http://www.nndc.bnl.gov/about/nndc.html#address

NNDC (2008) (c). National Nuclear Data Center, BNL, Upton NY 11973-5000; see the web site:

http://www.nndc.bnl.gov/qcalc

Nordberg, M.E. (1971). Cornell Report CLNS 138.

Norgett, M.J., Robinson, M.T. and Torrens, I.M. (1975). *Nucl. Engin. and Des.* **33**, 50.

Norlin, B. (2007). PhD Thesis, Sundsvall Mid Sweden University, Doctorate Thesis 26 ISBN 978-91-85317-55-4.

Normand, E. (1998). *IEEE Trans. on Nucl. Sci.* **45**, 2904.

Normand, E. and Baker, T.J. (1993). *IEEE Trans. on Nucl. Sci.* **40**, 1484.

Northcliffe, L.C. (1960). *Phys.Rev.* **120**, 1744.

Northcliffe, L.C. (1963). *Ann. Rev. of Nucl. Sci.* **13**, 67.

Norton, P., Braggins, T. and Levinstein, H. (1973). *Phys. Rev. B* **8 (no. 12)**, 5632.

Novov, V. et al. (1999). *Phys. Rev. Lett.* **83**, 1104.

Nowlin, R.N. et al. (1993). *IEEE Trans. on Nucl. Sci.* **40 (no. 6)**, 1686.

Nussbaum, A. (1973). *Phys. Status Solidi A* **19**, 441.

Nygard, E. et al. (1991). *Nucl. Instr. and Meth. in Phys. Res. A* **301**, 506.

Nymmik, R. (2000). *Adv. Space Res.* **11 (no. 11)**, 1875.

Nymmik, R. (2006). *Adv. Space Res.* **38**, 1182.

See also [ISO-15391 (2004)].

Nymmik, R.A., Panasyuk, M.I. and Suslov, A.A. (1992). *Adv. Space Res.* **17 (no. 2)**, (2)19.

Nymmik, R.A. and Suslov, A.A. (1996). *Radiat. Meas.* **26 (no. 3)**, 477.

Nymmik, R.A., Panasyuk, M.I., Pervaja, T.I. and Suslov, A.A. (1992). *Nucl. Tracks Radiat. Meas.* **20 (no. 3)**, 427.

See also the web site:

https://creme96.nrl.navy.mil

Oberg, D.L. et al. (1994). Measurement of Single Event Effects in the 87c51 Microcontroller, Workshop Record IEEE Radiation Effects Data Workshop, 43.

Oganessian, Yu. Ts. et al. (1999) (a). *Eur. Phys. J.* **5**, 63; *Nature* **400**, 242.

Oganessian, Yu. Ts. et al. (1999) (b). *Phys. Rev. Lett.* **83**, 3154.

O'Gorman, T.J. (1996). *IEEE Trans. Electron Devices* **ED-41**, 553.

O'Gorman, T.J. et al. (1996). *IBM J. Res. Develop.* **40**, 41.

Öğütt, S. and Chelikowsky, J.R. (2003). *Phys. Rev. Lett.* **91 (no. 23)**, 235503-1.

Olsen, H. (1955). *Phys. Rev.* **99**, 1335.

Onsager, L. (1938). *Phys. Rev.* **54**, 554.

Ordonez, C.E., Bolozdynya, A. and Chang, W. (1997) (a). IEEE Nuclear Science Symposium Conference Record, Vol. 2, 1122.

Ordonez, C.E., Bolozdynya, A. and Chang, W. (1997) (b). IEEE Nuclear Science Symposium Conference Record, Vol. 2, 1361.

Oreglia, M. et al. (1982). *Phys. Rev. D* **25**, 2259.

See also:

Nernst, R. et al. (1985). *Phys. Rev. Lett.* **54**, 2195.

Otterlund, I. (1977). Inclusive production on emulsion with accelerator beams, Proc. of the Topical meeting on Multiparticle Production on Nuclei at Very High Energies, Bellini, G., Bertocchi, L. and Rancoita, P.G. Editors, IAEA-SMR-21, 31.

Otterlund, I. (1980). High energy reactions on nuclei, *Nucl. Phys. A* **335**, 507–516.

Ougouag, A.M., Williams, G.J., Danjaji, M.B., Yang, S.Y. and Meason, J.L. (1990). *IEEE Trans. on Nucl. Sci.* **37**, 2219.

Øverbø, I. (1977). *Phy. Lett. B* **71**, 412.

Øverbø, I. (1979). *Physica Scripta* **19**, 299.

Øverbø, I., Mork, K.J. and Olsen, H.A. (1968). *Phy. Rev.* **175**, 1978; *Phy. Rev. A* **8**, 668 (1973).

Pacheco, A.F. and Strottman, D. (1989). *Phys. Rev. D* **40**, 2131.

Palmieri, V.G. et al. (1998). *Nucl. Instr. and Meth. in Phys. Res. A* **413**, 475.

Pan, L.K. and Wang, C.K.C. (1999). *Nucl. Instr. and Meth. in Phys. Res. A* **420**, 345.

Parker, E.N. (1957). *Astrophys. J. Supp.* **3**, 51; *Phys. Rev.* **107**, 924; *Phys. Rev.* **109**, 1874 (1958).

Parker, E.N. (1958). *Astrophys. J.* **128**, 664.

Parker, E.N. (1960). *Astrophys. J.* **132**, 821.

Parker, E.N. (1961) (a). *Astrophys. J.* **133**, 1014.

Parker, E.N. (1961) (b). *Astrophys. J.* **134**, 20.

Parker, E.N. (1963). Interplanetary Dynamical Process, Interscience, New York.

Parker, E.N. (1964). *J. Geophys. Res.* **69**, 1755.

Parker, E.N. (1965). *Planet. Space Sci.* **13**, 9.

Parker, E.N. (1966). *Planet. Space Sci.* **14**, 371.

Parker, E.N. (2007). Solar Wind, Handbook of the Solar-Rerrestrial Environment, Kamide, Y. and Chian Editors, Springer-Verlag, Berlin.

Parkin, D.M. and Coulter, C.A. (1977). *Trans. Am. Nucl. Soc.* **27**, 300.

Parkin, D.M. and Coulter, C.A. (1979). *J. Nucl. Mat.* **85&86**, 611.

Parks, G.K. (2004). Physics of Space Plasmas - 2nd Edition, Westview Press, Cambridge (MA).

Partridge, R. et al. (1980). *Phys. Rev. Lett.* **44**, 712.

Patterson, H.W. and Thomas, R.H. (1981). Accelerator Health Physics, Academic Press Inc., New York, 90.
See also:
Carpenter, J.M. and Yelon, W.B. (1986). Neutron Scattering, Methods of Experimental Physics 23 A, Academic Press Inc., New York, 99.
Kirkby, J. et al. (1988). Report of the Compact Detector Group, Proc. of the Workshop on Experiments, Detectors and Experimental Areas for the Supercollider (July 7–17,1987, Berkeley), World Scientific, Singapore, 388.

PDB (1980): Kelly, R.L. et al. (Particle Data Group) (1980). Review of Particle Physics, *Rev. of Mod. Phys.* **52**, S1.

PDB (1996 and 1998): Barnett, R.M. et al. (Particle Data Group) (1996). Review of Particle Physics, *Phys. Rev. D* **54**, 1.
See also:
PDB: Caso, C. et al. (Particle Data Group) (1998). Review of Particle Physics, *The E. Phys. Jou.* **3**, 1.

PDB (2000): Groom, D.E. et al. (Particle Data Group) (2000). Review of Particle Physics, *The Eur. Phys. Jou. C* **15**, 1.
See also Tables and Plots on the 2002 review [PDB (2002)].

PDB (2002): Hagiwara, K. et al. (Particle Data Group) (2002). Review of Particle Physics, *Phys. Rev. D* **66**, 010001.
Data files on cross sections (and their references) and available on the pdg group web site and are courtesy of the COMPAS Group, IHEP, Protvino, Russia:
http://pdg.lbl.gov/2002/contents_plots.html

PDB (2004): Eidelman, S. et al. (Particle Data Group) (2004). Review of Particle Physics,

Phys. Lett. B **592**, 1.

PDB (2008): Amsler, C. et al. (Particle Data Group) (2008). Review of Particle Physics, *Phys. Lett. B* **667**, 1.

The latest versions of tables, reference data and constants are available at the Particle Data Group web site: *http://pdg.lbl.gov/pdg.html*

PDG (2008). The Atomic and Nuclear Properties of Materials for more than 300 Substances are available at the web site:

http://pdg.lbl.gov/2007/AtomicNuclearProperties/index.html

Peigneux, J.P. et al. (1996). *Nucl. Instr. and Meth. in Phys. Res. A* **378**, 410.

Pensotti, S., Rancoita, P.G., Simeone, C., Vismara, L., Barbiellini, G. and Seidman, A. (1988). *Nucl. Instr. and Meth. in Phys. Res. A* **270**, 327.

Pensotti, S., Rancoita, P. G., Seidman, A. and Vismara, L. (1988). *Nucl. Instr. and Meth. in Phys. Res. A* **265**, 266.

Perkins, D.H. (1986). Introduction to High Energy Physics - 3rd Edition Addison-Wesley Publ. Comp., Inc. Menlo Park, CA.

Pernegr, J., Aebischer, B., Freudenreich, K., Frosch, R., Gentit, F.X., Mhlemann, P., Wetzel, W., Beusch, W., Chaloupka, V., Websdale, D., Astbury. P., Lee. G., Letheren, M., Bellini, G., Di Corato, M., Palombo, F., Rancoita, P.G., and Vegni, G. (1978). *Nucl. Phys. B* **134**, 436, doi: 10.1016/0550-3213(78)90458-3.

Perko, J.S. (1987). *Astron. Astrophys.* **184**, 119.

Petersen, E. (1981). *IEEE Trans. on Nucl. Sci.* **28**, 3981.

Petersen, E.L. (1992). *IEEE Trans. on Nucl. Sci.* **39**, 1960.

Pfeiffer, F. (2004). PhD Thesis, University of Erlangen-Nuremberg.

Physics Handbook (1972). American Institute of Physics Handbook, 3rd. ed., New York, McGraw-Hill.

Pickel, J.C. (1996). *IEEE Trans. on Nucl. Sci.* **43**, 483.

Pinkau, K. (1965). *Phys. Rev. B* **139**, 1548.

Pirollo, S. et al. (1999). *Nucl. Instr. and Meth. in Phys. Res. A* **426**, 126.

Pitzl, D. et al. (1992). *Nucl. Instr. and Meth. in Phys. Res. A* **311**, 98.

Pizzo, V. (1978). *J. Geophys. Res.* **83 (no. A12)**, 5563; *J. Geophys. Res.* **96 (no. A4)**, 5405 (1991).

Pneuman, G.W. and Kopp, R.A. (1971). *Solar Phys.* **18**, 258.

Pochodzalla, J. et al. (1995). *Phys. Rev. Lett.* **75**, 1040.

See also:

Bowman, D.R. et al. (1991). *Phys. Rev. Lett.* **67**, 1527.

Tke, J. et al. (1995). *Phys. Rev. Lett.* **75**, 2920.

Charity, R.J. et al. (1988). *Nucl. Phys. A* **483**, 371.

Barashenkov, V.S. et al. (1974). *Nucl. Phys. A* **222**, 204.

Poirier, R., Avalos, V., Dannefaer, S., Schiettekatte, F., Roorda, S. and Misra, S.K. (2003). *Physica B* **340–342**. 752.

Popovic, R.S. (2004). Hall Effect Devices, Institute of Physics Publishing, Bristol.

Pospisil, S. et al. (1993). *Rad. Prot. Dosim.* **46**, 115.

Potgieter, M.S. (1998). *Space Scie. Rev.* **83**, 147.

Potgieter, M.S. (2008). *J. of Atmosph. and Solar-Terr. Phys.* **70**, 207.

Potgieter, M.S., Le Roux, J.A., Burlaga, L.F. and McDonald, F.B. (1993). *Astrophys. J.* **403**, 760.

Potgieter, M.S., Le Roux, J.A., McDonald, F.B. and Burlaga, L.F. (1993). The causes of the 11-year and 22-year cycles in cosmuc rays modulation, Proceedings of the 23rd International Cosmic Ray Conf., Calgary, Canada (19–30/7/1993), 525.

Potgieter, M.S. and Moraal, H. (1985). *Astrophys. J.* **294**, 425.

Povh, B., Rith, K., Scholz, C. and Zetsche, F. (1995). Particles and Nuclei: an Introduction to the Physical Concepts, Springer-Verlag Publ., Berlin - Heidelberg - New York.

Powell, C.F., Fowler, C. and Perkins, C. (1959). The Study of Elementary Particles by Photographic Method, Pergamon Press, London.

Pratt, H.R. (1960). *Phys. Rev.* **119**, 1619; *Phys. Rev.* **120**, 1717.

Pratt, R.H., Tseng, H.K., Lee, C.M., Kissel, L., MacCallum, C. and Riley, M. (1977). *Atomic Data Nucl. Data Tables* **20**, 175; Errata in **26**, 477 (1981).

Pretzel, K. (2005). *J. Phys. G: Nucl. and Part. Phys.* **31**, R133–R149; Historical Review of Calorimeter Developments, Calorimetry in Particle Physics (Proc. of the 10th International Conference on Calorimetry in High Energy Physics, Pasadena, 25–29 March 2003), Zhu, R.-Y. Editor, World Sientific, Singapore, 3 (2003).

Price, W.J. (1964). Nuclear Radiation Detection - 2nd Edition, McGraw-Hill Book Company, New York.

Priest, E.R. (1985). Introduction to Solar Activity, Chapter 1 in Solar System Magnetic Fields, Priest, E.R. Editor, D. Reidel Publishing Company, Dordrecht.

Primakoff, H. (1951). *Phys. Rev.* **81**, 899.

Pripstein, M. (WALIC Collab.) (1991). Developments in warm liquid calorimetry, Proc. of the Symposium on Detector Research and Development for the Superconducting Supercollider, Fort Worth, Texas, 15–18 October 1990, Dombeck, T., Kelley, V. and Yost, G.P. Editors, World Scientific, Singapore, 105.
See also:
Alvarez-Taviel, F.J. et al. (UA1 Collab.) (1989). *Nucl. Instr. and Meth. in Phys. Res. A* **279**, 114.

Pritchard, R.L., Angell, J.B., Adler, R.B., Early, J.M. and Webster W.M. (1961). *Proc. IRE* **49**, 725.

Privitera, V., Coffa, S., Priolo, F. and Rimini, E. (1998). *La Riv. Del Nuovo Cimento* **21** **(no. 11)**, 1.

Prokoshkin, Yu D. (1980). Proceedings of Second ICFA Workshop on Possibilities and Limitations of Accelerators and Detectors, Les Diablerets, 4–10 October 1979, Amaldi, U. Editor, 405.

Putley, E.H. (1960). The Hall Effect and Related Phenomena, Butterworth & Co. Ltd., London.

Rakavy, G. and Ron, A. (1965). *Phys. Lett.* **19**, 207.
See also:
Phys. Rev. **159**, 50 (1967).

Raman, S., Houser, C.A., Walkiewicz, T.A. and Towner, I.S. (1978). *At. Data and Nucl. Tables* **21**, 567.

Ramo, S. (1939) *Proc. IRE* **27(9)**, 584.

Ramsey, C.E. and Vail, P.J. (1970). *IEEE Trans. on Nucl. Sci.* **6**, 310.

Rancoita, P.G. (1984). Silicon detectors and elementary particle physics, *J. Phys. G: Nucl. Phys.* **10**, 299, doi: 10.1088/0305-4616/10/3/007.

Rancoita, P.G. (2005). Systematic Investigation of Irradiated VLSI Bipolar Transistors for Space Environment Applications, *talk given at the Italian Physical Society Conference*, Catania 26 September–1 October.

Rancoita, P.G. (2008). Private communication.

Rancoita, P.G. and Seidman, A. (1982). *La Riv. del Nuovo Cimento* **Vol. 5, N. 7**.

Rancoita, P.G. and Seidman, A. (1984). *Nucl. Instr. and Meth. A* **226**, 369, doi: 10.1016/0168-9002(84)90051-2.
See also:
Barbiellini, G., Buksh, P., Cecchet, G., Hemery, J.Y., Lemeilleur, F., Rancoita,

P.G., Vismara, G. and Seidman (1985). *Nucl. Instr. and Meth. A* **235**, 216, doi: 10.1016/0168-9002(85)90556-X.

Randall, B.A. and Van Allen, J.A. (1986). *Geophys. Res. Lett.* **13**, 628.
    See also:
    Repko, J.S. (1987). *Astron. Astrophys.* **184**, 119.

Ranft, J. (1972). *Part. Accelerators* **3**, 129.
    See also:
    Gabriel, T.A. and Schmidt, W. (1976). *Nucl. Instr. and Meth.* **134**, 271.

Rao, U.R. et al. (1971). *Solar Phys.* **19**, 209.
    See also:
    Jokipii, J.R. (1966). *Astrophys. J.* **146**, 480.
    Fisk, L.A. and Axford, W.I. (1969). *Solar Phys.* **7**, 486.

Rasmussen, J.O (1965). Alpha Decay, in Alpha Beta and Gamma Ray Spectroscopy, Siegbahn, K. Editor, North Holland Publ. Co., Amsterdam.

RBS: Rutherford back-scattering experiment made at the Van de Graaff Tandem accelerator of the University of Montreal (2008). See:
    Bouchami, J. et al. (2008). MPX-ATLAS note 001 Universit de Montral, 28 May.

Ravi, K.V. (1981). Imperfections and Impurities in Semiconductor Silicon, John Wiley & Sons, New York.

Reed, R.A., Carts, M.A., Marshall, P.W., Marshall, C.J., Musseau, O., McNulty, P.J., Roth, D.R., Buchner, S., Melinger, J. and Corbiere, T. (1997). *IEEE Trans. on Nucl. Sci.* **44**, 2224.

Reinard, A.A. and Fisk, L.A. (2004). *Astrophys. J.* **608**, 533.

Reinecke, J.P.L. and Potgieter, M.S. (1994). *J. Geophys. Res.* **99 (no. A8)**, 14761.

Reitz, J.R. and Milford, F.J. (1970). Foundation of Electromagnetic Theory - 2nd Edition and 3rd printing, Addison-Wesley Publishing Company Inc., Reading.

Ressell, T. et al. (1993). *Phys. Rev. D* **48**, 5519.

Richter, W.A. et al. (1991). *Nucl. Phys. A* **523**, 325.

Riederer, S.J., Pelc N.J. and Chesler, D.A. (1978). *Phys. Med. Biol.* **23**, 446.

Ritson, D.M. (1961). Techniques of High Energy Physics, Interscience Publishers Inc., New York.

Riytz, A. (1991). *At. Data and Nucl. Tables* **47**, 205

Roberts, T. et al. (1979). *Nucl. Phys. B* **159**, 56.

Robinson, M.T. (1970). Nuclear Fission Reactors *Nuclear Fission Reactors*, British Nuclear Energy Society, London, 364.

Robinson, M.T. and Torrens, I.M. (1974). *Phys. Rev. B* **9**, 5008.

Rohrlich, F. and Carlson, B.C. (1954). *Phys. Rev.* **93**, 38.

Roldan, J.M., Ansley, W.E., Cressler, J.D. and Clark, S.D. (1997). *IEEE Trans. on Nucl. Sci.* **44 (no. 6)**. 1965.

Rollins, G. (1990). *IEEE Trans. on Nucl. Sci.* **37**, 1961.

Rode, D.L. (1973). *Phys. Status Solidi B* **55**, 687.

Roman, P. (1965). Advanced Quantum Theory, Addison-Wesley Publ. Comp., Inc. Reading MA.

Ros, E. (1991). *Nucl. Phys. B (Proc. Supp.)* **23 (Issue 1)**, 51.

Rose, M.E. and Korff, S.A. (1941). *Phys. Rev.* **59**, 850.

Rose Collab.: RD48 Collab. (2000). See web page:
    http://rd48.web.cern.ch/rd48/

Rose Collab.: Lindström, G. 2001 et al. (2001). *Nucl. Instr. and Meth. in Phys. Res. A* **466**, 308.

Rosen, J.L. (1977). Coherent Nuclear Production at Brookhaven and Fermilab Accelerator

Laboratories, Proceedings of the Topical Meeting on Multiparticle Production on Nuclei at Very High Energies, Bellini, G., Bertocchi, L. and Rancoita, P.G. Editors, IAEA-SMR-21, 305.

Rossi, B. (1964). High-Energy Particles, Englewood Cliffs, N.J., Prentice Hall (initially published in 1952).

Rossi, B. and Greisen, K. (1941). *Rev. Mod. Phys.* **13**, 240.

Roy, P. (1994). Étude de la réponse en courant de détecteurs silicium opérés en environnement de très hautes radiations, M.Sc. Thesis, Université de Montréal.

Roy, P. (2000). Études des caractéristiques électriques de détecteurs au silicium dans les conditions d'irradiation du LHC, Ph.D. Thesis, Université de Montréal.

Rudstam, G.Z. (1966). *Naturforsh.* **21a**, 1027.

Russ, J.S. (1990). Neutron energy deposition mechanisms in silicon calorimetry, Proceedings of the Workshop on Calorimetry for Supercollider, Tuscaloosa March 13–17 1989, Donaldson, R. and Gilchriese, M.G.D. Editors, World Scientific, Singapore, 531.

See also:

Bertrand, C. et al. (1990). Silicon Calorimetry for SSC, Proc. of the Workshop on Calorimetry for Supercollider, Tuscaloosa March 13–17 1989, Donaldson, R. and Gilchriese, M.G.D. Editors, World Scientific, Singapore, 489.

Furuno, K. et al. (1991). Neutron flux suppression with polyethylene moderators in the EMPACT silicon EM calorimeter, Proc. of the Symposium on Detector research and development for the Supercollider, Fort Worth October 15–18 1990, Dombeck, T., Kelly, V. and Yost, G.P. Editors, World Scientific, Singapore, 379.

Russ, J.S. et al. (1991). Neutron fluences in iron or lead calorimeters at the SSC, Proc. of the Symposium on Detector research and development for the Supercollider, Fort Worth October 15–18 1990, Dombeck, T., Kelly, V. and Yost, G.P. Editors, World Scientific, Singapore, 683.

Rutherford, E. (1911). *Phil. Mag. [6]* **21**, 669.

Sadrozinski, H.F.W. (2000). *IEEE Trans. Nucl. Sci.* **48**, 933.

Sah, C.T. and Reddi, V.G.K. (1964). *IEEE Trans. on Electron Devices* **11 (no. 7)**, 345.

Sah, C.T., Noyce, R.N. and Shockley, W. (1957). *Proc. of the IRE* **45 (no. 9)**, 1228.

Sakurai, J.J. (1994). Modern Quantum Mechanics - rev. Edition, Reading (Ma), Addison-Wesley.

Santocchia, A. et al. (2004). *Nucl. Instr. and Meth. in Phys. Res. A* **518**, 352.

Sanuki, T. et al. (2000). *Astrophys. J.* **545**, 1135.

Samara, G.A. (1988). *Phys. Rev. B* **37 (no. 14)**, 8523.

Sattler, A.R. (1965). *Phys. Rev. A* **138**, 1815.

Sauli, F. (1977). Principles of Operation of Multiwire Proportional and Drift Chamber, CERN 77-09.

Savolainen, V. et al. (2002). *J. Crystal. Growth* **243 (2)**, 243.

Schade, J. and Herrick, D. (1969). *Solid–State Elec.* **12**, 857.

Schimidt, D.M. et al. (1995). *IEEE Trans. on Nucl. Sci.* **42 (no. 6)**, 1541.

Schlickeiser, R. (2002). Cosmic Rays Astrophysics, Springer, Berlin.

Schneegans, M.A. et al. (1982). *Nucl. Instr. and Meth. in Phys. Res.* **193**, 445.

Schram, E. (1963). Organic Scintillation Detectors, New York, Elsevier.

Schroder, D.K. (1997). *IEEE Trans. on Electron Devices* **ED-44**, 160.

Schultz, W. (1971). *Solid–State Elec.* **14**, 227.

Schulz, M. (2007). Magnetosphere, Handbook of the Solar-Rerrestrial Environment, Kamide, Y. and Chian Editors, Springer-Verlag, Berlin.

Schwadron, N.A., McComas, D.J., Elliott, H.A., Gloeckler, G., Geiss, J. and von Steiger,

R. (2005). *J. Geophys. Res.* **110**, A04104, doi: 10.1029/2004JA010896.

Schwartz, S.J. (1985). Solar Wind and the Earth's Bow Shock, Chapter 8 of Solar System Magnetic Fields, Edited by Priest, E.R., Reidel Publishing Company, Dordrecht.

Segre, E. (1977). Nuclei and Particles, 2nd ed., W A Benjamin Inc., New York.

Séguinot, J. (1988). Les compteurs Cherenkov: applications et limites pour l'identification des particules, développements et perspectives, lecture given at l'Ecole Joliot-Curie, Maubuisson, France, 25–30 September, 249.

Séguinot, J. and Ypsilantis, T. (1977). *Nucl. Instr. and Meth.* **142**, 377.

Seide, W. et al. (2002). Dark Matter in Astro and Particle Physics, Klapdor-Kleingrothaus, H.V. and Viollier, R.D. Editors, Springer-Verlag, Berlin. 517.

Seitz, F. (1958). On the Theory of the Bubble Chambers, *The Physics of Fluids* **1**, 2.

Seitz, F. and Koehler, J.S. (1956). *Solid State Physics* **vol. 2**, edited by Seitz, F. and Turnbull, D., Academic Press Inc., New York.

Selesnick, R.S. et al. (1995). *J. Geophys. Res.* **100**, 9503.

Seltzer, S.M. (1993). *Rad. Res.* **136**, 147.

Seltzer, S.M. and Berger, M.J. (1964). Energy Loss Straggling of Protons and Mesons: Tabulation of the Vavilov Distribution from: Studies in Penetration of charged particles in matter, Publication 1133, National Academy of Sciences - National Research Council, Washington DC, 187.

Seltzer, S.M. and Berger, M.J. (1984). *Int. J. Appl. Radiat. Isot* **35 (no 7)**, 665.

Seltzer, S.M. and Berger, M.J. (1985). *Nucl. Instr. and Meth. in Phys. Res. B* **12**, 95.

Seltzer, S.M., Inokuti, M., Paul, H. and Bichsel, H. (2001). *Letter to the Editor, Radiat. Res.* **155**, 378.

Semico Research Corporation (2002). Gate Arrays Wane while Standard Cells Soar: ASIC Market Evolution Continues, **SC103-02**.

Sessom, A. et al. (1979). *Nucl. Instr. and Meth.* **161**, 371.
    See also:
    Bollini, D. et al. (1980). *Nucl. Instr. and Meth.* **171**, 237.
    Gabriel, T. et al. (1982). *Nucl. Instr. and Meth. in Phys. Res.* **195**, 461.

Shea, M.A., Smart, D.F. and McCracken, K.G. (1965). *J. Geophys. Res.* **70 (17)**, 4117.

Sheeley, N.R. Jr., Nash, A.G. and Wang, Y.-M. (1987). *Astrophys. J.* **319**, 481.

Shimano, Y. et al. (1989). *IEEE Trans. on Nucl. Sci.* **36**, 2344.

Shivakumar, P., Kistler, M., Keckler, S.W., Burger, D. and Alvisi, L. (2002). Modeling the Effect of Technology Trends on the Soft Error Rate of Combinational Logic, Proceedings of the 2002 International Conference on Dependable Systems and Networks, 389.

Shklee, K.L. and Nahory, R.E. (1970). *Phys. Rev. Lett.* **24**, 942.

Shockley, W. (1949). *Bell Syst. Tech. J.* **28**, 435.

Shockley, W. and Prim, R.C. (1953). *Phys. Rev.* **90**, 753.

Short, K.L. (1987). Microprocessors and Programmed Logic - 2nd Edition, Prentice Hall Inc., New Jersey.

Shulek, P. et al. (1966). *Yad. Fiz.* **4**, 564 [Shulek, P. et al. (1967). *Sov. Jou. Nucl. Phys.* **4**, 400].

Shur, M. (1996). Introduction to Electronic Devices, John Wiley & Sons, New York.

Shwartz, J.M. et al. (1966). *J. of App. Phys.* **37**, 745.

SICAPO Collab.: Barbiellini, G., Cecchet, G., Hemery, J.Y., Lemeilleur, F., Leroy, C., Levman, G., Rancoita, P.G. and Seidman, A. (1985) (a). *Nucl. Instr. and Meth. in Phys. Res. A* **235**, 55, doi: 10.1016/0168-9002(85)90245-1.

SICAPO Collab.: Barbiellini, G., Cecchet, G., Hemery, J.Y., Lemeilleur, F., Leroy, C., Levman, G., Rancoita, P.G. and Seidman, A. (1985) (b). *Nucl. Instr. and Meth. in*

Phys. Res. A **236**, 316, doi: 10.1016/0168-9002(85)90167-6.

SICAPO Collab.: Campanella, M., Croitoru, N., Groppi, F., Lemeilleur, F., Pensotti, S., Rancoita, P.G. and Seidman, A. (1986). *Nucl. Instr. and Meth. in Phys. Res. A* **243**, 93, doi: 10.1016/0168-9002(86)90826-0.

SICAPO Collab.: Barbiellini, G. Lemeilleur, F., Rancoita, P.G. and Seidman, A. (1987). *Nucl. Instr. and Meth. in Phys. Res. A* **257**, 543, doi: 10.1016/0168-9002(87)90959-4.

SICAPO Collab.: Ferri, G., Groppi, F., Lemeilleur, F., Pensotti, S., Rancoita, P.G., Seidman, A. and Vismara, L. (1988) (a). *Nucl. Instr. and Meth. in Phys. Res. A* **273**, 123, doi: 10.1016/0168-9002(88)90806-6.

See also:

Lemeilleur, F., Rancoita, P.G. and Seidman, A. (1987). *IEEE Trans. Nucl. Sci.* **34**, 538.

SICAPO Collab.: Furetta, C., Pensotti, S., Rancoita, P.G., Vismara, L., Barbiellini, G. and Seidman, A. (1988) (b). *IEEE Trans. Nucl. Sci.* **35**, 446.

SICAPO Collab.: Lemeilleur, F., Lamarche, F., Leroy, C., Paludetto, R., Pensotti, S., Rancoita, P.G., Vismara, L. and Seidman, A. (1989) (a). *Nucl. Instr. and Meth. in Phys. Res. A* **279**, 66, doi: 10.1016/0168-9002(89)91063-2.

SICAPO Collab.: Lemeilleur, F., Borchi, E., Fedder, I., Fretwurst, E., Lindstroem, G., Lamarche, F., Leroy, C., Furetta, C., Paludetto, R., Pensotti, S., Rancoita, P.G., Simeone, C., Vismara, L., Seidman, A., Barbiellini, G., Penzo, A., Giubellino, P., Ramello, L., and Riccati,L. (1989) (b). *Phys. Lett. B* **222**, 518.

SICAPO Collab.: Borchi, E., Bruzzi, M., Furetta, C., Giubellino, P., Lamarche, F., Lemeilleur, F., Leroy, C., Macii, R., Paludetto, R., Pensotti, S., Penzo, A., Ramello, L., Rancoita, P.G., Riccati, L., Seidman, A., Vismara, L. (1989) (c). Electromagnetic shower energy filtering effect. A way to achieve the compensation condition ($e/\pi=1$) in hadronic calorimetry, *Phys. Lett. B* **222**, 525–532.

SICAPO Collab.: Lemeilleur, F., Borchi, E., Fretwurst, E., Lindstroem, G., Lamarche, L., Leroy, C., Furetta, C., Pensotti, S., Rancoita, P.G., Vismara, L., Seidman, A., Giubellino, P., Ramello, L., Riccati, L., Barbiellini, G. and Penzo, A. (1989) (d). *IEEE Trans. Nucl. Sci.* **36**, 331.

SICAPO Collab.: Angelis, A.L.S., Borchi, E., Bruzzi, M., Furetta, C., Giubellino, P., Lamarche, F., Leroy, C., Macii, R., Manoukian-Bertrand, C., Mazzoni, S., Paludetto, R., Pensotti, S., Penzo, A., Ramello, L., Rancoita, P.G, Riccati, L., Seidman, A., Steni, R., Villari, A. and Vismara, L. (1990) (a). Evidence for the compensation condition in Si/U hadronic calorimetry by the local hardening effect. *Phys. Lett. B* **242**, 293–298.

SICAPO Collab.: Borchi, E., Bruzzi, M., Furetta, C., Giubellino, P., Lamarche, F., Lemeilleur, F., Leroy, C., Macii, R., Paludetto, R., Pensotti, S., Penzo, A., Ramello, L., Rancoita, P.G., Riccati, L., Seidman, A. and Vismara, L. (1990) (b). *IEEE, Trans. Nucl. Sci.* **37**, 1186.

SICAPO Collab.: Borchi, E., Furetta, C., Giubellino, P., Glaser, M., Lamarche, F., Lemeilleur, F., Leroy, C., Manoukian-Bertrand, C., Paludetto, R., Pensotti, S., Penzo, A., Rancoita, P.G., Ramello, L., Riccati, L. and Vismara, L. (1991) (a). *Nucl. Phys. B. (Proc. Suppl.)* **23 A**, 119.

SICAPO Collab.: Borchi, E., Furetta, C., Guibellino, P., Lamarche, F., Leroy, C., Macii, R., Manoukian-Bertrand, C., Paludetto, R., Pensotti, S., Penzo, A., Ramello, L., Rancoita, P.G., Riccati, L., Salvato, G., Seidman, A., Villari, A. and Vismara, L. (1991) (b). Evidence for compensation and study of lateral shower development in Si/U hadron calorimeters *IEEE Trans. Nucl. Sci.* **38**, 403–407, doi:

10.1109/23.289333.

SICAPO Collab.: Borchi, E., Furetta, C., Giubellino, P., Lamarche, F., Leroy, C., Macii, R., Manoukian-Bertrand, C., Paludetto, R., Pensotti, S., Penzo, A., Ramello, L., Rancoita, P.G., Riccati, L., Salvato, G., Seidman, A., Villari, A. and Vismara, L. (1991) (c). *Nucl. Phys. B (Proc. Supp.)* **23 A** 62.

See also:

Borchi, E., Leroy, C., Rancoita, P.G. and Seidman, A. (1991). Compensating Silicon Calorimetry for SSC/LHC colliders, Proc. of the Symposium on Detector Research and Development for the Superconducting Supercollider, Fort Worth, Texas, 15–18 October 1990, Dombeck, T., Kelley, V. and Yost, G.P. Editors, World Scientific, Singapore, 30.

Bertrand, C., Borchi, E., Brau, J.E., Bruzzi, M., Bugg, W.M., Furuno, K., Lamarche, F., Leroy, C., Lindstroem, G., Mazzoni, S., Ohsugi, T., Rancoita, P.G., Russ. J. and Vismara, L. (1990). Silicon Calorimeters for SSC, Proc. of the Workshop on Calorimetry for SSC, Tuscaloosa (USA), 13–17 March 1989, Donaldson, R. and Gilchriese, M.G.D. Editors, World Scientific, Singapore, 489.

SICAPO Collab.: Borchi, E., Macii, R., Mazzoni, S., Fedder, I., Lindstroem, G., Bertrand, C., Lamarche, F., Leroy, C., Villari,A., Bruzzi, M., Furetta, C., Paludetto, R., Pensotti, S., Rancoita, P.G., Simeone, C., Venturelli, L., Vismara, L., Brau, J.E., Croituro, N., Seidman, A., Berridge, S.C., Bugg, W.M., Giacomich, R., Penzo, A., Toppano, E., Giubellino, P., Ramello, L., Riccati, L., Pisani, M. and Steni, R. (1989). *Nucl. Instr. and Meth. in Phys. Res. A* **279**, 57, doi: 10.1016/0168-9002(89)91062-0.

Åkesson, T. et al. (1988). Report of the non magnetic Detector Group, Proc. of the Workshop on Experiments, Detectors and Experimental Areas for the Supercollider (July 7–17 1987, Berkeley), World Scientific, Singapore, 472.

SICAPO Collab.: Borchi, E., Bosetti, M., Furetta, C., Leroy, C., Macii, R., Manoukian-Bertrand, C., Paludetto, R., Pensotti, S., Pessina, G., Rancoita, P.G., Seidman, A., Terzi, G. and Vismara, L. (1991) (d). *Nucl. Phys. B (Proc. Supp.)* **23 A**, 352; Proc. ECFA Large Hadron Collider Workshop, Aachen, 1990 (CERN 90/10, Geneva), Vol. III, 721 (1990); oral presentation by Steni, R. at the ECFA Study Week on Instrumentation Technology for High-Luminosity Hadron Colliders, Barcelona, Spain (1989).

SICAPO Collab.: Borchi, E., Furetta, C., Giubellino, P., Lamarche, F., Leroy, C., Macii, R., Paludetto, R., Pensotti, S., Penzo, A., Pisani, M., Ramello, L., Rancoita, P.G., Riccati, L., Salvato, G., Seidman, A., Terzi, G., Villari, A. and Vismara, L. (1992) (a). Evidence for compensation in a Si/(Fe, Pb) hadron calorimeter by the filtering effect, *Phys. Lett. B* **280**, 169–174.

SICAPO Collab.: Angelis, A.L.S., Borchi, E., Furetta, C., Giubellino, P., Lamarche, F., Leroy, C., Macii, R., Manoukian-Bertrand, C., Paludetto, R., Pensotti, S., Penzo, A., Ramello, L., Rancoita, P.G., Riccati, L., Salvato, G., Seidman, A., Villari, A.and Vismara, L. (1992) (b). *Nucl. Inst. and Meth. in Phys. Res. A* **314**, 425.

SICAPO Collab.: Borchi, E., Bosetti, M., Leroy, C., Pensotti, S., Penzo, A., Rancoita, P.G., Rattaggi, M. and Terzi, G. (1993) (a). Evidence for compensation in a Si hadron calorimeter, *IEEE Trans. Nucl. Sci.* **40**, 508–515, doi: 10.1109/23.256610.

SICAPO Collab.: Borchi, E., Bosetti, M., Furetta, C., Lamarche, F., Leroy, C., Paludetto, R., Pensotti,S., Penzo, A., Rancoita, P.G., Rattaggi, M., Salvato, G., Terzi, G. and Vismara, L. (1993) (b). Systematic investigation of the electromagnetic filtering effect as a tool for achieving the compensation condition in silicon hadron calorimetry, *Nucl. Instr. and Meth. in Phys. Res. A* **332**, 85–90; *Nucl. Phys. B (Proc. Supp.)* **32**, 91.

SICAPO Collab.: Borchi, E., Lamarche, F., Leroy, C., Pensotti, S., Penzo, A. and Rancoita, P.G. (1993) (c). *Nucl. Phys. B (Proc. Supp.)* **32**, 47.

SICAPO Collab.: Bosetti M., Bosetti, M., Furetta, C., Leroy, C., Pensotti, S., Penzo, A., Rancoita, P.G., Rattaggi, M., Redaelli, M., Salvato, G. and Terzi, G. (1994) (a). Systematic investigation of the electromagnetic energy resolution on sampling frequency using silicon calorimeters, *Nucl. Instr. and Meth. in Phys. Res. A* **345**, 244–249.

SICAPO Collab.: Bosetti, M., Furetta, C., Leroy C., Pensotti, S., Rancoita, P.G., Rattaggi, M., Redaelli, M., Rizzatti, M., Seidman. A. and Terzi, G., (1994) (b). *Nucl. Instr. and Meth. in Phys. Res. A* **343**, 435, doi: 10.1016/0168-9002(94)90221-6.

SICAPO Collab.: Bosetti, M., Furetta, C., Leroy, C., Pensotti, S., Rancoita, P.G., Rattaggi, M., Redaelli, M., Rizzatti, M., Seidman, A. and Terzi, G. (1994) (c). *Nucl. Instr. and Meth. in Phys. Res. A* **345**, 250, doi: 10.1016/0168-9002(94)90998-9.

SICAPO Collab.: Furetta, C., Leroy, C., Pensotti, S., Penzo, A. and Rancoita, P.G. (1995) (a). *Nucl. Instr. and Meth. in Phys. Res. A* **361**, 149.

SICAPO Collab.: Furetta, C., Leroy, C., Pensotti, S., Penzo, A. and Rancoita, P.G. (1995) (b). *Nucl. Instr. and Meth. in Phys. Res. A* **357**, 64.

SICAPO Collab.: Bosetti, M., Croitoru, N., Furetta, C., Leroy, C., Pensotti, S., Rancoita, P.G., Rattaggi, M., Redaelli, M., Rizzatti, M. and Seidman, A. (1995) (c). *Nucl. Instr. and Meth. in Phys. Res. A* **361** 461 doi: 10.1016/0168-9002(95)00277-4

SICAPO Collab.: Bosetti, M., Croitoru, N., Furetta, C., Leroy, C., Pensotti, S., Rancoita, P.G., Rattaggi, M., Redaelli, M. and Seidman, A. (1995) (d). *Nucl. Instr. and Meth. in Phys. Res. B* **95**, 219, doi: 10.1016/0168-583X(94)00439-0.

SICAPO Collab.: Furetta, C., Gambirasio, A., Lamarche, F., Leroy, C., Pensotti, S., Penzo, A., Rattaggi, M. and Rancoita, P.G. (1996). *Nucl. Instr. and Meth. in Phys. Res. A* **368**, 378.

Sigmund, P. (1997). *Phys. Rev. A* **56**, 3781.

Sigmund, P. (1998). *Nucl. Instr. and Meth. in Phys. Res. B* **135**, 1.

Sigmund, P. (2006). Particle Penetration and Radiation Effects (General Aspects and Stopping of Swift Point Charges), Springer-Velarg, Berlin.

Sigmund, P. and Schinner, A. (2001) (a). *Eur. Phys. J. D* **15**, 165.

Sigmund, P. and Schinner, A. (2001) (b). *Nucl. Instr. and Meth. in Phys. Res. B* **174**, 535.

Sigmund, P. and Schinner, A. (2003). *Nucl. Instr. and Meth. in Phys. Res. B* **212**, 110.

Simon, M. and Heinbach, U. (1996). *Astrophys. J.* **456**, 519.
  See also: *Astrophys. J.* **441**, 209 (1995).

Simpson, J.A. (1983). Elemental and Isotopic Composition of the Galactic Cosmic Rays, *Ann. Rev. of Nucl. Part. Sci.* **33**, 323–381, doi: 10.1146/annurev.ns.33.120183.001543.

Sims, A.J. et al. (1994). *IEEE Trans. on Nucl. Sci.* **41**, 2361.

Singer, S.F. (1958). *Phys. Rev. Lett.* **1**, 181.

Smart, D.F. and Shea, M.A. (1985). Galactic Cosmic Radiation and Solar Energetic Particles, Chapter 6 of Handbook of Geophysics and the Space Environment, Editor Jursa, A.S., Air Force Geophysics Lab, Springfield.

Smart, D.F. and Shea, M.A. (1989). *J. Spacecraft and Rockets* **26**, 403.

Smith, E.C., Binder, D., Compton, P.A. and Wilbur, R.I. (1966). *IEEE Trans. on Nucl. Sci.* **13**, 11.

Smith, E.J. (1979). *Rev. Geophys. ans Space Phys.* **17 (no. 4)**, 610.

Smith, E.J. (2004). Magnetic Field in the Outer Heliosphere, Physics of the Outer Heliosphere, Proceedings of the 3rd International IGPP Conference (Riverside 8–13/2/2004), AIP Conference Proceedings **vol. 719**, Editors Florinski, V., Pogorelov, N.V. and Zank, G.P., American Institute of Physics, Melville, 213.

Smith, E.J. et al. (1995). *Space Scie. Rev.* **72**, 165.

Smith, E.J., Tsurutani, B.T. and Rosenberg, R.L. (1978). *J. Geophys. Res.* **83 (no. A2)**, 717.

Smith, F.M., Birnbaum, W. and Barkas, W.H. (1953). *Phy. Rev. (Letter to the Editor)* **91**, 765.

Smith, R.A. (1978). Semiconductors, Cambridge University Press, Cambridge.

Sokolsky, P. (1989). Introduction to Ultrahigh Energy Cosmic Ray Physics, Addison-Wesley, Redwood City, CA.

Sonnenblick, R. et al. (1991). *Nucl. Instr. and Meth. in Phys. Res. A* **310**, 189.

Sopko, B., Hazdra, P., Kohout, Z., Mrázek, D. and Pospíšil, S. (1997). Production of FZ silicon wafers by Polovodiče/Prague and MESA detector production, 2nd Workshop on Radiation Hardening of Silicon Detectors, CERN RD-48 *Collaboration internal report.*

Sorcini, B.B. and Brahme, A. (1994). *Phys. Med. Biol.* **39**, 795.

Sörenssen, A. (1965). *Nuovo Cim.* **38**, 745; **41**, 543 (1966).

Sowerby, B.D. (1971). *Nucl. Instr. and Meth.* **97**, 145.

Spjeldvik, W.N. and Rothwell, P.L. (1985). The Radiation Belts, Chapter 5 of Handbook of Geophysics and the Space Environment, Editor Jursa, A.S., Air Force Geophysics Lab, Springfield.

Sprawls, P. Jr. (1993). Physical Principles of Medical Imaging - 2nd Edition, Aspen Publishers Inc., Gaithersburg, 656.

Srinivasan, G.R., Tang, H.K. and Murley, P.C. (1994). Parameter-free, predictive modeling of single event upsets due to protons, neutrons, and pions in terrestrial cosmic rays, *IEEE Trans. on Nucl. Sci.* **vol. 41, issue 6, part 1,** 2063–2070, doi: 10.1109/23.340543.

Srour, J.R. (1973). *IEEE Trans. on Nucl. Sci.* **20**, 190.

Srour, J.R. and Hartman, R.A. (1989). *IEEE Trans. on Nucl. Sci.* **36 (no. 6)**, 1825.

Srour, J.R. and McGarrity, J.M. (1988). *Proc. of the IEEE* **76 (no. 11)**, 1443.

Srour, J.R., Long, D.M., Millward, D.G., Fitzwilson, R.L. and Chadsey, W.L. (1984). Radiation Effects on and Dose Enhancement of Electronic Materials, Noyes Publications, Park Ridge.

Srour, J.R., Marshall, C.I. and Marshall, P.W. (2003). *IEEE Trans. on Nucl. Sci.* **30 (no. 3)**, 653.

Srour, J.R., Othmer, S. and Chiu, K.Y. (1975). *IEEE Trans. on Nucl. Sci.* **22 (no. 6)**, 2656.

SSC Detector R&D at BNL (1990). Status Report, ed. by Yu, B. and Radeka, V., BNL 52244 UC-400, 14.

Stanev, T. (2004). High Energy Cosmic Rays, Springer-Praxis, Berlin.

Stapor, W.J., Meyers, J.P., Langworthy, J.B. and Petersen, E.L. (1990). *IEEE Trans. on Nucl. Sci.* **37**, 1966.

Stather, J.W. and Smith, H. (2005). Biological Effects of Ionizing Radiation, in Landolt–Börnstein, Group VIII: Advanced Materials and Technologies, Vol. 4: Radiological Protection, Martienssen, W. Editor in Chief, Springer-Verlag, Berlin.

Sternheimer, R.M. (1961). Interaction of Radiation with Matter, Methods of Experimental Physics Vol. 5 part A, Yuan, L.C.L. and Wu, C.-S. Editors, Academic Press, New York, 1.

Sternheimer, R.M. (1966). *Phys. Rev.* **145**, 247.

Sternheimer, R.M. (1971). *Phys. Rev. B* **3**, 3681.
> See also:
> *Phys. Rev.* **117**, 484 (1960).

Sternheimer, R.M. and Peierls, R.F. (1971). *Phys. Rev. B* **3**, 3681.
> See also:
> Sternheimer, R.M. (1960). *Phys. Rev.* **117**, 485.

Sternheimer, R.M., Berger, M.J. and Seltzer, S.M. (1984). *Atomic Data and Nucl. Data Tables* **30**, 261.

Stodolsky, L. (1966). *Phys. Rev.* **144**, 1145.
> See also:
> Koester, L.J. (1975). Scattering and Multipion Production Reaction leaving ^{12}C in the 2^+ State, in Proc. of the Topical Meeting on Hugh-Energy Collisions involving Nuclei, Bellini, G., Bertocchi, L. and Rancoita, P.G. Editors, Ed. Compositori, Bologna, 293.
> Frabetti, P.L. et al. (1979). *Nucl. Phys. B* **158**, 57.

Stoehr, U. (2005). Diploma Thesis, Albert-Ludwigs-University Freiburg.

Stoermer, K. (1930). *Z. Astroph.*, **1**, 237.

Stone, N.J. (2001). Table of Nuclear Magnetic Dipole and Electric Quadrupole Moments, Oxford Physics Clarendon Laboratory, Oxford U.K.; vailable on the web site:
> *http://www.nndc.bnl.gov/nndc/stone_moments/*
> see also [Tuli (2000)] and the web site:
> *http://www.nndc.bnl.gov/wallet/walletcontent.shtml*

Stone, S.L. et al. (1978). *Nucl. Instr. and Meth.* **151**, 387.

Storm, E. and Israel, H.I. (1967). LA-3753.
> See also Section 2.4.2 in [Hubbell (1969)].

Strom, D. (2008). Aspects of the SID detector design for the International Linear Collider, Proc. of the 10th ICATPP (8–12 October 2007, Como, Italy), Editors Barone, M., Gaddi, A., Leroy, C., Price, L., Rancoita, P.G. and Ruchti, R., World Scientific, Singapore, 511.
> See also:
> Brau, J.E., Breindenbach, M., Baltay, C., Frey, R.E. and Strom, D.M. (2007). *Nucl. Instr. and Meth. in Phys. Res. A* **579**, 567.

Suess, S.T. (1990). *Rev. of Geophys.* **28**, 1.

Suffert, M. (1988). Nouveaux scintillateurs et techniques associées, cours à l'École Joliot-Curie 1988, Instrumentation en Physique Nucléaire et en Physique des Particules, Maubuisson, France, 26–30 September, 49.

Summers, G.P., Burke, E.A., Shapiro, P., Messenger, S.R. and Walters, R.J. (1993). *IEEE Trans. on Nucl. Sci.* **40**, 1372.

Summers, G.P. et al. (1987). *IEEE Trans. on Nucl. Sci.* **34**, 1134.

Sumner, R. (1988). *Nucl. Instr. and Meth. in Phys. Res. A* **265**, 252.
> See also:
> Bakken, J.A. et al. (1987). *Nucl. Instr. and Meth. in Phys. Res. A* **254**, 535.

Swan, D.W. (1963). *Proc. Phys. Soc.* **82**, 74.

Swank, R.K. (1954). *Ann. Rev. Nucl. Sci.* **4**, 137.
> See also:
> Wright, G.T. (1955). *Proc. Phys. Soc. B* **68**, 929.
> Laval, M. et al. (1983). *Nucl. Instr. and Meth. in Phys. Res.* **206** 169.
> Blucher, E. et al. (1986). *Nucl. Instr. and Meth. in Phys. Res. A* **249**, 201.
> Kubota, S. et al. (1988). *Nucl. Instr. and Meth. in Phys. Res. A* **268**, 275.
> Adeva, B. et al. (1990). *Nucl. Instr. and Meth. in Phys. Res. A* **289**, 35.

Holl, I., Lorentz, E. and Mageras, G. (1988). *IEEE Trans. Nucl. Sci.* **35**, 105.

Symon, K.R. (1948). Fluctuation in Energy-Loss by High Energy Charged Particles in Passing through Matter, Thesis, Harward University; parts of this work appear in Section 2.7 of [Rossi (1964)].

Sze, S.M. (1965). Semiconductor devices, Physics and Technology Murray Hill, New Jersey.

Sze, S.M. (1981). Physics of Semiconductor Devices - 2nd Edition, John Wiley & Sons, New York.

Sze, S.M. (1985). Semiconductor Devices, Physics and Technology - 2nd Edition, John Wiley & Sons, New York.

Sze, S.M. (1988). VLSI Technology - 2nd Edition, McGraw–Hill, New York; (1983) 1st Edition.

Sze, S.M. and Ng, K.K. (2007). Physics of Semiconductor Devices - 3rd Edition, John Wiley & Sons, Hoboken, New Jersey.

Tabata, T., Andreo, P. and Ito, R. (1994). *Atomic Data and Nucl. Data Tables* **56**, 105, doi: 10.1006/adnd.1994.1003.

Taber, A. and Normand, E. (1993). *IEEE Trans. on Nucl. Sci.* **40**, 120.

Tagesen, S., Vonach, H. and Wallner, A. (2002). Evaluations of the fast neutron cross sections of ^{28}Si including complete covariance information, INDC, International Nuclear Data Committee, IAEA Nuclear Data Section, INDC(AUS)-018; available on the web site:

http://www-nds.iaea.org/indc_sel.html

Takeda, A. et al. (2003). *Phys. Lett. B* **572**, 145.

Tanenbaum, M. and Thomas, D.E. (1956). Diffused Emitter and Base Silicon Transistors, *Bell Syst. Techn. Journ.* **35**, 1.

Taroni, A. and Zanarini, G. (1969). *Nucl. Instr. and Meth.* **67**, 277.

Taylor, N.B. (1995). Guide for the Use of the International System of Units (SI), NIST Special Publication 811, National Institute of Standards and Technology, Gaithersburg, MD; available on the web site:

http://physics.nist.gov/Pubs/SP811/contents.html

Tesarek, R.J. et al. (2005). Single Event Effects and Their Mitigation for the Collider Detector at FERMILAB, *FERMILAB-CONF-05-511-E.*

Thomas, J. and Imel, D.A. (1987). *Phys. Rev A* **36**, 614.

Tiecke, H. (1989). *Nucl. Instr. and Meth. in Phys. Res. A* **277**, 42.

Timothy, A.F., Krieger, A.S. and Vaiana, G.S. (1975). *Solar Phys.* **42**, 135.

ToI (1996, 1998, 1999): Firestone, R.B., Shirley, V.S., Baglin, C.M. and Chu, S.Y.F., Zipkin, J. (1996). 8th Edition of the Table of Isotopes, John Wiley & Sons (Inc.) Publ., New York.

Firestone, R.B., Baglin. C.M. and Chu, S.Y.F. (1998). Update to the 8th Edition of the Table of Isotopes, John Wiley & Sons (Inc.) Publ., New York; and Update to the 8th Edition of the Table of Isotopes, John Wiley & Sons (Inc.) Publ., New York (1999).

Gamma-ray Energy and Intensity Standards (Appendix C) are available on the web site:

http://ie.lbl.gov/toipdf/eandi.pdf

as part of the 8th Edition of the Table of Isotopes, John Wiley & Sons (Inc.) Publ., New York.

α-Particles Energy and Intensity Standards (Appendix C) are available on the web site:

http://ie.lbl.gov/toipdf/alpha2.pdf

as part of the 8th Edition of the Table of Isotopes, John Wiley & Sons (Inc.) Publ.,

New York.

See also [Riytz (1991)].

Topkar, A. et al. (2006). Preshower silicon strip detectors for the CMS experiment at LHC, Proc. of the 9th ICATPP (17–21 October 2005, Como, Italy), ISBN 981-256-798-4, Editors Barone, M., Borchi, E., Gaddi, A., Leroy, C., Price, L., Rancoita, P.G. and Ruchti, R., World Scientific, Singapore, 411.

Toptygin, I.N. (1985). Cosmic Rays in Interplanetary Magnetic Fields, D.Reidel Publishing Company, Dordrecht.

Torrens, I.M. (1972). Interatomic Potentials, Academic Press, New York.

Torrens, I.M. and Robinson, M.T. (1972). Interatomic Potentials and Simulations of Lattice Defects, edited by Gehlen, P.C., Beeler, J.R. Jr. and Jaffee, R.I., Plenum Press, New York, 421–438.

Tousignant, O. et al. (1999). *Proc. SPIE* **3768**, 38.

Tove, P.A. and Seibt, W. (1967). *Nucl. Instr. and Meth. in Phys. Res. A* **51**, 261.

Tovey, D.R. et al. (2000). *Phys. Lett. B* **488**, 17.

Tsai, Y. (1974). *Rev. Mod. Phys.* **46**, 815.

Tseng, H.K. and Pratt, R.H. (1972). *Phys. Rev. A* **6**, 2049; *Phys. Rev. A* **21**, 454 (1980).

Tsveybak, I., Bugg, W., Harvey, J.A. and Walter, J. (1992). *IEEE Trans. on Nucl. Sci.* **NS-39 (no 6)**, 1720.

Tsyganenko, N.A. (1995). *J. Geophys. Res.* **100**, 5599.

Tsyganenko, N.A. and Stern, D.P. (1996). *J. Geophys. Res.* **101**, 27187.

Tuli, J.K. (2000). Nuclear Wallet Cards 6th Edition, National Nuclear Data Center for the U.S. Nuclear Data Program, BNL Upton, New York 11973-5000, USA.

Tyagi, M.S. (1991). Introduction to Semiconductor Materials and Devices, John Wiley & Sons, New York.

Tylka, A.J. et al. (1996). *Astron. Astrophys.* **17** (2), 47.

Uher, J. et al. (2007) (a). "Silicon Detectors for Neutron Imaging" in Nuclear Physics Methods and Accelerators in Biology and Medicine", Editors: Granja, C., Leroy, C. and Stekl, I., *AIP Conference Proceedings* **Vol. 958**, American Institute of Physics, New York, USA 101.

Uher, J. et al. (2007) (b). *Nucl. Instr. and Meth. in Phys. Res. A* **576**, 32.

Van Allen, L., Ludwig, G. and McIlwain, R. (1958). Observation of High Intensity Radiation by Satellites 1958 Alpha and Gamma, *IGY Satellite Series* (Natl. Acad. Sci. Washington D.C.) **3**, 73.

van der Pauw, L.J. (1958). *Philips Res. Rep.* **13**, 1.

van der Ziel, A. (1976). Physics of Semiconductor Devices - 3rd Edition, Prentice Hall, Englewood Cliffs.

Van Ginneken, A. (1989). Non Ionizing Energy Deposition in Silicon for Radiation Damage Studies, *FN-522*, Fermi National Accelerator Laboratory.

van Lint, V.A.J. and Leadon, R.E. (1966). *IEEE Trans. on Nucl. Sci.* **NS-13**, 11.

van Lint, V.A., Flanahan, T.M., Leadon, R.E., Naber, J.A. and Rogers, V.C. (1980). Mechanisms of Radiation Effects in Electronic Materials, Wiley Interscience, New York.

van Lint, V.A.J., Leadon, R.E. and Colweel, J.F. (1972). *IEEE Trans. on Nucl. Sci.* **NS-19**, 181.

Van Zeghbroeck, B. (2007). See the web site:
http://ece-www.colorado.edu/ bart/book/pndiode.htm

Vasilescu, A. (1997). The NIEL scaling hypothesis applied to neutron spectra of irradiation facilities in the ATLAS and CMS SCT, *ROSE/TN/97-2*.
See also:

Vasilescu, A. and Lindstroem, G. (2000). Notes on the fluence normalization based on the NIEL scaling hypothesis *ROSE/TN/2000-02.*

Vasilescu, A. and Lindstroem, G. (2000). Displacement damage in silicon (on-line compilation available at:

http://sesam.desy.de/members/gunnar/Si-dfuncs.html

Based for neutrons on:

Griffin, P.J. et al. (1993). SNL RML Recommended Dosimetry Cross Section Compendium *SAND92-0094* (Sandia National Lab.); (1996) private communication for $1.025 \times 10^{-10} < E_k^n < 19.95$ MeV.

Konobeyev, A. (1992). *J. Nucl. Mater.* **186**, 117 (for $20 < E_k^n < 800$ MeV).

Ref. [Huhtinen and Aarnio (1993)]; private communication for $8.05 < E_k^n < 8.995 \times 10^3$ MeV.

Based for protons on:

Ref. [Huhtinen and Aarnio (1993)]; private communication for $15 < E_k^p < 9.005 \times 10^3$ MeV.

Vavilov, P.V. (1957). *Zh. Eksp. Teor. Fiz.* **32**, 957 [Vavilov, P.V. (1957). *Sov. Phys. - JETP* **5**, 749].

Vavilov, V.S. and Ukhin, N.A. (1977). Radiation Effects in Semiconductors and Semiconductor Devices, Consultants Bureau (a division of Plenum Publishing Press), New York.

Videau, H. et al. (1984). *Nucl. Instr. and Meth. in Phys. Res. A* **225**, 481.

Virdee, T.S. (1999). Experimental Techniques, 1998 European School on High-Energy Physics - CERN-JINR School of Physics, 23 Aug – 5 Sep 1998, St Andrews, Scotland UK, Ellis, N. and March-Russell, J. Editors, CERN-99-04.

Viswanathan, C.R., Divakaruni, R. and Kizziar, J. (1991). *IEEE Trans. on Electron Device Lett.* **12**, 503.

von Weizsäcker, C.F. (1936). *Z. Physik* **96**, 96.

See also:

Bethe, H.A. (1936). *Rev. Mod. Phys.* **8**, 8.

Feenberg, E. (1947). *Rev. Mod. Phys.* **19**, 239.

Vykydal, Z., Jakubek, J. and Pospisil, S. (2006). *Nucl. Instr. and Meth. in Phys. Res. A* **563**, 112.

Wagener, J.L. and Milnes, A.G. (1964). *App. Phys. Lett.* **5**, 186.

Wagoner, D.E. et al. (1985). *Nucl. Instr. and Meth. in Phys. Res. A* **238**, 315.

Wahab, M.A. (1999). Solid State Physics, Structure and Properties of Materials, Narosa Publishing House, New Delhi.

Walenta, A.H., Fisher, J., Okuno, H. and Wang, C.L. (1979). *Nucl. Instr. and Meth.* **161**, 45.

Walker, W. and Sah, C.T. (1973). *Phys. Rev. B* **7**, 4587.

Walske, M.C. (1962). *Phys. Rev.* **101**, 940; *Phys. Rev.* **88**, 1283 (1952).

Wang, Y.-M., Sheeley, N.R. Jr., Nash, A.G. and Shampine, L.R. (1988). *Astrophys. J.* **327**, 427.

Wang, Y.-M. and Sheeley, N.R. Jr. (1993). *Astrophys. J.* **414**, 916.

Watkins, G.D. and Corbett, J.W. (1961) (b). *Phys. Rev.* **121**, 1001.

Watkins, G.D. and Corbett, J.W. (1961) (a). *Phys. Rev. Lett.* **7 (no. 8)**, 314.

Watkins, G.D. and Corbett, J.W. (1964). *Phys. Rev.* **134 (no 5A)**, 1359.

Watkins, G.D. and Corbett, J.W. (1965) (a). *Phys. Rev.* **138 (no 2A)**, 543.

Watkins, G.D. and Corbett, J.W. (1965) (b). *Phys. Rev.* **138 (no 2A)**, 555.

Watkins, G.D., Corbett, J.W., Chrenko, R.M. and McDonald, R.S. (1961). *Phys. Rev.* **121**, 1015.

Webber, W.R. and Lezniak, J.A. (1974). *Astrophys. Space Scie.* **30**, 361.

Webber, W.R. and Potgieter, M.S. (1989). *Astrophys. J.* **344**, 779.

Webster, W.M. (1954). *Proc. IRE* **42**, 914.

Weekes, T.C. (1988). *Phys. Reports (Review Section of Physics Letters)* **160**, 1.

Weilhammer, P. et al. (1995). Proceedings of the 2nd International Symposium on Development and Application of Semiconductor Tracking Detectors, Hiroshima.

Weisberg, L.R. and Schade, J. (1968). *J. of Appl. Phys.* **39**, 5149.

Weisskopf, V. (1937). *Phys. Rev.* **52**, 295.
    See also:
    Bohr, A. and Mottelson, B.R. (1969). Nuclear Structure, Vol. I, W.A. Benjamin Inc., New York, Chapter 2.
    Burcham, W. E. (1973). Nuclear Physics, Longman Group LTD, London, Chapter 15.

Werber, W.R. et al. (1991). *The Astrophys. J.* **380**, 230.

Wheeler, J.A. and Lamb, W.E. (1939). *Phys. Rev.* **55**, 858.

White, R.S. (1970). Space Physics (Chapter 5), Gordon and Breach, New York.

White, T.O. (1988). *Nucl. Instr. and Meth. in Phys. Res. A* **273**, 820.

White-Grodstein, G. (1957). NBS circular 583.

Wielunski, M. et al. (2004). *Nucl. Instr. and Meth. in Phys. Res. A* **517**, 240.

Wigmans, R. (1987). *Nucl. Instr. and Meth. in Phys. Res. A* **259**, 389.

Wigmans, R. (1988). *Nucl. Instr. and Meth. in Phys. Res. A* **265**, 273.

Wigmans, R. (1991). *Ann. Rev. Nucl. Part. Sci.* **41**, 133.

Wigmans, R. (2000). Calorimetry, Oxford, Clarendon Press.

Wikipedia: Charge number (2008) (a). See web site:
    http://en.wikipedia.org/wiki/Charge_number

Wikipedia: DRIE (2008) (b). See web site:
    http://en.wikipedia.org/wiki/DRIE

Wilkins, R.D. et al. (1971). *Nucl. Instr. and Meth.* **92** 381.

Williams, E.J. (1932). *Proc. Roy. Soc. A* **135**, 108.
    See also:
    Livingstone, M.S. and Bethe, H.A. (1937). *Rev. Mod. Phys.* **9**, 245.
    Uehling, E.A. (1954). *Ann. Rev. Part. Sci.* **4**, 315.

Williamson, C.F., Boujot, J.-P. and Picard, J. (1966). Tables of range and stopping power of chemical elements for charged particles of energy 0.05 to 500 MeV, Rapport CEA–R 3042, 321.

Wilson, R.R. (1947). *Phys. Rev.* **71**, 385.

Winzeler, H. (1965). *Nucl. Phys.* **69**, 661.

Wisshak, K. et al. (1989). *IEEE Trans. Nucl. Sci.* **36**, 101.

Wolf, H.F. (1969). Silicon Semiconductor Data, Pergamon Press, Oxford.

Wolf, H.F. (1971). Semiconductors Wiley-Interscience, New York.

Woolley, J.A., Lamar, L.E., Stradley, N.H. and Harshbarger, D.M. (1979). *IEEE Trans. on Components, Hybrids and manufacturing Technology* **CHMT-2**, 388.

Wright, A.G. (1974). *Jou. Phys. A* **7**, 2085.

Wu, A. et al. (1997). *IEEE Trans. on Nucl. Sci.* **44 (no. 6)**, 1914.

Xapsos, M.A. et al. (1994). *IEEE Trans. on Nucl. Sci.* **41**, 1945.

Yodh, G. (1987). High Energy Cosmic Rays and Astrophysics, Proceedings of the ICFAS School on instrumentation in elementary particle physics, Trieste, Italy, June 1987, Fabjan, C.W. and Plicher, J.E. Editors, World Scientific, Singapore, 298.

Young, G.R. et al. (1989). *Nucl. Instr. and Meth. in Phys. Res. A* **279**, 503.

Yuda, T. (1969). *Nucl. Instr. and Meth.* **73**, 301.

Yuda, T., Masaike, A., Kusumegi, A., Murata, Y., Ohta, I. and Nishimura, J. (1970). *Il Nuovo Cimento* **65** *A*, 205.

Zheng-Ming, L. and Brahme, A. (1993). *Radiat. Phys. Chem.* **41**, 673.

Zhu, R.-Y. (1993). Proceedings of the 4th International Conference on Calorimetry in High Energy Physics, La Biodola, Isola d'Elba, Italy, Menzione, A. and Scribano, A. Editors, World Scientific, Singapore, 301.

Ziegler, J. (2001). TRIM-98 available at:
http://www.srim.org/SRIM/SRIMLEGL.htm

Ziegler, J. (2006). SRIM.EXE version 2003 (©) 1984-2002 available at:
http://www.srim.org/SRIM/SRIMLEGL.htm

Ziegler, J.F. and Biersack, J.P. (2000). TRIM (Transport of Ions in Matter), version TRIM-96 in SRIM-2000 (The Stopping and Range of Ions in Matter), version 2000.39 coding by Mawrick, D.J. et al., ©1998, 1999 by IBM Co.

Ziegler, J.F., Biersack, J.P. and Littmark, U. (1985) (a). The Stopping Range of Ions in Solids, Vol. 1, Pergamon Press, New York.
See also [Ziegler and Biersack (2000)].

Ziegler, J.F., Biersack, J.P. and Littmark, U. (1985) (b). TRIM (Transport of Ions in Matter): version 95.4 in Ref. [Ziegler, Biersack and Littmark (1985a)].

Ziegler, J.F., Ziegler, M.D. and Biersack, J.P. (2008) (a). SRIM - The Stopping and Range of Ions in Matter, SRIM Co., Chester.
See also [Ziegler and Biersack (2000)].

Ziegler, J.F., Ziegler, M.D. and Biersack, J.P. (2008) (b). SRIM - The Stopping and Range of Ions in Matter, version 2008.03 available at:
http://www.srim.org/

Ziegler, J.F. et al. (1996). *IBM J. Res. Develop.* **40**, 51.

Zoutendyck, J.A., Edmonds, L.D. and Smith, L.S. (1989) (a). *IEEE Trans. on Nucl. Sci.* **36**, 2267.

Zoutendyck, J.A., Edmonds, L.D. and Smith, L.S. (1989) (b). *IEEE Trans. on Nucl. Sci.* **37**, 1844.

Zoutendyck, J.A., Schwartz, H.R. and Nevill, L.R. (1989). *IEEE Trans. on Nucl. Sci.* **35**, 1644.

Zurbuchen, T.H. (2007). *Annu. Rev. Astron. Astrophys.* **45**, 297.

Zurbuchen, T.H., Schwadron, N.A. and Fisk, L.A. (1997). *J. Geophys. Res.* **102**, 24172.

Index